Metabolic Adaptations in Plants During Abiotic Stress

Metabolic Adaptations in Plants During Abiotic Stress

Edited by
Akula Ramakrishna and Sarvajeet Singh Gill

CRC Press
Taylor & Francis Group
Boca Raton London New York

CRC Press is an imprint of the
Taylor & Francis Group, an **informa** business

CRC Press
Taylor & Francis Group
6000 Broken Sound Parkway NW, Suite 300
Boca Raton, FL 33487-2742

First issued in paperback 2021

ISBN 13: 978-1-03-209429-8 (pbk)
ISBN 13: 978-1-138-05638-1 (hbk)

Library of Congress Cataloging-in-Publication Data

Names: Ramakrishna, Akula, editor. | Gill, Sarvajeet Singh, editor.
Title: Metabolic adaptations in plants during abiotic stress / editor(s):
Akula Ramakrishna and Sarvajeet Singh Gill.
Description: Boca Raton, FL : CRC Press, Taylor & Francis Group, 2018.
Identifiers: LCCN 2018016983 | ISBN 9781138056381 (hardback : alk. paper)
Subjects: LCSH: Plants--Metabolism. | Plants--Effect of stress on. |
Plants--Adaptation.
Classification: LCC QK881 .M43 2018 | DDC 581.7--dc23
LC record available at https://lccn.loc.gov/2018016983

Visit the Taylor & Francis Web site at
http://www.taylorandfrancis.com

and the CRC Press Web site at
http://www.crcpress.com

Dr. Ravishankar Gokare

This book is dedicated to Dr. Ravishankar Gokare, a renowned plant biotechnologist

and food biotechnologist of international repute.

Ravishankar Gokare was born in 1954 in Bangalore. He completed his BSc in 1972 at Bangalore University and further studies, MSc and PhD, in botany at the M.S. University of Baroda. He was a PhD student of Professor Atul R. Mehta, a renowned plant tissue culturist trained in the lab of Professor H.E. Street at Swansea, who was known as the Father of Modern Plant Tissue Culture. During his PhD he worked on metabolic regulation of secondary pathways in plants. He continued his postdoctoral work at CSIR Laboratory at Jammu-Tawi working on medicinal plants. He joined the Central Food Technological Research Institute, Mysore, in 1984 as Scientist B, where he had the fortune of working with India's most vibrant Algal Research Group, which was started by Dr. Wolfgang Becker of University of Tubingen in association with the eminent algal biotechnologist, Dr. L.V. Venkataraman. He rose to the level of Chief Scientist and Head of Plant Cell Biotechnology Department, a position he held for over 18 years.

Ravishankar initiated plant biotechnology and food biotechnology research at the Central Food Technological Research Institute (CFTRI), where he established a strong research group working on aspects of micro-propagation, conservation of biodiversity, bioactive metabolites, biosynthetic pathways in plants and microbes, metabolic regulation of secondary metabolites, post-harvest technologies for value addition to foods, and downstream processing of metabolites of food applications, such as food pigments, non-nutritive sweeteners, flavor molecules, etc. He was a pioneer in using elicitor technology to enhance metabolite yield in a standing crop of various plantation crops, which followed as an agronomic practice in various countries. Ravishankar is a plant conservation expert responsible for developing the tissue culture technology for a banana variety, Nanjangud Rasabale, which was facing the brink of extinction. Similarly, he was responsible for the development of technology for *Decalepis hamiltonii* (swallow root), a rich herb for adding value as a potent antioxidant source and an anticancer agent with the development of several recipes for health food application, including health drinks. The technology of swallow root was combined with the development of a mass production system through tissue culture methodology and the agronomic practice of cultivation, allowing for the conservation and sustainable utilization of this endangered herb. The benefits of this technology were transferred to the Soliga Tribes of BR Hills who are custodians of the germplasm, exemplifying the sharing of Intellectual Property Rights. Under Ravishankar's leadership, this was acclaimed as a star project of the Department of Science and Technology of the Government of India. In 1992, he received the Indian Science Congress Award for Young Scientist from the then Prime Minister of India for these contributions.

Ravishankar moved from metabolic regulation of secondary pathways to metabolic engineering in several systems. His group pioneered in the initiation of coffee biotechnology research under the Jai Vigyan project of the Government of India, sponsored by the Department of Biotechnology, which resulted in the development of transgenic coffee plants with lowered caffeine. The technology is useful for improvement of coffee plants for various traits. Various institutions including the Coffee Board, which collaborated under the All India Coordinated Project, benefited from the transformation system developed by his group. Ravishankar and his team worked on metabolic engineering of carotenoids in algae, with the production of astaxanthin in *Dunaliella* spp., which is otherwise known for β-carotene only. This was achieved by cloning the genes of astaxanthin pathway from *Haematococcus pluvialis* into *Dunaliella*. Similarly, the metabolic engineering of fatty acid production to produce gamma linolenic acid in soybean by the expression of the Delta 6 desaturase gene of Spirulina is another example of his innovative research. He is also well known for his studies on genomics in higher plants and microalgal systems. These studies earned him a coveted award in new biology given by the Indian Science Congress Association.

Ravishankar is known for his contributions with Dr. L.V. Venkatraman in commercializing spirulina technology in India. This is a model for several other biotechnologists in India wherein the involvement of industry was realized from the beginning of the scale-up to the transfer of technology, culminating in Asia's largest spirulina production unit in Nanjangud by Bhallarpur Industries Limited. This contribution was recognized by the Government of India with a National Technology Day Award in 2003. Presently, the Government of Karnataka has taken up spirulina supplementation for malnourished children, under the mid-day meal program. His group have developed several value-added spirulina food products for meeting the micronutrient deficiencies caused by a deficiency of iron, gamma linolenic acid, and vitamin A in staple diets. Ravishankar is a great mentor, too. He has guided over 25 PhD students as a supervisor and mentored over 160 graduate students, who have occupied high positions in India and abroad in academia, R&D labs, and in industry. He has been a popular teacher who is known for motivating young minds. His love for the academics and student-centric activities have been well recognized at his present role as Vice President of Life Sciences and Biotechnology at Bangalore's pioneering research and innovation driven institution, namely Dayananda Sagar Institutions/Dayananda Sagar University at Bengaluru.

His innovative ideas, hard work, and dedication is evidenced by the publication of 340 research and review papers in peer-reviewed journals and over 45 patents awarded in India and abroad. His work has received over 13,500 citations with an h-index of 57 at present. Ravishankar is a fellow of several academies in India and abroad. He plays an active role in the International Union of Food Science and Technology and the Institute of Food Technologists of the USA, who have recognized him with the honor of fellow. He is widely traveled and has given invited talks in many international symposia and seminars. He has been a visiting professor to universities in Japan, Taiwan, and Russia.

This book is dedicated to Dr. Ravishankar Gokare A, for mentoring a large number of students including one of the authors, Dr. Ramakrishna, and several researchers, who are contributing immensely to the field of plant biotechnology for food application. His simplicity, humble behavior, kindness to colleagues, and positive outlook is exemplary. We sincerely hope that he will continue to make contributions of global relevance and to nurture young generation of researchers to be global leaders.

Contents

Section I Abiotic Stress Management and Its Impact on Plants

Section II Role of Major Plant Metabolites During Abiotic Stress Management

Section III Role of Specialized Proteins During Abiotic Stress Management

Section IV Role of Signaling Molecules Under Abiotic Stress Management

Section V   Biotechnological Applications to Improve the Plant Metabolic Pathways Towards Better Adaptations

Foreword

Not every scientist in the world has the basic skills to write and compile a comprehensive book on a challenging research area like *Metabolic Adaptations in Plants During Abiotic Stress.* Dr. Akula Ramakrishna and Dr. Sarvajeet Singh Gill completed this work remarkably. They have teamed up to cover all the burning topics and mechanics pertaining to metabolic adaptations of plants during the adverse and stressed conditions related to polyamines, indoleamines, antioxidative systems, adaptations in seeds under arid conditions, the roles of osmoregulators, secondary metabolites and miRNAs, kinase signaling, the impact of ozone and flavonoids, mechanisms and operating pathways during stresses, photoinhibitory responses, genomic response, antioxidant isoenzymes, management of heavy metal stress, pesticide stress, ATPases in ion homeostasis, genetic engineering approaches for abiotic stress tolerance in plants for sustainable agriculture, and food security. All these topics make this book wonderful and highly useful to the scientific community.

This book is worth reading for every student and scientist doing active research in agriculture and abiotic studies. Some of you may be wondering why I am praising this book. This book matters because our climate is changing, and the plant kingdom is seriously affected, which plays a significant role in the sustainability of human life and the whole ecosystem on Earth.

I strongly believe the content described here will be very useful for taking precautionary steps to sustain our agricultural system during upcoming adverse conditions.

This book provides a wide view of abiotic stress research materials to diversify each person's learning experiences. Students should cultivate new ideas by relating to the current research and knowledge described in simple and eloquent terms in each chapter. This thorough research should improve their knowledge and nurture their research productivity, as well stimulate incentives for further learning. This well-organized book provides an excellent opportunity for students, as well as for the whole scientific community.

I feel honored and privileged to have this opportunity to write a foreword to Dr. Ramakrishna and Dr. Gill's book. The challenges of abiotic stress in plant growth and development are evident, and the authors' dedication in this direction is appreciated. This is the area of future research that will help us all to thrive, along with plants.

Shashi Kumar, PhD
Group Leader, Metabolic Engineering Group,
International Center for Genetic Engineering and
Biotechnology (ICGEB),
Aruna Asaf Ali Marg, New Delhi - 110 067

Acknowledgments

Our special thanks and heartiest gratitude to all the contributors of this volume, who have provided an ample treatise on the topic. We understand that the contributors are active scientists and have agreed to share their findings and perspectives with the readers, who will benefit from learning about the implications and applications of the subject matter of the book as an emerging area of science.

We are grateful to Alice Oven (Senior Editor), Jennifer Blaise (Senior Editorial Assistant), Lara Silva McDonnell (Senior Project Manager), and the team from Taylor & Francis for their help in bringing out the volume in an attractive manner.

Both of us are thankful to our families who encouraged us to take up this task and permitted us to take the time off from our relationships.

Dr. Ramakrishna is thankful to Monsanto Crop Breeding Center, Bengaluru, for their support and also thankful to his mentors Dr. G.A. Ravishankar and Dr. P. Giridhar for their guidance. Dr. Ramakrishna also extends thanks to the staff and students of the Plant Cell Biotechnology Department, CFTRI, Mysore, India. Dr. Gill also acknowledges the unconditional support and guidance of Professor (Dr.) Narendra Tuteja, ICGEB, New Delhi.

Akula Ramakrishna
Monsanto Breeding Station, Bangalore, India

Sarvajeet Singh Gill
MD University, Rohtak, India

Editors

Dr. Ramakrishna Akula is currently a scientist at Monsanto Breeding Station, Bangalore, India. Dr. Ramakrishna holds a master's degree from Sri Krishna Devaraya University, Anantapur, India. He started his research career in 2005 at the Department of Plant Cell Biotechnology, CFTRI, Mysuru, in the research group of Dr. G.A. Ravishankar. He is a Senior Research Fellow of CSIR, New Delhi. He obtained his PhD in biochemistry from University of Mysore, Mysuru, in the area of development of high frequency somatic embryogenesis and regulation of secondary metabolites in *Coffea canephora*. He worked extensively on the role of serotonin, melatonin, and calcium-mediated signaling in plants. He has made significant contributions to metabolic engineering of secondary metabolites from plants and abiotic stress in plants. He has worked in the area of tissue culture, in *vitro* production, and regulation of plant secondary metabolites from food value plants that include natural pigment caffeine, steviosides, anthocyanins, and carotenoids. He is the author of three books, 12 peer-reviewed publications, two reviews, and eight chapters in books. His books include *Serotonin and Melatonin: Their Functional Role in Plants, Food, Phytomedicine, and Human Health*, *Metabolic Adaptations in Plants During Abiotic Stress* (CRC Press, 2016) and *Neurotransmitters in Plants: Perspectives and Applications* (CRC Press, 2018). He is a member of the Society for Biotechnologists (India). He is a fellow of the Society for Applied Biotechnology, India (2012), and has received the Global Vegetable Research Excellence Award (2017), three global technology recognition awards, a Rapid Recognition Award, Test Master, Asia Veg R&D quarterly recognitions, and special recognition from the Monsanto company. He attended the Fifth International Symposium on Plant Neurobiology held in 2009 in Florence, Italy. He also attended the Technical Community of Monsanto (TCM) held in 2016, in St. Louis, Missouri.

Dr. Sarvajeet Singh Gill is currently working as assistant professor at the Centre for Biotechnology, Maharshi Dayanand University, Rohtak, Haryana, India. In 2001, Dr. Gill completed his MSc in Botany from Aligarh Muslim University, Aligarh, India, with Gold Medal. Soon after, Dr. Gill started his research career (M.Phil. and PhD, 2001–2008) in plant stress physiology and molecular biology at AMU. Dr. Gill has made significant contributions towards abiotic stress tolerance in crop plants. Dr. Gill's research includes abiotic stress tolerance in crop plants, reactive oxygen species signaling and antioxidant machinery, gene expression, helicases, crop improvement, transgenics, nitrogen and sulfur metabolism, and plant fungal symbiotic interactions. Together with Dr. Narendra Tuteja at the International Centre for Genetic Engineering and Biotechnology (ICGEB), New Delhi, he worked on plant helicases for abiotic stress tolerance. He further explored the mechanism of stress tolerance by overexpressing PDH45 in tobacco and rice (*Plant Mol Biol* 82(1–2):1–22, 2013, and *PLoS One* 9(5):e98287, 2014). A novel function of plant MCM6 in salinity stress tolerance has also been reported that can will help to improve crop productivity at sub-optimal conditions (*Plant Mol Biol* 76(2011):19–34, 2014). Herbicide and salt stress tolerance (PDH45 + EPSPS) in plants has also been explored (*Front. Plant Sci.* 8:364, 2017). Dr. Gill helped to develop salinity-tolerant tobacco and rice plants, without affecting the overall yield. This research uncovers new pathways to plant abiotic stress tolerance and indicates the potential for improving crop production at sub-optimal conditions. A recipient of the INDIA Research Excellence & Citation Award 2017 from Clarivate Analytics (Web of Science), Sarvajeet Gill has edited several books with Springer, Wiley, Elsevier, CABI, and others, and has a number of research papers, review articles, and book chapters to his name.

Contributors

Rubén Alcázar
Department of Biology, Healthcare and Environment
University of Barcelona
Barcelona, Spain

Susana S. Araújo
Department of Biology and Biotechnology 'L. Spallanzani'
Universitá degli Studi di Pavia
Pavia, Italy
and
Plant Cell Biotechnology Laboratory
Instituto de Tecnologia Química e Biológica António Xavier
(ITQB-NOVA)
Oeiras, Portugal

Ariel D. Arencibia
Centre of Biotechnology in Natural Resources
Faculty of Agrarian and Forestry Sciences
Universidad Catolica del Maule
Talca, Chile

Bavita Asthir
Department of Biochemistry
Punjab Agricultural University
Ludhiana, India

Priyanka Babuta
Department of Botany
University of Delhi
Delhi, India

Navtej S. Bains
Department of Plant Breeding and Genetics
Punjab Agricultural University
Ludhiana, India

Alma Balestrazzi
Department of Biology and Biotechnology 'L. Spallanzani'
Universitá degli Studi di Pavia
Pavia, Italy

Hela Ben Ahmed
Laboratory of Plant Ecology
University of Tunis El Manar
Tunis, Tunisia

Renu Bhardwaj
Department of Environment and Botanical Sciences
Guru Nanak Dev University
Amritsar, India

Paramita Bhattacharjee
Department of Food Technology and Biochemical Engineering
Jadavpur University
Kolkata, India

Brijmohan Singh Bhau
Plant Sciences Division
Central University of Jammu
Jammu, India

Marco Biancucci
Department of Biology and Biotechnology
Sapienza University
Rome, Italy

Bitupon Borah
Academy of Scientific and Innovative Research (AcSIR)
CSIR–North East Institute of Science and Technology
Jorhat, India

Natasa Cerekovic
CNR-ISPA
National Research Council, Ecotekne
Lecce, Italy

Emilio Cervantes
IRNASA–CSIC
Salamanca, Spain

Sasanka Chakraborti
Department of Biochemistry and Central Research Cell
M M Institute of Medical Sciences & Research
M M (Deemed to be) University
Ambala, India

Soumi Chakraborty
Department of Food Technology and Biochemical Engineering
Jadavpur University
Kolkata, India

Abdellah Chalh
Laboratory of Plant Ecology
University of Tunis El Manar
Tunis, Tunisia

Belen Colavolpe
Unidad de Biotecnología 1
INTECH (CONICET-UNSAM)
Buenos Aires, Argentina

Mona G. Dawood
Botany Department
National Research Centre
Cairo, Egypt

Renu Deswal
Department of Botany
University of Delhi
Delhi, India

V.R. Devaraj
Department of Biochemistry
Central College Campus
Bangalore University
Bangalore, India

Nizar Dhaoui
Laboratory of Plant Ecology
University of Tunis El Manar
Tunis, Tunisia

Ashok Dhawan
Centre for Plant Sciences
Central University of Punjab
Bathinda, India

Allah Ditta
Department of Environmental Sciences
Shaheed Benazir Bhutto University Sheringal
Khyber Pakhtunkhwa, Pakistan

Sandeep Kumar Dixit
National Institute of Plant Genome Research
Aruna Asaf Ali Marg
New Delhi, India

Neha Dogra
Department of Botany
Punjabi University
Patiala, India

Myrene Dsouza
Department of Chemistry
Mount Carmel College, Autonomous
Vasanthnagar, India

Kathryn Dumschott
Faculty of Science
The University of Sydney
Sydney, Australia

Mohamed E. El-Awadi
Botany Department
National Research Centre
Cairo, Egypt

Fabiana Espasandin
Laboratorio de Biotecnología Aplicada y Genómica Funcional
Instituto de Botánica del Nordeste (IBONE-CONICET)
Universidad Nacional del Nordeste
Corrientes, Argentina

Parammal Faseela
Department of Botany
University of Calicut
Kerala, India

Nadia Fatnassi
GenXpro GmbH
Frankfurt Innovation Zentrum
Frankfurt am Main, Germany

Barbi Gogoi
Academy of Scientific and Innovative Research (AcSIR)
CSIR–North East Institute of Science and Technology
Jorhat, India

Carolina Gomes
Institute of Plant Genetics
Polish Academy of Sciences
Poznan, Poland

José Javier Martín Gómez
IRNASA-CSIC
Salamanca, Spain

Aarti Gupta
National Institute of Plant Genome Research
Aruna Asaf Ali Marg
New Delhi, India

Gurpreet
ICAR–Central Soil Salinity Research Institute
Karnal, India

Itzell Eurídice Hernández-Sánchez
Biología Molecular
Instituto Potosino de Investigación Científica y Tecnológica AC
San Luis Potosí, México

Ralph Horres
GenXpro GmbH
Frankfurt Innovation Zentrum,
Frankfurt am Main, Germany

Marine Hussain
Academy of Scientific and Innovative Research (AcSIR)
CSIR–North East Institute of Science and Technology
Jorhat, India

Satya N. Jena
Genetics and Plant Molecular Division
CSIR–National Botanical Research Institute
Lucknow, India

Shweta Jha
Department of Botany
J.N.V. University
Jodhpur, India

Juan Francisco Jiménez-Bremont
Biología Molecular
Instituto Potosino de Investigación Científica y Tecnológica AC
San Luis Potosí, Mexico

Babita Joshi
Academy of Scientific and Innovative Research (AcSIR)
CSIR–North East Institute of Science and Technology
Jorhat, India

Gurpreet Kaur
Department of Biochemistry
Punjab Agricultural University
Ludhiana, India

Shruti Kaushik
Department of Botany
Punjabi University
Patiala, India

Mahipal Singh Kesawat
Institute of Molecular Biology and Genetics
School of Biological Sciences
Seoul National University
Seoul, Republic of Korea

Tushar Khare
Department of Biotechnology
Modern College of Arts, Science and Commerce
S. P. Pune University
Ganeshkhind, India

Ashwani Kumar
ICAR–Central Soil Salinity Research Institute
Karnal, India

Bhumesh Kumar
ICAR–Directorate of Weed Research
Jabalpur, India

Manish Kumar
Department of Botany
Punjabi University
Patiala, India
and
Department of Environment and Botanical Sciences
Guru Nanak Dev University
Amritsar, India

Manu Kumar
Department of Bioindustry and Bioresource Engineering
Plant Engineering Research Institute
Sejong University
South Korea

Sandeep Kumar
Department of Botany
Punjabi University
Patiala, India
and
Department of Environment and Botanical Sciences
Guru Nanak Dev University
Amritsar, India

Pankaj Kumar
National Institute of Plant Genome Research
New Delhi, India

Sujit Kumar
ICAR–Directorate of Groundnut Research
Junagadh, India

Vinay Kumar
Department of Biotechnology
Modern College of Arts, Science and Commerce
S. P. Pune University
Ganeshkhind, India

Lucy Lalthafamkimi
Academy of Scientific and Innovative Research (AcSIR)
CSIR–North East Institute of Science and Technology
Jorhat, India

Anca Macovei
Department of Biology and Biotechnology 'L. Spallanzani'
Universitá degli Studi di Pavia
Pavia, Italy

Isha Madaan
Department of Botany
Punjabi University
Patiala, India

Santiago Maiale
Unidad de Biotecnología 1
INTECH (CONICET-UNSAM)
Buenos Aires, Argentina

Anita Mann
ICAR–Central Soil Salinity Research Institute
Karnal, India

Israel Maruri-Lopez
Biología Molecular
Instituto Potosino de Investigación Científica y Tecnológica AC
San Luis Potosí, Mexico

Roberto Mattioli
Department of Biology and Biotechnology
Sapienza University
Rome, Italy

Ibtissem Medyouni
Laboratory of Plant Ecology
Faculty of Sciences
University of Tunis El Manar
Tunis, Tunisia

Andrew Merchant
Faculty of Science
The University of Sydney
Sydney, Australia

Hajer Mimouni
Laboratory of Plant Ecology
Faculty of Sciences
University of Tunis El Manar
Tunis, Tunisia

Adra Mouellef
Department of Biology and Plant Ecology
University of Constantine 1
Constantine, Algeria

Soumya Mukherjee
Department of Botany
Jangipur College
University of Kalyani
West Bengal, India

Ruquia Mushtaq
Department of Botany
Punjabi University
Patiala, India

Kavya Naik
Department of Biochemistry
Central College Campus
Bangalore University
Bangalore, India

Maria Azucena Ortega-Amaro
Biología Molecular
Instituto Potosino de Investigación Científica y Tecnológica AC
San Luis Potosí, Mexico

Ankesh Pandey
Genetics and Plant Molecular Division
CSIR–National Botanical Research Institute
Lucknow, India

Parul Parihar
Ranjan Plant Physiology and Biochemistry Laboratory
Department of Botany
University of Allahabad
Allahabad, India

Kaninika Paul
Department of Food Technology and Biochemical Engineering
Jadavpur University
Kolkata, India

Jorge A.P. Paiva
Institute of Plant Genetics
Polish Academy of Sciences
Poznan, Poland

Palmiro Polronieri
CNR–ISPA
National Research Council, Ecotekne
Lecce, Italy

Pooja
ICAR–Regional Station
Sugarcane Breeding Institute
Karnal, India

Sheo Mohan Prasad
Ranjan Plant Physiology and Biochemistry Laboratory
Department of Botany
University of Allahabad
Allahabad, India

Jos Thomas Puthur
Department of Botany
University of Calicut
Kerala, India

Srinath Rao
Plant Tissue Culture and Genetic Engineering Laboratory
Gulbarga University, Kalburgi
Karnataka, India

Gokare A. Ravishankar
Dayananda Sagar Institutions
Bangalore, India

Alma Laura Rodriguez-Piña
Biología Molecular
Instituto Potosino de Investigación Científica y Tecnológica AC
San Luis Potosí, México

Rubal
Centre for Plant Sciences
Central University of Punjab
Bathinda, India

Oscar A. Ruiz
Unidad de Biotecnología 1
INTECH (CONICET-UNSAM)
Chascomús, Argentina
and
Instituto de Fisiología y Recursos Genéticos Vegetales
"Ing. Victorio S Trippi" (IFRGV-CIAP-INTA)
Córdoba, Argentina

Ezzeddine Saadaoui
National Institute of Research in Rural Engineering, Waters and Forests (INRGREF)
Regional Station of Gabès
University of Carthage
Tunis, Tunisia

H. Sandhya
Plant Tissue Culture and Genetic Engineering Laboratory
Gulbarga University Kalburgi
Karnataka, India

Pedro Sansberro
Laboratorio de Biotecnología Aplicada y Genómica Funcional
Instituto de Botánica del Nordeste (IBONE-CONICET)
Universidad Nacional del Nordeste
Corrientes, Argentina

Angelo Santino
CNR–ISPA
National Research Council, Ecotekne
Lecce, Italy

Harpreet Sekhon
Department of Botany
Punjabi University
Patiala, India
and
Department of Environment and Botanical Sciences
Guru Nanak Dev University
Amritsar, India

Muthappa Senthil-Kumar
National Institute of Plant Genome Research
Aruna Asaf Ali Marg
New Delhi, India

Samrin Shaikh
Department of Biotechnology
S. P. Pune University
Ganeshkhind, Pune, India

Sudhir Shukla
Genetics and Plant Molecular Division
CSIR–National Botanical Research Institute
Lucknow, India

Anil K. Singh
Indian Institute of Agricultural Biotechnology (Indian Council of Agricultural Research)
PDU Campus
Namkum, India

Anita Singh
Center of Advanced Study In Botany
Institute of Science
Banaras Hindu University
Varanasi, India

Rachana Singh
Ranjan Plant Physiology and Biochemistry Laboratory
Department of Botany
University of Allahabad
Allahabad, India

Sangeeta Singh
Amity Institute of Biotechnology
Amity University
Manesar, India

Asari Kandi Sinisha
Department of Botany
University of Calicut
Kerala, India

Geetika Sirhindi
Department of Botany
Punjabi University
Patiala, India

Millicent Smith
Faculty of Science
The University of Sydney
Sydney, Australia

Yaiphabi Sougrakpam
Department of Botany
University of Delhi
Delhi, India

Dinesh Kumar Srivastava
Dr Y.S. Parmar University of Horticulture and Forestry
Solan, India

Ajay Kumar Thakur
ICAR–Directorate of Rapeseed-Mustard Research
Bharatpur, India

A. Thilagavathy
Department of Biochemistry
Bangalore University
Bangalore, India

Thuruthummel Thomas Dhanya Thomas
Department of Botany
University of Calicut
Kerala, India

Antonio F. Tiburcio
Department of Biology, Healthcare and Environment
Faculty of Pharmacy
University of Barcelona
Barcelona, Spain

Santwana Tiwari
Ranjan Plant Physiology and Biochemistry Laboratory
Department of Botany
University of Allahabad
Allahabad, India

Maurizio Trovato
Department of Biology and Biotechnology
Sapienza University
Rome, Italy

Juan Manuel Vilas
Unidad de Biotecnología 1
INTECH (CONICET-UNSAM)
Buenos Aires, Argentina

Shabir H. Wani
Mountain Research Centre for Field Crops
Sher-e-Kashmir University of Agricultural Sciences and
Technology
Kashmir, India
and
Department of Plant, Soil and Microbial Sciences
Michigan State University
East Lansing, Michigan

Salma Wasti
Laboratory of Plant Ecology
University of Tunis El Manar
Tunis, Tunisia

Nadia Ykhlef
Department of Biology and Plant Ecology
University of Constantine 1
Constantine, Algeria

Section I

Abiotic Stress Management and Its Impact on Plants

1

Effects of Different Abiotic Stresses on Primary Metabolism

Belen Colavolpe*, Fabiana Espasandin*, Juan Manuel Vilas*, Santiago Maiale, Pedro Sansberro, and Oscar A. Ruiz

CONTENTS

Salinity Stress

General Considerations of Salt Stress as a Global Problem

Salt stress is one of the major abiotic stresses that seriously affect crop growth and yield. According to the Food and Agriculture Organization FAO (2009), there are at least 800 million hectares of land subjected to salinity in the world, accounting for as much as 6% of the world's total area. Although some of the salt-affected influences are the result of natural causes, the majority are derived from degraded cultivated agricultural land (Tang et al., 2014). It is estimated that 30% of the world's irrigated areas already suffer from salinity problems (UNESCO Water Portal, 2007). Each year, soil salinity increases worldwide, and it is predicted that it will cause a loss of up to 50% of the cultivatable land by 2050 (Henry et al., 2015). Expansion of agriculture to semi-arid and arid regions with the use of intensive irrigation will increase secondary salinization as a result of changes in the hydrologic balance of the soil between water applied (irrigation or rainfall) and water used by crops (transpiration). Moreover, the faster-than-predicted change in global climate (Intergovernmental Panel on Climate Change, 2007) and the different available scenarios for climate change suggest an increase

in aridity for the semi-arid regions of the globe in the near future. Together with overpopulation this will lead to an overexploitation of water resources for agriculture purposes and increased constraints on plant growth and survival; this will affect the ability to realize crop yield potential (Chaves et al., 2009).

The United States Department of Agriculture (USDA) Salinity Laboratory defines a saline soil as having an electrical conductivity of the saturation extract (ECc) of 4 dS m^{-1} or more. ECc is the electrical conductivity of the saturated paste extract, that is, of the solution extracted from a soil sample after being mixed with sufficient water to produce a saturated paste. The most widely accepted definition of a saline soil has been adopted from FAO (1996) as one that has an ECc of 4 dS m^{-1} or more and soils with ECcs exceeding 15 dS m^{-1} are considered strongly saline. Traditionally, four levels of soil salinity based on saline irrigation water have been distinguished: low salinity is defined by electrical conductivity of less than 0.25 dS m^{-1}; medium salinity, 0.25–0.75 dS m^{-1}; high salinity, 0.75–2.25 dS m^{-1}; and very high salinity with an electrical conductivity exceeding 2.25 dS m^{-1} (US Salinity Laboratory Staff, 1954). The common cations associated with salinity are Na$^+$, Ca^{2+} and Mg^{2+}, while the common anions are Cl$^-$, SO$_4^{2-}$ and HCO$_3^-$ (FAO: Land and Water Division, 2013).

Saline stress in plants refers to the presence of neutral salts such as NaCl or Na$_2$SO$_4$ in soil, whereas alkaline stress is only

---

* These authors contributed equally to this work.

related to the occurrence of alkaline salts: Na_2CO_3 or $NaHCO_3$. These conditions often co-occur in nature, with variable neutral to alkaline salt proportions according to the soil (Paz et al., 2014). Since Na^+ in particular causes deterioration of the physical structure of soil and along with Cl^- is toxic to plants, these are considered the most important ions. Historically soils were classified as saline, sodic or saline-sodic based on the total concentration of salt and the ratio of Na^+ to Ca^{2+} and Mg^{2+} in the saturated extract of the soil (Yadav et al., 2011).

According to the incapacity to grow on high salt medium, plants have also been classified as glycophytes or halophytes. Most plants are glycophytes and cannot tolerate salt stress and in a large proportion of crops cannot tolerate Na^+ concentrations at 50 mM or higher (Tang et al., 2014). The deleterious effects of salinity on plant growth are associated with: (1) low osmotic potential of soil solution (water stress), (2) nutritional imbalance, (3) specific ion effect (salt stress) or (4) a combination of these factors (Parvaiz and Satyawati, 2008). Early plant responses to water and salt stress have been considered mostly identical; drought and salinity share a physiological water deficit that affects, more or less intensely, all plant organs (Munns, 2002). However, under prolonged salt stress plants respond in addition to dehydration to hyper-ionic and hyper-osmotic stress (Fricke et al., 2006). Na^+ and Cl^- are taken up and, increasingly, displace mineral nutrients such as K^+, Ca^{2+} and nitrate (Campestre et al., 2016). Although accumulation of Na^+ and Cl^- causes osmotic and water potential adjustment of cells, it increases the risk of long-term ion toxicity. If Na^+ and Cl^- are not compartmentalized appropriately, exported or secreted, leaf tissue water deficit per se can be triggered not only by low soil water content but also by high vapor pressure deficit of the atmosphere (Chaves et al., 2016; Paz et al., 2014). Photosynthesis, together with cell growth, is among the primary processes to be affected by drought (Chaves, 1991) or by salinity (Munns et al., 2006). The effects can be direct, as the decreased CO_2 availability caused by diffusion limitations through the stomata and the mesophyll, and alterations of photosynthetic metabolism, or they can arise as secondary effects, namely oxidative stress. The latter are mostly present under multiple stress conditions and can seriously affect leaf photosynthetic machinery (Chaves et al., 2009).

In recent years, much attention has been devoted to the involvement of polyamines (PAs) as second messengers in the context of a variety of environmental stresses (Marina et al., 2008). These low-molecular-weight, aliphatic nitrogenous compounds, which are protonated at physiological pH, were originally thought to bind to anionic macromolecules, including proteins and nucleic acids, and thus to perform a structural role (Gárriz et al., 2003). The three commonest plant PAs, namely putrescine (Put), spermidine (Spd) and spermine (Spm), are now known; they protect salinity-stressed plants by aiding the accumulation of sugars, proline (Pro) and other osmolytes, and adjusting ion channels to maintain the plant's internal K^+ and Na^+ balance. They also serve to increase the activity of a range of antioxidant enzymes, thereby improving the plant's ability to control oxidative stress. The involvement of PAs in the response of higher plants to salinity stress has been widely reported (Zheng et al., 2016) and deserves to be mentioned, however, we focus in this chapter on plant primary metabolism and how different abiotic stresses affect it.

In recent years, tremendous advances have been achieved in salt stress studies. Plant breeders have fostered some salt-tolerant lines of crops by conventional breeding; moreover, a transgenic approach is employed to improve crop salt tolerance and a number of transgenic lines have been found to be effective under control conditions. Metabolomics is becoming a tool to understand the cellular mechanism of abiotic stress and acts as a viable option for the biotechnological improvement of halophytes (Tang et al., 2014). Nevertheless, the salt tolerance trait is a multigenic property and is related to physiological, biochemical and molecular processes, so genetic engineering to produce salt-tolerant crops is limited seriously in nature (Zhu, 2003). More energy, material and financial resources are needed to invest in research. Only in this way can the mechanisms and principles be discovered and corresponding solutions be proposed.

Growth and Crop Production under Salt Stress Conditions

The effect of salt on plant establishment has been assessed in maize, sorghum, rice, wheat and soybean. Osmotic stress resulting from drought or salinized soils can be disastrous for crop development (Westgate and Boyer, 1985). Both stresses have a common but not exclusive osmotic component. Salinity generates an immediate osmotic stress followed by later ion toxicity after continued exposure (Carillo et al., 2011). During stress conditions, plants need to maintain internal water potential below that of soil and maintain turgor and water uptake for growing. This requires an increase in osmosis, by uptake of soil solutes, synthesis of metabolic solutes or accommodation of the ionic balance in the vacuoles (Parvaiz and Satyawati, 2008). Growing or surviving in a saline soil imposes some costs: the cost of excluding salt by intracellular compartmentalization and of excreting it through salt glands. This cost, however, is relatively small in relation to that needed to synthesize organic solutes for osmotic adjustment (Munns, 2002).

As was mentioned before, plants have been classified as glycophytes or halophytes. Most plants are glycophytes and cannot tolerate salt stress (Parvaiz and Satyawati, 2008). Nevertheless, important differences exist in salt tolerance between species. In most plants, higher levels of the activity of antioxidant enzymes are considered as salt tolerance mechanisms (Ashraf, 2009; Zhang et al., 2014). Indeed, previous studies have shown that within the same species, salt-tolerant cultivars generally have higher constitutive or enhanced antioxidant enzyme activity under salt stress when compared with salt-sensitive cultivars. This has been demonstrated in numerous plant species such as cotton (*Gossypium herbaceum*), rice and pea. Moreover, the response of plant antioxidant enzymes to salinity has been shown to vary among plant species, tissues and subcellular localizations (Mittova et al., 2003). Dramatic variances are found between plant species production in saline versus control conditions over a prolonged period of time. For example, after some time in 200 mM NaCl, a salt-tolerant species such as sugar beet (*Beta vulgaris*) might have a reduction of only 20% in dry weight, a moderately tolerant species such as cotton might have a 60% reduction, and a sensitive species such as soybean might be dead (Greenway and Munns, 1980). In very salt-sensitive species, salt-specific effects can become visible after several days at high salinities. If the salinity is high, and if the plant has a poor ability to exclude NaCl, marked injury in older leaves might occur within days, as found for white lupin (*Lupinus albus*) once the salinity increased above 100 mM NaCl (Munns, 1988). *Medicago sativa* plants showed after 4 and 6 weeks of salt treatment a biomass reduction of about 40% and 50% with 150 mM NaCl, respectively (López-Gómez et al., 2014). At first the growth reduction is quickly apparent, and is due to the salt outside the roots. It is essentially a water stress or osmotic phase, for which there is surprisingly little genotypic variation (Munns, 2002). The growth reduction is presumably regulated by hormonal signals coming from the roots. Then, there is a second phase of growth reduction, which takes time to develop, and results from internal injury. This internal injury is due to Na^+ or Cl^- (or both) accumulating in transpiring leaves to excessive levels, exceeding the ability of the cells to compartmentalize these ions in the vacuole. Ions then build up rapidly in the cytoplasm and inhibit enzyme activity or they build up in the cell walls and dehydrate the cell (Flowers and Yeo, 1986). This process inhibits the growth of the younger leaves by reducing the supply of carbohydrates to the growing cells (Munns, 2002).

Evidence indicates that plants have two phases of growing response that have been shown clearly, for example, for maize and wheat cultivars. Two maize cultivars with two-fold differences in rates of Na^+ accumulation in leaves had the same growth reduction for 15 days in 80 mM NaCl. Furthermore, another two maize cultivars, again with two-fold differences in Na^+ accumulation, had the same growth reduction for 4 weeks in 100 mM NaCl, and it was not until 8 weeks of salinity that a growth difference was clearly observed (Munns, 2002).

Differences are evident between the phenotype given by salt-stressed maize plants versus control irrigated only with nutritive solution (Henry et al., 2015). Similar results were found in wheat (Munns et al., 1995). Maize plants are most susceptible during a period of 2 weeks around the time of silking, and kernel abortion is the limiting factor for yield (Boyer, 2010). Recent studies identify events as early as 1 day after pollination to be critical for determining whether the embryo will abort (Chaves et al., 2003). With rice, also, a clear distinction has been made between the initial effects of salinity, which are recoverable, and the long-term effects that result from the accumulation of salt in the expanded leaves (Yeo et al., 1991).

In general, in the first few seconds or minutes of salt stress exposure, plant cells lose water and shrink. Then, cells regain their original volume but cell elongation rates are reduced, leading to lower rates of leaf and root growth. Over a number of days, changes in cell elongation and cell division lead to slower leaf appearance and smaller final size, and leaf growth is usually more affected than root growth (Munns, 2002). Root ionic status does not increase with time, as in leaves, and they often have a lower Na^+ and Cl^- concentration than the external solution, which rarely happens in leaves. For example, in wheat growing in 150 mM NaCl, Na^+ in the roots was only 20–40 mM (Gorham et al., 1990). In plants with high salt uptake rates, the oldest leaf may start to show symptoms of injury. After months, differences between plants with high and low salt uptake rates become very apparent, with a large amount of leaf injury and complete death in some cases if the salinity level is high enough (Munns, 2002).

In maize, kernel abortion induced by osmotic stress correlates with reduced evapotranspiration and photosynthesis (Setter and Flannigan, 2001). Along with impaired photosynthesis in source leaves comes a reduction of seed sink strength. Abortion caused by osmotic stress correlates with depleted sucrose and reduced sugar levels, reduced sucrose degrading enzyme activity and transcript levels, and depletion of starch in the kernels (Henry et al., 2015). These events occur in a short period of time around pollination and can be partially prevented by stem sucrose feeding (Boyer, 2010). As a result of impaired photosynthesis and sink strength, sugar allocation to the reproductive organs is disrupted and the young embryo rapidly starves and aborts. Sugars not only serve as a source of carbon units and metabolic energy but also function as signaling molecules reporting carbon status within the cell (León and Sheen, 2003; Henry et al., 2015).

Seedling stage is the more vulnerable phase of durum wheat (*Triticum durum*) growth under salinity (Carillo et al., 2008). This species is more sensitive to salinity than bread wheat (*Triticum aestivum*) and poor yields on saline soil are partly due to the poor ability of durum wheat to exclude sodium (Annunziata et al., 2017). In particular, salinity greatly increases the levels of proline and glycine betaine (GB) in durum wheat (Munns, 2002; Carillo et al., 2008), as in other Poaceae. In many halophytes, leaf concentration of Pro, GB or both contributes to the osmotic pressure in the cell as a whole. In glycophytes, proline and GB have lower concentrations but, being partitioned exclusively to the cytoplasm, which makes up about 10% of the volume of the cell, they are able to determine significant osmotic pressure and balance the vacuolar osmotic potential (Annunziata et al., 2017).

The effect of salinity on the germination, vegetative growth and yield of cotton has also been reported (Ahmads et al., 2002; Guo et al., 2012). Cotton is classified as a salt-tolerant crop, but this tolerance is actually limited and varies according to the growth and developmental stages of the plant. Breeders have sought to make cotton more tolerant to salt through various methods, including traditional plant breeding and biotechnological approaches such as creating transgenic cotton (Chen et al., 2016). The growth rates of cotton plant roots and leaves decrease with increasing salt concentration, which may be a result of osmotic injury or specific ion toxicity (Meloni et al., 2001). Salt stress significantly reduces the growth rates in surface area, volume, average diameter of the cotton roots, and dry weights of roots and leaves (Zhang et al., 2014). This is accompanied by strong changes in carbohydrate metabolism owing to severe impairments in the photosynthetic and respiration apparatus (Chen et al., 2016).

On the other hand, legumes are classified as salt-sensitive crop species and their productivity is particularly affected because nodular nitrogenase activity markedly decreases upon exposure to mild saline conditions (Läuchli, 1984; López-Gómez et al., 2014). Salinity stress reduces seed germination, seedling growth, nodulation, biomass accumulation and seed yield (Essa, 2002). López-Gómez et al. (2014) have studied *M. sativa* and determined that nitrogenase activity is strongly inhibited by 150 mM NaCl 2 weeks after treatment initiation. The glycophyte *Lotus tenuis* (Waldst and Kit, syn. *L. glaber*; Kirkbride, 2006) is the best-adapted legume forage in the lowlands of the Buenos Aires Province (the most important cattle production region in Argentina) and is also affected by salt stress. The negative effect of NaCl on root length of *L. tenuis* is in agreement with previous results (Echeverria et al., 2008) along with important levels of growth inhibition on its shoots that significantly decrease persistence and yield (Mazzanti et al., 1986; Paz et al., 2014). Singleton and Bohlool (1983) showed that nodule function was relatively more resistant to salt stress than plant growth. High soybean yields require large amounts of nitrogen (N); the least expensive source of N for soybean is biological fixation of atmospheric N_2 by the symbiotic association between plant and soil bacteria belonging mainly to the genera *Bradyrhizobium* and *Sinorhizobium*, which are collectively called soybean rhizobia. Rhizobia infect the roots of legumes and induce formation of nodules, where nitrogen fixation takes place (Baghel et al., 2016). Similarly to the nitrogenase activity, nodule dry weight was also reduced by salt treatments particularly in the case of *M. sativa* in which 35% and 50% of reduction was obtained by 100 and 150 mM NaCl after 2 weeks (López-Gómez et al., 2014).

Owing to the high oil and protein content in its seeds, soybean is an important economic dicot crop and the demand for it is increasing continuously. However, as a salt-sensitive species, the growth and development are severely affected by salt stress. Exploiting resistant varieties and improving salt tolerance of soybean, therefore, became the goal of many researchers (Chen et al., 2011). In mature seedlings of cultivated soybean (*Glycine max*), the water potential was reduced 11.2 times and the relative water content dropped to 64.2% under 300 mM treatment. In addition, all the plants under this stress displayed symptoms of water loss and 30% died after the stress. For young seedlings the accumulation of Pro increased rapidly when the concentration of the NaCl reached 100 mM or higher. Under 300 mM, the contents of the Pro reached 34.4 times. Previous studies on the synthesis and degradation of Pro have proved a close relationship between Pro and salt resistance. GB was another important metabolite playing an important role in salt resistance and its behavior was similar to Pro (Wu et al., 2014). The Pro content in the cotyledons of the 4-day-old germinating soybeans increased in the plants subjected to NaCl stress as well as the gamma-aminobutyric acid (GABA) content (Yin et al., 2015).

Plant Chemical and Physical Response to Salt Stress

An efficient response to the environment is particularly important for plants, as sessile organisms. This means an ability of cells to quickly sense the surrounding environmental signals. Systemic signals generated by the tissue exposed to abiotic and biotic stress act in the co-ordination and execution of plant stress responses in terms of metabolic and developmental adjustments (Chaves et al., 2003). Under salt and drought, these responses are triggered by primary osmotic stress signals, which have an impact on both the source of carbohydrate (photosynthesis) and the mobilization/utilization of carbohydrate reserves (sink strength), or by secondary signal metabolites that generally increase or decrease in a transient mode. The latter include hormones (e.g. abscisic acid (ABA), ethylene, cytokinins), reactive oxygen species (ROS) and intracellular second messengers, e.g. phospholipids, sugars, etc. (Chaves et al., 2003). Salt stress signaling consists of ionic and osmotic detoxification and signaling to coordinate cell division and enzyme expansion. Several salt-responsive signaling pathways, such as salt overly sensitive (SOS), ABA, Ca^{2+} signal transduction, protein kinase, phospholipid, ethylene and

jasmonate acid (JA) induced signaling pathways, have been predicted (Zhang et al., 2012).

Stomata close in response to leaf turgor decline, high vapor pressure deficit in the atmosphere (Chaves et al., 2009) and early physiological events such as root-generated chemical signals. A close relationship is usually found between stomata conductance (Gs) and net CO_2 assimilation (A_N) (Flexas et al., 2004); photosynthesis rate is affected by salt stress directly due to CO_2 deficiency and Rubisco (RuBP) activity, and indirectly by reduced chlorophyll and total carotenoid content (Tang et al., 2014).

Since roots are the site of perception of salt in the environment, their responses and adaptive behavior form the first line of defense against stress damage (Ji et al., 2013). High concentrations of NaCl outside the roots reduce the water potential and make it more difficult for the root to extract water (Zhang et al., 2014). Sodium exclusion by root cells is the primary protecting response in plants that delays the toxic effects of high cytoplasmic Na^+. The comparison of unidirectional Na^+ uptake fluxes and the rates of net accumulation of Na^+ in roots indicate that the vast majority of the Na^+ taken up into the root symplast is extruded back to the apoplast and soil solution (Ji et al., 2013). Proper regulation of ion flux is necessary for cells to keep the concentrations of toxic ions low and to accumulate essential ions. Plant cells employ primary active transport, mediated by H^+-ATPases (Yokoi et al., 2002) that create a proton-motive force that drives the transport of all other ions and metabolites; and a secondary transport, mediated by channels and co-transporters, to maintain characteristically high concentrations of K^+ and low concentrations of Na^+ in the cytosol (Parvaiz and Satyawati, 2008).

The cytoplasm also accumulates low-molecular-mass compounds commonly called compatible solutes that do not interfere with normal biochemical reactions; rather, they replace water in biochemical reactions. While some compatible osmolytes are essential elemental ions (such as K^+) (Xiong et al., 2002), the majority are organic solutes (Zhu, 2003). However, the solutes that accumulate vary with the organism and even between plant species. A major category of organic osmotic solutes consists of simple sugars (mainly fructose and glucose), sugar alcohols (glycerol and methylated inositols) and complex sugars (trehalose, raffinose and fructans). Carbohydrates (glucose, fructose, sucrose, fructans) and starch accumulate under salt stress, playing a leading role in osmoprotection, osmotic adjustment, carbon storage and radical scavenging (Parvaiz and Satyawati, 2008). Others include quaternary amino acid derivatives (proline, glycine betaine, β-alanine betaine, proline betaine, tertiary amines 1,4,5,6-tetrahydro-2-methyl-4-carboxyl pyrimidine) and sulfonium compounds (choline O-sulfate, dimethyl sulfonium propionate) (Bohnert and Jensen, 1996; Yokoi et al., 2002).

Under mild stress, a small decline in stomata conductance may have protective effects against stress, by allowing plants to save water and improving plant water-use efficiency. The result of closing stomata is the reduction of CO_2 diffusion. There is an increasing body of evidence that shows g_m (internal leaf conductance to CO_2 diffusion) decreases in response to drought and salinity (Flexas et al., 2004). These changes in mesophyll conductance may be linked to physical alterations in the structure of the intercellular spaces due to leaf shrinkage or to alterations in the biochemistry (bicarbonate to CO_2 conversion) and/or membrane permeability (aquaporins) (Chaves et al., 2009).

The cell cortex is a specialized layer of the cytoplasm underlying the plasma-membrane that is composed of a network of microtubules and actin filaments. Cortical microtubules are highly dynamic and remodeled by numerous stimuli. Salt stress induces dynamic cytoskeletal changes, with initial depolymerization of microtubules at the onset of stress followed by repolymerization. Both depolymerization and reorganization of the cortical microtubules are important for the plant's ability to withstand salt stress (Ji et al., 2013).

Changes in leaf biochemistry that result in down-regulation of the photosynthetic metabolism may occur in response to lowered carbon substrate under prolonged stresses (Flexas et al., 2006). A deactivation of the carboxylating enzyme RuBP by low intercellular CO_2 has been observed along with other important photosynthetic proteins that are down-regulated by different mechanisms (carbonylation, phosphorylation/dephosphorylation and redox changes in thiol groups). These salt-induced alterations may induce great disturbances in photochemical activity, decreasing the growth and increasing salt plant sensitivity (Henry et al., 2015; Silveira and Carvalho, 2016). There is a large amount of data on initial RuBP activity and only one study in which nitrate reductase activity was followed concomitantly with Gs during a drought cycle. Even so, it seems clear that both enzymes share a common pattern on regulation with decreasing Gs. The fact that initial RuBP activity remains unaffected from maximum Gs down to 0.1 mol H_2O m^{-2} s^{-2} implies that within this range, photosynthesis is not impaired by the carboxylation capacity (Flexas et al., 2004). Early biochemical effects that involve alterations in photophosphorylation (a decrease in the amount of ATP leading to a decreased regeneration of Rubisco) have also been described (Tezara et al., 1999) and seem to be dependent on species showing different thresholds for metabolic down-regulation (Lawlor and Cornic, 2002).

Under salt stress, metabolic limitations of photosynthesis resulting from increased concentrations of Na^+ and Cl^{-2} in the leaf tissue (in general above 250 mM) do occur (Munns et al., 2006). As previously pointed out, the fast changes in gene expression following stress imposition that have been observed suggest that alterations in metabolism start very early. Although its role is not totally clear yet, photorespiration may also be involved in protecting the photosynthetic apparatus against light damage as suggested by its increase under drought observed in several species. Photorespiratory-produced hydroxide peroxide (H_2O_2) may also be responsible for signaling and acclimation under restricted CO_2 availability (Noctor et al., 2002). Several lines of evidence suggest that stomatal closure in moderately salt-stressed leaves leads to enhanced rates of photorespiration. The following parameters, all indicative of higher rates of photorespiration, have been shown to increase: the CO_2 compensation point, the light to dark ratio of CO_2 production, the stimulation of photosynthesis by lowering the O_2 concentration, the activity of glycolate oxidase and the formation of photorespiratory metabolites, such as glycine, serine and glycolate. The maintenance of considerable rates of electron transport in CO_2-free air also indicates a significant occurrence of photorespiration in salt-stressed leaves. In addition to sustained rates of electron transport due to photorespiration, the formation of zeaxanthin (one of the most common carotenoid alcohols found in nature) also mitigates against photo inhibitory damage, although this protection by zeaxanthin is not complete in high light (Wingler et al., 2000).

Salt-tolerant plants can not only regulate ion and water movements more efficiently but should also have a better antioxidant system for effective removal of ROS (Noctor et al., 2002). Molecular oxygen in its ground state, triplet oxygen, is essential to life on earth. It is a relatively stable molecule that does not directly cause damage to living cells. However, when triplet oxygen receives extra energy or electrons, it generates a variety of ROS that will cause oxidative damage to various components of living cells including lipids, proteins and nucleic acids (Abogadallah, 2010). Salt stress causes excessive generation of ROS: singlet oxygen ($^1O2$), hydrogen peroxide (H_2O_2), superoxide anions (O^{-2}) and hydroxyl radicals (OH•). To mitigate the oxidative damage initiated by ROS formed under salt stress, plants possess a complex antioxidant system, including non-enzymatic antioxidants such as ascorbic acid, glutathione, tocopherols and carotenoids; antioxidant enzymes such as superoxide dismutase, catalase, glutathione peroxidase and enzymes of the so-called ascorbate-glutathione cycle, including ascorbate peroxidase and glutathione reductase (Zhang et al., 2014). Plants produce ROS under normal conditions essentially from photosynthesis, photorespiration and respiration. The most common ROS generated under normal conditions are O^{-2} and H_2O_2 perhaps as a result of electron leakage from the photosynthetic and respiratory electron transport chains to oxygen. Another source of ROS (H_2O_2) is photorespiration resulting from the oxygenase activity of RuBP. Rates of photorespiration are basically controlled by the ratio of [CO_2] to [O_2] and temperature. In C_3 plants, photorespiration constitutes about 20%–30% of photosynthesis under the current atmospheric conditions at 25°C. In contrast, C_4 plants show lower rates of photorespiration (3.5%–6% of photosynthesis) under various environmental conditions due to their CO_2 concentrating mechanism. The role of ROS detoxification under salt stress may have resulted at least in part from (1) the technical inability to determine the major sources(s) of ROS under salt stress and therefore the appropriate antioxidant enzyme required and (2) the expectation that ROS scavenging enzymes, particularly catalases and peroxidases, perform similar functions (Abogadallah, 2010).

Salinity can affect growth in a number of ways. The first phase of the growth response is due to the osmotic effect of the salt in the soil solution and produces a suite of effects identical to those of water stress caused by drought. Later, there may be an additional effect on growth; if excessive amounts of salt enter the plant they will eventually rise to toxic levels in the older transpiring leaves, causing premature senescence. This will reduce the amount of assimilate that the plant can produce, and a reduction in assimilate transported to the growing tissues may further limit growth. This is the second phase of the growth response and is the phase that clearly separates species and genotypes that differ in the ability to tolerate saline soil (Munns, 2002). In summary, in order to guarantee survival under such detrimental circumstances, plants have evolved a series of biochemical and molecular processes to acclimatize themselves to the environment (Yan et al., 2013). The specific biochemical strategy contains: (1) ion regulation and compartmentalization, (2) induced biosynthesis of compatible solutes, (3) induction of antioxidant enzymes, (4) induction of plant hormones and (5) changes in photosynthetic pathway. The molecular mechanism includes: (1) the salt overly sensitive (SOS) pathway for ion homeostasis, (2) the protein kinase pathway for stress signaling, (3) the phytohormone signaling pathway under high salt stress and (4) the associated genes encoding salt-stress proteins, such as genes for photosynthetic enzymes, synthesis of compatible solutes, vacuolar-sequestering enzymes and for radical-scavenging enzymes (Tang et al., 2014).

The accumulated knowledge on physiological, cellular and molecular responses of plants to drought and salinity, including the signaling events occurring under both stresses, is already permitting great progress in crop management and breeding (Chaves et al., 2009). Some improvement in plant stress tolerance has been achieved by introducing stress-inducible genes into some model plants. To further understand the complexity of plant response to drought and salt, including the effects on photosynthesis, is important to strengthen multilevel genomics and physiological studies, covering different intensity and timing of imposition of the stresses in genotypes with different sensitivity to stress.

Cold Stress

Effect of Cold Stress on Primary Plant Metabolism

Climatic changes affect the normal performance of the plant. Low temperatures cause damage in different physiological stages (seedling, reproductive stage and grain filling), resulting in a reduction in crop yield. Low temperatures are a major environmental factor limiting the productivity and the geographic areas where agriculture can be developed.

Changes in environmental conditions make plants reprogramme metabolic fluxes because cold temperatures produce high metabolic requirements to maintain normal physiological processes. Chilling treatments of sensitive tissues increases the activity of invertase, catalase, pyruvate decarboxylase, glucose-6-phosphate dehydrogenase and phosphoenolpyruvate carboxykinase. Likewise, decreases have been observed in a number of systems including malate dehydrogenase and amylase (Lyon, 1973).

Plants that have evolved with mechanisms of tolerance and adaptation to cold have C_3 type photosynthesis and rapidly mobilize the reserves stored. Moreover, changes in starch metabolism and raffinose family oligosaccharide synthesis are all participants in the global response to cold stress (Janska et al., 2010).

Accumulations of soluble sugars from the hydrolysis of starch (Chen et al., 2014) and their roles in stabilizing biological components, particularly for raffinose family oligosaccharides (RFO) (Tarkowski and Van den Ende, 2015), are much-studied processes in plant physiology to understand the effects of cold and cold acclimation.

It has been reported that the accumulation of galactinol assimilates such as glucose, fructose and sucrose, as well as changes in the biosynthesis of protein lipids, are adopted strategies for cold stress tolerance in plants; these act as cryoprotective molecules (Rodziewicz et al., 2014).

In C_4 plants such as maize (*Zea mays*), the leaf structure minimizes photorespiration, which is a major source of ROS in C_3 plants (Foyer, 2002.)

Changes in environmental conditions such as cold temperatures associated with light energy result in an imbalance between the light energy absorbed through photochemistry versus the energy utilized through metabolism, which reflects the relative reduction state of the photosystem, generating ROS. This acts as a signal to influence the nuclear expression of a specific cold-acclimation gene (Huner et al., 1998).

Photosynthesis and Carbon Metabolism

Several authors have shown that the rate of photosynthesis decreases in higher plants during exposure under low temperatures. This reduction is due to stomatal closure and the consequent disability for CO_2 fixation. On the other hand, cold temperatures bring about photoinhibition of photosystem II (PSII) caused by the enhanced production of reactive oxygen species damaging the photosynthetic machine.

The key enzyme for CO_2 incorporation is ribulose-1,5-bisphosphate carboxylase/oxygenase (RuBP). It has been postulated that in seedlings subjected to cold, large RuBP subunits and RuBP-binding protein were downregulated, which may suggest chloroplast damage (Rodziewicz et al., 2014). Furthermore, it has been shown in rice (*Oryza sativa*) seedlings that synthesis of RuBP is drastically reduced after cold exposure (Hahn and Walbot, 1989). In wheat (*Triticum aestivum*) exposed to prolonged cold, stress fragmentation of RuBP was also evidenced in proteomics profile; it was suggested that RuBP proteolysis causes sugar accumulation during cold exposure (Rinalducci et al., 2011).

Regardless of the phenological state of the plant (vegetative or reproductive), cold-sensitive wheat is affected as photosystem proteins and protein related to electron transport are down-regulated, causing alterations in the flow of electrons in the chloroplast (Rinalducci et al., 2011; Xu et al., 2013). These proteins were enhanced in cold-tolerant wheat (Xu et al., 2013).

However, in rice seedlings, the proteins related to the generation of energy and sugar biosynthesis are activated under conditions of cold temperatures (Makoto Hashimoto and Komatsu, 2007).

The transcriptomics profile of rice seedling reveals that the genes involved in photosynthetic processes PSI and PSII were down-regulated (Kyonoshin Maruyama et al., 2014). RNA sequencing studies in the legume *Lotus japonicus* subjected to growth at low temperatures also show down-regulation in genes corresponding to energy metabolism that affect the photosynthetic process and chloroplast development (Calzadilla et al., 2016). In soybean (*Glycine max*), comparative proteomic analysis of seedling leaves of cold-tolerant and -sensitive cultivars indicated that cold stress also affects photosystem proteins, indicating the tolerance is caused by less energy depletion in the tolerant cultivar (Tian et al., 2015). The same result was found in maize seedling where a proteomics profile based in i-TRAQ revealed which proteins are involved in photosynthesis; light harvesting and light reaction are down-regulated (Wang et al., 2016).

On the other hand, the opposite results were obtained in cold-tolerant soybean seed germinates exposed to 4°C, where the proteomic profile revealed an increase in most proteins linked to energy metabolism (Cheng et al., 2010).

The Role of Carbohydrates in Response to Cold Stress

Plants produce biochemical changes in response to low temperatures as a mechanism of adaptation and defense. A transition occurs when the metabolism of starch is directed towards a metabolism aimed at accumulating oligosaccharides as cryo-protectants giving an osmotic adjustment response (Janska et al., 2010; Beck et al., 2007; Xiong and Zhu, 2002; Guy et al., 1992).

It could be said that this increase in the amount of sucrose, like a storage carbohydrate, is because it can be easily remobilized depending on the metabolic needs under stress conditions (Guy et al., 1992).

In wheat plants, seedling and reproductive stage exposed to chilling showed an increase in soluble carbohydrates (Savitch et al., 2000; Vargas et al., 2007; Rinalducci et al., 2011; Xu et al., 2013), especially sucrose, and an increase in the capacity for sucrose utilization through the biosynthesis of fructans, accompanied by an induction of the activity of the enzyme invertase, exhibiting a regulation mediated by free sugars in response to the cold (Savitch et al., 2000; Vargas et al., 2007). The same regulation was observed in legume nodules of *Lotus japonicus* in non-stressing conditions, indicating a possible role of invertases in producing hexoses for starch production when the metabolic requirements are high (Flemetakis et al., 2006).

RNA sequencing in seeds of tolerant indica rice detected high expression in enzymes involved in the synthesis of sucrose and breakdown of polysaccharides to generate simple sugars, such as glucose, compared to cold-sensitive indica rice (Dametto et al., 2015).

Comparisons between two soybean cultivars (tolerant and sensitive) in seedling stage reveal that cold stress affects the total amount of soluble carbohydrates, decreasing in both cultivars. At the same time, the protein profiles show a reduction in pentoses phosphate and glycolysis pathways for both cultivars (Tian et al., 2015). Other reports in seed soybean showed that proteins associated with carbohydrate enhanced starch metabolism as well as increasing the granule-bound starch gene (Cheng et al., 2010). In *Lotus japonicus* those genes of starch and sucrose metabolism also showed up-regulation (Calzadilla et al., 2016). In maize, most of the proteins associated with carbohydrate metabolism were accumulated after the chilling stress (Wang et al., 2016).

Metabolome analyses in rice seedling treated under low temperature revealed that levels of monosaccharides increased, while transcriptomics analyses indicated that several genes encoding enzymes involved in starch degradation, sucrose metabolism and the glyoxylate cycle are upregulated and that these changes are correlated with the accumulation of glucose, fructose and sucrose. In particular, high expression levels of genes encoding isocitrate lyase and malate synthase in the glyoxylate cycle correlate with increased glucose levels (Maruyama et al., 2014). In the reproductive stage of rice (booting stage) the starch and sucrose contents in the leaf blades and stems after panicle initiation did not undergo significant changes and carbon availability from source tissues is unlikely to affect chilling sensitivity in spikelet at the booting stage (Suzuki et al., 2015). Different results were reported in three maize inbred lines where a decrease in sucrose and glucose levels was found when analyzed by metabolism profile (Sun et al., 2016).

In another important forage legume, when two cultivars (*Medicago sativa* and *Medicago falcata*) were compared, both showed sugar accumulation (sucrose, galactinol and raffinose), although *M. sativa* displaying show accumulation of sugars. In *M. falcata* all sugars accumulated rapidly and stayed at high levels except for raffinose (Zhuo et al., 2013).

The raffinose family oligosaccharides (RFOs) are synthesized from sucrose by the subsequent addition of activated galactinol moieties donated by galactinol (ElSayed et al., 2014). Raffinose was accumulated in leaves accompanied by an increase in galactinol (Saito et al., 2011). It has also been observed that levels of transcripts of raffinose synthase decreased under cold stress but are increased under dehydration conditions (Maruyama et al., 2014).

Microarray analysis of rice plant at booting stage revealed that cold treatments during vegetative growth provoked up-regulation of a raffinose synthase gene, showing the same pattern as the heat shock protein genes (Suzuki et al., 2015). It has been proposed that higher levels of raffinose synthase, as well as raffinose, can assist in osmotic adjustments as well as in membrane and protein stabilization and to protect plants from oxidative stress caused by cold stress (Suzuki et al., 2015; ElSayed et al., 2014). Furthermore, there is evidence that in some plant species, change in RFOs is part of a mechanism for carbon storage rather than protection against environmental changes (ElSayed et al., 2014). Transcriptional up-regulation of the raffinose oligosaccharide pathway results in accumulation of monosaccharides and disaccharides, including glucose, fructose, sucrose, galactinol and raffinose (Jansko et al., 2010).

Higher levels of the transcript of the galactinol synthase encoded gene (MfGolS1) were induced and maintained in *M. falcata* than in *M. sativa* during cold acclimation, in accordance with the accumulation of sugars and the differential cold tolerance between *M. falcata* and *M. sativa* (Zhuo et al., 2013).

Galactinol synthase is the enzyme that catalyzes the first step of RFOs. Transgenic tobacco that overexpresses the exogenous gene MfGolS1 of *M. falcata* turned out to be more tolerant to freezing and chilling (Zhuo et al., 2013).

Trehalose is another nonreducing disaccharide that serves as a protector against stress. Trehalose synthesized via a phosphorylated intermediate, trehalose 6-phosphate (Tre6P), has been a reporter that the gene of trehalose-6-phosphate (TPS) increases in freezing stress (Song et al., 2016a). There is a strong correlation between sucrose and Tre6P. Parallel changes in their levels indicate that synthesis and degradation of Tre6P may be regulated by sucrose (Lunn et al., 2014). The demand for starch is regulated by sucrose-dependent changes in Tre6P that modulate the rate of starch breakdown according to the demand for sucrose (Lunn et al., 2014).

An accumulation of trehalose may act as a regulator of stress, giving greater tolerance under low temperatures. Transgenic rice plants which contained the exogenous gene trehalose biosynthetic (otsA and otsB) from *Escherichia coli* exhibited constant plant growth, less photo-oxidative damage and more favorable mineral balance under low-temperature stress conditions (Garg et al., 2002).

Fatty Acid Metabolism

Fatty acids are components of cellular membrane, thylakoid membrane and cutin waxes; trienoic fatty acids (TAs), hexadecatrienoic acid (16:3) and linolenic acid (18:3) are the major polyunsaturated fatty acid species in membrane lipids (Upchurch, 2008), and changes in their regulated plant response against low temperatures. Cold temperatures cause damage to the lipid membrane; these oxidative lipid injuries are generated by ROS resulting in lipid peroxidation and changes in lipid compositions.

In physiological conditions, the membrane lipid is in the liquid crystal phase, but under low-temperature stress, it is transformed from the liquid phase into the gel phase, thus increasing the permeability of the membrane (Lyons, 1973; Chen et al., 2014; Tarkowski and Van den Ende, 2015). This arrangement of membrane fluidity, with the transition to gel phase, is accompanied by an increase in the proportion of unsaturated fatty acids (Beck et al., 2007; Janska et al., 2010; Chen et al., 2014).

Releasing α-linolenic acid (18:3) is a way of modulating the membrane flow, in chloroplast saturated phosphatidylglycerol (PG), with saturated fatty acids (16:0, 18:0) playing a key role in acclimatization of plants under chilling. Increased TAs in the plasma membrane enhance low-temperature tolerance in plants during the early growth stage (Iba, 2002; Upchurch, 2008).

The chloroplast is a sensitive target of low temperatures; differences between lipid compositions in chloroplasts may be the cause of plant sensitivity or tolerance to cold stress. The content of PGs in chloroplasts is 16:0 or 16:1; in tolerant plants the levels of 16:0, 16:1 and 18:1 are less than in sensitive plants (Iba, 2002).

Alfalfa studies comparing tolerant and cold-sensitive varieties showed that low-temperature-resistant varieties contained a higher percentage of polyunsaturated fatty acids in chloroplast membranes (Peoples et al., 1978).

Willemot (1977) described that saturated fatty acids at low temperature in wheat are not related to frost hardiness. Furthermore, in alfalfa it has been shown that linoleic acid accumulates at low temperature while wheat accumulates linolenic acid. In rice there is a positive relationship between the unsaturated fatty acid composition of the chloroplast membrane and the photosynthetic tolerance to chilling (Zhu et al., 2007).

A metabolome study in maize showed that triacylglycerides and diacylglycerides produced from the glycerolipid metabolic pathway significantly decreased when plant seedlings were exposed to low temperatures (Sun et al., 2016). In soybean, protein profiles comparing a sensitive and tolerant variety show that the low-temperature-tolerant cultivar improves the biosynthesis of fatty acids (Tian et al., 2015).

RNA sequencing in rice and lotus reveals that in sensitive plants there is a down-regulated gene in the metabolic lipid pathway (Dametto et al., 2015; Calzadilla et al., 2016). This indicates this is a generalized response in plants against cold stress where a regulation metabolic process exists related to unsaturated fatty acid accumulation.

Summary

There is a general response for plants against cold stress. The metabolic changes focus on counteracting the oxidative stress in the chloroplast and lowering the photoinhibition, particularly regulating the photosynthesis and carbohydrate metabolism.

The biochemical response produced by cold damage is generally associated with the accumulation of unsaturated fatty acids in order to decrease lipid peroxidation. Sugar accumulation, seems to be a response to support growth under temperature stress, acting like photoprotective molecules, lowering levels of ROS, and like osmotic molecules, regulating the osmotic imbalance caused by the cold. This mechanism is summarized in Figure 1.1.

As a strategy for future breeding projects it is thought that a plant phenotype that is more energy efficient at the level of chloroplast would lower the levels of ROS produced by the cold and the consequent photoinhibition. This would make the plant more metabolically efficient where the biochemical processes are geared towards growth rather than repair of damage, resulting in a greater grain yield in crop species.

Water Deficit

Plants Water Balance

Plant water balance explains the behavior of plants in terms of how they control the hydration of their cells, which has important implications in the physiological and metabolic processes that determine the quantity and quality of plant growth (Passioura et al., 2010). The plant water balance is determined by the ratio between the fraction of water lost in evaporation (transpiration) to the atmosphere and its absorption from the soil. When transpiration exceeds absorption, the relative water content (RWC) and the cellular volume decrease, determining the loss of cell turgor and at the same time increasing the solute levels in the cell, so osmotic potential (π) and water potential (ψ) fall (Lawlor and Cornic, 2002). This situation could be modified by osmotic adjustment (OA) in species that have such a defense mechanism. OA has been and is well established as a major regulator of turgor and stomata conductance (Gs) in drought (Blum, 2016). Turner and Jones (1980) indicate that OA was effective in sustaining turgor, RWC and Gs at low leaf water potential in sorghum. A lower state of cell turgor and RWC causes slow growth and Gs. In this context two parameters describe the water status of plants: RWC and ψ (Nobel, 2009). RWC is a measure of relative change in cell volume; ψ is the result of cell turgor and π, and thus depends both on solute concentration and cell wall rigidity and does not relate directly to cell volume (Kaiser, 1987; Jones, 2007). OA together with cell wall rigidity regulate turgor under dehydration, where the former is generally more effective and prevalent than the latter (Bartlett et al., 2012). Small cells are generally more conducive to OA than large ones. It is recognized that turgor, Gs and growth can be regulated by hydraulic and/or hormonal signals, directly or indirectly (Blum, 2016).

Plant Strategies to Mitigate Damage from Dehydration

From an agricultural perspective, drought is ultimately defined in terms of its effects on yield, since this is the relevant issue when addressing the improvement of crop production under water-limited environments (Passioura, 2007). Most plants that are grown in field conditions experience water stress during part of their growing cycle, especially at midday and early afternoon in the summer or due to the scarcity or absence of rain for long periods of time. This threat affects phenology, carbon fixation, distribution of assimilates and plant reproduction (Fuad-Hassan et al., 2008); for example, legumes are highly sensitive to water deficit especially during the reproductive period, flowering and pod filling (Araújo et al., 2015). Consequently, the timing of water deficits during the season (e.g. sowing, crop establishment,

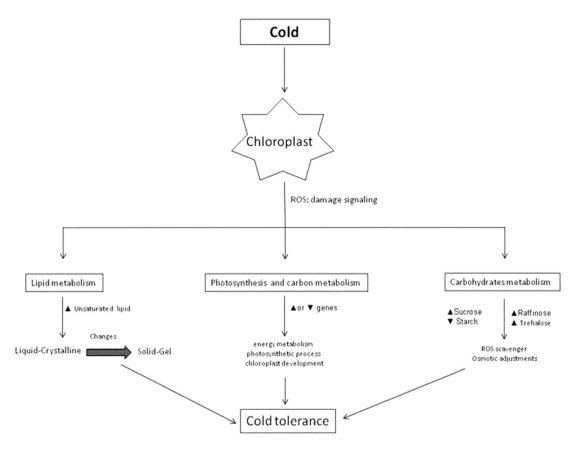

FIGURE 1.1 Schematic representation of how metabolic routes vary in response to cold.

flowering or grain filling) may have a much larger impact on yield than the intensity of drought per se (Pinheiro and Chaves, 2010). In this way, to maintain productivity, agriculture uses about 70% of water resources in the world. It is estimated that, by 2030, the requirement for irrigation will increase by 14% (FAO, 2009).

Mesophyte species that experience water stress modify or adjust their cell metabolism to the new conditions imposed by stress. When the regulation is inadequate and the metabolism is not adjusted to maintain its functions, they trigger a series of harmful effects that can cause the death of the vegetable. Regulation is strongly influenced by species and environment. Metabolic changes involve biochemical, physiological and structural modifications (Passioura, 2007). With some exceptions, the lack of water of tissues below a critical level is accompanied by irreversible changes in the structure and finally, the death of the plant (Azcón-Bieto and Talón, 2008). In nature, plants can either be subjected to slowly developing water shortage (days, weeks or months) or face short-term water deficits (hours to days). In the case of slow water deficits, plants can either escape dehydration by shortening their life cycle or optimize their resource gain in the long term through acclimation responses. In some modern maize hybrids the anthesis-silking interval under drought has become shorter and selection has possibly led to an increase in the growth of spikelets (Bänziger and Cooper, 2001). In the case of rapid dehydration, plants react to minimizing water loss or exhibiting metabolic protection (induced or constitutive) against the damaging effects of dehydration and co-developing oxidative stress (Chaves et al., 2003). Fast and slow desiccation can have totally different results in terms of physiological response or adaptation (McDonald and Davies, 1996), but the importance of time in shaping plant response may change dramatically according to genotype and environment (Chaves et al., 2003).

Plants can strengthen their response to a drought by avoiding tissue dehydration, maintaining the potential water as high as possible or by tolerating low tissue water potential using metabolic strategies. Although many genes are induced by water deficit, the molecular functions of all these are not known. In addition, the impact of a change in gene expression of a gene with a known function, such as an enzymatic function, on cellular homeostasis may not be exactly understood (Bray, 2007). Dehydration avoidance is associated with a variety of adaptive traits involving minimizing water loss and maximizing water uptake. Water loss is minimized by density or regulation stomatal (Bell et al., 2007), by reducing light absorbance or by decreasing leaf area due to less cell division and expansion rates in leaves. One of the earliest water-saving mechanisms, present in a great majority of plants, is reduced leaf growth (Aguirrezábal et al., 2006) or the earlier senescence of older leaves in the case of prolonged stress (Pinheiro and Chaves, 2010). This reduction in foliage dimension or lower stomata density leads to a decreased transpiration area (Quan et al., 2016). Water uptake is maximized by adjusting the growth pattern to increase growth of roots. This is the main resource to maximize water absorption. This response is mediated by hormones: abscisic acid (ABA), ethylene and their interactions (Wilkinson and Davies, 2002). Changes in the root/shoot ratio as well as temporary accumulation of reserves in the stem occur in several species under water deficit (Chaves et al., 2003). As the key process of primary metabolism, photosynthesis plays a central role in plant performance under drought (Pinheiro and Chaves, 2010). The decline observed in leaf net carbon assimilation (A) as a result of plant water deficits is followed by an alteration and partitioning of the photo-assimilates at the whole plant level, corresponding in general to an increase in the root/shoot ratio (Sharp, 2002).

Experience from drought-resistant cereal cultivation during a century of scientific breeding clearly indicates that drought resistance in crop plants under stress is mainly derived from their ability to sustain tissue hydration during drought (i.e. dehydration avoidance), rather than an ability to sustain biological function when tissues are dehydrated (i.e. dehydration tolerance) (Araus et al., 2012).

Response Mechanisms That Take Place in Leaves

In order to avoid excessive water loss in times of drought, regulation of stomata aperture together with the reduction of the leaf area are the immediate responses of plants, which could result in a dehydration cell that would produce xylem cavitation, causing plant death (Comstok, 2002). However, a stomatal aperture is essential in leaf cooling by latent heat exchange, and this characteristic is an objective of selection in plant breeding (Lu et al., 1997). For example, many species have been classified as isohydric or anisohydric depending on the sensitivity of stomata to soil dehydration. Isohydric plants are those that close their stomata when they sense a drop in soil water potential or an increase in atmospheric demand. On the other hand, anisohydric plants continue to transpire even when soil water content diminishes, because these plants are keeping their stomata open longer (Tardieu and Simonneau, 1998). The different response of stomata to water scarcity is also associated with different root/shoot ratios, with the anisohydric stomatal behavior often being associated with larger root systems and high capacity for osmoregulation that supports water uptake until soil water content is low. Osmoregulation requires a dehydration signal to be developed, more likely to take place in aniso- than in isohydric plants. The anisohydric strategy allows a closer match between water availability and consumption, with a positive impact on season growth (Chaves et al., 2016). Regardless of the mechanism of perception and signaling that causes stomatal closure, many plants show a tendency to increase water use efficiency (WUE) when the stress is moderate as a result of a non-linear relationship between carbon assimilation and Gs, by which the plant restricts the loss of water without excessively affecting photosynthesis (Blum, 2005; Taiz and Zeiger, 2003). When comparing C_3 and C_4 species it is apparent that C_4 plants exhibit higher WUE (A/E ratio) due to higher A and lower transpiration (E). In warm regions, where C_4 species evolved, photorespiration is stimulated considerably, as well as the demand for transpiration for cooling. By increasing CO_2 around the carboxylating enzyme Rubisco (RuBP), C_4 plants greatly enhanced carboxylation efficiency and were able to inhibit photorespiration (Sage, 2004). By producing smaller stomata density or reducing stomata aperture plants will function at low stomata conductance (Gs) and have lower E (Chaves et al., 2016).

However, in this case, the CO_2 concentration at the RuBP site in the mesophyll cells is limited (Perdomo et al., 2017). In this sense, Perdomo et al. (2017) observed that the limitations to photosynthesis under drought were mainly caused by a stomata closure (lower Gs) and increased leaf resistance to CO_2 transport from the atmosphere to the site of carboxylation (diffusive limitations) in C_3 and C_4 species. Conversely, under the combination of water deficit and heat stress, biochemical limitations have been observed with decreased RuBP activation and impaired ATP synthesis, which is due to a decrease in electron transport. Experimental studies on CO_2 assimilation of mesophytic C_3 plants under water stress revealed a decreased relative water content (RWC), actual rate of A and potential rate (A_{pot}). In this way, there are two general types of relation of A_{pot} to RWC, which are called Type 1 and Type 2. Type 1 has two main phases. As RWC decreases from 100% to 75%, A_{pot} is unaffected, but decreasing Gs results in smaller A, and lower CO_2 concentration inside the leaf (C_i) and in the chloroplast (C_c), the latter falling possibly to the compensation point. Below 75% RWC, there is metabolic inhibition of A_{pot}, inhibition of A then being partly reversible by elevated CO_2; Gs regulates A progressively less, and C_i and CO_2 compensation points rise. In the Type 2 response, A_{pot} decreases progressively at RWC 100% to 75%, with A being progressively less restored to the unstressed value by elevated CO_2. Decreased Gs leads to a lower C_i and C_c but they probably do not reach compensation point: Gs becomes progressively less important and metabolic limitations more important as RWC falls (Lawlor and Cornic, 2002). The primary effect of low RWC on A_{pot} is most probably caused by limited RuBP synthesis, RuBP activity or activation state, as a result of decreased ATP synthesis, either through inhibition of coupling factor activity or amount due to increased ion concentration or impaired activation of RuBP by Rubisco activase (Perdomo et al., 2017). Electron transport is maintained (but down-regulated) over a wide range of RWC (Sage et al., 2008; Taylor et al., 2011). Metabolic imbalance results in amino acid accumulation and decreased protein synthesis. These conditions profoundly affect cell functions and ultimately cause an excess of radiant energy that is not used for photosynthesis and that the plant must eliminate in order to avoid the overproduction of reactive chemical species that potentially can produce oxidative damages that would compromise the photosynthetic (Chaves et al., 2003). Autotrophic organisms have direct and indirect mechanisms that capture the excess of incident photons (photoreceptors, photochromes, neocromes, phytochromes, rhodopses and cryptochromes), variation of pH in the thylakoid lumen, changes in the oxidation-reduction state, production of reactive oxygen species (ROS) and accumulation of metabolites, which will induce changes in nuclear gene expression and gene expression in the plastid. These mechanisms attempt to decrease the damage caused by the adverse situation (Li et al., 2009). In order to avoid excess light, plants can reduce light absorbance by leaf rolling, a dense trichome layer increasing reflectance and steep leaf angles. They can divert absorbed light by photochemical reactions to other processes such as thermal dissipation (Pastenes et al., 2004; Baker, 2008). In fact, it is widely accepted that photosynthesis regulation in response to the environment is highly dynamic and is modulated in the short term by thermal energy dissipation. This increase in energy dissipation is linked (at least partially) to a parallel increase in photoprotective carotenoid levels. In this way, the "xanthophyll cycle" plays a primordial role in the thermal dissipation process, which consists of light-dependent interconversion of three xanthophylls (oxygenated carotenoids): violaxanthin, anteraxanthin and zeaxanthin. This is a cyclic reaction involving a de-epoxidation sequence of violaxanthin di-epoxide via anteroxanthine mono-epoxide to form zeaxanthin (without epoxy groups). The cycle is regulated by light; violaxanthin is de-epoxidated to anteraxanthin and zeaxanthin under conditions of high photon flux (or when photosynthetic activity is diminished by stress), while in the dark (or when the stress factor is absent), zeaxanthin is again epoxidized to violaxanthin (Demming-Adams and Adams, 1996; Nelson and Yocum, 2006). Moreover, other processes have significant photoprotective and stabilizing functions, including the antioxidant properties of vitamin E (α-tocopherol), a constitutive component of the lipid matrix of thylakoid membranes (Li et al., 2009). In species under water deficit, the use of absorbed light (photosynthesis or photorespiration) and thermal dissipation is not sufficient to retain excess energy; electrons in a highly excited state are transferred to molecular oxygen (O_2) that form the reactive oxygen species (ROS), such as H_2O_2, O_2^-, HO^- and $^1O_2$. All these molecules are highly toxic and can cause oxidative damage to proteins, DNA and lipids (Miller et al., 2010). In this situation, the plant has antioxidant molecules and enzymes that are in different cell compartments and can scavenge ROS (superoxide dismutase, catalases, ascorbate-glutathione cycle such as glutathione reductase and ascorbate peroxidases). The antioxidant activity depends on the degree of severity of stress, genotype and stage of development. The acclimatization of plants to drought is associated with increased activity of antioxidant enzymes that maintain the concentration of ROS at relatively low levels (Signorelli et al., 2013; Quan et al., 2016). On the other hand, seedlings subjected to water deficit show nitric oxide (NO) production (Arasimowicz-Jelonek et al., 2009). However, when there is a deregulated synthesis or overproduction of NO that may have toxic physiological consequences, it results in nitrosative stress due to reactive nitrogen species (RNS) formation. RNS are formed by the interaction of NO with free radicals and oxygen. RNS include NO^-, nitrogen dioxide (NO_2^-), S-nitrosothiols (SNOs) and peroxynitrite ($ONOO^-$) (Airaki et al., 2011). Water stress in *Lotus japonicus* results in an increase in NO levels and reduced S-nitrosoglutathione reductase activity (an enzyme that maintains low levels of RNS), especially in roots (Signorelli et al., 2013). Under environmental stress, an increase in ROS and RNS levels can cause damage in cells, but it must be pointed out that both ROS and RNS can also serve as secondary messengers in signaling for the activation of defense responses to pathogens, abiotic stress, programmed cell death and development (Suzuki and Mittler, 2006; Arasimowicz-Jelonek et al., 2009; Gill and Tuteja, 2010; Miller et al., 2010, Airaki et al., 2011; Signorelli et al., 2013).

Response Mechanisms That Take Place in Roots

Root systems are responsible for water and nutrient uptake and provide physical stability, and store nutrients and carbohydrates. In addition, roots may act as sensors for water deficit conditions and send signals to shoots above ground. It is well documented that species adapted to dry climatic regimes generally have higher root/shoot ratios and deeper root systems than species that are suited to humid climatic conditions (Brunner et al., 2015). Although the root/shoot ratio tends to increase, the biomass of fine roots in particular is often reduced as a consequence of reduced transpiration and respiration rates. The benefit, however, is that young roots are able to take up water more efficiently than older ones, and thus, root shedding and re-growth represent a more suitable acclimation of plants to reduced water supply if the plant can afford this strategy. In experiments with trees it was observed that whether a tree maintains old roots or sheds old roots and produces new ones in response to drought is determined by the benefit to cost ratio in terms of water uptake and carbon expenditure. Root shedding and the construction of new roots mean the investment of a considerable amount of energy in the process of root turnover (Eissenstat et al., 2000). Before this, plants cease growth and close the stomata to prevent water loss under drought conditions, which results in a simultaneous decrease in photosynthesis. However, maintenance respiration responds more slowly to drought than photosynthesis, resulting in a carbon deficit and forcing the plant to utilize stored carbohydrates. If the carbon deficit persists for a long time, carbohydrates will be depleted and plants will experience carbon starvation, resulting in plant mortality. In the roots of *Robinia*, Yan et al. (2017) observed an increase in soluble sugar but a decrease in starch, indicating that the roots began to utilize the starch, creating a carbon deficit in the roots. Regier et al. (2009) applied drought conditions to two contrasting *Populus nigra* clones and observed that the drought-adapted clone had significantly more starch but significantly less sucrose, glucose and fructose in the roots and conversely a drought-sensitive clone had a reduction of carbohydrate due to lower starch levels inducing degradation of starch to maintain root respiration. In soybean, under severe water deficit, there was an increase in transcripts associated with starch synthesis and a decrease in transcripts associated with starch degradation. Conversely, galactinol and raffinose synthase related genes were up-regulated, which might enhance osmotic tolerance in the root (Song et al., 2016b). In this way, the elevated level of proline found in alfalfa roots indicates the ability to regulate water deficit stress tolerance (Rahman et al., 2016). Generally, a decrease in soluble sugars is observed in the branch, stem, bark and root tissues after rewatering due to recovery of tissue growth (Brunner et al., 2015; Yan et al., 2017).

Summary

Cold, drought and salinity are those environmental factors which affect plants in many respects and which, due to their widespread occurrence, cause the most fatal economic losses in agriculture. All three forms of abiotic stress affect the water relations of a plant on the cellular as well as whole plant level, causing specific as well as unspecific damage (Figure 1.2) and responses. The plant's response to these stresses (drought, salt and cold) is through the response mechanisms that are raised at both molecular and cellular levels, as well as physiological and biochemical, which allow them to acclimate and survive stress. In general, the first step in switching on a molecular response in response to an environmental signal is its perception by specific receptors; these are physical signals (changes in the pressure of cell turgor, alterations in the walls and cellular volume) that are then converted to biochemical

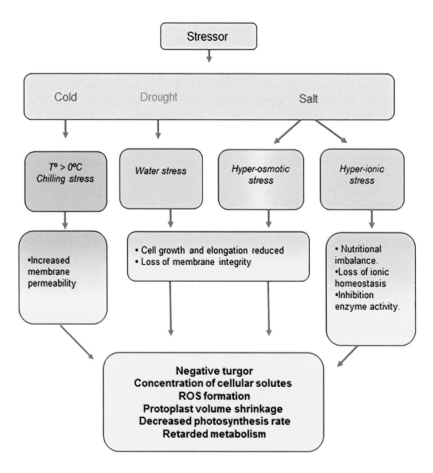

FIGURE 1.2   Similar (bold types) and dissimilar damages produced by cold, salt and drought in plant tissues.

signals, which are generated in response to stress (Bray, 2007). These include receptor-like kinases, receptor tyrosine kinases, G-protein coupled receptors, iontropic channel related receptors, histidine kinases, nuclear hormone receptors and extra- and intracellular Ca^{2+} sensors. Ca^{2+} influx and histidine kinases have been identified in response to cold. It has been suggested that in *Arabidopsis* a heterotrimeric G-protein is involved in ABA response in guard cells (Xiong and Ishitani, 2006). Then, followed by the generation of secondary messengers (inositol phosphates, ROS) that can modulate the intracellular Ca^{2+} level and initiate a signal transduction cascade that involves protein phosphorylation/dephosphorylation, where different Ca^{2+} sensors are included, a protein decodes their signals and has a Ca^{2+} binding motif, the so-called EF hand motif: calmodulins (CaMs), calcineurin B-like proteins (CBLs), Ca^{2+}-dependent protein kinases (CDPKs), mitogen-activated protein kinase (MAPKs) and phospholipid-cleaving enzymes (Shi, 2007). In salt stress, it is important to emphasize the ion-specific signaling pathway SOS (salt overlay sensitive), which responds to changes of Ca^{2+} levels in cytoplasm, described in *A. thaliana* (Zhou et al., 2007). Finally, these would reach a large number of genes, enzymes, hormones, and metabolites. The products of these genes can participate in the generation of regulatory molecules such as ABA, and ethylene and salicylic acid, which in turn can initiate a secondary signaling process (Xiong et al., 2002). The different environmental variables that determine osmotic stress constitute a complex of stimuli that possess different and sometimes related attributes in

which each one of them provides particular information to the cells. Given this multiplicity of signals, it is expected that the plant possesses multiple primary sensors that perceive the signal of stress. The increase of Ca^{2+} in cytoplasm, as a result of the influx from apoplast or the release of intracellular compartments, is an early response to osmotic stress (Shi, 2007). The products of inducible genes to osmotic stress can be classified into two groups: the first group includes functional proteins (which participate directly in the biochemical mechanisms of damage control, repair and acclimatization: aquaporins and some key enzymes of osmolytes biosynthesis, chaperones, antioxidants), whereas the second group comprises regulatory proteins involved in signal transduction and gene expression (e.g. transcription factors, protein kinases, protein phosphatases) as well as diverse effectors involved in signaling (e.g. calmodulin-binding protein, SOS pathways; Seki et al., 2003). Hundreds of genes are induced or repressed under osmotic stress. Many of these stress-inducible genes are activated by ABA, and their products may improve tolerance to stress (Xiong, 2007). It is not surprising that gene expression is regulated by this phytohormone, although not all genes that are induced by stress are regulated by ABA, indicating the existence of several signaling pathways (Ishitani et al., 1997). Among the most important pathways are ABA-dependent and ABA-independent, which isolate and characterize the *cis*-regulatory elements and the transcriptional factors responsible for its activation (Busk and Pages, 1998). The ABA-regulated promoters, called ABRE (ABA response elements), contain ACGT

TABLE 1.1

Promotors, *Cis*-Regulatory Elements and Transcription Factors in Response to Osmotic Stress

Promotors	*Cis* Elements	Transcription Factors	Stressor
RD29A	ABRE	CBF/DREB1	Cold
RD29A	DRE/CRT	DREB2	Salt, ABA and drought
RD29B	ABRE	bZIP	Salt and drought
RD22	MYB/MYC	MYBR/MYCR	Drought and ABA

nucleotides as the *core* of their sequence and form part of the elements called G-box. More than 20 ABREs have been described in genes that respond to dehydration and drought, which bind to trans- factors regulated by ABA, such as bZIP (Kobayashi et al., 2008). Also, other ABA-responsive transcription factors known as MYBR/MYCR constitute an indirect pathway for ABA regulation through the R22 promoter gene. On the other hand, ABA-independent genes are characterized by containing a conserved dehydration response element (DRE, TACCGACAT) in their promoter region, related to gene regulation by interaction with an ABA-independent signaling cascade and matched by the gene RD29B promoter. It is a faster response route than the previous one (Table 1.1). This was discovered through studies of the RD29A promoter gene, which is induced by both ABA-dependent and independent pathways (Xiong et al., 2002; Shinozaki and Yamaguchi-Shinozaki, 2007). The genes that are upregulated by ABA (a) encode enzymes that function in the biosynthesis of compatible solutes that could lower leaf water potential and facilitate water uptake and retention, (b) can directly detoxify ROS (Jiang and Zhang, 2002) and (c) encode polypeptides that may help to restore the native structure of abnormally folded proteins or could promote the degradation of unfolded proteins that cannot be repaired (Xiong et al., 2007). Although osmotic stress alone can activate these stress-responsive genes, ABA can synergistically enhance their expression, enhancing the ability of the plants to respond to stress (Tuteja et al., 2011). In addition, a complex network of response involving ABA and PAs and/or nitric oxide (Arasimowicz-Jelonek et al., 2009) has been identified and observed in different species, such as *Lotus tenuis* (Espasandin et al., 2018), *Coffea canephora* (Marracini et al., 2012) and *Lycopersicum esculentum* (Diao et al., 2017) exposed to salt, water and cold stress, respectively. Finally, the vegetal metabolism is altered, inducing biochemical and physiological responses that create tolerance (Figure 1.3).

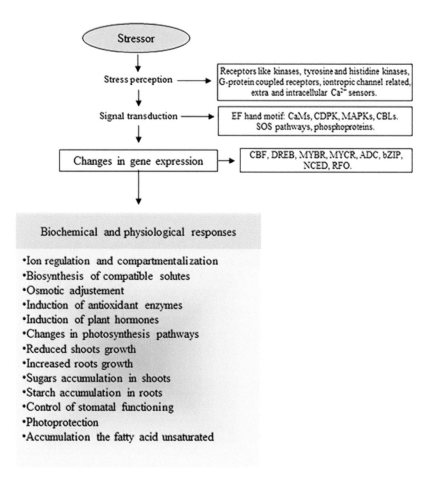

FIGURE 1.3   Biochemical and physiological responses to drought, salt and cold stress in plants.

REFERENCES

Abogadallah, G. M. 2010. Antioxidative defense under salt stress. *Plant Signaling and Behavior* 5(4):369–374.

Aguirrezábal, L., Bouchier-Combaud, S., Radziejwoski, A., Dauzat, M., Cookson, S.J., Granier, G. 2006. Plasticity to soil water deficit in *Arabidopsis thaliana*: Dissection of leaf development into underlying growth dynamic and cellular variables reveals invisible phenotypes. *Plant, Cell and Environment* 29:2216–2227.

Ahmads, S., Khan, N., Iqbal, M.Z. 2002. Salt tolerance of cotton (*Gossypium hirsutum L.*). *Asian Journal of Plant Science* 1:715–719.

Airaki, M., Sánchez-Moreno, L., Leterrier, M., Barroso, J.B., Palma, J.M., Corpas, F.J. 2011. Detection and quantification of S-nitrosoglutathione (GSNO) in pepper (*Capsicum annuum L.*) plant organs by LC-ES/MS. *Plant and Cell Physiology* 52:2006–2015.

Annunziata, M.G., Ciarmiello, L.F., Woodrow, P., Maximova, E., Fuggi, A., Carillo, P. 2017. Durum wheat roots adapt to salinity remodeling the cellular content of nitrogen metabolites and sucrose. *Frontiers in Plant Science* 7, Article 2035.

Arasimowicz-Jelonek, M., Floryszak-Wieczorek, J., Kubis, J. 2009. Interaction between polyamine and nitric oxide signaling in adaptive responses to drought in cucumber. *Journal of Plant Growth Regulation* 28:177–186.

Araújo, S.S., Beebe, S., Crespi, M., Delbreil, B., González, E.M., Gruber, V., Lejeune-Henaut, I., et al. 2015. Abiotic stress responses in legumes: Strategies used to cope with environmental challenges. *Critical Reviews in Plant Science* 34:237–280.

Araus, J.L., Serret, M.D., Edmeades, G.O. 2012. Phenotyping maize for adaptation to drought. *Frontiers in Physiology* 3:305.

Ashraf, M. 2009. Biotechnological approach of improving plant salt tolerance using antioxidants as markers. *Biotechnology Advance* 27:84–93.

Azcón-Bieto, J., Talón, M. 2008. *Fundamentos de Fisiología Vegetal*. 2nd edition. Interamericana-McGraw-Hill, Madrid.

Baghel, L., Kataria, S., Guruprasad, K.N. 2016. Static magnetic field treatment of seeds improves carbon and nitrogen metabolism under salinity stress in soybean. *Bioelectromagnetics* 37:455–470.

Baker, N. 2008. Chlorophyll fluorescence: A probe of photosynthesis *in vivo*. *Annual Review of Plant Biology* 59:89–113.

Bänziger, M., Cooper, M.E. 2001. Breeding for low input conditions and consequences for participatory plant breeding-examples from tropical maize and wheat. *Euphytica* 122:503–519.

Bartlett, M.K., Scoffoni, C., Sack, L. 2012. The determinants of leaf turgor loss point and prediction of drought tolerance of species and biomes: A global meta-analysis. *Ecology Letters* 15:393–405.

Beck, E.H., Fettig, S., Knake, C., Harting, K., Bhattarai, T. 2007. Specific and unspecific responses of plants to cold and drought stress. *Journal of Biosciences* 32:501–510.

Bell, L.W., Williams, A.H., Ryan, M.H., Ewing, M.A. 2007. Water relations and adaptations to increasing water deficit in three perennial legumes, *Medicago sativa, Dorycnium hirsutum* and *Dorycnium rectum*. *Plant and Soil* 290:231–243.

Blum, A. 2005. Drought resistance, water-use efficiency, and yield potential: Are they compatible, dissonant, or mutually exclusive? *Australian Journal of Agriculture Research* 56:1159–1168.

Blum, A. 2016. Osmotic adjustment is a prime drought stress adaptive engine in support of plant production. *Plant, Cell and Environment* 40:4–10.

Bohnert, H.J., Jensen, R.G. 1996. Strategies for engineering water-stress tolerance in plants. *Trends Biotechnology* 14:89–97.

Boyer, J.S. 2010. Drought decision making. *Journal of Experimental Botany* 61:3493–3497.

Bray, E.A. 2007. Molecular and physiological responses to water-deficit stress. In *Advances in Molecular Breeding toward Drought and Salt Tolerant Crops*. M.A. Jenks et al. (eds.). Springer, Dordrecht, the Netherlands, pp. 121–140.

Brunner, I., Herzog, C., Dawes, M.A., Arend, M., Sperisen, C. 2015. How tree roots respond to drought. *Frontiers in Plant Science* 6:547.

Busk, K., Pages, M. 1998. Regulation of abscisic acid-induced transcription. *Plant Molecular Biology* 37:425–435.

Calzadilla, P., Maiale, S., Ruiz, O., Escaray, F. 2016. Transcriptome response mediated by cold stress in *Lotus japonicus*. *Frontiers in Plant Science* 7:1–16.

Campestre, M.P., Castagno, L.N., Estrella, M.J., Ruiz, O.A. 2016. *Lotus japonicus* plants of the Gifu B-129 ecotype subjected to alkaline stress improve their Fe^{2+} bioavailability through inoculation with *Pantoea eucalypti* M91. *Journal of Plant Physiology* 192:47–55.

Carillo, P., Annunziata, M.G., Pontecorvo, G., Fuggi, A., Woodrow, P. 2011. Salinity stress and salt tolerance. In *Abiotic Stress in Plants: Mechanisms and Adaptations*. A. Shanker (ed.). InTech, Rijeka, Croatia, pp. 21–38.

Carillo, P., Mastrolonardo, G., Nacca, F., Parisi, D., Verlotta, A., Fuggi, A. 2008. Nitrogen metabolism in durum wheat under salinity: Accumulation of proline and glycine betaine. *Functional Plant Biology* 35:412–426.

Chaves, M.M. 1991. Effects of water deficits on carbon assimilation. *Journal of Experimental Botany* 42:1–16.

Chaves, M.M., Costa, J.M., Zarrouka, O., Pinheiro, C., Lopes, C.M., Pereira, J.S. 2016. Controlling stomatal aperture in semi-arid regions: The dilemma of saving water or being cool? *Plant Science* 251:54–64.

Chaves, M.M., Flexas, J., Pinheiro, C. 2009. Photosynthesis under drought and salt stress: Regulation mechanisms from whole plant to cell. *Annals of Botany* 103:551–560.

Chaves, M.M., Maroco, J.P., Pereira, J.S. 2003. Understanding plant responses to drought from genes to the whole plant. *Functional Plant Biology* 30:239–264.

Chen, H., He, H., Yu, D. 2011. Over expression of a novel soybean gene modulating Na^+ and K^+ transport enhances salt tolerance in transgenic tobacco plants. *Physiologia Plantarum* 141:11–18.

Chen, L.J., Xiang, H.Z., Miao, Y., Zhang, L., Guo, Z.F., Zhao, X.H., Lin, J.W., Li, T.L. 2014. An overview of cold resistance in plants. *Journal of Agronomy and Crop Science* 200:237–245.

Chen, T., Zhang, L., Shang, H., Liu, S., Peng, J., Gong, W., Shi, Y., et al. 2016. iTRAQ-based quantitative proteomic analysis of cotton roots and leaves reveals pathways associated with salt stress. *PLoS One* 11(2):e0148487.

Cheng, L., Gao, X., Li, S., Shi, M., Javeed, H., Jing, X., Yang, G., He, G. 2010. Proteomic analysis of soybean [*Glycine max* (L.) Meer.] seeds during imbibition at chilling temperature. *Molecular Breeding* 26:1–17.

Comstok, J. 2002. Hydraulic and chemical signaling in the control of stomatal conductance and transpiration. *Journal of Experimental Botany* 53:195–200.

Dametto, A., Sperotto R.A., Adamski, J.M., Blasi, É.A.R., Cargneluttie, D., De Oliveira, L.F.V., Ricachenevsky, F.K., et al. 2015. Cold tolerance in rice germinating seeds revealed by deep RNAseq analysis of contrasting indica genotypes. *Plant Science* 238:1–12.

Demming-Adams, B., Adams, W.W. 1996. The role of xanthophylls cycle carotenoids in the protection of photosynthesis. *Trends in Plant Science* 1:21–26.

Diao, Q., Song, Y., Shi, D., Qi, H. 2017. Interaction of polyamines, abscisic acid, nitric oxide, and hydrogen peroxide under chilling stress in tomato (*Lycopersicon esculentum* mill.) seedlings. *Frontiers in Plant Science* 8:203.

Echeverria, M., Scambato, A.A., Sannazzaro, A.I., Maiale, S., Ruiz, O.A., Menéndez, A.B. 2008. Phenotypic plasticity with respect to salt stress response by *Lotus glaber*: The role of its AM fungal and rhizobia symbionts. *Mycorrhiza* 18:317–329.

Eissenstat, D.M., Wells, C.E., Yanai, R.D., Whitbeck, J.L. 2000. Building roots in a changing environment: Implication for root longevity. *New Phytology* 147:33–42.

ElSayed, A.I., Rafudeen, M.S., Golldack, D. 2014. Physiological aspects of raffinose family oligosaccharides in plants: Protection against abiotic stress. *Plant Biology* 16:1–8.

Espasandin, F., Calzadilla, P., Maiale, S., Ruiz, O., Sansberro, P. 2018. Overexpression of the arginine decarboxylase gene improves tolerance to salt stress in *Lotus tenuis* plants. *Journal of Plant Growth Regulation* 37(1):156–165.

Essa, T.A. 2002. Effect of salinity stress on growth and nutrient composition of three soybeans (*Glycine max* L. Merrill) cultivars. *Journal of Agronomy and Crop Science* 188:86–93.

FAO: Land and Water Division (NRL). 2013. http://www.fao.org/ag/agl/agll/spush/intro.htm. Accessed 26 August 2013.

FAO. 1996. *The Production Yearbook*. Rome, Italy.

FAO. 2009. www.fao.org/askfao/topicsList.do?mainAreaId=20263.

Flemetakis, E., Efrose, R.C., Ott, T., Stedel, C., Aivalakis, G., Udvardi, M.K., Katinakis, P. 2006. Spatial and temporal organization of sucrose metabolism in *Lotus japonicus* nitrogen-fixing nodules suggests a role for the elusive alkaline/neutral invertase. *Plant Molecular Biology* 62:53–69.

Flexas, J., Bota, J., Loreto, F., Cornic, G., Sharkey, T.D. 2004. Diffusive and metabolic limitations to photosynthesis under drought and salinity in C3 plants. *Plant Biology (Stuttgart)* 6(3):269–79.

Flexas, J., Ribas-Carbó, M., Bota, J., Galmés, J., Henkle, M., Martínez-Cañellas, S. 2006. Decreased Rubisco activity during water stress is not induced by decreased relative water content but related to conditions of low stomatal conductance and chloroplast CO_2 concentration. *New Phytologist* 172:73–82.

Flowers, T.J., Yeo, A.R. 1986. Ion relations of plant under drought and salinity. *Australian Journal of Plant Physiology* 13:75–91.

Foyer, H.C., Vanacker, H., Gomez, L.D., Harbinson, J. 2002. Regulation of photosynthesis and antioxidant metabolism in maize leaves at optimal and chilling temperatures: Review. *Plant Physiology and Biochemistry* 40:659–668.

Fricke, W., Akhiyarova, G., Wei, W., Alexandersson, E., Miller, A., Kjellbom, P.O., Richardson, A., et al. 2006. The short-term growth response to salt of the developing barley leaf. *Journal of Experimental Botany* 57:1079–1095.

Fuad-Hassan, A., Tardieu, F., Turc, O. 2008. Drought-induced changes in anthesis-silking interval are related to silk expansion: A spatio-temporal growth analysis in maize plants subjected to soil water deficit. *Plant, Cell and Environment* 31:1349–1360.

Garg, A.K., Kim, J.K., Owens, T.G., Ranwala, A.P., Choi, Y.D., Kochian, L.V., Wu, R.J. 2002. Trehalose accumulation in rice plants confers high tolerance levels to different abiotic stresses. *Proceedings of the National Academy of Sciences* 99:15898–15903.

Gárriz, A., Dalmasso, M.C., Marina, M., Rivas, E.I., Ruiz, O.A., Pieckenstain, F.L. 2003. Polyamine metabolism during the germination of *Sclerotinia sclerotiorum* ascospores and its relation with host infection. *New Phytologist* 161:847–854.

Gill, S.S., Tuteja, N. 2010. Reactive oxygen species and antioxidant machinery in abiotic stress tolerance in crop plants. *Plant Physiology and Biochemistry* 48:909–930.

Gorham, J., Wyn Jones, R.G., Bristol, A. 1990. Partial characterization of the trait for enhanced K+-Na+ discrimination in the D genome of wheat. *Planta* 180:590–597.

Greenway, H., Munns, R. 1980. Mechanisms of salt tolerance in non-halophytes. *Annual Review of Plant Physiology* 31:149–190.

Guo, W.X., Mass, S.J., Bronson, K.F. 2012. Relationship between cotton yield and soil electrical conductivity, topography, and Landsat imagery. *Precision Agronomy* 13:678–692.

Guy, Ch., Huber, J.L.A., Huber, S.C. 1992. Sucrose phosphate synthase and sucrose accumulation at low temperature. *Plant Physiology* 100:502–508.

Hahn, M., Walbot, V. 1989. Effects of cold-treatment on protein synthesis and mRNA levels in rice leaves. *Plant Physiology* 91:930–938.

Hashimoto, M., Komatsu, S. 2007. Proteomic analysis of rice seedlings during cold stress. *Proteomics Journal* 7:1293–1302.

Henry, C., Bledsoe, S.W., Griffiths, C.A., Kollman, A., Paul, M.J., Sakr, S., Lagrimini, M. 2015. Differential role for trehalose metabolism in salt-stressed maize. *Plant Physiology* 169:1072–1089.

Huner, N.P.A., Öquist, G., Sarhan, F. 1998. Energy balance and acclimation to light and cold. *Trends in Plant Science* 3:224–230.

Iba, K. 2002. Acclimative response to temperature stress in higher plants: Approaches of gene engineering for temperature tolerance. *Annual Review Plant Biology* 53:225–245.

Intergovernmental Panel on Climate Change. 2007. http://www.ipcc.ch. Accessed 25 October 2007.

Ishitani, M., Xiong, L., Stevenson, B., Zhu, J.K. 1997. Genetic analysis of osmotic and cold stress signal transduction in *Arabidopsis*: Interactions and convergence of abscisic acid-dependent and abscisic acid-independent pathways. *The Plant Cell* 9:1935–1949.

Janska, P., Marsik, P., Zelenkova S., Ovesna, J. 2010. Cold stress and acclimation: What is important for metabolic adjustment?. *Plant Biology* 12:395–405.

Ji, H., Pardo, J.M., Batelli, G., Van Oosten, M.J., Bressan, R.A., Li, X. 2013. The salt overly sensitive (SOS) pathway: Established and emerging roles. *Molecular Plant* 6:275–286.

Jiang, M., Zhang, J. 2002. Water stress-induced abscisic acid accumulation triggers the increased generation of reactive oxygen species and up-regulates the activities of antioxidant enzymes in maize leaves. *Journal of Experimental Botany* 53:2401–2410.

Jones, H.G. 2007. Monitoring plant and soil water status: Established and novel methods revisited and their relevance to studies of drought tolerance. *Journal of Experimental Botany* 58:119–130.

Kaiser, W.M. 1987. Effect of water deficit on photosynthetic capacity. *Physiologia Plantarum* 71:142–149.

Kirkbride, J.H. 2006. The scientific name of narrow-leaf trefoil. *Crop Science* 46:2169–2170.

Kobayashi, F., Maeta, E., Terashima, A., Kawaura, K., Ogihara, Y., Takumi, S. 2008. Development of abiotic stress tolerance via bZIP-type transcription factor LIP19 in common wheat. *Journal of Experimental Botany* 59:891–905.

Läuchli, A. 1984. Salt exclusion: An adaptation of legumes for crops and pastures under saline conditions. In *Salinity Tolerance in Plants: Strategies for Crop Improvement*. R.C. Staples, G.H. Toennissen (eds.). Wiley, New York, pp. 171–187.

Lawlor, D.W., Cornic, G. 2002. Photosynthetic carbon assimilation and associated metabolism in relation to water deficits in higher plants. *Plant, Cell and Environment* 25:275–294.

León, P., Sheen, J. 2003. Sugar and hormone connections. *Trends in Plant Science* 8:110–116.

Li, Z., Wakao, S., Fischer, B., Niyogi, K. 2009. Sensing and responding to excess light. *Annual Review of Plant Biology* 60:239–60.

López-Gómez, M., Hidalgo-Castellanos, J., Iribarne, C., Lluch, C. 2014. Proline accumulation has prevalence over polyamines in nodules of *Medicago sativa* in symbiosis with *Sinorhizobium meliloti* during the initial response to salinity. *Plant and Soil* 374:149–159.

Lu, Z.M., Cheng, J.W., Percy, R.G., Zeiger, E. 1997. Photosynthetic rate, stomatal conductance and leaf area in two cotton species (*Gossypium barbadense* and *Gossypium hirsutum*) and their relation with heat resistance and yield. *Australian Journal of Plant Physiology* 24:693–700.

Lunn, J.E., Delorge, I., Figueroa, C.M., Van Dijck, P., Stitt, M. 2014. Trehalose metabolism in plants. *The Plant Journal* 79:544–567.

Lyons, J.M. 1973. Chilling injury in plants. *Annual Review Plant Physiology* 24:445–466.

Marina, M., Maiale, S., Rossi, F., Romero, M., Rivas, M., Gárriz, A., Ruiz, O., Pieckenstain, F. 2008. Apoplastic polyamine oxidation plays different roles in local responses of tobacco to infection by the necrotrophic fungus *Sclerotinia sclerotiorum* and the biotrophic bacterium *Pseudomonas*. *Plant Physiology* 147:2164–2178.

Marraccini, P., Vinecky, F., Alves, G., Ramos, H., Elbet, S., Vieira, N., Carneiro, F., et al. 2012. Differentially expressed genes and proteins upon drought acclimation in tolerant and sensitive genotypes of *Coffea canephora*. *Journal of Experimental Botany* 63:4191–4212.

Maruyama, K., Urano, K., Yoshiwara, K., Morishita, Y., Sakurai, N., Suzuki, H., Kojima, M., et al. 2014. Integrated analysis of the effects of cold and dehydration on rice metabolites, phytohormones, and gene transcripts. *Plant Physiology* 164:1759–1771.

Mazzanti, A., Darwich, N.A., Cheppi, C., Sarlangue, H. 1986. Persistencia de pasturas cultivadas en zonas ganaderas de la Pcia. de Buenos Aires. *Revista Argentina de Producción Animal* 6:65.

McDonald, A.J.S., Davies, W.J. 1996. Keeping in touch: Responses of the whole plant to deficits in water and nitrogen supply. *Advances in Botanical Research* 22:229–300.

Meloni, D.A., Oliva, M.A., Ruiz, H.A., Martinez, C.A. 2001. Contribution of proline and inorganic solutes to osmotic adjustment in cotton under salt stress. *Journal of Plant Nutrition* 24:599–612.

Miller, G., Suzuki, N., Ciftci-Yilmaz, S., Mittler, R. 2010. Reactive oxygen species homeostasis and signaling during drought and salinity stresses. *Plant, Cell and Environment* 33:453–467.

Mittova, V., Tal, M., Volokita, M., Guy, M. 2003. Up-regulation of the leaf mitochondrial and peroxisomal antioxidative systems in response to salt-induced oxidative stress in the wild salt-tolerant tomato species (*Lycopersicon pennellii*). *Plant Cell and Environment* 26:845–856.

Munns, R. 1988. Effect of high external NaCl concentrations of ion transport within the shoot of *Lupinus albus*. I. Ions in xylem sap. *Plant, Cell and Environment* 11:283–289.

Munns, R. 2002. Comparative physiology of salt and water stress. *Plant, Cell and Environment* 25:239–250.

Munns, R., James, R.A., Läuchli, A. 2006. Approaches to increasing the salt tolerance of wheat and other cereals. *Journal of Experimental Botany* 57:1025–1043.

Munns, R., Schachtman, D.P., Condon, A.G. 1995. The significance of a two-phase growth response to salinity in wheat and barley. *Australian Journal of Plant Physiology* 22:561–569.

Nelson, N., Yocum, C.F. 2006. Structure and function of photosystems I and II. *Annual Review of Plant Biology* 57:521–565.

Nobel, P. 2009. *Physicochemical and Environmental Plant Physiology*. 4th edition. Elsevier, Los Angeles, CA

Noctor, G., Veljovic-Jovanovic, S., Driscoll, S., Novitskaya, L., Foyer, C.H. 2002. Drought and oxidative load in the leaves of C3 plants: A predominant role for photorespiration? *Annals of Botany* 89:841–850.

Parvaiz, A., Satyawati, S. 2008. Salt stress and phyto-biochemical responses of plants: A review. *Plant Soil and Environment* 54:89–99.

Passioura, J. 2007. The drought environment: Physical, biological and agricultural perspectives. *Journal of Experimental Botany* 58:113–117.

Passioura, J.B. 2010. Scaling up: The essence of effective agricultural research. *Functional Plant Biology* 37:585–591.

Pastenes, C., Pimentel, P., Lillo, J. 2004. Leaf movements and photoinhibition in relation to water stress in field-grown beans. *Journal of Experimental Botany* 56:425–433.

Paz, R.C., Reinoso, H., Espasandin, F.D., Gonzalez Antivilo, F.A., Sansberro, P.A., Rocco, R.A., Ruiz, O.A., Menéndez, A.B. 2014. Alkaline, saline and mixed saline–alkaline stresses induce physiological and morph o-anatomical changes in *Lotus tenuis* shoots. *Plant Biology* 16:1042–1049.

Peoples, T.R., Koch, D.W., Smith, S.C. 1978. Relationship between chloroplast membrane fatty acid composition and photosynthetic response to a chilling temperature in four alfalfa cultivars. *Plant Physiology* 61:472–473.

Perdomo, J.A., Capó-Bauçà, S., Carmo-Silva, E., Galmés, J. 2017. Rubisco and rubisco activase play an important role in the biochemical limitations of photosynthesis in rice, wheat, and maize under high temperature and water deficit. *Frontiers in Plant Science* 8:1–15.

Pinheiro, C., Chaves, M.M. 2010. Photosynthesis and drought: can we make metabolic connections from available data? *Journal of Experimental Botany* 62:869–882.

Quan, W., Liu, X., Wang, H., Chan, Z. 2016. Comparative physiological and transcriptional analyses of two contrasting drought tolerant alfalfa varieties. *Frontiers in Plant Science* 6:1256–1272.

Rahman, Md. A., Yong-Goo, K., Iftekhar, A., Liu, G-Sh., Hyoshin, L., Jeung Joo, L., Byung-Hyun, L. 2016. Proteome analysis of alfalfa roots in response to water deficit stress. *Journal of Integrative Agriculture* 15:1275–1285.

Regier, N., Streb, S., Cocozza, C., Schaub, M., Cherubini, P., Zeeman, S.C., Frey, B. 2009. Drought tolerance of two black poplar (*Populus nigra* L.) clones: Contribution of carbohydrates and oxidative stress defense. *Plant, Cell and Environment* 32:1724–1736.

Rinalducci, S., Egidi, M.G., Karimzadeh, G., Jazii, F.R., Zolla, L. 2011. Proteomic analysis of a spring wheat cultivar in response to prolonged cold stress. *Electrophoresis* 32:1807–1818.

Rodziewicz, P., Swarcewicz, B., Chmielewska, K., Wojakowska, A., Stobiecki, M. 2014. Influence of abiotic stresses on plant proteome and metabolome changes. *Acta Physiologiae Plantarum* 36:1–19.

Sage, R.F. 2004. The evolution of C4 photosynthesis. *New Phytology* 161:341–370.

Sage, R.F., Way, D.A., Kubien, D.S. 2008. Photosynthetic carbon assimilation and associated metabolism in relation to water deficits in higher plants. *Journal of Experimental Botany* 59:1581–1595.

Saito, M., Yoshida, M. 2011. Expression analysis of the gene family associated with raffinose accumulation in rice seedlings under cold stress. *Journal of Plant Physiology* 168:2268–2271.

Savitch, L.V., Harney, T., Huner, N.P.A. 2000. Sucrose metabolism in spring and winter wheat in response to high irradiance, cold stress and cold acclimation. *Physiologia Plantarum* 108:270–278.

Seki, M., Kamei, A., Satou, M., Sakurai, T., Fujita, M., Oono, Y., Yamaguchi-Zhinozaki, K., Zhinozaki, K. 2003. Transcriptome analysis in abiotic stress conditions in higher plants. *Topics in Current Genetics* 4:271–295.

Setter, T.L., Flannigan, B.A. 2001. Water deficit inhibits cell division and expression of transcripts involved in cell proliferation and endoreduplication in maize endosperm. *Journal of Experimental Botany* 52:1401–1408.

Sharp, R.E. 2002. Interaction with ethylene: Changing views on the role of abscisic acid in root and shoot growth responses to water stress. *Plant, Cell and Environment* 25:211–222.

Shi, H. 2007. Integration of Ca^{2+} in plant drought and salt stress signal transduction pathways. In *Advances in Molecular Breeding toward Drought and Salt tolerant Crops*. M.A. Jenks, P.M. Hasegawa, S.M. Jain (eds.). Springer, Dordrecht, the Netherlands, pp. 141–182.

Shinozaki, K., Yamaguchi-Shinozaki, K. 2007. Gene networks involved in drought stress response and tolerance. *Journal of Experimental Botany* 58:221–227.

Signorelli, S., Corpas, F.J., Borsani, O., Barroso, J.B., Monza, J. 2013. Water stress induces a differential and spatially distributed nitro-oxidative stress response in roots and leaves of *Lotus japonicus*. *Plant Science* 201–202:137–146.

Silveira, A.G., Carvalho, F.E.L. 2016. Proteomics, photosynthesis and salt resistance in crops: An integrative view. *Journal of Proteomics* 143:24–35.

Singleton, P.W., Bohlool, B.B. 1983. The effect of salinity on the functional components of the soybean *Rhizobium japonicum* symbiosis. *Crop Science* 23:815–818.

Song, L., Jiang, L., Chen, Y., Shu, Y., Bai, Y., Guo, C. 2016a. Deep-sequencing transcriptome analysis of field-grown *Medicago sativa* L. crown buds acclimated to freezing stress. *Functional and Integrative Genomics* 16(5):495–511.

Song, L., Prince, S., Valliyodan, B., Joshi, T., Maldonado dos Santos, J.V., Wang, J., Lin, L et al. 2016b. Genome-wide transcriptome analysis of soybean primary root under varying water deficit conditions. *BMC Genomics* 15:57.

Sun, C.X., Gao, X.X., Li, M.Q., Fu, J.Q., Zhang, Y.L. 2016. Plastic responses in the metabolome and functional traits of maize plants to temperature variations. *Plant Biology* 18:249–261.

Suzuki, K., Aoki, N., Matsumura, H., Okamura, M., Ohsugi, R., Shimono, H. 2015. Cooling water before panicle initiation increases chilling-induced male sterility and disables chilling-induced expression of genes encoding OsFKBP65 and heat shock proteins in rice spikelets. *Plant, Cell and Environment* 38:1255–1274.

Suzuki, N., Mittler, R. 2006. Reactive oxygen species and temperature stresses: A delicate balance between signaling and destruction. *Physiologia Plantarum* 126:45–51.

Taiz, L., Zeiger, E. 2003. *Plant Physiology.* 3rd edition. Sinauer Associates Inc., Sunderland, MA.

Tang, X., Mu, X., Shao, H., Wang, H., Brestic, M. 2014. Global plant-responding mechanisms to salt stress: Physiological and molecular levels and implications in biotechnology. *Critical Reviews in Biotechnology.* Early Online: 1–13.

Tardieu, F., Simonneau, T. 1998. Variability among species of stomatal control under fluctuating soil water status and evaporative demand: Modelling isohydric and anisohydric behaviors. *Journal of Experimental Botany* 49:419–432.

Tarkowski, Ł.P., Van den Ende, W. 2015. Cold tolerance triggered by soluble sugars: A multifaceted countermeasure. *Frontiers in Plant Science* 6:1–7.

Taylor, S.H., Ripley, B.S., Woodward, F.I., Osborne, C.P. 2011. Drought limitation of photosynthesis differs between C3 and C4 grass species in a comparative experiment. *Plant, Cell and Environment* 34:65–75.

Tezara, W., Mitchell, V.J., Driscoll, S.D., Lawlor, D.W. 1999. Water stress inhibits plant photosynthesis by decreasing coupling factor and ATP. *Nature* 401:914–917.

Tian, X., Liu, Y., Huang, Z., Duan, H., Tong, J., He, X., Gu, W., Ma, H., Xiao, L. 2015. Comparative proteomic analysis of seedling leaves of cold-tolerant and -sensitive spring soybean cultivars. *Molecular Biology Reports* 42:581–601.

Turner, N.C., Jones, M.M. 1980. Turgor maintenance by osmotic adjustment: A review and evaluation. In *Adaptation of plants to water and high temperature stress*. N.C. Turner, P.J. Kramer (eds.). Wiley InterScience, New York, pp. 87–103.

Tuteja, N., Gill, S., Tuteja, R. 2011. Plant responses to abiotic stresses: Shedding light on salt, drought, cold, and heavy metal stress. In *Omics and Plant Stress Tolerance*. N. Tuteja, S. Gill, R. Tuteja (eds.). Bentham Science Publisher, Sharjah, UAE, pp. 39–64.

UNESCO Water Portal. 2007. http://www.unesco.org/water. Accessed 25.

Upchurch, R.G. 2008. Fatty acid unsaturation, mobilization, and regulation in the response of plants to stress. *Biotechnology Letters* 30:967–977.

USSL STAFF. 1954. Diagnosis and improvement of saline and alkali soils. In *Agriculture Handbook 60*. L.A. Richards (ed.). USDA, Washington, DC (Reprinted 1969).

Vargas, W.A., Pontis, H.G., Salerno, G.L. 2007. Differential expression of alkaline and neutral invertases in response to environmental stresses: Characterization of an alkaline isoform as a stress-response enzyme in wheat leaves. *Planta* 226:1535–1545.

Wang, X., Shan, X., Wu, Y., Su, S., Li, S., Liu, H., Han, J., Xue, C., Yuan, Y. 2016. iTRAQ-based quantitative proteomic analysis reveals new metabolic pathways responding to chilling stress in maize seedlings. *Journal of Proteomics* 146:14–24.

Westgate, M.E., Boyer, J.S. 1985. Carbohydrate reserves and reproductive development at low leaf water potentials in maize. *Crop Science* 25:762–769.

Wilkinson, S., Davies, J. 2002. ABA-based chemical signaling: The coordination of responses to stress in plants. *Plant, Cell and Environment* 25:195–210.

Willemot, C., Hope, H.J., Williams, R.J., Michaud, R. 1977. Changes in fatty acid composition of winter wheat during frost hardening. *Cryobiology* 14:87–93.

Wingler, A., Lea, P.J., Quick, W.P., Leegood, R.C. 2000. Photorespiration: Metabolic pathways and their role in stress protection. *Philosophical Transactions of the Royal Society of London B: Biological Sciences* 355:1517–1529.

Wu, G., Zhou, Z., Chen, P., Tang, X., Shao, H., Wang, H. 2014. Comparative ecophysiological study of salt stress for wild and cultivated soybean species from the Yellow River Delta, China. *The Scientific World Journal.* Article ID 651745.

Xiong, L. 2007. Abscisic acid in plant response and adaptation to drought and salt stress. In *Advances in Molecular Breeding toward Drought and Salt Tolerant Crops.* M. Jenks, P. Hasegawa, S. Jain (eds.). Springer, Dordrecht, the Netherlands, pp. 193–221.

Xiong, L., Ishitani, M. 2006. Stress signal transduction: Components, pathways and network integration. In *Abiotic Stress Tolerance in Plants.* A.K. Rai, T. Takabe (eds.). Springer, Dordrecht, the Netherlands, pp. 3–29.

Xiong, L., Schumaker, K.S., Zhu, J.K. 2002. Cell signaling during cold, drought, and salt stress. *The Plant Cell* 14:S165–S183.

Xiong, L., Zhu, J.K. 2002. Molecular and genetic aspects of plant responses to osmotic stress. *Plant, Cell and Environment* 25:131–139.

Xu, J., Li, Y., Sun, J., Du, L., Zhang, Y., Yu, Q., Liu, X. 2013. Comparative physiological and proteomic response to abrupt low temperature stress between two winter wheat cultivars differing in low temperature tolerance. *Plant Biology* 15:292–303.

Yadav, S., Irfan, M., Ahmad, A., Hayat, S. 2011. Causes of salinity and plant manifestations to salt stress: A review. *Journal of Environmental Biology* 32:667–685.

Yan, J., Wang, B., Jiang, Y., Cheng, L., Wu, T. 2013. GmFNSII-controlled soybean flavone metabolism responds to abiotic stresses and regulates plant salt tolerance. *Plant and Cell Physiology* 55:74–86.

Yan, W., Zhong, Y., Shangguan, Z. 2017. Rapid response of the carbon balance strategy in *Robinia pseudoacacia* and *Amorpha fruticosa* to recurrent drought. *Environmental and Experimental Botany* 138:46–56.

Yeo, A.R., Lee, K.S., Izard, P., Boursier, P.J., Flowers, T.J. 1991. Short- and long-term effects of salinity on leaf growth in rice (*Oryza sativa* L.). *Journal of Experimental Botany* 42:881–889.

Yin, Y., Yang, R., Han, Y., Gu, Z. 2015. Comparative proteomic and physiological analyses reveal the protective effect of exogenous calcium on the germinating soybean response to salt stress. *Journal of Proteomics* 113:110–126.

Yokoi, S., Quintero, F.J., Cubero, B., Ruiz, M.T., Bressan, R.A., Hasegawa, P.M., Pardo, J.M. 2002. Differential expression and function of *Arabidopsis thaliana* NHX Na^+/H^+ antiporters in the salt stress response. *Plant Journal* 30:529–539.

Zhang, H., Han, B., Wang, T., Chen, S., Li, H., Zhang, Y., Dai, S. 2012. Mechanisms of plant salt response: Insights from proteomics. *Journal of Proteome Research* 11:49–67.

Zhang, L., Ma, H., Chen, T., Pen, J., Yu, S., Zhao, X. 2014. Morphological and physiological responses of cotton (*Gossypium hirsutum* L.) plant to salinity. *PLoS One* 9:e112807.

Zheng, Q., Liu, J., Liu, R., Wu, H., Jiang, C., Wang, C., Guan, Y. 2016. Temporal and spatial distributions of sodium and polyamines regulated by brassinosteroids in enhancing tomato salt resistance. *Plant Soil* 400:147–164.

Zhou, F., Sosa, J., Feldmann, K.A. 2007. High throughput approaches for the identification of salt tolerance genes in plants. In *Advances in Molecular Breeding toward Drought and Salt Tolerant Crops.* M.A. Jenks, P.M. Hasegawa, S.M Jain. (eds.), Springer, Dordrecht, the Netherlands, pp. 359–379.

Zhu, J.K. 2003. Regulation of ion homeostasis under salt stress. *Current Opinion in Plant Biology* 6:441–445.

Zhu, S.-Q., Yu, C.-M., Liu, X.-Y., Ji, B.-H., Jiao, D.-M. 2007. Changes in unsaturated levels of fatty acids in thylakoid PSII membrane lipids during chilling-induced resistance in rice. *Journal of Integrative Plant Biology* 49:463–471.

Zhuo, Ch., Wang, T., Lu, S., Zhao, Y., Li, X., Guo, Z. 2013. A cold responsive galactinol synthase gene from *Medicago falcata* (MfGolS1) is induced by myo-inositol and confers multiple tolerances to abiotic stresses. *Physiologia Plantarum* 149:67–78.

2

Metabolic Adaptation and Allocation of Metabolites to Phloem Transport and Regulation Under Stress

Kathryn Dumschott, Andrew Merchant, and Millicent Smith

CONTENTS

Introduction

Plants must endure changes in their environment to promote both productivity and survival. Changes in the allocation of resources, particularly that of carbon amongst leaf-level soluble pools, represent a central mechanism by which downstream changes in structure and morphology are elicited. Allocation of carbon to specific compound classes in preparation for long-distance transport represents a major process underpinning plant-scale acclimation to changes in resource availability and development of adaptive traits. Phloem is the major conduit for long-distance transport in plants. Consequently, the processes of phloem transport are well characterised. Despite this, consideration of phloem-mediated transport of photoassimilates is less studied in the monitoring of plant health, most likely due to difficulties in the collection of phloem contents.

Long-distance movement of materials throughout the phloem stream varies in accordance with growth conditions. Among a range of potential consequences, unimpeded movement of photoassimilates is crucial to avoiding short-term damage associated with downstream starvation or upstream repression of the photosynthetic cycle (Ainsworth and Ort 2010). The phloem contents are therefore tightly regulated and contain dominant compound classes that are chemically reduced to avoid disruption to cellular processes. This phenomenon offers the potential for targeted analysis that reflects metabolic and physiological changes at the leaf level to changes in environmental conditions.

A select few compound classes dominate the phloem-transported pool of photoassimilates. Allocation to these compound classes is commonly closely associated with the primary reactions of photosynthesis (Stitt et al. 2010) and hence carry the capacity to impart both positive and negative influences over carboxylation rates (Ainsworth and Ort 2010). Shifts in allocation to metabolite classes combined with changes in the quantitative flux of compounds through the phloem transport stream have the capacity to significantly influence whole plant adaptive traits.

Here we outline major considerations in the utilisation of phloem sap contents in the evaluation of plant physiological performance. Following a brief introduction to the structure and function of phloem-mediated transport, we outline examples of how phloem sap analysis can be used to infer plant physiological status. Finally, we identify some important considerations for the sampling of phloem for those seeking to use phloem sap in future plant monitoring and improvement programs.

Phloem Function Communicates amongst Plant Organs

The process of phloem transport consists of a tightly regulated series of events. Phloem is the central conduit responsible for the transport of nutrients, defensive compounds, and information signals synthesised in leaves to the rest of the plant (Turgeon and Wolf 2009). Unlike dead xylem cells that facilitate the vast majority of water transport in plants, phloem is made up of living cells with a plasma membrane imparting regulation over solute movement. In brief, specialised cells termed 'sieve elements' form to combine a series of sieve tubes, which act as open channels for transport (Pritchard 2007). At the end of each sieve tube lie sieve plates containing pores of 0.5–1.5 μm in length and 1–15 μm in diameter (Pritchard 2007) which form the major resistance to photoassimilate flow along the sieve tube (Oparka and Turgeon 1999). Sieve tube elements are associated with one or more companion cells to form a sieve element-companion cell complex (SE-CCC). Companion cells have a metabolic function, supplying ATP and proteins to the sieve tube element and being involved in the rapid exchange of solutes through plasmodesmata (van Bel 2003). Plasmodesmatal apertures provide an uninterrupted and relatively unrestricted pathway for materials within the phloem (Turgeon 2010), as they allow cytoplasmic continuity between two adjacent cells by passing through their cell walls (Pritchard 2007).

Transport within the phloem stream is driven by a passive mechanism described by the pressure-flow model (Munch 1930), such that the mass flow of solutes is driven by a hydrostatic pressure gradient between sources and sinks of photoassimilates (Thompson 2006). The pressure-flow model is driven by the processes of phloem loading and unloading to maintain a downward gradient of pressure potential contributed to by the solutes (Thompson 2006). Loading of photoassimilates into the phloem therefore concentrating sugars in the sieve element, increases turgor pressure (e.g. Turgeon 2010) and is thought to be a major driver of phloem flow. Loading of sugars primarily occurs in minor vein networks, with major leaf veins and the remaining phloem pathway principally functioning as a long-distance transport conduit (van Bel et al. 1996; Lalonde et al. 2003). Phloem loading is recognised as being either apoplastic or symplastic, depending on the path of migration into the sieve element. In apoplastic loading, photoassimilates move by diffusion from the mesophyll cell into the apoplast, after which they are actively transported into the sieve element (Turgeon 2010). The transporter for this active step is specific to each compound. For instance, sucrose is transported via an ATP-dependent sucrose-H^+ symporter (Lalonde et al. 2003).

Symplastic loading involves the movement of sugars by diffusion from the mesophyll cell to the sieve element through plasmodesmata. Due to its independence from energy consumption, symplastic loading may lead to higher flux rates (Turgeon 2010). Evidence for symplastic phloem loading is largely drawn from the common occurrence of galactosylated sugars in the phloem sap and led to the proposal of the polymer trapping model for phloem loading (Turgeon 1991). This mechanism proposes that sucrose synthesised in the mesophyll cell is diffused into the bundle sheath and intermediary cells via plasmodesmata (Turgeon 2010). Within the intermediary cells, raffinose family oligosaccharides (RFOs) are synthesised from transported sucrose and galactinol. The resulting increased size of the RFO prevents diffusion back into the bundle sheath, and thus they move into the sieve element. Consequently, species using the polymer-trapping model of symplastic loading commonly contain RFOs and other galactosylated photoassimilates as constituents in their phloem.

For most other phloem components, very little is known about specific transporters, but the identification and characterisation of transporters is essential to understanding phloem functions (Dinant 2008). Due to the selectivity of apoplastic loading, transport rates may be modified, thus acting as a significant regulator of upstream and downstream metabolic processes. Tight control over this central process of plant growth and development may provide significant advantages for balancing resource use, a notion supported by its more common occurrence among plants living in marginal environments (Turgeon et al. 2001). Loading of photoassimilates into phloem may limit plant growth (Sowinski 1999; Turgeon 2000), and as such, the collective 'mobile pool' of photoassimilates provides valuable integration of plant physiological status. A corollary is that the mobile carbon pool helps plants adapt and acclimate to short- and medium-term changes in the environment.

Morphological, chemical, and now molecular tools have provided several lines of evidence for the existence of both apoplastic and symplastic transmembrane movement of solutes into the phloem (Turgeon 2000, 2006; Noiraud et al. 2001;

Ramsperger-Gleixner et al. 2004; Pommerenig et al. 2007) as well as the fact that this capacity may vary with plant age (e.g. Ramsperger-Gleixner et al. 2004). Whilst substantial evidence for the function of H+ symports via apoplastic pathways have been collected for the transport of sucrose (Knop et al. 2004) and polyols such as mannitol (Noiraud et al. 2001) and sorbitol (Ramsperger-Gleixner et al. 2004), to date, no alternative symplastic mechanism for the movement of cyclic polyols (apart from diffusion) has been suggested. This is the case even though oligomerisation of some cyclic polyols has been demonstrated in certain species via a mechanism closely related to that of raffinose synthesis (Peterbauer et al. 2003; Ma et al. 2005; Lahuta 2006). The functional significance of these mechanisms cannot be overstated and offer promising molecular targets for the identification and characterisation of the dynamics of metabolite transport via the phloem.

Phloem Contents Indicate Physiological Status

Metabolomic characterisation of phloem sap content is rare due to the significant challenges associated with unadulterated sample collection. While complete identification of solutes in phloem has been difficult, it is known that translocated solutes dissolved within water are a complex mix of organic and inorganic materials including, but not limited to, sugars, sugar alcohols, amino acids, hormones, ions, RNAs and proteins, and some secondary compounds involved in defence and protection (Turgeon and Wolf 2009; Turgeon 2010). Solute composition in the phloem is not constant, and variation has been found between plant species, developmental stages (Pate and Atkins 1983), and changes in environment (Pate et al. 1974; Smith and Milburn 1980). Phloem sap is also likely to be modified along the transport route by metabolism in companion cells, leakage from sieve elements, and retrieval of materials from the transpiration stream (Lalonde et al. 2003). As most of what we know about phloem composition is influenced by phloem sampling methods (Pritchard 2007) our understanding of the transported solute pool is not complete.

For recently fixed photoassimilates, sucrose is a ubiquitous constituent of all phloem sap (Lalonde et al. 2003) and in most plants is the major constituent (Dinant et al. 2010). Other sugars may also be translocated, including raffinose family oligosaccharides (RFO) such as raffinose, stachyose, verbascose, ajugose, and polyols, including mannitol and sorbitol (Lalonde et al. 2003; Turgeon and Wolf 2009; Dinant et al. 2010). More generally, carbohydrates are transported in phloem in the form of non-reducing sugars to prevent damage to the phloem stream (e.g. Raven 1991). Despite this, more recent studies suggest that in some species reducing compounds such as fructose and glucose are also transported in large amounts despite their likelihood of causing cellular damage (van Bel and Hess 2008).

Studies detecting changes in the composition of phloem metabolites in response to stress must be interpreted on the background of flux. For example, increases in the absolute concentrations of metabolites in the phloem are likely decoupled from photosynthetic rates, as concentrations may be influenced by reduced water supply or reduced flux rates. Changes in the ratio of solutes obtained from phloem sap over seasonal cycles have been observed in phloem-bled *Eucalyptus* (Merchant et al. 2010b) and found to be reversible under controlled conditions (Merchant et al. 2010a), suggesting that non-parametric approaches (such as the ratio of compounds within the phloem) may be used in evaluating physiological status. Another promising non-parametric approach is the use of naturally occurring carbon isotopes ($\delta^{13}C$), a commonly used modelling approach to determine aspects of plant water use and carbon fixation (Farquhar et al. 1982, 1989; Farquhar and Richards 1984; Seibt et al. 2008). Recent work has identified that sugars derived from the phloem sap are a good representation of recently fixed photoassimilates in leaves (Keitel et al. 2006). The movement of carbon into the phloem and subsequent distribution to other plant components (i.e. heterotrophic tissues) incorporates a number of steps under metabolic, diffusional, and enzymatic control (see for example Cernusak et al. 2009). The potential for fractionation of isotopes during this process warrants further investigation due to its likely influence over predictions of water use efficiency obtained from the use of carbon isotope abundance in heterotrophic tissues. Despite these limitations, several studies have used the naturally occurring isotopic abundance of phloem-derived metabolites to correlate strongly with leaf-level processes among a range of plant types (Gessler et al. 2008, 2009; Merchant et al. 2010b, 2011; Merchant 2012).

These findings illustrate the potential for phloem-derived metabolites to inform plant-scale physiological processes. At present, determining the reliability and functional significance of such patterns is difficult due to our inability to obtain 'pure' phloem sap. Sieve elements are under high internal turgor pressure. If damage occurs, the release of this pressure causes solutes within the sieve element to surge to the cut end, potentially leading to the release of phloem sap (Taiz and Zeiger 2010). Due to the connectivity of the phloem stream, to avoid extensive damage, sieve elements may be rapidly sealed through wounding mechanisms. Phloem structural proteins called P-proteins are found within many plant groups and appear to function by plugging sieve plate pores, essentially sealing damaged sieve elements (Clark et al. 1997). While this mechanism works for the short term, a longer-term solution is the production of callose. Callose is a glucose polymer that is synthesised in response to damage, stress, or developmental preparation such as dormancy, and when pressure within the sieve element is lost, callose is deposited on sieve plates, sealing the sieve element from surrounding tissue (van Bel 2003). Callose will remain in place until the sieve elements have been able to return to high internal pressure. Sampling techniques must therefore overcome these responses in order to enable the unadulterated sample collection of the transported pool.

Phloem Sampling

Phloem sap is difficult to sample due to the damage that occurs when pressurised sieve elements are wounded. Sudden pressure release can damage a series of cells, rupture organelles and anchoring molecules, disrupt sieve element reticulum, alter the configuration of phloem proteins, plug pores, and initiate wound reactions. It may also pull substances into sieve elements from surrounding cells, particularly companion cells, leading to contamination of sap (Turgeon and Wolf 2009). There are a number of methods of sampling phloem sap, each with particular strengths for ease, repeatability, and levels of contamination; however, no method gives a clear, complete result of the composition of moving phloem sap (Turgeon and Wolf 2009). Each study, therefore, must consider the strengths of each technique and select the most suitable technique for collection of the compounds of interest.

Feeding insects can be used to sample sieve-tube solute composition, as their stylets tap into a single sieve element (Dinant et al. 2010). This method is effective because insects take advantage of the high turgor pressure within the sieve element, and cell contents are forced through the stylet into the gut of the insect (Dinant et al. 2010). Stylectomy has been an important development in phloem research, as it avoids contamination from other cells, which is an artefact of alternative phloem sampling methods. However, this method can be technically difficult and is restricted to particular plant-insect combinations (Turgeon and Wolf 2009). It has been noted that stylet exudate from aphids may differ to insect-free exudate (Kennedy and Mittler 1953). Aphids may potentially alter sap chemistry, as insect saliva is exuded to inhibit sieve plate blocking and composition may be selected by aphid dietary preferences (Dinant et al. 2010).

Some plant species bleed exudate from severed phloem elements, encouraging pressure release of solutes (Pate et al. 1998). The advantage of this technique is that it collects the transported pool of phloem sap contents without the need for more invasive extraction techniques that will extract metabolites from the companion cell complexes. This method relies on sufficient pressure within sieve tubes to promote exudation and may not be compatible with some species or studies, particularly if plants are under a severe water deficit (Pate et al. 1998). Contamination of collected sap with solutes from surrounding tissues can also be problematic (Pate et al. 1974). For instance, incisions too deep into the cambium resulting in sap being sucked back and escaping into the xylem (Pate et al. 1998). Wounding mechanisms that are initiated when a plant is severed may also impact the effectiveness of exudation. The depth of the cut and the longevity of sap collection may therefore influence the identity and quantity of the harvested compounds.

The extraction of phloem sap by excision of the phloem material may be achieved by either retrieval of exudate from sieve tubes via osmosis or by centrifugation (see Gessler et al. 2004; Devaux et al. 2009). For osmosis, sections of stem or cambium tissue are usually cut from the plant and placed in a solution of distilled water (Devaux et al. 2009). In some cases the addition of a chelating agent to the solution may be necessary to avoid callose sealing sieve tubes and preventing exudation through osmosis (van Bel 2003). Adding a chelating agent such as EDTA (ethylenediaminetetraacetic acid) prevents callose production, which can block sieve plates and inhibit exudation (Gessler et al. 2004), although the additions of chelating agents that contain carbon, oxygen, or nitrogen can decrease the precision of studies measuring isotopic abundance (Gessler et al. 2004). Extraction methods usually yield exudate from the entire sieve element-companion cell complex, making it difficult to analyse the composition of the sieve element alone.

Summary

The use of phloem sap as an integrated reflection of plant physiological performance holds great promise. Phloem is the central conduit for the long-distance movement of photo-assimilates and thus has the capacity to integrate both spatially and temporally. Consequently, phloem metabolite and isotope content (particularly that of carbon) are closely coupled with leaf-level biochemistry. Using phloem-derived information to predict leaf-level processes is restricted by the consideration of flux and the challenges of obtaining 'pure' phloem sap. Several options for obtaining phloem sap exist, although their use depends on the species under investigation. Non-parametric measures of phloem contents provide useful information across a range of species, demonstrating the potential for phloem-sap diagnostic tools in the plant sciences.

REFERENCES

Ainsworth, E.A., Ort, D.R. (2010) How do we improve crop production in a warming world? *Plant Physiology* 154, 526–530.

Cernusak, L.A., Tcherkez, G., Keitel, C., Cornwell, W.K., Santiago, L.S., Knohl, A., Barbour, M.M., et al. (2009) Why are non-photosynthetic tissues generally ^{13}C enriched compared with leaves in C_3 plants? Review and synthesis of current hypotheses. *Functional Plant Biology* 36, 199–213.

Clark, A.M., Jacobsen, K.R., Bostwick, D.E., Dannenhoffer, J.M., Skaggs, M.I., Thompson, G.A. (1997) Molecular characterization of a phloem-specific gene encoding the filament protein, phloem protein 1 (PP1), from *Cucurbita maxima*. *Plant Journal* 12, 49–61.

Devaux, M., Ghashghaie, J., Bert, D., Lambrot, C., Gessler, A., Bathellier, C., Ogee, J., Loustau, D. (2009) Carbon stable isotope ratio of phloem sugars in mature pine trees throughout the growing season: Comparison of two extraction methods. *Rapid Communications in Mass Spectrometry* 23, 2511–2518.

Dinant, S. (2008) Phloem, transport between organs and long-distance signalling. *Comptes Rendus Biologies* 331, 334–346.

Dinant, S., Bonnemain, J.L., Girousse, C., Kehr, J. (2010) Phloem sap intricacy and interplay with aphid feeding. *Comptes Rendus Biologies* 333, 504–515.

Farquhar, G.D., Ehleringer, J.R., Hubick, K.T. (1989) Carbon isotope discrimination and photosynthesis. *Annual Review of Plant Physiology and Plant Molecular Biology* 40, 503–537.

Farquhar, G.D., O'Leary, M.H., Berry, J.A. (1982) On the relationship between carbon isotope discrimination and the intercellular carbon dioxide concentration in leaves. *Australian Journal of Plant Physiology* 9, 121–137.

Farquhar, G.D., Richards, R.A. (1984) Isotopic composition of plant carbon correlates with water use efficiency of wheat genotypes. *Australian Journal of Plant Physiology* 11, 539–552.

Gessler, A., Brandes, E., Buchmann, N., Helle, G., Rennenberg, H., Barnard, R.L. (2009) Tracing carbon and oxygen isotope signals from newly assimilated sugars in the leaves to the tree-ring archive. *Plant Cell and Environment* 32, 780–795.

Gessler, A., Rennenberg, H., Keitel, C. (2004) Stable isotope composition of organic compounds transported in the phloem of European beech—Evaluation of different methods of phloem sap collection and assessment of gradients in carbon isotope composition during leaf-to-stem transport. *Plant Biology* 6, 721–729.

Gessler, A., Tcherkez, G., Peuke, A.D., Ghashghaie, J., Farquhar, G.D. (2008) Experimental evidence for diel variations of the carbon isotope composition in leaf, stem and phloem sap organic matter in *Ricinus communis*. *Plant Cell and Environment* 31, 941–953.

Keitel, C., Matzarakis, A., Rennenberg, H., Gessler, A. (2006) Carbon isotopic composition and oxygen isotopic enrichment in phloem and total leaf organic matter of European beech (*Fagus sylvatica* L.) along a climate gradient. *Plant Cell and Environment* 29, 1492–1507.

Kennedy, J.S., Mittler, T.E. (1953) A method of obtaining phloem sap via the mouth-parts of aphids. *Nature* 171, 528–528.

Knop, C., Stadler, R., Sauer, N., Lohaus, G. (2004) AmSUT1, a sucrose transporter in collection and transport phloem of the putative symplastic phloem loader Alonsoa meridionalis. *Plant Physiology* 134, 204–214.

Lahuta, L.B. (2006) Biosynthesis of raffinose family oligosaccharides and galactosyl pinitols in developing and maturing seeds of winter vetch (Vicia villosa Roth.). *Acta Societatis Botanicorum Poloniae* 75, 219–227.

Lalonde, S., Tegeder, M., Throne-Holst, M., Frommer, W.B., Patrick, J.W. (2003) Phloem loading and unloading of sugars and amino acids. *Plant Cell and Environment* 26, 37–56.

Ma, J.M., Horbowicz, M., Obendorf, R.L. (2005) Cyclitol galactosides in embryos of buckwheat stem-leaf-seed explants fed D-chiro-inositol, myo-inositol or D-pinitol. *Seed Science Research* 15, 329–338.

Merchant, A. (2012) Developing phloem δ^{13}C and sugar composition as indicators of water deficit in *Lupinus angustifolius*. *Hortscience* 47, 691–696.

Merchant, A., Peuke, A.D., Keitel, C., Macfarlane, C., Warren, C., Adams, M.A. (2010a) Phloem sap and leaf δ^{13}C, carbohydrates and amino acid concentrations in *Eucalyptus globulus* change systematically according to flooding and water deficit treatment. *Journal of Experimental Botany* 61, 1785–1793.

Merchant, A., Tausz, M., Keitel, C., Adams, M.A. (2010b) Relations of sugar composition and δ^{13}C in phloem sap to growth and physiological performance of *Eucalyptus globulus* (Labill) *Plant Cell and Environment* 33, 1361–1368.

Merchant, A., Wild, B., Richter, A., Bellot, S., Adams, M.A., Dreyer, E. (2011) Compound-specific differences in (13)C of soluble carbohydrates in leaves and phloem of 6-month-old Eucalyptus globulus (Labill). *Plant Cell and Environment* 34, 1599–1608.

Munch, E. (1930) *Die Stoffbewegungen in der Pflanze.* Jena, Germany: Gustav Fischer Verlagsb.

Noiraud, N., Maurousset, L., Lemoine, R. (2001) Identification of a mannitol transporter, AgMaT1, in celery phloem. *Plant Cell* 13, 695–705.

Oparka, K.J., Turgeon, R. (1999) Sieve elements and companion cells—Traffic control centers of the phloem. *Plant Cell* 11, 739–750.

Pate, J.S., Atkins, C.A. (1983) Xylem and phloem transport and the functional economy of carbon and nitrogen of a legume leaf. *Plant Physiology* 71, 835–840.

Pate, J.S., Sharkey, P.J., Lewis, O.A.M (1974) Phloem bleeding form legume fruits—Technique for study of fruit nutrition. *Planta* 120, 229–243.

Pate, J., Shedley, E., Arthur, D., Adams, M. (1998) Spatial and temporal variations in phloem sap composition of plantation-grown *Eucalyptus globulus*. *Oecologia* 117, 312–322.

Peterbauer, T., Brereton, I., Richter, A. (2003) Identification of a digalactosyl ononitol from seeds of adzuki bean (Vigna angularis). *Carbohydrate Research* 338, 2017–2019.

Pommerenig, B., Papini-Terzi, F.S., Sauer, N. (2007) Differential regulation of sorbitol and sucrose loading into the phloem of *Plantago major* in response to salt stress. *Plant Physiology* 144, 1029–1038.

Pritchard, J. (2007) Solute transport in the phloem. In: *Plant Solute Transport*, A.R. Yeo, T.J. Flowers (Eds.). Oxford: Blackwell Publishing.

Ramsperger-Gleixner, M., Geiger, D., Hedrich, R., Sauer, N. (2004) Differential expression of sucrose transporter and polyol transporter genes during maturation of common plantain companion cells. *Plant Physiology* 134, 147–160.

Raven, J.A. (1991) Long-term functioning of enucleate sieve elements—Possible mechanisms of damage avoidance and damage repair. *Plant Cell and Environment* 14, 139–146.

Seibt, U., Rajabi, A., Griffiths, H., Berry, J.A. (2008) Carbon isotopes and water use efficiency: Sense and sensitivity. *Oecologia* 155, 441–454.

Smith, J.A.C, Milburn, J.A. (1980) Osmoregulation and the control of phloem-sap composition in Ricinus-communis L. *Planta* 148, 28–34.

Sowinski, P. (1999) Transport of photoassimilates in plants under unfavourable environmental conditions. *Acta Physiologiae Plantarum* 21, 75–85.

Stitt, M., Lunn, J., Usadel, B. (2010) Arabidopsis and primary photosynthetic metabolism—More than the icing on the cake. *Plant Journal* 61, 1067–1091.

Taiz, L., Zeiger, E. (2010) *Plant Physiology*, 5th edition. Sunderland: Sinauer Associates Inc.

Thompson, M.V. (2006) Phloem: The long and the short of it. *Trends in Plant Science* 11, 26–32.

Turgeon, R. (1991) Symplastic phloem loading and the sink-source transition in leaves: A model. In: *Recent Advances in Phloem Transport and Assimilate Compartmentation*, J.L. Bonnemain, S. Delrot, W.J. Lucas, J. Dainty (Eds.). Nantes, France: Quest Editions, pp. 18–22.

Turgeon, R. (2000) Plasmodesmata and solute exchange in the phloem. *Australian Journal of Plant Physiology* 27, 521–529.

Turgeon, R. (2006) Phloem loading: How leaves gain their independence. *Bioscience* 56, 15–24.

Turgeon, R. (2010) The role of phloem loading reconsidered. *Plant Physiology* 152, 1817–1823.

Turgeon, R., Medville, R., Nixon, K.C. (2001) The evolution of minor vein phloem and phloem loading. *American Journal of Botany* 88, 1331–1339.

Turgeon, R., Wolf, S. (2009) Phloem transport: Cellular pathways and molecular trafficking. *Annual Review of Plant Biology* 60, 207–221.

van Bel, A.J.E (2003) The phloem, a miracle of ingenuity. *Plant Cell and Environment* 26, 125–149.

van Bel, A.J.E, Hendriks, J.H.M, Boon, E.J.M.C., Gamalei, Y.V., vandeMerwe, A.P. (1996) Different ratios of sucrose/raffinose-induced membrane depolarizations in the mesophyll of species with symplasmic (*Catharanthus roseus, Ocimum basilicum*) or apoplasmic (*Impatiens walleriana, Vicia faba*) minor-vein configurations. *Planta* 199, 185–192.

van Bel, A.J.E, Hess, P.H. (2008) Hexoses as phloem transport sugars: The end of a dogma? *Journal of Experimental Botany* 59, 261–272.

3

Mechanism of Salt Stress Tolerance and Pathways in Crop Plants

Manu Kumar and Mahipal Singh Kesawat

CONTENTS

Abbreviations

ABA	Abscisic acid
CDPKs	Calcium-dependent protein kinases
MAPK	Mitogen activated protein kinase
NOXes	NADPH oxidases
POX	Peroxidase
RLKs	Receptor-like kinases
SOS	Salt overly sensitive

Introduction

Salt stress tolerance is a very complex phenomenon which involves many gene products. Rice plants can respond to this stress as a whole organism and as individual cells. Salt stress induces ionic stress and osmotic stress which increases Na^+ ion toxicity, disturbs ion homeostasis, and alters cellular metabolism. To cope with adversary conditions to growth, plants carry out osmotic and ionic signaling that makes them salt stress tolerant by maintaining osmotic and ion homeostasis. Osmotic stress affects the growth of plants mainly through ABA-dependent and ABA-independent pathways and plants overcome salt stress by the activation of many downstream genes. Ionic stress on the other hand activates the IP3 pathway and calcium signaling, which in turn activates the calmodulin pathway and salt overly sensitive (SOS) pathway to maintain ion homeostasis.

Osmotic Stress

Plant response to salt and drought stress is similar in nature. Both are involved in disruption of the ion and osmotic equilibrium of the plant, and therefore, the signaling pathways tend to be similar (Figure 3.1). Abscisic acid (ABA) is thought to have a role in regulating plant water balance and osmotic stress by regulating plant guard cells. ABA expression is increased by both salt and drought stress. Plant response to osmotic stress is to maintain osmotic homeostasis via an ABA-dependent pathway and an ABA-independent pathway (Ramakrishna et al. 2011, 2013; Kumar et al. 2013).

ABA-Dependent Pathway in Rice

Salt receptors perceive salt stress signals, which leads to the generation of many secondary signal molecules such as ABA, Ca^{2+} inositol phosphates, and reactive oxygen species (ROS). According to studies, no salt receptor has been reported yet. In the salinity stress response, ABA acts as a crucial signal (Fujita et al. 2011). Salinity stress increases the ABA content, which leads to the expression of numerous genes. Osmotic stress is critically mediated by an increase in the ABA level generated by rice. The biosynthesis of ABA is induced by osmotic stress; increased levels of ABA exert positive feedback on its own biosynthetic pathway. The expression of the phytoene synthase gene 9-cis-epoxycarotenoid dioxygenases (*OsNCED3*, *OsNCED4*, and *OsNCED5*) and *OsPSY3* in rice is very highly induced under salt stress, and their expression leads to the biosynthesis of ABA. In turn, ABA activates the expression of *OsPSY3* and *OsNCED*s genes in rice by positive feedback regulation (Welsch et al. 2008). This shows that the *OsPSY3* and *OsNCED*s genes act upstream of the ABA biosynthesis pathway. ABA acts as a regulator, which can initiate a second round of signaling. We will discuss here the transcription factor (TF) genes, heat shock protein genes, calcium-dependent protein kinase genes, and micro RNA that are involved in salt stress tolerance through the ABA-dependent pathway in rice plants.

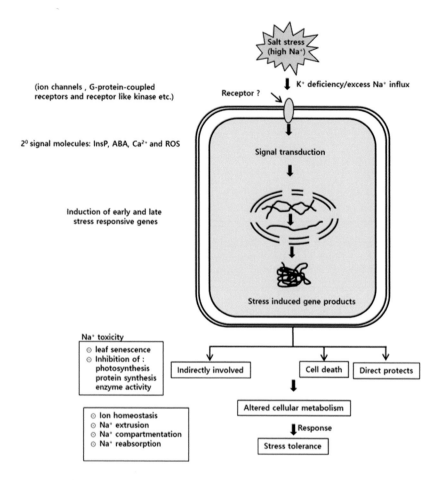

FIGURE 3.1 Functional demarcation of salt stress signaling pathways in rice. The inputs for osmotic and ionic and signaling pathways are osmotic (e.g., turgor) and ionic (excess NaCl) changes. Osmotic and ion homeostasis is the output of osmotic and ionic signaling. Growth and alteration of cellular metabolism under salt stress depend on osmotic and ionic signaling. (Adapted from Tuteja, N. and Sopory, S.K., *Plant Signal. Behav.*, 3, 79, 2008.)

The transcriptional regulatory network of *cis*-acting elements mediated by ABA-dependent TFs and *cis*-acting elements involved in salt stress gene expression is depicted in Figure 3.2. A promoter of stress-induced genes contains *cis*-regulatory elements such as ABRE, MYCRS/MYBRS, and NARC, which are regulated by various upstream TFs (Figure 3.2). The ABA-dependent salinity stress signaling activates *OsHsfC1b*, *OsARAB*, *MYC/MYB*, and *OsNAC/SNAC*, which bind to *ABRE*, *MYCRS/MYBRS*, NARC, and regulatory elements, respectively, to induce stress-responsive genes.

The promoter region of ABA-inducible genes has been identified. It contains a conserved *cis*-acting element named ABRE. ABRE-binding proteins include the *Arabidopsis* cDNAs that encode bZIP-type TFs (Shinozaki et al. 2006). Among these genes, *AREB2/ABF4*, *ABF3*, and *AREB1/ABF2* were reported to be induced by osmotic stress and ABA (Fujita et al. 2011). ABA-dependent posttranscriptional modification is important for the activation of *AREB1*. Salt stress induced one of the rice bZIP-types, *OsABF2* (Hossain et al. 2010). *OsABF2*, a T-DNA insertion mutant, showed salt stress sensitivity compared with control plants. The loss of function approach validates the *OsABF2* as a positively functioning TF. *OsABI5*, a rice bZIP-type TF, was isolated from rice panicles (Zou et al. 2008). Its expression was induced by high ABA and salt, but drought and cold stress reduced its expression in rice

seedlings. *OsABI5-Ox* plants were highly salt sensitive. In contrast, *OsABI5*-antisense plants were salt tolerant, but fertility was decreased (Zou et al. 2008). These findings indicate that *OsABI5* might regulate plant fertility and adaptation to stress. Nijhawan et al. (2008) analyzed the expression of 89 *OsbZIP* genes by surveying the rice genome for bZIP family proteins. Their microarray analysis indicated that *OsbZIP23* was upregulated by salt stress. Salt stress and ABA induced the expression of *OsbZIP23* (Xiang et al. 2008). Overexpressing *OsbZIP23* transgenic rice exhibited improved high-salinity stress tolerance. In contrast, salinity stress tolerance decreased in a null mutant for this gene. Microarray analyses of *OsbZIP23-Ox* and mutant revealed 37 possible *OsbZIP23*-specific target genes, which showed a reverse expression pattern in both types of *OsbZIP23* plants. Overexpression of the dehydrin gene, *OsDhn1*, improves drought and salt stress tolerance through scavenging of ROS in rice (Kumar et al. 2014). Ectopic expression of *OsSta2* enhances salt stress tolerance in rice (Kumar et al. 2017). These findings suggest that as in *Arabidopsis*, rice AREB homologs also regulate ABA and salt stress–inducible gene expression.

Heat shock factors (HSFs) play an important role in the response to abiotic stress. In *Arabidopsis*, *HsfA2* controls the response to osmotic stress, salt, submergence, and anoxia (Ogawa et al. 2007; Banti et al. 2010). In rice, the expression of *OsHsfC1b* was induced by

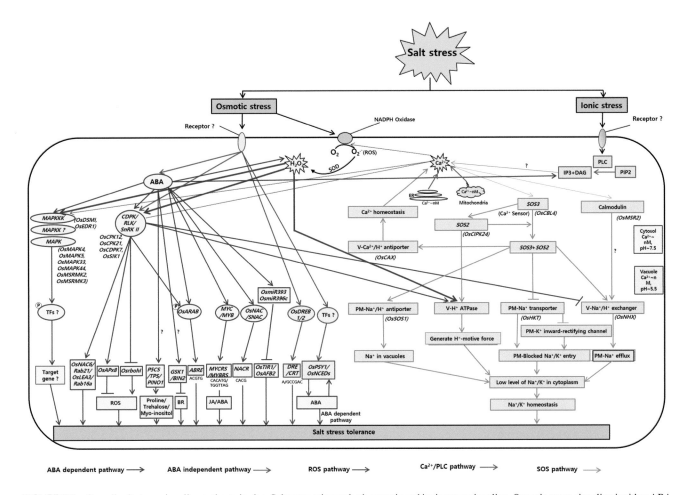

FIGURE 3.2 Overall salt stress signaling pathway in rice. Salt stress triggers both osmotic and ionic stress signaling. Osmotic stress signaling is either ABA dependent or ABA independent. ABA-dependent signaling includes MAP kinase cascades, CDPK, RLK, SnRK, OsRAB1, MYC/MYB, OsNAC/SNAC, and micro RNAs. ABA-independent signaling includes OsDREB1, OsDREB2, OsPSY1, and OsNCEDs. Ionic stress signaling is via the Ca²⁺/PLC, SOS, and CaM pathways. A blue arrow indicates an ABA-dependent pathway, a brown arrow shows an ABA-independent pathway, a blue-violet arrow shows an ROS pathway, a deep pink arrow shows a Ca²⁺/PLC pathway, and an orange-red arrow shows an SOS pathway. (Adapted from Kumar, K., et al., *Rice*, 6, 27, 2013.)

ABA, salt, and mannitol, whereas *hsfc1b* miRNA and mutant lines showed decreased osmotic and salt stress tolerance and increased sensitivity to salt-responsive genes involved in ion homeostasis and signaling and to external ABA (Schmidt et al. 2012). Three putative ABA-response elements were found in the promoter of *OsHsfC1b* (Mittal et al. 2009; Schmidt et al. 2012), and ABA induced their expression. These findings show the role of *OsHsfC1b* in ABA-dependent pathways. ABA hypersensitivity is often accompanied by hypersensitivity to osmotic stress and salt (Borsani et al. 2001; Pandey et al. 2005; Zhu et al. 2010). Thus, osmotic stress and the salt tolerance conferred by *OsHsfC1b* are most likely ABA dependent.

SNF1-related protein kinases (SnRKs) function as key regulators of salt stress and ABA signaling in plants and were found to be activated by ABA and diverse stress signals (Yoshida et al. 2002; Jossier et al. 2009). In rice, the entire SnRK2 family has been shown to be activated by hyperosmotic stress, and this activation involves phosphorylation (Kobayashi et al. 2004). *SAPK1* becomes active at NaCl concentrations higher than 300 mM, whereas *SAPK2* becomes active at lower concentrations. The C-terminal domain is responsible for these responses, as revealed by domain exchange experiments (Kobayashi et al. 2004). Several studies have shown that in response to salinity stress, *Arabidopsis* and rice SnRK2s are activated. Some are also regulated by ABA and some directly by

osmotic stress. Analysis of the C-terminal domain of SnRK2 showed that this region was responsible for ABA activation (Kobayashi et al. 2004). Furthermore, *Arabidopsis* SnRK2.6, which is activated by osmotic stress and is involved in stomatal closure in response to ABA, has two regulatory domains at its C-terminus (Mustilli et al. 2002; Yoshida et al. 2002). ABA-dependent activation was inhibited when part of Domain II was deleted, whereas osmotic-dependent activation remained unchanged. In the ABA-dependent activation of SnRK2.6, the PP2C encoded by ABI1 might play an important role, since the abi1-1 mutation inhibits its ABA-dependent activation, and the PP2C interacts directly with SnRK2.6 (Yoshida et al. 2006). Salt and osmotic stress induces kinase activity of *OSRK1/ SAPK6* in vitro and when transiently expressed in rice suspension cells (Kobayashi et al. 2004; Chae et al. 2007). Overexpressing *OSRK1* in rice under the control of CaMV 35S promoter indicates that *OSRK1* is an upstream regulator of stress signaling in rice (Nam et al. 2012). The direct phosphorylation of various downstream targets (*SLAC1*, *KAT1*, *AtRbohF*, and TFs required for the expression of numerous stress response genes) regulates the plant response to ABA via SnRK2s pathways (Kulik et al. 2011). The SnRK2 protein kinase family has evolved specifically for hyperosmotic stress signaling by modifying the C-terminal domain for ABA responsiveness (Kobayashi et al. 2004). SnRKs could affect stress signaling through ABA-response element binding proteins (AREBPs), a *bZIP* TF family that is unique to plants and regulates the expression of ABA-responsive genes. AREBPs contain highly conserved SnRK1 target sites, which are very good substrates for phosphorylation by SnRK1 (Zhang et al. 2008). SnRK2-type protein kinases phosphorylate AREBPs (Furihata et al. 2006; Kobayashi et al. 2005). *SAPK4* (one of the sucrose nonfermenting 1-related protein kinases), acts as a regulator of salt acclimation in rice that controls photosynthetic activity and ionic homeostasis and allows continued development and growth in high-salinity conditions. *SAPK4*-Ox rice plants show decreased Na$^+$ accumulation and reduced transcript amounts of *OsNHX1*, indicating that salt stress tolerance is caused by cellular Na$^+$ exclusion, not by vacuolar sequestration of the ion. The expression levels of the vacuolar ATPase increase and the Cl-channel *OsCLC1* decrease in transgenic and wild-type (WT) rice (Diedhiou et al. 2008; Nakamura et al. 2006). Transcription of SAPK4 was analyzed by Kobayashi et al. (2004) in the roots, leaf sheaths, and leaf blades of 30-DAG rice in normal conditions, and under ABA, mannitol, and NaCl treatment; increased transcription in blades and roots was found after NaCl and ABA treatment.

MYB proteins contain a highly conserved DNA-binding domain: the MYB domain. This group of 52 amino acids consists of up to four imperfect amino acid sequence repeats (R), which form three α-helices (Ogata et al. 1996). MYB proteins can be divided into different classes depending on the number of adjacent repeats. 4R-MYB has four repeats, 3R-MYB (R1R2R3-MYB) has three repeats, R2R3-MYB has two repeats, and the MYB-related type usually, but not always, has a single repeat (Dubos et al. 2010; Jin et al. 1999; Rosinski et al. 1998). In plants, the MYB TFs are involved in the control of several biological functions, including the regulation of salt stress and signal transduction via the ABA-dependent pathway. In *Arabidopsis*, overexpression of AtMYB44 and AtMYB15 can improve salinity stress tolerance (Jung et al. 2008; Agarwal et al. 2006; Ding et al. 2009). Salt stress up-regulates two soybean CaM isoforms (GmCaM1 and GmCaM4), which bind to the MYB CaMBD in a Ca^{2+}-dependent manner and mediate the DNA-binding activity of an R2R3-type MYB TF (MYB2), which is a transcriptional regulator of salt- and dehydration-responsive genes (Yoo et al. 2005; Galon et al. 2010). Rice *OsMYB3R-2* has three imperfect repeats in the DNA-binding domain, the same as in animal c-MYB proteins. Salt stress induces the expression of *OsMYB3R-2* along with cold and drought stress. *OsMYB3R-2* overexpression in *Arabidopsis* improves salt stress tolerance along with cold and drought stress tolerance (Dai et al. 2007). The expression of some downstream genes, such as DREB2A, COR15a, and RCI2A, was increased to a higher level in *OsMYB3R-2*-overexpressing plants than in wild type. These results indicate that *OsMYB3R-2* acts as a master switch in stress tolerance (Dai et al. 2007). *OsMYB2* expression is induced by salt stress along with cold and drought stress. The overexpression of *OsMYB2* in rice showed increased salt stress tolerance along with cold and drought stress tolerance without affecting the growth rate as compared with control plants. In the *OsMYB2*-overexpressing plants, the downstream stress-related genes *OsDREB2A*, *OsRab16A*, and *OsLEA3* were up-regulated (Yang et al. 2012). These findings suggest that *OsMYB2* encodes a stress-responsive MYB TF that plays a regulatory role in tolerance of rice to salt, cold, and drought stress (Table 3.1).

NAC transcription factors also regulate the expression of salinity stress-responsive genes through an ABA-dependent pathway, and they were isolated initially from *Arabidopsis*. Three NACs (ANAC019, ANAC055, and ANAC072) were isolated from *Arabidopsis* by the yeast one hybrid screening method as TFs that regulate the expression of a salt-induced gene, *ERD1* (Tran et al. 2004). Under salt stress conditions, the expression of all three NAC genes was enhanced. Microarray analysis of transgenic *Arabidopsis* plants overexpressing these NAC genes revealed that several stress-inducible genes were up-regulated. Stress tolerance was increased significantly by these up-regulated genes in the transgenic lines (Yamaguchi-Shinozaki and Shinozaki 2006). High-salinity stress in rice induced several NAC genes. *OsNAC6/SNAC2* is one of the NAC TFs in rice whose expression was induced by salt stress (Hu et al. 2008; Nakashima et al. 2007). Transgenic rice plants overexpressing *OsNAC6* showed low yield and growth retardation under normal conditions but still showed improved tolerance to high-salinity stress. A stress-inducible promoter in transgenic plants may negate the negative effects of low yield and growth retardation under normal conditions. Downstream genes of *OsNAC6*-Ox plants were revealed by microarray analyses, and these included many stress-inducible genes. In a transient transactivation assay, one of these genes, which was a gene that encoded peroxidase, was activated. *OsNAC6* functions as a transcriptional activator in both biotic stress and abiotic responses, as these results indicated. Moreover, *OsHDAC1*, a rice gene that encodes a histone deacetylase, epigenetically represses *OsNAC6* expression (Chung et al. 2009). Salt stress induced *OsNAC5* expression, as also did cold, drought, methyl jasmonate, and ABA treatments (Takasaki et al. 2010). The growth pattern of *OsNAC5*-Ox plants was similar to that of control plants under normal conditions. Transgenic rice showed better salt tolerance on *OsNAC5* overexpression. *OsLEA3* is a downstream gene found after microarray analyses using *OsNAC5*-Ox. In the *OsLEA3* promoter region, *OsNAC5* binds to the NAC recognition core sequence (CACG) as shown by gel mobility shift assays. *OsNAC5* expression was reduced in RNA interference (RNAi) transgenic rice plants, which showed less tolerance to salt stress than control plants, whereas salt stress tolerance was increased by overexpression of *OsNAC5* (Song et al. 2011). *OsNAC5* expression level also correlated positively with the accumulation of proline and soluble sugars. In rice, SNAC1 was identified as a key component that functions in stress responses (Hu et al. 2006). Salt stress induced the expression of *SNAC1*, and its overexpression led to improved salt tolerance. Growth and productivity were unaffected by the *SNAC1* overexpression. These results indicate that for the transcriptional networks of salinity stress response in plants, NAC (NAM, ATAF, and CUC) TFs are also important, along with DREB (dehydration responsive element binding) and AREB (abscisic acid responsive elements binding) TFs.

Zinc finger proteins are among the most abundant proteins in eukaryotic genomes. They are functionally extraordinarily diverse and involved in DNA recognition, RNA packaging, transcriptional activation, regulation of apoptosis, and protein folding. The zinc finger was first recognized 27 years ago as a repeated zinc-binding motif, containing conserved cysteine and histidine ligands, in *Xenopus* transcription factor IIIA (TFIIIA) (Miller et al. 1985). Since that time, a number of other

TABLE 3.1

Salt Stress–Related Genes Described in This Chapter

Gene Name	Salt Stress Response	Source	Role in Stress Response	Reference
Osmotic Stress				
ABA-Dependent Pathway in Rice				
OsPSY3/OsNCED3/OsNCED4/OsNCED5	Up-regulated	*Oryza sativa*	Regulate salt stress tolerance through ABA biosynthesis	Welsch et al. (2008)
OsABF2	Down-regulated	*Oryza sativa*	Gene expressed modulation through an ABA-dependent pathway	Hossain et al. (2010)
OsABI5	Up-regulated	*Oryza sativa*	Transgenic rice plants overexpressing *OsABI5* showed high sensitivity to salt stress	Zou et al. (2008)
OsbZIP23	Up-regulated	*Oryza sativa*	Regulation of expression of genes involved in stress response and tolerance	Xiang et al. (2008)
OsHsfC1b	Up-regulated	*Oryza sativa*	Positively regulate salt stress tolerance	Schmidt et al. (2012)
SAPK1/SAPK2	Up-regulated	*Oryza sativa*	Regulation of expression of genes involved in stress response and tolerance	Kobayashi et al. (2004)
OSRK1/SAPK6	Up-regulated	*Oryza sativa*	Regulation of expression of genes involved in stress response and tolerance	Kobayashi et al. (2004) and Chae et al. (2007)
SLAC1/KAT1/AtRbohF	Up-regulated	*Arabidopsis thaliana*	*OSRK1* downstream genes	Kulik et al. (2011)
SAPK4	Up-regulated	*Oryza sativa*	Maintain Na^+ and Cl^- homeostasis by reducing *NHX* and *CLC1* expression	Diedhiou et al. (2008) and Nakamura et al. (2006)
OsMYB3R-2	Up-regulated	*Oryza sativa*	Regulation of stress-responsive gene expression	Dai et al. (2007)
OsMYB2	5 h (Up-regulated) 24 h (Down-regulated)	*Oryza sativa*	Overexpression of *OsMYB2* in rice seedlings functions as positive regulator of salt, cold, and dehydration stress tolerance	Yang et al. (2012)
OsLEA3/OsRab16A/OsDREB2A	Up-regulated	*Oryza sativa*	*OsMYB2* downstream genes	Yang et al. (2012)
OsNAC6/SNAC2	Up-regulated	*Oryza sativa*	Regulation of stress-responsive gene expression	Nakashima et al. (2007) and Hu et al. (2008)
OsNAC5	Up-regulated	*Oryza sativa*	Regulation of OsLEA3 expression	Takasaki et al. (2010) and Song et al. (2011)
SNAC1	Up-regulated	*Oryza sativa*	Play important role in enhancing salt stress tolerance	Hu et al. (2006)
ONAC045	Up-regulated	*Oryza sativa*	Regulation of expression of genes involved in salt and drought stress response and tolerance	Zheng et al. (2009)
OsNUC1	Up-regulated	*Oryza sativa*	Regulation of stress-responsive gene expression	Thanaruksa et al. (2012)
ZFP252	Up-regulated	*Oryza sativa*	Accumulation of sugars and proline	Xu et al. (2008)
OSISAP1	Up-regulated	*Oryza sativa*	Plays an important role in enhancing salt stress tolerance	Mukhopadhyay et al. (2004)
DST	Down-regulated	*Oryza sativa*	Negative regulation of H_2O_2-induced stomatal closure	Huang et al. (2009b)
ZFP179	Up-regulated	*Oryza sativa*	Accumulation of sugars and proline	Sun et al. (2010)
OsTZF1	Up-regulated	*Oryza sativa*	Plays an important role in enhancing salt and drought stress tolerance	Jan et al. (2013)
OsCPK21	Up-regulated	*Oryza sativa*	Positively regulates the signaling pathways that are involved in the response to ABA and salt stress tolerance	Asano et al. (2011)

(Continued)

TABLE 3.1 (CONTINUED)

Salt Stress–Related Genes Described in This Chapter

Gene Name	Salt Stress Response	Source	Role in Stress Response	Reference
OsNAC6/Rab21/OsLEA3	Up-regulated	Oryza sativa	OsCPK21 downstream genes	Asano et al. (2011) and Duan et al. (2012)
OsDCPK7	Up-regulated	Oryza sativa	Plays an important role in enhancing salt stress tolerance	Saijo et al. (2000)
Rab16a/SalT/wsi18	Up-regulated	Oryza sativa	OsCDPK7 downstream genes	Saijo et al. (2000)
OsSIK1	Up-regulated	Oryza sativa	Plays an important role in enhancing salt stress tolerance	Ouyang et al. (2010)
OsDSM1	Up-regulated	Oryza sativa	Early signaling component in regulating responses to osmotic stress by regulating scavenging of ROS	Ning et al. (2010)
OsEDR1	Up-regulated	Oryza sativa	An essential positive regulator of tolerance to salt stress	Kim et al. (2003)
OsMAPK4/OsMAPK5	Up-regulated	Oryza sativa	Positively regulate salt stress tolerance	Fu et al. (2002) and Xiong et al. (2003)
OsMAPK33/OsMAPK44	Up-regulated	Oryza sativa	Play an important role in salt stress tolerance through unfavorable ion homeostasis	Lee et al. (2011) and Jeong et al. (2006)
OsMSRMK2/OsMSRMK3	Up-regulated	Oryza sativa	Play a role in salt stress signaling pathway in rice	Agrawal et al. (2002, 2003)
OsmiR396c/OsmiR393	Up-regulated	Oryza sativa	Overexpression shows sensitive phenotype in salt stress	Gao et al. (2010, 2011)
OsTIR1/OsAFB2	Down-regulated	Oryza sativa	OsmiR393 overexpression plays an important role that leads to less tolerance to salt and drought	Xia et al. (2012)
OsWRKY45-2	Up-regulated after 4 days	Oryza sativa	Positively regulates ABA signaling and negatively regulates rice response to salt stress	Tao et al. (2011)
OsTIFY11a	Up-regulated	Oryza sativa	Regulation of seed germination	Ye et al. (2009)
OsGSK1	Down-regulated	Oryza sativa	Negatively regulates salt stress tolerance	Koh et al. (2007)
OsP5CR/OsP5CS1	Up-regulated	Oryza sativa	Gene expressed modulation through an ABA-dependent pathway	Sripinyowanich et al. (2010)
OsNADP-ME2	Up-regulated	Oryza sativa	Plays an important role in enhancing tolerance of plants to salt and osmotic stress	Liu et al. (2007)
OsDhn1	Up-regulated	Oryza sativa	Plays an important role in enhancing salt and drought stress tolerance	Kumar et al. (2014)
OsERF922	Down-regulated	Oryza sativa	Negatively regulates salt stress tolerance	Liu et al. (2012)
OsSKIPa	Up-regulated	Oryza sativa	Positively regulates salt and drought stress tolerance	Hou et al. (2009)
c-GS2	Up-regulated	Oryza sativa	Accumulation of glutamine leads to salt stress tolerance	Hoshida et al. (2000)
OsSta2	Up-regulated	Oryza sativa	Positively regulates salt stress tolerance and increases yield	Kumar et al. (2017)
ABA-Independent Pathway in Rice				
OsDREB1A	Up-regulated	Oryza sativa	Regulation of stress-responsive gene expression and accumulation of proline	Ito et al. (2006)
OsDREB1B	Up-regulated	Oryza sativa	Induction of TF expression	Ito et al. (2006)
OsDREB1F	Up-regulated	Oryza sativa	ABA-responsive regulation of gene expression	Wang et al. (2008)
OsDREB2A	Up-regulated	Oryza sativa	Regulation of salt stress tolerance	Mallikarjuna et al. (2011)

(Continued)

TABLE 3.1 (CONTINUED)

Salt Stress–Related Genes Described in This Chapter

Gene Name	Salt Stress Response	Source	Role in Stress Response	Reference
OsCPK12	Not regulated at the transcriptional level in response to salt stress	*Oryza sativa*	An essential positive regulator of tolerance to salt stress	Asano et al. (2012)
Osrbohl/OsAPx2/OsAPx8	Regulated by *OsCPK12*	*Oryza sativa*	Plays an important role in H_2O_2 homeostasis	Asano et al. (2012)
OsSPL1	Up-regulated	*Oryza sativa*	Decreases the tolerance of plants to salt stress	Zhang et al. (2012)
Salt Stress–Related Genes from Organisms Other Than Rice				
coda	Up-regulated	*Arthrobacter globiformis*	Positively regulates salt and cold stress tolerance in rice	Sakamoto et al. (1998)
P5CS	Up-regulated	*Vigna aconitifolia* L.	Plays an important role in enhancing salt and drought stress tolerance	Zhu et al. (1998)
otsA + otsB	Up-regulated	*Escherichia coli*	Positively regulates salt and drought stress tolerance through the accumulation of trehalose	Garg et al. (2002)
hva1	Up-regulated	*Hordeum vulgare*	Accumulation of HVA1 leads to salt and drought stress tolerance	Xu et al. (1996)
Mn-SOD	Up-regulated	*Saccharomyces cerevisiae*	Transgenic plants showed enhanced tolerance to salt	Tanaka et al. (1999)
katE	Up-regulated	*Escherichia coli*	Decomposes H_2O_2 and increases salt stress tolerance	Motohashi et al. (2010)
TPS + TPP	Up-regulated	*Escherichia coli*	Transgenic plants showed increased tolerance to salt stress	Jang et al. (2003)
Ionic Stress				
OsMSR2	Up-regulated	*Oryza sativa*	A novel rice gene encoding a calmodulin-like protein consisting of three predicted Ca^{2+} binding sites	Xu et al. (2011)
OsCIPK15	Up-regulated	*Oryza sativa*	An essential positive regulator of tolerance to salt stress	Xiang et al. (2007)
OsSOS1/OsCIPK24/OsCBL4	Up-regulated	*Oryza sativa*	Important in calcium signaling	Atienza et al. (2007)
CNAtr	Up-regulated	*Mus musculus*	Overexpression of calcineurin gene in rice seedlings increases salt stress tolerance	Ma et al. (2005)
AgNHX1	Up-regulated	*Atriplex gmelini*	Plays an important role in enhancing salt stress tolerance by regulating vacuolar-type Na+/H+ antiporter	Ohta et al. (2002)
nhaA	Up-regulated	*Escherichia coli*	Transgenic rice grew faster as compared with the wild types under salt stress conditions	Wu et al. (2005)
SOD2	Up-regulated	*Saccharomyces pombe*	Plays an important role in enhancing salt stress tolerance	Zhao et al. (2006a)
OsNHX1	Up-regulated	*Oryza sativa*	Transgenic plants showed increased tolerance to salt stress	Fukuda et al. (2004, 2011)
SsNHX1	Up-regulated	*Suaeda salsa*	Transgenic rice carrying a vacuolar Na+/H+ antiporter gene showed salt tolerance	Zhao et al. (2006b)
OsAKT1	Up-regulated	*Oryza sativa*	Plays an important role in enhancing salt stress tolerance	Golldack et al. (2003)
SaVHAc1	Up-regulated	*Spartina alterniflora*	Plays an important role in enhancing salt stress tolerance	Baisakh et al. (2012)

zinc-binding motifs have been identified and designated as zinc fingers. Zinc finger structures are as diverse as their functions. The classical Cys2His2 zinc finger family is among the most abundant in eukaryotic genomes. A few members of this family were up-regulated by abiotic stress in *Arabidopsis*. In eukaryotes, the Cys-2/His-2-type zinc finger is one of the best-characterized DNA-binding motif TFs. This motif is represented by about 30 amino acids and two pairs of conserved Cys and His bound tetra-hedrally to a zinc ion, the signature CX2-4CX3FX5LX2HX3-5H (Sakamoto et al. 2004; Pabo et al. 2001). The salt and drought stress–responsive TFIIIA-type zinc finger protein gene ZFP252 was identified in rice. During the functional analysis of *ZFP252*, the overexpression of *ZFP252* in rice elevated the expression of stress defense genes and enhanced rice tolerance to salt and drought stresses as compared with *ZFP252* antisense and non-transgenic plants; it also increased the amount of free proline and soluble sugars (Xu et al. 2008). This shows that in salt and drought stress, *ZFP252* plays a crucial role in rice. Another gene from rice, *OSISAP1*, encodes a zinc finger protein induced by salt, cold, submergence, and heavy metal stress, and so on. ABA also induced this gene. Its overexpression in transgenic tobacco conferred tolerance to salt stress, cold, and dehydration without affecting growth as in a normal WT plant in normal conditions (Mukhopadhyay et al. 2004). *OSISAP1* also seems to be an important player in determining the stress response in plants. Huang et al. (2009) isolated the *DST* (drought and salt tolerance) gene from rice. As the name suggests, it regulates salt and drought stress tolerance through the opening and closing of stomata. *DST* negatively regulates stomatal closure and direct modulation of genes related to H_2O_2 homeostasis. A *DST* mutant enhanced salt and drought tolerance in rice by increasing stomatal closure and reducing stomatal density. Another salt-responsive zinc finger protein gene, *ZFP179*, was identified through microarray analysis from rice. *ZFP179* was highly expressed due to salt stress and ABA. Thus, the overexpression of *ZFP179* in rice leads to increased salt tolerance. The levels of free proline and soluble sugars were also increased in transgenic plants compared with WT plants under salt stress (Sun et al. 2010). The *ZFP179* transgenic rice showed high expression levels of a number of stress-related genes, including *OsDREB2A*, *OsP5CS*, *OsProT*, and *OsLea3*, under salt stress (Sun et al. 2010). This suggests that *ZFP179* plays a crucial role in the plant response to salt stress. *OsTZF1* is a CCCH-type zinc finger gene in rice. Its expression was induced by high-salt stress, drought, and hydrogen peroxide. The overexpression of *OsTZF1* driven by a maize ubiquitin promoter exhibited delayed seed germination, growth retardation at the seedling stage, and delayed leaf senescence: exactly opposite to the phenotype shown by RNAi plants. The overexpressing plants showed improved tolerance to salt and drought stress, and vice versa for *OsTZF1*-RNAi plants (Jan et al. 2013). Based on these findings, *OsTZF1* may play an important role in the improvement of stress tolerance in various plants through the control of RNA metabolism of stress-responsive genes.

In plant salt stress signal transduction pathways, calcium acts as a messenger. Calcium-dependent protein kinases (CDPKs) play important roles in regulating downstream components in calcium signaling pathways. In rice under high-salinity conditions, the survival rate of *OsCPK21*-Ox plants is higher than that of WT plants (Asano et al. 2011). ABA treatment increases the inhibition of seedling growth in the *OsCPK21*-Ox plants more than in WT plants. *OsCPK21*-Ox plants induce several ABA- and high-salinity stress–inducible genes such as *OsNAC6*, *Rab21*, and *OsLEA3* more than WT plants (Asano et al. 2011; Nakashima et al. 2007; Mundy and Chua 1988; Duan and Cai 2012). This shows that *OsCPK21* is involved in the ABA-dependent pathway in response to salt stress. Salt stress also induces another CDPK, *OsCDPK7*. Overexpression of *OsCDPK7* in rice leads to salt and drought tolerance. *OsCDPK7*-Ox plants enhance the induction of some stress-responsive genes in response to salt and drought, such as *Rab16a*, *SalT*, and *wsi18* (Saijo et al. 2000). The fact that these stress-inducible genes are also up-regulated by ABA shows that *OsCDPK7* has a role in salt stress tolerance via the ABA-dependent pathway.

Receptor-like kinases (RLKs) play an essential part in development, plant growth, and salt stress tolerance. Salt, H_2O_2, drought, and ABA treatments induce the expression of an RLK gene, *OsSIK1*. *OsSIK1*-Ox rice plants show higher tolerance to drought and salt stresses than WT plants, and the mutants *sik1-1* and *sik1-2*, as well as RNAi plants, are sensitive to salt and drought stresses (Ouyang et al. 2010). This shows that *OsSIK1* plays an important role in salt stress tolerance in rice through the ABA-mediated pathway.

The mitogen-activated protein kinase (MAP kinase) cascade plays a crucial role in salt stresses. Plants use the MAP kinase pathway to translate external stimuli into cellular responses by the activation of the mitogen-activated protein kinase (MAPK) cascade. MAP kinase cascades are highly conserved in eukaryotes and consist of three consecutively acting protein kinases, an MAP kinase kinase kinase (MEKK), an MAP kinase kinase (MKK), and an MAPK (MPK), which are linked in various ways with upstream receptors and downstream targets (Jonak et al. 2002). Upstream signals activate MAPKKKs, which then phosphorylate MAPKKs; MAPKKs in turn activate a specific MAPK. The downstream targets of MAPKs can be TFs (Lin et al. 1993; Jonak et al. 2002).

A drought-hypersensitive mutant, *dsm1*, of a putative MAPKKK gene in rice (*Oryza sativa*) shows more sensitivity to drought stress than WT plants at both seedling and panicle development stages. In DSM1-RNA interference lines, the *DSM1* gene was induced by salt, drought, and ABA, but not by cold. The dsm1 mutant showed more sensitivity to oxidative stress due to an increase in ROS damage caused by the reduced peroxidase (POX) activity (Ning et al. 2010). This indicates that *DSM1* may be a novel MAPKKK functioning as an early signaling component in regulating responses to osmotic stress by regulating the scavenging of ROS in rice. The expression of another MAPKKK, *OsEDR1*, a single copy gene from rice, is up-regulated by high salt, drought, sugar, heavy metals, protein phosphatase inhibitors, the fungal elicitor chitosan, cutting, and treatment with jasmonic acid (JA), salicylic acid, ethylene, ABA, and hydrogen peroxide (Kim et al. 2003). These results suggest that *OsEDR1* plays an important role in stress signaling pathways and development. There are no reports of MKK in rice playing any role in salt stress. The up-regulation of *Oryza sativa* MAPK4 (*OsMAPK4*) expression level under high salt, cold, and sugar starvation shows that *OsMAPK4* functions in an osmotic stress signaling pathway (Fu et al. 2002). The kinase activity of MAPK gene (*OsMAPK5*) from rice was inducible by various abiotic (salt, drought, cold, and wounding) and biotic (pathogen infection) stresses as well as ABA. The suppression of *OsMAPK5* expression and its kinase activity in the dsRNAi transgenic plants shows a reduction in salt, drought, and cold tolerance. By contrast, *OsMAPK5*-Ox lines exhibited increased kinase activity and increased tolerance to salt, drought, and cold stresses (Xiong et al. 2003). This indicates that *OsMAPK5* can positively regulate salt, drought, and cold tolerance. Another rice MAPK, *OsMAPK33*, is mainly induced by drought stress, but *OsMAPK33*-Ox lines showed higher sodium uptake into cells, resulting in a lower K^+/Na^+ ratio inside the cell than in WT plants and *OsMAPK33* mutant lines (Lee et al. 2011). This result indicates that *OsMAPK33* may play a negative role in salt stress tolerance through unfavorable ion homeostasis. A putative rice MAPK gene, *OsMAPK44*, encodes a protein of 593 amino acids that has the MAPK family signature, and its expression is highly induced by salt and drought stress but not by cold stress. Exogenous ABA and H_2O_2 also induced the expression of *OsMAPK44*. *OsMAPK44*-Ox transgenic rice plants show greater Na^+/K^+ than *OsMAPK44* mutant transgenic lines (Jeong et al. 2006). This indicates that *OsMAPK44* may have a role in ion balance in the presence of salt stress. *OsMSRMK2* in JA-treated rice seedlings was induced within 15 min by high salt, sucrose, drought, signaling molecules, protein phosphatase inhibitors, ultraviolet irradiation, fungal elicitors, heavy metals, cutting, JA, salicylic acid, and ethylene, suggesting a role for *OsMSRMK2* in rice defense/stress response pathways (Agrawal et al. 2002). The expression of *OsMSRMK3*, a novel rice MAPK, was induced by high salt/sugar, ABA, hydrogen peroxide, cutting, JA, salicylic acid, ethylene, protein phosphatase inhibitors, chitosan, and heavy metals (Agrawal et al. 2003). This indicates a role of *OsMSRMK3* in the salt stress signaling pathway in rice.

MicroRNAs (miRNAs) are endogenous small RNAs of ~21 nucleotides, which are ubiquitous regulators of gene expression and play a key role in the down-regulation of gene expression at the posttranscriptional level in plants and animals. Numerous miRNAs are either up-regulated or down-regulated by stress treatment. *osa-MIR396c* shows a dramatic transcription change following salt stress, alkali stress, and ABA in rice. Rice and *Arabidopsis* transgenic plants overexpressing *osa-MIR396c* showed reduced salt and alkali stress tolerance compared with that of WT plants (Gao et al. 2010). The expression level of another microRNA, *osa-MIR393*, also changed. Transgenic rice and *Arabidopsis thaliana* that overexpress *osa-MIR393* show more sensitivity to salt stress as compared with WT plants by reducing the expression of two auxin receptor genes *OsTIR1* and *OsAFB2* (Gao et al. 2011). This shows that *osa-MIR393* and *osa-MIR396c* play an important role in the salt stress signaling pathway in rice.

Plant NADPH oxidases (NOXes) produce ROS and are important in plant innate immunity. The rice respiratory burst oxidase homolog B (*OsRbohB*) gene encodes a NOX, the regulatory mechanisms of which are largely unknown. Takahashi et al. (2012) reported that *OsRbohB* shows ROS-producing activity. Treatment with a Ca^{2+} ionophore (ionomycin) and a protein phosphatase inhibitor (calyculin A) activated ROS-producing activity. Thus, *OsRbohB* is activated by both Ca^{2+} and protein phosphorylation.

Salt stress tolerance was reduced in *OsWRKY45-2*-overexpressing lines, which also showed increased ABA sensitivity. In contrast, *OsWRKY45-2*-suppressing lines showed increased salt stress tolerance and reduced ABA sensitivity (Tao et al. 2011; Kumar et al. 2017). This shows that *OsWRKY45-2* positively regulates ABA signaling and negatively regulates rice

response to salt stress. *OsTIFY11a* is one of the stress-inducible genes. Overexpression of this gene resulted in increased salt and dehydration stress tolerance (Ye et al. 2009). *OsGSK1* knockout plants showed enhanced tolerance to salt, cold, heat, and drought stresses when compared with non-transgenic (NT) plants. *OsGSK1* overexpressers showed stunted growth (Koh et al. 2007). Therefore, stress-responsive *OsGSK1* may have physiological roles in stress signal transduction pathways and developmental processes. *OsP5CS1* gene expression is induced by salt stress. *OsP5CR* up-regulation increased proline accumulation and salt resistance induced by topical ABA application indirectly or directly from OsP5CS1 (Sripinyowanich et al. 2010). In a rice screen, *NADP-ME2* showed increased transcription during 72-h exposure to $NaHCO_3$, NaCl, and polyethylene glycol stresses. These results suggest that rice *NADP-ME2* responds to salts and osmotic stresses. Overexpressing *NADP-ME2* *Arabidopsis* plants show salt stress tolerance as compared with WT (Liu et al. 2007). These results suggest that NADP-ME2 has a role in enhancing salt tolerance in plants. *OsDhn1*-overexpressing rice plants showed salt and drought stress tolerance, indicating that *OsDhn1* plays an important role in tolerance to salt and drought stress (Kumar et al. 2014). The rice AP2/ERF type TF *OsERF922* is strongly induced by salt treatments and ABA. *OsERF922*-overexpressing lines exhibited decreased tolerance to salt stress with an increased Na^+/K^+ ratio in the shoots. The ABA levels were found to be increased in the overexpressing lines and decreased in the RNAi plants (Liu et al. 2012). *OsSKIPa* is a rice homolog of human Ski-interacting protein, whose expression is induced by various abiotic stresses and phytohormone treatments. *OsSKIPa*-overexpressing transgenic rice exhibited significantly improved growth performance due to salt, ABA, and drought resistance at both the seedling and reproductive stages. Transcription levels of many stress-related genes, including *SNAC1* and rice homologs of CBF2, PP2C, and RD22, increase under drought stress conditions in *OsSKIPa*-overexpressing rice (Hou et al. 2009). The GS2 gene from rice also showed salt stress tolerance when overexpressed as compared with a control plant by retaining more than 90% activity of Photosystem II in comparison to the complete loss of Photosystem II in the control plant (Hoshida et al. 2000). Thus, the enhancement of photorespiration leads to improved salt stress tolerance in rice plants.

ABA-Independent Pathway

There are several salt stress–inducible genes that do not respond to ABA treatment; they work in ABA-independent pathway.

TFs such as *DREB1A*, *DREB1B*, *DREB1F*, and *DREB2A* transactivate the DRE *cis* element of osmotic stress genes and thereby maintain the osmotic stress equilibrium of the cell. *DREB1A*, *DREB1B*, *DREB1F*, and *DREB2A* are DREB-type genes identified as TFs from *Arabidopsis*. They have a conserved ERF/AP2 domain that binds to a *cis*-acting element, *DRE/CRT*. During high-salinity stress in *Arabidopsis*, the *DRE/CRT* regulates gene expression through the core sequence A/GCCGAC (Shinozaki et al. 2005). The expression of *DREB1A*, *DREB1B*, *DREB1F*, and *DREB2A* was induced by salt stress, and the four encoded DREB proteins were major transcriptional activators required for salt-inducible gene expression. Transgenic rice overexpressing DREB-type genes driven by the cauliflower mosaic virus (CaMV) 35S promoter displayed strong salt tolerance. Growth retardation was minimized by the use of a stress-inducible promoter. The rice genome contains at least 14 DREB-type genes, among which *DREB1A*, *DREB1B*, *DREB1F*, and *DREB2A* are induced by salt stress (Ito et al. 2006; Wang et al. 2008; Mallikarjuna et al. 2011). Overexpression of *OsDREB1A* and *OsDREB1B* induced strong expression of stress-responsive genes in transgenic rice plants, which resulted in not only improved stress tolerance to high salinity but also growth retardation of plants under non-stress growth conditions (Ito et al. 2006). Osmoprotectants, such as proline and a variety of sugars, were accumulated in the transgenic rice plants under control conditions. Transgenic *Arabidopsis* overexpressing *DREB1A* and *DREB2A* also showed growth retardation under non-stress conditions and improved salinity tolerance (Dubouzet et al. 2003; Mizoi et al. 2011). For the activation of *DREB2A*, the protein stability mediated by ubiquitin E3 ligases is important (Qin et al. 2007). This result showed similarity between rice *OsDREB1A*, *OsDREB2A* and *Arabidopsis DREB1A*, *DREB2A*. Microarray analysis revealed that the target stress-inducible genes of *OsDREB1A* and *OsDREB2A* encode proteins thought to function in stress tolerance in the plants, which is similar to the target genes of DREB1 and DREB2 proteins in *Arabidopsis* (Sakuma et al. 2006). These observations showed that the *DREB/CBF* TFs are conserved in rice, and *DREB1*- and *DREB2*-type genes are useful for the improvement of salt stress tolerance in transgenic rice.

Rice *OsCPK12*, which belongs to group IIa of the CDPK family, enhances tolerance to salt stress by reducing the accumulation of ROS. Furthermore, *OsCPK12*-Ox rice plants show increased sensitivity to exogenously applied ABA. Under high salt conditions, the accumulation of H_2O_2 in *OsCPK12*-Ox plants was lower than that in WT plants, whereas in *oscpk12* and *OsCPK12* RNAi plants, the accumulation was higher (Asano et al. 2012). This shows that *OsCPK12* confers salt stress tolerance by repressing ROS accumulation rather than by affecting ABA-mediated salt stress signaling. *OsCPK12* negatively regulates the expression of *OsrbohI* and positively regulates ROS detoxification by controlling the expression of *OsAPx2* and *OsAPx8* under conditions of high salinity (Asano et al. 2012). These results also suggest that *OsCPK12* functions in salt stress signaling pathways and is an essential positive regulator of tolerance to salt stress.

Sphingolipids function as signaling mediators to regulate stress response and plant growth and development. Functional analysis of a rice S1P lyase gene *OsSPL1* explored its possible involvement in abiotic stress responses in transgenic tobacco plants. Salt and oxidative stress tolerance decreased with overexpression of *OsSPL1* in transgenic tobacco compared with the wild type. This suggests that rice *OsSPL1* plays an important role in abiotic stress responses.

The *codA* gene for choline oxidase from the soil bacterium *Arthrobacter globiformis* was genetically engineered into rice to synthesize glycinebetaine. Transgenic plants showed an improvement in salt stress tolerance, as they recovered their growth after salt stress as compared with WT plants (Sakamoto et al. 1998). This suggests that *codA* has a role in salt stress tolerance. A *P5CS* cDNA from mothbean (*Vigna aconitifolia* L.) was introduced into the rice genome. *P5CS* expression under the control of a stress-inducible promoter led to an increase in biomass under salt and water stress conditions as compared with the nontransformed control plants. Proline accumulation was also high as compared with the wild type (Zhu et al. 1998). This showed that P5CS has

a significant role in salt and drought stress tolerance. Trehalose functions as a compatible solute in the stabilization of biological structures under abiotic stress in bacteria, fungi, and invertebrates, but its level in plants is not high enough. The *Escherichia coli* trehalose biosynthetic genes *otsA* and *otsB* were overexpressed as a fusion gene to manipulate abiotic stress tolerance in rice. Transgenic lines exhibited sustained plant growth, less photo-oxidative damage, and more favorable mineral balance under salt, drought, and low-temperature stress conditions (Garg et al. 2002). These findings demonstrate that the overproduction of trehalose increased tolerance of abiotic stress and enhanced productivity. *HVA1*, a late embryogenesis abundant (LEA) protein gene from barley (*Hordeum vulgare* L.), was introduced into rice. Expression of the barley *HVA1* gene regulated by the rice actin 1 gene promoter at a high level resulted in a significant increase in salt and drought tolerance and higher growth rates than in nontransformed control plants under stress conditions (Xu et al. 1996). This supports the idea that LEA proteins play an important role in the protection of plants under salt or water stress conditions.

Superoxide dismutase (SOD) plays a potential role in protection against salt stress. The coding region of the mitochondrial *Mn-SOD* gene from *Saccharomyces cerevisiae* was fused with the chloroplast targeting signal of the glutamine synthetase gene. Under salt stress conditions, the SOD activity decreased in both transgenic and WT plants, but the decrease was faster in the control plants. At high salinity, the ascorbate peroxidase activity of the overexpressing plants was about 1.5-fold higher than that in the wild type (Tanaka et al. 1999). This suggests that for salt resistance in rice, an increased level of ascorbate peroxidase in the overexpressing plants is an important factor. *katE* is a gene derived from *Escherichia coli* encoding catalase, which decomposes H_2O_2, one of many ROS produced by salt stress. Transgenic rice showed catalase activities of *katE* about 1.5–2.5-fold higher than those of NT plants. Transgenic plants showed greater salt stress tolerance than the wild type without affecting the growth of the plant (Motohashi et al. 2010). This showed that *katE* plays a crucial role in salt stress tolerance. Trehalose has a crucial role in stress tolerance in plants. Transgenic rice plants producing trehalose showed increased tolerance to drought, salt, and cold stress. Transgenic rice was generated by the engineering of a gene encoding a trehalose-6-P phosphatase (*TPP*) of *Escherichia coli* and a bifunctional fusion (*TPSP*) of the trehalose-6-phosphate synthase (*TPS*) driven by the maize ubiquitin promoter (Jang et al. 2003). This suggests that trehalose synthesis plays an important role in salt stress tolerance.

Ionic Stress

Rice undergoes ionic stress conditions under salt stress, and the plant responds to this by maintaining ionic homeostasis. To maintain ionic homeostasis, rice induces many genes by different pathways. These pathways include the Ca^+/phospholipid C (PLC) signaling pathway and the salt overlay sensitive (SOS) pathway.

Ca^+/Phospholipid C (PLC) Signaling Pathway

The salt stress tolerance mechanism is a complex signaling network that involves multiple signal transduction pathways, allowing plants to respond properly and rapidly to salt stress (Zhu 2001). PLC signaling is one of these. In plant signaling, the network role of the calcium is important in providing salinity tolerance. Cytosolic Ca^{2+} increases due to high salinity, which triggers the stress signal transduction pathway for stress tolerance. Ca^{2+} may primarily be from extracellular sources, because the addition of EGTA or BAPTA blocks calmodulin-mediated activity (Guo et al. 2006). The activation of phospholipase C may also cause Ca^{2+} release, leading to the hydrolysis of phosphatidylinositol bisphosphate to inositol trisphosphate and the subsequent release of Ca^{2+} from intracellular Ca^{2+} stores such as mitochondria and the endoplasmic reticulum. The phosphatidic acid (PA) level increases after salt stress, even though its proportion in the cell is small (Testerink and Munnik 2005). Phospholipase D (PLD) can form PA, but the indirectly sequential action of diacylglycerol (DAG) kinase (DGK) and PLC can also produce PA. Polyphosphoinositides, such as phosphatidylinositol-4,5-bisphosphate (PI(4,5)P2) and phosphatidylinositol 4-phosphate (PI4P), are hydrolyzed by PLC into inositol bisphosphate (IP_2) or inositol triphosphate (IP_3) and DAG. IP_3 always releases Ca^{2+} from intracellular stores, while DAG remains in the membrane, and DGK rapidly phosphorylates DAG to PA. PLCs are involved in intracellular calcium release. Salt stress leads to an increase of IP_3 after 3 h of treatment. PLC activity hydrolyzes PI(4,5)P2 into two second messengers: IP_3 and DAG (Yotsushima et al. 1993; Parre et al. 2007). IP_3 generated by PLCs is an essential second messenger in Pro accumulation in response to salt treatment. IP_3 releases calcium from intracellular plant stores (Alexandre et al. 1990; Allen et al. 1995; Gilroy et al. 1990), but genes encoding IP_3-gated calcium channels have not been reported in plants.

SOS Pathway

A calcium sensor (calcium-binding protein) can provide an additional level of regulation in calcium signaling. These sensors recognize and decode the information in calcium signatures and pass it to the downstream to initiate a phosphorylation cascade, which leads to the regulation of gene expression. Three genes, *SOS1*, *SOS2*, and *SOS3*, were identified in *Arabidopsis* by screening of mutants that were oversensitive to salt stress (Wu et al. 1996; Liu et al. 2000; Halfter et al. 2000). The SOS pathway is conserved in rice. The rice transporter *OsSOS1* helped Na^+/H^+ exchange in plasma membrane vesicles of yeast (*Saccharomyces cerevisiae*) cells and reduced the total Na^+ content in the cell. Putative rice homologs of the *Arabidopsis* protein kinase SOS2 and its Ca^{2+}-dependent activator SOS3 were identified as *OsCIPK24* and *OsCBL4* (Atienza et al. 2007). The SOS3 gene encodes a calcineurin B-like (CBL) protein (calcium sensor). This is a calcium-binding protein that senses the cytosolic Ca^{2+} concentration and transmits the signal downstream. Figure 3.2 shows the SOS pathway. *SOS2* encodes a novel serine/threonine protein kinase known as the CBL-interacting protein kinase. *SOS2* protein kinase physically interacts with and is activated by the calcium-binding protein *SOS3* (Halfter et al. 2000). Analysis of the *sos1* mutant of *Arabidopsis* gave the first target of the SOS2–SOS3 pathway (Ding et al. 1997). The salt-hypersensitive *sos1* mutant of *Arabidopsis thaliana* has reduced sodium uptake, and *SOS1* is an Na^+/H^+ antiporter (Sahi et al. 2000). *SOS1*, *SOS2*, and *SOS3* work in a common pathway under salinity conditions, as

proved by genetic analysis (Zhu et al. 2002; Zhang et al. 2004). The SOS2–SOS3 kinase complex phosphorylates SOS1 directly (Figures 3.2 and 3.3). SOS pathways also seem to play an important role in removing excess Na^+ ions from the cell and help in maintaining cellular homeostasis.

Salt stresses elicit changes in intracellular calcium levels that serve to convey information and activate adaptive responses. Ca^{2+} signals are perceived by different Ca^{2+} sensors, and calmodulin (CaM) is one of the best-characterized Ca^{2+} sensors in eukaryotes. *OsMSR2* is a novel calmodulin-like protein gene isolated from rice. Its expression was strongly up-regulated by stresses, including salt, cold, drought, and heat, in different tissues at different developmental stages of rice. The expression of *OsMSR2* conferred enhanced tolerance to high salt and drought in *Arabidopsis* (Xu et al. 2011). CBL protein-interacting protein kinases (CIPKs) are signaling molecules that are initiated by plants in stress conditions. *OsCIPK15* overexpression resulted in significantly improved tolerance to salt stress, cold, and drought (Xiang et al. 2007).

Genomic Overview of Transporters in Rice Salt Tolerance

Salt stress tolerance is a very complex phenomenon. Various components are involved in imparting salinity tolerance to plants. Higher salt stress causes an imbalance of sodium ions. Various pumps, ions, and Ca^{2+} sensors and their downstream genes play important roles in overcoming that unbalance. The roles of different ion pumps/channels are shown in Figure 3.3. Some channels have more selectivity to K^+ over Na^+. The K inward-rectifying channel mediates the influx of K^+ on the plasma membrane, and it selectively accumulates K^+ over Na^+ on the plasma membrane hyperpolarization. The low-affinity Na^+ transporter, histidine kinase transporter (HKT), blocks the entry of Na^+ into the cytosol. Platten et al. (2006) give the nomenclature of HKT, which is a key determinant of salinity tolerance. The voltage-independent channel, which acts as an entry gate for Na^+ into the plant cell, is a nonspecific cation channel. The K^+ outward-rectifying channel opens during depolarization of the plasma membrane and mediates the efflux of K^+ and the influx of Na^+ ion, leading to accumulation of Na^+ ions in the cytosol. The vacuolar Na^+/H^+ exchanger (NHX) pushes excess Na^+ ions into vacuoles. The electrochemical gradient generated by H^+-ATPase helps in the extrusion of Na^+ ions from the plant cell. H^+-ATPase permits the H^+ movement inside along the electrochemical gradient and extrusion of Na^+ out of the cytosol. H^+/Ca^{2+} antiporter (CAX1) is the pump that helps in Ca^{2+} ion homeostasis (Figure 3.3).

A low-affinity Na^+ transporter, *OsHKT1*, mediates the entry of Na^+ ions into root cells during salt stress in rice. It blocks the Na^+ ion entry (Figure 3.3), hence contributing to Na^+ ion homeostasis. *SOS2* also interacts with and activates the NHX (Garciadeblás et al. 2003). This causes excess Na^+ ion to be sequestered and pushed into vacuoles. Other calcium-binding proteins, such as calnexin and calmodulin, interact and activate the NHX to maintain the Ca^{2+} level in the cell. *SOS2* also targets *CAX1* to maintain the cytosolic Ca^{2+} level. Components of the SOS pathway cross talk with each other by interacting with other components

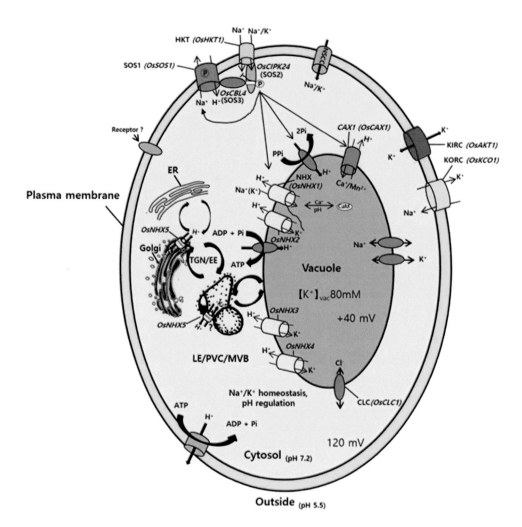

FIGURE 3.3 Representation of a plant cell showing the regulation of ion homeostasis by various ion pumps. The salt stress signal is perceived by a plasma membrane–situated receptor or salt sensor. This signal is responsible for activating the SOS pathway, the components of which help in regulating some of these pumps. The various pumps/channels are the K^+ inward-rectifying channels (KIRC), HKT, nonspecific cation channels (NSCC), K^+ outward-rectifying channel (KORC), Na^+/H^+ antiporters (SOS1), vacuolar Na^+/H^+ exchangers (NHX), and H^+/Ca^+ antiporter (CAX1). Na^+ extrusion from plant cells is powered by the electrochemical gradient generated by H+-ATPases, which permits the Na^+/H^+ antiporters to couple the passive movement of H^+ inside along the electrochemical gradient and the extrusion of Na+ out of cytosol. ER, endoplasmic reticulum; LE/PVC/MVB, late endosome/pre-vacuolar compartment/ multivesicular body; RE, recycling endosome. The salt stress signal is sensed by SOS3, which activates SOS2, which leads to the activation of SOS1, as shown in detail in Figure 3.2. (Adapted from Kumar, K., et al., *Rice*, 6, 27, 2013.)

to maintain cellular ion homeostasis, which helps in improving stress tolerance.

Na^+/H^+ and K^+/H^+ exchange activity is mediated by *NHX1* (Zhang and Blumwald 2001; Apse et al. 2003), which is found on the vacuole membrane. In *Arabidopsis*, the overexpression of the tomato *LeNHX2* also resulted in an increased endosomal K^+/H^+ exchange and demonstrated that the regulation of K^+ homeostasis by endosomal compartments is also important for plant growth and development (Venema et al. 2003; Rodriguez-Rosales et al. 2008) *NHX2* is also found on the vacuole membrane and maintains K^+/H^+ homeostasis. Using reverse genetics, direct evidence for the role of vacuolar *NHX1* and *NHX2* in K^+ homeostasis was recently provided (Bassil et al. 2011b; Barragan et al. 2012).

An excess of K^+ does not affect the expression of other vacuolar NHX isoforms (*OsNHX3*), possibly because particular isoforms may alter their expression only under specific conditions. *NHX3* and *NHX4* are both found on the vacuole membrane, where they maintain K^+/H^+ homeostasis, but are less significant

to vacuolar Na^+ uptake. *AtNHX3* and *AtNHX4* isoforms are expressed differently from tissue to tissue, and the growth of *Arabidopsis* improved under external Na^+, but not K^+, in knockout mutants lacking all four vacuolar NHX members (i.e., *NHX1–NHX4*) (Bassil et al. 2012). Bassil et al. (2011a) showed that the *Arabidopsis* intracellular Na^+/H^+ antiporters *NHX5* and *NHX6* are endosome associated and necessary for plant growth and development. It was not possible to assess whether *NHX5* and *NHX6* were required for Na^+ uptake into the endosomes or whether their main role was associated with vesicular trafficking and pH homeostasis. In rice, only *NHX5* is reported. Two other divergent members localized to the plasma membrane, *AtNHX7* and *AtNHX8*, have been reviewed previously (Chinnusamy et al. 2005).

Inward-rectifying K^+ channels, such as *AtAKT1* and *AtKAT1* from *Arabidopsis* and *OsAKT1* from rice, have been functionally characterized (Sentenac et al. 1992; Anderson et al. 1992; Müller-Röber et al. 1995; Golldack et al. 2003). Plant inward-rectifying

K^+ channels show high selectivity for K^+ over other monovalent cations, and the channels are reputed to specifically mediate K^+ uptake and transport in plant cells (Schachtman et al. 1992). Sodium homeostasis in plants is controlled by Na^+/H^+ antiporters in the plasma membrane and tonoplast. In *Arabidopsis thaliana*, the Na^+/H^+ antiporter SOS1 regulates the long-distance transport of sodium from roots to shoots and sodium efflux into the roots. SOS1 activity is regulated through phosphorylation by the protein kinase SOS2 together with the calcium-sensing regulatory subunit SOS3. Functional homologs of the SOS genes have been identified in rice. *OsSOS1*, *OsCIPK24*, and *OsCBL4* are the rice orthologues of *Arabidopsis SOS1*, *SOS2*, and *SOS3* genes, respectively (Atienza et al. 2007). They suppressed the salt sensitivity of the corresponding mutants of *Arabidopsis*. Reverse genetics demonstrates the importance of the SOS system in the salt tolerance of rice plants. Mutant lines in public rice collections bearing gene disruptions in *OsSOS1, OsCBL24,* and *OsCBL4*, created by insertion of T-DNA or retrotransposons, have been analyzed for salt tolerance. Our data indicate that mutants with reduced activity of OsSOS1 or OsCIPK24 are indeed salt sensitive. The gene expression pattern of OsSOS1 will be analyzed using promoter::GUS transcriptional fusions to corroborate a role in long-distance transport and sodium efflux in rice plants. SOS1 proteins contain self-inhibitory domains located at their carboxy terminal. The deletion of this carboxy terminal inhibitory domain of *OsSOS1* resulted in a greater transport activity and enhanced salt tolerance in yeast cells, which were both independent of *CIPK24/CBL4*.

Conclusion

As described earlier, numerous studies toward understanding plant salt stress response mechanisms have been reported in the last decades. However, there is still no clear-cut picture. For example, many candidate genes that regulate the signal cascade during salt stress are still to be elucidated. Furthermore, it would be interesting to know which of the transport processes explained in this chapter could enhance plant performance. Advanced biotechnology tools should help scientists to develop salt stress–tolerant crops. In this context, important genes for enhancing salt tolerance could be transporters, ROS scavengers, and so on, which could have an important role in ion homeostasis and positively influence the plant to cope with salt stress.

Acknowledgments

MK acknowledges Nisha Thakur for proofreading.

REFERENCES

Agarwal M, Hao Y, Kapoor A, Dong CH, Fujii H, Zheng X, Zhu JK (2006) A R2R3 type MYB transcription factor is involved in the cold regulation of CBF genes and in acquired freezing tolerance. *Journal of Biological Chemistry* 281:37636–37645.

Agrawal GK, Agrawal SK, Shibato J, Iwahashi H, Rakwal R (2003) Novel rice MAP kinases *OsMSRMK3* and *OsWJUMK1* involved in encountering diverse environmental stresses and developmental regulation. *Biochemical and Biophysical Research Communications* 300:775–783.

Agrawal GK, Rakwa R, Iwahashi H (2002) Isolation of novel rice (*Oryza sativa* L.) multiple stress responsive MAP kinase gene, *OsMSRMK2*, whose mRNA accumulates rapidly in response to environmental cues. *Biochemical and Biophysical Research Communications* 294:1009–1016.

Alexandre J, Lassalles J, Kado R (1990) Opening of Ca^{2+} channels in isolated red beet root vacuole membrane by inositol 1,4,5-trisphosphate. *Nature* 343:567–570.

Allen GJ, Muir SR, Sanders D (1995) Release of Ca^{2+} from individual plant vacuoles by both InsP3 and cyclic ADP-ribose. *Science* 268:735–737.

Anderson JA, Huprikar SS, Kochian LV, Lucas WJ, Gaber RF (1992) Functional expression of a probable *A. thaliana* potassium channel *in S. cerevisiae*. *Proceedings of the National Academy of Sciences of the United States of America* 89:3736–3740.

Apse MP, Sottosanto JB, Blumwald E (2003) Vacuolar cation/H^+ exchange, ion homeostasis, and leaf development are altered in a T-DNA insertional mutant of *AtNHX1*, the Arabidopsis vacuolar Na^+/H^+ antiporter. *Plant Journal* 36:229–239.

Asano T, Hakata M, Nakamura H, Aoki N, Komatsu S, Ichikawa H, Hirochika H, Ohsugi R (2011) Functional characterisation of *OsCPK21*, a calcium-dependent protein kinase that confers salt tolerance in rice. *Plant Molecular Biology* 75:179–191.

Asano T, Hayashi N, Kobayashi M, Aoki N, Miyao A, Mitsuhara I, Ichikawa H, et al. (2012) A rice calcium-dependent protein kinase *OsCPK12* oppositely modulates salt-stress tolerance and blast disease resistance. *Plant Journal* 69:26–36.

Atienza JM, Jiang X, Garciadeblas B, Mendoza I, Zhu JK, Pardo JM, Quintero FJ (2007) Conservation of the salt overly sensitive pathway in rice. *Plant Physiology* 143:1001–1012.

Baisakh N, Rao MVR, Rajasekaran K, Subudhi P, Janda J, Galbraith D, Vanier C, Pereira A (2012) Enhanced salt stress tolerance of rice plants expressing a vacuolar H^+-ATPase subunit c1 (*SaVHAc1*) gene from the halophyte grass *Spartina alterniflora* Löisel. *Plant Biotechnology Journal* 10:453–464.

Banti V, Mafessoni F, Loreti E, Alpi A, Perata P (2010) The heat-inducible transcription factor HsfA2 enhances anoxia tolerance in Arabidopsis. *Plant Physiology* 152:1471–1483.

Barragán V, Leidi EO, Andrés Z, Rubio L, De Luca A, Fernández JA, Cubero B, Pardo JM (2012) Ion exchangers *NHX1* and *NHX2* mediate active potassium uptake into vacuoles to regulate cell turgor and stomatal function in Arabidopsis. *Plant Cell* 24:1127–1142.

Bassil E, Coku A, Blumwald E (2012) Cellular ion homeostasis: Emerging roles of intracellular *NHX* Na^+/H^+ antiporters in plant growth and development. *Journal of Experimental Botany* 63:5727–5740. doi:10.1093/jxb/ers250

Bassil E, Ohto MA, Esumi T, Tajima H, Zhu Z, Cagnac O, Belmonte M, Peleg Z, Yamaguchi T, Blumwald E (2011a) The Arabidopsis intracellular Na^+/H^+ antiporters *NHX5* and *NHX6* are endosome associated and necessary for plant growth and development. *Plant Cell* 23:224–239.

Bassil E, Tajima H, Liang YC, Ohto MA, Ushijima K, Nakano R, Esumi T, Coku A, Belmonte M, Blumwald E (2011b) The Arabidopsis Na^+/H^+ antiporters NHX1 and NHX2 control vacuolar pH and K^+ homeostasis to regulate growth, flower development, and reproduction. *Plant Cell* 23:3482–3497.

Borsani O, Cuartero J, Fernández JA, Valpuesta V, Botella MA (2001) Identification of two loci in tomato reveals distinct mechanisms for salt tolerance. *Plant Cell* 13:873–887.

Chae MJ, Lee JS, Nam MH, Cho K, Hong JY, Yi SA, Suh SC, Yoon IS (2007) A rice dehydration-inducible SNF1-related protein kinase 2 phosphorylates an abscisic acid responsive element-binding factor and associates with ABA signaling. *Plant Molecular Biology* 63:151–169.

Chinnusamy V, Jagendorf A, Zhu JK (2005) Understanding and improving salt tolerance in plants. *Crop Science* 45:437–448.

Chung PJ, Kim YS, Jeong JS, Park SH, Nahm BH, Kim JK (2009) The histone deacetylase OsHDAC1 epigenetically regulates the OsNAC6 gene that controls seedling root growth in rice. *Plant Journal* 59:764–776.

Dai X, Xu Y, Ma Q, Xu W, Wang T, Xue Y, Chong K (2007) Overexpression of an R1R2R3 MYB gene, *OsMYB3R-2*, increases tolerance to freezing, drought, and salt stress in transgenic Arabidopsis. *Plant Physiology* 143:1739–1751.

Diedhiou CJ, Popova OV, Dietz KJ, Golldack D (2008) The SNF1-type serine-threonine protein kinase *SAPK4* regulates stress-responsive gene expression in rice. *BMC Plant Biology* 8:49.

Ding L, Zhu JK (1997) Reduced sodium uptake in the salt-hypersensitive *sos1* mutant of *Arabidopsis thaliana*. *Plant Physiology* 113:795–799.

Ding Z, Li S, An X, Liu X, Qin H, Wang D (2009) Transgenic expression of MYB15 confers enhanced sensitivity to abscisic acid and improved drought tolerance in *Arabidopsis thaliana*. *Journal of Genetics and Genomics* 36:17–29.

Duan J, Cai W (2012) *OsLEA3-2*, an abiotic stress induced gene of rice plays a key role in salt and drought tolerance. *PLOS One* 7(9):e45117. doi:10.1371/journal.pone.0045117

Dubos C, Stracke R, Grotewold E, Weisshaar B, Martin C, Lepiniec L (2010) MYB transcription factors in Arabidopsis. *Trends in Plant Science* 15:573–581.

Dubouzet JG, Sakuma Y, Ito Y, Kasuga M, Dubouzet EG, Miura S, Seki M, Shinozaki K, Yamaguchi-Shinozaki K (2003) OsDREB genes in rice, *Oryza sativa* L., encode transcription activators that function in drought-, high-salt- and cold-responsive gene expression. *Plant Journal* 33:751–763.

Fu SF, Chou WC, Huang DD, Huang HJ (2002) Transcriptional regulation of a rice mitogen-activated protein kinase gene, *OsMAPK4*, in response to environmental stresses. *Plant and Cell Physiology* 43:958–963.

Fujita Y, Fujita M, Shinozaki K, Yamaguchi-Shinozaki K (2011) ABA-mediated transcriptional regulation in response to osmotic stress in plants. *Journal of Plant Research* 124(4): 509–525. doi: 10.1007/s10265-011-0412-3.

Fukuda A, Nakamura A, Hara N, Toki S, Tanaka Y (2011) Molecular and functional analyses of rice NHX-type Na^+/H^+ antiporter genes. *Planta* 233:175–188.

Fukuda A, Nakamura A, Tagiri A, Tanaka H, Miyao A, Hirochika H, Tanaka Y (2004) Function, intracellular localization and the importance in salt tolerance of a vacuolar $Na^{(+)}/H^{(+)}$ antiporter from rice. *Plant and Cell Physiology* 45:146–159.

Furihata T, Maruyama K, Fujita Y, Umezawa T, Yoshida R, Shinozaki K, Yamaguchi-Shinozaki K (2006) Abscisic acid dependent multisite phosphorylation regulates the activity of a transcription activator *AREB1*. *Proceedings of the National Academy of Sciences of the United States of America* 103:1988–1993.

Galon Y, Finkler A, Fromm H (2010) Calcium-regulated transcription in plants. *Molecular Plant* 3:653–669.

Gao P, Bai X, Yang L, Lv D, Li Y, Cai H, Ji W, Guo D, Zhu Y (2010) Over-expression of osa-MIR396c decreases salt and alkali stress tolerance. *Planta* 231:991–1001.

Gao P, Bai X, Yang L, Lv D, Pan X, Li Y, Cai H, Ji W, Chen Q, Zhu Y (2011) *osa-MIR393*: A salinity- and alkaline stress-related microRNA gene. *Molecular Biology Reports* 38:237–242.

Garciadeblás B, Senn ME, Bañuelos MA, Rodríguez-Navarro A (2003) Sodium transport and HKT transporters: The rice model. *Plant Journal* 34:788–801.

Garg AK, Kim JK, Owens TG, Ranwala AP, Choi YD, Kochian LV, Wu RJ (2002) Trehalose accumulation in rice plants confers high tolerance levels to different abiotic stresses. *Proceedings of the National Academy of Sciences of the United States of America* 99:15898–15903.

Gilroy S, Read ND, Trewavas AJ (1990) Elevation of cytoplasmic calcium by caged calcium or caged inositol triphosphate initiates stomatal closure. *Nature* 346:769–771.

Golldack D, Quigley F, Michalowski CB, Kamasani UR, Bohnert HJ (2003) Salinity stress-tolerant and -sensitive rice (*Oryza sativa* L.) regulate AKT1-type potassium channel transcripts differently. *Plant Molecular Biology* 51:71–81.

Guo J, Duff HJ (2006) Calmodulin kinase II accelerates L-type Ca2+ current recovery from inactivation and compensates for the direct inhibitory effect of [Ca2+]i in rat ventricular myocytes. *The Journal of Physiology* 574(2):509–518.

Halfter U, Ishitani M, Zhu JK (2000) The Arabidopsis SOS2 protein kinase physically interacts with and is activated by the calcium-binding protein SOS3. *Proceedings of the National Academy of Sciences of the United States of America* 97:3735–3740.

Hoshida H, Tanaka Y, Hibino T, Hayashi Y, Tanaka A, Takabe T, Takabe T (2000) Enhanced tolerance to salt stress in transgenic rice that overexpresses chloroplast glutamine synthetase. *Plant Molecular Biology* 43:103–111.

Hossain MA, Cho JI, Han M, Ahn CH, Jeon JS, An G, Park PB (2010) The ABRE binding bZIP transcription factor *OsABF2* is a positive regulator of abiotic stress and ABA signaling in rice. *Journal of Plant Physiology* 167:1512–1520.

Hou X, Xie K, Yao J, Qi Z, Xiong L (2009) A homolog of human ski-interacting protein in rice positively regulates cell viability and stress tolerance. *Proceedings of the National Academy of Sciences of the United States of America* 106:6410–6415.

Hu H, Dai M, Yao J, Xiao B, Li X, Zhang Q, Xiong L (2006) Overexpressing a NAM, ATAF, and CUC (NAC) transcription factor enhances drought resistance and salt tolerance in rice. *Proceedings of the National Academy of Sciences of the United States of America* 103:12987–12992.

Hu H, You J, Fang Y, Zhu X, Qi Z, Xiong L (2008) Characterization of transcription factor gene *SNAC2* conferring cold and salt tolerance in rice. *Plant Molecular Biology* 67:169–181.

Huang XY, Chao DY, Gao JP, Zhu MZ, Shi M, Lin HX (2009) A previously unknown zinc finger protein, DST, regulates drought and salt tolerance in rice via stomatal aperture control. *Genes & Development* 23:1805–1817.

Ito Y, Katsura K, Maruyama K, Taji T, Kobayashi M, Seki M, Shinozaki K, Shinozaki KY (2006) Functional analysis of rice DREB1/CBF-type transcription factors involved in cold responsive gene expression in transgenic rice. *Plant and Cell Physiology* 47:141–153.

Jan A, Maruyama K, Todaka D, Kidokoro S, Abo M, Yoshimura E, Shinozaki K, Nakashima K, Shinozaki KY (2013) *OsTZF1*, a CCCH-tandem zinc finger protein, confers delayed senescence and stress tolerance in rice by regulating stress-related genes. *Plant Physiology* 161:1202–1216. doi:10.1104/pp.112.205385.

Jang IC, Oh SJ, Seo JS, Choi WB, Song SI, Kim CH, Kim YS, et al. (2003) Expression of a bifunctional fusion of the *Escherichia coli* genes for trehalose-6-phosphate synthase and trehalose-6-phosphate phosphatase in transgenic rice plants increases trehalose accumulation and abiotic stress tolerance without stunting growth. *Plant Physiology* 131:516–524.

Jeong MJ, Lee SK, Kim BG, Kwon TR, Cho WS, Park YT, Lee JO, Kwon HB, Byun MO, Park SC (2006) A rice (*Oryza sativa* L.) MAP kinase gene, *OsMAPK44*, is involved in response to abiotic stresses. *Plant Cell, Tissue and Organ Culture* 85:151–160.

Jin H, Martin C (1999) Multifunctionality and diversity within the plant MYB-gene family. *Plant Molecular Biology* 41:577–585.

Jonak C, Okresz L, Bogre L, Hirt H (2002) Complexity, cross talk and integration of plant MAP kinase signaling. *Current Opinion in Plant Biology* 5:415–424.

Jossier M, Bouly JP, Meimoun P, Arjmand A, Lessard P, Hawley S, Grahame Hardie D, Thomas M (2009) SnRK1 (SNF1-related kinase 1) has a central role in sugar and ABA signalling in *Arabidopsis thaliana*. *Plant Journal* 59:316–328.

Jung C, Seo JS, Han SW, Koo YJ, Kim CH, Song SI, Nahm BH, Choi YD, Cheong JJ (2008) Over-expression of AtMYB44 enhances stomatal closure to confer abiotic stress tolerance in transgenic Arabidopsis. *Plant Physiology* 146:623–635.

Kim JA, Agrawal GK, Rakwal R, Han KS, Kim KN, Yun CH, Heu S, Park SY, Lee YH, Jwa NS (2003) Molecular cloning and mRNA expression analysis of a novel rice (*Oryza sativa* L.) MAPK kinase kinase, *OsEDR1*, an ortholog of Arabidopsis *AtEDR1*, reveal its role in defense/stress signalling pathways and development. *Biochemical and Biophysical Research Communications* 300(4):868–876.

Kobayashi Y, Murata M, Minami H, Yamamoto S, Kagaya Y, Hobo T, Yamamoto A, Hattori T (2005) Abscisic acid-activated SNRK2 protein kinases function in the gene-regulation pathway of ABA signal transduction by phosphorylating ABA response element binding factors. *Plant Journal* 44:939–949.

Kobayashi Y, Yamamoto S, Minami H, Kagaya Y, Hattori T (2004) Differential activation of the rice sucrose nonfermenting1-related protein kinase2 family by hyperosmotic stress and abscisic acid. *Plant Cell* 16:1163–1177.

Koh S, Lee SC, Kim MK, Koh JH, Lee S, An G, Choe S, Kim SR (2007) T-DNA tagged knockout mutation of rice *OsGSK1*, an orthologue of Arabidopsis BIN2, with enhanced tolerance to various abiotic stresses. *Plant Molecular Biology* 65:453:466.

Kulik A, Wawer I, Krzywinska E, Bucholc M, Dobrowolska G (2011) SnRK2 protein kinases—key regulators of plant response to abiotic stresses. *OMICS: A Journal of Integrative Biology* 15:859–872.

Kumar K, Kumar M, Kim SR, Ryu H, Cho YG (2013) Insights into genomics of salt stress response in rice. *Rice* 6:27. doi:10.1186/1939-8433-6-27.

Kumar M (2013) Crop plants and abiotic stresses. *Journal of Biomolecular Research and Therapeutics* 3:e125. doi: 10.4172/2167-7956.1000e125.

Kumar M, Choi J, An G, Kim S-R (2017) Ectopic expression of *OsSta2* enhances salt stress tolerance in rice. *Frontiers in Plant Science* 8:316. doi:10.3389/fpls.2017.00316.

Kumar M, Gho YS, Jung KH, Kim SR (2017) Genome-wide identification and analysis of genes, conserved between *japonica* and *indica* rice cultivars, that respond to low-temperaturestress at the vegetative growth stage. *Frontiers in Plant Science* 30;8:1120. https://doi.org/10.3389/fpls.2017.01120.

Kumar M, Lee SC, Kim JY, Kim SJ, Aye SS, Kim SR (2014). Over-expression of dehydrin gene, *OsDhn1*, improves drought and salt stress tolerance through scavenging of reactive oxygen species in rice (*Oryza sativa* L.). *Journal of Plant Biology* 57:383–393. doi:10.1007/s12374-014-0487-1.

Lee SK, Kim BG, Kwon TR, Jeong MJ, Park SR, Lee JW, Byun MO, et al. (2011) Overexpression of the mitogen-activated protein kinase gene *OsMAPK33* enhances sensitivity to salt stress in rice (*Oryza sativa* L.). *Journal of Biosciences* 36:139–151.

Lin LL, Wartmann M, Lin AY, Knopf JL, Seth A, Davis RJ (1993) cPLA2 is phosphorylated and activated by MAP kinase. *Cell* 72:269–278.

Liu D, Chen X, Liu J, Ye J, Guo Z (2012) The rice ERF transcription factor *OsERF922* negatively regulates resistance to *Magnaporthe oryzae* and salt tolerance. *Journal of Experimental Botany* 63:3899–3912.

Liu J, Ishitani M, Halfter U, Kim CS, Zhu JK (2000) The *Arabidopsis thaliana* SOS2 gene encodes a protein kinase that is required for salt tolerance. *Proceedings of the National Academy of Sciences of the United States of America* 97:3730–3734.

Liu S, Cheng Y, Zhang X, Guan Q, Nishiuchi S, Hase K, Takano T (2007) Expression of an NADP-malic enzyme gene in rice (*Oryza sativa*. L) is induced by environmental stresses; over-expression of the gene in Arabidopsis confers salt and osmotic stress tolerance. *Plant Molecular Biology* 64:49–58.

Ma X, Qian Q, Zhu D (2005) Expression of a calcineurin gene improves salt stress tolerance in transgenic rice. *Plant Molecular Biology* 58:483–495.

Mallikarjuna G, Mallikarjuna K, Reddy MK, Kaul T (2011) Expression of *OsDREB2A* transcription factor confers enhanced dehydration and salt stress tolerance in rice (*Oryza sativa* L.). *Biotechnology Letters* 33:1689–1697.

Miller J, McLachlan AD, Klug A (1985) Repetitive zinc-binding domains in the protein transcription factor IIIA from *Xenopus* oocytes. *The EMBO Journal* 4:1609–1614.

Mittal D, Chakrabarti S, Sarkar A, Singh A, Grover A (2009). Heat shock factor gene family in rice: Genomic organization and transcript expression profiling in response to high temperature, low temperature and oxidative stresses. *Plant Physiology and Biochemistry* 47(9):785–795. doi: 10.1016/j.plaphy.2009.05.003.

Mizoi J, Shinozaki K, Yamaguchi-Shinozaki K (2011) AP2/ERF family transcription factors in plant abiotic stress responses. *Biochimica et Biophysica Acta* 1819:86–96.

Motohashi T, Nagamiya K, Prodhan SH, Nakao K, Shishido T, Yamamoto Y, Moriwaki T, et al. (2010) Production of salt stress tolerant rice by overexpression of the catalase gene, *katE*, derived from *Escherichia coli*. *Asia Pacific Journal of Molecular Biology and Biotechnology* 18:1.

Mukhopadhyay A, Vij S, Tyagi AK (2004) Overexpression of a zinc-finger protein gene from rice confers tolerance to cold, dehydration, and salt stress in transgenic tobacco. *Proceedings of the National Academy of Sciences of the United States of America* 101:6309–6314.

Müller-Röber B, Ellenberg J, Provart N, Willmitzer L, Busch H, Becker D, Dietrich P, Hoth S, Hedrich R (1995) Cloning and electrophysiological analysis of *KST1*, an inward rectifying K+ channel expressed in potato guard cells. *The EMBO Journal* 14:2409–2416.

Mundy J, Chua NH (1988) Abscisic acid and water-stress induce the expression of a novel rice gene. *The EMBO Journal* 7:2279–2286.

Mustilli AC, Merlot S, Vavasseur A, Fenzi F, Giraudat J (2002) Arabidopsis *OST1* protein kinase mediates the regulation of stomatal aperture by abscisic acid and acts upstream of reactive oxygen species production. *Plant Cell* 14:3089–3099.

Nakamura A, Fukuda A, Sakai S, Tanaka Y (2006) Molecular cloning, functional expression and subcellular localization of two putative vacuolar voltage-gated chloride channels in rice (*Oryza sativa* L.). *Plant and Cell Physiology* 47:32–42.

Nakashima K, Tran LS, Van Nguyen D, Fujita M, Maruyama K, Todaka D, Ito Y, Hayashi N, Shinozaki K, Shinozaki KY (2007) Functional analysis of a NAC-type transcription factor *OsNAC6* involved in abiotic and biotic stress-responsive gene expression in rice. *Plant Journal* 51:617–630.

Nam MH, Huh SM, Kim KM, Park WJ, Seo JB, Cho K, Kim DY, Kim BG, Yoon IS (2012) Comparative proteomic analysis of early salt stress-responsive proteins in roots of SnRK2 transgenic rice. *Proteome Science* 10:25.

Nijhawan A, Jain M, Tyagi AK, Khurana JP (2008) Genomic survey and gene expression analysis of the basic leucine zipper transcription factor family in rice. *Plant Physiology* 146:333–350.

Ning J, Li X, Hicks LM, Xiong L (2010) A raf-like MAPKKK gene *DSM1* mediates drought resistance through reactive oxygen species scavenging in rice. *Plant Physiology* 152:876–890.

Ogata K, Kanei-Ishii C, Sasaki M, Hatanaka H, Nagadoi A, Enari M, Nakamura H, Nishimura Y, Ishii S, Sarai A (1996) The cavity in the hydrophobic core of Myb DNA binding domain is reserved for DNA recognition and trans-activation. *Nature Structural and Molecular Biology* 3:178–187.

Ogawa D, Yamaguchi K, Nishiuchi T (2007) High level overexpression of the Arabidopsis *HsfA2* gene confers not only increased thermotolerance but also salt/osmotic stress tolerance and enhanced callus growth. *Journal of Experimental Botany* 58:1–11.

Ohta M, Hayashi Y, Nakashima A, Hamada A, Tanaka A, Nakamura T, Hayakawa T (2002) Introduction of a Na+/H+ antiporter gene from *Atriplex gmelini* confers salt tolerance to rice. *FEBS Letters* 532:279–282.

Ouyang SQ, Liu YF, Liu P, Lei G, He SJ, Ma B, Zhang WK, Zhang JS, Chen SY (2010) Receptor-like kinase *OsSIK1* improves drought and salt stress tolerance in rice (*Oryza sativa*) plants. *Plant Journal* 62:316–329.

Pabo CO, Peisach E, Grant RA (2001) Design and selection of novel Cys2His2 zinc finger proteins. *Annual Review of Biochemistry* 70:313–340.

Pandey GK, Grant JJ, Cheong YH, Kim BG, Li L, Luan S (2005) ABR1, an APETALA2-domain transcription factor that functions as a repressor of ABA response in Arabidopsis. *Plant Physiology* 139:1185–1193.

Parre E, Ghars MA, Leprince AS, Thiery L, Lefebvre D, Bordenave M, Richard L, Mazars C, Abdelly C, Savouré A. (2007) Calcium signaling via phospholipase C is essential for proline accumulation upon ionic but not nonionic hyperosmotic stresses in Arabidopsis. *Plant Physiology* 144:503–512.

Platten JD, Cotsaftis O, Berthomieu P, Bohnert H, Davenport RJ, Fairbairn DJ, Horie T, et al. (2006) Nomenclature for *HKT* transporters, key determinants of plant salinity tolerance. *Trends in Plant Science* 11:372–374.

Qin F, Kakimoto M, Sakuma Y, Maruyama K, Osakabe Y, Tran LS, Shinozaki K, Yamaguchi-Shinozaki K (2007) Regulation and functional analysis of ZmDREB2A in response to drought and heat stresses in *Zea mays* L. *Plant Journal* 50:54–69.

Ramakrishna A, Ravishankar GA (2011) Influence of abiotic stress signals on secondary metabolites in plants. *Plant Signaling & Behavior* 6:1720–1731.

Ramakrishna A, Ravishankar GA (2013) Role of plant metabolites in abiotic stress tolerance under changing climatic conditions with special reference to secondary compounds. In: *Climate Change and Plant Abiotic Stress Tolerance*, N. Tuteja and S. S. Gill (Eds.). Wiley-VCH Verlag GmbH& Co. KGaA, Weinheim, Germany.

Rodriguez-Rosales MP, Jiang XY, Galvez FJ, Aranda MN, Cubero B, Venema K (2008) Overexpression of the tomato K+/H+ antiporter *LeNHX2* confers salt tolerance by improving potassium compartmentalization. *New Phytologist* 179:366–377.

Rosinski JA, Atchley WR (1998) Molecular evolution of the Myb family of transcription factors: Evidence for polyphyletic origin. *Journal of Molecular Evolution* 46:74–83.

Schachtman DP, Munns R (1992) Sodium accumulation in leaves of triticum species that differ in salt tolerance. *Australian Journal of Plant Physiology* 25:239–250.

Saijo Y, Hata S, Kyozuka J, Shimamoto K, Izui K (2000) Overexpression of a single Ca^{2+} dependent protein kinase confers both cold and salt/drought tolerance on rice plants. *Plant Journal* 23:319–327.

Sakamoto A, Alia, Murata N (1998) Metabolic engineering of rice leading to biosynthesis of glycinebetaine and tolerance to salt and cold. *Plant Molecular Biology* 38:1011–1019.

Sakamoto H, Maruyama K, Sakuma Y, Meshi T, Iwabuchi M, Shinozaki K, Shinozaki KY (2004) Arabidopsis Cys2/His2-type zinc-finger proteins function as transcription repressors under drought, cold, and high-salinity stress conditions. *Plant Physiology* 136:2734–2746.

Sakuma Y, Maruyama K, Osakabe Y, Qin F, Seki M, Shinozaki K, Yamaguchi-Shinozaki K (2006) Functional analysis of an Arabidopsis transcription factor, DREB2A, involved in drought-responsive gene expression. *Plant Cell* 18:1292–1309.

Schmidt R, Schippers JHM, Welker A, Mieulet D, Guiderdoni E, Roeber BM (2012) Transcription factor *OsHsfC1b* regulates salt tolerance and development in *Oryza sativa* ssp. Japonica. *AoB Plants* doi:10.1093/aobpla/pls011.

Sentenac H, Bonneaud N, Minet M, Lacroute F, Salmon JM, Gaymard F, Grignon C (1992) Cloning and expression in yeast of a plant potassium ion transport system. *Science* 256:663–665.

Shi H, Ishitani M, Kim C, Zhu JK (2000) The *Arabidopsis thaliana* salt tolerance gene SOS1 encodes a putative Na+/H+ antiporter. *Proceedings of the National Academy of Sciences of the United States of America* 97:6896–6901.

Shinozaki KY, Shinozaki K (2005) Organization of cis-acting regulatory elements in osmotic- and cold-stress-responsive promoters. *Trends in Plant Science* 10:88–94.

Song SY, Chen Y, Chen J, Dai XY, Zhang WH (2011) Physiological mechanisms underlying *OsNAC5*-dependent tolerance of rice plants to abiotic stress. *Planta* 234:331–345.

Sripinyowanich S, Klomsakul P, Boonburapong B, Bangyeekhun T, Asami T, Gu H, Buaboocha T, Chadchawan S (2010) Exogenous ABA induces salt tolerance in indica rice (*Oryza sativa* L.): The role of *OsP5CS1* and *OsP5CR* gene expression during salt stress. *Environmental and Experimental Botany* 86:94–105.

Sun SJ, Guo SQ, Yang X, Bao YM, Tang HJ, Sun H, Huang J, Zhang HS (2010) Functional analysis of a novel Cys2/His2-type zinc finger protein involved in salt tolerance in rice. *Journal of Experimental Botany* 61:2807–2818.

Takahashi S, Kimura S, Kaya H, Iizuka A, Wong HL, Shimamoto K, Kuchitsu K (2012) Reactive oxygen species production and activation mechanism of the rice NADPH oxidase *OsRbohB*. *The Journal of Biochemistry* 152:37–43.

Takasaki H, Maruyama K, Kidokoro S, Ito Y, Fujita Y, Shinozaki K, Shinozaki KY, Nakashima K (2010) The abiotic stress-responsive NAC-type transcription factor *OsNAC5* regulates stress-inducible genes and stress tolerance in rice. *Molecular Genetics and Genomics* 284:173–183.

Tanaka Y, Hibin T, Hayashi Y, Tanaka A, Kishitani S, Takabe T, Yokota S, Takabe T (1999) Salt tolerance of transgenic rice overexpressing yeast mitochondrial *Mn-SOD* in chloroplasts. *Plant Science* 148:131–138.

Tao Z, Kou Y, Liu H, Li X, Xiao J, Wang S (2011) *OsWRKY45* alleles play different roles in abscisic acid signalling and salt stress tolerance but similar roles in drought and cold tolerance in rice. *Journal of Experimental Botany* 62:4863–4874.

Testerink C, Munnik T (2005) Phosphatidic acid: A multifunctional stress signaling lipid in plants. *Trends in Plant Science* 10:368–375.

Thanaruksa R, Sripinyowanich S, Udomchalothorn T, Chadchawan S (2012) Effects of *OsNUC1* motif on transgenic *Arabidopsis thaliana* L. growth under salt stress condition. *Thai Journal of Botany* 4:145–157.

Tran LSP, Nakashima K, Sakuma Y, Simpson SD, Fujita Y, Maruyama K, Fujita M, Seki M, Kazuo Shinozaki K, Shinozaki KY (2004) Isolation and functional analysis of Arabidopsis stress-inducible NAC transcription factors that bind to a drought-responsive *cis*-element in the early responsive to dehydration stress 1 promoter. *Plant Cell* 16:2481–2498.

Tuteja N, Sopory SK (2008) Plant signaling in stress: G-protein coupled receptors, heterotrimeric G-proteins and signal coupling via phospholipases. *Plant Signaling & Behavior* 3:79–86.

Venema K, Belver A, Marin-Manzano MC, Rodriguez-Rosales MP, Donaire JP (2003) A novel intracellular K⁺/H⁺ antiporter related to Na⁺/H⁺ antiporters is important for K⁺ ion homeostasis in plants. *Journal of Biological Chemistry* 278:22453–22459.

Wang Q, Guan Y, Wu Y, Chen H, Chen F, Chu C (2008) Overexpression of a rice *OsDREB1F* gene increases salt, drought, and low temperature tolerance in both Arabidopsis and rice. *Plant Molecular Biology* 67:589–602.

Welsch R, Wust F, Bar C, Babili SA, Beyer P (2008) A third phytoene synthase is devoted to abiotic stress-induced abscisic acid formation in rice and defines functional diversification of phytoene synthase genes. *Plant Physiology* 147:367–380.

Wu L, Fan Z, Guo L, Li Y, Chen ZL, Qu LJ (2005) Overexpression of the bacterial *nhaA* gene in rice enhances salt and drought tolerance. *Plant Science* 168:297–302.

Wu SJ, Ding L, Zhu JK (1996) *SOS1*, a genetic locus essential for salt tolerance and potassium acquisition. *Plant Cell* 8:617–627.

Xia K, Wang R, Ou X, Fang Z, Tian C, Duan J, Wang Y, Zhang M (2012) *OsTIR1* and *OsAFB2* downregulation via OsmiR393 overexpression leads to more tillers, early flowering and less tolerance to salt and drought in rice. *PLoS One* 7(1):e30039. doi:10.1371/journal.pone.0030039.

Xiang Y, Huang Y, Xiong L (2007) Characterization of stress-responsive *CIPK* genes in rice for stress tolerance improvement. *Plant Physiology* 144:1416–1428.

Xiang Y, Tang N, Du H, Ye H, Xiong L (2008) Characterization of *OsbZIP23* as a key player of the basic leucine zipper transcription factor family for conferring abscisic acid sensitivity and salinity and drought tolerance in rice. *Plant Physiology* 148:1938–1952.

Xiong L, Yang Y (2003) Disease resistance and abiotic stress tolerance in rice are inversely modulated by an abscisic acid–inducible mitogen-activated protein kinase. *Plant Cell* 15:745–759.

Xu D, Duan X, Wang B, Hong B, Ho THD, Wu R (1996) Expression of a late embryogenesis abundant protein gene, *hva1*, from barley confers tolerance to water deficit and salt stress in transgenic rice. *Plant Physiology* 110:249–257.

Xu DQ, Huang J, Guo SQ, Yang X, Bao YM, Tang HJ, Zhang HS (2008) Overexpression of a TFIIIA-type zinc finger protein gene *ZFP252* enhances drought and salt tolerance in rice (*Oryza sativa* L.). *FEBS Letters* 582:1037–1043.

Xu GY, Rocha PSCF, Wang ML, Xu ML, Cui YC, Li LY, Zhu YX, Xia X (2011) A novel rice calmodulin-like gene, *OsMSR2*, enhances drought and salt tolerance and increases ABA sensitivity in Arabidopsis. *Planta* 234:47–59.

Yamaguchi-Shinozaki K, Shinozaki K (2006) Transcriptional regulatory networks in cellular responses and tolerance to dehydration and cold stresses. *Annual Review of Plant Biology* 57:781–803.

Yang A, Dai X, Zhang WH (2012) A R2R3-type MYB gene, *OsMYB2*, is involved in salt, cold, and dehydration tolerance in rice. *Journal of Experimental Botany* 63:2541–2556.

Ye H, Du H, Tang N, Li X, Xiong L (2009) Identification and expression profiling analysis of TIFY family genes involved in stress and phytohormone responses in rice. *Plant Molecular Biology* 71:291–305.

Yoo JH, Park CY, Kim JC, Do Heo W, Cheong MS, Park HC, Kim MC, et al. (2005). Direct interaction of a divergent CaM isoform and the transcription factor, MYB2, enhances salt tolerance in Arabidopsis. *Journal of Biological Chemistry* 280:3697–3706.

Yoshida R, Hobo T, Ichimura K, Mizoguchi T, Takahashi F, Aronso J, Ecker JR, Shinozaki K (2002) ABA-activated SnRK2 protein kinase is required for dehydration stress signaling in Arabidopsis. *Plant and Cell Physiology* 43:1473–1483.

Yoshida R, Umezawa T, Mizoguchi T, Takahashi S, Takahashi F, Shinozaki K (2006) The regulatory domain of SRK2E/OST1/SnRK2.6 interacts with ABI1 and integrates abscisic acid (ABA) and osmotic stress signals. *Journal of Biological Chemistry* 281:5310–5318.

Yotsushima K, Mitsui T, Takaoka T, Hayakawa T, Igaue I (1993) Purification and characterization of membrane-bound inositol phospholipid-specific phospholipase C from suspension-cultured rice (*Oryza sativa* L.) cells. *Plant Physiology* 102:165–172.

Zhang H, Zhai J, Mo J, Li D, Song F (2012) Overexpression of rice sphingosine-1-phosphate lyase gene *OsSPL1* in transgenic tobacco reduces salt and oxidative stress tolerance. *Journal of Integrative Plant Biology* 54:652–662.

Zhang HX, Blumwald E (2001) Transgenic salt-tolerant tomato plants accumulate salt in foliage but not in fruit. *Nature Biotechnology* 19:765–768.

Zhang JZ, Creelman RA, Zhu JK (2004) From laboratory to field. Using information from Arabidopsis to engineer salt, cold, and drought tolerance in crops. *Plant Physiology* 135:615–621.

Zhang Y, Andralojc PJ, Hey SJ, Primavesi LF, Specht M, Koehler J, Parry MAJ, Halford NG (2008) Arabidopsis sucrose non-fermenting-1-related protein kinase-1 and calcium-dependent protein kinase phosphorylate conserved target sites in ABA response element binding proteins. *Annals of Applied Biology* 153:401–409.

Zhao F, Guo S, Zhang H, Zhao Y (2006a) Expression of yeast *SOD2* in transgenic rice results in increased salt tolerance. *Plant Science* 170:216–224.

Zhao F, Wang Z, Zhang Q, Zhao Y, Zhang H (2006b) Analysis of the physiological mechanism of salt-tolerant transgenic rice carrying a vacuolar Na$^+$/H$^+$ antiporter gene from *Suaeda salsa*. *Journal of Plant Research* 119:95–104.

Zheng X, Chen B, Lu G, Han B (2009) Overexpression of a NAC transcription factor enhances rice drought and salt tolerance. *Biochemical and Biophysics Research Communications* 379:985–989.

Zhu B, Su J, Chang M, Verma DPS, Fan YL, Wu R (1998) Overexpression of a Δ^1-pyrroline-5-carboxylate synthetase gene and analysis of tolerance to water- and salt-stress in transgenic rice. *Plant Science* 139:41–48.

Zhu JK (2001) Cell signaling under salt stress, water and cold stresses. *Current Opinion in Plant Biology* 5:401–406.

Zhu JK (2002) Salt and drought stress signal transduction in plants. *Annual Review of Plant Biology* 53:247–273.

Zhu Q, Zhang J, Gao X, Tong J, Xiao L, Li W, Zhang H (2010) The Arabidopsis AP2/ERF transcription factor RAP2.6 participates in ABA, salt and osmotic stress responses. *Gene* 457:1–12.

Zou M, Guan Y, Ren H, Zhang F, Chen F (2008) A bZIP transcription factor, *OsABI5*, is involved in rice fertility and stress tolerance. *Plant Molecular Biology* 66:675–683.

4

Recent Advances on the Modulatory Role of ATPases toward Salt Tolerance in Plants

Soumya Mukherjee

CONTENTS

Abbreviations

ABA	Abscisic acid
CaM	Calmodulin
ER	Endoplasmic reticulum
NO	Nitric oxide
OU	Ouabain

Introduction

Soil salinity is one of the major abiotic stress factors known to affect plant growth and crop productivity. Salinity stress annually leads to infertility of 2–3 Mha of agricultural land worldwide (Sekmen et al. 2010). Sodium chloride (NaCl) is a predominant salt in soils, which causes soil salinity and hyperosmotic stress in plants. Salt-tolerant plants (halophytes) comprise 0.25% of known angiosperms and include 350 species among Caryophyllales, Alismatales, Malpighiales, Poales, and Lamiales (Flowers et al. 2010). Glycophytes range from salt-susceptible to salt-sensitive types, depending on Na+ sequestration or exclusion ability. Glycophytes and halophytes have strong differences in terms of their evolutionary and adaptational backgrounds. Salt stress signaling involves a series of physiological events coordinated to maintain ion homoeostasis in plants (Tables 4.1 and 4.2). Salt tolerance in plants depends on the following factors: (1) accumulation and sequestration of Na+ and Cl−; (2) maintenance of K+ and Ca2+ pools as essential macronutrients in the presence of high Na+ levels; (3) regulation of long-distance Na+ transport and its accumulation in different aerial organs following transpiration; (4) accumulation of suitable osmolytes to

prevent desiccation; and (5) modulation of efficient antioxidant machineries. A change in the cytosolic calcium level is one of the immediate signaling events evident as an early response to salt stress. Plasma membrane Na+/H+ transporters (SOS 1) and H+-ATPases are associated sensors regulating sodium efflux in salt-stressed cells (Zhu 2000). Changes in cytosolic and vacuolar pH occur as a result of sodium sensing (Kader et al. 2007). This is followed by the activation of several hormones and enzymes (Gao et al. 2004). The regulation of Na+ transport and its efflux from the cells is modulated by a set of SOS genes (SOS 1–SOS 6) (Zhu 2002). Sodium influx into the root cells and its sequestration or long-distance transport are highly regulated. The influx of sodium ions into the root cells mostly occurs by passive diffusion, mediated by nonspecific cation channels (NSCCs) in the plasma membrane and some high-affinity potassium transporters (HKTs; Davenport and Tester 2000; Mäser et al. 2002). The entry of Na+ into the epidermis of the root cells is followed by its apoplastic and symplastic movement to the cortical zone, endodermis, and xylem parenchyma cells (Munns and Tester 2008). The development of Casparian strips is initiated in the differentiated zone of the root endodermis, thus directing the flow of water and restricting toxic solutes from entering the stelar cells. The elongation zone of the roots is highly susceptible to sodium influx due to the higher absorption capability in this region. Sodium, being a cation, alters the resting membrane potential of the cells (Kader 2006). Osmotic stress pertains to the loss of turgor pressure and water deficiency caused by the decreased water potential of saline soil. The electric potential of cortical cells and of xylem parenchyma varies from −120 to −60 mV, thus causing higher activity of Na+/H+ antiporters in xylem cells to exclude sodium ions (Munns and Tester 2008).

TABLE 4.1

Early Signaling Events Associated with Salt Stress in Plants

Molecules/Proteins	Site of Action	Reference
Calcium	Increase in $[Ca^{2+}]_{cyt}$ with concentration range >200 nM	Zhu (2001)
IP$_3$ (Inositol triphosphate)	Membrane-bound signaling molecule, released in cytosol to trigger stored calcium release from organelles	Pical et al. (1999)
Phosphatidic acid	Produced in membranes and appears in cytosol	Park et al. (2003)
Phospholipases C and D	PLC and PLD provide salt tolerance by producing phosphatidic acid as signaling molecule	Hong et al. (2008)
Protein kinases	Localized in cytosol or organelles to carry out protein phosphorylation. Histidine kinases are major osmosensors of salt-stress	Hermann and Jung (2010)
SOS 1	Early sensors of NaCl stress, localized in cell and vacuolar membranes facilitating Na$^+$ efflux (Na$^+$/H$^+$ antiport)	Munns and Tester (2008)
SOS 2/SOS 3	Ser/Thr protein kinase (SOS 2) works in coordination with myristoylated Ca^{2+}-binding protein (SOS 3) to phosphorylate SOS 1	Zhu (2002)
SOS 4	Pyridoxal kinase has a role in root hair development	Shi and Zhu (2002)
SOS 5	Cell surface adhesion protein, Na$^+$ sensor	Shi et al. (2003)
SOS 6	Cellulose–synthase-like protein, provides osmotic tolerance	Zhu et al. (2010)
Nitric oxide (NO)[a]	Root tips and elongation zone, cotyledons, oil body surface (rapid response)	David et al. (2010)

[a] Transiently produced molecule, but may stabilize within the cells to provide late signaling effects.

TABLE 4.2

Late Signaling Events associated with Salt Stress in Plants

Molecules/Proteins	Site of Action	Reference
ABA (abscisic acid)	Synthesis triggered in roots and leaves on desiccation or water stress	Zhang and Davis (1990)
Glutathione[a]	Roots and leaves	Magdalena et al. (2008)
Tyrosine-nitrated proteins	Growing region of roots; oil body surface of cotyledons	David (2012)
Protein phosphorylation	Reported in leaves	Lv et al. (2014)
Reactive oxygen species[b]	Meristematic zone of roots	David (2012)
Peroxidase	Roots	Jiang et al. (2007)
Superoxide dismutase	Roots	Jiang et al. (2007)
Catalase	Roots	Hernandez et al. (2010)

[a] Interacts with nitric oxide leading to signaling cross talk.

[b] May also be involved in early signaling event.

H$^+$-ATPase Regulation and SOS Signaling Pathway Provide Tolerance toward Salt Stress

The various perspectives associated with salt stress tolerance and ion homoeostasis have been discussed with respect to H$^+$-ATPase activity and its regulation. Investigations have been reported in various crop and non-crop model systems that have been used for H$^+$-ATPase activity regulation during salt stress. The dominant H$^+$-ATPase isoforms present in plant membranes serve as the major enzyme regulating electrochemical potential in the cell. The process of ion homoeostasis mostly involves coordination of the Na$^+$/K$^+$ ratio and regulation of their transport across the membrane. The regulation of H$^+$-ATPase largely depends on the membrane composition of the cells. It has been suggested that abiotic stress factors (including salt stress) alter the fatty acid composition and permeability of plant membranes. Low water potential and osmotic stress are likely to alter the fatty acid composition and degree of unsaturated fatty acids. H$^+$-ATPase activity has been reported to be regulated by such factors in vivo (Kasamo 2003; Hernandez et al. 2002). The plasma membrane–localized

H$^+$-ATPase confers salt tolerance by exhibiting higher activity in salt-tolerant plants (Chen et al. 2007; Sahu and Shaw 2009). However, negative regulation or salt-insensitive H$^+$-ATPase activity has also been reported in the leaves of some plants (Pitan et al. 2009; Wakeel et al. 2010). Salt tolerance in resistant genotypes of maize has also been found to be independent of H$^+$-ATPase activity (Pitan et al. 2009). Salt stress–induced elevation in H$^+$-ATPase activity pertains to both transcriptional and post-transcriptional modifications (Janicka-Russak and Klobus 2007; Janicka-Russak et al. 2013). Salt tolerance has been reported to coincide with increased accumulation of H$^+$-ATPase mRNA (Janicka-Russak and Klobus 2007; Sahu and Shaw 2009). The temporal and spatial expression of various isoforms of H$^+$-ATPase is involved in the salt stress–induced regulation of ion homoeostasis. The salt-induced isoforms help to cope with the Na$^+$ efflux substantiated by the H$^+$ electrochemical gradient across the cell membranes. This temporal regulation of stress-induced H$^+$-ATPase activity is attributed to the molecular cross-talk events between calcium, Ca^{2+}-calmodulin-dependent kinases and H$^+$-ATPase enzyme. The calcium-dependent protein kinase (CDPK)–dependent phosphorylation of various enzymatic proteins, including

H⁺-ATPase, pertains to salt tolerance in plants (Camoni et al. 2000; Klobus and Janicka-Russak 2004). Investigations in rice revealed that the activation of NaCl stress–induced H⁺-ATPase activity is mediated by phospholipase Dα expression. The phospholipase gene knock-down effect was manifested by a decrease in transcript levels of NaCl stress–induced H⁺-ATPase (Shen et al. 2011). Sahu and Shaw (2009) reported the correlation of salt-tolerant rice cultivars with higher levels of H⁺-ATPase transcripts in comparison with that in salt-sensitive cultivars. The regulation of an inducible form of H⁺-ATPase in salt tolerance has been reported by Gewaudant et al. (2007). Transgenic tobacco (PMA4) plants transformed with the gene for the autoinhibitory domain of H⁺-ATPase exhibited better growth in saline conditions. Wild-type plants that possessed the constitutive form of H⁺-ATPase isoform showed growth reduction in saline conditions. Similar investigations of the expression of salt stress–induced isoforms of H⁺-ATPase have been reported in rice and cucumber (Sahu and Shaw 2009; Janicka-Russak et al. 2013). The SOS pathway of the sodium efflux system is energized by H⁺-ATPase activity. The hydrogen efflux mechanism causes a proton-mediated electrochemical gradient, which drives the Na⁺/H⁺ antiport system (Blumwald et al. 2000). Halophytes exhibit constitutive expression of the Na⁺/H⁺ antiport system. The three SOS pathways (SOS 1, SOS 2, and SOS 3) are effective for salt tolerance in Arabidopsis. The SOS 1 antiporter system has been reported to possess sodium efflux and sensing activities during salt stress (Zhu 2003; Conde et al. 2011). Shi et al. (2002) reported an important function of SOS 1 in sodium uptake and transport through xylem vessels. The SOS 1 study has put forward important information regarding sodium sensing and its transport during salt stress. Thus,

H⁺-ATPase regulates important functions in terms of sodium mobility in plants (Figure 4.1). Overexpressing transgenic Arabidopsis with increased activity of SOS 1 exhibited less sodium accumulation and better tolerance of salt stress (Shi et al. 2003). The function of SOS 1 is further coordinated by SOS 2 and SOS 3. The sodium sensing mechanism is carried out by the SOS 2 and SOS 3 components. This signaling event is mediated by an increase in calcium levels in the cell. The activity of inositol triphosphate and diacylglycerol has been reported in this context (Tuteja 2007). The perception of salt stress elevates cytosolic calcium levels. This, in turn, activates the SOS 3-calcium sensor. SOS 3, after binding to calcium, changes its conformation and interacts with the FISL motif of SOS 2 (Ser/Thr protein kinase), thus activating its substrate phosphorylation. The activated SOS 2–SOS 3 complex induces the sodium antiport activity of SOS 1 (Chinnusamy et al. 2005). Aquaporin water channels have, however, been found to be independent of H⁺-ATPase activity during salt stress in capsicum (Martinez-Ballesta et al. 2003). Evidence for a sodium effluxing isoform of H⁺-ATPase (AHA4) has been reported in Arabidopsis root endodermis (Vitart et al. 2001). This isoform confers salt tolerance in Arabidopsis. Increased activity of H⁺-ATPase in *Spartina patens* has been reported to be associated with the mechanism of salt tolerance (Wu and Seliskar 1998). The spatial regulation of vacuolar V-H⁺-ATPase has been investigated in response to NaCl stress in *Broussonetia papyrifera* (Zhang et al. 2012). Root vacuolar H⁺-ATPase activity (E subunit) exhibited higher increase in the roots in comparison with leaves. NaCl stress–induced morphological changes in leaf growth coincided with higher H⁺-ATPase activity in *Aeluropus littoralis* (Olfatmiri et al. 2014).

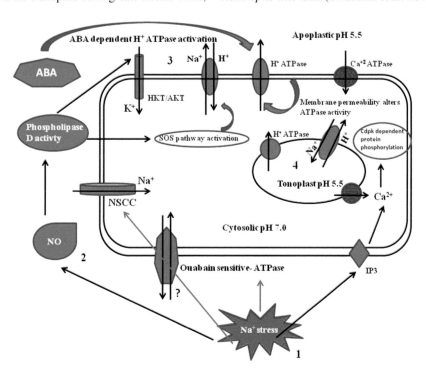

FIGURE 4.1 Regulation of H⁺-ATPase and Ca²⁺-ATPase activity during salt stress in plants. 1: Perception of Na⁺ stress and activation of IP₃-mediated signal cascade, induction of OU-sensitive ATPase activity. 2: Nitric oxide–mediated SOS pathway activation, mediated by phospholipase D and retention of potassium ion in cytosol. 3: ABA-dependent H⁺-ATPase activity coupled to sodium efflux and ion homoeostasis. 4: Sodium sequestration in vacuole, proton gradient maintenance, and calcium signature maintained by Ca²⁺-ATPase.

Nitric Oxide, ABA, and Ethylene-Mediated Crosstalk Regulates H⁺-ATPase Activity During Salt Stress

Nitric oxide–mediated salt stress tolerance operates through altered sodium uptake and its intracellular sequestration. Among various mechanisms of salt tolerance mediated by NO, the regulation of H^+-ATPase activity is a primary signaling event operative in plants. Investigations in *Avicennia marina* exhibited increased salt secretion from salt glands in response to NO treatments (Chen et al. 2010). The application of 100 μM sodium nitroprusside (SNP) as an NO donor positively regulated sodium efflux activity from salt glands. This resulted in lowering of the Na^+/K^+ ratio in hypodermal and epidermal cells of the leaves. Furthermore, the effect of 100 μM SNP was manifested as increased activity of plasma membrane–bound H^+-ATPase and the vacuolar Na^+/H^+ antiport system. The positive regulation of NO-mediated increase in their activity was confirmed by immune-blot and mRNA analysis using reverse transcription polymerase chain reaction. The reversal of positive effects of NO was obtained using NO scavengers. Interesting investigations in barley seedling roots revealed the coordinated effect of hydrogen sulfide and NO in ion homoeostasis (Chen et al. 2015). The mechanism of this regulation operated through elevation in the activity of H^+-ATPase and the Na^+/H^+ antiporter system (Figure 4.1). The effect of hydrogen sulfide was found to be similar to that of SNP, and thus, it acted as an NO donor. The effect of H_2S was manifested by an alteration in the potassium efflux mechanism and upregulation of an inward rectifying K^+ channel (HvAKT1). Nitric oxide–mediated salt tolerance has been reported in wheat seedling roots. The effect of SNP application was analyzed in terms of Na^+/K^+ ratio, cellular K^+ and Na^+ levels, and plasma membrane–localized H^+-ATPase activity (Ruan Hai-hua et al. 2004). The application of SNP in the absence and presence of NaCl stress led to increased acidification of the external medium, thus suggesting higher proton extrusion activity of H^+-ATPase. Furthermore, the application of SNP increased the cellular K^+ content and thus lowered the Na^+/K^+ ratio. However, a correlation of increased H^+-ATPase activity and increased Na^+/K^+ ratio as an effect of NO has been reported in *Phragmites* sp. (Zhao et al. 2004). NO thus functions as a potential secondary messenger during salt stress and elevates the expression and activity of H^+-ATPase in the plasma membranes. Similar events of NO-mediated modulation of H^+-ATPase and H^+-pyrophosphatase (PPase) activity have been reported in maize (Zhang et al. 2006). Nitric oxide enhances salt tolerance in maize seedlings by increasing the activity of the proton pump and Na^+/H^+ antiport in the tonoplast. The mechanism of ATPase activity modulation by NO was reported to be mediated by phospholipase D activity. This resulted in increased vacuolar H^+-ATPase activity, proton extrusion, and subsequent Na^+/H^+ exchange activity during salt stress. Abscisic acid (ABA)–mediated regulation of plasma membrane H^+-ATPase activity has been reported by Janicka-Russak and Klobus (2007). This suggests that the cross talk between ABA and H^+-ATPase operates through ABA-responsive genes and increased transcript levels of H^+-ATPase. NaCl stress–induced inhibition of H^+-ATPase and H^+-PPase activity has been reported to be alleviated by exogenous nitric oxide application in

cucumber roots (Shi et al. 2007). A similar observation has also been reported in tomato seedlings during copper stress by Zhang et al (2009). Interesting observations in Arabidopsis report the coordinated effect of ethylene and NO in promoting H^+-ATPase during NaCl stress (Wang et al. 2009). The application of exogenous 1-aminocyclopropane-1-carboxylic acid (ACC, an ethylene precursor) produced effects on H^+-ATPase activity and NaCl tolerance similar to that of SNP. Furthermore, ethylene-insensitive mutant etr1–3 callus of Arabidopsis was observed to be more sensitive to salt stress in comparison with that of the wild type. This report suggests that there is cross talk between NO and ethylene in regulating H^+-ATPase activity during salt stress. Ethylene has been suggested to act downstream of NO during such events of salt stress signaling.

Calcium-Calmodulin Signaling Is Associated with Salt Stress–Induced Regulation of Ca²⁺-ATPase Activity

Calcium is a major macronutrient involved in several functions of growth and development in plants (White and Broadley 2003). The versatility of calcium as a major signaling molecule lies in its chemical nature, which renders it suitable for interaction with other biomolecules. The valence state, ionization potential, and molecular structure of calcium ions make it possible to bind with proteins, lipids, and other biomolecules (Jaiswal 2001). Ca^{2+} has a high affinity for the carboxylate oxygen of acidic amino acids, namely aspartate and glutamate, which are abundant in cellular proteins. Calcium has an ionic radius of 100–200 pm, favorable for its site-specific binding to a variety of biomolecules (Jaiswal 2001). Cells at normal state have $[Ca^{2+}]_{cyt}$ of 10–200 nM, while the intracellular calcium reserves of the cell wall, vacuole, apoplast, mitochondria, and endoplasmic reticulum harbor calcium stores of concentration 1–10 mM. This partitioning of calcium ion imparts scope for its involvement in signaling responses initiated by rapid release of calcium ions from organelles to the cytosol of cells (Reddy et al. 2011). Fluctuations in cytosolic calcium levels provide a diverse yet specific set of signaling clues, termed calcium signatures. The calcium signature depends on the rate of calcium influx mediated by ion channels and their cellular localization (Tuteja 2009). An NaCl stress–induced transient increase in $[Ca^{2+}]_{cyt}$ has been reported in *Arabidopsis* sp. (Knight et al. 1997). Calcium supplementation recovers Na^+-induced osmotic imbalance and preferably maintains K^+/Na^+ ratio in salt-stressed plant cells (Cramer and Jones 1996; Sun et al. 2009). The effect of calcium on root development, salt stress amelioration, and ion flux was earlier investigated in salt-stressed barley roots (Shabala et al. 2003). Salinity stress induces displacement of membrane-bound Ca^{2+} by Na^+, thus resulting in leakage or efflux of K^+ from the cytosol (Cramer et al. 1987). Root tips exhibit calcium influx, and its cytosolic distribution is necessary for tip-oriented growth. The growing region of roots involves Ca^{2+} binding to growing cell wall polysaccharides, cytoskeletal functions, organellar transport, and Ca^{2+}-calmodulin functions (Schiefelbein et al. 1992). The localization and analysis of Ca^{2+} dynamics by confocal laser scanning microscopy has recently been reported in Arabidopsis

(Tanaka et al. 2010; Zhu et al. 2013). Root tips exhibiting fluorescence due to increased Ca^{2+} imply the involvement of calcium as an early signaling molecule in the actively growing meristematic cells induced by osmotic stress. X-ray microanalysis of ion distribution revealed lower accumulation of Na^+ in the epidermis and endodermis in comparison with the cortex of primary roots of sunflower subjected to NaCl stress (Ebrahimi and Bhatla 2012). Analysis of Ca^{2+} dynamics in roots of Arabidopsis using aequorin-based luminescence imaging has recently shown asynchronous oscillations of Ca^{2+} in response to osmotic stress (Zhu et al. 2013). The endodermis and pericycle of Arabidopsis roots undergo prolonged oscillations of Ca^{2+} distinct from other cells in response to 220 mM NaCl (Kiegle et al. 2000).

Calmodulin (CaM) is a major calcium sensor in cells, which further propagates the signaling response to several calmodulin-binding proteins, namely CDPK, mitogen-activated protein kinase (MAPK), calcineurins, and so on. Salt stress–induced increase in CaM levels has been reported to positively regulate the activity of superoxide dismutase (SOD), ascorbate peroxidase (APX), and catalase (CAT) in *Bruguiera* sp. (Li et al. 2009). Saeng-ngam et al. (2012) reported salt stress–induced overexpression of calmodulin gene-*OsCam1-1*, which confers resistance to salt stress in two rice cultivars. This mechanism of salt tolerance was further mediated by a CaM-induced trigger in ABA biosynthesis. Thus, calmodulin is a major sensor of salt stress, which imparts tolerance or avoidance to Na^+-mediated physiological imbalances. CaM-induced leaf senescence and H_2O_2 elevation in the presence of salt stress have recently been reported in sweet potato (Chen et al. 2012). This CaM-induced indirect response, although deleterious, was stated to be mediated by additional components. Thus, CaM participates in several mechanisms of direct and indirect association to downstream proteins, providing a varied physiological response in plants. Reports suggest that apocalmodulin (unbound to calcium) interacts with various proteins in animals, plants, and fungi (Rhoads and Friedberg 1997; Jurado et al. 1999). A wide range of proteins associated with diverse physiological functions can bind to apocalmodulin. These include actin-binding proteins (actin-activated ATPase), enzymes (phosphorylase b kinase, adenyl cyclase, and nitric oxide synthase), and inositol triphosphate (IP3)–activated receptors and calcium channels (Jurado et al. 1999). However, plant-based apocalmodulin activity requires further investigations in response to abiotic stress. The above-mentioned proteins exhibiting binding affinity to apoCaM also function in plants and are expected to show modulation induced by abiotic stress.

Ca^{2+}-ATPases Regulate Brief Calcium Signatures and Ion Homoeostasis During Salt Stress

The P type Ca^{2+}-ATPases are comprised of two subgroups: IIA and IIB. The two subgroups of Ca^{2+}-ATPases represent ER-localized (ECA; IIA) and autoinhibitory types (ACA; IIB), respectively (Geisler et al. 2000). The regulation of Ca^{2+}-ATPases involves calcium–CaM interaction and protein phosphorylation activities. Calcium plays a pivotal role in ion homoeostasis and stress amelioration in adverse conditions. Salt stress–induced calcium signatures are associated with calcium

exchange between tonoplast, organellar reserves, and cytoplasm. This signaling cascade is initiated by increased Ca^{2+}-ATPase activity and associated events. Ca^{2+}-ATPase activity has been reported to be associated with increased salt tolerance. This has been attributed to increased cytosolic calcium ions, decreased Na^+ accumulation, and better stress amelioration. Ca^{2+}-ATPase possesses different CaM binding domains, which are specific for different isoforms of CaM. CaM binds to Ca^{2+}-ATPase in a dose-dependent manner. The abundance of Ca^{2+}-ATPase isoforms relate to differential expression in various plant tissues. Salt stress–induced Ca^{2+}-ATPase activity pertains to the activation of specific isoforms. Ca^{2+}-ATPase-mediated regulation of cytosolic calcium ion can be mediated by both CaM-dependent and CaM-independent pathways (Tuteja and Mahajan 2007). Some Ca^{2+}-ATPase isoforms, however, operate independently of CaM activity. Ca^{2+}-ATPases are less abundant in comparison to H^+-ATPase and comprise 0.1% of membrane proteins (Tuteja and Mahajan 2007). Calcium sequestration in tonoplast is mediated by Ca^{2+}/H^+ antiporters. This antiporter is energized by the proton gradient developed by proton efflux due to H^+-ATPase activity (Tuteja and Mahajan 2007). The presence of Ca^{2+}-ATPases and Ca^{2+}/H^+ maintains a low concentration of Ca^{2+} in the cytosol. The differential distribution of Ca^{2+}-ATPases in the various regions of root cells provides physiological significance for ion homoeostasis (Felle et al. 1992). An abundance of apoplastic activity of Ca^{2+}-ATPase has been reported in and around the root vessels, thus suggesting the crucial regulation of calcium transport. This regulation might provide new insights into the context of salt stress–induced calcium uptake or homoeostasis. Ca^{2+}-ATPases of Type IIA and IIB have been reported to be involved in salt stress–induced calcium signaling (Geisler et al. 2000). Wimmers et al. (1992) suggested a marked increase in the mRNA levels of Ca^{2+}-ATPase in tomato plants subjected to 50 mM NaCl stress. Tobacco cell cultures exhibited Ca^{2+}-ATPase gene expression induced by a higher intensity of NaCl (428 mM) stress (Perez-Prat et al. 1992). The transcript level analysis of salt stress–induced expression of Ca^{2+}-ATPase revealed higher accumulation within 1 hour, which decreased within 6 hours of stress induction (Chung et al 2000). This suggests a rapid signaling event of Ca^{2+}-ATPase activation and the calcium signature associated with salt stress. Gene knockout analysis of vacuolar ACA in moss has revealed the physiological significance of Ca^{2+}-ATPase in providing NaCl stress tolerance (Qudeimat et al. 2008). The investigation was substantiated by the observation of NaCl stress–induced rapid elevation in cytosolic calcium levels. Significant investigations in *Oryza sativa* have revealed the role of the Ca^{2+}-ATPase gene (OsACA 6) in modulating salt stress tolerance (Huda et al 2013). Overexpression transgenic lines for *OsACA 6* revealed increased transcript levels in response to salt stress. The tolerance behavior was affirmed by a plethora of effects such as increased photosynthetic ability, malonaldehyde content, reduction in electrolytic leakage, reduced reactive oxygen species (ROS) generation, and greater proline accumulation (Huda et al 2013). The beneficial role of salt stress–induced calcium signatures has been reported to be mediated by Type IIB (PCA1) Ca^{2+}-ATPase in *Physcomitrella patens* (Qudeimat et al 2008). A transgene construct tagged with reporter green fluorescent protein exhibited the localization of this Ca^{2+}-ATPase in the vacuolar membranes. Mutant lines deficient in functional PCA1 were observed to alter calcium responses. Thus, PCA 1 was stated to have beneficial roles in modulating salt stress tolerance by evoking transient and rapid calcium signatures (Qudeimat et al. 2008). A similar investigation on the role of Ca^{2+}-ATPase (ACA 2) in yeast has been reported by Anil et al. (2008). The calcium pump (ACA2) operates to regulate salt stress–induced ion homoeostasis. This has been further attributed to adaptive Na^+ uptake regulation. Interesting observations exhibited that the activity of ACA 2 was regulated by CaM-mediated phosphorylation. The beneficial role of ACA 2 during salt stress was elucidated by upregulation of NHX 1 activity involved in vacuolar sequestration of Na^+ ions (Anil et al 2008). The *ACA4* gene for Type IIB Ca^{2+}-ATPase from Arabidopsis also conferred salt stress tolerance in yeast (Geisler et al. 2000).

Ouabain-Sensitive ATPases in Plants: A Physiological Enigma

Over the past few decades, plant scientists have been assuming that plants possess ouabain (OU)-sensitive ATPases, which might share some functional similarity with Na^+/K^+-ATPase in animals (Minorsky 2002). However, further investigations on this ATPase eventually declined in the late 1990s due to lack of a suitable methodology. Genomic sequence data confirmed the absence of Na^+/K^+-ATPase in plants. This, however, did not rule out the possibility of finding some other type of OU-sensitive ATPases in plants having functional homology to Na^+/K^+-ATPase. Recent findings in sunflower revealed that salinity stress affects seed germination and seedling growth (David et al. 2010). Seedlings subjected to NaCl stress (120 mM) for 48 h exhibit inhibition of hypocotyl extension and a significant reduction in root proliferation after 4 and 6 days of germination. The Na^+/K^+ ratio in roots increases with the progress of germination, and it is accompanied by a significant mobilization of sodium ions into the cotyledons. Among various mechanisms likely to block Na^+ influx into the plant cells, NaCl stress is likely to enhance OU-sensitive ATPase activity, which has earlier been reported to facilitate sodium efflux in plants, thereby maintaining ion homoeostasis (Vakhmistrov et al. 1982). OU, a well-known cardiac glycoside, is known to be a specific inhibitor of Na^+/K^+-ATPase activity in animals (Fortes 1977). Earlier investigations have demonstrated the physiological effects of OU in plant processes, such as Na^+ fluxes, stomatal opening, pulvinar function, flowering, and transpiration (Cram 1968; Thomas 1970; Watanabe 1971; Oota 1974; Morant-Avice et al. 1997). Plants such as *Hordeum* sp. and *Halocnemum* sp. have been reported to possess Na^+, K^+, and Mg^{2+}-activated ATPases, and their activity is affected by OU (Vakhmistrov et al. 1982). The OU sensitivity of similar kinds of ATPases in sugarbeet roots has also been investigated (Lindberg 1982). OU inhibits Na^+ efflux and causes a subsequent increase in the intracellular sodium ion concentration in excised barley and corn roots (Nassery and Baker 1972; Davis and Jaworski 1979). OU-sensitive sodium efflux is an active process in plants and is likely to be similar to the sodium pumps operative in animals and giant algal coenocytes (Cram 1968). Tikhaya and Mishutina (1981) reported the presence of OU-sensitive ATPases

in some glycophytes and halophytes. Dose–response curves of OU in plants are similar to those found in animal cells (Brown et al. 1964; Thomas 1970). Two forms of OU-sensitive ATPases have been reported to exist in plants, with pH for optimal activity ranging between 5.5–6.0 and 8.0 (Lindberg 1982).

Ouabain-Sensitive ATPase Activity Modulation Plays a Significant Role in NaCl Stress Tolerance in Plants

Among the various mechanisms likely to block Na^+ influx into the plant cells subjected to NaCl stress, enhancement of OU-sensitive ATPase activity has been reported to facilitate sodium efflux in plants, thereby maintaining ion homoeostasis (Vakhmistrov et al. 1982). OU-sensitive sodium efflux is an active process in plants and is likely to be similar to the sodium pumps operative in animals and giant algal coenocytes (Cram 1968). Two forms of OU-sensitive ATPases have been reported to exist in plants, with pH for optimal activity ranging between 5.5–6.0 and 8.0 (Lindberg 1982). Reports on the calcium sensitivity of ATPase activity in plants have revealed elevation of H^+-ATPase activity in unstressed seedling roots by calcium treatment (Klobus and Janicka-Russak 2004). OU-sensitive ATPase activity is reported to be regulated through the modulation of calcium binding proteins, namely calmodulin and calnaktin (Yingst et al. 1992; Wang et al. 1998). 9-anthroylouabain (a fluorescent derivative of OU known to localize OU-sensitive ATPase activity in animal cells) has been used in a plant system for the first time (Mukherjee and Bhatla 2014). OU, a glycoside, is a well-known inhibitor of Na^+/K^+-ATPase in animals. The localization of OU-sensitive ATPase activity in the nuclear membrane of sunflower seedling root protoplasts suggests its possible role in ion homoeostasis between the cytoplasm and the nucleus. The presence of OU-sensitive ATPases has been reported in glycophytes and halophytes (Tikhaya and Mishutina 1981; Vakhmistrov et al. 1982). Intracellular sodium concentrations have been reported to regulate Na^+/K^+-ATPase activity in animals (Barlet-Bas et al. 1990; Senatorov et al. 2000; Sun et al. 2011). In resting cells, the enzyme has been reported to be less active (Senatorov et al. 2000). OU-sensitive ATPase activity in *Halocnemum* sp. is stimulated by 238% due to sodium induction (Vakhmistrov et al. 1982). The rise in sodium concentration induces active units of the enzyme to function in the membranes and cytosolic compartments, where 110% enhancement in the enzyme activity has been observed in rat thalamic neurons on treatment with Na^+ ionophore (Senatorov et al. 2000). Observations in sunflower seedling roots substantiate the fact that OU affects ion transport in plant cells by modulating OU-sensitive ATPase activity through its binding with OU receptors (Mukherjee and Bhatla 2014). This highlights the evolutionary significance of OU-sensitive ATPases in plants, associated with sodium exchange activity. Although further work is necessary for the characterization of the putative enzyme, these findings provide promising scope for further work on understanding the mechanism of action of OU-sensitive ATPase in plants, which in turn regulates sodium accumulation during salt stress.

Conclusion

The role of various ATPases and their biochemical regulation has helped in the current understanding of salt stress tolerance and ion homoeostasis in plants. The process of sodium efflux and its sequestration appears to be the primary approach in various plants to help combat oxidative stress. Sensitivity to salt stress and inhibition of growth rate vary among different plants ranging from halophytes to glycophytes. The halophytes possess higher ATPase activity in comparison with glycophytes, thus suggesting their adaptive nature. A transgenic approach has been adopted in various crop species to regulate salt stress–induced growth and productivity. Such methods have involved the overexpression of ATPase genes in candidate crops. Further investigations are necessary in terms of the regulation of H^+-ATPase and Ca^{2+}-ATPase activity. Pathways of hormonal cross talk with ATPase and deciphering the steps of the signaling cascade are also prime requirements for a better understanding of salt stress signaling. The characterization of ouabain-sensitive ATPase in plants will provide deeper knowledge on the role of sodium and potassium regulation by plants operative during salt stress.

Acknowledgments

The author is grateful to Professor S.C. Bhatla (Department of Botany, University of Delhi) for providing necessary guidance in writing the present review. The author is also thankful to Dr. Akula Ramakrishna for his assistance in the initiation of this work.

REFERENCES

Anil VS, Rajkumar P, Kumar P, Mathew MK. A plant Ca^{2+} pump, ACA2, relieves salt hypersensitivity in yeast. Modulation of cytosolic calcium signature and activation of adaptive Na^+ homeostasis. *J. Biol. Chem.* 2008; 283: 3497–3506.

Barlet-Bas C, Khadouri C, Marsy S, Doucet A. Enhanced intracellular sodium concentration in kidney cells recruits a latent pool of Na-K-ATPase whose size is modulated by corticosteroids. *J. Biol. Chem.* 1990; 265: 7799–7803.

Blumwald E, Aharon GS, Apse MP. Sodium transport in plant cells. *Biochim. Biophys. Acta* 2000; 1465: 140–151.

Brown HD, Jackson RT, Dupuy HJ. Transport of sugar in *Allium*: Effects of inhibitors and ethylene. *Nature* 1964; 202: 722–723.

Camoni L, Iori V, Marra M, Aducci P. Phosphorylation-dependent interaction between plant plasma membrane H^+-ATPase and 14-3-3 proteins. *J. Biol. Chem.* 2000; 275: 9919–9923.

Chen HJ, Lin HW, Huang GJ, Lin YH. Sweet potato calmodulin SPCAM is involved in salt stress-mediated leaf senescence, H_2O_2 elevation and senescence-associated gene expression. *J. Plant Physiol.* 2012; 169: 1892–1902.

Chen J, Wang WH, Wu FH, He EM, Liu X, Shangguan ZP, Zheng HL. Hydrogen sulfide enhances salt tolerance through nitric oxide-mediated maintenance of ion homeostasis in barley seedling roots. *Sci. Rep.* 2015; 5: 12516. doi:10.1038/srep12516.

Chen J, Xiao Q, Wu F, Dong X, He J, Pei Z, Zheng H. Nitric oxide enhances salt secretion and Na^+ sequestration in a mangrove plant, *Avicennia marina*, through increasing the expression of H^+-ATPase and Na^+/H^+ antiporter under high salinity. *Tree Physiol.* 2010; 30: 1570–1586.

Chen J, Xiong DY, Wang WH, Hu WJ, Simon M, Xiao Q, Liu TW, Liu X, Zheng HL. Nitric oxide mediates root K^+/Na^+ Balance in a mangrove plant, *Kandelia obovata*, by enhancing the expression of AKT1-Type K^+ channel and Na^+/H^+ antiporter under high salinity. *PLoS One* 2013; 8(8): e71543. doi:10.1371/journal.pone.0071543.

Chen Z, Pottosin II, Cuin TA, Fuglsang AT, Tester M, Jha D, Zepeda-Jazo I, et al. Root plasma membrane transporters controlling K^+/Na^+ homeostasis in salt-stressed barley. *Plant Physiol.* 2007; 145: 1714–1725.

Chinnusamy V, Jagendorf A, Zhu JK. Understanding and improving salt tolerance in plants. *Crop Sci.* 2005; 45: 437–448.

Chung WS, Lee SH, Kim JC, Do Heo W, Kim MC, Park CY, Park HC, et al. Identification of a calmodulin-regulated soybean Ca^{2+}-ATPase (SCA1) that is located in the plasma membrane. *Plant Cell* 2000; 12: 1392–1407.

Conde A, Chaves MM, Gerós H. Membrane transport, sensing and signaling in plant adaptation to environmental stress. *Plant Cell Physiol.* 2011; 52: 1583–1602.

Cram WJ. The effects of ouabain on sodium and potassium fluxes in excised roots of carrot. *J. Exp. Bot.* 1968; 19: 611–616.

Cramer GR, Jones RL. Osmotic stress and abscisic acid reduce cytosolic calcium activities in roots of *Arabidopsis thaliana*. *Plant Cell Environ.* 1996; 19: 1291–1298.

Cramer GR, Lynch J, Lauchli A, Epstein E. Influx of Na^+, K^+, and Ca^{2+} into roots of salt-stressed cotton seedlings. *Plant Physiol.* 1987; 83: 510–516.

Davenport RJ, Tester M. A weakly voltage-dependent, nonselective cation channel mediates toxic sodium influx in wheat. *Plant Physiol.* 2000; 122: 823–834.

David A. Involvement of nitric oxide and associated biomolecules in sunflower seedling growth in response to salt stress. PhD Thesis. 2012; University of Delhi, Delhi, India.

David A, Yadav S, Bhatla SC. Sodium chloride stress induces nitric oxide accumulation in root tips and oil body surface accompanying slower oleosin degradation in sunflower seedlings. *Physiol. Plant.* 2010; 140: 342–354.

Davis RF, Jaworski AZ. Effects of ouabain and low temperature on the sodium efflux pump in excised corn roots. *Plant Physiol.* 1979; 63: 940–946.

Ebrahimi R, Bhatla SC. Ion distribution measured by electron probe X-ray microanalysis in apoplastic and symplastic pathways in root cells in sunflower plants grown in saline medium. *J. Biosci.* 2012; 37: 713–721.

Felle HH, Tretyn A, Wagner G. The role of the plasma-membrane Ca^{2+}-ATPase in Ca^{2+} homeostasis in *Sinapis alba* root hairs. *Planta* 1992; 188: 306–313.

Flowers TJ, Galal HK, Bromham DL. Evolution of halophytes: Multiple origins of salt tolerance in land plants. *Funct. Plant Biol.* 2010, 37: 604–612.

Fortes PAG. Anthroylouabain: A specific fluorescent probe for the cardiac glycoside receptor of the Na-K ATPase. *Biochemistry* 1977; 16: 531–540.

Gao D, Knight MR, Trewavas AJ, Sattelmacher B, Plieth C. Self-reporting *Arabidopsis* expressing pH and [Ca^{2+}] indicators unveil ion dynamics in the cytoplasm and in the apoplast under abiotic stress. *Plant Physiol.* 2004; 134: 898–908.

Geisler M, Axelsen KB, Harper JF, Palmgren MG. Molecular aspects of higher plant P-type Ca^{2+}-ATPases. *Biochim. Biophys. Acta* 2000; 1465: 52–78.

Géwaudant F, Duby G, Stedingk E, Zhao R. Expression of a constitutively activated plasma membrane H$^+$-ATPase alters plant development and increases salt tolerance. *Plant Physiol.* 2007; 144(August): 1763–1776.

Heermann R, Jung K. Stimulus perception and signaling in histidine kinases. In: *Bacterial Signaling*, Kramer R and Jung K (eds.). Wiley-VCH Verlag GmbH, Weinheim, 2010; pp. 135–152.

Hernandez A, Cooke D, Clarkson D. In vivo activation of plasma membrane H$^+$-ATPase hydrolytic activity by complex lipid-bound unsaturated fatty acids in *Ustilago maydis*. *Eur. J. Biochem.* 2002; 269: 1006–1011.

Hernandez M, Fernandez-Garcia N, Diaz-Vivancos P, Olmos E. A different role for hydrogen peroxide and the antioxidative system under short and long salt stress in *Brassica oleracea* roots. *J. Exp. Bot.* 2010; 61: 521–535.

Hong Y, Pan X, Welti R, Wang X. The effect of phospholipase Dα_3 on *Arabidopsis* response to hyperosmotic stress and glucose. *Plant Signal. Behav.* 2008; 3: 1099–1100.

Huda KM, Banu MS, Garg B, Tula S, Tuteja R, Tuteja N. OsACA6, a P-type IIB Ca2+ ATPase promotes salinity and drought stress tolerance in tobacco by ROS scavenging and enhancing the expression of stress-responsive genes. *Plant J.* 2013; 76, 997–1015. doi:10.1111/tpj.12352.

Jaiswal JK. Calcium—how and why? *J. Biosci.* 2001; 26: 357–363.

Janicka-Russak M, Kabala K, Wdowikowska A, Kłobus G. Modification of plasma membrane proton pumps in cucumber roots as an adaptation mechanism to salt stress. *J. Plant Physiol.* 2013; 170: 915–922.

Janicka-Russak M, Kłobus G. Modification of plasma membrane and vacuolar H$^+$-ATPase in response to NaCl and ABA. *J. Plant Physiol.* 2007; 164: 295–302.

Jiang Y, Yang B, Harris NS, Deyholos MK. Comparative proteomic analysis of NaCl stress-responsive proteins in *Arabidopsis* roots. *J. Exp. Bot.* 2007; 58: 3591–3607.

Jurado LA, Chockalingam PS, Jarrett HW. Apocalmodulin. *Physiol. Rev.* 1999; 79: 661–682.

Kader A. Salt stress in rice: Adaptive mechanisms for cytosolic sodium homeostasis. PhD Thesis. 2006; Faculty of natural resources and agricultural sciences, Department of Plant Biology and Forest Genetics. Uppsala, Sweden.

Kader MA, Lindberg S, Seidel T, Golldack D, Yemelyanov V. Sodium sensing induces different changes in free cytosolic calcium concentration and pH in salt-tolerant and salt-sensitive rice (*Oryza sativa* L.) cultivars. *Physiol. Plant.* 2007; 130: 99–111.

Kasamo K. Regulation of plasma membrane H$^+$-ATPase activity by the membrane environment. *J. Plant Res.* 2003; 116: 517–523.

Kiegle E, Moore CA, Haseloff J, Tester MA, Knight MR. Cell type specific calcium responses to drought, salt and cold in the *Arabidopsis* root. *Plant J.* 2000; 23: 267–278.

Klobus G and Janicka-Russak M. Modulation by cytosolic components of proton pump activities in plasma membrane and tonoplast from *Cucumis sativus* roots during salt stress. *Physiol. Plant.* 2004; 121: 84–92.

Knight H, Trewavas AJ, Knight MR. Calcium signalling in *Arabidopsis thaliana* responding to drought and salinity. *Plant J.* 1997; 12: 1067–1078.

Li N, Li C, Chen S, Chang Y, Zhang Y, Wang R, Shi Y, Zheng X, Fritz E, Hüttermann A. Abscisic acid, calmodulin response to short term and long term salinity and the relevance to NaCl-induced antioxidant defense in two mangrove species. *Open For. Sci. J.* 2009; 2: 48–58.

Lindberg S. Sucrose and ouabain effects on the kinetic properties of a membrane bound (Na$^+$ + K$^+$ + Mg^{2+}) ATPase in sugar beet roots. *Physiol. Plant.* 1982; 54: 455–460.

Lv DW, Li X, Zhang M, Gu AQ, Zhen SM, Wang C, Li XH, Yan YM. Large-scale phosphoproteome analysis in seedling leaves of *Brachypodium distachyon* L. *BMC Genom.* 2014; 15: 375. doi:10.1186/1471-2164-15-375.

Magdalena G, Maria S, Barbara G. Effect of short- and long-term salinity on the activities of antioxidative enzymes and lipid peroxidation in tomato roots. *Acta Physiol. Plant.* 2008; 30: 11–18.

Martinez-Ballesta MC, Martinez V, Carvajal M. Aquaporin functionality in relation to H$^+$-ATPase activity in root cells of *Capsicum annuum* grown under salinity. *Physiol. Plant.* 2003; 117(4): 413–420.

Mäser P, Gierth M, Schroeder JI. Molecular mechanisms of potassium and sodium uptake in plants. *Plant Soil* 2002; 247: 43–54.

Minorsky PV. News from the archives: Do plants have ouabain (OU)-sensitive ATPases? *Plant Physiol.* 2002; 130: 4–5.

Morant-Avice AA, Jurvilliers P, Tremblin G, Coudret A. Effect of ouabain on stomatal movements and transpiration rate of *Secale cereale*. *Biol. Plant* 1997; 39: 235–242.

Mukherjee S, Bhatla SC. A novel fluorescence imaging approach to monitor salt stress-induced modulation of ouabain-sensitive ATPase activity in sunflower seedling roots. *Physiol. Plant.* 2014; 150: 540–549.

Munns R, Tester M. Mechanisms of salinity tolerance. *Annu. Rev. Plant Biol.* 2008; 59: 651–681.

Nassery H, Baker DA. Extrusion of sodium ions by barley roots I. Characteristics of the extrusion mechanism. *Ann. Bot.* 1972; 36: 881–887.

Olfatmiri H, Alemzadeh A, Zakipour Z. Up-regulation of plasma membrane H$^+$-ATPase under salt stress may enable *Aeluropus littoralis* to cope with stress. *Mol. Biol. Res. Commun.* 2014; 3. 67–75.

Oota Y. Removal of the sugar inhibition of flowering in *Lemna gibba* G$_3$ by catecholamines. *Plant Cell Physiol.* 1974; 15: 63–68.

Park KY, Jung JY, Park J, Hwang JU, Kim YW, Hwang I, Lee Y. A role for phosphatidylinositol 3-phosphate in abscisic acid-induced reactive oxygen species generation in guard cells. *Plant Physiol.* 2003; 132: 92–98.

Perez-Prat E, Narashimhan ML, Binzel ML, Botella MA, Chen Z, Valpuesta V, Bressan RA, Hasegawa PM. Induction of a putative Ca^2-ATPase mRNA in NaCl adapted cells. *Plant Physiol.* 1992; 100: 1471–1478.

Pical C, Westergren T, Dove SK, Larsson C, Sommarin M. Salinity and hyperosmotic stress induce rapid increases in phosphatidylinositol 4,5-biphosphate, diacylglycerol pyrophosphate, and phosphatidyl choline in *Arabidopsis thaliana* cells. *J. Biol. Chem.* 1999; 274: 38232–38240.

Pitan B, Schubert S, Mühling K. Decline in leaf growth under salt stress is due to an inhibition of H$^+$ pumping activity and increase in apoplastic pH of maize leaves. *J. Plant Nutr. Soil Sci.* 2009; 172: 535–543.

Qudeimat E, Faltusz AMC, Wheeler G, Lang D, Brownlee C, Reski R, Frank W. A PIIB-type Ca^{2+} ATPase is essential for stress adaptation in *Physcomitrella patens*. *Proc. Natl. Acad. Sci. USA* 2008; 105: 19555–19560.

Reddy ASN, Ali GS, Celesnik H, Day IS. Coping with stresses: Roles of calcium and calcium/calmodulin regulated gene expression. *Plant Cell* 2011; 23: 2010–2032.

Rhoads AR, Friedberg F. Sequence motifs for calmodulin recognition. *FASEB J.* 1997; 11: 331–340.

Ruan HH, Shen WB, Xu LL. Nitric oxide modulates the activities of plasma membrane ATPase and PPase in wheat seedling roots and promotes the salt tolerance against salt stress. *Acta Bot. Sin.* 2004; 46: 415–422.

Saeng-ngam S, Takpirom W, Buaboocha T, Chadchawan S. The role of the OsCam1-1 salt stress sensor in ABA accumulation and salt tolerance in rice. *J. Plant Biol.* 2012; 55: 198–208.

Sahu B, Shaw B. Salt-inducible isoform of plasma membrane H$^+$-ATPase gene in rice remains constitutively expressed in natural halophyte, *Suaeda maritima*. *J. Plant Physiol.* 2009; 166: 1077–1089.

Schiefelbein JW, Shilpey A, Rowse P. Calcium influx at the tip of growing root-hair cells of *Arabidopsis thaliana*. *Planta* 1992; 187: 455–459.

Sekmen AH, Bor M, Ozdemir F, Turkan I. Current concepts about salinity and salinity tolerance in plants. In: *Climate Change and Plant Abiotic Stress Tolerance*, Tuteja N and Gill SS (eds.), Volume 1. Wiley Blackwell, Singapore, 2010; pp. 163–188.

Senatorov VV, Stys PK, Hu B. Regulation of Na$^+$, K$^+$-ATPase by persistent sodium accumulation in adult rat thalamic neurones. *J. Physiol.* 2000; 525: 343–353.

Shabala S, Shabala L, Volkenburgh EV. Calcium mediates root K$^+$/Na$^+$ homeostasis in poplar species differing in salt tolerance. *Funct. Plant Biol.* 2003; 30: 507–514.

Shen P, Wang R, Zhang W. Rice phospholipase Dα is involved in salt tolerance by the mediation of H$^+$-ATPase activity and transcription. *J. Integr. Plant Biol.* 2011; 53(4): 289–299.

Shi H, Kim YS, Guo Y, Stevenson B, Zhu JK. The *Arabidopsis* SOS5 locus encodes a putative cell surface adhesion protein and is required for normal cell expansion. *Plant Cell* 2003; 15: 19–32.

Shi H, Lee B, Wu S-J, Zhu JK. Overexpression of a plasma membrane Na$^+$/H$^+$ antiporter gene improves salt tolerance in *Arabidopsis thaliana*. *Nat. Biotechnol.* 2003; 21: 81–85.

Shi H, Quintero FJ, Pardo JM, Zhu JK. The putative plasma membrane Na$^+$/H$^+$ antiporter SOS1 controls long-distance Na$^+$ transport in plants. *Plant Cell* 2002; 14: 465–477.

Shi H, Zhu JK. SOS4, a pyridoxal kinase gene, is required for root hair development in *Arabidopsis*. *Plant Physiol.* 2002; 129: 585–593.

Shi Q, Ding F, Wang X, Wei M. Exogenous nitric oxide protect cucumber roots against oxidative stress induced by salt stress. *Plant Physiol. Biochem.* 2007; 45: 542–550.

Sun H, Zhang L, Ren C, Chen C, Fan S, Xia J, Lin H, Hu C. The expression of Na$^+$, K$^+$-ATPase in *Litopenaeus vannamei* under salinity stress. *Mar. Biol. Res.* 2011; 7: 623–628.

Sun J, Dai S, Wang R, Chen S, Li N, Zhou X, Lu C, et al. Calcium mediates root K$^+$/Na$^+$ homeostasis in poplar species differing in salt tolerance. *Tree Physiol.* 2009; 29: 1175–1186.

Tanaka K, Swanson SJ, Gilroy S, Stacey G. Extracellular nucleotides elicit cytosolic free calcium oscillations in *Arabidopsis*. *Plant Physiol.* 2010; 154: 705–719.

Thomas DA. The regulation of stomatal aperture in tobacco leaf epidermis strips: II. The effect of ouabain. *Aust. J. Biol. Sci.* 1970; 23: 981–989.

Tikhaya NI, Mishutina NE. Comparison of some membrane-bound ATPases of glycophytes and halophytes. *Plant Soil* 1981; 63: 25–26.

Tuteja N. Integrated calcium signaling in plants. In: *Signaling and Communication in Plants*, Baluška F and Mancuso S (eds.). 2009; Springer, pp. 29–49.

Tuteja N. Mechanisms of high salinity tolerance in plants. *Methods Enzymol.* 2007; 428: 419–438.

Tuteja N, Mahajan S. Calcium signaling network in plants: An overview. *Plant Signal. Behav.* 2007; 2: 79–85; PMID: 19516972.

Vakhmistrov DB, Tikhaya NI, Mishutina NE. Characterization and comparison of membrane-bound Na,K,Mg-ATPase from tissues of *Hordeum vulgare* L. and *Halocnemum strobilaceum* L. *Physiol. Plant.* 1982, 55: 155–160.

Vitart V, Baxter I, Doerner P, Harper JF. Evidence for a role in growth and salt resistance of a plasma membrane H$^+$-ATPase in the root endodermis. *Plant J.* 2001; 27: 191–201.

Wakeel A, Hanstein S, Pitann B, Schubert S. Hydrolytic and pumping activity of H$^+$-ATPase from leaves of sugar beet (*Beata vulgaris* L.) as affected by salt stress. *J. Plant Physiol.* 2010; 167: 725–731.

Wang H, Liang X, Wan Q, Wang X, Bi Y. Ethylene and nitric oxide are involved in maintaining ion homeostasis in Arabidopsis callus under salt stress. *Planta* 2009; 230: 293–307.

Wang J, Adachi M, Rhoads DE. A calnaktin-like inhibitor of Na$^+$, K$^+$-ATPase in rat brain: Regulation of α_1 and α_2 isozymes. *Comp. Biochem. Physiol.* 1998; 119B: 241–246.

Watanabe S. Ouabain and IAA effects on *Mimosa pudica*. *Artes Liberals Iwate Univ.* 1971; 8: 75–80.

White PJ, Broadley MR. Calcium in plants. *Ann. Bot.* 2003; 92: 487–511.

Wimmers LE, Ewing NN, Bennett AB. Higher plant Ca^{2+}-ATPase: Primary structure and regulation of mRNA abundance by salt. *Proc. Natl. Acad. Sci. USA* 1992; 89: 9205–9209.

Wu J, Seliskar DM. Salinity adaptation of plasma membrane H$^+$-ATPase in the salt marsh plant *Spartina patens*: ATP hydrolysis and enzyme kinetics. *J. Exp. Bot.* 1998; 49: 1005–1013.

Yingst DR, Ye-Hu J, Chen H, Barret V. Calmodulin increases Ca-dependent inhibition of the Na,K-ATPase in human red blood cells. *Arch. Biochem. Biophys.* 1992; 295: 49–54.

Zhang J, Davies WJ. Changes in the concentration of ABA xylem sap as a function of changing soil water status will account for changes in leaf conductance. *Plant Cell Environ.* 1990; 13: 277–285.

Zhang M, Fang Y, Liang Z, Huang L. Enhanced expression of vacuolar H$^+$-ATPase subunit E in the roots is associated with the adaptation of *Broussonetia papyrifera* to salt stress. *PLoS One* 2012; 7(10): e48183.

Zhang YK, Han XJ, Chen XL, Hong J, Cui XM. Exogenous nitric oxide on antioxidative system and ATPase activities from tomato seedlings under copper stress. *Sci. Hortic.* 2009; 123: 217–223.

Zhang YY, Wang LL, Liu YL, Zhang Q, Wei QP, Zhang WH. Nitric oxide enhances salt tolerance in maize seedlings through increasing activities of proton-pump and Na1/H1 antiport in the tonoplast. *Planta* 2006; 224: 545–555.

Zhao L, Zhang F, Guo J, Yang Y, Li B, Zhang L. Nitric oxide functions as a signal in salt resistance in the calluses from two ecotypes of reed. *Plant Physiol.* 2004; 134: 849–857. doi:10.1104/pp.103. 030023.

Zhu J, Lee BH, Dellinger M, Cui X, Zhang C, Wu S, Nothnage EA, Zhu JK. A cellulose synthase-like protein is required for osmotic stress tolerance in *Arabidopsis*. *Plant J.* 2010; 63: 128–140

Zhu JK. Genetic analysis of plant salt tolerance using *Arabidopsis*. *Plant Physiol.* 2000; 124: 941–948.

Zhu JK. Plant salt tolerance. *Trend. Plant Sci.* 2001; 6: 66–71.

Zhu JK. Regulation of ion homeostasis under salt stress. *Curr. Opin. Plant Biol.* 2003; 6: 441–445.

Zhu JK. Salt and drought stress signal transduction in plants. *Annu. Rev. Plant Biol.* 2002; 53: 247–273.

Zhu X, Feng Y, Liang G, Liu N, Zhu JK. Aequorin-based luminescence imaging reveals stimulus- and tissue-specific Ca^{2+} dynamics in *Arabidopsis* plants. *Mol. Plant.* 2013; 6: 444–455.

5

Physiological and Phenological Responses of Crop Plants under Heat Stress

Allah Ditta

CONTENTS

Abbreviations

AFLP	Amplified fragment length polymorphism
AOX	Alternative oxidase
At-NhaK2	$Na^+(K^+)/H^+$ antiporter
ATPase	Adenosine triphosphatase
CO$_2$	Carbon dioxide
COX	Cytochrome c oxidase
CTD	Canopy temperature depression
DNA	Deoxy ribonucleic acid
GFD	Grain filling duration
GS	Genomic selection
GTPase Arl1	Arf-like guanosine triphosphate
GWAS	Genome-wide association studies
H$^+$	Hydrogen ion
IPCC	Intergovernmental Panel on Climate Change
K$^+$	Potassium ion
MAS	Marker-assisted selection
MARS	Marker-assisted recurrent selection
Na$^+$	Sodium ion
QTLs	Quantitative trait loci
R	Leaf respiration
ROS	Reactive oxygen species
SRES	Special Report on Emissions Scenarios
WGS	Whole-genome shotgun sequencing

Introduction

Climatic changes due to global warming have been a great challenge for sustainable agriculture (IPCC, 2014). Crop adaptation to changing environments will ensure food security in the future, as climate change models around the world have predicted with great certainty the occurrence of heat stress episodes that will badly impact crop productivity (Battisti and Naylor, 2009; Teixeira et al., 2013; Challinor et al., 2014). During this century, heat stress is expected to be a defining environmental change in many regions of the world. Moreover, disruption of climates and increased frequency of drought and storm events with an increase in temperature are expected to reduce crop productivity (Figure 5.1).

Many factors have been found to be responsible for this occurrence of frequent and increased heat stress episodes. For example, the concentration of CO$_2$ in the atmosphere has increased from 280 to 400 ppm since the pre-industrial period and is expected to reach 700 ppm, resulting in a temperature increase of 2°C–6°C in Europe (IPCC, 2013). Frequent and increased heat stress episodes are likely to occur and would result in the shifting or even complete extinction of many species (Field et al., 2014).

Plant growth and development can be severely affected under unfavorable environmental conditions, since plants are immobile. Frequent and increased heat stress episodes in plants are likely to occur in the form of daily and seasonal fluctuations in temperature. These fluctuations in temperature are thought to be exogenous signals for the plants to develop properly through many physiological and phenological adaptations (Balasubramanian et al., 2006; Stavang et al., 2009; Wigge, 2013) (Figure 5.2).

Physiologically, heat stress can result in denaturation of proteins and fluidization of membranes, accumulation of reactive oxygen species (ROS), which ultimately reduce plant growth and survivability, aggregation and cell structural disruption, increased ion leakage, and elevated malondialdehyde content (Suzuki and Mittler, 2006; Wahid et al., 2007; Ahuja et al., 2010;

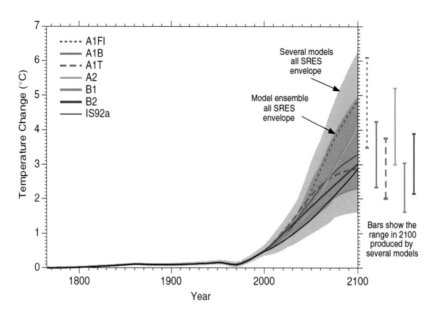

FIGURE 5.1 Global average surface temperature change. (From IPCC. Climate Change 2014: Synthesis Report. Contribution of Working Groups I, II and III to the Fifth Assessment Report of the Intergovernmental Panel on Climate Change.) SRES, Special Report on Emissions Scenarios.

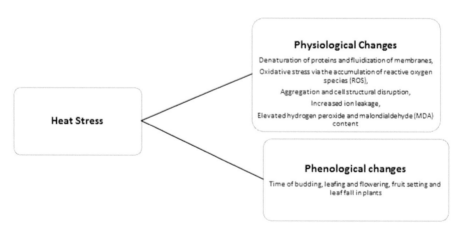

FIGURE 5.2 Effect of heat stress on physiological and phenological processes of crop plants.

Xu et al., 2013). Under heat stress, accumulation of ROS causes oxidative stress and results in lipid peroxidation and electrolyte leakage (Suzuki and Mittler, 2006). High temperature results in irreversible damage to the cellular membranous structure and ultimately results in the loss of cellular function. Moreover, these frequent and increased heat stress episodes may weaken the photosynthetic capacity, resulting in decreased net photosynthetic rate and ultimately, decreased growth and yield of crop plants (Ogweno et al., 2008; Shanmugam et al., 2013). To cope with increased and frequent heat stress, plants have evolved certain unique mechanisms. For example, the production of different enzymatic and non-enzymatic antioxidants relieves oxidative stress (Apel and Hirt, 2004). Moreover, plants have evolved many morphological and metabolic adjustments to cope with heat stress (Wang et al., 2015).

At both genotypic and species level, phenological sensitivity is considered a good indicator of success at increased temperatures (Springate and Kover, 2014). Under short- and long-term climate changes, plant phenology has been one of the most sensitive and easily observable indicators for the assessment and development of

crop models for climate variability (Cleland et al., 2007). A significant advancement in spring phenological events, that is, budding, leafing, and flowering of plants, and, with less certainty, changes in timing of autumn phenological events, that is, fruit setting and leaf fall, have been observed in various studies (Schwartz et al., 2006; Pudas et al., 2008; Julien and Sobrino, 2009; Gordo and Sanz, 2010). These phenological responses under short- and long-term climate changes vary among species (Cleland et al., 2012). This chapter gives a comprehensive and critical analysis of the response mechanisms of crop plants, especially from the physiological and phenological point of view, at elevated temperature.

Acquired Thermotolerance and Transgenerational Memory

The threshold temperature at which this irreversible change starts can vary among individuals of the same species under different climatic conditions experienced prior to heat stress (Barua et al., 2008). When the plants have undergone a gradual increase in heat stress or have undergone prior exposure to moderately high temperature, greater thermotolerance is observed (Larkindale and Vierling, 2008). This reprogramming in gene expression is termed *acquired thermotolerance* or *acclimation*, partly dependent on the rapid expression of heat shock proteins (Wang et al., 2004). Moreover, regions at high altitude have been found to be most sensitive to increased and frequent heat stress (IPCC, 2014).

For better survival under heat stress, plants have evolved certain mechanisms—for example, acquired thermotolerance—to maintain cellular homeostasis (Senthil-Kumar and Udayakumar, 2004; Wang et al., 2004). Through acquired thermotolerance, better adaptive capacity against heat stress has been observed in many crops (Hong et al., 2003; Rizhsky et al., 2004; Jagadish et al., 2011; Lin et al., 2014; Raju et al., 2014). Many reasons have been found for the better adaptive capacity of crop plants under heat stress through acquired thermotolerance. For example, in many plant species, greater production of heat shock proteins during the seedling and vegetative stages has been found in pre-treated plants in comparison to those without pre-treatment due to acquired thermotolerance (Wollenweber et al., 2003; Wang et al., 2012; Raju et al., 2014). On the other hand, transgenerational memory has been found to be a responsible factor for acquired thermotolerance. In transgenerational memory, offspring adaptation is dependent on parental response toward a stress (Molinier et al., 2006; Suter and Widmer, 2013). Recently, transgenerational response under heat stress was investigated in rice (Shi et al., 2016). Rice seeds were taken from previous generations that had shown lower fertility during the flowering stage under heat stress. Pre-acclimation to heat stress during the vegetative and initial reproductive stages resulted in better response only in heat-tolerant genotypes at anthesis under severe post-treatment heat stress (40°C). In future, similar investigations in other crops regarding transgenerational response under heat stress need to be performed, as rice is extremely sensitive during the flowering stage.

Physiological Changes

Leaf respiration (*R*) in crop plants is sensitive to heat stress and has the ability to acclimate against it. Plants from the same species will have similar leaf *R* due to acclimation when grown at different temperatures. The acclimation ability of crop plants varies among different species. This is also the reason why leaf *R* values measured at a common temperature are higher in the case of cold-acclimated plants compared with plants exposed to higher temperature and acclimated (Atkin et al., 2005). According to Guy and Vanlerberghe (2005), the higher values of leaf *R* are due to the increased activity of enzymes involved in the mito-chondrial electron transport chain. The cytochrome *c* oxidase (COX) and alternative oxidase (AOX) are the O_2 consumption sites in plants. In plants with intact leaves, there is a heterogeneous population of mitochondria among contrasting cell types. Recently, this behavior was observed in *Vicia faba* leaves, showing different mitochondrial populations between photosynthetic and non-photosynthetic cell types, and hence responding differently at different temperatures, as in the case of warm- and cold-developed leaves of *V. faba* (Long et al., 2015). Moreover, it was also found that the contribution of non-photosynthetic cell types to leaf *R* is increased at low temperature.

Heat stress causes transient changes in the intracellular pH of microbes, which induces the synthesis of heat shock proteins (Coote et al., 1991; Weitzel et al., 1987). Increased activity of membrane ATPase, responsible for intracellular pH homeostasis, was observed in *Saccharomyces cerevisiae* under heat stress (Coote et al., 1991). Moreover, intracellular GTPase Arl1 and organellar cation/H^+ antiporters in *S. cerevisiae* play a critical role during abiotic stresses, including heat stress (Marešová and Sychrová, 2010; Ditta, 2013; Ramakrishna and Ravishankar, 2011). These antiporter systems are critical in the regulation of intracellular pH (Soemphol et al., 2011). Recently, the At-NhaK2 gene in *Acetobacter tropicalis* SKU1100 was reported, which is essential for K^+ and pH homeostasis under heat stress, as it functions as an $Na^+(K^+)/H^+$ antiporter (Soemphol et al., 2015).

Phenological Changes

Heat stress can occur during different growth and developmental stages of crop plants, and if this stress, even for a short time, coincides with the critical developmental stages of the crop plants such as flowering, it can reduce spikelet fertility and seed set (Jagadish et al., 2010; Rang et al., 2011). Also, the impact of heat stress during gametogenesis is irreversible, and hence, shorter and longer periods of heat stress have the same impact during critical stages of growth and development. According to Jagadish et al. (2013), heat stress can reduce pollen viability and spikelet fertility during gametogenesis. To reduce the susceptibility of heat-sensitive cultivars, significant research projects have identified heat-tolerant donors that can be integrated into molecular breeding programs and ultimately, reduce the anticipated damage caused (Jagadish et al., 2010; Ye et al., 2012).

Overall Phenology

During the last decade, considerable shifts in grapevine phenology, such as shortening of the growth intervals and earlier onset of phenological stages due to heat stress, have been observed in many vineyard regions of the world (Jones, 2006; Bock et al., 2011; Sadras and Petrie, 2011; Tomasi et al., 2011; Webb et al., 2011; Daux et al., 2012). Although early maturation results in higher vintage ratings, it results in the loss of wine typicity due to the changes in aromatic profiles and balance between sugar content and acidity in grapes at harvest (Bock et al., 2011).

Earlier analysis of temperature (1981–2007) and phenological (1986–2011) data showed that average and heat-related extreme temperature indices were high; the greatest rate of change was observed during flowering and the smallest during the ripening period. Ripening was observed at high temperature due to earlier flowering and veraison. Moreover, an advancement in the beginning of budburst (3.6 days), the beginning of flowering (3.1 days), the beginning of veraison (5.2 days), and harvest (7.4 days) was estimated with each degree rise in temperature (Ruml et al., 2016).

Recently, the impact of soil warming and a long growing season on the vegetative and reproductive phenology of eight species of shrub suggested that in the tundra, an early onset of the growing season due to decreased snow cover would not result in greater productivity of shrubs (Khorsand et al., 2015).

Yield Parameters

Under heat stress, time to anthesis in wheat under a particular set of growing conditions has been found to be a useful trait (Snape et al., 2001). Its variability can be used to determine fine-tuned growth and development patterns under prevailing environmental conditions (Blum, 2011). During the flowering stage, heat stress severely affects crop yields due to reduced pollen viability, pollen tube growth and pollen production per flower, fertilization failure, and embryo abortion and reduced pod set in various crops (Guilioni et al., 2003; Young et al., 2004; Salem et al., 2007; Petkova et al., 2009; Devasirvatham et al., 2012; Sadras et al., 2012; Jiang et al., 2015).

Crop Strategies for Heat Stress Tolerance

Usually, heat tolerance is characterized by a smaller effect on essential metabolic processes, and a heat stress–tolerant genotype is usually characterized by higher photosynthetic rates (Prasad et al., 2008; Scafaro et al., 2010). This character is a multigenic trait, so multiple opportunities exist for crop improvement under this stress. Under field conditions, it is a big challenge for scientists to screen heat-tolerant genotypes due to interaction with other environmental factors. A variety of screenable traits have been discovered for the selection of heat-tolerant genotypes under field conditions, and this may also be conducted under controlled conditions (Hall, 2011). However, as well as being costly, this process does not allow the natural selection of genotypes under field conditions due to the environmental interaction mentioned earlier (Souza et al., 2012).

Although the method has been used for screening, the main objective of the crop breeders has been to find a set of effective thermotolerance markers in various crop species. To cope with heat stress, crop plants adopt various morphological, physiological, and biochemical mechanisms. In this regard, the analysis of plant phenology provides a better understanding of plant responses and facilitates the characterization of various other heat tolerance traits at the molecular level (Wahid et al., 2007). Advanced plant phenomics methodologies have enabled us to study plant performance under controlled and field conditions. Through these methodologies, certain genes related to the traits of interest have been identified through the establishment of relationships between genetics and the associated phenotypes (Sivasankar et al., 2012).

Moreover, changes in growth and development of crop plants under heat stress not only vary between plant tissues and growth stages but also depend on the temperature and heat stress duration. However, it is critical to choose an appropriate phenotype due to the variation of the function of response genes under heat stress across plant tissues and growth stages. The need of the hour is to adopt a systematic approach that includes a range of heat stress conditions and identification of the functions of potential heat stress response genes. For this purpose, efficient screening techniques/procedures, for example, molecular genetic markers, would be required to identify key donors and reveal their inheritance pattern through molecular genetics. In modern plant genetic analysis, sequence-based micro-array analysis and amplified fragment length polymorphism (AFLP) have been successfully employed (Vos et al., 1995). Also, the identification of such markers would reduce the costs of DNA sequencing. Latest association studies, genetic linkage analyses, marker-assisted selection (MAS), and trait mapping could be employed for the introduction of these markers (Duran et al., 2010).

Despite the characterization of heat stress response genes, molecular markers for direct trait selection have still been a challenge for scientists. Therefore, the selection of indirect physiological traits would be a workable alternative in genetic engineering. For example, early maturation is closely related to smaller yield loss. Genetic variability under heat stress could also be used to screen certain germplasms with earlier flowering and with a large number of fertile spikelets/panicles (Adams et al., 2001; Singh et al., 2010). At the cellular level, electron transport rate, membrane integrity, chlorophyll accumulation, and enzyme viability have also been found to be useful traits for the selection of heat-tolerant genotypes (Cottee et al., 2010; Selvaraj et al., 2011).

As an example, membranous organelles are damaged under heat stress, resulting in the loss of chlorophyll. Therefore, detection of chlorophyll concentration could be a useful approach for the selection of heat-tolerant genotypes (Shah and Paulsen, 2003). Similarly, pollen viability and seed set are also considered important traits for the selection of heat-tolerant genotypes in rice. Actually, there is a strong correlation among pollen production and viability, anther dehiscence, and seed set, which would result in easier dehiscence in the anthers of heat-tolerant genotypes compared with those of susceptible ones under heat stress (Prasad et al., 2006; Jagadish et al., 2010).

Molecular markers associated with phenotypic traits are searched through methods associated with segregation mapping, genomic introgression, and association mapping (Morgante and Salamini, 2003). In the case of major food crops that have an important role in food security, re-sequencing would be a good method for the identification of markers that co-segregate under heat stress. The marker data sets collected through whole genome sequencing (WGS) would be helpful in genome-wide association studies (GWAS) and for quantitative trait locus (QTL) mapping, which has become the latest method for the identification of specific chromosome segments containing candidate genes for heat tolerance (Argyris et al., 2011; Zhang et al., 2012).

As mentioned in the previous section, QTL mapping has wide applications and is widely accepted. Several studies have identified various QTLs under different environmental stresses, including heat stress. For example, QTLs associated with acquired thermotolerance in a heat-sensitive *Arabidopsis* mutant and multiple loci for thermotolerance in wheat and maize have been identified (Hong et al., 2003; Bai, 2011; Paliwal et al., 2012). QTLs associated with heat stress have been identified in rice at the flowering stage, but their confirmation and fine mapping have not been reported yet (Ye et al., 2012). In lettuce, a major QTL for germination and an additional one having a smaller effect have also been identified (Argyris et al., 2008). Similarly, different traits such as thousand grain weight, GFD (final grain yield divided by the average rate of grain dry weight accumulation during the linear period of grain formation), CTD (difference between the ambient air temperature and the canopy temperature) yield, and senescence related have also been used for QTL identification under heat stress (Pinto et al., 2010; Vijayalakshmi et al., 2010).

These identified QTLs could be combined for tolerance to various abiotic stresses prevailing under field conditions (Hirayama and Shinozaki, 2010; Roy et al., 2011).

For the identification of QTLs in several crops under abiotic stress, association genetics and high-throughput marker genotyping platforms have played an important role, as association or linkage disequilibrium mapping is cost effective and a high-resolution methodology (Varshney et al., 2009; Ahuja et al., 2010). The candidate QTLs could be further introgressed after their identification in the elite line through MAS techniques (Lopes and Reynolds, 2010; Thomson et al., 2010). Using this technique, there might be problems in developing superior genotypes, as the candidate traits might be controlled by small-effect QTLs or several epistatic QTLs (Collins et al., 2008). In this regard, pyramiding several QTLs from large populations in the same genetic background, marker-assisted recurrent selection (MARS), or genome selection (GS) approaches could be employed (Tester and Langridge, 2010).

In comparison to the MAS technique, a MARS or GS method predicts breeding values based on the data derived from a vast number of molecular markers with a high coverage of the genome and greater chances of recombination due to successive crossing (Ramalho et al., 2005). Moreover, this method is novel, cost effective, repetitive, and accumulative, which makes it more reliable, as it uses all data for prediction and delivers more accurate predictions with an increased rate of genetic gain (Habier et al., 2009; Heffner et al., 2009). For example, this method has been successfully used at the Federal University of Lavras to develop heat stress–tolerant potato genotypes with expressive gains in quality and yield of tuber (Benites and Pinto, 2011). This has been confirmed through simulation studies, which have predicted a high correlation between predicted and true breeding values over several generations without the need to re-phenotype (Doná et al., 2012). Using this approach, only field trials are conducted under field conditions for the ultimate selection of heat stress–tolerant genotypes and their yield performance (Mason et al., 2010). Precise phenotypic protocols would be required for the detection of small yet significant phenotypic changes caused by introducing a single gene into breeding material (Cattivelli et al., 2008). The most limiting factor under this approach is the lack of a precise phenotyping protocol, which is a pre-requisite for the assessment of complex genetic networks associated with QTLs under field conditions.

In this regard, a modern genetic modification approach could also be employed for the transfer of advantageous traits to a cultivar. This approach includes the identification of the responsible gene for the desired trait, and there is no barrier to transferring the required genes across different species within the plant kingdom or even from animals. Transformation protocols for the potential introgression of genes from non-plant species are available for most of the food crop species, and a combination of useful genes could be transferred into the same plant. However, the efficiency of transformation protocols is low in the case of major crops such as wheat (*Triticum aestivum*), barley, and rice (*Oryza sativa*), and there are regulatory restrictions on the use of transgenic plants, which makes this approach economically impractical (Takeda and Matsuoka, 2008). Nevertheless, there are reports of the use of spatially and temporally restricted transgene expression via tissue-specific and stress-inducible promoters to combine stress tolerance with high yield potential while avoiding the negative effects of a stress gene on plant growth under favorable

conditions (Nakashima et al., 2007). Thus, the future commercial success of transgenic breeding will depend on public acceptance of genetically modified plants and their produce and on the development of clearly defined and scientifically based regulatory frameworks in this regard (Godfray et al., 2010).

Carbon Dioxide Enrichment Alleviates Heat Stress

During the last two centuries, elevated atmospheric CO_2 concentrations and increased surface temperatures have resulted in global climate change (Xu et al., 2013). Although the atmospheric concentration of CO_2 has already increased from 0.028% to more than 0.039%, it is predicted to reach 0.07% by the end of this century (IPCC, 2007). Similarly, the surface temperature will increase by 1.1–6.4°C on average by the end of the current century compared with the end of the last century. Commercially, supplemental CO_2 is being used in greenhouses during summer around the world to enhance the productivity and quality of crops under heat stress (Li et al., 2013). According to Yu et al. (2014), this situation would cause abiotic stress due to high temperature and elevated concentration of CO_2 in crop plants and their interactive effects on plant growth. Under natural conditions, the productivity and quality of crop plants are severely affected under heat stress.

Earlier, it was found that elevated concentration of CO_2 promotes plant growth and productivity under heat stress, and this effect might be compounded by high temperature (Qaderi et al., 2006; Xu et al., 2013; Yu et al., 2014). In a study under both ambient temperature and heat stress in tall fescue, an elevated concentration of CO_2 (800 ppm) enhanced the photosynthetic rate, which was suggested to be due to an abundance of proteins involved in photosynthetic light reactions, electron transport, carbon assimilation, and antioxidant metabolism (Ramakrishna and Ravishankar, 2013; Yu et al., 2014).

Conclusion

Heat stress tolerance is regarded as a polygenic character, which is usually measured through the crop yield parameter. This parameter is also dependent on many processes and mechanisms related to plants, soil, and other environmental factors. Therefore, it is the need of the hour to devise a comprehensive and useful approach, which could use both classical and modern genetic engineering tools. Using these modern and state-of-the-art tools, it would be possible to find a denser genome-wide marker that has coverage for almost all crop species. Moreover, we need an accelerated discovery of QTLs and their transfer by GS, which would ultimately result in an increased flexibility in the interplay between phenotypic evaluation and selection. These tools are expected to reduce genotyping costs. Overall, public acceptance of genetically modified plants and their produce would play an important role in the future commercial success of transgenic breeding. Success would also be equally dependent on the development of a clearly defined and scientifically based regulatory structure.

REFERENCES

Adams S, Cockshull K, Cave C. Effect of temperature on the growth and development of tomato fruits. *Ann. Bot.* 2001;88:869–877.

Ahuja I, De Vos RCH, Bones AM, Hall RD. Plant molecular stress responses face climate change. *Trends Plant Sci.* 2010;15(12):664–674.

Apel K, Hirt H. Reactive oxygen species: Metabolism, oxidative stress, and signal transduction. *Ann. Rev. Plant Biol.* 2004;55:373–399.

Argyris J, Dahal P, Hayashi E, Still DW, Bradford KJ. Genetic variation for lettuce seed thermoinhibition is associated with temperature-sensitive expression of abscisic acid, gibberellin, and ethylene biosynthesis, metabolism, and response genes. *Plant Physiol.* 2008;148:926–947.

Argyris J, Truco MJ, Ochoa O, McHale L, Dahal P, VanDeynze A, Michelmore RW, Bradford KJ. A gene encoding an abscisic acid biosynthetic enzyme (LsNCED4) collocates with the high temperature germination locus Htg6.1 in lettuce (*Lactuca* sp.). *Theor. Appl. Genet.* 2011; 122:95–108.

Atkin OK, Bruhn D, Hurry VM, Tjoelker MG. Evans Review No. 2: The hot and the cold: Unravelling the variable response of plant respiration to temperature. *Funct. Plant Biol.* 2005;32(2):87–105.

Bai J. Genetic variation of heat tolerance and correlation with other agronomic traits in a maize (*Zea mays* L.) recombinant inbred line population. Available at: http://hdl.handle.net/2346/13572. 2011.

Balasubramanian S, Sureshkumar S, Lempe J, Weigel D. Potent induction of *Arabidopsis thaliana* flowering by elevated growth temperature. *Plos Genet.* 2006;2:980–989.

Barua D, Heckathorn SA, Coleman JS. Variation in heat-shock proteins and photosynthetic thermotolerance among natural populations of *Chenopodium album* L. from contrasting thermal environments: Implications for plant responses to global warming. *J. Integr. Plant Biol.* 2008;50:1440–1451.

Battisti DS, Naylor RL. Historical warnings of future food insecurity with unprecedented seasonal heat. *Science* 2009;323:240–244.

Benites FRG, Pinto CABP. Genetic gains for heat tolerance in potato in three cycles of recurrent selection. *Crop Breed. Appl. Biotechnol.* 2011;11:133–140.

Blum A. Drought resistance–Is it really a complex trait? *Functional Plant Biology* 2011;38(10):753–757.

Cattivelli L, Rizza F, Badeck FW, Mazzucotelli E, Mastrangelo AM, Francia E, Marè C, Tondelli A, Stanca AM. Drought tolerance improvement in crop plants: An integrated view from breeding to genomics. *Field Crops Res.* 2008;105:1–14.

Challinor AJ, Watson J, Lobell DB, Howden SM, Smith DR, Chhetri N. A meta-analysis of crop yield under climate change and adaptation. *Nat. Clim. Change* 2014;4(4):287.

Cleland EE, Allen JM, Crimmins TM, Dunne JA, Pau S, Travers SE, Zavaleta ES, Wolkovich EM. Phenological tracking enables positive species responses to climate change. *Ecology* 2012;93:1765–1771.

Cleland EE, Chuine I, Menzel A, Mooney HA, Schwartz MD. Shifting plant phenology in response to global change. *Trends Ecol. Evol.* 2007;22(7):357–365.

Collins NC, Tardieu F, Tuberosa R. Quantitative trait loci and crop performance under abiotic stress: Where do we stand? *Plant Physiol.* 2008;144:469–486.

Coote PJ, Cole MB, Holyoak C. Thermal inactivation of *Listeria monocytogenes* during a process simulating temperatures achieved during microwave heating. *J. Appl. Bacteriol.* 1991;70(6):489–494.

Cottee N, Tan D, Bange M, Cothren J, Campbell L. Multi-level determination of heat tolerance in cotton (*Gossypium hirsutum* L.) under field conditions. *Crop Sci.* 2010;50:2553–2564.

Daux V, Garcia De Cortazar-Atauri I, Yiou P, Chuine I, Garnier E, Le Roy Ladurie E, Mestre O, Tardaguila J. An open-access database of grape harvest dates for climate research: Data description and quality assessment. *Clim. Past* 2012;8:1403–1418.

Devasirvatham V, Tokachichu RN, Trethowan RM, Tan DKY, Gaur PM, Mallikarjuna N. Effect of high temperature on the reproductive development of chickpea genotypes under controlled environments. *Funct. Plant Biol.* 2012;39:1009–1018.

Ditta A. Salt tolerance in cereals: Molecular mechanisms and applications. In: *Molecular Stress Physiology of Plants*, Rout GR, Das AB (Eds.). Springer, India; 2013, pp. 133–154.

Doná AA, Miranda GV, DeLima RO, Chaves LG, eGama EEG. Genetic parameters and predictive genetic gain in maize with modified recurrent selection method. *Chilean J. Agric. Res.* 2012;72:1:33.

Duran C, Eales D, Marshall D, Imelfort M, Stiller J, Berkman PJ, Clark T, et al. Future tools for association mapping in crop plants. *Genome* 2010;53(1):1017–1023.

Field CB, Barros VR, Mach K, Mastrandrea M. Climate change 2014: Impacts, adaptation, and vulnerability, vol. V. New York: Cambridge University Press Cambridge; 2014.

Godfray HCJ, Beddington JR, Crute IR, Haddad L, Lawrence D, Muir JF, Pretty J, Robinson S, Thomas SM, Toulmin C. Food security: The challenge of feeding 9 billion people. *Science* 2010;327:812–818.

Gordo O, Sanz JJ. Impact of climate change on plant phenology in Mediterranean ecosystems. *Global Change Biol.* 2010;16:1082–1106.

Guilioni L, Wéry J, Lecoeur J. High temperature and water deficit may reduce seed number in field pea purely by decreasing plant growth rate. *Funct. Plant Biol.* 2003;30:1151–1164.

Guy RD, Vanlerberghe GC. Partitioning of respiratory electrons in the dark in leaves of transgenic tobacco with modified levels of alternative oxidase. *Physiol. Plant.* 2005;125(2):171–180.

Habier D, Fernando RL, Dekkers JCM. Genomic selection using low-density marker panels. *Genetics* 2009;182:343–353.

Hall AE. Breeding cowpea for future climates. In: *Crop Adaptation to Climate Change*, Yadav SS, Redden R, Hatfield JL, Lotze-Campen H, Hall AJW (Eds.). John Wiley & Sons, Hoboken; 2011.

Heffner EL, Sorrells ME, Jannink JL. Genomic selection for crop improvement. *Crop Sci.* 2009;49:1–12.

Hirayama T, Shinozaki K. Research on plant abiotic stress responses in the post-genome era: Past, present and future. *Plant J.* 2010;61:1041–1052.

Hong SW, Lee U, Vierling E. *Arabidopsis* hot mutants define multiple functions required for acclimation to high temperatures. *Plant Physiol.* 2003;132:757–767.

IPCC. Climate Change 2007: The Physical Science Basis. Contribution of Working Group I to the Fourth Assessment Report of the Intergovernmental Panel on Climate Change. Solomon S, Qin D, Manning M, Chen Z, Marquis M, Averyt KB, Tignor M, Miller HL (Eds.). Cambridge University Press, Cambridge, UK; 2007.

IPCC. Climate Change 2014: Synthesis Report. Contribution of Working Groups I, II and III to the Fifth Assessment Report of the Intergovernmental Panel on Climate Change [Core Writing Team, Pachauri RK, and Meyer LA (Eds.)]. IPCC, Geneva, Switzerland; 2014, p. 151.

IPCC. Working Group I Contribution to the IPCC Fifth Assessment Report on Climate Change 2013: The Physical Science Basis, Summary for Policymakers. 2013. www.climatechange2013. org/images/report/WG1AR5_SPM_FINAL.pdf

Jagadish SVK, Craufurd PQ, Shi W, Oane R. A phenotypic marker for quantifying heat stress impact during microsporogenesis in rice (*Oryza sativa* L.). *Funct. Plant Biol.* 2013;41:48–55.

Jagadish SVK, Muthurajan R, Oane R, Wheeler TR, Heuer S, Bennett J, Craufurd PQ. Physiological and proteomic approaches to address heat tolerance during anthesis in rice (*Oryza sativa* L.). *J. Exp. Bot.* 2010;61:143–156.

Jagadish SVK, Muthurajan R, Rang ZW, Malo R, Heuer S, Bennett J, Craufurd PQ. Spikelet proteomic response to combined water deficit and heat stress in rice (*Oryza sativa* cv. N22). *Rice* 2011;4:1–11.

Jiang Y, Lahlali R, Karunakaran C, Kumar S, Davis AR, Bueckert RA. Seed set, pollen morphology and pollen surface composition response to heat stress in field pea. *Plant Cell Environ.* 2015;38(1):2387–2397.

Jones GV. Climate change and wine: Observations, impacts and future implications. *Aust. N. Z. Wine Ind. J.* 2006;21:21–26.

Julien Y, Sobrino JA. Global land surface phenology trends from GIMMS database. *Int. J. Remote Sens.* 2009;30:3495–3513.

Khorsand RR, Oberbauer SF, Starr G, Puma IPL, Pop E, Ahlquist L, Baldwin T. Plant phenological responses to a long-term experimental extension of growing season and soil warming in the tussock tundra of Alaska. *Global Change Biol.* 2015;21(12):4520–4532.

Larkindale J, Vierling E. Core genome responses involved in acclimation to high temperature. *Plant Physiol.* 2008;146:748–761.

Li X, Zhang GQ, Sun B, Zhang S, Zhang YQ, Liao YWK, Zhou YH, Xia XJ, Shi K, Yu JQ. Stimulated leaf dark respiration in tomato in an elevated carbon dioxide atmosphere. *Sci. Rep.* 2013;3:1–8.

Lin MY, Chai KH, Ko SS, Kuang LY, Lur HS, Charng YY. A positive feedback loop between Heat Shock Protein101 and Heat Stress-Associated 32-Kd Protein modulates long-term acquired thermotolerance illustrating diverse heat stress responses in rice varieties. *Plant Physiol.* 2014;164:2045–2053.

Long BM, Bahar NHA, Atkin OK. Contributions of photosynthetic and non-photosynthetic cell types to leaf respiration in *Vicia faba* L. and their responses to growth temperature. *Plant Cell Environ.* 2015;38:2263–2276.

Lopes MS, Reynolds MP. Partitioning of assimilates to deeper roots is associated with cooler canopies and increased yield under drought in wheat. *Funct. Plant Biol.* 2010;37:147–156.

Marešová L, Sychrová H. Genetic interactions among the Arl1GTPase and intracellular Na⁺/H⁺ antiporters in pH homeostasis and cation detoxification. *FEMS Yeast Res.* 2010;10:802–811.

Mason RE, Mondal S, Beecher FW, Pacheco A, Jampala B, Ibrahim AM, Hays DB. QTL associated with heat susceptibility index in wheat (*Triticum aestivum* L.) under short-term reproductive stage heat stress. *Euphytica* 2010;174, 423–436.

Molinier J, Ries G, Zipfel C, Hohn B. Transgeneration memory of stress in plants. *Nature* 2006;442:1046–1049.

Morgante M, Salamini F. From plant genomics to breeding practice. *Curr. Opin. Biotechnol.* 2003;14:214–219.

Nakashima K, Tran LSP, Van Nguyen D, Fujita M, Maruyama K, Todaka D, Ito Y, Hayashi N, Shinozaki K, Yamaguchi-Shinozaki K. Functional analysis of a NAC-type transcription factor OsNAC6 involved in abiotic and biotic stress-responsive gene expression in rice. *Plant J.* 2007;51:617–630.

Ogweno J, Song X, Shi K, Hu W, Mao W, Zhou Y, Yu J, Nogues S. Brassinosteroids alleviate heat-induced inhibition of photosynthesis by increasing carboxylation efficiency and enhancing antioxidant systems in *Lycopersicon esculentum*. *J. Plant Growth Regul.* 2008;27:49–57.

Paliwal R, Röder MS, Kumar U, Srivastava J, Joshi AK. QTL mapping of terminal heat tolerance in hexaploid wheat (*T. aestivum* L.). *Theor. Appl. Genet.* 2012;125:561–575.

Petkova V, Nikolova V, Kalapchieva SH, Stoeva V, Topalova E, Angelova S. Physiological response and pollen viability of *Pisum sativum* genotypes under high temperature influence. *Acta Hort.* 2009;830:665–674.

Pinto RS, Reynolds MP, Mathews KL, McIntyre CL, Olivares-Villegas JJ, Chapman SC. Heat and drought adaptive QTL in a wheat population designed to minimize confounding agronomic effects. *Theor. Appl. Genet.* 2010;121:1001–1021.

Prasad P, Boote K, Allen L, Sheehy J, Thomas J. Species, ecotype and cultivar differences in spikelet fertility and harvest index of rice in response to high temperature stress. *Field Crops Res.* 2006;95:398–411.

Prasad P, Pisipati S, Mutava R, Tuinstra M. Sensitivity of grain sorghum to high temperature stress during reproductive development. *Crop Sci.* 2008;48:1911–1917.

Pudas E, Leppälä M, Tolvanen A, Poikolainen J, Venäläinen A, Kubin E. Trends in phenology of *Betula pubescens* across the boreal zone in Finland. *Int. J. Biometeorol.* 2008;52(4):251–259.

Qaderi MM, Kurepin LV, Reid DM. Growth and physiological responses of canola (*Brassica napus*) to three components of global climate change: Temperature, carbon dioxide and drought. *Physiol. Plant.* 2006;128:710–721.

Raju BR, Narayanaswamy BR, Mohankumar MV, Sumanth KK, Rajanna MP, Mohanraju B, Udayakumar M, Sheshshayee MS. Root traits and cellular level tolerance hold the key in maintaining higher spikelet fertility of rice under water limited conditions. *Funct. Plant Biol.* 2014;41:930–939.

Ramakrishna A, Ravishankar GA. Influence of abiotic stress signals on secondary metabolites in plants. *Plant Signal. Behav.* 2011;6:1720–1731.

Ramakrishna A, Ravishankar GA. Role of plant metabolites in abiotic stress tolerance under changing climatic conditions with special reference to secondary compounds. In: *Climate Change and Plant Abiotic Stress Tolerance*, Wiley-VCH Verlag GmbH & Co. KGaA, Weinheim, Germany; 2013, pp. 705–726.

Ramalho MAP, Abreu ÂFB, Santos JB. Genetic progress after four cycles of recurrent selection for yield and grain traits in common bean. *Euphytica* 2005;144:23–29.

Rang ZW, Jagadish SVK, Zhou QM, Craufurd PQ, Heuer S. Effect of heat and drought stress on pollen germination and spikelet fertility in rice. *Environ. Exp. Bot.* 2011;70:58–65.

Rizhsky L, Liang H, Shuman J, Shulaev V, Davletova S, Mittler R. When defense pathways collide: The response of *Arabidopsis* to a combination of drought and heat stress. *Plant Physiol.* 2004;134:1683–1696.

Roy SJ, Tucker EJ, Tester M. Genetic analysis of abiotic stress tolerance in crops. *Curr. Opin. Plant Biol.* 2011;14:232–239.

Ruml M, Korać N, Vujadinović M, Vuković A, Ivanišević D. Response of grapevine phenology to recent temperature change and variability in the wine-producing area of Sremski Karlovci, Serbia. *J. Agric. Sci.* 2016;154(2):186–206.

Sadras VO, Lake L, Chenu K, McMurray LS, Leonforte A. Water and thermal regimes for field pea in Australia and their implications for breeding. *Crop Pasture Sci.* 2012;63:33–44.

Sadras VO, Petrie PR. Climate shifts in southeastern Australia: Early maturity of Chardonnay, Shiraz and Cabernet-Sauvignon is associated with early onset rather than faster ripening. *Aust. J. Grape Wine Res.* 2011;17:199–205.

Salem MA, Kakani VG, Koti S, Reddy KR. Pollen-based screening of soybean genotypes for high temperatures. *Crop Sci.* 2007;47:219–231.

Scafaro AP, Haynes PA, Atwell BJ. Physiological and molecular changes in *Oryza ameridionalis* Ng., a heat-tolerant species of wild rice. *J. Exp. Bot.* 2010;61:191–202.

Schwartz MD, Ahas R, Aasa A. Onset of spring starting earlier across the Northern Hemisphere. *Global Change Biol.* 2006;12:343–351.

Selvaraj MG, Burow G, Burke JJ, Belamkar V, Puppala N, Burow MD. Heat stress screening of peanut (*Arachis hypogaea* L.) seedlings for acquired thermotolerance. *Plant Growth Regul.* 2011;65:83–91.

Senthil-Kumar M, Udayakumar M. Development of thermotolerant tomato (*Lycopersicon esculentum* Mill.) lines: An approach based on mutagenesis. *India J. Plant Biol.* 2004;31:139–148.

Shah N, Paulsen G. Interaction of drought and high temperature on photosynthesis and grain-filling of wheat. *Plant Soil* 2003;257:219–226.

Shanmugam S, Kjaer KH, Ottosen CO, Rosenqvist E, Kumari SD, Wollenweber B. The alleviating effect of elevated CO_2 on heat stress susceptibility of two wheat (*Triticum aestivum* L.) cultivars. *J. Agron. Crop Sci.* 2013;199:340–350.

Shi W, Lawas LMF, Raju BR, Jagadish SVK. Acquired thermo-tolerance and trans-generational heat stress response at flowering in rice. *J. Agron. Crop Sci.* 2016;202(4):309–319.

Singh RK, Redoña E, Refuerzo L. Varietal improvement for abiotic stress tolerance in crop plants: Special reference to salinity in rice. In: *Abiotic Stress Adaptation in Plants*, Gonzalez Fontes A (Ed.). Studium Press Llc, Houston; 2010, pp. 387–415.

Sivasankar S, Williams RW, Greene TW. Abiotic stress tolerance in plants: An industry perspective. In: *Improving Crop Resistance to Abiotic Stress*, Tuteja N, Gill SS, Tiburcio AF, Tuteja R (Eds.). Wiley-Blackwell, Hoboken; 2012, pp. 27–47.

Snape J, Butterworth K, Whitechurch E, Worland AJ. Waiting for fine times: Genetics of flowering time in wheat. In: *Wheat in a Global Environment. Developments in Plant Breeding*, Bedö Z, Láng L (Eds.) vol 9. Springer, Dordrecht; 2001, pp. 67–74.

Soemphol W, Deeraksa A, Matsutani M, Yakushi T, Toyama H, Adachi O, Yamada M, Matsushita K. Global analysis of the genes involved in the thermotolerance mechanism of thermotolerant *Acetobacter tropicalis* SKU1100. *Biosci. Biotechnol. Biochem.* 2011;75:1921–1928.

Soemphol W, Tatsuno M, Okada T, Matsutani M, Kataoka N, Yakushi T, Matsushita K. A novel Na+(K+)/H+ antiporter plays an important role in the growth of *Acetobacter tropicalis* SKU1100 at high temperatures via regulation of cation and pH homeostasis. *J. Biotechnol.* 2015;211:46–55.

Souza MA, Pimentel AJB, Ribeiro G. Breeding for heat-stress tolerance. In: *Plant Breeding for Abiotic Stress Tolerance*, Fritsche-Neto R, Borém A (Eds.). Springer, Berlin; 2012, pp. 137–156.

Springate DA, Kover PX. Plant responses to elevated temperatures: A field study on phenological sensitivity and fitness responses to simulated climate warming. *Global Change Biol.* 2014;20(2):456–465.

Stavang JA, Gallego-Bartolome J, Gomez MD, Yoshida S, Asami T, Olsen JE, Garcia-Martinez JL, Alabadi D, Blazquez MA. Hormonal regulation of temperature-induced growth in *Arabidopsis. Plant J.* 2009;60:589–601.

Suter L, Widmer A. Environmental heat and salt stress induce transgenerational phenotypic changes in *Arabidopsis thaliana*. *PLoS One* 2013;8:e60364.

Suzuki N, Mittler R. Reactive oxygen species and temperature stresses: A delicate balance between signaling and destruction. *Physiol. Plant.* 2006;126:45–51.

Takeda S, Matsuoka M. Genetic approaches to crop improvement: Responding to environmental and population changes. *Nat. Rev. Genet.* 2008;9:444–457.

Teixeira EI, Fischer G, van Velthuizen H, Walter C, Ewert F. Global hot-spots of heat stress on agricultural crops due to climate change. *Agric. For. Meteorol.* 2013;170:206-215.

Tester M, Langridge P. Breeding technologies to increase crop production in a changing world. *Science* 2010;327:818–822.

Thomson MJ, deOcampo M, Egdane J, Rahman MA, Sajise AG, Adorada DL, Tumimbang-Raiz E, et al. Characterizing the Saltol quantitative trait locus for salinity tolerance in rice. *Rice* 2010;3:148–160.

Tomasi D, Jones GV, Giust M, Lovat L, Gaiotti F. Grapevine phenology and climate change: Relationships and trends in the Veneto Region of Italy for 1964–2009. *Am. J. Enol. Vitic.* 2011;62:329–339.

Varshney RK, Nayak SN, May GD, Jackson SA. Next-generation sequencing technologies and their implications for crop genetics and breeding. *Trends Biotechnol.* 2009;27:522–530.

Vijayalakshmi K, Fritz AK, Paulsen GM, Bai G, Pandravada S, Gill BS. Modeling and mapping QTL for senescence-related traits in winter wheat under high temperature. *Mol. Breed.* 2010;26:163–175.

Vos P, Hogers R, Bleeker M, Reijans M, vandeLee T, Hornes M, Friters A, et al. AFLP: A new technique for DNA finger printing. *Nucl. Acids Res.* 1995;23:4407–4414.

Wahid A, Gelani S, Ashraf M, Foolad MR. Heat tolerance in plants: An overview. *Environ. Exper. Bot.* 2007;61:199–223.

Wang W, Vinocur B, Shoseyov O, Altman A. Role of plant heat-shock proteins and molecular chaperones in the abiotic stress response. *Trends Plant Sci.* 2004;9:244–252.

Wang X, Cai J, Liu F, Jin M, Yu H, Jiang D, Wollenweber B, Dai T, Cao W. Pre-anthesis high temperature acclimation alleviates the negative effects of postanthesis heat stress on stem stored carbohydrates remobilization and grain starch accumulation in wheat. *J. Cereal Sci.* 2012;55:331–336.

Wang X, Dinler BS, Vignjevic M, Jacobsen S, Wollenweber B. Physiological and proteome studies of responses to heat stress during grain filling in contrasting wheat cultivars. *Plant Sci.* 2015;230:33–50.

Webb LB, Whetton PH, Barlow EWR. Observed trends in winegrape maturity in Australia. *Global Change Biol.* 2011;17:2707–2719.

Weitzel G, Pilatus U, Rensing L. The cytoplasmic pH, ATP content and total protein synthesis rate during heat-shock protein inducing treatments in yeast. *Exp. Cell Res.* 1987;170:64–79.

Wigge PA. Ambient temperature signalling in plants. *Curr. Opin. Plant Biol.* 2013;16:661–666.

Wollenweber B, Porter JR, Schellberg J. Lack of interaction between extreme high-temperature events at vegetative and reproductive growth stages in wheat. *J. Agron. Crop Sci.* 2003;189:142–150.

Xu J, Duan X, Yang J, Beeching JR, Zhang P. Enhanced reactive oxygen species scavenging by overproduction of superoxide dismutase and catalase delays postharvest physiological deterioration of cassava storage roots. *Plant Physiol.* 2013;161(3):1517–1528.

Xu Z, Shimizu H, Yagasaki Y, Ito S, Zheng Y, Zhou G. Interactive effects of elevated CO_2, drought, and warming on plants. *J. Plant Growth Regul.* 2013;32:692–707.

Ye C, Argayoso MA, Redoña ED, Sierra SN, Laza MA, Dilla CJ, Mo Y, et al. Mapping QTL for heat tolerance at flowering stage in rice using SNP markers. *Plant Breed.* 2012;131:33–41.

Young LW, Wilen RW, Bonham-Smith PC. High temperature stress of *Brassica napus* during flowering reduces micro- and mega-gametophyte fertility, induces fruit abortion, and disrupts seed production. *J. Exp. Bot.* 2004;55:485–495.

Yu J, Yang Z, Jespersen D, Huang B. Photosynthesis and protein metabolism associated with elevated CO_2 mitigation of heat stress damages in tall fescue. *Environ. Exp. Bot.* 2014;99:75–85.

Zhang WB, Jiang H, Qiu PC, Liu CY, Chen FL, Xin DW, Li CD, Hu GH, Chen QS. Genetic overlap of QTL associated with low-temperature tolerance at germination and seedling stage using BILs in soybean. *Can. J. Plant Sci.* 2012;92:1–8.

6

Biochemical and Molecular Mechanisms of High-Temperature Stress in Crop Plants

Gurpreet Kaur, Bavita Asthir, and Navtej S. Bains

CONTENTS

Introduction

Abiotic stresses are often interrelated. Either individually or in combination, they cause morphological, physiological, biochemical, and molecular changes that adversely affect plant growth and productivity and ultimately, yield. Heat, drought, cold, and salinity are the major abiotic stresses that induce severe cellular damage in plant species, including crop plants (Ahmad and Prasad 2012; Ramakrishna and Ravishankar 2011, 2013). Heat stress (HS) caused by global warming is one of the major concerns for losses of yield in temperate climates (Hancock et al. 2014). The global surface temperature change is projected to likely exceed 1.5°C by the end of the 21st century (IPCC 2014). Plant growth and development involve numerous biochemical reactions that are sensitive to temperature. A delicate balance exists between multiple pathways residing in different organelles of plant cells, known as *cellular homeostasis* (Kocsy et al. 2013). This coordination between different organelles may be disrupted during temperature stresses due to variation in temperature optima in different pathways within cells. A direct result of stress-induced cellular changes is the overproduction of reactive oxygen species (ROS) in plants. These are produced in such a way that they are confined to a small area and also by a specific pattern of biological responses. The production of ROS is an inevitable consequence of aerobic metabolism during stressful conditions (Bhattacharjee 2012). ROS are highly reactive and toxic, affecting various cellular functions in plant cells through damage to nucleic acids, protein oxidation, and lipid peroxidation, eventually resulting in cell death. Major ROS-scavenging mechanisms include enzymatic systems consisting of superoxide dismutase (SOD), catalase (CAT), peroxidases (POD), ascorbate peroxidase (APX), and glutathione reductase (GR) and nonenzymatic systems comprising ascorbic acid (AsA) and glutathione (GSH) (Foyer and Noctor 2012; Das and Roychoudhury 2014).

Indirect or slower heat injuries include inactivation of enzymes in chloroplasts and mitochondria, inhibition of protein synthesis, protein degradation, and loss of membrane integrity. Under heat-stress conditions, plants accumulate different metabolites such as antioxidants, osmoprotectants, heat shock proteins (HSPs), and metabolites from different pathways to enable correcting the equilibrium of the plant cellular redox state and balancing fluctuations in plant cells (Bokszczanin and Fragkostefanakis 2013). At the molecular level, HS causes alterations in the expression of genes involved in direct protection from HS (Hartl et al. 2011). These include genes responsible for the expression of osmoprotectants, detoxifying enzymes, transporters, and regulatory proteins. In recent times, exogenous applications of protectants in the form of osmoprotectants (proline, glycine betaine, trehalose, etc.), phytohormones (abscisic acid [ABA], gibberellic acids, jasmonic acids [JA], brassinosteroids, salicylic acid, etc.), signaling molecules (e.g., nitric oxide [NO]), polyamines (putrescine, spermidine, and spermine), trace elements, and nutrients have been found effective in mitigating HS-induced damage in plants.

Breeding for selecting genotypes with improved thermotolerance using an assortment of genetic approaches is, therefore, one of the most vital objectives in crop improvement programs. For this reason, a thorough understanding of the physiological responses of plants to high temperature, mechanisms of heat tolerance, and possible strategies for improving crop thermotolerance is crucial.

Plant Response to Heat Stress

HS has an independent mode of action on the physiology and metabolism of plant cells. It affects all aspects of plant processes, such as germination, growth, development, reproduction, and yield. Various physiological injuries have been observed at elevated temperatures, such as scorching of leaves and stems, leaf abscission and senescence, shoot and root growth inhibition, or fruit damage, which consequently lead to decreased plant productivity. HS exerts a negative effect on germination percentage and plant emergence, and results in abnormal seedlings, poor seedling vigor, and reduced radicle and plumule growth of germinated seedlings in various cultivated plant species. HS results in reductions in tillers, spike, and floret numbers per plant, as well as spikelets per spike (Dawson and Wardlaw 1989; Allen et al. 1995; Prasad et al. 2006a). Within a floret, anthers and pollen are more sensitive to HS than ovules, and floret sterility at temperatures ≥ 30 °C has been correlated with diminished anther dehiscence (Saini and Aspinall 1982; Matsui et al. 2000), production of fewer pollen grains (Prasad et al. 2006a,b), pollen sterility (Saini and Aspinall 1982; Saini et al. 1984; Sakata et al. 2000; Prasad et al. 2006a), and reduced in vivo pollen germination (Jagadish et al. 2010). Yield components that are negatively affected by HS after fertilization include grain numbers and weight (Prasad et al. 2006a,b; Farooq et al. 2011).

Heat stress may adversely affect vital physiological processes, such as photosynthesis, respiration, water relations, and membrane stability, and also modulate levels of hormones and primary and secondary metabolites. The photochemical modifications in the carbon flux of the chloroplast stroma and those of the thylakoid membrane system are considered the primary sites of heat injury (Wise et al. 2004), as photosynthesis and the enzymes of the Calvin–Benson cycle, including ribulose 1,5-bisphosphate carboxylase (Rubisco) and Rubisco activase, are very sensitive to increased temperature and are severely inhibited even at low levels of HS. A specific detrimental effect of high stress occurs in the thylakoid membranes, which typically show swelling, increased leakiness, physical separation of the chlorophyll light harvesting complex II from the PSII core complex, and disruption of PSII-mediated electron transfer (Ristic et al. 2009). According to Kaushal et al. (2011), oxidative damage, measured as lipid peroxidation and hydrogen peroxide concentration, increased with HS (45/40 °C); pertinently, lipid peroxidation was found to increase to a greater extent, indicating membrane injury. It has been suggested that HS might uncouple enzymes and metabolic pathways, causing the accumulation of unwanted and harmful ROS: most commonly singlet oxygen ($^1O_2$), superoxide radical (O_2^{-}), hydrogen peroxide (H_2O_2), and hydroxyl radical ($OH^{.}$), which are responsible for oxidative stress. ROS have potentially toxic effects, as they can induce protein oxidation, DNA damage, lipid peroxidation of membranes, and destruction of pigments (Apel and Hirt 2004; Xu et al. 2006; Hasanuzzaman et al. 2012a); however, plants have evolved a variety of responses to extreme temperatures that help to minimize damages and provide cellular homeostasis (Kotak et al. 2007). Oxidative damage due to ROS production during long-term exposure to high temperature led to changes in malondialdehyde (MDA) content and O_2^- production, which were observed at two growth stages: the early (4 days old) and late stages (7 days old) of wheat (*Triticum aestivum*) seedling development (Savicka and Skute 2010; Cossani and Reynolds 2012). In another study on wheat, increased MDA concentration was observed in the first leaf of wheat seedlings during HS conditions, which is due to the increased production of superoxide radical (O_2^-) (Bohnert et al. 2006). According to Kumar et al. (2012c), a high temperature of 40/35 °C (day/night

temperature) resulted in a 1.8-fold and 1.2–1.3-fold increase of MDA content in rice and maize genotypes, respectively, over the control treatment. A direct link exists between ROS scavenging and plant stress tolerance under temperature-stress conditions, which is often related to enhanced activities of antioxidative defense enzymes that confer stress tolerance to HS.

HS modifies the activities of carbon metabolism enzymes, starch accumulation, and sucrose synthesis by down-regulating specific genes in carbohydrate metabolism. Many plant species accumulate other osmolytes as well, such as sugar alcohols (polyols) or tertiary and quaternary ammonium compounds (Sairam and Tyagi 2004). Osmolyte production under HS is thought to increase protein stability and stabilize the structure of the membrane bilayer (Sung et al. 2003; Mirzaei et al. 2012). Secondary metabolites such as phenolics, including flavonoids, anthocyanins, and plant steroids, are also significantly involved in plant responses under HS and usually play roles in abiotic stress responses generally associated with tolerance to heat.

Mechanism of Heat Stress Tolerance

Heat tolerance is the ability of the plant to grow and produce an economic yield under HS conditions. Plants have evolved various mechanisms for flourishing under higher prevailing temperatures. They include short-term avoidance/acclimation mechanisms and long-term evolutionary adaptations such as changing leaf orientation, transpirational cooling, and alteration of membrane lipid composition. The closure of stomata and reduced water loss, increased stomatal and trichomatous densities, and larger xylem vessels are common heat-induced features in plants (Bita and Girates 2013). In many crop plants, early maturation is closely correlated with smaller yield losses under HS, which may be attributed to the engagement of an escape mechanism. Some major tolerance mechanisms, including ion transporters, late embryogenesis abundant (LEA) proteins, osmoprotectants, antioxidant defense, and factors involved in signaling cascades and transcriptional control, are essentially significant in counteracting the stress effects.

Biochemical Mechanism of Heat Stress Tolerance

Plants have evolved a complex antioxidative defense system to counteract the injurious effects of ROS produced due to HS conditions. It includes antioxidant enzymes, such as SOD, CAT, guaiacol peroxidase (GPX), APX, dehydroascorbate reductase (DHAR), GR, and glutathione S-transferase (GST), and nonenzymatic antioxidants, such as flavonoids, anthocyanin, carotenoids, and ascorbic acid (AA) (Suzuki et al. 2012) (Figure 6.1). The enzyme SOD converts O_2^- to H_2O_2, whereas CAT and peroxidases dismutate H_2O_2. Catalase eliminates H_2O_2 by breaking it down to H_2O and O_2, but peroxidases require reducing equivalents to scavenge H_2O_2. GPX requires a phenolic compound, guaiacol, as an electron donor to decompose H_2O_2, while APX uses a reduced form of ascorbate (AsA) to protect cells against the damaging effects of H_2O_2 (Tripathy and Oelmüller 2012). The oxidized form of AsA produced by the action of APX is regenerated via the AsA-glutathione cycle or the Halliwell–Asada pathway involving monodehydroascorbate reductase (MDHAR) and

DHAR, and finally, the oxidized glutathione (GSSG) is reduced by GR using the reducing power of NADPH (Figure 6.1). GSTs are a collection of multifunctional proteins that are found essentially in all organisms. AsA and reduced glutathione (GSH) are potent nonenzymatic antioxidants within the cell. AsA scavenges the most dangerous forms of ROS, that is, $^{\cdot}OH$, O_2^-, and H_2O_2, through the action of APX, while glutathione participates in maintaining the cellular AsA pool in the reduced state through the Halliwell–Asada pathway as well as serving as a major thiol disulfide redox buffer in plants.

The water–water cycle, or the Mehler peroxidase reaction, involves the leakage of electrons from the photosynthetic electron transport chain to oxygen with the generation of superoxide, which is further dismutated by SOD, forming H_2O_2. ROS that escape this cycle undergo detoxification by SOD and the stromal AsA-glutathione cycle. GPX is also involved in H_2O_2 removal. The activities of different antioxidant enzymes are temperature sensitive, and activation occurs at different temperature ranges. Chakraborty and Pradhan (2011) observed that CAT, APX, and SOD showed an initial increase before declining at 50 °C, whereas POX and GR activities declined at all temperatures ranging from 20°C to 50°C. In addition, total antioxidant activity was at a maximum of 35°C–40°C in the tolerant varieties and at 30°C in the susceptible ones. The activity of the enzymes GST, APX, and CAT was more enhanced in the cultivar that showed better tolerance to HS and protection against ROS

production (Suzuki and Mittler 2006; Goyal and Asthir 2010; Ahmad and Prasad 2012). Pearl millet plantlets showed a significant increase in SOD, CAT, and peroxidase activities during HS (Tikhomirova 1985). In a similar fashion, the exposure of a thermo-tolerant (BPR5426) and a thermo-sensitive (NPJ119) Indian mustard (*Brassica juncea*) genotype to high temperature (45°C) revealed higher SOD, CAT, APX, and GR activities in tolerant genotypes (Rani et al. 2013). The protection mechanism against HS in wheat varieties appeared to be correlated with the antioxidant level, though changes in activity were observed for different antioxidant enzymes.

The synchronized action of ascorbate, α-tocopherol, and glutathione results in the detoxification of ROS and limits oxidative stress in plants (Hameed et al. 2012). Ascorbate is distributed in almost all the plant parts; it is synthesized in the mitochondria and transported to other parts of the plant (Foyer 2015). Ascorbate is used as a substrate by APX to reduce H_2O_2 to H_2O in the AsA–glutathione cycle and generate monodehydroascorbate, which further dissociates to AA and dehydroascorbate. α-Tocopherol, along with other antioxidants, scavenges lipid peroxyl radical. It acts as a lipophilic antioxidant and interacts with polyunsaturated acyl groups of lipids to reduce the deleterious effects of ROS (Tripathy and Oelmüller 2012). α-Tocopherol stabilizes membranes and also acts as a substance that modulates signal transduction. Glutathione is a nonprotein thiol that plays a key role in H_2O_2 detoxification. It has been reported that the

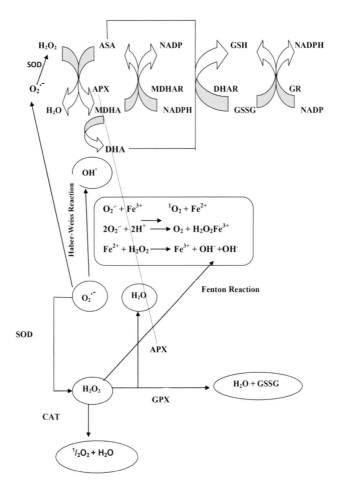

FIGURE 6.1 ROS production and its scavenging.

conversion ratio of reduced GSH to its oxidized form GSSG during the detoxification of H_2O_2 is the indicator of cellular redox balance via the glutathione peroxidase cycle (Goyal and Asthir 2010). These events were widely reported in plants under various abiotic stresses. Glutathione and ascorbate are now considered as important components of redox signaling in plants (Suzuki et al. 2012).

The accumulation of osmolytes such as proline, glycinebetaine, and trehalose is a well-known adaptive mechanism in plants against HS. In recent decades, the exogenous application of protectants such as osmoprotectants, phytohormones, signaling molecules, trace elements, and so on has shown a beneficial effect on the tolerance of plants grown under HS, as these protectants have growth-promoting and antioxidant capacity (Asthir 2015a,b). Supplementation with proline and glycinebetaine considerably reduced H_2O_2 production, improved the accumulation of soluble sugars, and protected the developing tissues from HS effects. Exogenous proline and glycinebetaine application also improved the K^+ and Ca^{2+} contents and increased the concentrations of free proline, glycinebetaine, and soluble sugars, which rendered the buds more tolerant to HS (Hasanuzzaman et al. 2014). Identically, exogenous applications of several phytohormones were found to be effective in mitigating HS in plants. Chhabra et al. (2009) studied the phytohormone-induced amelioration of heat tolerance stress in *B. juncea* and found that soaking seeds in 100 µM indole acetic acid (IAA), 100 µM GA, 50 and 100 µM kinetin, and 0.5 and 1 µM ABA was effective in mitigating the effect of HS ($47 \pm 0.5°C$). The significant observation was that both growth-promoting and growth-retarding hormones were effective in mitigating HS effects. Chen et al. (2006) treated grape seedlings with 50 µM JA solution and observed that JA could reduce the effect of stress under HS (42 °C) by upregulating antioxidant enzyme (SOD, CAT, and POD) activity. Kumar et al. (2010) investigated the effect of different concentrations of 24-epibrassinolide (24-EBL) on the growth and antioxidant enzymes of mustard (*B. juncea*) seedlings. Polyamine provided plants with protection from HS by maintaining the thermostability of thylakoid membranes and increasing photosynthetic efficiency, which rendered the plants more tolerant to heat-induced oxidative stress (Pathak et al. 2014).

Molecular Aspects of Heat Stress Tolerance

Plants are capable of adapting to HS by reprogramming their transcriptome, proteome, and metabolome and even by activating cell death mechanisms, leading to entire plant death (Qi et al. 2011; Sánchez-Rodríguez et al. 2011). The most important characteristic of thermotolerance is the massive production of HSPs (Asthir 2015a,b); however, as heat tolerance is a multigenic character, numerous biochemical and metabolic traits are also involved in the development and maintenance of thermotolerance: antioxidant activity, membrane lipid unsaturation, gene expression and translation, protein stability, and accumulation of compatible solutes (Figure 6.2).

Different types of HSP include Hsp100/ClpB, Hsp90/HtpG, Hsp70/DnaK, Hsp60/GroEL, and small HSP (sHSP) proteins (Timperio et al. 2008). HSPs act as molecular chaperones and stabilize several cellular proteins under temperature stress, which has been reported to be a highly conserved response

(Wang et al. 2004). Due to their thermotolerant nature, the expression of HSPs can be induced by heat treatment in the presence of conserved heat shock elements (HSEs) in the promoter region of heat shock genes, which trigger transcription in response to heat (Driedonks et al. 2015). HSEs consist of alternating units of pentameric nucleotides (50-nGAAn-30) that serve as the binding site for Hsf. Efficient Hsf binding requires at least three alternating units (50-nGAAnnTTCnnGAAn-30). *Arabidopsis* has 21 HSF genes belonging to three major classes—HsfA, HsfB, and HsfC—based on structural differences (Evrard et al. 2013). HsfAs appear to be the major factor(s) responsible for heat-induced activation of heat shock genes. HsfBs apparently lack the heat-inducible transactivation function, despite having normal DNA binding function, and might act as co-activators of transcription with HsfAs. In tomato, HsfA1a, HsfA2, and HsfB1 seem to form a regulatory network that is responsible for the expression of HS-responsive genes (Chan-Schaminet et al. 2009). HsfA1a is constitutively expressed and regulates the HS-induced expression of HsfA2 and HsfB1. HsfA1a was therefore defined as a master regulator of HSR in tomato, whereas HsfA2 is the major HSF in thermotolerant cells. HsfA1a also functions as a nuclear retention factor and co-activator of HsfA2 by forming an HsfA1a–HsfA2 heterooligomeric complex. In *Arabidopsis*, analysis of HsfA1a, HsfA1b, and HsfA2 knockout mutants suggests that HsfA1a and HsfA1b are important for the initial phase of HS-responsive gene expression, and that HsfA2 controls expression under prolonged HS and recovery conditions (Liu and Charng 2013). Lee et al. (1995) successfully altered the expression level of HSPs by making a change in the transcription factor (*AtHSF1*) responsible for HSPs in *Arabidopsis* plants and able to produce transgenic HS-tolerant *Arabidopsis*. Malik et al. (1999) reported an increase in thermotolerance in transgenic carrot cell lines and other plants by constitutive expression of carrot Hsp17.7gene–driven *CaMV35S* promoter. It has already been reported that mitochondrial small HSP (*MT-sHSP*) in tomato has a molecular chaperone function in vitro (Sanmiya et al. 2004).

LEA proteins, ubiquitin, and dehydrins have also been found to play important roles in protection from HS (Tolleter et al. 2010; Hincha and Thalhammer 2012). LEA proteins prevent aggregation and protect citrate synthase (involved in ATP production) from desiccating conditions during HS (Battaglia and Covarrubias 2013). Wang et al. (2008) investigated an LEA gene from Tamarix, expressed in yeast, which showed enhanced tolerance to high temperature. Similarly, ubiquitin and conjugated-ubiquitin synthesis during the first 30 min of exposure emerged as an important mechanism of heat tolerance in mesquite and soybean experiencing HS (Kaushal 2016). In *Arabidopsis*, the dehydration-responsive element binding protein (DREB) of the AP2/ERF family, DREB 2A, has a negative regulatory domain in the central region, which plays an important role in HS. The overexpression of DREB2A CA in transgenic *Arabidopsis* enhanced tolerance to HS (Sato et al. 2014). In rice, OsDREB 2B has been found to encode DREB2-type transcription factors to improve tolerance to HS (Matsukura et al. 2010). An in vitro study by Brini et al. (2010) indicated that wheat dehydrin (DHN-5) protein preserved the enzyme activities of glucosidase and glucose oxidase/peroxidase from adverse effects induced by heating in vitro. The OsLEA3-1 gene from *Oryza sativa* has

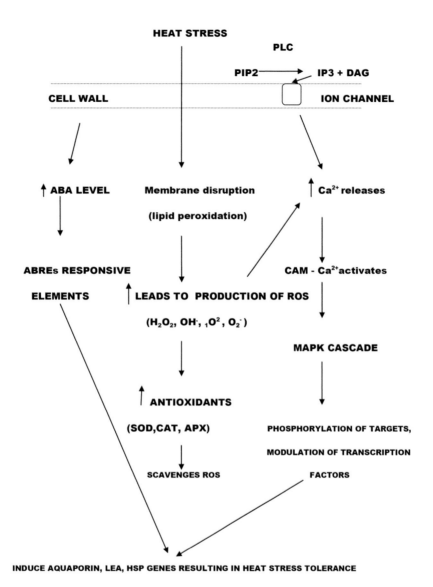

FIGURE 6.2   Mechanism of heat stress tolerance.

been found to play an important role in conferring tolerance to a wide range of environmental stresses, such as salt, heavy metal ion, hyperosmotic, heat, and UV radiation, which suggests that the OsLEA3-1 gene may contribute to the ability of plants to adapt to stressful environments (Duan and Cai 2012).

Strategies for Crop Improvement for Heat Stress Tolerance

To date, conventional breeding schemes have been extensively deployed for uncovering the HS-relevant gene(s) and their inheritance patterns, thus illuminating the causal molecular mechanism (Farooq et al. 2011). To this end, advances in DNA marker discovery and genotyping assays have permitted the accurate determination of the chromosomal position of the quantitative trait loci (QTLs) responsible for HS in different crops (Bonneau et al. 2013). The application of QTL mapping has elucidated the genetic relationships among tolerance to various stresses. Molecular marker technology has identified and characterized QTL with significant effects on stress protection during different

stages of plant development to find genetic relationships among different stresses (Fooland 2005). In *Arabidopsis*, four genomic loci (QTLs) determining its capacity to acquire thermotolerance were identified. The use of restriction fragment length polymorphism revealed mapping of 11 QTLs for pollen germination and pollen tube growth under heat stress in maize. QTL mapping studies for heat tolerance have been conducted on various rice populations at flowering stages. However, confirmation and fine mapping of the identified QTLs for heat tolerance have not been reported yet (Ye et al. 2012). Multiple loci for heat tolerance have been identified in wheat (Paliwal et al. 2012) and maize (Bai 2011). In rice, a marker-based approach of advanced backcross was used to develop introgression lines in the background of "Teqing" for the detection of the heat-sensitive line "YIL106" (Lei et al. 2013). Similarly, heat-tolerant (XN0437T) and heat-sensitive (XN0437S) introgression lines were also recovered from another backcross inbred population derived from the cross: Xieqingzao B 9 N22 and 9 Xieqingzao B (Jianglin et al. 2011). Near isogenic lines (NILs) created by introducing the qWB6-allele from "Hana-echizen" into the background of

"Niigata-wase" showed considerable reduction in the incidence of heat-induced injuries such as white-back kernels (Kobayashi et al. 2013).

Microarray-based expression analysis of rice pollens exposed to high temperature revealed the down-regulation of some important tapetum-specific genes under HS; however, other genes, such as Osc6, OsRAFTIN, and TDR genes, remained unaffected during high temperature (Endo et al. 2009). In addition, by combining microarray and complementary DNA amplified fragment length polymorphism (cDNA-AFLP) techniques, Bita et al. (2011) demonstrated the transcriptomic response of meiotic anthers in heat-tolerant as well as heat-sensitive tomato genotypes. The application of Affymetrix 22K Barley 1 Gene Chip microarray illustrated the expression patterns of 958 induced genes and 1122 repressed genes during caryopsis development in barley under early high-temperature stress (Mangelsen et al. 2011). The involvement of different HSPs with ROS, hormones, and sugars was tested in tomato by Affymetrix tomato genome array and cDNA-AFLP-based transcriptome profiling (Frank et al. 2009). Likewise, genome-wide expression patterns of genes were analyzed in sensitive and tolerant wheat lines using a GeneChip wheat genome array, thereby enabling access to the HS-responsive genes (Qin et al. 2008). A similar transcriptome analysis in potato at high soil temperature uncovered a suite of stress-related genes that encode for HSPs in periderm (Ginzberg et al. 2011).

Conclusion

Extreme temperatures, resulting from current climate change, are considered to be major abiotic stresses for crop plants. During temperature stresses, the overproduction of ROS can be a major risk factor to plant cells and may also enhance the expression of ROS-detoxifying and -scavenging enzymes. Under HS conditions, plants accumulate different metabolites (such as antioxidants, osmoprotectants, heat shock proteins, etc.), and different metabolic pathways and processes are activated. These changes emphasize the importance of physiological and molecular studies to reveal the mechanisms underlying HS responses. Molecular approaches that uncover the response and tolerance mechanisms will pave the way for engineering plants capable of tolerating HS and could be the basis for the development of crop varieties capable of producing economic yields under HS.

REFERENCES

Ahmad, P., Prasad, M.N.V. 2012. *Environmental Adaptations and Stress Tolerance of Plants in the Era of Climate Change.* New York, NY: Springer, pp. 297–324.

Allen, L.H., Baker, J.T., Albrecht, S.L., Boote, K.J., Pan, D., Vu, J.C.V. 1995. Carbon dioxide and temperature effects on rice. In: Peng S, Ingram KT, Neue H-U, Ziska LH, eds. *Climate Change and Rice.* Berlin: Springer, pp. 258–277.

Apel, K., Hirt, H. 2004. Reactive oxygen species: Metabolism, oxidative stress, and signal transduction. *Annu. Rev. Plant Biol.* 55:373–399.

Asthir, B. 2015a. Mechanisms of heat tolerance in crop plants. *Biol. Plant.* 59:620–628.

Asthir, B. 2015b. Protective mechanisms of heat tolerance in crop plants. *J. Plant Interact.* 10:202–210.

Bai, J. 2011. Genetic variation of heat tolerance and correlation with other agronomic traits in a maize (*Zea mays* L.) recombinant inbred line population. Available from: http://hdl.handle.net/2346/13572.

Battaglia, M, Covarrubias, A.A. 2013. Late embryogenesis abundant (LEA) proteins in legumes. *Front. Plant Sci.* 4:190.

Bhattacharjee, S. 2012. The language of reactive oxygen species signalling in plants. *J. Bot.* 22:985298.

Bita, C.E, Gerats, T. 2013. Plant tolerance to high temperature in a changing environment: Scientific fundamentals and production of heat stress-tolerant crops. *Front. Plant Sci.* 4:273.

Bita, C.E., Zenoni, S., Vriezen, W.H., Mariani, C., Pezzotti, M., Gerats, T. 2011. Temperature stress differentially modulates transcription in meiotic anthers of heat-tolerant and heat-sensitive tomato plants. *BMC Genom.* 12:384.

Bohnert, H.J., Gong, Q., Li, P., Ma, S. 2006. Unravelling abiotic stress tolerance mechanisms—getting genomics going. *Curr. Opin. Plant Biol.* 9:180–188.

Bokszczanin, K.L, Fragkostefanakis, S. 2013. Perspectives on deciphering mechanisms underlying plant heat stress response and thermotolerance. *Front. Plant Sci.* 4:315.

Bonneau, J., Taylor, J., Parent, B. 2013. Multi-environment analysis and improved mapping of a yield-related QTL on chromosome 3B of wheat. *Theor. Appl. Genet.* 126:747–761.

Brini, F., Saibi, W., Amara, I., Gargouri, A., Masmoudi, K., Hanin, M. 2010. The wheat dehydrin DHN-5 exerts a heat-protective effect on beta-glucosidase and glucose oxidase activities. *Biosci. Biotech. Biochem.* 74:1050–1054.

Chakraborty, U., Pradhan, D. 2011. High temperature-induced oxidative stress in *Lens culinaris*, role of antioxidants and amelioration of stress by chemical pre-treatments. *J. Plant Interact.* 6:43–52.

Chan-Schaminet, K.Y., Baniwal, S.K., Bublak, D., Nover, L., Scharf, K.D. 2009. Specific interaction between tomato HsfA1 and HsfA2 creates hetero-oligomeric superactivator complexes for synergistic activation of heat stress gene expression. *J. Biol. Chem.* 284:20848–20857.

Chen, P., Yu, S., Zhan, Y., Kang, X.L. 2006. Effects of jasmonate acid on thermotolerance of grape seedlings. *J. Shihezi Univ. Nat. Sci.* 24:87–91.

Chhabra, M.L., Dhawan, A., Sangwan, N., Dhawan, K., Singh, D. 2009. Phytohormones induced amelioration of high temperature stress in *Brassica juncea* (L.) Czern & Coss. Proceedings of 16th Australian Research Assembly on Brassicas, September 10–14, Ballarat, Australia.

Cossani, C.M., Reynolds, M.P. 2012. Physiological traits for improving heat tolerance in wheat. *Plant Physiol.* 160:1710–1718.

Das, K., Roychoudhury, A. 2014. Reactive oxygen species (ROS) and response of antioxidants as ROS-scavengers during environmental stress in plants. *Front. Environ. Sci.* 2:10.

Dawson, I.A., Wardlaw, I.F. 1989. The tolerance of wheat to high temperatures during reproductive growth III. Booting to anthesis. *Aust. J. Agric. Res.* 40:965–980.

Driedonks, N., Xu, J., Peters, J.L., Park, S., Rieu, I. 2015. Multi-level interactions between heat shock factors, heat shock proteins, and the redox system regulate acclimation to heat. *Front. Plant Sci.* 6:999.

Duan, J., Cai, W. 2012. *OsLEA3–2*, an abiotic stress induced gene of rice plays a key role in salt and drought tolerance. *PLoS One* 7:e45117.

Endo, M., Tsuchiya, T., Hamada, K., Kawamura, S., Yano, K., Ohshima, M., Higashitani, A., Watanabe, M., Kawagishi-Kobayashi, M. 2009. High temperatures cause male sterility in rice plants with transcriptional alterations during pollen development. *Plant Cell Physiol.* 50:1911–1922.

Evrard, A., Kumar, M., Lecourieux, D., Lucks, J., von Koskull-Döring, P., Hirt, H. 2013. Regulation of the heat stress response in *Arabidopsis* by MPK6-targeted phosphorylation of the heat stress factor HsfA2. *Peer J.* 1:e59.

Farooq, M., Bramley, H., Palta, J.A., Siddique, K.H.M. 2011. Heat stress in wheat during reproductive and grain-filling phases. *Crit. Rev. Plant Sci.* 30:491–507.

Fooland, M.R. 2005. Breeding for abiotic stress tolerances in tomato. In: Ashraf, M., Harris, P.J.C., eds. *Abiotic Stresses: Plant Resistance through Breeding and Molecular Approaches.* New York, NY: The Haworth Press, pp. 613–684.

Foyer, C.H. 2015. Redox homeostasis: Opening up ascorbate transport. *Nat. Plants* 1:1401.

Foyer, C.H., Noctor, G. 2012. Managing the cellular redox hub in photosynthetic organisms. *Plant Cell Environ.* 35:199–201.

Frank, G., Pressman, E., Ophir, R., Althan, L., Shaked, R., Freedman, M., Shen, S., Firon, N. 2009. Transcriptional profiling of maturing tomato (*Solanum lycopersicum* L) microspores reveals the involvement of heat shock proteins, ROS scavengers, hormones, and sugars in the heat stress response. *J. Exp. Bot.* 60:3891–3908.

Ginzberg, I., Barel, G., Ophir, R., Tzin, E., Tanami, Z., Muddarangappa, T., de Jong, W., Fogelman, E. 2011. Transcriptomic profiling of heatstress response in potato periderm. *J. Exp. Bot.* 60:4411–4421.

Goyal, M., Asthir, B. 2010. Polyamine catabolism influences antioxidative defense mechanism in shoots and roots of five wheat genotypes under high temperature stress. *Plant Growth Regul.* 60:13–25.

Hameed, A., Goher, M., Iqbal, N. 2012. Heat stress-induced cell death, changes in antioxidants, lipid peroxidation and protease activity in wheat leaves. *J. Plant Growth Regul.* 31:283–291.

Hancock, R.D., Morris, W.L., Ducreux, L.J., Morris, J.A., Usman, M., Verrall, S.R., Fuller, J., Simpson, C.G., Zhang, R., Hedley, P.E. 2014. Physiological, biochemical and molecular responses of the potato (*Solanum tuberosum* L.) plant to moderately elevated temperature. *Plant Cell Environ.* 37:439–450.

Hartl, F.U., Bracher, A., Hayer-Hartl, M. 2011. Molecular chaperones in protein folding and proteostasis. *Nature* 475:324–332.

Hasanuzzaman, M., Alam, M.M., Rahman, A., Hasanuzzaman, M., Nahar, K., Fujita, M. 2014. Exogenous proline and glycine betaine mediated upregulation of antioxidant defense and glyoxalase systems provides better protection against salt-induced oxidative stress in two rice (*Oryza sativa* L.) varieties. *BioMed. Res Int.* 2014:757219.

Hasanuzzaman, M., Hossain, M.A., Teixeira da Silva, J.A., Fujita, M. 2012a. Plant responses and tolerance to abiotic oxidative stress: Antioxidant defense is a key factor. In: Bandi, V., Shanker, A.K., Shanker, C., Mandapaka, M., eds. *Crop Stress and Its Management: Perspectives and Strategies.* Berlin: Springer, pp. 261–316.

Hincha, D.K., Thalhammer, A. 2012. LEA proteins: IDPs with versatile functions in cellular dehydration tolerance. *Biochem. Soc. Trans.* 40:1000–1003.

Intergovernmental Panel on Climate Change (IPCC). 2014. Climate Change 2014: Synthesis Report. Contribution of Working Groups I, II, III to the Fifth Assessment Report of the Intergovernmental Panel on Climate Change.

Jagadish, S.V.K., Muthurajan, R., Oane, R., Wheeler, T.R., Heuer, S., Bennett, J., Craufurd, P.Q. 2010. Physiological and proteomic approaches to address heat tolerance during anthesis in rice (*Oryza sativa* L.). *J. Exp. Bot.* 61:143–156.

Jiang-lin, L., Hong-yu, Z., Xue-lian, S., Ping-an, Z., Ying-jin, H. 2011. Identification for heat tolerance in backcross recombinant lines and screening of backcross introgression lines with heat tolerance at milky stage in rice. *Rice Sci.* 18:279–286.

Kaushal, N., Bhandari, K., Siddique, K.H.M., Nayyar, H. 2016. Food crops face rising temperatures: An overview of responses, adaptive mechanisms, and approaches to improve heat tolerance. *Cogent Food Agric.* 2:1134380.

Kaushal, N., Gupta, K., Bhandhari, K., Kumar, S., Thakur, P., Nayyar, H. 2011. Proline induces heat tolerance in chickpea (*Cicer arietinum* L.) plants by protecting vital enzymes of carbon and antioxidative metabolism. *Physiol. Mol. Biol. Plants* 17:203–213.

Kobayashi, A., Sonoda, J., Sugimoto, K., Kondo, M., Iwasawa, N., Hayashi, T., Tomita, K., Yano, M., Shimizu, T. 2013. Detection and verification of QTLs associated with heat-induced quality decline of rice (*Oryza sativa* L.) using recombinant inbred lines and near-isogenic lines. *Breed. Sci.* 63:339–346.

Kocsy, G., Tari, I., Vankova, R., Zechmanne, B., Gulyas, Z., Poor, P. 2013. Redox control of plant growth and development. *Plant Sci.* 211:77–91.

Kotak, S., Larkindale, J., Lee, U., von Koskull-Doring, P., Vierling, E., Scharf, K.D. 2007. Complexity of the heat stress response in plants. *Curr. Opin. Plant Biol.* 10:310–316.

Kumar, S., Gupta, D., Nayyar, H. 2012c. Comparative response of maize and rice genotypes to heat stress: Status of oxidative stress and antioxidants. *Acta Physiol. Planta.* 34:75–86.

Kumar, M., Sirhindi, G., Bhardwaj, R., Kumar, S., Jain, G. 2010. Effect of exogenous H2O2 on antioxidant enzymes of *Brassica juncea* L. seedlings in relation to 24-epibrassinolide under chilling stress. *Indian J. Biochem. Biophys.* 47:378–382.

Lee, J.H., Hubel, A., Schoffl, F. 1995. Derepression of the activity of genetically engineered heat shock factor causes constitutive synthesis of heat shock proteins and increased thermotolerance in transgenic *Arabidopsis*. *Plant J.* 8:603–612.

Lei, D., Tan, L., Liu, F., Chen, L., Sun, C. 2013. Identification of heat-sensitive QTL derived from common wild rice (*Oryza rufipogon* Griff.). *Plant Sci.* 201–202:121–127.

Liu, H.C., Charng, Y.Y. 2013. Common and distinct functions of *Arabidopsis* class A1 and A2 heat shock factors in diverse abiotic stress responses and development. *Plant Physiol.* 163:276–290.

Malik, M.K., Slovin, J.P., Hwang, C.H., Zimmerman, J.L. 1999. Modified expression of a carrot small heat shock protein gene, Hsp17.7, results in increased or decreased thermotolerance. *Plant J.* 20:89–99.

Mangelsen, E., Kilian, J., Harter, K., Jansson, C., Wanke, D., Sundberg, E. 2011. Transcriptome analysis of high-temperature stress in developing barley caryopses: Early stress responses and effects on storage compound biosynthesis. *Mol. Plant.* 4:97–115.

Matsui, T., Omasa, K., Horie, T. 2000. High temperature at flowering inhibits swelling of pollen grains, a driving force for thecae dehiscence in rice. *Plant Prod. Sci.* 3:430–434.

Matsukura, S., Mizoi, J., Yoshida, T., Todaka, D., Ito, Y., Maruyama, K. 2010. Comprehensive analysis of rice DREB2-type genes that encode transcription factors involved in the expression of abiotic stress-responsive genes. *Mol. Genet. Genom.* 283:185–196.

Mirzaei, M., Pascovici, D., Atwell, B.J., Haynes, P.A. 2012. Differential regulation of aquaporins, small GTPases and V-ATPases proteins in rice leaves subjected to drought stress and recovery. *Proteomics* 12:864–877.

Paliwal, R., Röder, M.S., Kumar, U., Srivastava, J., Joshi, A.K. 2012. QTL mapping of terminal heat tolerance in hexaploid wheat (*T. aestivum* L.). *Theor. Appl. Genet.* 125:561–575.

Pathak, M.R., Teixeira da Silva, J.A., Wani, S.H. 2014. Polyamines in response to abiotic stress tolerance through transgenic approaches. *GM Crops Food* 5:87–96.

Prasad, P.V.V., Boote, K.J., Allen, L.H.J.R. 2006a. Adverse high temperature effects on pollen viability, seed-set, seed yield and harvest index of grain-sorghum [*Sorghum bicolor* (L.) Moench] are more severe at elevated carbon dioxide due to higher tissue temperature. *Agric. For. Meteor.* 139:237–251.

Prasad, P.V.V., Boote, K.J., Allen, L.H.J.R, Sheehy, J.E., Thomas, J.M.G. 2006b. Species, ecotype and cultivar differences in spikelet fertility and harvest index of rice in response to high temperature stress. *Field Crops Res.* 95:398–411.

Qi, Y., Wang, H., Zou, Y., Liu, C., Liu, Y., Wang, Y. 2011. Overexpression of mitochondrial heat shock protein 70 suppresses programmed cell death in rice. *FEBS Lett.* 585:231–239.

Qin, D., Wu, H., Peng, H., Yao, Y., Ni, Z., Li, Z., Zhou, C., Sun, Q. 2008. Heat stress-responsive transcriptome analysis in heat susceptible and tolerant wheat (*Triticum aestivum* L) by using Wheat Genome Array. *BMC Genom.* 9:432.

Ramakrishna, A., Ravishankar, G.A. 2011. Influence of abiotic stress signals on secondary metabolites in plants. *Plant Signal. Behav.* 6:1720–1731.

Ramakrishna, A., Ravishankar, G.A. 2013. Role of plant metabolites in abiotic stress tolerance under changing climatic conditions with special reference to secondary compounds. In: Ramakrishna, A., Ravishankar, G.A., eds. *Climate Change and Plant Abiotic Stress Tolerance*, Wiley-VCH Verlag GmbH & Co. KGaA, pp. 705–726.

Rani, T., Yadav, R.C., Yadav, N.R., Rani, A., Singh, D. 2013. A genetic transformation in oilseed brassicas—A review. *Indian J. Agric. Sci.* 83:367–373.

Ristic, Z., Momèiloviæ, I., Bukovnik, U., Prasad, P., Fu, J., DeRidder, B.P. 2009. Rubisco activase and wheat productivity under heat-stress conditions. *J. Exp. Bot.* 60:4003–4014.

Saini, H.S., Aspinall, D. 1982. Abnormal sporogenesis in wheat (*Triticum aestivum* L.) induced by short periods of high-temperature. *Ann. Bot.* 49:835–846.

Saini, H.S, Sedgley, M., Aspinall, D. 1984. Developmental anatomy in wheat of male-sterility induced by heat-stress, water deficit or abscisic acid. *Aust. J. Plant Physiol.* 11:243–253.

Sairam, R.K., Tyagi, A. 2004. Physiology and molecular biology of salinity stress tolerance in plants. *Curr. Sci.* 86:407–421.

Sakata, T., Takahashi, H., Nishiyama, I., Higashitani, A. 2000. Effects of high temperature on the development of pollen mother cells and microspores in barley *Hordeum vulgare* L. *J. Plant Res.* 113:395–402.

Sánchez-Rodríguez, E., Moreno, D.A., Ferreres, F., Rubio-Wilhelmi, M.M., Ruiz, J.M. 2011. Differential responses of five cherry tomato varieties to water stress: Changes on phenolic metabolites and related enzymes. *Phytochemistry* 72:723–729.

Sanmiya, K., Suzuki, K., Egawa, Y., Shono, M. 2004. Mitochondrial small heat-shock protein enhances thermotolerance in tobacco plants. *FEBS Lett.* 557:265–268.

Sato, H., Mizoi, J., Tanaka, H. 2014. *Arabidopsis* DPB3-1, a DREB2A interactor, specifically enhances heat stress-induced gene expression by forming a heat stress-specific transcriptional complex with NF-Y subunits. *Plant Cell* 26:4954–4973.

Savicka, M., Skute, N. 2010. Effects of high temperature on malondialdehyde content, superoxide production and growth changes in wheat seedlings (*Triticum aestivum* L.). *Ekologija* 56:26–33.

Sung, D.Y., Kaplan, F., Lee, K.J., Guy, C.L. 2003. Acquired tolerance to temperature extremes. *Trends Plant Sci.* 8:179–187.

Suzuki, N., Koussevitzky, S., Mittler, R., Miller, G. 2012. ROS and redox signalling in the response of plants to abiotic stress. *Plant Cell Environ.* 35:259–270.

Suzuki, N., Mittler, R. 2006. Reactive oxygen species and temperature stresses: A delicate balance between signaling and destruction. *Physiol. Plant.* 126:45–51.

Tikhomirova, E.V. 1985. Changes of nitrogen metabolism in millet at elevated temperatures. *Field Crops Res.* 11:259–264.

Timperio, A.M., Egidi, M.G., Zolla, L. 2008. Proteomics applied on plant abiotic stresses: Role of heat shock proteins (HSP). *J. Proteomics* 71:391–411.

Tolleter, D., Hincha, D.K., Macherel, D. 2010. A mitochondrial late embryogenesis abundant protein stabilizes model membranes in the dry state. *Biochim. Biophys. Acta* 1798:1926–1933.

Tripathy, B.C., Oelmüller, R. 2012. Reactive oxygen species generation and signaling in plants. *Plant Signal. Behav.* 7:1621–1633.

Wang, B.F, Wang, Y., Zhang, D. 2008 Verification of the resistance of a LEA gene from *Tamarix* expression in *Saccharomyces* to abiotic stresses. *J. For. Res.* 19:58–62.

Wang, W., Vinocur, B., Shoseyov, O., Altman, A. 2004. Role of plant heat-shock proteins and molecular chaperones in the abiotic stress response. *Trends Plant Sci.* 9:244–252.

Wise, R., Olson, A., Schrader, S., Sharkey, T. 2004. Electron transport is the functional limitation of photosynthesis in field-grown pima cotton plants at high temperature. *Plant Cell Environ.* 27:717–724.

Xu, S., Li, J., Zhang, X., Wei, H., Cui, L. 2006. Effects of heat acclimation pre-treatment on changes of membrane lipid peroxidation, antioxidant metabolites, and ultrastructure of chloroplasts in two cool-season turf grass species under heat stress. *Environ. Exp. Bot.* 56:274–285.

Ye, C., Argayoso, M.A., Redoña, E.D., Sierra, S.N, Laza, M.A., Dilla, C.J. 2012. Mapping QTL for heat tolerance at flowering stage in rice using SNP markers. *Plant Breed.* 131:33–41.

7

Profiles of Antioxidant Isoenzymes and Physiological Behavior of Tomato Exposed to NaCl Stress and Treated with Salicylic Acid

Salma Wasti, Nizar Dhaoui, Ibtissem Medyouni, Hajer Mimouni, Hela Ben Ahmed, and Abdellah Chalh

CONTENTS

Abbreviations

$^1O_2$	Singlet oxygen
CAT	Catalases
DNA	Deoxyribonucleic acid
DW	Dry weight
FW	Fresh weight
GSH	Glutathione
H_2O_2	Hydrogen peroxide
O_2^-	Superoxide
OH·	Hydroxyl radical
PAGE	Polyacrylamide gel electrophoresis
POX	Guaiacol peroxidase
ROS	Reactive oxygen species
SA	Salicylic acid
SOD	Superoxide dismutase

Introduction

Physiological Approaches to Enhance Plant Salinity Tolerance

Enhancing stress tolerance in plants has major implications for agriculture and horticulture. Different methods have been developed. Serotonin, melatonin, and salicylic acid possess antioxidative and growth-inducing properties, thus proving beneficial for biotic and abiotic stress (Kaur et al., 2015; Wasti et al., 2017). These compounds have been found to generate a wide range of metabolic and physiological responses in plants by affecting their growth and development. For example, they improve plant growth, crop yields, photosynthesis, and ion uptake and transport (Wei et al., 2015; Mimouni et al., 2016). In this chapter, we are interested in one of these compounds: salicylic acid, a phenolic compound naturally occurring in plants in very low amounts. It may be considered as a potential growth regulator to improve plant salinity stress resistance.

Plant Responses to Salt Stress and Salicylic Acid

Salinization in arid and semi-arid ecosystems results from natural evaporation (Munns et al., 2006) and from irregular and insufficient rainfall (Mezni et al., 2002). Salinization also comes from irrigation, which is often poorly controlled (Bennaceur et al., 2001). These issues require reflection on the strategies to be undertaken to understand the mechanisms that enable plants to adapt to new environmental conditions and to maintain their growth and productivity (Trinchant et al., 2004). Depending on the degree of salinity in the environment, glycophytes in particular are exposed to changes in their morphophysiological (Bennaceur et al., 2001), biochemical (Grennan, 2006), and mineral behavior (Martinez et al., 2005). In Tunisia, nearly 100,000 ha of irrigated perimeters are deeply affected by the importance of salinization. Seventy-five percent of the soils are in a range from moderately to highly sensitive to salinization (Hachicha, 2007). The soils affected by salts cover about 1.5 million hectares, about 10% of the surface of the country.

They are encountered throughout the territory, but it is especially in the center and the south that the aridity of the climate is causing their expansion (Hachicha, 2007). Saline ground is constantly growing, so the development of methods to improve crop tolerance to salinity is essential and is receiving considerable attention.

Tomato (*Solanum lycopersicum*) has an important place in the human diet, since it is consumed worldwide throughout the year. It ranks as the world's leading producer of fruits, with a production of about 127 million tons in 2007 (Food and Agriculture Organization, 2008). It is considered a healthy food because it is low in calories and rich in minerals and contains many antioxidants. For all these reasons, it is targeted by many improvement programs. Tomato salt tolerance could be improved not only by genetic selection but also through the use of adapted physiological tools. The small phenolic molecule salicylic acid (SA) plays a key role in plant defense. It is required for the recognition of pathogen-derived components and subsequent establishment of local resistance in the infected region as well as systemic resistance at the whole plant level (Hua, 2009), so it is known as a signal molecule that induces responses to environmental stresses (Szalai et al., 2013). Its effectiveness has been tested not only for saline stress (Wasti et al., 2012; Mimouni et al., 2016) but also for other abiotic stresses such as heavy metals (Zhou et al., 2009), drought (Lakzayi, 2014), oxidative stress (Krantev et al., 2008), heat (Chakraborty and Tongden, 2005), and cold stress (Tasgin et al., 2003). The exogenous application of SA improves the growth of several species subjected to different types of abiotic stresses such as saline stress. According to Sakhabutdinova et al. (2003), the increase in growth probably results from the effect of SA on root apical meristems. Indeed, SA stimulates cell division at the level of these meristems, which results in the growth of the plant. Also, the effect of SA on mineral nutrition has been studied in several cases of abiotic stress. In tomato, the application of a dose of 0.01 mM improves K^+ and Ca^{2+} absorption in plants under saline conditions (Wasti et al., 2012; Manaa et al., 2014).

Salinity induces severe plant metabolic disturbances, as it generates reactive oxygen species (ROS), which disturb the cellular redox system in favor of oxidized forms, thereby creating oxidative stress, which can damage DNA, inactivate enzymes, and cause lipid peroxidation (Molassiotis et al., 2006). Therefore, their levels must be carefully monitored and controlled in cells. Plants possess antioxidant defense systems, both enzymatic and non-enzymatic, which maintain the ROS balance within the cell. They use a diverse array of enzymes, such as superoxide dismutases (SOD), catalases (CAT), and peroxidases, as well as low–molecular mass antioxidants such as ascorbate and reduced glutathione (GSH), to scavenge ROS (Fahad and Bano, 2012). Khan et al. (2014) reported that SA application alleviated the adverse effects of salt stress in mungbean by improving plant growth and also enhancing the antioxidant system. However, the improvement of salt tolerance by exogenous SA depends on genotype and the concentration of SA used. This chapter focuses on the influence of exogenous SA on the behavior of a cultivated glycophyte, *Solanum lycopersicum* var. Golden Sunrise, under salt stress and studies the effect of SA on the antioxidant enzyme profiles.

Tomato Plants and Culture

Tomato seeds were surface-sterilized with 20% sodium hypochlorite solution for 10 min and rinsed with demineralized sterile water. Seeds were then germinated in the dark at 25 °C on filter paper wetted with distilled water. After 10 days, the germinated seedlings were transferred to nutrient solution. This solution was composed of macroelements (mM): K^+ 3.0; Ca^{2+} 3.5; Mg^{2+} 1.0; NO_3^- 8.0; SO_4^{2-} 1.0; $PO_4H_2^-$ 1.5, and microelements (ppm): Fe 1.4; Mn 0.25; B 0.16; Cu 0.03; Zn 0.03; and Mo 0.01. The experiment was conducted in an air-conditioned room, under artificial light (150 µmol m^{-2} s^{-1}; 16 h photoperiod), at 25°C day/20°C night, and with air humidity 60%–80% day/night.

Ten days after being transferred, the plants were at the third leaf stage. Treatments were designed as follows: 1) Control (only nutrient solution); 2) NaCl 100 mM; 3) NaCl 100 mM plus 0.01 mM SA. The pH of the nutrient solution was adjusted with KOH (1.0 N). After 21 days of treatment, for the determination of the dry weight (DW), plants were then separated into roots, stems, and leaves, and oven-dried at −80°C (for 3 days). Besides, fresh shoot and root samples from each plant were immediately frozen in liquid nitrogen and stored at −80°C until the biochemical analysis was performed.

Growth

For plant growth and ion analysis, 12 independent dry matter measurements and ion analyses were performed on separated leaves, stems, and roots.

Ion Analysis

For the measurement of cations, plant material was dried at 80°C and digested with nitric acid (1% [v/v] HNO_3) according to the method of Wolf (1982). K^+ and Na^+ were analyzed by flame emission using an Eppendorf spectrophotometer. Cl^- was quantified by a colorimetric method using a HaakeBuchler digital chloridometer (Buchler instruments Inc., New Jersey, United States).

Proline Dosage

The proline content was determined using the method of Bates et al. (1973). Proline was extracted from organ samples of 200 mg fresh weight (FW) with 10 mL of 3% sulfosalicylic acid at 70°C for 30 min. Extracts were filtered through filter paper. After the addition of acid ninhydrin and glacial acetic acid to the extracts, the mixture was heated at 90 °C for 1 h in a water bath. The reaction was then stopped by using an ice bath. The mixture was extracted with toluene, and the absorbance of the fraction with toluene aspired from the liquid phase was spectrophotometrically determined at 520 nm. Proline concentration was determined using a calibration curve as µmol proline g^{-1} FW.

Protein Extraction and Quantification

An aliquot of frozen leaf and root material was ground to a fine powder with liquid nitrogen and extracted (100 mg FW, 300 mL) at 4 °C in 100 mM Tris-HCl buffer (pH 8.0) containing 10 mM EDTA, 50 mM KCl, 20 mM MgCl$_2$, 0.5 mM phenylmethylsulfonyl fluoride (PMSF), 1 mM dichlorodiphenyltrichloroethane (DDT), 0.1% (v/v) Triton X-100, and 10% (w/w) polyvinylpyrrolidone (PVP). The homogenate was centrifuged at 13,000×g for 40 min at 4 °C, and the supernatant was clarified by filtration. The filtrate was used for the determination of antioxidative enzyme activities. Soluble protein concentration in enzyme extract was estimated according to Bradford (1976), using Sigma reagent (B6916) and bovine serum albumin as standard. Three replicates per treatment were used.

Native Gel Electrophoresis and Enzyme Activity Staining

Polyacrylamide gel electrophoresis (PAGE) was performed to separate the different enzyme isoforms using a discontinuous gel system under nondenaturing conditions, essentially as described by Tewari et al. (2008).

Samples of tomato leaf or root extracts were separated by gel electrophoresis in 10% (w/v) polyacrylamide slab gel at pH 8.9 under native conditions, according to Davis (1964). Staining for SOD activity was carried out as described by Beauchamp and Fridovich (1971). The gel was first soaked in 50 mM sodium phosphate (pH 7.5) containing 4.8 mM 3-(4,5-dimethylthiazol-2-yl)-2,5-diphenyltetrazolium bromide (MTT) in darkness for 20 min, followed by soaking in 50 mM sodium phosphate (pH 7.5) containing 0.4% (v/v) N,N,N0,N0-tetramethylethylenediamine (TEMED) and 26 mM riboflavin, and subsequently illuminated for 10 min. The three types of SOD, Fe-SOD, Mn-SOD, and Cu-Zn SOD, were identified

using inhibitors. Mn-SOD was visualized by its insensitivity to 5 mM H$_2$O$_2$ and 2 mM potassium cyanide (KCN), while Cu-Zn SOD was sensitive to 2 mM KCN. Fe-SOD was inhibited by 5 mM H$_2$O$_2$ (Navari-Izzo et al., 1998).

Peroxidase (POD) isoforms were visualized on gels according to Vallejos (1983). POD was localized by incubating the gel for 20 min in a reaction mixture containing 20 ml 0.1 M phosphate buffer (pH 7.0), 4 mL 0.01% H$_2$O$_2$, and 4 mL 0.5% guaiacol or 3,3′-diaminobenzidine. The gel was then rinsed with distilled water and scanned immediately.

Physiological Responses of Tomato to Salt and Salicylic Acid

Agricultural production throughout the world can be limited by a diversity of abiotic stresses, especially salinity (Munns and Tester, 2008). Salinity affects plant growth, crop development, and productivity (Khan et al., 2014). It also induces the production of ROS (Khan et al., 2014) and triggers the inhibition of photosynthesis and mineral absorption by disrupting nutritional (Mimouni et al., 2016) and metabolic behavior (Munns and Tester, 2008). As shown in Figure 7.1a, the presence of NaCl 100 mM in the growth medium reduced DW production of the whole plant by 42% compared with control plant. The magnitude of salt response, however, varied according to the plant organ considered; the roots are much less sensitive to salt than the shoots. The decrease in plant DW was 53%, 41%, and 27%, respectively, in stems, leaves, and roots (Figure 7.1b–d).

The immediate response of a plant exposed to biotic or abiotic stress is the activation of molecules inducing a set of transduction

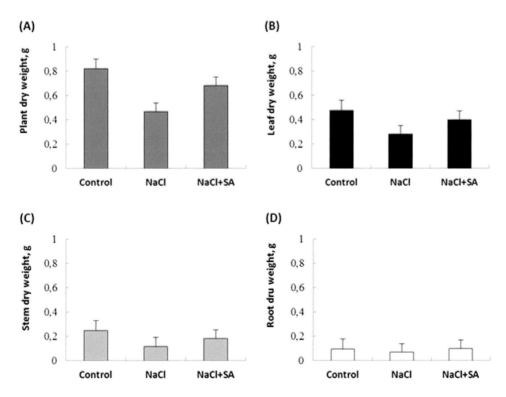

FIGURE 7.1 Dry weight of (A) whole plant, (B) leaves, (C) stems and (D) roots of tomato plants grown for 21 days in nutrient solution with 100 mM NaCl±0.01 mM salicylic acid. Data are means of 12 replicates±standard error (P=0.05).

pathways. Several signaling molecules have been identified in plants, such as jasmonic acid, ethylene, and SA. The role of SA as a defense signal has been established in plants (Ganesan and Thomas, 2001). Exogenous SA has a protective effect in plants subjected to different abiotic stresses (Wasti et al., 2012; Mimouni et al., 2016). In this chapter, the application of 0.01 mM SA attenuated the depressive effect of the salt. NaCl+SA significantly improved plant DW in Golden Sunrise by 42%, 53%, and 44% in leaves, stems, and roots, respectively, as compared with saline conditions (NaCl). These results are in agreement with those of (Wasti et al., 2012; Manaa et al., 2014; Mimouni et al., 2016).

The presence of salt in root medium causes an alteration in mineral nutrition, limiting the absorption and transport of essential cations such as K^+ and Ca^{2+} (Sairam et al., 2002; Wasti et al., 2012). Under salt stress conditions, we observed an accumulation of Na^+ and Cl^- in the different organs of tomato (Figure 7.2b,c). Wasti et al. (2017) also found that NaCl salinity increased Na^+ content in plant tissue of tomato. Simultaneously, a decrease in the K^+ concentrations in leaves, stems, and roots was observed (Figure 7.2a). The extent of the response depends, however, on the plant organs considered. For example, K^+ content significantly decreased under salt treatment (NaCl); the reduction was about 52%, 46%, and 41%, respectively, in leaves, stems, and roots

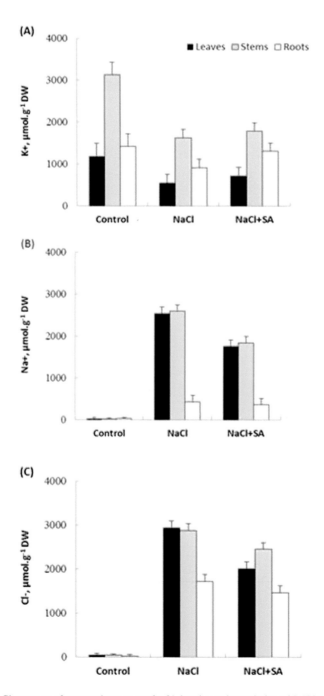

FIGURE 7.2 (A) K^+, (B) Na^+, and (C) Cl^- contents of tomato plants grown for 21 days in nutrient solution with 100 mM NaCl ± 0.01 mM salicylic acid. Data are means of 12 replicates ± standard error ($P = 0.05$).

compared with control. The application of SA decreased the Na^+ and Cl^- content, whereas the K^+ content showed an increase over that of saline conditions (Figure 7.2). Thus, SSA increases tolerance to NaCl, limiting the transport of Na^+ to the leaves and improving selectivity in favor of K^+. Gunes et al. (2007) also find that pretreatment of barley by 1 mM of SA improves K/Na selectivity in the presence of 200 mM NaCl. Our data are also in agreement with the results of Fahad and Bano (2012), who demonstrated that SA significantly decreased the Na^+, Ni^{+3}, Pb^{+4}, Zn^{+2}, and Na^+/K^+ content of roots, while the Co^{+3}, Mn^{+2}, Cu^{+3}, Fe^{+2}, K^+, and Mg^{+2} content increased under salinity stress.

The synthesis of proline and its accumulation in the cytosol is one of the most important adaptations of plants to salinity and water deficiency (Kumar et al., 2000). In addition to its role as an osmolyte for osmotic adjustment, proline contributes to stabilizing cell structures and regulating redox potential under stress conditions (Hsu et al., 2003). In the tomato var. Golden Sunrise, 100 mM NaCl induces a strong increase in proline content in the shoots and roots, about eight times higher than in the control (Figure 7.3). Several studies have demonstrated the accumulation of proline in roots and leaves of tomato in salt conditions (Kafi et al., 2011, Byordi, 2012; El Sayed et al., 2015). The application of SA induced the accumulation of proline in Golden Sunrise. This induction was about 10 and 19 times more than the control in leaves and roots, respectively (Figure 7.3). These results are in agreement with those of Bybordi (2014) on wheat. Indeed, the accumulation of proline in plants exposed to salt stress combined with SA treatment may be a crucial element in plant adaptation to salinity and in the maintenance of the water balance induced by osmotic stress (Li et al., 2014).

Proteins that accumulate in plants under saline conditions may provide a form of nitrogen storage that is reused later and may play a role in osmotic adjustment (Amini and Ehsanpour, (2005). Our data indicated that salt increases protein content in tomato

var. Golden Sunrise (Figure 7.4). The increase was about 37% and 27% in leaves and roots, respectively, compared with plants treated with NaCl. Several proteins induced by salt stress have been identified in tomatoes. Pareek et al. (1997) have suggested that these proteins can be used as important molecular markers for the improvement of salt tolerance. Moreover, these proteins are synthesized according to the needs of the plant, and their accumulation is not generally carried out with the same intensity in different organs. Moreover, our study shows that SA treatment stimulates protein synthesis in tomato, especially in leaves, with a 64% increase compared with control and 20% compared with those treated with NaCl only, but has no effect on the roots (Figure 7.4). Several studies have highlighted the possible roles of SA in the stimulation of protein synthesis. In tomato var. Rio Fuego, SA stimulates the synthesis of proteins that provide hormonal balance (Tari et al., 2002). In barley, SA pretreatment stimulates the synthesis of proteins that are included in the response to oxidative stress and decrease lipid peroxidation, thus protecting the cell membrane (El-Tayeb, 2005).

Salt and Salicylic Acid–Induced Changes in Antioxidative Enzymes

Salt stress also generates ROS such as superoxide (O_2^-), hydrogen peroxide (H_2O_2), hydroxyl radical (OH•), and singlet oxygen ($^1O_2$) (Brosché et al., 2010). High levels of these ROS are responsible for lipid peroxidation and can damage proteins, chlorophylls, and nucleic acids (Schutzengel and Poole, 2002). However, nature has equipped all plants with protective antioxidant systems to counter the oxidative damage caused by ROS (Apel and Hirt, 2004). There are several antioxidants found in plants, belonging to one of two categories: enzymatic and non-enzymatic systems (Ndhala et al., 2010). SOD and peroxidase (POX) are among the major antioxidant enzymes involved in scavenging ROS (Askari

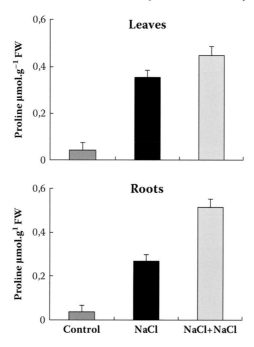

FIGURE 7.3 Proline content of tomato plants grown for 21 days in nutrient solution with 100 mM $NaCl \pm 0.01$ mM salicylic acid. Data are means of six replicates $\pm$ standard error ($P = 0.05$).

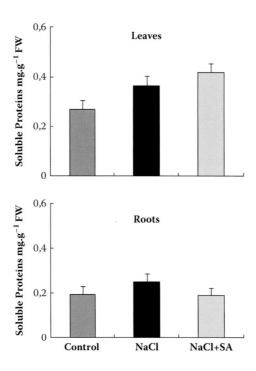

FIGURE 7.4   Soluble protein content of tomato plants grown for 21 days in nutrient solution with 100 mM NaCl ± 0.01 mM salicylic acid. Data are means of six replicates ± standard error (*P* = 0.05).

et al., 2006). In our study, we explored the impact of NaCl on the antioxidant isozyme profiles. Figure 7.5 shows that the gel does not reveal any activity of SOD in control plants. However, salt stimulates the activity of SOD, in the presence or in the absence of SA, in leaves as well as in roots. The gel also revealed a new isoenzyme stimulated by SA in leaves as compared with roots. This isoenzyme is another form of SOD (Figure 7.5). Indeed, there are three types of superoxide dismutase: Mn-SOD, Fe-SOD, and CuZn-SOD (Beauchamp and Fridovich, 1971). The increase in SOD activity is due to the induction of its isozymes (Sairam et al., 2005).

The activity of POX, which essentially eliminates oxygen peroxide, was revealed on polyacrylamide gel (PAGE) in the presence of sodium chloride. The activity of POX in control tomato roots was represented by two isoenzymes. Also, native PAGE showed two bands of two identical isoenzymes for the other treatments. However, in the presence of SA, the intensity and the size of moving bands increased compared with control and NaCl treatment (Figure 7.6).

In leaves, three identical isoenzymes were revealed. The activity of POX was attenuated for both treatments, NaCl and SA+NaCl, compared with control (Figure 7.6). According to Del Rio et al. (2002), the level of the intervention of this enzyme depends on the intensity of the presence of ROS: it is normal to have high ROS at the roots, since these organs are in direct contact with the salt.

The absence of the fourth band in leaves can be explained by two factors. First, the activity of this isozyme is inhibited by the salt, since it is maintained in control plants. The application of SA had no effect on this activity. Second, it is due to the duration of the application of SA. Indeed, El-Khallal et al. (2009) showed that in maize, SA induced stimulation of this enzyme when the plants were subjected to NaCl (100 mM), after 5 weeks and not after 3 weeks.

In summary, two main conclusions can be drawn from these findings. First, a depressive effect of NaCl on the growth response of tomato var Golden Sunrise was observed after 21 days of treatment. Second, the application of SA enhances salt tolerance in tomato and increases plant growth. This

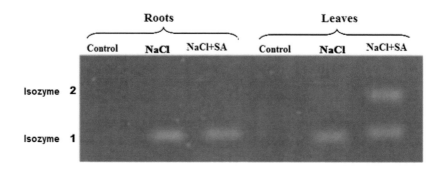

FIGURE 7.5   Activity staining of SOD isozymes, in roots and leaves, after a native PAGE of NaCl-treated tomato plants.

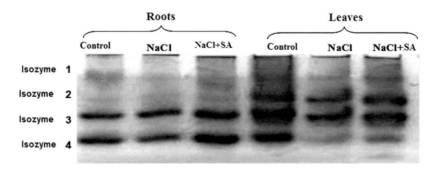

FIGURE 7.6 Activity staining of POX isozymes, in roots and leaves, after a native PAGE of NaCl-treated tomato plants.

enhancement could be associated with a stimulation of the supply of essential elements, essentially K^+, a decrease in the accumulation of Na^+ and Cl^- in plant organs, the synthesis and accumulation of organic solutes, and also an increase in the antioxidant enzymes SOD and POX, preventing the toxic accumulation of ROS.

REFERENCES

Amini F, Ehsanpour AA. Soluble proteins, proline, carbohydrates and Na^+/K^+ changes in two tomato (*Lycopersicon esculentum* Mill.) cultivars under in vitro salt stress. *Am. J. Biochem. Biotech.* 2005; 1: 212–216.

Apel K, Hirt H. Reactive oxygen species: Metabolism, oxidative stress, and signal transduction. *Annu. Rev. Plant Physiol. Plant Mol. Biol.* 2004; 55: 373–399.

Askari H, Edqvist J, Hajheidari M, Kafi M, Salekdeh GH. Effects of salinity levels on proteome of *Suaeda aegyptiaca* leaves. *Proteomics* 2006; 6: 2542–2554.

Bates L, Waldren R, Teare I. Rapid determination of free proline for water-stress studies. *Plant Soil* 1973; 39: 205–207.

Beauchamp C, Fridovich I. Superoxide dismutase: Improved assays and an assay applicable to acrylamide gels. *Anal. Biochem.* 1971; 44: 276–287.

Bennaceur M, Rahmoun C, Sdiri H, Medahi M, Selmi M. Effet du stress salin sur la germination, la croissance et la production de grains de blé. *Sécheresse* 2001; 12: 167–174.

Bradford M. A rapid and sensitive method for quantitation of microgram quantities of protein utilizing the principle of protein-dye-binding. *Anal. Biochem.* 1976; 72: 248–254.

Brosché M, Overmyer K, Wrzaczek M, Kangasjärvi J, Kangasjärvi S. Chapter 5: Stress signaling III: Reactive oxygen species (ROS). In: Pareek SA, Sopory SK, Bohnert HJ and Govindjee HJ (eds.). *Abiotic Stress Adaptation in Plants: Physiological, Molecular and Genomic Foundation*, 2010; pp. 91–102.

Bybordi A. Effect of ascorbic acid and silicium on photosynthesis, antioxidant enzyme activity, and fatty acid contents in canola exposure to salt stress. *J. Integr. Agric.* 2012; 11: 1610–1620.

Bybordi A. Interactive effects of silicon and potassium nitrate in improving salt tolerance of wheat. *J. Integr. Agric.* 2014; 13: 1889–1899.

Chakraborty U, Tongden C. Evaluation of heat acclimation and salicylic acid treatments as potent inducers of thermotolerance in *Cicer arietinum* L. *Curr. Sci.* 2005; 89: 384–389.

Davis BJ. Disc electrophoresis. II. Method and application to human serum proteins. *Ann. N Y Acad. Sci.* 1964; 121: 404–427.

Del Rio LA, Corpas FJ, Sandalio LM, Palma JM, Gomez M, Barroso JB. Reactive oxygen species, antioxidant systems and nitric oxide in peroxisomes. *J. Exp. Bot.* 2002; 53: 1255–1272.

El-Khallal SM, Hathout TA, Ashour AEA, Kerrit AA. Brassinolide and salicylic acid induced growth, biochemical activities and productivity of maize plants grown under salt stress. *Res. J. Agric. Biol. Sci.* 2009; 5: 380–390.

El Sayed Hameda EA, Baziad Salih AM, Basaba Reem AAS. Alleviation of adverse effects of salinity stress on tomato (*Solanum lycopersicum* L.) plants by exogenous application of ascorbic acid. *Eur. J. Acad. Essays* 2015; 2: 23–33.

El-Tayeb M. Response of barley grains to the interactive effect of salinity and salicylic acid. *Plant Growth Regul.* 2005; 45: 215–224.

Fahad S, Bano A. Effect of salicylic acid on physiological and biochemical characterization of maize grown in saline area. *Pak. J. Bot.* 2012; 44: 1433–1438.

FAO. (Food and Agricultural Organization). Land and plant nutrition management service. 2008. www.fao.org/ag/agl/agll/spush.

Ganesan V, Thomas G. Salicylic acid response in rice: Influence of salicylic acid on H_2O_2 accumulation and oxidative stress. *Plant Sci.* 2001; 160: 1095–1106.

Grennan AK. High impact abiotic stress in rice. An "omic" approach. *Plant Physiol.* 2006; 140: 1139–1141.

Gunes A, Inal A, Alpaslan M, Eraslan F, Bagci EG, Cicek N. Salicylic acid induced changes on some physiological parameters symptomatic for oxidative stress and mineral nutrition in maize (*Zea mays* L.) grown under salinity. *J. Plant Physiol.* 2007; 164: 728–736.

Hachicha M. Les sols salés et leur mise en valeur en Tunisie. *Sécheresse* 2007; 18: 45–50.

Hsu SY, Hsu YT, Kao CH. The effect of polyethylene glycol on proline accumulation in rice leaves. *Biol. Plant* 2003; 46: 73–78.

Hua L. Dissection of salicylic acid-mediated defense signaling networks. *Plant Signal Behav.* 2009; 4(8): 713–717.

Kafi M, Nabati J, Masoumi A, Mehrgerd MZ. Effect of salinity and silicon application on oxidative damage of sorghum [*Sorghum bicolor* (L.) moench.]. *Pak. J. Bot.* 2011; 43: 2457–2462.

Kaur H, Mukherjee S, Baluska F, Bhatla SC. Regulatory roles of serotonin and melatonin in abiotic stress tolerance in plants. *Plant Signal. Behav.* 2015; 10(11): e1049788. doi:10.1080/15592324.2015.1049788.

Khan MIR, Asgher M, Khan NA. Alleviation of salt-induced photosynthesis and growth inhibition by salicylic acid involves glycinebetaine and ethylene in mungbean (*Vigna radiata* L.). *Plant Physiol. Biochem.* 2014; 80: 67–74.

Krantev A, Yordanova R, Janda T, Szalai G, Popova L. Treatment with salicylic acid decreases the effect of cadmium on photosynthesis in maize plants. *J. Plant Physiol.* 2008; 165: 920–931.

Kumar P, Jyothi LN, Mani VP. Interactive effects of salicylic acid and phytohormones on photosynthesis and grain yield of soybean (*Glycine max* L. Merrill). *Plant Physiol. Plant Mol. Biol.* 2000; 6: 179–186.

Lakzayi M, Sabbagh E, Rigi K, Keshtehgar A. Effect of salicylic acid on activities of antioxidant enzymes, flowering and fruit yield and the role on reduce of drought stress. *Int. J. Farm. Allied Sci.* 2014; 3: 980–987.

Li T, Hu Y, Du X, Tang H, Shen C, Wu J. Salicylic acid alleviates the adverse effects of salt stress in *Torreya grandis* cv. Merrillii seedlings by activating photosynthesis and enhancing antioxidant systems. *PLoS One* 2014; 9(10): e109492. doi:10.1371/journal.pone.0109492.

Manaa A, Gharbi E, Mimouni H, Wasti S, Aschi-Smiti S, Lutts S, Ben Ahmed H. Simultaneous application of salicylic acid and calcium improves salt tolerance in two contrasting tomato (*Solanum lycopersicum*) cultivars. *S. Afr. J. Bot.* 2014; 95: 32–39.

Martínez JP, Kinet JM, Bajji M, Lutts S. NaCl alleviates polyethylene glycol-induced water stress in the halophyte species *Atriplex halimus* L. *J. Exp. Bot.* 2005; 6: 2421–2431.

Mezni M, Albouchi A, Bizid E, Hamza M. Effet de la salinité des eaux d'irrigation sur la nutrition minérale chez trois variétés de luzerne pérenne (*Medicago sativa* L.). *Agronomie* 2002; 22: 283–291.

Mimouni H, Wasti S, Manaa A, Chalh A, Vandoorne B, Lutts S, Ben Ahmed H. Does salicylic acid (SA) improve tolerance to salt stress in plants? A study of SA effects on tomato plant growth, water dynamics, photosynthesis, and biochemical parameters. *Omics* 2016; 20: 180–190.

Molassiotis A, Sotiropoulos T, Tanou G, Diamantidis G, Therios I. Boron induced oxidative damage and antioxidant and nucleolytic responses in shoot tips culture of the apple rootstock EM9 (*Malus domestica Borkh*). *Environ. Exp. Bot.* 2006; 56: 54–62.

Munns R, James RA, Lauchli A. Approaches to increasing the salt tolerance of wheat and other cereals. *J. Exp. Bot.* 2006; 57: 1025–1043.

Munns R, Tester M. Mechanisms of salinity tolerance. *Ann. Rev. Plant Biol.* 2008; 59: 651–681.

Navari-Izzo F, Quartacci MF, Pinzino C, Dalla vecchia F, Sgherri CLM. Thylakoid-bound and stromal antioxidative enzymes in wheat treated with excess copper. *Physiol. Plant.* 1998; 104: 630–638.

Ndhala AR, Moyo M, Staden J. Natural antioxidants: Fascinating or mythical biomolecules? *Molecules* 2010; 15: 6905–6930.

Pareek A, Singla SL, Grover A. Salt responsive proteins/genes in crop plants. In: Jaiwal PK, Singh RP, Gulati A (eds.). *Strategies for Improving Salt Tolerance in Higher Plants.* Science Publishers, USA, 1997; pp. 365–382.

Sairam RK, Rao KV, Srivastava GC. Differential response of wheat genotypes to long term salinity stress in relation to oxidative stress, antioxidant activity and osmolyte concentration. *Plant Sci.* 2002; 163: 1037–1046.

Sairam RK, Srivastava GC, Agarwal S, Meena RC. Differences in antioxidant activity in response to salinity stress in tolerant and susceptible wheat genotypes. *Biol. Plant.* 2005; 49: 85–91.

Sakhabutdinova AR, Fatkhutdinova DR, Bezrukova MV, Shakirova FM. Salicylic acid prevents the damaging action of stress factors on wheat plants. *Bulg. J. Plant Physiol.* 2003; 314–319.

Schutzendubel A, Polle A. Plant responses to abiotic stresses: heavy metal-induced oxidative stress and protection by mycorrhization. *J. Exp. Bot.* 2002; 53: 1351–1365.

Szalai G, Krantev A, Yordanova R, Popova LP, Janda T. Influence of salicylic acid on phytochelatin synthesis in *Zea mays* during Cd stress. *Turk. J. Bot.* 2013; 37: 708–714.

Tari I, Csiszar J, Szalai G, Horvath F, Pecsvaradi A, Kiss G, Szepesi A, Szabo M, Erdei L. Acclimation of tomato plants to salinity stress after a salicylic acid pre-treatment. *Acta Biol. Szeged* 2002; 46: 55–56.

Tasgin E, Atici O, Nalbantoglu B. Effects of salicylic acid and cold on freezing tolerance in winter wheat leaves. *Plant Growth Regul.* 2003; 41: 231–236.

Tewari RK, Kim S, Hahn EJ, Paek KY. Involvement of nitric oxide-induced NADPH oxidase in adventitious rootgrowth and antioxidant defense in *Panax ginseng. Plant Biotechnol. Rep.* 2008; 2: 113–122.

Trinchant JC, Boscari A, Spennato G, Van de Sype G, Le Rudulier D. Proline betaine accumulation and metabolism in alfalfa plants under sodium chloride stress. Exploring its compartmentalization in nodules. *Plant Physiol.* 2004; 135: 1583–1594.

Vallejos CE. Enzyme activity staining. In: Tanksley SD, Orton TJ (eds.). *Isozymes in Plant Genetics and Breeding. Part A.* Elsevier, Amsterdam, The Netherlands, 1983; pp. 469–516.

Wasti S, Manaa A, Mimouni H, Nasairi A, Medyouni I, Gharbi E, Gautier H, Ben Ahmed H. Exogenous application of calcium silicate improves salt tolerance in two contrasting tomato (*Solanum lycopersicum*) cultivars. *J. Plant Nutr.* 2017; 4: 673–684.

Wasti S, Mimouni H, Smiti S, Zid E, Ben Ahmed H. Enhanced salt tolerance of tomatoes by exogenous salicylic acid applied through rooting medium. *Omics* 2012; 16: 200–207.

Wei W, Li QT, Chu YN, Reiter RJ, Yu XM, Zhu DH, Zhang WK, et al. Melatonin enhances plant growth and abiotic stress tolerance in soybean plants. *J. Exp. Bot.* 2015; 66(3): 695–707.

Wolf B. A comprehensive system of leaf analyses and its use for diagnosing crop nutrient status. *Commun. Soil Sci. Plant Anal.* 1982; 13: 1035–1059.

Zhou ZS, Guo K, Elbaz AA, Yang ZM. Salicylic acid alleviates mercury toxicity by preventing oxidative stress in roots of *Medicago sativa. Environ. Exp. Bot.* 2009; 65: 27–34.

8

Toxicity of Heavy Metal and Its Mitigation Strategies Through Application of Nutrients, Hormones, and Metabolites

Rachana Singh, Parul Parihar, Anita Singh, and Sheo Mohan Prasad

CONTENTS

Abbreviations

APX	Ascorbate peroxidase
AsA	Ascorbate
BRs	Brassinosteroids
CAT	Catalase
CK	Cytokinin
GA	Gibberellic acid
GSH	Glutathione
JA	Jasmonic acid
MDA	Malondialdehyde
MeJA	Methyl jasmonate
NP-SH	Non-protein thiols
PC	Phytochelatin
POD	Peroxidase
ROS	Reactive oxygen species
SOD	Superoxide dismutase
SOR	Superoxide radical

Introduction

Over the past 200 years, the emission of heavy metals in the environment has risen immensely and metal contamination issues have now become common all over the world. Heavy metals constitute an ill-defined group of inorganic chemical hazards, which includes any metallic element that has a comparatively high density and is toxic or poisonous even at its lower concentration (Lenntech Water Treatment and Air Purification, 2004). These criteria apply to metals and metalloids that have an atomic density higher than 4 g/cm^3, or 5 times the density of water (Hawkes, 1997). There are about 40 elements that come into this category. The most common heavy metals found on contaminated sites are arsenic (As), cadmium (Cd), chromium (Cr), copper (Cu), mercury (Hg), nickel (Ni), lead (Pb) and zinc (Zn) (GWRTAC, 1997). 'Stress' can be simply defined as any material that can cause objectionable effects, thereby impairing the environmental condition, reducing the quality of life and sooner or later possibly causing death. These types of materials should not be present in the environment beyond the tolerance limit. The main sources of heavy metals are found in dispersed form in rock formations. On the other hand, rapid industrialization and urbanization have increased the anthropogenic contribution of heavy metals to the atmosphere/biosphere. The largest availability of heavy metals is in the soil and water ecosystems, whereas the atmosphere, in comparison, has a smaller proportion in the form of particulates or vapours. From soil and water, heavy metals enter into plants, where toxicity varies depending on the plant species, composition of soil and pH and specific metal and its concentration, as several heavy metals are essential for plant growth. For instance, both Cu and Zn either function as cofactors and activators of the enzymes of important reactions or act as a prosthetic group in metalloproteins (catalytic property).

On the basis of their coordination chemistry, heavy metals have been categorized as class B metals that fall under non-essential trace elements, which are extremely toxic, such as silver (Ag), Hg, Ni and Pb (Nieboer and Richardson, 1980; Tchounwou et al., 2012). These heavy metals are bioaccumulative, neither readily broken down nor easily metabolized, but remain persistent in the environment. These kinds of heavy metals accumulate in crops (Table 8.1) and thereby enter the food chain through uptake at the primary producer level and then after through consumption at the consumer levels of the ecological system. Whereas in aquatic ecosystems, the whole plant body is in direct contact with these metals that are directly absorbed on the leaf surface. Heavy metals directly enter the human body during either inhalation or ingestion. On the other hand, agricultural and industrial activities, urbanization and traffic, mining and waste incineration have significantly increased the entry of heavy metals through inhalation in the human body. Due to their toxic impact on plants and the human body, several techniques are being employed to detoxify them at the initial level. Among several techniques, the application of protectants like mineral nutrients, phytohormones and metabolites is cost effective and quite efficient. These techniques are a sustainable approach for enhancing productivity. In this chapter we will briefly discuss heavy metals with special reference to their sources, impact and mechanism of toxicity mediation. In later sections, we will also give an overview of recent techniques that are being deployed to alleviate the toxicity induced by these heavy metals and how these protectants are regulating the cellular metabolism in order to reduce the damage induced by heavy metals.

Sources of Heavy Metals

Heavy metals are released into the environment via different sources such as (i) agricultural sources, (ii) domestic effluent, (iii) industrial sources, (iv) atmospheric sources, (v)natural sources, etc. (Figure 8.1). Heavy metals are commonly released into the environment from both natural and anthropogenic sources. Agricultural, mining and smelting activities have contaminated large areas of the world, for instance, Cd, Cu and Pb in North Greece (Zanthopolous et al., 1999), Cr, Cu, Cd, Ni, Pb and Zn in Australia (Smith, 1996) and mainly by heavy metals such as Cu, Cd and Zn in China, Indonesia, Hawaii and Japan (Herawati et al., 2000; Wan et al., 2016; Tam et al., 2016).

Heavy metals originate naturally in the soil environment from the geochemical weathering of parent rocks or rock outcroppings, i.e. the pedogenetic processes. The geochemical materials mainly have Cd, cobalt (Co), Cr, Cu, Hg, manganese (Mn), Ni, Pb, tin (Sn) and Zn. Mostly sedimentary rocks take part in soil formation but generally these are not easily weathered, so they contribute only a small source of heavy metals. On the other hand, igneous rocks such as augite, hornblende and olivine contribute considerable amounts of Co, Cu, Mn, Ni and Zn to the soil. Volcanoes have been reported to be a source of aluminium (Al), Cu, Hg, Mn, Ni, Pb and Zn (Seaward and Richardson, 1990). Wind dust that originates from the Sahara desert has high concentrations of Cr, iron (Fe), Mn, Ni, Pb and Zn (Ross, 1994). Some principal eruptions have more extensive effects, such as emissions from Mount Etna, Sicily, which comprise 10×10^6 kg year^{-1} of Cd, along with Cr, Cu, Mn and Zn (Climino and Ziino, 1983). This volcanic activity also drastically increased Hg content in the soil and plants of the surrounding area (Barghiani et al., 1987). Forest and prairie fires significantly contribute to the

airborne emissions of heavy metals (Ross, 1994; Kristensen and Taylor, 2012).

Agricultural and Other Common Sources of Heavy Metals

The main sources of heavy metal in agriculture are the different types of fertilizers, including irrigation waters, liming and pesticides and sewage sludge. There is a considerable amount of Cr, Cd, Ni, Pb and Zn that has been reported in inorganic fertilizers, fungicides and phosphate fertilizers. Limes, manure and sewage sludge are the biggest source of Cd in soil after their application (Nriagu, 1988; Yanqun et al., 2005). Although the concentrations which are being added by these fertilizers in agricultural soil are low, the frequent use of phosphate fertilizer and the long persistence time of metals may result in high accumulations of some metals, which can be fatal (Verkleji, 1993; Hansell et al., 2006). Table 8.2 shows concentrations of different heavy metals in the agricultural soil of heavy metal–affected countries. Further, Verkleji (1993) has suggested that various pesticides that are formulated to control the diseases of fruit, vegetable and grain crops are adding enormous amounts of heavy metals to the soil system. In Canadian fruit orchards, lead arsenate pesticide was frequently used to control parasitic insects for more than six decades, which contaminated the orchard soil with a high amount of heavy metals such as As, Pb and Zn (Ross, 1994).

Natural and man-made processes have created metal-containing airborne particulates. In the prevailing climatic conditions, these particulates may be wind-blown over long distances. Another cause of atmospheric heavy metal increase is high temperature anthropogenic sources. According to Robock (2000), volcanic eruptions like geothermal sources cause significant atmospheric pollution. In lakes and rivers, domestic effluents possibly comprise the largest source of elevated metal values. The use of detergents is a potential pollution hazard; Nielsen and Skagerlind (2007) reported that most enzyme detergents commonly contain significant amounts of boron (B), Cr, Co, Fe, Mn, strontium (Sr) and Zn like elements. With regard to the pollution caused by urban areas, Bradford (1997) revealed that urban storm water runoff is a major source of pollutants to surface waters. Table 8.2 shows concentrations of different heavy metals in water reservoirs of heavy metal–affected countries.

Industrial Sources of Heavy Metals

Mining, milling and refinement are the major sources of industrial heavy metals. During mining and tailing operations, larger and heavier particles that settle down to the bottom of the flotation unit are directly discharged into environments like onsite wetlands, which lead to elevated concentrations (DeVolder et al., 2003). For instance, coalmines are rich sources of As, Fe and Cd, etc. that directly or indirectly enhances deposition of heavy metals in the soil around the coalfield.

In 1997, Lacerda reported that gold mines are the major source of Hg, as Hg is utilized in gold mines and the mobilization of high amounts of Hg from old mines had significantly moved

TABLE 8.1

Different Heavy Metal Concentrations in Various Food Stuffs

Metal	Country	Food Stuff(s)	Concentration	References
Cadmium	Finland	Green vegetables, potatoes and berries	1–21, 6 and 14 µg/kg	Tahvonen and Kumpalainen (1995)
	Great Britain	Bread, flour, green vegetables, potatoes and milk	30, 6, 30, 2 and 1 µg/kg	Ysart et al. (1999)
	Belgium	Celery and beans (unpolluted area), celery and beans (polluted area)	680 and 150 µg/kg (unpolluted areas), 2430 and 420 µg/kg (polluted areas)	Staessen et al. (1994)
	Germany	Bread/cereals, green vegetables, fresh fruit and milk	1–50, <1–10, <1–2, <1–1 µg/kg	Muller et al. (1998)
	France	Apple, carrot, beans, tomatoes, spinach, lettuce, grains, milk and egg	5, 9, 5, 5, 35.5, 21, 20–70, 1 and 2 µg/kg	Malmauret et al. (2002)
	Denmark	Bread, carrot, spinach and potatoes	30–42, 20, 65 and 21 µg/kg	Larsen et al. (2002)
	Sweden	Wheat flour and rye flour	29 and 17 µg/kg	Jorhem et al. (2001)
	India	Milk and baby food	0.1 and 0.5–18 (GM) µg/kg	Tripathi et al. (1999)
	Japan	Polished raw rice (polluted area)	210 µg/kg	Cai et al. (2001)
	China	Pulses, soy beans and cereals	56, 74 and 9 µg/kg	Zhang et al. (1998)
Chromium	USA	Meat, fish, fruit and vegetables	<10 to 1300 µg/kg	Ministry of Agriculture, Fisheries and Food (1985); Agency for Toxic Substances and Disease Registry (1989)
Arsenic	India	Rice	153.11 µg/kg	EFSA (2014)
		Vegetables	90–3900 µg/kg	Das et al. (2004)
		Pulses	1300 µg/kg	Santra et al. (2013)
		Chicken meat	286 µg/kg	Islam et al. (2013)
		Fish	3000 µg/kg	Lin and Liao (2008)
Lead	Finland	Rye flour, wheat-based cereals, muesli, pasta and bread	16, 22, 34,18 and 8–19 µg/kg	Tahvonen and Kumpalainen (1994)
	Sweden	Wheat bran	27 µg/kg	Jorhem et al. (2001)
	China	Cereal, foxtail millet, maize grain, wheat flour, rice flour, pulses, kidney beans, soy beans	31, 54, 35, 29, 23, 26, 25 and 31 µg/kg	Watanabe et al. (1998)
	Brazil	Milk (near lead factory)	40 (MV) µg/kg	Okada et al. (1997)
	USA	Milk	0.83 µg/kg	Schaum et al. (2003)
	Turkey	Milk	0.2–0.5 µg/kg	Simsek et al. (2000)
	Spain	Milk-free infant cereals and Milk-added infant cereals	36–306 and 54–598 µg/kg	Roca de Togores (1999)
	Australia	Infant formula and Beikost	1.6 (GM) and 2.9 µg/kg	Gulson et al. (2001)
	India	Milk and baby food	2–3 (GM) and 40–78 µg/kg	Tripathi et al. (1999)
	United Kingdom	Green vegetables, other vegetables, flour and potatoes	10, 20, 20 and 10 µg/kg	Ysart et al. (2000)
	France	Apple, carrot, beans, tomatoes, spinach, lettuce, grains and egg	5, 9, 5, 5, 35.5, 21, 20, 1 and 2 µg/kg	Malmauret et al. (2002)
	Poland	Vegetables, fruits and cereals	9–1044, 13–144 and nd-760 µg/kg	Szymszak et al. (1993)
Mercury	Poland	Wheat/rye, vegetables and fruits	2.4, 0.5 and 1.1 µg/kg	Jedrzejczak (2002)
	Egypt	Rice grain (imported)	3.2 µg/kg	Al Saleh and Shinwari (2001)
	Japan	Toothed whales and dolphins liver (boiled) and kidney (boiled)	370,000 and 40,500 µg/kg	Endo et al. (2002)
	Sweden	Pig meat, liver, kidney and bovine meat, liver and kidney	9, 15, 19, 5, 6 and 10 µg/kg	Jorhem et al. (1991)
	Spain	Calf meat, liver, kidney and cow meat, liver and kidney	0.4 (GM), 0.9, 12.2, 0.4, 1 and 10.8 µg/kg	López Alonso et al. (2000)
	The Netherlands	Sheep meat, liver and kidney	1, 4 and 9 µg/kg	Vos et al. (1990)
	Germany	River fish	16–812 µg/kg	Falter and Scholer (1994)
	USA (North Mississippi lakes)	Bass, crappie, gar, carp and catfish	1400, 1690, 1890, 630 and 820 µg/kg	Huggett et al. (2001)
	Italy	Cephalopods (2 species)	70 and 270 µg/kg	Storelli and Marcotrigiano (1999)

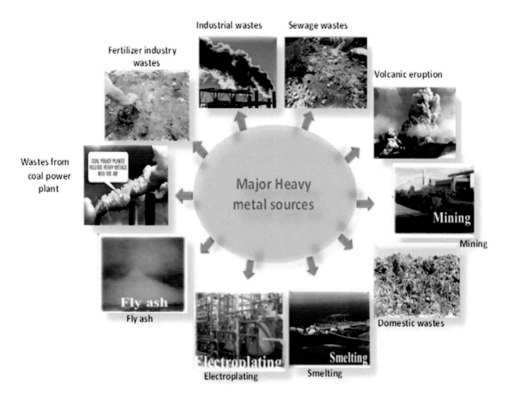

FIGURE 8.1   Different sources of heavy metals.

this pollutant into the environment up to 1960. Presently, Hg is speedily spreading throughout tropical regions, mainly in Asia and Latin America, affecting over 10 million people (Lacerda, 1997). Soil and water bodies are also being contaminated with dust produced during crude ore transportation, runoff from the erosion of mine wastes, metal corrosion and leaching of heavy metals to groundwater and soil (Table 8.3). Soil is also being contaminated by the processing of refineries. Petroleum combustion, coal-burning power plants, high-tension lines and nuclear power stations all significant add B, Cd, Cu, caesium (Cs), Ni, Se and Zn to the environment (Verkleji, 1993; Xiao et al., 2010).

Besides the above-mentioned sources, other sources of heavy metals include landfills, refuse incineration and transportation including aircraft, automobiles and diesel-powered vehicles. Major anthropogenic sources are the corrosion of commercial waste products and coal burning that produces fly ash, which adds Cu, Cr, Pb and Zn into the environment (Al-Hiyaly et al., 1988). According to Verkleji (1993) and Gill (2014), coal burning contributes Al, Cd, Fe, Hg, Ni, Mn and titanium (Ti) into soils while oil burning adds Fe, Ni, Pb and vanadium (V) to the environment. During the transportation of vehicles, metal emission contains Al from catalysts, Ni and Zn from tires, Ni and Zn from aerosol emissions and Cd and Cu from diesel engines.

Symptoms of Metal Stress in Plants

Plants are sensitive to both the excess availability and deficiency of some heavy metal ions. For instance, essential micronutrients are essential at low concentrations, while at higher concentrations, they are strongly deleterious to metabolic activities. In agricultural soil, the contamination with heavy metals has become a significant environmental concern due to their possible adverse ecological effects. They enter into plants and ultimately cause health hazards in humans (Singh et al., 2015).

As is a toxic metalloid; one of its forms, arsenate (AsV), is a chemical analogue of inorganic phosphate (Pi). That is why it competes for the same carriers for uptake in the root plasmalemma of plants (Meharg and Macnair, 1992; Farias et al., 2017). According to Meharg and Macnair (1992), the suppression of a high-affinity Pi/As uptake system leads to tolerance in grasses. Further, Meharg (1994) suggested that this suppression in the Pi/As uptake system lessens As influx up to a level, which can then easily be detoxified by the plant system. However, assimilation of As over the life history of plants may result in As accumulation in very high concentrations such as in *Holcus lanatus* 560 mg/g As and in *Agrostis tenuis* 3,470 mg/g As; this was reported by Porter and Peterson (1975). According to Meharg (1994), As may undergo a transformation to form other less phytotoxic As forms within plant cells. In plants, As may transform into arsenite (AsIII), mono methyl arsenic acid (MMA) and dimethylarsinic acid (DMA). These methylated forms of As are metabolized into arsenosugars and organo phospholipids (Phillips, 1990).

The uptake and distribution of Co are species specific in plants and controlled by different mechanisms (Li et al., 2004; Bakkaus et al., 2005). Plants are able to accumulate a small amount of Co from the soil. Li et al. (2009) have shown that excess Co can cause an adverse effect on biomass and shoot growth of tomato (*Lycopersicon esculentum* L.), oilseed rape (*Brassica napus* L.) and barley (*Hordeum vulgare* L.) seedlings. High concentrations of Co can reduce the catalase (CAT), chlorophyll, protein and Fe concentration in cauliflower plants. Excessive Co also decreased water potential and transpiration rate (Chatterjee and Chatterjee, 2000).

TABLE 8.2

Concentrations of Different Heavy Metals (μg/g) in Soil (Agricultural and Urban) of the Heavy Metal–Affected Countries

Location (Country)	Metal								References
Agricultural Soil	*Arsenic (As)*	*Cadmium (Cd)*	*Chromium (Cr)*	*Copper (Cu)*	*Mercury (Hg)*	*Nickel (Ni)*	*Lead (Pb)*	*Zinc (Zn)*	
Beijing, China	–	0.18	75.74	28.05	–	–	18.48	81.10	Liu et al. (2005)
India	–	0.05	1.23	2.62	–	0.14	2.82	4.65	Prajapati and Meravi (2014)
Iran	–	0.34	10.36	9.62	–	11.28	5.17	11.56	Sayyed and Sayadi (2011)
Korea	0.78	0.12	–	2.98	0.05	–	5.25	4.78	Kim and Kim (1999)
Slovakia	–	–	–	65.00	–	29.00	139.00	140.00	Wilcke et al. (2005)
Spain	–	1.42	63.48	107.65	–	34.75	213.93	427.80	Zimakowska-Gnoinska et al. (2000)
USA	–	13.5	48.5	48	–	29	55	88.5	Jean-Philippe et al. (2012)
Urban Soil		*Cadmium (Cd)*	*Chromium (Cr)*	*Copper (Cu)*	*Nickel (Ni)*	*Lead (Pb)*	*Zinc (Zn)*		
Beijing, China		0.15	35.60	23.70	27.80	28.60	65.60		Zheng et al. (2008)
France		0.53	42.08	20.06	14.47	43.14	43.14		Hernandez et al. (2003)
Iran		1.53	63.79	60.15	37.53	46.59	94.09		Sayadi and Rezaei (2014)
Syria		–	57.00	34.00	39.00	17.00	103.00		Moller et al. (2005)
Spain		3.76	–	57.01	–	1505.45	596.09		Rodríguez et al. (2009)
Turku, Finland		0.17	59.00	23.00	24.10	17.00	90.00		Salonen and Korkka-Niemi (2007)

TABLE 8.3

Concentrations of Different Heavy Metals in Water Reservoir of the Heavy Metal–Affected Countries

Metal	Location (Country)	Concentration	References
Chromium (Cr)	Antarctic lake	<0.6 to 30 µg/l	Masuda et al. (1988)
	USA (surface water)	84 µg/l	Office of Drinking Water (1987)
	Central Canada (surface water)	0.2–44 µg/l	Inland Waters Directorate, Environment Canada (1985)
	Rhine	Below 10 µg/l	RIWA (1989)
	India	Below 2 µg/l	Handa (1988)
Arsenic (As)	Argentina, Pampa, Cordoba (groundwater)	100–3810 µg/l	Nicolli et al. (1989)
	Calcutta, India (groundwater)	50–23,080 µg/l	Mandal et al. (1996)
	Nakhon Si Thammarat Province, Thailand (groundwater)	1.25–5114 Shallow µg/l	Williams et al. (1996)
	Western USA (drinking water)	1–48,000 µg/l	Welch et al. (1988)
	Mono Lake, California, USA	10,000–20,000 µg/l	Maest et al. (1992)
	Uranouchi Inlet, Japan	22.0–32.0 µg/l	Hasegawa (1996)
	Mole River, NSW, Australia	110–600 (up to 13900) µg/l	Ashley andLottermoser (1999)
Nickel (Ni)	Baltic water	0.09–1.08 µg Ni dm^{-3}	Szefer (2002)
	River water	0.7 µg dm^{-3}	Gaillardet et al. (2003)
	Poland (drinking water)	17 µg dm^{-3}	Kocjan et al. (2002a)
	Poland (bottled mineral waters)	0.71–3.20 µg dm^{-3}	Długaszek et al. (2006)
Mercury (Hg)	Idrija, Slovenia (rainwater)	3.15–24.4 ng/l	Kocman et al. (2011)
	Wujiang, Guizhou, China (rainwater)	0.19–36 ng/l	Guo et al. (2008)
	San Joaquin, Qro., Mexico (rainwater)	1.5–339 ng/l	Martinez-Trinidada et al. (2012)
Iron (Fe)	Odra (Viadua, Europe)	250 µg/l	Adamiec and Helios-Rybicka (2002)
	Gomti (India)	0.176 µg/l	Singh et al. (2005)
	Euphrates (Turkey, Syria and Iraq)	105.60 µg/l	Hassan et al. (2010)
	Tsurumi (Japan)	0.241 µg/l	Mohiuddin et al. (2010)
Copper (Cu)	Odra (Viadua, Europe)	8.24 µg/l	Adamiec and Helios-Rybicka (2002)
	Gomti (India)	3.13×10^{-3} µg/l	Singh et al. (2005)
	Keritis (Chania)	3.75 µg/l	Papafilippaki et al. (2008)
	Euphrates (Turkey, Syria and Iraq)	2.48 µg/l	Hassan et al. (2010)
	Tsurumi (Japan)	0.510 µg/l	Mohiuddin et al. (2010)
Zinc (Zn)	Odra (Viadua, Europe)	55.4 µg/l	Adamiec and Helios-Rybicka (2002)
	Gomti (India)	0.02272 µg/l	Singh et al. (2005)
	Keritis (Chania)	21.5 µg/l	Papafilippaki et al. (2008)
	Euphrates (Turkey, Syria and Iraq)	10.50 µg/l	Hassan et al. (2010)
Lead (Pb)	Odra (Viadua, Europe)	1.77 µg/l	Adamiec and Helios-Rybicka (2002)
	Gomti (India)	0.02118 µg/l	Singh et al. (2005)
	Keritis (Chania)	1.44 µg/l	Papafilippaki et al. (2008)
	Euphrates (Turkey, Syria and Iraq)	0.10 µg/l	Hassan et al. (2010)
	Tsurumi (Japan)	0.038 µg/l	Mohiuddin et al. (2010)
Cadmium (Cd)	Odra (Viadua, Europe)	0.140 µg/l	Adamiec and Helios-Rybicka (2002)
	Gomti (India)	2.6 10–4 µg/l	Singh et al. (2005)
	Keritis (Chania)	0.012 µg/l	Papafilippaki et al. (2008)
	Euphrates (Turkey, Syria and Iraq)	2.14 µg/l	Hassan et al. (2010)
	Rio Rimao, Peru	100 µg/l	WHO/UNEP (1989)
	Global (surface fresh water)	0.05 µg/kg	Nriagu (1980)
	Global (groundwater)	0.1 µg/kg	Nriagu (1980)
	Global (ocean)	0.06 µg/kg	Nriagu (1980)

Plants grown in Cd-contaminated soil exhibit symptoms like browning of root tips, chlorosis, growth inhibition and ultimately death (Mohanpuria et al., 2007; Guo et al., 2008). Das et al. (1997) have reported that Cd can interfere with the uptake and transport of calcium (Ca), potassium (K), magnesium (Mg) and phosphorous (P). In roots, Cd inhibits Fe (III) reductase, thereby causing Fe(II) deficiency and finally affecting photosynthesis (Alcantara et al., 1994). Cd decreased nitrate absorption and its transportation from roots to shoots, thereby inhibiting nitrate reductase activity in shoots (Hernandez et al., 1996; Irfan et al., 2014). According to Fodor et al. (1995), Cd reduces the ATPase activity in the plasma membranes of wheat and sunflower roots. Cd also alters the functionality of membranes by inducing lipid peroxidation and disturbance in chloroplast metabolism, thereby inhibiting the biosynthesis of chlorophyll and reducing the activity of enzymes participating in carbon dioxide (CO_2) fixation (De Filippis and Ziegler, 1993).

Copper (Cu) is a micronutrient that plays an important role in CO_2 assimilation and ATP synthesis in plants. It is a necessary constituent of the plastocyanin protein of photosynthetic machinery and cytochrome oxidase of the respiratory electron transport chain (Demirevska-kepova et al., 2004). According to Lewis et al. (2001), high concentrations of Cu in soil causes injury to plants that may lead to a decrease in plant growth and leaf chlorosis. Excessive Cu in plants produces oxidative stress and reactive oxygen species (ROS) that disturb the metabolic pathways and damage important macromolecules (Stadtman and Oliver, 1991; Hegedus et al., 2001; Gill and Tuteja, 2010).

Zinc (Zn) is a vital micronutrient that involves numerous metabolic processes of the plant. An excessive amount of Zn in the soil suppresses growth and development and metabolism, thereby increasing oxidative damage in plants such as *Brassica juncea* and *Phaseolus vulgaris* (Cakmak and Marshner, 1993; Prasad et al., 1999). High levels of Zn cause chlorosis and Cu and Mn deficiencies in plant shoots (Ebbs and Kochian, 1997). Lee et al. (1996) have reported that excessive Zn also causes P deficiency, which appears in the form of a purplish-red color in affected leaves.

Mercury (Hg) exists in many forms in arable land, i.e. Hg°, Hg^{+2}, HgS and methyl-Hg; although, Hg^{+2} (ionic form) is predominant in agricultural soil (Han et al., 2006). Hg in the soil remains in a solid phase through adsorption onto clay particles, organic matter and sulphides. Excessive Hg^{+2} can be extremely toxic and may cause injuries and physiological disorders in plant bodies (Zhou et al., 2007). Zhang and Tyerman (1999) reported that Hg^{+2} binds with water channel proteins that induce leaf stomata closing and causes hindrance in water flow in plants. High Hg^{+2} concentration hinders mitochondrial activities and induces the production of ROS, which leads to the interruption of lipid biomembranes and cellular metabolism in plants (Messer et al., 2005; Cargnelutti et al., 2006).

Cr is a highly toxic metal that has a detrimental effect on plant growth and development. Cr slows down the process of seed germination, which might be due to the depressive effect of Cr on amylases activity and on the consequent transport of sugars to the embryo axes (Zeid, 2001). Zeid (2001) explained that increased Cr concentration might increase protease activity, which might reduce the germination of Cr-treated seeds. Cr greatly affects the photosynthetic mechanism in terms of electron transport, CO_2 fixation, photophosphorylation and specific enzyme activities (Clijsters and Van Assche, 1985). High levels of Cr may cause three potential metabolic modifications in plants: (i) changes in pigment production, i.e. chlorophyll, anthocyanin, which are involved in the life provisions of plants, (ii) increases in the production of glutathione (GSH) and ascorbic acid, which may damage plants, (iii) changes in the metabolic pool to channelize the new biochemically related metabolic products, i.e. histidine and phytochelatins (PCs), which confer tolerance or resistance to excessive Cr (Boonyapookana et al., 2002; Shanker et al., 2003b; Schmfger, 2001).

Pb is one of the abundant toxic elements present in the soil. It adversely affects the growth, morphology and photosynthetic mechanism of plants. Pb has been reported to inhibit seed germination in *Pinus helipensis* and *Spartiana alterniflora* thereby interfering with important enzymes (Morzck and Funicclli, 1982; Nakos, 1979). Pb has been reported to inhibit root and stem elongation and leaf expansion in barley, *Allium* and *Raphanus sativas* (Gruenhage and Jager, 1985; Juwarkar and Shende, 1986). Pb^{+2} shows an inhibitory effect on growth and biomass production that may derive from effects on the metabolic processes of plants (Van Assche and Clijsters, 1990). Excessive Pb causes alterations in membrane permeability, water imbalance, disturbances in mineral nutrition and inhibition of enzyme activities (Sinha et al., 1988a,b; Sharma and Dubey, 2005). Pb inhibits enzyme activity thereby reacting with their sulfhydryl groups. A high level of Pb also increases ROS production in plants and leads to oxidative stress (Reddy et al., 2005).

Nickel (Ni) is an important micronutrient and found in natural soil in trace concentrations. In polluted soil, Ni^{+2} concentration may vary from 200 to 26,000 mg/kg, i.e. 20- to 30-fold higher than the overall range (10–1,000 mg/kg) found in natural soil (Izosimova, 2005). A high level of Ni^{+2} alters the physiological processes and causes toxicity like chlorosis and necrosis in plants (Pandey and Sharma, 2002; Rahman et al., 2005). High Ni^{+2} concentration in the soil impairs nutrient balance and leads to the disorder of cell membrane (H-ATPase activity) functions, as reported in shoots of *Oryza sativa* (Ros et al., 1992). Excess Ni^{+2} also increases malondialdehyde (MDA) concentration in wheat plants that might disturb membrane fluidity/functionality and ion homeostasis in the cytoplasm, especially of K^+. Excess Ni^{+2} may interfere with the water uptake in both monocot and dicot plants, and this decrease in water uptake indicates the severity of Ni^{+2} toxicity in plants (Pandey and Sharma, 2002; Gajewska et al., 2006).

Manganese (Mn) is an important micronutrient for plants but excess Mn accumulation in leaves reduces the photosynthetic rate of plants (Kitao et al., 1997a,b; Srivastava and Dubey, 2011). After its uptake by roots, Mn is readily transported from root to shoot via the transpiration stream but not readily remobilized to other organs after reaching the leaves (Loneragan, 1988). According to Wu (1994), the common symptoms of Mn toxicity are necrotic brownish spots on stems, petioles and leaves. 'Crinkle leaf' is also an important symptom that commonly occurs in stems, petioles and young leaves. Roots with Mn toxicity are brown in color and sometimes crack (Foy et al., 1995; Ei-Jaoual and Cox, 1998). Excessive Mn inhibits chlorophyll biosynthesis thereby blocking Fe process (Clarimont et al., 1986; Huang et al., 2016).

Iron (Fe) is an essential element for the growth and development of plants as it plays an important role in several biological processes such as chlorophyll biosynthesis, chloroplast development and photosynthesis. It is an important constituent of heme proteins such as CAT, peroxidase (POD), cytochromes, leghemoglobin and iron sulphur proteins, i.e. aconitase, ferredoxin and superoxide disumutase (SOD) (Marschner, 1995; Chandramouli et al., 2007). Fe toxicity symptoms may only be expressed during flooded conditions that involve the microbial reduction of insoluble Fe^{+3} in soluble Fe^{+2} (Becker and Asch, 2005). High Fe^{+2} uptake by roots and its transportation via the transpiration stream to leaves causes iron toxicity in plants. Excessive Fe^{+2} in leaves causes oxidative stress and generates ROS, which impairs cellular structure and damages membranes, DNA and proteins (Arora et al., 2002; de Dorlodot et al., 2005). According to Sinha et al. (1997), Fe toxicity causes reduction in photosynthesis and an increase in ascorbate peroxidase (APX) activity and oxidative stress.

Mitigation Strategies Employed under Heavy Metal Stress

Among the several strategies for mitigating heavy metal toxicity, nutrient management, hormone application and metabolite application are emerging as efficient tools (Figure 8.2). Several studies have been performed showing the ameliorative role of nutrients, hormones and metabolites under heavy metal toxicity. The detailed role of these mitigants in ameliorating toxicity induced by heavy metal will be discussed in the following sections.

Mineral Nutrients

Mineral nutrients *viz.*, potassium (K), nitrogen (N), calcium (Ca), sulphur (S), phosphorus (P), and trace metals *viz.*, iron (Fe), copper (Cu), manganese (Mn), molybdenum (Mo), cobalt(Co) and zinc (Zn) have been shown to alleviate heavy metal stress. Mineral nutrients as well as trace elements are needed for the proper growth and development of plants and they also play a major role in alleviating metal-induced toxicity in crop plants. Herein, we have summarized the role of several nutrients in ameliorating heavy metal stress:

The most vital nutrient, which is a component of nucleic acid, proteins and vitamins, is nitrogen (N). Apart from being an important component, it has an important role in maintaining the nitrogen ratio in plants. It has also been found to be involved in mitigating damage induced by heavy metals by increasing the biosynthesis of chlorophyll, photosynthetic activity and production of enzymatic and non-enzymatic antioxidants like proline and GSH, etc. (Sharma and Dietz, 2006; Lin et al., 2011). It has been suggested that application of N at 7.5 mM (optimal value) could reduce Cd-induced damage in *Helianthus* seedlings by increasing the activity of RuBisCO and antioxidant enzymes, and protein content (Pankovic et al., 2000). Some studies have also been performed to confirm the role of N by analyzing the impact on seedlings grown under N-deficient as well as N-supplemented conditions (Polesskaya et al., 2004; Zhao et al., 2005; Chaffei et al., 2008). In addition, enhanced N application improves the protein amount in crops cultivated in Cd-contaminated areas (Wångstrand et al., 2007). In a recent study on poplar plants it was shown that nitrogen supplementation to Cd-stressed plants up-regulated the expression of γ-glutamylcysteine synthetase (γ-GCs), glutathione synthetase (ECGs) and PC synthetase genes, thereby enhancing GSH and PC synthesis (Lin et al., 2016).

Phosphorus (P) is another important mineral nutrient, which is a major constituent of nucleic acid and is also important for the phosphorylation process. A study by Sarwar et al. (2010) suggested that P alleviates Cd-induced toxicity by forming a complex between P and Cd, thereby decreasing the uptake and mobility of Cd. In a recent study by Manikandan et al. (2016) on vetiver plants, they showed that the growth of seedlings exposed to Cd at 50 and 150 mg L^{-1} declined due to Cd accumulation and enhanced H_2O_2 and MDA contents; however, P supplementation improved the Cd-induced decline in growth by enhancing antioxidant activity (SOD, CAT and POD). Another mineral nutrient, K is also an important component required for stimulating cell elongation and the osmoregulation process. Moreover, it is also required for stomatal movement, photosynthetic processes and synthesis of carbohydrate, proteins and nitrogen compounds (Zhao et al., 2005; Siddique et al., 2012; Zorb et al., 2014). Apart from this, it is also a cofactor of some enzymes and regulates their activity by maintaining the pH between 7 and 8 (Mengel, 2007; Siddique et al., 2012). Exogenous application of K at 60 mg kg^{-1} soil enhanced the non-enzymatic antioxidant, i.e. ascorbate (AsA) and GSH contents, thereby improving the Cd-induced toxicity (Shen et al., 2000). Similar to this, Zhao et al. (2005)

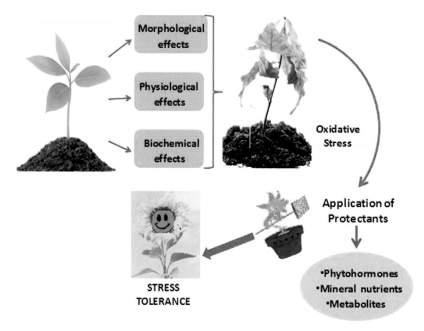

FIGURE 8.2   Schematic diagram showing effect of heavy metal stress and its mitigation using protectants.

performed work on two varieties of *Triticum* seedlings and showed that when test seedlings were exposed to Cd at different concentrations, i.e. from 0 to 55 mg kg^{-1} soil, decline in growth was noticed; however, when seedlings were supplemented with different sources of K, alleviation in Cd-induced toxicity was noticed. In a recent study by Ahmad et al. (2016), the role of potassium in ameliorating Cd toxicity was investigated and the study suggested that K application minimized the Cd toxicity by enhancing organic acids, metabolites, mineral nutrients and antioxidant enzymes (SOD, CAT, APX and GR).

Furthermore, another mineral nutrient involved in regulating growth and development processes in plants is sulphur (S), as it is an important component of coenzymes, vitamins and amino acids. Apart from constituting an important component, it also plays an important role in alleviating heavy metal stress by enhancing the formation of thiols like non-protein thiols (NP-SH), PCs and GSH (Noctor et al., 2011). Studies have shown that S supplementation enhances ATP-sulfurylase (ATPS), serine acetyl transferase (SAT) and O-acetyl serine (OAS) thiol lyase activity (Howarth et al., 2003; Wangeline et al., 2004; Astolfi et al., 2004; Khan et al., 2007). The enhancement in the activity of these enzymes suggests increased S-assimilation and thiol production, conferring protection against metal stress (Zhang et al., 2010). Apart from stimulating thiol contents, it is also involved in ethylene production and its signalling, thereby aiding plants under metal-induced stress (Masood et al., 2012). In a recent study it was suggested that S acts as a pool for synthesis of thiol and PCs and also enhances the antioxidant system, which is important for combating heavy metal stress (Bashir et al., 2015).

Ca is essential for maintaining the structure and permeability of the cell membrane and is also an important component of cell division and elongation translocation of carbohydrate contents (White, 2000; El-Beltagi and Mohamed, 2013). Ca is also involved in signal transduction pathways as it is the secondary messenger and has importance in cell metabolism, nutrient absorption and their movements across membranes (Talukdar, 2012; El-Beltagi and Mohamed, 2013). Several studies have shown its involvement in alleviating heavy metal toxicity (Siddique et al., 2012; El-Beltagi and Mohamed, 2013; Ahmad et al., 2016). In a recent study, Ca was shown to enhance the growth of *Cicer* seedlings by increasing the antioxidant system, and levels of organic solutes and metabolites (Ahmad et al., 2016).

Apart from the role of macronutrients in regulating important growth and developmental processes, trace metals also play an important role as they are cofactors of several enzymes and redox reactions (Sanita di Toppi and Gabbrielli, 1999). The impact of trace metals under heavy metal stress has been analyzed in some studies (Singh et al., 2016). The mechanism of tolerance mediated by trace metals is either direct or indirect (Sarwar et al., 2010). The direct mechanism includes lowering the availability of heavy metals by changing their solubility, competing with the heavy metal for the same transporter and sequestrating the toxic metal in vacuoles by enhancing PCs (Baszynski et al., 1980; Salt and Rauser, 1995; Qiu et al., 2005; Hart et al., 2005; Shi et al., 2005; Zaccheo et al., 2006; Matusik et al., 2008). The indirect mechanism includes dilution by enhancing the biomass accumulation and antioxidant system (Hassan et al., 2010; Suzuki, 2005; Jalloh et al., 2009). It has been reported that Zn efficiently reduces the toxicity induced by Cd in *Thlapsi violacea*

(Street et al., 2010). Similar to Zn, Fe has also been reported to reduce the uptake of heavy metals and their further translocation to shoots. Fe stabilizes the thylakoid complex and reduces the impact of Cd on the photosynthesis process (Qureshi et al., 2010). The overall tolerance mechanisms of trace metals include enhancement of enzymatic and non-enzymatic antioxidants, improved chlorophyll biosynthesis and protection of the thylakoid structure (Zornoza et al., 2010; Tkalec et al., 2014).

Phytohormones

Plant hormones or phytohormones have been considered as an effective tool for alleviating metal-induced toxicity in plants. The role of classical hormones *viz.*, auxins, cytokinin (CK), gibberellic acid (GA), ethylene and abscisic acid (ABA), has been performed on several crop plants. Apart from this, studies on other groups of hormones *viz.*, jasmonic acid (JA), brassinosteroids (BRs), indoleamines (serotonin and melatonin) and salicylic acid, have also been performed. All these studies have shown that plant growth hormones are involved in alleviating metal-induced toxicity. In the following section, we have discussed the role of some hormones and their mechanism of action for alleviating metal-induced toxicity:

Auxin is the growth regulating hormone involved in processes of apical dominance, tropism movement and root and shoot formation (Vanneste and Friml, 2009). Apart from regulating growth and development processes endogenously, exogenous application of auxin stimulates defence processes in plants under conditions of abiotic stress. Jain and Khurana (2009) reported that auxin induces changes in the expression of several important genes for regulating growth and development processes in plants. Similarly, it has also been shown that auxin enhances the antioxidant potential of the plant system under stress conditions (Bashri et al., 2015). A study conducted on *Trigonella* seedlings grown under Cd stress showed a considerable decline in growth, pigment contents, and photosynthetic (O$_2$ evolution), PSII photochemistry and carbonic anhydrase (CA) activity, while the exogenous application of auxin ameliorated the negative impact on all the above studied parameters (Bashri and Prasad, 2016). Similarly, He et al. (2017) reported that auxin is involved in enhancing the antioxidant property of the hyperaccumulator, *Pteris vitata*, when grown under arsenic stress. Auxin-mediated signalling is regulated by nitric oxide (NO) and its involvement in regulating metal toxicity has been investigated in several studies (Peto et al., 2011; Yuan and Huang, 2016).

Another important growth regulator is CK, which plays an important role in regulating the process of cell division and differentiation, root and shoot initiation, formation of buds, expansion of leaves and synthesis of chlorophyll. Apart from this, CK also delays senescence and alleviates the toxicity induced by salinity, low temperature, drought and heavy metals (Letham et al., 1978; Barciszewski et al., 2000; Pospisilova et al., 2000; Werner et al., 2001; Dodd, 2003; Singh and Prasad, 2014). The toxicity alleviation capability of CK is due to its stimulatory effect on the antioxidant system, chlorophyll synthesis and improvement in nitrogen metabolism processes (Al-Hakimi, 2007; Gangwar et al., 2010).

Gibberellins are the group of hormones whose precursors are terpenoids, and are highly involved in stimulating (i) root and shoot growth, (ii) initiation and development of the floral organ, (iii) hydrolysis enzymes of the aleurone layer in maize and other cereal grains and (iv) mitotic division in the cells (Matsouka, 2003). Several studies have suggested its role in alleviating metal stress (Hisamatsu et al., 2000; Iqbal et al., 2014; Zhu et al., 2012). Hamayun et al. (2010) reported that exogenous application of GA in soybean plants enhanced the level of two important components *viz.*, daidzein and genistein, involved in alleviating stress conditions. Another study, performed by Gangwar et al. (2010), suggested that exogenous application of GA alleviated the negative impact of Cr in *Pisum* seedlings, by improving the activity of ammonia assimilating enzymes. In addition to this, Zhu et al. (2012) suggested that GA at 5 µM concentration alleviated the Cd-induced toxicity by reducing uptake of Cd and reducing lipid peroxidation. Reduction in Cd uptake could be subjected to decline in the expression of the IRT1 gene (Fe transporter gene associated with Cd absorption). The overall mechanism of alleviation by GA is governed by the enhancement in antioxidant activity, decreased uptake of the toxic metals, enhancement in important growth regulating processes, i.e. photosynthesis and nitrogen metabolism.

Another group of hormones, Brassinosteroids (BRs) that are steroidal have been found to play an important role in growth and development processes. BRs help in improving the enzymatic as well as non-enzymatic antioxidant system thereby improving the tolerance capability of plants. Several studies have been conducted to study the impact on plants grown under heavy metal stress (Anuradha and Rao, 2007a,b; Hayat et al., 2010; Hasan et al., 2011; Ramakrishna and Rao, 2013, 2015). In a study on beetroot, authors have shown a reduction in Pb content after application of 24-epibrassinolide, a BR, which suggests that it helps plants in overcoming metal-induced toxicity by reducing the metal uptake (Khripach et al., 1999). Another study, performed on *Brassica* seedlings, suggested its role in improving shoot growth and thereby improving the biomass of the seedling (Sharma et al., 2007). BRs have also been proven to improve the chlorophyll content and sugar and protein amount in plants and also enhance the synthesis of PCs (Bajguz, 2000, 2002). Another type of BR is homobrassinoloid (HBL), which has been shown to reduce the toxic effect of metal on several crop plants like *Brassica*, *Zea mays*, *Raphanus* and *Triticum aestivum* (Anuradha and Rao, 2007b; Bhardwaj et al., 2007; Hayat et al., 2007; Sharma et al., 2010). HBL is known to be involved in stimulating the enzyme activity of the photosynthesis process. In a recent study, seeds pre-soaked in HBL showed a considerable decline in the uptake of Cr and thereby improved the growth of test seedlings (Sharma et al., 2011, 2016). Methyl jasmonate (MeJA) is a methyl ester of JA, which is a derivative of fatty acid and involved in regulating physiological and biochemical processes of plants (Norastehnia et al., 2007). MeJA, when applied at low concentrations (10^{-7} to 10^{-5}), stimulates the biosynthetic process of chlorophyll, PSII photochemistry and photosynthetic process (Maksymiec and Krupa, 2002; Kováčik et al., 2011). Apart from stimulating these processes, MeJA also enhances antioxidant capacity and reduces the formation of peroxides, products of aldehydes and H_2O_2 contents (Keramat et al., 2009; Kováčik et al., 2011; Noriega et al., 2012; Singh and Shah, 2014). On the contrary, studies have also shown that higher concentrations of MeJA tend to pose negative impacts like inhibiting stomatal opening, the photosynthesis process and

growth (Anjum et al., 2011; Yan et al., 2015). Moreover, MeJA-induced tolerance in plants is mediated by regulation in genes involved in GSH biosynthesis, as an increase in mRNA level was reported in *Arabidopsis* (Xiang and Oliver, 1998). In a recent study, increased GSH content was reported in *Glycine max* and *Zea mays*, when applied with JA (Noriega et al., 2012; Singh and Shah, 2014). In a recent study on *Brassica*, performed by Per et al. (2016), it was shown that the decline in growth induced by Cd was alleviated by MeJA, and this protective mechanism was mediated by enhanced S- assimilation and thereby increased GSH production that aided the plants and protected the photosynthetic apparatus from damage. A recent study on faba bean seedlings grown under Cd stress showed a decline in growth, leaf water content and pigment content, and application of JA relieved the impact by decreasing Cd accumulation, H_2O_2 and MDA equivalents content and enhancing the production of osmolytes and activity of antioxidants (SOD, CAT, APX and GR) (Ahmad et al., 2017). Overall studies suggested that JA and its methyl ester, MeJA, play an important role as pathogenesis-related molecules and a detoxification agent (through the action of antioxidants) and maintaining the redox status of cells (Farmer, 2007; Cao et al., 2009; Soares et al., 2010; Guo et al., 2013; Gill et al., 2013).

Metabolites

Metabolites like ascorbate (AsA), glutathione(GSH) and phenols, etc., are an important component of the cell system and are involved in protecting cells against conditions of oxidative stress, for example, AsA and GSH are mainly involved in maintaining the redox status of cells and phenols are of a photo-protective nature. Herein, we have briefly summarized the role of some metabolites and their mechanisms of action under stress conditions.

Ascorbate (AsA) is an antioxidant and is also an important component that regulates the growth of plants. AsA regulates growth by (i) controlling the hydroxyproline-rich protein synthesis, which is important for cell division, (ii) formatting cross-links between glycoproteins and polymers of cell walls and (iii) participating in redox reactions of elongation of plasmalemma (Zhang, 2013). Research has suggested that it acts as a cofactor for phytohormones (ethylene, GA and abscisic acid) biosynthesis as well as an important component for signalling pathways of these phytohormones (Pastori et al., 2003). Apart from its important regulatory function, it is also involved in alleviating heavy metal stress. Under metal stress condition, AsA has been shown to inhibit the uptake of metal and also emit metals from the cell system. Battke et al. (2005) suggested that AsA, due to its electrochemical potential, was capable of reducing Hg^{+2} into Hg and volatilizing the toxic metal out of the cell. It has also been shown that the AsA level increases when plants are exposed to heavy metals like Cd, Pb and Hg, which suggests its protective role under stress (Huang et al., 2010). AsA is also an important part of the AsA-GSH cycle, where it is involved in maintaining the activity of APX, which is involved in the detoxification of H_2O_2.

Glutathione (GSH) is a tripeptide thiol, involved in maintaining the redox status of the cells; it is a component of every cellular compartment like the chloroplast, cytosol, endoplasmic reticulum and vacuole. The thiol group of GSH makes it reactive and therefore it is involved in many cellular reactions. It protects the plant system under heavy metal stress by regulating the metabolic process. GSH controls the H_2O_2 level by maintaining the redox pathways of the AsA-GSH cycle and the oxidized form is GSSG. The ratio of GSH/GSSG is very important as it indicates the redox status of the cell (Millar et al., 2003; Foyer and Noctor, 2005; Shao et al., 2008). Apart from its direct role in detoxifying oxidative stress generated by heavy metals, it is also involved in chelating the toxic metal/radicals by its conjugation with glutathione S-transferase (GST; Edwards et al., 2000; Edwards and Dixon, 2005). These GSH-GST-metal/radical complexes are transferred into the vacuole; there they are sequestrated and thereby protect the cell from damage (Klein et al., 2006; Yazaki, 2006). GSH is found to be implicated at a very early stage of development in the embryo of *Arabidopsis*, as it was seen that loss of function of the GSH biosynthetic gene, i.e. AtGSH1, resulted in a lethal effect on the embryo (Cairns et al., 2006). In addition, GSH is a precursor of PCs that are also peptides that have the capability of binding metals and thereafter detoxifying them in vacuoles. Due to their chelating nature, these are considered important in maintaining cell homeostasis (Hirata et al., 2005).

Proline (Pro) is an osmolyte and component of cell walls that acts as a radical scavenger and a macromolecule stabilizer (Seregin et al., 2003; Zarei et al., 2012). It has a multifarious role like regulating seed development, adapting under stress conditions and maintaining the osmotic balance within the cell as well as signalling (Szabados and Savoure, 2010; Burritt, 2012; reviewed by Gill et al., 2014). Its potentiality under heavy metal stress has been proved in many works. Pro accumulation takes place in the cell as a result of an imbalance in water content; therefore it is also known as an osmoregulator or osmoprotectant. Mourato et al. (2012) reported that Pro is involved in scavenging/quenching the hydroxyl radical and oxygen (singlet state) that are generated as a resultant of metal stress. Under heavy metal stress, it enhances the activity of antioxidant enzymes, maintains osmotic balance of the cell and intracellular pH (Rastgoo et al., 2011; Mourato et al., 2012). Anjum et al. (2014) reported that Pro and GSH work in a coordinated manner, as there is an increase in ROS level under metal stress and the available GSH in the cell shifts to GSSG. This increase in GSSG increases ROS, especially H_2O_2 accumulation, which in turn enhances Pro accumulation (Anjum et al., 2010, 2012; Noctor et al., 2012). The accumulated Pro generates the reducing power, i.e. NADPH, which again maintains GSH level via glutathione reductase (Anjum et al., 2010, 2012; Noctor et al., 2012). Apart from this, it has been suggested that the interaction of Pro and GSH stimulates the phytohormone production that is involved in defence, thereby conferring tolerance to plants (Mhamdi et al., 2010; Ghanta et al., 2014; Iqbal et al., 2014; Anjum et al., 2014).

Conclusions

At present, heavy metal pollution in soil and water is one of the major emerging problems facing mankind as it is continuously causing lethal effects, either directly by polluting drinking water or indirectly by affecting the agricultural productivity of plants. The ability of plants to survive in this critical condition depends on the type of plant. Mineral nutrients, phytohormones and metabolites seem to play a major role in determining plant tolerance. This chapter answers many questions like how heavy metal stress is mediated and what are the basic symptoms or impacts on plant systems; it looks at the major processes affected in cellular systems, as well as the role of these protectants in mitigating the metal stress via their mechanisms of action. An effort has also been made to show how two metabolites interact and regulate the tolerance mechanism in plants under heavy metal stress.

Acknowledgments

The University Grants Commission, New Delhi, is acknowledged for providing financial support to Rachana Singh and Parul Parihar in the form of a D.Phil. research scholarship to carry out this work. The author A. Singh is thankful to the CSIR [No. 13(8913-A)/2017-pool] for providing funding as a senior research associate. The authors also acknowledge the University of Allahabad for providing necessary facilities.

REFERENCES

Adamiec E, Helios-Rybicka E. Distribution of pollutants in the Odra river system part IV. Heavy metal distribution in water of the upper and middle Odra River, 1998–2000. *Pol. J. Environ. Stud.* 2002;11:669–673.

Agency for Toxic Substances and Disease Registry. Toxicological profile for chromium. US Public Health Service, Washington, DC, 1989 (ATSFDR/TP 88/10).

Ahmad P, Abdel Latef AA, AbdAllah EF, Hashem A, Sarwat M, Anjum NA, Gucel S. Calcium and potassium supplementation enhanced growth, osmolyte secondary metabolite production, and enzymatic antioxidant machinery in cadmium-exposed chickpea (*Cicer arietinum* L.). *Front. Plant Sci.* 2016;7:513.

Ahmad P, Alyemeni MN, Wijaya L, Alam P, Ahanger MA, Alamri SA. Jasmonic acid alleviates negative impacts of cadmium stress by modifying osmolytes and antioxidants in faba bean (*Vicia faba* L.). *Arch. Agron. Soil Sci.* 2017;1:1–11.

Al Saleh I, Shinwari N. Report on the levels of cadmium, lead and mercury in imported rice grain samples. *Biol. Trace Elem. Res.* 2001;83:91–96.

Alcantara E, Romera FJ, Canete M, De La Guardia MD. Effects of heavy metals on both induction and function of root Fe(III) reductase in Fe-deficient cucumber (*Cucumis sativus* L.) plants. *J. Exp. Bot.* 1994;45:1893–1898.

Al-Hakimi AMA. Modification of cadmium toxicity in pea seedlings by kinetin. *Plant Soil Environ.* 2007;53:129–135.

Al-Hiyaly SA, McNeilly T, Bradshaw AD. The effect of zinc concentration from electricity pylons-evolution in replicated situation. *New Phytol.* 1988;110:571–580.

Angino EE, Magnuson LM, Waugh TC, Galle OK, Bredfeldt J. Arsenic in detergents: Possible danger and pollution hazard. *Science* 1970;168:389–392.

Anjum NA, Ahmad I, Mohmood I, Pacheco M, Duarte AC, Pereira E, Umar S, Ahmad A, Khan NA, Iqbal M, Prasad MNV. Modulation of glutathione and its related enzymes in plants' responses to toxic metals and metalloids—A review. *Environ. Exp. Bot.* 2012;75:307–324.

Anjum NA, Ibrahim MA, Armando CD, Eduarda P, Ahmad I, Iqbal M. Glutathione and proline can coordinately make plants withstand the joint attack of metal(loid) and salinity stresses. *Front. Plant Sci.* 2014;5:662.

Anjum NA, Umar S, Chan MT. *Ascorbate-Glutathione Pathway and Stress Tolerance in Plants.* Springer, Dordrecht, 2010.

Anjum SA, Xie XY, Farooq M, Wang LC, Xue LI, Shahbaz M, Salhab J. Effect of exogenous methyl jasmonate on growth, gas exchange and chlorophyll contents of soybean subjected to drought. *Afr. J. Biotechnol.* 2011;24:9647–9656.

Anuradha S, Rao SSR. Effect of 24-epibrassinolide on the growth and antioxidant enzyme activities in radish seedlings under lead toxicity. *Indian J. Plant Physiol.* 2007a;12:396–400.

Anuradha S, Rao SSR. The effect of brassinosteroids on radish (*Raphanus sativus* L.) seedlings growing under cadmium stress. *Plant Soil Environ.* 2007b;5:465–472.

Arora A, Sairam RK, Srivastava GC. Oxidative stress and antioxidative system in plants. *Curr. Sci.* 2002;82:1227–1338.

Astolfi S, Zuchi S, Passera C. Role of sulphur availability on cadmium-induced changes of nitrogen and sulphur metabolism in maize (*Zea mays* L.) leaves. *J. Plant Physiol.* 2004;61:795–802.

Bajguz A. Blockade of heavy metals accumulation in *Chlorella vulgaris* cells by 24-epibrassinolide. *Plant Physiol. Biochem.* 2000;38:797–801.

Bajguz A. Brassinosteroids and lead as stimulators of phytochelatins synthesis in *Chlorella vulgaris. J. Plant Physiol.* 2002;159:321–324.

Bakkaus E, Gouget B, Gallien JP, Khodja H, Carrot H, Morel JL, Collins R. Concentration and distribution of cobalt in higher plants: The use of micro-PIXE spectroscopy. *Nucl. Instr. Methods B* 2005;231:350–356.

Barciszewski J, Siboska G, Rattan SIS, Clark BFC. Occurrence, biosynthesis and properties of kinetin (N6-furfurylade- nine). *Plant Growth Regul.* 2000;32:257–265.

Barghiani C, Gloffre D, Bargali R. Mercury content in *Pinus* sp of the Mt. In: *Etna Volcanic Area, in Heavy Metals in the Environment,* 2. Lindberg JE, Hutchinson TC (Eds.). New Orleans, 1987; 51.

Bashir H, Ibrahim MM, Bagheri R, Ahmad J, Arif IA, Baig MA, Qureshi MI. Influence of sulfur and cadmium on antioxidants, phytochelatins and growth in Indian mustard. *AoB PLANTS* 2015;7.

Bashri G, Prasad SM. Indole acetic acid modulates changes in growth, chlorophyll *a* fluorescence and antioxidant potential of *Trigonella foenum graecum* L. grown under cadmium stress. *Acta Physiol. Plant.* 2015;37:49.

Battke F, Ernst D, Halbach S. Ascorbate promotes emission of mercury vapour from plants. *Plant Cell Environ.* 2005;28:1487–1495.

Becker M, Asch F. Iron toxicity in rice—Conditions and management concepts. *J. Plant Nutr. Soil Sci.* 2005;168:558–573.

Bhardwaj R, Arora N, Sharma P, Arora HK. Effects of 28- homobrassinolide on seedling growth, lipid peroxidation and antioxidative enzyme activities under nickel stress in seedlings of *Zea mays* L. *Asian J. Plant Sci.* 2007;6:765–772.

Boonyapookana B, Upatham ES, Kruatrachue M, Pokethitiyook P, Singhakaew S. Phytoaccumulation and phytotoxicity of cadmium and chromium in duckweed *Wolffia globosa. Int. J. Phytoremed.* 2002;4:87–100.

Bradford WI. Urban storm water pollutant loadings a statistical summary through. *JWPCF* 1997;49:610–613.

Burritt DJ. Proline and the cryopreservation of plant tissues: Functions and practical applications. In: *Current Frontiers in Cryopreservation.* Katkov I (Eds.), InTech, Rijeka, Croatia, 2012; 415–426.

Cai Y, Aoshima K, Katoh T, Teranishi H, Kasuya M. Renal tubular dysfunction in male inhabitants of a cadmium-polluted area in Toyama, Japan – An eleven year follow-up study. *J. Epidemiol.* 2001;18:180–189.

Cairns NG, Pasternak M, Wachter A, Cobbett CS, Meyer AJ. Maturation of *Arabidopsis* seeds is dependent on glutathione biosynthesis within the embryo. *Plant Physiol.* 2006;141:446–455.

Cakmak I, Marshner H. Effect of zinc nutritional status on superoxide radical and hydrogen peroxide scavenging enzymes in bean leaves. In: *Plant Nutrition—From Genetic Engineering Field Practice.* Barrow NJ (Eds.). Kluwer, the Netherlands, 1993;133–137.

Cao S, Zheng Y, Wang K, Jin P, Rui H. Methyl jasmonate reduces chilling injury and enhances antioxidant enzyme activity in postharvest loquat fruit. *Food Chem.* 2009;115:1458–1463.

Cargnelutti D, Tabaldi LA, Spanevello RM, Jucoski GO, Battisti V, Redin M, Linares CEB, Dressler VL, Flores MM, Nicoloso FT, Morsch VM, Schetinger MRC. Mercury toxicity induces oxidative stress in growing cucumber seedlings. *Chemosphere* 2006;65:999–1106.

Chaffei C, Gouia H, Debouba M, Ghorbel MH. Differential toxicological response to cadmium stress of bean seedlings grown with NO_3^- or NH_4^+ as nitrogen source. *Int. J. Bot.* 2008;4:14–23.

Chandramouli K, Unciuleac MC, Naik S, Dean DR, Huynh BH, Johnson MK. Formation and properties of [4Fe__4S] clusters on the IscU scaffold protein. *Biochemistry* 2007;46:6804–6811.

Chatterjee J, Chatterjee C. Phytotoxicity of cobalt, chromium and copper in cauliflower. *Environ. Pollut.* 2000;109:69–74.

Clarimont KB, Hagar WG, Davis EA. Manganese toxicity to chlorophyll synthesis in tobacco callus. *Plant Physiol.* 1986;80:291–293.

Clijsters H, Van Assche F. Inhibition of photosynthesis by heavy metals. *Photosynth. Res.* 1985;7:31–40.

Climino G, Ziino M. Heavy metal pollution part VII emissions from Etna volcanic. *Geophys. Res. Lett.* 1983;10:31–38.

Das HK, Mitra AK, Sengupta PK, Hossain A, Islam F, Rabbani GH. Arsenic concentrations in rice, vegetables, and fish in Bangladesh: A preliminary study. *Environ. Int.* 2004;30:383–387.

Das P, Samantaray S, Rout GR. Studies on cadmium toxicity in plants: A review. *Environ. Pollut.* 1997;98:29–36.

de Dorlodot S, Lutts S, Bertin P. Effects of ferrous iron toxicity on the growth and mineral composition of an inter specific rice. *J. Plant Nutr.* 2005;28:1–20.

De Filippis LF, Ziegler H. Effect of sublethal concentrations of zinc, cadmium and mercury on the photosynthetic carbon reduction cycle of Euglena. *J. Plant Physiol.* 1993;142:167–172.

Demirevska-Kepova K, Simova-Stoilova L, Stoyanova Z, Holzer R, Feller U. Biochemical changes in barley plants after excessive supply of copper and manganese. *Environ Exp Bot.* 2004;52:253–266.

DeVolder PS, Brown SL, Hesterberg D, Pandya K. Metal bioavailability and speciation in a wetland tailings repository amended with biosolids compost, wood ash, and sulphate. *J. Environ. Qual.* 2003;32:851–864.

Dodd IC. Hormonal interactions and stomatal responses. *J. Plant Growth Regul.* 2003;22:32–46.

EFSA. Dietary exposure to inorganic arsenic in the European population. *EFSA J.* 2014;12:3597.

Ebbs SD, Kochian LV. Toxicity of zinc and copper to *Brassica* species: Implications for phytoremediation. *J. Environ. Qual.* 1997;26:776–781.

Edwards R, Dixon D. Plant glutathione transferases. *Methods Enzymol.* 2005;401:169–186.

Edwards R, Dixon DP, Walbot V. Plant glutathione S-transferases: Enzymes with multiple functions in sickness and in health. *Trends Plant Sci.* 2000;5:193–198.

Ei-Jaoual T, Cox DA. Manganese toxicity in plants. *J. Plant Nutr.* 1998;21:353–386.

Endo T, Haraguchi K, Sakata M. Mercury and selenium concentrations in internal organs of toothed whales and dolphins marketed for human consumption in Japan. *Sci. Total Environ.* 2002;300:15–22.

Eshleman A, Siegel SM, Siegel BZ. Is mercury from Hawaiian volcanoes a natural source of pollution? *Nature* 1971;223:471–475.

Falter R, Scholer HF. Determination of methyl-, ethyl-, phenyl and total mercury in Neckar River fish. *Chemosphere* 1994;29:1333–1338.

Farias JG, Bernardy K, Schwalbert R, del frari BK, Meharg A, Carey M, Marques ACR, Signes-pastor A, Sausen D, Schorr MRW, Tavares MS, Nicoloso FT. Effect of phosphorus on arsenic uptake and metabolism in rice cultivars differing in phosphorus use efficiency and response. *An. Acad. Bras. Ciênc.* 2017;89:163–174.

Farmer EE. Plant biology: Jasmonate perception machines. *Nature* 2007;448:659–660.

Foy CD, Weil RR, Coradetti CA. Differential manganese tolerances of cotton genotypes in nutrient solution. *J. Plant Nutr.* 1995;18:685–706.

Foyer CH, Noctor G. Redox homeostasis and antioxidant signaling: A metabolic interface between stress perception and physiological responses. *Plant Cell* 2005;17:1866–1875.

Gajewska E, Sklodowska M, Slaba M, Mazur J. Effect of nickel on antioxidative enzymes activities, proline and chlorophyll contents in wheat shoots. *Biol. Planta.* 2006;50:653–659.

Gangwar S, Singh VP, Prasad SM, Maurya JN. Modulation of manganese toxicity in *Pisum sativum* L. seedlings by kinetin. *Sci. Hortic.* 2010;126:467–474.

Ghanta S, Datta R, Bhattacharyya D, Sinha R, Kumar D, Hazra S, Mazumdar AB, Chattopadhyay S. Multistep involvement of glutathione with salicylic acid and ethylene to combat environmental stress. *J. Plant Physiol.* 2014;171:940–950.

Gill M. Heavy metal stress in plants: A review. *Inter. J. Adv. Res.* 2014;2:1043–1055.

Gill SS, Anjum, NA, Hasanuzzaman M, Gill R, Trivedi DK, Ahmad I, Pereira E, Tuteja, N. Glutathione and glutathione reductase: A boon in disguise for plant abiotic stress defense operations. *Plant Physiol. Biochem.* 2013;70:204–212.

Gill SS, Gill R, Anjum NA. Target osmoprotectants for abiotic stress tolerance in crop plants—Glycine betaine and proline. In: *Plant Adaptation to Environmental Change: Significance of Amino Acids and Their Derivatives.* Anjum NA, Gill SS, Gill R (Eds.). CAB International, Wallingford, CT, 2014;97–108.

Gill SS, Tuteja N. Reactive oxygen species and antioxidant machinery in abiotic stress tolerance in crop plants. *Plant Physiol. Biochem.* 2010;48:909–930.

Gruenhage L, Jager HJ. Effect of heavy metals on growth and heavy metals content of *Allium Porrum* and *Pisum sativum.* *Angew Bot.* 1985;59:11–28.

Gulson BL, Mizon KJ, Palmer JM, Patison N, Law AJ, Korsch MJ, Mahaffey KR, Donnelly JB. Longitudinal study of daily intake and excretion of lead in newly born infants. *Environ. Res.* 2001;85:232–245.

Guo J, Chen YZ, Li MS, Shi L, Yan XF. Does MYC2 really play a negative role in jasmonic acid-induced indolic glucosinolate biosynthesis in *Arabidopsis thaliana*? *Russ. J. Plant Physiol.* 2013;60:100–107.

Guo J, Dai X, Xu W, Ma M. Over expressing GSHI and AsPCSI simultaneously increases the tolerance and accumulation of cadmium and arsenic in *Arabidopsis thaliana.* *Chemosphere* 2008;72:1020–1026.

GWRTAC. Remediation of metals-contaminated soils and groundwater. Tech. Rep. TE-97-01, GWRTAC, Pittsburgh, PA, 1997, GWRTAC-E Series.

Hamayun M, Shin JH, Khan SA, Ahmad B, Shin DH, Khan AL, Lee IJ. Exogenous gibberellic acid reprograms soybean to higher growth and salt stress tolerance. *J. Agric. Food Chem.* 2010;58:7226–7232.

Han FX, Su Y, Monts DL, Waggoner AC, Plodinec JM. Binding distribution, and plant uptake of mercury in a soil from Oak Ridge, Tennessee, USA. *Sci. Total Environ.* 2006;368:753–768.

Handa BK. Occurrence and distribution of chromium in natural waters of India. *Adv. Environ. Sci. Technol.* 1988;20:189–214.

Hansell AL, Horwell CJ, Oppenheimer C. The health hazards of volcanoes and geothermal areas. *Occup. Environ. Med.* 2006;63:149–156.

Hasan SA, Hayat S, Ahmad A. Brassinosteroids protect photosynthetic machinery against the cadmium induced oxidative stress in two tomato cultivars. *Chemosphere* 2011;84:1446–1451.

Hassan FM, Saleh MM, Salman JM. A study of physicochemical parameters and nine heavy metals in the Euphrates River, Iraq. *E-J Chem.* 2010;7:685–692.

Hawkes JS. Heavy metals. *J. Chem. Educ.* 1997;74:1369–1374.

Hayat S, Ali B, Hassan SA, Ahmad A. Brassinosteroid enhanced the level of antioxidants under cadmium stress in *Brassica juncea.* *Environ. Exp. Bot.* 2007;60:33–41.

Hayat S, Hasan SA, Hayat Q, Ahmad A. Brassinosteroids protect *Lycopersicon esculentum* from cadmium toxicity applied as shotgun approach. *Protoplasma* 2010;239:3–14.

Hegedus A, Erdei S, Horvath G. Comparative studies of H_2O_2 detoxifying enzymes in green and greening barley seedings under cadmium stress. *Plant Sci.* 2001;160:1085–1093.

Herawati N, Suzuki S, Hayashi K, Rivai IF, Koyoma H. Cadmium, copper and zinc levels in rice and soil of Japan, Indonesia and China by soil type. *Bull. Environ. Contam. Toxicol.* 2000;64:33–39.

Hernandez L, Probst A, Probst JL, Ulrich E. Heavy metal distribution in some French forest soils: Evidence for atmospheric contamination. *Sci. Total Environ.* 2003;312:195–219.

Hernandez LE, Carpena-Ruiz R, Garate A. Alterations in the mineral nutrition of pea seedlings exposed to cadmium. *J. Plant Nutr.* 1996;19:1581–1598.

Hirata K, Tsuji N, Miyamoto K. Biosynthetic regulation of phytochelatins, heavy metal-binding peptides. *J. Biosci. Bioeng.* 2005;100:593–599.

Howarth JR, Domínguez-Solís JR, Gutíerrez-Alcalá G, Wray JL, Romero LC, Gotor C. The serine acetyl transferase gene family in *Arabidopsis thaliana* and the regulation of its expression by cadmium. *Plant Mol. Biol.* 2003;51:589–598.

Huang GY, Wang YS, Sun CC, Dong JD, Sun ZX. The effect of multiple heavy metals on ascorbate, glutathione and related enzymes in two mangrove plant seedlings (*Kandelia candel* and *Bruguiera gymnorrhiza*). *Oceanol. Hydrobiol. Stud.* 2010;39:11–25.

Huang YL, Yang S, Long GX, Zhao ZK, Li XF, Gu MH. Manganese toxicity in sugarcane plantlets grown on acidic soils of Southern China. *PLoS ONE* 2016;11:e0148956.

Huggett DB, Steevens JA, Allgood JC, Lutken CB, Grace CA, Benson WH. Mercury in sediment and fish from North Mississippi Lakes. *Chemosphere* 2001;42:923–929.

Iqbal N, Umar S, Khan NA, Khan MIR. A new perspective of phytohormones in salinity tolerance: Regulation of proline metabolism. *Environ. Exp. Bot.* 2014;100:34–42.

Irfan M, Ahmad A, Hayat S. Effect of cadmium on the growth and antioxidant enzymes in two varieties of *Brassica juncea*. *Saudi J. Biol. Sci.* 2014;21:125–131.

Islam MS, Awal MA, Mostofa M, Begum F, Myenuddin M. Detection of arsenic in chickens and ducks. *Int. J. Sci. Res. Manage.* 2013;1:56–62.

Izosimova A. Modelling the interaction between calcium and nickel in the soil-plant system. *FAL Agric. Res.*, Special issue 2005;288:99.

Jain M, Khurana JP. Transcript profiling reveals diverse roles of auxin-responsive genes during reproductive development and abiotic stress in rice. *FEBS Lett.* 2009;276:3148–3162.

Jean-Philippe SR. Labbé N, Franklin JA, Johnson A. Detection of mercury and other metals in mercury contaminated soils using mid-infrared spectroscopy. *Proc. Int. Acad. Ecol. Environ. Sci.* 2012;2:139–149.

Jedrzejczak R. Determination of total mercury in foods of plant origin in Poland by cold vapour atomic absorption-spectrometry. *Food Addit. Contam.* 2002;19:996–1002.

Jorhem L, Slorach S, Sundsstorm B, Ohlin B. Lead, cadmium, arsenic and mercury in meat, liver and kidney in Swedish pigs and cattle in 1984-88. *Food Addit. Contam.* 1991;8:201–211.

Jorhem L, Sundsstorm B, Engman J. Cadmium and other metal in Swedish wheat and rye flours: Longitudinal study 1983–1997. *J. AQAC Int.* 2001;84:1984–1992.

Juwarkar AS, Shende GB. Interaction of Cd-Pb effect on growth yield and content of Cd, Pb in barley. *Ind. J. Environ. Health* 1986;28:235–243.

Keramat B, Kalantari KM, Arvin MJ. Effects of methyl jasmonate in regulating cadmium induced oxidative stress in soybean plant (*Glycine max* L.). *Afr. J. Microbiol. Res.* 2009;31:240–244.

Khan NA, Samiullah, Singh S, Nazar R. Activities of antioxidative enzymes, sulphur assimilation, photosynthetic activity and growth of wheat (*Triticum aestivum*) cultivars differing in yield potential under cadmium stress. *J. Agron. Crop Sci.* 2007;193:435–444.

Khripach VA, Zhabinskii VN. De Groot AE. *Brassinosteroids: A New Class of Plant Hormones*. Academic Press, San Diego, 1999.

Kim KH, Kim SH. Heavy metal contamination of agricultural soils in central regions of Korea. *Water Air Soil Contam.* 1999;111:109–122.

Kitao M, Lei TT, Koike T. Effects of manganese in solution culture on the growth of five deciduous broad-leaved tree species with different successional characters from northern Japan. *Photosynthetica* 1997b;36:31–40.

Kitao M, Lei TT, Koike T. Effects of manganese toxicity on photosynthesis of white birch (*Betula platyphylla* var. *japonica*) seedlings. *Physiol. Plant.* 1997a;101:249–256.

Klein M, Burla B, Martinoia E. The multidrug resistance-associated proteins (MRP/ABCC) subfamily of ATP-binding cassette transporters in plants. *FEBS Lett.* 2006;580:1112–1122.

Kovácik J, Klejdus B, Štork F, Hedbavny J, Backor M. Comparison of methyl jasmonate and cadmium effect on selected physiological parameters in *Scenedesmus quadricauda* (chlorophyta, chlorophyceae). *J. Phycol.* 2011;47:1044–1049.

Kristensen LJ, Taylor MP. Fields and forests in flames: Lead and mercury emissions from wildfire pyrogenic activity. *Environ. Health Perspect.* 2012;120:a56–a57.

Lacerda LD. Global mercury emissions from gold and silver mining. *Water Air Soil Pollut.* 1997;97:209–221.

Larsen EH, Andersen NL, Moller A, Petersen A, Mortensen GK, Petersen J. Monitoring the content and intake of trace elements from food in Denmark. *Food Addit. Contam.* 2002;19:33–46.

Lee CW, Choi JM, Pak CH. Micronutrient toxicity in seed geranium (*Pelargonium* x *hortorum* Bailey). *J. Am. Soc. Hortic. Sci.* 1996;121:77–82.

Leigh RA, Jones RGW. A hypothesis relating critical potassium concentrations for growth to the distribution and functions of this ion in the plant-cell. *New Phytol.* 1984;97:1–13.

Lenntech Water Treatment and Air Purification. Water treatment. Lenntech, Rotterdameseweg, the Netherlands, 2004 (http://www. excelwater.com/thp/filters/Water-Purification.htm).

Letham, DS, Goodwin PB, Higgins TJV. Phytohormones and related compounds: A comprehensive treatise. In: *Phytohormones and the Development of Higher Plants*, vol. 2. Letham DS, Goodwin PB, Higgins TJV (Eds.). Elsevier North-Holland Biomedical Press, Amsterdam, 1978.

Lewis S, Donkin ME, Depledge MH. Hsp 70 expression in *Enteromorpha intestinalis* (Chlorophyta) exposed to environmental stressors. *Aqua Toxicol.* 2001;51:277–291.

Li HF, Gray C, Mico C, Zhao FJ, McGrath SP. Phytotoxicity and bioavailability of cobalt to plants in a range of soils. *Chemosphere* 2009;75:979–986.

Li Z, McLaren RG, Metherell AK. The availability of native and applied soil cobalt to ryegrass in relation to soil cobalt and manganese status and other soil properties. *N. Z. J. Agric. Res.* 2004;47:33–43.

Lin MC, Liao CM. Assessing the risks on human health associated with inorganic arsenic intake from groundwater-cultured milk fish in southwestern Taiwan. *Food Chem. Toxicol.* 2008;46:701–709.

Lin T, Wan X, Zhang F. The short-term responses of glutathione and phytochelation synthetic pathways genes to additional nitrogen under cadmium stress in poplar leaves. *Russ. J. Plant Physiol.* 2016;63:754.

Lin T, Zhu X, Zhang F, Wan X. The detoxification effect of cadmium stress in *Populus yunnanensis*. *Res. J. Bot.* 2011;4:13–19.

Liu WH, Zhao JZ, Ouyang ZY, Söderlund L, Liu GH. Impacts of sewage irrigation on heavy metal distribution and contamination in Beijing, China. *Environ. Inter.* 2005;31:805–812.

Loneragan JF. Distribution and movement of manganese in plants. In: *Manganese in soils and plants*. Graham RD, Hannam RJ, Uren NC (Eds.). Kluwer, Dordrecht, 1988;113–124.

López Alonso M, Benedito JL, Miranda M, Castillo C, Hernandez J, Shore RF. Toxic and trace elements in liver, kidney and meat from cattle slaughtered in Galicia (NW Spain). *Food Addit. Contam.* 2000;17:447–457.

Maksymiec W, Krupa Z. The in vivo and in vitro influence of methyl jasmonate on oxidative processes in *Arabidopsis thaliana* leaves. *Acta Physiol. Plant.* 2002;24:351–357.

Malmauret L, Parent-Massin D, Hardy JL, Verger P. Contaminants in organic and conventional foodstuffs in France. *Food Addit. Contam.* 2002;19:524–532.

Manikandan R, Ezhili N, Venkatachalam P. Phosphorus supplementation alleviation of the cadmium-induced toxicity by modulating oxidative stress mechanisms in vetiver grass (*Chrysopogon zizanioides* (L.) Roberty. *J. Environ. Eng.* 2016;142:C4016003. doi:10.1061/(ASCE)EE.1943-7870.0001112#sthash.

Marschner H. *Mineral Nutrition of Higher Plants*, 2nd edn. Academic Press, Toronto, 1995.

Masood A, Iqbal N, Khan NA. Role of ethylene in alleviation of cadmium-induced photosynthetic capacity inhibition by sulfur in mustard. *Plant Cell Environ.* 2012;35:524–533.

Masuda N, Nakaya S, Burton HR, Torii T. Trace element distributions in some saline lakes of the Vestfold Hills, Antarctica. *Hydrobiologia* 1988;165:103–114.

Matsuoka M. Gibberellins signaling: How do plant cells respond to GA signals?. *J. Plant Growth Regul.* 2003;22:123–125.

Meharg AA, Macnair MR. Suppression of the high affinity phosphate uptake system; a mechanism of arsenate tolerance in *Holcus lanatus* L. *J. Exp. Bot.* 1992;43:519–524.

Meharg AA. Integrated tolerance mechanisms-constitutive and adaptive plant-response to elevated metal concentrations in the environment. *Plant Cell Environ.* 1994;17:989–993.

Messer RL, Lockwood PE, Tseng WY, Edwards K, Shaw M, Caughman GB, Lewis JB, Wataha JC. Mercury (II) alters mitochondrial activity of monocytes at sublethal doses via oxidative stress mechanisms. *J. Biomed. Mat. Res. B* 2005;75:257–263.

Mhamdi A, Hager J, Chaouch S, Queval G, Han Y, Taconnat L, Saindrenan P, Gouia H, Issakidis-Bourguet E, Renou JP, Noctor G. *Arabidopsis* glutathione reductase 1 plays a crucial role in leaf responses to intracellular H_2O_2 and in ensuring appropriate gene expression through both salicylic acid and jasmonic acid signaling pathways. *Plant Physiol.* 2010;153:144–1160.

Millar AH, Mittova V, Kiddle G, Heazlewood JL, Bartoli CG, Theodoulou FL, Foyer CH. Control of ascorbate synthesis by respiration and its implications for stress responses. *Plant Physiol.* 2003;133:443–447.

Ministry of Agriculture, Fisheries and Food. Survey of aluminium, antimony, chromium, cobalt, indium, nickel, thallium and tin in food. 15. Report of the Steering Group on Food Surveillances; The Working Party on the Monitoring of Foodstuffs for Heavy Metals. Her Majesty's Stationery Office, London, 1985.

Mohanpuria P, Rana NK, Yadav SK. Cadmium induced oxidative stress influence on glutathione metabolic genes of *Camella sinensis* (L.). O Kuntze. *Environ. Toxicol.* 2007;22:368–374.

Mohiuddin KM, Zakir HM, Otomo K, Sharmin S, Shikazono N. Geochemical distribution of trace metal pollutants in water and sediments of downstream of an urban river. *Int. J. Environ. Sci. Tech.* 2010;7:17–28.

Moller A, Müller HW, Abdullah A, Abdelgawad G, Utermann J. Urban soil contamination in Damascus, Syria: Concentrations and patterns of heavy metals in the soils of the Damascus Ghouta. *Geoderma* 2005;124:63–71.

Morzck E Jr, Funicclli NA. Effect of lead and on germination of *Spartina alterniflora* Losiel seeds at various salinities. *Environ. Exp. Bot.* 1982;22:23–32.

Mourato M, Reis R, Martins LL. Characterization of plant antioxidative system in response to abiotic stresses: A focus on heavy metal toxicity. In: *Advances in Selected Plant Physiology Aspects.* Montanaro G, Dichio B (Eds.), InTech, Vienna, Austria, 2012;23–44.

Muller M, Anke M, Illing-Gunther H, Thiel HC. Oral cadmium exposure of adults in Germany. 2: Basket calculations. *Food Addit. Contam.* 1998;15:135–141.

Nakos G. Lead pollution: Fate of lead in soil and its effects on *Pinus haplenis*. *Plant Soil* 1979;50:159–161.

Nieboer E, Richardson DHS. The replacement of the nondescript term heavy metals by a biologically and chemistry significant classification of metal ions. *Environ. Pollut. Series B.* 1980;1:3–26.

Nielsen PH, Skagerlind P. Cost-neutral replacement of surfactants with enzymes. *Household Pers. Care Today* 2007;4:3–7.

Noctor G, Mhamdi A, Chaouch S, Han YI, Neukermans J, Marquez–Garcia B Queval G, Foyer CH. Glutathione in plants: An integrated overview. *Plant Cell Environ.* 2012;35:454–484.

Noctor G, Queval G, Mhamdi A, Chaouch S, Foyer CH. Glutathione. The *Arabidopsis* Book 2011;9:e0142.

Norastehnia A, Sajedi RH, Nojavan-Asghari M. Inhibitory effects of methyl jasmonate on seed germination in maize (*Zea mays*): Effect on α-amylase activity and ethylene production. *Gen. Appl. Plant Physiol.* 2007;33:13–23.

Noriega G, Santa Cruz D, Batlle A, Tomaro M, Balestrasse K. Heme oxygenase is involved in the protection exerted by jasmonic acid against cadmium stress in soybean roots. *J. Plant Growth Regul.* 2012;31:79–89.

Nriagu JO. A silent epidemic of environmental metal poisoning? *Environ. Pollut.* 1988;50:139–161.

Office of Drinking Water. *Health Advisory—Chromium.* US Environmental Protection Agency, Washington, DC, 1987.

Okada IA, Sukuma AM, Maio FD, Dovidauskas S, Zenebon A. Evaluation of cadmium and lead levels in milk in environmental contamination in the Vale do Paraiba region, southeastern Brazil. *Rev. Saude Publica.* 1997;31:140–143.

Pandey N, Sharma CP. Effect of heavy metals Co^{2+}, Ni^{2+} and Cd^{2+} on growth and metabolism of cabbage. *Plant Sci.* 2002;163:753–758.

Pankovic D, Plesnicar M, Arsenijeevic-Maksimovic I, Petrovic N, Sakac Z, Kastori R. Effects of nitrogen nutrition on photosynthesis in Cd-treated sunflower plants. *Ann. Bot.* 2000;86:841–847.

Papafilippaki AK, Kotti ME, Stavroulakis GG. Seasonal variations in dissolved heavy metals in the Keritis River, Chania, Greece. *Global Nest J.* 2008;10:320–325.

Pastori GM, Kiddle G, Antoniw J, Bernard S, Veljovic-Jovanovic S, Verrier PJ, Noctor G, Foyer CH. Leaf vitamin C contents modulate plant defense transcripts and regulate genes that control development through hormone signaling. *Plant Cell.* 2003;15:939–951.

Per TS, Khan NA, Masood A, Fatma M. Methyl Jasmonate alleviates cadmium-induced photosynthetic damages through increased S-assimilation and glutathione production in Mustard. *Front Plant Sci.* 2016;7:1933.

Peto A, Lehotai N, Lozano-Juste J, León J, Tari I, Erdei L, Kolbert Z. Involvement of nitric oxide and auxin in signal transduction of copper-induced morphological responses in *Arabidopsis* seedlings. *Ann. Bot.* 2011;108:449–457.

Phillips DJH. Arsenic in aquatic organisms: A review of emphasizing chemical speciation. *Aqua Toxicol.* 1990;16:151–186.

Polesskaya OG, Kashirina EI, Alekhina ND. Changes in the activity of antioxidant enzymes in wheat leaves and roots as a function of nitrogen source and supply. *Russ. J. Plant Physiol.* 2004;51:615–620.

Porter EK, Peterson PJ. Arsenic accumulation by plants on mine waste (United Kingdom). *Environ. Pollut.* 1975;4:365–371.

Pospisilova J, Synkova H, Rulcova J. Cytokinins and water stress. *Biol. Planta.* 2000;43:321–328.

Prajapati SK, Meravi N. Heavy metal speciation of soil and *Calotropis procera* from thermal power plant area. *Proc. Int. Acad. Ecol. Environ. Sci.* 2014;4:68–71.

Prasad KVSK, Pardha saradhi P, Sharmila P. Concerted action of antioxidant enzyme and curtailed growth under zinc toxicity in *Brassica juncea*. *Environ. Exp. Bot.* 1999;42:1–10.

Rahman H, Sabreen S, Alam S, Kawai S. Effects of nickel on growth and composition of metal micronutrients in barley plants grown in nutrient solution. *J. Plant Nutr.* 2005;28:393–404.

Ramakrishna B, Rao SSR. Foliar application of brassinosteroids alleviates adverse effects of zinc toxicity in radish (*Raphanus sativus* L.) plants. *Protoplasma* 2015;252:665–667.

Ramakrishna B, Rao SSR. Preliminary studies on the involvement of glutathione metabolism and redox status against zinc toxicity in radish seedlings by 28-homobrassinolide. *Environ. Exp. Bot.* 2013;96:52–58.

Rastgoo L, Alemzadeh A, Afsharifar A. Isolation of two novel isoforms encoding zinc- and copper-transporting P1BATPase from Gouan (*Aeluropus littoralis*). *Plant Omics J.* 2011;4:377–383.

Reddy AM, Kumar SG, Jyotsnakumari G, Thimmanayak S, Sudhakar C. Lead induced changes in antioxidant metabolism of horsegram (*Macrotyloma uniflorum* (Lam.) Verdc.) and bengalgram (*Cicer arietinum* L.). *Chemosphere* 2005;60:97–104.

RIWA. De samenstelling van het Rijnwater in 1986–1987. Composition of the water of the Rhine in 1986 and 1987. Amsterdam, 1989.

Robock A. Volcanic eruptions and climate. *Rev. Geophys.* 2000;38:191–219.

Roca de Togores M, Ferre R, Frigola AM. Cadmium and lead in infants cereals-electrothermal-atomic-absorption-spectroscopic determination. *Sci. Total Environ.* 1999;23:197–201.

Rodríguez L, Ruiz E, Alonso-Azcárate J, Rincon J. Heavy metal distribution and chemical speciation in tailings and soils around a Pb–Zn mine in Spain. *J. Environ. Manage.* 2009;90:1106–1116.

Ros R, Cook David T, Picazo C, Martinez-Cortina I. Nickel and cadmium-related changes in growth, plasma membrane lipid composition, atpase hydrolytic activity and proton pumping of rice (*Oryza sativa* L. cv. Bahia) shoots. *J. Exp. Bot.* 1992;43:1475–1481.

Ross SM. *Toxic Metals in Soil–plant Systems*. Wiley, Chichester, 1994;469.

Salonen VP, Korkka-Niemi K. Influence of parent sediments on the concentration of heavy metals in urban and suburban soils in Turku, Finland. *Appl. Geochem.* 2007;22:906–918.

Santra SC, Samal AC, Bhattacharya P, Banerjee S, Biswas A, Majumdar J. Arsenic in food chain and community health risk: A study in gangetic West Bengal. *Proc. Environ. Sci.* 2013;18:2–13.

Sarwar N, Saifullah, Malhi SS, Zia MH, Naeem A, Bibi S, Farida G. Role of mineral nutrition in minimizing cadmium accumulation by plants. *J. Sci. Food Agric.* 2010;90:925–937.

Sayadi MH, Rezaei MR. Impact of land use on the distribution of toxic metals in surface soils in Birjand city, Iran. *Proc. Int. Acad. Ecol. Environ. Sci.* 2014;4:18–29.

Sayyed MRG, Sayadi MH. Variations in the heavy metal accumulations within the surface soils from the Chitgar industrial area of Tehran. *Proc. Int. Acad. Ecol. Environ. Sci.* 2011;1:36–46.

Schaum J, Schuda L, Wu C, Sears R, Ferrario J, Andrews K. A national survey of persistent, bioaccumulative, and toxic (PBT) pollutants in the United States supply. *J. Expo. Anal. Environ. Epidemiol.* 2003;13:177–186.

Schmfger MEV. Phytochelatins: Complexation of metals and metalloids, studies on the phytochelatin synthase. PhD Thesis, Munich University of Technology (TUM), Munich, 2001.

Seaward MRD, Richardson DHS. Atmospheric sources of metal pollution and effects on vegetation. In: *Heavy Metal Tolerance in Plants Evolutionary Aspects*. Shaw AJ (Eds.). CRC Press, Boca Raton, 1990;75–94.

Seregin IV, Kozhevnikova AD, Kazyumina EM, Ivanov VB. Nickel toxicity and distribution in maize roots. *Russ. J. Plant Physiol.* 2003;50:711–717.

Shanker AK, Djanaguiraman M, Pathmanabhan G, Sudhagar R, Avudainayagam S. Uptake and phytoaccumulation of chromium by selected tree species. In: Proceedings of the International Conference on Water and Environment held in Bhopal, India, 2003.

Shao HB, Chu LY, Lu ZH, Kang CM. Primary antioxidant free radical scavenging and redox signaling pathways in higher plant cells. *Int. J. Biol. Sci.* 2008;4:8–14.

Sharma I, Pati PK, Bhardwaj R. Effect of 28-homobrassinolide on antioxidant defence system in *Raphanus sativus* L. under chromium toxicity. *Ecotoxicology* 2011;20:862–874.

Sharma I, Pati PK, Bhardwaj R. Regulation of growth and antioxidant enzyme activities by 28-homobrassinolide in seedlings of *Raphanus sativus* L. under cadmium stress. *Indian J. Biochem. Biophys.* 2010;47:172–177.

Sharma P, Bhardwaj R, Arora N, Arora HK. Effect of 28- homobrassinolide on growth, zinc metal uptake and antioxidative enzyme activities in *Brassica juncea* L. seedlings. *Braz. J. Plant Physiol.* 2007;19:203–210.

Sharma P, Dubey RS. Lead toxicity in plants. *Braz. J. Plant Physiol.* 2005;17:35–52.

Sharma P, Kumar A, Bhardwaj R. Plant steroidal hormone epibrassinolide regulate—Heavy metal stress tolerance in *Oryza sativa* L. by modulating antioxidant defense expression. *Environ. Exp. Bot.* 2016;122:1–9.

Sharma SS, Dietz KJ. The significance of amino acids and amino acid-derived molecules in plant responses and adaptation to heavy metal stress. *J. Exp. Bot.* 2006;57:711–726.

Shen W, Nada K, Tachibana S. Involvement of polyamines in the chilling tolerance of cucumber cultivars. *Plant Physiol.* 2000;124:431–439.

Shujuan H, Yongjun H, Hongbin W, Haijuan W, Qinchun L. Effects of indole-3-acetic acid on arsenic uptake and antioxidative enzymes in *Pteris cretica* var. nervosa and *Pteris ensiformis*. *Int. J. Phytoremed.* 2017;19:231–238.

Simsek O, Gultekin R, Oksuz O, Kurultay S. The effect of environmental pollution on the heavy metal content of raw milk. *Nahrung* 2000;44:360–363.

Singh I, Shah, K. Exogenous application of methyl jasmonate lowers the effect of cadmium-induced oxidative injury in rice seedlings. *Phytochemistry* 2014;31:57–66.

Singh R, Singh S, Parihar P, Singh VP, Prasad SM. Arsenic contamination, consequences and remediation techniques: A review. *Ecotoxicol. Environ. Saf.* 2015;112:247–270.

Singh S, Parihar P, Singh R, Singh VP, Prasad SM. Heavy metal tolerance in plants: Role of transcriptomics, proteomics, metabolomics, and ionomics. *Front. Plant Sci.* 2016;6:1143.

Singh VK, Singh KP, Mohan D. Status of heavy metals in water and bed sediments of River Gomti—A tributary of the Ganga River, India. *Environ. Monit. Assess.* 2005;105:43–67.

Sinha S, Gupta M, Chandra P. Oxidative stress induced by iron in *Hydrilla verticillata* (i.f) Royle: Response of antioxidants. *Ecotoxicol. Environ. Saf.* 1997;38:286–291.

Sinha SK, Srinivastava HS, Mishra SN. Effect of lead on nitrate reductase activity and nitrate assimilation in pea leaves. *Bot. Pollut.* 1988;57:457–463.

Sinha SK, Srinivastava HS, Mishra SN. Nitrate assimilation in intact and excised maize leaves in the presence of lead. *Bull. Environ. Contam. Toxicol.* 1988a;41:419–422.

Slooff W, Cleven RFMJ, Janus JA, van der Poel P, van Beelen P, Boumans LJM, Canton JH, Eerens HC, Krajnc EI, de Leeuw FAAM, Matthijsen AJCM, van de Meent D, van der Meulen A, Mohn GR, Wijland GC, de Bruijn PJ, van Keulen A, Verburgh JJ, van der Woerd KF. *Integrated Criteria Document Chromium*. National Institute of Public Health and Environmental Protection, Bilthoven, the Netherlands, 1989.

Smith SR. *Agricultural Recycling of Sewage Sludge and the Environment*. CAB International, Wallingford, UK, 1996.

Srivastava S, Dubey RS. Manganese-excess induces oxidative stress, lowers the pool of antioxidants and elevates activities of key antioxidative enzymes in rice seedlings. *Plant Growth Regul.* 2011;64:1–16.

Stadtman ER, Oliver CN. Metal-catalyzed oxidation of proteins. Physiological consequences. *J. Biol. Chem.* 1991;266:2005–2008.

Staessen JA, Lauwerys RR, Ide G, Roels HA, Vyncke G, Amery A. Renal function and historical environmental cadmium pollution from zinc smelters. *Lancet* 1994;343:1523–1527.

Storelli MM, Marcotrigiano GO. Cadmium and total mercury in some Cephalopods from the South Adriatic Sea (Italy). *Food Addit. Contam.* 1999;16:261–265.

Szabados L, Savoure A. Proline: A multifunctional amino acid. *Trends Plant Sci.* 2010;15:89–97.

Szymszak J, Ilow R, Regulska-Ilow B. Levels of cadmium and lead in vegetables, fruits, cereals and soil from areas differing in the degree of industrial pollution and from green houses. *Rocz. Panstw. Zakl. Hig.* 1993;44:331–346.

Tahvonen R, Kumpalainen J. Lead and cadmium content in Finnish breeds. *Food Addit. Contam.* 1994;11:621–631.

Tahvonen R, Kumpalainen J. Lead and cadmium in some berries on the Finnish market in 1991–1993. *Food Addit. Contam.* 1995;12:789–798.

Tam E, Miike R, Labrenz S, Sutton AJ, Elias T, Davis J, Chen YL, Tantisira K, Dockery D, Avo E. Volcanic air pollution over the Island of Hawai'i: Emissions, dispersal, and composition. *Environ. Inter.* 2016;92–93:543–552.

Tchounwou PB, Yedjou CG, Patlolla AK, Sutton DJ. Heavy metals toxicity and the environment. *EXS* 2012;101:133–164.

Tripathi RM, Raghunath R, Sastry VN, Krishnamoorthy VN. Daily intake of heavy metals in infants through milk and milk products. *Sci. Total Environ.* 1999;227:229–235.

Van Assche F, Clijsters H. Effects of metals on enzyme activity in plants. *Plant Cell Environ.* 1990;13:195–206.

Vanneste S, Friml J. Auxin: A trigger for change in plant development. *Cell* 2009;136:1005–1016.

Verkleji JAS. The effects of heavy metals stress on higher plants and their use as bio monitors. In: *Plant as Bioindicators: Indicators of Heavy Metals in the Terrestrial Environment*. Markert B (Eds.). VCH, New York, 1993;415–424.

Vos G, Lammers H, Kan CA. Cadmium and lead in muscles tissue and organs of broilers, turkeys and spent hens and in mechanically deboned poultry meat. *Food Addit. Contam.* 1990;7:83–91.

Wan D, Han Z, Yang Z, Yang G, Liu X. Heavy metal pollution in settled dust associated with different urban functional areas in a heavily air-polluted city in North China. *Int. J. Environ. Res. Public Health* 2016;13:1119.

Wangeline AL, Burkhead JL, Hale KL, Lindblom SD, Terry N, Pilon M, Pilon-Smits EA. Over-expression of ATP sulfurylase in Indian mustard: Effects on tolerance and accumulation of twelve metals. *J. Environ. Qual.* 2004;33:54–60.

Wångstrand H, Eriksson J, Öborn I. Cadmium concentration in winter wheat as affected by nitrogen fertilization. *Eur. J. Agron.* 2007;26:209–214.

Watanabe T, Shimbo S, Higashikawa K, Ikeda M. Lead and cadmium content in cereals and pulses in north-eastern China. *Sci. Total Environ.* 1998;220:137–145.

Werner T, Motyka V, Strnad M., Schmulling T. Regulation of plant growth by cytokinin. *Plant Biol.* 2001;98:10487–10492.

Wilcke W, Krauss M, Kobza J. Concentrations and forms of heavy metals in Slovak soils. *J. Plant Nutr. Soil Sci.* 2005;168:676–686.

Wu S. Effect of manganese excess on the soybean plant cultivated under various growth conditions. *J. Plant Nutr.* 1994;17:993–1003.

Xiang C, Oliver DJ. Glutathione metabolic genes coordinately respond to heavy metals and jasmonic acid in *Arabidopsis*. *Plant Cell* 1998;1539–1550.

Xiao Y, Jalkanen H, Yang Y, Mambote C, Boom R. Ferrovanadium production from petroleum fly ash and BOF flue dust. *Min. Eng.* 2010;23:1155–1157.

Yan Z, Zhang W, Chen J, Li X. Methyl jasmonate alleviates cadmium toxicity in *Solanum nigrum* by regulating metal uptake and antioxidative capacity. *Biol. Plant*, 2015.59:373–381.

Yanqun Z, Yuan L, Jianjun C, Haiyan C, Li Q, Schratz C. Hyper accumulation of Pb, Zn and Cd in herbaceous grown on lead-zinc mining area in Yunnan, China. *Environ. Int.* 2005;31:755–762.

Yazaki K. ABC transporters involved in the transport of plant secondary metabolites. *FEBS Lett.* 2006;580:1183–1191.

Ysart G, Miller P, Crews H, Robb P, Baxter M, de L'Argy C, Lofthouse S, Sargent C, Harrison N. Dietary exposure estimates of 30 elements from the UK Total Diet Study. *Food Addit. Contam.* 1999;16:391–403.

Ysart G, Miller P, Croasdale M, Crews H, Robb P, Baxter M, de l'Argy C, Harrison N. UK total diet study. Dietary exposures to aluminium, arsenic, cadmium, chromium, copper, lead, mercury, nickel, selenium, tin and zinc. *Food Addit. Contam.* 2000;17:775–786.

Yuan H, Huang X. Inhibition of root meristem growth by cadmium involves nitric oxide-mediated repression of auxin accumulation and signalling in *Arabidopsis. Plant Cell Environ.* 2016;39:120–135.

Zanthopolous N, Antoniou V, Nikolaidis E. Copper, zinc, cadmium and lead in sheep grazing in North Greece. *Bull. Environ. Contam. Toxicol.* 1999;62:691–699.

Zarei S, Ehsanpour AA, Abbaspour J. The role of over expression of P5CS gene on proline, catalase, ascorbate peroxidase activity and lipid peroxidation of transgenic tobacco (*Nicotiana tabacum* L.) plant under in vitro drought stress. *J. Cell Mol. Res.* 2012;4:43–49.

Zeid IM. Responses of *Phaseolus vulgaris* to chromium and cobalt treatments. *Biol. Planta.* 2001;44:111–115.

Zhang WH, Tyerman SD. Inhibition of water channels by HgCl₂ in intact wheat root cells. *Plant Physiol.* 1999;120:849–857.

Zhang Y. Biological role of ascorbate in plants. In: *Ascorbic Acid in Plants*, Zhang Y (Eds.). Springer Briefs in Plant Science, New York, 2013.

Zhang ZC, Chen BX, Qiu BS. Phytochelatin synthesis plays a similar role in shoots of the cadmium hyperaccumulator *Sedum alfredii* as in non-resistant plants. *Plant Cell Environ.* 2010;33:1248–1255.

Zhang ZW, Watanabe T, Shimbo S, Higashikawa K, Ikeda M. Lead and cadmium content in cereals and pulses in north-eastern China. *Sci. Total Environ.* 1998;220:137–145.

Zhao DL, Reddy KR, Kakani VG, Reddy VR. Nitrogen deficiency effects on plant growth, leaf photosynthesis, and hyperspectral reflectance properties of sorghum. *Eur. J. Agron.* 2005;22:391–403.

Zheng Y, Chen T, He J. Multivariate geostatistical analysis of heavy metals from Beijing, China. *J. Soils Sedim.* 2008;8:51–58.

Zhou ZS, Huang SQ, Guo K, Mehta SK, Zhang PC, Yang ZM. Metabolic adaptations to mercury-induced oxidative stress in roots of *Medicago sativa* L. *J. Inorg. Biochem.* 2007;101:1–9.

Zimakowska-Gnoinska D, Bech J, Tobias FJ. Assessment of the heavy metal contamination effects on the soil respiration in the Baix Llobregat (Catalonia, NE Spain). *Environ. Monit. Asses.* 2000;61:301–313.

9

Regulation of Pesticide Stress on Metabolic Activities of Plant

Santwana Tiwari, Anita Singh, and Sheo Mohan Prasad

CONTENTS

Abbreviations

APX	Ascorbate peroxide
ATP	Adenosine triphosphate
BR	Brassinosteroid
CAT	Catalase
DHAR	Dehydroascorbate reductase
EBR	Epibrassinosteroid
GR	Glutathione reductase
GSH	Glutathione
GST	Glutathione-s-transferase
MDA	Malondialdehyde
PSI	Photosystem I
PSII	Photosystem II
ROS	Reactive oxygen species
SA	Salicylic acid
SOD	Superoxide dismutase
WHO	World Health Organization

Introduction

Day by day population explosion is becoming a great barrier for the development of different countries, especially for the developing countries like India. Due to the increase in population, the problem of starvation is increasing as well. To diminish this problem, it is essential to increase food production (Carvalho, 2006). Several agriculture practices have been applied to increase the production of food crops. Use of agrochemicals is an important strategy to enhance the yield of agricultural crops. These agrochemicals are categorized in two large groups: chemical fertilizers and pesticides (Dhaliwal and Arora, 1996). However, these agrochemicals have both positive and negative effects on environmental factors as well as on human health (Essumang et al., 2009; Dankyi et al., 2014). Pesticides are commonly used in agricultural fields to control the numbers of insects, bacteria, fungi, algae, grass, and weeds (Gilden et al., 2010). In 2500 BC, pesticides were used intentionally for the first time, by Sumerians. They rubbed their bodies with foul-smelling sulfur compounds that protected them from insects and mites (Parween et al., 2016). There is no doubt that the application of pesticides increases agricultural efficiency but with more frequent use the chance of their misapplication and accidental exposure is also increased. It has been observed that only <0.1% of total pesticides reach to the site of action and the remaining volume get lost via spray drift, off-target deposition, run-off, photodegradation, and so on. Due to this large proportion of losses, pesticide application has two sides: it plays a beneficial role by increasing agricultural efficiency, but when applied in excess there are deleterious effects on non-target plants. When pesticides are applied in the field, they directly or indirectly interact with vital physiological processes and may disrupt the normal genetic makeup of plants resulting in inhibition of growth (Carvalho, 2006; Akhtar et al., 2009). These indications may be in the form of visible physiological or biochemical effects, including early senescence or generation of reactive oxygen species, etc. Along with this, application of pesticides on a large scale increases pest resistance to pesticides, their accumulation in fruit and vegetable crops, and also leads to loss of biodiversity and natural habitats (Baig et al., 2012).

In order to overcome this situation plants endogenously produce several enzymes to reduce the level of pesticide toxicity (Cherian and Oliveira, 2005). At first the enzymes cytochrome P450 monooxygenases (P450), peroxidases, and carboxylesterases activate the pesticides and then glutathione-S-transferase (GST) and UDP-glycosyltransferase help in the conjugation of activated pesticides with glutathione (GSH) and glucose in order to release less toxic and more soluble metabolites. Thereafter, they accumulate in vacuoles or in the apoplast. However, the endogenous productions of these enzymes are not able to overcome the loss caused by the excessive application of pesticides. So, to deal with this situation, several other remedies are also adapted (Pandya and Saraf, 2010; Channapatna, 2001) and among them the application of phytohormones is one of the sustainable ways to solve this problem. Sharma et al. (2017) have reported that application of 24-epibrassinolide (100 nM EBR L^{-1}) enhances the growth and antioxidant activity of *Brassica juncea* treated with imidacloprid pesticide. A similar effect of salicylic acid on *Pisum sativum* was also reported, by Singh et al. (2016). Amelioration of pesticide stress by brassinosteroids was also reported by Sharma et al. (2015) on *Oryza sativa*. Similarly, many reports suggest that the application of phytohormones is one of the better ways to reduce pesticide stress (Figure 9.1) (Singh and Prasad, 2014; Soaresa et al., 2016). The present chapter reviews different types of pesticides and their harmful effects on plants. It contains information related to the mechanism behind the regulation of physiological attributes of plants in response to pesticide stress. Moreover, the strategy involved in mitigation of pesticide stress particularly by phytohormones is also discussed in detail (Table 9.1).

Types of Pesticides

Pesticides are the important chemicals recurrently applied in agriculture to protect crop plants from harmful pests like fungi, insects, mites, and rodents (Freedman, 1995) and from competition with abundant but unwanted plant species to obtain higher yields in less time, crops of high quality, and also to reduce the labor. They are used to prevent disease in plants as well as in human beings. They include a wide range of chemicals including

insecticides, herbicides, fungicides, rodenticides, plant growth regulators, and synthetic insecticides like carbamates, pyrethroids, organophosphates, and pyrethroids (Casida and Quistad, 2004). They are classified as follows:

1. Chemical pesticides:
   a. Organophosphates: These chemicals are produced by the reaction between phosphoric acid and alcohol. They cause blockage of the nervous system by inhibiting the activity of enzyme acetylcholinesterase, resulting in its accumulation causing overstimulation of muscles. Insecticides, nerve gas herbicides, etc., are the main examples of this type.
   b. Carbamates: They are esters of carbamic acid. Their mode of action is similar to that of organophosphates, i.e., by inhibiting acetylcholinesterase, but here bond formation is reversible. Mainly insecticides are included in this category.
   c. Organochlorines: These are chlorinated hydrocarbon derivatives. They mainly disrupt the hormonal system by disrupting the endocrine system of the body by behaving just like normal hormones and thus causing adverse health problems.
   d. Pyrethroids: They cause paralysis by acting as a neuro-poison disrupting the endocrine system. They are a synthetic form of natural pyrethrin obtained from the flower Chrysanthemum. These are esters of ketoalcohol which are derivatives of chrysanthemic and pyrethroic acids. They are more stable than pyrethrin in the presence of sunlight. As they easily pass through the exoskeleton of insects, they have become very popular. Some examples that are used regularly are deltamethrin, cypermethrin, etc.

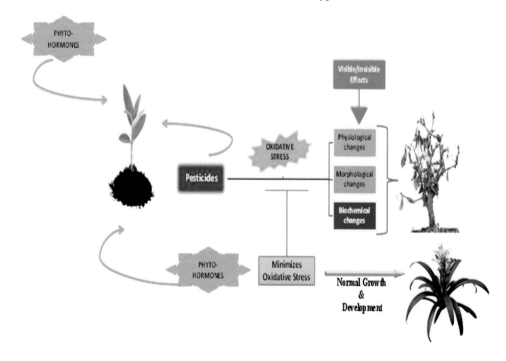

FIGURE 9.1   Impact of pesticide and phytohormones application on physiological aspects of plants.

TABLE 9.1

Impact of Pesticide Application on Physiological Activities of Plants

S. No	Name of Pesticide	Model System	Concentration	Effect	Phytohormones	Reference
1.	Chlorothalonil, Carbendazim	*Solanum lycopersicum*	Commercial recommended dose	Increase in glutathione content, GSTs activity, and glutathione reductase along with gene expression of GSTs, GR, GSH, and GPX in tomato leaves but this increase was not shown by carbendazim.	Not specified	Wang et al. (2010)
2.	Captan Maneb Zineb Carbaryl Diazinon Malathion Methomyl Permethrin Bentazon Sethoxydim PCNB Trifluralin	*Glycine max, Phaseolus lunatus*	106.5 g/L 8.1 g/L 7.2 g/L 15.6 g/L 115.2 g/L 11.7 g/L 18.8 g/L 30.0 g/L 2.8 g/L 29.4 g/L 15.5 kg/ha 2.8 L/ha	Reduction in N_2 fixation, decrease in enzyme activity involved in the N_2-fixation process.	Not specified	Yueh et al. (1993)
3.	Pretilachlor	*Azolla pinnata, Azolla microphylla*	5, 10, and 20 ppm	Biomass accumulation, photosynthetic pigments; chlorophyll a, b and carotenoids, photosynthetic oxygen yield, and photosynthetic electron transport activities, i.e. photosystem II (PS II) and photosystem I (PS I) in both the species declined with the increasing doses of pretilachlor. The lower dose mainly damages the oxidation site of PS II while higher doses damage the reduction site of PS II. A significant increase in respiration also noticed. Activity of nitrate reductase, nitrite reductase, glutamine synthetase, and glutamate synthase was also severely affected.	Not specified	Prasad et al. (2005)
4.	Chlorpyrifos	*Oryza sativa*	0.02%, 0.04%, and 0.06%	Adverse effect on growth and protein content of seedlings whereas leads to enhancement in the level of MDA and proline content. The activity of antioxidant enzymes increased.	24-epibrassinolide	Sharma et al. (2012)
5.	Paraquat Fluazifop-*p*-butyl Haloxyfop Flusilazole Cuproxat Cyazofamidl Imidacloprid Chlorpyrifos Abamectin	*Cucumis sativus* L. cv. Jinyan No. 4	2.76 g/L^{-1} 0.20 g/L^{-1} 0.14 g/L^{-1} 0.04 g/L^{-1} 0.54 g/L^{-1} 0.22 g/L^{-1} 0.02 g/L^{-1} 0.48 g/L^{-1} 0.005 g/L^{-1}	Plant shows severest phytotoxic symptoms with the highest reduction in net photosynthesis, decrease in quantum yield.	24-epibrassinolide	Xia et al. (2006)

(Continued)

TABLE 9.1 (CONTINUED)

Impact of Pesticide Application on Physiological Activities of Plants

S. No	Name of Pesticide	Model System	Concentration	Effect	Phytohormones	Reference
6.	Imidacloprid	*Brassica juncea* L. *Oryza sativa*	0, 250, 300, and 350 mg/kg 0.01%, 0.015%, and 0.02%	Relatively less increase in biosynthesis of non-enzymatic antioxidants. Adverse effect on growth and protein content of seedlings whereas an enhancement in the level of MDA and proline content was evinced. The activities of antioxidant enzymes such as superoxide dismutase (SOD), ascorbate peroxidase (APX), catalase (CAT), glutathione reductase (GR), and monodehydroascorbatereductase (MDHAR) increased after treatment with IMI.	Epibrassinolid	Sharma et al. (2013, 2016)
7.	Isoproturon	*Pisum sativum*	10 mM	Inhibition in growth variables like shoot and root height, fresh and dry biomass. Chlorophyll, carotenoid and protein content was decreased. Protein and activity of nitrate reductase was also inhibited this is mainly because of induced oxidative stress. Activity of SOD, APX, and CAT increased while GPX activity is decreased.	Salicylic acid	Singh et al. (2016)
8.	Fluroxypyr	*Chlamydomonas reinherdetii*	0.05–1.0 mgL^{-1}	Degradation of DNA due to increase in oxidative stress.	Not specified	Zhang et al. (2010)
9.	Carbendazim	*Cucumis sativus* L.	0, 5, 50 and 100 mg kg^{-1}	Increased antioxidant activity for detoxification of activated oxygen species.	Not specified	Zhang et al. (2007)
10.	Chlorpyrifos Propiconazole	*Hordeum vulgare* L.	0.05%, 0.1%, and 0.5%	Induced genotoxicity, reduced morphological parameters, chromosomal aberrations and decreased in photosynthetic pigments at high concentration.	Not specified	Dubey et al. (2016)

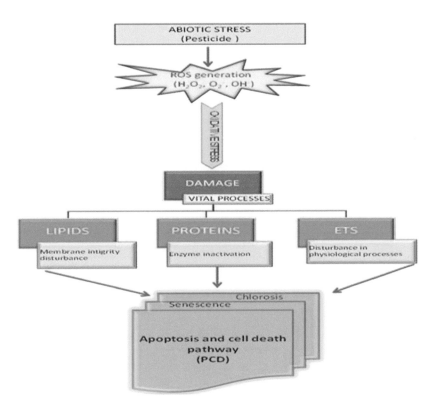

FIGURE 9.2 Disturbance in the physiological functioning of plant under pesticide stress.

2. Biopesticides: They are naturally occurring materials derived from living organisms or their metabolites, like bacteria, fungi, plants, etc. They act against bacteria, fungi, or viruses.

3. Insecticides: act especially against insects.

4. Algicides: mainly control or kill the growth of algae.

5. Herbicides: act against weeds.

6. Bactericides: mainly act against bacteria.

7. Fungicide: act against fungi.

8. Rodenticides: kill/control rodents.

9. Larvicides: act against larvae.

10. Virucides: act against viruses.

Although pesticides were designed to protect crops or human health from vector-borne diseases, they tend to pollute the environment. When they are used in high volumes they can be lethal. According to a World Health Organization (WHO) estimation, about 3,000,000 pesticides are responsible for causing 220,000 deaths worldwide (WHO, 2007). Among different kinds of pesticides, some biopesticides are found to be less harmful compared to chemical pesticides (Kumar and Singh, 2015). However, their constant application may lead to a disturbance in the physiological functioning and consequent death of a plant cell (Figure 9.2).

Effect of Pesticide Stress on Physiological Attributes of Plants

In agriculture, pesticides were introduced to fulfill the need for increased food, which is the prime demand for a growing population; thus the excessive use of pesticides has become a necessary evil. They mainly affect photosynthesis, respiration, growth, cell/nuclear division, and the synthesis of proteins, carotenoids, or lipids (DeLorenzo et al., 2001; Jaga and Dharmani, 2003). Undifferentiating use of various pesticides in current scenarios has led to a serious environmental problem. Their major impact can be seen on the physiological and morphological attributes of plants (Rio et al., 2012; Kilic et al., 2015).

Effect on Photosynthetic Machinery and Growth of Plant

Pesticide stress causes visible as well as invisible effects on plants. Visible effects include stunting of plants, reduction in leaf area, shoot distortion, and chlorosis, which are prominent indications of deleterious effects in a dose-dependent manner. These effects are also coupled with a decrease in photosynthetic pigments (Murthy et al., 2005; Kaushik and Inderjit, 2006). A decrease in the biosynthesis of chlorophyll *a*, chlorophyll *b*, carotenoids, and total pigment content caused by the application of benomyl was reported in *Helianthus annuus* (Dias, 2012). Reduction in pigment content leads to yellowing, browning, or drying of some parts of maize leaves by the application of clethodim (Radwan, 2012) and almost complete retardation of growth by the application of monocrotophos was also reported by Saraf and Sood (2002). These pesticides disrupt the basic process by inhibiting electron transport of photosystem II, which results in the interruption of the photosynthesis process by binding dimethoate directly to the different sites of the photosynthetic electron transport chain that decrease photosynthetic oxygen yield in the plant *Vigna unguiculata* (Mishra et al., 2008). Bader and Abdel-Basset (1999) reported that the application of triforine, a fungicide, strongly inhibits the electron transport system in chloroplast. After the application of fungicide, a detailed study on the cellular level is available describing the damage to the photosynthetic apparatus (Saladin et al., 2003; Untiedt and Blanke, 2004; Petit et al., 2008). The growth factor is mainly accomplished by photosynthetic rate and mobilization of carbohydrate reserves (Petit et al., 2008) any alteration in these factors disrupts the growth of the plant adversely (Dias, 2012). Bashir et al. (2007) have reported a reduction in growth of *Lens culinaris* L. by mancozeb due to inactivitation of the enzyme 4-hydroxyphenylpyruvate dioxygenase, which is mainly required for the division of meristematic tissues (Mishra et al., 2008). According to Stevens et al. (2008), continuous exposure of imidacloprid reduces the germination and growth of rice seeds and seedlings, respectively. The reduction in growth may be due to the direct interference of pesticide on the activity of photosynthetic pigment and photosynthetic activity and this effect is also reported in *Glycine max* L. (Panduranga et al., 2005) and *Vigna unguiculata* L. (Mishra et al., 2008). The adverse effect of pretilachlor was seen in photosynthetic activity in *Azolla* sp. (Prasad et al., 2016). Several studies reveal that the chlorophyll *a* fluorescence activity is modified negatively by the application of fungicides. Alteration in several parameters of chlorophyll *a* fluorescence concluded that the light reaction of photosynthesis is quite sensitive to fungicide application (Saladin et al., 2003; Xia et al., 2006). The effective quantum yield of PSII (Φ_{PSII}) as well as the maximum quantum efficiency of PSII (Fv/Fm) is adversely affected by the application of benzimidazoles and triazole, and dithiocarbonate also (Nason et al., 2007). Blockage at the Qi site of cytochrome *f* during electron transport between PSII and PSI was seen by the application of the fungicide strobilurin by Nason et al. (2007). It is well known that vitamin synthesis takes place in the chloroplast, and any fluctuation in the synthesis of photosynthetic pigment substances may also cause a decline in the nutrition value of the pepper plant. The effect of herbicide treatments on chlorophyll content in foliage at flowering (45 DAS) steadily declined with increasing rates of herbicides but was significant ($p < 0.05$) only at 400 mg active ingredient (a.i.) kg^{-1} of atrazine, isoproturon, and metribuzin (Khan et al., 2006). It is well-reported that the process of photosynthesis is inhibited by the application of different pesticides and a plant's physiological state may also be affected. Therefore, biochemical parameters like ATP-formation, CO_2-fixation, and O_2-evolution which is a key factor of photosynthesis, have been used as reliable indicators for pesticides and other pollutant toxicity. In contrast to photosynthesis, respiration was significantly increased in *Azolla* sp. by the application of pretilachlor (Prasad et al., 2016). Raja et al. (2012) also reported that the respiration rate is increased by the application of endosulfan in *Azolla microphylla*. This enhancement may be for the fulfillment of energy costs of the cell, which is associated with the repair of different cellular organelles (Prasad et al., 2016). Nitrogen is regarded as one of the most essential elements and its availability in the soil is one of the key factors to determine plant growth and productivity. It was reported that nitrogen-containing compounds and other primary metabolites of plants may also contribute to plant resistance (Haukioja et al., 1991; Berenbaum, 1995). Any significant

alteration by pesticides in the nitrogen uptake and the activity of key nitrogen assimilatory enzymes would severely affect the growth and productivity of the crops. Prasad et al. (2016) have reported that pretilachlor affects nitrogen assimilatory enzymes negatively, binding with their sulfhydryl (-SH) group. Another reason might be the damage of the photosynthetic electron transport chain which donates the electron for the reduction of NO_2^- via ferredoxin. Additionally, a decrease in the activity of these enzymes may also be associated with the inhibition at transcriptional level.

Effect of Pesticides on Biochemical Activities of Plants

Apart from the visible effect of toxicity of pesticides in plants, many studies illustrate the indirect effects of the abundant use of pesticides on targeted as well as non-targeted organisms. A common sign of isoproturon toxicity is a disproportionate accumulation of ROS (reactive oxygen species) (Gill and Tuteja, 2010; Singh et al., 2016). It is well-reported that pesticides or herbicides are related to the overproduction of free radicals and oxidative stress. ROS at high concentrations are extremely harmful to living organisms. They are generated by stepwise reduction of molecular oxygen (O_2) by electron-transfer reactions leading to the production of the highly reactive ROS (Gill and Tuteja, 2010; Takagi et al., 2017). One of the consequences is the excessive generation or accumulation of malondialdehyde (MDA) content (Sergiev et al., 2006), which is the cause of severe disturbances in the membrane integrity, which may lead to senescence and ultimately death (Ogweno et al., 2009; Xu et al., 2006; Li et al., 2004; Guo et al., 2006). Sergiev et al. (2006) reported that glyphosate induces stress in maize plants. Imidacloprid application also causes stress in plants (Zhou et al., 2015), including generation of ROS. It also includes free radicals such as superoxide anion ($O_2 \bullet -$), hydroxyl radical ($\bullet OH$), as well as non-radical molecules like hydrogen peroxide (H_2O_2), singlet oxygen ($_1O^2$), and so on. Oxyfluorfen is capable of decreasing growth indirectly by generating singlet oxygen ($_1O^2$), superoxide radical ($O_2 \bullet^-$), hydrogen peroxide (H_2O_2), and hydroxyl radicals (OH). According to Mittler et al. (2004) enhanced production of ROS by cell organelles like chloroplast and mitochondria hampers carbon fixation in plants which also results in the formation of SOR and H_2O_2 by the photosystem (Foyer and Nector, 2005; Gomes et al., 2017). The presence of acifluorfen and endosulfan in lower organisms leads to the production of lipid hydroperoxides that disturb the integrity of different cell organelle's membranes (Prasad et al., 2005; Bashri and Prasad, 2016). Lipid peroxidation and free radical generation are mainly considered as agents for leaf senescence. When the level of ROS increases, it damages the chlorophyll, leading to chlorosis (Dubey et al., 2016). Another adverse effect of pesticides is that they may alter the competitive interactions among species, and species dominance, richness, and distribution, which can lead to further adverse effects in the ecosystem (Kleijn and Snoeijing, 1997). Several studies have explored how heavy use of pesticide severely affects the soil microorganisms, including cyanobacteria *Nostoc*, *Anabaena*, etc., which are essential components of soil responsible for its fertility.

When an increased level of ROS exceeds the defense mechanisms, a cell is said to be in a state of "oxidative stress." The toxicity of pesticides affects plants at different levels, either disrupting biochemical characteristics or by disturbing signaling mechanisms (Fufezan et al., 2002; Ramal et al., 2007; Liu et al., 2009; Ramel et al., 2009; Unver et al., 2010). The excessive generation or accumulation of ROS leads to the destruction of functional protein, inactivation of enzymes, oxidation of lipids, and changing conformation of nucleic acids (Ryter et al., 2007; Franco et al., 2009, 2010; Tebourbi et al., 2011). In order to reduce the level of pesticide, the activity of antioxidants is increased by the application of chlorpyrifos in rice plants (Sharma et al., 2012). The molecules generated by stress are mainly responsible for production of stress proteins as well as up-regulation of antioxidants. There are numerous receptors present on the cell surface that sense the external foreign bodies to develop the cascade of events leading to neutralization and elimination of such compounds (Terbourbi et al., 2011). Plants possess many antioxidative enzymes which mainly scavenge ROS and minimize its deleterious impact. These are the critical weapons of a plant cell which maintain the appropriate internal mechanism to regulate the correct functioning of the plant (Dubey et al., 2016). The antioxidant systems are mainly categorized in two types, i.e., enzymatic and non-enzymatic antioxidants. Enzymatic antioxidants are mainly superoxide dismutase (SOD), catalase (CAT), ascorbate peroxides (APX), dehydroascorbate reductase (DHAR), glutathione reductase (GR), and glutathione-s-transferase (GST). Non-enzymatic antioxidants, mainly ascorbate and glutathione, are effectively involved in detoxifying the ROS effect in plants (Godim et al., 2012; Thounaojam et al., 2012). SOD is mainly considered as the first line of defense against ROS. SOR ($O_2{}^{\bullet-}$), which is produced by the photosynthetic and respiratory process, is very reactive. Its activity increases by the interference of chlorpyrifos in rice plants (Sharma et al., 2012). SOD dismutates SOR into the less damaging species H_2O_2 (Dubey et al., 2016). This process leads to protection against oxidative stress. Further, catalase (CAT) is mainly found in peroxisome to catalyze the conversion of hydrogen peroxide (H_2O_2) into water and oxygen. APX also scavenges H_2O_2 and is essential for the protection of chloroplast and other cell components from damage caused by H_2O_2 and hydroxyl radicals ($\bullet OH$). DHAR is basically located in the chloroplast and cytosol of higher plants. It plays an important role in the ascorbate-glutathione cycle in higher plants. It utilizes glutathione (GSH) to reduce dehydroascorbate (DHA) and thus regenerates reduced ascorbate (AsA). Ascorbate is oxidized to DHA by means of successive reversible single electron transfers producing monodehydroascorbate (MDHA) as a free radical intermediate. MDHA spontaneously disproportionates to DHA because MDHA radicals have a relatively short lifetime. DHA is reduced to AsA by DHAR in a reaction requiring reduced GSH. By liberating protons that are utilized in the recycling reaction of AsA, DHAR is converted into oxidized glutathione (GSSG) (Dubey et al., 2016). Ascorbate-glutathione is also known as the Asada-Halliwell pathway, which plays a crucial role in detoxification of ROS. In this cycle, glutathione reductase (GR) is a very important enzyme, which belongs to the flavoenzyme family. It mainly catalyzes reduction of reduced GSSG into GSH and in turn NADP$^+$ is converted into NADPH. It maintains the level of glutathione by providing it in the plant cell for scavenging the ROS by other enzymes. Therefore, its involvement helps the plant cell for self-defense. Glutathione-s-transferase plays a key role in the cellular detoxification metabolism in different organisms including plants. Plant GST not only inactivates toxic compounds but also plays an important role in the defense system. GST helps in the electrophilic addition of the reduced form of glutathione (GSH; γ-glutamylcysteine-glycine) in varieties of hydrophobic compounds (Dubey et al., 2016). The resulting molecule, i.e., glutathione-s-conjugates are less reactive and more polar than the xenobiotic chemicals. Detoxification by GSH conjugation is widely reported, which is a very effective system in plants that metabolizes various pesticides such as triazines, thiocarbamates, and chloroacetanilides. Maracacci et al. (2006) reported the effectiveness of GSH conjugation for detoxifying atrazine in *Chrysopogon zizanioides* Nash, a species resistant to herbicide. Apart from enzymatic antioxidants, non-enzymatic antioxidants participate simultaneously during the process of detoxification against pesticides. Among these, ascorbate (AsA) and glutathione (GSH) play crucial roles in pesticide metabolism. They protect the cell from the harmful effect of ROS by scavenging directly or indirectly through activation of the defense mechanism. Many studies have reported that their concentration in cells are between 10–300 mM and 0.1–25 mM, respectively (Ball et al., 2004; Shao et al., 2008). Ascorbate and glutathione are also responsible for redox signaling through the maintenance of the ratio between reduced and oxidized forms. Zhang et al. (2007) have reported a strong correlation between antioxidant production and fungicide carbendazim concentration. They observed that in the roots, stem, and leaves of cucumber the activity of SOD, CAT, and GPX enzymes increases in order to maintain a balance between the formation and detoxification of activated oxygen species. Wu et al. (2002) have reported significant enhancement in SOD and CAT activity after the exposure of the fungicides azoxystrobin and epoxiconazole. It was also observed that applications of paraquat resulted in increased membrane permeability and MDA levels and a decrease in unsaturated fatty acids levels that consequently lead to enhancement of the lipid peroxidation reactions (Parween et al., 2016). Catalase enzymes also provide tolerance toward herbicide stress and enhancement in their activity during herbicide exposure (Radetski et al., 2000; Jung et al., 2006). Further, to maintain the physiological functioning of plants in response to pesticide stress, some other strategies were adapted that are discussed in detail in the following section.

Regulation of Pesticide Stress by Phytohormone Application

Many compounds are present in plants that directly or indirectly influence the developmental processes and enable the plant system to rapidly adapt in response to any alternation in the internal or external environment. These compounds are commonly called phytohormones. Their endogenous levels are altered according to the external environmental stress condition (Wang et al., 2005; Ashraf et al., 2008; Ramakrishna and Ravishankar, 2011, 2013). They are one of the important parts of the internal system of the plant which integrate metabolic and developmental events in the form of response against external factors and are essential for many processes which influence the quality and yield of the crops. Out of several strategies adopted by plants to neutralize the adverse effect of different stresses, phytohormones provide signals to plants for their survival under stress conditions. They show impact on growth, development, nutrient allocation, and source or sink transitions. These changes may be due to biotic or abiotic factors. Abiotic factors like salinity, heat, cold, drought, flooding, and soil infertility cause crop loss all over the world (Ramakrishna and Ravishankar, 2011). Due to all these factors, the quality and quantity of crop productions are adversely affected, which are the main emerging problems for farmers. This is mainly because of the frequent use of pesticides in the agricultural field. Pesticides directly affect the physiological as well as other aspects of the plant. Several pesticides are responsible for causing toxicity to the plant that consequently results in retardation of the growth rate of the plants (Sharma et al., 2015; Singh et al., 2016). They are mainly used for betterment of crop production against biotic stresses but in turn they hamper the growth and metabolism of the plants (Sharma et al., 2013). For these reasons, the requirement of food cannot be fulfilled according to the need of the population. To fight against these conditions, several strategies are advised to the farmers to improve agriculture conditions but they are not very efficient or cost effective. In this era of fast-growing techniques and strategies for the improvement of agricultural practice, sustainable agriculture is the biggest challenge (Dhaubhadel et al., 2002; Zhang et al., 2007). In recent times, phytohormones have been focused on by researchers to reduce pesticide stress (Table 9.1). Auxins are growth promoting hormones reported to impart resistance to the toxicity of several herbicides. In *Zea mays* benzyladenine shows a protective effect on chlorophyll, carotenoid, and ascorbic acid contents of leaves against paraquat by inducing the protein kinase cascade activated by mitogen, which reduces oxidative damage. Its pretreatment significantly increases SOD activity (Durmus et al., 2015). Phenylurea cytokinin reduced the toxicity of 10 mM glyphosate by increasing the potentiality of antioxidant systems (Sergeiv et al., 2006). In *Oryza sativa* plants, abscisic acid reduces the toxicity of phosphinothricin and also reduces the accumulation of ammonium ion in leaves (Hsu and Kao, 2004).

Along with the traditional hormones, there are certain signaling molecules like salicylic acid (SA) which are reported for their protective action against pesticide toxicity (Sakhabutdinova et al., 2003). SA is a plant phenolic compound that is commonly present in plant tissues, which is capable of regulating plant growth and development (Gallego-Giraldo et al., 2011; Vicente and Plasencia, 2011; Berkowitz et al., 2016). Popova et al. (2004) also reported that SA is involved in plant response to herbicide stress. Salicylic acid status and ROS production in plants subjected to biotic/abiotic stresses are well established (Kawano, 2003). SA induces adaptability to herbicide toxicity in some crop plants including wheat, oat, maize, rapeseed, and sunflower (Cui et al., 2010; Liang et al., 2012; Radwan, 2012; Kaya and Yigit, 2014; Yigit et al., 2016). SA reversed the effect of isoproturon in pea seedlings and enhanced the growth and accumulation of biomass (Singh et al., 2016). The increase in growth is directly related to the increase in photosynthesis and chlorophyll content. Enhancement in the photosynthesis and chlorophyll content could be attributed to an increase in the activity of certain

enzymes with the application of salicylic acid (Hayat et al., 2005). Pretreatment of barley plants with SA before paraquat application protected them against paraquat-induced damage by inducing the antioxidant potential of the plant (Popova et al., 2002; Ananievaa et al., 2004; Apel and Hirt, 2004). Salicylic acid facilitates paraquat tolerance in *Hordeum vulgare* (Ananieva et al., 2002) and *Zea mays* seedlings. It had improved growth of *Brassica napus* treated by napropamide and reduced its levels in the plants. Its application also decreased the number of ROS and increased the activities of POD, guaiacol peroxidase (GPX), and GST (Cui et al., 2010). It prevents herbicides from entering into the cells and consequently the cell membrane integrity is improved. Similarly, Radwan (2012) reported SA-induced alleviation of oxidative stress caused by clethodim in maize (*Zea mays* L.). Wang and Zhang (2017) reported that at low concentrations (1, 2, and 4 mg L^{-1}) SA significantly alleviates the toxicity of chlorpyrifos on wheat seedlings.

Brassinosteroids (BRs) are plant steroids usually occurring in almost every part of the plant, such as pollen grain, floral buds, fruits, and seeds as well as in vascular cambium, leaves, shoots, and roots, which are well-known for increasing the capability of plants to resist against various abiotic stresses like pesticides (Hayat et al., 2010, 2014; Sharma et al., 2012, 2013, 2015; Bajguz and Hayat, 2009). It has been reported that application of 24-epibrassinolide (EBR) externally to plants decreases the pesticide residues (Xia et al., 2009; Zhou et al., 2015; Sharma et al., 2016a,b). Its application may also change gene expression in plants, which increases the potentiality of plants against pesticide stress (Xia et al., 2009; Sharma et al., 2015; Zhou et al., 2015). Treating the seed by soaking with EBR before sowing has been reported to ameliorate the pesticide toxicity in plants (Sharma et al., 2012, 2016a). Moreover, seed soaking with EBR leads to the recovery of the growth of *B. juncea* seedlings treated with Imidacloprid (IMI) pesticide. The improvement in seedling growth after the application of EBR might be due to the ability of BRs to modulate the biosynthesis of cellulose, along with cell division and cell elongation (González-García et al., 2011; Xie et al., 2011). Sharma et al. (2015) reported the increase in growth parameters after the exogenous application of BRs in rice seedlings. It has also been reported by Honnerova et al. (2010) that the transcription and translation process is being up-regulated by BRs during the biosynthesis and also reduces the degree of degradation of chlorophyll. A study by Pinol and Simon (2009) reveals the protective action of brassinosteroids against photosynthesis-inhibiting herbicides. Their study shows that epibrassinolide induced the recovery of chlorophyll fluorescence and photosynthetic CO_2 assimilation from damage caused by terbutryn in *Vicia faba* plants. Brassinosteroids alleviate the harmful effect of s-triazine herbicide, which is analyzed by chlorophyll fluorescence together with the measures of photosynthetic CO_2 assimilation and plant growth (Pinol et al., 2011). BR treatment induced alleviation of the damage in rice caused by the application of simazine, butachlor, or pretilachlor (Sasse, 2003). Xia et al. (2009) studied the phytotoxic effect of nine pesticides – three herbicides, three fungicides and three insecticides – on cucumber leaves. They reported that 24 epibrassinolide (24-epiBL) increased the resistance of plants to pesticides probably by enhancing CO_2 assimilation. BRs have also been reported as effective in reducing the damage caused by pesticides. Sharma et al. (2016a) worked out that application of epibrassinolide (EBR) increased the length of seedlings, dry weight, pigment contents, polyphenols, total phenols, and organic acids under Imidacloprid (IMI) toxicity (Sharma et al., 2013). The application of IMI tends to degrade the protein due to enhanced activity of proteases or by autophagy to degrade oxidized proteins under stress conditions (Xiong et al., 2007). The application of EBL in the presence of IMI led to an enhancement in the level of proteins which may be due to the well-documented effect of BRs on activation of total protein synthesis and induced de novo polypeptide synthesis (Dhaubhadel, 2002). Proline accumulation under stress acts as an osmoprotectant and helps in the stabilization of enzymes and membrane structures (Bandurska, 2001)and its accumulation mainly happens in response to various stresses, which helps in stress regulation of cellular acidity and ROS scavenging (Rodriguez and Radman, 2005). It is also considered as a signaling molecule that can modulate mitochondrial function and influence gene expression in the nucleus and thereby increases plants' ability to tolerate stress (Szabados, 2010). Thus, EBL-induced proline accumulation under IMI stress may be perceived as a stress protection strategy. EBL application also decreases the lipid peroxidation resulting in the amelioration of IMI stress, as demonstrated by Sarma et al. (2013). The damaging effect of pesticides is mainly diminished by the up-regulation of antioxidants like SOD, POD, etc. Banerjee et al. (2001) and Chaudhary et al. (2012) have also reported that the expressions of some key antioxidant genes (Cu/Zn-SOD, Fe-SOD, Mn-SOD, APX, CAT, and GR) were found to be up-regulated in rice seedlings under various treatments of IMI and EBL (Gomez et al., 2004).

Conclusion

Pesticide application increases plant yield by reducing the attack of pests, but when applied in excess they can have deleterious effects on plant physiology and contaminate other natural resources. Pesticides can also result in food chain contamination through their accumulation in food crops and cause serious health hazards. To deal with this problem, several innovative approaches are being taken. Among them, the application of phytohormones is one of the sustainable strategies to reduce the negative impact of pesticides on plants. Exogenous application of plant hormones like auxin, salicylic acid, and brassinosteroids plays an important role in alleviating pesticide toxicity in plants. These hormones help in the regulation of pesticide stress by up- and down-regulating the expression of specific genes involved in processes related to detoxification. Further, to elucidate ameliorative strategies in the future, some proteomic and genomic studies should be done to understand the mechanism behind the regulation of physiological functioning of plants under pesticide stress.

Acknowledgments

The University Grants Commission, New Delhi, is acknowledged for providing financial support to Santwana Tiwari as a D.Phil. research scholar. The author A. Singh is thankful to the CSIR [No. 13(8913-A)/2017-pool] for providing funds as a senior research associate. The authors also acknowledge the University of Allahabad for providing necessary facilities.

REFERENCES

Akhtar MW, Sengupta D, Chowdhury A. Impact of pesticides use in agriculture: Their benefits and hazards. *Interdiscip Toxicol.* 2009; 2: 1–12.

Ananieva AE, Alexieva VS, Popova LP. Treatment with salicylic acid decreases the effects of paraquat on photosynthesis. *J. Plant Physiol.* 2002; 159(7): 685–693.

Ananieva AE, Christova KN, Popova LP. Exogenous treatment with salicylic acid leads to increased antioxidant capacity in leaves of barley plants exposed to Paraquat. *J. Plant Physiol.* 2004; 161: 319–328.

Apel K, Hirt H. Reactive oxygen species: Metabolism, oxidative stress, and signal transduction. *Annu. Rev. Plant Biol.* 2004; 55: 373–399.

Ashraf M, Athar HR, Harris PJC, Kwon TR. Some prospective strategies for improving crop salt tolerance. *Adv. Agron.* 2008; 97: 45–110.

Bader KP, Abdel-Basset R. Adaptation of plants to anthropogenic and environmental stresses. In: *The Effects of Air Constituents and Plant-Protective Chemicals, Handbook of Plant and Crop Stresses*, Pessarakli M (Ed.). New York, NY: Marcel Dekker, 1999, pp. 973–1010.

Baig KJ, Wagner P, Ananjeva NB, Bohme W. A morphology-based taxonomic revision of *Laudakia* Gray, 1845 (Squamata: Agamidae). *Vertebr. Zool.* 2012; 62: 213–260.

Bajguz A, Hayat S. Effects of brassinosteroids on the plant responses to environmental stresses. *Plant Physiol Biochem.* 2009; 47: 1–8.

Ball L, Accotto GP, Bechtold U, Creissen G, Funck D, Jimenez A, Kular B, et al. Evidence for a direct link between glutathione synthesis and stress defence gene expression in *Arabidopsis*. *Plant Cell* 2004; 16: 2448–2462.

Bandurska H. Does proline accumulated in leaves of water deficit stressed barley plants confine cell membrane injuries? II. Proline accumulation during hardening and its involvement in reducing membrane injuries in leaves subjected to severe osmotic stress. *Acta Physiol. Plant* 2001; 23(4): 483–490.

Banerjee BD, Seth V, Ahmed RS. Pesticide-induced oxidative stress: Perspectives and trends. *Rev. Environ. Health.* 2001; 16: 1–40.

Bashir F, Siddiqi TO, Mahmooduzzafar, Iqbal M. Effects of different concentrations of mancozeb on the morphology and anatomy of *Lens culinaris* L. *Ind. J. Environ. Sci.* 2007; 11: 71–74.

Bashri G, Prasad SM. Exogenous IAA differentially affects growth, oxidative stress and antioxidants system in Cd stressed Trigonella foenum-graecum L. seedlings: Toxicity alleviation by up-regulation of ascorbate glutathione Cycle. *Ecotoxicol. Environ. Saf.* 2016; 132: 329–338.

Berenbaum, May R. The chemistry of defense: Theory and practice. *Proc. Natl. Acad. Sci.* 1995; 92(1): 2–8.

Berkowitz O, De Clercq I, Van Breusegem F, Whelan J. Interaction between hormonal and mitochondrial signalling during growth, development and in plant defence responses. *Plant, Cell Environ.* 2016; 39: 1127–1139.

Carvalho FP. Agriculture, pesticides, food security and food safety. *Environ. Sci. Policy* 2006; 9: 685–692.

Casida JE, Quistad GB. Organophosphate toxicology: Safety aspects of on acetylcholinesterase secondary targets. *Chem Res Toxicol.* 2004; 17: 983–998.

Channapatna SP. The genetically modified crop debate in the context of agricultural evolution. *Plant Physiol.* 2001; 126: 8–15.

Chaudhary SP, Kanwar M, Bhardwaj R, Yu JQ, Tran LP. Chromium stress mitigation by polyamine-brassinosteroid application involves phytohormonal and physiological strategies in Raphanus sativus L. *PLoS One* 2012; 7: 33210.

Cherian S, Oliveira MM. Transgenic plants in phytoremediation: Recent advances and new possibilities. *Environ. Sci. Technol.* 2005; 39: 9377–9390.

Cui J, Zhang R, Wu GL, Zhu HM, Yang H. Salicylic acid reduces napropamide toxicity by preventing its accumulation in rapeseed (*Brassica napus* L.). *Arch. Environ. Contam. Toxicol.* 2010; 59: 100–108.

Dankyi E, Gordon C, Carboo D, Fomsgaard IS. Quantification of neonicotinoid insecticide residues in soils from cocoa plantations using a QuEChERS extraction procedure and LC-MS/ MS. *Sci. Total Environ.* 2014; 499: 276–283.

DeLorenzo ME, Scott GI, Ross PE. Toxicity of pesticides to aquatic microorganisms: A review. *Environ. Toxicol. Chem.* 2001; 20: 84–98.

Dhaliwal GS, Arora R. *Principles of Insect Pest Management.* National Agricultural Technology Information Centre, 1996.

Dhaubhadel S, Browning KS, Gallie DR, Krishna P. Brassinosteroid functions to protect the translational machinery and heat-shock protein synthesis following thermal stress. *Plant J.* 2002; 29: 681–691.

Dias MC. Phytotoxicity: An overview of the physiological responses of plants exposed to fungicides. *J. Bot.* 2012. doi:10.1155/2012/135479

Dubey G, Mishra N, Prasad SM. Metabolic responses of pesticides in plants and their ameliorative processes. In: *Plant Responses to Xenobiotics*, Singh A, Singh RP, Prasad SM (Eds.), Springer Nature Singapore Pte Ltd., 2016.

Durmuş N, Bekircan T. Pretreatment with polyamines alleviate the deleterious effects of diuron in maize leaves. *Acta Biol. Hung.* 2015; 66: 52–65.

Essumang DK, Togoh GK, Chokky L. Pesticide residues in the water and fish (Lagoon Tilapia) samples from Lagoons in Ghana. *Bull. Chem. Soc. Ethiop.* 2009; 23: 19–27.

Foyer CH, Noctor G. Redox homeostasis and antioxidant signalling: A metabolic link between stress perception and physiological responses. *Plant Cell* 2005; 17: 1866–1875.

Franco R, Li S, Rodriguez-Rocha H, Burns M, Panayiotidis MI. Molecular mechanisms of pesticide-induced neurotoxicity: Relevance to Parkinson's disease. *Chem. Biol. Interact.* 2010; 188: 289–300.

Franco R, Sanchez-Olea R, Reyes-Reyes EM, Panayiotidis MI. Environmental toxicity, oxidative stress and apoptosis: Menagea trois. *Mutat. Res.* 2009; 674: 3–22.

Freedman B. *Environmental Ecology: The Ecological Effects of Pollution, Disturbance, and Other Stresses.* San Diego, CA: Academic Press, 1995.

Fufezan C, Rutherford AW, Krieger-Liszkay A. Singlet oxygen production in herbicide-treated photosystem II. *FEBS Lett.* 2002; 532: 407–410.

Gallego-Giraldo L, Escamilla-Trevino L, Jackson LA, Dixon RA. Salicylic acid mediates the reduced growth of lignin down-regulated plants. *Proc. Natl. Acad. Sci. U.S.A.* 2011; 108: 20814–20819.

Gilden RC, Katie H, Barbara S. Pesticides and health risks. *J. Obst. Gynecol. Neonat. Nurs.* 2010; 39(1): 103–110.

Gill SS, Tuteja N. Reactive oxygen species and antioxidant machinery in abiotic stress tolerance in crop plants. *Plant Physiol. Biochem.* 2010; 48: 909–930.

Gomes T, Xie L, Brede O, Solhaug KA, Salbu B, Tollefsen KE. Sensitivity of the green algae *Chlamydomonas reinhardtii* to gamma radiation: Photosynthetic performance and ROS formation. *Aqua Toxicol.* 2017; 183: 1–10.

Gomez JM, Jimenez A, Olmos E, Sevilla F. Location and effects of long-term NaCl stress on superoxide dismutase and ascorbate peroxidase isoenzymes of pea (*Pisum sativum* cv. Puget) chloroplasts. *J. Exp. Bot.* 2004; 55: 119–130.

González-García MP, Vilarrasa-Blasi J, Zhiponova M, Divol F, Mora-García S, Russinova E. Brassinosteroids control meristemsize by promoting cell cycle progression in Arabidopsis roots. *Development* 2011; 138: 849–859.

Guo YP, Zhou HF, Zhang LC. Photosynthetic characteristics and protective mechanisms against photooxidation during high temperature stress in two citrus species. *Sci. Hortic.* 2006; 108: 260–267.

Haukioja, Erkki. Induction of defenses in trees. *Annu. Rev. Entomol.* 1991; 36(1): 25–42.

Hayat S, Fariduddin Q, Ali B, Ahmad A. Effect of salicylic acid on growth and enzyme activities of wheat seedlings. *Acta Agron. Hung.* 2005; 53: 433–437.

Hayat Q, Hayat S, Irfan M, Ahmad A. Effect of exogenous salicylic acid under changing environment: A review. *Environ. Exp. Bot.* 2010; 68: 14–25.

Hayat S, Khalique G, Wani AS, Alyemeni MN, Ahmad A. Protection of growth in response to 28-homobrassinolide under the stress of cadmium and salinity in wheat. *Int. J. Biol. Macromol.* 2014; 64: 130–136.

Honnerova J, Rothova O, Hola D, Koccova M, Kohout L, Kvasnica M. The exogenous application of brassinosteroids to Zea mays (L.) stressed by long-term chilling does not affect the activities of photosystem 1 or 2. *J. Plant Growth Regul.* 2010; 29(4): 500–505.

Hsu YT, Kao CH. Phosphinothricin tolerance in rice *Oryza sativa* L. seedlings is associated with elevated abscisic acid in the leaves. *Bot. Bull. Acad. Sin.* 2004; 45: 41–48.

Jaga K, Dharmani C. Sources of exposure to and public health implications of organophosphate pesticides. *Pan Am. J. Public Health.* 2003; 14: 171–185.

Jung S, Chon S, Kuk Y. Differential antioxidant responses in catalase-deficient maize mutants exposed to norflurazon. *Biol. Plant* 2006; 50: 383–388.

Kaushik S, Inderjit S. Phytotoxicity of selected herbicides to mung bean (*Phaseolus aureus* Roxb.). *Environ. Exp. Bot.* 2006; 55: 41–48.

Kawano T. Roles of the reactive oxygen species-generating peroxidase reactions in plant defense and growth induction. *Plant Cell Rep.* 2003; 21: 829–837.

Kaya A, Yigit E. The physiological and biochemical effects of salicylic acid on sunflowers (*Helianthus annuus*) exposed to flurochloridone. *Ecotox. Environ. Saf.* 2014; 106: 232–238.

Khan MS, Chaudhary P, Wani PA, Zaidi A. Biotoxic effects of the herbicides on growth, seed yield and grain protein of green gram. *J. Appl. Sci. Environ. Manage.* 2006; 10: 141–146.

Kilic S, Duran RE, Coskun Y. Morphological and physiological responses of maize (*Zea mays* L.) seeds grown under increasing concentrations of chlorantraniliprole insecticide. *Pol. J. Environ. Stud.* 2015; 3: 1069–1075.

Kleijn G, Snoeijing IJ. Field boundary vegetation and the effects of agrochemical drift: Botanical change caused by low levels of herbicide and fertilizer. *J. Appl. Ecol.* 1997; 34: 1413–1425.

Kumar S, Singh A. Biopesticides: Present status and the future prospects. *J. Biofertil. Biopest.* 2015; 6: 129.

Li J, Zhao X, Matsui S, Maezawa S. Changes in antioxidative levels in aging cucumber (*Cucumis sativus* L.) leaves. *J. Jpn. Soc. Hort. Sci.* 2004; 73: 491–495.

Liang Lu, Yan Li Lu, Hong Yang. Toxicology of isoproturon to the food crop wheat as affected by salicylic acid. *Environ. Sci. Pollut. Res.* 2012; 19(6): 2044–2054.

Liu H, Weisman D, Ye YB, Cui B, Huang YH, Colon-Carmona A, Wang ZH. An oxidative stress response to polycyclic aromatic hydrocarbon exposure is rapid and complex in *Arabidopsis thaliana*. *Plant Sci.* 2009; 176: 375–382.

Marcacci S, Raveton M, Ravenel P, Schwitzguebel J. "Conjugation of atrazine in vetiver (Chrysopogon zizanioides Nash) grown in hydroponics. *Environ. Exp. Bot.* 2006; 56(2): 205–215.

MRivas-San Vicente, J Plasencia. Salicylic acid beyond defence: Its role in plant growth and development. *J. Exp. Bot.* 2011; 62: 3321–3338.

Mishra V, Srivastava G, Prasad SM, Abraham G. Growth, photosynthetic pigments and photosynthetic activity during seedling stage of cowpea (*Vigna unguiculata*) in response to UV-B and dimethoate. *Pest. Biochem. Physiol.* 2008; 92: 30–37.

Mittler R, Vanderauwera S, Gollery M, Van Breusegem F. Reactive oxygen gene network of plants. *Trends Plant Sci.* 2004; 9: 490–498.

Murthy GP, Prasad GM, Sudarshana MS. Toxicity of different imbibition periods of dimethoate on germination, chlorophyll *a/b* and dry matter of *Glycine max* (L.) Merrill. cv. KHSB-2, during early seedling growth. *J. Physiol. Res.* 2005; 18: 199–201.

Nason MA, Farrar J, Bartlett D. Strobilurin fungicides induce changes in photosynthetic gas exchange that do not improve water use efficiency of plants grown under conditions of water stress. *Pest. Manage. Sci.* 2007; 63: 1191–1200.

Ogweno JO, Song XS, Hu WH, Shi K, Zhou YH, Yu JQ. Detached leaves of tomato differ in their photosynthetic physiological response to moderate high and low temperature stress. *Sci. Hortic.* 2009; 123: 17–22.

Panduranga G, Prasad GM, Sudarshana MS. Toxicity of different imbibition periods of dimethoate on germination, chlorophyll a/b and dry matter of Glycine max (L.) Merrill. cv. KHSB-2, during early seedling growth. *J. Physiol. Res.* 2005; 18: 199–201.

Pandya U, Saraf M. Application of fungi as a biocontrol agent and their biofertilizer potential in agriculture. *J. Adv. Dev. Res.* 2010; 1(1): 90–99.

Parween T, Jan S, Mahmooduzzafar S, Fatma T, Siddiqui ZH. Selective effect of pesticides on plant—A review. *Crit. Rev. Food Sci. Nut.* 2016; 56(1): 160–179.

Petit N, Fontaine F, Clement C, Vaillant-Gaveau N. Photosynthesis limitations of grape vine after treatment with the fungicide fludioxonil. *J. Agric. Food Chem.* 2008; 56: 6761–6767.

Pinol R, Esther S. *Protective Effects of Brassinosteroids Against Herbicides. Brassinosteroids: A Class of Plant Hormone.* The Netherlands: Springer, 2011; pp. 309–344.

Pinol R, Simon E. Effect of 24-epibrassinolide on chlorophyll fluorescence and photosynthetic CO2 assimilation in *Vicia faba* plants treated with the photosynthesis inhibiting herbicide terbutryn. *J. Plant Growth Regul.* 2009; 28: 97–105.

Popova LP, Ananieva EA, Alexieva VS. Treatment with salicylic acid decreases the effects of paraquat on photosynthesis. *J. Plant Physiol.* 2002; 159: 685–693.

Popova LP, Ananieva EA, Christov KN. Exogenous treatment with salicylic acid leads to increased antioxidant capacity in leaves of barley plants exposed to Paraquat. *J. Plant Physiol.* 2004; 61: 319–328.

Prasad SM, Dwivedi R, Zeeshan M. Growth, photosynthetic electron transport, and antioxidant responses of young soybean seedlings to simultaneous exposure of nickel and UV-B stress. *Photosynthetica* 2005; 43: 177–185.

Prasad SM, Kumar S, Parihar P, Singh A, Singh R. Evaluating the combined effects of pretilachlor and UV-B on two Azolla species. *Pest. Biochem Physiol.* 2016; 128: 45–56. doi:10.1016/j.pestbp.2015.10.006

Radetski CM, Sylvie C, Jean-François F. Classical and biochemical endpoints in the evaluation of phytotoxic effects caused by the herbicide trichloroacetate. *Environ. Exp. Bot.* 2000; 44(3): 221–229.

Radwan DEM. Salicylic acid induced alleviation of oxidative stress caused by clethodim in maize (*Zea mays* L.) leaves. *Pest. Biochem. Physiol.* 2012; 102: 182–188.

Raja W, Rathaur P, John SA, Ramteke PW. Endosulfan induced changes in growth rate, pigment composition and photosynthetic activity of mosquito fern *Azolla microphylla*. *J. Stress Physiol. Biochem.* 2012; 8: 98–109.

Ramakrishna A, Ravishankar GA. Influence of abiotic stress signals on secondary metabolites in plants. *Plant Sig. Behav.* 2011; 6: 1720–1731.

Ramakrishna A, Ravishankar GA. Role of plant metabolites in abiotic stress tolerance under changing climatic conditions with special reference to secondary compounds. In: *Climate Change and Plant Abiotic Stress Tolerance*, Tuteja N, Gill SS (Eds.), Wiley-VCH Verlag GmbH & Co. KGaA, 2013; pp. 705–726.

Ramel F, Sulmon C, Bogard M, Couee I, Gouesbet G. Differential dynamics of reactive oxygen species and antioxidative mechanisms during atrazine injury and sucrose-induced tolerance in *Arabidopsis thaliana* plantlets. *BMC Plant Biol.* 2009; 9: 28.

Ramel F, Sulmon C, Cabello-Hurtado F, Taconnat L, Martin-Magniette ML, Renou JP, El Amrani A, Couee I, Gouesbet G. Genome-wide interacting effects of sucrose and herbicide-mediated stress in *Arabidopsis thaliana*: Novel insights into atrazine toxicity and sucrose-induced tolerance. *BMC Genom.* 2007; 8: 450.

Rio AD, Bamberg J, Centeno-Diaz R, Salas A, Roca W, Tay D. Effects of the pesticide furadan on traits associated with reproduction in wild potato species. *Am. J. Plant Sci.* 2012; 3: 1608–1612.

Rodriguez R, Redman R. Balancing the generation and elimination of reactive oxygen species. *Proc. Natl. Acad. Sci.* 2005; 102: 3175–3176.

Ryter SW, Kim HP, Hoetzel A, Park JW, Nakahira K, Wang X, Choi AM. Mechanisms of cell death in oxidative stress. *Antioxid. Redox Signal.* 2007; 9: 49–89.

Sakhabutdinova AR, Fatkhutdinova DR, Bezrukova MV, Shakirova FM. Salicylic acid prevents damaging action of stress factors on wheat plants. *Bulg. J. Plant Physiol.* 2003; 314–319.

Saladin G, Magne C, Clement C, Effects of fludioxonil and pyrimethanil, two fungicides used against Botrytis cinerea, on carbohydrate physiology in *Vitis vinifera* L. *Pest. Manage. Sci.* 2003; 59: 1083–1092.

Saraf M, Sood N. Influence of monocrotophos on growth oxygen uptake and exopolysaccharide production of *Rhizobium* NCIM 2771 on chickpea. *J. Indian Bot. Soc.* 2002; 82: 157–164.

Sasse JM. Physiological actions of brassinosteroids: An update. *J. Plant Growth Regul.* 2003; 22: 276–288.

Sergiev IG, Alexieva VS, Ivanov SV, Moskova II, Karanov EN. The phenyl urea cytokinin 4PU-30 protects maize plants against glyphosate action. *Pest. Biochem. Physiol.* 2006; 85: 139–146.

Shao H-B, Chu L-Y, Lu Z-H, Kang C-M. Primary antioxidant free radical scavenging and redox signaling pathways in higher plant cells. *Int. J. Biol. Sci.* 2008; 4: 8–14.

Sharma A, Kumar V, Singh R, Thukral AK, Bhardwaj R. Effect of seed pre-soaking with 24-epibrassinolide on growth and photosynthetic parameters of *Brassica juncea* L. in imidacloprid soil. *Ecotoxicol. Environ. Saf.* 2016a; 133: 195–201.

Sharma A, Kumar V, Thukral AK, Bhardwaj R. Epibrassinolide imidacloprid interaction enhances non-enzymatic antioxidants in *Brassica juncea* L. *Ind. J. Plant Physiol.* 2016b; 21: 70–75. doi:10.1007/s40502-016-0203-x

Sharma A, Thakur S, Kumar V, Kesavan AK, Thukral AK, Bhardwaj R. 24-epibrassinolide stimulates imidacloprid detoxification by modulating the gene expression of *Brassica juncea* L. *BMC Plant Biol.* 2017; 17: 56.

Sharma I, Bhardwaj R, Pati PK. Exogenous application of 28-homobrassinolide modulates the dynamics of salt and pesticides induced stress responses in an elite rice variety Pusa Basmati-1. *J. Plant Growth Regul.* 2015; 34: 509–518.

Sharma I, Bhardwaj R, Pati PK. Mitigation of adverse effects of chlorpyrifos by 24-epibrassinolide and analysis of stress markers in a rice variety Pusa Basmati-1. *Ecotoxicol. Environ. Saf.* 2012; 85: 72–81.

Sharma I, Bhardwaj R, Pati PK. Stress modulation response of 24-epibrassinolide against imidacloprid in an elite indica rice variety Pusa Basmati-1. *Pest. Biochem. Physiol.* 2013; 105: 144–153.

Singh H, Singh NB, Singh A, Hussain I, Yadav V. Physiological and biochemical effects of salicylic acid on *Pisum sativum* exposed to isoproturon. *Arch. Agron. Soil Sci.* 2016; 62: 1425–1436.

Singh S, Prasad SM. Growth, photosynthesis and oxidative responses of Solanum melongena L. seedlings to cadmium stress: Mechanism of toxicity amelioration by kinetin. *Sci. Hort.* 2014; 176: 1–10.

Soaresa C, Alexandra de Sousa A, Pinto A, Azenha M, Teixeira J, Antunes Azevedoc R, Fidalgo F. Effect of 24-epibrassinolide on ROS content, antioxidant system, lipid peroxidation and Ni uptake in *Solanum nigrum* L. under Ni stress. *Environ. Exp. Bot.* 2016; 122: 115–125.

Stevens MM, Reinke RF, Coombes NE, Helliwell S, Mo J. Influence of imidacloprid seed treatments on rice germination and early seedling growth. *Pest. Manage. Sci.* 2008; 64: 215–222.

Szabados L, Arnould S. Proline: A multifunctional amino acid. *Trends Plant Sci.* 2010; 15(2): 89–97.

Takagi D, Ishizaki K, Hanawa H, Mabuchi T, Shimakawa G, Yamamoto H, Miyake C. Diversity of strategies for escaping reactive oxygen species production within photosystem I among land plants: P700 oxidation system is prerequisite for alleviating photoinhibition in photosystem I. *Physiol. Plant.* 2017; 161: 56–74. doi:10.1111/ppl.12562

Tebourbi O, Mohsen S, Khemais BR. Molecular mechanisms of pesticide toxicity. *Pesticides in the Modern World-Pests Control* and *Pesticides Exposure* and *Toxicity Assessment*. InTech, 2011.

Thounaojam TC, Panda P, Mazumdar P, Kumar D, Sharma GD, Sahoo L. Excess copper induced oxidative stress and response of antioxidants in rice. *Plant Physiol. Biochem.* 2012; 53: 33–39.

Untiedt R, Blanke MM. Effects of fungicide and insecticide mixtures on apple tree canopy photosynthesis, dark respiration and carbon economy. *Crop Prot.* 2004; 23: 1001–1006.

Unver T, Bakar M, Shearman RC, Budak H. Genome-wide profiling and analysis of *Festuca arundinacea* miRNAs and transcriptomes in response to foliar glyphosate application. *Mol. Genet. Genom.* 2010; 283: 397–413.

Wang C, Zhang Q. Exogenous salicylic acid alleviates the toxicity of chlorpyrifos in wheat plants (*Triticum aestivum*). *Ecotoxicol. Environ. Saf.* 2017; 137: 218–224.

Wang P, Du Y, Li Y, Ren D, Song CP. Hydrogen peroxide-mediated activation of MAP kinase 6 modulates nitric oxide biosynthesis and signal transduction in *Arabidopsis*. *Plant Cell* 2010; 22: 2981–2998.

Wang Y, Ying J, Kuzma M, Chalifoux M, Sample A, McArthur C, Uchacz T, Sarvas C, Wan J, Dennis DT. Molecular tailoring of farnesylation for plant drought tolerance and yield protection. *Plant J.* 2005; 43: 413–424.

WHO. Joint FAO/WHO Expert Standards Program. Codex Alimentation Commission, Geneva, Switzerland, 2007. Available online http://www.who.int_[accessed 10/09/2012].

Wu XY, Von Tiedemann A. Impact of fungicides on active oxygen species and antioxidant enzymes in spring barley (Hordeum vulgare L.) exposed to ozone. *Environ Pollut.* 2002; 116: 37–47.

Xia XJ, Huang YY, Wang L, Huang LF, Yu YL, Zhou YH. Pesticides-induced depression of photosynthesis was alleviated by 24-epibrassinolide pretreatment in *Cucumis sativus* L. *Pest. Biochem. Physiol.* 2006; 86: 42–48.

Xia XJ, Zhang Y, Wu JX, Wang JT, Zhou YH, Shi K. Brassinosteroids promote metabolism of pesticides in cucumber. *J. Agric. Food Chem.* 2009; 57: 8406–8413.

Xie L, Yang C, Wang X. Brassinosteroids can regulate cellulose biosynthesis by controlling the expression of CESA genes in *Arabidopsis*. *J. Exp. Bot.* 2011; 62: 495–506.

Xiong, Yan. Degradation of oxidized proteins by autophagy during oxidative stress in Arabidopsis. *Plant Physiol.* 2007; 143(1): 291–299.

Xu S, Li J, Zhang X, Wei H, Cui L. Effects of heat acclimation pretreatment on changes of membrane lipid peroxidation, antioxidant metabolites, and ultrastructure of chloroplasts in two cool-season turfgrass species under heat stress. *Environ. Exp. Bot.* 2006; 56: 274–285.

Yigit E, Akbulut GB, Bayram D, Kaya A, Gok Y. Effect of salicylic acid and selenium on antioxidant system of Avena sativa L. under fenoxaprop-p-ethyl stress. *Fresen Environ. Bull.* 2016; 25: 874–884.

Yueh, LY, David L. Hensley. Pesticide effect on acetylene reduction and modulation by soybean and lima bean. *J. Am. Soc. Hortic. Sci.* 1993; 118(1): 73–76.

Zhang ZB, Xu P, Jia JZ, Zhou RH. Quantitative trait loci for leaf chlorophyll fluorescence traits in wheat. *Aust. J. Crop Sci.* 2010; 4: 571–579.

Zhang ZL, Wei N, Wu QX, Ping ML. Antioxidant response of *Cucumis sativus* L. to fungicide carbendazim. *Pest. Biochem. Physiol.* 2007; 89: 54–59.

Zhou Y, Xia X, Yu G, Wang J, Wu J, Wang M, Yang Y, Shi K, Yu Y, Chen Z, Gan J, Yu J. Brassinosteroids play a critical role in the regulation of pesticide metabolism in crop plants. *Sci. Rep.* 2015; 5: 9018.

10

Oxidative Stress and Its Management in Plants During Abiotic Stress

Parammal Faseela, Asari Kandi Sinisha, Thuruthummel Thomas Dhanya Thomas, and Jos Thomas Puthur

CONTENTS

Abbreviations

$^1O_2$	singlet oxygen
ABA	abscisic acid
APX	ascorbate peroxidase
AsA	ascorbate
BR	brassinosteroid
CAT	catalase
CK	cytokinin
DHAR	dehydroascorbate reductase
ET	ethylene
GA	gibberellic acid
GPX	guaiacol peroxidase
GR	glutathione reductase
GSH	reduced glutathione
H_2O_2	hydrogen peroxide
$HO_2{}^{\bullet}$	perhydroxyl radical
JA	jasmonic acid
MDHAR	monodehydroascorbate reductase
$O_2{}^{\bullet}-$	superoxide anion
OH•	hydroxyl radical
ROS	reactive oxygen species

SA   salicylic acid
SL   strigolactone
SOD   superoxide dismutase

Introduction

Abiotic stress is a limiting factor for plants that affects crop productivity and food production throughout the world. Food security is a major issue pertaining to the majority of the population in the world, mainly due to a rapid rise in population and extreme changes in the climate (Lesk et al. 2016). Abiotic stresses cause major problems for plants in terms of growth, development, and yield. There are different factors which cause abiotic stress in plants including temperature [high and low (chilling, freezing)], light (high and low), UV-B radiation, water (drought, flooding, and submergence), chemical factors (heavy metals and pH), salinity, a deficiency or an excess of essential nutrients, gaseous pollutants (ozone, sulfur dioxide), mechanical factors, and other less frequently occurring stressors (Ramakrishna and Ravishankar 2011). Plants tolerate different abiotic stress factors in different ways, which involves various physiological, molecular, metabolic, and morphological phenomena that may affect growth and development of plants (Hernandez et al. 2010; Wang and Frie 2011; Ramakrishna and Ravishankar 2013; Fahad et al. 2017).

High-temperature stress (heat stress) and drought stress are the main effects of climate change that adversely affect crop productivity. It was reported that from 1980 to 2015 there were 21% and 40% reductions in yields of wheat (*Triticum aestivum* L.) and maize (*Zea mays* L.), respectively, due to drought stress (Daryanto et al. 2016). Drought and salinity stress changes the osmotic balance in plant cells by reducing water availability, leading to loss of cell turgor and the generation of reactive oxygen species (ROS) that are detrimental to plant growth and development. These stresses also cause a reduction in plant growth, biomass, and leaf chlorophyll content and an increase in total proline content. In addition to these, a significant increase in hydrogen peroxide content and lipid peroxidation is recorded in plants exposed to stress(es) (Nxele et al. 2017). Low temperature (0–15°C) is another abiotic stress factor in plants which induces more damage to chloroplast by the activation of galactolipase followed by the liberation of free fatty acids which are more prone to peroxidation by lipoxygenases and ROS (Kaniuga 2008).

The increased dependence of agriculture on chemical fertilizers and rapid industrialization has added toxic metals including heavy metals to environment systems. Heavy metal stress is generally caused by Zn, Cd, Pb, Cu, etc. Cd is a toxic heavy metal that will remain in soil persistently due to its higher half-life and easily soluble properties (Ali et al. 2015). According to Girotto et al. (2013), excess concentration of these metals increases the production of ROS in living systems, thus leading to oxidative stress. It causes impairment in source and sink capacities, reduction in total chlorophyll and also lowers the concentration of some trace elements such as Zn (Jinadasa et al. 2016). When supplied above the optimum level, essential heavy metals such as Cu and Zn cause toxicity to plants. A high concentration of Cu alters the properties of membranes and ion channels, and decreases the

photosynthetic rate, thus affecting the electron transport chain (Cambrollé et al. 2013, 2015).

High-intensity light and UV-B radiation, both inherent parts of sunlight, are major abiotic stress factors in tropical areas. UV-B induces damages to DNA, proteins and membrane lipids by the production of ROS like H_2O_2 and hydroxyl radicals. Even at low doses of UV-B, ROS are produced either by metabolic disorders or due to the high activity of membrane-bound NADPH-oxidases and cell wall-bound peroxidases and this can harm the electron transport chain in chloroplast and mitochondria (Hideg et al. 2013; Czegeny et al. 2014; Muller-Xing et al. 2014).

Production of ROS is a common feature in plants when various abiotic stresses are encountered. ROS are endogenously produced mainly in chloroplasts, mitochondria, and peroxisomes as the by-products of various metabolic pathways and act as cellular messengers and redox regulators in plant cells. Over-accumulation of ROS causes oxidative stress, leading to protein denaturation, lipid peroxidation, and oxidation of DNA, which result in cellular damage and finally cell death (Gill and Tuteja 2010; Raja et al. 2017).

Sources of ROS

Major sources of ROS are chloroplast, peroxisomes, and mitochondria. Chloroplast is a major organelle that carries out photosynthesis in plants and algae. Photosynthesis is the process in which light energy is captured in the extensive membrane of the thylakoid and converted into high energy molecules such as ATP and NADPH. In the subsequent steps using ATP and NADPH, carbon dioxide is reduced to sugars (Foyer et al. 2012). The photosynthetic electron transport chain is a source of ROS production and induces oxidative stress and results in oxidation of DNA, proteins, and peroxidation of lipids (Gill and Tuteja 2010; Woodson 2016).

Peroxisomes are single membrane-bound subcellular organelles, consisting of two major enzymatic units such as catalases and H_2O_2-producing flavin-containing oxidases. Other H_2O_2-producing enzymes in peroxisomes are glycolate oxidase, nitric oxide synthase, and acyl Co A oxidases (Bonekamp et al. 2009; Rio and Huertas 2016; Deb and Nagotu 2017). Photorespiratory glycolate oxidase reaction, fatty acid β-oxidation, the enzymatic reaction of flavin oxidases, and the disproportionation of O_2 radicals are the main metabolic processes that generate H_2O_2 in peroxisomes (Foyer and Noctor 2003; Gill and Tuteja 2010).

Mitochondria or energy factories of the cell are another source of ROS. Under normal conditions, cytochrome C oxidase or complex III of ETC react with O_2 along with the transfer of four electrons, and as a result water is produced. Rarely other ETC components can react with O_2 and result in the production of superoxide anion ($O_2 \bullet^-$). Mitochondria can produce ROS from any of the four complexes and complex I and III are more common (Sweetlove and Foyer 2004). Reverse electron flow occurs from complex II to complex I when substrates for complex I are limited and this process generates ROS at complex I (Turrens 2003). In addition to this, completely reduced ubiquinone donates an electron to complex III and as a result ubisemiquinone radical will be formed and the electron will move to O_2 and thus ROS are generated (Murphy 2009). In the

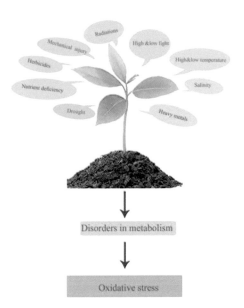

FIGURE 10.1 Various abiotic stress factors, leading to subsequent generation of various types of ROS species such as $^1O_2$ (singlet oxygen), $O_2^{\bullet-}$ (superoxide anion radical), peroxynitrite ($OONO^-$), perhyroxyl radical ($HO_2\bullet$), hydrogen peroxide (H_2O_2), hydroxyl radical ($OH\bullet$), and finally oxidative stress in plants.

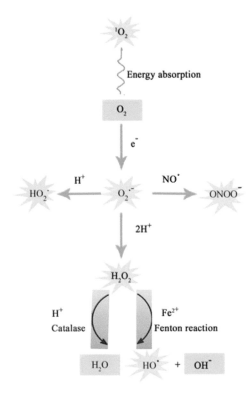

FIGURE 10.2 Schematic representation of the formation of various ROS such as $^1O_2$ (singlet oxygen), $O_2^{\bullet-}$ (superoxide anion radical), nitric oxide radical ($NO\bullet$), peroxynitrite ($OONO^-$), perhyroxyl radical ($HO_2\bullet$), hydrogen peroxide (H_2O_2), and hydroxyl radical ($OH\bullet$) from molecular oxygen (O_2).

presence of Fe and Cu, H_2O_2 produce the most active hydroxyl radical, which can move through the membrane and results in membrane peroxidation (Rhoads et al. 2006). Figure 10.1 represents various abiotic stress factors, and the subsequent generation of various types of ROS species, which will, finally, lead to oxidative damage in plants.

ROS: Types

Oxygen is a diatomic molecule having two free electrons in outer orbit with the same spin quantum number, which makes it accept only one electron at a time and results in the formation of reactive oxygen species. There are different types of ROS such as hydroxyl radical, the superoxide anion radical, hydrogen peroxide, singlet oxygen, nitric oxide radical, hypochlorite radical, and various lipid peroxides. When molecular oxygen accepts a single electron, less reactive $O_2^{\bullet-}$ is formed, which is the first step in the formation of ROS. Since it is a charged molecule, it can pass only through anion channels. Under acidic conditions, two such molecules undergo dismutation by superoxide dismutase and H_2O_2 is produced. As H_2O_2 is uncharged, it is freely permeable through membranes (Newsholme et al. 2012).

Hydrogenation of $O_2^{\bullet-}$ results in the formation of the perhydroxyl radical ($HO_2\bullet$). Superoxide radical ($O_2^{\bullet-}$) also reacts with nitric oxide radical ($NO\bullet$) and thus, another ROS, peroxynitrite ($OONO^-$), is formed. In addition to this $O_2^{\bullet-}$ gives its electron to (Fe^{3+}) and reduces it to Fe^{2+}. Ferrous ion (Fe^{2+}) then reduces H_2O_2 to hydroxyl radical ($OH\bullet$), which is the most reactive ROS species. Hydroxyl radical is the major ROS which causes oxygen toxicity *in vivo* in plants. Finally, ferrous ion is oxidized to ferric ion by H_2O_2. This process is known as the Haber-Weiss reaction, the final step of which is referred to as the Fenton reaction.

Further reduction of hydroxyl radical will result in the formation of water. Another ROS, singlet oxygen ($^1O_2$), can be formed by photoexcitation of chlorophyll under low CO_2 concentrations in the cell (Gill and Tuteja 2010; Newsholme et al. 2012; Yadav and Sharma 2016). Schematic representation of different ROS formation from molecular oxygen in response to abiotic stress is shown in Figure 10.2.

ROS-Induced Oxidative Damage

Damage to Lipids

Lipid peroxidation is one of the damaging effects of ROS. It involves three major steps: initiation, progression, and termination. Presence of transition metals such as Fe and Cu is involved in the initiation step. ROS such as $O_2^{\bullet-}$, H_2O_2, and highly active $OH\bullet$ induce the peroxidation of polyunsaturated fatty acids (PUFA) by removing hydrogen and producing lipid aldehydes, alkenals, and hydroxyalkenals (HAEs), such as 4-hydroxy-2-nonenal (HNE) and malondialdehyde (Winger et al. 2005). The products of the initiation step behave in the same manner and a series of reactions take place, ultimately reacting with proteins and nucleic acids. Smaller, lipid-derived reactive electrophile species such as key oxylipins are produced during lipid peroxidation (Almeras et al. 2003).

Various abiotic stress factors such as Cd in *Brassica juncea* cultivars (Mobin and Khan 2007), elevated levels of O_3 in rice cultivars (Wang et al. 2013), drought stress in *Phaseolus vulgaris* (Zlatev et al. 2006), low level of salinity stress (Ashraf et al. 2010), and UV-B irradiated seedlings of *Cassia auriculata* (Rodriguez et al. 2010) were found to enhance the rate of lipid peroxidation.

Damage to Proteins

ROS induces covalent modifications to proteins which are referred to as protein oxidation. Most of them are irreversible but a few which involve sulfur-containing amino acids are reversible. Irreversible modifications include carbonylation, nitrosylation, and reactions with products of PUFA oxidation (Moller et al. 2007). Under chilling stress, ROS such as $O_2^{\bullet-}$ will be formed as the product of the Fenton reaction in Fe-S centers, which will cleave the large subunit of Rubisco into five major fragments and also Fe-S centers; in addition, PSI B is degraded, resulting in the leakage of Fe. Carbonyl group accumulation under oxidative stress can be correlated with the protein oxidation. Treatment of pea leaves with Cd showed twofold increases in the level of carbonyl groups over the control (Puertas et al. 2002).

Damage to DNA

Methylation, oxidation, depurination, and deamination are the reactions occurring in DNA damage due to ROS in plants. This damage results in single-stranded and double-stranded breaks in DNA. Highly active OH• and $^1O_2$ induce damage to DNA by reacting with guanine nucleotide and finally 8-hydroxy guanine will be formed. ROS also induces the changes in methylation of cytosine which brings alterations in gene expression (Kim et al. 2011). In addition to this, indirect oxidation also occurs, which involves a reaction between guanine and products of lipid peroxidation such as MDA, and extra ring structures will be formed (Jeong et al. 2005).

ROS Defense Machinery in Plants

The production and elimination of ROS in plants at the intracellular level are generally balanced and regulated by production of efficient enzymatic and non-enzymatic antioxidant defense systems. Antioxidants can fu nction as electron donors in plants and thus they reduce ROS to less harmful molecules. Recently, many studies have established that the induction of the cellular antioxidant machinery in plants is mandatory for protection against various abiotic stresses and investigations on disturbances or alterations in this system are an excellent strategy to explain the different ROS signaling pathways. Enzymatic systems include superoxide dismutase (SOD), catalase (CAT), ascorbate peroxidase (APX), guaiacol peroxidase (GPX), glutathione reductase (GR), monodehydroascorbate reductase (MDHAR), and dehydroascorbate reductase (DHAR). Non-enzymatic components consist of ascorbic acid, glutathione, proline, α-tocopherol, carotenoids, phenolic acids, flavonoids, sugars, proline, etc. (Kasote et al. 2015; Caverzan et al. 2016).

Enzymatic Antioxidants

It is well known that plant cells employ antioxidant defense systems to protect themselves against ROS-induced oxidative stress. The enzymes localized in the different subcellular compartments, which comprise the antioxidant machinery, include guaiacol peroxidase (GPX), superoxide dismutase (SOD), catalase (CAT), and enzymes of the ascorbate-glutathione (AsA-GSH) cycle such as ascorbate peroxidase (APX), monodehydroascorbate reductase (MDHAR), dehydroascorbate reductase (DHAR), and glutathione reductase (GR) (Gill and Tuteja 2010). The activities of many enzymes of the antioxidant defense system in plants have been found to increase significantly to combat the oxidative stress induced by various environmental stresses and the maintenance of a balance between antioxidant enzymes and the toxic ROS has been connected to increased tolerance of plants to various environmental stresses (Sharma et al. 2012; You and Chan 2015).

Guaiacol Peroxidase (GPX, 1.11.1.7)

GPX is a heme-containing protein, which can function as an effective quencher of reactive intermediary forms of O_2 and peroxy radicals under stressed conditions. It oxidizes aromatic electron donors such as guaiacol and pyragallol by consuming H_2O_2. It belongs to class III or the 'secreted plant peroxidases' and these enzymes have a capacity to undertake a second cyclic reaction known as a hydroxylic reaction, which is distinct from the peroxidative reaction (Foyer and Shigeoka 2011). There are various reports about the changes in peroxidase activity under various stresses and such studies revealed that peroxidases could be employed as sensitive and accurate stress markers. The activity of GPX highly varies depending upon plant species and stress type and its level was enhanced in plants under various stressful conditions of the environment (Caverzan et al. 2016; Jisha and Puthur 2014). GPX is not only associated with defense against abiotic and biotic stresses, but also has a role in the biosynthesis of lignin in the cell wall, biosynthesis of ethylene, wound healing, and degradation of auxin (Gaspar et al. 1991).

Superoxide Dismutase (SOD, 1.15.1.1)

Various environmental stresses lead to the increased generation of ROS, where SOD has been proposed to be important in plant stress tolerance and provide the first line of defense against the toxic effects of elevated levels of ROS. The SOD scavenges $O_2^{\bullet-}$ by catalyzing its dismutation in two different ways, one being its reduction to H_2O_2 and the other its oxidization to O_2. Based on the metal co-factor used by the enzyme, SODs are classified into three types and these SODs are located in different compartments of the cell. Cu/Zn-SODs are present in chloroplast and cytosol, which requires Cu and Zn as cofactors to catalyze the dismutation of superoxide radicals into H_2O_2. Mn-SOD is found in the mitochondrial matrix and utilizes Mn as the cofactor and the Fe-SODs are distributed in chloroplast, which requires Fe as a cofactor (Feng et al. 2016). There are various reports about the enhancement of SOD activity in plants when exposed to various environmental stresses (Ren et al. 2016; Shackira and Puthur 2017; Wang et al. 2016b).

Catalase (CAT, 1.11.1.6)

CAT, the first enzyme to be discovered and characterized, is a tetrameric heme-containing enzyme that serves as an efficient ROS scavenging system to reduce the oxidative damage in plants induced by various stresses. It catalyzes the dismutation of two molecules of H_2O_2 into water and oxygen and it has high specificity for H_2O_2. CATs are unique as they do not require any cellular reducing equivalent like other types of scavenging enzymes. CAT scavenges H_2O_2 produced in peroxisomes arising from the oxidation of the photorespiratory products and in the glyoxysomes generated by oxidases involved in the β-oxidation of fatty acids (Sharma et al. 2012). These proteins are localized to the cytosol, mitochondria, chloroplasts, and most abundantly in peroxisomes. Environmental stresses cause either an increase or decrease of CAT activity in plants, depending on the type, intensity, and duration of the treatment (Pandey et al. 2015; Caverzan et al. 2016).

Enzymes of Ascorbate-Glutathione (AsA-GSH) Cycle

The AsA-GSH cycle or Halliwell-Asada pathway involves successive oxidation and reduction of AsA, GSH, and NADPH for detoxification of H_2O_2 and it also has a role in AsA and GSH regeneration. The AsA-GSH cycle is catalyzed by the enzymes APX, MDHAR, DHAR, and GR and is present in the cytosol, chloroplast, mitochondria, and peroxisomes. In the AsA-GSH cycle, ascorbate is oxidized to monodehydroascorbate (MDHA), which is then reduced to AsA by MDHAR, or spontaneously disproportionated to AsA and dehydroascorbate (DHA). DHA is then reduced to ascorbate through a GSH-dependent reaction by DHAR and oxidized glutathione (GSSG) is reduced by GR using NADPH as the electron donor (Foyer and Shigeoka 2011). The AsA-GSH cycle plays a key role in scavenging ROS and protecting plants from oxidative stress induced by various environmental stresses (Kapoor et al. 2015). Figure 10.3 illustrates the AsA-GSH cycle in response to abiotic stress in plants.

Ascorbate Peroxidase (APX, 1.1.11.1)

APX is a member of the Class I superfamily of heme peroxidases and it plays a significant role in the reduction of H_2O_2 into H_2O by using two molecules of AsA as electron donors and generates two molecules of MDHA. APX has a higher affinity for H_2O_2 than other antioxidant enzymes and reduces H_2O_2 to H_2O after utilizing ascorbate as a specific electron donor in chloroplasts, cytosol, mitochondria, and peroxisomes and the apoplastic space also (Sofo et al. 2015). In plants, there are five different APX isoforms based on amino acid sequences and they are distributed in various organelles, such as chloroplast stromal (sAPX), thylakoidal (tAPX), mitochondrial (mitAPX), peroxisomal, and glyoxisomal (mAPX) and cytosolic form (cAPX) (Racchi 2013).

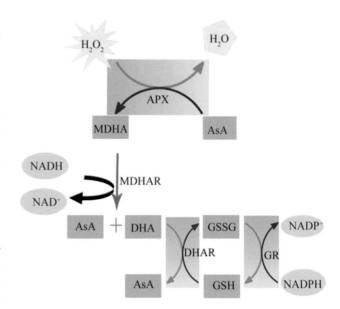

FIGURE 10.3 The ascorbate-glutathione cycle in response to abiotic stress. Ascorbate peroxidise (APX), momodehydroascorbate reductase (MDHAR), dehydroascorbate reductase (DHAR), glutathione reductase (GR), ascorbate (AsA), monodehyrdoascorbate (MDHA), dehydroascorbate (DHA), reduced glutathione (GSH), and oxidized glutathione (GSSG).

Due to its higher affinity for H_2O_2, APX has a predominant role in plants when exposed to different environmental stress. It has been reported that APX activity was increased on exposure to UV, drought, salt, cold, heavy metal, heat, pathogen infection, wound stress, and other biotic or abiotic stresses. According to Koussevitzky et al. (2008) cytosolic APX1 activity was enhanced in *Arabidopsis thaliana* when subjected to the combined stress of drought and heat.

Monodehydroascorbate Reductase (MDHAR, 1.6.5.4)

MDHAR functions in maintaining the reduced pools of ascorbate by recycling the oxidized form of ascorbate. It is a flavin adenine dinucleotide (FAD) enzyme present in chloroplasts, cytosol, peroxisomes, and mitochondria and it plays an important role in maintaining the AsA pool. MDHA produced by APX is reduced back to AsA by MDHAR preferring NADH rather than NADPH as the electron donor (Park at al. 2016). Several authors have reported increased activity of MDHAR in plants when subjected to various stresses (Mohseni et al. 2015; Wang et al. 2016a).

Dehydroascorbate Reductase (DHAR, 1.8.5.1)

In plants, DHAR and MDHAR are the two primary enzymes concerned in the recycling of AsA. DHAR regenerates the reduced AsA for detoxification of ROS by catalyzing the reduction of oxidized ascorbate (DHA) using glutathione (GSH) as a hydrogen donor before the irreversible spontaneous hydrolysis of DHA to form 2,3-diketogulonic acid. MDHA produced by APX can also lead to the formation of DHA which is further reduced to AsA by the enzyme DHAR. Due to the necessity for the maintenance of a reduced pool of AsA, DHAR is of prime importance in oxidative stress tolerance (Kapoor et al. 2015). Mostofa et al. (2015) reported the enhancement in DHAR activity in rice seedlings under Cu stress. Do et al. (2016) studied the crystal structures of DHAR from *Oryza sativa* L. japonica (OsDHAR) in the native, ascorbate-bound, and GSH-bound forms for elucidating the molecular basis of DHAR catalysis.

Glutathione Reductase (GR, 1.6.4.2)

GR is a homodimeric FAD-containing NADPH-dependent oxidoreductase which catalyzes the reduction of oxidized glutathione (GSSG) to reduced glutathione (GSH) and it protects cells against oxidative damage. This enzyme maintains a high GSH/GSSG ratio in cells and is predominantly found in chloroplast, mitochondria, cytosol, and peroxisomes (Edwards et al. 1990). The GSH pool maintained by GR is important for active protein function, and so both GR and GSH play an important role in determining the tolerance of a plant under various environmental stresses. Gill et al. (2013) discussed the significance of GSH and GR in crop plants exposed to various abiotic stresses such as salinity, drought, chilling, and heavy metals. Previous studies have revealed that the GR activity was enhanced under salinity stress in maize roots (AbdElgawad et al. 2016) and in wheat under drought stress (Devi et al. 2012).

Non-Enzymatic Antioxidants

The non-enzymatic antioxidants protect the cell from damage by directly detoxifying the ROS by donating electrons or hydrogen or they can reduce substrates for antioxidant enzymes (Kasote et al. 2015). These molecules include ascorbate (AsA) and glutathione (GSH), α-tocopherol, carotenoids, phenolic compounds, flavonoids, and amino acid cum osmolyte proline (Sharma et al. 2012; Kasote et al. 2015).

Ascorbate (AsA)

Ascorbate or vitamin C, the most abundant and water-soluble antioxidant, is one of the most important and extensively studied antioxidants in plants (Arrigoni and De Tullio 2000). It is considered as a most powerful ROS scavenger in plants under various stresses that are formed from photosynthetic and respiratory processes because of its ability to donate electrons in a number of enzymatic and non-enzymatic reactions. It plays a significant role in the conversion of hydrogen peroxide and superoxide radicals to water by the enzyme APX. AsA protects the entire photosynthetic machinery by acting as an alternative electron donor of photosystem II (PSII) in leaves with inactive oxygen-evolving complex; thereby it protects PSII against donor side photoinhibition (Tóth et al. 2011). AsA acts as a cofactor of violaxanthin de-epoxidase and thus can dissipate the excess excitation energy from chloroplasts. AsA functions in tocopherol regeneration and also acts as a substrate for synthesis of important organic acids such as L-oxalic acids, L-tartaric, and L-threonic acid (Debolt et al. 2007).

Reduced Glutathione (GSH)

Glutathione is a low-molecular-weight thiol (γ-glutamyl-cysteinyl-glycine) and it functions as a reductant in scavenging of ROS in plant cells (Gill and Tuteja 2010). As it is one of the most important metabolites of plants, it plays a significant role in the regulation of cellular metabolism (Zagorchev et al. 2013). It is present in almost all cellular compartments such as cytosol, mitochondria, chloroplasts, endoplasmic reticulum, vacuoles, peroxisomes, and apoplast. In plants, glutathione is mainly present in a reduced state and when exposed to lethal conditions of various environmental stresses the oxidized form of glutathione (GSSG) is increased in cells; thus, under normal conditions, a high GSH:GSSG ratio is maintained in plants (Noctor et al. 2011). In stressed conditions, GSH along with ascorbate play a central role in scavenging of ROS. Glutathione plays a central role in metal homeostasis by acting as phytochelatins due to the presence of sulfhydryl group in cysteine, enabling GSH to chelate metals and participate in redox cycling (Jozefczak et al. 2012). Various reports revealed that the concentrations of GSH increased in plants when exposed to various stresses such as drought and salt in *Arabidopsis* (Cheng at el. 2015), high light in *Arabidopsis* (Heyneke et al. 2013), and heavy metal and ultraviolet in pea (Saleh 2007).

Tocopherols (Vitamin E)

There are four main isomers of tocopherols (α, β, γ, δ) in plants and among them α-tocopherol is the most active, a naturally occurring lipophilic antioxidant. Due to the presence of three methyl groups, it is the largest scavenger of peroxyl radicals in lipid bilayers (Munné-Bosch 2005). α-Tocopherol is present only in green parts of plants, and it is located in the chloroplast envelope, thylakoid membranes, and plastoglobuli. It plays a major role in the chloroplastic antioxidant component in plants by the deactivation of photosynthesis-derived ROS in thylakoid membranes (Kasote et al. 2015). α-Tocopherol is known to protect the lipids and other membrane components by quenching the singlet oxygen and hydroxyl radicals during photoinhibition and thus helping the maintenance of PSII structure and function (Miret and Munné-Bosch 2015). α-Tocopherol has been shown to increase in plants and imparts tolerance to various stresses such as salinity (Ellouzi et al. 2013), lead, copper, cadmium, and mercury (Zengin and Munzuroglu 2005).

Carotenoids

Carotenoids are also one of the lipophilic antioxidants and they play a multitude of functions in plant metabolism including oxidative stress tolerance. Moreover, carotenoids function as accessory light-harvesting pigments in photosynthetic organisms. As an antioxidant, carotenoids have a primary role in photoprotection by quenching triplet state and excited chlorophyll molecules and scavenging singlet oxygen formed within the photosynthetic apparatus (Lu and Li 2008). Zeaxanthin, a carotenoid, is mainly involved in the dissipation of harmful excess excitation energy under stress conditions in plants (Young 1991). Carotenoids have a structural role also in the PSI assembly and the stabilization of light harvesting complex proteins and the thylakoid membrane. Many studies have described the physiological role of carotenoids in plants under various stresses (Zhou et al. 2017; Kang et al. 2017).

Proline

The accumulation of cellular osmolytes in plants under various stresses is a common process. Among them, proline, an imino acid, is increasingly accumulated in many plants and it has multifunctional roles in response to a wide range of biotic and abiotic stresses (Szabados and Savoure 2009). In addition to functioning as a compatible osmolyte, it plays a significant role in scavenging of ROS, stabilizing subcellular structures, maintaining intracellular redox homeostasis (e.g., ratio of $NADP^+/NADPH$ and GSH/GSSG), metal chelation, and also as a signaling molecule under stress conditions (Verbruggen and Hermans 2008; Hayat et al. 2012). During osmotic stress, the balance between the increased production and reduced degradation determines the osmoprotective and other functions of proline. Numerous studies revealed that the proline content increases in plants under various environmental stresses such as salt (Gharsallah et al. 2016) and heavy metal (Shackira et al. 2017).

Sugars

Sugars such as glucose, sucrose, fructose, and trehalose are important regulatory molecules with both signaling and ROS scavenging functions in response to a number of stresses in plants (Smeekens et al. 2010). They can function as signals useful for the plant in sensing and controlling the photosynthetic activity and also for maintaining the cellular redox potential (Keunen et al. 2013). Sugars have a dual role in plants with respect to ROS. Firstly they are directly involved or linked with ROS-producing metabolic pathways such as mitochondrial respiration or photosynthesis. In reverse, they can also feed NADPH-producing metabolism such as the oxidative pentose-phosphate pathway, which helps plants in ROS scavenging processes (Keunen et al. 2013). According to ElSayed et al. (2014), sucrose and sucrosyl oligosaccharides (including fructans and raffinose family oligosaccharides) either directly detoxify ROS in chloroplasts and vacuoles or are indirectly involved in stimulating the activity of the other common antioxidants. The synergistic interaction of sugars and phenolic compounds functions as an integrated redox system in plants, quenching ROS and contributing to stress tolerance (Bolouri-Moghaddam et al. 2010). There are various reports that trehalose serves as a signaling molecule and is involved in plant responses to abiotic and biotic stresses (Mostofa et al. 2015; Govind et al. 2016).

Phenolic Compounds

Plant phenolics are the most widely distributed secondary metabolites in plants which possess one or more aromatic rings with one or more hydroxyl groups. They are mainly classified into five major groups, phenolic acids, flavonoids, lignins, stilbenes, and tannins, and they are distinguished by the number of constitutive carbon atoms in conjugation with the structure of the basic phenolic skeleton (Michalak 2006). Phenolic compounds are abundantly present in plant cells and have the ability to reduce reactive oxygen species and also function as metal chelators (Ksouri et al. 2008). The antioxidant capacity of phenolics will increase with the number of free hydroxyls and conjugation of side chains to the aromatic rings (Bergmann et al. 1994). They are involved in the response to stress by acting synergistically with other physiological antioxidants such as ascorbate or tocopherol and finally amplify their biological effects. Previously, various authors reported the high antioxidant capacity of plants under various stresses and correlated it to their total phenolic content (Gharibi et al. 2016; Mirshad and Puthur 2016).

Flavonoids and phenolic acids are the largest classes of plant phenolics, which were found to have excellent antioxidant activity in both *in vitro* and *in vivo* studies (Kasote et al. 2015). The ROS scavenging properties of flavonoids are restricted to few structures, namely, dihydroxy B-ring substituted flavonoids, but not to their monohydroxy B-ring substituted counterparts. Flavonoids have a role in antioxidant systems and are found in the chloroplast, which scavenges ROS and stabilizes the chloroplast envelope. Flavonoids protect plants from different abiotic stresses; also they can function as anti-herbivory agents in plants under biotic stress. They are UV screening compounds or UV filters and effectively scavenge singlet oxygen generated after exposure to excess UV radiation (Agati et al. 2012). According to Dai and Mumper (2010), phenolics and flavonoids have great antioxidant property and are more effective than Vitamin C, E, and carotenoids.

Gene Activation of ROS Scavenging Antioxidant Enzymes

The molecular control mechanisms for abiotic stress tolerance are based on the activation and regulation of stress-related genes. These genes are involved in stress responses, such as signaling, transcriptional control, and free-radical and toxic-compound scavenging (Wang et al. 2003). Signaling pathways are induced in response to abiotic stresses, and recent molecular and genetic studies have revealed that these pathways involve diverse responses in plants (Sreenivasulu et al. 2007). Signal recognition activates signal transduction pathways which result in altered gene expression of different antioxidant enzymes, which results in better acclimatization of the plant towards abiotic stress conditions. Genes of FeSOD are *FSD1*, *FSD2*, and *FSD3*, Cu/ZnSOD genes are *CSD1*, *CSD2*, and *CSD3*, and MnSOD gene is *MSD1* (Kliebenstein et al. 1999).

Isoforms of CAT are *CAT1*, *CAT2*, and *CAT3*, which are found on separate chromosomes and are regulated independently. CAT1 and CAT2 are located in peroxisomes and cytosol, and *CAT3* in mitochondria (Skadsen et al. 1995). The APX mRNA is located in cytosol APX1, APX2, chloroplast APX5, APX6, APX7, and APX8, and peroxisomal APX3 and APX4 (Hong et al. 2007).

The GR gene is *GR1* and *GR2*, and cytosolic GR is GR2 (Bashir et al. 2007). Gene families of GST are large and highly diverse with 25 members reported in soybean, 42 in maize, and 54 in *Arabidopsis* (Sappl et al. 2004). These are cytoplasmic proteins, but microsomal, plastidic, nuclear, and apoplastic isoforms have also been reported (Frova 2003). Seven related proteins of

GPX are identified in the cytosol, chloroplast, mitochondria, and endoplasmic reticulum, named AtGPX1–AtGPX7 in *Arabidopsis* (Millar et al. 2003).

Role of Phytohormones in Activating Antioxidant Machinery

Plant hormones play an important role in regulating developmental processes and signaling networks in response to a wide range of abiotic stresses. Hormones such as salicylic acid (SA), jasmonates (JA), ethylene (ET), abscisic acid (ABA), auxin, gibberellic acid (GA), cytokinin (CK), brassinosteroids (BR), and strigolactones (SLs) have a role in activating the antioxidant machinery in plants (Bari et al. 2009).

ROS also act as a secondary messenger in hormone-mediated cellular responses in plants such as ABA transduction pathway in guard cells. ABA is responsible for plant stress responses, such as changes in physiology, proteins, and the closure of stomata (Raja et al. 2017). ABA-induced H_2O_2 is a signal in mediating stomatal closure which reduces water loss through the activation of calcium-permeable channels in the plasma membrane. ABA also upregulates synthesis of osmoprotectants and antioxidant enzymes thereby conferring desiccation tolerance (Chaves et al. 2003). According to Zhang et al. (2006), the ABA concentration was increased when plants were exposed to salinity. GA has an important role in abiotic stress response and adaptation (Munteanu et al. 2014). In dormant barley grains, GA signaling and ROS content are low, whereas ABA signaling is high, resulting in dormancy. Exogenous H_2O_2 does not appear to alter ABA biosynthesis and signaling, but has a more prominent effect on GA signaling, inducing a change in hormonal balance that results in germination (Bahin et al. 2011).

Auxins, cytokinins, and gibberellins enhanced the antioxidants such as superoxide dismutase, ascorbate peroxidase, catalase, ascorbate, and glutathione activities (Piotrowska-Niczyporuk et al. 2012). Auxin is involved in defense responses via regulation of numerous genes and mediation of crosstalk between abiotic and biotic stress responses (Fahad et al. 2015). ROS generation is stimulated by auxin, which mediates gravitropism. ROS production in abiotic stress alters auxin gradients that in turn decrease auxin-mediated signaling (Raja et al. 2017). Auxins and cytokinins elevate the activity of antioxidant enzymes under stress conditions. Cytokinins are active oxygen scavengers, which enhances the detoxification of active oxygen species (Szechyńska-Hebda et al. 2007).

JA activates plant defense responses to abiotic stresses including drought, salinity, and low temperature (Dar et al. 2015). It activates the antioxidant machinery when plants are exposed to heavy metal stress (Yan et al. 2013). Lignin is produced in plants as a response to abiotic stress and lignin biosynthesis is regulated through the interactions between JA and ROS (Denness et al. 2011).

SA plays an important role in the regulation of plant growth, fruit ripening, development, and abiotic stresses tolerance (Hara et al. 2012). Low concentrations of SA enhance the antioxidant capacity of plants, but high concentrations of SA cause cell death or susceptibility to abiotic stresses (Jumali et al. 2011). SA with

ABA is involved in the regulation of drought response (Miura and Tada 2014). The increased osmotic stress tolerance of transgenic *Arabidopsis* is through the overexpression of the salicylate hydroxylase (NahG) gene, resulting from decreased SA-mediated ROS generation (Borsani et al. 2001). In stressed plants, SA is directly or indirectly involved in signaling pathways and interplays with ROS and GSH (Raja et al. 2017). The BR have an important role in plants subjected to various abiotic stresses like high temperature, chilling, soil salinity, light, drought, flooding, metals/metalloids, and organic pollutants (Wani et al. 2016). Increased BR levels induce prolonged ROS accumulation, which causes a stress response including ABA biosynthesis and stomatal closure (Raja et al. 2017).

Phytohormone Interaction in Activating Antioxidation Machinery

Plant hormones interact together to fine-tune the defense against abiotic stresses (Harrison 2012). Major phytohormones including ABA, auxin, BRs, GAs, JA, and ET are the biosynthetic and core signaling components in plants (Singh and Jwa 2013). In drought stress condition, ABA interacts with JA and nitric oxide (NO) to stimulate stomatal closure, and regulates the gene expression, which includes the induction of genes associated with response to ethylene, cytokinin, or auxin. Stress-induced JA production interacts with ABA-mediated stomatal closure by stimulating the influx of extracellular Ca^{2+} and activating H_2O_2/NO signaling. JA interacts with ABA-mediated stomatal closure by enhancing Ca^{2+} influx, which stimulates calcium-dependent protein kinase (CDPK) production and the resulting signal cascade (Harrison 2012). The activated ABA, ET, GA, and CK signals are critical in determining plant stress responses through the enhancement of antioxidation machinery (Stamm and Kumar 2013). Figure 10.4 illustrates that abiotic stresses in plants upregulate the signal transduction pathways, leading to activation of antioxidants for ROS scavenging (enzymatic, non-enzymatic, and phytohormones) to neutralize the harmful effects of ROS and finally the plant can regain healthy growth and development.

ROS Signaling Pathway

The ROS signaling network contains kinases, phosphatase, and ROS-responsive transcription factors. H_2O_2 activates ANP1, a mitogen-activated protein kinase (MAPK; MAP kinase kinase kinase [MAPK3]), and, through an unidentified intermediate kinase, which activates two downstream MAPKs, AtMPK3 and AtMPK6, which eventually upregulate *GST6* (glutathione-S-transferases) and *HSP18.2* genes. The overexpression of ANP1 in transgenic plants resulted in enhanced tolerance to heat shock, freezing, and salt stress. The serine/threonine kinase OXI1 (oxidative signal-inducible1) is another essential element of H_2O_2 signaling. H_2O_2 and abiotic stresses activate OXI1, which in turn activates AtMPK3 and AtMPK6 (Rentel et al. 2004). OMTK1 (oxidative stress-activated MAP triple-kinase 1) is another H_2O_2-inducible kinase, which activates the downstream MAP kinase

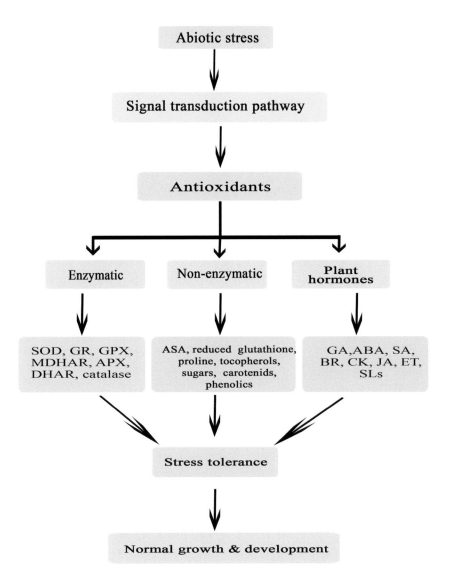

FIGURE 10.4 Abiotic stress in plants upregulates the signal transduction pathway, leading to activation of antioxidants for ROS scavenging (enzymatic, non-enzymatic and phytohormones) to neutralize the harmful effects of ROS and finally the plant can regain healthy growth and development.

MMK3. The MMK3 is in turn activated by ethylene and ROS signaling (Nakagami et al. 2004).

H_2O_2 also increases the expression of NDPK2 (nucleotide diphosphate kinase 2). Overexpression of NDPK2 reduces the accumulation of H_2O_2 and results in enhanced tolerance to cold, salt, and oxidative stress. NDK1 (nucleotide diphosphate kinase 1) interacts with catalases, resulting in enhanced ability to detoxify H_2O_2. Several members of the ERF and Myb family transcription factors are induced by $^1O_2$ (Gadjev et al. 2006). Heat-shock transcription factors are H_2O_2 sensors and the downstream genes are involved in defense against a variety of stresses, including oxidative stress. Two genes encoding WRKY-family transcription factors and two zinc-finger transcription factors ZAT11 and ZAT12 are upregulated by $O_2^{\bullet-}$, $^1O_2$, and H_2O_2. ZAT12 has a role in abiotic stress signaling as ZAT12 overexpressors have elevated transcript levels of oxidative- and light stress-responsive transcripts and ZAT12-deficient plants are more sensitive to H_2O_2-induced oxidative stress (Suzuki et al. 2012). Figure 10.5 illustrates the ROS signaling pathways activated in stress tolerance mechanisms of plants when subjected to various abiotic stresses.

Bioengineering Plants for Enhancing Antioxidation Potential

Abiotic stress tolerance significantly increases in transgenic plants through the overexpression of enzymatic and non-enzymatic antioxidants. Enhanced salt tolerance was achieved in transgenic *Arabidopsis* plants via the overexpression of Mn-SOD, Cu/Zn-SOD, Fe-SOD, CAT, and POD (Wang et al. 2004). Overexpressing Cu/Zn-SOD transgenic tobacco plants and rice showed multiple stress tolerance (Prashanth et al. 2008). The expression of Mn-SOD in *Triticum aestivum* transgenic plants exhibited photooxidative stress tolerance, lower oxidative damage, and a significant increase in SOD and GR activities (Melchiorre et al. 2009). The overexpression of Mn-SOD

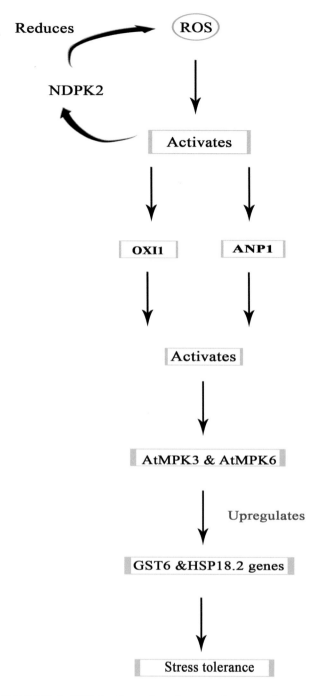

FIGURE 10.5 ROS signaling pathway in stress tolerance mechanism of plants when subjected to various abiotic stresses. ROS activation ANP1 (MAPK3) and OX11 (oxidative signal-indicuble1), which activates AtMPK3 and AtMPK6 (mitogen-activated protein kinase). AtMPK3 and AtMPK6 upregulate *GST6* (glutathione-S-transferase6) and HSP18.2 (heat shock protein 18.2) genes. ROS also activates the expression of NDPK2 (nucleotide diphosphate kinase2) and the overexpression of NDPK2 reduces the production of ROS. Finally, these pathways enhance the stress tolerance mechanism in plants.

in *Arabidopsis* resulted in enhancement in activities of other SODs (SOD, Cu/Zn-SOD, Fe-SOD), CAT, and POD. As a final outcome of this, *Arabidopsis* showed enhanced salt tolerance (Wang et al. 2010). Overexpression of CAT genes in transgenic plants like *Oryza sativa* (*CAT*) enhanced the low-temperature

stress tolerance (Matsumura et al. 2002), in *Nicotiana tabacum* (*CAT3*) exhibited Cd stress tolerance (Gichner et al. 2004), and in *Nicotiana tabacum (katE)* showed salt stress tolerance (Al-Taweel et al. 2007).

The APX gene overexpression exhibited UV-B, heat, drought, and chilling stress tolerance in *Lycopersicon esculentum* (*cAPX*) (Wang et al. 2005, 2006), water deficit tolerance in *Nicotiana tabacum* (*APX3*) (Yan et al. 2003), salt stress tolerance in *Arabidopsis thaliana* (*APX1*) (Xu et al. 2008), and NaCl stress tolerance in *Nicotiana tabacum* (*swpa4*) (Kim et al. 2008). Transgenic *Triticum aestivum* (Melchiorre et al. 2009) and *Gossypium hirsutum* (Mahan et al. 2009) plants showed chilling stress tolerance and photoprotection via the expression of GR. *Arabidopsis thaliana* (*parB*) exhibited salt stress tolerance (Katsuhara et al. 2005); *Gossypium hirsutum* (*Nt107*) and *Oryza sativa* (*GST*) (Yu et al. 2003) showed salt and paraquat stress tolerance; and *Nicotiana tabacum* (*GST+GPX*) transgenic plants showed enhanced stress tolerance via GST overexpression (Jogeswar et al. 2006). According to Yoshimura et al. (2004), enhanced GPX in *Nicotiana tabacum* transgenic plants is responsible for their stress tolerance.

Overexpression of MDAR1 in *Nicotiana tabacum* transgenic plants showed tolerance to ozone, salt, and PEG stress. In *Nicotiana tabacum*, drought, salt, ozone, low temperature, and NaCl stress tolerance was enhanced via the overexpression of DHAR (Eltayeb et al. 2007). The overexpression of DHAR, AsA, and GSSG genes in transgenic tobacco plants also showed oxidative stress tolerance (Lee et al. 2007).

The P5CS [Proline P5CS (D1-Pyrroline-5-carboxylate-synthetase)] gene transfer to *Triticum aestivum* exhibited drought tolerance (Vendruscolo et al. 2007), to *Saccharum* spp. showed drought tolerance by proline accumulation (Molinari et al. 2007), and in *Petunia hybrida* showed drought tolerance and high proline content (Yamada et al. 2005). The transfer of the P5CR gene [Proline P5CR (D1-pyrroline-5-carboxylate reductase)] to *Arabidopsis thaliana* exhibited salt tolerance (Ma et al. 2008) and drought stress tolerance in *Glycine max* (Simon-Sarkadi et al. 2006). According to Czuj et al. (2009), the Met25 [O-acetylhomoserine-oacetylserine (OAH-OAS) sulfhydrylase] gene transfer to *Linum ussitatissimum* showed increased cysteine and methionine biosynthesis, which resulted in a significant increase in glutathione.

Conclusion

Nowadays, multiple abiotic stress factors such as salt, drought, temperature, light, and heavy metals adversely affect various aspects of plant growth, development, and crop productivity and these will remain a challenge to the natural environment and agriculture. Various environmental stresses lead to excessive production of ROS, causing damage to cell machinery in plants. Plants respond to unfavorable environmental conditions using highly efficient strategies to cope with and adapt to different types of abiotic stress factors. Efficient scavenging of ROS produced during various abiotic stresses requires the action of several enzymatic as well as non-enzymatic antioxidants present in plants. Several genes responsible for abiotic stress tolerance have been identified which code for antioxidants, phytohormones, or enzymes that synthesize important stress response compounds. A fruitful strategy for the identification of stress tolerance genes and their molecular signaling pathways in many plant species has been achieved. Transgenic plants with some of these genes have been produced and found to be an emerging and effective measure to counteract the negative impact of abiotic stresses on crop production.

REFERENCES

AbdElgawad, H., Zinta, G., Hegab, M., Pandey, R., Asard, H., and Abuelsoud, W. 2016. High salinity induces different oxidative stress and antioxidant responses in maize seedling organs. *Front. Plant Sci.* 7:276. doi:10.3389/fpls.2016.00276.

Agati, G., Azzarello, E., Pollastri, S., and Tattini, M. 2012. Flavonoids as antioxidants in plants: Location and functional significance. *Plant Sci.* 196:67–76.

Ali, B., Deng, X., Hu, X., et al. 2015. Deteriorative effects of cadmium stress on antioxidant system and cellular structure in germinating seeds of *Brassica napus* L. *J. Agric. Sci. Technol.* 17:63–74.

Almeras, E., Stolz, S., Vollenweider, S., et al. 2003. Reactive electrophile species activate defense gene expression in *Arabidopsis*. *Plant J.* 34:205–16.

Al-Taweel, K., Iwaki, T., Yabuta, Y., Shigeoka, S., Murata, N., and Wadano, A. 2007. A bacterial transgene for catalase protects translation of D1 protein during exposure of salt-stressed tobacco leaves to strong light. *Plant Physiol.* 145:258–65.

Arrigoni, O., and De Tullio, M. C. 2000. The role of ascorbic acid in cell metabolism: Between gene-directed functions and unpredictable chemical reactions. *J. Plant Physiol.* 157:481–88.

Ashraf, M. A., Ashraf, A., and Ali, Q. 2010. Response of two genetically diverse wheat cultivars to salt stress at different growth stages: Leaf lipid peroxidation and phenolic content. *Pak. J. Bot.* 42:559–65.

Bahin, E., Bailly, C., Sotta, B., Kranner, I., Corbineau, F., and Leymarie, J. 2011. Crosstalk between reactive oxygen species and hormonal signalling pathways regulates grain dormancy in barley. *Plant Cell Environ.* 34:980–93.

Bari, R., Jonathan, D., and Jones, G. 2009. Role of plant hormones in plant defence responses. *Plant Mol. Biol.* 69:473–88.

Bashir, K., Nagasaka, S., Itai, R. N., et al. 2007. Expression and enzyme activity of glutathione reductase is upregulated by Fe-deficiency in graminaceous plants. *Plant Mol. Biol.* 65:277–84.

Bergmann, H., Leinhos, V., and Machelett, B. 1994. Increase of stress resistance in crop plants by using phenolic compounds. *Acta Hort.* 381:390–97.

Bolouri-Moghaddam, M. R., Roy, K. L., Xiang, L., Rolland, F., and Van den Ende, W. 2010. Sugar signalling and antioxidant network connections in plant cells. *FEBS J.* 277:2022–37.

Bonekamp, N. A., Volkl, A., Fahimi, H. D., and Schrader, M. 2009. Reactive oxygen species and peroxisomes: Struggling for balance. *Biofactors* 35:346–55.

Borsani, O., Díaz, P., Agius, M. F., Valpuesta, V., and Monza, J. 2001. Water stress generates an oxidative stress through the induction of a specific Cu/Zn superoxide dismutase in *Lotus corniculatus* leaves. *Plant Sci.* 161:757–63.

Cambrollé, J., García, J. L., Figueroa, M. E., and Cantos, M. 2015. Evaluating wild grapevine tolerance to copper toxicity. *Chemosphere* 120:171–78.

Cambrolle, J., Garcia, J. L., Ocete, R., Figueroa, M. E., and Cantos, M. 2013. Growth and photosynthetic responses to copper in wild grapevine. *Chemosphere* 93:294–301.

Caverzan, A., Casassola, A., and Brammer, S. P. 2016. Reactive oxygen species and antioxidant enzymes involved in plant tolerance to stress. In *Abiotic and Biotic Stress in Plants: Recent Advances and Future Perspectives*, ed. A. K. Shanker, and C. Shanker, 463–80. Rijeka, Croatia: InTech.

Chaves, M. M., Maroco, J. P., and Pereira, J. S. 2003. Understanding plant responses to drought-from genes to the whole plant. *Funct. Plant Biol.* 30:239–64.

Cheng, M. C., Ko, K., Chang, W. L., Kuo, W. C., Chen, G. H., and Lin, T. P. 2015. Increased glutathione contributes to stress tolerance and global translational changes in *Arabidopsis*. *Plant J.* 83:926–39.

Czegeny, G., Wu, M., Der, A., Eriksson, L. A., Strid, A., and Hideg, E. 2014. Hydrogen peroxide contributes to the ultraviolet-B (280–315 nm) induced oxidative stress of plant leaves through multiple pathways. *FEBS Lett.* 588:2255–61.

Czuj, T., Zuk, M., Starzycki, M., Amir, R., and Szopa, J. 2009. Engineering increases in sulfur amino acid contents in flax by overexpressing the yeast Met25 gene. *Plant Sci.* 177:584–92.

Dai, J., and Mumper, R. 2010. Plant phenolics: Extraction, analysis and their antioxidant and anticancer properties. *Molecules* 15:7313–52.

Dar, T. A., Uddin, M., Khan, M. M. A., Hakeem, K. R., and Jaleel, H. 2015. Jasmonates counter plant stress: A review. *Environ. Exp. Bot.* 115:49–57.

Daryanto, S., Wang, L., and Jacinthe, P. A. 2016. Global synthesis of drought effects on maize and wheat production. *PLoS One* 11:1–11.

Deb, R., and Nagotu, S. 2017. Versatility of peroxisomes: An evolving concept. *Tissue Cell* 49:209–26.

Debolt, S., Melino, V., and Ford, C. M. 2007. Ascorbate as a biosynthetic precursor in plants. *Ann. Bot.* 99:3–8.

Denness, L., McKenna, J. F., Segonzac, C., et al. 2011. Cell wall damage-induced lignin biosynthesis is regulated by a reactive oxygen species- and jasmonic acid-dependent process in *Arabidopsis*. *Plant Physiol.* 156:1364–74.

Devi, R., Kaur, N., and Gupta, A. K. 2012. Potential of antioxidant enzymes in depicting drought tolerance of wheat (*Triticum aestivum* L.). *Indian J. Biochem. Biophys.* 49:257–65.

Do, H., Kim, I. S., Jeon, B. W., et al. 2016. Structural understanding of the recycling of oxidized ascorbate by dehydroascorbate reductase (OsDHAR) from *Oryza sativa* L. Japonica. *Sci. Rep.* 6:19498. doi:10.1038/srep19498.

Edwards, E. A., Rawsthorne, S., and Mullineaux, P. M. 1990. Subcellular distribution of multiple forms of glutathione reductase in leaves of pea (*Pisum sativum* L.). *Planta* 180:278–84.

Ellouzi, H., Hamed, K. B., Cela, J., Müller, M., Abdelly, C., and Munné-Bosch, S. 2013. Increased sensitivity to salt stress in tocopherol-deficient *Arabidopsis* mutants growing in a hydroponic system. *Plant Signal. Behav.* 8:e23136. doi:10.4161/psb.23136.

ElSayed, A. I., Rafudeen, M. S., and Golldack, D. 2014. Physiological aspects of raffinose family oligosaccharides in plants: Protection against abiotic stress. *Plant Biol.* 16:1–8.

Eltayeb, A. E., Kawano, N., Badawi, G. H., et al. 2007. Overexpression of monodehydroascorbate reductase in transgenic tobacco confers enhanced tolerance to ozone, salt and polyethylene glycol stresses. *Planta* 225:1255–64.

Fahad, S., Bajwa, A. A., Nazir, U., et al. 2017. Crop production under drought and heat stress: plant responses and management options. *Front. Plant Sci.* 8:1–16.

Fahad, S., Hussain, S., Matloob, A., et al. 2015. Phytohormones and plant responses to salinity stress: A review. *Plant Growth Regul.* 75:391–404.

Feng, K., Yu, J., Cheng, Y., et al. 2016. The *SOD* gene family in tomato: Identification, phylogenetic relationships, and expression patterns. *Front. Plant Sci.* 7:1279. doi:10.3389/fpls.2016.01279.

Foyer, C. H., Neukermans, J., Queval, G., Noctor, G., and Harbinson, J. 2012. Photosynthetic control of electron transport and the regulation of gene expression. *J. Exp. Bot.* 63:1637–61.

Foyer, C. H., and Noctor, G. 2003. Redox sensing and signaling associated with reactive oxygen in chloroplasts, peroxisomes and mitochondria. *Physiol. Plant* 119:355–64.

Foyer, C. H., and Shigeoka, S. 2011. Understanding oxidative stress and antioxidant functions to enhance photosynthesis. *Plant Physiol.* 155:93–100.

Frova, C. 2003. The plant glutathione transferase gene family: Genomic structure, functions, expression and evolution. *Physiol. Plant* 119:469–79.

Gadjev, I., Vanderauwera, S., Gechev, T., et al. 2006. Transcriptomic footprints disclose specificity of reactive oxygen species signaling in *Arabidopsis*. *Plant Physiol.* 141:434–45.

Gaspar, T., Penel, C., Hagege, D., and Greppin, H. 1991. Peroxidase in plant growth, differentiation and developmental processes. In *Biochemical, Molecular and Physiological Aspects of Plant Peroxidases*, ed. J. Lobarzewski, H. Greppin, C. Pennel, and T. Gaspar, 249–280. Switzerland: Geneva University Press.

Gharibi, S., Tabatabaei, B. E. S., Saeidi, G., and Goli, S. A. H. 2016. Effect of drought stress on total phenolic, lipid peroxidation, and antioxidant activity of Achillea species. *Appl. Biochem. Biotechnol.* 178:796–09.

Gharsallah, C., Fakhfakh, H., Grubb, D., and Gorsane, F. 2016. Effect of salt stress on ion concentration, proline content, antioxidant enzyme activities and gene expression in tomato cultivars. *AoB Plants* 8:plw055. doi:10.1093/aobpla/plw055.

Gichner, T., Patkova, Z., Szakova, J., and Demnerova, K. 2004. Cadmium induces DNA damages in tobacco roots, but no DNA damage, somatic mutations or homologous recombinations in tobacco leaves. *Mutat. Res. Genet. Toxicol. Environ. Mut.* 559:49–57.

Gill, S. S., Anjum, N. A., Hasanuzzaman, M., et al. 2013. Glutathione and glutathione reductase: A boon in disguise for plant abiotic stress defense operations. *Plant Physiol. Biochem.* 70:204–12.

Gill, S. S., and Tuteja, N. 2010. Reactive oxygen species and antioxidant machinery in abiotic stress tolerance in crop plants. *Plant Physiol. Biochem.* 48:909–30.

Girotto, E., Ceretta, C. A., Rossato, L. V., et al. 2013. Triggered antioxidant defense mechanism in maize grown in soil with accumulation of Cu and Zn due to intensive application of pig slurry. *Ecotox. Environ. Saf.* 93:145–55.

Govind, S. R., Jogaiah, S., Abdelrahman, M., Huntrike, S. S., and Tran, L. S. 2016. Exogenous trehalose treatment enhances the activities of defense-related enzymes and triggers resistance against downy mildew disease of pearl millet. *Front. Plant Sci.* 7:1593. doi:10.3389/fpls.2016.01593.

Hara, M., Furukawa, J., Sato, A., Mizoguchi, T., and Miura, K. 2012. Abiotic stress and role of salicylic acid in plants. In *Abiotic Stress Responses in Plants*, ed. A. Parvaiz, and M. N. V. Prasad, 235–351. New York: Springer.

Harrison, M. A. 2012. Cross-talk between phytohormone signaling pathways under both optimal and stressful environmental conditions. In *Phytohormones and Abiotic Stress Tolerance in Plants*, ed. N. A. Khan, R. Nazar, N. Iqbal, and N. A. Anjum, 49–76. Berlin/Heidelberg: Springer.

Hayat, S., Hayat, Q., Alyemeni, M. N., Wani, A. S., Pichtel, J., and Ahmad, A. 2012. Role of proline under changing environments. *Plant Signal. Behav.* 7:1456–66.

Hernandez, M. N., Fernandez, G. P. D., and Olmos, E. 2010. A different role for hydrogen peroxide and the antioxidative system under short and long salt stress in *Brassica oleracea* roots. *J. Exp. Bot.* 61:521–35.

Heyneke, E., Luschin-Ebengreuth, N., Krajcer, I., Wolkinger, V., Müller, M., and Zechmann, B. 2013. Dynamic compartment specific changes in glutathione and ascorbate levels in *Arabidopsis* plants exposed to different light intensities. *BMC Plant Biol.* 13:104–23.

Hideg, E., Jansen, M. A. K., and Strid, A. 2013. UV-B exposure, ROS, and stress: Inseparable companions or loosely linked associates? *Trends Plant Sci.* 18:107–15.

Hong, C. Y., Hsu, Y. T., Tsai, Y. C., and Kao, C. H. 2007. Expression of ascorbate peroxidase 8 in roots of rice (*Oryza sativa* L.) seedlings in response to NaCl. *J. Exp. Bot.* 58:3273–83.

Jeong, Y. C., Nakamura, J., Upton, P. B., and Swenberg, J. A. 2005. Pyrimido[1,2-α]-purin-10(3H)-one, M1G, is less prone to artifact than base oxidation. *Nucl. Acids Res.* 33:6426–34.

Jinadasa, N., Collins, D., Holford, P., Milham, P. J., and Conroy, J. P. 2016. Reactions to cadmium stress in a cadmium-tolerant variety of cabbage (*Brassica oleracea* L.): Is cadmium tolerance necessarily desirable in food crops. *Environ. Sci. Pollut. Res.* 23:5296–306.

Jisha, K. C., and Puthur, J. T. 2014. Halopriming of seeds imparts tolerance to NaCl and PEG induced stress in *Vigna radiata* (L.) Wilczek varieties. *Physiol. Mol. Biol. Plants* 20:303–12.

Jogeswar, G., Pallela, R., Jakka, N. M., et al. 2006. Antioxidative response in different sorghum species under short-term salinity stress. *Acta Physiol. Plant* 28:465–75.

Jozefczak, M., Remans, T., Vangronsveld, J., and Cuypers, A. 2012. Glutathione is a key player in metal-induced oxidative stress defenses. *Int. J. Mol. Sci.* 13:3145–75.

Jumali, S. S., Said, I. M., Ismail, I., and Zainal, Z. 2011. Genes induced by high concentration of salicylic acid in *Mitragyna speciosa*. *Aust. J. Crop Sci.* 5:296–303.

Kang, L., Park, S., Ji, C. Y., Kim, H. S., Lee, H., and Kwak, S. 2017. Metabolic engineering of carotenoids in transgenic sweet potato. *Breed. Sci.* 67:27–34.

Kaniuga, Z. 2008. Chilling response of plants: Importance of galactolipase, free fatty acids and free radicals. *Plant Biol.* 10:171–84.

Kapoor, D., Sharma, R., Handa, N., et al. 2015. Redox homeostasis in plants under abiotic stress: Role of electron carriers, energy metabolism mediators and proteinaceous thiols. *Front. Environ. Sci.* 3:13. doi:10.3389/fenvs.2015.00013.

Kasote. D. M., Katyare. S. S., Hegde. M. V., and Bae. H. 2015. Significance of antioxidant potential of plants and its relevance to therapeutic applications. *Int. J. Biol. Sci.* 11:982–91.

Katsuhara, M., Otsuka, T., and Ezaki, B. 2005. Salt stress-induced lipid peroxidation is reduced by glutathione S-transferase, but this reduction of lipid peroxides is not enough for a recovery of root growth in *Arabidopsis*. *Plant Sci.* 169:369–73.

Keunen, E., Peshev, D., Vangronsveld, J., Van Den Ende, W., and Cuypers, A. 2013. Plant sugars are crucial players in the oxidative challenge during abiotic stress: Extending the traditional concept. *Plant Cell Environ.* 36:1242–55.

Kim, J. H., Moon, J. Y., Park, E. Y., Lee, K. H., and Hong, Y. C. 2011. Changes in oxidative stress biomarker and gene expression levels in workers exposed to volatile organic compounds. *Ind. Health* 49:8–14.

Kim, Y. H., Kim, C. Y., Song, W. K., et al. 2008. Overexpression of sweetpotato swpa4 peroxidase results in increased hydrogen peroxide production and enhances stress tolerance in tobacco. *Planta* 227:867–81.

Kliebenstein, D. J., Dietrich, R. A., Martin, A. C., Last, R. L., and Dangl, J. L. 1999. LSD1 regulates salicylic acid induction of copper zinc superoxide dismutase in *Arabidopsis thaliana*. *Mol. Plant Microbe Interact.* 12:1022–26.

Koussevitzky, S., Suzuki, N., Huntington, S., et al. 2008. Ascorbate peroxidase 1 plays a key role in the response of *Arabidopsis thaliana* to stress combination. *J. Biol. Chem.* 283:34197–203.

Ksouri, R., Megdiche, W., Falleh, H., et al. 2008. Influence of biological, environmental and technical factors on phenolic content and antioxidant activities of Tunisian halophytes. *C. R. Biol.* 331:865–73.

Lee, Y. P., Kim, S. H., Bang, J. W., Lee, H. S., Kwak, S. S., and Kwon, S. Y. 2007. Enhanced tolerance to oxidative stress in transgenic tobacco plants expressing three antioxidant enzymes in chloroplasts. *Plant Cell Rep.* 26:591–98.

Lesk, C., Rowhani, P., and Ramankutty, N. 2016. Influence of extreme weather disasters on global crop production. *Nature* 529:84–87.

Lu, S., and Li, L. 2008. Carotenoid metabolism: Biosynthesis, regulation, and beyond. *J. Integr. Plant Biol.* 50:778–85.

Ma, L., Zhou, E., Gao, L., Mao, X., Zhou, R., and Jia, J. 2008. Isolation, expression analysis and chromosomal location of P5CR gene in common wheat (*Triticum aestivum* L.). *S. Afr. J. Bot.* 74:705–12.

Mahan, J. R., Gitz, D. C, Payton, P. R., and Allen, R. 2009. Overexpression of glutathione reductase in cotton does not alter emergence rates under temperature stress. *Crop Sci.* 49:272–80.

Matsumura, T., Tabayashi, N., Kamagata, Y., Souma, C., and Saruyama, H. 2002. Wheat catalase expressed in transgenic rice can improve tolerance against low temperature stress. *Physiol. Plant* 116:317–27.

Melchiorre, M. G., Robert, V., Trippi, R., and Racca, H. R. 2009. Lascano, Superoxide dismutase and glutathione reductase overexpression in wheat protoplast: Photooxidative stress tolerance and changes in cellular redox state. *Plant Growth Regul.* 57:57–68.

Michalak, A. 2006. Phenolic compounds and their antioxidant activity in plants growing under heavy metal stress. *Polish J. Environ. Stud.* 15:523–30.

Millar, A. H., Mittova, V., Kiddle, G., et al. 2003. Control of ascorbate synthesis by respiration and its implication for stress responses. *Plant Physiol.* 133:443–47.

Miret, J. A., and Munné-Bosch, S. 2015. Redox signaling and stress tolerance in plants: A focus on vitamin E. *Ann. N.Y. Acad. Sci.* 1340:29–38.

Mirshad, P. P., and Puthur, J. T. 2016. Arbuscular mycorrhizal association enhances drought tolerance potential of promising bioenergy grass (*Saccharum arundinaceum* retz.). *Environ. Monit. Assess.* 188:425. doi:10.1007/s10661-016-5428-7.

Miura, K., and Tada, Y. 2014. Regulation of water, salinity, and cold stress responses by salicylic acid. *Front Plant Sci.* 5:4. doi:10.3389/fpls.2014.00004.

Mobin, M. and Khan, N. A. 2007. Photosynthetic activity, pigment composition and anti- oxidative response of two mustard (*Brassica juncea*) cultivars differing in photosynthetic capacity subjected to cadmium stress. *J. Plant Physiol.* 164:601–10.

Mohseni, A., Nematzadeh, G. A., Dehestani, A., Shahin, B., and Soleimani, E. 2015. Isolation, molecular cloning and expression analysis of *Aeluropus littoralis* Monodehydroascorbate reductase (MDHAR) gene under salt stress. *J. Plant Molec. Breeding* 3:72–80.

Molinari, H. B. C., Marur, C. J., Daros, E., et al. 2007. Evaluation of the stress-inducible production of proline in transgenic sugarcane (*Saccharum* spp.): Osmotic adjustment, chlorophyll fluorescence and oxidative stress. *Physiol. Plant* 130:218–29.

Moller, I. M., Jensen, P. E. and Hansson, A. 2007. Oxidative modifications to cellular components in plants. *Annu. Rev. Plant Biol.* 58:459–81.

Mostofa, M. G., Hossain, M. A., Fujita, M., and Tran, L. S. P. 2015. Physiological and biochemical mechanisms associated with trehalose-induced copper-stress tolerance in rice. *Sci. Rep.* 5:11433. doi:10.1038/srep11433.

Muller-Xing, R., Xing, Q., and Goodrich, J. 2014. Footprints of the sun: Memory of UV and light stress in plants. *Front. Plant Sci.* 5:474. doi:10.3389/fpls.2014.00474.

Munné-Bosch, S. 2005. The role of α-tocopherol in plant stress tolerance. *J. Plant Physiol.* 162:743–48.

Munteanu, V., Gordeev, V., Martea, R., and Duca, M. 2014. Effect of gibberellin cross talk with other phytohormones on cellular growth and mitosis to endoreduplication transition. *Int. J. Adv. Res. Biol. Sci.* 1:136–53.

Murphy, M. P. 2009. How mitochondria produce reactive oxygen species. *Biochem. J.* 417:1–13.

Nakagami, H., Kiegerl, S., and Hirt, H. 2004. OMTK1, a novel MAPKKK, channels oxidative stress signaling through direct MAPK interaction. *J. Biol. Chem.* 279:26959–66.

Newsholme, P., Rebelato, E., Abdulkader, F., Krause, M., Carpinelli, A., and Curi, R. 2012. Reactive oxygen and nitrogen species generation, antioxidant defenses, and β-cell function: A critical role for amino acids. *J. Endocrinol.* 21:411–20.

Noctor, G., Mhamdi, A., Chaouch, S., et al. 2011. Glutathione in plants: An integrated overview. *Plant Cell Environ.* 35:454–84.

Nxele, X., Klein, A., and Ndimba, B. K. 2017. Drought and salinity stress alters ROS accumulation, water retention, and osmolyte content in sorghum plants. *S. Afr. J. Bot.* 108:261–66.

Pandey, P., Muthappa, V., and Senthil-Kumar, M. 2015. Shared and unique responses of plants to multiple individual stresses and stress combinations: Physiological and molecular mechanisms. *Front. Plant Sci.* 6:723. doi:10.3389/fpls.2015.00723.

Park, A. K., Kim, I. S., Do, H., et al. 2016. Structure and catalytic mechanism of monodehydroascorbate reductase, MDHAR, from *Oryza sativa* L. Japonica. *Sci. Rep.* 6:33903. doi:10.1038/srep33903.

Piotrowska-Niczyporuk, A., Bajguz, A., Zambrzycka, E., and Godlewska-Żyłkiewicz, B. 2012. Phytohormones as regulators of heavy metal biosorption and toxicity in green alga *Chlorella vulgaris* (Chlorophyceae). *Plant Physiol. Biochem.* 52:52–65.

Prashanth, S. R., Sadhasivam, V., and Parida, A. 2008. Over expression of cytosolic copper/zinc superoxide dismutase from a mangrove plant *Avicennia marina* in indica rice var Pusa Basmati-1 confers abiotic stress tolerance. *Transgenic Res.* 17:281–91.

Puertas, M. C. R., Palma, J. M., Gomez, M., Rio, L. A. D. and Sandalio, L. M. 2002. Cadmium causes the oxidative modification of proteins in pea plants. *Plant Cell Environ.* 25:677–86.

Racchi, M. L. 2013. Antioxidant defenses in plants with attention to *Prunus* and *Citrus* spp. *Antioxidants* 2:340–69.

Raja, V., Majeeda, U., Kang, H., Andrabi, K. I., and John, R. 2017. Abiotic stress: Interplay between ROS, hormones and MAPKs. *Environ. Exp. Bot.* 137:142–57.

Ramakrishna, A., and Ravishankar, G. A. 2011. Influence of abiotic stress signals on secondary metabolites in plants. *Plant Signal. Behav.* 6:1720–31.

Ramakrishna, A., and Ravishankar, G. A. 2013. Role of plant metabolites in abiotic stress tolerance under changing climatic conditions with special reference to secondary compounds. In *Climate Change and Plant Abiotic Stress Tolerance*, ed. N. Tuteja and S. S. Gill, 705–26. Weinheim, Germany: Wiley-VCH Verlag GmbH & Co. KGaA.

Ren, J., Sun, L. N., Zhang, Q. Y., Song, X. S. 2016. Drought tolerance is correlated with the activity of antioxidant enzymes in *Cerasus humilis* seedlings. *Biomed. Res. Int.* 2016:1–9.

Rentel, M. C., Lecourieux, D., Ouaked, F., et al. 2004. OXI1 kinase is necessary for oxidative burst-mediated signalling in *Arabidopsis*. *Nature* 427:858–61.

Rhoads, D. M., Umbach, A. L., Subbaiah, C. C. and Siedow, J. N. 2006. Mitochondrial reactive oxygen species contribution to oxidative stress and interorganellar signaling. *Plant Physiol.* 141:357–66.

Rio, L. A., and Huertas, E. L. 2016. ROS generation in peroxisomes and its role in cell signaling. *Plant Cell Physiol.* 57:1364–76.

Rodriguez, E. S., Wilhelmi, M. M. R., Cervilla, L. M., et al. 2010. Genotypic differences in some physiological parameters symptomatic for oxidative stress under moderate drought in tomato plants. *Plant Sci.* 178:30–40.

Saleh, A. A. H. 2007. Influence of UV_{A+B} radiation and heavy metals on growth, some metabolic activities and antioxidant system in Pea (*Pisum sativum*) plant. *Am. J. Plant Physiol.* 2:139–54.

Sappl, P. G., Onate-Sanchez, L., Singh, K. B., and Millar, A. H. 2004. Proteomic analysis of glutathione S-transferases of *Arabidopsis thaliana* reveals differential salicylic acid-induced expression of the plant-specific phi and tau classes. *Plant Mol. Biol.* 54:205–19.

Shackira, A. M., and Puthur, J. T. 2017. Enhanced phytostabilization of cadmium by a halophyte-*Acanthus ilicifolius* L. *Int. J. Phytoremed.* 19:319–26.

Shackira, A. M., Puthur, J. T., and Salim, N. 2017. *Acanthus ilicifolius* L. a promising candidate for phytostabilization of zinc. *Environ. Monit. Assess.* 189:282. doi:10.1007/s10661-017-6001-8.

Sharma. P., Jha. A. B., Dubey. R. S., and Pessarakli. M. 2012. Reactive oxygen species, oxidative damage, and antioxidative defense mechanism in plants under stressful conditions. *J. Bot.* 2012:1–26. doi:10.1155/2012/217037.

Simon-Sarkadi, L., Kocsy, G., Várhegyi, Á., Galiba, G., and De Ronde, J. A. 2006. Stress induced changes in the free amino acid composition in transgenic soybean plants having increased proline content. *Biol. Plant* 50:793–96.

Singh, R., and Jwa, N. S. 2013. Understanding the responses of rice to environmental stress using proteomics. *J. Proteome Res.* 12:4652–69.

Skadsen, R. W., Schulz-Lefert, P., and Herbt, J. M. 1995. Molecular cloning, characterization and expression analysis of two classes of catalase isozyme genes in barley. *Plant Mol. Biol.* 29:1005–14.

Smeekens, S., Ma, J., Hanson, J., and Rolland, F. 2010. Sugar signals and molecular networks controlling plant growth. *Curr. Opin. Plant Biol.* 13:274–79.

Sofo, A., Scopa, A., Nuzzaci, M., and Vitti, A. 2015. Ascorbate peroxidase and catalase activities and their genetic regulation in plants subjected to drought and salinity stresses. *Int. J. Mol. Sci.* 16:13561–78.

Sreenivasulu, N., Sopory, S. K., and Kavi-Kishor, P. B. 2007. Deciphering the regulatory mechanisms of abiotic stress tolerance in plants by genomic approaches. *Gene* 388:1–13.

Stamm, P., and Kumar, P. P. 2013. Auxin and gibberellin responsive *Arabidopsis* SMALL AUXIN UP RNA36 regulates hypocotyls elongation in the light. *Plant Cell Rep.* 32:759–69.

Suzuki, N., Koussevitzky, S., Mittler, R., and Miller, G. 2012. ROS and redox signalling in the response of plants to abiotic stress. *Plant Cell Environ.* 35:259–70.

Sweetlove, L. J., and Foyer, C. H. 2004. Roles for reactive oxygen species and antioxidants in plant mitochondria. In *Plant Mitochondria: From Genome to Function, Advances in Photosynthesis and Respiration*, ed. D. A. Day, A. H. Millar, and J. Whelan, 307–20. The Netherlands: Kluwer Acad. Press.

Szabados, L., and Savoure, A. 2009. Proline: A multifunctional amino acid. *Trends Plant Sci.* 15:89–97.

Szechyńska-Hebda, M., Skrzypek, E., Dąbrowska, G., Biesaga-Kościelniak, J., Filek, M., and Wędzony, M. 2007. The role of oxidative stress induced by growth regulators in the regeneration process of wheat. *Acta Physiol. Plant* 29:327–37.

Tóth, S. Z., Nagy, V., Puthur, J. T., Kovács, L., and Garab, G. 2011. The physiological role of ascorbate as photosystem II electron donor: Protection against photoinactivation in heat-stressed leaves. *Plant Physiol.* 156:382–92.

Turrens, J. F. 2003. Mitochondrial formation of reactive oxygen species. *J. Physiol.* 552:335–44.

Vendruscolo, E. C. G., Schuster, I., Pileggi, M., et al. 2007. Stress-induced synthesis of proline confers tolerance to water deficit in transgenic wheat. *J. Plant Physiol.* 164:1367–76.

Verbruggen, N., and Hermans, C. 2008. Proline accumulation in plants: A review. *Amino Acids* 35:753–59.

Wang, J., Zeng, Q., Zhu, J., Liu, G., and Tang, H. 2013. Dissimilarity of ascorbate–glutathione (AsA–GSH) cycle mechanism in two rice (*Oryza sativa* L.) cultivars under experimental free-air ozone exposure. *Agric. Ecosyst. Environ.* 165:39–49.

Wang, W., Vinocur, B., and Altman, A. 2003. Plant responses to drought, salinity and extreme temperatures: Towards genetic engineering for stress tolerance. *Planta* 218:1–14.

Wang, X., Cai, X., Xu, C., Wang, Q., and Dai, S. 2016a. Drought-responsive mechanisms in plant leaves revealed by proteomics. *Int. J. Mol. Sci.* 17:1706–36.

Wang, X., Zhang, H., Gao, Y., and Zhang, W. 2016b. Characterization of Cu/Zn-SOD enzyme activities and gene expression in soybean under low nitrogen stress. *J. Sci. Food Agric.* 96:2692–97.

Wang, Y., and Frei, M. 2011. Stressed food—The impact of abiotic environmental stresses on crop quality. *Agric. Ecosyst. Environ.* 141:271–86.

Wang, Y., Wisniewski, M., Meilan, R., Cui, M., and Fuchigami, L. 2006. Transgenic tomato (*Lycopersicon esculentum*) overexpressing cAPX exhibits enhanced tolerance to UV-B and heat stress. *J. Appl. Horticult.* 8:87–90.

Wang, Y., Wisniewski, M., Meilan, R., Cui, M., Webb, R., and Fuchigami, L. 2005. Overexpression of cytosolic ascorbate peroxidase in tomato confers tolerance to chilling and salt stress. *J. Am. Soc. Hortic. Sci.* 130:167–73.

Wang, Y., Ying, Y., Chen, J., and Wang, X. C. 2004. Transgenic *Arabidopsis* overexpressing Mn-SOD enhanced salt-tolerance. *Plant Sci.* 167:671–77.

Wang, Y. C., Qu, G. Z., Li, H. Y., et al. 2010. Enhanced salt tolerance of transgenic poplar plants expressing a manganese superoxide dismutase from *Tamarix androssowii*. *Mol. Biol. Rep.* 37:1119–24.

Wani, S. H., Kumar, V., Shriram, V., and Sahd, S. K., 2016. Phytohormones and their metabolic engineering for Abiotic stress tolerance in crop plants. *Crop J.* 4:162–76.

Winger, A. M., Millar, A. H., and Day, D. A. 2005. Sensitivity of plant mitochondrial terminal oxidases to the lipid peroxidation product 4-hydroxy-2-nonenal (HNE). *Biochem. J.* 387:865–70.

Woodson, J. D. 2016. Chloroplast quality control—Balancing energy production and stress. *New Phytol.* 212:36–41.

Xu, W. F., Shi, W. M., Ueda, A., and Takabe, T. 2008. Mechanisms of salt tolerance in transgenic Arabidopsis thaliana carrying a peroxisomal ascorbate peroxidase gene from barley. *Pedosphere* 18:486–95.

Yadav, N. and Sharma, S. 2016. Reactive oxygen species, oxidative stress and ROS scavenging system in plants. *J. Chem. Pharm. Res.* 8:595–604.

Yamada, M., Morishita, H., Urano, K., et al. 2005. Effects of free proline accumulation in petunias under drought stress. *J. Exp. Bot.* 56:1975–81.

Yan, J., Wang, J., Tissue, D., Holaday, A. S., Allen, R., and Zhang, H. 2003. Photosynthesis and seed production under water-deficit conditions in transgenic tobacco plants that overexpress an arabidopsis ascorbate peroxidase gene. *Crop Sci.* 43:1477–83.

Yan, Z., Chen, J., and Li, X. 2013. Methyl jasmonate as modulator of Cd toxicity in *Capsicum frutescens* var. *fasciculatum* seedlings. *Ecotoxicol. Environ. Saf.* 98:203–09.

Yoshimura, K., Miyao, K., Gaber, A., et al. 2004. Enhancement of stress tolerance in transgenic tobacco plants overexpressing *Chlamydomonas* glutathione peroxidase in chloroplasts or cytosol. *Plant J.* 37:21–33.

You, J., and Chan, Z. 2015. ROS regulation during abiotic stress responses in crop plants. *Front. Plant Sci.* 6:1092. doi:10.3389/fpls.2015.01092.

Young, A. J. 1991. The photoprotective role of carotenoids in higher plants. *Physiol. Plant* 83:702–8.

Yu, T., Li, Y. S., Chen, X. F., Hu, J., Chang, X., and Zhu, Y. G. 2003. Transgenic tobacco plants overexpressing cotton glutathione S-transferase (GST) show enhanced resistance to methyl viologen. *J. Plant Physiol.* 160:1305–11.

Zagorchev, L., Seal, C. E., Kranner, I., and Odjakova, M. 2013. A central role for thiols in plant tolerance to abiotic stress. *Int. J. Mol. Sci.* 14:7405–32.

Zengin, F. K., and Munzuroglu, O. 2005. Effects of some heavy metals on content of chlorophyll, proline and some antioxidant chemicals in bean (*Phaseolus vulgaris* L.) seedlings. *Acta Biol. Cracov. Ser. Bot.* 47:157–64.

Zhang, J., Jia, W., Yang, J., and Ismail, A. M. 2006. Role of ABA in integrating plant responses to drought and salt stresses. *Field Crops Res.* 97:111–19.

Zhou, R., Yu, X., Ottosen, C., et al. 2017. Drought stress had a predominant effect over heat stress on three tomato cultivars subjected to combined stress. *BMC Plant Biol.* 17:24. doi:10.1186/s12870-017-0974-x.

Zlatev, Z. S., Lidon, F. C., Ramalho, J. C., and Yordanov, I. T. 2006. Comparison of resistance to drought of three bean cultivars. *Biol. Plant* 50:389–94.

11

Plant Genome Response Related to Phenylpropanoid Induction under Abiotic Stresses

Ariel D. Arencibia

CONTENTS

Abbreviations

ACC	1-aminocyclopropane-1-carboxylic acid
APX	Ascorbate peroxidase
CAT	Catalase
DHAR	Dehydroascorbate reductase
GPX	Guaiacol peroxidases
GPX	Glutathione peroxidase
GR	Glutathione reductase
JA	Jasmonic acid
LEA	Embryogenesis abundant-like
MDA	Proline malondialdehyde
MDHAR	Monodehydroascorbate reductase
MGT	Mean germination time
PAL	L-phenyloalanine ammonia-lyase
PEG	Polyethylene-glicol
POX	Peroxidase
PPO	Polyphenol oxidase
ROS	Reactive oxygen species
SOD	Superoxide dismutase
TE	Transposable element
TGP	Transgenerational plasticity
TIB	Temporary Immersion Bioreactors
UV-B	Ultraviolet radiation

Introduction

Phenylpropanoid Induction and Oxidative Stress

As stated by Boudet (2007), it is impossible to imagine plant species without polyphenols because these compounds are a characteristic of plants related to the adaptation and colonization of land. These secondary metabolites display key functional aspects of plant life, such as: (a) structural roles in different supporting or protective tissues; (b) involvement in defense strategies; and (c) signaling properties in interactions between plants and their environment. Higher plants synthesize several thousand different phenolic compounds and the number is continually increasing. Vascular plant leaves contain hydroxycinnamic acid esters, amides and glycosides, glycosylated flavonoids, especially flavonols and proanthocyanidins, and their derivates. Lignin, suberin, and pollen sporopollenin are examples of phenolics containing polymers. Some soluble phenolics are widely distributed, e.g., chlorogenic acid, but the distribution of many other structures is restricted to specific genera or families, making them convenient biomarkers for taxonomic studies (Boudet, 2007).

Phenolic compounds are one of the most important groups of secondary metabolites. Phenolics are characterized by at least one aromatic ring (C6) bearing one or more hydroxyl groups. They are mainly synthesized from cinnamic acid, which is formed from phenylalanine by the action of L-phenylalanine ammonia-lyase (PAL; EC 4.3.1.5), the branch point enzyme between the primary (shikimate pathway) and secondary (phenylpropanoid) metabolism. The importance of the phenylpropanoid pathway could be explained by the fact that in normal growth conditions this route originates approximately 20% of the carbon fixed by plants (Díaz et al., 2001).

Phenolic compounds have been correlated with an array of functions in plants. An induction of phenylpropanoid metabolism and the production of phenolics can be evidenced in plants under different environmental factors and stress conditions (Sakihama and Yamasaki, 2002). The synthesis of isoflavones, and some other flavonoids, is induced when plants are infected or injured (Takahama and Oniki, 2000), or are under low temperatures and have a deficit of nutrients (Ruiz et al., 2003). Plants accumulate UV-absorbing flavonoids and other phenolic compounds mainly in vacuoles of epidermal cells in order to prevent the penetration of UV-B into the deepest plant tissues (Kondo and Kawashima, 2000). Additionally, flavonoids secreted from legume roots activate the genes of root nodule bacteria (Winkel-Shirley, 2002).

According to Hernández et al. (2009), an antioxidant compound must be a molecule that: (i) could donate electrons or hydrogen atoms (i.e., low reduction potential); (ii) yields an antioxidant-derived radical that (iii) is efficiently quenched by other electron or hydrogen sources in order to prevent cellular damage; and (iv) whose properties are spatially and temporally correlated with oxidative stress events. Usually, both ascorbate (vitamin C) and α-tocopherol (vitamin E) are used as a control of the antioxidant capacities because of the well-known *in planta* antioxidant activities (Rice-Evans et al., 1997).

Severe stress conditions might inactivate antioxidant enzymes, while up-regulating the biosynthesis of flavonoids (Fini et al., 2012). Stress-responsive dihydroxy B-ring-substitutedflavonoids have great potential to inhibit the generation of reactive oxygen species (ROS) and reduce the levels of ROS once they are formed, i.e., to perform antioxidant functions. These flavonoids are located within or in the proximity of ROS generation centers in severely stressed plants. "Antioxidant" flavonoids are found in the chloroplast, which suggests a scavenger role of singlet oxygen and stabilizers of the chloroplast outer envelope membrane. Dihydroxy B-ring-substituted flavonoids are present in the nucleus of mesophyll cells and may inhibit ROS generation, making complexes with Fe and Cu ions. Vacuolar dihydroxy B-ring flavonoids have been reported to serve as co-substrates for vacuolar peroxidases in order to reduce the H_2O_2 escape from the chloroplast, following the depletion of the ascorbate peroxidase activity. Antioxidant flavonoids may successfully control key steps of cell growth and differentiation, therefore regulating the development of the whole plant as well as individual organs (Agati et al., 2012).

Furthermore, Agati et al. (2013) discuss the role of the different functions potentially provided by flavonoids in photoprotection, mainly their ability to scavenge ROS and control plant development. Firstly, the authors suggest a model in which chloroplast-located flavonoids scavenge H_2O_2 and singlet oxygen produced under excess light stress, consequently avoiding programmed cell death. Secondly, they propose that vacuolar flavonoids, in conjunction with peroxidases and ascorbic acid, comprise a secondary antioxidant system designed to detoxify H_2O_2. Additionally, Agati et al. (2013) hypothesize about the crucial functions of flavonols as regulators during the evolution of land-plants, based on their ability to modulate auxin movement and auxin catabolism. These regulatory functions of flavonoids, which are shared by plants and animals, are fully accomplished in the nM concentration range, as it is likely to occur in early land-plants. Summarizing, Agati et al. (2013) conclude that functions of flavonoids as antioxidants and/or developmental regulators might contribute to photoprotection, suggesting, from an evolutive point of view, that UV-B screening was just one of the multiple functions served by flavonoids when the plants responded to an increase in sunlight irradiance.

In order to determine the effect of the activity on antioxidant enzymes and phenylpropanoid biosynthesis, the excess of light stress, in combination with the different levels of drought stress, was studied in *Fraxinus ornus* plants (Fini et al., 2012). The results led to a hypothesis about the zeaxanthin function

as a chloroplast antioxidant. Reductions in ascorbate peroxidase and catalase activities, as drought-stress augmented, were correlated by superior productions of esculetin and quercetin 3-O-glycosides, both phenylpropanoids displaying efficient capability for scavenging H_2O_2. The accumulation of esculetin and quercetin 3-O-glycosides in the vacuoles of the mesophyll cells was consistent with their putative functions as reducing agents for H_2O_2 in lightly stressed leaves. It was hypothesized that vacuolar phenylpropanoids could constitute a secondary antioxidant system, activated upon the depletion of primary antioxidant defenses, and aimed at keeping whole-cell H_2O_2 within a sub-lethal concentration range (Fini et al., 2012).

The differential gene expression in leaves of *in vitro* cultured *Vitis vinifera* cv. Malbec plants was determined under UV-B radiation at rates of 16 h at $\cong 8.25$ μW cm^{-2} or 4 h at $\cong 33$ μW cm^{-2} (Pontin et al., 2010). Results revealed that genes modulated by a high rate of UV-B doubled the number of genes modulated by low fluency UV-B. General protective responses, the induction of pathways regulating the synthesis of phenylpropanoids, the induction of antioxidant defense systems, and the activation of pathogen defense and abiotic stress responses could be crucial in inducing grapevine responses against UV-B radiation. In contrast, a low UV-B fluence rate regulated the expression of the auxin and abscisic acids, as well as the modification of the cell walls which can be associated with UV-B acclimation-like processes (Pontin et al., 2010).

The effect of H_2O_2 on salt stress acclimation has been studied in relation to plant growth, lipid peroxidation, and the activity of antioxidative enzymes of a salt-sensitive maize genotype. Pretreatment with 1 μM H_2O_2 for a period of 2 days, induced an increase in salt tolerance during the successive exposure to salt stress. This was confirmed by plant growth, lipid peroxidation, and antioxidative enzyme activities. The variations in lipid peroxidation and antioxidative enzymes SOD, APX, GPX, GR, and CAT activities, in both leaves and roots, suggest that differences in the antioxidative enzyme activities between acclimated and un-acclimated plants could clarify the improved tolerance of the acclimated plants to salt stress, and that the H_2O_2 metabolism was involved as a signal in the processes of maize salt acclimation (Dias et al., 2005).

In a related paper, exploring how to improve salt tolerance in wheat, the H_2O_2 treatment was evaluated as a priming procedure. Results revealed the activation of antioxidants and H_2O_2 scavenging in seeds after 5 h. Thus, H_2O_2-treated seeds, for a period of 8 h, germinate in a saline medium and reduce the mean germination time (MGT) in comparison to the water control level of H_2O_2. On the other hand, seedlings from H_2O_2-treated seeds were grown under salinity, suggesting the activation of the antioxidant system. These seedlings displayed superior photosynthetic capacity, which can be explained by improving the leaf gas exchange due to the stomatal component. Additionally, the H_2O_2 treatment also improved leaf water relations and displayed greater tissue K$^+$, Ca^{2+}, NO$_3^-$, PO$_4^{3-}$ levels, and improved the K$^+$:Na$^+$ ratio. Altogether, the results proposed that H_2O_2 signals the induction of antioxidants in seeds, which continued in the seedlings in order to offset the ion-induced oxidative damage (Wahid et al., 2007).

The consequences of increasing NaCl stresses were studied on two cultivars of *Sesamum indicum*. The results indicated that growth parameters, lipid peroxidation, antioxidative enzyme activities, and proline accumulation differ depending on the cultivar, where growth parameters, lipid peroxidation, and proline accumulation were in correlation when the cultivar was tolerant to salts (Koca et al., 2007).

Caper (*Capparis ovata* Desf.) is a perennial shrub (xerophyte) displaying drought resistance which is well adapted to the Mediterranean ecosystem. In this case, parameters indicative of oxidative stress, as well as antioxidant enzymes, were studied in relation to their tolerance to polyethylene glycol-mediated drought stress. For induction of drought stress, the seedlings were subjected to PEG 6000 of osmotic potential −0.81 MPa for a period of 14 days. As a result, lipid peroxidation increased in PEG stressed seedlings as compared to non-stressed seedlings. Moreover, PEG treatment caused a decrease in shoot fresh and dry weights, relative water content, and chlorophyll fluorescence. The total activity of antioxidative enzymes SOD, APX, POX, CAT, and GR was higher in stressed seedlings. The authors concluded that increased drought tolerance was correlated with the diminishing oxidative injury because it makes the antioxidant system function at higher rates under drought stress (Ozku et al., 2009).

In order to understand the adaptability of alfalfa (*Medicago sativa* L.) to environmental stresses, six cultivars were studied under 200 mM NaCl or 35% PEG artificial stresses. The results demonstrated that the tolerance of a cultivar to salt or drought stresses is related to an enhanced activity of antioxidant enzymes SOD, POD, APX, and CAT in its shoots and roots. The study highlights the role of antioxidant enzymes in the tolerance of alfalfa seedlings to drought and salinity conditions, typical of desertification.

The molecular response of oak to drought stress and its capacity to recover were studied using the proteomic approach. Twenty-three different spots were observed when comparing the two-dimensional electrophoresis profile to both drought affected and recovered plants. The identified proteins belong to the photosynthesis, carbohydrate and nitrogen metabolism, as well as the stress-related protein functional categories (Echevarría et al., 2009). On the other hand, the identification of the differentially expressed proteins and phosphoproteins induced by drought was studied in rice using proteomic approaches. While the embryogenesis abundant (LEA)-like protein and chloroplast Cu–Zn superoxide dismutase (SOD) were up-regulated by drought, the Rieske Fe–S precursor protein was down-regulated. Ten drought-responsive phosphoproteins were identified: NAD-malate dehydrogenase, OSJNBa0084K20.14 protein, abscisic acid- and stress-inducible protein, ribosomal protein, drought-induced S-like ribonuclease, ethylene-inducible protein, guanine nucleotide-binding protein, beta subunit-like protein, r40c1 protein, OSJNBb0039L24.13 protein, and germin-like protein 1. Seven of these phosphoproteins have not been previously reported to be involved in rice drought stress. Altogether, these results provide a comprehensive approach to the regulatory mechanism of drought-induced proteins and implicate new proteins in response to drought stress (Ke et al., 2009).

Physical wounding, as caused by plant culture manipulation, activates several important changes in plant metabolism, such as the accumulation and oxidation of phenolic compounds causing tissue-browning. Wounding induces an increase in the activity of the phenylalanine ammonia-lyase (PAL; EC 4.3.1.5), a key enzyme

in the synthesis of phenylpropanoid compounds. PAL activity is correlated with the accumulation of phenolic compounds that are oxidized by polyphenol oxidases (PPOs) and peroxidases (POXs). PPO enzymes are responsible for the typical browning of plant extracts and damaged tissues caused by spontaneous polymerization and cross-linking of θ-quinones. In order to understand this process, the genomic characterization of sugarcane plants was determined by suppressing key genes in the phenylpropanoid pathway; as a result, a new function of phenolic metabolites has been characterized during micropropagation in temporary immersion bioreactors (TIBs). Genes related to cell metabolism and development (10), plant defenses (9), phenylpropanoids (7), methyl jasmonate response (5), ethylene (5), oxidative burst (3), and auxin (3) pathways, among others (8) were induced in sugarcane plants multiplicating in TIBs with phenolic metabolites. Results support that phenylpropanoids could act as elicitor molecules of other biochemical pathways. Additionally, during adaptation, plants with the highest levels of phenolics exhibited an increased number of functional roots, a high growth rate, and an early ability to be colonized by the natural sugarcane endophytic *Gluconacetobacter diazotrophicus* (Arencibia et al., 2008).

Integrated metabolome and transcriptome analysis have revealed that many important metabolic pathways are regulated at the transcriptional level. However, there are also many metabolic pathways that are only regulated at the post-transcriptional level, for example, by RNA processing, translational, post-translational regulation, or feedback mechanisms. In this way, integrated "omics" analyses are essential in recognizing the spectrum of functions of metabolite regulatory networks during genomic responses to abiotic stresses (Urano et al., 2010).

Induced Phenylpropanoids as Elicitors

The phenolic profiles of higher plants are considered a result of long and complex processes of evolution and coevolution. The regulation of their synthesis involves sophisticated mechanisms that drive the precise control of the induction of the precise molecules to the right place and the right time or in response to environmental signals. These integrated responses represent good examples of systems biology (Boudet, 2007).

In this context, phenylpropanoid metabolites could be considered signal molecules. As an example, four phenylpropanoids were isolated from the leaves of *Piper sarmentosum* and the identified metabolites were characterized as 1-allyl-2,6-dimethoxy-3,4-methylenedioxybenzene, 1-allyl-2,4,5-trimethoxybenzene, 1-(1-E-propenyl)-2,4,5-trimethoxybenzene, and 1-allyl-2-methoxy-4,5-methylenedioxybenzene, which showed antimicrobial activities against *Escherichia coli* and *Bacillus subtilis* (Masuda et al., 1991). Furthermore, another new metabolite complex was isolated from the fruits of *Piper sarmentosum*, where, for the first time, some of them demonstrated antituberculosis and antiplasmodial activities (Rukachaisirikul et al., 2004). Additionally, a strong antifeedant and some toxicity effects from the essential oils of *Piper sarmentosum* were demonstrated on *Brontispa longissima*, with Myristicin (65.22%) and trans-caryophyllene (13.89%) as its major components. The high antifeedant and contact toxicity effects were observed in the 1st–2nd instar larvae (Qin et al., 2010). At the same time, the aqueous extract of *Piper sarmentosum*-AEPS exhibits opioid-mediated anti-nociceptive activity at the peripheral and central levels, as well as anti-inflammatory activity, which confirmed the traditional uses of the plant in the treatment of pain- and inflammatory-related ailments (Zakaria et al., 2010).

Development of synthetic biosystems could be an alternative for the production of high-value plant metabolites. As an example, Xiao et al. (2013) investigated metabolite biosynthesis for biotechnological applications in non-model plants and established a data-mining framework employing next-generation sequencing and computational algorithms. As direct search results obtained from public databases, functional information to over 800,000 putative transcripts was assigned by the authors, as well as the selection of biosynthetic gene candidates associated with six specialized metabolic pathways.

In this sense, some progress reconstituting a plant's natural pathways in *Escherichia coli* or yeast (*Saccharomyces cerevisiae*) was cited by Xiao et al. (2013). As examples, some pathways involved: (a) taxadiene, a key isoprenoid intermediate in taxol biosynthesis; (b) amorphadiene, the sesquiterpene olefin precursor to artemisinin; (c) artemisinic acid, the immediate precursor to artemisinin; (d) Diterpene fragrance precursors cis-abienol and sclareol; and (e) reticuline, a key intermediate in the biosynthesis of codeine and morphine. However, in order to obtain appropriate catalytic efficiencies, some limitations need to be solved, such as: (a) the isolation and functional characterization of numerous unknown biosynthetic genes; (b) the integrated management of specific catalytic characteristics of each enzyme for the efficient operation of the metabolic pathway in a heterologous system (Ajikumar et al., 2010).

Integration of plant genomics and yeast functional genomics has exposed some unexpected challenges for the selection of appropriate promoters and transcription terminators to promote efficient recombinant gene expression for optimized pathway flux. Additionally, key parameters affecting the steady-state levels of foreign proteins in yeast as well as post-transcriptional issues are not well understood (Facchini et al., 2012). In this way, efforts in interdisciplinary research in plant genomics, microbial engineering, and synthetic biology are expected to pave the way for the acceleration of the discovery of new enzymes. It is expected that the management of plant enzymes, coupled with their deployment in metabolically optimized microbes could be a target for getting a high-throughput genomic tool and a novel type of biology engineering (Facchini et al., 2012).

In the case of plant phenylpropanoids, the current developments in enzymology and molecular biology give perspectives as to how to engineer plant biomass (Vassão et al., 2008). For example, the content of lignin biopolymers could be reduced and the carbon flow redirected toward the production of related phenylpropanoids, such as the more valuable allyl/propenyl phenols. The characterization of the biochemical pathway of allyl/propenyl phenols (e.g., eugenol, chavicol, estragole, and anethole), which are important components of plant spice aromas and flavors, originate an approach or a tool to engineer the phenylpropanoid metabolism. The proteins implicated in the latter step are homologous to well-characterized phenylpropanoid reductases (pinoresinol-lariciresinol, isoflavone, phenylcoumaran-benzylic ether, and leucoanthocyanidin reductases), with similar catalytic mechanisms being

operative. The approach could also facilitate wood processing in pulp/paper industries and offer sources of renewable plant-derived biofuels, intermediate chemicals in polymer industries, or specialty chemicals in cosmetology and nutraceutical industries (Vassão et al., 2008).

Eugenol has well-known antioxidant, antimicrobial, and anti-inflammatory properties. It is commonly used as a pesticide and fumigant, as well as used to protect foods from microorganisms during storage. Furthermore, it has the property of enhancing the skin permeation of various drugs and is widely used as an analgesic in dentistry (Vassão et al., 2010). Transgenic poplars for the biochemical pathway that produces eugenol, chavicol, p-anol, and visoeugenol, which use the available substrates of coniferyl and p-coumaryl alcohols (Lu et al., 2017), were obtained. Results from field trials showed that transgenic poplar lines generated eugenol and chavicol glucosides in foliage tissue, as well as their corresponding aglycones in trace amounts, and several transformed lines, displayed unexpected precocious flowering after a 4-year field trial growth. Authors voted for additional biotechnological approaches in order to produce sterile plant lines, and, at the same time, exploit the possibility of reaching a significant increase in these metabolites in foliage and stems via systematic deployment of numerous "omics", systems biology, synthetic biology, and metabolic flux modeling approaches (Lu et al., 2017). In a previous related study, phenylethanol was produced by introducing a two-enzyme step process from Phe in poplar foliage tissue (Costa et al., 2013). However, in the research of Lu et al. (2017), the availability of substrates, namely monolignols coniferyl and p-coumaryl alcohols, were required much further downstream in the phenylpropanoid pathway.

As an alternative to synthetic biology, plant cultures in photomixotrophic cultures surface as a promising tool to induce phenylpropanoids as a model of secondary metabolites. For example, the induction of phenolic metabolites was standardized during sugarcane micropropagation in temporary immersion bioreactors (TIBs). In TIBs supplemented with CO_2 enrichment, the transcript levels of both phenylalanine ammonia-lyase (PAL EF189195) and ribulose-1,5-biphosphate carboxylase/oxygenase (Rubisco CF576750) were correlated to an increase in the phenolic levels in the culture medium. When the phenolics were sprayed onto the tomato plants infected with *R. solanacearum*, an early defense signaling mechanism was induced, resulting in the protection of the plant against the tomato bacterial wilt disease. In this case, an application of plant phenolic metabolites as elicitors of resistance was identified in the tomato bacterial wilt *Solanum lycopersicum–Ralstonia solanacearum* pathosystem (Yang et al., 2010).

Another example is plant phenolics as elicitors for tolerance to abiotic stress. A significant induction of phenolic compounds and plant survival were confirmed in sugarcane plants treated with 5 mM H_2O_2. In comparison with the control treatments, sugarcane plants treated with H_2O_2 demonstrated a difference in the micropropagation efficiency when multiplied in TIBs supplemented with PEG 20%. The expressions of selected genes related to photosynthesis, ethylene, auxins, oxidative burst, and defense pathways were confirmed during the entire PEG 20% stress in the plants that came from the 5 mM H_2O_2 treatment. Transcript expression analysis sustained the hypothesis that H_2O_2 induced the oxidative burst and the phenylpropanoid pathways, resulting in the elicitation and/or maintenance of the defensive response mechanism in micropropagated plants (Arencibia et al., 2012).

Plant Response to Heavy Metal Stress

The contamination of soil and water with heavy metals has produced a major environmental problem, leading to significant losses in crop productivity and harmful health effects. Thus, contact with toxic metals can induce the ROS which are constantly produced in plants cells. Some of the ROS species are highly toxic and must be detoxified by cellular responses in stress-tolerant plants which have the capacity to survive and grow (Gratão et al., 2005). However, several pathways take advantage of ROS in a good way, i.e., (a) ROS are involved in lignin formation in cell walls; (b) ROS are direct protectants against invading pathogens and also activate signals of HR-hypersensitive response or phytoalexin biosynthesis (Michalak, 2006).

Plants have homeostatic mechanisms that allow them to maintain accurate concentrations of vital metal ions in cellular compartments and minimize the negative effects of an overload of nonessential ones. One of the adverse effects of metal ions is the generation of harmful ROS which originates plant oxidative stress. Plant damage originates when the capability of antioxidant processes and detoxification mechanisms are lower than the quantity of induced ROS. In addition to the systems of low-molecular antioxidants and specific enzymes, in the last few years an increased number of papers highlight the role of flavonoids, phenylpropanoids, and phenolic acids as efficient antioxidant elements. Phenolic compounds can act as metal chelators and, on the other hand, can directly scavenge molecular active oxygen species (Michalak, 2006).

Induction of phenolic compound biosynthesis was observed in wheat, in response to nickel toxicity (Díaz et al., 2001) and in maize, in response to aluminum (Winkel-Shirley, 2002). *Phaseolus vulgaris* exposed to Cd^{2+} accumulate soluble and insoluble phenolics, while *Phyllanthus tenellus* leaves contain more phenolics than the control plants after being sprayed with copper sulfate (Diaz et al., 2001). A rise in soluble phenolics such as lignin biosynthesis precursors may explain the characteristic variations induced by stressors: for example, an augment or increase in the cell wall endurance and the formation of physical barriers, avoiding the destructive action of heavy metals (Diaz et al., 2001).

Antioxidant activity of phenolic compounds could be explained by their ability to chelate metals. Phenolics possess hydroxyl and carboxyl groups which are able to bind metal ions such as iron and copper (Jung et al., 2003). In some plant species, the roots in contact with heavy metals exude high levels of phenolics. They may inactivate ions by chelating and suppressing the superoxide-driven Fenton reaction, as one of the most important sources of ROS (Rice-Evans et al., 1997). As an example, tannin-rich plants such as tea are protected by the direct chelation of Mn. Direct chelation or binding to polyphenols was also observed with methanol extracts of rhizome polyphenols from *Nympheae* for Cr, Pb, and Hg (Lavid et al., 2001).

It was also shown that ethylene could be associated with the inhibitory action of Cu on the roots and leaves of the dicotyledon plant (Maksymiec and Krupa, 2007), including the induction of ethylene production in plants exposed to Cd, Cu, Fe, and Zn (Maksymiec, 2007). Results reveal that both Cu and Cd induce ethylene production by inducing the ACC synthase activity and the expression of its genes. Additionally, the relationship between the ethylene increase and the lipoxygenase activity increase has been demonstrated (Gora and Clijsters, 1989). Altogether, outcomes reveal that heavy metals stimulate the jasmonate pathway and/or lipoxygenase-mediated reactive oxygen species (ROS) formation, and that exogenous JA also augments ethylene concentration, particularly by stimulating the activity of ACC synthase and ACC oxidase (Kruzmane et al., 2002; Saniewski et al., 2003). A review from Maksymiec (2007) points to an indirect heavy metal action through the induction of signaling pathways, mainly those linked with jasmonate, ethylene, and H_2O_2. Signal pathways affected by heavy metals could respond to stress via a decline in plant growth (increasing resistance to stress) or accelerate senescence after a longer exposure to stress (decreasing plant resistance).

Additionally, metal ions play a key function in the antioxidant activity, because they are vital cofactors of a large amount of antioxidant enzymes. As an example, the isoforms of super oxide dismutase (SOD) contain bound heavy metal ions. Cu and Zn constitute the cofactor of the Cu/ZnSOD associated with chloroplast, glyoxisomes contain Mn-SOD, and the Fe-SOD has been found in the chloroplast of some plants (Nagajyoti et al., 2010). The ROS produced in leaf cells are removed by the complex enzymes CAT, APX, GPX, SOD, and GR of antioxidant systems (Prasad et al., 2001).

The relationship between multiple heavy metal stress and the activity of antioxidative enzymes and lipid peroxidation was reported in mangrove plants *Kandelia candel* and *Bruguiera gymnorrhiza*. In the heavy metal-stressed plants, SOD and POX activities showed different levels, while CAT activity augmented with stress levels in *K. candel*, but remained unchanged in *B. gymnorrhiza*. However, the SOD, CAT, and POX activities in the roots of heavy metal-stressed plants first increased and then decreased. Results indicate that for monitoring purposes, POX activity in roots and leaves might serve as a biomarker of heavy metal stress in *K. candel*, while lipid peroxidation might serve as a biomarker in *B. gymnorrhiza* (Zhang et al., 2007).

The impacts of lead, copper, and zinc over a concentration gradient of 0.05–0.20 mg/L on MDA and SOD were investigated in the cyanobacterium *Spirulina platensis*-S5. Results demonstrated a reduction in growth while its MDA, SOD, and proline increased under stress. The increased amount of MDA was indicative of free radical formation under heavy metals, while the increase of both SOD and proline levels supported the incidence of a scavenging mechanism (Choudhary et al., 2007).

In this way, the behavior of the aquatic bryophyte *Fontinalis antipyretica* Hedw was studied on different heavy metal (Cd, Cu, Pb, Zn) stress. Stressed plants showed increased levels of lipid peroxidation and enzyme activities SOD, CAT, GR, APX, and GPX. Authors suggested that plants have antioxidant enzymes which were induced either non-specifically (SOD and APX) or depending on the nature of the contamination. Results confirmed the important role of antioxidant defenses in the mechanisms of plant resistance to heavy metal stress (Dazy et al., 2009).

The relationship of tomato (*Lycopersicon esculentum*) seedlings stressed with four cadmium (Cd) levels were analyzed regarding the expression of the antioxidative enzymes, and the interaction with Mn, Zn, Cu, and Fe microelements. A significant increase in MDA, SOD, and POX activities was demonstrated which indicates that the Cd induces an oxidative stress response in tomato plants. A tissue-dependent response was found in four microelement concentrations to Cd stress in the tomato leaves, stems, and roots (Dong et al., 2006).

The effects of salt stress on the activity of antioxidative enzymes and lipid peroxidation have been studied in leaves and roots of contrasting maize genotypes for salt tolerance. In leaves of salt-stressed plants, SOD, APX, GPX, and GR activities increased with time, which was more evident in the salt-tolerant than in the salt-sensitive genotype. Salt stress had no significant effect on the CAT activity in the salt-tolerant one, but it was reduced significantly in the salt-sensitive genotype. Data revealed that CAT and GPX enzymes had the greatest H_2O_2 scavenger activity in both leaves and roots. Altogether, results indicate that oxidative stress could play a significant role in salt-stressed maize plants, at least in part, through the maintenance and/or increase of the antioxidant enzyme activity (Dias et al., 2006).

The accumulation of ROS, antioxidant enzyme activities, as well as gene expression patterns of antioxidant enzymes were determined in Kentucky bluegrass (*Poa pratensis* L) under drought stress and during recovery. Results confirmed that drought stress induced the leaf activities of enzymes such as ascorbate peroxidase (APX, EC 1.11.1.11), monodehydroascorbate reductase (MDHAR, EC 1.6.5.4), and dehydroascorbate reductase (DHAR, EC 1.8.5.1), and the root activities of catalase (CAT, EC 1.11.1.6), glutathione reductase (GR, EC 1.6.4.2), and MDHAR, while reducing the root activities of superoxide dismutase (SOD, EC 1.15.1.1) and DHAR, respectively. In the case of leaves, the expression of iron SOD (*Fe-SOD*), cytosolic copper/zinc SOD (*Cu/ZnSOD*), chloroplastic *Cu/ZnSOD*, and *DHAR* were down-regulated by drought stress but recovered

to the control level after rewatering; whereas the expressions of *GR* and *MDHAR* were up-regulated and remained at those levels after recovery. In the case of the roots, the expressions of cytosolic *Cu/ZnSOD*, manganese SOD (*Mn-SOD*), cytosolic *APX*, *GR*, and *DHAR* were down-regulated under drought stress. Notably, no differences in the *CAT* transcript abundance were noted among the treatments. The conclusion highlighted that antioxidant enzymes and their gene expressions may be differentially or cooperatively involved in the defense mechanisms in the leaves and roots of Kentucky bluegrass exposed to drought stress and during recovery (Bian and Jiang, 2009).

Tolerance to metal stress was studied in the seedlings of three pine species (*Pinus radiata*, *P. pinaster*, and *P. canariensis*), where the cones were randomly collected at two sites, characterized by contrasting environmental conditions (Arencibia et al., 2016). Results showed large differences among provenances in seedling tolerance to $CuSO_4$ and $AlCl_3$ in terms of survival and growth. *P. pinaster* showed the highest activity level for POX, SOD, and CAT enzymes, while *P. canariensis* and *P. radiata* had the intermediate and lowest values, respectively. Differential gene expression among pine seedlings under metal stress for two genes (Cu-Zn-superoxide dismutase and RuBisCo) confirmed *P. pinaster* as the most tolerant species (Arencibia et al., 2016). The principal component analysis of the three pine populations under $CuSO_4$ and $AlCL_3$ stress treatments is shown in Figure 11.1.

Components C1 and C2 account for 87.67% of the experimental variability. Component C1 (X axis: 59.23%) was reliably represented by the interactions between the variables of the enzyme (POX, SOD, CAT) activities. Meanwhile, the component C2 (Y axis: 28.44%) was characterized by the interactions between the phenotypic variables (NDP, NSP) with stress treatment (salts concentration). Furthermore, the image shows that some plants plot in overlapping positions. Overall, the plot of these principal

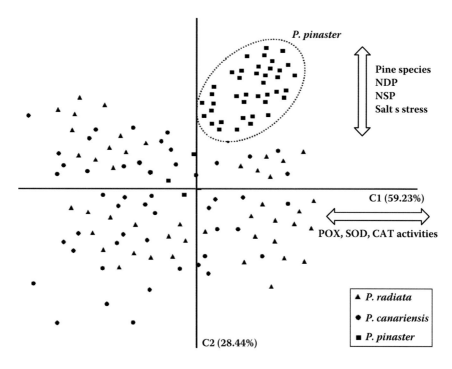

FIGURE 11.1 Plot of the principal component analysis of pine populations treated with $CuSO_4$ (100 mM; 200 mM; 300 mM) or $AlCL_3$ (50 mM; 100 mM).

TABLE 11.1

Canonical Analysis of *Pinus spp* Populations under Abiotic Stress with $CuSO_4$ and $AlCl_3$

Population	*P. radiata*	*P. canariensis*	*P. pinaster*	Misclassification (%)
P. radiate	25	17	4	65.2
P. canariensis	41 **(0.367)**	37	5	48.6
P. pinaster	6 **(0.783)**	18 **(0.485)**	63	12.5
TOTAL	**72**	**72**	**72**	

Note: In brackets: Mahalanobis test distances.

components showed that *P. radiata* and *P. canariensis* plants were interspersed, while *P. pinaster* plants were grouped in the upper-right quadrant.

Canonical analysis of the metal tolerance data was used to evaluate the variability within and among the three studied pine populations (Table 11.1). The verified level of misclassification in the *P. radiata* population (65.2%) could be explained by its high susceptibility to the studied metal concentrations. Moreover, the misclassification level (48.6%) of the *P. canariensis* plants can corroborate intermediate tolerance closer to the *P. radiata* plants. The lowest misclassification level (12.5%) of the *P. pinaster* plants strengthens the conclusion about its tolerance to the experimental treatments with $CuSO_4$ or $AlCL_3$.

The results of the Mahalanobis test (Table 11.1) showed greater distances between *P. pinaster* and the populations of *P. radiata* and *P. canariensis*, whereas reduced but significant distance was demonstrated between *P. radiata* and *P. canariensis* populations. This confirms the tolerance of *P. pinaster* to the studied $CuSO_4$ or $AlCL_3$ stresses and the reliability of the selected variables (phenotypic and molecular) to characterize the pine response to metals concentrations.

Abiotic Stress and Transgenerational Plasticity

Epigenetics refers to the changeable states of the epigenome of a cell, consisting of the entirety of epigenetic marks in the form of histone modifications, histone variants, small RNAs, and DNA methylation (Jablonka and Raz, 2009), thereby modulating the transcriptional activity of the genome (Deal and Henikoff, 2011). Changes in the epigenome are of greater significance through development and in response to environmental conditions, where epigenetic modifications could be transmitted to the progeny and can occur spontaneously. Examples of how epigenetic modifications could contribute directly or indirectly to the diversification of populations during evolution are: (a) when affecting a protein-coding gene, this can directly lead to a variant phenotype; (b) moreover, if the consequence is the release of TEs for gene silencing, this could produce genetic mutations (Becker and Weige, 2012).

Changes in the epigenetic regulation of gene expression induced by environmental exposure can persist in the sexual generation of stressed plants, where these transgenerational effects could be inherited over successive generations (Molinier et al., 2006). One possible mechanism of stress adaptation can be explained by the skill plants have to modify their physiology, which results in stress causing less damage and can also lead to changes in DNA methylation (Boyko et al., 2010).

Plants have the capacity to react and adapt to abiotic stresses in order to survive under a range of environmental conditions using stress tolerance mechanisms. In this way, the tolerance mechanisms result from different processes involving physiological and biochemical changes that originate in adaptive or morphological changes. Novel approaches in genome-wide analysis have revealed an array of networks that control global gene expression, protein modification, and metabolite production. Genetic regulation and epigenetic regulation, including changes in nucleosome distribution, histone modification, DNA methylation, and npcRNAs (non-protein-coding RNA), play a key role in abiotic stress gene regulation. Transcriptomics, metabolomics, bioinformatics, and high-throughput DNA sequencing have noticeably improved our understanding of global plant systems in response and adaptation to stress conditions (Urano et al., 2010). Remarkably, both forward and reverse genetic approaches have elucidated genes and gene products that are involved in gene expression, signal transduction, and stress tolerance (Yamaguchi and Shinozaki, 2006).

The transgenerational effects may comprise the adaptive traits of agronomic relevance. Boyko et al. (2010) proved that the exposure of *Arabidopsis* plants to abiotic stresses, including salt, ultraviolet light, cold, heat, and flood, caused higher homologous recombination frequencies and genome methylation, which resulted in tolerance to stresses in the untreated progenies. In summary, authors hypothesize that stress-induced transgenerational responses in *Arabidopsis* depend on altered DNA methylation and smRNA silencing pathways.

For a long time, transgenerational plasticity was considered to transmit only negative effects, caused by the stressful parental environment, for example, the production of smaller seeds. However, new evolutionary studies have shown that these inherited environmental effects can comprise specific adjustments that are functionally adaptive to the parental conditions. It is accepted that adaptive effects can be transmitted by means of diverse mechanisms, including epigenetic modifications such as DNA methylation, which could be transmitted across multiple generations, influencing the ecological and evolutionary dynamics of plants, as well as the spread of species, including invasive plants (Herman and Sultan, 2011).

Plasticity could be considered as an evolutionary change which may be necessary for plant establishment and spread in stressful areas. Dyer et al. (2010) explain that transgenerational plasticity (TGP) can originate a pre-adapted progeny that displays traits related to an increased performance under stress conditions. For example, in *Cyperus esculentus*, maternal plants growing under nutrient deficiency can place a disproportional number of propagules into nutrient-rich patches. Additionally, studying the invasive grass *Aegilops triuncialis*, authors demonstrate that maternal response to soil conditions confers stress tolerance in seedlings by increasing photosynthetic efficiency. Dyer et al. (2010) also suggest that the maternal environment can have profound effects on offspring success and that TGP may play a significant role in some plants.

A study exploring if the induced changes could be followed by heritable effects was conducted on offspring phenotypes in dandelion (Verhoeven and van Gurp, 2012). Results revealed the effects of parent plants treated with jasmonic acid on their offspring's specific traits, such as root–shoot biomass ratio, tissue P-content, and leaf morphology. Furthermore, the effects could enhance the offspring's ability to cope with the same stress as their parents. Results give evidence of the occurrence of the transgenerational effects in apomictic dandelions. In parallel, a zebularine treatment affected the generation response to nutrient stress, pointing to the role of DNA methylation in phenotypic plasticity towards diverse nutrient environments (Verhoeven and van Gurp, 2012).

Transgenerational adaptive plasticity means that plants can sense varying growth conditions and modify the phenotype of their progeny in order to enhance their behavior under new environments. Boyko and Igor (2011) reviewed some examples of abiotic stress: (a) the light environment of the maternal plant influenced its offspring's flowering pattern in *Campanulastrum americanum* in an adaptive manner when plants grew in a light environment that was similar to their maternal light environment, which led to higher survival rates of seeds and germination; (b) the maternal photoperiod and temperature had positive adaptive effects on phenology and frost hardiness in Norway spruce (*Picea abies*) progeny; and (c) exposure of *A. thaliana* plants to cold during bolting and seed maturation improved the recovery of the photosynthetic yield in their immediate progeny under chilling and freezing conditions.

Research on *Arabidopsis* and *Pinus silvestris* growing in the surrounding area of the Chernobyl reactor revealed that the augmented genome methylation was correlated with genome stability and stress tolerance in response to irradiation and, in the case of pine, after a month of seasonal hardening, some trees can establish tolerance to temperatures as low as −70°C (Boyko et al., 2010). In addition to physiological adaptation related to the accumulation of metabolites and proteins in the seeds of the stressed parental plants, both the genome rearrangement and the epigenetic modifications are proposed as the potential mechanisms for originating the transgenerational plasticity (Boyko and Kovalchuk, 2011)

Continuing with the *Pinus* species, Arencibia et al. (2016) demonstrated a higher number of seedlings tolerant to $CuSO_4$ or $AlCL_3$ when these were originated from seeds collected from mother trees that were growing in a more severe environment. In this way, the number of metal-tolerant seedlings recovered from the Llico area was higher (87.4%) than those from Huilquilemu (12.6%). Both of these localities are in the same region (Maule, Chile), but are separated by a coastal mountain range which determines the highly contrasting environmental conditions between the two sites in terms of salt exposure, rain regime, soil type, etc. In addition, Llico is a secular agroforestry system located along the coast, where pines were planted as protection barriers, while Huilquilemu is an experimental field where the germplasm of different plant species is under maintenance and characterization. The results sustain the hypothesis that plant phenotype depends on genotype–environment interactions, which are specific to the area where they grow, but at the same time it could also be determined by the environment experienced by the parent trees. The canonical variate analysis (CVA) supported the occurrence of a parental effect on seedling tolerance to Cu and Al, independently from the species studied (Arencibia et al., 2016). In this way, different activities of the POX, SOD, and CAT enzymes, as well as the differential expression of gene coding for Cu-Zn-superoxide dismutase and Ribulose bisphosphate carboxylase (RuBisCo), were demonstrated mainly in plantlets originated from seeds coming from Llico, the most stressed area. Altogether, the results corroborate that photosynthetic activity and oxidative stress might be considered as sensitive biological indicators of heavy metal stress (Schutzendubel and Polle, 2002). Additionally, evidence pointed to the role of these enzymatic systems and their connection with the phenylpropanoid pathways, of great significance to plant development in parallel to the adaptative response to adverse environmental conditions, which probably contribute, directly or indirectly, to the diversification of populations during evolution (Boudet, 2007; Becker and Weige, 2012).

Summary

Phenylpropanoid compounds are an essential group of secondary metabolites that, under standard conditions, originate approximately 20% of the carbon fixed by plants, which are correlated with several essential functions in plants. In the case of severe abiotic stress conditions, phenylpropanoids might activate antioxidant enzymes, while up-regulating the biosynthesis of flavonoids inhibits and/or reduces the generation and the levels of ROS. Antioxidant flavonoids are found in the chloroplast, which proposes a possible function as scavengers of singlet oxygen and stabilizers of the chloroplast outer envelope membrane.

Integrated "omics" analysis should be essential in recognizing the array of functions of metabolite regulatory networks during genomic responses to abiotic stresses. Moreover, in parallel, the development of synthetic biosystems could be an alternative for the production of high-value plant metabolites. In the meantime, induction of phenylpropanoids in controlled *in vitro* conditions demonstrates that it could be a robust alternative to eliciting a plant defense system for their protection from both abiotic and biotic stresses.

The role of phenylpropanoids has been confirmed as an efficient antioxidant element in achieving plant tolerance to heavy metal stress. Phenolic compounds can act as metal chelators and, on the other hand, can directly scavenge molecular species of active oxygen. Furthermore, metal ions are vital cofactors of a large amount of antioxidant enzymes.

Evidence points to the role of phenylpropanoids, connected to other enzymatic pathways, as being essential for the development of plants and their adaptive responses to adverse environmental conditions, which could contribute to the diversification of populations during the evolution process.

REFERENCES

Agati G, Azzarello E, Pollastri S, Tattini M. Flavonoids as antioxidants in plants: Location and functional significance. *Plant Science* 2012; 196: 67–76.

Agati G, Brunetti C, Di Ferdinando M, Ferrini F, Pollastri S, Tattini M. Functional roles of flavonoids in photoprotection: New evidence, lessons from the past. *Plant Physiology and Biochemistry* 2013; 72: 35–45.

Ajikumar PK, Xiao WH, Tyo KE, Wang Y, Simeon F, Leonard E, Mucha O, Phon TH, Pfeifer B, Stephanopoulos G. Isoprenoid pathway optimization for taxol precursor overproduction in *Escherichia coli*. *Science* 2010; 330: 70–74.

Arencibia AD, Bernal A, Yang L, Cortegaza L, Carmona ER, Pérez A, Hu C-J, Li Y-R, Zayas CM, Santana I. New role of phenylpropanoid compounds during sugarcane micropropagation in Temporary Immersion Bioreactors (TIBs). *Plant Science* 2008; 175: 487–496.

Arencibia AD, Bernal A, Zayas C, Carmona E, Cordero C, González G, García R, Santana I. Hydrogen peroxide induced phenylpropanoids pathway eliciting a defensive response in plants micropropagated in Temporary Immersion Bioreactors (TIBs). *Plant Science* 2012; 195: 71–79.

Arencibia AD, Rodriguez C, Roco L, Vergara C, González N; García R. Tolerance to heavy metal stress in seedlings of three pine species from contrasting environmental conditions in Chile. *IForest-Biogeosciences and Forestry* 2016; 9: 937–945.

Becker C, Weige D. Epigenetic variation: Origin and transgenerational inheritance. *Current Opinion in Plant Biology* 2012; 15: 562–567.

Bian S, Jiang Y. Reactive oxygen species, antioxidant enzyme activities and gene expression patterns in leaves and roots of Kentucky bluegrass in response to drought stress and recovery. *Scientia Horticulturae* 2009; 120: 264–270.

Boudet AM. Evolution and current status of research in phenolic compounds. *Phytochemistry* 2007; 68: 2722–2735.

Boyko A, Blevins T, Yao Y, Golubov A, Bilichak A, Ilnytskyy Y, Hollander J, Meins FJ, Kovalchuk I. Transgenerational adaptation of Arabidopsis to stress requires DNA methylation and the function of dicer-like proteins. *PLoS One* 2010; 5(3): e9514. doi:10.1371/journal.pone.0009514.

Boyko A, Kovalchuk I. Genome instability and epigenetic modification — heritable responses to environmental stress? *Current Opinion in Plant Biology* 2011; 14: 260–266.

Choudhary M, Jetley UK, Khan MA, Zutsh S, Fatma T. Effect of heavy metal stress on proline, malondialdehyde, and superoxide dismutase activity in the cyanobacterium *Spirulina platensis*-S5. *Ecotoxicology and Environmental Safety* 2007; 66: 204–209.

Costa MA, Marques JV, Dalisay DS, Herman B, Bedgar DL, Davin LB, Lewis NG. Transgenic hybrid poplar for sustainable and scalable production of the commodity/specialty chemical, 2-phenylethanol. *PLoS One* 2013; 8: e83169. doi.org/10.1371/.

Dazy M, Masfaraud J-F, Férard J-F. Induction of oxidative stress biomarkers associated with heavy metal stress in *Fontinalis antipyretica* Hedw. *Chemosphere* 2009; 7: 297–302.

Deal RB, Henikoff S. Histone variants and modifications in plant gene regulation. *Current Opinion in Plant Biology* 2011; 14: 116–122.

Dias AN, Tarquinio PJ, Eneas FJ, Braga ACE, Gomes FE. Effect of salt stress on antioxidative enzymes and lipid peroxidation in leaves and roots of salt-tolerant and salt-sensitive maize genotypes. *Environmental and Experimental Botany* 2006; 56: 87–94.

Dias AN, Tarquinio PJ, Eneas FJ, Rolim JVM, Gomes FE. Hydrogen peroxide pre-treatment induces salt-stress acclimation in maize plants. *Journal of Plant Physiology* 2005; 162: 1114–1122.

Díaz J, Bernal A, Pomar F, Merino F. Induction of shikimate dehydrogenase and peroxidase in pepper (*Capsicum annum* L.) seedlings in response to copper stress and its relation to lignification. *Plant Science* 2001; 161: 179–188.

Dong J, Wu F, Zhang G. Influence of cadmium on antioxidant capacity and four microelement concentrations in tomato seedlings (*Lycopersicon esculentum*). *Chemosphere* 2006; 64: 1659–1666.

Dyer AR, Brown CS, Espeland EK, McKay JK, Meimberg H, Rice KJ. The role of adaptive trans-generational plasticity in biological invasions of plants. *Evolutionary Applications* 2010; 3: 179–192.

Echevarría ZS, Ariza D, Jorge I, Lenz C, Del Campo A, Jorrín JV, Navarro RM. Changes in the protein profile of *Quercus ilex* leaves in response to drought stress and recovery. *Journal of Plant Physiology* 2009; 166: 233–245.

Facchini PJ, Bohlmann J, Covello PS, De Luca V, Mahadevan R, Page JE, Ro DK, Sensen CW, Storms R, Martin VJJ. Synthetic biosystems for the production of high-value plant metabolites. *Trends in Biotechnology* 2012; 30: 127–131.

Fini A, Guidi L, Ferrini F, Brunetti C, Di Ferdinando M, Biricolti S, Pollastri S, Calamai L, Tattini M. Drought stress has contrasting effects on antioxidant enzymes activity and phenylpropanoid biosynthesis in *Fraxinus ornus* leaves: An excess light stress affair? *Journal Plant Physiology* 2012; 169: 929–939.

Gora L, Clijsters H. Effect of copper and zinc on the ethylene metabolism in *Phaseolus vulgaris* L. In: Clijsters H (ed.). *Biochemical and Physiological Aspects of Ethylene Production in Lower and Higher Plants*. Kluwer, Dordrecht, 1989, pp. 219–228.

Gratão PL, Polle A, Lea PJ. Azevedo RA. Making the life of heavy metal-stressed plants a little easier. *Functional Plant Biology* 2005; 32: 481–494.

Herman JJ, Sultan SE. Adaptive transgenerational plasticity in plants: Case studies, mechanisms, and implications for natural populations. *Frontiers in Plant Sciences* 2011; 27. doi:10.3389/fpls.2011.00102.

Hernández I, Alegre L, Van Breusegem F, Munné-Bosch S. How relevant are flavonoids as antioxidants in plants? *Trends in Plant Sciences* 2009; 14: 125–132.

Jablonka E, Raz G. Transgenerational epigenetic inheritance: Prevalence, mechanisms, and implications for the study of heredity and evolution. *Quarterly Review of Biology* 2009; 84: 131–176.

Jung C, Maeder V, Funk F, Frey B, Sticher H, Frossard E. Release of phenols from *Lupinus albus* L. roots exposed to Cu and their possible role in Cu detoxification. *Plant Soils* 2003; 252: 301–312.

Ke Y, Han G, He H, Li J. Differential regulation of proteins and phosphoproteins in rice under drought stress. *Biochemical and Biophysical Research Communications* 2009; 379: 33–138.

Koca H, Bor M, Ozdemi F, Türkan İ. The effect of salt stress on lipid peroxidation, antioxidative enzymes and proline content of sesame cultivars. *Environmental and Experimental Botany* 2007; 60: 344–351.

Kondo N, Kawashima M. Enhancement of tolerance to oxidative stress in cucumber (*Cucumis sativus* L.) seedlings by UV-B irradiation: Possible involvement of phenolic compounds and antioxidant enzymes. *Journal of Plant Research* 2000; 113: 311–317.

Kruzmane D, Jankevica L, Ievinsh G. Effect of regurgitant from *Leptinotarsa decemlineata* on wound responses in *Solanum tuberosum* and *Phaseolus vulgaris*. *Physiologia Plantarum* 2002; 115: 577–584.

Lavid N, Schwartz A, Yarden O, Telor E. The involvement of polyphenols and peroxidase activities in heavy metal accumulation by epidermal glands of waterlily (Nymphaeceaea). *Planta* 2001; 212: 323.

Lu D, Yuan X, Kim SJ, Marques JV, Chakravarthy PP, Moinuddin SGA, Luchterhand R, Herman B, Davin LB, Lewis NG. Eugenol specialty chemical production in transgenic poplar (*Populus tremula x P.alba*) field trials. *Plant Biotechnology Journal* 2017; 15: 970–981. doi:10.1111/pbi.12692.

Maksymiec W. Signaling responses in plants to heavy metal stress. *Acta Physiologia Plantarum* 2007; 29: 177–187.

Maksymiec W, Krupa Z. Effects of methyl jasmonate and excess copper on root and leaf growth. *Biologia Plantarum* 2007; 51: 322–326.

Masuda T, Inazumi A, Yamada Y, Padolina WG, Kikuzàk H, Nakatani N. Antimicrobial phenylpropanoids from *Piper sarmentosum*. *Phytochemistry* 1991; 30: 3227–3228.

Michalak A. Phenolic compounds and their antioxidant activity in plants growing under heavy metal stress. *Polish Journal of Environmental Studies* 2006; 15: 523–530.

Molinier J, Ries G, Zipfel C, Hohn B. Transgeneration memory of stress in plants. *Nature* 2006; 442: 1046–1049.

Nagajyoti PC, Lee KD, Sreekanth TVM. Heavy metals, occurrence and toxicity for plants: A review. *Environmental Chemical Letters* 2010; 8: 199–216.

Ozkur O, Ozdemir F, Bor M, Turkan I. Physiochemical and antioxidant responses of the perennial xerophyte *Capparis ovata* Desf. to drought. *Environmental and Experimental Botany* 2009; 66: 487–492.

Pontin MA, Piccoli PN, Francisco R, Bottini R, Martinez-Zapater JM, Lijavetzky D. Transcriptome changes in grapevine (*Vitis viniferaL.*) cv. Malbec leaves induced by ultraviolet-B radiation. *BMC Plant Biology* 2010; 10: 224.

Prasad MNV, Greger M, Landberg T. *Acacia nilotica* L. bark removes toxic elements from solution: Corroboration from toxicity bioassay using *Salix viminalis* L. in hydroponic system. *International Journal of Phytoremediation* 2001; 3: 289–300.

Qin W, Huang S, Li C, Chen S, Peng Z. Biological activity of the essential oil from the leaves of *Piper sarmentosum* Roxb. (Piperaceae) and its chemical constituents on *Brontispa longissima* (Gestro) (Coleoptera: Hispidae). *Pesticide Biochemistry and Physiology* 2010; 96: 132–139.

Rice-Evans C, Miller N, Paganga G. Antioxidant properties of phenolic compounds. *Trends Plant Sciences* 1997; 2: 152–159.

Ruiz JM, Rivero RM, López-Cantanero I, Romero L. Role of Ca^{2+} in metabolism of phenolic compounds in tobacco leaves (*Nicotiana tabacum* L.). *Plant Growth Regulators* 2003; 41: 173–177.

Rukachaisirikul T, Siriwattanakit P, Sukcharoenphol K, Wongvein C, Ruttanaweang P, Wongwattanavuch P, Suksamrarn A. Chemical constituents and bioactivity of *Piper sarmentosum*. *Journal of Ethnopharmacology* 2004; 93: 173–176.

Sakihama Y, Yamasaki H. Lipid peroxidation induces by phenolics in cinjunction with aluminium ions. *Biologia Plantarum* 2002; 45: 249–254.

Saniewski M, Ueda J, Miyamoto K, Urbanek H. Interaction between ethylene and other plant hormones in regulation of plant growth and development in natural conditions and under abiotic and biotic stresses. In: Vendrell M, Klee H, Pech JC, Romojaro F (eds.). *Biology and Biotechnology of the Plant Hormone Ethylene III*. IOS Press, 2003, pp. 263–270.

Schutzendubel A, Polle A. Plant responses to abiotic stresses: Heavy metal induced oxidative stress and protection by mycorrhization. *Journal of Experimental Botany* 2002; 53: 1351–1365.

Takahama U, Oniki T. Flavonoid and some other phenolics as substrates of peroxidase: physiological significance of the redox reactions. *Journal of Plant Research* 2000; 113: 37.

Urano K, Kurihara Y, Seki M, Shinozaki K. 'Omics' analyses of regulatory networks in plant abiotic stress responses. *Current Opinion in Plant Biology* 2010; 13: 132–138.

Vassão DG, Davin LB, Lewis NG. Metabolic engineering of plant Allyl/propenyl phenol and lignin pathways: Future potential for biofuels/bioenergy, polymer intermediates, and specialty chemicals? *Advances in Plant Biochemistry and Molecular Biology* 2008; 1: 385–428.

Vassão DG, Kim KW, Davin LB, Lewis NG. Lignans (neolignans) and allyl/propenyl phenols: Biogenesis, structural biology, and biological/human health considerations. In: Mander L, Liu HW (eds.). *Comprehensive Natural Products II. Chemistry and Biology*. Vol. 1: Structural Diversity I. Oxford, UK: Elsevier, 2010, pp. 815–928.

Verhoeven KJ, van Gurp TP. Transgenerational effects of stress exposure on offspring phenotypes in apomictic dandelion. *PLOS One* 2012; 7: e38605.

Wahid A, Perveen M, Gelani S, Basra SA. Pretreatment of seed with H_2O_2 improves salt tolerance of wheat seedlings by alleviation of oxidative damage and expression of stress proteins. *Journal of Plant Physiology* 2007; 164: 283–294.

Winkel-Shirley B. Biosynthesis of flavonoids and effects of stress. *Current Opinion Plant Biology* 2002; 5: 218–223.

Xiao M, Zhanga Y, Chenc X, Leec EJ, Barber C, Chakrabartyc R, Desgagné-Penixc I, et al. Transcriptome analysis based on next-generation sequencing of non-model plants producing specialized metabolites of biotechnological interest. *Journal of Biotechnology* 2013; 166: 122–134.

Yamaguchi SK, Shinozaki K. Transcriptional regulatory networks in cellular responses and tolerance to dehydration and cold stresses. *Annual Review Plant Biology* 2006; 57: 781–803.

Yang L, Zambrano Y, Hu C-J, Carmona ER, Bernal A, Pérez A,Li Y-R; Guerra A, Santana I, Arencibia AD. Sugarcane metabolites produced in CO_2-rich Temporary Immersion Bioreactors (TIBs) induce tomato (*Solanum lycopersicum*) resistance against bacterial wilt (*Ralstonia solanacearum*). *In Vitro Cell Development-Plan* 2010; 46: 558–568.

Zakaria ZA, Patahuddin H, Mohamad AS, Israf DA, Sulaiman MR. *In vivo* anti-nociceptive and anti-inflammatory activities of the aqueous extract of the leaves of *Piper sarmentosum*. *Journal of Ethnopharmacology* 2010; 128: 42–48.

Zhang F, Wang YS, Lou ZP, Dong JD. Effect of heavy metal stress on antioxidative enzymes and lipid peroxidation in leaves and roots of two mangrove plant seedlings (*Kandelia candel* and *Bruguiera gymnorrhiza*). *Chemosphere* 2007; 67: 44–50.

12

Metabolic Control of Seed Dormancy and Germination: New Approaches Based in Seed Shape Quantification in Desert Plants

Emilio Cervantes, José Javier Martín Gómez and Ezzeddine Saadaoui

CONTENTS

Abbreviations

ABA	abscisic acid
ctr	constitutive triple response
ein	ethylene insensitive
etr	ethylene resistant
mRNA	messenger RNA

Introduction

Deserts and semi-deserts are the most extensive of the world's land biomes, occupying more than 30% of the Earth's surface (Lioubimtseva, 2004). Based on the precipitation and mean temperatures they may be classified as Hot-Dry or Cold Deserts. Most Hot-Dry Deserts are near the tropics and have very little rainfall, about 10–15 cm a year on average (FAO, 1989), and concentrated in short periods between long rainless periods. Nevertheless, total rainfall in Hot-Dry Deserts is highly variable and may reach up to 60 cm a year. Mean air temperature ranges from 20 to 25°C with maximum values from 43.5 to 49°C. Solar radiation is intense, with values of 710–1050 $Kj \times cm^{-2} \times year^{-1}$ (Baskin and Baskin, 2001). Here we use the term "desert" in its strict sense to classify land where vegetation is virtually absent, except for watercourses. Ephemeral grasses and herbs can appear after infrequent rain showers. In contrast, vegetation in semi-desert regions is a mixture of grasses, herbs, and small, short trees and shrubs up to 2 metres in height, interspersed with bare areas. The rainfall in these regions varies from 100 to 300 millimetres; most of this is unreliable and confined to several months, occurring as local storms or scattered rain showers (FAO, 1989). The soils of the desert are poorly developed with low amounts of nitrogen and organic matter. Non-sandy soils may be saline. In addition, calcium carbonate and gypsum are abundant and may form hardpans. The soils are poor, sandy, salt or saline-sodic, with pebbles, high wind erosion and poorly evolved (Mtimet, 2001). Trees are scarce, being usually limited to the edges of rivers that may be dry most of the time, and their growth is limited to a few metres in height. The predominant forms of vegetation include shrubs and chamaephytes or dwarf-shrubs with sizes decreasing in increased aridity. Perennial grasses may be abundant such as *Stipa*, *Panicum* or *Aristida* as well as annual grasses (Houérou, 2008; Ayyad and Ghabbour, 1986).

The Sahara is the world's largest hot desert, covering an important part of North Africa from the Atlantic Ocean to the Red Sea, with a total surface area of 9 million km². Its flora comprises around 2800 species of vascular plants. Approximately a quarter of these are endemic and about half of these species are common with the flora of the Arabian deserts (Houérou, 2008). In Tunisia, one-third of the territory is in danger of desertification. In particular, the loss of biodiversity caused by overgrazing is a serious issue of concern in southern regions. Because of the extremely dry soil, only 1–10% of the ground is covered by vegetation and the dry biomass of perennial plants is 100 kg/y × Ha (Gamoun et al., 2012).

Central Sahara is estimated to include five hundred species of plants, which is extremely low considering the huge extent of the area. A number of species of plants in the Sahara belong to families typical of warm arid or xeric environments such as the Aizoaceae and Zygophillaceae. Others belong to families more frequent in the tropical and sub-tropical regions such as the Capparaceae. Finally, other species of plants in the Sahara belong to families of cosmopolitan distribution that include species frequent in temperate climates such as Amaranthaceae, Asteraceae, Boraginaceae, Brassicaceae, Cariophyllaceae, Fabaceae, Geraniaceae, Poaceae, Plumbaginaceae and others.

Succulent plants are less frequent in the Sahara and Australia than in other Hot-Dry Deserts. Semi-desert and desert climate regions are important habitats for plants that have adapted to dry conditions. The main species in these regions are *Acacia tortilis* subsp. *Raddiana* and *Vachellia* sp., both in the Fabaceae; *Stipagrostis pungens*, in the Gramineae and *Peganum harmala*, in the family Nitrariaceae (You et al., 2016). This work presents the possibilities of seed shape quantification as a method to study adaptations to the Sahara climate in diverse plant families. Our interest is to search for new models that may be useful for the description of mechanisms involved in the metabolic control of seed dormancy and germination.

Dormancy and Germination in Seeds: Generalities

Seeds are the products of sexual reproduction in plants. They contain the zygote-derived, new generation organism: the embryo. The structure of the embryo includes the root and shoot primordia and meristematic cells. Surrounding the embryo are the seed covers. The testa and pericarp are layers of variable cellular structure and size that protect the embryo and, in particular cases, also contain specialized formations for transport in air, water or through animals including wings, spines and others. Embryo development must be arrested to avoid plants growing on or close to the mother-plant. Thus seed dispersal may occur either in time or in space, or both. While specific structures guarantee dispersal in space, regulatory mechanisms are responsible for delayed germination and dormancy.

Embryo dormancy may be due to the chemical and mechanical effect of the surrounding structures, testa and pericarp. This is considered as physiologically non-deep dormancy and embryos germinate after artificial isolation. This is the case in Arabidopsis, where dormancy may be broken by chilling and after-ripening (Bentsink and Koornneef, 2008). In other cases, the growth potential of the embryo is restricted and dormancy mechanisms may become more complex and difficult to overcome.

Germination occurs when the growth of the radicle in the embryo causes protrusion through the testa and pericarp surrounding the seed. The balance between dormancy and germination is regulated by a combination of endogenous and environmental factors in a way that is particular for each plant species. Seeds of model plants may have very complex processes regulating germination and, in this sense, model plants are not the best models for the study of seed germination (Cervantes, 2008). It may be interesting to find other models based on ecological criteria.

Model plants belong to the type r strategy (Begon et al., 2006). They were chosen as models because of their small size, short life cycle and absence of special adaptations, but all these characteristics are incompatible with a lack of dormancy and easy germination.

Hormone Metabolism in Seed Dormancy and Germination

Among the hormones, ABA has been known for a long time as a major regulator of seed dormancy and germination. ABA added exogenously maintains dormancy and blocks germination. Ethylene is involved in the control of growth and developmental processes from germination to senescence, and in germination ethylene interacts with ABA. Ethylene insensitive mutants such as *ein2* and *etr1-1* are hypersensitive to ABA (Beaudoin et al., 2000; Ghassemian et al., 2000), suggesting that ethylene negatively regulates seed dormancy by inhibiting ABA action (Beaudoin et al., 2000). In addition, there is a cross-talk between sugar signalling and ethylene suggested by the sugar insensitive phenotype of constitutive triple response ethylene mutant *ctr1* (Gibson et al., 2001) and the sugar hypersensitive phenotype of *etr1-1* (Zhou et al., 1998). Thus ABA, ethylene and sugar signalling strongly interact at the level of germination and early seedling growth in *Arabidopsis* (Bentsink and Koornneef, 2008).

Gibberellic acid was known for a long time to be an important regulator of germination in cereals and in *Arabidopsis*, and clear evidence of a role for auxin in germination has been obtained from an analysis of the regulation of Auxin Response Factor10 (ARF10) by microRNA (miRNA) miR160 (Liu et al., 2007). This work reveals the importance of microRNA in the regulation of seed germination.

Synthesis of Germination Proteins

Seeds are structures specialized in the storage of materials and information. They contain all the molecules required to complete the developmental program of a new plant, but this is progressive and depends on the environmental conditions, thus seeds must also have mechanisms that may respond differently to changing environmental conditions.

At the beginning of development, the process of seed germination consists of cell elongation more than cell division (Barrôco et al., 2005), and it depends more on translation than on transcription (Holdsworth et al., 2008). An important question concerns the strategies for stability in seed RNA. Diverse RNA molecules are stored in seeds and the proteins required for germination and the early steps of development may be rapidly synthesized

(Rajjou et al., 2004). These RNA may have sequences in 3′ responsible for their stability.

Mitochondria exist in the dry seeds as large structures (promitochondria). Upon imbibition, these elongate and produce the mature mitochondria. Mitochondrial synthesis is coupled with progression in the cell cycle (Rodríguez et al., 2015). The mRNA encoding prohibitin, a mitochondrial protein, is present in the dry seeds of *Arabidopsis thaliana* (de Diego et al., 2007). A role for prohibitin has been described in the control of the cell cycle as well as in the integrity and functionality of the mitochondrial inner membrane.

Plant Adaptations to the Desert

General adaptations of plants to arid conditions include aspects in their structure and function directed to efficiently collect and store water and to reduce water loss. Adaptations to arid conditions of the desert may involve changes in the whole organism or in particular organs. Morphological adaptations include reduced leaf size and surface, thickened shoots to store water and the presence of spickles and long root systems. Physiological adaptations include waxy coating, shoot photosynthesis and water accumulation in tissues. Adaptations may also imply general changes in life cycle, such as, for example, shortened cycles; or particular aspects of the life cycle such as growth arrest at the vegetative stage, particular types of seed dormancy or rapid root elongation.

In a comparison of the desert plant species with their relatives from more temperate climates, both morphological and physiological adaptations are frequently found in the former. Examples can be found in a variety of acacia trees, palms, spiny shrubs and grasses. Reduced shoot growth avoids water loss by strong winds. This may be accompanied by storing water in their thick stems to use in dry periods, by having long roots that travel horizontally to reach the maximum area of water and to find any surface moisture and by small thick leaves or needles to prevent water loss by evapo-transpiration. Plant leaves may dry out totally and then recover.

In the Fabaceae, *Retama raetam*, a typical plant of the Sahara, has an acclimation strategy of partial plant dormancy in order to survive the dry season. There are two different populations of stems in this plant: those of the upper canopy, exposed to direct sunlight, and those of the lower canopy, protected from direct sunlight. During the dry season, stems of the upper canopy have a reduced number of proteins, and dormancy is evident by the post-transcriptional suppression of gene expression, as well as the suppression of photosynthesis. Translation is induced specifically in stems of the upper canopy which protect the lower canopy by shading (Mittler et al., 2001).

Other desert species such as *Atriplex* sp. (Amaranthaceae) present high adaptation to saline soils and high water-use efficiency, mainly linked to the C4 metabolism. Species of the genus *Atriplex* also have a high ability to absorb nitrogen from the soil and can benefit from the action of nitrogen-fixing microorganisms (Mulas and Mulas, 2004).

The majority of plants in the arid zone are non-succulent perennials. These are hardy plants, including grasses, woody herbs, shrubs and trees that withstand the stress of the arid zone environment. Many non-succulent perennials have "hard" seeds that do not readily germinate (FAO, 1989).

Seeds in the Desert

Seed populations in the desert are complex and their densities may reach several thousand per square metre. Strategies for survival involve reduced size, the presence of specialized cuticles, production of mucilages and special types of dormancy. The ecology of seeds and their germination patterns can determine limits in the geographical range, adaptation to environmental variation, species diversity, and community responses to climate change. Huang et al. (2016) studied the seeds of 13 species of the Arizona desert and registered a relationship between ecological conditions and the physiology of germination, and also with the diversity of seeds. These traits act in the dynamics of the population for each species.

In Tunisia, *C. spinosa* subsp. *rupestris* exists in different regions, but in the Sahara, it shows specific characteristics: the presence of hair, a high number of stamens, and small size and high number of seeds; these characteristics are adaptations to aridity (Saadaoui et al., 2011). Chemical composition of seeds varied, for example populations of desert *C. spinosa* subsp. *rupestris* show low protein composition (Tlili et al., 2011). Its fixed oils are rich in myristic and palmitic acids and low in linoleic and palmitoleic acids (Saadaoui et al., 2015). Desert ephemeral plants are basicarpic species, producing flowers immediately above ground level, and retain their seeds in hard, lignified structures for a long period of time after the mother-plant dies, releasing seeds during rainfall events (Martínez-Berdeja, 2014).

Morphological seed characteristics, such as shape and size, can be used to distinguish species and varieties and predict seed persistence in soil (Arman and Gholipour, 2013; Mahdavi et al., 2012; Saadaoui et al., 2013). A relationship between seed size and soil persistence has been described such that smaller seeds are more persistent (Peco et al., 2003). Concerning shape and the specific role that it may have in the adaptation to a given environment, it is important to establish methods that allow seed shape quantification, such as those based in geometric models (Cervantes et al., 2016).

Geometric Models in the Description and the Quantification of Seed Shape

Seed shape is an important characteristic for plant description, but the useful application of seed shape in taxonomy, development and evolution requires quantification. A practical way to quantify seed shape consists in the comparison of seed images with geometric figures. Geometric figures used as models for the quantification of seed shape include the cardioid, ellipse and ovoid (Cervantes et al., 2016). J index values represent the percentage of the resemblance of seed images to these figures, and they must be around 90 or higher to permit comparisons between subspecies or varieties or to study the effect of climatic factors on seed development. Quantification of seed shape is a required step in the search for trends between subspecies, species and genera of the same family, as well as to study general aspects of seed morphology and its relationship with seed survival, dispersal and plant evolution. In previous work, we have investigated morphological variation in the seeds of some Saharan plants. The method used for seed size quantification involves the comparison of seed shape with a geometric figure and has been applied to the seeds of *Capparis spinosa* (Capparaceae; Saadaoui et al., 2013), *Rhus tripartita* (Anacardiaceae; Saadaoui et al., 2017a), *Ricinus communis* (Euphorbiaceae; Martín Gómez et al., 2016a), *Nerium oleander* (Apocynaceae; Saadaoui et al., submitted) and *Olea europaea* (Oleaceae; Hannachi et al., 2017). Similar methods could be applied to plant species belonging to other families. To explore this and evaluate the possibility of a general extension of seed shape comparison to the family level, we have classified the plant families according to their seed shape into two preliminary groups, based on the similarity of their seeds to a cardioid or similarity to other geometric figures.

Seed Images Resembling a Cardioid

The seeds of species from diverse taxonomic groups resemble a cardioid or a modified cardioid. This is the case for the majority of species in some families such as the Brassicaceae, Capparaceae, Malvaceae and Fabaceae.

In other families, seeds of some species resemble cardioids but the seeds from other species adjust better to other models. This is the case in the Resedaceae, Aizoaceae, Cariophyllaceae, Menispermaceae and Solanaceae. We review these two groups separately.

The Cardioid Model Applies to All Species in a Family

Seed images of most species in the family Brassicaceae resemble a cardioid or a modified cardioid. In the model plant, *Arabidopsis thaliana*, seed images adjust very well to a cardioid elongated by

a factor of Phi (Cervantes *et al.*, 2010) and the similarity increases in the course of germination, reaching values of J index close to 95% in the first hour of imbibition (Martín Gómez et al, 2014). Differences in seed shape were found between wild-type seeds and mutants in the ethylene signal transduction pathway as well as mutants in cellulose biosynthesis, with reduced values indicating shape alteration for mutant seeds (Cervantes *et al.*, 2010; Martín Gómez et al, 2014).

Seeds of Saharan plants in the family Brassicaceae also resemble cardioids such as species of the genera *Crambe*, *Didesmus*, *Diplotaxis*, *Eruca*, *Henophyton* (syn. *Oudneya*), *Farsetia*, *Lobularia*, *Matthiola*, *Morettia*, *Schouwia* and *Sisymbrium* (for examples in *Didesmus aegyptius*, *Diplotaxis harra*, *Eruca sativa*, *Farsetia aegyptia* and *Morettia philaeana* see Tantawy et al., 2004; for *Henophyton* see Gorai et al., 2014; for *Schouwia*, El Nagar and Soliman, 1999; see Figure 12.1: Brassicaceae).

The family Capparaceae is related to the Brassicaceae (Hall et al., 2002). Seeds of *Capparis spinosa*, *Cleome arborea* and other genera of the Capparaceae resemble a cardioid. This allows the quantification of seed shape in varieties or during development as well as in diverse populations and in response to environmental stress (Figure 12.2: Capparaceae).

In *Capparis spinosa*, subspecies *rupestris* grows in the desert, while subspecies *spinosa* is reduced to the sub-humid regions in the north of Tunisia. The comparison between both subspecies reveals smaller seed size in subspecies *rupestris*. Also differences in J index values were detected between subspecies. Division of the seed image (and the cardioid) in four quadrants allows us to define different seed types, demonstrating that morphological variation (amount of seed types) is higher in subspecies *rupestris* (Saadaoui et al., 2013; Martín Gómez et al., 2016b). The comparison between seed populations from diverse origins reveals that seeds from plants grown in the desert have more morphological diversity than seeds obtained from other populations (Martín Gómez et al., 2016b).

The images of seeds of genera in the Malvaceae also present similarity with cardioids (Figure 12.3; Malvaceae:

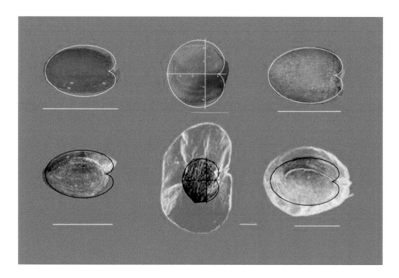

FIGURE 12.1   Seeds of the Brassicaceae adjust well to a cardioid or to modified cardioids: Above: *Sisimbrium sp.*, *Eruca sp*, *Diplotaxis erucoides*. Below: *Didesmus bipinnatus*, *Farsetia aegyptia*, *Lobularia libyca* (from Tantawy *et al.*, 2004). Scale bar = 1 mm. *Diplotaxis erucoides* image is from the Chinese National Herbarium (PE). Images of *Didesmus bipinnatus*, *Farsetia aegyptia*, *Lobularia lybica* are adapted from Tantawy et al. (2004).

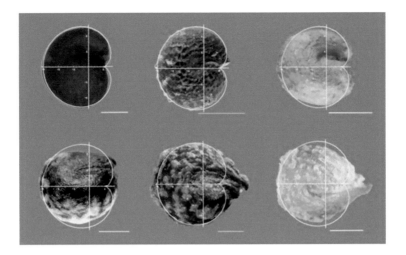

FIGURE 12.2 Seeds of the Capparaceae adjust well to a cardioid: Above: *Capparis spinosa, Cleome gynandra, Cleome spinosa*. Below: *Cleome platy-carpa, Cleome serrulata and Cleome lutea*. Scale bar = 1 mm. Images of *Cleome gynandra, Cleome spinosa, Cleome Platycarpa, Cleome serrulata, Cleome lutea* are from USDA.

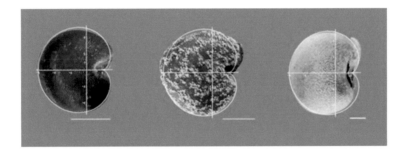

FIGURE 12.3 Malvaceae: *Malva sylvestris, Abutilon lignosum* and *Abelmoschus esculentus*. Scale bar = 1 mm. *Abutilon lignosum* and *Abelmoschus esculentus* are from USDA.

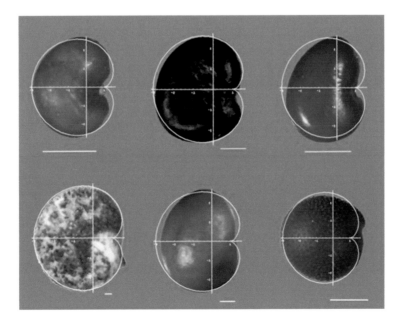

FIGURE 12.4 Fabaceae. Above: *Astragalus botanensis, Colutea arborescens, Crotalaria lanceolata*. Below: *Lupinus digitatus, Retama raetam, Ononis arvensis*. Scale bar = 1 mm. *Astragalus bothanensis, Colutea arborescens, Crotalaria lanceolata, Retama raetam, Ononis avensis* are from Chinese National Herbarium (PE). *Lupinus digitatus* image is from the database of Kew Royal Botanic Gardens.

FIGURE 12.5   Resedaceae: *Reseda alba*, *R. complicata* and *R. lutea*. Scale bar=1 mm. *Reseda alba* and *Reseda lutea* are from USDA. *Reseda complicata*, from the Chinese National Herbarium (PE).

Malvasylvestris, *Abutilon lignosum* and *Abelmoschus esculentus*). It may be interesting to compare seed shape in populations from these species to study the range of shape variation and also to test the hypothesis that morphological variation may be increased in the desert.

In the large and complex family of the Fabaceae, seeds of most species adjust well to a cardioid or modified cardioid, while there is some variation in shape for others (Figure 12.4). The model legumes *Lotus japonicus* and *Medicago truncatula* resemble cardioids and modified cardioids respectively (Cervantes et al., 2012). Similar to *Arabidopsis thaliana*, differences in seed shape

were found between wild-type seeds and mutants in the ethylene signal transduction pathway, with reduced values indicating shape alteration in mutant seeds (Cervantes *et al.*, 2012).

The Cardioid Model Applies to Some but Not All Species or Genera in a Family

In the Resedaceae (Figure 12.5: *Reseda alba*, *R. complicata* and *R. lutea*), seeds resemble a cardioid in *Reseda alba* and *R. complicata* but the images of seeds of *R. lutea* adjust better to an ovoid. Shape quantification in a variety of populations from

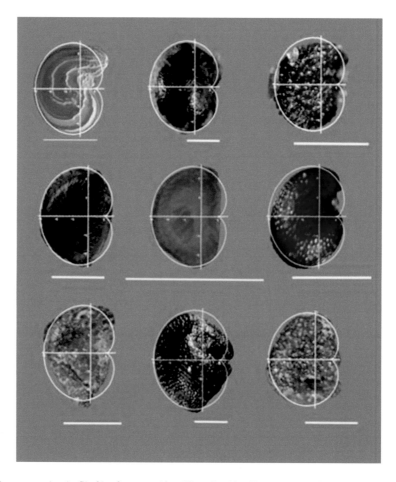

FIGURE 12.6   Aizoaceae: *Aizoon canariensis*, *Gisekia pharmaceoides*, *Glinus lotoides*, *Hesperoyucca whipplei*, *Mollugo cerviana*, *Mollugo pentaphylla*, *Mesembryanthemum nodifolium*, *Mesembryanthemum crystalinum*, *Mesembryanthemum pomeridianum*. Scale bar=0.5 mm. *Aizon canariensis* is from the Department of Horticulture and Crop Science. The Ohio State University. *Hesperoyucca whipplei* is from USDA. *Gisekia pharmaceoides*, *Glinus lotoides*, *Mollugo penthaphylla*, *Mesembriaterium nodifolia*, *Mesembriaterium cristalinum*, *Mesembriaterium pomeridianum* are from the Chinese National Herbarium (PE).

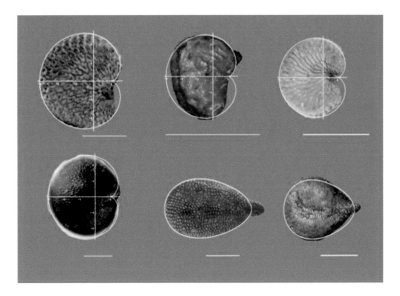

FIGURE 12.7 Cariophyllaceae. Above: *Lychnis coronaria, Polycarpaea arida, Silene armeria*. Below: *Spergula arvensis, Dianthus armeria, Tunica saxifraga*. Scale bar = 0.5 mm. *Lychnis coronaria, Silene armería, Spergula arvensis, Dianthus armería* are from USDA. *Tunica saxifraga* from the Chinese National Herbarium (PE).

diverse origins may help to understand the relationships between shape and taxonomical categories as well as changes in shape in response to environmental conditions.

In the Aizoaceae, seeds resemble a cardioid in *Aizoanthemum, Aizoon, Cypsalea*, and *Sesubium* (Hassan et al, 2005). See, for example, *Aizoon canariensis* (Figure 12.6). The similarity to a cardioid is variable in *Gisekia pharmaceoides, Glinus lotoides, Mollugo cerviana, Mollugo pentaphyla, Mesembryanthemum nodifolium, Mesembryanthemum crystalinum*, and *Mesembryanthemum pomeridianum* as well as in seeds of *Yucca* and *Hesperoyucca* species (Figure 12.6). *Gunniopsis*, a genus of South Australia, resembles *Mesembryanthemum* in that both genera present an interesting variation in shape with seeds resembling a cardioid in some species, but not in others.

In the Caryophyllaceae, seeds resemble a cardioid in *Gypsophila, Lychnis, Spergula* and *Silene* as well as in some species, but not all, of *Dianthus*. Other species of *Dianthus* and *Tunica*, for example *Dianthus armeria* and *Tunica saxifraga*, resemble more an ovoid (Figure 12.7). In the Menispermaceae, seeds of *Cocculus, Gisekia* and *Glinus* resemble cardioids but not in other species such as *Stephania japonica* (Figure 12.8). In the Solanaceae *Anthocercis, Atropa, Capsicum, Datura, Hyosciamus* and *Mandragora* seeds resemble cardioids but in *Nicotiana* some species resemble cardioids and others don't (Figure 12.8).

The Anacardiaceae family presents an interesting case in which the cardioid model is valid only for particular species. In this family, only one genus has been described in the desert flora: *Rhus*. While in the North American desert (Sonoran, Chihuahuan, Mojave and Great Basin Deserts) we find *Rhus ovata* and *Rhus trilobata*, *Rhus tripartita* is the species present in the Sahara. Seeds of all three desert species present a high similarity with the cardioid (Figure 12.9), while other species of the Anacardiaceae don't. In this family, it is thus interesting to remark on the relationship of seed shape (cardioid) with reduced plant size (rapid life cycle) and life in the desert. In Tunisia,

southern populations of *Rhus tripartita* (desert) are characterized by small seeds and lower J index values compared to populations of the semi-arid area (Saadaoui et al., 2017a).

Other Geometric Figures as Models for Seed Shape Analysis

Seeds Resembling an Ellipse

Images of the seeds of the genera *Jatropha* and *Ricinus* in the Euphorbiaceae adjust well to an ellipse (Figure 12.10). Seed images of other species of the genus also resemble this figure, but in some species the similarity between seed images and the

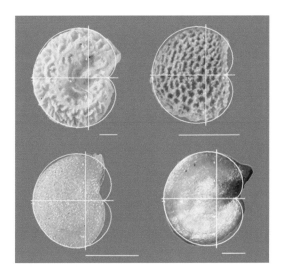

FIGURE 12.8 Menispermaceae and Solanaceae: Above: *Cocculus carolinus* and *Hyosciamus niger*. Below: *Solanum nigrum* and *Capsicum annuum*. Scale bar = 1 mm. *Cocculus carolinus, Hyosciamus niger, Solanum nigrum, Capsicum annuum* are from USDA.

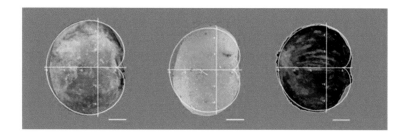

FIGURE 12.9 Anacardiaceae: *Rhus tripartita*, *Rhus ovata* and *Rhus trilobata*. Scale bar = 1 mm. *Rhus ovata is* from USDA. *Rhus trilobata* from the Chinese National Herbarium (PE).

ellipse may be reduced (for example see *Euphorbia dendroides* in Figure 12.10). Our analysis of seed shape in *Ricinus* plants grown spontaneously in different regions of Tunisia revealed reduced size and J index in seeds obtained from the populations grown in the desert (Martín Gómez et al., 2016a). The culture in experimental conditions of plants from seeds grown in the wild resulted in increased J index in the first generation (Saadaoui et al., 2017b).

Nerium oleander (Apocynaceae) seeds adjust well to an ellipse (not shown). Seeds from the semi-arid and Saharan climatic regions had higher values of J index and lower roundness than seeds from humid regions (Saadaoui et al., submitted). Similar to observations in *Ricinus* seeds, lower values of J index were found in the spontaneous than in the cultivated plants. In this case, seeds from the Lower Arid and Upper Saharan climatic regions had higher J index values and lower roundness than seeds from the Lower Humid region (Saadaoui et al., 2017a).

Olea europea seeds adjust well to an ellipse. Hannachi et al. (2017) studied stone characteristics of spontaneous *Olea europea* in different climatic regions of Tunisia and registered high size and low circularity and roundness index in arid regions in comparison with sub-humid and semi-arid bioclimates.

Seeds Resembling an Ovoid

Seed shape models based on the ovoid may be used for the Asteraceae (*Helianthus annuus* L.), Rutaceae (*Citrus reticulata* Blanco), Cucurbitaceae (*Ecballium elaterium* L. and *Citrullus colocynthis*), Caryophyllaceae and Asclepidaceae (*Calotropis procera*) (see Figure 12.11).

Conclusion

In the Fabaceae, seeds of species in the genera *Acacia*, *Colutea* and *Pueraria* also resemble cardioids but in many cases it is possible to observe a diversity where some species of a given genus resemble a cardioid more than others, for example in *Ebenus* (Bayrakdar et al., 2010). In these cases it may be interesting to analyze the morphology of seeds in relation to the climatic region of origin as well as to particular micro-environmental conditions. The classification of seeds in morphological types may provide interesting clues in taxonomy. Accurate seed shape quantification may help to investigate to what extent seed shape is a fixed taxonomical trait or variations in shape may be associated with plant micro-environmental, climatic or soil conditions.

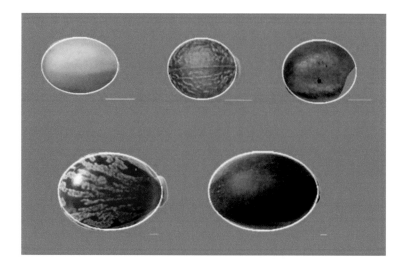

FIGURE 12.10 Euphorbiaceae: *Euphorbia esula*, *Euphorbia epithymoides*, *Euphorbia dendroides*, *Ricinus comunis*, *Jatropha curcas*. Scale bar = 1 mm. *Euphorbia esula* is from the Department of Horticulture and Crop Science. The Ohio State University. *Euphorbia epithymoides* and *Euphorbia dendroides* are from the Chinese National Herbarium (PE).

FIGURE 12.11   Seed shape models based on the ovoid may be used for the Asteraceae (*Helianthus annuus* L.), Rutaceae (*Citrus reticulata* Blanco), Cucurbitaceae (*Ecballium elaterium* L. and *Citrullus colocynthis*), Caryophyllaceae (*Dianthus armeria*), and Asclepidaceae (*Calotropis procera*). Scale bar = 1 mm. *Citrullus collocynthis* and *Calotropis procera* are from USDA.

Acknowledgments

The images used were obtained from different sources as indicated in the legend to each figure: USDA-NRCS PLANTS Database (http://plants.usda.gov, 11 July 2016); National Plant Data Team, Greensboro, NC 27401-4901 USA; The National Herbarium (PE), Institute of Botany Chinese Academy of Sciences 20 Nanxincun, Xiangshan, Beijing 100093, P.R. China (http://pe.ibcas.ac.cn/sptest/SeedSearch.aspx, 11 July 2016); Kew Royal Botanic Gardens; Department of Horticulture and Crop Science, The Ohio State University. We also thank the South Australia Conservation Centre for images of *Polycarpaea arida* and *Mollugo cerviana*. The rest of the images are from our laboratory.

Other Useful Web Sites Consulted (Sites Ordered by the Frequency of Utilization)

List of plant families of the Sahara: http://www.sahara-nature.com/liste_famille.htm?aff=famille

Inventory of plants native to Algeria: http://algerianativeplants.net/html/plante-algerie-inventaire.php?page=115

Herbary of IPE-CSIC (Jaca, Spain): http://herbario.ipe.csic.es/es/listado-imagenes.php?galeria=3&p=13

South Australian Seed Conservation Centre: http://saseedbank.com.au/species_information.php?rid=2197

B&T World seed directory: https://b-and-t-world-seeds.com/directory.html

World checklist of selected plant families at Kew Gardens: http://apps.kew.org/wcsp/reportbuilder.do;jsessionid=61744FA9E0B800B49E1E0BA91DC03A81

Seed images in Wikipedia: https://commons.wikimedia.org/wiki/Seeds

The collection of seed images of Miliza Tischler (1900–1990): http://semina-tischler.de/fam/-cru1.htm

Seed Bank of Korea: http://www.seedbank.re.kr/detail.php?seed=1600.2

Hyppaweed science of INRA-DIJON: https://www2.dijon.inra.fr/hyppa/hyppa-a/dipmu_ah.htm

REFERENCES

Arman M, Gholipour A. Seed morphology diversity in some Iran endemic silene (Caryophyllaceae) species and their taxonomic significance. *Acta Biol. Szegediensis* 2013; 57(1): 31–37.

Ayyad MA, Ghabbour SI. Hot deserts of Egypt and Sudan. In: *Ecosystems of the World* (M. Evenari, Y. Noy-Meir, and D. W. Godall, eds.). Elsevier, Amsterdam, 1986; vol. 12 B, 149–202.

Barrôco RM, Van Poucke K, Bergervoet JH, De Veylder L, Groot SP, Inzé D, Engler G. The role of the cell cycle machinery in resumption of postembryonic development. *Plant Physiol.* 2005; 137(1): 127–140.

Baskin CC, Baskin JM. *Seeds: Ecology, Biogeography, and Evolution of Dormancy and Germination.* Academic Press, San Diego, 2001.

Bayrakdar F, Aytaç Z, Suludere Z, Candan S. Seed morphology of *Ebenus* L. species endemic to Turkey. *Turk J. Bot.* 2010; 34: 283–289.

Beaudoin N, Serizet C, Gosti F, Giraudat J. Interactions between abscisic acid and ethylene signaling cascades. *Plant Cell* 2000; 12(7): 1103–1115.

Begon M, Townsend CA, Harper JL. *Ecology: From Individuals to Ecosystems.* 4th ed. Blackwell, Maldon, MA, 2006.

Bentsink L, Koornneef M. Seed dormancy and germination. *Arabidopsis Book* 2008; 6: e0119. https://www.ncbi.nlm.nih.gov/pmc/articles/PMC3243337/

Cervantes E. Challenging arabidopsis thaliana as the ultimate model species: Can seed germination be the achiles' heel? *Eur. J. Plant Sci. Biotechnol.* 2008; 2(1): 106–109.

Cervantes E, Martin JJ, Ardanuy R, de Diego JG, Tocino A. Modeling the *Arabidopsis* seed shape by a cardioid: Efficacy of the adjustment with a scale change with factor equal to the Golden Ratio and analysis of seed shape in ethylene mutants. *J. Plant Physiol.* 2010; 167: 408–410. doi:10.1016/j.jplph.2009.09.013.

Cervantes E, Martin JJ, de Diego JG, Chan PK, Gresshoff P, Tocino A. Seed shape in model legumes: Approximation by a cardioid reveals differences between *Lotus* and *Medicago*. *J. Plant Physiol.* 2012; 169(14): 1359–1365. doi:10.1016/j.jplph.2012.05.019.

Cervantes E, Martín Gómez JJ, Saadaoui, E. Updated methods for seed shape analysis. *Scientifica (Hyndawi)* 2016. doi:10.1155/2016/5691825.

de Diego JG, David Rodríguez F, Rodríguez Lorenzo JL, Cervantes E. The prohibitin genes in *Arabidopsis thaliana*: Expression in seeds, hormonal regulation and possible role in cell cycle control during seed germination. *J. Plant Physiol.* 2007; 164(3): 371–373.

El Nagar SM, Soliman MA. Biosystematic studies on Schouwia DC. (Brassicaceae) in Egypt. *Flora Mediterr.* 1999; 9: 175–183.

FAO (Food and Agriculture Organization). *Arid Zone Forestry: A Guide for Field Technicians.* Publications Division, Food and Agriculture Organization of the United Nations, Via delle Terme di Caracalla, Rome, Italy, 1989.

Gamoun M, Hanchi B, Neffati M. Dynamic of plant communities in Saharan rangelands of Tunisia. *Arid Ecosyst.* 2012; 2: 105–110.

Ghassemian M, Nambara E, Cutler S, Kawaide H, Kamiya Y, McCourt P. Regulation of abscisic acid signaling by the ethylene response pathway in *Arabidopsis*. *Plant Cell* 2000; 12(7): 1117–1126.

Gibson SI, Laby RJ, Kim D. The sugar-insensitive1 (*sis1*) mutant of Arabidopsis is allelic to *ctr1*. *Biochem. Biophys. Res. Commun.* 2001; 280(1): 196–203.

Gorai M, El Aloui W, Yang X, Neffati M. Toward understanding the ecological role of mucilage in seed germination of a desert shrub Henophytondeserti: Interactive effects of temperature, salinity and osmotic stress. *Plant Soil* 2014; 374: 727–738.

Hall JC, Sytsma KJ, Iltis HH. Phylogeny of Capparaceae and Brassicaceae based on chloroplast sequence data. *Am. J. Bot.* 2002; 89(11): 1826–1842.

Hannachi H, Martín JJ, Saadaoui E, Cervantes E. Stone diversity in wild and cultivated olive trees (*Olea europaea* L.). *Dendrobiology* 2017; 77: 19–32.

Hassan, NMS, Meve U, Liede-Schumann S. Seed coat morphology of Aizoaceae–Sesuvioideae, Gisekiaceae and Molluginaceae and its systematic significance. *Bot. J. Linnean Soc.* 2005; 148: 189–206.

Holdsworth MJ, Finch-Savage WE, Grappin P, Job D. Post-genomics dissection of seed dormancy and germination. *Trends Plant Sci* 2008; 13(1): 7–13.

Houérou HN. *Bioclimatology and Biogeography of Africa.* Springer Verlag, Berlin. 2008; ISBN 978-3-540-85192-9.

Huang Z, Liu S, Bradford KJ, Huxman TE, Venable DL. The contribution of germination functional traits to population dynamics of a desert plant community. *Ecology* 2016; 97: 250–261.

Lioubimtseva E. Climate change in arid environments: Revisiting the past to understand the future. *Prog. Phys. Geogr.* 2004; 28(4): 502–530.

Liu PP, Montgomery TA, Fahlgren N, Kasschau KD, Nonogaki H, Carrington JC. Repression of AUXIN RESPONSE FACTOR10 by microRNA160 is critical for seed germination and post-germination stages. *Plant J.* 2007; 52(1): 133–146.

Mahdavi M, Assadi M, Fallahian F, Nejadsattari T. The systematic significance of seed micro-morphology in *Stellaria* L. (Caryophyllaceae) and its closes relatives in Iran. *Iran. J. Bot.* 2012; 18(2): 302–310.

Martín Gómez JJ, Saadaoui E, Cervantes E. Climatic effects in seed shape of castor bean (*Ricinus communis* L.) grown in Tunisia. *J. Agric. Ecol. Res. Int.* 2016a; 8(1): 1–11. doi:10.9734/JAERI/2016/23934.

Martín JJ, Saadaoui E, Cervantes E. Seed shape quantification in Capparis spinosa L.: Effect of subspecies and geographic regions. *SCIREA J. Agric.* 2016b; 1(1): 79–90.

Martín JJ, Tocino A, Ardanuy R, de Diego JG, Cervantes E. Dynamic analysis of *Arabidopsis* seed shape reveals differences in cellulose mutants. *Acta Physiol Plant.* 2014; 36: 1585–1592.

Martínez-Berdeja A. Rainfall variability in deserts and the timing of seed release in chorizanthe rigida a serotinous winter desert annual. Thesis in Biology. University of California, Riverside, CA, 2014.

Mittler R, Merquiol E, Hallak-Herr E, Rachmilevitch S, Kaplan A, Cohen M. Living under a 'dormant' canopy: A molecular acclimation mechanism of the desert plant Retama raetam. *Plant J.* 2001; 25(4): 407–416.

Mtimet A. Soils of Tunisia. In: *Soil Resources of Southern and Eastern Mediterranean Countries* (P. Zdruli, P. Steduto, C. Lacirignola, and L. Montanarella, eds.). Options Méditerranéennes: Série B. Etudes et Recherches; n. 34. Bari: CIHEAM, 2001, 243–262.

Mulas M, Mulas G. The strategic use of *Atriplex* and *Opuntia* to combat desertification. University of Sassari, Italy, 2004.

Peco B, Traba J, Levassor C, Sánchez AM, Azcárate EM. Seed size, shape and persistence in dry Mediterranean grass and scrublands. *Seed Sci. Res.* 2003; 13: 87–95.

Rajjou L, Gallardo K, Debeaujon I, Vandekerckhove J, Job C, Job D. The effect of alpha-amanitin on the *Arabidopsis* seed proteome highlights the distinct roles of stored and neosynthesized mRNAs during germination. *Plant Physiol.* 2004; 134(4): 1598–1613.

Rodríguez JL, De Diego JG, Rodríguez FD, Cervantes E. Mitochondrial structures during seed germination and early seedling development in *Arabidopsis thaliana*. *Biologia* 2015; 70(8): 1019–1025.

Saadaoui E, Guetat A, Massoudi Ch, Tlili N, Khaldi A. Wild Tunisian *Capparis spinosa* L.: Subspecies and seed fatty acids. *Int. J. Curr. Res. Acta. Rev.* 2015; 3(1): 315–327.

Saadaoui E, Guetat A, Tlili N, El Gazzah M, Khaldi A. Subspecific variability of Tunisian wild populations of *Capparis spinosa* L. *J. Med. Plants Res.* 2011; 5(17): 4339–4348.

Saadaoui E, Martín Gómez JJ, Cervantes E. Intraspecific variability of seed morphology in *Capparis spinosa* L. *Acta Biol. Cracovensia (Sect Botanica)* 2013; 55: 1–8. doi:10.2478/abcsb-2013-0027.

Saadaoui E, Martín Gómez JJ, Cervantes E. Effect of climate in seed diversity of wild Tunisian *Rhus tripartita* (Ucria) Grande. *J. Adv. Biol. Biotechnol.* 2017a; 13(4): 1–10.

Saadaoui E, Martín Gómez JJ, Ghazel N, Ben Yahia K Tlili N, Cervantes, E. Genetic variation and seed yield in Tunisian castor bean (*Ricinus communis* L.) *Bot. Sci.* 2017b; 95(2): 271–281.

Saadaoui E, Martin Gómez JJ, Kaouther ben Yahia K, Cervantes E. Seed morphology and germination in spontaneous and cultivated populations of *Nerium oleander* var. Villa Romaine grown in Tunisia, submitted.

Tantawy ME, Khalifa SF, Hassan SA, Al-Rabiai, GT. Seed Exomorphic Characters of Some Brassicaeae (LM and SEM Study). *Int. J. Agri. Biol.* 2004; 6(5): 821–830.

Tlili N, Saadaoui E, Sakouhi F, Elfalleh W, El Gazzah M, Triki S, Khaldi A. Morphology and chemical composition of Tunisian caper seeds: Variability and population profiling. *Afr. J. Biotechnol.* 2011; 10(10): 2112–2118.

You H, Jin H, Khaldi A, Kwak M, Lee T, Khaine I, Jang J, et al. Plant diversity in different bioclimatic zones in Tunisia. *J. Asia-Pac. Biodivers.* 2016; 9(1): 56–62.

Zhou L, Jang JC, Jones TL, Sheen J. Glucose and ethylene signal transduction crosstalk revealed by an Arabidopsis glucose-insensitive mutant. *Proc. Natl. Acad. Sci. USA.* 1998; 956(1): 10294–10299.

13

Plant Ionomics: An Important Component of Functional Biology

Anita Mann, Sangeeta Singh, Gurpreet, Ashwani Kumar, Pooja, Sujit Kumar, and Bhumesh Kumar

CONTENTS

Abbreviations

AAS	Atomic absorption spectroscopy
FAAS	Flame atomic absorption spectrometry
GFAAS	Graphite furnace atomic absorption spectrometry
ICP-AES	Inductively coupled plasma atomic emission spectrometry
ICP-MS	Inductively coupled plasma mass spectrometry
INAA	Instrumental neutron activation analysis
LA-ICP-MS	Laser-ablation inductively coupled plasma mass spectrometry
PIXE	Proton-induced X-ray emission
XF	X-ray fluorescence

Introduction: Definition of Ionomics

Elements are essential building blocks of living systems and are involved in many processes like the regulation of electrochemical balance of cellular compartments; they act as cofactors in biochemical reactions, and as structural components in biological molecules and complexes. Many elements can also have deleterious effects within an organism if they accumulate above the critical limit, hence an efficient compartmentalization is essential to maintain the ionic homeostasis according to

the need of the functional status of an organism. Therefore, the study of physiological and genetic networks controlling uptake, transport, accumulation, and regulation of elements (ionome) is a challenging avenue for researchers and has been the focus of many research groups. As plants take up all elements from the soil environment except carbon and oxygen, regulation of the uptake and compartmentalization of elements as well as interaction with other environmental factors is extremely crucial for survival and functioning. The level and distribution of an element in a given plant species can be affected by several factors like changes in the soil chemical environment; changes in the environment due to biotic or abiotic factors; different morphology at growth stages of the plant; presence and accumulation of chelators; availability of ion transporters/channels; and extent of ion compartmentalization.

The functional life of plants is supported and sustained by four basic pillars, namely transcriptome, proteome, metabolome, and ionome, which represent the sum of all the expressed genes, proteins, metabolites, and elements within an organism. The dynamic response and interaction of these 'omes' together can be regarded as 'system biology', which defines how a living system functions in totality. Studies on the functional connections between the genome and the transcriptome (Martzivanou and Hampp, 2003; Becher et al., 2004; Leonhardt et al., 2004), proteome (Koller et al., 2002), and metabolome (Fiehn et al., 2000) have been undertaken in many plant species. However, in contrast, the study of the ionome still needs to be explored (Lahner et al., 2003; Hirschi, 2003; Rea, 2003) as the majority of genes and gene networks involved in its regulation are still unknown. The ionome is involved in a vast array of important biological processes (White et al., 2012) including electrophysiology, signalling, enzymology, osmoregulation, and transport and is also used for phylogenetic analysis; thus, its study promises to provide new insight into dissecting the system biology of a given plant species.

Huge variation in soil nutrient composition is common in agriculture systems; it may range from an extreme deficiency of nutrients to excess nutrients. Plant response to the deprivation of essential mineral nutrients is fairly well described, but an understanding of how plants sense and signal changes (at the cellular- or whole-plant level) in the availability or deficiency of nutrients is lacking. For maximum benefits to farmers, nutrients need to be used efficiently and in an eco-friendly manner to avoid negative environmental impacts. It will be of immense significance to gain a better understanding of how crop plants can be designed to grow more efficiently in environments with minimum nutrient inputs. The advent and availability of modern high-throughput ionome analysis techniques provide us with knowledge about the effect of these changes on the whole ionome and help in understanding the functional status of plants and their regulation at gene level.

The basic biology of nutrient ion homeostasis in plants was described in the 20th century (Marschner, 1995). One of the key advances enabling organisms to survive was the evolution of ion transport systems which have been the primary focus of work involved in characterizing the ionome in plants. In the past two decades, transporters for many different ions have been characterized (Maser et al., 2001), suggesting the involvement of multiple genes and even multiple gene families in transport. Considering that transport across different membranes within plants exhibits diverse kinetics and dynamics (Rausch, 2002; Sanders, 2002; Curie and Briat, 2003; Very and Sentenac, 2003; Zhu, 2003), understanding of the ionome seems to be essential. Analysis of the *Arabidopsis* genome revealed that 5% of its approximately 25,000 genes are involved in regulating the ionome (Lahner, 2003). Plants contain relatively few different cell types (Martin, 2001) and these cell types are thought to perform many physiological functions and, hence, possess a unique ionome (Punshon et al., 2009; Karley and Philip, 2009). Current interest is primarily focused on quantifying and differentiating the ionome of different plant species, biotypes, and mutants (Watanabe et al., 2007; Baxter, 2009; Chen et al., 2009; White and Broadley, 2009; Broadley et al., 2010). The ionome is also expected to

be influenced significantly by developmental and environmental factors, thus making the study of the ionome more complicated, but at the same time, more relevant too.

Useful Techniques in Ionomics

To maintain cellular homeostasis, concentrations, chemical speciation, and localization of mineral nutrients and other trace elements need to be regulated efficiently (Zhao et al., 2014). Study of the above aspects warrants sophisticated techniques and instrumentation. A range of techniques are available for mapping elements at various spatial scales which are useful in mapping the cellular distributions of macro/micro nutrients and metals; however, the technique must be selective, keeping in view the experimental plan and demands. The most commonly used analytical methods in ionomics since the conception of the term to date are summarized here.

Atomic Absorption Spectroscopy (AAS)

Atomic absorption spectroscopy was first used as a spectroanalytical technique by Robert Wilhelm Bunsen and Gustav Robert Kirchhoff during the second half of the 19th century at the University of Heidelberg, Germany. The technique is based on the principle of determining the concentration of a particular element in a sample using the absorption of optical radiation by free atoms in the gaseous state. In order to analyze a sample, first it has to be atomized, which can be done using flames and electro thermal (graphite tube) atomizers. The atoms should then be irradiated using a radiation source and passed through a monochromator in order to separate the element-specific radiation from any other radiation emitted by the radiation source (Figure 13.1). After passing through a monochromator, element-specific radiation flux without a sample and with a sample in the atomizer is measured using a detector, and the ratio between the two values (the absorbance) is converted to analyte concentration or mass using the Beer-Lambert Law. Based on the method of atomization, the following two types of atomic absorption spectrometry are being used.

Flame Atomic Absorption Spectrometry (FAAS)

In FAAS, atomization is achieved by a flame generated using a mixture of either air/acetylene or N_2O/acetylene to evaporate the solvent and to dissociate the sample into its component atoms. The use of a flame limits the excitation temperature reached by a sample to a maximum of approximately 2,600°C (with the N_2O/

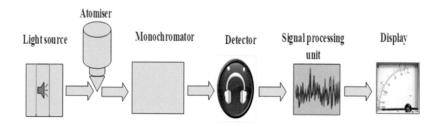

FIGURE 13.1   Line diagram of a typical atomic absorption spectrophotometer.

acetylene flame), which is good enough for many plant mineral elements and heavy metals such as lead or cadmium, and transition metals like manganese or nickel with a detection in the sub-micromolar range. However, there are a number of refractory elements like V, Zr, Mo, and B which do not perform well with a flame source because the maximum temperature reached using a flame is insufficient to atomize these elements, yielding poor sensitivity.

Graphite Furnace Atomic Absorption Spectrometry (GFAAS)

Limitations of the FASS technique led to the advent of the refined technique 'GFAAS', which is essentially the same except that the flame is replaced by a small, electrically heated graphite tube or cuvette. The graphite tube can easily be heated to a temperature up to 3,000°C to generate the cloud of atoms. The higher atom density and longer retention time in the tube improve furnace AAS detection limits a thousand times as compared to FAAS with a detection limit in the sub-nanomolar range. Still, because of the temperature limitation, use of graphite cuvettes for detection of refractory elements (V, Zr, Mo, and B) is somewhat limited due to inefficient performance.

The graphite furnace has several advantages over a flame furnace. First it accepts solutions, slurries, or solid samples. Second, it is a much more efficient atomizer than a flame furnace and it can directly accept very small absolute quantities of sample. It also provides a reducing environment for easily oxidized elements. There have been many studies measuring concentrations of toxic metals such as Ag, As, Cd, Cr, Cu, Hg, Ni, and Pb in rainwater and their deposition into surface waters and on soils using AAS. A sequential extraction procedure for Se(IV) and Se(0) independently was developed and analyzed by GFAAS, whereas only total selenium could be analyzed by AAS (Sammut et al., 2010). This research group also proposed the analysis of beryllium species using GFAAS. Sodium, K^+, Ca^{2+}, and Mg^{2+} were analyzed with a double beam atomic absorption spectrophotometer (Báez et al., 2007; García et al., 2009) in atmospheric aerosols, metals, and ions that play an important role in the content of chemical species and of many elements in atmospheric ecosystem interfaces. The eight inorganic elements (Ca, P, Mn, Zn, Ni, Fe, K, and Mg) have been detected in medicinal plants *B. monosperma*, *C. fistula*, *C. toona*, *T. cordifolia*, and *Q. infectoria* using AAS (Vermani et al., 2010).

Proton-induced X-ray Emission (PIXE)

PIXE is an X-ray spectrographic technique that can be used for the non-destructive and multiple elemental analysis of solid, liquid, or aerosol-filtered samples without a prerequisite need of digestion. Exciting the inner shell electrons in the target atoms is induced by energetic protons leading to the expulsion of these inner shell electrons. Expulsion of inner shell electrons produced a unique X-ray spectrum characteristically specific to the element. The number of X-rays emitted is proportional to the mass of that corresponding element in the sample being analyzed. The PIXE technique offers advantages over the previously described techniques in terms of time and precision.

X-ray Fluorescence (XF)

The use of an X-ray beam to excite fluorescent radiation from the sample was first proposed by Glocker and Schreiber (1928). Since then, the technique has been widely used successfully for elemental and chemical analysis of metals, glass, ceramics and building materials, and for research in geochemistry, forensic science, and archaeology. Emission of characteristic fluorescence can be induced by bombarding with high-energy short wavelength rays (X-rays or gamma rays). If an atom is exposed to radiation with an energy greater than its ionization potential, ionization of the atom may take place concomitantly with the ejection of one or more electrons from the atom. X-rays and gamma rays possess energy high enough to expel tightly held electrons from the inner orbitals of the atom, thus making the atom electronically unstable. To fill the electron's deficiency in inner orbitals, an electron from higher orbitals falls into inner orbitals during which energy is released in the form of a photon, which is equal to the energy difference of the two orbitals involved. Detection of the element is achieved by mapping emission radiation having the characteristic energy of the atoms present in that element. In the early 1990s, before the ionome or ionomics had been defined, Delhaize and co-workers (Delhaize et al., 1993) applied XRF for the successful multi-element screening of more than 100,000 mutagenized *A. thaliana* seedlings to identify mutants with altered ionomes.

X-ray Photoelectron Spectroscopy (XPS)

XPS is a surface-sensitive quantitative spectroscopic technique that measures composition, empirical formula, chemical speciation, and electronic state of the elements. This technique requires high-vacuum to ultra-high-vacuum (10^{-8} to 10^{-9} millibar) conditions; however, recent advances allow operation at ambient pressure. XPS spectra are obtained by irradiation of a sample with a beam of X-rays or gamma rays and mapping the kinetic energy and number of electrons that escape from the top 0 to 10 nm of the material being analyzed. XPS showed improved potential over techniques and can detect all elements with an atomic number of 3 or more; however, it is not fully capable of detecting elements with an atomic number of 1 (hydrogen) and 2 (helium). Extended XPS provides detailed information about the bond lengths between the element of interest and the element(s) it is coordinated with, along with qualitative information on the identity of the ligating atoms. These approaches were first applied to plants in the mid-1990s by Salt and colleagues (Salt et al., 1995) and Kramer and colleagues (1996) for the chemical speciation of Cd and Ni respectively in bulk plant tissues. Quantification and speciation of Se in the vascular tissue of the plants using X-ray absorbance spectroscopy and high-resolution quantitative chemically specific imaging were also used to determine the localization and chemical speciation of As in the As-hyperaccumulating fern *Pteris vittata* (Pickering et al., 2006). This type of imaging revealed in exquisite detail the amount and localization of arsenite, arsenate, and As(III) coordinated with thiol ligands in several tissues in the sporophyte and the gametophyte of *P. vittata*. Gamma-ray sources are inherently radioactive, and therefore do not require large power supplies and can be used in small portable XRF instruments, which are useful for rapid ionomic analyses in the greenhouse or the field. For example, plant tissue and soil could be analyzed directly in the field, and decisions about the collection of fresh tissue for genotyping of a segregating mapping population can be made. Such experiments could be very valuable for the identification of quantitative trait loci (QTL) involved in ionomic adaptation to particular soil conditions.

Inductively Coupled Plasma Atomic Emission Spectrometry (ICP-AES)

ICP-AES, commonly referred to as ICP, uses an inductively coupled plasma source to atomize and ionize the sample and exciting them to a level where they emit light of a characteristic wavelength. A detector measures the intensity of the emitted light and calculates the concentration of that particular element in the sample. During ICP analysis, the sample is exposed to extremely high temperature (as high as 10,000°C), enabling atomization of even the most refractory elements with a greater efficiency than other techniques. The technique offers several benefits like multi-element analysis (60 elements in a single run), lower detection limit (1–10 ppb), less time consumption ($\leq$1 min per element), and high precision.

Inductively Coupled Plasma Mass Spectrometry (ICP-MS)

As indicated by its name, ICP-MS also uses an inductively coupled plasma source to dissociate the sample into constituent atoms or ions; however, unlike ICP-AES, the ions themselves are detected instead of the light emitted by them. Ions are extracted from the plasma and passed through the mass spectrometer and separated based on their atomic mass-to-charge ratio (m/z) and detected by quadrupole or magnetic sector instrumentation. This technique is highly efficient with a detection limit in the ppt (parts per trillion) range; multi-element low backgrounds provide the best detection limits, capable of performing multi-element analysis and high precision, and capable of discriminating at isotopic level.

Using inductively coupled plasma spectroscopy, Lahner et al. (2003) had quantified 18 elements, including essential macro- and micronutrients and various nonessential elements, in shoots of 6,000 mutagenized M2 *Arabidopsis thaliana* plants. Recently, metal contaminants have been analyzed using inductively coupled plasma mass spectrometer (ICP-MS) in water-soluble and gulal colors (Debnath et al., 2015). Punshon and colleagues (2004) analyzed U and Ni on the surface of leaves using ICP-MS.

Laser-Ablation Inductively Coupled Plasma Mass Spectrometry (LA-ICP-MS)

This technique has the potential of mapping the distribution of elements in solid samples. A sample is first ablated by the laser beam to induce aerosol, which is transported with a carrier gas (Ar or He) into the inductively coupled plasma ion source. After ionization, charged ions are detected by the mass spectrometer based on their m/z.

Instrumental Neutron Activation Analysis (INAA)

This technique is useful for multi-analysis of trace and major elements in different geological, environmental, and biological matrices. A sample, when subjected to a neutron flux, produces radioactive nuclides, which emit gamma rays having characteristic energy. This energy is then mapped and compared with standards. Elemental analysis of plant samples by INAA dates back to at least the mid-1960s. Multi-element analysis of plant samples by INAA was used to identify and monitor areas of heavy metal pollution across broad geographical areas over multiple years (Barandovski et al., 2008). INAA was also used to perform multi-element quantification on plant samples collected within and across broad phylogenetic groupings for the identification of trends in mineral nutrient and trace element accumulation in plants across taxa (Watanabe et al., 2007). INAA has been applied to perform ionomics in the study of breast cancer, colorectal cancer, and brain cancer; in these studies the ionome was shown to be perturbed in the diseased tissues or organisms.

Super-SIMS

Super-sims (secondary ion mass spectrometry) is an ultrasensitive analytical technique capable of detecting stable elements and isotopes up to the parts per trillion (ppt) level. An energetic primary ion beam removes particles from the top surface and converts them into secondary ions, which in turn are analyzed in a mass spectrometer to give information about the elemental or molecular distribution within the sample. The only limitation to this method is that it uses only solid and vacuum stable samples; hence this technique finds more application in physics, material science, and polymer chemistry.

In this view, availability of a broad range of techniques undoubtedly offers choices to users. These techniques have their own advantages and limitations too, which need to be accounted during the selection of a technique. Detection limits, sample throughput, dynamic range, precision, and capability of isotopic discrimination might be major criteria for selection; however, cost and time taken per sample also need to be considered. A brief comparative analysis of various ionomic techniques has been summarized in Table 13.1.

Link Between Ionomics and Functional Biology

Numerous studies based on the elemental profile in a variety of plant systems are available in literature, which could provide ample information about total concentration, uptake, and translocation within a tissue, organ, or whole plant. Ionomics, being a high-throughput phenotyping platform, offers the possibility of rapidly generating large ionomics data sets on many thousands of individual plants. Utilization of such a phenotyping platform to screen mapping populations with available modern genetic tools provides a very powerful approach for the identification of genes and gene networks that regulate the ionome. Baxter et al. (2008) established a multivariable shoot ionomic signature, consisting of manganese (Mn), cobalt (Co), zinc (Zn), molybdenum (Mo), and cadmium (Cd), that is indicative of a plant's Fe nutritional status by evaluating the *Arabidopsis* shoot ionome in plants grown under different Fe nutritional conditions. This signature has been validated against known Fe response genes (*IRT1* and *FRO2*). Using a logistic regression

TABLE 13.1

Summary of Most Commonly Used Elemental Analysis Techniques

	Flame AAS	GFAAS	XRF	NAA	ICP-AES	ICP-MS
Detection limits	Very good for some elements	Excellent for some elements	Very good for some elements	Very good for multi elements	Very good for most elements	Excellent for most elements
Sample throughput	10–15 sec per element	3–4 min per element	30 sec – several minutes, depends on no of elements measured	0.1 to 1×10^6 ng g^{-1}	1–60 elements per minute	All elements in <1 minute
Dynamic range	10^3	10^2	10^{-8} to 10^{-9}	1–10^7	10^6	10^8
Precision Short term Long term	0.1%–1.0% 2-beam 1–2& 1-beam <10%	0.5%–5% 1%–10% (tube lifetime)	NA	5% 0.1%	0.1%–2% 1%–5%	0.5%–2% 2%–4%
Dissolved solids in solution	0.5%–5%	>20% (slurries)	Powdered or fused into glass	Sample encapsulated in a vial made of polyethylene or quartz	0%–20%	0.1%–0.4%
Elements analyzed	68+	50+	15–20	74	73	82
Sample volume required	Large	Very small	Extensive	Very small	Medium	Very small to medium
Isotope analysis	No	No	No	Yes	No	Yes
Disadvantage	Flame insufficient to atomize few refractory elements	Detection of refractory elements is limited	Difficult to quantify elements lighter than Z=11	Sample remains radioactive for long	None except running cost is slightly higher	It introduces a lot of interfering species: argon from the plasma, component gasses of air that leak through the cone orifices, and contamination from glassware and the cones

model (LRM), trained on this multivariable ionomic signature, Fe response state of individual plants can be successfully classified. This model can detect alterations in the Fe response status of a plant. In another work (Baxter et al., 2012), the elemental composition (ionome) of a set of 96 wild accessions of the genetic model plant *Arabidopsis thaliana* grown in hydroponic culture and soil have been defined using inductively coupled plasma mass spectrometry (ICP-MS). Significant genetic effects were detected for all 19 elements analyzed, which suggested that the ionome of a plant tissue is variable, yet tightly controlled by genes and gene× environment interactions. The ionomic data set provides a valuable resource for mapping studies to identify genes regulating elemental accumulation. Indeed, the ionomics approach has been successfully used to clone genes responsible for natural variation (Baxter et al., 2010) in Na, Co, Mo, S, and Cu homeostasis in *A. thaliana*. EMS- and FN-mutagenized populations have been successfully used for the identification of various ionomic mutants, including mutants with perturbations in single elements, such as P (Delhaize and Randall, 1995) and Na (Nublat et al., 2001), and mutants with multiple ionomic changes. Lahner et al. (2003) and Chen et al. (2009) performed large (>2,000 plants) ionomic screens of mutagenized populations of *Arabidopsis thaliana* grown on soil and *Lotus japonica* grown in liquid culture, respectively. Punshon et al. (2013) have

provided methodological details on using elemental imaging to aid or accelerate gene functional characterization by narrowing down the search for candidate genes to the tissues in which elemental distributions are altered. They have used synchrotron X-ray microprobes as a technique of choice, which can now be used to image all parts of an *Arabidopsis* plant in a hydrated state. Elemental images of leaves, stem, root, siliques, and germinating hypocotyls have been produced by the workers. Strong evidence for the coordination of networks with ionome came from the work on *Arabidopsis*, which suggested that out of the 50 ion-profile mutants identified (Lahner et al., 2003), few (11%) showed changes in only one element. Such evidence necessitates the urgency to unravel the links among four pillars of functional biology; transcriptomics, proteomics, metabolomics, and ionomics. It is well evident that several genes, proteins, and metabolites showed differential regulation in response to nutrient availability or deficiency (Negishi et al., 2002; Maathuis et al., 2003; Wang et al., 2003), suggesting a possible link among the regulatory networks. To establish and integrate a link among these regulatory networks is a challenging and fascinating avenue for plant scientists. As a significant step, a high-throughput ion-profiling strategy has been developed based on robust inductively coupled plasma mass spectroscopy (ICP-MS) technology (Lahner et al., 2003), which is quite useful for genomic scale profiling of nutrient and trace elements. Further, clues about the regulation of element or ionic homeostasis have been provided by many researchers (Gojon et al., 2009; Luan et al., 2009; Pilon et al., 2009; Tejada et al., 2009; Puig and Penarrubia, 2009).

Transport of elements across the different membranes is mediated by transport proteins (transporters), which in some cases may be very specific at the functional and regulatory level; however, in other cases, these transporters may not be specific to confer membrane flexibility for transport. It has been suggested that metallo-chaperones can help in delivering metal micronutrients to specific proteins and transporters, activating transcription factors, or acting as transcriptional regulators themselves (Ramesh et al., 2003). A number of transporters involved in the homeostasis of metal ions (Fe^{2+}, Zn^{2+}, Mn^{2+}, and Cu^{2+}) have been identified in the rice genome (Koike et al., 2004; Narayanan et al., 2007; Sperotto et al., 2010); however, the specific function of many of these transport proteins remains unclear. Furthermore, a functional model depicting the rate-limiting step for the transport of ions to the plant parts of economic importance like seeds, grains or fruits, is not available. Access to such information coupled and supported by transcriptomics, proteomics, and metabolomics may help in the bio-fortification of edible parts (Norton et al., 2010) and to increase the tolerance to toxic metals (Induri et al., 2012). Integration of ionomics with other 'omics' can contribute much more to understanding the role of a particular gene/protein/metabolite in ion transport (Figure 13.2). Once ionomics QTL have been identified, genomic tools available for *A. thaliana*, and to some extent rice and maize, can be used to locate the genes that underlie these QTL and, thus, describe the traits at a molecular level. With the advent and access of the new robust technologies and protocols, potential candidate genes involved in mineral acquisition, transport, distribution, and storage can be identified. Success in doing so will definitely pave the way for the development of designer crops having higher mineral content as part of the bio-fortification program, either through genetic engineering or through molecular breeding approaches (White and Broadley, 2009; Singh et al., 2013), and may help in ensuring nutritional security globally.

Ionomic Database

The availability and easy accessibility of databases are essential in any 'omics' to facilitate potential *in silico* exploration of data among researchers for designing novel hypotheses and performing experiments. Scientists at Purdue University (USA) have developed a searchable database containing ionomics information on more than 22,000 plants. The database can be searched online for mutants altered in a specific element or set of elements along with the gene name and AGI gene codes for reverse genetics used in T-DNA insertional lines. This ionomics database can be found at http://hort.agriculture.purdue.edu/Ionomics/database.asp. Another ionomics database on ionomic mutants (Lahner et al., 2003) is available through the ABRC website (http://www.biosci.ohio-state.edu). Genes identified in the T-DNA insertional ionomics database can be further characterized in detail using the SIGnAL database (http://signal.salk.edu/cgibin/tdnaexpress). The Purdue Ionomics Information Management System (PiiMS) is an open access web-enabled integrated workflow, which provides data storage, high-throughput analysis, and data acquisition, along with other useful tools for hypothesis development. PiiMS contains information related to concentrations of major, minor, and trace elements in more than 60,000 shoot tissue samples of *Arabidopsis thaliana*. These samples include mutants, natural accessions, and populations of recombinant inbred lines from more than 800 independent experiments (Baxter et al., 2007). These databases can be regarded as novel user-friendly *in silico* laboratories that provide the integration of ionomic data with other available genomic, biochemical, and physiological data.

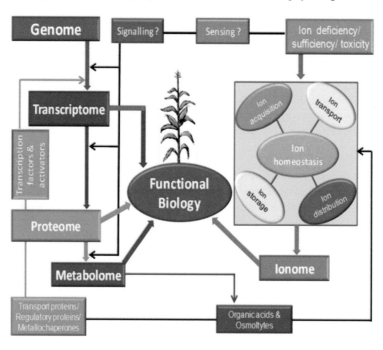

FIGURE 13.2 Possible linkage among genomics, transcriptomics, proteomics, metabolomics, and ionomics.

Conclusion

In adjusting nutrient acquisition to the needs of the plant, there is a coordination of regulatory networks and cross-talk between homeostatic pathways regulating different nutrients. If this is so, then a robust coordination must exist among pathways that warrant the need of a holistic approach such as 'system biology' integrating the available 'omics'. We still have to explore the plant ionome and the mechanisms that regulate it. Considering the multifaceted role of nutrients and other elements in plant biology, ionomics has to play a very important part in the study of systems biology. Various techniques may be fruitfully used to measure elemental composition. In practice, the total elemental composition of an organism is rarely determined. Ionomics alone can ensure an eco-friendly and economical use of nutrients for achieving sustainable food production in years to come. Establishment of a robust link among available data through different 'omics' approaches would further pave the way towards bio-fortification of harvestable plant parts and may help in achieving an optimistic goal of 'food and nutritional security'. Questions in physiology, ecology, evolution, and many other fields can be investigated using ionomics, often coupled with bioinformatics and other genetic tools. High-throughput ionomic phenotyping has created the need for data management systems to collect, organize, and share the collected data with researchers worldwide. Lots of research has already been done; now it's time to explore and integrate the assets of plant biology available to us to find what we are missing, and, finally, with some more effort, to fill in the blanks.

REFERENCES

Báez, A., Belmont, R., García, R.M., Padilla, H., and Torres, M.C. 2007. Chemical composition of rainwater collected at a southwest site of Mexico City, Mexico. *Rev. Atmos. Res.* 86: 61–75.

Barandovski, L., Cekova, M., Frontasyeva, M.V., Pavlov, S.S., Stafilov, T., Steinnes, E., and Urumov, V. 2008. Atmospheric deposition of trace element pollutants in Macedonia studied by the moss biomonitoring technique. *Environ. Monit. Assess.* 138: 107–118. doi:10.1007/s10661-007-9747-6.

Baxter, I. 2009. Ionomics: Studying the social network of mineral nutrients. *Curr. Opin. Plant Biol.* 12: 381–386.

Baxter, I., Brazelton, J.N., Yu, D., Huang, Y.S., Lahner, B., Yakubova, E., Li, Y., et al. 2010. A coastal cine in sodium accumulation in *Arabidopsis thaliana* is driven by natural variation of the sodium transporter *AtHKT1;1. PLoS Genet.* 6: e1001193.

Baxter, I., Christian, H., Lahner, B., Elena, Y., Tikhonova, M., Verbruggen, N., Chao, D., and Salt, D.E. 2012. Biodiversity of mineral nutrient and trace element accumulation in *Arabidopsis thaliana. PLoS ONE* 7(4): e35121.

Baxter, I., Ouzzani, M., Orcun, S., Kennedy, B., Jandhyala, S.S., and Salt, D.E. 2007. Purdue ionomics information management system. An integrated functional genomics platform. *Plant Physiol.* 143(2): 600–611.

Baxter, I.R., Vitek, O., Lahner, B., Muthukumar, B., Borghi, B., Morrissey, J., Lou Guerinot, M.L., and Salt, D.E. 2008. The leaf ionome as a multivariable system to detect a plant's physiological status. *Proc. Natl. Acad. Sci. USA* 105(33): 12081–12086.

Becher, M., Talke, I.N., Krall, L., and Kramer, U. 2004. Cross-species microarray transcript profiling reveals high constitutive expression of metal homeostasis genes in shoots of the zinc hyperaccumulator *Arabidopsis halleri. Plant J.* 73: 251–268.

Broadley, M.R., Hammond, J.P., White, P.J. and Salt, D.E. 2010. An efficient procedure for normalizing ionomics data for *Arabidopsis thaliana. New Phytol.* 186: 270–274.

Chen, Z., Shinano, T., Ezawa, T., Wasaki, J., Kimura, K., Osaki, M. and Zhu, Y. 2009. Elemental interconnections in Lotus japonicus: A systematic study of the effects of elements additions on different natural variants. *Soil Sci. Plant Nutr.* 55(1): 91–101.

Curie, C., and Briat, J.F. 2003. Iron transport and signaling in plants. *Annu. Rev. Plant Biol.* 54: 183–206.

Debnath, M., Lal, P., and Jain, R. 2015. Inductively coupled plasma-mass spectrometric profiling of metal contamination in Holi colours. *Curr. Sci.* 108(8): 1424–1427.

Delhaize, E., and Randall, P.J. 1995. Characterization of a phosphate-accumulator mutant of *Arabidopsis thaliana. Plant Physiol.* 107: 207–213.

Delhaize, E., Randall, P.J., Wallace, P.A., and Pinkerton, A. 1993. Screening *Arabidopsis* for mutants in mineral nutrition. *Plant Soil* 155/156: 131–134.

Fiehn, O., Kopka, J., Dormann, P., Altmann, T., Trethewey, R.N., and Willmitzer, L. 2000. Metabolite profiling for plant functional genomics. *Nat. Biotech.* 8: 1157–1161.

García, R., Belmont, R., Padilla, H., Torres, M.C., and Báez, A. 2009. Trace metals and inorganic ion measurements in rain from Mexico City and a nearby rural area. *Rev. Chem. Ecol.* 25(2): 71–86.

Glocker, R., and Schreiber, H. 1928. Quantitative röntgenspektralanalyse mit kalterregung des spektrums. *Ann. Physik.* 85: 1089.

Gojon, A., Philippe, N., and Davidian, J.C. 2009. Root uptake regulation: A central process for NPS homeostasis in plants. *Curr. Opin. Plant Biol.* 12(3): 328–338.

Hirschi, K.D. 2003. Striking while the ionome is hot: Making the most of plant genomic advances. *Trends Biotechnol.* 21: 520–521.

Induri, B.R., Ellis, D.R., Slavov, G.T., Yin, T., Zhang, X., Muchero, W., Tuskan, G.A., and DiFazio, S.P. 2012. Identification of quantitative trait loci and candidate genes for cadmium tolerance in Populus. *Tree Physiol.* 32(5): 626–638.

Karley, A.J., and Philip, J.W. 2009. Moving cationic minerals to edible tissues: Potassium, magnesium, calcium. *Curr. Opin. Plant Biol.* 12(3): 291–298.

Koike, S., Inoue, H., Mizuno, D., Takahashi, M., Nakanishi, H., Mori, S., and Nishizawa, N.K. 2004. OsYSL2 is a rice metal-nicotianamine transporter that is regulated by iron and expressed in the phloem. *Plant J.* 39: 415–424.

Koller, A., Washburn, M.P., Lange, B.M., Andon, N.L., Deciu, C., and Haynes, P.A. 2002. Proteomic survey of metabolic pathways in rice. *Proc. Natl. Acad. Sci. USA* 99: 11969–11974.

Kramer, U., Cotter-Howells, J.D., Charnock, J.M., Baker, A.J.M., and Smith, J.A.C. 1996. Free histidine as a metal chelator in plants that accumulate nickel. *Nature* 379: 635–38.

Lahner, B., Gong, J., Mahmoudian, M., Smith, E.L., Abid, K.B., Rogers, E.E., Guerinot, M.L., Harper, J.F., Ward, J.M., and McIntyre, L. 2003. Genomic scale profiling of nutrient and trace elements in *Arabidopsis thaliana. Nat. Biotech.* 21: 1215–1221.

Leonhardt, N., Kwak, J.M., Robert, N., Waner, D., Leonhardt, G., and Schroeder, J.I. 2004. Microarray expression analysis of Arabidopsis guard cells and isolation of a recessive abscisic acid hypersensitive protein phosphatase 2C mutant. *Plant Cell* 16: 596–615.

Luan, S., Wenzhilan, S., and Chul, L. 2009. Potassium nutrition, sodium toxicity, and calcium signaling: Connections through the CBL-CIPK network. *Curr. Opin. Plant Biol.* 12(3): 339–346.

Maathuis, F.J., Filatov, V., Herzyk, P., Krijger, G.C., Axelsen, K.B., Chen, S., Green, B.J., Li, Y., Madagan, K.L., and Sanchez-Fernandez, R. 2003. Transcriptome analysis of root transporters reveals participation of multiple gene families in the response to cation stress. *Plant J.* 35: 675–692.

Marschner, H. 1995. *Mineral Nutrition of Higher Plants.* 2nd edn. London: Academic Press.

Martin, C., Kiran, B., and Kim, B. 2001. Shaping in plant cells. *Curr. Opin. Plant Biol.* 4(6): 540–549.

Martzivanou, M., and Hampp, R. 2003. Hyper-gravity effects on the Arabidopsis transcriptome. *Physiol. Plant* 118: 221–231.

Maser, P., Thomine, S., Schroeder, J.I., Ward, J.M., Hirschi, K., Sze, H., Talke, I.N., Amtmann, A., Maathuis, F.J.M., and Sanders, D. 2001. Phylogenetic relationships within cation-transporter families of *Arabidopsis thaliana. Plant Physiol.* 126: 1646–1667.

Narayanan, N., Narayanan, M.W., Vasconcelos, M., and Grusk, A. 2007. Expression profiling of *Oryza sativa* metal homeostasis genes in different rice cultivars using a cDNA macroarray. *Plant Physiol. Biochem.* 45(5): 277–286.

Negishi, T., Nakanishi, H., Yazaki, J., Koshimoto, N., Fujii, F., Shimbo, K., Yamamoto, K., Sakata, K., Sasaki, T., and Kikuchi, S. 2002. cDNA microarray analysis of gene expression during Fe-deficiency stress in barley suggests that polar transport of vesicles is implicated in phytosiderophore secretion in Fe-deficient barley roots. *Plant J.* 30: 83–94.

Norton, G.J., Deacon, C.M., Xiong, L., Huang, S., Meharg, A.A., and Price, A.H. 2010. Genetic mapping of the rice ionome in leaves and grain: Identification of QTLs for 17 elements including arsenic, cadmium, iron and selenium. *Plant Soil* 329: 139–153.

Nublat, A., Desplans, J., Casse, F., and Berthomieu, P. 2001. sas1, an Arabidopsis mutant overaccumulating sodium in the shoot, shows deficiency in the control of the root radial transport of sodium. *Plant Cell* 13: 125–137.

Pickering, I.J., Gumaelius, L., Harris, H.H., Prince, R.C., Hirsch, G., Banks, J.A., Salt, D.E., and George, G.N. 2006. Localizing the biochemical transformations of arsenate in a hyperaccumulating fern. *Environ. Sci. Technol.* 40: 5010–5014.

Pilon, M., Christopher, M.C., Karl, R., Salah E. Abdel-Ghany, and Frederic, G. 2009. Essential transition metal homeostasis in plants. *Curr. Opin. Plant Biol.* 12(3): 347–357.

Puig, S., and Penarrubia, L. 2009. Placing metal micronutrients in context: Transport and distribution in plants. *Curr. Opin. Plant Biol.* 12(3): 299–306.

Punshon, T., Guerinot, M.L., and Lanzirotti, A. 2009. Using synchrotron X-ray fluorescence microprobes in the study of metal homeostasis in plants. *Ann. Bot.* 103: 665–672.

Punshon, T., Jackson, B.P., Bertsch, P.M., and Burger, J. 2004. Mass loading of nickel and uranium on plant surfaces: Application of laser ablation-ICP-MS. *J. Environ. Monit.* 6: 153–159.

Punshon, T., Ricachenevsky, F.K., Hindt, M., Socha, A.L., and Zuber, H. 2013. Methodological approaches for using synchrotron X-ray fluorescence (SXRF) imaging as a tool in ionomics: Examples from *Arabidopsis thaliana. Metallomics* 5(9): 1133–1145. doi:10.1039/c3mt00120b.

Ramesh, S.A., Shin, R., Eide, D.J., and Schachtman, D.P. 2003. Differential metal selectivity and gene expression of two zinc transporters from rice. *Plant Physiol.* 133: 126–134.

Rausch, C., and Bucher, M. 2002. Molecular mechanisms of phosphate transport in plants. *Planta* 216: 23–37.

Rea, P.A. 2003. Ion genomics. *Nat. Biotech.* 21: 1149–1151.

Salt, D.E., Prince, R.C., Pickering, I.J., and Raskin, I. 1995. Mechanisms of cadmium mobility and accumulation in Indian mustard. *Plant Physiol.* 109: 1427–1433.

Sammut, M.L., Noack, Y., Rose, J., Hazemann, J.L., Proux, O., Depoux, M., Ziebel, A., and Fiani, E. 2010. Speciation of Cd and Pb in dust emitted from sinter plant. *Chemosphere* 78: 445.

Sanders, D., Pelloux, J., Brownless, C., and Harper, J.F. 2002. Calcium at the crossroads of signaling. *Plant Cell* 14: S410–S417.

Singh, U.M., Sareen, P., Sengar, R.S., and Kumar, A. 2013. Plant ionomics: A newer approach to study mineral transport and its regulation. *Acta Physiol. Plant.* 35: 2641–2653.

Sperotto, R.A., Tatiana, B., Guilherme, L., Duarte, L.S., Santos, M., Grusak, A., and Janette P.F. 2010. Identification of putative target genes to manipulate Fe and Zn concentrations in rice grains. *J. Plant Physiol.* 167: 1500–1506.

Tejada-Jimenez, M., Galvan, A., Fernandez, E., and Llamas, A. 2009. Homeostasis of the micronutrients Ni, Mo and Cl with specific biochemical functions. *Curr. Opin. Plant Biol.* 12: 358–363.

Vermani, A., Navneet, P., and Chauhan, A. 2010. Physico-chemical analysis of ash of some medicinal plants growing in Uttarakhand. *India. Nat. Sci.* 8 (6): 88–91.

Very, A.A., and Sentenac, H. 2003. Molecular mechanisms and regulation of K^+ transport in higher plants. *Annu. Rev. Plant Biol.* 54: 575–603.

Wang, R., Okamoto, M., Xing, X., and Crawford, N.M. 2003. Microarray analysis of the nitrate response in Arabidopsis roots and shoots reveals over 1,000 rapidly responding genes and new linkages to glucose, trehalose-6-phosphate, iron, and sulfate metabolism. *Plant Physiol.* 132: 556–567.

Watanabe, T., Broadley, M.R., and Jansen, S. 2007. Evolutionary control of leaf element composition in plants. *New Phytol.* 174: 516–523.

White, P.J., and Broadley, M.R. 2009. Biofortification of crops with seven mineral elements often lacking in human diets-iron, zinc, copper, calcium, magnesium, selenium and iodine. *New Phytol.* 182: 49–84.

White, P.J., Broadley, M.R., Thompson, J.A., McNicol, J.W., Crawley, M.J., Poulton, P.R., and Johnston, A.E. 2012. Testing the distinctness of shoot ionomes of angiosperm families using the Rothamsted Park Grass Continuous Hay experiment. *New Phytol.* 196(1): 101–109.

Zhao, Fang-J., Katie, L., Moore, L.E., and Zhu, Y.G., Imaging element distribution and speciation in plant cells. *Trends Plant Sci.* 19(3): 183–192.

Zhu, J.K. 2003. Regulation of ion homeostasis under salt stress. *Curr. Opin. Plant Biol.* 6: 441–445.

Section II

Role of Major Plant Metabolites During Abiotic Stress Management

14

Role of Glutamate-Derived Amino Acids under Stress Conditions: The Case of Glutamine and Proline

Marco Biancucci, Roberto Mattioli, Adra Mouellef, Nadia Ykhlef, and Maurizio Trovato

CONTENTS

Abbreviations

CRISPR/Cas9	Clustered regularly interspaced short palindromic repeats/Cas 9
GABA	γ-amino butyric acid
GOGAT	Glutamine 2-oxoglutarate aminotransferase
GS	Glutamine synthetase
NADH-GDH	NADH-dependent glutamate dehydrogenase
nat-siRNA	Natural small interfering antisense RNA
NiR	Nitrite reductase
NO	Nitric oxide
NR	Nitrate reductase
NUE	Nitrogen use efficiency
P5CDH	Pyrroline-5-carboxylate dehydrogenase
P5CR	Pyrroline-5-carboxylate reductase
P5CS	Pyrroline-5-carboxylate synthetase
ProDH	Proline dehydrogenase
QTL	Quantitative trait locus
rbcS	Rubisco small subunit
RFLP	Restriction fragment length polymorphism
ROS	Reactive oxygen species
SNP	Single nucleotide polymorphism
SSR	Single sequence repeat

The Glutamate Family of Amino Acids and Its Role in Stress Response

In addition to their primary function as building blocks for protein synthesis, the amino acids derived from glutamate (Glu), notably glutamine (Gln), proline (Pro), arginine (Arg), and their close derivatives γ-amino butyric acid (GABA) and nitric oxide (NO), are considered signaling molecules and master regulators of plant development. The metabolism of these amino acids, particularly Glu, Gln, and Arg, is central for nitrogen assimilation, storage, and redistribution, and links and integrates nutritional and environmental clues (Okumoto et al., 2016). Glu and Gln are considered entry points and master regulators of nitrogen metabolism in higher plants and play a critical role in the regulation and integration of carbon and nitrogen metabolism, while Arg is regarded as the major storage form of organic nitrogen in plants because of its high nitrogen to carbon ratio (Winter et al., 2015). Asp, usually considered only as the precursor of the aspartate-family of amino acids – a group of amino acids including the essential amino acids Methionine (Met), Lysine (Lys), Threonine (Thr), and Isoleucine (Ile) – might be included in this family because it is derived from glutamate through the action of the enzyme aspartate aminotransferase (EC 2.6.1.1), which catalyzes the transamination between glutamate and oxaloacetate to produce aspartate and 2-oxoglutarate (Lea and Azevedo, 2007). Glu, Gln, Pro, and GABA, have all been assigned multiple functions and, to different extents, seem to have critical roles as signaling molecules involved in plant development, particularly in the reproductive phase (Trovato et al., 2008; Biancucci et al., 2015). A unifying feature of the glutamate-derived amino acids is their importance in the response of plants to different types of stresses. From a meta-analysis of publicly available microarray data, Less and Galili (2008) found that abiotic stress caused extensive transcriptional regulation of the biosynthetic enzymes involved in proline and arginine metabolism, contrary to other amino acids, which showed no or little alterations in their biosynthetic metabolism. As another example, in a study conducted on the resurrection plant *Sporobolus stapfianus* subjected to dehydration stress, Martinelli et al. (2007) found that Glu, Gln, Asp,

Asn, and Ala together constituted 80% of the total free amino acid pool in unstressed plants, while the most striking and meaningful differences in stressed plants were found in Pro, Asn, Asp, Arg, Val, and GABA. Although the accumulation of these amino acids is not a universal response to stress conditions, the role played by this particular class of amino acid is of central importance during stress conditions and will be reviewed in this chapter with particular emphasis on glutamine and proline and their possible use for crop improvement.

Glutamine Metabolism: A Key Component of Nitrogen Use Efficiency (NUE) Involved in Stress Responses

In higher plants nitrogen is obtained from nitrogen-fixing bacteria, soil inorganic nitrate, photorespiration, phenylpropanoid biosynthesis, and amino acid catabolism. The inorganic NO_3^- found in the soil cannot directly be used as a substrate for protein synthesis in plants and has to be reduced to NO_2^- by nitrate reductase (NR; EC 1.6.6.1-3) and to $NH4^+$ by nitrite reductase (NiR; EC 1.7.2.1). Regardless of its source, nitrogen is ultimately used as NH_4^+, which, because of its inherent toxicity, is readily assimilated into the amino acids glutamine and glutamate by the combined action of the enzymes glutamine synthetase (GS; EC 6.3.1.2) and glutamine 2-oxoglutarate aminotransferase (glutamate synthase, GOGAT; EC 1.4.1.13) in the so-called GS/GOGAT cycle (Temple et al., 1998). As an alternative route, NADH-dependent glutamate dehydrogenase (NADH-GDH; EC 1.4.1.2) can catalyze the direct incorporation of ammonium to glutamate by reductive amination of 2-oxoglutarate (Miflin and Lea, 1980; Cammaerts and Jacobs, 1985). Both Gln and Glu can subsequently serve as nitrogen donors for the biosynthesis of essentially all other amino acids and nitrogen compounds (Coruzzi, 2003).

The key enzyme of glutamine metabolism in higher plants is the enzyme glutamine synthetase, which catalyzes the ATP-dependent condensation of glutamine starting from glutamate and ammonia (Figure 14.1). Two major GS isoforms are present in plants: A cytosolic isoform (GS1), which is mainly found in roots where it plays a major role in $NH4^+$ assimilation from the soil, and a plastidic isoform (GS2), mainly found in chloroplasts and mitochondria, where it actively scavenges the $NH4^+$ produced during photorespiration (Taira et al., 2004). In most species examined, *GS1* is coded for by a small multigene family composed of a variable number of members depending on the species, while *GS2* seems to always be encoded by a single gene. As an example, three *GS1* genes have been identified in rice, seven in wheat, five in maize, five in *Arabidopsis*, and as many as sixteen in *Brassica napus* (Orsel et al., 2014).

Being involved in so many aspects of nitrogen metabolism, it is no wonder that GS is a pivotal component of NUE, a complex quantitative trait related to the crop capacity to use available nitrogen and maximize productivity. Accordingly, GS has been extensively studied to understand how it is regulated and how it regulates nitrogen metabolism in plants (Bernard and Habash, 2009; Lea and Miflin, 2010; Thomsen et al., 2014). Indeed, several authors have shown a direct correlation between the activity

of GS and either the yield or the biomass of plants transgenics for *GS* (Hirel et al., 1997; Habash et al., 2001; Oliveira et al., 2002). According to Habash et al. (2001), for example, the heterologous expression in wheat of a *Phaseolus vulgaris GS1* driven by a rice rbcS promoter resulted in plants with enhanced nitrogen accumulation and, at least in one line, with significantly more roots and higher grain yield than untransformed controls. In addition, knockout mutants of specific *GS1* isogenes were reported to be severely impaired in growth and yield (Tabuchi et al., 2005; Martin et al., 2006; Funayama et al., 2013).

In addition to its central role in NUE and plant yield, glutamine metabolism is also involved in plant tolerance to drought, cold, and saline conditions. GS gene transcription and/or enzyme activity has been reported to increase in response to saline conditions in different species, such as potato, (Teixeira and Fidalgo, 2009), barley (Kant et al., 2007), foxtail millet (Veeranagamallaiah et al., 2007), tomato (Cramer et al., 1999), and ryegrass (Sagi et al., 1998). Furthermore, transgenic plants overexpressing GS have been reported to be drought-tolerant (Hoshida et al., 2000; el-Khatib et al., 2004). GS1 is thought to be indirectly involved in stress tolerance as a major source of proline in the phloem (Brugiére et al., 1999) and because it is involved in the remobilization of nitrogen during prolonged water stress (Bauer et al., 1997). As to the latter point, Bauer et al. (1997) reported that water stress stimulates tomato cytosolic *GS1* gene expression, while the expression of plastidic *GS2* remains unchanged during drought. The overexpression of *GS1* likely serves to sustain the production of glutamine, which, in turn, represents a vehicle to remobilize nitrogen in tomato leaves in response to chronic water stress. In the end, the role of GS in plants under environmental stress conditions might be twofold: On one hand, the increase in GS activity during stress may be correlated with, and important for a corresponding increase in glutamate availability, an essential pre-requisite to support the massive accumulation of proline typically observed in stressed plants (Bauer et al., 1997), and, on the other hand, increased nitrogen assimilation and transport is known to be critically important for stress tolerance. In support of the former point, we note that an increase in both GS and GOGAT activity has been described in response to salt and water stress and that glutamate is typically found along with glutamine accumulation upon salt and water stress (Berteli et al., 1995; Bauer et al., 1997; Borsani et al., 1999; Díaz et al., 2005). Accordingly, we note that nitrogen, supplied in different forms, is generally known to alleviate the detrimental effects of water and salt stress (Gimeno et al., 2009; Orsel et al., 2014; Rashid et al., 2014). Indeed, the difficulty of the root system to uptake inorganic nitrogen from the soil against a water potential, coupled with a reduced intake of organic nitrogen produced by a severe reduction in photorespiration, biosynthetic cellular reactions, and plant transpiration, cause a dramatic reduction in usable nitrogen (Rashid et al., 2014), which can be alleviated by a better NUE. Another stress-related aspect GS is involved in is related to the metabolism of plant phenolics, a widely-distributed class of secondary metabolites that are involved in stress responses (Cheynier et al., 2013). Flavonoids, in particular, are known to respond to almost all kinds of unfavorable environmental conditions providing protection against UV light and pathogens (Winkel-Shirley, 2002) and exerting high antioxidative activity *in vitro*, and probably *in vivo* (Nakabayashi and Saito, 2015).

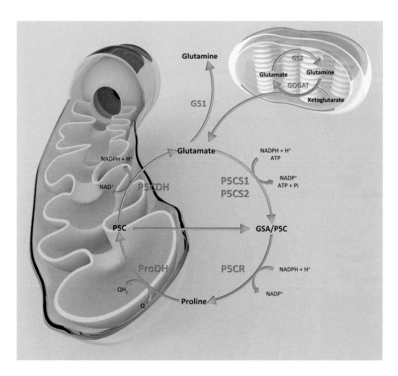

FIGURE 14.1 Proline and glutamine metabolic pathways. Simplified scheme of proline and glutamine pathways. The main route of proline synthesis starts from glutamate, which is reduced to proline in the cytosol (or in the chloroplast under stress conditions) by the sequential action of P5CS1/2 and P5CR. Proline is transported into the mitochondrion where it is oxidized back to glutamate by ProDH and P5CDH, respectively. When in excess, P5C can exit the mitochondrion and feed proline synthesis in the cytosol. Multiple isozymes of GS are localized either in chloroplasts and mitochondria or in the cytosol. A single GS2 isozyme is involved, in the chloroplast, in recycling the ammonium produced by photorespiration, while multiple GS1 isozymes are mainly responsible for the assimilation of ammonium derived from nitrate reduction.

Consistently, increased tolerance to oxidative stress and drought stress has been reported in transgenic plants that over accumulated flavonoids (Nakabayashi et al., 2014). Furthermore, by transcriptomic and metabolomic analysis, García-Calderón et al., (2015) reported that in *Lotus japonicus* GS2 activity affects phenolic metabolism, altering different aspects of phenolic metabolism in response to different types of stress. Overall, the essential role of GS in improving NUE in plants appears to be closely intertwined with stress tolerance and phenolic metabolism. This opens up a fascinating perspective to improve NUE as well as stress tolerance and phenolic metabolism by engineering only one enzyme.

Proline: Multi-Functional Amino Acid Involved in Stress and Development

A rapid and reversible increase of the intracellular free proline concentration is a typical response to stress conditions shared by many organisms, including protozoa (Kaneshiro et al., 1969; Poulin et al., 1987), eubacteria (Csonka, 1989), marine invertebrates (Burton, 1991), algae (Brown and Hellebust, 1978), and a large number of plant species (Savouré et al., 1995; Yoshiba et al., 1995; Vendruscolo et al., 2007; Yamchi et al., 2007; Trovato et al., 2008; Kumar et al., 2010; Szabados and Savouré, 2010; Huang et al., 2013). Proline accumulation can be triggered by numerous abiotic stresses and a few biotic stresses (Trovato et al., 2008), but water deficit, as caused by drought, cold, and

excessive salinity, is by far the most effective inducer of proline accumulation in plants.

The accumulation of proline during stress relies on the combined action of synthesis and degradation at different levels of expression, particularly at the transcriptional level, where proline accumulation is accounted for by the coordinated upregulation of the genes involved in proline synthesis (Savouré et al., 1995; Yoshiba et al., 1995; Peng et al., 1996; Igarashi et al., 1997; Strizhov et al., 1997) and by the downregulation of the genes involved in proline catabolism (Yoshiba et al., 1995; Kiyosue et al., 1996; Miller et al., 2009; Sharma et al., 2011). However, transport (Girousse et al., 1996), post-transcriptional regulation (Borsani et al., 2005), post-translational modifications (Bhaskara et al., 2015), and regulation of enzyme activities are also known to play an important role in determining the actual levels of intracellular proline. As an example, the accumulation of P5CS1, the enzyme coding for the rate-limiting step of proline synthesis in plants, is feedback-inhibited by proline in bacteria (Csonka et al., 1988; Dandekar and Uratsu, 1988) and in *V. aconitifolia* (Hu et al., 1992; Zhang et al., 1995). In the P5CS enzyme from *V. aconitifolia* a conserved phenylalanine, at position 129, was found to be necessary for feedback inhibition by proline (Zhang et al., 1995). Accordingly, a tobacco mutant with a mutated version of P5CS1 carrying a phenylalanine to alanine substitution at position 129 (F129A) exhibited decreased feedback inhibition and high levels of proline accumulation (Zhang et al., 1995). A post-transcriptional regulatory mechanism based on a natural small interfering antisense RNA (nat-siRNA), has been reported to fine-tune the levels of *P5CDH* mRNA, the gene coding for the second step of proline catabolism (Borsani et al., 2005). A 24 bp-long nat-siRNA was detected in Arabidopsis under salt stress conditions and was shown to be correlated with *P5CDH* degradation. The generation of this nat-siRNA is guided by the formation of a complementary pair of transcripts between *P5CDH* and *SRO5* – a partially overlapping gene of unknown function – induced under salt stress and has been found to be dependent on the proteins DICER-LIKE 2 (DCL2), RNA-DEPENDENT RNA POLYMERASE 6 (RDR6), SUPPRESSOR OF GENE SILENCING 3 (SGS3), and NUCLEAR RNA POLYMERASE D 1a (NRPD1A). As for post-translational mechanisms, recent proteomic studies have confirmed that protein levels do not always match transcript levels particularly in the case of transcriptionally downregulated genes (Vélez-Bermúdez and Schmidt, 2014). According to Bhaskara et al. (2012), this may be the case for *ProDH1*, the gene coding for the first step of proline catabolism. Indeed, the protein levels of ProDH1, as detected by specific anti-*ProDH1* antibodies, were found to be high at 24 h after stress treatment, although the corresponding mRNA levels had been dramatically downregulated as early as 10 h after stress treatment.

Interestingly, proline accumulation also occurs under non-stressed conditions in specific phases of plant development (Mattioli et al., 2009a), such as flower transition (Mattioli et al., 2008), embryo development (Székely et al., 2008; Mattioli et al., 2009b), pollen development (Funck et al., 2012; Mattioli et al., 2012), and root elongation (Trovato et al., 1997; Verslues and Sharp, 1999; Biancucci et al., 2015). Although the functions and the molecular mechanisms of proline in plant development are still unclear, at least in some cases the accumulation of proline occurring during normal development may be related to natural forms of stress. The massive proline accumulation observed in pollen grains, which in tomato may represent up to 80% of the total amino acidic pool (Schwacke et al., 1999), is consistent with this hypothesis. According to Chiang and Dandekar (1995), the role of proline in pollen grains is to protect cellular structures from the denaturation caused by the process of natural dehydration to which pollen grains are naturally subjected.

Proline Metabolism

In higher plants, proline is synthesized in the cytoplasm, and probably in the chloroplast, from glutamate (Figure 14.1), which is converted in glutamic-semialdehyde (GSA) by the bi-functional enzyme δ1-pyrroline-5-carboxylate synthetase (P5CS; EC 2.7.2.11/1.2.1.41), which catalyzes both the ATP-dependent phosphorylation of glutamate to γ-glutamyl-phosphate and its NADPH-dependent reduction to glutamic-γ-semi-aldehyde (GSA). This latter intermediate molecule spontaneously cyclizes into δ1-pyrroline-5-carboxylate (P5C), which, in turn, is reduced to proline by the action of P5C reductase (P5CR; EC 1.5.1.2), using NADPH as the hydrogen donor once again (Figure 14.1). The subcellular localization of the P5CS proteins is generally believed to be cytosolic, consistent with the lack of recognizable motifs in their amino acid sequence. Székely et al. (2008), however, found that in stressed leaves the fusion protein P5CS1-GFP relocalizes from the cytoplasm to the chloroplasts, raising the possibility that under stress conditions proline synthesis may take place in the chloroplast.

Proline synthesis can also proceed from ornithine, either through ornithine δ-aminotransferase (δ-OAT; EC 2.6.1.13) under conditions of high nitrogen availability (da Rocha et al., 2012) or by direct deamination of ornithine catalyzed by the enzyme ornithine cyclodeaminase (OCD; EC 4.3.1.12) (Trovato et al., 2001). However, the relevance of these two alternative pathways for proline synthesis has been questioned (Funck et al., 2008; Sharma et al., 2013).

Subsequently, proline is translocated to and catabolized in the mitochondrion by the enzyme proline dehydrogenase (ProDH; EC 1.4.3), which oxidizes proline to pyrroline-5-carboxylate (P5C), which, in turn, is oxidized to glutamate by the enzyme pyrroline-5-carboxylate dehydrogenase (P5CDH; EC 1.5.1.12) (Figure 14.1).

The enzyme catalyzing the first and rate-limiting step of proline synthesis in higher plants, P5CS, is coded for in most plant species by *P5CS1* and *P5CS2*, two paralog genes sharing extensive sequence homology (Strizhov et al., 1997), while only a single gene – *P5CR* – is known to encode P5CR. Similarly, the enzyme catalyzing the rate-limiting step of proline catabolism in Arabidopsis plants, ProDH, is encoded by the two paralog genes *ProDH1* and *ProDH2* (Funck et al., 2010), with a single gene, *P5CDH*, coding for the enzyme catalyzing the second step of proline catabolism, from P5C to Glu. The two couples of paralog genes, *P5CS* and *ProDH*, are differentially regulated in most plant species with only one of the paralog genes responding to environmental cues. In Arabidopsis compelling evidence, based on GUS analysis (Yoshiba et al., 1999), expression analysis (Strizhov et al., 1997), and analysis of T-DNA insertional mutants (Székely et al., 2008), indicates that *P5CS1* is induced by stress and is responsible for stress tolerance. *P5CS2*, on the contrary, is not involved in stress responses (Székely et al., 2008) and is expressed at low levels in all tissues and at high levels in floral tissues and embryos (Székely et al., 2008; Mattioli et al., 2009b) and in dividing cells from callus and cell cultures (Strizhov et al., 1997). Similar to *P5CS1* and *P5CS2*, the expression of *ProDH1* and *ProDH2* also exhibits a non-redundant but partially overlapping expression pattern. In particular, the expression of *ProDH1*, which represents the dominant isoform, and *P5CDH*, the gene coding for the enzyme that catalyzes the conversion of P5C to Glu, are repressed by osmotic stress and upregulated by Pro (Verbruggen and Hermans, 2008). *ProDH2*, in contrast, is expressed at lower levels, especially in the vascular system, and is induced by proline and salt but repressed by sugars (Funck et al., 2008).

Proline Transport and Signaling

The strict compartmentalization of proline metabolism between cytosol and organelles, and the massive redistribution of proline between tissues and organs in the course of development or under stress (i.e. towards roots during stress or towards reproductive organs in non-stressed plants after floral transition) implies the existence of amino acid carriers for proline. During non-stressed conditions, for example, the dramatic increase of free proline found in Arabidopsis reproductive tissues after floral transition

(Chiang and Dandekar, 1995) may be in part accounted for by proline transport, since proline has been detected in both phloem and xylem sap of several plant species and long-distance transport trough phloem vessels has been well-documented (Girousse et al., 1996; Mäkelä et al., 1996). In addition, Verslues and Sharp (1999) reported that, under low water potential, the accumulation of proline in the elongation zone of the primary maize root depends on transport and not on synthesis.

Plant transporters that can move proline across the plasma membrane have been found both in the amino acid transporter (ATF) or amino acid/auxin permease (AAAP) family and in the APC (amino acid–polyamine–choline) family (Rentsch et al., 2007).

In the ATF/AAAP family, proline transporters have been found in different subfamilies, namely in the amino acid permease (AAP) family, the lysine-histidine transporter (LHT) family, and the proline transporter (ProT) family. Unlike other transporters that can recognize and transport a large number of different amino acids, the ProT family (ProT1, ProT2, and ProT3) is highly specific and can transport only proline and the related molecules GABA and glycine betaine, although at different affinities (Lehmann et al., 2010).

Because proline accumulation in response to stress is dependent on *de novo* protein synthesis (Stewart et al., 1986; Verbruggen et al., 1993), the accumulation of this amino acid is generally believed to be a secondary response to stress regulated by a signaling cascade (Hare et al., 1999). Indeed, *P5CS1* expression is upregulated by high salinity, dehydration, cold (Delauney and Verma, 1990; Hu et al., 1992; Yoshiba et al., 1995, 1997, 1999; Hong et al., 2000), light (Hayashi et al., 2000), and nitric oxide (Zhao et al., 2009) and downregulated by brassinosteroids (Abraham et al., 2003) and different signaling pathways are likely to control proline accumulation. The best-characterized signaling pathway related to proline metabolism has been linked to abscisic acid (ABA) and osmotic stress (Savouré et al., 1997; Strizhov et al., 1997; Abraham et al., 2003). It is well known that osmotic stresses such as salinity, dehydration, and cold induce a rapid surge of ABA, which, in turn, leads to *P5CS1* upregulation and proline accumulation. According to Yoshiba et al. (1995), exogenous ABA treatment results in *P5CS* upregulation as early as 1 h after hormonal treatment. Consistently, salt stress-induced proline accumulation is inhibited in ABA-deficient *aba1* and ABA-insensitive *abi1* Arabidopsis mutants (Savouré et al., 1997; Strizhov et al., 1997; Abraham et al., 2003). *P5CS2*, on the contrary, does not seem to be induced by ABA (Strizhov et al., 1997; Abraham et al., 2003). In agreement, bioinformatic analysis of the *P5CS1* and *P5CS2* promoters showed the presence of cis-acting ABA-responsive elements (ABRE) in the former but not in the latter (Hare et al., 1999). The relationship between ABA and proline synthesis under stress, however, is not straightforward, as witnessed by the fact that ABA, by itself, cannot induce proline accumulation in absence of stress (Sharma and Verslues, 2010). Moreover, an ABA-independent regulation of proline synthesis was revealed by the partial expression of *P5CS1* in the ABA-deficient *aba2-1*, and ABA- and proline-deficient *p5cs1-1 aba2-1* double mutants (Sharma and Verslues, 2010). A signaling cascade relying on phospholipase-induced second messengers, has been proposed to mediate stress-induced proline synthesis (Knight et al., 1997; Thiery et al., 2004; Parre et al., 2007). In plants, as in animals, extra-cellular stimuli can activate

phospholipase C (PLC) that specifically hydrolyzes phosphatidylinositol 4,5-biphosphate to generate inositol 1,4,5-triphosphate (IP3) and 1-2-diacylglycerol (DAG). These latter two molecules act as second messengers to release intracellular caged Ca^{++} and activate protein kinase C (PKC), respectively.

The Role of Proline in Stress Tolerance

It is widely accepted that proline accumulation is a common reaction of plant cells to most environmental stresses, especially drought, cold, and salinity stress, but its actual function, its mechanism of action, and its real effectiveness in improving stress tolerance are much less clear.

In spite of the prevailing opinion that sees proline accumulation as an adaptive mechanism to improve stress tolerance and although a clear-cut correlation between proline accumulation and osmotolerance is well established in bacteria (Csonka, 1989), the correlation between proline and stress tolerance in plants is still an object of debate. A rich scientific literature is available on this topic but is of little help in addressing this issue because of the co-existence of papers reporting either positive (Rhodes et al., 1986; Kohl et al., 1991; Chiang and Dandekar, 1995; Kavi Kishor et al., 1995; Nanjo et al., 1999; Hong et al., 2000; Kant et al., 2006; Székely et al., 2008; Miller et al., 2009) or inconsistent (Liu and Zhu, 1997; Xin and Browse, 1998; Maggio et al., 2002; Mani et al., 2002; Chen et al., 2007; Widodo et al., 2009; Bhaskara et al., 2012) correlations between proline accumulation and stress tolerance. Proline accumulation during stress has been proposed to behave as a compatible osmolyte, a ROS scavenger, a redox balancer, a cytosolic pH buffer, a stabilizing agent for macromolecules and subcellular structures, and a source of energy to be used in energy-demanding cellular processes or during recovery from stress (for reviews, see Hare and Cress, 1997; Kavi Kishor et al., 2005; Trovato et al., 2008; Verbruggen and Hermans, 2008; Szabados and Savouré, 2010).

Because of its exceptionally high solubility in water (14 kg/L; Le Rudulier et al., 1984), its zwitterionic nature, neutral pH, and good compatibility with cellular reactions, the most obvious function proposed for proline is to be a compatible osmolyte, which can accumulate at high concentrations as an osmoregulator, increasing the plant capacity to retain water or reducing water losses in adverse conditions (Szabados and Savouré, 2010; Kavi Kishor and Sreenivasulu, 2014; Signorelli et al., 2014). The actual levels of proline accumulation depend on the different plant species and may reach concentrations up to 100 times higher than unstressed controls (Verbruggen and Hermans, 2008) up to 230 mM (Büssis and Heineke, 1998; Ashraf and Foolad, 2007). Proline is not the only compatible osmolyte that accumulates in plants under stress and different proteinogenic and non-proteinogenic amino acids (e.g. Asn, Asp, Arg, Ala, GABA, glycine betaine), sugars (e.g. sorbitol and trehalose), polyols, and polyamines can substitute for or complement the action of proline. However, under stress conditions, proline is probably the most widely-used compatible osmolyte in the plant kingdom, possibly because of the many different functions that this multi-functional amino acid can accomplish, and its metabolism has been the object of intense research to understand its underlying molecular mechanisms and, ultimately, to improve drought tolerance in crops

(Lehmann et al., 2010; Szabados and Savouré, 2010; Verslues and Sharma, 2010; Kavi Kishor and Sreenivasulu, 2014).

Proline can also act as an effective stabilizer of macromolecular structures and membranes (Low, 1985; Kandpal and Rao, 1985), a cryoprotectant (Xin and Browse, 1998; Takagi et al., 2000), an antioxidant, and a ROS scavenger (Floyd and Nagy, 1984; Smirnoff and Cumbes, 1989; Smirnoff, 1993). The capacity to accumulate at high concentrations in the cytoplasm without interfering with cellular structures is at the basis of proline action under limited water content. In these conditions, proline might behave as a water substitute capable of stabilizing cellular structures by means of hydrophilic interactions and hydrogen bonds. Proline is proposed to stabilize enzymes and membranes by inducing preferential hydration of proteins and by forming hydrophilic colloids in water with hydrophobic backbones interacting with proteins (Rajendrakumar et al., 1994).

The action of proline as an antioxidant and ROS scavenger is particularly important because of the dramatic effect of ROS on the plant cell. It is well known that essentially all biotic and abiotic stresses cause an overproduction of reactive oxygen species, such as singlet oxygens ($^1O_2$), superoxide anion radicals ($\bullet O_2^-$), hydrogen peroxide (H_2O_2), and hydroxyl radicals ($OH\bullet$). All these molecules cause severe and irreversible damage to proteins, lipids, carbohydrates, DNA, and membranes and can eventually lead to cell death. Furthermore, it is well established that the plant ability to control oxidant levels by scavenging these hyper-reactive molecules is well correlated with stress tolerance (Loggini et al., 1999; Torres-Franklin, 2008). More recently the accumulation of nitric oxide-derived molecules called reactive nitrogen species (RNS) have been found to follow, and contribute to ROS accumulation in inducing cell damage and cell death (Corpas and Barroso, 2013).

As early as the end of the last century, proline has been proposed to have a beneficial role under stress by acting as a scavenger of reactive oxygen species (ROS) forming stable adducts with the latter (Floyd and Nagy, 1984; Smirnoff and Cumbes, 1989; Smirnoff, 1993). Accordingly, transgenic plants overexpressing *P5CS1* have been found to be more tolerant to salt stress (Kavi-Kishor et al., 1995) and Arabidopsis *p5cs1* knockout mutants, with reduced levels of proline and incapable of accumulating proline under stress, have been reported to be hypersensitive to stress and to exhibit ROS accumulation under drought or salt stress (Székely et al., 2008). In addition, exogenous proline supplementation has been reported to alleviate the negative effects of stress in different plant systems by inducing antioxidant enzymes (Saranga et al., 1992; Khedr et al., 2003; Hoque et al., 2007).

While a beneficial role of proline for plant cells under stress situations is generally accepted, the underlying mechanism of action is uncertain, and it is still controversial whether proline can act by inducing antioxidant and detoxifying enzymes or by directly reacting with reactive species.

Indeed, under stress conditions, a number of antioxidant enzymes, such as superoxide dismutase (Islam et al., 2009; Xu et al., 2009), catalase (Hoque et al., 2007; Islam et al., 2009; Xu et al., 2009), Glutathione S-transferase, and ascorbate peroxidase (Hoque et al., 2007; Yan et al., 2011) have been reported to increase their activity in cell cultures treated with exogenous proline, in agreement with the hypothesis that proline stimulates antioxidant activity.

In addition, or alternatively, because of the high reactivity of ROS and RNS molecules, a direct action of proline as a scavenger has been hypothesized, but at present, not clearly demonstrated. The capacity of proline to scavenge the $OH\bullet$ radical is generally accepted by the scientific community, although a convincing demonstration is still lacking. Chemically, proline reacts with $OH\bullet$ under hydrogen abstraction by forming a more stable radical, as shown by Rustgi et al. (1977), and a direct scavenging of proline to the $OH\bullet$ radical, was proposed as early as 1989 by Smirnoff et al., but only indirectly demonstrated (Smirnoff and Cumbes, 1989). Recently, Signorelli et al. (2014) suggested that proline could scavenge the $OH\bullet$ radical within a pro-pro cycle without proline consumption. In this cycle, the $OH\bullet$ radical attacks proline by H-abstraction and the resulting pro$\bullet$ radical reacts again with the $OH\bullet$ radical to form P5C, which is the direct substrate of proline synthesis (Signorelli et al., 2014). A role for proline in scavenging the single singlet oxygen $^1O_2$ – the molecular oxygen in its lowest energy electronically excited state – has been also hypothesized but remains controversial, since the quenching of the singlet oxygen by proline has been demonstrated *in vitro* by Alia et al. (2001) but recently it was questioned by Signorelli et al. (2013) who provided evidence that proline cannot quench $^1O_2$ in aqueous buffer.

A different school of thought, however, proposes that stress tolerance may be correlated with proline metabolism rather proline itself. This point of view is supported by a number of instances in which a correlation between proline accumulation and stress tolerance is weak or lacking. The Arabidopsis mutant *pdh1-2*, for instance, carries a mutation in the *ProDH1* gene and accumulates proline. Nevertheless, under low water potential, the mutant exhibited a severe growth reduction and high stress susceptibility as a *p5cs1-4* proline-deficient mutant (Bhaskara et al., 2015). Similar growth reduction observed under drought conditions in *p5cs1-4* and *pdh1-2* mutants, in spite of the striking differences in proline accumulation, has been interpreted by the authors as proof that proline accumulation is not correlated, by itself, with drought tolerance (Bhaskara et al., 2015). Another argument against the correlation between proline and stress tolerance comes from the study of natural Arabidopsis accessions, among which a 10-fold variation in proline accumulation is observed under drought conditions (Kesari et al., 2012). No clear-cut correlation between proline and drought was found among these accessions and, strangely enough, accessions from drier regions had lower proline accumulation than accessions from less dry or temperate regions (Kesari et al., 2012). The conclusion reached by Bhaskara et al. (2015) is that proline accumulation contributes to, but is not essential for, drought tolerance and plant cells may rely on different anatomical, physiological, and metabolic strategies to cope with stress. A reduced metabolic flux through the cycle of proline synthesis and catabolism, however, can be a crucial factor limiting their growth.

The hypothesis that changes in proline metabolic flux may be more important for stress tolerance than free proline content was first hypothesized by Phang (1985) in animal systems and confirmed in plants by Hare and Cress (1997) and Hare et al. (1998) in two seminal review articles. The strict compartmentation between proline anabolism, that takes place in the cytoplasm or in the chloroplast, and proline catabolism, that takes place in mitochondria, as well as the cycling of proline and P5C, an intermediate molecule of both synthesis and catabolism, is central in this system. The cycling of proline and P5C through PRODH and P5CR results in the passage of reducing equivalents from cytosol to mitochondria (Phang et al., 1985; Miller et al., 2009), buffering the NADP+/NADPH level in the cytosol and driving the oxidative pentose phosphate pathway (Phang et al., 1985). Thanks to the production of NADP+ coupled with the reduction of glutamate to proline, this amino acid may reduce stress-induced cellular acidification and sustain the oxidative pentose phosphate pathway, which is dependent on NADP+ and inhibited by NADPH. Increased flux through the oxidative pentose phosphate pathway is believed, in turn, to support purine nucleotide biosynthesis and phenolic metabolism during stress recovery (Hare and Cress, 1997; Kohl et al., 1988). In addition, during relief from stress, proline catabolism in the mitochondrion may prime oxidative respiration providing energy to the cell during the phase of stress recovery: Electrons are transferred to the mitochondrial electron transport chain directly from PRODH via ubiquinone. Overall, the oxidation of one molecule of proline yields 30 ATP equivalents (Atkinson, 1977) and is therefore well-suited to sustain high energy-requiring processes, such as root elongation (Mauro et al., 1996; Verslues and Sharpe, 1999; Trovato et al., 2001), bolting (Samach et al., 2000; Mattioli et al., 2008), or, in animal systems, during the earliest and more expensive stages of flight of the honeybee (Micheu et al., 2000). Evidence for proline metabolic flux influencing the NADP+/NADPH ratio in plants has been provided by different groups (Kohl et al., 1991; Hare and Cress, 1997; Sharma et al., 2011; Giberti et al., 2014), including Bhaskara e al. (2015), who ascribed the similar drought-dependent reduction in growth of *p5cs1-4* and *pdh1-2* to a similar reduction in the NADP+/NADPH ratio, compared to wild type (Bhaskara et al., 2015). Moreover, Giberti et al. (2014) reported that the activity of P5CR is regulated by proline and chloride ions in a different way depending on the use of either NADH or NADPH as a co-factor, consistent with the importance of the NADP+/NADPH ratio in proline metabolism. In addition, proline accumulation has been reported to balance the NADP+/NADPH ratio during photoinduced stress. Since photosynthesis is insensitive to stress, under stressful conditions plants may be exposed to light intensities exceeding those that can be used for carbon assimilation, resulting in NADPH imbalance, photoinhibition, and ROS production. Under such conditions upregulation of proline synthesis has been reported to keep the NADP+/NADPH ratio at normal levels (Hare and Cress, 1997; Szabados and Savouré, 2010) further supporting the importance of the redox regulation in proline-induced stress tolerance (Hare et al., 1998).

Role of GS and Proline in Crop Improvement

In spite of the complex debate on the functions and benefits of glutamine and proline accumulation during stress, these amino acids have long been studied as potential tools to improve both yield and stress tolerance in agricultural crops. The negative impact of environmental stresses on agricultural crops is particularly severe in arid or semi-arid countries, especially in large regions of Southeast Asia and in Saharan and sub-Saharan countries. In Algeria, for example, where only 3.4% of the land is arable and 84% is desert, although over 60% of the arable land is cultivated with cereals, especially wheat, cereal yields are poor and unreliable and the country has become one of the world's leading importers of wheat. On a worldwide scale, the negative impact of these adverse conditions is dramatic and brings heavy economic losses and severe environmental and societal damage and, because of global warming and increasing population, their negative effects are predicted to increase. According to the Intergovernmental Panel on Climate Change (IPCC), the global surface temperature may rise from 2.7 to 7.8°C by 2100, depending on the model used, and recent authoritative studies (Karl et al., 2015) claim that trends may be even worse than those predicted by the IPCC. These rapid changes prompt the development of new stress-tolerant and high-productivity crops capable of withstanding drought, salinity, and temperature stress and maintaining or improving current levels of productivity.

Although it is clear that plants counteract drought and environmental stresses by a combination of different types of adaptation, including morphologic, metabolic, and physiologic changes, the possibility to increase stress tolerance by raising the accumulation of only one, or just a few metabolites, is tempting and has drawn the interest of many laboratories worldwide. Indeed, a number of agricultural crops, including wheat, rice, maize, sugarcane, carrot, potato, and pigeon pee, have been transformed either with genes coding for P5CS (Zhu et al., 1998; Sahawel and Hassan, 2002; Han and Hwang, 2003; Hmida-Sayari et al., 2005; Su and Wu, 2004; Molinari, 2007; Vendruscolo et al., 2007; Hayano-Kanashiro et al., 2009; Kumar et al., 2010; Surekha et al., 2013) or GS (Gallardo et al., 1999; Hoshida et al., 2000; Fuentes et al., 2001; Oliveira et al., 2002; El-Khatib et al., 2004;) resulting in improved stress tolerance or NUE. Therefore, these genes, although not necessary, may be sufficient to confer, or contribute to, either plant stress tolerance or higher yields, at least in pilot conditions (Sawahel and Hassan, 2002; Vendruscolo et al., 2007). In addition, all the genes and the regulations involved in these metabolic pathways are well known allowing a facile biotechnology approach based on the generation of transgenic or cisgenic plants. As a note of caution, however, it must be said that in some cases (Singh et al., 1972; Liu and Zhu, 1997; Xin and Browse, 1998; Chen et al., 2007; Widodo et al., 2009; Maggio et al., 2002; Mani et al., 2002; Bhaskara et al., 2012; Fan et al., 2015) no correlation was found between the accumulation of these amino acids and stress tolerance, implying the existence of complex regulations still poorly understood. In addition, there is limited data derived from experiments run under field conditions, where environmental variability and plant plasticity may introduce more complex scenarios, prompting the analysis of the metabolic fluxes of these metabolites under stress conditions and in field conditions.

The recent development of powerful genome editing techniques, such as the CRISPR/Cas9 technology (Kozano-Juste and Cutler, 2014), opens up fascinating possibilities to fine-tune the metabolism of these amino acids to improve the crop resilience to environmental stresses and gain superior yields, provided a sufficient knowledge of the metabolism of these amino acids is available.

As an alternative approach to transgene technologies, allelic variants of these genes associated with useful traits, may be used as genetic tags in marker-assisted breeding programs.

Conventional breeding has the power to generate novel plant cultivars and germplasm but needs a long time to select and stabilize inbred lines, subjective evaluations, and empirical selection methods, particularly in crops with large and complex genomes such as durum (*Triticum durum*) or bread wheat (*Triticum aestivum*) with allotetraploid or, respectively, allohexaploid genomes. Marker-assisted selection (Tanksley and McCouch, 1997) offers the possibility to accelerate and extend the classical breeding techniques by exploiting molecular markers reliably associated with quantitative trait loci (QTL) such as disease resistance, abiotic stress tolerance, crop yield, and nitrogen use efficiency (Slafer et al., 2005). A number of molecular markers, such as Single Sequence Repeats (SSRs), Restriction Fragment Length Polymorphism (RFLPs), Single Nucleotide Polymorphisms (SNPs), Random Amplification of Polymorphic DNAs (RAPDs), and Amplified Fragment Length Polymorphisms (AFLPs) can be used to track genes or genomic regions to be associated, by genetic linkage analysis, with useful QTLs. In a study tailored to assess the usefulness of molecular markers in wheat breeding programs, for example, 26 SSR markers were used and found able to unambiguously discriminate between as many as 40, closely related, different wheat genotypes including drought, semi-tolerant, and non-drought tolerant genotypes (Bousba et al., 2013).

Since 1979, proline accumulation has been recognized as a heritable trait (Hanson et al., 1979) and has been associated with drought stress tolerance in several crop plants where stress-tolerant ecotypes accumulate proline in significantly higher amounts than in stress-sensitive ecotypes (Ashraf and Foolad, 2007). Examples of this stringent correlation can be found, for example, in creeping bentgrass (Ahmad et al., 1981), tobacco (van Rensburg et al., 1993), wheat (van Heerden and de Villiers, 1996), alfa alfa (Petrusa and Winicov, 1997), rice (Hsu et al., 2003), cowpea (Hamidou et al., 2007), maize (Moussa and Abdel-Aziz, 2008), chickpea (Mafakheri et al., 2010), and soybean (Masoumi et al., 2011). Accordingly, genes related to proline metabolism, especially *P5CS* and *ProDH*, coding for the rate-limiting steps of proline synthesis and, respectively, catabolism, have been used as reliable markers to detect QTLs associated with different types of stress, particularly drought and high salinity (Sairam et al., 2002; Sayed et al., 2012; Wehner et al., 2016; Mwenye et al., 2016).

As for GS, Hirel et al. (2001) and Masclaux et al. (2001) reported a positive correlation between NUE and GS, by analyzing, in maize inbred lines, a number of NUE-related traits including GS activity. The locations on the maize genome of the QTLs affecting yield parameters turned out to nicely coincide with those of cytosolic GS (GS1). A robust correlation between GS and QTLs for NUE and grain yield was confirmed in maize by Gallais and Hirel (2004). In particular, a precise co-location was found between the gene encoding cytosolic GS (*gln4* locus) and a major grain yield QTL on chromosome 5. In addition, two GS genes (*gln1* and *gln2*) on chromosome 1 and the GS gene (*gln3*) on chromosome 4 were found to be associated with QTLs for NUE. Moreover, Habash et al. (2007) investigated nitrogen use in wheat by characterizing QTLs for 21 traits associated with growth, yield, and leaf nitrogen assimilation during grain fill in wheat. All QTLs for GS activity were co-localized together with those for grain N, with increased activity associated with higher grain N, but little or no activity on grain yield components. In particular, one QTL cluster for GS activity was found to co-localize with a *GS2* gene mapped on chromosome 2A, and another with the *GSr* isogene on chromosome 4A. Co-localization of the activity of, or abundance of GS – particularly the cytosolic isozyme GS1 – with QTLs for N remobilization, grain size, growth, grain fill, and nitrogen recycling was additionally reported in maize (Obara et al., 2001), rice (Tabuchi et al., 2005), and sugarcane (Whan et al., 2010) to further confirm the usefulness of GS as QTL marker.

Conclusion and Future Perspectives

The progressive worsening of environmental conditions caused by global warming, the increasing demand for natural resources, and the continuous rise of the world population, are generating dramatic losses on crop yields, prompting the need for the creation of a new generation of high-yield, stress-tolerant crops.

In spite of the conflicting positions that divide the scientific community on the role of glutamate-derived amino acids under stress, it is clear that in most plant species these amino acids are associated with, and exert a beneficial effect on, the plant capacity to use and recycle nitrogen and tolerate stress injuries. The metabolism of glutamine and proline, in particular, because of the preeminent role of these amino acids in nitrogen use efficiency and, respectively, stress tolerance, can potentially be modified or followed to develop plants with superior performances.

As plant crops often acquire drought and salinity resistance at the expense of yield potential (Blum, 2005), it will be of major importance to modify both glutamine and proline metabolism to generate drought-tolerant crops of high productivity.

The exploitation of powerful novel techniques of genome editing, such as CRISPR/Cas9, to precisely engineer genes involved in glutamine and proline metabolism, as well as the use of new molecular markers, such as SNPs and genotyping-by-sequencing (GBS) markers (Yang et al., 2012; He et al., 2014) associated with both GS and proline traits, is expected to boost and accelerate this process in the next future.

A sound knowledge of the molecular and biochemical details of these metabolic pathways, is, of course, a preliminary requirement of these biotechnological advances and much has still to be done to achieve a full comprehension of glutamine and proline metabolism and their underlying molecular regulations. Post-transcriptional and post-translational regulations, in particular, have been recently discovered to significantly contribute to the overall regulations of glutamine and proline metabolism (Seabra and Carvalho, 2015; Bhaskara et al., 2015) and need to be further investigated. Equally important is understanding whether the metabolism of these amino acids rather than their accumulation is important for stress tolerance, and to unveil the role of, and the intricate connections among, the multiple functions these peculiar amino acids bring about.

REFERENCES

Abraham E, Rigo G, Székely G, Nagy R, Koncz C, Szabados L. Light-dependent induction of proline biosynthesis by abscisic acid and salt stress is inhibited by brassinosteroid in Arabidopsis. *Plant Mol. Biol.* 2003; 51: 363–372.

Ahmad I, Wainwright SJ, Stewart GR. The solute and water relations of *Agrostis stolonifera* ecotypes differing in their salt tolerance. *New Phytol.* 1981; 87: 615–629.

Alia A, Mohanty P, Matysik J. Effect of proline on the production of singlet oxygen. *Amino Acids* 2001; 21: 195–200.

Ashraf M, Foolad MR. Roles of glycinebetaine and proline in improving plant abiotic stress tolerance. *Environ. Exp. Bot.* 2007; 59: 206–216.

Atkinson DE. *Cellular Energy Metabolism and Its Regulation.* New York: Academic Press, 1977.

Bauer D, Biehler K, Fock H, Carrayol E, Migge A, Becker TW. A role for cytosolic glutamine synthetase in the remobilization of leaf nitrogen during water stress. *Physiol. Plant.* 1997; 99: 241–248.

Bernard SM, Habash D. The importance of cytosolic glutamine synthetase in nitrogen assimilation and recycling. *New Phytol.* 2009; 182: 608–620.

Berteli F, Corrales E, Guerrero C, Ariza MJ, Pilego F, Valpuesta V. Salt stress increases ferredoxin-dependent glutamate synthase activity and protein level in the leaves of tomato. *Physiol. Plant.* 1995; 93: 259–264.

Bhaskara GB, Nguyen TT, Verslues PE. Unique drought resistance functions of the Highly ABA-Induced clade A protein phosphatase 2Cs. *Plant Physiol.* 2012; 160: 379–395.

Bhaskara GB, Yang TH, Verslues P. Dynamic proline metabolism: Importance and regulation in water limited environments. *Front. Plant Sci.* 2015; 6: 484.

Biancucci M, Mattioli R, Forlani G, Funck D, Costantino P, Trovato M. Role of proline and GABA in sexual reproduction of angiosperms. *Front. Plant Sci.* 2015; 6: 680.

Blum, A. Drought resistance, water-use efficiency, and yield potential—Are they compatible, dissonant, or mutually exclusive? *Aust. J. Agric. Res.* 2005; 56. 1159–1168.

Borsani O, Díaz P, Monza J. Proline is involved in water stress responses of *Lotus corniculatus* nitrogen fixing and nitrate fed plants. *J. Plant Physiol.* 1999; 155: 269–273.

Borsani O, Zhu J, Verslues PE, Sunkar R, Zhu JK. Endogenous siRNAs derived from a pair of natural cis-antisense transcripts regulate salt tolerance in Arabidopsis. *Cell* 2005; 123: 1279–1291.

Bousba R, Baum M, Jighly A, Djekoune A, Lababidi S, Benbelkacem A, Labhilili M, Gaboun F, Ykhlef N. Association analysis of genotypic and phenotypic traits using SSR marker in durum wheat. *Online Int. Interdiscip. Res. J.* 2013; III: 60–79.

Brown LM, Hellebust JA. Sorbitol and proline as intracellular osmotic solutes in the green alga *Stichococcus bacillaris*. *Can. J. Bot.* 1978; 56: 676–679.

Brugiére N, Dubois F, Limami AM, Lelandais M, Roux Y, Sangwan RS, Hirel B. Glutamine synthetase in the phloem plays a major role in controlling proline production. *Plant Cell* 1999; 11: 1995–2011.

Burton RS. Regulation of proline synthesis in osmotic response: Effects of protein synthesis inhibitors. *J. Exp. Zool.* 1991; 259: 272–277.

Büssis D, Heineke D. Acclimation of potato plants to polyethylene glycol-induced water deficit. *J. Exp. Bot.* 1998; 49: 1361–1370.

Cammaerts D, Jacobs M. A study of the role of glutamate dehydrogenase in the nitrogen metabolism of Arabidopsis thaliana. *Planta* 1985; 163: 517–526.

Chen Z, Cuin TA, Zhou M, Twomey A, Naidu BP, Shabala S. Compatible solute accumulation and stress-mitigating effects in barley genotypes contrasting in their salt tolerance. *J. Exp. Bot.* 2007; 58: 4245–4255.

Cheynier V, Comte G, Davies KM, Lattanzio V, Martens S. Plant phenolics: recent advances on their biosynthesis, genetics, and ecophysiology. *Plant Physiol. Biochem.* 2013; 72: 1–20.

Chiang HH, Dandekar AM. Regulation of proline accumulation in Arabidopsis during development and in response to desiccation. *Plant Cell Environ.* 1995; 18: 1280–1290.

Corpas FJ, Barroso JB. Nitro-oxidative stress vs oxidative or nitrosative stress in higher plants. *New Phytol.* 2013; 199: 633–635.

Coruzzi GM. *Primary N-assimilation into Amino Acids in Arabidopsis. The Arabidopsis Book* 2003; 2: e0010.

Cramer MD, Gao ZF, Lips SH. The influence of dissolved inorganic carbon in the rhizosphere on carbon and nitrogen metabolism in salinity-treated tomato plants. *New Phytol.* 1999; 142: 441–453.

Csonka LN. Physiological and genetic responses of bacteria to osmotic stress. *Microbiol. Rev.* 1989; 53: 121–147.

Csonka LN, Gelvin SB, Goodner BW, Orser CS, Siemieniak D, Slightom JL. Nucleotide sequence of a mutation in the proB gene of Escherichia coli that confers proline overproduction and enhanced tolerance to osmotic stress. *Gene* 1988; 64: 199–205.

da Rocha IM, Vitorello VA, Silva JS, Ferreira-Silva SL, Viégas RA, Silva EN, Sileira JA. Exogenous ornithine is an effective precursor and the δ-ornithine amino transferase pathway contributes to proline accumulation under high N recycling in salt-stressed cashew leaves. *J. Plant Physiol.* 2012; 169: 41–49.

Dandekar AM, Uratsu SL. A single base pair change in proline biosynthesis genes causes osmotic stress tolerance. *J. Bacteriol.* 1988; 170: 5943–5945.

Delauney AJ, Verma DPS. A soybean gene encoding Δ1-pyrroline-5-carboxylate reductase was isolated by functional complementation in *Escherichia coli* and is found to be osmoregulated. *Mol. Gen. Genet.* 1990; 221: 299–305.

Díaz P, Borsani O, Márquez A, Monza J. Osmotically induced proline accumulation in *Lotus corniculatus* leaves is affected by light and nitrogen source. *Plant Growth Regul.* 2005; 46: 223–232.

El-khatib RT, Hamerlynck E, Gallardo F, Kirby E. Transgenic poplar characterized by ectopic expression of a pine cytosolic glutamine synthetase gene exhibits enhanced tolerance to water stress. *Tree Physiol.* 2004; 24: 729–736.

Fan Y, Shabala S, Ma Y, Xu R, Zhou M. Using QTL mapping to investigate the relationships between abiotic stress tolerance (drought and salinity) and agronomic and physiological traits. *BMC Genom.* 2015; 16: 43.

Floyd RA, Nagy ZS. Formation of long-lived hydroxyl free radical adducts of proline and hydroxyproline in a Fenton reaction. *Biochem. Biophys. Acta* 1984; 790: 94–97.

Fuentes SI, Allen DJ, Ortiz-Lopez A, and Hernández G. Overexpression of cytosolic glutamine synthetase increases photosynthesis and growth at low nitrogen concentrations. *J. Exp. Bot.* 2001; 52: 1071–1081.

Funayama K, Kojima S, Tabuchi-Kobayashi M, Sawa Y, Nakayama Y, Hayakawa T, Yamaya T. Cytosolic glutamine synthetase1;2 is responsible for the primary assimilation of ammonium in rice roots. *Plant Cell Physiol.* 2013; 54: 934–943.

Funck D, Eckar S, Müller G. R Non-redundant functions of two proline dehydrogenase isoforms in Arabidopsis. *BMC Plant Biol.* 2010; 10: 70.

Funck D, Stadelhofer B, Koch W. Ornithine-δ-aminotransferase is essential for arginine catabolism but not for proline biosynthesis. *BMC Plant Biol.* 2008; 8: 40.

Funck D, Winter G, Baumgarten L, Forlani G. Requirement of proline synthesis during Arabidopsis reproductive development. *BMC Plant Biol.* 2012; 12: 191.

Gallais A, Hirel B. An approach to the genetics of nitrogen use efficiency in maize. *J. Exp. Bot.* 2004; 55: 295–306.

Gallardo F, Fu J, Cantón FR, García-Gutiérrez A, Cánovas FM, Kirby EG. Expression of a conifer glutamine synthetase gene in transgenic poplar. *Planta* 1999; 210: 19–26.

García-Calderón M, Pons-Ferrer T, Mrázova A, Pal'ove-Balang P, Vilková M, Pérez-Delgado CM, Vega JM, Eliášová A, Repčák M, Márquez AJ, Betti M. Modulation of phenolic metabolism under stress conditions in a *Lotus japonicus* mutant lacking plastidic glutamine synthetase. *Front. Plant Sci.* 2015; 6: 760.

Giberti S, Funck D, Forlani G. Δ1-pyrroline-5-carboxylate reductase from *Arabidopsis thaliana*: stimulation or inhibition by chloride ions and feedback regulation by proline depend on whether NADPH or NADH acts as cosubstrate. *New Phytol.* 2014; 202: 911–919.

Gimeno V, Syvertsen JP, Nieves M, Simón I, Martínez V, García-Sánchez F. Additional nitrogen fertilization affects salt tolerance of lemon trees on different rootstocks. *Sci. Hortic.* 2009; 121: 298–305.

Girousse C, Bournoville R, Bonnemain JL. Water deficit-induced changes in concentrations in proline and some other amino acids in the phloem sap of alfalfa. *Plant Physiol.* 1996; 111: 109–113.

Habash DZ, Bernard S, Schondelmaier J, Weyen J, Quarrie SA. The genetics of nitrogen use in hexaploid wheat: N utilisation, development and yield. *Theor. Appl. Genet.* 2007; 114: 403–419.

Habash DZ, Massiah AJ, Rong HL, Wallsgrove RM, Leigh RA. The role of cytosolic glutamine synthetase in wheat. *Ann. Appl. Biol.* 2001; 138: 83–89.

Hamidou F, Zombre G, Braconnier S. Physiological and biochemical responses of cowpea genotypes to water stress under glasshouse and field conditions. *J. Agron. Crops* 2007; 193: 229–237.

Han KH, Hwang CH. Salt tolerance enhanced by transformation of a P5CS gene in carrot. *J. Plant Biotechnol.* 2003; 5: 149–53.

Hanson AD, Nelsen CE, Pedersen AR, Everson EH. Capacity for proline accumulation during water stress in barley and its implications for breeding for drought resistance. *Crop Sci.* 1979; 19: 489–493.

Hare PD, Cress WA. Metabolic implications of stress-induced proline accumulation in plants. *Plant Growth Regul.* 1997; 21: 79–102.

Hare PD, Cress WA, Van Staden J. Dissecting the roles of osmolyte accumulation during stress. *Plant Cell Environ.* 1998; 21: 535–553.

Hare PD, Cress WA, Van Staden J. Proline synthesis and degradation: A model system for elucidating stress-related signal transduction. *J. Exp. Bot.* 1999; 50: 413–434.

Hayano-Kanashiro C, Calderon-Vazquez C, Ibarra-Laclette E, Herrera-Estrella L, Simpson J. Analysis of gene expression and physiological responses in three Mexican maize landraces under drought stress and recovery irrigation. *PLoS ONE* 2009; 4: e7531.

Hayashi F, Ichino T, Osanai R, Wada K. Oscillation and regulation of proline content by *P5CS* and *ProDH* gene expressions in the light/dark cycles in *Arabidopsis thaliana* L. *Plant Cell Physiol.* 2000; 41: 1096–1101.

He J, Zhao X, Laroche A, Lu ZX, Liu H, Li Z. Genotyping-by-sequencing (GBS), an ultimate marker-assisted selection (MAS) tool to accelerate plant breeding. *Front. Plant Sci.* 2014; 5: 484.

Hirel B, Bertin P, Quillere I, Bourdoncle W, Attagnant C, Dellay C, Gouy A, Cadiou S, Retailliau C, Falque M, Gallais A. Towards a better understanding of the genetic and physiological basis for nitrogen use efficiency in maize. *Plant Physiol.* 2001; 125: 1258–1270.

Hirel B, Phillipson B, Murchie E, Suzuki A, Kunz C, Ferrario S, Limami A, Chaillou S, Deleens E, Brugière N, Chaumont-Bonnet M, Foyer C, Morot-Gaudry J-F. Manipulating the pathway of ammonium assimilation in transgenic non-legumes and legumes. *J. Plant Nutr. Soil Sci.* 1997; 160: 283–290.

Hmida-Sayari A, Gargouri-Bouzid R, Bidani A, Jaoua L, Savouré A, Jaoua S. Overexpression of Δ1-pyrroline-5-carboxylate synthetase increases proline production and confers salt tolerance in transgenic potato plants. *Plant Sci.* 2005; 169: 746–52.

Hong Z, Lakkineni K, Zhang Z, Verma DPS. Removal of feedback inhibition of Δ1-pyrroline-5-carboxylate synthetase results in increased proline accumulation and protection of plants from osmotic stress. *Plant Physiol.* 2000; 122: 1129–1136.

Hoque A, Okuma E, Banu NA, Nakamura Y, Shimoishi Y, Murata Y. Exogenous proline mitigate the detrimental effect of salt stress more than exogenous betaine by increasing antioxidant enzyme activities. *Plant Physiol.* 2007; 164: 553–561.

Hoshida H, Tanaka Y, Hibino T, Hayashi Y, Tanaka A, Takabe T, Takabe T. Enhanced tolerance to salt stress in transgenic rice that overexpresses chloroplast glutamine synthetase. *Plant Mol. Biol.* 2000; 43: 103–111.

Hsu SY, Hsu YT, Kao CH. The effect of polyethylene glycol on proline accumulation in rice leaves. *Biol. Plant.* 2003; 46: 73–78.

Hu CAA, Delauney AJ, Verma DPS. A bifunctional enzyme Δ1-pyrroline-5- carboxylate synthetase catalyzes the first two steps in proline biosynthesis in plants. *Proc. Natl. Acad. Sci. USA* 1992; 89: 9354–9358.

Huang Z, Zhao L, Chen D, Liang M, Liu Z, Shao H, Long X. Salt stress encourages proline accumulation by regulating proline biosynthesis and degradation in Jerusalem artichoke plantlets. *PLoS ONE* 2013; 8: 4.

Igarashi Y, Yoshiba Y, Sanada Y, Yamaguchi-Shinozaki K. Characterization of the gene for Δ1-pyrroline-5-carboxylate synthetase and correlation between the expression of the gene and salt tolerance in *Oryza sativa* L. *Plant Mol. Biol.* 1997; 33: 857–865.

Islam MM, Hoque MA, Okuma E, Banu MN, Shimoishi Y, Nakamura Y, Murata Y. Exogenous proline and glycinebetaine increase antioxidant enzyme activities and confer tolerance to cadmium stress in cultured tobacco cells. *J. Plant Physiol.* 2009; 166: 1587–1597.

Kandpal RP, Rao NA. Alterations in the biosynthesis of proteins and nucleic acids in finger millet (*Eleucine coracana*) seedlings during water stress and the effect of proline on protein biosynthesis. *Plant Sci.* 1985; 40: 73–79.

Kaneshiro ES, Holz GG Jr, Dunham PB. Osmoregulation in a marine ciliate, *Miamiensis avidus*. II. Regulation of intracellular free amino acids. *Biol. Bull.* 1969; 137: 161–169.

Kant S, Kant P, Lips H, Barak S. Partial substitution of NO3− by NH4+ fertilization increases ammonium assimilating enzyme activities and reduces the deleterious effects of salinity on the growth of barley. *J. Plant Physiol.* 2007; 164: 303–311.

Kant S, Kant P, Raveh E, Barak S. Evidence that differential gene expression between the halophyte, *Thellungiella halophila*, and *Arabidopsis thaliana* is responsible for higher levels of the compatible osmolyte proline and tight control of Na + uptake in *T. halophila*. *Plant Cell Environ.* 2006; 29: 1220–1234.

Karl TR, Arguez A, Huang B, Lawrimore JH, McMahon JR, Menne MJ, Peterson TC, Vose RS, Zhang HM. Possible artifacts of data biases in the recent global surface warming hiatus. *Science* 2015; 348: 1469–1472.

Kavi Kishor PB, Hong Z, Miao GH, Hu CAA, Verma DPS. Overexpression of Δ1-pyrroline-5-carboxylate synthetase increases proline production and confers osmo-tolerance in transgenic plants. *Plant Physiol.* 1995; 108: 1387–1394.

Kavi Kishor PB, Sangam S, Amrutha RN, Sri Laxmi P, Naidu KR, Rao KRSS, Sreenath Rao, Reddy KJ, Theriappan P, Sreenivasulu N. Regulation of proline biosynthesis, degradation, uptake and transport in higher plants: Its implications in plant growth and abiotic stress tolerance. *Curr. Sci.* 2005; 88: 424–438.

Kavi Kishor PB, Sreenivasulu N. Is proline accumulation per se correlated with stress tolerance or is proline homeostasis a more critical issue? *Plant Cell Environ.* 2014; 37: 300–311.

Kesari R, Lasky JR, Villamor JG, Marais DLD, Chen YJC, Liu TW, Juenger TE, Verslues PE. Intron-mediated alternative splicing of *Arabidopsis P5CS1* and its association with natural variation in proline and climate adaptation. *Proc. Natl. Acad. Sci. USA* 2012; 109: 9197–9202.

Khedr HA, Abbas MA, Wahid AAA, Quick PW, Abogadallah GM. Proline induces the expression of salt-stress-responsive proteins and may improve the adaptation of *Pancratium maritimum* L. to salt-stress. *J. Exp. Bot.* 2003; 392: 2553–2562.

Kiyosue T, Yoshiba Y, Yamaguchi-Shinozaki K, Shinozaki K. A nuclear gene encoding mitochondrial proline dehydrogenase, an enzyme involved in proline metabolism, is upregulated by proline but downregulated by dehydration in *Arabidopsis*. *Plant Cell* 1996; 8: 1323–1335.

Knight H, Trewavas AJ, Knight MR. Calcium signalling in *Arabidopsis* responding to drought and salinity. *Plant J.* 1997; 12: 1067–1078.

Kohl DH, Kennelly EJ, Zhu Y, Schubert KR, Shearer G. Proline accumulation, nitrogenase (C2 H2 reducing) activity and activities of enzymes related to proline metabolism in drought-stressed soybean nodules. *J. Exp. Bot.* 1991; 240: 831–837.

Kohl DH, Schubert KR, Carter MB, Hagedorn CH, Shearer G. Proline metabolism in N2-fixing root nodules: Energy transfer and regulation of purine synthesis. *Proc. Natl. Acad. Sci. USA* 1988; 85: 2036–2040.

Kozano-Juste, Cutler SR. Plant genome engineering full bloom. *Trends Plant Sci.* 2014; 19: 284–287.

Kumar V, Shriram V, Kavi Kishor PB, Jawali N, Shitole MG. Enhanced proline accumulation and salt stress tolerance of transgenic indica rice by over-expressing *P5CSF129A* gene. *Plant Biotech. Rep.* 2010; 4: 37–48.

Le Rudulier D, Strom AR, Dandekar LT Smith, Valentine RC. Molecular biology of osmoregulation. *Science* 1984; 224: 1064–1068.

Lea PJ, Azevedo RA. Nitrogen use efficiency. 2. Amino acid metabolism. *Ann. Appl. Biol.* 2007; 151: 269–275.

Lea PJ, Miflin BJ. Nitrogen assimilation and its relevance to crop improvement. *Annu. Plant Rev.* 2010; 42: 1–40.

Lehmann S, Funck D, Szabados L, Rentsch D. Proline metabolism and transport in plant development. *Amino Acids* 2010; 39: 949–962.

Less H, Galili G. Principal transcriptional programs regulating plant amino acid metabolism in response to abiotic stresses. *Plant Physiol.* 2008; 147: 316–330.

Liu J, Zhu JK. Proline accumulation and salt-stress-induced gene expression in a salt-hypersensitive mutant of Arabidopsis. *Plant Physiol.* 1997; 114: 591–596.

Loggini B, Scartazza A, Brugnoli E, Navari-Izzo F. Antioxidative defense system, pigment composition, and photosynthetic efficiency in two wheat cultivars subjected to drought. *Plant Physiol.* 1999; 119: 1091–1100.

Low PS. Molecular basis of the biological compatibility of nature's osmolytes. In: Gilles R, Gilles-Baillien M. (eds.). *Transport Processes, Iono- and Osmoregulation*, 469–477. Berlin: Springer-Verlag, 1985.

Mafakheri A, Siosemardeh A, Bahramnejad B, Struik, PC, Sohrabi Y. Effect of drought stress on yield, proline and chlorophyll contents in three chickpea cultivars. *Aust. J. Crop Sci.* 2010; 4: 580–585.

Maggio A, Miyazaki S, Veronese P, Fuijita T, Ibeas JI, Damsz B, Narasmhan ML, Hasegawa PM, Joly RJ, Bressan RA. Does proline accumulation play an active role in stress-induced growth reduction? *Plant J.* 2002; 31: 699–712.

Mäkelä P, Peltonen-Sainio P, Jokinen K, Pehu E, Setälä H, Hinkkanen R, Somersalo S: Uptake and translocation of foliar-applied glycinebetaine in crop plants. *Plant Sci.* 1996: 121: 221–230.

Mani S, Van De Cotte B, Van Montagu M, Verbruggen N. Altered levels of proline dehydrogenase cause hypersensitivity to proline and its analogs in Arabidopsis. *Plant Physiol.* 2002; 128: 73–83

Martin A, Lee J, Kichey T, Gerentes D, Zivy M, Tatout C, Dubois F, Balliau T, Valot B, Davanture M, Tercé-Laforgue T. Two cytosolic glutamine synthetase isoforms of maize are specifically involved in the control of grain production. *Plant Cell* 2006; 18: 3252–3274.

Martinelli T, Whittaker A, Bochicchio A, Vazzana C, Akira Suzuki, Masclaux-Daubresse C. Amino acid pattern and glutamate metabolism during dehydration stress in the 'resurrection' plant *Sporobolus stapfianus*: A comparison between desiccation-sensitive and desiccation-tolerant leaves. *J. Exp. Bot.* 2007; 58: 3037–3046.

Masclaux C, Quillere I, Gallais A, Hirel B. The challenge of remobilization in plant nitrogen economy. A survey of physio-agronomic and molecular approaches. *Ann. Appl. Biol.* 2001; 138: 69–81.

Masoumi H, Darvish F, Daneshian J, Nourmohammadi G, Habibi D. Chemical and biochemical responses of soybean (*Glycine max* L.) cultivars to water deficit stress. *Aust. J. Crop Sci.* 2011; 5: 544–553.

Mattioli R, Biancucci M, Lonoce C, Costantino P, Trovato M. Proline is required for male gametophyte development in Arabidopsis. *BMC Plant Biol.* 2012; 12: 236.

Mattioli R, Costantino P, Trovato M: Proline accumulation in plants:not only stress. *Plant signal Behav.* 2009a; 4: 1016–1018.

Mattioli R, Falasca G, Sabatini S, Costantino P, Altamura MM, Trovato M. The proline biosynthetic genes P5CS1 and P5CS2 play overlapping roles in Arabidopsis flower transition but not in embryo development. *Physiol. Plant.* 2009b; 137: 72–85.

Mattioli R, Marchese D, D'Angeli S, Altamura MM, Costantino P, Trovato M. Modulation of intracellular proline levels affects flowering time and inflorescence architecture in Arabidopsis. *Plant Mol. Biol.* 2008; 66: 277–288.

Mauro ML, Trovato M, De Paolis A, Gallelli A, Costantino P, Altamura MM. The plant oncogene *rolD* stimulates flowering in transgenic tobacco plants. *Dev. Biol.* 1996; 180: 693–700.

Micheu S, Crailsheim K, Leonhard B. Importance of proline and other amino acids during honeybee flight (Apis mellifera carnica POLLMANN). *Amino Acids* 2000; 18: 157–175.

Miflin BJ, Lea PJ. Ammonia assimilation. In: Miflin BJ (ed.). *Biochemistry of Plants*, 169–202. New York: Academic Press, 1980.

Miller G, Honig A, Stein H, Suzuki N, Mittler R, Zilberstein A. Unraveling Δ1-pyrroline-5-carboxylate-proline cycle in plants by uncoupled expression of proline oxidation enzymes. *J. Biol. Chem.* 2009; 284: 26482–26492.

Molinari HBC, Marur CJ, Daros E, Campos MKF, De Carvalho JFRP, Filho JCB, Pereira LFP, Vieira LGE. Evaluation of the stress-inducible production of proline in transgenic sugarcane (*Saccharum* spp.): Osmotic adjustment, chlorophyll fluorescence and oxidative stress. *Physiol. Plant.* 2007; 130: 218–229.

Moussa HR, Abdel-Aziz SM. Comparative response of drought tolerant and drought sensitive maize genotypes to water stress. *Aust. J. Crop Sci.* 2008; 1: 331–36.

Mwenye OJ, van Rensburg L, van Biljon A, van der Merwe R. The role of proline and root traits on selection for drought-stress tolerance in soybeans: A review. *S. Afr. J. Plant Soil* 2016; 34: 1–12.

Nakabayashi R, Saito K. Integrated metabolomics for abiotic stress responses in plants. *Curr. Opin. Plant Biol.* 2015; 24: 10–16.

Nakabayashi R, Yonekura-Sakakibara K, Urano K, Suzuki M, Yamada Y, Nishizawa T, Matsuda F, Kojima M, Sakakibara H, Shinozaki K, Michael AJ. Enhancement of oxidative and drought tolerance in Arabidopsis by overaccumulation of antioxidant flavonoids. *Plant J.* 2014; 77: 367–379.

Nanjo T, Kobayashi M, Yoshiba Y, Kakubari Y, Yamaguchi-Shinozaki K, Shinozaki K. Antisense suppression of proline degradation improves tolerance to freezing and salinity in *Arabidopsis. FEBS Lett.* 1999; 461: 205–210.

Obara M, Kajiura M, Fukuta Y, Yano M, Hayashi M, Yamaya T, Sato T. Mapping of QTLs associated with cytosolic glutamine synthetase and NADH-glutamate synthase in rice (*Oryza sativa* L.). *J. Exp. Bot.* 2001; 52: 1209–1217.

Okumoto S, Funck D, Trovato M, Forlani G. Editorial: Amino acids of the glutamate family: Functions beyond primary metabolism. *Front. Plant Sci.* 2016; 7: 318.

Oliveira IC, Brears T, Knights TJ, Clark A, Coruzzi GM. Overexpression of cytosolic glutamate synthetase. Relation to nitrogen, light, and photorespiration. *Plant Physiol.* 2002; 129: 1170–1180.

Orsel M, Moison M, Clouet V, Thomas J, Leprince F, Canoy AS, Just J, Chalhoub B, Masclaux-Daubresse C. Sixteen cytosolic glutamine synthetase *genes* identified in the *Brassica napus* L. genome are differentially regulated depending on nitrogen regimes and leaf senescence. *J. Exp. Bot.* 2014; 65: 3927–3947.

Parre E, Ghars MA, Leprince AS, Thiery L, Lefebvre D, Bordenave M, Richard L, Mazars C, Abdelly C, Savouré A. Calcium signaling via phospholipase C is essential for proline accumulation upon ionic but not nonionic hyperosmotic stresses in *Arabidopsis. Plant Physiol.* 2007; 144: 503–512.

Peng Z, Lu Q, Verma DP. Reciprocal regulation of Δ1-pyrroline-5-carboxylate synthetase and proline dehydrogenase genes controls proline levels during and after osmotic stress in plants. *Mol. Gen. Genet.* 1996; 253: 334–341.

Petrusa LM, Winicov I. Proline status in salt tolerant and salt sensitive alfalfa cell lines and plants in response to NaCl. *Plant Physiol. Biochem.* 1997; 35: 303–310.

Phang JM. The regulatory functions of proline and pyrroline-5-carboxylic acid. *Curr. Top Cell Regul.* 1985; 25: 91–132.

Poulin R, Larochelle J, Hellebust JA. The regulation of amino acid metabolism during hyperosmotic stress in *Acanthamoetla castellanii. J. Exp. Zool.* 1987; 243: 365–378.

Rajendrakumar SC, Reddy BV, Reddy AR. Proline–protein interactions: Protection of structural and functional integrity of M4 lactate dehydrogenase. *Biochem. Biophys. Res. Commun.* 1994; 201: 957–963.

Rashid A, Waraich EA, Ashraf MY, Shamim A, Aziz T. Does nitrogen fertilization enhance drought tolerance in sunflower? A review. *J. Plant Nutr.* 2014; 37: 942–963.

Rentsch D, Schmidt S, Tegeder M. Transporters for uptake and allocation of organic nitrogen compounds in plants. *FEBS Lett.* 2007; 581: 2281–2289.

Rhodes D, Handa S, Bressan RA. Metabolic changes associated with adaptation of plant cells to water stress. *Plant Physiol.* 1986; 82: 890–903.

Rustgi S, Joshi A, Moss H, Riesz P. ESR of spin trapped radicals of aqueous solutions of amino acids: Reactions of the hydroxyl radical. *Int. J. Rad. Biol. Rel. Stud. Phys. Chem. Med.* 1977; 31: 415–440.

Sagi M, Dovrat A, Kipnis T, Lips SH. Nitrate reductase, phosphoenolpyruvate carboxylase and glutamine synthetase in annual ryegrass as affected by salinity and nitrogen. *J. Plant Nutr.* 1998; 21: 707–723.

Sairam RK, Rao KV, Srivastava GC. Differential response of wheat genotypes to long term salinity stress in relation to oxidative stress, antioxidant activity and osmolyte concentration. *Plant Sci.* 2002; 163: 1037–1046.

Samach A, Onouchi H, Gold SE, Dittha GS, Schwarz-Sommer S, Yanofsky MF, Coupland G. Distinct roles of *CONSTANS* target genes in reproductive development of *Arabidopsis. Science* 2000; 288: 1613–1616.

Saranga Y, Rhodes D, Janick J. Changes in amino acid composition associated with tolerance to partial desiccation of celery somatic embryos. *J. Am. Soc. Hort. Sci.* 1992; 117: 337–341.

Savouré A, Hua XJ, Bertauche N, Van Montagu M, Verbruggen N. Abscisic acid-independent and abscisic acid-dependent regulation of proline biosynthesis following cold and osmotic stresses in Arabidopsis thaliana. *Mol. Gen. Genet.* 1997; 254: 104–109.

Savouré A, Jaoua S, Hua XJ, Ardiles W, VanMontagu M, Verbruggen N. Isolation and characterization, and chromosomal location of a gene encoding the Δ1-pyrroline- 5-carboxylate synthetase in *Arabidopsis. FEBS Lett.* 1995; 372: 13–19.

Sawahel W, Hassan AH. Generation of transgenic wheat plants producing high levels of the osmoprotectant proline. *Biotechnol. Lett.* 2002; 24: 721.

Sayed MA, Schumann H, Pillen K, Naz AA, Léon J. AB-QTL analysis reveals new alleles associated to proline accumulation and leaf wilting under drought stress conditions in barley (*Hordeum vulgare* L.). *BMC Genet.* 2012; 13: 61.

Schwacke R, Grallath S, Breitkreuz KE, Stransky H, Frommer WB, Rentsch D. LeProT1, a transporter for proline, glycine betaine, and γ -amino butyric acid in tomato pollen. *Plant Cell* 1999; 11: 377–391.

Seabra AR, Carvalho HG. Glutamine synthetase in *Medicago truncatula*, unveiling new secrets of a very old enzyme. *Front. Plant Sci.* 2015; 6: 578.

Sharma S, Shinde S, Verslues PE. Functional characterization of an ornithine cyclodeaminase-like protein of Arabidopsis thaliana. *BMC Plant Biol.* 2013; 13: 182.

Sharma S, Verslues PE. Mechanisms independent of abscisic acid (ABA) or proline feedback have a predominant role in transcriptional regulation of proline metabolism during low water potential and stress recovery. *Plant Cell Environ.* 2010; 33: 1838–1851.

Sharma S, Villamor JG, Verslues PE. Essential role of tissue-specific proline synthesis and catabolism in growth and redox balance at low water potential. *Plant Physiol.* 2011; 157: 292–304.

Signorelli S, Arellano JB. Melo, T.B.; Borsani, O; Monza, J. Proline Does Not Quench Singlet Oxygen: Evidence to Reconsider Its Protective Role in Plants. *Plant Physiol. Biochem.* 2013; 64, 80-83.

Signorelli S, Coitiño EL, Borsani O, Monza J. Molecular mechanisms for the reaction between .OH radical and proline: Insights on the role as reactive oxygen species scavenger in plant stress. *J. Phys. Chem.* 2014; 118: 37–47.

Singh TN, Aspinall D, Paleg LG. Proline accumulation and varietal adaptability to drought in barley: A potential metabolic measure of drought resistance. *Nat. New Biol.* 1972; 236: 188–190.

Slafer GA, Araus JL Royo C, Del Moral LFG. Promising eco-physiological traits for genetic improvement of cereal yields in Mediterranean environments. *Ann. Appl. Biol.* 2005; 146: 61–70.

Smirnoff N. The role of active oxygen in the response of plants to water deficit and desiccation. *New Phytol.* 1993; 125: 27–58.

Smirnoff N, Cumbes QJ. Hydroxyl radical scavenging activity of compatible solutes. *Phytochemistry* 1989; 28: 1057–1060.

Stewart CR, Voetberg G, Rapayati PJ. The effects of benzyladenine, and cordycepin on wilting-induced abscisic acid and proline accumulation and abscisic acid- and salt induced proline accumulation in barley leaves. *Plant Physiol.* 1986; 82: 703–707.

Strizhov N, Ábrahám E, Ökresz L, Blickling S, Zilberstein A, Schell J, Koncz C, Szabados L. Differential expression of two P5CS genes controlling proline accumulation during salt-stress requires ABA and is regulated by ABA1, ABI1 and AXR2 in *Arabidopsis*. Plant J. 1997; 12: 557–569.

Su J, Wu R. Stress-inducible synthesis of proline in transgenic rice confers faster growth under stress conditions than that with constitutive synthesis. *Plant Sci.* 2004; 166: 941–948.

Surekha CH, Nirmala Kumari K, Aruna LV, Suneetha G, Arundhati A, Kavi Kishor PB. Expression of the *Vigna aconitifolia* P5CSF129A gene in transgenic pigeon pea enhances proline accumulation and salt tolerance. *Plant Cell Tissue Organ Cult.* 2013; 116: 27–36.

Szabados L, Savouré A. Proline: A multifunctional amino acid. *Trends Plant Sci.* 2010; 15: 89–97.

Székely G, Ábrahám E, Cséplo Á, Rigo G, Zsigmond L, Csiszár J, Ayaydin F, Strizhov N, Jásik J, Schmelzer E, Koncz C, Szabados L: Duplicated P5CS genes of Arabidopsis play distinct roles in stress regulation and developmental control of proline biosynthesis. *Plant J.* 2008; 53: 11–28.

Tabuchi M, Sugiyama K, Ishiyama K, Inoue E, Sato, T, Takahashi H, Yamaya T. Severe reduction in growth rate and grain filling of rice mutants lacking OsGS1;1, a cytosolic glutamine synthetase1;1. *Plant J.* 2005; 42: 641–651.

Taira M, Valtersson U, Burkhardt B, Ludwig RA. *Arabidopsis thaliana* GLN2-encoded glutamine synthetase is dual targeted to leaf mitochondria and chloroplasts. *Plant Cell* 2004; 16: 2048–2058.

Takagi H, Sakai K, Morida K, Nakamori S. Proline accumulation by mutation or disruption of the proline oxidase gene improves resistance to freezing and desiccation stresses in *Saccharomyces cerevisiae*. *FEMS Microbiol. Lett.* 2000; 184: 103–108.

Tanksley SD, McCouch SR. Seed banks and molecular maps: Unlocking genetic potential from the wild. *Science* 1997; 227: 1036–1066.

Teixeira J, Fidalgo F. Salt stress affects glutamine synthetase activity and mRNA accumulation on potato plants in an organ-dependent manner. *Plant Physiol. Biochem.* 2009; 47: 807–813.

Temple SJ, Vance CP, Gantt SJ. Glutamate synthase and nitrogen assimilation. *Trends Plant Sci.* 1998; 3: 51–56.

Thiery L, Leprince AS, Lefebvre D, Ghars MA, Debarbieux E, Savouré A. Phospholipase D is a negative regulator of proline biosynthesis in *Arabidopsis thaliana*. *J. Biol. Chem.* 2004; 279: 14812–14818.

Thomsen HC, Eriksson D, Møller IS, Schjoerring JK. Cytosolic glutamine synthetase: A target for improvement of crop nitrogen use efficiency? *Trends Plant Sci.* 2014; 19: 656–663.

Torres-Franklin ML, Contour-Ansel D, Zuily-Fodil Y, Pham-Thi AT. Molecular cloning of glutathione reductase cDNAs and analysis of GR gene expression in cowpea and common bean leaves during recovery from moderate drought stress. *J. Plant Physiol.* 2008; 165: 514–21.

Trovato M, Maras B, Linhares F, Costantino P. The plant oncogene *rolD* encodes a functional ornithine cyclodeaminase. *Proc. Natl. Acad. Sci. USA* 2001; 98: 13449–13454.

Trovato M, Mattioli R, Costantino P. Multiple roles of proline in plant stress tolerance and development. *Rend. Lincei* 2008; 19: 325–346.

Trovato M, Mauro ML, Costantino P, Altamura MM. Agrobacterium rhizogenes *rolD* gene is developmentally regulated in transgenic tobacco. *Protoplasma* 1997; 197: 111–120.

van Heerden PDR, de Villiers OT. Evaluation of proline accumulation as an indicator of drought tolerance in spring wheat cultivars. *S. Afr. J. Plant Soil* 1996; 13: 17–21.

van Rensburg L, Kruger GHJ, Kruger H. Proline accumulation as drought tolerant criterion: Its relationship to membrane integrity and chloroplast ultrastructure in *Nicotiana tabacum* L. *J. Plant Physiol.* 1993; 141: 188–194.

Veeranagamallaiah G, Chandraobulreddy P, Jyothsnakumari G, Sudhakar C. Glutamine synthetase expression and pyrroline-5-carboxylate reductase activity influence proline accumulation in two cultivars of foxtail millet (*Setaria italica* L.) with differential salt sensitivity. *Environ. Exp. Bot.* 2007; 60: 239–244.

Vélez-Bermúdez, I. C., Schmidt, W. The conundrum of discordant protein and mRNA expression. Are plants special? *Front. Plant Sci.* 2014; 5: 619.

Vendruscolo ECG, Schuster I, Pileggi M, Scapim CA, Molinari HBC, Marur CJ, Vieira LG. Stress-induced synthesis of proline confers tolerance to water deficit in transgenic wheat. *J. Plant Physiol.* 2007; 164: 1367–1376.

Verbruggen N, Hermans C. Proline accumulation in plants: A review. *Amino Acids* 2008; 35: 753–759.

Verbruggen N, Villarroel R, Van Montagu M. Osmoregulation of a pyrroline-5-carboxylate reductase gene in *Arabidopsis*. *Plant Physiol*. 1993; 103: 771–781.

Verslues PE, Sharma S. Proline metabolism and its implications for plant-environment interaction. *Arabidopsis Book* 2010; 8: e0140.

Verslues PE, Sharp RE. Proline accumulation in maize (*Zea mays* L.) primary roots at low water potentials. II. Metabolic source of increased proline deposition in the elongation zone. *Plant Physiol*. 1999; 119: 1349–1360.

Wehner G, Balko C, Humbeck K, Zyprian E, Ordon F. Expression profiling of genes involved in drought stress and leaf senescence in juvenile barley. *BMC Plant Biol*. 2016; 16: 3.

Whan A, Robinson N, Lakshmanan P, Schmidt S, Aitken K. A quantitative genetics approach to nitrogen use efficiency in sugarcane. *Funct. Plant Biol*. 2010; 37: 448–454.

Widodo, Patterson JH, Newbigin Ed, Tester M, Bacic A, Roessner U. Metabolic responses to salt stress of barley (*Hordeum vulgare* L.) cultivars, Sahara and Clipper, which differ in salinity tolerance. *J. Exp. Bot*. 2009; 60: 4089–4103.

Winkel-Shirley B. Biosynthesis of flavonoids and effect of stress. *Curr. Opin. Plant Biol*. 2002; 5: 218–223.

Winter G, Todd CD, Trovato M, Forlani G, Funck D. Physiological implications of arginine metabolism in plants. *Front. Plant Sci*. 2015; 6: 534.

Xin Z, Browse J. *eskimo1* mutants of *Arabidopsis* are constitutively freezing-tolerant. *Proc. Natl. Acad. Sci. USA* 1998; 95: 7799–7804.

Xu J, Yin H, Li X. Protective effects of proline against cadmium toxicity in micropropagated hyperaccumulator, Solanum nigrum L. *Plant Cell Rep*. 2009; 28: 325–333.

Yamchi A, Jazii FR, Mousavi A, Karkhane AA. Proline accumulation in transgenic tobacco as a result of expression of Arabidopsis Δ1-Pyrroline-5-carboxylate synthetase (P5CS) during osmotic stress. *J. Plant Biochem. Biotech*. 2007; 16: 9–15.

Yang H, Tao Y, Zheng Z, Li C, Sweetingham MW, Howieson JG. Application of next-generation sequencing for rapid marker development in molecular plant breeding: A case study on anthracnose disease resistance in *Lupinus angustifolius* L. *BMC Genom*. 2012; 13: 1.

Yian Z, Guo S, Shu S, Sun J, Tezuka T. Effects of proline on photosynthesis, root reactive oxygen species (ROS) metabolism in two melon cultivars (*Cucumis melo* L.) under NaCl stress. *Afr. J. Biotech*. 2011; 10: 18381–18390.

Yoshiba Y, Kiyosue T, Katagiri T, Ueda H, Wada K, Harada Y, Shinozaki K. Correlation between the induction of a gene for Δ1-pyrroline-5-carboxylate synthetase and the accumulation of proline in *Arabidopsis* under osmotic stress. *Plant J*. 1995; 7: 751–760.

Yoshiba Y, Kiyosue T, Nakashima K, Yamaguchi-Shinozaki K, Shinozaki K. Regulation of levels of proline as an osmolyte in plants under water stress. *Plant Cell Physiol*. 1997; 38: 1095–1102.

Yoshiba Y, Nanjo T, Miura S, Yamaguchi-Shinozaki K, Shinozaki K. Stress-responsive and developmental regulation of Δ1-pyrroline-5-carboxylate synthetase 1 (P5CS1) gene expression in *Arabidopsis thaliana*. *BBRC* 1999; 261: 766–772.

Zhang CS, Lu Q, Verma DPS. Removal of feedback inhibition of Δ1-pyrroline-5-carboxylate synthetase, a bifunctional enzyme catalyzing the first two steps of proline biosynthesis in plants. *J. Biol. Chem*. 1995; 270: 20491–20496.

Zhao MG, Chen L, Zhang LL, Zhang WH. Nitric reductase-dependent nitric oxide production is involved in cold acclimation and freezing tolerance in Arabidopsis. *Plant Physiol*. 2009; 151: 755–767.

Zhu B, Su J, Chang M, Verma DPS, Fan YL, Wu R. Overexpression of a Δ^1-pyrroline-5-carboxylate synthetase gene and analysis of tolerance to water- and salt-stress in transgenic rice. *Plant Sci*. 1998; 139: 41–48.

15

Role of Glycinebetaine and Trehalose as Osmoregulators During Abiotic Stress Tolerance in Plants

Mona G. Dawood and Mohamed E. El-Awadi

CONTENTS

Abbreviations

APX	ascorbate peroxidase
BADH	betaine aldehyde dehydrogenase
CAT	catalase
CDH	choline dehydrogenase enzyme
CMO	choline monooxygenase
COD	choline oxidase
CodA	choline oxidase A
GB	glycinebetaine
MDA	malondialdehyde
POX	peroxidase
ROS	reactive oxygen species
SOD	superoxide dismutase
TPP	trehalose 6-phosphate phosphatase
TPS	trehalose 6-phosphate synthase
Tre	trehalose
UDP-glucose	uridine diphosphoglucose

Introduction

Nowadays, the world is marked by environmental pollution, global scarcity of water resources, and increased salinity of soils and waters. Moreover, desertification covers more and more of the world's terrestrial area and is one of the main causes of crop loss (Vinocur and Altman, 2005). Water stresses (mainly drought and salinity) are a major problem in reduced agricultural productivity and are widespread in many regions especially in tropical, semi-arid, and arid regions, and are expected to cause serious salinity of more than 50% of all arable lands by the year 2050 (Ashraf, 1994). Water deficits result from low and erratic rainfall, poor soil water storage, and when the rate of transpiration exceeds water uptake by plants. The cellular water deficits result in the concentration of solutes, loss of turgor, change in cell volume, disruption of water potential gradients, change in membrane integrity, denaturation of proteins, and changes in several physiological and molecular components (Griffiths and Parry, 2002; Lawlor and Cornic,

2002; Parry et al., 2002; Raymond and Smirnoff, 2002; Bartels and Souer, 2003). In addition, the effect of water deficits on plants is complex, and its adverse effects include ion toxicity, nutrient imbalance, and deficiencies. Abiotic stress effects depend on the degree and duration of the stress, developmental stage of the plant, genotypic capacity of species, and environmental interactions.

Plant Adaptation to Abiotic Stresses

Under abiotic stresses, plants have developed a wide variety of highly sophisticated and efficient mechanisms to sense, respond to, and adapt to a wide range of environmental changes. Plants exhibit various responses to these stresses at the molecular, cellular, and whole plant levels (Raamakrishna et al., 2011, 2013). Generally, a range of adaptive responses including morphological (Jaleel et al., 2009), physiological (Harb et al., 2010), and biochemical changes (Ahmadi et al., 2010) may contribute to increased plant tolerance to the stresses. Plants are able to tolerate and survive such adverse conditions by sending signals to change the metabolism for the activation/synthesis of defense mechanisms in different parts of the plant (Siopongco et al., 2009) and by adjusting the membrane system and cell wall architecture, altering the cell cycle and rate of cell division, and also by metabolic tuning (Atkinson and Urwin, 2012).

Plant adaptation to environmental stresses is dependent upon the activation of cascades of molecular networks involved in stress perception, signal transduction, and the expression of specific stress-related genes and metabolites such as the production of stress proteins, up-regulation of antioxidants, and accumulation of compatible solutes (osmolytes).

Proteins involved in the biosynthesis of osmoprotectant compounds, detoxification enzyme systems, proteases, transporters, and chaperones are among the multiple protein functions triggered as the first line of direct protection from stress (Wang et al., 2009; Krasensky et al., 2012).

At the same time, abiotic stresses affect the biosynthesis, concentration, transport, and storage of primary and secondary metabolites. Metabolic adjustments in response to abiotic stressors involve fine adjustments in amino acid, carbohydrate, and amine metabolic pathways. Moreover, there are more than 200,000 known plant secondary metabolites, representing a vast reservoir of diverse functions. When the environment is adverse and plant growth is affected, metabolism of secondary metabolites is profoundly involved in signaling, physiological regulation, and defense responses. This chapter presents an overview on glycinebetaine (GB) and trehalose (Tre) accumulation in plants exposed to abiotic stress as well as their role in the enhancement of plant tolerance to abiotic stress.

Osmolyte Accumulation

Natural accumulation of osmoprotectants varies in plants and it often increases several-fold during exposure to osmotic stress (Rhodes and Hanson, 1993; Bohnert et al., 1995). The accumulation of osmolyte (compatible solutes) in response to various stresses is a widely distributed phenomenon in the plant kingdom and regarded as a basic strategy for the protection and survival of plants under abiotic stress by lowering water potential without decreasing actual water contents and leading to osmotic adjustment (Serraj and Sinclair, 2002; Zhonghua et al., 2007).

Osmotic adjustment is the key adaptation of plants at the cellular level to minimize the effects of water stress–induced damage in crop plants (Blum, 2005) and helps plants under drought in two ways: (1) it helps to maintain leaf turgor to improve stomatal conductance for efficient intake of CO_2 (Kiani et al., 2007) and (2) it promotes the root's ability to uptake more water (Chimenti et al., 2006). In plant cells, osmoprotectants are normally restricted to the cytosol, chloroplasts, and other cytoplasmic compartments. The changes in these metabolites at the cellular level are thought to be associated with protecting cellular function or with maintaining the structure of cellular components. Several genes that are involved in the metabolism of these osmolytes have been found to increase the tolerance of abiotic stress in plants. However, many plants lack the ability to synthesize the special metabolites that are naturally accumulated in the stress-tolerant organisms (Bartels and Sunkar, 2005). Therefore, it was hoped that crop stress tolerance could be improved by introducing one or many genes implicated in the synthesis of a specific osmoprotectant. Hence, metabolic engineering of synthetic pathways to osmoprotectants has been vigorously pursued during the last decade. Understanding the metabolism, transport, and roles of the osmoprotectants during stress is vital in developing plants for stress tolerance. There is an urgent need to identify the signaling components related to the osmolytes' biosynthesis and degradation under stress and at recovery.

Osmoprotectants

These osmolytes (compatible solutes) are highly soluble, carry no net charge, and are non-toxic at high concentrations (McNeil and Nuccio, 1999). They do not cause any disruption to plant metabolism or have any detrimental effects on membranes, enzymes, and other macromolecules, even at higher concentrations (Cechin et al., 2006; Kiani et al., 2007). Compatible solutes are hydrophilic, so they can replace water at the surface of proteins, complex protein structures, and membranes, which explain their action as osmoprotectants (Hasegawa et al., 2000). These solutes possess a multitude of hydroxyl (-OH) groups that facilitate hydrogen bonds with water molecules in the cytoplasm and assist in keeping functional macromolecules in solution. The chemical nature of these small molecular weight organic osmoprotectants is diverse. These compounds fall into several groups: amino acids (e.g. proline), quaternary ammonium compounds (GB), polyols, and sugars (mannitol, D-ononitil, Tre, sucrose, and fructan) (Banuet al., 2010; Krasenskyet al., 2012). It is worthy to mention that the increase of osmoprotectants is achieved either by altering metabolism (increasing biosynthesis and/or decreasing degradation) or by transport (increased uptake and/or decreased export) (Kramerand Morbach, 2004; Takabe et al., 2005).

Physiological Role of Osmoprotectants

These metabolites accumulate under stress and can maintain turgor pressure, protect macromolecules of cells from the damaging effects of reactive oxygen species (ROS) (Farooq et al., 2009), protect cell membrane against oxidation (Ashraf and Foolad, 2007), stabilize proteins, enzymes, and cell structures (Martinezet al., 2004; Bartelsand Sunkar, 2005), and act as antioxidants (Rhodes and Hanson, 1993; Ashraf and Foolad, 2007; Ben Hassine et al., 2008),thermo stabilizers (Akashi et al., 2001; Kaushik and Bhat, 2003), or scavengers that help plants to avoid and/or tolerate stresses (Bartels and Sunkar, 2005). Scavenging of ROS to restore redox metabolism, preservation of cellular turgor by the restitution of osmotic balance, and associated protection and stabilization of proteins and cellular structures are among the multiple protective functions of compatible osmoprotectants during environmental stress (Rathinasabapathi, 2000; Yancey, 2005).

It was found that the protective effects of osmolytes are a result of enhancing the structural stability of native protein. Osmolytes can protect proteins from the unfolding and aggregation induced by extreme environmental stress (Bolen and Baskakov, 2001). The effects of osmolytes on preventing proteins against aggregation are due to the preferential increase of free energy of the activated complex (unfolded protein) which shifts the equilibrium between the native state and the activated complex to favor the native state. The effect of osmolytes on the free energy of the native protein is small. The protection of enzyme activity by osmolytes is due to the enhancement of the structural stability of the whole protein rather than the active site.

There is evidence showing that compatible solutes stabilize enzymes. For example, ribulose-1,5-bisphosphate carboxylase/oxygenase (Rubisco) activity is suppressed by high concentrations of NaCl, but GB and proline can protect this enzyme against inhibition (Solomon et al., 1994; Nomura et al., 1998). Fructans, another compatible solute, have the ability to stabilize phosphatidylcholine liposomes during freeze (Hincha et al., 2000).

High concentrations of compatible solutes can increase cellular osmotic pressure (Delauney and Verma, 1993), and because of their high hydrophilic property, they maintain the turgor pressure and water content of cells, and may replace water molecules around nucleic acids, proteins, and membranes during water shortages (Hoekstra et al., 2001). Cell-water deficits cause an increase in the concentration of ions that destabilize macromolecules. Compatible solutes may prevent interaction between these ions and cellular components by replacing the water molecules around these components, thereby, protecting against destabilization during drought.

Some compatible solutes function as scavengers of hydroxyl radicals (Shen et al., 1997; Akashi et al., 2001). It has been reported that levels of free radicals are decreased in tobacco plants that have been transformed due to the accumulation of more proline (Hong et al., 2000).

Glycinebetaine (GB)

GB (N,N,N-trimethylglycine) is a quaternary ammonium compound, an extremely efficient compatible solute that occurs naturally in a wide variety of plants, animals, and microorganisms. GB is an amphoteric compound that is electrically neutral over a wide range of physiological pH values. It is extremely soluble in

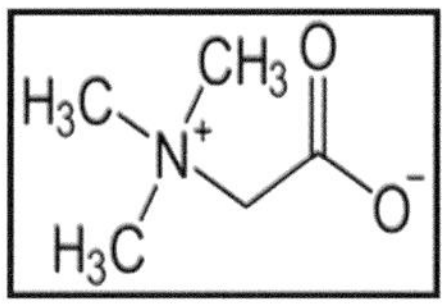

FIGURE 15.1 Chemical structure of GB.

water but includes a non-polar hydrocarbon moiety that consists of three methyl groups. The molecular features of GB allow it to interact with both hydrophilic and hydrophobic domains of macromolecules, such as enzymes and protein complexes (Sakamoto and Murata, 2002) (Figure 15.1).

Accumulation and Translocation of Glycinebetaine

The accumulation of GB depends not only on the type of plant species (Moghaieb et al., 2004), plant varieties (Cha-Um et al., 2007; Ben Hassine et al., 2008), and plant organelles (Zhu et al., 2003), but also on environmental factors, such as salinity, drought, alkaline stress, and extreme temperatures, etc. (Saneoka et al., 2001; Girija et al., 2002; Longstreth et al., 2004; Cui et al., 2008; Zhang et al., 2008).

GB content was markedly different among cotyledons, between roots, stems, leaves, and flowers (including seeds) (Wang et al., 2004). The GB content of these organs were very low during the earlier stages of plant development and increased as the plant developed. Roots accumulated a small amount of GB at all stages of plant development (Rezae et al., 2012).

GB is translocated easily to the developing organs immediately after application (Yang and Lu, 2005). The stability of GB, in terms of the practical utility in crop production, is sufficient and GB remains un-metabolized up to 17 days after application (Makela et al., 1996a). When [14C] GB solution was exogenously applied to the foliage of summer turnip rape (*Brassica rapa* L. ssp. oleifera), [14C] GB was translocated to roots within two hours of application. One day after application, labeled GB was translocated to all plant parts of turnip rape plants and plants were able to translocate foliar-applied GB from their leaves to other organs, indicating that the use of surfactants accelerates the penetration of foliar-applied GB (Makela et al., 1996a).

GB biosynthesis has been reported to increase in most crop species under water deficit conditions (Martinez et al., 2005; Hessine et al., 2009), while in others the natural synthesis of GB is considerably lower than the required level to protect the plants from the stress conditions (Subbarao et al., 2001). Plant species vary in their capacity to synthesize GB (Ashraf and Foolad, 2007). Spinach (*Spinacia oleracea*) and barley (*Hordeum vulgare*) produce high levels of GB in their chloroplasts, which allow the crops to sustain growth despite stressful conditions (Sakamoto and Murata, 2002), while others, such as *Arabidopsis* and tobacco, do not synthesize this compound. Maize is not a significant producer of GB and maize cultivars differ in their capacity to synthesize GB (Rhodes and Rich, 1988; Brunk et al., 1989). Yang and Lu (2006) stated that maize has the ability to absorb and accumulate high levels of exogenous foliar-applied GB. Makela et al. (1996a) reported that it can be translocated from shoot to root immediately after application in summer turnip rape.

The fact that many agronomically important crops, such as rice and potato, are betaine-deficient means that it might be possible to increase their stress tolerance by genetic manipulation and would allow non-accumulators or low-level accumulators to accumulate betaine at protective levels (McCue and Hanson, 1990). There are different methods that could be used to introduce a GB synthetic system into non-GB-accumulating plants to improve their stress tolerance (Chen and Muata, 2008, 2011).

Biosynthesis of Glycinebetaine

In known biological systems, GB is synthesized via two distinct pathways from two distinct substrates: oxidation of choline and N-methylation of glycine, respectively, as shown in Figure 15.2 (Chenet al., 2002; Sakamoto and Murata, 2002). In plants, the enzyme choline monooxygenase (CMO) first converts choline into betaine aldehyde and then an NAD^+ dependent enzyme, betaine aldehyde dehydrogenase (BADH) produces GB. These enzymes are mainly found in chloroplast stroma and their activity is increased in response to stress. In *E. coli*, GB is synthesized by the choline dehydrogenase enzyme (CDH) along with BADH. Whereas. in soil bacterium (*Arthrobacter globiformis*), choline oxidaseA (codA) converts choline into GB and H_2O_2 in a single step.

Transgenic Plant and Glycinebetaine Biosynthesis

Genetic engineering has allowed the introduction of GB-biosynthetic pathways into GB-deficit species (Sakamoto and Murata, 2002). One of the methods was the introduction of the BADH gene, which has been frequently introduced into a variety of plants including tomato (Jia et al., 2002), tobacco (Yang et al., 2005; Ci et al., 2007; Zhou et al., 2008), wheat (Guo et al., 2000), and potato (Zhang et al., 2009a) for enhanced tolerance to salt, drought, or extreme temperatures. The other was COD (choline oxidase), which itself does not exist in higher plants at all. Previous reports showed that the

COD gene was also introduced into *Arabidopsis* (Sulpice et al., 2003; Waditee et al., 2005), tobacco (Huang et al., 2000), rice (Konstantinova et al., 2002; Mohanty et al., 2002; Kathuria et al., 2009), tomato (Park et al., 2007; Goel et al., 2011; Li et al., 2011), maize (Quan et al., 2004), potato (Ahmad et al., 2008), and *Eucalyptus globules* (Matsunaga et al., 2012) to improve their stress tolerance.

GB synthesizing enzymes have been targeted to cytosol, mitochondria, and chloroplast as mentioned by Giri (2011). In transgenic rice with chloroplast-targeted GB accumulation, it was found that protection of photosynthetic machinery against salt and cold stress was better than in plants with cytosolic GB accumulation, even though GB accumulation was 5-fold higher in later plants. Plants with chloroplast-targeted GB synthesis, even though they accumulated the least amount of GB, showed better seedling growth following chilling treatment. These results suggested that GB accumulation in chloroplast is a better strategy for engineering abiotic stress tolerance in plants (Su et al., 2006; Park et al., 2007; Ahmad et al., 2008; Wang et al., 2010a,b; Peel et al., 2010) (Figure 15.3).

GB accumulates at a high concentration (4–40 μmol g^{-1} FW) in naturally accumulating plants like spinach and sugar beet, and acts as an osmoregulator in abiotic stress conditions. However, GB synthesizing genes carrying transgenic plants produced a much reduced amount of GB (0.05–5 μmol g^{-1} FW) (Chen and Murata, 2002; Chen and Murata, 2011; Giri, 2011). Holmström et al. (2000) stated that GB at a concentration of 0.035 μmol g^{-1} fresh weight could impart cold and salinity stress tolerance in transgenic tobacco.

Physiological Role of Glycinebetaine

Exogenous application of GB is known to have beneficial effects on the growth and final yield of a number of crops, particularly those which do not normally accumulate significant amount of GB under abiotic stresses (Park et al., 2006; Ashraf and Foolad, 2007; Hoque et al., 2007a; Raza et al., 2007; Chen and Murata,

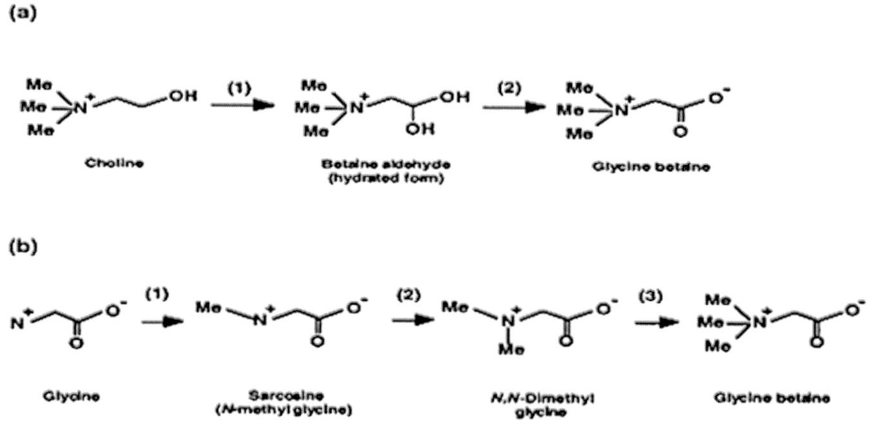

(a)

(b)

FIGURE 15.2 The two main pathways for the synthesis of GB. (a) The synthesis of GB from choline. In this pathway, choline is oxidized to GB via a two-step reaction [indicated by (1) and (2)], in which one or two enzymes are involved. In higher plants, reaction (1) is catalyzed by CMO and reaction (2) is catalyzed by BADH. In mammals and some bacteria (such as *E. coli*), the conversion of choline to GB resembles that in plants, with the exception that CDH replaces CMO in reaction (1). Both reactions are catalyzed by COD in certain bacteria, such as *Arthrobacter globiformi*. (b) The synthesis of GB from glycine. GB is generated as the result of three successive N -methylations of glycine (Sakamoto and Murata, 2002). The enzymes GSMT and SDMT have partially overlapping functions, catalyzing the first two [indicated as (1) and (2)] and last two [indicated as (2) and (3)] methylations, respectively. In higher plants, the reactions are catalyzed by CMO and BADH, both of which are localized in the stroma of chloroplasts.

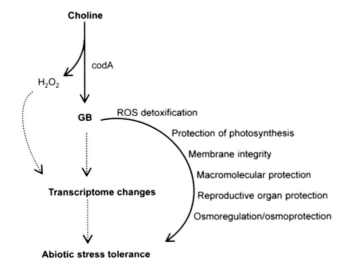

FIGURE 15.3 Model for mechanisms of abiotic stress tolerance in codA expressing plants. Dotted arrows indicate possible involvement of H_2O_2 or GB in transcriptome changes and subsequent stress tolerance. codA-mediated conversion of choline into GB, releases H_2O_2 as byproduct. The H_2O_2 might activate stress-related transcripts in transgenic plants and enhance stress tolerance. Such gene regulation can also be the result of GB accumulation. Given the limited accumulation of GB in transgenic plants, stress-related transcriptome changes might contribute to the observed effects of GB on stress tolerance (Giri, 2011).

2008; Cha-Um and Kirdmanee, 2010; Wang et al., 2010a,b; Mahouachi et al., 2012; Dawood and Sadak, 2014).

Exogenous application of GB has been reported to enhance water stress tolerance in sorghum (Agboma et al., 1997a), sunflower (Hussain et al., 2008), common beans (Xing and Rajashekar, 1999), soybean (Agboma et al., 1997b), and canola (Dawood and Sadak, 2014). On the other hand, there are a few reports suggesting a lack of such positive effects or even apparent negative effects of exogenous GB on plant growth under stress conditions. Foliar application of GB did not affect the yield components or endogenous levels of GB in cotton plants grown under drought stress (Meek et al., 2003) and was not effective on wheat growth (Agboma et al., 1997a). The effectiveness of exogenous application of GB depends on the type of species, the developmental stage of the plant, the application level, and the number of applications, etc. (Ashraf and Foolad, 2005, 2007; Ashraf et al., 2008).

GB affords osmoprotection for plants and protects cell components from harsh conditions by the maintenance of the water potential equilibrium in the cell, which in turn maintains the turgor pressure during water deficit conditions (Makela et al., 1998; Xing and Rajashekar, 1999).

Gibon et al. (1997) showed that application of exogenous GB increased internal GB concentration that caused osmotic adjustment and improved cell growth.

In addition to its role in osmotic adjustment, many studies suggest that GB plays a vital role in the protection of thylakoid membranes and photosynthetic apparatus (Allakhverdiev et al., 2003; Ma et al., 2006; Zhao et al., 2007; Wang et al., 2010a), stabilization of the complex proteins and membranes, protection of transcriptional and translational machineries, and as a molecular chaperone in the refolding of enzymes such

as Rubisco (Sakamoto and Murata, 2000; Chen and Murata, 2002), maintaining enzyme activity, stabilizing membrane structure of photosystem II (PSII) complex, preventing oxidative damage to membranes, and enhancing an antioxidative defense system under osmotic stress (Ma et al., 2006; Raza et al., 2007; Yang et al., 2007; Chen and Murata, 2008; Ben Hassine et al., 2008).

GB protects photosynthetic machinery (Zhao et al., 2001; Allakhverdiev et al., 2003) by preventing photoinhibition (Ma et al., 2006), partially preserving the net photosystem-II efficiency and enhancing the tolerance of the photosynthetic apparatus of plants subjected to various stress conditions (Zhao et al., 2001; Cherian et al., 2006; Demiral and Turkan, 2006; Wang et al., 2010a,b). GB efficiently protects various components of the photosynthetic machinery, such as Rubisco and the oxygen-evolving PSII complex from stress (Murata et al., 2007). It preserves the normal cellular turgor pressure and plays an important role in respiration and photosynthesis. Photosynthesis depends on the application rate of exogenous GB; low concentrations of GB (2–20 mM) lead to enhanced photosynthesis and growth while higher concentrations (>20 mM) lead to reduced stomatal conductance, photosynthetic capacity, and growth (Agboma et al., 1997a; Yang and Lu, 2006). The ameliorating effect of GB on photosynthesis can be attributed to the increase in stomatal conductance augmented by increased turgor pressure in guard cells. Exogenous applications of GB appear to increase internal precursor choline in leaves, prevent chlorophyll degradation, and inhibit the activity of chlorophylase enzyme. Harinasut et al. (1996) showed that exogenous application of GB on rice protected PSII from salt stress damage and increased chlorophyll concentration. Moreover, the ability of GB to act as a molecular chaperon protecting photosystem II against oxidative stress has recently been shown in the halophyte species *Chenopodium quinoa* (Shabala et al., 2012).

GB application protects the photosynthetic machinery when exposed to water deficit conditions in maize (Xing and Rajashekar, 1999), soybean (Agboma et al., 1997b), and canola (Dawood and Sadak, 2014).

The GB maintained the integrity of cell membranes, thylakoid membrane, and electron flow through it (Allakhverdiev et al., 2003; Groppa and Benavides, 2008), and enhanced leaf turgor potential by anchoring enzymes, functional proteins, and lipids against different stressors. Many proteins are prone to aggregation under stress, thereby losing their native structure and activity. The osmolytes have been shown to stabilize protein native structures and antagonize the inhibition of protein biosynthesis (Sakamoto and Murata, 2002; Al-Taweel et al., 2007; Takahashi and Murata, 2008). Two osmolytes, GB and proline, have been shown to destabilize the double-helix DNA and lower the melting temperature of DNA *in vivo*. This would make GB a candidate to regulate gene expression under abiotic stress by activating replication and transcription (Giri, 2011).

Glycinebetaine and Salt Tolerance

Through studies of both plant physiology and genetics, it has been demonstrated that the level of accumulated GB is correlated with the degree of salt tolerance (Saneoka et al., 1995). Exogenous supply of GB increased the salt tolerance of some plants that are

TABLE 15.1

Chlorophyll a (Chl a), Chlorophyll b (Chl b) (mg kg^{-1}), Membrane Permeability (%), and Proline Content (µmol g^{-1} fresh weight) of Maize Plants Grown Under High Saline Conditions in the Presence of GB Applied Foliarly

Treatments	Chl a	Chl b	Membrane Permeability	Proline
C	1276 [a]	898 a	13.23 d	0.48 c
C+GB1	1287 a	908 c	13.45 d	0.49 c
C+GB2	1265 a	902 b	14.45 d	0.42 d
S	898 d	587 d	36.54 a	0.74 b
S+GB1	980 c	698 c	27.68 b	0.86 a
S+GB2	1078 b	756 b	24.38 c	0.89 a
Interaction S×GB	**	*	*	*

Notes: Values followed by different letters in the same column are significantly different at $P \leq 0.05$.

*$P < 0.05$; **$P < 0.01$; C: Control treatment (nutrient solution alone); S: 100 mM NaCl; GB1 and GB2: 25 and 50 mM GB, respectively (Kaya et al., 2013).

unable to accumulate GB (Harinasut et al., 1996; Hayashi et al., 1998). The accumulation of GB *in vivo* enhanced the ability of plants to tolerate high concentrations of salt; studies were made of the physiological consequences of the transgenic engineering of GB synthesis via the over expression of two choline-oxidizing enzymes, namely COD and CDH, proving the property of GB as an osmoregulator (Hayashi et al., 1997, 1998; Holmström et al., 2000). Transgenic *Arabidopsis* plants that produced COD in their chloroplasts not only acquired resistance to high concentrations of NaCl during germination but also were able to tolerate high levels of salt during the subsequent growth of seedlings and mature plants (Hayashi et al., 1997, 1998). In addition, *Brassica juncea* and the Japanese persimmon (*Diospyros kaki*) have been successfully transformed to tolerate salt stress through the introduction and over expression of a gene for COD (Prasad et al., 2000; Gao et al., 2000). GB improves salt tolerance in tobacco BY-2 cells by increasing the activity of enzymes involved in the antioxidant defense system (Hoque et al., 2007b). GB must overcome the deleterious effects of oxidative stress by activating or stabilizing ROS-scavenging enzymes and/or repressing the production of ROS by an unknown mechanism (Chen and Murata, 2008).

Kaya et al. (2013) stated that exogenous application of GB increased chlorophyll and proline levels and maintained membrane permeability by lowering the electrolyte leakage of maize plants under salinity stress (Table 15.1).

GB treatments significantly reduced Na$^+$ concentration and increased calcium (Ca^{2+}) and potassium (K$^+$) concentrations in the maize seedling tissues. 50 mM GB mitigated some of the deleterious effects of salt stress by maintaining membrane permeability and enhancing antioxidant enzyme activities. Hasanuzzaman et al. (2014) suggest that exogenous application of GB increased rice seedling tolerance to salt-induced oxidative damage by up-regulating their antioxidant defense system.

Glycinebetaine and Drought Tolerance

Foliar application of GB could increase its internal content in soybean plants up to 60 µmol/g dry weights, leading to an improvement in photosynthesis activity, nitrogen fixation,

leaf-area development, and seed yield of both well-irrigated and drought-stressed soybean plants (Makela et al., 1996b). Agboma et al. (1997a) reported that 6 kg ha^{-1} exogenously applied GB increased maize grain yield from 4.2 to 5.0 and 3.2 to 4.3 kg ha^{-1} under well-watered and deficit conditions, respectively. In the view of Naidu et al. (1998), cotton crop yield could be improved from 10% to 50% with foliar application of GB, even under mild field stress conditions. Exogenous application of GB in turnip rape plants improved net photosynthesis and reduced photorespiration under drought and salt stress (Makela et al., 1998). Foliar spray of GB mitigated the unpleasant drought stress effects on sunflower achene weight (Iqbal et al., 2005), improved gas exchange characteristics and biomass production in sunflower (Iqbal et al., 2009), and increased dry matter, grain yield, and osmolytes in maize (Zhang et al., 2009b). Under water stress conditions, exogenous applications of 100 mM GB improved the 1,000-grain weight and number of grains per spike through increasing photosynthesis and maintaining leaf turgor potential. Raza et al. (2014) concluded that the application of GB and K (100 mM and 1.5%, respectively) is the best strategy to ameliorate the drought impact on wheat at milking stage and to improve yield production. Miri and Armin (2013) concluded that spraying GB at stem elongation and presowing stages showed great positive effects on alleviating the harmful effects of drought on corn plants. Its usage is affected by the time of application, concentration, and stress severity. In addition, concentration of 150 ppm GB was superior to other concentrations and presowing spraying was better than stem elongation. GB can protect yields during periods of water deficit and high temperatures, even under field conditions with seasonal variability. Foliar application of GB applied weekly or alternated weekly increased maize grain yield by 13% and 6%, respectively, relative to the control. Increased yields are attributed to stress alleviation by GB, which resulted in higher photosynthesis and total plant biomass (Reddy et al., 2014). Shahbaz et al. (2011) mentioned that foliar-applied 50 mM GB was the most effective concentration in enhancing various growth attributes and grain yield, as well as the levels of some key metabolites of wheat cultivars under drought stress conditions. In addition, Aldesuquy (2014) mentioned that exogenous application of GB on wheat plants could counteract the adverse effects of drought by increasing growth vigor of root and shoot, leaf area, concentration of osmoprotectants, retention of pigment content, and keeping out the polysaccharides concentration and/or stabilization of essential proteins. GB application was found to be effective in mitigating the harmful effects of water deficit conditions on the photosynthetic capacity of wheat plants, possibly due to its role in preventing photoinhibition (Ma et al., 2006), protecting the Rubisco enzyme and lipids of the photosynthetic apparatus, and maintaining electron flow through thylakoid membranes, thereby maintaining photosynthetic efficiency (Allakhverdiev et al., 2003; Shahbaz et al., 2011). The GB-treatment reduced the malondialdehyde (MDA) contents which led to cell membrane stability by reducing ROS (Farooq et al., 2010), and substantially ameliorating the impact of drought on membrane integrity and stability in the maize plants (Anjum et al., 2012). This reduction in MDA contents could be attributed to the putative role of osmolytes in alleviating the deleterious effects of stress on the structure of cell membranes and on the activities of different enzymes, as well as their role in reducing the generation of highly destructive free radicals (Smirnoff and Cumbes, 1989; Ali, 2011). GB protects membranes and proteins against the destabilizing effects of dehydration during abiotic stress, and it has an ability to scavenge free radicals and accelerate the activities of antioxidant enzymes (Ashraf and Foolad, 2007). Wang et al. (2010a,b) suggested that GB induced an increase in osmotic adjustments for drought tolerance by improving the antioxidative defense system including antioxidant enzymes in wheat crop. Ma et al. (2006) found that GB-treated wheat plants increased superoxide dismutase (SOD) and ascorbate peroxidase (APX) activities and showed higher photosynthetic activity and water stress tolerance. Furthermore, GB-treated rice plants exhibit increases in SOD, peroxidase (POX) and catalase (CAT) activities, indicating a more efficient quenching of ROS (Farooq et al., 2010). Compatible solutes, such as GB and proline, could enhance the levels of phenolic compound (Ali and Ashraf, 2011) and flavonoids in drought-stressed maize plants (Ali et al., 2013). A positive correlation in seed oil antioxidant activity and different antioxidant compounds under the effect of water deficit conditions and exogenous application of organic osmolytes has already been reported in some earlier studies in maize (Ali et al., 2010; Ali and Ashraf, 2011). In addition, GB had the capacity to scavenge free radicals, which is more important than their role as a mere osmolyte (Ashraf and Foolad, 2007; Ali, 2011). Compatible solutes (GB or proline) improved the oil quantity and quality due to their protective effect on cellular structures during fatty oil biosynthesis and storage, which occurs in liposomes or oleosomes in seeds during the seed filling stage (Taiz and Zeiger, 2006; Ali et al., 2013). Ali (2011) stated that exogenous GB improved the quality of maize oil by decreasing the un-saponifiable matter and increasing oil saponification and iodine values, the measure of oil unsaturation. Dawood and Sadak (2014) stated that 20 mMGB was the most pronounced and effective treatment in alleviating the deleterious effect of moderate or severe drought stress on canola plants as shown in Tables 15.2 and 15.3.

Glycinebetaine and Cold Tolerance

Cold tolerance has been associated with the accumulation of GB in several plant species (Kishitani et al., 1994). In addition, exogenous application of GB has been reported to induce cold tolerance (Chen et al., 2000; Sakamoto et al., 2000; Xing and Rajashekar, 2001). Cold tolerance induced by GB appeared to be related to an elevation in relative leaf water content, chlorophyll, and sucrose, and a decrease in abscisic acid, as well as in active oxygen species (MDA and hydrogen peroxide). GB protects the activities of enzymes and proteins, and stabilizes membranes (Rhodes and Hanson, 1993) and photosynthetic apparatus (Lee et al., 1997; McNeil et al., 1999) under chilling and freezing temperatures.

Chen et al. (2000) reported that GB caused a reduction in lipid peroxidation and membrane damage in chilling-stressed maize plants.

Nayyar et al. (2005) treated chickpea plants with 1 mM GB for 3 consecutive days prior to exposure to cold stress. They found that chilling injury (assessed as electrolyte leakage) was reduced by 63% while cellular respiration increased by 69%. Moreover, application of GB at the bud stage resulted in improvement in flower functioning in terms of pollen germination (*in vitro* and *in vivo*), pollen viability, pollen tube growth, stigma receptivity, and ovule viability. The floral retention, pod set, and pod retention were increased by 47%, 38%, and 23%, respectively, over the control. Treatment with GB at the pod-filling stage caused 30%, 37%, 46%, and 9% enhancement in seed yield/plant, number of seeds/100 pods, single-seeded pods/plant, and individual seed weight, respectively, while the number of double seeded pods was not affected significantly.

Regarding plant genetic engineering, the seeds of transgenic plants were more tolerant to low temperatures during imbibition and germination with higher frequencies and accelerated rates of germination, respectively, than controls (Alia et al., 1998a). The production of biomass by both young and mature plants of

TABLE 15.2

Effect of GB On MDA, Hydrogen Peroxide and Antioxidant Enzymes of Fresh Canola Leaves at Different Levels of Drought Stress (D) (Means of Two Successive Seasons)

Treatments		MDA	H2O2	POX	PPO	SOD	CAT	APX	NR
		μmol/g fresh wt		Unit/min/g fresh wt					(nM NO/g fresh wt/h)
D0	GB0	8.11	2.69	333	18.30	18.35	18.75	0.413	300
	GB1	4.41	1.95	402	26.13	20.82	28.62	0.432	342
	GB2	4.52	1.11	592	33.95	24.30	36.65	0.451	362
	GB3	4.31	0.91	672	41.00	27.20	46.60	0.481	384
D1	GB0	14.11	11.74	474	24.1	23.45	38.90	0.482	280
	GB1	1.90	7.83	558	35.45	24.70	43.80	0.512	292
	GB2	10.77	7.44	693	39.10	30.40	58.35	0.542	311
	GB3	9.36	6.49	720	45.53	35.50	60.60	0.560	335
D2	GB0	23.45	13.98	529	33.30	33.60	49.20	0.528	242
	GB1	6.34	11.44	647	38.75	40.55	59.85	0.552	280
	GB2	13.01	10.30	805	47.15	45.75	64.45	0.560	301
	GB3	7.97	9.36	893	60.62	53.40	72.85	0.580	320
LSD 5%		0.27	0.27	17.29	2.13	2.65	2.87	0.023	10.25

Notes: D0 (95% FC); D1 (75% FC); D2 (50% FC); GB0 (0 mM); GB1 (10 mM); GB2 (15 mM); GB3 (20 mM) (Dawood and Sadak, 2014).

TABLE 15.3

Effect of GB on Seed Yield, Chemical Composition and Antioxidant Activity of the Yielded Canola seeds at Different Levels of Drought Stress (D) (Means of Two Successive Seasons)

Treatments		Seed yield/plant (g)	Oil (%)	Carbohydrate (%)	Protein (%)	Phenolic Content (mg/g)	Tannins (mg/100 g)	Flavonoids (mg/g)	Antioxidant Activity (%)
D0	GB0	3.15	43.92	19.50	18.66	15.95	0.82	0.37	31.0
	GB1	3.32	44.47	20.65	22.45	19.05	0.97	0.44	32.0
	GB2	3.35	45.68	21.45	22.49	19.30	1.20	0.42	34.0
	GB3	4.07	45.01	23.92	22.98	22.30	2.34	0.55	37.0
D1	GB0	2.37	40.20	15.52	21.83	14.70	0.80	0.36	26.0
	GB1	2.84	42.48	19.06	22.31	17.15	1.11	0.41	28.5
	GB2	2.95	43.20	17.50	23.58	17.20	1.22	0.42	30.0
	GB3	3.10	43.09	20.66	24.54	17.60	1.39	0.44	31.0
D2	GB0	1.41	32.99	13.13	22.67	8.45	0.39	0.32	20.5
	GB1	1.48	35.08	14.77	23.87	14.25	0.58	0.33	22.5
	GB2	1.64	36.45	17.66	26.11	15.15	0.94	0.35	23.5
	GB3	2.26	38.05	17.84	27.29	17.90	1.11	0.47	24.8
LSD 5%		0.18	1.07	0.86	0.50	1.09	0.16	0.09	1.51

Notes: D0 (95% FC); D1(75% FC); D2 (50% FC); GB0 (0 mM); GB1 (10 mM); GB2 (15 mM); GB3 (20 mM) (Dawood and Sadak, 2014).

the transgenic strain of *Arabidopsis* was also enhanced at low temperatures, as compared with controls (Hayashi et al., 1997; Alia et al., 1998a). Sakamoto et al. (2000) showed that the transformation of *Arabidopsis* with the *codA* gene for COD enhanced freezing tolerance significantly when the tolerance was evaluated in terms of viability and the retention of intracellular ions after freezing treatments.

Glycinebetaine and Heat Tolerance

Heat stress is often defined as the rise in temperature beyond a threshold level for a period of time sufficient to cause irreversible damage to plant growth and development. In general, a transient elevation in temperature, usually 10–15°C above ambient, is considered heat shock or heat stress. However, heat stress is a complex function of intensity (temperature in degrees), duration, and rate of increase in temperature. The extent to which it occurs in specific climatic zones depends on the probability and period of high temperatures occurring during the day and/or the night (Wahid et al., 2007).

Heat tolerance is generally defined as the ability of the plant to grow and produce economic yield under high temperatures. Early experiments *in vitro* indicated that GB protects some enzymes and protein complexes from heat-induced destabilization. Therefore, it has been postulated that GB increased the resistance to high-temperature stress. More recent experiments showed that transformed *Arabidopsis* that accumulated GB exhibited enhanced tolerance to high temperatures during the imbibition and germination of seeds, as well as during the growth of young seedlings (Alia et al., 1998b). It also seems likely that GB might alleviate the effects of heat shock because the extent of the induction of heat-shock proteins was significantly reduced in these transgenic plants.

Trehalose (Tre)

Tre is a non-reducing disaccharide of glucose that is not easily hydrolyzed by acid, and the glycosidic bond is not cleaved by

a-glucosidase (Bartels and Sunkar, 2005). Tre is highly soluble but chemically un-reactive, making it compatible with cellular metabolism even at high concentrations (Figure 15.4).

Occurrence of Trehalose

Tre is present in significant concentrations in several bacteria and fungi but rare in vascular plants (Fernandez et al., 2010; Lunnet al., 2014). Almost all members of the plant kingdom do not seem to accumulate detectable amounts of Tre. It is present in plants in extremely small quantities, just close to detection level, but it plays a significant role in metabolic processes associated with abiotic stress tolerance (Bae et al., 2005; Aghdasi et al., 2008; Luo et al., 2010). Tre is only present in a few species: seventy Selaginella species (notably *S. lepidophylla*— the "rose of Jericho", a resurrection plant), *Botrychium lunaria* (a fern), and the spermatophytes: *Echinopspersicus, Carex brunescens*, and *Fagus silvatica* (Elbein, 1974). Muller et al. (1995a) found Tre in the grass *Sporobolus stapfianus* and in *Ophioglossum vulgatum*. Moreover, Ghasempour et al. (1998) detected Tre in minute but detectable amounts in desiccation-tolerant plants from several families and in diverse locations: *Borya constricta* (Liliaceae, Western Australia), *Coleochloa setifera* (Cyperceae, Mpumalanga, South Africa), *Eragrostiella nardoides* (Poaceae, Northern India), *Eragrostis nindensis* (Poaceae, Namibia), *Microchloa kunthii* (Poaceae, Zimbabwe), *Tripogon jacqemontii* (Poaceae, India), *Ramonda Myconi* (Gesneriaceae, Pyrenees-Spain), and *Sporobolus pyramidalis* (Poaceae, Northern Cape, South Africa). According to Muller et al. (2001), Tre can be detected in *Arabidopsis thaliana*.

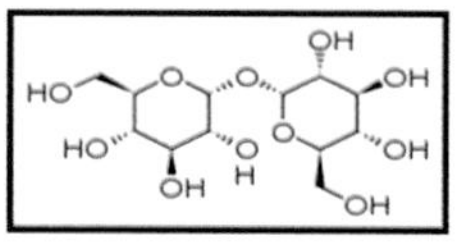

FIGURE 15.4 Structure of Tre.

Trehalose Biosynthesis

The main pathway by which Tre is synthesized in microorganisms is composed of two steps. First, trehalose 6-phosphate synthase (TPS) catalyzes the synthesis of trehalose 6-phosphate (T6P) from glucose-6-phosphate and uridine diphosphoglucose (UDP-glucose), and the second, trehalose 6-phosphate phosphatase (TPP) catalyzes the dephosphorylation of T6P to Tre. In plants, T6P level is highly regulated by enzymes that either directly metabolize it or by trehalase, which breaks down Tre (Figure 15.5).

Other plant species that do not accumulate Tre have been shown to contain trehalase like alfalfa, black Mexican sweet corn, jack pine, wheat, sunflower, white spruce, and canola, as described by Kendall et al. (1990).

Distribution and activity of trehalase have been studied in *Arabidopsis*, where it seems to be present throughout the plant, although higher activity was detected in flower organs, particularly in anthers (Muller et al., 2001). Vogel et al. (2001) showed that the Tre content in *Arabidopsis* increased when trehalase inhibitors were added to growth medium. Plants that accumulate Tre, such as *S. lepidophylla*, have trehalase but with lower activity (Muller et al., 1995a). In soybean, it has been demonstrated that a gene responsible for the synthesis of trehalase is expressed in many tissues, with particular relevance to the nodules (Aeschbacher et al., 1999). Contrasting with the difficulty detecting Tre, trehalase activity is found in most of the higher plants across all major taxonomic groups (Goddijn and van Dun, 1999).

T6P is the intermediate product of Tre biosynthesis (Eastmond et al., 2003). This intermediate has also been reported to be involved in the signaling, through sugar-mediated responses and carbohydrate allocation and metabolism (Kolbe et al., 2005). T6P was found to regulate the utilization of sugars for storage starch synthesis by promoting the reductive activation of AGPase in the plastid (Kolbe et al., 2005). Recently, T6P has been found to act as an inhibitor of SnRK1 (SNF-related kinase), a down regulator of genes involved in biosynthetic reactions that affect the transcript abundances of approximately 1,000 genes in *Arabidopsis* which play a central role in the response to starvation (Baena-Gonzalez et al., 2007).

The study of genetically engineered *Arabidopsis* plants with low and high levels of T6P accumulation showed that this sugar controls carbohydrate utilization and growth (Schluepmann et al., 2004a,b). T6P promoted the redox activation of ADP-glucose pyrophosphorylase and subsequently starch biosynthesis (Lunn et al., 2006). Additionally, Tre induced expression and activity of Suc:Suc-1-fructosyltransferase and Suc:fructan-6-fructosyltransferase, the enzymes involved in fructan biosynthesis, in barley, although glucose or mannitol is required in the presence of Tre to induce fructan accumulation (Wagner et al., 1986; Muller et al., 2000).

Transgenic Plant and Trehalose Biosynthesis

Several attempts have been carried out to increase Tre accumulation through genetic engineering in both model and crop plants using genes of bacterial and yeast origin (Pellny et al., 2004; Wang et al., 2005). Moreover, homologous genes for Tre biosynthesis have been discovered in several wild and crop plants (Goddijnvan Dun, 1999; Zentella et al., 1999), which make them attractive candidates for gene transfer along with genes isolated from prokaryotes. Tre biosynthesis genes from plants such as *A. thaliana* were also used for plant transformation with encouraging results. Genetic engineering of plants with plant-derived genes may probably receive better consumer acceptance than those with genes of microbial origin.

The discovery of homologous genes for Tre biosynthesis in *Selaginella lepidophylla*, *Arabidopsis thaliana*, and several crop plants suggests that the ability to synthesize Tre may be widely distributed in the plant kingdom (Goddijn and van Dun, 1999). The introduction of Tre biosynthetic genes into several plants improved their drought tolerance without having any detrimental effects on plant growth or grain yield (Pilon-Smits et al., 1999; Abebe et al., 2003).

An alternative strategy for the accumulation of Tre seems to be the blocking of trehalase, the enzyme involved in Tre breakdown. In fact, the use of trehalase inhibitors has led to high levels of accumulation of Tre (Goddijn et al., 1997).

Numerous experiments have been carried out to create stress-tolerant plants by introducing microbial Tre biosynthetic genes

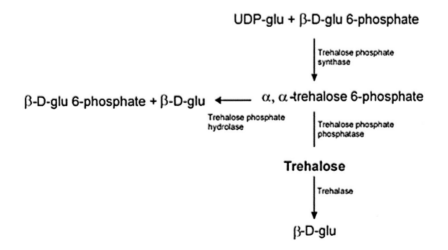

FIGURE 15.5 Tre synthesis and metabolism.

in tobacco (Holmström et al., 1996; Goddijn et al., 1997; Romero et al., 1997; Pilon-Smits et al., 1998; Lee et al., 2003; Han et al., 2005), rice (Garg et al., 2002; Jang et al., 2003), tomato (Cortina and Culianez-Macia, 2005), potato (Goddijn et al., 1997), and *Arabidopsis* (Karim et al., 2007; Miranda et al., 2007).

Tre genes of *Escherichia coli* and *Saccharomyces cerevisiae* have been expressed in tobacco and potatoes, though only in very small quantities. However, it was observed that these small amounts of Tre increase the drought resistance of the transformed plants (Willmitzer et al., 1997). Jang et al. (2003) stated that the transgenic rice expressing the bifunctional enzyme TPSP derived from an in-frame fusion of TPS and TPP from *E. coli*, resulted in the accumulation of Tre and the tolerance of plants to abiotic stress without exerting any negative effects on plant growth.

Holmström et al. (1996) described the Agrobacterium-mediated transformation of tobacco using the TPS gene (TPS1) from *S. cerevisae* under the control of theats1A gene (Rubisco small unit) promoter from *Arabidopsis*. When detached transgenic leaves were air-dried, they had a higher Tre content and slower water loss than the leaves from wild-type plants and remained fresh even after 24 h. Similarly, when plantlets were air-dried, the transgenic lines were able to withstand desiccation better than wild-type plants.

A TPS enzyme was two to three times higher in transgenic plants than in wild type (Almeida et al., 2007a). This enzyme was predominantly present in the vacuoles and in the cell wall, and to a lesser extent in the cytosol.

Tre levels increased significantly in both leaf and seed extracts of transgenic plants (up to 1.076 mg/g fresh weight), and Tre accumulation, in transgenic plants, showed no growth inhibitions or any of the other usually reported phenotypes (lancet-shaped, yellowish leaves, abnormal root system, and sterility) (Jang et al., 2003). Moreover, transgenic plants showed a marked resistance to multiple abiotic stresses, such as drought, salt, and cold (Garget al., 2002; Jang et al., 2003). Wang et al. (2005) reported that transgenic plants of sugarcane (*Saccharum officinarum* L.) showed an enhancement in Tre production and increased shoot and root biomass as compared with non-transgenic plants under drought stress.

The low levels of Tre in transgenic plants can be explained by specific trehalase activity, which degrades Tre; it might be possible to increase Tre accumulation and hence increase abiotic stress tolerance by down regulating trehalase activity (Penna, 2003).

Transgenic tobacco, expressing a bacterial TPP (*OtsB*) gene, showed better growth under drought stress (Pilon-Smits et al., 1995). In addition, transgenic tobacco plants possessing the yeast T6P synthetase gene (*TPS1*) exhibited improved drought tolerance (Romeroet al., 1997). Potato plants expressing genes for the Tre synthesis showed unanticipated morphological changes and modification of structural carbohydrates (Yeo et al., 2000).

In transgenic rice which received the otsA and otsB genes (TPS and TPP in higher plants) from *Escherichia coli*, Tre accumulated 3–10 fold higher when compared to the wild type, and the overproduction of Tre increased tolerance to abiotic stresses (Garg et al., 2002).

These findings demonstrate the feasibility of engineering rice for increased tolerance to abiotic stress, and enhanced productivity through tissue-specific or stress-dependent overproduction of Tre (Garg et al., 2002).

Compared with non-transgenic rice, several independent transgenic lines exhibited sustained plant growth, less photo-oxidative damage, and more favorable mineral balance under salt, drought, and low-temperature stress conditions. The transformants accumulated higher levels of products of Tre compared to many other known transgenic plants (400-fold higher than tobacco co-transformed with *Escherichia coli TPS* and *TPP*, two-fold higher than rice transformed with a bi functional fusion gene (*TPSP*) of TPS and TPP of *E. coli*, and 12-fold higher than tobacco transformed with yeast (*TPS1* gene)) (Djilianovet al., 2005). These results suggest that transgenic plants transformed with the *TSase* gene can accumulate high levels of Tre and have enhanced tolerance to drought and salt.

Transgenic plants that express the TPS and/or TPP genes from microorganisms not only exhibit an increase in drought tolerance; they also show strong developmental alterations. Many of the transgenic plants reported are on dicot plants, which generally produce very low levels of Tre with growth stunting (Holmström et al., 1996; Goddijn et al., 1997; Romero et al., 1997; Pilon-Smits et al., 1998). Interestingly, rice plants appear to be more tolerant to Tre than dicot plants, since the exogenous application of Tre produced no growth inhibition or visible changes in the appearance of rice (Garcia et al., 1997). A gene that encodes a bifunctional fusion enzyme TPSP, produced from the TPS and TPP in *Escherichia coli*, was introduced into rice and produced Tre levels up to 0.1% of the fresh weight without any visible growth inhibition, which was coupled with increased tolerance to drought, salt, and cold stresses (Garg et al., 2002; Jang et al., 2003).

Gaff (1996) reported that tobacco plants transformed for the gene Tps1, which enables transgenic plants to synthesize the disaccharide Tre, exhibit increased drought tolerance compared to the wild type due to improving water retention and desiccation tolerance through stomatal closing, preserve the stability of the chloroplast envelope, and maintain the osmotic potential of the chloroplast at milder drought stress.

Physiological Role of Trehalose

It has been reported that exogenously applied Tre is readily absorbed by leaf tissues and roots and transported to other parts of the plant (Smith and Smith, 1973; Luo et al., 2010).

Tre is considered one of the potential osmoprotectants that stabilize dehydrated enzymes, proteins, and lipid membranes efficiently, as well as protects biological structures from damage during desiccation (Garg et al., 2002).

It can also stabilize proteins and membranes of plants when exposed to stress by replacing hydrogen bonding through polar residues, preventing protein denaturation, and fusion of membranes (Iturriaga et al., 2009). Exogenously applied Tre significantly ameliorated the drought stress–induced inhibitory effects on plant–water relation parameters; this may be due to the role of Tre in plant osmotic adjustment, as reported earlier in different studies, (Rodrıguez et al., 1998; Grennan, 2007; Zeid, 2009) and lowering of the cell osmotic potential, which consequently enabled the plants to absorb water from soil by generating osmotic-driven force to increase the overall plant water contents (Grennan, 2007; Kaya et al., 2007). Besides their possible roles in osmotic adjustment and in stabilizing membranes upon water stress (Lokhande and Suprasanna, 2012), Tre is an essential component of the mechanisms that coordinate metabolism with plant growth adaptation and development (Paul, 2007), and accumulate in response to abiotic stresses (Bae et al., 2005; Aghdasi et al., 2008; Yuanyuan et al., 2009; Luo et al., 2010).

Upon salt stress, Fougere et al. (1991) reported that Tre concentrations in roots and bacteroides of alfalfa (*Medicago sativa* L.) increased significantly (3.5-fold and 4.4-fold respectively), but they were still too low to account for an osmoprotectant role for Tre. Garcia et al. (1997) showed that Tre accumulated in small amounts in rice roots 3 days following salt stress. External applications of low concentration (up to 5 mM) of Tre reduced Na^+ accumulation and growth inhibition, and higher concentration (10 mM) decreased chlorophyll loss in leaf blades and preserved root integrity.

Tre serves as a carbohydrate storage molecule and plays a major role in plant carbohydrate metabolism (Eastmond et al., 2003), acts as an energy or carbon reserve, and protects biomolecules from water stress, heat, cold, and oxygen radicals (Elbein et al., 2003). Tre accumulation influences the alteration of sugar metabolism leading to an osmoprotectant effect under stress (Djilianov et al., 2005). Tre can serve as a transport sugar, similar to the function of sucrose (Muller et al., 1999). Indeed, Garg et al. (2002) reported that the increased Tre accumulation in rice correlates with higher soluble carbohydrate levels and an elevated capacity for photosynthesis under both stressed and unstressed conditions. Over the years, the role of Tre in the acquisition of plant tolerance to stress has been linked to sugar signaling rather than an action as an osmoprotectant alone (Iturriaga et al., 2009). Tre not only sustains the growth of sink tissues, but also affects sugar-sensing systems that regulate the expression, either positively or negatively, of a variety of genes involved in photosynthesis, respiration, and the synthesis and degradation of starch and sucrose (Hare et al., 1998). Moreover, Fructan biosynthesis in barley leaves is affected by Tre (Muller et al., 2001).

Carbohydrate contents of soybean's roots and nodules are influenced by the presence of Tre accumulation (Muller et al., 1995b). Tre feeding to soybean plants affects sucrose metabolism in the root (Muller et al., 1998), as well as the starch biosynthesis in the cotyledons and leaves of *Arabidopsis* seedlings (Wingler et al., 2000). Accumulation of Tre through trehalase inhibition causes a strong reduction of starch and sucrose contents (Muller et al., 2001), thus suggesting a role in carbon allocation (Rolland et al., 2002).

Similarly, Zeid (2009) also reported that exogenous application of Tre significantly mitigated the adverse effects of salt stress on maize plants in terms of biomass production and protecting the photosynthetic apparatus. Paul et al. (2001) and Zeid (2009) suggest that improved photosynthetic performance can be achieved by sugar signaling mechanisms and Tre metabolism through its interaction with sugar-signaling pathways that can enhance photosynthetic capacity.

Other mechanisms by which Tre protect plants from stress include maintaining the activities of different antioxidant enzymes (Muller et al., 1995a) as well as reducing the generation

of highly destructive free radicals (Fillinger et al., 2001; Hincha et al., 2002). Furthermore, Tre has been reported to be very effective in increasing lipid bilayer fluidity (Crowe et al., 1984) and in preserving enzyme stability in different cultivars of bean (Colaco et al., 1992).

Ali and Ashraf (2011) concluded that foliar-applied Tre ameliorated the adverse effects of drought stress on the growth of maize plants by increasing plant photosynthetic activity; adjusting plant water status, and increasing the plant defense mechanism against oxidative stress by enhancing the activities of some key antioxidant enzymes (POD and CAT) coupled with enhanced production of non-enzymatic antioxidants (phenolics and tocopherols) (Avonce et al., 2004; Zeid, 2009).

Zeid (2009) mentioned that 10 mMTre pretreatment alleviated the adverse effects of salinity stress on the metabolic activity of maize seedlings by increasing Hill-reaction activity, photosynthetic pigments, nucleic acids content, and the ratio of K/Na ions, and stabilization of the plasma membranes via decreasing the rate of ion leakage and the rate of lipid peroxidation of the root cells. Moreover, Tre pretreatment of winter wheat protected thylakoid membranes from heat damage, maintained cell membrane integrity, and reduced ROS accumulation (Luo et al., 2010).

Exogenous Tre reduced proline accumulation in two maize cultivars under drought stress and increased biomass production, improved plant–water relations and some key photosynthetic attributes (Ali and Ashraf, 2011).

Rodriguez et al. (1998) determined the presence of Tre in 9 bean (*Phaseolus vulgaris*) cultivars and its correlation with resistance to drought stress. They found that bean cultivars exhibiting high nodule Tre levels and/or a high degree of Tre stimulation in response to water stress, exhibited a high leaf relative water content and the most resistance to water stress.

Tre preserves the integrity and native state of proteins and lipid bilayers (Crowe et al., 1984; Colacoet al., 1992), and maintains the pumps needed to exclude excess NaCl from the photosynthetic organelles. The reducing effect of Tre on lipid peroxidation may indicate that Tre is an antioxidant agent, or at least plays a role in this respect (Azevedo et al., 2006; Stepien and Klobus, 2005; Zeid, 2009). However, other roles have been proposed for Tre and TPS: regulation of plant growth and development, especially the regulation of embryo maturation (Eastmond et al., 2002; Eastmond and Graham, 2003), implicit transition to flowering (Van Dijken et al., 2004), seedling development (Schluepmann et al., 2004a,b), inflorescence architecture (Satoh-Nagasawa et al., 2006), and stress-signaling-mediated glucose and abscisic acid (Avonce et al., 2004).

The Mechanisms by Which Trehalose Protects Biological Molecules

Water replacement, glass transformation, and chemical stability (Almeida et al., 2007b). These three mechanisms are not mutually exclusive and all may contribute to the stabilizing effects of Tre (Richards et al., 2002). The water replacement theory states that macromolecules are normally stabilized by water that forms hydrogen bonds around these molecules. Tre has greater flexibility in the glycosidic linkage between the two D-glucose molecules compared to other disaccharides such as sucrose, allowing Tre to conform to the irregular polar groups of those same macromolecules. According to the "glass transformation" theory, sugars in solution can transform into or maintain a glassy state instead of crystallizing. Tre is unique in that it forms a non-hygroscopic glass, stable at high temperatures or desiccation, holding molecules in a form that allows them to return to a native state and function upon rehydration (Sola-Penna and Meyer-Fernandes, 1998). Tre is one of the most chemically stable sugars, as the 1,1-glycosidic linkage makes it essentially non-reducing. It is highly resistant to hydrolysis and in general chemically inert in its interactions with proteins.

On the other hand, Tre has toxic effects on some plants, namely *Cuscutare flexa*, a parasitic plant (dodder), and *Lemna minor* (duckweed), and this effect was associated with low trehalase activity (Veluthambi et al., 1981). Because these two plants do not undergo any symbiosis with microorganisms that would produce Tre, it has been suggested that trehalase, in symbiotic plants, could play a very important role in the detoxification of Tre produced by symbiotic microorganisms (Muller et al., 1999) and in the prevention of parasitic infections by fungus requiring Tre for their metabolism such as *Magnaporthe grisea* (Foster et al., 2003).

Tre production in dicot plants has resulted in morphological growth defects or altered metabolism (Yeo et al., 2000; Cortina and Culianez-Macia, 2005; Miranda et al., 2007; Suarez et al., 2009).

Seedlings of *Arabidopsis* cultured in MS medium supplemented with 100 mM Tre failed to develop primary leaves and primary roots (Aghdasi et al., 2010). Schluepmann et al. (2004a,b) summarized that exogenously supplied Tre resulted in T6P accumulation, which is a growth inhibitor. Although exogenous Pro and Tre did not clearly show protective roles during the salt stress period, they obviously furnished the plants with an enhanced ability to recover as compared with stressed plants without the osmoprotectants.

Conclusion

Osmolytes (GB and Tre) are non-toxic and accumulate to significant levels without disrupting plant metabolism. Exogenous application of GB and Tre is known to have beneficial effects on the growth and final yield of a number of crops under abiotic stresses. The effectiveness of exogenous application of GB and Tre depends on the type of species, the developmental stage of the plant, the application level, and the number of applications, etc.

GB maintains the turgor pressure, protects thylakoid membranes and photosynthetic apparatus, stabilizes the complex proteins and membranes, protects transcriptional and translational machineries, prevents oxidative damage to membranes, and enhances the antioxidative defense system under stress.

Tre stabilizes the proteins and membranes of plants, prevents protein denaturation, and lowers the cell osmotic potential, which consequently enables the plants to absorb water from the soil by generating osmotic-driven force to increase the overall plant water contents. Tre serves as a carbohydrate storage molecule and plays a major role in plant carbohydrate metabolism by acting as an energy or carbon reserve. Tre not only sustains the growth of sink tissues, but also affects sugar-sensing systems that regulate the expression, either positively or negatively, of a variety of genes involved in photosynthesis, respiration, and the synthesis and degradation of starch and sucrose.

REFERENCES

Abebe T, Guenzi AC, Martin B, Cushman JC. Tolerance of mannitol-accumulating transgenic wheat to water stress and salinity. *Plant Physiol*. 2003; 131: 1748–1755.

Aeschbacher RA, Müller J, Boller T, Wiemkem A. Purification of the trehalase GMTRE1 from soybean nodules and cloning of its cDNA. GMTRE is expressed at low levels in multiple tissues. *Plant Physiol*. 1999; 119: 489–495.

Agboma PC, Jones MGK, Peltonen Sainio P, Rita H, Pehu E. Exogenous glycine betaine enhances grain yield of maize, sorghum and wheat grown under two supplementary watering regimes. *J. Agron. Crop. Sci*. 1997a; 178(1): 29–37.

Agboma PC, Sinclair TR, Jokinen K, Peltonen-Sainio P, Pehu E. An evaluation of the effect of exogenous glycinebetaine on the growth and yield of soybean: Timing of application, watering regimes and cultivars. *Field Crops Res*. 1997b; 54(1): 51–64.

Aghdasi M, Schluepmann H, Smeekens S. Characterization of *Arabidopsis* seedlings growth and development under trehalose feeding. *J. Cell Mol. Res*. 2010; 2: 1–9.

Aghdasi M, Smeekens S, Schluepman H. Microarray analysis of gene expression patterns in *Arabidopsis* seedlings under trehalose, sucrose and sorbitol treatment. *Int. J. Plant Prod*. 2008; 2: 309–320.

Ahmad R, Kim MD, Back KH, Kim HS, Lee, HS, Kwon SY, Murata N, Chung WI, Kwak SS. Stress-induced expression of choline oxidase in potato plant chloroplasts confers enhanced tolerance to oxidative, salt and drought stresses. *Plant Cell Rep*. 2008; 27: 687–698.

Ahmadi A, Emam Y, Pessarakli M. Biochemical changes in maize seedlings exposed to drought stress conditions at different nitrogen levels. *Plant Nut*. 2010; 33(4): 541–556.

Akashi K, Miyake C, Yokota A. Citrulline, a novel compatible solute in drought tolerant wild watermelon leaves, is an efficient hydroxyl radical scavenger. *FEBS Lett*. 2001; 508: 438–442.

Aldesuquy HS. Glycine betaine and salicylic acid induced modification in water relations and productivity of drought wheat plants. *J. Stress Physiol. Biochem*. 2014, 10: 55–73.

Ali Q. Exogenous use of some potential organic osmolytes in enhancing drought tolerance in maize (*Zae mays* L.). A thesis submitted in partial fulfillment of the requirements for the degree of doctor of philosophy. Botany Department, Faculty of Science, University of Agriculture, Faisalabad, Pakistan, 2011; p. 312.

Ali Q, Ashraf M. Exogenously applied glycine betaine enhances seed and seed oil quality of maize (*Zea mays* L.) under water deficit conditions. *Environ. Exp. Bot*. 2011; 71: 249–259.

Ali Q, Ashraf M, Anwar F. Seed composition and seed oil antioxidant activity of maize under water stress. *J. Am. Oil Chem. Soc*. 2010; 87: 1179–1187.

Ali QF, Anwar M, Ashraf Saari N, Perveen R. Ameliorating effects of exogenously applied proline on seed composition, seed oil quality and oil antioxidant activity of maize (*Zea mays* L.) under drought stress. *Int. J. Mol. Sci*. 2013, 14: 818–835.

Alia HH, Chen THH, Murata N. Transformation with a gene for choline oxidase enhances the cold tolerance of *Arabidopsis* during germination and early growth. *Plant Cell Environ.* 1998a; 21: 232–239.

Alia HH, Sakamoto A, Murata N. Enhancement of the tolerance of *Arabidopsis* to high temperatures by genetic engineering of the synthesis of glycine betaine. *Plant J.* 1998b; 16: 155–161.

Allakhverdiev SI, Hayashi H, Nishiyama Y, Ivanov AG, Aliev JA, Klimov VV, Murata N, Carpentier R. Glycinebetaine protects the D1/D2/Cytb559 complex of photosystem II against photo-induced and heat-induced inactivation. *J. Plant Physiol.* 2003; 160: 41–49.

Almeida AM, Cardoso LA, Santos DM, Torné JM, Fevereiro PS. Trehalose and its applications in plant biotechnology. *In Vitro Cell Dev. Biol. Plant* 2007b; 43: 167–177.

Almeida AM, Santos M, Villalobos E, Araújo SS, van Dijck P, Leyman B, Cardoso LA, Santos D, Fevereiro PS, Torné JM. Immunogold localization study of trehalose-6-phosphate synthase in leaf segments of wild type and transgenic tobacco plants expressing the AtTPS1 gene from *Arabidopsis thaliana*. *Protoplasma* 2007a; 230(1): 41–49.

Al-Taweel K, Iwaki T, Yabuta Y, Shigeoka S, Murata N, Wadano A. A bacterial transgene for catalase protects translation of D1 protein during exposure of salt stressed tobacco leaves to strong light. *Plant Physiol.* 2007; 145: 258–265.

Anjum SA, Saleem MF, Wang L, Bilal MF, Saeed A. Protective role of glycine betaine in maize against drought-induced lipid peroxidation by enhancing capacity of antioxidative system. *Aust. J. Crop Sci.* 2012; 6: 576–583.

Ashraf M. Breeding for salinity tolerance in plants. *Crit. Rev. Plant Sci.* 1994; 13: 17–42.

Ashraf M, Athar HR, Harris PJC, Kwon TR. Some prospective strategies for improving crop salt tolerance. *Adv. Agron.* 2008; 97: 45–110.

Ashraf M, Foolad MR. Pre-sowing seed treatment-a shot gun approach to improve germination, plant growth, and crop yield under saline and non-saline conditions. *Adv. Agron.* 2005; 88: 223–271.

Ashraf M, Foolad MR. Roles of glycinebetaine and proline in improving plant abiotic stress resistance. *Environ. Exp. Bot.* 2007; 59: 206–216.

Atkinson NJ, Urwin PE. The interaction of plant biotic and abiotic stresses: From genes to the field. *J. Exp. Bot.* 2012; 63(10): 3523–3543.

Avonce N, Leyman B, Mascorro-Gallardo JO, Van Dijck P, Thevelein JM, Iturriaga G. The Arabidopsis trehalose-6-P synthase AtTPS1 gene is a regulator of glucose, abscisic acid, and stress signaling. *Plant Physiol.* 2004; 136: 3649–3659.

Azevedo NAD, Prisco JT, Eneas FJ, Abreu CEB, Gomes FE. Effect of salt stress on antioxidative enzymes and lipid peroxidation in leaves and roots of salt-tolerant and salt-sensitive maize genotypes. *Environ. Exp. Bot.* 2006; 56: 87–94.

Bae H, Herman E, Bailey B, Bae HJ, Sicher R. Exogenous trehalose alters *Arabidopsis* transcripts involved in cell wall modification, abiotic stress, nitrogen metabolism, and plant defense. *Physiol Plant.* 2005; 125: 114–126.

Baena-Gonzalez E, Rolland F, Thevelein JM, Sheen J. A central integrator of transcription networks in plant stress and energy signaling. *Nature.* 2007; 449: 938–942.

Banu MN, Hoque MA, Watanabe-Sugimoto M, Islam MM, Uraji M, Matsuoka K, Nakamura Y, Murata Y. Proline and glycinebetaine ameliorated NaCl stress via scavenging of hydrogen peroxide and methylglyoxal but not superoxide or nitric oxide in tobacco cultured cells. *Biosci. Biotechnol. Biochem.*, Research Support, Non-U.S. Gov't. 2010; 74(10): 2043–2049.

Bartels D, Souer E. Molecular responses of higher plants to dehydration. In: *Plant Responses to Abiotic Stress. Topics in Current Genetics.* Berlin: Springer, 2003; 4: 9–38.

Bartels D, Sunkar R. Drought and salt tolerance in plants. *Crit. Rev. Plant Sci.* 2005; 24: 23–58.

Ben Hassine A, Ghanem ME, Bouzid S, Lutts S. An inland and a coastal population of the Mediterranean xero-halophyte species *Atriplex halimus* L. differ in their ability to accumulate proline and glycine betaine in response to salinity and water stress. *J. Environ. Bot.* 2008; 59: 1315–1326.

Blum A. Drought resistance, water-use efficiency, and yield potential are they compatible, dissonant, or mutually exclusive? *Aust. J. Agric. Res.* 2005; 56: 1159–1168.

Bohnert HJ, Nelson DE, Jensen RG. Adaptations to environmental stresses. *Plant Cell.* 1995; 7: 1099–1111.

Bolen DW, Baskakov IV. The osmophobic effect: Natural selection of a thermodynamic force in protein folding. *J. Mol. Biol.* 2001; 310: 955–963.

Brunk DG, Rich PJ, Rhodes D. Genotypic variation for glycine betaine among public in breds of maize. *Plant Physiol.* 1989; 91(3): 1122–1125.

Cechin I, Rossi SC, Oliveira VC, Fumis TF. Photosynthetic responses and proline content of mature and young leaves of sunflower plants under water deficit. *Photosynthetica* 2006; 44: 143–146.

Cha-Um S, Kirdmanee C. Effect of glycine betaine on proline, water use, and photosynthetic efficiencies, and growth of rice seedlings under salt stress. *Turk. J. Agric. For.* 2010; 34: 517–527.

Cha-Um S, Supaibulwatana K, Kirdmanee C. Glycine betaine accumulation, physiological characterizations and growth efficiency in salt-tolerant and salt-sensitive lines of indica rice (Oryza *sativa* L. ssp. indica) in response to salt stress. *J. Agron. Crops Sci.* 2007; 193: 157–166.

Chen TH, Murata N. Enhancement of tolerance of abiotic stress by metabolic engineering of betaines and other compatible solutes. *Curr. Opin. Plant Biol.* 2002; 5: 250–257.

Chen TH, Murata N. Glycine betaine: An effective protectant against abiotic stress in plants. *Trends Plant Sci.* 2008; 13: 499–505.

Chen TH, Murata N. Glycine betaine protects plants against a biotic stress: Mechanisms and biotechnological applications. *Plant Cell Environ.* 2011; 34: 1–20.

Chen WP, Li PH, Chen TH. Glycine betaine increases chilling tolerance and reduces chilling-induced lipid peroxidation in *Zea mays* L. *Plant Cell Environ.* 2000; 23: 609–618.

Cherian S, Reddy M, Ferreira R. Transgenic plants with improved dehydration-stress tolerance: Progress and future prospects. *Biol. Plant.* 2006; 50: 481–495.

Chimenti CA, Marcantonio M, Hall AJ. Divergent selection for osmotic adjustment results in improved drought tolerance in maize (*Zea mays*L.) in both early growth and flowering phases. *Field Crop Res.* 2006; 95: 305–315.

Ci H, Zhang N, Wang D. Enhancement of tobacco drought and salt-tolerance with introduced BADH gene. *Acta Agron. Sin.* 2007; 33: 1335–1339.

Colaco C, Sen S, Thangavelu M, Pinder S, Roser B. Extraordinary stability of enzymes dried in trehalose: Simplified molecular biology. *Nature Biotech.* 1992; 10: 1007–1011.

Cortina C, Culiáñez-Macià FA. Tomato abiotic stress enhanced tolerance by trehalose biosynthesis. *Plant Sci.* 2005; 169: 75–82.

Crowe JH, Crowe LM, Chapman D. Preservation of membranes in anhydrobiotic organisms: The role of trehalose. *Science* 1984; 223: 701–703.

Cui XH, Hao FS, Chen H, Chen J, Wang XC. Expression of the *Vicia faba*VfPIP1 gene in *Arabidopsis thaliana* plants improves their drought resistance. *J. Plant Res.* 2008; 121(2): 207–214.

Dawood MG, Sadak M Sh. Physiological role of glycinebetaine in alleviating the deleterious effects of drought stress on canola plants (*Brassica napus* L.). *Middle East J. Agric. Res.* 2014; 3(4): 943–954.

Delauney AJ, Verma DPS. Proline biosynthesis and osmoregulation in plants. *Plant J.* 1993; 4: 215–223.

Demiral T, Türkan I. Exogenous glycine betaine affects growth and proline accumulation and retards senescence in two rice cultivars under NaCl stress. *Environ. Exp. Bot.* 2006; 56: 72–79.

Djilianov D, Georgieva T, Moyankova D, Atanassov A, Shinozaki K, Smeeken SCM, Verma DPS. Murata N. Improved abiotic stress tolerance in plants by accumulation of osmoprotectants-gene transfer approach. *Biotechnol. Biotechnol. Equip.* 2005; 19(sup3): 63–71.

Eastmond PJ, Graham IA. Trehalose metabolism: A regulatory role for trehalose-6-phosphate? *Curr Opin Plant Biol.* 2003; 6: 231–235.

Eastmond PJ, Li Y, Graham IA. Is trehalose-6-phosphate a regulator of sugar metabolism in plants? *J. Exp. Bot.* 2003; 54: 533–537.

Eastmond PJ, Van Dijken A, Spielman M, Kerr A, Tissier A, Dickinson, HG, Jones JD, Smeekens SC, Graham IA. Trehalose-6-phosphate synthase 1, which catalyses the first step in trehalose synthesis, is essential for *Arabidopsis* embryo maturation. *Plant J.* 2002; 29: 225–235.

Elbein AD. The metabolism of a,a- Trehalose. *Adv. Carbohydr. Chem. Biochem.* 1974; 30: 227–256.

Elbein AD, Pan YT, Pastuszak I. Carrol D. New insights on trehalose: A multifunctional molecule. *Glycobiology* 2003; 13: 17R–27R.

Farooq M, Wahid A, Kobayashi N, Fujita D, Basra SMA. Plant drought stress: Effects, mechanisms and management. *Agron. Sustain. Dev.* 2009; 29: 185–212.

Farooq M, Wahid A, Lee DJ, Cheema SA, Aziz T. Comparative time course action of the foliar applied glycinebetaine, salicylic acid, nitrous oxide, brassinosteroids and spermine in improving drought resistance of rice. *J. Agron. Crop Sci.* 2010; 196: 336–345.

Fernandez O, Béthencourt L, Quero A, Sangwan RS, Clement C. Trehalose and plant stress responses: Friend or foe? *Trends Plant Sci.* 2010; 15: 409–417.

Fillinger S, Chaveroche MK, van Dijck P, de Vries R, Ruijter G, Thevelein T, d'Enfert C. Trehalose is required for the acquisition of tolerance to a variety of stresses in the filamentous fungus *Aspergillus nidulans*. *Microbiology* 2001; 147: 1851–1862.

Foster AJ, Jenkinson JM, Talbot NJ. Trehalose synthesis and metabolism are required at different stages of plant infection by *Magnaporthe grisea*. EMBO J. 2003; 22: 225–235.

Fougere F, Le Rudulier D, Streeter JG. Effects of salt stress on amino acid, organic acid, and carbohydrate composition of roots, bacteroids, and cytosol of alfalfa (*Medicago sativa* L.). *Plant Physiol.* 1991; 96: 1228–1236.

Gaff D. Tobacco-plant desiccation tolerance. *Nature (London)* 1996; 382(6591): 502.

Gao M, Sakamoto A, Miura K, Murata N, Sugiura A, Tao R. Transformation of Japanese persimmon (*Diospyroskaki* Thunb.) with a bacterial gene for choline oxidase. *Mol. Breed.* 2000; 6: 501–510.

Garcia AB, Engler J, Iyer S, Gerats T, Van Montagu M, Caplan AB. Effects of osmoprotectants upon NaCl stress in rice. *Plant Physiol.* 1997; 115(1): 159–169.

Garg AK, Kim JK, Owens TG, Ranwala AP, Choi YD, Kochian LV, Wu RJ. Trehalose accumulation in rice plants confers high tolerance levels to different abiotic stresses. *Proc. Nat. Acad. Sci. USA* 2002; 99: 15898–15903.

Ghasempour HR, Gaff DF, Williams R, Gianello RD. Contents of sugars in leaves of drying desiccation tolerant flowering plants, particularly grasses. *Plant Growth Regul.* 1998; 24: 185–191.

Gibon Y, Bessieres MA, Larher F. Is glycine betaine a non-acompatible solute in higher plants that do not accumulate it? *Plant Cell Environ.* 1997; 20: 329–340.

Giri J. Glycinebetaine and abiotic stress tolerance in plants. *Plant Signal. Behav.* 2011; 6(11): 1746–1751.

Girija C, Smith BN, Swamy PM. Interactive effects of sodium chloride and calcium chloride on the accumulation of proline and glycine betaine in peanut (*Arachis hypogaea* L.). *Environ. Exp. Bot.* 2002; 47: 1–10.

Goddijn OJM, Van Dun K. Trehalose metabolism in plants. *Trends Plant Sci.* 1999; 4(8): 315–319.

Goddijn OJ, Verwoerd TC, Voogd E, Krutwagen RW, de Graaf PT, van Dun K, Poels J, Ponstein AS, Damm B, Pen J. Inhibition of trehalase activity enhances trehalose accumulation in transgenic plants. *Plant Physiol.* 1997; 113: 181–190.

Goel D, Singh AK, Yadav V, Babbar SB, Murata N, Bansal KC. Transformation of tomato with a bacterial codA gene enhances tolerance to salt and water stresses. *J. Plant Physiol.* 2011; 168: 1286–1294.

Grennan AK. The role of trehalose biosynthesis in plants. Plant Physiol. 2007; 144: 3–5.

Griffiths H, Parry MAJ. Plant responses to water stress. *Ann. Bot.* 2002; 89: 801–802.

Groppa MD, Benavides MP. Polyamines and abiotic stress: Recent advances. *Amino Acids* 2008; 34: 35–45.

Guo B, Zhang Y, Lee H, Du L, Lee Y, Zhang J, Chen S, Zhu Z. The transformation and expression of wheat with Betaine aldehyde hydrogenase (BADH) gene. *Chin. Bull Bot.* 2000; 42: 279–283.

Han SE, Park SR, Kwon HB, Yi BY, Lee GB, Byun MO. Genetic engineering of drought-resistant tobacco plants by introducing the trehalose phosphorylase (TP) gene from *Pleurotussajorcaju*. *Plant Cell Tissue Organ Cult.* 2005; 82: 151–158.

Harb A, Krishnam A, Ambavaram MMR, Pereira A. Molecular and physiological analysis of drought stress in *Arabidopsis* reveals early responses leading to acclimation in plant growth. *Plant Physiol.* 2010; 154(3): 1254–1271.

Hare PD, Cress WA, Van Staden J. Dissecting the roles of osmolyte accumulation during stress. *Plant Cell Environ.* 1998; 21: 535–553.

Harinasut P, Tsutsui K, Takabe T, Nomura M, Takabe T, Kishitani S. Exogenous glycinebetaine accumulation and increased salt-tolerance in rice seedlings. *Biosci. Biotecha. Biochem.* 1996; 60: 366–368.

Hasanuzzaman M, Alam MM, Rahman A, Hasanuzzaman M, Nahar K, Fujita M. Exogenous proline and glycinebetaine mediated upregulation of antioxidant defense and glyoxalase systems provides better protection against salt-induced oxidative stress in two rice (*Oryza sativa* L.) varieties. *BioMed. Res. Int.* 2014; Article ID 757219, 17pp.

Hasegawa PM, Bressan RA, Zhu JK, Bohnert HJ. Plant cellular and molecular responses to high salinity. *Annu. Rev. Plant Physiol. Plant Mol. Biol.* 2000; 51: 463–499.

Hayashi H, Mustardy AL, Deshnium P, Ida M, Murata N. Transformation of *Arabidopsis thaliana* with the codA gene for choline oxidase: Accumulation of glycine betaine and enhanced tolerance to salt and cold stress. *Plant J.* 1997; 12: 133–142.

Hayashi H, Sakamoto A, Nonaka H., Chen THH, Murata N. Enhanced germination under high-salt conditions of seeds of transgenic *Arabidopsis* with a bacterial gene (codA) for choline oxidase. *J. Plant Res.* 1998; 111: 357–362.

Hessine K, Martínez JP, Gandour M, Albouchi A, Soltani A, Abdelly C. Effect of water stress on growth, osmotic adjustment, cell wall elasticity and water use efficiency in *Spartina alterniflora*. *Environ. Exp. Bot.* 2009; 67: 312–319.

Hincha DK, Hellwege EM, Heyey AG, Crowe JH. Plant fructans stabilize phosphatidylcholine liposomes during freeze-drying. *Eur. J. Biochem.* 2000; 267: 535–540.

Hincha DK, Zuther E, Hellwege EM, Heyer AG. Specific effects of fructo- and gluco-oligosaccharides in the preservation of liposomes during drying. *Glycobiology* 2002; 12: 103–110.

Hoekstra FA, Golovina EA, Buitink J. Mechanisms of plant desiccation tolerance. *Trends Plant Sci.* 2001; 6: 431–438.

Holmström KO, Mantyla E, Wellin B, Mandal A, Palva ET. Drought tolerance in tobacco. *Nature* 1996; 379: 683–684.

Holmström KO, Somersalo S, Mandal A, Palva TE, Welin B. Improved tolerance to salinity and low temperature in transgenic tobacco producing glycine betaine. *J. Exp. Bot.* 2000; 51: 177–185.

Hong Z, Lakkineni K, Zhang Z, Verma DP. Removal of feedback inhibition of A1-pyrroline-5-carboxylase synthetase results in increased proline accumulation and protection of plants from osmotic stress. *Plant Physiol.* 2000; 122: 1129–1136.

Hoque MA, Banu MN, Okuma E, Amako K, Nakamura Y, Shimoishi Y, Murata Y. Exogenous proline and glycine betaine increases NaCl-induced ascorbate glutathione cycle enzyme activities and proline improves salt tolerance more than glycinebetaine in tobacco Bright yellow-2suspension- cultured cells. *J. Plant Physiol.* 2007a; 164(11): 1457–1468.

Hoque MA, Okuma E, Banu MNA, Nakamura Y, Shimoishi Y, Murata Y. Exogenous proline mitigates the detrimental effects of salt stress more than exogenous betaine by increasing antioxidant enzyme activities. *J. Plant Physiol.* 2007b; 164: 553–561.

Huang J, Hirji R, Adam L, Rozwadowski K, Hammerlindl J, Keller W, Selvaraj G. Genetic engineering of glycinebetaine production toward enhancing stress tolerance in plants: Metabolic limitations. *Plant Physiol.* 2000; 122: 747–756.

Hussain M, Malik MA, Farooq M, Ashraf MY, Cheema MA. Improving drought tolerance by exogenous application of glycinebetaine and salicylic acid in sunflower. *J. Agron. Crop Sci.* 2008; 194(3): 193–199.

Iqbal N, Ashraf MY, Ashraf M. Influence of water stress and exogenous glycinebetaine on sunflower achene weight and oil percentage. *Int. J. Sci. Technol.* 2005; 2: 155–160.

Iqbal N, Ashraf MY, Ashraf M. Influence of exogenous glycine betaine on gas exchange and biomass production in sunflower (*Helianthus annuus* L.) under water limited conditions. *J. Agron. Crop Sci.* 2009; 195: 420–426.

Iturriaga G, Suarez R, Nova-Franco B. Trehalose metabolism: From osmoprotection to signalling. *Int. J. Mol. Sci.* 2009; 10: 3793–3810.

Jaleel CA, Manivannan P, Wahid A, Farooq M, Al-Juburi HJ, Somasundaram R, Panneerselvam R. Drought stress in plants: A review on morphological characteristics and pigments composition. *Int. J. Agric. Biol.* 2009; 11(1): 100–105.

Jang IC, Oh SJ, Seo JS, Choi WB, Song SI, Kim CH, Kim YS, Seo HS, Choi YD, Nahm BH, Kim JK. Expression of a bifunctional fusion of the *Escherichia coli* genes for trehalose-6-phosphate synthase and trehalose-6-phosphate phosphatase in transgenic rice plants increases trehalose accumulation and abiotic stress tolerance without stunting growth. *Plant Physiol.* 2003; 131: 516–524.

Jia G, Zhu Z, Chang F, Li Y. Transformation of tomato with the BADH gene from Atriplex improves salt tolerance. *Plant Cell Rep.* 2002; 21: 141–146.

Karim S, Aronsson H, Ericson, H, Pirhonen M, Leyman B, Welin B, Mäntylä E, Palva E, Dijck P, Holmström KO. Improved drought tolerance without undesired side effects in transgenic plants producing trehalose. *Plant Mol. Biol.* 2007; 64: 371–386.

Kathuria H, Giri J, Nataraja KN, Murata N, Udayakumar M, Tyagi A. K. Glycine betaine induced water-stress tolerance in codA-expressing transgenic indica rice is associated with up-regulation of several stress responsive genes. *Plant Biotechnol. J.* 2009; 7: 512–526.

Kaushik JK, Bhat R. Why is trehalose an exceptional protein stabilizer? An analysis of the thermal stability of proteins in the presence of the compatible osmolyte trehalose. *J. Biol. Chem.* 2003; 278: 26458–26465.

Kaya C, Sönmez O, Aydemir S, Dikilitaş M. Mitigation effects of glycinebetaine on oxidative stress and some key growth parameters of maize exposed to salt stress. *Turk. J. Agric. For.* 2013; 37: 188–194.

Kaya C, Tuna AL, Ashraf M., Altunlu H. Improved salt tolerance of melon (*Cucumis melo* L.) by the addition of proline and potassium nitrate. *Environ. Exp. Bot.* 2007; 60: 397–403.

Kendall EJ, Adams RP, Kartha KK. Trehalase activity in plant tissue cultures. *Phytochemistry* 1990; 29: 2525–2528.

Kiani SP, Talia P, Maury P, Grieu P, Heinz R, Perrault A, Nishinakamasu V, Hopp E, Gentzbittel L, Paniego N, Sarrafi A. Genetic analysis of plant water status and osmotic adjustment in recombinant inbred lines of sunflower under two water treatments. *Plant Sci.* 2007: 172: 773–787.

Kishitani S, Watanabe K, Yasuda S, Arakawa K, Takabe T. Accumulation of glycine betaine during cold accumulation and freezing tolerance in leaves of winter and spring barley plants. *Plant Cell Environ.* 1994; 17: 89–95.

Kolbe A, Tiessen A, Schluepmann H, Paul Ulrich MS, Geigenberger P. Trehalose-6-phodphate regulates starch synthesis via post-translation redox activation of ADP-glucose pyrophosphorylase. *Proc. Natl. Acad. Sci. USA* 2005; 102: 11118–11123.

Konstantinova T, Parvanova D, Atanassov A, Djilianov D. Freezing tolerant tobacco, transformed to accumulate osmoprotectants. *Plant Sci.* 2002; 163: 157–164.

Kramer R, Morbach S. BetP of *Corynebacterium glutamicum*, a transporter with three different functions: Betaine transport, osmosensing, and osmoregulation. *Biochim. Biophys. Acta* 2004; 1658, 31–36.

Krasensky J, Jonak C. Drought, salt, and temperature stress-induced metabolic rearrangements and regulatory networks. *J. Exp. Bot.* 2012; 63(4): 1593–1608.

Lawlor DW, Cornic G. Photosynthetic carbon assimilation and associated metabolism in relation to water deficits in higher plants. *Plant Cell Environ.* 2002; 25: 275–294.

Lee CB, Hayashi H, Moon BY. Stabilization by glycinebetaine of photosynthetic oxygen evolution by thylakoid membranes from *Synechococcus* PCC7002. *Mol. Cells* 1997; 7: 296–299.

Lee SB, Kwon HB, Kwon SJ, Park SC, Jeong MJ, Han SE, Byun MO, Daniell H. Accumulation of trehalose within transgenic chloroplasts confers drought tolerance. *Mol Breed.* 2003; 11: 1–13.

Li S, Li F, Wang J, Zhang W, Meng Q, Chen TH, Murata N, Yang X. Glycine betaine enhances the tolerance of tomato plants to high temperature during germination of seeds and growth of seedlings. *Plant Cell Environ.* 2011; 34: 1931–1943.

Lokhande VH, Suprasanna P. Prospects of halophytes in understanding and managing abiotic stress tolerance. In: Ahmad P, Prasad MNV, eds. *Environmental Adaptations and Stress Tolerance of Plants in the Era of Climate Change.* New York: Springer, 2012; 29–56.

Longstreth DJ, Burrow GB, Yu G. Solutes involved in osmotic adjustment to increasing salinity in suspension cells of *Alternanthera philoxeroides* Griseb. *Plant Cell Tissue Organ Cult.* 2004; 78: 225–230.

Lunn JE, Delorge I, Figueroa CM, Dijck VP, Stitt M. Trehalose metabolism in plants. *Plant J.* 2014; 79: 544–567.

Lunn J, Feil R, Hendriks JHM, Gibon Y, Morcuende R, Osuna D, Scheible WR, Carillo P, Hajirezaei MR, Stitt M. Sugar-induced increases in trehalose6-phosphate are correlated with redox activation of ADP glucose pyrophosphorylase and higher rates of starch synthesis in *Arabidopsis thaliana*. *Biochem. J.* 2006; 397: 139–148.

Luo Y, Li F, Wang GP, Yang XH, Wang W. Exogenously-supplied trehalose protects thylakoid membranes of winter wheat from heat-induced damage. *Biol. Plant* 2010; 54: 495–501.

Ma QQ, Wang W, Li YH, D. Li, Zou QQ. Alleviation of photo-inhibition in drought-stressed wheat (*Triticum aestivum*) by foliar-applied glycinebetaine. *J. Plant Physiol.* 2006; 163: 165–175.

Mahouachi J, Argamasilla R, Gomez-Cadenas A. Influence of exogenous glycinebetaine and abscisic acid on papaya in responses to water-deficit stress. *J. Plant Growth Regul.* 2012; 31: 1–10.

Makela P, Jokinen K, Kontturi M, Peltonen-Sainio P, Pehu E, Somersalo S. Foliar application of glycine betaine: A novel product from sugar beet as an approach to increase tomato yield. *Ind. Crops Prod.* 1998; 7: 139–148.

Makela P, Mantila J, Hinkkanen R, Pehu E, Peltonen-Sainio P. Effect of foliar applications of glycine betaine on stress tolerance, growth, and yield of spring cereals and summer turnip rape in Finland. *J. Agron. Crop Sci.* 1996a; 176(4): 223–234.

Makela P, Peltonensainio P, Jokinen K, Pehu E, Setala H, Hinkkanen R, Somersalo S. Uptake and translocation of foliar-applied glycine betaine in crop plants. *Plant Sci.* 1996b; 121: 221–234.

Martinez AT, Speranza M, Ruiz-Dueñas FJ, Ferreira P, Camarero S, Guillén F, Martínez MJ, Gutiérrez A, Del Río JC. Biodegradation of lignocellulosics: Microbial, chemical, and enzymatic aspects of the fungal attack of lignin. *Int. Microbiol.* 2005; 8: 195–204.

Martinez J, Lutts S, Schanck A, Bajji M, Kinet JM. Is osmotic adjustment required for water stress resistance in the Mediterranean shrub *Atriplex halimus* L? *J. Plant Physiol.* 2004; 161: 1041–1051.

Matsunaga E, Nanto K, Oishi M, Ebinuma H, Morishita Y, Sakurai N, Suzuki H, Shibata D. Agrobacterium-mediated transformation of *Eucalyptus globulus* using explants with shoot apex with introduction of bacterial choline oxidase gene to enhance salt tolerance. *Plant Cell Rep.* 2012; 31: 225–235.

McCue K, Hanson A. Drought and salt tolerance: Towards understanding and application. *Trends Biotechnol.* 1990; 8: 358–362.

McNeil SD, Nuccio ML, Hanson AD. Betaines and related osmoprotectants. Targets for metabolic engineering of stress resistance. *Plant Physiol.* 1999; 120: 945–949.

Meek C, Oosterhuis D, Gorham J. Does foliar applied glycine betaine affect endogenous betaine levels and yield in cotton? *Crop Manage.* 2003. Available online. doi:10.1094/CM-2003-0804-02-RS.

Miranda JA, Avonce N, Suárez R, Thevelein JM, Van Dijck P, Iturriaga G. A bifunctional TPS–TPP enzyme from yeast confers tolerance to multiple and extreme abiotic-stress conditions in transgenic *Arabidopsis*. *Planta* 2007; 226: 1411–1414.

Miri HR, Armin M. The interaction effect of drought and exogenous application of glycine betaine on corn (*Zea mays* L.). *Eur. J. Exp. Biol.* 2013; 3(5): 197–206.

Moghaieb REA, Saneoka H, Fujita K. Effect of salinity on osmotic adjustment, glycine betaine accumulation and the betaine aldehyde dehydrogenase gene expression in two halophytic plants, *Salicornia europaea* and *Suaedamaritima*. *Plant Sci.* 2004; 166: 1345–1349.

Mohanty A, Kathuria H, Ferjani A, Sakamoto A, Mohanty P, Murata N, Tyagi A. Transgenics of an elite indica rice variety Pusa Basmati 1 harbouring the codA gene are highly tolerant to salt stress. *Theor. Appl. Genet.* 2002; 106: 51–57.

Muller J, Aeschbacher RA, Sprenger N, Boller T, Wiemken A. Disaccharide-mediated regulation of sucrose:fructan-6-fructosyltransferase, a key enzyme of fructan synthesis in barley leaves. *Plant Physiol.* 2000; 123: 265–274.

Muller J, Aeschbacher R, Wingler A, Boller T, Wiemken A. Trehalose and trehalase in *Arabidopsis*. *Plant Physiol.* 2001; 125: 1086–1093.

Muller J, Boller T, Wiemkem A. Trehalose and trehalase in plants: recent developments. *Plant Sci.* 1995a; 112: 1–9.

Muller J, Boller T, Wiemkem A. Effects of validamycin A, a potent trehalose inhibitor, and phytohormones on trehalose metabolism in roots and root nodules of soybean and cowpea. *Planta* 1995b; 197: 362–368.

Muller J, Boller T, Wiemkem A. Trehalose affects sucrose synthase and invertase activities in soybean (*Glycine max* [L.] Merr) roots. *J. Plant Physiol.* 1998; 153: 255–257.

Muller J, Wiemken R, Aeschbacher R. Trehalose metabolism in sugar sensing and plant development. *Plant Sci.* 1999; 147: 37–47.

Murata N, Takahashi S, Nishiyama Y, Allakhverdiev SI. Photoinhibition of photosystem II under environmental stress. *Biochim. Biophys. Acta (BBA)-Bioenerg.* 2007; 1767: 414–421.

Naidu BP, Cameron DF, Konduri SV. Improving drought tolerance of cotton by glycine betaine application and selection. In: Michalk DL, Pratley JE, eds. *Proceedings of the 9th*

Australian Agronomy Conference; Wagga Wagga, Australia. Serpentine, Australia: Australian Society of Agronomy, 20–23 July 1998; 1–5.

Nayyar H, Bains T, Kumar S. Low temperature induced floral abortion in chickpea: Relationship with abscisic acid and cryoprotectants in reproductive organs. *Environ. Exp. Bot.* 2005; 53: 39–47.

Nomura M, Hibino T, Takabe T, Sugiyama T, Yokota A, Miyake H, Takabe T. Transgenically produced glycinebetaine protects ribulose 1,5-bisphosphate carboxylase/oxygenase from inactivation in Synechococcus sp. PCC7942 under salt stress. *Plant Cell Physiol.* 1998; 39: 425–432.

Park E, Jeknic JZ, Chen TH. Exogenous application of glycine betaine increases chilling tolerance in tomato plants. *Plant Cell Physiol.* 2006; 47: 706–714.

Park EJ, Jeknic Z, Pino MT, Murata N, Chen TH. Glycine betaine accumulation is more effective in chloroplasts than in the cytosol for protecting transgenic tomato plants against abiotic stress. *Plant Cell Environ.* 2007; 30: 994–1005.

Parry MAJ, Andraloje PJ, Khan S, Lea PJ, Keys AJ. Rubisco activity: Effects of drought stress. *Ann. Bot.* 2002; 89: 833–839.

Paul M. Trehalose 6-phosphate. *Curr. Opin. Plant Biol.* 2007; 10: 303–309.

Paul M, Pellny T, Goddijn O. Enhancing photosynthesis with sugar signals. *Trends Plant Sci.* 2001; 6: 197–200.

Peel GJ, Mickelbart MV, Rhodes D. Choline metabolism in glycinebetaine accumulating and non-accumulating near-isogenic lines of *Zea mays* and *Sorghum bicolor. Phytochemistry* 2010; 71: 404–414.

Pellny TK, Ghannoum O, Conroy JP, Schluepmann H, Smeekens S, Andraloje J, Krause KP, Goddijn O, Paul MJ. Genetic modification of photosynthesis with *E. coli* genes for trehalose synthesis. *Plant Biotechnol.* 2004; 2: 71–82.

Penna S. Building stress tolerance through over-producing trehalose in transgenic plants. *Trends Plant Sci.* 2003; 8: 355–357.

Pilon-Smits EAH, Ebskamp MJM, Paul MJ, Jeuken MJW, Weisbeek PJ, Smeekens S CM. Improved performance of transgenic fructan-accumulating tobacco under drought stress. *Plant Physiol.* 1995; 107: 125–130.

Pilon-Smits EAH, Terry N, Sears T, Dun KV. Enhanced drought resistance in fructan-producing sugar beet. *Plant Physiol. Biochem.* 1999; 37: 313–317.

Pilon-Smits EAH, Terry N, Sears T, Kim H, Zayed A, Hwang S, van Dun, K, Voogd E, Verwoerd TC, Krutwagen RH, Goddijn OJ. Trehalose-producing transgenic tobacco plants show improved growth performance under drought stress. *J. Plant Physiol.* 1998; 152: 525–532.

Prasad KVSK, Sharmila P, Kumar PA, Saradhi P. Transformation of *Brassica juncea* (L.) Czern with a bacterial codA gene enhances its tolerance to salt stress. *Mol. Breed.* 2000; 6: 489–499.

Quan R, Shang M, Zhang H, Zhao Y, Zhang J. Engineering of enhanced glycine betaine synthesis improves drought tolerance in maize. *Plant Biotechnol. J.* 2004; 2: 477–486.

Ramakrishna A, Ravishankar GA. Influence of abiotic stress signals on secondary metabolites in plants. *Plant Signal. Behav.* 2011; 6: 1720–1731.

Ramakrishna A, Ravishankar GA. Role of plant metabolites in abiotic stress tolerance under changing climatic conditions with special reference to secondary compounds. In:

Tuteja N, Gill SS, eds. *Climate Change and Abiotic Stress Tolerance.* Weinheim: Wiley-VCH, 2013. ISBN 978-3-527-33491-9.

Rathinasabapathi B. Metabolic engineering for stress tolerance: Installing osmoprotectant synthesis pathways. *Ann. Bot.* 2000; 86(4): 709–716.

Raymond MJ, Smirnoff N. Proline metabolism and transport in maize seedlings at low water potential. *Ann Bot.* 2002; 89: 813–823.

Raza MAS, Saleem MF, Shah GM, Khan IH, Raza A. Exogenous application of glycinebetaine and potassium for improving water relations and grain yield of wheat under drought. *J. Soil Sci. Plant Nutr.* 2014; 14(2): 348–364.

Raza SH, Athar HR, Ashraf M, Hameed A. Glycine betaine induced modulation of antioxidant enzymes activities and ion accumulation in two wheat cultivars differing in salt tolerance. *Environ. Exp Bot.* 2007; 60: 368–376.

Reddy KR, Seepaul R, Henry WB, Gajanayake B, Lokhande S, Brand D. Maize (*Zea mays* L.) yield and aflatoxin accumulation responses to exogenous glycinebetaine application. *Int. J. Plant Prod.* 2014; 8(2): 271–290.

Rezaei MA, Kaviani B, Masouleh AK. The effect of exogenous glycine betaine on yield of soybean [*Glycine max* (L.) Merr.] in two contrasting cultivars Pershing and DPX under soil salinity stress. *Plant Omics J.* 2012; 5(2): 87–93.

Rhodes D, Hanson AD. Quaternary ammonium and tertiary sulfonium compounds in higher plants. *Annu. Rev. Plant Physiol. Plant Mol. Biol.* 1993; 44: 357–384.

Rhodes D, Rich PJ. Preliminary genetic studies of the phenotype of betaine deficiency in *Zea mays* L. *Plant Physiol.* 1988; 88(1): 102–108.

Richards AB, Krakowka S, Dexter LB, Schmid H. Wolterbeek APM, Waalkens-Berendsen DH, Shigoyuki M, Kurimoto AM. Trehalose: A review of properties, history of use and human tolerance, and results of multiple safety studies. *Food Chem. Toxicol.* 2002; 40: 871–898.

Rodriguez FR, Mellor RB, Arias C, Cabriales PJJ. The accumulation of trehalose in nodules of several cultivars of common bean (*Phaseolus vulgaris*) and its correlation with resistance to drought stress. *Physiol Plant.* 1998; 102: 353–359.

Rolland F, Moore B, Heen SJ. Sugar sensing and signaling in plants. *Plant Cell* 2002; 14(Suppl. 1): S185–S205.

Romero C, Belles JM, Vaya JL, Serrano R, Culianez-Macia FA. Expression of the yeast trehalose-6-phosphate synthase gene in transgenic tobacco plants: Pleiotropic phenotypes include drought tolerance. *Planta* 1997; 201: 293–297.

Sakamoto A, Alia RV, Chen THH, Murata N. Transformation of *Arabidopsis* with the codA gene for choline oxidase enhances freezing tolerance of plants. *Plant J.* 2000; 22(5): 449–453.

Sakamoto A, Murata N. Genetic engineering of glycine-betaine synthesis in plants: Current status and implications for enhancement of stress tolerance. *J. Exp. Bot.* 2000; 51: 81–88.

Sakamoto A, Murata N. The role of glycine betaine in the protection of plants from stress: Clues from transgenic plants. *Plant Cell Environ.* 2002; 25: 163–171.

Saneoka H, Ishiguro S, Moghaieb RE. Effect of salinity and abscisic acid on accumulation of glycine betaine and betaine aldehyde dehydrogenase mRNA in sorghum leaves (*Sorghum bicolor*). *J. Plant Physiol.* 2001; 158: 853–859.

Saneoka H, Nagasaka C, Hahn DT, Yang WJ, Premachandra GS, Joly RJ, Rhodes D. Salt tolerance of glycinebetaine-deficient and -containing maize lines. *Plant Physiol.* 1995; 107: 631–638.

Satoh-Nagasawa N, Nagasawa N, Malcomber S, Sakai H, Jackson D. A trehalose metabolic enzyme controls inflorescence architecture in maize. *Nature* 2006; 441: 227–230.

Schluepmann H, Pellny T, van Dikken A, Smeekens S, Paul M. Trehalose 6-phosphate is indispensable for carbohydrate utilization and growth in *Arabidopsis thaliana. Proc. Natl. Acad. Sci. USA.* 2004a; 100(11): 6849–6854.

Schluepmann H, Van Dikken A, Aghdasi M, Wobbes B, Paul M, Smeekens S. Trehalose mediated growth inhibition of *Arabidopsis* seedlings is due to trehalose-6-phosphate accumulation. *Plant Physiol.* 2004b; 135: 879–890.

Serraj R, Sinclair TR. Osmolyte accumulation: Can it really help increase crop yield under drought conditions? *Plant Cell Environ.* 2002; 25: 333–341.

Shabala L, Mackay A, Tian Y, Jacobsen SE, Zhou D, Shabala S. Oxidative stress protection and stomatal patterning as components of salinity tolerance mechanism in quinoa (*Chenopodium quinoa*). *Physiol. Plant* 2012; 146: 26–38.

Shahbaz M, Masood Y, Parveen S, Ashraf M. Is foliar applied glycinebetaine effective in mitigating the adverse effects of drought stress on wheat (*Triticum aestivum* L.)? *J. Appl. Bot. Food Tech.* 2011; 84: 192–199.

Shen B, Jensen RG, Bohnert HJ. Mannitol protects against oxidation by hydroxyl radicals. *Plant Physiol.* 1997; 115: 527–532.

Siopongco J, Sekiya K, Yamauchi A, Egdane J, Ismail AM, Wade LJ. Stomatal responses in rainfed lowland rice to partial soil drying; comparison of two lines. *Plant Prod. Sci.* 2009; 12: 17–28.

Smirnoff N, Cumbes QJ. Hydroxyl radical scavenging activity of compatible solutes. *Phytochemistry* 1989; 28: 1057–1060.

Smith SE, Smith FA. Uptake of glucose, trehalose and mannitol by leaf slices of the orchid *Bletilla hyacinthina. New Phytol.* 1973; 72: 957–964.

Sola-Penna M, Meyer-Fernandes JR. Stabilization against thermal inactivation promoted by sugars on enzyme structure and function: Why is trehalose more effective than other sugars? *Arch. Biochem. Biophys.* 1998; 360: 10–14.

Solomon A, Beer S, Waisel Y, Jones GP, Paleg LG. Effects of NaCl on the carboxylating activity of Rubisco from *Tamarix jordanis* in the presence and absence of proline- water stress related compatible solutes. *Plant Physiol.* 1994; 90: 198–204.

Stepien P, Klobus G. Antioxidant defense in the leaves of C3 and C4 plants under salinity stress. *Physiol. Plant* 2005; 125: 31–40.

Su J, Hirji R, Zhang L, He C, Selvaraj G, Wu R. Evaluation of the stress-inducible production of choline oxidase in transgenic rice as a strategy for producing the stress-protectant glycine betaine. *J. Exp. Bot.* 2006; 57: 1129–1135.

Suarez R, Calderon C, Iturriaga G. Improved tolerance to multiple abiotic stresses in transgenic alfalfa accumulating trehalose. *Crop Sci.* 2009; 49: 1791–1799.

Subbarao GV, Wheeler RM, Levine LH, Stutte GW. Glycine betaine accumulation, ionic and water relations of red-beet at contrasting levels of sodium supply. *J. Plant Physiol.* 2001; 158: 767–776.

Sulpice R, Tsukaya H, Nonaka H, Mustardy L, Chen T, Murata N. Enhanced formation of flowers in salt stressed *Arabidopsis* after genetic engineering of the synthesis of glycine betaine. *Plant J.* 2003; 36: 165–176.

Taize L, Zeiger E. *Plant Physiology*, 4th ed. Sunderland, MA: Sinauer Associates, Inc., 2006.

Takabe T, Rai V, Hibino T. Osmotic stresses; Metabolic engineering of glycinebetaine. Section V: P.135. In: Rai AK, Takabe T, eds. *Abiotic Stress Tolerance in Plants Toward the Improvement of Global Environment* and *Food.* Dordrecht, the Netherlands: Springer, 2005.

Takahashi S, Murata N. How do environmental stresses accelerate photoinhibition? *Trends Plant Sci.* 2008; 13: 178–182.

Van Dijken A, Schluepmann H, Smeekens SCM. *Arabidopsis* trehalose-6-phosphate synthase 1 is essential for normal vegetative growth and transition to flowering. *Plant Physiol.* 2004; 135: 969–977.

Veluthambi K, Mahadevan S, Maheshwari R. Trehalose toxicity in *Cuscuta reflexa. Plant Physiol.* 1981; 68: 1369–1374.

Vinocur B, Altman A. Recent advances in engineering plant tolerance to abiotic stress: Achievements and limitations. *Curr. Opin. Biotechnol.* 2005; 16(2): 123–132.

Vogel G, Fiehn O, Jean-Richard-dit-Bressel L, Boller T, Wiekem A, Aeschbacher RA, Wingler A. Trehalose metabolism in *Arabidopsis:* Occurrence of trehalose and molecular cloning and characterization of trehalose-6-phosphate synthase homologues. *J. Exp. Bot.* 2001; 52: 1817–1826.

Waditee R, Bhuiyan M, Nazmul H, Rai V, Aoki K, Tanaka Y, Hibino T, Suzuki S, Takano J, Jagendorf A. Genes for direct methylation of glycine provide high levels of glycinebetaine and abiotic-stress tolerance in *Synechococcus* and *Arabidopsis. Proc Natl Acad Sci.* 2005; 102: 1318.

Wagner W, Wiemken A, Matile P. Regulation of fructan metabolism in leaves of barley (*Hordeumvulgare* L. Cv. Gerbel). *Plant Physiol.* 1986; 81: 444–447.

Wahid A, Gelani S, Ashraf M, Foolad MR. Heat tolerance in plants: An overview. *Environ. Exp. Bot.* 2007; 61: 199–223.

Wang GP, Hui Z, Li F, Zhao MR, Zhang J, Wang W. Improvement of heat and drought on photosynthetic tolerance in wheat by accumulation of glycine betaine. *Plant Biotech. Rep.* 2010a; 4: 212–222.

Wang GP, Zhang XY, Li F, Luo Y, Wang W. Over accumulation of glycine betaine enhances tolerance to drought and heat stress in wheat leaves in the protection of photosynthesis. *Photosynthetica* 2010b; 48: 117–126.

Wang YM, Meng YL, Nii N. Changes in glycine betaine and related enzyme contents in *Amaranthus tricolor* under salt stress. *J. Plant Physiol. Molbiol.* 2004; 30(5): 496–502.

Wang Z, Zhang S, Yang B, Li Y. Trehalose synthase gene transfer mediated by *Agrobacterium tumefaciens* enhances resistance to osmotic stress in sugarcane. *Sugar Tech.* 2005; 7: 49–54.

Willmitzer L, Heyer A, Kossmann J. Production of new or modified carbohydrates in transgenic plants. *Proceedings of the 5th Symposium on Renewable Resources*, 1997.

Wingler A, Fritzius T, Wiemken A, Boller T, Aeschbacher R. Trehalose induces the ADP-glucose pyrophosphorylase gene, ApL3, and starch synthesis in *Arabidopsis. Plant Physiol.* 2000; 124: 105–114.

Xing W, Rajashekar CB. Alleviation of water stress in beans by exogenous glycine betaine. *Plant Sci.* 1999; 148: 185–195.

Xing W, Rajashekar CB. Glycine betaine involvement in freezing tolerance and water stress in *Arabidopsis thaliana. Environ. Exp. Bot.* 2001; 46: 21–28.

Yancey PH. Organic osmolytes as compatible, metabolic and counteracting cytoprotectants in high osmolarity and other stresses. *J. Exp. Biol.* 2005; 208(Pt 15): 2819–2830.

Yang X, Lu C. Photosynthesis is improved by exogenous glycine betaine in salt-stressed maize plants. *Physiol Plant.* 2005; 124(3): 343–352.

Yang X, Lu C. Effects of exogenous glycine betaine on growth, CO_2 assimilation, and photosystem II photochemistry of maize plants. *Physiol Plant.* 2006; 127(4): 593–602.

Yang XH, Liang Z, Lu CM. Genetic engineering of the biosynthesis of glycine betaine enhances photosynthesis against high temperature stress in transgenic tobacco plants. *Plant Physiol.* 2005; 138: 2299–2309.

Yang XH, Wen XG, Gong HM, Lu QT, Yang ZP, Tang YL, Liang Z, Lu CM. Genetic engineering of the biosynthesis of glycine betaine enhances thermo tolerance of photosystem II in tobacco plants. *Planta* 2007; 225: 719–733.

Yeo ET, Kwon HB, Han SE, Lee JT, Ryu JC, Byu MO. Genetic engineering of drought resistant potato plants by introduction of the trehalose-6-phosphate synthase (TPS1) gene from *Saccharomyces cerevisiae. Mol Cell.* 2000; 30: 263–268.

Yuanyuan M, Yali Z, Jiang L, Hongbo S. Roles of plant soluble sugars and their responses to plant cold stress. *Afr. J. Biotech.* 2009; 8: 2004–2010.

Zeid IM. Trehalose as osmoprotectant for maize under salinity-induced stress. *Res. J. Agric. Biol. Sci.* 2009; 5(5): 613–622.

Zentella R, Mascorro-Gallardo JO, Van Dijck P, Folch-Mallol J, Bonini B, Van Vaeck C, Gaxiola R, Covarrubias AA, Nieto-Sotelo J, Thevelein JM, Iturriaga G. A *Selaginella lepidophylla* trehalose-6-phosphate synthase complements growth and stress-tolerance defects in a yeast tps1 mutant. *Plant Physiol.* 1999; 119(4): 1473–1482.

Zhang H, Zhang S, Yang J, Zhang J, Wang Z. Postanthesis moderate wetting drying improves both quality and quantity of rice yield. *Agron. J.* 2008; 100: 726–734.

Zhang LX, Li SX, Liang ZS. Differential plant growth and osmotic effects of two maize (*Zea mays* L.) cultivars to exogenous glycinebetaine application under drought stress. *Plant Growth Regul.* 2009b; 8: 297–305.

Zhang N, Ci H, Li L, Yang T, Zhang C, Wang D. Enhanced drought tolerance of potato with introduced BADH gene. *Acta Agron. Sin.* 2009a; 35: 1146–1150.

Zhao XX, Ma QQ, Liang C, Fang Y, Wang YQ, Wang W. Effect of Glycine betaine on function of thylakoid membranes in wheat flag leaves under drought stress. *Biol. Plant* 2007; 51: 584–588.

Zhao Z, Chen G, Zhang C. Interaction between reactive oxygen species and nitric oxide in drought-induced abscisic acid synthesis in root tips of wheat seedlings. *Aust. J. Plant Physiol.* 2001; 28: 1055–1061.

Zhonghua Ch, Tracey A, Cuin M, Amanda T, Bodapati P, Shabala S. Compatible solute accumulation and stress- mitigating effects in barley genotypes contrasting in their salt tolerance. *J. Exp. Bot.* 2007; 58(15/16): 4245–4255.

Zhou S, Chen X, Zhang X, Li Y. Improved salt tolerance in tobacco plants by co-transformation of a betaine synthesis gene BADH and a vacuolar Na+/H+ antiporter gene SeNHX1. *Biotechnol. Lett.* 2008; 30: 369–376.

Zhu MY, Ahn SJ, Matsumoto H. Inhibition of growth and development of root border cells by Al. *Physiol. Plant.* 2003; 117: 359–367.

16

Polyamine Metabolism and Abiotic Stress Tolerance in Plants

Rubén Alcázar and Antonio F. Tiburcio

CONTENTS

Abbreviations

1,3-DAP	1,3-diaminopropane
AAC (s/ox)	1-amino-cyclopropane-1-carboxylic acid (synthase/oxidase)
ABA	abscisic acid
ADC	arginine decarboxylase
AI	agmatine; iminohydrolase
CuAO	copper amine oxidase
CPA	N-carbamoylputrescine amidohydrolase
DAO	diamine oxidase
GABA	γ-aminobutyric acid
ODC	ornithine decarboxylase
PAO	polyamine oxidase
Polyamines	PAs
Putrescine	Put
SPDS	spermidine synthase
Spermidine	Spd
SPMS	spermine synthase
Spermine	Spm
SAM	S-adenosylmethionine

Introduction

Polyamines (PAs) have been known to exist since Antonie van Leeuwenhoek first made an observation of a crystalline structure he described as glittering and translucent during the late 16th century (van Leeuwenhoek 1678). However, the structure of what he had seen was not deciphered until 1924, when Otto Rosenheim managed to synthesize the three major PAs, the diamine Putrescine (Put) $[NH_2(CH_2)_4NH_2]$, the triamine Spermidine (Spd) $[NH_2(CH_2)_3NH(CH_2)_4NH_2]$ and the tetraamine Spermine (Spm) $[NH_2(CH_2)_3NH(CH_2)_4NH(CH_2)_3NH_2]$, or the Böttcher crystals as they were called (Rosenheim 1924). Rosenheim's discovery was based on samples from human seminal fluid, and no one at the time could possibly know of the wide distribution of these amines in the branches of life (Bachrach 2010). Since then, the number of PA structures identified has increased, including the identification of isomers and conjugated forms.

Today we know that PAs can be found in almost every living organism analyzed (Pegg and Michael 2010). Although the small aliphatic PA molecules are known to chelate with cationic transition metals (Paoletti 1984) and conjugate with hydroxycinnamic acids (Martin-Tanguy 1997) and proteins, free PAs are fully protonated at pH 7.4 and interact electrostatically with negatively charged molecules. Interactions with DNA, RNA and proteins (Aikens et al. 1983; Venkiteswaran et al. 2005; Alcázar et al. 2010a) and associations with negatively charged cell wall constituents (Geny et al. 1997) have been found. How many protons a PA can use in reactions with other molecules depends partly on its charge and the surrounding pH (Aikens et al. 1983). However, not only is the number of charges relevant for PA functions but so is their distribution in the molecule. PAs are important for plant growth and development, senescence and survival (Tiburcio et al. 2014). Some of the processes known to be involved include cell division and differentiation, programmed cell death, pollen development, flowering, fruit ripening, seed development and germination, leaf abscission, and grain yield, as well as protection against different types of abiotic and biotic stress (Alcázar et al. 2010a; Moschou and Roubelakis-Angelakis 2013; Tiburcio et al. 2014; Shabala and Pottosin 2014; Anwar et al. 2015; Liu et al. 2015; Ramakrishna and Ravishankar 2011, 2013).

Through years of research, it has become clear that it is of great importance to study PAs in relation to crop stress tolerance. With increasing environmental fluctuations and a rapidly expanding human population that needs to be fed, the question of how to increase crop quality and production has become one of the greatest challenges of the 21st century (FAO 2009; Manning 2015). This review will examine PA metabolism in relation to some of the major abiotic stress factors that affect crop production in agriculture today, notably high salinity and water stress. Putative mechanisms of action for PA mediated stress tolerance will be summarized, including recent findings on PA crosstalk with the ABA signalling pathway during environmental stress.

Polyamine Metabolism

The biosynthesis and breakdown of PAs have been extensively studied and it is well understood in plants (Figure 16.1) (Tavladoraki et al. 2011; Marco et al. 2012). Put has the simplest molecular structure and can be synthesized from ornithine, a nitrogen rich product from the urea cycle, by the action of ornithine decarboxylase (ODC). Even though the ODC pathway is the most common pathway in plants, animals and fungi, no *ODC* gene has been discovered in the plant model species *Arabidopsis thaliana* (Hanfrey et al. 2001). A second biosynthesis pathway exists in plants where arginine decarboxylase (ADC) reacts with arginine to produce agmatine. One of the most recently characterized enzymes in plant PA biosynthesis, agmatine imino-hydrolase (AIH) (Janowitz et al. 2003), subsequently converts agmatine into N-carbamoylputrescine. Putrescine is subsequently produced by the hydrolytic activity of N-carbamoylputrescine amidohydrolase (CPA) that form Put and the by-products carbon dioxide and ammonia. Put acts as a precursor of Spd and Spm, which are produced by the addition of aminopropyl groups

(Marco et al. 2012). This means that homeostasis and the production of other PAs is rate limited by the levels of free Put in the cell (Casero and Marton 2007). The aminopropyl groups needed to convert Put to Spd and Spm are donated by decarboxylated S-adenosylmethionine (dcSAM), synthesized through decarboxylation of SAM by SAM decarboxylase (SAMDC). The addition of aminopropyl group(s) to Put or Spd is then mediated by Spd synthase (SPDS) and Spm synthase (SPMS) activities, to yield Spd and Spm, respectively. The decarboxylated SAM is the only known molecule that can donate the aminopropyl group and SAM is also a precursor for the production of the important phytohormone ethylene. These are two reasons why this step constitutes a second rate limiting step in PA biosynthesis (Ge et al. 2006). Other mechanisms of cellular PA regulation depend on conjugation to hydroxycinnamic acids (Martin-Tanguy 1997) and transport. Transcriptional regulation of PA biosynthetic genes by upstream ORFs has been reported. A gene encoding a heptapeptide uORF that regulate translation of *ADC* in carnation (*Dianthus caryophyllus* L.) and a pentapeptide of an *ODC* gene in tomato (*Lycopersicon esculentum* Mill.) were among the first to be discovered in plants (Chang et al. 2000; Kwak and Lee 2001). More recent examples include uORFs for the genes *AdoMetDC1-3* that encode SAMDC and *AtPAO1* and *AtPAO2* that encode polyamine oxidase (PAO) 1 and 2 respectively (as extensively reviewed by Jorgensen and Dorantes-Acosta (2012).

The PAO enzymes are responsible for the back conversion of Spm to Spd and the catabolic degradation of both, while Put is degraded by diamine oxidase (DAO) or copper amine oxidase (CuAO). DAO degradation of Put produces ammonia (NH₃), hydrogen peroxide (H₂O₂) and Δ1-pyrroline, a shared product that forms from the previous step of both Put and Spd degradation. Δ1-pyrroline is further degraded into γ-aminobutyric acid (GABA), another signalling molecule (Ramesh et al. 2015). The other products from Spd oxidation are 1,3-diaminopropane (1,3-DAP) and H₂O₂. Spm on the other hand is degraded into 3-aminopropanal and H₂O₂ or 1-(3-aminopropanal)-pyrroline,

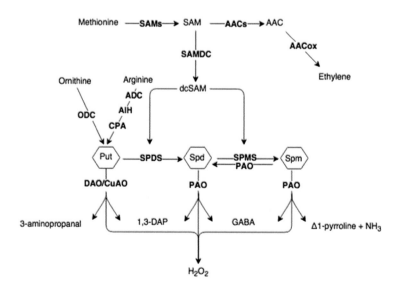

FIGURE 16.1 Plant PA biosynthetic pathway. SAM(s/dc), S-adenosylmethionine (synthase/decarboxylase); dcSAM, decarboxylated SAM; AAC (s/ox), 1-amino-cyclopropane-1-carboxylic acid (synthase/oxidase); ADC, arginine decarboxylase; AIH, agmatine iminohydrolase; CPA, N-carbamoylputrescine amidohydrolase; ODC, ornithine decarboxylase; SPDS, spermidine synthase; SPMS, spermine synthase; PAO, polyamine oxidase; DAO, diamine oxidase; 1,3-DAP, 1,3-diaminopropane; CuAO, copper amine oxidase; GABA, γ-aminobutyric acid. Key enzymes are emphasized in bold type.

1,3-DAP and H_2O_2. Note that degradation of all three major PAs result in H_2O_2 formation (Figure 16.1). More on the importance of H_2O_2 production can be found under the heading discussing PA and ABA crosstalk.

Polyamine Transport

Little is known about how plant PAs are transported between cells or to their subcellular localization. However, L-amino acid transporters (LATs) have been shown to be involved in transport of small amino acids, paraquat (PQ) and PAs (Fujita and Shinozaki 2014).

The first LAT protein with high affinity for PAs was found in rice (*Orysa sativa*) by Mulangi et al. (2011). The gene encoding the transporter, named *polyamine uptake transporter 1* (*PUT1*), was expressed in all tissues and organs except seeds and roots of developing plants and the protein imported radiolabelled Spd at a rate of ~12 pmol $\times$ mg^{-1} protein $\times$ min^{-1}. Further characterization showed that precursor uptake (of ornithine and agmatine) had a negligible effect on the rate of Spd uptake but competitive inhibition of up to 70% could be measured in the presence of Spm, methionine, glutamine or asparagine (Mulangi et al. 2011). This evidence demonstrates a broad specificity of this amino acid transporter. The same research group identified five more Spd transporters in the same gene cluster as *PUT1* in *Arabidopsis* and rice the following year (Mulangi et al. 2012).

The gene *resistant to methyl viologen1* (*RMV1*) (also known as *LAT1*) encodes a LAT PQ transporter that is localized to the plasma membrane. During a study on PQ resistance in plants by Fujita et al. (2012) the researchers observed that the *RMV1* overexpressing mutant was hypersensitive to PQ but also showed elevated PA uptake in cells. Exogenous PA treatment also lowered PQ toxicity suggesting that there may be a competition between PAs and PQ for RMV1 transportation (Fujita et al. 2012). Further research by Li et al. (2013) about genes encoding proteins structurally similar to *RMV1* found another putative LAT family member *paraquat resistant 1* (*PAR1*). *PAR1* is expressed in the whole plant with the highest mRNA accumulation in seeds and were shown to be localized intracellularly to the Golgi apparatus and responsible for PQ (and possibly PA) accumulation in chloroplasts (Li et al. 2013). It is likely that the three LATs (PUT1, RMV1 and PAR1) are responsible for part of both the inter/intra cell PA transportation mechanism since they all localize in different parts of the cell. The precise biochemical function of PA transporters and PA transportation in the cell remains unknown.

Polyamine Metabolism in Response to Abiotic Stress

As a plant encounters a stressful environment, transcriptional reprogramming occurs that induces production of the hormone abscisic acid (ABA), which in turn activates ABA-dependent signalling pathways to cope with the stress. The PAs Put, Spm and Spd have been implicated in a wide range of stress responses, both biotic and abiotic. Examples include fungal and viral infection (Walters 2000), salt, drought, heavy metal, mineral, oxidative, UV light and extreme temperature stress (Groppa and

Benavides 2007; Alcázar et al. 2011; Bitrián et al. 2012; Gupta et al. 2013). Recent advances for the functions of PAs during abiotic stress, mechanisms of action and signalling crosstalk with ABA will be reviewed in this section.

Polyamines and Saline Stress

High salt content in the soil is a major environmental stress factor for crop production. Saline environments have three main effects on the plant: (1) the osmotic potential of the plant cells is decreased, (2) ions such as Na^+ can reach toxic concentrations as salts dissociate in contact in water and (3) oxidative damage is caused to e.g. photosystem II (Pospíšil and Prasad 2014) through accumulation of reactive oxygen species (ROS) (Shabala and Pottosin 2014). As salt content increases so do some PA levels, but is this accumulation a response to stress or as a consequence of some unrelated effect?

Several clues come from studying exogenous PA application to stressed plants and/or manipulation of genes encoding PA synthesis enzymes. Urano et al. (2004) used an *Arabidopsis* ADC mutant (*adc2-1*) that produces 25% of the Put levels that wild type (WT) control plants and subjugated them to 150 mM NaCl treatment *in vitro*. They observed that the *adc2-1* plants were more sensitive to a saline environment than WT plants and importantly, exogenous application of Put rescued the phenotype, showing that PAs can have an effect on short-term acclimation to excess salt. Further research in other species has since reproduced this result. Shi et al. (2008) exposed a salt sensitive cucumber (*Cucumis sativus* L.) variety to 100 mM salt and found that exogenous application of Put diminished the observed detrimental effects on the rate of photosynthesis and leaf water potential. They hypothesized that Put may have a protective role based on their results. However, many research groups have found Put levels to decrease in several species, while Spm and Spd levels increase, suggesting that Put may act as a precursor to other PAs that ameliorate salt stress symptoms rather than Put itself (Zapata et al. 2004; Yamaguchi et al. 2006; Alet et al. 2011; Puyang et al. 2016). Kasukabe et al. (2004) produced evidence that complement this result because they could correlate overexpression of *SPDS* (leading to Spd accumulation) from *Cucurbita ficifolia* in *Arabidopsis* to heightened tolerance to several abiotic stressors, including salt stress.

Puyang (2016) performed a long term (28 days) salinity stress test where they added Put as a pre-treatment to Kentucky blue grass (*Poa pratensis* L.) and measured activities of ADC, ODC, SAMDC and PAO and the levels of Put, Spd and Spm as well as molecular phenotypes of salt stress. Authors defined a salt stress phenotype baseline by gradually adding up to 200 mM NaCl as a foliar spray treatment. They found decreased chlorophyll content, K^+/Na^+ ratio and Ca^{2+} and Mg^{2+} contents. K^+/Na^+ ratio decreases upon addition of Spd have further support from other studies such as Zhao et al. (2007) and Edwards et al. (2015). NaCl also increased the levels of electrolyte leakage (EL) from leaves and increased proline, Na^+ and free Put, Spd and Spm levels in the cell. The activities of ADC, ODC, SAMDC and PAO also increased. With the baseline established in the control plants they added 1 mM Spd to others and found this to be sufficient to improve all health measurements taken. At day 28 the Spd treated plants had a 58% higher chlorophyll content and a decreased EL by 43% compared to non-Spd treated plants. Na+ content decreased by 40%, and the levels of K+, Ca2+ and Mg2+, and the K+/Na+ ratio and proline content were further increased compared to non-treated plants in a similar fashion. Furthermore, the application of Spd decreased Put levels and increased Spm and Spd (Puyang et al. 2016). Previous studies have shown similar results in EL, ion and chlorophyll content in plants upon addition of PAs in rice (Chattopadhayay et al. 2002) and that ADC, ODC, SPDS, SPMS, SAMDC and DAO enzymes are induced by salt stress and reduce symptoms through their function in various species (Alcázar et al. 2006b, 2010a; Zhang et al. 2014; Yin et al. 2014). It seems that Spd is the main PA that alleviates salt stress symptoms and while Spm is also induced, it may only confer a subtle effect on salt tolerance in rice (Maiale et al. 2004) and *A. thaliana* (Alet et al. 2012) while Put is canalized into the production of higher PAs as noted earlier. Correlations between SPMS activity and Spm accumulation and salt stress tolerance have also been found while studying an *acaulis5/spms* double mutant in *Arabidopsis* and in several salt sensitive cultivars of rice (Do et al. 2014). Furthermore Do et al. (2014) found the PA levels to be quite similar in comparisons between salt and drought induced stress in rice, except for a decrease in Spd during drought stress while salt stress Spd levels were unchanged. During both types of stress, Put and Spm levels were dominant across 21 cultivars with varying sensitivity. These results indicate that the puzzle of PA mediated salt tolerance is far from completely characterized (Yamaguchi et al. 2006).

Polyamines and Drought Stress

PAs have been implicated in water deficiency stress in several species. Some evidence includes results from a transcriptomic profiling of water stressed *A. thaliana* plants. Alcázar et al. (2006b) showed that genes for Put, Spd and Spm synthesis (*ADC2*, *SPDS1* and *SPMS*) are highly induced in WT plants when water availability drops (Alcázar et al. 2006b). *SAMDC2* and *AIH* have also been found to be specifically drought induced in rice (Do et al. 2014). Furthermore, an *Arabidopsis* mutant (*pao1 pao5*) that had a 62% reduced cytoplasmic PAO activity, and as such lost PA degradation activity, produced significantly less ROS, had a higher water content and lived longer compared to WT plants after water supply was cut off (Sagor et al. 2016).

Since genes encoding *SPDS* (for Spd synthesis) have been shown to be upregulated during drought stress, could Spd mitigate drought stress as well as salt stress? Depending on species this could be the case. Research by Yin et al. (2014) showed that exogenous application of Spd and/or Spm to plants of the Chinese dwarf cherry (*Cerasus humili*) was enough to mitigate the oxidative damage that results from drought stress in WT plants. Plants treated with Spd or Spm had enhanced water content and prevented water deficiency induced lipid membrane peroxidation. They could also show that while there was an accumulation of both endogenous Spd and Spm, Spd application was more effective than the addition of Spm in relieving stress symptoms. The transgenic *C. ficifolia SPDS* overexpressing *Arabidopsis* mutant made by Kasukabe et al. (2004) also seemed to confer tolerance in their water deficient plants. However, cellular Spd content was not measured. Alcázar et al. (2011) found a decrease in Spd levels in dehydrated *A. thaliana* plants even though SPDS activity was

high. It is possible the *SPDS* from *C. ficifolia* may behave differently in the *Arabidopsis* genetic background, such as not being exposed to posttranslational regulation.

During onset of drought Put levels increase in several species, including *A thaliana*, *Craterostigma plantagineum* (Alcázar et al. 2011), rice and *Datura stramonium* (Capell et al. 2004). As time progresses these levels deplete while Spm and Spd levels rise to higher levels in all above species except for *Arabidopsis*. This canalization resulted in an eightfold increase in Spm levels in *C. plantagineum* and correlated with water deficiency tolerance (Alcázar et al. 2011). This result is in compliance with a later study using rice that showed Put levels to decrease and Spm to become the dominantly accumulated PA in long term water stressed rice plants (Do et al. 2013). Again, this seems to indicate that the pool of Put is depleted in favour for synthesis of higher PAs that have a role in protection during drought stress in several species, as it is the case during salt stress. Alcázar et al. (2010b) used an *ADC2* overexpressing *A. thaliana* mutant in one of their drought stress experiments where they measured water content, transpiration rate and modulation of Put, Spd and Spm synthesis. They could conclude that high Put content alone was enough to confer drought resistance because the higher tolerance mutant accumulated high levels of Put while there was a sharp decrease in Spd and Spm levels. Similar results have been reported using a stress inducible oat *ADC* transgenic variety of *Lotus tenuis* (Espasandin et al. 2014). A likely explanation for high enzyme activity but low PA accumulation in *Arabidopsis* was suggested by Alcázar et al. (2011) that could show that SPDS (and SPMS) activity was involved in canalization of Put to Spd (and Spd to Spm) during early drought onset, followed by Spm to Put back conversion. This activity would provide an effective means of metabolic regulation of PA levels as needed while *A. thaliana* acclimatizes to water deficiency (Alcázar et al. 2011).

Other Abiotic Stresses

As mentioned, saline and drought are not the only abiotic stress factors that PAs can induce a degree of tolerance against. PA induction of a carotenoid (lycopene) has been reported in postharvest cold stress-injured tomato fruit (Neily et al. 2011). And NO treated postharvest bananas (*Musa* spp.) have shown alleviation of cold stress due to accumulation of PAs as well as the PA breakdown products GABA and proline (Wang et al. 2016), which are all known to be involved in tolerance to cold stress (Gupta et al. 2013). PAs are also included in the response against high temperatures in ripening tomato fruits as reported by Cheng et al. (2012). Heavy metal toxicity mitigation has also been reported in wheat (*Triticum* ssp.) seedlings in response to Cadmium (Rady and Hemida 2015). PAs have implications for tolerance against less obvious stresses like mechanical wounding (Perez-Amador et al. 2002) and ozone stress (Sharma et al. 1996) as well. Too little ozone is not good either; reports have been published regarding UV-damage due to ozone depletion. Notably light in the UV-A through UV-C spectra (400–320 nm, 320–280 nm and 280–200 nm) as reviewed by (Fariduddin et al. 2013). This article deals a lot with stress resulting from low water availability, but a lot of water can also be a problem for plants. Flooding tolerance has been reported in submerged rice during the tillering stage (lateral shoot emergence) through

Spd application by the same mechanisms described below (e.g. restored chlorophyll damage, inhibition of lipid peroxidation and antioxidation maintenance) (Liu et al. 2015).

Polyamine Mechanisms of Action

Modulation of PA metabolism during different types of stress provide an interesting insight into what is happening inside the cell as a plant experiences stress, but it can't explain PA function alone. To be able to annotate biological meaning to the modulations of PA metabolism, biochemical and physiological response measurements, as well as looking into PA metabolism in response to the well-known stress and senescence hormone ABA, can provide answers. In this section a summary of contemporary work is presented and evaluated. A simplified summary can be viewed in Figure 16.2.

Ion Channel Interactions, Protein/Membrane Stabilization Properties During Salt Stress

Ion toxicity, electrolyte leakage, loss of osmotic potential and chlorophyll are some of the effects salt stress has on plants as previously described. Since PAs can ameliorate these detrimental symptoms, the question of how they do it comes to mind. As mentioned, PAs can interact with negatively charged molecules. This also includes negative residues in membrane proton pumps (Zhao et al. 2007; Alcázar et al. 2010a; Puyang et al. 2016). It has been shown that PAs can block fast acting vacuolar cation channels in barley (*Hordeum vulgare*) and thus lower positively charged ion flux across the vacuolar tonoplast in a charge dependent manner (Spm, $^{4+}$ > Spd $^{3+}$ >> Put $^{2+}$) (Brüggemann et al. 1998). This by itself could prevent loss of K^+ ions during salt stress and contributes to Na^+/K^+ homeostasis (Alcázar et al. 2010a). Additionally, if Spm synthesis is knocked out, as in the case of the *A. thaliana acl5/spms* mutant, plants become hypersensitive to salt. Complementation with exogenous Spm also cured the salt stress phenotype implicating the polycationic Spm in salt stress (Yamaguchi et al. 2006). The charge dependent action of PAs was also reported by Zhao et al. (2007) that applied extracellular PAs to barley plants and measured root epidermal- and cortical- and xylem parenchyma-cell ion content. They showed that application of Spm significantly blocked inward Na^+ and K^+ channels and the block effect increased as the PA levels increased. Interestingly, this effect was not observed when PAs were applied intracellularly, indicating that the target site on the

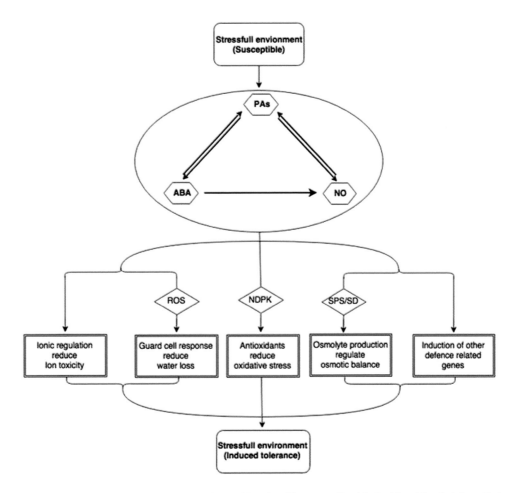

FIGURE 16.2 Schematic illustration of the effects PA metabolism and PA signalling crosstalk with abscisic acid and nitric oxide has on physiological acclimatisation during abiotic stress. PA, polyamine; ABA, abscisic acid; NO, nitric oxide; ROS, reactive oxygen species (H_2O_2); NDPK, nucleoside diphosphate kinase; SPS/SD, sucrose phosphate synthase and sorbitol dehydrogenase. Hexagons symbolize PAs and their crosstalk with signalling molecules and the rhombus symbolize examples downstream effect mediating enzymes/molecules.

ion transporter is located on the apoplastic part of the cell membrane. They also found shoot outward K$^+$ channels were blocked. Together their results show that PAs help maintain cellular Na$^+$/ K$^+$ homeostasis systemically by repression of Na$^+$ influx and K$^+$ efflux, diminishing ion toxicity to a tolerable level (Zhao et al. 2007). Plasma membrane H$^+$/Na$^+$ antiporter activity has been reported to be inhibited in salt sensitive cultivars of rice as well. Roy et al. (2005) measured plasma membrane H$^+$-ATPase activity in salt sensitive and tolerant cultivars and found that sensitive varieties had a much lower ATPase activity (and NADPH oxidase activity) and that addition of Spd could dramatically restore this enzyme activity as well as lower the damage to the plasma membrane. They further showed that tolerant cultivars were naturally rich in both Spd and Spm while sensitive cultivars were rich in Put only. In cucumber however it seems that a lower level of PAs (and thus lower levels of Plasma membrane and vacuolar H$^+$-ATPase blockage) is what seems to give a beneficial effect (Janicka-Russak et al. 2010).

Not only are PAs involved in protection against salt-induced electrolyte/osmolyte loss through ion channel activity. Reports about PAs binding to negatively charged phosphate head groups in the plasma membrane (Chattopadhayay et al. 2002; Roy et al. 2005) and microsomal membranes (Legocka and Kluk 2005; Groppa and Benavides 2007) have been published as well. By binding and thus stabilizing membranes, monovalent cations remain inside the cell and osmoregulation activity remain intact. Chattopadhayay et al. (2002) measured internal concentrations of Na$^+$, K$^+$, Ca^{2+} and Mg^{2+} and amino acid content in roots and shoots of the same rice cultivars as Zhao et al. (2007) and subjected them to 150 mM NaCl stress with or without PA addition. 1 mM Spd and Spm addition reduced leakage of the measured cations significantly due to stabilization properties of PAs and also mediated chloroplastic gene transcription (*rbcL*, *psbA*, *psbB*). Thus indicating their role in reducing loss of photosynthetic activity during salt stress. Ion channel interaction, together with plasma membrane stabilization effects of Spd and Spm, can thus explain reduction in salt-induced loss of electrolytes, osmotic potential and reduce ion toxicity.

PA Modulation of Osmolyte Metabolism

Except for maintaining turgor pressure by regulating ion content through vacuolar- and plasma- membrane stabilization, PAs protect plants from drought related stress by influencing osmotic potential through modulating carbohydrate and amino acid metabolism (Li et al. 2015; Li et al. 2016). A proteomic and physiologic study aimed to investigate the protective effect of exogenous PA application during salt and drought stress in Bermudagrass (*Cynodon dactylon*) by Shi et al. (2013) identified 36 protein coding genes that were regulated by Put, Spd and Spm by way of time of flight mass spectrum analysis. The functions of the identified genes were related to electron transport chains, ROS metabolism and antioxidant production. Among the genes were one encoding nucleoside diphosphate kinase (NDPK), which is an important enzyme involved in multiple stress responses including drought and carcinogenic polycyclic hydrocarbon stress (Zadražnik et al. 2013; Liu et al. 2014). Several were, however, involved in glycolysis and the Calvin cycle (two and nine respectively), which is related to the synthesis and

breakdown of glucose, which, when joined together, can act as an osmoregulator (Ackerson 1981). Further testing revealed significant differences in osmolyte accumulation between Put, Spm or Spd pre-treated plants and non-treated plants during salt and drought stress, mainly proline, sucrose and total soluble sugars. This points to the direct effects in osmolyte accumulation that enhanced PA levels have during stress (Shi et al. 2013).

Another study using a drought tolerant and a drought sensitive cultivar of white clover (*Trifolium repens*) could also identify changes in osmolyte production due to exogenous PA treatment (Li et al. 2015). Key enzyme activity for the water-soluble carbohydrates sucrose and sorbitol (sucrose phosphate synthase and sorbitol dehydrogenase) was significantly increased (but sucrose synthase activity was lowered). Furthermore, glucose, sucrose, fructose and sorbitol content was increased in cells after treatment with Spm during drought stress. Spm application resulted additionally in higher dehydrin production, which, as the name suggests, are proteins produced during dehydration. This activity was induced earlier in the drought susceptible cultivar, but both cultivars had enhanced dehydrin production following Spm addition, and both were able to maintain cellular turgor pressure (Li et al. 2015). Additional work by the same research group using physiological and an isobaric labelling proteomics method took this work further but treated white clover plants with Spd instead of Spm. White clover plants that were either treated or non-treated (with 50 μM Spd) were grown with 20% polyethylene glycol 6000 dissolved in Hoagland's medium to induce drought stress. The subsequent physiological analysis revealed that treated plants with higher endogenous Spd levels demonstrated a delay in drought responses, i.e. water use efficacy, lower oxidative damage symptoms and continuously high photosynthetic rate. The ensuing proteome analysis revealed that the basis for Spd mediated drought stress could be due to an ability to maintain higher expression of proteins involved in amino acid synthesis, protein biosynthesis, carbon metabolism, antioxidant and stress defence proteins (ascorbate peroxidase, glutathione peroxidase, and dehydrins) and proteins involved in ABA signalling pathways (Li et al. 2016). These results provide a glimpse into the many cellular processes that PAs modulate during stress and the power of "omics" approaches in elucidating complex networks and interactions.

Polyamine and ABA, NO Crosstalk: Stomata Regulation and ROS Scavenging Activity

Spm has been implicated in stomata closure during drought. As drought stress sets in, the Spm synthesis incapable *Arabidopsis* mutant *acl4/spms* lose more water faster due to a higher rate of transpiration than WT plants (Yamaguchi et al. 2007). Liu et al. (2000) has also found Spm to block KAT1-like inward K$^+$ channels in guard cells, which resulted in stomata closure in *V. faba*. However, Spm is not the most widely studied PA in regard to stomata activity and water loss retention. An extensive amount of evidence has been produced that indicates that PAs, ABA and NO signalling pathways overlap during abiotic stress (Liu et al. 2000; Alcázar et al. 2006b; An et al. 2008; Rosales et al. 2011; Bitrián et al. 2012; Espasandin et al. 2014; Dong et al. 2016). ABA function during drought and salt stress includes mediation of water loss reductions through stomata regulation. Exogenous application of ABA has also been found to modulate PA synthesis (Alcazar 2006a), and PA synthesis to modulate ABA production, at least in response to cold (Cuevas et al. 2008). Further research on the topic by Espasandin et al. (2014) showed that Put induced ABA synthesis through modulation of ABA at the transcriptional level by influencing the activity of the *9-cis-epoxycarotenoid dioxygenase* (*NCED*) gene in *Lotus tenuis*. These results point to a positive feedback loop between ABA and PAs that reinforces tolerance to low water availability (Figure 16.2). Alcázar et al. (2010b) used an *Arabidopsis* mutant that constitutively expresses *ADC2*, giving a high accumulation of Put with no changes in Spd or Spm, and investigated the effects of this during drought stress. They found that there might exist a threshold for Put mediated drought tolerance and, as the *ADC2* overexpressing mutant overcame it, stomata apertures got smaller and stomata conductance and transpiration rate was lowered. Similarly, *ADC2*, *SPDS1* and *SPMS* gene expression was later found to be induced by drought in an ABA-dependent manner because these activities were abolished in ABA deficient and insensitive mutants (*aba2-3* and *aba1-1*) (Alcázar et al. 2006a). Furthermore Alcázar et al. (2006a) could pinpoint a possible reason for this effect. They could identify ABA responsive elements (ABRE) and ABRE-related motifs in the promoter region of *ADC2 and SPMS*. A possible mode of action could thus be related to abruption of signalling crosstalk between PAs and ABA if ABREs normally induce transcription of PAs through this mechanism.

Not only do PAs themselves participate in water loss reductions during water stress. The production of H_2O_2 by PA oxidation also has implications for stomata guard cell activity. An et al. (2008) found that H_2O_2 production was stimulated in fava beans (*Vicia faba*) by way of apoplastic CuAO enzymes that in turn were induced by exogenous ABA application. This correlated with reduced transpiration rates and stomata closure. Application of exogenous Put gave a similar response. The authors attributed this response to Put mediated CuAO production of H_2O_2 that participate in ABA-related stomata closure. Nitric oxide (NO) signalling has also been implicated in stomata closure, which is furthermore dependent on H_2O_2 for this effect (Neill et al. 2008). NO is mainly produced through the enzymatic action of nitric oxide synthase (NOS) and nitrate reductase (NR) (Crawford 2006), but PAs are also involved in NO synthesis (Rosales et al. 2011) and NO induces PA synthesis (at least during cold stress in postharvest fruit) (Wang et al. 2016) (see Figure 16.2). Even though NR activity was reduced by Spm and Spd treatment during the first hours of incubation, NR activity was raised 64% by Put and 114% by Spm and Spd treatment after 21 hours of incubation in wheat (Rosales et al. 2011). The exact mechanism and whether it is a major pathway for NO production or not is unclear (Rosales et al. 2011). PAs have been shown to be responsible for inducing NO production in the elongation zone and root tip of *Arabidopsis* as well (Tun et al. 2006). Tanou et al. (2014) could also infer that application of PAs could restore nitrate reductase function and other NO-related genes during salt stress. Taken together these results point towards the likelihood that a PA, ABA, ROS and NO crosstalk is needed for guard cell response regulation and NO/ABA-related defence gene expression during water deficiency stress or cold stress.

Several types of abiotic stress have a connection to oxidative metabolism and ROS generation, namely hydroxyl radicals

($\cdot OH^-$), super oxide ($\cdot O_2^-$) and H_2O_2, which can be detrimental to plant cells. Plants have, as such, a protective antioxidative system where enzymatic- (superoxide dismutases (SODs), catalase (CAT), peroxidases (POD), glutathione reductases (GRs) and ascorbate peroxidase (APX)) and non-enzymatic- (such as carotenoids and tocopherol) reactions occur to render these radicals inert, reviewed by Kar (2015). As DAO/CuAO degrade Put, and PAO degrades Spd and Spm, H_2O_2 production occurs (Figure 16.1). These molecules are generally produced during stress and can cause oxidative damage to plasma membranes increasing EL and reducing the amount of chlorophyll (as noted earlier) if they accumulate. This speaks against PA function as antioxidants. However, exogenous Put treatment has been correlated to lower ROS production and enhance salt tolerance (Zhang et al. 2014). By reducing the activity of PAO, Sagor et al. (2016) found ROS production to be lowered as well as an induction of other defence genes. Reports have been published that implicate PAs as direct acting antioxidants, including Yin et al. (2014), which showed, among other findings, that Spd and Spm prevented drought induced lipid peroxidation and reduced the amount of free radicals ($\cdot O_2^-$) in *C. humili* seedlings, and Zhang et al. (2014), that mention that the protonation state of PAs at physiological pH enables them to directly scavenge free radicals. One may however postulate that these results seem contradictory compared to the fact that PA oxidation produce ROS. However, the discussion on how PA accumulation mediates lower oxidative stress is currently focused on results showing that Put and other PAs act as indirect ROS scavengers by the induction of antioxidant enzymes (Neill et al. 2008; Shi et al. 2013; Wi et al. 2014; Zhang et al. 2014; Yin et al. 2014; Tanou et al. 2014; Sánchez-Rodríguez et al. 2016), as well as carotenoids (Neily et al. 2011; Espasandin et al. 2014) in various species. Wi et al. (2014) used a *Capsicum annuum SAMDC* overexpressing transgenic *Arabidopsis* mutant that they subjugated to drought stress and measured expression of various antioxidant-related genes. Measurements of the $\cdot O_2^-$ producing NADPH oxidase genes *RbohD* and *RbohF*, and the genetic activity of the mitochondrial ROS detoxifying enzymes *manganese-superoxide dismutase* (*MnSODmi*) and *catalase 1* (*CAT1*), showed significantly increased mRNA accumulation in the mutant compared to WT plants. This activity also resulted in less drought stress symptoms occurring. Zhang et al. (2014) could prove that the addition of Put to the growth medium ameliorated the detrimental effects of salt stress on plant growth and biomass production in Soybean (*Glycine max* L.). Put was canalized into Spm and Spd, which increased in concentration as free and conjugated forms in the roots. Antioxidant activity was reduced in non-treated salt stressed plants while the plants with extra Spd and Spm got restored activities of SOD, POD, CAT and APX. Greater ROS scavenging efficacy by the action that higher PAs exert on the antioxidation enzymes SOD and CAT has also been reported in tomato during drought (Sánchez-Rodríguez et al. 2016), in Bermudagrass with the addition of enhanced POD activity after salt and drought treatments (Shi et al. 2013) and in Seville orange (*Citrus aurantium* L) following salt treatment with the addition of enhanced GR, Dehydroascorbate reductase (DHAR), mono-DHAR and ascorbate oxidase. The exact molecular mechanism that PAs use to induce antioxidant defence activation and carotenoid production however remains unknown.

Conclusion

This review focused on the interaction between the PAs Put, Spd and Spm during salt and drought stress. These are, however, not the only PAs that may have implications in abiotic stress. Recent work has proposed that thermospermine, an isomer of Spm, is involved in various stress responses including correct root xylem differentiation during simulated stress (Ghuge et al. 2015), as well as ameliorating salt stress (Zarza et al. 2016). PA metabolism often changes in response to abiotic stress and this correlates to increased tolerance through a number of mechanisms. Stress is often accompanied by a rise in Put levels followed by Put to Spm and Spd conversion resulting in an accumulation of higher PAs. These act to stabilize phospholipid membranes and proteins to prevent EL and interact with ion channels to regulate ion homeostasis, stomata guard cells to prevent water loss, carbohydrate metabolism enzymes to increase osmotic potential and ABA to induce a signal cascade that leads to activation of other defence genes.

In the recent history of PA research, most progress has been made using forward and reverse genetics and measuring the metabolic and physiologic effects of exogenous application of PAs while studying one or a few components at a time. This approach has taught us a lot but there are many gaps of knowledge left. How the different PAs are transported remains an elusive problem as well as the exact signalling pathway used by PAs in cross talk with hormones and other signalling molecules. Although PAs are low molecular weight substances, PA protection against stress is not free. A recent report shows that there may be a PA-dependent trade-off between plant growth and protection against stress (Mellidou et al. 2016). The advent of "omics" approaches such as hypothesis driven metabolomics and proteomics may help to shed light on complex interactions and quantitative traits such as biomass accumulation during stress. By studying the systems holistically and viewing the problem from a systems biology view, we can begin to puzzle together the sum of the parts to create a more complete picture of the effects of PA metabolism. It is clear however that we need to study PAs further because this knowledge may help to future proof our crop production systems against environmental changes.

REFERENCES

Ackerson RC. 1981. Osmoregulation in cotton in response to water stress II. Leaf carbohydrate status in relation to osmotic adjustment. *Plant Physiol.* 67: 489–493.

Aikens D, Bunge S, Onasch F, Parker R, Hurwitz C, Clemans S. 1983. The interactions between nucleic acids and polyamines. *Biophys. Chem.* 17: 67–74.

Alcázar R, Altabella T, Marco F, Bortolotti C, Reymond M, Koncz C, Carrasco P, Tiburcio AF. 2010a. Polyamines: molecules with regulatory functions in plant abiotic stress tolerance. *Planta* 231: 1237–1249.

Alcázar R, Bitrián M, Bartels D, Koncz C, Altabella T, Tiburcio AF. 2011. Polyamine metabolic canalization in response to drought stress in Arabidopsis and the resurrection plant *Craterostigma plantagineum*. *Plant Signal. Behav.* 6: 243–250.

Alcázar R, Cuevas JC, Patron M, Altabella T, Tiburcio AF. 2006a. Abscisic acid modulates polyamine metabolism under water stress in *Arabidopsis thaliana*. *Physiol. Plantarum* 128: 448–455.

Alcázar R, Marco F, Cuevas JC, Patron M, Ferrando A, Carrasco P, Tiburcio AF, Altabella T. 2006b. Involvement of polyamines in plant response to abiotic stress. *Biotechn. Lett.* 28: 1867–1876.

Alcázar R, Planas J, Saxena T, Zarza X, Bortolotti C, Cuevas J, Bitrián M, Tiburcio AF, Altabella T. 2010b. Putrescine accumulation confers drought tolerance in transgenic Arabidopsis plants overexpressing the homologous *ARGININE DECARBOXYLASE* 2 gene. *Plant Physiol. Biochem.* 48: 547–552.

Alet AI, Sánchez DH, Cuevas JC, Marina M, Carrasco P, Altabella T, Tiburcio AF, Ruiz OA. 2012. New insights into the role of spermine in *Arabidopsis thaliana* under long-term salt stress. *Plant Sci.* 182: 94–100.

Alet AI, Sánchez DH, Ferrando A, Tiburcio AF, Alcazar R, Cuevas JC, Altabella T, et al. 2011. Homeostatic control of polyamine levels under long-term salt stress in *Arabidopsis*. *Plant Signal. Behav.* 6: 237–242.

An Z, Jing W, Liu Y, Zhang W. 2008. Hydrogen peroxide generated by copper amine oxidase is involved in abscisic acid-induced stomatal closure in *Vicia faba*. *J. Exp. Bot.* 59: 815–825.

Anwar R, Mattoo AK, Handa AK. 2015. Polyamine interactions with plant hormones: Crosstalk at several levels. In: Kusano T, Suzuki H (eds.). *Polyamines*, pp. 267–302. Springer, Japan, Tokyo.

Bachrach U. 2010. The early history of polyamine research. *Plant Physiol. Biochem.* 48: 490–495.

Bitrián M, Zarza X, Altabella T, Tiburcio AF, Alcázar R. 2012. Polyamines under abiotic stress: metabolic crossroads and hormonal crosstalks in plants. *Metabolites* 2: 516–528.

Brüggemann LI, Pottosin II, Schönknecht G. 1998. Cytoplasmic polyamines block the fast-activating vacuolar cation channel. *Plant J.* 16: 101–105.

Capell T, Bassie L, Christou P. 2004. Modulation of the polyamine biosynthetic pathway in transgenic rice confers tolerance to drought stress. *Proc. Nat. Acad. Sci. USA* 101: 9909–9914.

Casero RA, Marton LJ. 2007. Targeting polyamine metabolism and function in cancer and other hyperproliferative diseases. *Nat. Rev. Drug Discovery* 6: 373–390.

Chang KS, Lee SH, Hwang SB, Park KY. 2000. Characterization and translational regulation of the arginine decarboxylase gene in carnation (*Dianthus caryophyllus* L.). *Plant J.* 24: 45–56.

Chattopadhayay MK, Tiwari BS, Chattopadhyay G, Bose A, Sengupta DN, Ghosh B. 2002. Protective role of exogenous polyamines on salinity-stressed rice (*Oryza sativa*) plants. *Physiol. Plantarum* 116: 192–199.

Cheng L, Sun R, Wang F, Peng Z, Kong F, Wu J, Cao J, Lu G. 2012. Spermidine affects the transcriptome responses to high temperature stress in ripening tomato fruit. *J. Zhejiang Univ. Sci. B* 13: 283–297.

Crawford NM. 2006. Mechanisms for nitric oxide synthesis in plants. *J. Exp. Bot.* 57: 471–478.

Cuevas JC, López-Cobollo R, Alcázar R, Zarza X, Koncz C, Altabella T, Salinas J, Tiburcio AF, Ferrando A. 2008. Putrescine is involved in *Arabidopsis* freezing tolerance and cold acclimation by regulating abscisic acid levels in response to low temperature. *Plant Physiol.* 148: 1094–1105.

Do PT, Degenkolbe T, Erban A, Heyer AG, Kopka J, Köhl KI, Hincha DK, Zuther E. 2013. Dissecting rice polyamine metabolism under controlled long-term drought stress. *PLoS One* 8: e60325.

Do PT, Drechsel O, Heyer AG, Hincha DK, Zuther E. 2014. Changes in free polyamine levels, expression of polyamine biosynthesis genes, and performance of rice cultivars under salt stress: A comparison with responses to drought. *Front. Plant Sci.* 5: 182. doi 10.3389/fpls.2014.00182.

Dong S, Hu H, Wang Y, Xu Z, Zha Y, Cai X, Peng L, Feng S. 2016. A *pqr2* mutant encodes a defective polyamine transporter and is negatively affected by ABA for paraquat resistance in *Arabidopsis thaliana*. *J. Plant Res.* 129: 899–907.

Edwards D, Oldroyd G, Chunthaburee S, Sanitchon J, Pattanagul W, Theerakulpisut P. 2015. Agriculture and climate change— Adapting crops to increased uncertainty (AGRI 2015) Application of exogenous spermidine (Spd) improved salt tolerance of rice at the seedling and reproductive stages. *Proc. Environ. Sci.* 29: 134.

Espasandin FD, Maiale SJ, Calzadilla P, Ruiz OA, Sansberro PA. 2014. Transcriptional regulation of *9-CIS-EPOXYCAROTENOID DIOXYGENASE* (*NCED*) gene by putrescine accumulation positively modulates ABA synthesis and drought tolerance in *Lotus tenuis* plants. *Plant Physiol. Biochem.* 76: 29–35.

Fariduddin Q, Varshney P, Yusuf M, Ahmad A. 2013. Polyamines: potent modulators of plant responses to stress. *J. Plant Interact.* 8: 1–16.

Food and Agriculture Administration of the United Nations (FAO). 2009. How to feed the world in 2050. Available online at: http://www.fao.org/fileadmin/templates/wsfs/docs/ expert_paper/How_to_Feed_the_World_in_2050.pdf. Accessed 11/05/2015.

Fujita M, Fujita Y, Iuchi S, Yamada K, Kobayashi Y, Urano K, Kobayashi M, Yamaguchi-Shinozaki K, Shinozaki K. 2012. Natural variation in a polyamine transporter determines paraquat tolerance in *Arabidopsis*. *Proc. Nat. Acad. Sci. USA* 109: 6343–6347.

Fujita M, Shinozaki K. 2014. Identification of polyamine transporters in plants: paraquat transport provides crucial clues. *Plant Cell Physiol.* 55: 855–861.

Ge C, Cui X, Wang Y, Hu Y, Fu Z, Zhang D, Cheng Z, Li J. 2006. *BUD2*, encoding an S-adenosylmethionine decarboxylase, is required for *Arabidopsis* growth and development. *Cell Res.* 16: 446–456.

Geny L, Broquedis M, Martin-Tanguy J, Bouard J. 1997. Free, conjugated, and wall-bound polyamines in various organs of fruiting cuttings of *Vitis vinifera* L. cv. cabernet sauvignon. *Am. J. Enol. Vit.* 48: 80–84.

Ghuge SA, Tisi A, Carucci A, Rodrigues-Pousada RA, Franchi S, Tavladoraki P, Angelini R, Cona A. 2015. Cell wall amine oxidases: new players in root xylem differentiation under stress conditions. *Plants* 4: 489–504.

Groppa MD, Benavides MP. 2007. Polyamines and abiotic stress: recent advances. *Amino Acids* 34: 35–45.

Gupta K, Dey A, Gupta B. 2013. Plant polyamines in abiotic stress responses. *Acta Physiol. Plantarum* 35: 2015–2036.

Hanfrey C, Sommer S, Mayer MJ, Burtin D, Michael AJ. 2001. *Arabidopsis* polyamine biosynthesis: absence of ornithine decarboxylase and the mechanism of arginine decarboxylase activity. *Plant J.* 27: 551–560.

Janicka-Russak M, Kabała K, Młodzińska E, Kłobus G. 2010. The role of polyamines in the regulation of the plasma membrane and the tonoplast proton pumps under salt stress. *J. Plant Physiol.* 167: 261–269.

Janowitz T, Kneifel H, Piotrowski M. 2003. Identification and characterization of plant agmatine iminohydrolase, the last missing link in polyamine biosynthesis of plants. *FEBS Lett.* 544: 258–261.

Jorgensen RA, Dorantes-Acosta AE. 2012. Conserved peptide upstream open reading frames are associated with regulatory genes in angiosperms. *Plant Genet. Genom.* 3: 191.

Kar RK. 2015. Interactive role of polyamines and reactive oxygen species in stress tolerance of plants. In: Chakraborty U, Chakraborty B (eds.). *Abiotic Stresses in Crop Plants*, pp. 212–221. CABI, Wallingford.

Kasukabe Y, He L, Nada K, Misawa S, Ihara I, Tachibana S. 2004. Overexpression of Spermidine Synthase enhances tolerance to multiple environmental stresses and up-regulates the expression of various stress-regulated genes in transgenic *Arabidopsis thaliana*. *Plant Cell Physiol.* 45: 712–722.

Kwak S-H, Lee SH. 2001. The regulation of Ornithine Decarboxylase gene expression by sucrose and small upstream open reading frame in tomato (*Lycopersicon esculentum* Mill). *Plant Cell Physiol.* 42: 314–323.

Legocka J, Kluk A. 2005. Effect of salt and osmotic stress on changes in polyamine content and arginine decarboxylase activity in Lupinus luteus seedlings. *J. Plant Physiol.* 162: 662–668.

Li J, Mu J, Bai J, Fu F, Zou T, An F, Zhang J, et al. 2013. PARAQUAT RESISTANT1, a Golgi-localized putative transporter protein, is involved in intracellular transport of paraquat. *Plant Physiol.* 162: 470–483.

Li Z, Jing W, Peng Y, Zhang XQ, Ma X, Huang LK, Yan Y. 2015. Spermine alleviates drought stress in white clover with different resistance by influencing carbohydrate metabolism and dehydrins synthesis. *PLoS One* 10: e0120708.

Li Z, Zhang Y, Xu Y, Zhang X, Peng Y, Ma X, Huang L, Yan Y. 2016. Physiological and iTRAQ-based proteomic analyses reveal the function of spermidine on improving drought tolerance in white clover. *J. Prot. Res.* 15: 1563–1579.

Liu H, Weisman D, Tang L, Tan L, Zhang W, Wang Z, Huang Y, Lin W, Liu X, Colón-Carmona A. 2014. Stress signaling in response to polycyclic aromatic hydrocarbon exposure in Arabidopsis thaliana involves a nucleoside diphosphate kinase, NDPK-3. *Planta* 241: 95–107.

Liu K, Fu H, Bei Q, Luan S. 2000. Inward potassium channel in guard cells as a target for polyamine regulation of stomatal movements. *Plant Physiol.* 124: 1315–1326.

Liu M, Chu M, Ding Y, Wang S, Liu Z, Tang S, Ding C, Li G. 2015. Exogenous spermidine alleviates oxidative damage and reduce yield loss in rice submerged at tillering stage. *Front. Plant Sci.* 6: 919. doi 10.3389/fpls.2015.00919.

Maiale S, Sánchez DH, Guirado A, Vidal A, Ruiz OA. 2004. Spermine accumulation under salt stress. *J. Plant Physiol.* 161: 35–42.

Manning DAC. 2015. How will minerals feed the world in 2050? *Proc. Geol. Assoc.* 126: 14–17.

Marco F, Alcázar R, Altabella T, Carrasco P, Gill SS, Tuteja N, Tiburcio AF. 2012. Polyamines in developing stress-resistant crops. In: Tuteja N, Gill SS, Tiburcio AF, Tuteja R (eds.). *Improving Crop Resistance to Abiotic Stress*, pp. 623–635. Wiley-VCH Verlag GmbH & Co. KGaA, Weinhem.

Martin-Tanguy J. 1997. Conjugated polyamines and reproductive development: biochemical, molecular and physiological approaches. *Physiol. Plantarum* 100: 675–688.

Mellidou I, Moschou PN, Ioannidis NE, Pankou C, Gémes K, Valassakis C, Andronis EA, et al. 2016. Silencing S-Adenosyl-L-Methionine Decarboxylase (SAMDC) in *Nicotiana tabacum* points at a polyamine-dependent trade-off between growth and tolerance responses. *Front. Plant Sci.* 7: 379. doi 10.3389/fpls.2016.00379.

Moschou PN, Roubelakis-Angelakis KA. 2013. Polyamines and programmed cell death. *J. Exp. Bot.* 65: 1285–1296; ert373.

Mulangi V, Chibucos MC, Phuntumart V, Morris PF. 2012. Kinetic and phylogenetic analysis of plant polyamine uptake transporters. *Planta* 236: 1261–1273.

Mulangi V, Phuntumart V, Aouida M, Ramotar D, Morris P. 2011. Functional analysis of OsPUT1, a rice polyamine uptake transporter. *Planta* 235: 1–11.

Neill S, Barros R, Bright J, Desikan R, Hancock J, Harrison J, Morris P, Ribeiro D, Wilson I. 2008. Nitric oxide, stomatal closure, and abiotic stress. *J. Exp. Bot.* 59: 165–176.

Neily MH, Matsukura C, Maucourt M, Bernillon S, Deborde C, Moing A, Yin Y-G, et al. 2011. Enhanced polyamine accumulation alters carotenoid metabolism at the transcriptional level in tomato fruit over-expressing spermidine synthase. *J. Plant Physiol.* 168: 242–252.

Paoletti P. 1984. Thermochemistry of metal-polyamine complexes. In: Silva MAVR da (ed.). *Thermochemistry and Its Applications to Chemical and Biochemical Systems*, pp. 339–352. Springer, Netherlands.

Pegg AE, Michael AJ. 2010. Spermine synthase. *Cel. Mol. Life Sci.* 67: 113.

Perez-Amador MA, Leon J, Green PJ, Carbonell J. 2002. Induction of the arginine decarboxylase *ADC2* gene provides evidence for the involvement of polyamines in the wound response in *Arabidopsis*. *Plant Physiol.* 130: 1454–1463.

Pospíšil P, Prasad A. 2014. Formation of singlet oxygen and protection against its oxidative damage in Photosystem II under abiotic stress. *J. Photochem. Photobiol. B, Biol.* 137: 39–48.

Puyang X, An M, Xu L, Han L, Zhang X. 2016. Protective effect of exogenous spermidine on ion and polyamine metabolism in Kentucky bluegrass under salinity stress. *Hort. Environ. Biotechnol.* 57: 11–19.

Rady MM, Hemida KA. 2015. Modulation of cadmium toxicity and enhancing cadmium-tolerance in wheat seedlings by exogenous application of polyamines. *Ecotox. Environ. Saf.* 119: 178–185.

Ramakrishna A, Ravishankar GA. 2011. Influence of abiotic stress signals on secondary metabolites in plants. *Plant Signal. Behav.* 6: 1720–1731.

Ramakrishna A, Ravishankar GA. 2013. Role of plant metabolites in abiotic stress tolerance under changing climatic conditions with special reference to secondary compounds. In: *Climate Change and Plant Abiotic Stress Tolerance*, pp. 705–726. Wiley-VCH Verlag GmbH & Co. KGaA, Weinheim.

Ramesh SA, Tyerman SD, Xu B, Bose J, Kaur S, Conn V, Domingos P, et al. 2015. GABA signalling modulates plant growth by directly regulating the activity of plant-specific anion transporters. *Nat. Comm.* 6: 7879.

Rosales EP, Iannone MF, Groppa MD, Benavides MP. 2011. Polyamines modulate nitrate reductase activity in wheat leaves: involvement of nitric oxide. *Amino Acids* 42: 857–865.

Rosenheim O. 1924. The isolation of spermine phosphate from semen and testis. *Biochem. J.* 18: 1253–1262.

Roy P, Niyogi K, SenGupta DN, Ghosh B. 2005. Spermidine treatment to rice seedlings recovers salinity stress-induced damage of plasma membrane and PM-bound H+-ATPase in salt-tolerant and salt-sensitive rice cultivars. *Plant Sci.* 168: 583–591.

Sagor GHM, Zhang S, Kojima S, Simm S, Berberich T, Kusano T. 2016. Reducing cytoplasmic polyamine oxidase activity in arabidopsis increases salt and drought tolerance by reducing reactive oxygen species production and increasing defense gene expression. *Plant Physiol.* 7: 214.

Sánchez-Rodríguez E, Romero L, Ruiz JM. 2016. Accumulation of free polyamines enhances the antioxidant response in fruits of grafted tomato plants under water stress. *J. Plant Physiol.* 190: 72–78.

Shabala S, Pottosin I. 2014. Regulation of potassium transport in plants under hostile conditions: implications for abiotic and biotic stress tolerance. *Physiol. Plantarum* 151: 257–279.

Sharma YK, Léon J, Raskin I, Davis KR. 1996. Ozone-induced responses in *Arabidopsis thaliana*: the role of salicylic acid in the accumulation of defense-related transcripts and induced resistance. *Proc. Nat. Acad. Sci. USA* 93: 5099–5104.

Shi H, Ye T, Chan Z. 2013. Comparative proteomic and physiological analyses reveal the protective effect of exogenous polyamines in the bermudagrass (*Cynodon dactylon*) response to salt and drought stresses. *J. Proteome Res.* 12: 4951–4964.

Shi K, Huang YY, Xia XJ, Zhang YL, Zhou YH, Yu JQ. 2008. Protective role of putrescine against salt stress is partially related to the improvement of water relation and nutritional imbalance in cucumber. *J. Plant Nutr.* 31: 1820–1831.

Tanou G, Ziogas V, Belghazi M, Christou A, Filippou P, Job D, Fotopoulos V, Molassiotis A. 2014. Polyamines reprogram oxidative and nitrosative status and the proteome of citrus plants exposed to salinity stress. *Plant Cell Environ.* 37: 864–885.

Tavladoraki P, Cona A, Federico R, Tempera G, Viceconte N, Saccoccio S, Battaglia V, Toninello A, Agostinelli E. 2011. Polyamine catabolism: target for antiproliferative therapies in animals and stress tolerance strategies in plants. *Amino Acids* 42: 411–426.

Tiburcio AF, Altabella T, Bitrián M, Alcázar R. 2014. The roles of polyamines during the lifespan of plants: from development to stress. *Planta* 240: 1–18.

Tun NN, Santa-Catarina C, Begum T, Silveira V, Handro W, Floh EIS, Scherer GFE. 2006. Polyamines induce rapid biosynthesis of nitric oxide (NO) in *Arabidopsis thaliana* seedlings. *Plant Cell Physiol.* 47: 346–354.

Urano K, Yoshiba Y, Nanjo T, Ito T, Yamaguchi-Shinozaki K, Shinozaki K. 2004. Arabidopsis stress-inducible gene for arginine decarboxylase AtADC2 is required for accumulation of putrescine in salt tolerance. *Biochem. Biophys. Res. Comm.* 313: 369–375.

van Leeuwenhoek A. 1678. Observationes D. Anthonii Leeuwenhoek, de natis e semine genitali animalculis. *Philos. Trans. R. Soc. Lond.* 12: 1040–1043.

Venkiteswaran S, Vijayanathan V, Shirahata A, Thomas T, Thomas TJ. 2005. Antisense Recognition of the HER-2 mRNA: Effects of Phosphorothioate Substitution and Polyamines on DNA·RNA, RNA·RNA, and DNA·DNA Duplex Stability. *Biochemistry* 44: 303–312.

Walters DR. 2000. Polyamines in plant–microbe interactions. *Physiol. Mol. Plant Pathol.* 57: 137–146.

Wang Y, Luo Z, Mao L, Ying T. 2016. Contribution of polyamines metabolism and GABA shunt to chilling tolerance induced by nitric oxide in cold-stored banana fruit. *Food Chem.* 197, Part A: 333–339.

Wi SJ, Kim SJ, Kim WT, Park KY. 2014. Constitutive S-adenosylmethionine decarboxylase gene expression increases drought tolerance through inhibition of reactive oxygen species accumulation in Arabidopsis. *Planta* 239: 979–988.

Yamaguchi K, Takahashi Y, Berberich T, Imai A, Miyazaki A, Takahashi T, Michael A, Kusano T. 2006. The polyamine spermine protects against high salt stress in Arabidopsis thaliana. *FEBS Lett.* 580: 6783–6788.

Yamaguchi K, Takahashi Y, Berberich T, Imai A, Takahashi T, Michael AJ, Kusano T. 2007. A protective role for the polyamine spermine against drought stress in Arabidopsis. *Biochem. Biophys. Res. Comm.* 352: 486–490.

Yin ZP, Li S, Ren J, Song XS. 2014. Role of spermidine and spermine in alleviation of drought-induced oxidative stress and photosynthetic inhibition in Chinese dwarf cherry (*Cerasus humilis*) seedlings. *Plant Growth Regul.* 74: 209–218.

Zadražnik T, Hollung K, Egge-Jacobsen W, Meglič V, Šuštar-Vozlič J. 2013. Differential proteomic analysis of drought stress response in leaves of common bean (*Phaseolus vulgaris* L.). *J. Proteom.* 78: 254–272.

Zapata PJ, Serrano M, Pretel MT, Amorós A, Botella MÁ. 2004. Polyamines and ethylene changes during germination of different plant species under salinity. *Plant Sci.* 167: 781–788.

Zarza X, Atanasov KE, Marco F, Arbona V, Carrasco P, Kopka J, Fotopoulos V, Munnik T, Gómez-Cadenas A, Tiburcio AF, Alcázar R. 2016. Polyamine oxidase 5 loss-of-function mutations in *Arabidopsis thaliana* trigger metabolic and transcriptional reprogramming and promote salt stress tolerance. *Plant Cell Environ.* 40: 527–542. doi:10.1111/pce.12714.

Zhang G, Xu S, Hu Q, Mao W, Gong Y. 2014. Putrescine plays a positive role in salt-tolerance mechanisms by reducing oxidative damage in roots of vegetable soybean. *J. Int. Agric.* 13: 349–357.

Zhao F, Song C-P, He J, Zhu H. 2007. Polyamines improve K+/Na+ homeostasis in barley seedlings by regulating root ion channel activities. *Plant Physiol.* 145: 1061–1072.

17

Plant Glycine-Rich Proteins and Abiotic Stress Tolerance

Juan Francisco Jiménez-Bremont, Maria Azucena Ortega-Amaro, Itzell Eurídice Hernández-Sánchez,
Alma Laura Rodriguez-Piña, and Israel Maruri-Lopez

CONTENTS

Introduction

Plants are amazing organisms: although they are sessile, plants are able to obtain nutrients, capture water and light, and can tolerate or even adapt to diverse stress conditions of their changing environment. Throughout the year, plants may suffer abiotic stresses such as salinity, drought, and extreme temperatures, among others, which affect the development, establishment, and survival of wild-plants and crops.

In the last decades, we have been observing how the planet is deteriorated by anthropogenic activities, thus affecting all organisms including plants. However, plants have evolvedmechanisms that allow them to survive adverse conditions. These stresses induce a wide variety of plant physiological and molecular responses, resulting in changes in gene expression patterns and, hence, in cell metabolism, inducing mechanisms of osmotic adjustment, repair and protection systems mediated by chaperones, production of late embryogenesis abundant (LEA) proteins, aquaporins, and glycine-rich proteins, among many others. All of these responses influence the distribution of plants, their survival, and crop yield. Therefore, one of the most important challenges of modern agriculture is to maintain and increase crop yields even under adverse environmental conditions.

The Glycine-Rich Proteins (GRPs) and Glycine-Rich Domain Proteins (GRDPs)

Plant glycine-rich proteins (GRPs) are a large family of proteins with a high glycine content, more than 20% of the total amino acids of the protein, which encodes diverse motifs or domains through the polypeptide chain (Sachetto-Martins et al., 2000; Mangeon et al., 2010). The study of GPRs began with the isolation and characterization of these genes in plants such as petunia (Condit and Meagher, 1986), common bean (Keller et al., 1988), maize (Gómez et al., 1988), and tobacco (van Kan et al., 1988). Although the first genes that encoded GRPs were discovered in plants, proteins with glycine signatures have been isolated in diverse species, such as animals, fungi, and cyanobacteria (Sachetto-Martins et al., 2000). Throughout the genome of *Arabidopsis thaliana* (Flores-Fusaro and Sachetto-Martins, 2007), *Saccharum officinarum* (Fusaro et al., 2001), and *Eucalyptus* (Bocca et al., 2005) more than 150 genes encoding GRP proteins have been identified; in contrast only 51 genes have been identified in the medicinal plant *Curcuma longa L.* (Kar et al., 2012).

Plant GRPs classification is based on the presence of characteristic domains and motifs, as well as the arrangement of the glycine repeats, and up to now, four classes have been proposed (Mangeon et al., 2009). Class I of GRPs contains a signal peptide followed by a glycine-rich region; the presence of this signal peptide indicates that these proteins can be synthesized in a different cellular compartment from where they carry out their function. GRPs of Class II contain a signal peptide and have a characteristic cysteine-rich C-terminal region. In the case of Class III, it presents a signal peptide and an N-terminal hydrophobic oleosin domain. Compared to other GRP classes, Class III has a lower overall glycine content. Classes I and II are related to a structural function in cell walls and plant development, while Class III is related to pollen hydration. Class IV of GRPs comprises those proteins with an RNA binding motif, such as RNA recognition motif (RRM) or a cold-shock domain (CSD), and may also contain CCHC zinc-fingers. Class IV of GRPs is subdivided into four subclasses: IVa containing one RRM motif, IVb has one RRM and a CCHC zinc-finger, IVc shows a cold-shock domain and two or more zinc-fingers, and IVd contains two RRMs (Mangeon et al., 2009). The RNA-binding GRPs play a role in biotic and abiotic stress responses (Kim et al., 2007a, 2008; Vermel et al., 2002).

Different signatures on the glycine-rich region have been identified, such as (I) GGX, (II) GGXXXGG, (III) GXGX, and (IV) GGX/GXGX, where frequently the X corresponds to alanine, serine or cysteine (Mangeon et al., 2010). Usually, the glycine repeats are interspersed along the entire protein; nevertheless, in some proteins, glycine repeats may form a short domain. Recently, these types of GRPs have been named glycine-rich domain proteins (GRDPs). GRDPs have been reported in *Eucalyptus* and *Arabidopsis* (Bocca et al., 2005; Rodríguez-Hernández et al., 2014; Ortega-Amaro et al., 2015).

In plants, genes encoding GRP and GRDP proteins have been involved in growth and development and are also modulated under abiotic and biotic stresses (Reddy and Poovaiah, 1987; Urbez et al., 2006; Long et al., 2013; Rodríguez-Hernández et al., 2014; Ortega-Amaro et al., 2015, 2016). Also, differences in transcript accumulation of some GRPs and GRDPs were found in response to phytohormone treatments, such as abscisic acid and indole-3-acetic acid (Kim et al., 2005, 2007b; Rodríguez-Hernández et al., 2014; Ortega-Amaro et al., 2015).

Glycine-rich proteins play important roles in signal transduction, transcriptional and post-transcriptional regulation, and in protein-protein interactions. In this chapter, we describe the role of GRPs and GRDPs under abiotic stress, and we will cover three sections: 1) cell wall glycine-rich proteins, 2) RNA-binding proteins associated with glycine-rich regions (RBP-GRPs), and 3) glycine-rich domain proteins (GRDPs) (Figure 17.1).

The Structural Glycine-Rich Proteins (Class I–III)

It has been reported that some GRPs play an important role in plant cell wall composition. The Classes I–III GRPs form a group of structural components of the cell wall, in addition to extensions and proline-rich proteins. Furthermore, the glycine content of these structural proteins is 60% or 70%. Several studies have revealed that these types of proteins are located in the cell wall of the protoxylem (Ringli et al., 2001a).

AtGRP3 a Cell Wall Glycine-Rich Protein

The plant cell wall provides the structural framework of plants and acts as the first line of defense. Additionally, the cell wall is the first step in signal transduction pathway when the plant is exposed to stress conditions (Houston et al., 2005). GRPs have been associated with signal transduction; such is the case of *Arabidopsis thaliana* GRP3. AtGRP3 is an extracellular glycine-rich protein belonging to the Class II of the GRP family (Flores-Fusaro and Sachetto-Martins, 2007; Park et al., 2001). This GRP is structurally organized into several domains, having a potential signal peptide, a glycine-rich region, and a cysteine-rich C-terminus (Park et al., 2001). AtGRP3 is a cell wall protein that binds and negatively regulates the plant wall-associated kinase, WAK1 (Park et al., 2001). WAK1 is involved in cell expansion and activation of defense responses against fungi and bacteria, and it functions as a serine/threonine receptor-like kinase (RLK) (Brutus et al., 2010; Mangeon et al., 2016). The *Atgrp3* null mutant line phenotype shows longer roots with augmented cell expansion in root cells. Conversely, the AtGRP3 overexpression lines have shorter roots (Mangeon et al., 2017). The *Atgrp3* null mutant also exhibits other phenotypes, such as enhanced aluminum tolerance (Mangeon et al., 2016). Consequently, the WAK1 overexpression line also exhibits tolerance to aluminum, confirming the negative regulation of the glycine-rich protein AtGRP3 against WAK1 (Sivaguru et al., 2003).

AtGRP5, AtGRDP23, and AtGRDP9 Are Involved in Stress Response

The *Arabidopsis thaliana* glycine-rich protein AtGRP5 was initially reported by de Oliveira et al. (1990). Based on its domain arrangement, AtGRP5 is classified in Class I of the GRP family (Mangeon et al., 2010). Its N-terminal contains the signal peptide consensus sequence, and it has a 66% glycine content. Promoter analysis using GUS expression assays revealed that *AtGRP5* is expressed in reproductive organs, epidermal tissues, and fruits of *Arabidopsis*. In addition, the *AtGRP5* is expressed in epidermal cells of leaves and stems, and also in guard cells and trichomes (Sachetto-Martins et al., 1995). *AtGRP5* gene is specifically induced by phytohormones, such as abscisic acid, ethylene, and salicylic acid (de Oliveria et al., 1990). In addition, AtGRP5 was found to be associated with somatic embryo development (Magioli et al., 2001).

The glycine-rich protein AtGRP23 is closely related to AtGRP5 in structure and contains 63% glycine. The enhanced expression of AtGRP23 and AtGRP5 by 16-hydroxypalmitic acid (HPA), one component of cutin, suggests that these genes could participate in plant defense against pathogens. Both genes are induced by abscisic acid and salicylic acid (Park et al., 2008).

Another glycine-rich protein belonging to Class I is AtGRP9, which codes for a protein of 127 amino acids with 43.3% glycine. It has been reported that NaCl and ABA treatments induce the *AtGRP9* transcript in *Arabidopsis*. Interestingly, a yeast two-hybrid analysis revealed that AtGRP9 protein interacts with *A. thaliana* AtCAD5 (Chen et al., 2007). Cinnamyl alcohol dehydrogenase (CAD) is a key enzyme in lignin biosynthesis, suggesting that AtGRP9 may participate in lignin synthesis in response to salt stress through interaction with AtCAD5 (Chen et al., 2007).

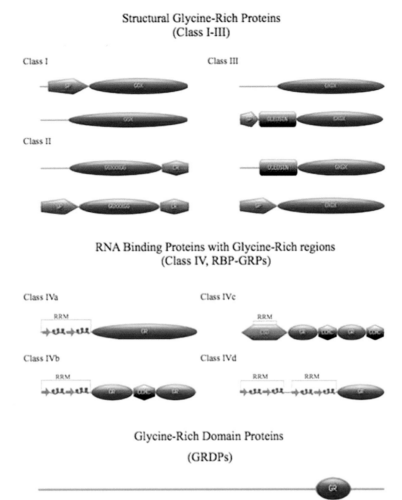

FIGURE 17.1 Diagrammatic representation of the classes of Glycine-rich proteins (GRPs). GRPs were subdivided into several classes such as structural GRPs that comprise classes I-III, RNA Binding Proteins associated with glycine-rich region (RBP-GRPs, Class IV), and Non-canonical Glycine-rich domain proteins (GRDPs). SP, signal peptide; Glycine-rich regions GR; CR, cysteine-rich domain; Oleosin, Oleosin domain; RRM, RNA-binding motif; CCHC, zinc-finger; CSD, Cold-shock domain. The schematic representations were performed using MyDomains tool (Hulo et al., 2008). The classes of GRPs were taken and modified as reported in Mangeon et al. (2010).

In other plants, glycine-rich proteins have also been reported as structural components of cell walls. The GRP-1 of Petunia and GRP-1.8 of *Phaseolus vulgaris* perform a structural role in cell walls as a part of a repair system (Condit et al., 1990). PvGRP-1.8 is classified as Class I and is localized in the primary cell wall involved in protoxylem development. It has been proposed that PvGRP-1.8 has a structural role in cell walls as a part of a repair system of the protoxylem and that this GRP connects lignin rings leading to cell wall fortification (Ryser and Ryser, 1992; Ryser et al., 1997; Ringli et al., 2001b). In tobacco, the NtCIG1 glycine-rich protein, belonging to Class II of the GRP family, is involved in the vascular tissue and enhances callose deposition in the cell wall (Ueki and Citovsky, 2002).

The AtGRP17 Is a Pollen-Specific Oleosin-Like Protein

The GRPs that belong to Class III present a GXGX pattern, and oleosin domains. These GRPs show a lower content of glycine residues than other classes of GRPs (Mangeon et al., 2010). In *A. thaliana*, five pollen membrane proteins have been reported

GRP19, GRP16, GRP17, GRP18, and GRP14 (de Oliveira et al., 1993; Mayfield et al., 2001). The AtGRP17 was identified as the most abundant pollen coat protein in *Arabidopsis* (de Oliveira et al., 1990; Mayfield et al., 2001). The AtGRP17 has an oleosin domain, which is usually present in proteins that are associated with oil bodies. In *Arabidopsis*, the *AtGRP17* transcript accumulation was observed in anthers in the later stages of flower development, and it displays a reduction before flower opening (de Oliveira et al., 1993). It has been reported that AtGRP17 protein is involved in pollen hydration and recognition. Therefore, the mutant *grp17-1* showed a severe delay in the commencement of pollen hydration, derived from a failure to interact with the stigma (Mayfield and Preuss, 2000).

The Glycine-Rich RNA-Binding Proteins (GRPs) Are Involved in the Abiotic Stress Response

Plant stress responses depend widely on transcriptional, post-transcriptional and translational regulation. In particular, post-transcriptional regulation of RNA implicates processes such as capping, splicing, polyadenylation, transport, and translation. Crucial components of RNA metabolism are the RNA-binding proteins, which regulate the RNA biogenesis, stability, transport, and localization, through interactions with RNA.

In particular, RNA-binding proteins contain at least one RNA-binding domain at the N-terminal, and the C-terminal may contain different types of assistant motifs, such as glycine-rich or arginine-rich motifs, CCHC-type finger motifs, and RD/SR-repeat motifs, among others (Ambrosone et al., 2012). The Class IV GRPs or RNA-binding glycine-rich protein (RB-GRP) plays a role in post-transcriptional gene regulation in response to some abiotic and biotic stresses (Kwak et al., 2016). It has been reported that the GR region of RB-GRPs can form glycine loops and ß-pleated sheets to facilitate the RNA-binding activity (Condit and Meagher, 1986; Steinert et al., 1991; Hanano et al., 1996). Two types of domains have been reported for the RNA binding sequence of this class (RB-GRP), the RNA recognition motif (RRM) or the cold-shock domain (CSD), to result in subclasses IVb and IVc, respectively (Mangeon et al., 2010). The RRM motif or RNA-binding domain is approximately 90 amino acids in length and contains two conserved sequences, RNP-1 and RNP-2, essential for binding activity (Adam et al., 1986). In addition to the β-sheet formation by the RNP motifs, the external β-strands and the loops, or even more RRM domains, allow high affinity and specific recognition (Maris et al., 2005). The CSD has been found in proteins with an RNA chaperone activity that works during cold stress, both in prokaryotes and eukaryotes (Karlson and Imai, 2003).

AtGRP2/GR-RBP2 Is Involved in Abiotic Stress Response

In *A. thaliana*, AtGRP2 (GR-RBP2; At4g13850) is a glycine-rich RNA-binding protein belonging to Class IV of the GRP family, which contains an N-terminal RNA recognition motif (RRM), and a C-terminal glycine-rich domain. It has been reported that AtGRP2 is located in the mitochondria, and it is induced by cold stress. Vermel et al. (2002) propose that AtGRP2 could participate in the post-transcriptional processing of mitochondrial gene expression, and in response to cold stress. Accordingly, *Arabidopsis* AtGRP2 overexpression lines showed germination and seedling growth tolerance under low and freezing temperatures as well as salt stress, suggesting that AtGRP2 is involved in these abiotic stress responses (Kim et al., 2007a). The capacity of AtGRP2 melting RNA structures and the RNA chaperone activity under cold treatments were demonstrated through heterologous expression of *AtGRP2* gene in *E. coli* RL211 strain and BX04 mutant strain (Kim et al., 2007a). Transgenic rice that overexpresses the AtGRP2 or AtGRP7 displayed a tolerant phenotype under drought stress conditions, achieving major chlorophyll contents and grain yield in comparison to wild-type plants (Yang et al., 2014).

AtGRP4 Has Protein Chaperone Activity

The designated AtGRP4 (At3g23830) is the smallest RNA-binding GRP from the *Arabidopsis* GR-RBP family. The AtGRP4 is a mitochondrial protein with 136 amino acids, containing only 12 glycine residues at the C-terminus (Elo et al.,

2003; Kwak et al., 2005). *In silico* structural prediction revealed that the N-terminal conserved regions that harbor the RRM domains, could adopt α-helical and β-strand conformations (Han et al., 2013).

The *AtGRP4* gene is modulated under abiotic stress conditions; it is induced by cold treatments, but down-regulated under high salinity or dehydration conditions. However, *Arabidopsis* transgenic lines that overexpress the *AtGRP4* gene did not confer cold tolerance, and when these lines were subjected to dehydration and salt stress, a negative impact in seed germination and seedling growth was observed (Kwak et al., 2005). Furthermore, the heterologous expression of AtGRP4 did not complement the cold sensitivity phenotype of BX04 mutant *E. coli* (Kim et al., 2007b). Despite the fact that AtGRP4 does not display RNA chaperone activity in *E. coli* (Kim et al., 2007b), AtGRP4 exhibits protein chaperone activity (Kim et al., 2007b). The ability of AtGRP4 to function as a protein chaperone is related to its capacity to form homo-oligomeric complexes of high molecular weight, where the first 25 amino acids of the N-terminal of the AtGRP4 are indispensable for the formation of these complexes, and for protein chaperone activity (Han et al., 2013).

AtGRP7 Is Involved in Cold Stress Response

The glycine-rich RNA-binding domain AtGRP7 (At2g21660), known as GR-RBP7, has been related to several important plant processes, such as the role in abiotic and biotic stress tolerance (Kim et al., 2008). Remarkably, the expression of *AtGRP7* is regulated by diverse abiotic stresses, and it is also controlled by the circadian clock (Carpenter et al., 1994; Heintzen et al., 1997). *AtGRP7* heterologous expression rescues the cold sensitivity phenotype of *E. coli* BX04 mutant (Kim et al., 2007b), which is attributed to the RNA chaperone activity. It has been reported that the ability of AtGRP7 to augment cold tolerance is related to its RNA chaperone activity (Kim et al., 2008).

Interestingly, it has been reported that the *AtGRP7* gene negatively auto-regulates the steady-state abundance of its own mRNA through a post-transcriptional mechanism (Staiger et al., 2003). The increased abundance of AtGRP7 protein in 35S::AtGRP7 overexpression lines results in a repression of the endogenous *Atgrp7* transcript (Heintzen et al., 1997). The AtGRP7 protein binds to its own transcript to auto-regulate itself (Streitner et al., 2012). This occurs by partial retention of its own intron, where an alternative 5´ splice site is used to retain part of the *Atgrp7* intron (Staiger et al., 2003). This alternative splicing generates a premature stop codon, and its degradation through the nonsense-mediated decay pathway (NMD, Schöning et al., 2008). In addition to alternative splicing of pre-mRNAs mediated by AtGRP7, this protein has other functions in RNA metabolism such as regulating mRNA export from the nucleus to the cytosol under cold treatments (Kim et al., 2008), and recently its participation in primary microRNA processing (Köster et al., 2014).

The AtGRP8 Gene Is Involved in Cold Stress Response

AtGRP8 (At4g39260) is paralogous to AtGRP7. It contains a putative RNA-binding domain and a carboxyl-terminal domain composed of stretches of glycine and serine with interspersed hydrophilic residues (Van Nocker and Vierstra, 1993). This RNA-binding GRP gene was found to be regulated by the circadian rhythm, and induced in response to cold treatment (4°C) and oxidative stress, suggesting functional roles in these responses (Schmidt et al., 2010; Carpenter et al., 1994). Proteomic analyses of *Arabidopsis* and GFP translational fusion have revealed that AtGRP8 is a nuclear/cytoplasmic protein that could be imported by AtTRN1, an importin-β family member (Pendle et al., 2005; Ziemienowicz et al., 2003). This dual localization suggests that AtGRP8 may have functions in both cellular compartments.

As described above, AtGRP7 can regulate itself, in turn, AtGRP8 also regulates its own transcripts. During cold stress, AtGRP8 promotes the use of a 5′ splice site of its own transcript, generating an alternative nonfunctional splice variant that is targeted by NMD mechanism (Schöning et al., 2008). Both AtGRP7 and AtGRP8 proteins were able to bind to their own mRNAs and each other *in vitro*. In addition, AtGRP8 was negatively regulated in the AtGRP7 overexpression transgenic lines, and also AtGRP8 regulates AtGRP7 transcript, demonstrating that AtGRP8/AtGRP7 are reciprocally regulated (Heintzen et al., 1997; Schöning et al., 2008). Using *Arabidopsis* independent RNA interference lines with reduced levels of AtGRP7 and AtGRP8, it was demonstrated that AtGRP7 and AtGRP8 proteins regulate flowering transition in *A. thaliana* (Streitner et al., 2008).

Camelina sativa CsGRP7 Genes and Sorghum SbGRBP Encoding to Glycine-Rich RNA-Binding Proteins

The functional characterization of proteins that shows homology to glycine-rich RNA-binding proteins is increasing in crop plants. Such is the case of SbGRBP, a gene of *sorghum* that responds to heat stress (Singh et al., 2017). SbGRBP localized in the nucleus and cytosol, and, in the presence of Ca^{2+} ions, it can interact with calmodulin. SbGRBP is also able to interact with nucleic acids (Singh et al., 2017).

Kwak and col. (2016) isolated three glycine-rich RNA-binding proteins from *Camelina sativa*, an oilseed crop that is important for biodiesel production. These three GRP genes, *CsGRP7a*, *CsGRP7b*, and *CsGRP7c*, were induced by cold stress, and their respective proteins were localized in the cytosol and nucleus. The authors demonstrated that the *Camelina* CsGRP7 (a, b, and c) proteins have RNA chaperone activity in *E. coli* systems, in which these GRPs had transcription anti-termination activities, and also they complemented the cold sensitivity of *E. coli* BX04 strain. Overexpression of *CsGRP7a* in *Camelina* transgenic lines induced enhanced root growth under cold stress, in comparison to *Camelina* parental plantlets (Kwak et al., 2016).

Class IVc, Glycine-Rich Protein with Cold-Shock Domain

The GRPs' Class IVc differs from the others' plant glycine-rich proteins because they possess a cold-shock domain (CSD) localized at the first N-terminal residues (Sasaki and Imai, 2011). The CSD is crucial for RNA and single- and double-stranded DNA binding (Graumann and Marahiel, 1996, 1998; Kloks et al., 2002). Some Class IVc GRP proteins have CCHC-type zinc-fingers (Kim et al., 2007b; Yang and Karlson, 2013).

The first characterized plant CSD protein was the wheat cold shock protein 1 (WCSP1), which was accumulated in crown tissue during cold acclimation (Karlson et al., 2002). The WCSP1 protein contains a glycine-rich region interspersed with three C-terminal CCHC zinc-fingers and has the ability to dissolve double-stranded RNA/DNA molecules and complement a cold-sensitive phenotype of an *E. coli* BX04 mutant strain (Karlson et al., 2002; Nakaminami et al., 2006). In *A. thaliana*, there are four genes encoding CSD proteins (AtCSP1-4) (Sasaki and Imai, 2011).

The AtCSP1 Gene Is Induced by Cold

The AtCSP1 (At4g36020) contains a CSD localized at the first 77 N-terminal residues and exhibits eight glycine-rich regions interspersed with seven CCHC-type zinc fingers throughout the rest of protein (Kim et al., 2007b). It is known that the *AtCSP1* gene is induced during cold stress in *A. thaliana* plants (4°C), and its heterologous expression was able to suppress the cold sensitivity of *E. coli* BX04 cells (Kim et al., 2007). Moreover, the AtCSP1 protein disclosed both DNA melting and RNA chaperone activity during cold adaptation process (Kim et al., 2007b). A subcellular localization analysis using an N-terminal GFP fusion revealed a dual nucleus and cytosol localization for AtCSP1 protein in onion epidermal cells (Yang and Karlson, 2013). Remarkably, the overexpression of *AtCSP1* in the *Arabidopsis grp7* mutant background rescued its cold sensibility phenotype.

The AtCSP2 Is Regulated by Abiotic Stress

The AtCSP2 (At4g38680) protein is composed of a nucleic acid-binding CSD domain, two glycine-rich domains interspersed with two CCHC zinc-fingers (Fusaro et al., 2007; Sasaki et al., 2007). *AtCSP2* transcript accumulation was noticed under cold stress in *Arabidopsis* plants under 6°C for 48 h (Fusaro et al., 2007). Also, Sasaki et al. (2007) showed that AtCSP2 mRNA and protein levels increased during cold acclimation at 4°C. In this sense, *AtCSP2* heterologous expression in *E. coli* BX04 mutant shows RNA chaperone activity (Sasaki et al., 2007; Park et al., 2009). Conversely, the overexpression of *AtCSP2* decreases freezing tolerance under cold-acclimated conditions possibly by negative regulation of CBFs and *cor* genes (Sasaki et al., 2013). The retarded germination phenotype in AtCSP2-overexpressing lines was related to higher ABA contents; this phenotype might be related to the AtCSP2 negative regulation of ABA catabolic gene CYP707A2 (Sasaki et al., 2015a). Since the *AtCSP2* gene is induced by salinity, assays of salt stress tolerance were performed with the double mutant (atcsp2/atcsp4) and AtCSP2-overexpressing lines; a clear stress tolerance was observed in this mutant and a sensitivity in the overexpressing lines. It has been proposed that AtCSP2 acts as a negative regulator of salt stress tolerance in *Arabidopsis* due to the alteration of the expression of genes critical for the abiotic stress responses (Sasaki et al., 2015b). Instead, the overexpression of *AtCSP2* in the *grp7* mutant rescued the cold susceptibility phenotype, which was detected through a higher survival rate at freezing temperatures relative to the *grp7* mutant (Park et al., 2009).

The AtCSP3 Is Involved in Abiotic Stress Tolerance

The AtCSP3 (At2g17870) possesses an N-terminal CSD and seven glycine-rich regions interspersed with seven CCHC zinc-fingers throughout protein (Kim et al., 2009). The role of AtCSP3 in abiotic stress tolerance acquisition has been demonstrated (Kim et al., 2009, 2013a). It is known that AtCSP3 complemented the *E. coli* BX04 strain and disclosed *in vitro* nucleic acid melting activity (Kim et al., 2009). Moreover, the *AtCSP3* gene was induced in *A. thaliana* seedlings during cold, salt, drought, and ABA treatments. The analysis of an *AtCSP3* promoter-GUS reporter line disclosed a shoot and root apical tissue-specific *AtCSP3* expression (Kim et al., 2009). Furthermore, the atcsp3 knockout mutant line was more sensitive to freezing (−2°C for 14 h), drought (without water for 5 d), and salt (200 mM NaCl) stresses. In contrast, the *AtCSP3* overexpression conferred enhanced tolerance to *A. thaliana* plants under these stresses in comparison to wild-type and the *atcsp3* mutant (Kim et al., 2009, 2013a). In this regard, AtCSP3 protein has an essential role in abiotic stress adaptation through the regulation of abiotic stress responsive genes; in particular, during salt and drought stress some genes, which were down-regulated in the *atcsp3* mutant, showed an increased expression in the AtCSP3 overexpressing plants, such is the case of *peroxidase 62* (At5g39580), *nitrate transporter 2* (At1g08090), and *WRKY18* (At4g31800) (Kim et al., 2013a).

Furthermore, a yeast two-hybrid analysis indicates that AtCSP3 forms several complexes with RNA associated proteins: the decapping protein 5 (At1g26110), three putative nuclear poly(A)-binding proteins (PABN1, At5g51120; PABN 2, At5g10350; PABN 3, At5g65260), three 60S ribosomal proteins (RPL36aB, At3g23390; RPL26A, At3g49910; RPL40A, At2g36170), and two members of the DEAD/DEAH-box helicases (RH15, At5g11200 and PRH75, At5g62190). A bimolecular fluoresce complementation approach performed on onion epidermal cells confirmed the *in vivo* interaction of AtCSP3 with these RNA associated proteins, which take place in both nuclear and cytoplasmic compartments (Kim et al., 2013b). These interactions of AtCSP3 with RNA associated proteins might suggest that this GRP may be involved in several processes of RNA metabolism.

The AtCSP4 Has an Important Role in Late Embryo Development

The AtCSP4 (At2g21060) protein contains a nucleic acid-binding CSD domain, two glycine-rich domains interspersed with two CCHC zinc-fingers (Sasaki and Imai, 2011), and a putative redundant function has been suggested with its paralogue. The subcellular localization of AtCSP4 was examined by a GFP C-terminal fusion having a nuclear and cytoplasmic localization in onion epidermal cells (Yang and Karlson, 2011). Both qRT-PCR and promoter analyses of AtCDS4 revealed that this gene was expressed in multiple tissues but accumulated preferentially in apical meristem and reproductive tissues, especially in dehiscence area, valves, and siliques with mature seeds (Yang and Karlson, 2011). Overexpression of *AtCSP4* diminishes silique length and caused embryo lethality (Yang and Karlson, 2011). However, the *atcsp4* mutant shows no apparent phenotypic abnormalities, possibly due to functional redundancy with its *AtCSP2*

paralogue. In plants that overexpress the *AtCSP4* gene, several MADS-box genes (*AP1, CAL, AG*, and *SHP2*) are induced during early stages of silique development (Yang and Karlson, 2011).

Glycine-rich Domain Proteins (GRDPs)

Since the identification of genes encoding proteins with a glycine-rich domain in the *Eucalyptus* and rice genome, a group of GRDPs has been studied recently (Bocca et al., 2005; Mangeon et al., 2010). In common bean, the *P. vulgaris* glycine-rich domain protein 1 (*PvGRDP1*) gene was isolated from a subtractive cDNA library generated under salt stress conditions. The *PvGRDP1* expression levels increase in common bean leaves after 2 and 5 days of salt treatment (Hernández-Lucero et al., 2014). Two orthologous genes of *PvGRDP1* were identified in *A. thaliana*, *AtGRDP1* and *AtGRDP2* (Ortega-Amaro et al., 2015; Rodriguez-Hernández et al., 2014). The structural features of AtGRDP1 and AtGRDP2 proteins are very similar; they present a domain of unknown function 1399 (DUF1399) at the N-terminal, a potential RNA-binding domain, and a glycine-rich domain at the carboxy-terminus. In the glycine-rich region, AtGRDP1 has five repeats of the consensus region XSGCGXXCXGXCG, while AtGRDP2 possess four consensus regions [CG]GGGCGG[GC] interspersed only with cysteine.

AtGRDP1 gene has an important role in abiotic stress tolerance and plays a regulatory role in ABA signaling. The 35S::AtGRDP1 overexpression lines showed a tolerant phenotype in germination and development of green cotyledons under salt and osmotic stress, also a marked resistance to ABA at high concentrations (Rodríguez-Hernández et al., 2014). Whereas the *Atgrdp1* mutant showed ABA hypersensitivity, mimicking the ABI3-overexpression phenotype. The ABA signaling central regulators, *ABI3* and *ABI5* transcripts, were found to be oppositely regulated in the *AtGRDP1* overexpression and mutant background (Rodríguez-Hernández et al., 2014). Further studies revealed that *AtGRDP1* transcript levels are important for the normal silique and seed development (Rodríguez-Hernández et al., 2017).

The *AtGRDP2* is involved in *Arabidopsis* growth and development, and also in response to salt stress. The growth rate was increased in the *Arabidopsis* 35S::AtGRDP2 overexpression lines, in contrast to the slow growth phenotype of the *Atgrdp2* mutants (Ortega-Amaro et al., 2015). Such a feature is possible since *ARF6, ARF8*, and *miR167* expression levels were altered in the different genetic backgrounds of AtGRDP2, also indole-3-acetic acid accumulation was higher in the AtGRDP2 overexpression lines (Ortega-Amaro et al., 2015). AtGRDP2 overexpression in lettuce showed alterations in the growth rate and time to flowering, similar to *Arabidopsis* AtGRDP2 overexpression lines. Also, the AtGRDP2 overexpression in *Arabidopsis* and lettuce display higher tolerance to salinity (Ortega-Amaro et al., 2015). In addition, *AtGRDP2* overexpression in common bean hairy-roots, generated by *Agrobacterium rhizogenes,* shows an increment in the biomass, an accelerated growth of hairy-roots, and an increased tolerance to salt stress (Ortega-Amaro et al., 2016). Altogether these reports show that plant GRDPs are involved in growth rate regulation and improvement of abiotic stress tolerance.

Conclusion

The up-to-date understanding of plant GRP proteins has increased significantly. The first GRPs were identified as proteins that had accumulated under different stress conditions, suggesting a potential role in stress tolerance. However, many other important roles have been elucidated in the recent years: i) GRPs with structural functions associated to plant cell walls, ii) GRPs with RNA-binding domains involved in post-transcriptional regulation. The usage of "Omics" technologies will offer the generation of novel data on GRPs, prompting the discovery of novel functions and the mechanisms of action of these glycine-rich proteins.

REFERENCES

Adam, S. A., Nakagawa, T., Swanson, M. S., Woodruff, T. K., and Dreyfuss, G. (1986). mRNA polyadenylate-binding protein: gene isolation and sequencing and identification of a ribonucleoprotein consensus sequence. *Molecular and Cellular Biology, 6*(8), 2932–2943.

Ambrosone, A., Costa, A., Leone, A., and Grillo, S. (2012). Beyond transcription: RNA-binding proteins as emerging regulators of plant response to environmental constraints. *Plant Science, 182*, 12–18.

Bocca, S. N., Magioli, C., Mangeon, A., Junqueira, R. M., Cardeal, V., Margis, R., and Sachetto-Martins, G. (2005). Survey of glycine-rich proteins (GRPs) in the Eucalyptus expressed sequence tag database (ForEST). *Genetics and Molecular Biology, 28*(3), 608–624.

Brutus, A., Sicilia, F., Macone, A., Cervone, F., and De Lorenzo, G. (2010). A domain swap approach reveals a role of the plant wall-associated kinase 1 (WAK1) as a receptor of oligogalacturonides. *Proceedings of the National Academy of Sciences, 107*(20), 9452–9457.

Carpenter, C. D., Kreps, J. A., and Simon, A. E. (1994). Genes encoding glycine-rich *Arabidopsis thaliana* proteins with RNA-binding motifs are influenced by cold treatment and an endogenous circadian rhythm. *Plant Physiology, 104*(3), 1015–1025.

Condit, C. M., McLean, B. G., and Meagher, R. B. (1990). Characterization of the expression of the petunia glycine-rich protein-1 gene product. *Plant Physiology, 93*(2), 596–602.

Condit, C. M., and Meagher, R. B. (1986). A gene encoding a novel glycine-rich structural protein of petunia. *Nature, 323*, 178–181.

Chen, A. P., Zhong, N. Q., Qu, Z. L., Wang, F., Liu, N., and Xia, G. X. (2007). Root and vascular tissue-specific expression of glycine-rich protein AtGRP9 and its interaction with AtCAD5, a cinnamyl alcohol dehydrogenase, in *Arabidopsis thaliana*. *Journal of Plant Research, 120*(2), 337–343.

de Oliveira, D. D., Franco, L. O., Simoens, C., Seurinck, J., Coppieters, J., Botterman, J., and Montagu, M. V. (1993). Inflorescence-specific genes from *Arabidopsis thaliana* encoding glycine-rich proteins. *The Plant Journal, 3*(4), 495–507.

de Oliveira, D. E., Seurinck, J., Inzé, D., Van Montagu, M., and Botterman, J. (1990). Differential expression of five *Arabidopsis* genes encoding glycine-rich proteins. *The Plant Cell, 2*(5), 427–436.

Elo, A., Lyznik, A., Gonzalez, D. O., Kachman, S. D., and Mackenzie, S. A. (2003). Nuclear genes that encode mitochondrial proteins for DNA and RNA metabolism are clustered in the *Arabidopsis* genome. *The Plant Cell, 15*(7), 1619–1631.

Flores Fusaro, A., and Sachetto-Martins, G. (2007). Blooming time for plant glycine-rich proteins. *Plant Signaling and Behavior, 2*(5), 386–387.

Fusaro, A., Mangeon, A., Junqueira, R. M., Rocha, C. A. B., Coutinho, T. C., Margis, R., and Sachetto-Martins, G. (2001). Classification, expression pattern and comparative analysis of sugarcane expressed sequences tags (ESTs) encoding glycine-rich proteins (GRPs). *Genetics and Molecular Biology, 24* (1–4), 263–273.

Fusaro, A. F., Bocca, S. N., Ramos, R. L. B., Barrôco, R. M., Magioli, C., Jorge, V. C., Coutinho, T. C., et al. (2007). AtGRP2, a cold-induced nucleo-cytoplasmic RNA-binding protein, has a role in flower and seed development. *Planta, 225*(6), 1339–1351.

Gómez, J., Sánchez-Martínez, D., Stiefel, V., Rigau, J., Puigdomènech, P., and Pagès, M. (1988). A gene induced by the plant hormone abscisic acid in response to water stress encodes a glycine-rich protein. *Nature, 334*, 262–264.

Graumann, P., and Marahiel, M. A. (1996). Some like it cold: response of microorganisms to cold shock. *Archives of Microbiology, 166*(5), 293–300.

Graumann, P. L., and Marahiel, M. A. (1998). A superfamily of proteins that contain the cold-shock domain. *Trends in Biochemical Sciences, 23*(8), 286–290.

Han, J. H., Jung, Y. J., Lee, H. J., Jung, H. S., Lee, K. O., and Kang, H. (2013). The RNA chaperone and protein chaperone activity of *Arabidopsis* glycine-rich RNA-binding protein 4 and 7 is determined by the propensity for the formation of high molecular weight complexes. *The Protein Journal, 32*(6), 449–455.

Hanano, S., Sugita, M., and Sugiura, M. (1996). Isolation of a novel RNA-binding protein and its association with a large ribonucleoprotein particle present in the nucleoplasm of tobacco cells. *Plant Molecular Biology, 31*(1), 57–68.

Heintzen, C., Nater, M., Apel, K., and Staiger, D. (1997). AtGRP7, a nuclear RNA-binding protein as a component of a circadian-regulated negative feedback loop in *Arabidopsis thaliana*. *Proceedings of the National Academy of Sciences, 94*(16), 8515–8520.

Hernández-Lucero, E., Rodríguez-Hernández, A. A., Ortega-Amaro, M. A., and Jiménez-Bremont, J. F. (2014). Differential expression of genes for tolerance to salt stress in common bean (Phaseolus vulgaris L.). *Plant Molecular Biology Reporter, 32*(2), 318–327.

Houston, N. L., Fan, C., Schulze, J. M., Jung, R., and Boston, R. S. (2005). Phylogenetic analyses identify 10 classes of the protein disulfide isomerase family in plants, including single-domain protein disulfide isomerase-related proteins. *Plant Physiology, 137*(2), 762–778.

Hulo, N., Bairoch, A., Bulliard, V., Cerutti, L., Cuche, B. A., de Castro, E., Lachaize, C., Langendijk-Genevaux, P. S., and Sigrist, C. J. A. (2008). The 20 years of PROSITE. *Nucleic Acids Research, 36*(Database issue), D245–D249.

Kar, B., Nayak, S., and Joshi, R. K. (2012). Classification and comparative analysis of Curcuma longa L. expressed sequences tags (ESTs) encoding glycine-rich proteins (GRPs). *Bioinformation, 8*(3), 142.

Karlson, D., and Imai, R. (2003). Conservation of the cold shock domain protein family in plants. *Plant Physiology*, *131*(1), 12–15.

Karlson, D., Nakaminami, K., Toyomasu, T., and Imai, R. (2002). A cold-regulated nucleic acid-binding protein of winter wheat shares a domain with bacterial cold shock proteins. *Journal of Biological Chemistry*, *277*(38), 35248–35256.

Keller, B., Sauer, N., and Lamb, C. J. (1988). Glycine-rich cell wall proteins in bean: gene structure and association of the protein with the vascular system. *The EMBO Journal*, *7*(12), 3625.

Kim, J. S., Jung, H. J., Lee, H. J., Kim, K., Goh, C. H., Woo, Y., Oh, S. H., Han, Y. S., and Kang, H. (2008). Glycine-rich RNA-binding protein7 affects abiotic stress responses by regulating stomata opening and closing in *Arabidopsis thaliana*. *The Plant Journal*, *55*(3), 455–466.

Kim, J. Y., Park, S. J., Jang, B., Jung, C. H., Ahn, S. J., Goh, C. H., Cho, K., Han, O., and Kang, H. (2007a). Functional characterization of a glycine-rich RNA-binding protein 2 in *Arabidopsis thaliana* under abiotic stress conditions. *The Plant Journal*, *50*(3), 439–451.

Kim, J. S., Park, S. J., Kwak, K. J., Kim, Y. O., Kim, J. Y., Song, J., Jang, B., Jung, C. H., and Kang, H. (2007b). Cold shock domain proteins and glycine-rich RNA-binding proteins from *Arabidopsis thaliana* can promote the cold adaptation process in Escherichia coli. *Nucleic Acids Research*, *35*(2), 506–516.

Kim, M.-H., Sasaki, K., and Imai, R. (2009). Cold shock domain protein 3 regulates freezing tolerance in *Arabidopsis thaliana*. *The Journal of Biological Chemistry*, *284*(35), 23454–23460. doi:10.1074/jbc.M109.025791

Kim, M.-H., Sato, S., Sasaki, K., Saburi, W., Matsui, H., and Imai, R. (2013a). COLD SHOCK DOMAIN PROTEIN 3 is involved in salt and drought stress tolerance in Arabidopsis. *FEBS Open Bio*, *3*, 438–442. doi:10.1016/j.fob.2013.10.003

Kim, M.-H., Sonoda, Y., Sasaki, K., Kaminaka, H., and Imai, R. (2013b). Interactome analysis reveals versatile functions of *Arabidopsis* COLD SHOCK DOMAIN PROTEIN 3 in RNA processing within the nucleus and cytoplasm. *Cell Stress and Chaperones*, *18*(4), 517–525. doi:10.1007/s12192-012-0398-3

Kim, Y. O., Kim, J. S., and Kang, H. (2005). Cold-inducible zinc finger-containing glycine-rich RNA-binding protein contributes to the enhancement of freezing tolerance in *Arabidopsis thaliana*. *The Plant Journal*, *42*(6), 890–900.

Kloks, C. P., Spronk, C. A., Lasonder, E., Hoffmann, A., Vuister, G. W., Grzesiek, S., and Hilbers, C. W. (2002). The solution structure and DNA-binding properties of the cold-shock domain of the human Y-box protein YB-1. *Journal of Molecular Biology*, *316*(2), 317–326.

Köster, T., Meyer, K., Weinholdt, C., Smith, L. M., Lummer, M., Speth, C., Grosse, I., Weigel, D., and Staiger, D. (2014). Regulation of pri-miRNA processing by the hnRNP-like protein AtGRP7 in *Arabidopsis*. *Nucleic Acids Research*, *42*(15), 9925–9936; gku716.

Kwak, K. J., Kim, H. S., Jang, H. Y., Kang, H., and Ahn, S. J. (2016). Diverse roles of glycine-rich RNA-binding protein 7 in the response of camelina (*Camelina sativa*) to abiotic stress. *Acta Physiologiae Plantarum*, *38*(5), 1–11.

Kwak, K. J., Kim, Y. O., and Kang, H. (2005). Characterization of transgenic *Arabidopsis* plants overexpressing GR-RBP4 under high salinity, dehydration, or cold stress. *Journal of Experimental Botany*, *56*(421), 3007–3016.

Long, R., Yang, Q., Kang, J., Zhang, T., Wang, H., Li, M., and Zhang, Z. (2013). Overexpression of a novel salt stress-induced glycine-rich protein gene from alfalfa causes salt and ABA sensitivity in *Arabidopsis*. *Plant Cell Reports*, *32*(8), 1289–1298.

Magioli, C., Barrôco, R. M., Rocha, C. A. B., de Santiago-Fernandes, L. D., Mansur, E., Engler, G., Margis-Pinheiro, M., and Sachetto-Martins, G. (2001). Somatic embryo formation in *Arabidopsis* and eggplant is associated with expression of a glycine-rich protein gene (Atgrp-5). *Plant Science*, *161*(3), 559–567.

Mangeon, A., Junqueira, R. M., and Sachetto-Martins, G. (2010). Functional diversity of the plant glycine-rich proteins superfamily. *Plant Signaling and Behavior*, *5*(2), 99–104.

Mangeon, A., Magioli, C., Menezes-Salgueiro, A. D., Cardeal, V., de Oliveira, C., Galvão, V. C., Margis, R., Engler, G., and Sachetto-Martins, G. (2009). AtGRP5, a vacuole-located glycine-rich protein involved in cell elongation. *Planta*, *230*(2), 253–265.

Mangeon, A., Menezes-Salgueiro, A. D., and Sachetto-Martins, G. (2017). Start me up: revision of evidences that AtGRP3 acts as a potential switch for AtWAK1. *Plant Signaling and Behavior*, *12*(2), e1191733.

Mangeon, A., Pardal, R., Menezes-Salgueiro, A. D., Duarte, G. L., de Seixas, R., Cruz, F. P., Cardeal, V., et al. (2016). AtGRP3 is implicated in root size and aluminum response pathways in *Arabidopsis*. *PloS One*, *11*(3), e0150583.

Maris, C., Dominguez, C., and Allain, F. H. T. (2005). The RNA recognition motif, a plastic RNA-binding platform to regulate post-transcriptional gene expression. *The FEBS Journal*, *272*(9), 2118–2131.

Mayfield, J. A., Fiebig, A., Johnstone, S. E., and Preuss, D. (2001). Gene families from the *Arabidopsis thaliana* pollen coat proteome. *Science*, *292*(5526), 2482–2485.

Mayfield, J. A., and Preuss, D. (2000). Rapid initiation of *Arabidopsis* pollination requires the oleosin-domain protein GRP17. *Nature Cell Biology*, *2*(2), 128–130.

Nakaminami, K., Karlson, D. T., and Imai, R. (2006). Functional conservation of cold shock domains in bacteria and higher plants. *Proceedings of the National Academy of Sciences*, *103*(26), 10122–10127.

Nocker, S., and Vierstra, R. D. (1993). Two cDNAs from *Arabidopsis thaliana* encode putative RNA binding proteins containing glycine-rich domains. *Plant Molecular Biology*, *21*(4), 695–699.

Ortega-Amaro, M. A., Rodríguez-Hernández, A. A., Rodríguez-Kessler, M., Hernández-Lucero, E., Rosales-Mendoza, S., Ibáñez-Salazar, A., Delgado-Sánchez, P., and Jiménez-Bremont, J. F. (2015). Overexpression of AtGRDP2, a novel glycine-rich domain protein, accelerates plant growth and improves stress tolerance. *Frontiers in Plant Science*, *5*, 782.

Ortega-Amaro, M. A., Rodríguez-Kessler, M., Rodríguez-Hernández, A. A., Becerra-Flora, A., Rosales-Mendoza, S., and Jiménez-Bremont, J. F. (2016). Overexpression of ArGRDP2 gene in common bean hairy roots generates vigorous plants with enhanced salt tolerance. *Acta Physiologiae Plantarum*, *38*, 66.

Park, A. R., Cho, S. K., Yun, U. J., Jin, M. Y., Lee, S. H., Sachetto-Martins, G., and Park, O. K. (2001). Interaction of the *Arabidopsis* receptor protein kinase Wak1 with a glycine-rich protein, AtGRP-3. *Journal of Biological Chemistry*, *276*(28), 26688–26693.

Park, J. H., Suh, M. C., Kim, T. H., Kim, M. C., and Cho, S. H. (2008). Expression of glycine-rich protein genes, AtGRP5 and AtGRP23, induced by the cutin monomer 16-hydroxypalmitic acid in *Arabidopsis thaliana*. *Plant Physiology and Biochemistry, 46*(11), 1015–1018.

Park, S. J., Kwak, K. J., Oh, T. R., Kim, Y. O., and Kang, H. (2009). Cold shock domain proteins affect seed germination and growth of *Arabidopsis thaliana* under abiotic stress conditions. *Plant and Cell Physiology, 50*, 869–878.

Pendle, A. F., Clark, G. P., Boon, R., Lewandowska, D., Lam, Y. W., Andersen, J., Mann, M., Lamond, A. I., Brown, J. W., and Shaw, P. J. (2005). Proteomic analysis of the *Arabidopsis* nucleolus suggests novel nucleolar functions. *Molecular Biology of the Cell, 16*(1), 260–269.

Reddy, A. S. N., and Poovaiah, B. W. (1987). Accumulation of a glycine rich protein in auxin-deprived strawberry fruits. *Biochemical and Biophysical Research Communications, 147*(3), 885–891.

Ringli, C., Hauf, G., and Keller, B. (2001b). Hydrophobic interactions of the structural protein GRP1. 8 in the cell wall of protoxylem elements. *Plant Physiology, 125*(2), 673–682.

Ringli, C., Keller, B., and Ryser, U. (2001a). Glycine-rich proteins as structural components of plant cell walls. *Cellular and Molecular Life Sciences, 58*(10), 1430–1441.

Rodríguez-Hernández, A. A., Muro-Medina, C. V., Ramírez-Alonso, J. I., and Jiménez-Bremont, J. F. (2017). Modification of AtGRDP1 gene expression affects silique and seed development in *Arabidopsis thaliana*. *Biochemical and Biophysical Research Communications, 486*(2), 252–256.

Rodríguez-Hernández, A. A., Ortega-Amaro, M. A., Delgado-Sánchez, P., Salinas, J., and Jiménez-Bremont, J. F. (2014). AtGRDP1 Gene encoding a glycine-rich domain protein is involved in germination and responds to ABA signalling. *Plant Molecular Biology Reporter, 32*(6), 1187–1202.

Ryser, U., and Ryser, B. (1992). Ultrastructural localization of a bean glycine-rich protein in unlignified primary walls of protoxylem cells. *The Plant Cell, 4*(7), 773–783.

Ryser, U., Schorderet, M., Zhao, G. F., Studer, D., Ruel, K., Haul, G., and Keller, B. (1997). Structural cell-wall proteins in protoxylem development: evidence for a repair process mediated by a glycine-rich protein. *The Plant Journal, 12*(1), 97–111.

Sachetto-Martins, G., Fernandes, L. D., Felix, D. B., and de Oliveira, D. E. (1995). Preferential transcriptional activity of a glycine-rich protein gene from *Arabidopsis thaliana* in protoderm-derived cells. *International Journal of Plant Sciences, 156*(4), 460–470.

Sachetto-Martins, G., Franco, L. O., and de Oliveira, D. E. (2000). Plant glycine-rich proteins: a family or just proteins with a common motif?. *Biochimica et Biophysica Acta (BBA)-Gene Structure and Expression, 1492*(1), 1–14.

Sasaki, K., and Imai, R. (2011). Pleiotropic roles of cold shock domain proteins in plants. *Frontiers in Plant Science, 2*, 116. doi:10.3389/fpls.2011.00116

Sasaki, K., Kim, M. H., and Imai, R. (2007). *Arabidopsis* COLD SHOCK DOMAIN PROTEIN2 is a RNA chaperone that is regulated by cold and developmental signals. *Biochemical and Biophysical Research Communications, 364*(3), 633–638.

Sasaki, K., Kim, M. H., and Imai, R. (2013). *Arabidopsis* COLD SHOCK DOMAIN PROTEIN 2 is a negative regulator of cold acclimation. *New Phytologist, 198*(1), 95–102.

Sasaki, K., Kim, M. H., Kanno, Y., Seo, M., Kamiya, Y., and Imai, R. (2015a). *Arabidopsis* COLD SHOCK DOMAIN PROTEIN 2 influences ABA accumulation in seed and negatively regulates germination. *Biochemical and Biophysical Research Communications, 456*(1), 380–384.

Sasaki, K., Liu, Y., Kim, M. H., and Imai, R. (2015b). An RNA chaperone, AtCSP2, negatively regulates salt stress tolerance. *Plant Signaling and Behavior, 10*(8), e1042637. doi:10.1080/15592324.2015.1042637

Schmidt, F., Marnef, A., Cheung, M. K., Wilson, I., Hancock, J., Staiger, D., and Ladomery, M. (2010). A proteomic analysis of oligo (dT)-bound mRNP containing oxidative stress-induced *Arabidopsis thaliana* RNA-binding proteins ATGRP7 and ATGRP8. *Molecular Biology Reports, 37*(2), 839–845.

Schöning, J. C., Streitner, C., Meyer, I. M., Gao, Y., and Staiger, D. (2008). Reciprocal regulation of glycine-rich RNA-binding proteins via an interlocked feedback loop coupling alternative splicing to nonsense-mediated decay in *Arabidopsis*. *Nucleic Acids Research, 36*(22), 6977–6987.

Singh, S., Virdi, A. S., Jaswal, R., Chawla, M., Kapoor, S., Mohapatra, S. B., Manoj, N., Pareek, A., Kumar, S., and Singh, P. (2017). A temperature-responsive gene in sorghum encodes a glycine-rich protein that interacts with calmodulin. *Biochimie, 137*, 115–123.

Sivaguru, M., Ezaki, B., He, Z. H., Tong, H., Osawa, H., Baluška, F., Volkmann, D., and Matsumoto, H. (2003). Aluminum-induced gene expression and protein localization of a cell wall-associated receptor kinase in *Arabidopsis*. *Plant Physiology, 132*(4), 2256–2266.

Staiger, D., Zecca, L., Kirk, D. A. W., Apel, K., and Eckstein, L. (2003). The circadian clock regulated RNA-binding protein AtGRP7 autoregulates its expression by influencing alternative splicing of its own pre-mRNA. *The Plant Journal, 33*(2), 361–371.

Steinert, P. M., Mack, J. W., Korge, B. P., Gan, S. Q., Haynes, S. R., and Steven, A. C. (1991). Glycine loops in proteins: their occurence in certain intermediate filament chains, loricrins and single-stranded RNA binding proteins. *International Journal of Biological Macromolecules, 13*(3), 130–139.

Streitner, C., Danisman, S., Wehrle, F., Schöning, J. C., Alfano, J. R., and Staiger, D. (2008). The small glycine-rich RNA binding protein AtGRP7 promotes floral transition in *Arabidopsis thaliana*. *The Plant Journal, 56*(2), 239–250.

Streitner, C., Köster, T., Simpson, C. G., Shaw, P., Danisman, S., Brown, J. W., and Staiger, D. (2012). An hnRNP-like RNA-binding protein affects alternative splicing by in vivo interaction with transcripts in *Arabidopsis thaliana*. *Nucleic Acids Research, 40*(22), 11240–11255.

Ueki, S., and Citovsky, V. (2002). The systemic movement of a tobamovirus is inhibited by a cadmium-ion-induced glycine-rich protein. *Nature Cell Biology, 4*(7), 478–486.

Urbez, C., Cercós, M., Perez-Amador, M. A., and Carbonell, J. (2006). Expression of PsGRP1, a novel glycine rich protein gene of Pisum sativum, is induced in developing fruit and seed and by ABA in pistil and root. *Planta, 223*(6), 1292–1302.

Van Kan, J. A., Cornelissen, B. J., and Bol, J. F. (1988). A virus-inducible tobacco gene encoding a glycine-rich protein shares putative regulatory elements with the ribulose bisphosphate carboxylase small subunit gene. *Molecular Plant-Microbe Interactions, 1*(3), 107–112.

Vermel, M., Guermann, B., Delage, L., Grienenberger, J. M., Maréchal-Drouard, L., and Gualberto, J. M. (2002). A family of RRM-type RNA-binding proteins specific to plant mitochondria. *Proceedings of the National Academy of Sciences*, *99*(9), 5866–5871.

Yang, D. H., Kwak, K. J., Kim, M. K., Park, S. J., Yang, K. Y., and Kang, H. (2014). Expression of *Arabidopsis* glycine-rich RNA-binding protein AtGRP2 or AtGRP7 improves grain yield of rice (Oryza sativa) under drought stress conditions. *Plant Science*, *214*, 106–112.

Yang, Y., and Karlson, D. (2013). AtCSP1 regulates germination timing promoted by low temperature. *FEBS Letters*, *587*(14), 2186–2192.

Yang, Y., and Karlson, D. T. (2011). Overexpression of *AtCSP4* affects late stages of embryo development in *Arabidopsis*. *Journal of Experimental Botany*, *62*(6), 2079–2091. doi: 10.1093/jxb/erq400

Ziemienowicz, A., Haasen, D., Staiger, D., and Merkle, T. (2003). *Arabidopsis* transportin1 is the nuclear import receptor for the circadian clock-regulated RNA-binding protein AtGRP7. *Plant Molecular Biology*, *53*(1–2), 201–212.

18

Compatible Solutes and Abiotic Stress Tolerance in Plants

Vinay Kumar, Tushar Khare, Samrin Shaikh, and Shabir H. Wani

CONTENTS

Abbreviations

BADH	Betaine aldehyde dehydrogenase
CMO	Choline monooxygenase
codA	Choline oxidase
GABA	γ-aminobutyric acid
GB	Glycine betaine
PAs	Polyamines
P5CDH	Δ^1-pyrroline-5-carboxylate dehydrogenase
P5CR	Δ^1-pyrroline-5-carboxylate reductase
P5CS	Δ^1-pyrroline-5-carboxylate synthetase
P2C	pyrroline 2-carboxylate
Pro	Proline
ProDH	Proline dehydrogenase
PS	Photosystem
Put	Putrescine
RFOs	Raffinose family oligosaccharides
ROS	Reactive oxygen species
SAMDC	*S*-adenosylmethionine decarboxylase
Spd	Spermidine
Spm	Spermine
TPSP	Trehalose-6-phosphate synthase/phosphatase
Tre	Trehalose
δ-OAT	δ-aminotransferase

Introduction

As sessile organisms, plants are often challenged by inimical environments of various biotic and abiotic stresses, leading to the prohibition of ideal cellular and developmental processes in the plants. To combat the situation of most of hostile abiotic stress conditions, including drought, salinity, heavy metal toxicity, anoxia, high or low temperature, nutrient imbalance, radiations and water logging; plants have evolved numerous responses, as well as adaptive mechanisms (Abebe et al. 2003; Groppa and Benavides 2008; Sanghera et al. 2011; Ramakrishna and Ravishankar 2011, 2013; Shriram et al. 2016). One of the adaptive mechanisms is the synthesis and accumulation of small organic molecules, which are known as compatible osmolytes, for the protection of cellular machinery against the unwanted hazardous effects of the mentioned abiotic environmental stresses (Chen and Jiang 2010; Liang et al. 2013; Wani et al. 2016a). Compatible solutes are small molecules that act as osmolytes and help organisms survive extreme osmotic stress conditions.

The properly regulated, composite sets of different biochemical and molecular mechanisms have been developed by different plant species for better survival (Munns and Tester 2008; Wani et al. 2016b). Most abiotic stress conditions share common effects and responses, such as a reduction in plant growth and development, photosynthetic rate, hampered flower production and grain filling, hormonal changes, proliferation of reactive oxygen species (ROS) and hence the oxidative damage (Khare et al. 2015). These changes are usually the result of tissue dehydration, sodium toxicity, ionic imbalance/deficiency and photoinhibition (Bohnert and Shen 1998). To combat against these deficient and unfavourable environments, the complex plant metabolism has been further changed in many diverse ways including compatible solute production. It helps to stabilize the cellular turgor by maintaining osmotic balance across the tissues as well as it improves the redox homeostasis and helps to scavenge excessively produced reactive oxygen species (Chinnusamy and Zhu 2009; Krasensky and Jonak 2012). Osmotic adjustment is considered as a central strategy by most of the plants, which is tailed by the accumulation of the compatible osmolytes like proline, glycine betaine and sugars (Chen and Jiang 2010; Wani and Gosal 2010). The most common classes of compatible solutes include betaines (b-alanine betaine, glycine betaine, proline betaine), sugars (glucose, fructose, fructan, raffinose and trehalose), sulfonium compounds (choline sulfate, dimethyl sulfonium propionate), polyols (mannitol, glycerol, sorbitol), polyamines (spermine, spermidine and putrescine) and amino acids (proline, glutamine) (Chen and Murata 2002; Cramer et al. 2011; Chen and Jiang 2010; Wani et al. 2013; Slama et al. 2015; Pathak and Wani 2015; Sah et al. 2016). Many of the mentioned low molecular weight entities have been used for improved abiotic stress tolerance in various plants mainly including crops; via application of diverse tools of genetic engineering. The present chapter is an attempt to provide an update on different compatible solutes, their differential accumulation under various stress environments and transgenic technology implements to improve abiotic stress tolerance in plants.

Compatible Solutes: Protective Role under Abiotic Stress

Under stress conditions, plants synthesize and accumulate compatible solutes, also termed as osmo-protectants, which are low molecular weight, neutral, non-lethal molecules which show high solubility in solutes. On the basis of their chemical nature these molecules can be categorized in three main classes as: (i) betaines and related compounds, (ii) amino acids (iii) polyols and non-reducing sugars and (iv) polyamines. Various abiotic stress responses and compatible solutes mediated stress tolerance mechanisms are depicted in Figure 18.1.

Betaines

Betaines are compounds comprising a carboxylic acid group, which are classified as quaternary ammonium compounds. They are considered as fully methylated amino or imino acids. Diverse forms of betaine occur depending upon the plant species and environmental stress; including glycine betaine, proline betaine, β-alanine betaine, choline-o-sulfate and 3-dimethylsulfoniopropionate (Nahar et al. 2016). Among the mentioned types glycine betaine (GB) (N,N,N-trimethylglycine), which was first discovered in *Beta vulgaris*, is the most common entity which is diversely distributed not only in plants but also in microbes and animals (Ashraf and Foolad 2007). GB is the most abundant quaternary ammonium compound which occurs during dehydrating conditions, which plays a protective role in the thylakoid membrane for osmotic adjustments in the chloroplast. In GB-accumulator plant species, tolerant genotypes usually show higher GB accumulation than sensitive genotypes in response to stress. However, the distribution of GB in plants is sporadic as some plants species including *Arabidopsis*, rice and tobacco are categorized as non-GB accumulators (Kumar et al. 2016; Nahar et al. 2016).

Synthesis of GB takes place via two distinctive pathways from two substrates: choline and glycine, respectively. Choline is converted to GB via two-enzyme pathway occurring naturally in several plants, animals and microorganisms. In this pathway, GB is formed in two-step oxidation of choline by the toxic intermediate compound betaine aldehyde. In higher plants, the processes are catalyzed in the stroma of chloroplast by choline monooxygenase (CMO) and NAD$^+$-dependent betaine aldehyde dehydrogenase (BADH). Such synthesis of GB is stress-responsive and the concentration of GB *in vivo* ranges amid plant species, from 40 to 400 µmol per gram of dry weight (Rhodes and Hanson 1993). GB is involved in the prevention of excess accumulation of ROS produced due to ionic and osmotic imbalance in the abiotic environment. It leads to the protection of photosynthetic machinery, membrane protection, activation of stress-induced genes; enhancement of the repair of photosystem-II (PSII) and preservation of enzymatic activities by protecting the quaternary structure of proteins (Al-Taweel et al. 2007; Chen and Murata 2008, 2011; Kumar et al. 2016; Takahashi and Murata 2008). One of the notable beneficial characteristics of GB is that it renders variation in facilitating the osmotic alteration by Na$^+$/K$^+$ discrimination, thus defending enzyme induction and maintaining membrane stability, which significantly contribute to the reduction in sensitivity against abiotic stress like salinity and drought

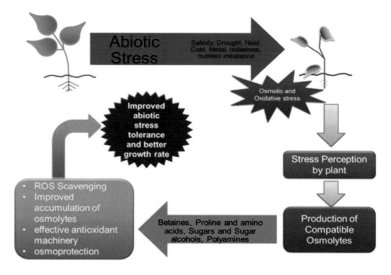

FIGURE 18.1 Abiotic stress responses and compatible solutes-mediated abiotic stress tolerance in plants.

(Ashraf and Foolad 2007). There are reports which indicate the abiotic stress-induced higher accumulation of GB in many crops including rice, barley, ground nut and some other higher plants (Jagendorf and Takabe 2001; Sawahel 2004; Ranganayakulu et al. 2013).

Amino Acids

Improved cellular amassing of the amino acids in plants under abiotic stress have been recorded earlier (Lugan et al. 2010). Many of the conventional, as well as non-conventional, amino acids including proline (Pro), alanine, glycine, glutamate, aspargine, γ-aminobutyric acid (GABA), pipecolic acid, citrulline, ornithine, etc., play roles as osmolytes by acting as precursors for most of the osmo-protectants. They also help in preventing membrane damage and ionic-toxicity. They are widely distributed in many of the angiosperm plant families including some important crop families like Poaceae, Brassicaceae and Fabaceae (Suprasanna et al. 2016).

Amongst amino acids, Pro has been credited as the most vital amino acid molecule which plays a highly beneficial role in the plants exposed to the variety of abiotic stresses. Pro has been recorded to serve as an osmo-protective agent, which plays diverse roles including metal chelation, antioxidant defense molecule and as a signal molecule. The stress environment results in the overproduction of the Pro in most of the plants, which is closely connected with improved stress tolerance via maintaining cellular turgor, membrane stability and effective controlling over the production and scavenging of ROS, which eventually protects the plant cells from oxidative damage (Kumar et al. 2010; Ramkrishna and Ravishankar 2011; Kumar and Khare 2015; Kumar et al. 2015; Wani et al. 2016a). Accumulation of Pro is a result of either *de novo* synthesis or declined degradation or both (dos Reis et al. 2012). In many plants Pro synthesis takes place via two pathways, namely from glutamate or ornithine pathways. However, glutamate-mediated Pro synthesis is more abundant in the plants under the osmotic stress (Lv et al. 2011; Witt et al. 2012). The potentiality of the Pro production mainly depends on the activity and expression of the catalytic enzymes involved in the synthetic pathway. These enzymes are Δ^1-pyrroline-5-carboxylate synthetase (P5CS) and Δ^1-pyrroline-5-carboxylate reductase (P5CR). Whereas, in mitochondria, Pro catabolism occurred via Pro dehydrogenase (ProDH) and Δ^1-pyrroline-5-carboxylate dehydrogenase (P5CDH), which metabolize Pro to glutamate. On the other hand, in the Ornithin pathway, a different set of enzymes, including δ-aminotransferase (δ-OAT), works to produce pyrroline 2-carboxylate which finally converted into Pro via pyrroline 2-carboxylate reductase (P2CR). The overall Pro level is dependent on the biosynthesis, catabolism and transport of the synthesized Pro in organelle and cellular level. Hence the biosynthesis of enzymes in the appropriate cellular region (P5CS and P5CR in cytosol and ProDH, P5CDH, OAT in mitochondria) is vital for Pro level maintenance in stress conditions (Szabados and Savoure 2010). During retrieval from stress, Pro is converted to P5C by PDH and then to glutamate by P5CDH. Improved Pro accumulation is observed under a variety of abiotic stresses including salinity, drought, heavy metal, plant pathogens and low temperature (Kumar and Khare 2015; Slama et al. 2015) in

many plant species including wheat, rice, barley, maize, tomato, brassica and *Arabidopsis*.

GABA is also a non-protein coding amino acid with α-ketoglutarate acting as its precursor. GABA is a stress-induced amino acid, which is formed from glutamate by the action of enzymes including glutamate dehydrogenase and glutamate decarboxylase. Under the abiotic environment, it plays a key role in osmotic and pH regulation, as well as in nitrogen metabolism and prevents the excess ROS accumulation (Barbosa et al. 2010; Renault et al. 2010). Accumulation of GABA has been marked in various adverse environments like salinity, flooding, cold, heat, as well as some transient environmental factors (Kinnersley and Turano 2000; Renault et al. 2010; Zhang et al. 2011).

Polyols and Non-Reducing Sugars

Sugars are primarily known as carbon suppliers and, hence, are considered an energy source for, not only plants, but also for most other organisms. Yet, sugars and sugar alcohols are widely accepted as osmoprotectants, which control overall osmotic adjustments and ROS scavenging in the plants growing under stress conditions (Ahmad and Sharma 2008; Livingston et al. 2009; Van den Ende and Valluru 2009; Koyro et al. 2012). Reduced forms of the sugars, such as glucose and fructose, or sugars, such as sucrose or fructans, have been reported as osmoprotectants under salinity and drought (Murakeozy et al. 2003; Kumar et al. 2007). Fructans are a known storage carbohydrate source which accumulates in roots or tubers. Pilon-Smits et al. (1995) described that fructan defends plants against severe drought. Some raffinose family oligosaccharides (RFOs) have been associated with mitigating the effects of abiotic stresses, such as cold, heat or dehydration (El Sayed et al. 2014). RFOs protect cellular integrity in desiccation and/or imbibition, encompassing longevity in the arid state and providing substrates for energy generation during germination.

Trehalose (Tre) is a non-reducing disaccharide consisting of two glucose units with α-α-(1-1) bonding. The Tre molecule shows unique bonding properties as both reducing ends of the molecule form glycosidic bonds. Hence, Tre shows higher stability as it is resistant to acidic hydrolysis (Richards et al. 2002). Tre bears a water replacement mechanism where it changes water by forming hydrogen bonds with a membrane or macromolecules during dehydration or freezing. Hence, Tre is known as a membrane or molecule stabilizer (Crowe 2007). It is the only sugar molecule which can endure a glass-like state when completely dehydrated (Richards et al. 2002; Elbein et al. 2003). In most of the higher plants Tre biosynthesis is achieved by an OtsA-OtsB pathway (Nahar et al. 2016).

Sugar alcohols (polyols) can be classified in two distinct types on the structural basis. They can be cyclic (myoinositol and pinitol) or linear (sorbitol, mannitol, xylitol and ribitol). Polyols are products of the reduction of aldoses or their phosphate esters. Most of the polyols are water soluble in nature (Tari et al. 2010). Polyols control the osmotic balance, enable the water adjustment in cytoplasm, facilitate the sodium sequestration in vacuole or apoplast and hence provide the membrane stability in abiotic stress conditions (Kanayama et al. 2006; Li et al. 2011). Another sugar alcohol, mannitol, originates by the action of enzyme mannose-6-phosphatase, in the reaction where mixture of glucose and fructose get interchanged into gluconic acid and mannitol. Likewise, in the sorbitol biosynthesis, it shares the usual hexose phosphate group with sucrose production in the cytosol.

Polyamines

Polyamines (PAs) are characterized as low molecular weight aliphatic nitrogen containing molecules. They are positively charged at a physiological level and act as cellular pH regulators (Groppa and Benavides 2008). PAs have higher affinity to bind with nucleic acids (DNA, RNA) and proteins via electrostatic linkages (Kusano et al. 2007). Predominantly studied aliphatic amine phytohormones including putrescine (Put), spermidine (Spd), and spermine (Spm) are reported in many of the plants. These polyamines are synthesized from methionine, ornithine and arginine by a series of reactions involving different sets of enzymes. In plants, diverse roles are performed by PAs with respect to different vital physiological and biochemical procedures such as senescence, development, cell proliferation, signal transduction as well as the regulation of gene expression in response to various stresses in eukaryotic and prokaryotic cells (Alcazar et al. 2006; Kusano et al. 2007). Under the course of various biotic and abiotic stresses, accumulation of PAs is recorded in many plants (Alcazar et al. 2006; Hussain et al. 2011).

Genetic Engineering Approaches for Synthesis and Accumulation of Compatible Solutes for Abiotic Stress Tolerance

One of the most important characteristic features of osmoprotectant is that their beneficial role is usually not species specific. This opens up a wide scope for the bioengineering of alien genes from other species into the new host plant species to prevent them from destructive damage due to one or more abiotic stresses. Due to the latest biotechnological advancements, as well as the characterization of whole genomes of many organisms, now it is very much possible to manipulate the genome of the desired plant with the necessary osmoprotectant gene. Elevated accumulation and production of these compatible osmolytes, along with improved radical scavenging and other transgene products, has been successfully confirmed with stress tolerance (Vinocur and Altman 2005; Gosal et al. 2009; Wani and Gosal 2011; Kumar et al. 2010,). Genes encoding for many of the earlier discussed osmoprotectants have been successfully employed to create more tolerant versions of different plants to abiotic stresses. Some of the examples have been summarized in Table 18.1.

As discussed earlier, some plants have been classed as non-GB accumulators and hence have been targeted for genetic engineering for osmotolerance by producing GB in them (McCue and Hanson 1990). GB is produced either by the oxidation/dehydrogenation of choline or by the *N*-methylation of glycine (Chen and Murata 2002). Hence, genes encoding GB-biosynthetic enzymes have been cloned from several microbes and plants to produce transgenic with higher GB accumulation and exhibiting enhanced tolerance to a variety of abiotic stresses.

Transgenic sweet potato plants that were raised expressing the *BADH* gene from spinach showed higher tolerance to a variety of abiotic stresses including salinity, oxidative stress as well as low temperature (Fan et al. 2012). The choline oxidase gene (*codA*) is expressed in transgenic *Brassica* plants by Wang et al. (2010). The resultant transgenic plants showed higher accumulation of the GB along with resistance to multiple abiotic stresses. The overexpression of betaine by manipulating the *BADH* gene through genetic engineering in chloroplast proved to be an imperative approach to induce salinity tolerance in desired crops (Kumar et al. 2004; Fitzgerald et al. 2009). Transgenic lines with higher accumulation of GB showed more tolerance to drought stress than wild-type plants. Transgenic lines showed higher GB accumulation, which provided greater protection to maintain the integrity of the cell membrane and higher activity of enzymes. Transgenic lines showed a better germination rate and an early vegetative stage as well as high grain yield. Similarly, the *beta* gene was inserted into wheat to enhance the salinity tolerance of the cultivar. Transformed wheat lines showed reduced membrane damage, a higher photosynthetic rate and a lower Na/K ratio under a higher salinity level (He et al. 2010). Transgenic cotton was produced by introducing the CMO gene (*AhCMO*) from *Atriplex hortensis*. Transgenic seedling showed higher accumulation of GB under 150 mM of salinity treatment. Hence the improved salinity tolerance in transgenic lines was observed with reduced osmotic potential, membrane leakage and lipid peroxidation and improved photosynthetic activity (Zhang et al. 2009).

In other studies, trehalose pathway genes have also been targeted in order to genetically engineer the plants for abiotic stress tolerance improvement. The genes from *Escherichia coli* namely *otsA* and *otsB* have been implemented to raise transgenic tobacco plants. Though transformed plants showed some morphological changes, drought tolerance level was improved than the control plants. To surpass such morphological changes, trehalose transgenics of rice were raised, by using the stress inducible or tissue specific expression of bifunctional trehalose-6-phosphate synthase/phosphatase (*TPSP*) fusion gene consisting of the *E. coli* trehalose biosynthetic genes *otsA and otsB* along with TPS and TPP activity, respectively. Transgenic *Saccharum officinarum* showed the improved drought stress tolerance with enhanced trehalose production, compared to non-transformed controls (Wang et al. 2005).

The behavior of the drought-sensitive potato cultivar (White Lady) and its drought-tolerant *TPS1* transgenic variant was observed under extended drought. Potato cultivar expressing the yeast *TPS1* gene exhibited better drought tolerance. Transgenic plants displayed augmented fructose, galactose and glucose accumulation, which added to the amplified accumulation of Pro, inositol and raffinose. The transgenic potato plant showed better drought tolerance, compared to the wild type. Similarly, potato tubers expressing *TPS1* showed higher accumulation of fructose, lucose, mannose, sucrose, aspergine, phenylalanine, abscisic acid, jasmonic acid, indole acetic acid and salicylic acid under stress conditions; either exposed in fresh or storage period (Juhász et al. 2014). Transgenic tomato plants overexpressing yeast Tre synthesis genes displayed greater tolerance to drought, salt and oxidative stresses (Cortina and Culianez-Macia 2005). Arabidopsis plants overexpressing *AtTPS1* were more drought tolerant and

TABLE 18.1

Summarized List of Few Compatible Solute Pathway-genes Targeted for Engineering Abiotic Stress Tolerance in Transgenic Plants

Gene	Source Organism	Transformed Plants	Improved Character	Reference
P5CS	*Vigna aconitifolia*	*Triticum aestivum* L.	Drought tolerance via reduction in oxidative stress	Vendruscolo et al. (2007)
P5CS	*Vigna aconitifolia*	*Saccharum* spp.	Drought tolerance, Pro accumulation, improved antioxidative defense	Molinari et al. (2007)
P5CS	*Arabidopsis thaliana* and *Oryza sativa*	*Petunia hybrid*	Higher Pro content, improved growth under drought stress	Yamada et al. (2005)
P5CSF129A	*Vigna aconitifolia*	*Oryza sativa*	Salinity tolerance, increased proline production	Kumar et al. (2010)
betA	*E. coli*	*Triticum aestivum*	Increased GB and chl content, low Na^+/K^+ ratio, solute potential, reduced membrane damage, high photosynthesis rate, higher yield under salt stress	He et al. (2010)
betA	*E. coli*	*Gossypium hirsutum*	Higher RWC and Photosynthesis rate, reduced ion leakage and lipid membrane peroxidation under drought stress	
AhCMO	*Atriplex hortensis*	*Gossypium hirsutum*	High GB synthesis, lowered osmotic potential, electrolyte leakage and lipid peroxidation, higher photosynthetic capacity under salt stress	Zhang et al. (2009)
BADH	*Spinacia oleracea*	*Nicotiana tabacum*	Improved CO_2 assimilation and seedling growth under salt stress	Yang et al. (2008)
otsA and *otsB*	*E. coli*	*Oryza sativa*	Higher Tre level, less photo-oxidative damage, improved mineral balance under salt and drought and low-temp. condition	
TPS1	*Pichia angusta*	*Solanum tuberosum'*	Increased yield under drought stress, Increased metabolite production including sugar, proteins, osmolytes, and hormones	Juhász et al. (2014)
mtlD	*E. coli*	*Triticum aestivum*	Greater drought and salt tolerance with better biomass production	Abebe et al. (2003)
SAMDC	*Tritordeum*	*Oryza sativa*	Increased seedling growth, spermine and spermidine production and salt tolerance.	Roy and Wu (2002)

showed glucose- and abscisic acid-insensitive phenotypes (Avonce et al. 2004). Very few attempts with focus on other small sugar molecules have been made to improve abiotic stress tolerance. Overexpression of the myo-inositol phosphate synthase gene *Medicago falcata* from (*MfMIPS1*) in tobacco showed improved MIPS activity and myoinositol, galactinol and raffinose conferring resistance to chilling, drought and salt stresses in transgenic plants (Tan et al. 2013).

Pro biosynthetic pathway genes are also one of the heavily targeted genes for the production of multiple abiotic stress tolerant plant lines. The pyrroline-5-carboxylate synthetase gene from *Arabidopsis thaliana* and *Oryza sativa* were transformed into *Petunia hybrid*, which showed higher Pro accumulation, improved growth as well as drought tolerance (Yamada et al. 2005). When wheat plants were transformed with the *Vigna aconitifolia* P5CS gene, significant Pro accumulation was observed along with improved enhanced tolerance against drought due to improved mechanisms for osmotic adjustments and oxidative stress (Vendruscolo et al. 2007). Similarly, the *P5CS* gene from *Vigna aconitifolia*, was successfully introduced in the *indica* rice cultivar ADT 43 via *Agrobacterium* mediated transformation by Karthikeyan et al. (2011). The transgenic plants were able to withstand 200 mM of NaCl with amended Pro production while non-transformed plants died in ten days under the same conditions. A similar strategy was implemented by Kumar et al. (2010), who overexpressed the *P5CSF129A* gene in the rice cultivar Karjat-3

under the control of *CaMV 35S* promoter. Under salt stress, an increased Pro accumulation and root growth was observed in transgenics (Kumar et al. 2010). Improved drought stress tolerance was observed in *Glycine max* cultivar Ibis, when it was transformed with the Pro *P5CR* gene, which also showed higher Pro accumulation and affected amino acid metabolism (Simon-Sarkadi et al. 2006).

Improved abiotic stress tolerance has also been achieved by increased amassing of sugar alcohols like mannitol. The *mtlD* (mannitol-1-phospho dehydrogenase) gene from *E. coli*, mannitol biosynthetic pathway gene, was successfully transformed into in *Oryza sativa* to raise several putative transgenic rice lines (cv. PB-1) (Punji et al. 2007). Transgenic plants exhibited excess mannitol accumulation, which can be linked to the improved drought and salinity stress environments (Punji et al. 2007). Similarly, the ectopic expression of a *mltD* gene from *E. coli* in *Triticum aestivum* was achieved by transformation and the transformed calli were cultivated under water deficit with 100 mM NaCl. Under the provided abiotic stress conditions, transformed plants showed better growth rate compared to the non-transformed plants (Abebe et al. 2003).

The cDNA encoding for SAMDC (*S*-adenosylmethionine decarboxylase), a key enzyme from polyamine biosynthetic pathway, from *Tritordeum* was introduced in a *japonica* rice cultivar (Roy and Wu 2002). Transgenic plants showed three to four times greater production of Spm and Spd with high seedling growth and salt tolerance level.

Conclusion

Increased biosynthesis and accumulation of compatible solutes is an effective strategy evolved by plants in response to and to offset the deleterious effects of abiotic stresses. Due to their multifunctional roles during stress and recovery phases, compatible solutes present potential candidates for genetic engineering aimed to impart abiotic stress tolerance in resultant transgenics. Several successful attempts have been made to overexpress various compatible solute pathway gene(s) for abiotic stress tolerance. However, efforts are needed to produce transgenic plants with enhanced tolerance against combined abiotic stress and to further commercialize the transgenic technologies.

Acknowledgments

The research activities in the laboratory of VK are supported by the funds under the Science and Engineering Research Board, Government of India, in the form of a Young Scientist Project [SERB-OYS; grant number SR/FT/LS-93/2011], and University Grants Commission Major Research Project [F. No. 41-521/2012 (SR)]. The use of the facilities created under the FIST program of the Department of Science and Technology (DST), Government of India and Star College Scheme of Department of Biotechnology (DBT), Government of India is gratefully acknowledged. SHW thanks University Grants Commission, New Delhi, India, for the Raman Postdoctoral Fellowship.

REFERENCES

Abebe T, Guenzi AC, Martin B, Cushman JC (2003) Tolerance of mannitol-accumulating transgenic wheat to water stress and salinity. *Plant Physiol.* 131:1748–55.

Ahmad P, Sharma S (2008) Salt stress and phytobiochemical responses of plants. *Plant Soil Environ.* 54:89–99.

Alcazar R, Marco F, Cuevas JC, Patron M, Ferrando A, Carrasco P, Tiburcio AF, Altabella T (2006) Involvement of polyamines in plant response to abiotic stress. *Biotechnol. Lett.* 28:1867–1876.

Al-Taweel K, Iwaki T, Yabuta Y, Shigeoka S, Murata N, Wadano A (2007) A bacterial transgene for catalase protects translation of d1 protein during exposure of salt-stressed tobacco leaves to strong light. *Plant Physiol.* 145:258–265.

Ashraf M, Foolad MR (2007) Roles of glycine betaine and proline in improving plant abiotic stress resistance. *Environ. Exp. Bot.* 59:206–216.

Avonce N, Leyman B, Mascorro-Gallardo JO, Van Dijck P, Thevelein JM, Iturriaga G (2004) The *Arabidopsis* trehalose-6-P synthase *AtTPS1* gene is a regulator of glucose, abscisic acid, and stress signaling. *Plant Physiol.* 136:3649–3659.

Barbosa JM, Singh NK, Cherry JH, Locy RD (2010) Nitrate uptake and utilization is modulated by exogenous-aminobutyric acid in Arabidopsis thaliana seedlings. *Plant Physiol. Biochem.* 48:443–450.

Bohnert HJ, Shen BO (1998) Transformation and compatible solutes. *Sci. Hortic.* 78:237–260.

Chen H, Jiang J-G (2010) Osmotic adjustment and plant adaptation to environmental changes related to drought and salinity. *Environ. Rev.* 18:309–319.

Chen TH, Murata N (2002) Enhancement of tolerance of abiotic stress by metabolic engineering of betaines and other compatible solutes. *Curr. Opin. Biotechnol.* 5:250–257.

Chen TH, Murata N (2008) Glycinebetaine: An effective protectant against abiotic stress in plants. *Trends Plant Sci.* 13:499–505.

Chen THH, Murata N (2011) Glycine betaine protects plants against abiotic stress: Mechanisms and biotechnological applications. *Plant Cell Environ.* 34:1–20.

Chinnusamy V, Zhu JK (2009) Epigenetic regulation of stress responses in plants. *Curr. Opin. Plant Biol.* 12:133–139.

Cortina C, Culianez-Macia FA (2005) Tomato abiotic stress enhanced tolerance by trehalose biosynthesis. *Plant Sci.* 169:75–82.

Cramer GR, Urano K, Delrot S, Pezzotti M, Shinozaki K (2011) Effects of abiotic stress on plants: A systems biology perspective. *BMC Plant Biol.* 11:163.

Crowe JH (2007) Trehalose as a "chemical chaperone": Fact and fantasy. *Adv. Exp. Med. Biol.* 594:143–158.

dos Reis SP, Lima AM, de Souza CRB (2012) Recent molecular advances on downstream plant responses to abiotic stress. *Int. J. Mol. Sci.* 13:8628–8647.

El Sayed AI, Rafudeen MS, Golldack D (2014) Physiological aspects of raffi nose family oligosaccharides in plants: Protection against abiotic stress. *Plant Biol.* 16:1–8.

Elbein AD, Pan YT, Pastuszak I, Carroll D (2003) New insights on trehalose: A multifunctional molecule. *Glycobiology* 13:17–27.

Fan W, Zhang M, Zhang H, Zhang P (2012) Improved tolerance to various abiotic stresses in transgenic sweet potato (Ipomoea batatas) expressing spinach betaine aldehyde dehydrogenase. *PLoS One* 7(5):e37344.

Fitzgerald TL, Waters DLE, Henry RJ (2009) Betaine aldehyde dehydrogenase in plants. *Plant Biol.* 11(2):119–130.

Gosal SS, Wani SH, Kang MS (2009) Biotechnology and drought tolerance. *J. Crop. Improv.* 23:19–54.

Groppa MD, Benavides MP (2008) Polyamines and abiotic stress: Recent advances. *Amino Acids* 34:35–45.

He C, Yang A, Zhang W, Gao Q, Zhang J (2010) Improved salt tolerance of transgenic wheat by introducing *betA* gene for glycine betaine synthesis. *Plant Cell Tiss. Organ Cult.* 101:65–78.

Hussain SS, Ali M, Ahmad M, Siddique KHM (2011) Polyamines: Natural and engineered abiotic and biotic stress tolerance in plants. *Biotechnol. Adv.* 29:300–311.

Jagendorf AT, Takabe T (2001) Inducers of glycine betaine synthesis in barley. *Plant Physiol.* 127:1827–1835.

Juhász Z, Balmer D, Sós-Hegedűs A, Vallat A, Mauch-Mani B, Bánfalvi Z (2014) Effects of drought stress and storage on the metabolite and hormone contents of potato tubers expressing the yeast trehalose-6-phosphate synthase 1. *Gene J. Agric. Sci.* 6:142–166.

Kanayama Y, Watanabe M, Moriguchi R, Deguchi M, Kanahama K, Yamaki S (2006) Effects of low temperature and abscisic acid on the expression of the sorbitol-6-phosphate dehydrogenase gene in apple leaves. *J. Jpn. Soc. Hortic. Sci.* 75:20–25.

Karthikeyan A, Pandian SK, Ramesh M (2011) Transgenic *indica* rice cv. ADT 43 expressing a Δ^1-pyrroline-5-carboxylate synthetase (*P5CS*) gene from *Vigna aconitifolia* demonstrates salt tolerance. *Plant Cell Tiss. Organ Cult.* 107:383–395.

Khare T, Kumar V, Kavi Kishor PB (2015) Na$^+$ and Cl$^-$ ions show additive effects under NaCl stress on induction of oxidative stress and the responsive antioxidative defense in rice. *Protoplasma.* 252:1149–1165. doi:10.1007/s00709-014-0749-2.

Kinnersley AM, Turano FJ (2000) Gamma aminobutyric acid (GABA) and plant responses to stress. *Crit. Rev. Plant Sci.* 19:479–509.

Koyro HW, Ahmad P, Geissler N (2012) Abiotic stress responses in plants: An overview. In: Ahmad P, Prasad MNV (eds.). *Environmental Adaptations and Stress Tolerance of Plants in the Era of Climate Change.* Springer, New York, pp. 1–28.

Krasensky J, Jonak C (2012) Drought, salt, and temperature stress-induced metabolic rearrangements and regulatory networks. *J. Exp. Bot.* 63:1593–1608.

Kumar S, Dhingra A, Daniell H (2004) Plastid-expressed betaine aldehyde dehydrogenase gene in carrot cultured cells, roots, and leaves confers enhanced salt tolerance. *Plant Physiol.* 136:2843–2854.

Kumar V, Khare T (2015) Individual and additive effects of Na$^+$ and Cl$^-$ ions on rice under salinity stress. *Arch. Agron. Soil Sci.* 61:381–395.

Kumar V, Shriram V, Hoque TS, Hasan MM, Burritt DJ, Hossain MA (2016) Glycinebetaine mediated abiotic oxidative-stress tolerance in plants: Physiological and biochemical mechanisms. In: Sarwat M, Ahmad A, Abdin MZ, Ibrahim MM (eds.). *Stress Signaling in Plants: Genomics and Proteomics Perspective*, Vol. 2. Springer International Publishing, Switzerland.

Kumar V, Shriram V, Hossain MA, Kishor PK (2015) Engineering proline metabolism for enhanced plant salt stress tolerance. In: Wani SH, Hossain MA (eds.). *Managing Salt Tolerance in Plants: Molecular and Genomic Perspectives.* CRC Press, Boca Raton, pp. 353–372.

Kumar V, Shriram V, Jawali N, Shitole MG (2007) Differential response of indica rice genotypes to NaCl stress in relation to physiological and biochemical parameters. *Arch. Agron. Soil Sci.* 53:581–592.

Kumar V, Shriram V, Kavi Kishor PB, Jawali N, Shitole MG (2010). Enhanced proline accumulation and salt stress tolerance of transgenic indica rice by over-expressing P5CSF129A gene. *Plant Biotechnol. Rep.* 4:37–48.

Kusano T, Yamaguchi K, Berberich T, Takahashi Y (2007) Advances in polyamine research in 2007. *J. Plant Res.* 120:345–350.

Li F, Lei HJ, Zhao XJ, Tian RR, Li TH (2011) Characterization of three sorbitol transporter genes in micropropagated apple plants grown under drought stress. *Plant Mol. Biol. Rep.* 30:123–130.

Liang X, Zhang L, Natarajan SK, Becker DF (2013) Proline mechanisms of stress survival. *Antioxid. Redox Signal.* 19:998–1011.

Livingston DP, Hincha DK, Heyer AG (2009) Fructan and its relationship to abiotic stress tolerance in plants. *Cell Mol. Life Sci.* 66:2007–2023.

Lugan R, Niogret MF, Leport L, Guégan JP, Larher FR, Savouré A, Kopka J, Bouchereau A (2010) Metabolome and water homeostasis analysis of *Thellungiella salsuginea* suggests that dehydration tolerance is a key response to osmotic stress in this halophyte. *Plant J.* 64:215–229.

Lv WT, Lin B, Zhang M, Hua XJ (2011) Proline accumulation is inhibitory to Arabidopsis seedlings during heat stress. *Plant Physiol.* 15:1921–1933.

McCue KF, Hanson AD (1990) Drought and salt tolerance: Towards understanding and application. *Trends Biotechnol.* 8:358–362.

Molinari HB, Marur CJ, Daros E, De Campos MK, De Carvalho JF, Pereira LF, Vieira LG (2007) Evaluation of the stress-inducible production of proline in transgenic sugarcane (*Saccharum* spp.): Osmotic adjustment, chlorophyll fluorescence and oxidative stress. *Physiol. Plant.* 130:218–229.

Munns R, Tester M (2008) Mechanism of salinity tolerance. *Annu. Rev. Plant Biol.* 59:651–681.

Murakeozy EP, Nagy Z, Duhaze C, Bouchereau A, Tuba Z (2003) Seasonal changes in the levels of compatible osmolytes in three halophytic species of inland saline vegetation in Hungary. *J. Plant Physiol.* 160:395–401.

Nahar K, Hasanuzzaman M, Fujita M (2016) Roles of osmolytes in plant adaptation to drought and salinity. In: Iqbal N, Nazar R, Khan NA (eds.). *Osmolytes and Plants Acclimation to Changing Environment: Emerging Omics Technologies.* Springer India, New Delhi, pp. 37–68.

Pilon-Smits E, Ebskamp M, Paul MJ, Jeuken M, Weisbeek PJ, Smeekens S (1995) Improved performance of transgenic fructan-accumulating tobacco under drought stress. *Plant Physiol.* 107:125–130.

Punji D, Chaudhary A, Rajam MV (2007). Increased tolerance to salinity and drought in transgenic indica rice by mannitol accumulation. *J. Plant Biochem. Biotech.* 16:1–7.

Ramakrishna A, Ravishankar GA (2011). Influence of abiotic stress signals on secondary metabolites in plants. *Plant Signal. Behav.* 6:1720–1731.

Ramakrishna A, Ravishankar GA (2013). Role of plant metabolites in abiotic stress tolerance under changing climatic conditions with special reference to secondary compounds. In: Tuteja N. Gill SS (eds.). *Climate Change and Plant Abiotic Stress Tolerance.* Wiley-VCH Verlag GmbH & Co. KGaA, Weinheim, Germany, pp. 705–726.

Ranganayakulu GS, Veeranagamallaiah G, Sudhakar C (2013) Effect of salt stress on osmolyte accumulation in two groundnut cultivars (*Arachis hypogaea* L.) with contrasting salt tolerance. *Afr. J. Plant Sci.* 12:586–592.

Renault H, Roussel V, Amrani E, Arzel M, Renault D, Bouchereau A, Deleu C (2010) The Arabidopsis pop2-1 mutant reveals the involvement of GABA transaminase in salt stress tolerance. *BMC Plant Biol.* 10:1–16.

Rhodes D, Hanson AD (1993) Quaternary ammonium and tertiary sulfonium compounds in higher-plants. *Annu. Rev. Plant Physiol. Plant Mol. Biol.* 44:357–384.

Richards AB, Krakiwka S, Dexter LB, Schid H, Wolterbeek APM, Waalkens-Berendsen DH, Shigoyuki A, Kurimoto M (2002) Trehalose: A review of properties, history of use and human tolerance, and results of multiple studies. *Food Chem. Toxicol.* 40:871–898.

Roy M, Wu R (2002) Overexpression of S-adenosylmethionine decarboxylase gene in rice increases polyamine level and enhances sodium chloride-stress tolerance. *Plant Sci.* 163:987–992.

Sah SK, Kaur G, Wani SH (2016) Metabolic engineering of compatible solute trehalose for abiotic stress tolerance in plants. In: Iqbal N, Nazar R, Khan NA (eds.). *Osmolytes and Plants Acclimation to Changing Environment: Emerging Omics Technologies.* Springer India, New Delhi, pp. 83–96.

Sanghera GS, Wani SH, Hussain W, Singh NB (2011) Engineering cold stress tolerance in crop plants. *Curr. Gen.* 14:30–43.

Sawahel W (2004) Improved performance of transgenic glycine betaine-accumulating rice plants under drought stress. *Biol. Plant.* 47:39–44.

Shriram V, Kumar V, Devarumath RM, Khare T, Wani SH (2016) MicroRNAs as potent targets for abiotic stress tolerance in plants. *Front. Plant Sci.* 7:817. doi:10.3389/fpls.2016.00817.

Simon-Sarkadi L, Kocsy G, Várhegyi Á, Galiba G, De Ronde JA (2006) Stress induced changes in the free amino acid composition in transgenic soybean plants having increased proline content. *Biol. Plant.* 50:793–796.

Slama I, Abdelly C, Bouchereau A, Flowers T, Savouré A (2015) Diversity, distribution and roles of osmoprotective compounds accumulated in halophytes under abiotic stress. *Ann. Bot.* 115:433–447.

Suprasanna P, Nikalje GC, Rai AN (2016) Osmolyte accumulation and implication in plant abiotic stress tolerance. In: Iqbal N, Nazar R, Khan NA (eds.). *Osmolytes and Plants Acclimation to Changing Environment: Emerging Omics Technologies,* Springer India, New Delhi, pp. 1–12.

Szabados L, Savoure A (2010) Proline: A multifunctional amino acid. *Trends Plant Sci.* 15:89–97.

Takahashi S, Murata N (2008) How do environmental stresses accelerate photoinhibition? *Trends Plant Sci.* 13:178–182.

Tan JL, Wang CY, Xiang B, Han R, Guo Z (2013) Hydrogen peroxide and nitric oxide mediated cold- and dehydration induced myo-inositol phosphate synthase that confers multiple resistances to abiotic stresses. *Plant Cell Environ.* 36:288–299.

Tari I, Kiss G, Deer AK, Csiszár J, Erdei L, Gallé Á, Gemes K, et al. (2010) Salicylic acid increased aldose reductase activity and sorbitol accumulation in tomato plants under salt stress. *Biol. Plant.* 54:677–683.

Van den Ende W, Valluru R (2009) Sucrose, sucrosyl oligosaccharides, and oxidative stress: Scavenging and salvaging? *J. Exp. Bot.* 60:9–18.

Vendruscolo ECG, Schuster I, Pileggi M, Scapim CA, Molinari HBC, Marur CJ, Vieira LG (2007) Stress-induced synthesis of proline confers tolerance to water deficit in transgenic wheat. *J. Plant Physiol.* 164:1367–1376.

Vinocur B, Altman A (2005) Recent advances in engineering plant tolerance to abiotic stress: Achievements and limitations. *Curr. Opin. Biotechnol.* 16:123–132.

Wang GP, Li F, Zhang J, Zhao MR, Hui Z, Wang W (2010) Overaccumulation of glycine betaine enhances tolerance of the photosynthetic apparatus to drought and heat stress in wheat. *Photosynthetica.* 48:30–41.

Wang Z, Zhang S, Yang B, Li Y (2005) Trehalose synthase gene transfer mediated by *Agrobacterium tumefaciens* enhances resistance to osmotic stress in sugarcane. *Sugar Tech.* 7:49–54.

Wani SH, Singh NB, Haribhushan A, Mir JI (2013) Compatible solute engineering in plants for abiotic stress tolerance—Role of glycine betaine. *Curr. Gen.* 14(3):157–165.

Pathak MR, Wani SH (2015) Salinity stress tolerance in relation to polyamine metabolism in plants. In: Wani SH, Hossain MA (eds.) *Managing Salt Tolerance in Plants: Molecular and Genomic Perspectives.* CRC Press, Boca Raton, pp. 241–250.

Wani SH, Gosal SS (2010) Genetic engineering for osmotic stress tolerance in plants-role of proline. *IUP J. Genet. Evol.* 3(4):14–25.

Wani SH, Gosal SS (2011) Introduction of OsglyII gene into Indica rice through particle bombardment for increased salinity tolerance. *Biol. Plant.* 55(3):536–540.

Wani SH, Kumar V, Shriram V, Sah SK. (2016a) Phytohormones and their metabolic engineering for abiotic stress tolerance in crop plants. *Crop J.* 4(3):162–176. doi:10.1016/j.cj.2016.01.010.

Wani SH, Sah SK, Hussain MA, Kumar V, Balachandra SM (2016b) Transgenic approaches for abiotic stress tolerance in crop plants. In: Al-Khayri JM, Jain SM, Johnson DV (eds.). *Advances in Plant Breeding Strategies,* Vol. 2: *Agronomic, Abiotic and Biotic Stress Traits.* Springer International Publishing, Switzerland.

Witt S, Galicia L, Lisec J, Cairns J, Tiessen A, Araus JL, Palacios-Rojas N, Fernie AR (2012) Metabolic and phenotypic responses of greenhouse-grown maize hybrids to experimentally controlled drought stress. *Mol. Plant.* 5:401–417.

Yamada M, Morishita H, Urano K, Shiozaki N, Yamaguchi-Shinozaki K, Shinozaki K, Yoshiba Y (2005) Effects of free proline accumulation in petunias under drought stress. *J. Exp. Bot.* 56:1975–1981.

Yang X, Liang Z, Wen X, Lu C (2008) Genetic engineering of the biosynthesis of glycinebetaine leads to increased tolerance of photosynthesis to salt stress in transgenic tobacco plants. *Plant Mol. Biol.* 66:73–86.

Zhang H, Dong H, Li W, Sun Y, Chen S, Kong X (2009) Increased glycine betaine synthesis and salinity tolerance in AhCMO transgenic cotton lines. *Mol. Breed.* 23(2):289–298.

Zhang J, Zhang Y, Du Y, Chen S, Tang H (2011) Dynamic metabolomics responses of Tobacco (*Nicotiana tabacum*) plants to salt stress. *J. Proteome Res.* 10:1904–1914.

19

Protective Role of Indoleamines (Serotonin and Melatonin) During Abiotic Stress in Plants

Ramakrishna Akula, Sarvajeet Singh Gill, and Gokare A. Ravishankar

CONTENTS

Abbreviations

CBFs	C-repeat-binding factors
DREBs	Drought response element-binding factors
HSFA1s	Heat-shock factors A1
MEL	Melatonin
SER	Serotonin

Introduction

Accumulation of secondary metabolites often occurs in plants exposed to different stresses, various elicitors, and by direct signal molecules action. It is widely known that secondary metabolites are essential for a plant's interaction with its environment, and for adaptation and defense against stress conditions (Bennett and Wallsgrove 1994; Ramakrishna and Ravishankar 2011, 2013). Serotonin (SER) and Melatonin (MEL) are two major indoleamines derived from tryptophan. They have been reported in various plant species and their physiological role has also been demonstrated in plants (Odjakova et al. 1997; Murch et al. 2001; Roshchina 2001; Ramakrishna et al. 2009, 2011a,b, 2012a,b,c; Ramakrishna 2015; Ravishankar and Ramakrishna 2016; Kang et al. 2007). SER, with its high antioxidant activity, protects plants from the oxidative damage caused by the process of senescence (Kang et al. 2007). Additionally, SER in plants has been found to vary widely with environmental factors including light levels and wavelength, abiotic and biotic stress and growth conditions, and location (Erland et al. 2015). Elevated levels of MEL probably help plants to protect against environmental stress caused by water and soil pollutants. MEL has been reported to protect plants against a variety of abiotic

and biotic stresses (Zuo et al. 2014; Liang et al. 2015; Shi et al. 2015a; Wang et al. 2015; Li et al. 2015; Xu et al. 2016). The role of MEL in stress management is mainly attributable to higher photosynthesis, improvement of oxidative stress, enhancement of cellular redox homeostasis, and regulation of the expression of stress responsive genes involved in signal transduction. All stresses including salt, drought, cold, heat, and chemical stress can cause up-regulation of MEL production in various plants, indicating the possible role of MEL as an important messenger in plant stress responses (Li et al. 2012; Bajwa et al. 2014; Zhang et al. 2014, 2015). The protective effect of exogenous MEL in plants has also been used to demonstrate its possible role as cellular antioxidant. Earlier reports suggest that MEL protects germ and reproductive tissues of plants from oxidative damage due to environmental factors (Tan et al. 2007; Paredes et al. 2009; Hardeland et al. 2007; Kaur et al. 2015). This chapter describes the protective role of SER and MEL in plants during various abiotic stresses. Tables 19.1 and 19.2 provide some examples of studies related to the influence of SER and MEL in stress tolerance mechanisms in plants.

Role of Indoleamines During Drought Stress

Drought limits plant growth due to photosynthetic decline. The harmful effects of high salt result from both a water shortage caused by osmotic stress and the interference of excess sodium ions with key biochemical processes (Zhang and Blumwald 2001). MEL up regulates the expression of C-repeat-binding factors (CBFs)/drought response element-binding factors (DREBs). Drought priming-induced MEL accumulation enhanced the antioxidant capacity in both chloroplasts and mitochondria.

TABLE 19.1

Protective Role of Melatonin During Abiotic Stress in Plants

Plant Species	Function	Reference
Daucus carota	Antiapoptotic effect	Lei et al. (2004)
Glycyrrhiza uralensis	Tolerance to photodamage and UV-B	Afreen et al. (2006)
Eichhornia crassipes	Radiation stress	Tan et al. (2007)
Brassica juncea	Radiation stress, drought, temperature and chemical pollutants	Tan et al. (2007)
Brassica oleracea	Protect seedlings against toxic copper ions	Posmyk et al. (2008)
Hardeum vulgare	Protect chlorophyll degradation during senescence	Arnao and Hernandez-Ruiz (2009)
Cucumis sativus	Chilling stress	Posmyk et al. (2009)
Datura metel	Protective role in developing flower buds	Murch et al. (2009)
Rhodiola crenulata	Improves the survival of cryopreserved callus	Zhao et al. (2011)
Arabidopsis	Cold stress	Bajwa et al. (2014) and Shi and Chan (2014)
Arabidopsis	Leaf senescence	Shi et al. (2015d)
Arabidopsis	Thermotolerance	Shi et al. (2015e)
Arabidopsis	Salt and drought stress	Shi et al. (2015c)
Arabidopsis	Oxidative stress	Weeda et al. (2014)
Arabidopsis (Transgenic plant)	Drought tolerance	Zuo et al. (2014)
Bermuda grass	Salt, drought and cold stresses	Shi et al. (2015b)
Bermuda grass	Oxidative stress	Shi et al. (2015f)
Rice (Transgenic plants)	Salt and cold stresses	Kang et al. (2010)
Rice	Cadmium stress	Byeon et al. (2015)
Rice	Leaf senescence and salt stress	Liang et al. (2015)
Malu shupehensis	Salt stress	Li et al. (2012)
Malushupehensis	Drought stress	Li et al. (2015)
Malushupehensis	Senescence	Wang et al. (2012, 2014)
Cucumber	Salt stress tolerance	Zhang et al. (2014)
Cucumber	Drought stress tolerance	Zhang et al. (2014)
Tomato	Drought stress tolerance	Wang et al. (2014)
Citrullus lanatus L	Cold stress tolerance	Li et al. (2017)
Vigna radiata L.	Chilling stress	Szafrańska et al. (2014)
Zea mays	Drought tolerance	Lin et al. (2015)
Red cabbage	Metal stress	Posmyk et al. (2008)
Malus baccata	Waterlogging tolerance	Zheng et al. (2017)
Nicotiana tabacum	Protect against ozone	Dubbles et al. (1995)Kolar et al. (1997)
Chenopodium rubrum	Damage photoperiodic events, circadian rhythms	
Hypericum perforatum	Reproductive physiology and flower development	Murch and Saxena (2002)
Wheat	Drought stress	Cui et al. (2017)

Mitigation potential of melatonin in drought stress and certain mechanisms of MEL-induced glutathione (GSH) and ascorbic acid (AsA) accumulation, and MEL treatment remarkably increased drought tolerance of wheat seedlings, as evidenced by increased antioxidant capacity and decreased endogenous Reactive oxygen species (ROS) level (Cui et al. 2017). Moreover, MEL application into maize and cucumber seeds had a positive effect on seedling development and on the yield of plants that had grown from them, especially those subjected to water-stress (Zhang et al. 2013). The MEL enriched transgenic _Arabidopsis_ had a greater tolerance to drought stress compared to the wild types (Zuo et al. 2014). The exogenous MEL also improved the tolerance of tomato plants against alkaline stress (Liu et al. 2015). Moreover, application of MEL to tomato plants enhances their root vigor, mitigates their stress-related damage to PSII reaction centers, minimizes the negative impact of drought by regulating the antioxidant system, and reduces the cellular content of toxic substances of the plants (Liu et al. 2015). MEL application

to apple leaves alleviated the drought-induced inhibition of photosynthesis (Wang et al. 2012). The protective effects of MEL on chlorophyll (Chl) decay, photosynthetic capacity, and stomata configurations have been reported previously in other stressors such as in drought and hot (Wang et al. 2012; Xu et al. 2016).

Role of Indoleamines During Cold Stress

Cold stress is one of the key abiotic stresses that limits crop growth, yield, and geographical distribution of plants, particularly in temperate zones and high-elevation environments (Andaya and Mackill 2003). Exposure of plants to cold stress results in changes in several physiological, biochemical, molecular, and metabolic processes including fluctuations of membrane fluidity, enzyme activities, and metabolism homeostasis (Bajwa et al. 2014). Recent studies show that exogenous MEL confers cold tolerance to plants. An increased accumulation of MEL accompanied by an induction

TABLE 19.2

Possible Physiological Functions of Serotonin in Plants

Plant Species	Function	Reference
Cowhage (*Mucuna pruriens*) Tree nettle (*Utrica dioica*)	Protective agent against predation	
Walnut seeds (*Juglans regia*)	Detoxification	
Barley (*Hordeum vulgare*)	Plant seed development	
Corn (*Zea mays*)	Photo morphogenetic responses	
Albizzia julibrissin Garden pea (*P. sativum*) Wild maracuja (*Passiflora quadrangularis*)	Adaptation to environmental changes	Odjakova and Hadjiivanova (1997)
Oat (*Avena sativa*)	Root growth and development	Roshchina (2001)
Walnut seeds (*Juglans regia*)	Secondary plant product	
St. John's wort (*Hypericum perforatum*)	Auxinic like effect & regulation of organogenesis, toot growth and development	Murch et al. (2001)
Radish seeds (*Hippeastrum hybridum*)	Stimulation of seeds germination	Roshchina (2001)
Rice (*Oryza sativa*)	Ion permeability & antioxidant	Kang et al. (2007)
Rice (*Oryza sativa*)	Maintain the cellular integrity of xylem parenchyma and companion cells and delay senescence	
Coffea Canephora	Induce somatic embryogenesis	Ramakrishna et al. (2012a,b)

in antioxidant enzyme activity in distant untreated tissues alleviated cold-induced oxidative stress. The effects of foliar and rhizospheric MEL pretreatment on the cold stress tolerance in untreated leaves and roots, respectively (Li et al. 2016). MEL applied to the seeds of *Vigna radiata* can improve the resistance mechanisms of plants during cold acclimation by increasing the activity of Phenylalanine ammonia lyase (PAL) after re-warming. A cold stress-responsive gene, *COR15a*, a gene encoding the expression of a transcription factor involved in freezing stress tolerance, *CAMTA1*, and transcription activators of ROS-related antioxidant genes, *ZAT10* and *ZAT12*, following cold stress (Bajwa et al. 2014). Cucumber seeds treated with MEL showed an improved germination rate during chilling stress (Posmyk et al. 2009). An interesting effect of MEL as a protector of cold-induced apoptosis in carrot suspension cells has been reported (Lei et al. 2004). Pretreatment with MEL diminished apoptosis induced by cold temperature in cultured carrot suspension cells (Lei et al. 2004). Cold stress-induced shrinkage and disruption of carrot cell plasma membranes were almost completely alleviated by MEL treatment. These findings suggest that MEL is helpful in managing harsh environments by maintaining membrane integrity. Lupin plants grown in a cold environment (6°C) show a 2.5-fold increase in MEL levels compared with control plants grown at 24°C. Moreover, Zhao et al. (2011) have reported that MEL improves the survival of cryopreserved callus of *Rhodiola crenulate*. The survival rate of the cryopreserved *Rhodiola crenulata* callus is ~60%. When the callus was pre-treated with 0.1 μM MEL prior to freezing in liquid nitrogen, a significant improvement in survival rate was observed. MEL significantly enhanced the recovery of cryopreserved shoot tips of American elm (*Ulmus americana* L.) (Uchendu et al. 2013). Shoot explants grown in MEL-enriched media showed increased regrowth. Shi et al. (2015a) found that cold stress activated the synthesis of MEL in Bermuda grass, and exogenous MEL improved its cold stress tolerance by scavenging ROS directly and improving the antioxidative enzymes activities. Previous reports that MEL application can enhance cold tolerance of *Arabidopsis* (Bajwa et al. 2014), wheat (Hulya et al. 2014),

and Bermuda grass (Hu et al. 2016). The possible role of MEL in growth, photosynthesis, and the response to cold stress in rice have been reviewed recently (Han et al. 2017). MEL pretreatment alleviated the detrimental effects of cold stress and accelerated the recovery, mainly by enhancing photosynthesis and antioxidant capacity in melon leaves (Zhang et al. 2017). These results indicate that both ABA-dependent and ABA-independent pathways may contribute to MEL-induced cold tolerance in *E. nutans*. Exogenous MEL up-regulated components of the cold-stress signaling pathways following specific time intervals of cold exposure. Transcripts of *EnCBF9* and *EnCBF14* accumulated immediately after 1 and 3 h of cold treatment (Fu et al. 2017).

Role of Indoleamines During Heat Stress

Extreme temperatures impair plant growth or even cause death by affecting membrane fluidity and enzyme activity. Endogenous MEL level in *Arabidopsis* leaves was significantly induced by heat stress treatment, and exogenous MEL treatment conferred improved thermo tolerance in *Arabidopsis*, indicating the involvement of *HSFA1s*-activated heat-responsive genes in MEL-mediated thermo tolerance in *Arabidopsis* (Shi et al. 2015a). Mel induces class A1 heat-shock factors (HSFA1s) and their possible involvement of thermo tolerance in *Arabidopsis*. MEL also promotes cellular protein protection through the induction of heat-shock proteins (HSPs) and autophagy to refold or degrade denatured proteins under heat stress in tomato plants. Moreover, HsfA1a up regulates MEL biosynthesis to confer cadmium tolerance in tomato plants (Cai et al. 2017).

Role of Indoleamines During Salt Stress

Salinity interferes with plant growth as it leads to physiological drought and ion toxicity. Salt stress-induced the regulation of endogenous SER and MEL accumulation, which had a direct correlation with seedling growth in sunflower. Exogenous SER and MEL treatments lead to a variable effect on hypocotyl elongation and root growth under NaCl stress. High salt stress (100 m*M*) has been reported to inhibit primary root length and lateral root primordia development results due to the inhibition of auxin accumulation in the emerging lateral root primordia (Zolla et al. 2010). The effect of NaCl stress on endogenous SER and MEL accumulation, and their differential spatial distribution in sunflower (*Helianthus annuus*) seedling roots and cotyledons, has been reported (Mukhrjee et al. 2014). Moreover, MEL and nitric oxide regulate sunflower seedling growth under salt stress accompanying differential expression of Cu/ZnSOD and Mn SOD (Arora and Bhatla 2017).

Several studies have shown that salt stress can increase MEl content in roots (Arnao and Hernández-Ruiz 2009; Mukherjee et al. 2014). Exogenous application of MEL enhances salt stress tolerance in *Malus hupehensis* and *Helianthus annuus* (Li et al. 2012; Mukherjee et al. 2014). MEL's role in regulating growth, ion homeostasis, and the response to oxidative stress in *Malus hupehensis* Rehd. under high-salinity conditions has been reported. The results suggested that the MEL enhanced maize salt tolerance in maize were most likely due to the improvement of photosynthetic capacity, antioxidative capacity, and ion homeostasis in leaves (Jiang et al. 2016).

Role of Indoleamines During Metal Stress

Heavy metals such as copper and zinc are essential for normal plant growth, but excess heavy metals are toxic. In addition, heavy metals cause oxidative deterioration of biomolecules by initiating free radical-mediated chain reactions resulting in lipid peroxidation, protein oxidation, and oxidation of nucleic acids. All the abiotic stresses cause an enhanced generation of ROS (Gill and Tuteja 2010). MEL and its precursors can bind to several toxic metals: aluminum with MEL, tryptophan, and SER; cadmium with MEL and tryptophan; copper with MEL and SER; Fe^{3+} with MEL and SER; Fe^{2+} with tryptophan only; lead with MEL, tryptophan, and SER; and zinc with MEL and tryptophan (Limson et al. 1998). Electrochemical studies show that MEL can bind to both Cu^{2+} and Cu^{1+}, which cause free radical damage (Parmar et al. 2002). It is documented that MEL affects biological systems not only through the direct quenching of free radicals but also *via* chelation of toxic metals (Flora et al. 2013). The potential relationships between MEL supplementation and environmental tolerance in plants have been reported. In pea plants treated with high levels of copper in the soil, the tolerance of pea plants to copper contamination is reportedly enhanced to a significant extent by the addition of MEL to the soil, raising the chances of their survival (Tan et al. 2007). The incubation of *Brassica oleracea* L. seeds with exogenous MEL solutions has a protective effect against copper, a highly toxic metal (Posmyk et al. 2008). The chemical stress induces the biosynthesis of MEL in barley roots, enhancing significantly their MEL content. Such an increase in MEL probably plays an important antioxidative role in the defence against chemically induced stress and other abiotic/biotic stresses (Arnao and Hernandez-Ruiz 2009). Cai et al. (2017) demonstrate that transcription factor heat-shock

factor A1a (HsfA1a) conferred cadmium (Cd) tolerance to tomato plants, in part through its positive role in inducing MEL biosynthesis under Cd stress. Moreover, HsfA1a confers Cd tolerance by activating the transcription of the *COMT1* gene and inducing the accumulation of melatonin that partially up regulates expression of HSPs. A novel mechanism in which RBOH activity and H_2O_2 signaling are important components of the MEL-induced stress tolerance in tomato plants. SER has also been found to be effective in mitigating the effects of heavy metal exposure in plants, though indirectly. SER has a relatively high capacity for binding cadmium to form stable complexes, and experiments in rice found that SER biosynthetic pathway genes, including TDC and T-5-H, were up regulated in response to cadmium.

Role of Indoleamines During Radiation Stress

Radiation stress can damage macromolecules, such as DNA and proteins, generate ROS, and impair cellular processes. MEL can significantly reverse the inhibitory effects of light and high temperature on the germination of photosensitive and thermo sensitive *Phacelia tanacetifolia*.

Benth seeds (Tiryaki and Keles 2012). MEL in sensing and responding to ultraviolet (UV) light. Though at high levels UV induces stress and requires defensive roles for MEL, which include increased pigment production and antioxidant action, at low levels UV is a useful signal to indicate the direction of sunlight (Afreen et al. 2006; Byeon et al. 2013). MEL may also mediate this light sensing process, possibly through COP receptors, particularly as this pathway has also been found to interact with the UV receptor UVR8 (Sanchez-Barcelo et al. 2016). This is a plausible proposition as the chromophore of UVR8 comprises several conserved tryptophan residues, suggesting a potential link to MEL, though this idea has yet to be explored. Also, tolerance to photodamage and UV-B of MEL in *Glycyrrhiza* (Afreen et al. 2006) and *Eichhornia* (Tan et al. 2007) has been reported.

Conclusion

Further studies are necessary to elucidate the interrelation between MEL/SER and other signaling molecules in response to abiotic stresses. More studies are necessary to elucidate the interaction of MEL/SER-induced regulation of calcium-dependent signaling under abiotic stress in plants. However, the molecular network that operates during different abiotic stresses mediated by MEL and SER remains to be determined. Transcriptomic and metabolomic changes in response to exogenous application of MEL/SER and their roles in abiotic stress tolerance need to be studied. Moreover, further studies are necessary to characterize SER and MEL receptors (if any) and their involvement in stress tolerance mechanisms.

REFERENCES

Afreen F, Zobayed S, Kozai T (2006) Melatonin in Glycyrrhiza uralensis: Response of plant roots to spectral quality of light and UV-B radiation. *J. Pineal Res.* 41:108–115.

Andaya V, Mackill D (2003) QTLs conferring cold tolerance at the booting stage of rice using recombinant inbred lines from a japonica × indica cross. *Theor. Appl. Genet.* 106:1084–1090.

Arnao MB, Hernandez-Ruiz A (2009) Protective effect of melatonin against chlorophyll degradation during the senescence of barley leaves. *J. Pineal Res.* 46:58–63.

Arora D, Bhatla SC (2017) Melatonin and nitric oxide regulate sunflower seedling growth under salt stress accompanying differential expression of Cu/Zn SOD and Mn SOD. *Free Rad. Biol. Med.* 106: 315.

Bajwa VS, Shukla MR, Sherif SM, Murch SJ, Saxena PK (2014) Role of melatonin in alleviating cold stress in *Arabidopsis thaliana*. *J. Pineal Res.* 56:238–245.

Bennett RN, Wallsgrove RM (1994) Secondary metabolites in plant defense mechanisms. *New Phytol.* 127:617–633.

Byeon Y, Back KW (2013) Melatonin synthesis in rice seedlings in vivo is enhanced at high temperatures and under dark conditions due to increased serotonin N-acetyltransferase and N-acetylserotonin methyltransferase activities. *J. Pineal Res.* 56:189–195.

Byeon Y, Lee HY, Hwang OJ, Lee HJ, Lee K, Back K (2015). Coordinated regulation of melatonin synthesis and degradation genes in rice leaves in response to cadmium treatment. *J Pineal Res.* 58:470–478. doi:10.1111/jpi.12232.

Cai S-Y, Zhang Y, Xu Y-P, Qi ZY, Li MQ, Ahammed GJ, et al. (2017) HsfA1a up regulates melatonin biosynthesis to confer cadmium tolerance in tomato plants. *J. Pineal Res.* 62:e12387.

Cui G, Zhao X, Liu S, Sun F, Zhang C, Xi Y (2017) Beneficial effects of melatonin in overcoming drought stress in wheat seedlings. *Plant Physiol. Biochem.* 118: 138e149.

Dubbels R, Reiter RJ, Klenke E, Goebel A, Schnakenberg E, Ehlers C, et al. (1995) Melatonin in edible plants identified by radioimmunoassay and by high performance liquid chromatography–mass spectrometry. *J. Pineal Res.* 18:28–31.

Erland LAE, Murch SJ, Reiter RJ, Saxena PK (2015) A new balancing act: The many roles of melatonin and serotonin in plant growth and development. *Plant Signal. Behav.* 10:e1096469–15.

Flora SJS, Shrivastava R, Mittal M (2013) Chemistry and pharmacological properties of some natural and synthetic antioxidants for heavy metal toxicity. *Curr. Med. Chem.* 20:4540–4574.

Fu J, Wu Y, Miao Y, Xu Y, Zhao E, Wang J, et al. (2017) Improved cold tolerance in *Elymus nutans* by exogenous application of melatonin may involve ABA-dependent and ABA-independent pathways. *Sci. Rep.* 7:39865. doi:10.1038/srep39865.

Gill SS, Tuteja N (2010) Reactive oxygen species and antioxidant machinery in abiotic stress tolerance in crop plants. *Plant Physiol. Biochem.* 48:909–930.

Han QH, Huang B, Ding CB, Zhang ZW, Chen YE, Hu C, et al. (2017) Effects of melatonin on anti-oxidative systems and Photosystem II in cold-stressed rice seedlings. *Front. Plant Sci.* 8: 785.

Hardeland R, Pandi-Perumal SR, Poeggeler B (2007) Melatonin in plants—Focus on a vertebrate night hormone with cytoprotective properties. *Funct. Plant Sci. Biotech.* 1:32–45.

Hu ZR, Fan JB, Xie Y, Amombo E, Liu A, Gitau MM, et al. (2016) Comparative photosynthetic and metabolic analyses reveal mechanism of improved cold stress tolerance in bermudagrass by exogenous melatonin. *Plant Physiol. Biochem.* 100:94–104. doi:10.1016/j.plaphy.2016. 01.008.

Hulya T, Serkan E, Mucip G, Okkes A, Yavuz D, Derya Y (2014) The regulatory effect of melatonin on physiological, biochemical and molecular parameters in cold-stressed wheat seedlings. *Plant Growth Regul.* 74:139–152.

Jiang C, Cui Q, Feng K, Xu D, Li C, Zheng Q (2016) Melatonin improves antioxidant capacity and ion homeostasis and enhances salt tolerance in maize seedlings. *Acta Physiol. Plant.* 38:1–9.

Kang K, Lee K, Park S, Kim YS, Back K (2010) Enhanced production of melatonin by ectopic overexpression of human serotonin N-acetyltransferase plays a role in cold resistance in transgenic rice seedlings. *J. Pineal Res.* 49:176–182.

Kang S, Kang K, Lee K, Back K (2007) Characterization of tryptamine 5-hydroxylase and serotonin synthesis in rice plants. *Plant Cell Rep.* 26:2009–2015.

Kaur H, Mukherjee S, Baluska F, Bhatla SC (2015) Regulatory roles of serotonin and melatonin in abiotic stress tolerance in plants. *Plant Signal. Behav.* 10:e1049788. doi:10.1080/15592324.20 15.1049788.

Kolar J, Machackova I, Eder J, Prinsen E, van Dongen W, van Onckelen H, Illnerova H (1997) Melatonin: Occurrence and daily rhythm in *Chenopodium rubrum*. *Phytochemistry* 8:1407–1413.

Lei X-Y, Zhu R-Y, Zhang G-Y, Dai Y-R (2004) Attenuation of cold-induced apoptosis by exogenous melatonin in carrot suspension cells: The possible involvement of polyamines. *J. Pineal Res.* 36:126–131.

Li H, Chang J, Chen H, Wang Z, Gu X, Wei C, et al. (2017) Exogenous melatonin confers salt stress tolerance to watermelon by improving photosynthesis and redox homeostasis. *Front. Plant Sci.* 8:295.

Li H, Chang J, Zheng J, Dong Y, Liu Q, Yang X, et al. (2017) Local melatonin application induces cold tolerance in distant organs of *Citrullus lanatus* L. via long distance transport. *Sci. Rep.* 7:40858.

Li C, Wang P, Wei Z, Liang D, Liu C, Yin L, et al. (2012) The mitigation effects of exogenous melatonin on salinity-induced stress in *Malus hupehensis*. *J. Pineal Res.* 53:298–306.

Li C, Tan DX, Liang D, Chang C, Jia D, Ma F (2015) Melatonin mediates the regulation of ABA metabolism, free-radical scavenging and stomatal behavior in two *Malus* species under drought stress. *J. Exp. Bot.* 66:669–680.

Li X, Tan DX, Jiang D, Liu F (2016) Melatonin enhances cold tolerance in drought-primed wild-type and abscisic acid-deficient mutant barley. *J. Pineal Res.* 61:328–339.

Liang C, Zheng G, Li W, Wang Y, Hu B, Wang H, et al. (2015) Melatonin delays leaf senescence and enhances salt stress tolerance in rice. *J. Pineal Res.* 59:91–101. doi:10.1111/jpi.12243.

Limson J, Nyokong T, Daya S (1998) The interaction of melatonin and its precursors with aluminum, cadmium, copper, iron, lead and zinc: An adsorptive voltammetric study. *J. Pineal Res.* 24:15–21.

Lin Q, Zhang JS, Chen SY (2015) Melatonin enhances plant growth and maize (*Zea mays* L.) seedling drought tolerance by alleviating drought-induced photosynthetic inhibition and oxidative damage. *Acta Physiol. Plant* 38:48.

Liu J, Wang W, Wang L, Sun Y (2015) Exogenous melatonin improves seedling health index and drought tolerance in tomato. *J. Plant Growth Regul.* 77:317–326.

Mukherjee S, David A, Yadav S, Baluška F, Bhatla SC (2014) Salt stress-induced seedling growth inhibition coincides with differential distribution of serotonin and melatonin in sunflower seedling roots and cotyledons. *Physiol. Plant* 152:714–28.

Murch SJ, Alan AR, Cao J, Saxena, PK (2009) Melatonin and serotonin in flowers and fruits of Datura metel L. *J. Pineal Res.* 47:277–283.

Murch SJ, Campbell SSB, Saxena PK (2001) The role of serotonin and melatonin in plant morphogenesis: Regulation of auxin induced root organogenesis in in vitro-cultured explants of St. John's wort (Hypericum perforatum L.). *In Vitro Cell Dev. Biol. Plant* 37:786–793.

Murch SJ, Saxena PK (2002) Melatonin: A potential regulator of plant growth and development? *In Vitro Cell Dev. Biol. Plant* 38:531–536.

Odjakova M, Hadjiivanova C (1997) Animal neurotransmitter substances in plants. *Bulg. J. Plant Physiol.* 23:94–102.

Paredes SD, Korkmaz A, Manchester LC, Tan DX, Reiter RJ (2009) Phytomelatonin: A review. *J. Exp. Bot.* 60:57–69.

Parmar P, Limson J, Nyokong T, Daya S (2002) Melatonin protects against copper-mediated free radical damage. *J. Pineal Res.* 32:237–242.

Posmyk MM, Kuran H, Marciniak K, Janas KM (2008) Presowing seed treatment with melatonin protects red cabbage seedlings against toxic copper ion concentrations. *J. Pineal Res.* 45:24–31.

Posmyk MM, Bałabusta M, Wieczorek M, Sliwinska E, Janas KM (2009) Melatonin applied to cucumber (Cucumis sativus L.) seeds improves germination during chilling stress. *J. Pineal Res.* 46:214–223.

Ramakrishna A (2015) Indoleamines in edible plants: Role in human health effects. In: *Indoleamines: Sources, Role in Biological Processes and Health Effects*, Angel Catalá (ed.). Biochemistry Research Trends Series. Nova Publishers. Biochemistry Research Trends, p. 279. ISBN: 978-1-63482-097-4.

Ramakrishna A, Dayananda C, Giridhar P, Rajasekaran T, Ravishankar GA (2011a). Photoperiod influence endogenous indoleamines in cultured green alga *Dunaliella bardawil*. *Indian J. Exp. Biol.* 49:234–240.

Ramakrishna A, Giridhar P, Ravishankar GA (2011b) Phytoserotonin: A review. *Plant Signal. Behav.* 6:800–809.

Ramakrishna A, Giridhar P, Ravishankar GA (2009) Indoleamines and calcium channels influence morphogenesis in *in vitro* cultures of *Mimosa pudica* L. *Plant Signal. Behav.* 12:1136–1141.

Ramakrishna A, Giridhar P, Udaya Sankar K, Ravishankar GA (2012a) Endogenous profiles of indoleamines: Serotonin and melatonin in different tissues of *Coffea canephora* P ex Fr. as analyzed by HPLC and LC-MS-ESI. *Acta Physiol. Plant* 34:393–396.

Ramakrishna A, Giridhar P, Udaya Sankar K, Ravishankar GA (2012b) Melatonin and serotonin profiles in beans of *Coffea* sps. *J. Pineal Res.* 52:470–476.

Ramakrishna A, Giridhar P, Jobin M, Paulose CS, Ravishankar GA (2012c) Indoleamines and calcium enhance somatic embryogenesis in cultured tissues of *Coffea canephora* P ex Fr. *Plant Cell Tissue Organ Cult.* 108:267–278.

Ramakrishna A, Ravishankar GA (2011) Influence of abiotic stress signals on secondary metabolites in plants. *Plant Signal Behav.* 6:1720–1731.

Ramakrishna A, Ravishankar GA (2013) Role of plant metabolites in abiotic stress tolerance under changing climatic conditions with special reference to secondary compounds. In: *Climate Change and Abiotic Stress Tolerance*, Tuteja N, Gill SS (eds). Wiley-VCH, Weinheim. ISBN 978-3-527-33491-9.

Ravishankar GA, Ramakrishna A (2016) *Serotonin and Melatonin: Their Functional Role in Plants, Food, Phytomedicine, and Human Health.* Taylor & Francis, CRC Press, Boca Raton, p. 568. ISBN 9781498739054.

Roshchina VV (2001) *Neurotransmitters in Plant Life.* Science Publishers, Enfield, pp. 4–81. ISBN 9781578081424.

Sanchez-Barcelo EJ, Mediavilla MD, Vriend J, Reiter RJ (2016) Constitutive photo morphogenesis protein 1 (COP1) and COP9 signal some, evolutionarily conserved photomorphogenic proteins as possible targets of melatonin. *J. Pineal Res.* 61:41–51. doi:10.1111/jpi.12340.

Shi H, Chan Z (2014) The cysteine2/histidine2-type transcription factor Zinc Finger of *Arabidopsis thaliana* 6-activated C-repeat-Binding factor pathway is essential form melatonin-mediated freezing stress resistance in *Arabidopsis*. *J. Pineal Res.* 57:185–191.

Shi H, Tan D-X, Reiter RJ, Ye T, Yang F, Chan Z (2015a) Melatonin induces class A1 heat-shock factors (HSFA1s) and their possible involvement of thermo tolerance in *Arabidopsis*. *J Pineal Res.* 58:335–342.

Shi H, Jiang C, Ye T, Tan DX, Reiter RJ, Zhang H, et al. (2015b). Comparative physiological, metabolomic, and transcriptomic analyses reveal mechanisms of improved abiotic stress resistance in bermudagrass [Cynodon dactylon (L). Pers.] by exogenous melatonin. *J. Exp. Bot.* 66:681–694. doi:10.1093/jxb/eru373.

Shi H, Qian Y, Tan DX, Reiter RJ, He C (2015c) Melatonin induces the transcripts of CBF/DREBs and their involvement in abiotic and biotic stresses in Arabidopsis. *J. Pineal Res.* 59:334–342. doi:10.1111/jpi.12262.

Shi H, Reiter RJ, Tan DX, Chan Z (2015d) INDOLE-3-ACETIC ACID INDUCIBLE 17 positively modulates natural leaf senescence through melatonin-mediated pathway in Arabidopsis. *J. Pineal Res.* 58:26–33. doi:10.1111/jpi.12188.

Shi H, Tan DX, Reiter RJ, Ye T, Yang F, Chan Z (2015e) Melatonin induces class A1heat-shock factors (HSFA1s) and their possible involvement of thermo tolerance in Arabidopsis. *J. Pineal Res.* 58:335–342.

Shi H, Wang X, Tan DX, Reiter RJ, Chan Z (2015f) Comparative physiological and proteomic analyses reveal the actions of melatonin in the reduction of oxidative stress in Bermuda grass (Cynodon dactylon (L). Pers.). *J. Pineal Res.* 59:120–131. doi:10.1111/jpi.12246.

Szafrańska K, Szewczyk R, Janas K (2014) Involvement of melatonin applied to *Vigna radiata* L. seeds in plant response to chilling stress. *Cent. Eur. J. Biol.* 9:1117–1126.

Tan DX, Manchester LC, Helton P, Reiter RJ (2007) Phytoremediative capacity of plants enriched with melatonin. *Plant Signal. Behav.* 2:514–516.

Tiryaki I, Keles H (2012) Reversal of the inhibitory effect of light and high temperature on germination of Phacelia tanacetifolia seeds by melatonin. *J. Pineal Res.* 52:332–339.

Uchendu EE, Shukla MR, Reed BM, Saxena PK (2013) Melatonin enhances the recovery of cryopreserved shoot tips of American elm (*Ulmus americana* L.). *J. Pineal Res.* 55:435–442.

Wang P, Sun X, Wang N, Tan DX, Ma F (2015) Melatonin enhances the occurrence of autophagy induced by oxidative stress in *Arabidopsis* seedlings. *J. Pineal Res.* 58:479–489. doi:10.1111/jpi.12233.

Wang P, Sun X, Xie Y, Li M, Chen W, Zhang S, et al. (2014) Melatonin regulates proteomic changes during leaf senescence in Malushupehensis. *J. Pineal Res.* 57:291–307. doi:10.1111/jpi.12169.

Wang P, Yin L, Liang D, Li C, Ma F, Yue Z (2012) Delayed senescence of apple leaves by exogenous melatonin treatment: Toward regulating the ascorbate-glutathionecycle. *J. Pineal Res.* 53:11–20. doi:10.1111/j.1600-079X.2011. 00966.x

Weeda S, Zhang N, Zhao X, Ndip G, Guo Y, Buck GA, et al. (2014) Arabidopsis transcriptome analysis reveals key roles of melatonin in plant defense systems. *PLoS One* 9:e93462. doi:10.1371/journal.pone.0093462.

Xu W, Cai S-Y, Zhang Y, Wang Y, Ahammed GJ, Xia X-J, et al. (2016) Melatonin enhances thermotolerance by promoting cellular protein protection in tomato plants. *J. Pineal Res.* 61:457–469.

Zhang HX, Blumwald E (2001) Transgenic salt-tolerant tomato plants accumulate salt in foliage but not in fruit. *Nat. Biotech.* 19:765–768.

Zhang N, Sun Q, Zhang H, Cao Y, Weeda S, Ren S, et al. (2015) Roles of melatonin in abiotic stress resistance in plants. *J. Exp. Bot.* 66:647–656.

Zhang YP, Yang SJ, Chen YY (2017) Effects of melatonin on photosynthetic performance and antioxidants in melon during cold and recovery. *Biol. Plant* 61:571–578.

Zhang HJ, Zhang N, Yang RC, Wang L, Sun QQ, Li DB, et al. (2014) Melatonin promotes seed germination under high salinity by regulating antioxidant systems, ABA and GA interaction in cucumber (*Cucumis sativus* L.). *J. Pineal Res.* 57:269–279.

Zhang N, Zhao B, Zhang HJ, Weeda S, Yang C, Yang ZC, et al. (2013) Melatonin promotes water-stress tolerance lateral root formation and seed germination in cucumber (*Cucumis sativus* L.). *J. Pineal Res.* 54:15–23.

Zhao Y, Qi LW, Wang WM, Saxena PK, Liu CZ (2011) Melatonin improves the survival of cryopreserved callus of *Rhodiola crenulata*. *J. Pineal Res.* 50:83–88.

Zheng XD, Zhou JZ, Tan DX, Wang N, Wang L, Shan DQ, et al. (2017) Melatonin improves waterlogging tolerance of *Malus baccata* (Linn.) Borkh. Seedlings by maintaining aerobic respiration, photosynthesis and ROS migration. *Front. Plant Sci.* 8:483. doi:10.3389/fpls.2017.00483.

Zolla G, Heimer YM, Barak S (2010) Mild salinity stimulates a stress-induced morphogenic response in *Arabidopsis thaliana* roots. *J. Exp. Bot.* 61:211–224.

Zuo B, Zheng X, He P, Wang L, Lei Q, Feng C, et al. (2014) Overexpression of MzASMT improves melatonin production and enhances drought tolerance in transgenic *Arabidopsis thaliana* plants. *J. Pineal Res.* 57:408–417.

20

Flavonoid Accumulation as Adaptation Response in Plants during Abiotic Stresses

Rubal, Ashok Dhawan, and Vinay Kumar

CONTENTS

Abbreviations

ANR	anthocyanin reductase
APX	ascorbate peroxidase
CAT	catalase
CHI	chalcone isomerase
CHS	chalcone synthase
DFR	dihydroflavonol reductase
EC	epicatechins
FLS	flavonol synthase
GR	glutathione reductase
GT	glycosyl trnasferase
NADH	Nicotinamide Adenine Dinucleotide Hydrate
PHE	phenylalanine
PAL	phenylalanine ammonia lyase
ROS	Reactive Oxidant Species
UBGAT	UDP-glucuronate: Baicalein 7-O-Glucuronosyltransferase
UV	ultraviolet

Introduction

Flavonoids are widespread throughout the plant kingdom. Several different classes of flavonoids, including anthocyanins, flavonols, and flavan-3-ols, are known to contribute in many ways to the growth and survival of plants (Dixon and Sumner 2003). Anthocyanins are known to facilitate the interaction between pollen and stigma and to attract pollinators and fruit dispersal agents in angiosperms. Flavonols may act as UV protectors and natural auxin transport regulators in plants (Mahajan et al. 2011a; Schenke et al. 2011). The flavan-3-ols are accumulated in various tissues of plants to provide protection against predation. At the same time, flavan-3-ols impart astringency and flavors to beverages such as wines, fruit juices, and tea (Dixon et al. 2005). The flavan-3-ols are also major quality factors for forage crops and are increasingly recognized as compounds beneficial to health. The other functions of flavonoids include protection against UV exposure, pathogenic fungi, and herbivores (Harborne and Grayer 1993; Tanner et al. 2003; Lee et al. 2008; Yuan et al. 2012). The flavonoids are also involved in establishing the interaction between plants and mutualistic mycorrhizal fungi (Aron and Kennedy 2008). The antioxidant property of flavonoids is likely due to the scavenging of free radicals (Hernández et al. 2004). A number of studies have been carried out on the structure and antioxidant activity relationship of flavonoids (Bros et al. 1990; Rice-Evans et al. 1996) to establish their role in adaptation to environmental stress conditions. Different types of environmental conditions, especially abiotic stress conditions, stimulate the accumulation of flavonoids in the vegetative tissues and organs of plants. Water deficiency has been reported to up-regulate the flavonoid content, adjusting it according to stress conditions. Similarly, flavonoid levels have been found to increase in plant responses to many abiotic stresses, including cold, drought, and salinity. In addition, salt stress is reported to cause alteration of flavonoids in plants. In short, the accumulation of flavonoids often occurs in plants subjected to various stress conditions. This brief review summarizes updates on the protective roles of flavonoids against various different abiotic factors in plants.

The focus of this review is updates on the role of flavonoids as an adaptation response to various abiotic stress factors in plants. The details of flavonoid biosynthetic pathways, with their regulation, role in protection against environmental constraints, and possible strategies for improving flavonoid content in plants, are presented and are subsequently used for drafting an effective approach for crop improvement programs under abiotic stress conditions.

Flavonoid Biosynthetic Pathway

Flavonoids in plants are either preformed or induced compounds. Preformed compounds are synthesized during the normal development of plant tissue and are accumulated in different parts of the normal cell, while induced flavonoids are synthesized by a plant in response to fungal attack, wounding, and different type of environmental stresses.

The flavonoid biosynthetic pathway is a major branch of the phenylpropanoid pathway and synthesizes diverse types of phenyl benzopyran molecules performing various tasks depending on the plant's demands. The general outline of the flavonoid biosynthetic pathway with the major focus on the flavan-3-ol pathway in plants is shown in Figure 20.1. The key precursors for all the classes of flavonoids are malonyl-CoA and p-coumaroyl-CoA. The first step of the flavonoid biosynthetic pathway

is catalyzed by chalcone synthase (CHS), which synthesizes a yellow-colored product known as chalcone. CHS catalyzes the condensation of three acetate units from malonyl-CoA with p-coumaroyl-CoA to form chalcone. Then, chalcone undergoes several enzymatic conversions and synthesizes other classes of flavonoids, except isoflavonoids and isoflavones. Chalcone is isomerized into a flavanone, naringenin, by the chalcone isomerase enzyme. Naringenin is then hydroxylated at the C-3 position by flavanone-3-hydroxylase (F3H) to form dihydroflavonols and eriodictyol. The dihydroflavonols comprise dihydrokaempferol (DHK), dihydroquercetin (DHQ), and dihydromyricetin (DHM). DHK is further hydroxylated, either at the 3' position or at both the 3' and 5' positions of the B ring, to form DHQ and DHM by F3'H and F3'5'H enzymes, respectively. The DHQ and DHM are further used for the biosynthesis of cyanidin and delphinidin, respectively. Stereospecific reduction of dihydroflavonols is catalyzed by dihydroflavonol 4-reductase (DFR) using NADPH to synthesize leucoanthocyanidins (flavan-3,4-diols). The formation of anthocyanidins from leucoanthocyanidins occurs by the action of anthocyanidin synthase. UDP-glucose:flavonoid

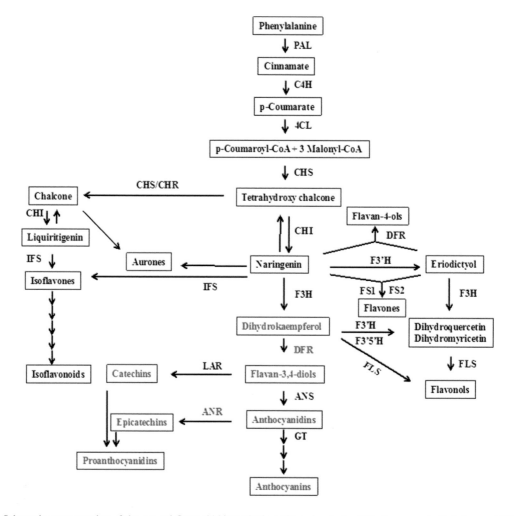

FIGURE 20.1 Schematic representation of the general flavonoid biosynthetic pathway in plants. 4CL, 4-coumarate:CoA ligase; ANR, anthocyanidin reductase; ANS, anthocyanidin synthase; C4H, chalcone 4-hydrolase; CHI, chalcone isomerase; CHR, chalcone reductase; CHS, chalcone synthase; DFR, dihydroflavonol 4-reductase; F3H, flavanone 3-hydroxylase; F3'H, flavanone-3'-hydroxylase; F3',5'H, flavonoid 3',5'-hydroxylase; FLS, flavonol synthase; FS1, flavone synthase 1; FS2, flavone synthase 2; GT, glycosyltransferase; IFS, isoflavone synthase; LAR, leucoanthocyanidin reductase; PAL, phenylalanine ammonia-lyase. The blue colored text shows the specific flavan-3-ol/PA biosynthetic pathway. The red colored text shows the enzymes that catalyze rate limiting steps of the flavan-3-ol/PA biosynthetic pathway. (Figure designed by author.)

3-O-glucosyltransferase catalyzes the transfer of the glucose moiety from UDP-glucose to the hydroxyl group at the 3′ position of the C ring of anthocyanidin. This interaction increases the stability and aqueous solubility of the final products in a vacuole. Interestingly, different classes of dihydroflavonols, such as DHK, DHQ, and DHM, are used as substrates for the formation of flavonols (kaempferol, quercetin, and myrecetin) by the action of flavonol synthase. Catechins (Cat) and epicatechins (EC) are synthesized in two different steps. Cat is produced by the action of leucoanthocyanidin reductase from the leucoanthocyanidin substrate, while anthocyanidin reductase (ANR) catalyzes the reduction of anthocyanidin to produce EC (Xie et al. 2004). Proanthocyanidins (PAs) are synthesized by the condensation of either similar monomers (either Cat or EC) or different monomers (both Cat and EC).

All other classes of flavonoids originate from this basic pathway. Stilbenes are synthesized by stilbene synthase, which shares the same substrate (p-coumaroyl-CoA) with CHS (Schroder 1999). The aurones class is formed by the action of aureusidin synthase using tetra-hydroxychalcone and penta-hydrochalcone as substrates (Nakayama et al. 2000). In legume plants, chalcone reductase (CHR) acts in association with CHS, producing the precursor for the synthesis of isoflavones and isoflavonoids. For example, deoxy-flavonoids include isoliquiritigenin, which is a major constituent of *Glycyrrhiza uralensis* (licorice) and is synthesized by the co-action of CHS and CHR. The isoliquiritigenin is further converted into 5-deoxy-isoflavonoids such as daidzein through the stepwise action of CHI and isoflavone synthase (IFS). The flavones are synthesized from the naringenin precursor by either cytochrome P450 or deoxygenase enzyme. In maize, the branch from the common pathway is determined by the activity of F3H, which leads to the production of 3-deoxy-anthocyanins and phlobaphenes from flavan-4-ol substrates. The flavan-4-ols are produced from flavanones by the action of DFR (Grotewold et al. 1998). The F3H activity determines the moment of carbon flux toward either anthocyanins or 3-deoxy-anthocyanins and phlobaphenes.

Flavonoids are modified into aglycones by a number of different reactions, such as hydroxylation, glycosyl transfer, methylation, acyl transfer, and hydroxylation, in each class. Naringenin and dihydrokaempferol are hydroxylated by the enzymes flavonoid 3′-hydroxylase and flavonoid 3′,5′-hydroxylase at the 3′ and 3′,5′ positions, respectively. Glycosyltransferase (GT) classes include flavonoid 5-O-glycosyl transferase (5GT) and flavonoid 7-O-glycosyl transferase (7GT). In addition to C3, glycosylation also occurs at C5, C7, and C3′ of anthocyanin by UDP-glucose: anthocyanin glucosyltransferase. 5GT has been isolated from *Perilla*, *Petunia*, and *Torenia*, while 7GT and 3GT have been isolated from *Scutellaria baicalensis* and *Gentiana triflora*, respectively (Mizutani et al. 2003). Methyl transferases, especially O-methyltransferase, are known to methylate flavonoids. Glycosylated flavonoids are also modified by acylation via flavonoid-specific acyltransferases such as malonyl-CoA; anthocyanin 3-O-glucoside-6′′-O-malonyltransferase from *Dahlia variabilis* (Suzuki et al. 2002). This modification increases the water solubility of glucosides and provides protection against degradation by glucosidase. A flavonol sulfotransferase catalyzing the transfer of the sulfate group to the 3-hydroxyl group of flavonol aglycones has been characterized from *Flaveria*

chloraefolia (Gidda and Varin 2006). The hydrolytic enzyme isoflavone 7-O-glucoside β-glucosidase, from *Cicer arientinum*, has been purified and characterized. This enzyme is known for its involvement in the two-step hydrolysis of isoflavone conjugates (Barz and Welle 2012). However, the modification of flavonoids by acetylation is not well understood, although it has been suggested that acyl groups stabilize the anthocyanin chromophore by addition at 2 and 4 positions (Winkel 2006). In addition to modification that occurs in the cytoplasm, additional modifications can occur at the final destination, in the cell wall or the vacuole, in the form of oxidation by peroxidase enzymes. In conclusion, the diverse flavonoid compounds show different roles in plants, which makes them an interesting area of research. The ability of flavonoids to influence various cellular processes in plants gives support for their protective roles under adverse conditions. A very tight transcriptional regulation of flavonoid biosynthetic pathway genes by regulatory proteins has been established for the accumulation of flavonoids under abiotic stress conditions.

Regulation of Flavonoid Biosynthesis

In plants, the flavonoid biosynthetic pathway is the most explored with respect to its regulation (Yu and McGonigle 2005). In addition to tissue-specific regulation, environmental factors also influence the regulation of flavonoid biosynthesis. Both abiotic and biotic stresses are known to influence the flavonoid biosynthetic pathway. Transcriptional regulation acts as the primary level of the regulatory mechanism for the biosynthesis of flavonoids (Lepiniec et al. 2006; Czemmel et al. 2012; Henry-Kirk 2012). To understand how flavonoid biosynthesis is regulated in plants, several transcriptional factors controlling the expression of known flavonoid biosynthetic genes have been isolated and studied recently. These regulatory transcription factors belong to six different families: Myc transcription factors (encoding basic helix-loop-helix proteins [bHLH]), Myb transcription factors, WD40-like proteins, WRKY transcription factors, MADS homeodomain genes, and TFIIIA-like proteins "WIP."

In all species analyzed to date, members of the protein family R2R3-MYB domains act as the common denominator in the regulation of flavonoid biosynthetic pathway genes. However, Myc, Myb, and WD40 are collectively known to control the biosynthesis of a wide range of anthocyanins and PAs. In *Arabidopsis thaliana*, the ternary MYB-BHLH-WDR complex formed from tt2 (Myb family), tt8 (Myc family), and ttg1 (WD-like protein) is known to activate the expression of *ANR* (Gonzalez et al. 2008). The overexpression of genes encoded by tt2, tt8, and ttg1 proteins from *A. thaliana* into *Fragaria ananassa* has been reported to up-regulate the expression of *F3′H*, *ANS*, *ANR*, and *LAR* genes (Schaart et al. 2012). Basic helix-loop-helix transcription factor (tt8) is also reported to activate *DFR* and *ANR* genes in *A. thaliana* (Nesi et al. 2000). Ku-like and HBF1 transcription factors have been reported to activate *CHS* gene in soya beans (Yu and McGonigle 2005). In maize, *C1* gene has regulated the flavonoid biosynthetic pathway by binding to the consensus cis element of the promoters of many flavonoid biosynthetic genes (Lepiniec et al. 2006). C1 belongs to the Myb transcription protein family.

It has been reported to interact with a transcription factor R gene (basic helix-loop-helix type) and activate the transcription of the entire flavonoid biosynthetic pathway (Grotewold et al. 2000). The C1 homolog from a distinct subfamily has been characterized for its role in regulating anthocyanin production in various species. The NtAn1 and NtAn2 transcription factors are known as flavonoid-related basic helix-loop-helix type and are predominantly expressed in tobacco flowers. A transgenic tobacco overexpressing *NtAn1a* and *NtAn1b* has been reported to up-regulate the expression of *CHS*, *CHI*, *F3H*, *DFR*, and *ANS* genes (Bai et al. 2011).

WD40 repeats from the b unit of heterotrimeric G protein activate flavonoid-related Myb-like transcription factors, such as AN11 from *Petunia hybrida* and TTG1 from *A. thaliana* (Sompornpailin et al. 2002; Hichri et al. 2011). These components directly regulate the action of transcription factors related to the flavonoid biosynthetic pathway. AN11 has been shown to regulate anthocyanin biosynthesis indirectly by regulating the Myb-like transcription factor anthocyanin2. TTG1 has been reported to regulate anthocyanin biosynthesis after binding with Myb-like transcription factor GLABROUS1. Besides regulating anthocyanin biosynthesis, TTG1 has also been reported to regulate trichome development and root epidermal cell patterning (Hichri et al. 2011).

Transcription factors from the WRKY family, such as *A. thaliana* ttg2, are also known to regulate flavonoid biosynthesis (Johnson et al. 2002; Ishida et al. 2007). WRKY transcription factor (ttg2) is a zinc finger–like protein and acts downstream of the WD40-like protein. Hence, WD40 is regulated by tt2 and Myb transcription factors and MADS homeodomain genes. Further, ttg2 protein directly regulates *BAN* (*banyuls*) by acting upstream of other regulatory genes. *Banyuls* gene encodes an ANR enzyme in *A. thaliana* seed coat. Similarly, WIP protein also shows interaction with *BAN* gene and influences the flavonoid biosynthetic pathway. The regulation of flavonols in *A. thaliana* is governed by a separate set of genes from those of the anthocyanin biosynthetic pathway, such as *AtMYB11* and *AtMYB12*. They have been shown to coordinate the regulation of *CHS*, *CHI*, *F3H*, and *FLS* expression in response to light in different plant tissues (Mehrtens et al. 2005; Stracke et al. 2007).

The MADS-box family of transcription factors are also known for their role in the regulation of the flavonoid biosynthetic pathway. The *SQUAMOSA*-class MADS-box transcription factor *VmTDR4* has been reported to be associated with anthocyanin biosynthesis in bilberry (Jaakola et al. 2010). A sweet potato (*Ipomoea batatas*) SQUA transcription factor, IbMADS10, has shown expression in tight correlation with anthocyanin biosynthesis, especially in the red root (Lalusin et al. 2006). The *A. thaliana* TT1 factor of the WIP subfamily of plant zinc finger proteins has been reported to be involved in PA biosynthesis (Sagasser et al. 2002).

Besides these major regulatory families, some other regulatory genes are also known to have either direct or indirect influence on flavonoid biosynthesis. *Anthocyaninless2* (*ANL2*) gene from *A. thaliana*, a homeobox gene of the homeodomain-leucine zipper (HD-Zip IV) family, is one such example. This gene codes for a homeodomain protein of the HD-GL2 group and is most likely to play a key role in anthocyanin accumulation and root development (Vernoud et al. 2009). Some negative regulators

from the MYB superfamily have also been found to down-regulate flavonoid biosynthesis. Such regulators share a conserved motif in their C- terminal end (Vom Endt et al. 2002). These suppressor genes encode MYB repressors/activators, which compete with endogenous MYB-related activators and down-regulate flavonoid biosynthesis. The overexpression of FaMYB1 transcription factor has been reported to down-regulate flavonoid biosynthesis in tobacco and also suppress the accumulation of anthocyanins and flavonols (Aharoni et al. 2001). The phenylpropanoid biosynthetic pathway in tobacco is also repressed by overexpression of AmMYB308 and AmMYB330 MYB transcription factors isolated from *Antirrhinum* (Tamagnone et al. 1998). Some components of the signal transduction pathway, which are present upstream of the transcription factors and can bind to the promoters of flavonoid biosynthetic pathway genes, have also been identified. One such example is the *Arabidopsis* ICX1 (increased chalcone synthase expression 1) mutant. The mutants have been reported to induce the expression of *CHS* and other flavonoid biosynthesis genes in response to several stimuli (Wade et al. 2003).

Thus, there is strong evidence regarding the regulation of flavonoid accumulation at the transcriptional level. The regulation of expression of flavonoid biosynthetic pathway genes by different types of transcriptional factor is well controlled in a plant species–specific fashion. This information may be useful in setting benchmarks for future research on the regulatory network controlling flavonoid biosynthesis and diversity.

Flavonoid Accumulation as Adaptation Response against Environmental Constraints

Plants have a complex antioxidant system to protect them against oxidative damage caused by environmental constraints. Flavonoids are an integral part of plant antioxidant system due to their free radical–scavenging property and antioxidant activity. The flavonoid concentration in plant cells is often over 1 mM, which helps to provide protection against oxidative stress. Flavonoids show antioxidant activity against oxidizable compounds and act as ideal scavengers of H_2O_2 due to their reduction potential. Flavonoids also act as inhibitors of the enzyme lipoxygenase, which converts polyunsaturated fatty acids (PUFA) to oxygen-containing derivatives (Sadik et al. 2003). So, they act as effective inhibitors of lipid peroxidation. Flavonoids are reported to be induced by the exposure of plants to different stresses, and their biosynthetic genes are generally up-regulated in response to one or many stresses, such as strong UV radiation, temperature extremes, drought, metal toxicity, nutrient deprivation (nitrogen and phosphorus deficiencies), and bacterial as well as fungal infections. Flavonoids are also considered as a stress symptom or part of the mechanisms to mitigate the effects of stresses. During exposure to biotic and abiotic stress, the activation of secondary metabolism as well as antioxidant metabolism is an integral part of plant adaptation when encountering such stresses (Oh et al. 2009).

UV-B radiation is highly energetic, and a small increase in it might modify plant growth. To protect against the harmful effects of UV light, plants have developed mechanisms to reduce UV penetration into the tissues. Under such stress, plants are reported to synthesize UV-protective compounds such as total phenolics and flavonoids (Ryan and Hunt 2005). Flavonoids absorb radiation in the UV region of the spectrum and act as a buffer by attenuating the excess energy. Flavonoid-deficient mutants of maize and *A. thaliana* have been reported to be sensitive to UV-B radiation. This has documented that flavonoids provide protection to the plants against UV-B light (Guo et al. 2008). Additionally, the flavonoid biosynthetic pathway is also known to be activated by UV exposure in many plant species (Gould and Lister 2006). UV light and high-intensity white light caused an increase in the transcription of flavonoid biosynthetic genes as well as their enzyme activities. Flavonoids have been reported to accumulate in the epidermal tissue of plants, where exposure to UV-B radiation is highest. Flavonoids provide a line of defense from UV-B-induced damage to the internal tissues of leaves and stems (Lee et al. 2008). Flavonoid accumulation has also influenced the photosynthetic ability of leaves during exposure to UV-B or UV-C radiation. Interestingly, the photosynthetic capacity of the green-leafy variety of coleus was reduced compared with the red-leafy variety during exposure to UV-B and UV-C light (Burger and Edwards 1996). Flavonoids have also been reported to protect DNA from the mutagenic effects of UV light both *in vivo* and *in vitro* (Mpoloka 2008). Flavonoids prevent peroxidation of lipids by scavenging reactive oxygen species (ROS). But prolonged accumulation of anthocyanin due to UV exposure might induce detrimental effects (Hada et al. 1998).

In watermelon, temperature variation has induced the biosynthesis of flavonoids and suppressed the oxidation of phenolics in response to damage caused by thermal stress (Rivero et al. 2001). In rose and sugarcane, anthocyanins have been reported to accumulate in leaf tissues during heat stress (Wahid and Ghazanfar 2006). Anthocyanins play a significant role during such environmental stresses. Higher anthocyanin accumulation is known to decrease leaf osmotic potential, which is linked to an increased uptake of water and a reduction in water loss (Chalker-Scott 2002). A temperature below the optimum for plant growth and development has also induced anthocyanin biosynthesis in seedlings of many plants, such as *A. thaliana*, maize, sorghum, and poncirus. Some commonality has been reported between anthocyanin biosynthesis and freezing (Chalker-Scott 1999). During chilling temperatures, induced anthocyanins have protective functions, such as cold-hardiness induction and an increase in leaf temperature. Accumulated anthocyanins in epidermal vacuoles have been known to prevent them from freezing at cold temperature. This function has also protected deciduous leaves from early frost fall. Cold hardiness induced by anthocyanins has further provided cross resistance to another stress during winters in plants (Chalker-Scott 1999).

Flavonoids have provided protection against drought stress by increasing the concentrations of EC, epicatechin gallate (ECG), and epigallocatechin gallate (EGCG) with other antioxidants in *Cistus clusii* (Hernández et al. 2004). Drought stress has induced anthocyanin pigments in cowpea seedlings and resurrection plants. A pretty purple cultivar of pepper containing higher levels of anthocyanins has been reported to show more resistance to drought stress than green cultivars of pepper (Bahler et al. 1991). Accumulated anthocyanins in ornamental shrubs have also been documented to provide tolerance against drought stress.

As compared with the fully hydrated state, resurrection plants have been reported to contain high amounts of anthocyanins to protect against injury caused by drought stress (Chalker-Scott 2002). In angiosperm evergreen species, the association between winter anthocyanins and drought stress has also been well documented (Hughes et al. 2010). During drought stress, maize seedlings with induced anthocyanins have been reported to show more tolerance (Kaliamoorthy and Rao 1994).

Besides being radical scavengers, flavonoids have been found to act as chelators for metals, depending on their molecular structure (Michalak 2006). Anthocyanin molecules have provided tolerance against heavy metal stresses (Gould 2004). In response to $CuSO_4$ treatment, much greater accumulation of flavonoids has occurred in *Ginkgo biloba* cell cultures. In *Ononis arvensis*, the flavonoid level was increased on exposure to $CuSO_4$ and $CdCl_2$ stress. Exposure to aluminum has stimulated the exudation of flavonoids in maize seedlings, which provided protection against Al-induced toxicity. Treatment with naringenin and quercetin has been reported to restore partial growth during metal stresses in *A. thaliana* plants. These observations have suggested the metal-binding activity of flavonoids and their role in heavy metal stress tolerance.

During nitrogen (N) and phosphorus (P) starvation, flavonoid accumulation has provided better adaptation or tolerance to plants (Peng et al. 2008). In N-depleted growth, flavonols have been reported to accumulate in the leaves of tomato (Løvdal et al. 2010). In *A. thaliana*, the expression of regulatory genes of flavonoid biosynthesis, such as *PAP1/2* and *GL3*, was induced by N-depletion in seedlings and leaves at the rosette stage (Lillo et al. 2008; Feyissa et al. 2009). Phosphate deficiency has also been found to alter isoflavonoid (naringenin and daidzein) content in the roots of bean plants (*Phaseolus vulgaris* L.) (Juszczuk et al. 2004). Deficiency of nitrogen and phosphorus has increased anthocyanin accumulation in plants, which helped them to maintain the integrity of cellular function by mobilizing inorganic minerals. Flavonoids isolated from white lupin roots have been observed to mobilize inorganic phosphate ion and play a significant role in an efficient phosphate acquisition strategy (Uhde-Stone et al. 2005; Tomasi et al. 2008).

Potential Strategies for Improving Flavonoid Content in Plants

A number of strategies have been adopted for improving the flavonoid content of agriculturally important crops. The conventional breeding approach has its own limitations and requires robust screening of lines and the time-consuming process of releasing tolerant varieties. On the other hand, metabolic engineering offers a reliable approach to the investigation of the protective roles of improved flavonoid content in plants for drafting a strategy for developing abiotic stress–tolerant crops.

Flavonoid metabolic engineering has highlighted two strategies for the generation of transgenic plants with higher levels of flavonoids. To achieve higher accumulation of flavan-3-ols, it is necessary to up-regulate the pathway leading to flavan-3-ol production or down-regulate the competing pathways (Griesser et al. 2008; Han et al. 2012a,b; Mahajan et al. 2012; Yuan et al. 2012). Transgenic plants have been developed using the introduction of either biosynthetic or regulatory genes of the flavonoid biosynthetic pathway from diverse plants to increase the flavan-3-ol level (Yoshida et al. 2008; Han et al. 2012a,b; Hancock et al. 2012; Yuan et al. 2012). The engineering of the flavonoid biosynthetic pathway by coexpression of a biosynthetic gene (*ANR*) and a regulatory gene (*PAP1*) has also provided evidence regarding the use of a combinatorial approach for enhancing flavan-3-ol content (Xie et al. 2006). For the down-regulation of genes corresponding to competing key enzymes of the flavan-3-ol biosynthetic pathway, different strategies such as antisense, sense suppression, and RNA interference (RNAi) have been applied (Hoffmann et al. 2006; Nakamura et al. 2006; Jagtap et al. 2011; Jiang et al. 2012; Mahajan et al. 2011b).

A. thaliana contains four *PAL* genes: *PAL1*, *PAL2*, *PAL3*, and *PAL4*. Among these genes, *PAL1* and *PAL2* have a redundant role in anthocyanin as well as flavan-3-ol biosynthesis, as confirmed by mutant analysis (Huang et al. 2010). Transgenic tomato overexpressing either *PhCHS* or *PhCHI* has been reported to possess modified flavonoid content as well as increased antioxidant potential (Verhoeyen et al. 2002). The suppression of *CHS* gene has been documented to result in increased antioxidant capacity of *Linum usitatissimum* via increasing PA accumulation (Zuk et al. 2012). The single gene overexpression or simultaneous expression of genes encoding CHS, CHI, and DFR enzymes in potato has resulted in a significant elevation of the anthocyanin level with improved antioxidant capacity (Lukaszewicz et al. 2002). The overexpression of *DFR* gene from *Medicago trancatula* has been studied in tobacco and rice with respect to the accumulation of flavan-3-ol/PA content (Xie et al. 2004; Takahashi et al. 2006). The overexpression of *DFR* genes (*DFR1* and *DFR2*) from *Populus trichocarpa* in tobacco and *P. tomentosa* has resulted in a greater accumulation of anthocyanins and PAs (Huang et al. 2012). RNAi-mediated silencing of *F3H* gene in strawberry has been reported to alter flavonoid content (Jiang et al. 2012). The ectopic expression of apple *F3'H* gene has also resulted in altered flavonoids in the *A. thaliana tt7* mutant grown under nitrogen stress (Han et al. 2012b). Transgenic tobacco with silenced *FLS* gene has been documented to have a higher level of flavan-3-ols (Mahajan et al. 2011b, 2012). Transgenic rice (mutant *Nootripathu*) overexpressing the *ANS* gene has been reported to have higher accumulation

of a mixture of flavonoids with increased antioxidant potential (Reddy et al. 2007). Transgenic poplar (*Populus tomentosa* Carr.) overexpressing *PtrLAR3* has also been reported to have higher accumulation of flavan-3-ols (Yuan et al. 2012). However, transgenic tobacco overexpressing *MtLAR* has shown a lack of flavan-3-ol production (Pang et al. 2007). The *ANR* from *M. trancatula* and *Malus domestica* has also been engineered for flavan-3-ol content in transgenic tobacco (Xie et al. 2003; Han et al. 2012a). Soybean grain with higher anthocyanin and lower PA content has been developed by the redirection of metabolic flux with the suppression of endogenous *ANR* genes *ANR1* and *ANR2*. The redirection of the flavonoid biosynthetic pathway toward the biosynthesis of flavan-3-ols in ripening strawberry fruit has been achieved through the down-regulation of *anthocyanidin GT* gene (Griesser et al. 2008). In addition to metabolic engineering of flavonoid biosynthetic pathway in plants, metabolic engineering of this pathway in *Escherichia coli* as well as *Saccharomyces cerevisiae* has also been carried out to produce flavonoids for use in the pharmaceutical industry (Chemler et al. 2007, 2008; Wang et al. 2011). Since *E. coli* and *S. cerevisiae* do not possess an endogenous flavonoid pathway, the transformation of production strains of these microbes with plant genes encoding metabolic enzymes is essential for *de novo* biosynthesis or bioconversion. Table 20.1 summarizes genetically modified plants with improved a special class of flavonoids after the manipulation of specific genes to provide tolerance against abiotic stresses.

Apart from the manipulation of flavonoid content using genetic engineering, another potential approach is the use of genome editing approaches, including CRISPR-Cas9 technology. The activation and silencing of regulatory gene(s) of the flavonoid biosynthetic pathway could be carried out using the CRISPR-Cas9 technology. A complete set of genes could be targeted to analyze their influence on flavonoid contents under abiotic stress factors. Similarly, the generation of a genomic library of important crops using the CRISPR-Cas9 technology would assist in the identification of novel regulators of the flavonoid biosynthetic pathway and help to provide in-depth knowledge of plant adaptation response against numerous abiotic stress factors by the manipulation of flavonoid contents. Enhancing flavonoid production by tuning the central metabolic pathway using a

TABLE 20.1

List of Latest Reports with Flavonoid Biosynthetic Pathway Genes That Are Exploited for Improving the Flavonoid Content in Response to Abiotic Stress Factors

S. No.	Name of Genes	Type of Flavonoid Accumulated	Abiotic Stress Factors	References
1	*FLS*	Flavonol	Salt, Drought UV-B	Nguyen et al. (2016)
2	*PAL*	Not available	Drought, UV	Huang et al. (2010)
3	*IbDFR*	Anthocyanins	Cold	Wang et al. (2013)
4	*SlF3HL*	Flavonoid 3-hydroxylase	Cold	Meng et al. (2015)
5	*LDOX*	Anthocyanin	Salt stress	Oosten et al. (2013)
6	*RsF3H*	Anthocyanin	Drought, UV	Lui et al. (2013)
7	*RrANR*	Proanthocyanidins	Oxidative stress	Luo et al. (2016)
8	*LcF3H*	Flavan-3-ols	Drought	Song et al. (2016)
9	*AtMTB12*	Total flavonoids content	Salt and drought stress	Wang et al. (2016)
10	*CsDFR* and *CsANR*	Flavan-3-ols	Alluminium Stress	Kumar and Yadav (2014)
11	*CsF3H*	Flavan-3-ols	Salinity	Mahajan and Yadav (2015)

CRISPR-interference system in *E. coli* also opens a new avenue for the production of flavonoids in plants (Wu et al. 2015).

Conclusion

The diverse array of flavonoid molecules contributes toward improving the antioxidant potential of plants under adverse conditions. The anthocyanins, flavonols, and flavan-3-ols have better antioxidant capacity than vitamin C and scavenge the free radicals generated by the oxidative stress injury mediated by almost all adverse environmental conditions. The genetic regulatory pathway has been documented to activate a dedicated biosynthetic pathway for the accumulation of a specialized class of flavonoids of interest according to urgent requirements for protection against toxicity mediated by adverse abiotic factors, ensuring better plant adaptability under stress conditions. However, the use of emerging technologies, including genome scale transcriptome analysis, will definitely accelerate the investigation of accumulation of flavonoids as an adaptive response against a specialized abiotic stress factor. Genome-wide association of the metabolic network will also provide an overall picture of the regulation of metabolic biosynthesis with accumulation of products in different compartments in plants. In addition, new emerging technology will increase the relevant information in the days to come. Overall, highly focused organized information relevant to flavonoid accumulation as an adaptive response contributes toward the crop improvement strategy for sustainable agriculture.

Acknowledgments

The authors express sincere thanks to the vice chancellor of Central University of Punjab for providing the basic facility for conducting this work.

REFERENCES

Aharoni A, De Vos CHR, Wein M, Sun Z, Greco R, Kroon A, Mol JN, O'Connell AP. The strawberry FaMYB1 transcription factor suppresses anthocyanin and flavonol accumulation in transgenic tobacco. *Plant J.* 2001;28:319–332.

Aron PM, Kennedy JA. Flavan-3-ols: Nature, occurrence and biological activity. *Mol. Nutr. Food Res.* 2008;52:79–104.

Bahler BD, Steffen KL, Orzolek MD. Morphological and biochemical comparison of a purple-leafed and a green-leafed pepper cultivar. *HortScience* 1991;26:736.

Bai S, Kasai A, Yamada K, Li T, Harada T. A mobile signal transported over a long distance induces systemic transcriptional gene silencing in a grafted partner. *J. Exp. Bot.* 2011;62:4561–4570.

Barz W, Welle R. Biosynthesis and metabolism of isoflavones and pterocarpan phytoalexins in chickpea, soybean and phytopathogenic fungi. In: Stafford H, Ibrahim R. (eds.) *Phenolic Metabolism in Plants.* Plenum Press, New York, pp. 139–164 *Chemistry* 2012;53:3313–3320.

Bros W, Heller W, Michel C, Saran M. Flavonoids as antioxidant, determination of radical-scavenging efficiency. *Methods Enzymol.* 1990;186:343–355.

Burger J, Edwards GE. Photosynthetic efficiency, and photo damage by UV and visible radiation, in red versus green leaf of coleus varieties. *Plant Cell Physiol.* 1996;37:395–399.

Chalker-Scott L. Do anthocyanins function as osmoregulators in leaf tissues? *Adv. Bot. Res.* 2002;37:104–129.

Chalker-Scott L. Environmental significance of anthocyanins in plant stress responses. *J. Photochem. Photobiol.* 1999;70:1–9.

Chemler JA, Koffas MAG. Metabolic engineering for plant natural product biosynthesis in microbes. *Curr. Opin. Biotechnol.* 2008;19:597–605.

Chemler JA, Lock LT, Koffas MAG, Tzanakakis ES. Standardized biosynthesis of flavan-3-ols with effects on pancreatic beta-cell insulin secretion. *Appl. Microbiol. Biotechnol.* 2007;77:797–807.

Czemmel S, Heppel SC, Bogs J. R2R3 MYB transcription factors: Key regulators of the flavonoid biosynthetic pathway in grapevine. *Protoplasma* 2012;249:S109–S118.

Dixon RA, Sumner LW. Role of anthocyanidin reductase, encoded by *BANYULS* in plant flavonoid biosynthesis. *Plant Physiol.* 2003;131:878–885.

Dixon RA, Xie DY, Sharma SB. Proanthocyanidins—a final frontier in flavonoid research? *New Phytol.* 2005;165:9–28.

Feyissa DM, Løvdal T, Olsen KM, Slimestad R, Lillo C. The endogenous GL3, but not EGL3, gene is necessary for anthocyanin synthesis as induced by nitrogen depletion in *Arabidopsis* rosette stage leaves. *Planta* 2009;230:747–754.

Gidda SK, Varin L. Biochemical and molecular characterization of flavonoid 7-sulfotransferase from *Arabidopsis thaliana. Plant Physiol. Biochem.* 2006;44:628–636.

Gonzalez A, Zhao M, Leavitt JM, Lloyd AM. Regulation of the anthocyanin biosynthetic pathway by the TTG1/bHLH/Myb transcriptional complex in *Arabidopsis* seedlings. *Plant J.* 2008;53:814–827.

Gould KS. Nature's Swiss army knife: The diverse protective roles of anthocyanins in leaves. *J. Biomed. Biotechnol.* 2004;2004: 314–320.

Gould KS, Lister C. Flavonoid functions in plants. In: Andersen ØM, Markham KR (eds.). *Flavonoids. Chemistry, Biochemistry, and Applications.* CRC Press, Boca Raton, 2006, pp. 397–441.

Griesser M, Hoffmann T, Bellido ML, Rosati C, Fink B, Kurtzer R, Aharoni A, Munoz-Blanco J, Schwab W. Redirection of flavonoid biosynthesis through the down-regulation of an anthocyanidin glucosyltransferase in ripening strawberry fruit. *Plant Physiol.* 2008;146:1528–1539.

Grotewold E, Chamberlin M, Snook M, Siame B, Butler L, Swenson J, Maddock S, St Clair G, Bowen B. Engineering secondary metabolism in maize cells by ectopic expression of transcription factors. *Plant Cell* 1998;10:721–740.

Grotewold E, Sainz MB, Tagliani L, Hernandez JM, Bowen B, Chandler VL. Identification of the residues in the Myb domain of maize C1 that specify the interaction with the bHLH cofactor R. *Proc. Natl. Acad. Sci. USA* 2000;97:13579–13584.

Guo J, Han W, Wang MH. Ultraviolet and environmental stresses involved in the induction and regulation of anthocyanin biosynthesis: A review. *Afr. J. Biotechnol.* 2008;7:4966–4972.

Hada M, Hashimoto T, Nikaido O, Shin M. UVB-induced DNA damage and its photorepair in nuclei and chloroplasts of *Spinacia oleracea* L. *J. Photochem. Photobiol.* 1998;68:319–322.

Han Y, Vimolmangkang S, Soria-Guerra RE, Korban SS. Introduction of apple *ANR* genes into tobacco inhibits expression of both *CHI* and *DFR* genes in flowers, leading to loss of anthocyanin. *J. Exp. Bot.* 2012a;63:2437–2447.

Han Y, Vimolmangkang S, Soria-Guerra RE, Rosales-Mendoza S, Zheng D, Lygin AV, Korban SS. Ectopic expression of apple F3′H genes contributes to anthocyanin accumulation in the *Arabidopsis* tt7 mutant grown under nitrogen stress. *Plant Physiol.* 2012b;153:806–820.

Hancock KR, Collette V, Fraser K, Greig M, Hong X, Richardson K, Jones C, Rasmussen S. Expression of the R2R3-MYB transcription factor TaMYB14 from *Trifolium arvense* activates proanthocyanidin biosynthesis in the legumes *Trifolium repens* and *Medicago sativa*. *Plant Physiol.* 2012;159:1204–1220.

Harborne JB, Grayer RJ. Flavonoids and insects. In: Harborne JB (ed.). *The Flavonoids: Advances in Research since 1986.* Chapman and Hall, London, 1993, pp. 589–618.

Henry-Kirk RA, McGhie TK, Andre CH, Hellens RP, Allan AC. Transcriptional analysis of apple fruit proanthocyanidin biosynthesis. *J. Exp. Bot.* 2012;63:5437–5450.

Hernández I, Alegre L, Munné-Bosch S. Drought-induced changes in flavonoids and other low molecular weight antioxidants in *Cistus clusii* under Mediterranean filed conditions. *Tree Physiol.* 2004;24:1303–1311.

Hichri I, Barrieu F, Bogs J, Kappel C, Delrot S, Lauvergeat V. Recent advances in the transcriptional regulation of the flavonoid biosynthetic pathway. *J. Exp. Bot.* 2011;62:2465–2483.

Hoffmann T, Kalinowski G, Schwab W. RNAi-induced silencing of gene expression in strawberry fruit (*Fragaria x ananassa*) by agrofiltration: A rapid assay for gene function analysis. *Plant J.* 2006;48:818–826.

Huang J, Gu M, Lai Z, Fan B, Shi K, Zhou YH, Yu JQ, Chen Z. Functional analysis of the *Arabidopsis* PAL gene family in plant growth, development, and response to environmental stress. *Plant Physiol.* 2010;153:1526–1538.

Huang Y, Gou J, Jia Z, Yang L, Sun Y, Xiao X, Song F, Luo K. Molecular cloning and characterization of two genes encoding dihydroflavonol-4-reductase from *Populus trichocarpa*. PLoS One 2012;7:e30364.

Hughes NM, Reinhardt K, Field TS, Gerardi AR, Smith WK. Association between winter anthocyanin production and drought stress in angiosperm evergreen species. *J. Exp. Bot.* 2010;61:1699–1709.

Ishida T, Hattori S, Sano R, Inoue K, Shirano Y, Hayashi H, Shibata D, et al. *Arabidopsis TRANSPARENT TESTA GLABRA2* is directly regulated by R2R3 MYB transcription factors and is involved in regulation of *GLABRA2* transcription in epidermal differentiation. *Plant Cell* 2007;19:2531–2543.

Jaakola L, Poole M, Jones MO, Kämäräinen-Karppinen T, Koskimäki JJ, Hohtola A, Häggman H, et al. A SQUAMOSA MADS box gene involved in the regulation of anthocyanin accumulation in bilberry fruits. *Plant Physiol.* 2010;153:1619–1629.

Jagtap UB, Gurav RG, Bapat VA. Role of RNA interference in plant improvement. *Naturwissenschaften* 2011;98:473–492.

Jiang F, Wang JY, Jia HF, Jia WS, Wang HQ, Xiao M. RNAi-mediated silencing of the flavanone 3-hydroxylase gene and its effect on flavonoid biosynthesis in strawberry fruit. *J. Plant Growth Regul.* 2012;32:182–190. doi:10.1007/s00344–012-9289–1.

Johnson CS, Kolevski B, Smyth DR. *TRANSPARENT TESTA GLABRA2*, a trichome and seed coat development gene of *Arabidopsis*, encodes a WRKY transcription factor. *Plant Cell* 2002;14:1359–1375.

Juszczuk IM, Wiktorowska A, Malusa E, Rychter AM. Changes in the concentration of phenolic compounds and exudation induced by phosphate deficiency in bean plants (*Phaseolus vulgaris* L.). *Plant Soil* 2004;267:41–49.

Kalimoorthy S, Rao AS. Effect of salinity on anthocyanin accumulation in the root of maize. *Ind. J. Plant Physiol.* 1994;37:169–170.

Kumar V, Yadav SK. Overexpression of CsDFR and CsANR enhanced flavonoids accumulation and antioxidant potential of roots in tobacco. *Plant Root* 2013;7:65–76.

Lalusin A, Nishita K, Kim S, Ohta M, Fujimura T. A new MADS-box gene (IbMADS10) from sweet potato (*Ipomoea batatas* (L.) Lam) is involved in the accumulation of anthocyanin. *Mol. Genet. Genomics* 2006;275:44–54.

Lee XZ, Liang YR, Chen H, Lu JL, Liang HL, Huang FP, Mamati EG. Alleviation of UV-B stress in *Arabidopsis* using tea catechins. *Afr. J. Biotechnol.* 2008;7:4111–4115.

Lepiniec L, Debeaujon I, Routaboul J, Baudry A, Pourcel L, Nesi N, Caboche M. Genetics and biochemistry of seed flavonoids. *Annu. Rev. Plant Biol.* 2006;57:405–430.

Lillo C, Lea US, Ruoff P. Nutrient depletion as a key factor for manipulating gene expression and product formation in different branches of the flavonoid pathway. *Plant Cell Environ.* 2008;31:587–601.

Liu M, Li X, Liu Y, Cao B. Regulation of flavanone 3-hydroxylase gene involved in the flavonoid biosynthesis pathway in response to UV-B radiation and drought stress in the desert plant, *Reaumuria soongorica*. *Plant Physiol. Biochem.* 2013;73:161–167.

Løvdal T, Olsen KM, Slimestad R,   M, Lillo C. Synergetic effects of nitrogen depletion, temperature, and light on the content of phenolic compounds and gene expression in leaves of tomato. *Phytochemistry* 2010;71:605–613.

Lukaszewicz M, Matysiak-Kara I, Skala J, Fecka I, Cisowski W, Szopa J. Antioxidant capacity manipulation in transgenic potato tuber by changes in phenolic compounds content. *J. Agric. Food Chem.* 2002;52:1526–1533.

Luo P, Shen Y, Jin S, Huang S, Cheng X, Wang Z. Overexpression of *Rosa rugosa* anthocyanin reductase enhances tobacco tolerance to abiotic stress through increased ROS scavenging and modulation of ABA signaling. *Plant Sci.* 2016;245:35–49.

Mahajan M, Ahuja PS, Yadav SK. Post-transcriptional silencing of flavonol synthase mRNA in tobacco leads to fruits with arrested seed set. *PLoS One* 2011b;6:e28315.

Mahajan M, Joshi R, Gulati A, Yadav SK. Increase in flavan-3-ols by silencing flavonol synthase mRNA affects the transcript expression and activity levels of antioxidant enzymes in tobacco. *Plant Biol.* 2012;14:725–733. doi:10.1111/j.1438-8677.2011.00550.x.

Mahajan M, Kumar V, Yadav SK. Effect of flavonoid-mediated free IAA regulation on growth and development of *in vitro*-grown tobacco seedlings. *Intern. J. Plant Dev. Biol.* 2011;5:42–48.

Mahajan M, Yadav SK. Overexpression of a tea flavanone 3-hydroxylase gene confers tolerance to salt stress and *Alternaria solani* in transgenic tobacco. *Plant Mol. Biol.* 2014;85:551–573.

Mehrtens F, Kranz H, Bednarek P, Weisshaar B. The *Arabidopsis* transcription factor MYB12 is a flavonol-specific regulator of phenylpropanoid biosynthesis. *Plant Physiol.* 2005;138:1083–1096.

Meng C, Zhang S, Deng YS, Wang GD, Kong FY. Overexpression of a tomato flavanone 3-hydroxylase-like protein gene improves chilling tolerance in tobacco. *Plant Physiol. Biochem.* 2015;96:388–400.

Michalak A. Phenolic compounds and their antioxidant activity in plants growing under heavy metal stress. *Pol. J. Environ. Stud.* 2006;15:523–530.

Mizutani M, Tsuda S, Suzuki K, Nakamura N, Fukui Y, Kusumi T, Tanaka Y. Evaluation of post transcriptional gene silencing methods using flower color as the indicator. *Plant Cell Physiol.* 2003;44:S122.

Mpoloka SW. Effects of prolonged UV-B exposure in plants. *Afr. J. Biotechnol.* 2008;7:4874–4883.

Nakamura N, Fukuchi-Mizutani M, Suzuki K, Miyazaki K, Tanaka Y. RNAi suppression of the anthocyanidin synthase gene in *Torenia hybrida* yields white flowers with higher frequency and better stability than antisense and sense suppression. *Plant Biotechnol.* 2006;23:13–17.

Nakayama T, Yonekura-Sakakibara K, Sato T, Kikuchi S, Fukui Y, Fukuchi-Mizutani M, Ueda T, et al. Aureusidin synthase: A polyphenol oxidase homolog responsible for flower coloration. *Science* 2000;290:1163–1166.

Nesi N, Debeaujon I, Jond C, Pelletier G, Caboche M, Lepiniec L. The TT8 gene encodes a basic helix-loop-helix domain protein required for expression of *DFR* and *BAN* genes in *Arabidopsis* siliques. *Plant Cell* 2000;12:1863–1878.

Nguyen NH, Kim JH, Kwon J, Jeong CY, Lee W, Lee D, Hong SW, Lee H. Characterization of *Arabidopsis thaliana* flavonol synthase 1 (FLS1)-overexpression plants in response to abiotic stress. *Plant Physiol. Biochem.* 2016;103:133–142.

Oh MM, Trick HN, Rajashekar CB. Secondary metabolism and antioxidants are involved in environmental adaptation and stress tolerance in lettuce. *J. Plant Physiol.* 2009;166: 180–191.

Pang YP, Peel GJ, Wright E, Wang Z, Dixon RA. Early steps in proanthocyanidin biosynthesis in the model legume *Medicago truncatula. Plant Physiol.* 2007;145:601–615.

Peng M, Hudson D, Schofield A, Tsao R, Yang R, Gu H, Bi YM, Rothstein SJ. Adaptation of *Arabidopsis* to nitrogen limitation involves induction of anthocyanin synthesis which is controlled by the *NLA* gene. *J. Exp. Bot.* 2008;59:2933–2944.

Reddy AM, Reddy VS, Scheffler BE, Wienand U, Reddy AR. Novel transgenic rice overexpressing anthocyanin synthase accumulates a mixture of flavonoids leading to an increased antioxidant potential. *Metab. Eng.* 2007;9:95–111.

Rice-Evans CA, Miller NJ, Paganga G. Structure-antioxidant activity relationships of flavonoids and phenolic acids. *Free Radic. Biol. Med.* 1996;20:933–956.

Rivero RM, Ruiz JM, Garcia PC, Lopez-Lefebre LR, Sanchez E, Romero L. Resistance to cold and heat stress: Accumulation of phenolic compounds in tomato and watermelon plants. *Plant Sci.* 2001;160:315–321.

Ryan KG, Hunt JE. The effect of UVB radiation on temperate southern hemisphere forests. *Environ. Pollut.* 2005;137:415–427.

Sadik CD, Sies H, Schewe T. Inhibition of 15-lipoxygenases by flavonoids: Structure-activity relations and mode of action. *Biochem. Pharmacol.* 2003;65:773–781.

Sagasser M, Lu GH, Hahlbrock K, Weisshaar B. A. *thaliana* TRANSPARENT TESTA 1 is involved in seed coat development and defines the WIP subfamily of plant zinc finger proteins. *Genes Dev.* 2002;16:138–149.

Schaart JG, Dubos C, De La Fuente IR, van Houwelingen AM, De Vos RC, Jonkar HH, Xu W, Routaboul JM, Lepiniec L, Bovy AG. Identification and characterization of MYB-BHLH-WD40

regulatory complexes controlling proanthocyanidin biosynthesis in strawberry (*Fragaria* x *ananassa*) fruits. *New Phytol.* 2012;197:454–467.

Schenke D, Böttcher C, Scheel D. Crosstalk between abiotic ultraviolet-B stress and biotic (flg22) stress signaling in *Arabidopsis* prevents flavonol accumulation in favor of pathogen defense compound production. *Plant Cell Environ.* 2011;34: 1849–1864.

Schroder J. Probing plant polyketide biosynthesis. *Nat. Struct. Biol.* 1999;6:714–716.

Sompornpailin K, Makita Y, Yamazaki M, Saito K. A WD-repeat-containing putative regulatory protein in anthocyanin biosynthesis in *Perilla frutescens. Plant Mol. Biol.* 2002;50:485–495.

Song X, Diao J, Ji J, Wang G, Guan C, Jin C. Molecular cloning and identification of a flavanone 3-hydroxylase gene from *Lycium chinense*, and its overexpression enhances drought stress in tobacco. *Plant Physiol. Biochem.* 2016;98:89–100.

Stracke R, Ishihara H, Huep G, Barsch A, Mehrtens F, Niehaus K, Weisshaar B. Differential regulation of closely related R2R3-MYB transcription factors controls flavonol accumulation in different parts of the *Arabidopsis thaliana* seedling. *Plant J.* 2007;50:660–677.

Suzuki H, Nakayama T, Yonekura-Sakakibara K, Fukui Y, Nakamura N, Yamaguchi MA, Tanaka Y, Kusumi T, Nishino T. cDNA cloning, heterologous expressions, and functional characterization of malonyl-coenzyme A: Anthocyanidin 3-O-glucoside-6''-O-malonyltransferase from dahlia flowers. *Plant Physiol.* 2002;130:2142–2151.

Takahashi H, Hayashi M, Goto F, Sato S, Soga T, Nishioka T, Tomita M, Yamada M, Uchimiya H. Evaluation of metabolic alteration in transgenic rice overexpressing dihydroflavonol-4-reductase. *Ann. Bot.* 2006;98:819–825.

Tamagnone L, Merida A, Stacey N, Plaskitt K, Parr A, Chang CF, Lynn D, Dow JM, Roberts K, Martin C. Inhibition of phenolic acid metabolism results in precocious cell death and altered cell morphology in leaves of transgenic tobacco plants. *Plant Cell* 1998;10:1801–1816.

Tanner GJ, Francki KT, Abrahams S, Watson JM, Larkin PJ, Ashton AR. Proanthocyanidin biosynthesis in plants: Purification of legume leucoanthocyanidin reductase and molecular cloning of its cDNA. *J. Biol. Chem.* 2003;278:31647–31656.

Tomasi N, Weisskopf L, Renella G, Landi L, Pinton R, Varanini Z, Nannipieri P, Torrent J, Martinola E, Cesco S. Flavonoids of white lupin roots participate in phosphorus mobilization from soil. *Soil Biol. Biochem.* 2008;40:1971–1974.

Uhde-Stone C, Liu J, Zinn KE, Allan DL, Vance CP. Transgenic proteoid roots of white lupin: A vehicle for characterizing and silencing root genes involved in adaptation to P stress. *Plant J.* 2005;44:840–853.

Van Oosten MJ, Sharkhuu A, Batelli G, Bressan RA, Maggio A. The *Arabidopsis thaliana* mutant air1 implicates SOS3 in the regulation of anthocyanins under salt stress. *Plant Mol. Biol.* 2013;83:405–415.

Verhoeyen ME, Bovy A, Collins G, Muir S, Robinson S, De Vos CHR, Colliver S. Increasing antioxidant levels in tomatoes through modification of the flavonoid biosynthetic pathway. *J. Exp. Bot.* 2002;53:2099–2106.

Vernoud V, Laigle G, Rozier F, Meeley R, Perez P, Rogowsky PM. The HD-ZIP IV transcription factor OCL4 is necessary for trichome patterning and anther development in maize. *Plant J.* 2009;59:883–894.

Vom Endt D, Kijne JW, Memelink J. Transcription factors controlling plant secondary metabolism: What regulates the regulators? *Phytochemistry* 2002;61:107–114.

Wade HK, Sohal AK, Jenkins GJ. *Arabidopsis ICX1* is a negative regulator of several pathways regulating flavonoid biosynthesis genes. *Plant Physiol.* 2003;131:707–715.

Wahid A, Ghazanfar A. Possible involvement of some secondary metabolites in salt tolerance of sugarcane. *J. Plant Physiol.* 2006;163:723–730.

Wang F, Kong W, Wong G, Fu L, Peng R, Li Z. AtMYB12 regulates flavonoids accumulation and abiotic stress tolerance in transgenic *Arabidopsis thaliana. Mol. Genet. Genom.* 2016;291:1545–1559.

Wang H, Fan W, Li H, Yang J, Huang J, Zhang P. Functional characterization of dihydroflavonol-4-reductase in anthocyanin biosynthesis of purple sweet potato underlies the direct evidence of anthocyanins function against abiotic stresses. *PLoS One* 2013;8:e78484.

Wang Y, Chen S, Oliver Yu. Metabolic engineering of flavonoids in plants and microorganisms. *Appl. Microbiol. Biotechnol.* 2011;91:949–956.

Winkel BSJ. The biosynthesis of flavonoids. In: Grotewold E (ed.). *The Science of Flavonoids*. Springer Science and Business Media, New York, 2006, pp. 71–96.

Wu J, Du G, Chen J, Zhou J. Enhancing flavonoid production by systematically tuning the central metabolic pathways based on a CRISPR interference system in *Escherichia coli. Sci. Rep.* 2015;5:13477. doi:10.1038/srep13477.

Xie DY, Sharma SB, Dixon RA. Anthocyanidin reductase from *Medicago truncatula* and *Arabidopsis thaliana. Arch. Biochem. Biophys.* 2004;422:91–102.

Xie DY, Sharma SB, Wright E, Wang ZY, Dixon RA. Metabolic engineering of proanthocyanidins through co-expression of anthocyanidin reductase and the PAP1 MYB transcription factor. *Plant J.* 2006;45:895–907.

Xie DY, Sharma SB, Paiva NL, Ferreira D, Dixon RA. Role of anthocyanidin reductase, encoded by *BANYULS* in plant flavonoid biosynthesis. *Science* 2003;299:396–399.

Yoshida K, Iwasaka R, Kaneko T, Sato S, Tabata S, Sakuta M. Functional differentiation of *Lotus japonicus* TT2s, R2R3-MYB transcription factors comprising a multigene family. *Plant Cell Physiol.* 2008;49:157–169.

Yu O, McGonigle B. Metabolic engineering of isoflavone biosynthesis. *Adv. Agron.* 2005;86:147–190.

Yuan L, Wang L, Han Z, Jiang Y, Zhao L, Liu H, Yang L, Luo K. Molecular cloning and characterization of *PtrLAR3*, a gene encoding leucoanthocyanidin reductase from *Populus trichocarpa*, and its constitutive expression enhances fungal resistance in transgenic plants. *J. Exp. Bot.* 2012;63:2513–2524.

Zuk M, Prescha A, Stryczewska M, Szopa J. Engineering flax plants to increase their antioxidant capacity and improve oil composition and stability. *J. Agric. Food Chem.* 2012;60:5003–5012.

21

The Role of Gamma Aminobutyric Acid (GABA) During Abiotic Stress in Plants

Paramita Bhattacharjee, Sasanka Chakraborti, Soumi Chakraborty, and Kaninika Paul

CONTENTS

Introduction

4-Aminobutyrate or γ-aminobutyric acid (GABA), a four-carbon non-protein amino acid, is ubiquitous in the plant kingdom, and its concentration in plant tissues is often similar to that of normal protein amino acids. It was discovered in plants more than half a century ago and is a significant component of the free amino acid pool. Typically, GABA levels in plant tissues are in the range of 0.03 to 2.00 μmol g^{-1} (fresh weight) but increase several-fold in response to biotic and abiotic stresses.

Metabolic Synthesis of GABA: The GABA Shunt

To evaluate the roles of GABA as a metabolic adapter in plants during abiotic stresses, it is necessary to understand metabolic synthesis of GABA in plants. In plants, GABA is mainly metabolized via a short pathway known as the GABA shunt, so called

because it bypasses two steps of the tricarboxylic acid (TCA) cycle. GABA has an amino group on the γ-carbon (Figure 21.1). This pathway converts glutamate to succinate via GABA (Aurisano et al. 1995; Bouché and Fromm 2004; Shelp et al. 1999; Kinnersley and Turano 2000) (Figure 21.2). The aforesaid pathway is composed of three enzymes, namely the cytosolic enzyme L-glutamate decarboxylase (GAD) and the mitochondrial enzymes GABA transaminase (GABA-T) and succinic semialdehyde dehydrogenase (SSADH). The first step of the shunt is the direct and irreversible α-decarboxylation of glutamate by glutamate decarboxylase (GAD, EC 4.1.1.15). The second enzyme involved in the GABA shunt, GABA-T (EC 2.6.1.19), catalyzes the reversible conversion of GABA to succinic semialdehyde using either pyruvate or α-ketoglutarate as amino acceptors (Figure 21.2). *In vitro* GABA-T activity appears to prefer pyruvate as amino receptor rather than α-ketoglutarate. Either activity exhibits a broad pH optimum range (8–10). The last step of the GABA shunt is catalyzed by SSADH

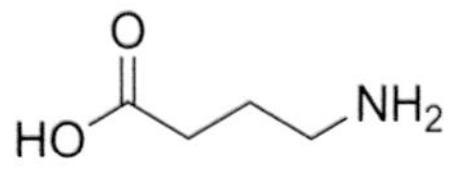

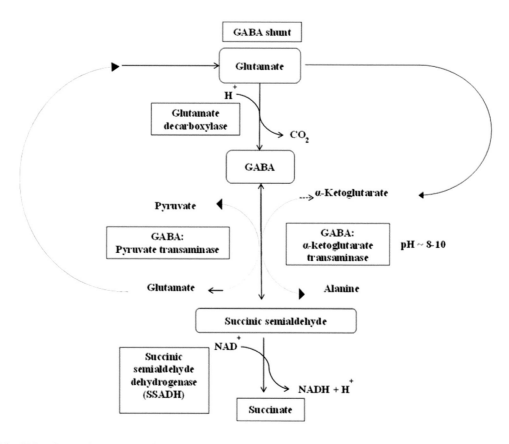

FIGURE 21.1   Structure of γ-aminobutyric acid (GABA).

(EC 1.2.1.16) which irreversibly oxidizes succinic semialdehyde to succinate (Aurisano et al. 1995; Bouché and Fromm 2004; Shelp et al. 1999; Kinnersley and Turano 2000).

Regulation of GABA Synthesis

The above synthetic pathway (GABA shunt) of GABA production is regulated by stress conditions, both biotic and abiotic. Plant tissues exposed to abiotic stresses such as acidosis, cold, anoxia, heat, salt, and drought promotes rapid accumulation of GABA.

A breakthrough in the study of GABA in plants came from molecular studies, particularly following the cloning of the Ca^{2+}/calmodulin (CaM)-regulated GABA-synthesizing enzyme GAD (Baum et al. 1993). It was the first direct evidence for Ca^{2+}-stimulated GAD activity obtained in *Petunia* where Ca^{2+}-dependent binding of calmodulin occurred to a 58 kDa protein. A similar phenomenon was also observed for a 62 kDa protein in fava beans (Shelp et al. 1999). Further evidence of Ca^{2+}/calmodulin (CaM) activation of GAD was obtained using Ca^{2+} channel-blockers and calmodulin antagonists in rice roots (Aurisano et al. 1995). These studies have established that plant

GADs are stimulated by Ca^{2+}/Ca-calmodulin, but not by Ca^{2+} or CaM individually. Further breakthrough studies of GABA in plants involved transgenic plants (Baum et al. 1996) and functional genomics in *Arabidopsis* which investigated mutants of the GABA metabolic pathway (Bouché and Fromm 2004; Palanivelu et al. 2003; Fait et al. 2005). Increased cytosolic Ca^{2+} concentration as well as cytosolic acidity independently governs GABA synthesis by stimulating GAD activity. Thus, stimulation of GAD activity has been divided into two phases based on Ca^{2+}/ CaM-dependent (Phase I) or acidic pH-dependent (Phase II) activity. Results from several investigations support a working model for the *in vivo* biphasic stimulation of GAD and control of GABA accumulation in plants exposed to environmental stresses (Figure 21.3). The interrelation, if any, between these phases inside the cells has not been clearly established. Each phase may occur simultaneously or both phases may occur as distinct events in no predetermined order, depending on the type, duration, and severity of stresses in plants (Shelp et al. 1999; Kinnersley and Turano 2000; Crawford et al. 1994).

Ca^{2+}/CaM-Dependent (Phase I) Activity

Thermal shocks (both heat and cold), salt level increase, lack of water, mechanical and osmotic changes in solute concentrations outside the cells, and depletion of oxygen (anoxia or hypoxia) all lead to an increase in cytosolic concentrations of Ca^{2+} or H^+.

FIGURE 21.2   The GABA shunt pathway (adapted from Shelp et al. 1999).

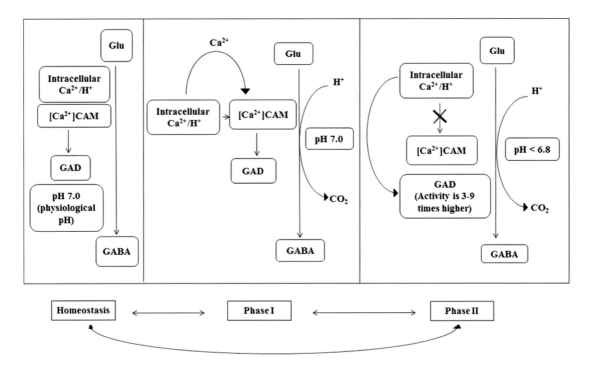

FIGURE 21.3 Biphasic regulation of GAD activity (adapted from Kinnersley and Turano 2000).

Acidic pH-Dependent (Phase II) Activity

Increased GABA levels result from a reduction in intracellular pH. Hypoxia (lack of oxygen in cells) is reported to reduce cytosolic pH by 0.4 to 0.8 pH units. Significantly, GAD exhibits a pH optimum of approximately 6.0 and is located in the cytosol. Since GAD activity is a proton consumer (Figure 21.3), a stress-induced reduction in intracellular pH from normal physiological values elevates GABA levels by stimulating GAD. This mechanism is part of the metabolic pH-stat. Thus, GABA accumulation ameliorates cytosolic acidification and tries to maintain pH homeostasis under stress (Shelp et al. 1999; Kinnersley and Turano 2000; Crawford et al. 1994).

Functional Roles of GABA in Plants as a Metabolic Adapter during Abiotic Stresses

The various roles of GABA in plants as a means to combat abiotic stresses are elaborated in the following sections.

GABA as an Alternate Substrate during Decreased Respiration (Anoxia) through Krebs Cycle Bypass

When glutamate enters the Krebs cycle as α-ketoglutarate, its conversion to succinate requires NAD (Figure 21.2). Alternatively, glutamate might enter as succinate via the GABA shunt, bypassing the dehydrogenase or transaminase reaction and the α-ketoglutarate dehydrogenase of the Krebs cycle. Therefore, during certain conditions such as hypoxia, decreases in respiration, and the resultant decrease in the [NAD] to [NADH] ratio, the NAD-dependent SSADH reaction and the entry of carbon into the Krebs cycle is limited, thereby causing GABA to accumulate. Thus, GABA provides an immediate substrate upon recovery from stress.

The highest decline in cytosolic pH occurs under oxygen deprivation (anoxia or hypoxia), which is the primary stress factor in flooded soils, and induces the greatest accumulation of GABA. The synthesis of GABA consumes a proton and raises pH (Kinnersley and Turano 2000). Depending on the duration and/or severity of the stress, such as severe anoxia, where pH approaches 6.0 after exposure to a bacterial elicitor that decreases cytosolic pH, the likelihood of metabolic dysfunction, membrane damage, and degradation of cellular components, chiefly disruption of vacuolar membranes may release acids in the cytosol, lowering pH< 6.8. Under these circumstances, GAD activity would be stimulated in a Ca^{2+}/CaM-independent manner. Below pH 6.8, acidic pH activates GAD with a pH optimum of 5.8, at which GAD activity is 3–9 times higher than that at 7.3 in the presence of Ca^{2+}/CaM. GAD activity increases with reduced pH and declines as pH recovers. These facts establish that acid-stimulated GABA synthesis does not involve Ca^{2+} flux (Shelp et al. 1999).

GABA as an Inducer of Stress Ethylene

Stress generally promotes ethylene production in plants (Morgan and Drew 1997). The response is so evident that the term 'stress ethylene' was coined by Abeles (1973) to describe this phenomenon. The association between GABA and ethylene has received little attention despite the fact that nearly all signals that induce GABA accumulation (Table 21.1) also increase production of ethylene. This includes chilling, high temperature, salinity, flooding, anoxia, drought, and mechanical damage (Abeles et al. 1992; Morgan and Drew 1997). Oxygen deprivation (anoxia) is the primary stress factor in flooded soils. In waterlogged soils,

TABLE 21.1

Abiotic Stress Factors Influencing Production of GABA in Plants

Common Name of Plant Source	Scientific Name of the Plant Source	Abiotic Stress Factor	References
Asparagus cells	*Asparagus sprengeri* Regel cells	Acidosis	Crawford et al. (1994)
Asparagus cells	*Asparagus sprengeri* Regel cells	Acidosis	Cholewa et al. (1997)
Soybean leaves	*Glycine max* [L.] Merr. cv Corsoy 79	Mechanical	Ramputh and Bown (1996)
Soybean leaves	*Glycine max* [L.] Merr. cv Corsoy 79	Mechanical	Wallace et al. (1984)
Arabidopsis	*Arabidopsis thaliana* Columbia-0 wild-type and mutant plants (tpc1-2 and pop2-5)	Mechanical	Scholz et al. (2017)
Radish leaves	*Raphanus sativus* [L.] var. Champion	Anoxia	Streeter and Thompson (1972)
Tea leaves	*Camellia sinensis* cv. Yabukita	Anoxia	Tsushida and Murai (1987)
Rice root and shoot	*Oryza sativa* [L.] cv. Arborio	Anoxia	Aurisano et al. (1995)
Bean leaves	*Phaseolus vulgaris* [L.] cv. Topcrop	Drought	Raggi (1994)
Turnip leaves	*Brassica rapa* [L.] var. Shogoin	Drought	Thompson et al. (1966)
Durum wheat plants	*Triticum durum* Desf. cv. Ofanto	Salinity	Woodrow et al. (2017)
Leaves and roots of wild tomato seedlings	*Lycopersicon esculentum* Mill cv. P-73	Salinity	Bolarin et al. (1995)
	Lycopersicon pennellii (Correll) D'arcy accession PE-47	Salinity	Bolarin et al. (1995)
Cowpea cells	*Vigna unguiculata* cv. *CB5*	Thermal	Mayer et al. (1990)
Creeping bentgrass	*Agrostis stolonifera* cv. Penncross	Thermal	Li et al. (2016)
Rice	*Oryza sativa* [L.] seeds (PR118)	Thermal	Nayyar et al. (2014)
Kentucky bluegrass	*Poa Pratensis* 'Midnight'	Thermal	Du et al. (2011)
Hybrid Bermuda grass	*Cynodon transvaalensis*×*Cynodon dactylon* 'Tifdwarf'	Thermal	Du et al. (2011)
Asparagus cells	*Asparagus sprengeri* Regel mesophyll cells	Cold	Cholewa et al. (1997)
Soybean leaves	*Glycine max* [L.] Merr. cv Corsoy 79	Cold	Wallace et al. (1984)
Sesame seed	*Sesamum indicum* [L.] var. Cumhuriyet	Metal	Bor et al. (2009)

promotion of adventitious rooting by ethylene provides a replacement root system for roots damaged by extreme oxygen deficiency (Morgan and Drew 1997).

GABA as an Adapter of Temperature Stimuli

Low temperature is one of the most harmful abiotic stresses that affect temperate plants. During cold shock, elevated Ca^{2+} level stimulates GABA synthesis and rapidly increases cytosolic Ca^{2+} but not H^+ levels. In the absence of cold shock, a Ca^{2+} ionophore stimulates cytosolic Ca^{2+} levels and GABA synthesis. Ca^{2+} channel-blockers or calmodulin antagonists inhibit cold shock-stimulated GABA synthesis, prevent stimulation by Ca^{2+}/calmodulin (but do not inhibit GABA synthesis in response to cytosolic acidification). It has been found that a monoclonal antibody specific for the 26 amino acid, calmodulin-binding, C-terminal region fully activates GAD in the absence of Ca^{2+}/calmodulin. These findings demonstrate that Ca^{2+}/calmodulin or an antibody binding to an auto-inhibitory domain activates GAD in plant cells under cold shock (Shelp et al. 1999).

Role of GABA in heat stress mitigation has been investigated in detail, although the mechanism is yet not fully established. Pysiological studies suggested that GABA could alleviate heat damage in photosynthesis by suppressing leaf senescence and balancing water relations by improving osmotic adjustment and water usage efficiency (Li et al. 2016). GABA-induced stress tolerance could also be acquired by an increase in antioxidant ability in various plant species under heat stress and is also effective for mitigating chill and drought stresses. GABA reportedly enhances the activity of several key enzymes involved in the ascorbate-glutathione (AsA-GSH) cycle and the accumulation of non-enzymatic antioxidants ascorbic acid (AsA) and glutathione (GSH), while it had no significant effects on activating superoxide dismutase (SOD), catalase (CAT), and peroxidase (POD). As a result, the reactive oxygen species (ROS) level and lipid peroxidation decline appreciably in GABA-treated plants under heat stress (Gill and Tuteja 2010; Li et al. 2016).

Carbohydrates are among the most abundant metabolites in plants, which play essential roles in plant tolerance to abiotic stresses, such as serving as energy sources, osmoregulants, and signaling molecules. Sugar alcohols such as mannitol and myo-inositol exhibit roles in signal transduction and as scavengers of ROS against abiotic stress. Metabolic profiling found that exogenous application of GABA leads to increased accumulations of sugars (sucrose, fructose, glucose, galactose, and maltose), amino acids (glutamic acid, aspartic acid, alanine, threonine, serine, and valine), organic acids (aconitic acid, malic acid, succinic acid, oxalic acid, and threonic acid) and sugar alcohols (mannitol and myo-inositol). These findings suggest that GABA-induced heat tolerance perhaps involves the enhancement of photosynthesis and ascorbate-glutathione (AsA-GSH) cycle, the maintenance of osmotic adjustment, and the increase in GABA shunt. The increased GABA shunt could supply intermediates to feed the TCA cycle of respiration metabolism during a long-term heat stress, thereby maintaining metabolic homeostasis (Li et al. 2016).

Proline is a stress-related amino acid, and its level varies with the severity of stress and the level of stress tolerance of plant species. Based on the analysis of metabolic pathways, increased

GABA shunt could be the supply of pyruvate and succinic semi-aldehyde to feed the TCA cycle, instead of being channelized for proline metabolism (Li et al. 2016).

GABA as an Adapter of Mechanical Stimulus

Mechanical stimulation or damage resulting from activities of phytophagous insects also stimulates GABA accumulation, which in turn is useful in preventing insect growth and development. Thus, GABA accumulation may act as a defense mechanism against phytophagous insects. Rapid GABA accumulation may represent one of many chemical defense systems that immobile plants deploy against phytophagous invertebrate pests (Ramputh and Bown 1996).

The term thigmomorphogenesis was introduced by Jaffe (1973) to describe the changes in plant development induced by the action of mechanical stimuli such as wind or rubbing on plant stems. A linkage between a mechanical stress-induced increase in Ca^{2+}, GABA, ethylene, and the resulting thigmo-morphogenetic responses remains to be fully established. Nevertheless, the indications are that thigmomorphogenesis could provide an early example of how stress-induced changes in Ca^{2+} are coupled with physiological responses. Within 1 s of mechanical stimulation to bean plants, there is a dramatic drop in the electrical resistance of the tissue, indicating an increase in membrane permeability to electrolytes (Jaffe and Biro 1979). The earliest thigmomorphogenetic events seem to involve membrane changes that allow Ca^{2+} to act as a secondary messenger, probably via CaM. Besides, GABA is metabolically positioned to link stress-induced changes in Ca^{2+} to thigmomor-phogenetic responses through an effect on ethylene biosynthesis (Kinnersley and Turano 2000).

Anaplerotic Role of GABA in Stress-related Metabolism

Anaplerotic reactions (or filling up reactions coined by Kornberg [1966]) are metabolic processes that ensure that the intramito-chondrial concentrations of the TCA cycle intermediates remain constant with time. The GABA shunt performs an anaplerotic role by providing carbon to the cycle through GABA catabo-lism to succinic acid. GABA metabolism is thus a major route for succinate production in roots and helps to buffer the TCA cycle activity. GABA is therefore a major player in central carbon metabolism/adjustment in roots under salt stress. These findings also suggest a role of GABA in respiration under low sugar con-ditions (Kornberg 1966). This amino acid accumulates in plants in response to drought and cold stress and disappears rapidly when the stress is removed. Thus, GABA serves as a metabolic homeostatic regulator. Literature also suggests that GABA pro-duced during stress can act as a source of carbon skeletons for the TCA cycle long after the stress factors have been removed (Renault et al. 2013).

GABA as a Biochemical pH-stat

In plants, GAD is activated by acidic pH and GABA accumulates in response to cytosolic acidification, so it is possible that GAD could participate in regulating the cytosolic pH of plants (Bouché

and Fromm 2004). This mechanism has been elaborated in the section on regulation of GABA synthesis (Crawford et al. 1994; Shelp et al. 1999; Kinnersley and Turano 2000).

GABA as an Osmotic Regulator

It has also been suggested that GABA could function as an osmolyte and mitigate water stress. A common factor in stresses caused by salinity, drought, and freezing is cellular dehydration, and as a result, the concentration of cellular con-stituents may increase to levels that cause membrane damage and ultimately cell death. A minimum of 28%–30% water is required in plant cells for the maintenance of functional integrity of the membrane structure. Cellular accumulation of GABA balances the decrease in water potential that occurs during cellular dehydration and thus protects plants during water stress (Kinnersley and Turano 2000; Ramakrishna and Ravishankar 2011, 2013).

Examples of GABA Synthesis under Specific Abiotic Stresses in Plants

GABA has been extensively reviewed by Shelp et al. (1999), Snedden and Fromn (1999), Kinnersley and Turano (2000), Bouché et al. (2003), and Bouché and Fromm (2004). From the aforesaid discussion, it is clear that plant adaptation to abiotic stresses is dependent on comprehensive responses associated with physiological and metabolic changes. GABA production enhances several folds in response to abiotic stress such as in drought-stressed cotton, heat-stressed cowpea cells, and soy-beans when subjected to cold stress and mechanical damage. It is opined that less time would be needed to elicit response dur-ing frostbite and during flooding vis-à-vis stresses from heat, drought, and salt. Stress-related kinetics of GABA accumula-tion are shown in Table 21.2 (Kinnersley and Turano 2000). The data shows time of accumulation of GABA in response to various abiotic stresses and percentage increase over its corre-sponding levels in unstressed (control) samples. It is opined that GABA synthesis is a common response to different stressors, but the ability of the plant in coping with these stresses affects the rate of GABA synthesis. While the minimum time required for GABA accumulation is 15 s in asparagus in response to acidosis, it is 5 days in tomato leaves consequent to salt stress (Kinnersley and Turano 2000). Figure 21.4 presents the abi-otic stress factors in plants which trigger GABA accumula-tion. Examples of GABA synthesis as a metabolic adaptation to these abiotic stress factors have been illustrated in the fol-lowing sections.

Thermal Stress (Heat and Cold Shock) in Regulation of GABA Synthesis

The patterns of GABA accumulation as a response to temperature fluctuations was found to differ between heat-shocked cowpea cells (Figure 21.5) and cold-shocked asparagus cells (Figure 21.6) with respect to time. While GABA accumulation in the cold-shocked asparagus cells occurred in minutes, it lasted for hours in heat-shocked cowpea cells. Besides, GABA accumulation in heat-shocked (26°C–42°C) cowpea cell cultures resulted in a fivefold increase

TABLE 21.2

Stress-Related Kinetics of GABA Accumulation in Plants

Stress	Plant	Time for GABA Accumulation*	% GABA Accumulation in the Said Time Relative to Unstressed Plants*
Acidosis	Asparagus cells	15 s	300
Mechanical damage	Soybean leaves	1 min	1800
	Soybean leaves	5 min	2700
Cold (6°C)	Soybean leaves	5 min	2000
Cold (10°C)	Asparagus cells	15 min	200
Anoxia	Radish leaves	4 h	10000
	Tea leaves	12 h	4000
	Rice root	24 h	750
	Rice shoot	24 h	1000
Drought	Bean leaves	3 days	200
	Turnip leaves	3 days	1000
Salt	Tomato root	4 days	200
	Tomato leaves	5 days	300
Heat	Cowpea cells	24 h	1800

* For each stress, the time to reach the greatest reported GABA accumulation relative to unstressed plants has been presented (adapted from Kinnersley and Turano 2000).

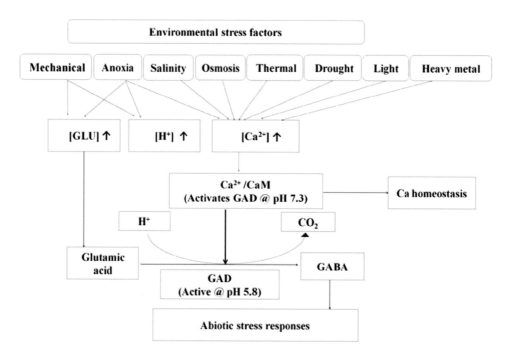

FIGURE 21.4 Proposed roles of GABA in plant stress responses (adapted from Kinnersley and Turano 2000)

in GABA levels in 1 h and 18-fold increase within 24 h (Shelp et al. 1999). These temporal differences in GABA accumulation could be correlated with severity of thermal stresses (Kinnersley and Turano 2000).

GABA-induced heat tolerance in creeping bentgrass *(Agrostis stolonifera)* involved enhancement of AsA-GSH cycle and an increase in GABA shunt. Under prolonged periods (28 days) of heat stress (35/30°C, day/night) in bentgrass, GABA enhanced production of antioxidants AsA and GSH, contributing to low ROS levels and increased protection to plants. These changes associated with GABA accumulation contributed to the maintenance of metabolic homeostasis (discussed in section 2.4.3) in

bentgrass (Li et al. 2016). Figure 21.7 exhibits a proposed model for GABA-induced heat tolerance in bentgrass associated with physiological and metabolic changes.

A similar role of GABA in antioxidant protection has also been demonstrated in leaves of rice *(Oryza sativa)* when exposed to short-term (10 days) high temperature (35/30°C, day/night) (Li et al. 2016). Nayyar et al. (2014) reported that exogenous GABA resulted in increases in SOD, CAT, ascorbate peroxidase (APX), and glutathione reductase (GR) activities in rice seedlings after a relatively short-term (10 days) heat stress (35/30°C, day/ night), indicating its protective role in scavenging heat-induced ROS. However, the mechanism of regulation of enzymatic and

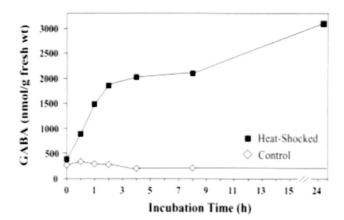

FIGURE 21.5 The time course of GABA accumulation in heat-shocked cowpea cell cultures (Kinnersley and Turano 2000).

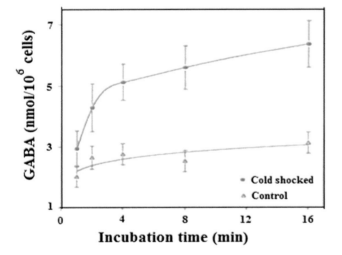

FIGURE 21.6 The time course of GABA accumulation in cold-shocked asparagus cell cultures (Kinnersley and Turano 2000).

non-enzymatic antioxidant pathways by GABA during long-term heat stress has not yet been elucidated (Nayyar et al. 2014).

Excess accumulation of proline could attenuate heat tolerance in *Arabidopsis* seedlings. Study of Du et al. (2011) implied that higher increases in proline content reflected greater damages in Bermuda grass and Kentucky bluegrass during early periods of heat stress (Du et al. 2011). This was also true in creeping bentgrass, wherein lower proline content of GABA-treated plants correlated well with lesser heat-induced damages in creeping bentgrass (Li et al. 2016). Heat shock also increased GABA levels several-fold in cowpea cells and in *Sesamum indicum* L. (cultivated variety Cumhuriyet) (Bor et al. 2009).

Drought Stress in Regulation of GABA Synthesis

Levels of GABA have been found to be more than threefold higher than levels of proline in leaves of cotton and bean, exposed to drought (caused by water deficit accompanied by high temperature and solar radiation) stress (Kinnersley and Turano 2000).

Among drought, salt, heavy metal, and high-temperature stresses given to sesame plants, GABA accumulation was found to be the highest in Se-treated groups followed by the plants under drought stress. GABA accumulation was also reported in the leaves of bean, turnip, and soybean under drought stress (Bor et al. 2009).

Salinity Stress in Regulation of GABA Synthesis

Both temporal variation and variation in the pattern of accumulation of GABA levels are seen between salt-sensitive and salt-tolerant cultivars under stress. Treatment with 140 mM NaCl showed significantly greater GABA accumulation in leaves and roots of a salt-sensitive cultivar of tomato (*Lycopersicon esculentum*), from the beginning of the salt treatment, compared with the salt tolerant relative cultivar (*L. pennellii*), possibly owing to metabolic differences between them (Bolarín et al. 1995; Kinnersley and Turano 2000).

Renault et al. (2013) found that GABA transaminase deficient *Arabidopsis* mutants are oversensitive to NaCl treatment (Renault et al. 2013). Therefore, the accumulation of GABA, in particular under salt stress, could depend not only on the rate of its degradation but also on the backward reaction from succinic semialdehyde to GABA, both reactions catalyzed by the same enzyme. Salt stress combined with Ca^{2+} treatment ameliorated inhibition of growth of foxtail millet (*Setaria italica* L.) and increased GAD activity and GABA accumulation, establishing that GABA accumulation under salt stress is regulated by Ca^{2+} (Bai et al. 2013).

GABA synthesis under salinity could contribute both as a biochemical pH-stat and as temporary nitrogen storage to decrease, together with asparagine, excess ammonium accumulated under stress. During simultaneous freezing and salt stress, GABA has a cryoprotective and scavenging efficiency exceeding that of proline (Woodrow et al. 2017).

Light Stress in Regulation of GABA Synthesis

GABA significantly accumulates in response to simultaneous high light and salinity in durum wheat shoots. Salinity increased GABA content by 1.5- and 5.8-fold in shoots of wheat plants exposed to low light and high light conditions compared to their respective controls, but its content was 43.7 and 165.7 fold higher under high light and salinity compared with low light and salinity (Woodrow et al. 2017). GABA accumulation has also been related to nitrogen starvation and light stress in green algae *Haematococcus pluvialis* (Recht et al. 2014).

Osmotic Stress in Regulation of GABA Synthesis

The mechanism of osmotic regulation in plants under stressed conditions has been discussed in the section on functional roles of GABA. It has been found that the common bean (*Phaseolus vulgaris* L.) is sensitive to water stress and exhibits reduced stomatal conductance under moderate water stress and osmotic shock, wherein the increase in GABA was significantly higher in the water-stressed leaves of the plants vis-à-vis that in non-stressed leaves (Raggi 1994). High levels of GABA have been found in osmotically stressed wheat seedlings (Bor et al. 2009). As zwitterion, GABA could act as an

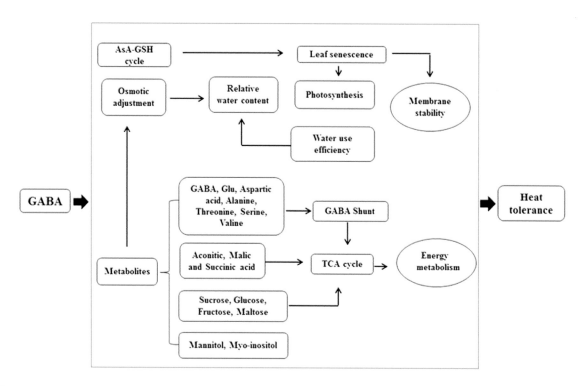

FIGURE 21.7   A proposed model for GABA-induced heat tolerance in creeping bentgrass associated with physiological and metabolic changes (adapted from Li et al. 2016).

osmolyte accumulating in cells under high light and salinity stresses without causing toxic effects (Woodrow et al. 2017).

Anoxia Stress in Regulation of GABA Synthesis

The mechanism of GABA accumulation under anoxia has been elaborated earlier. Excised leaves of plants incubated under anaerobic conditions showed increased accumulation of GABA (Thompson et al. 1966). GABA accumulation (synthesis mainly from glutamate) reached its maximum in rice roots within 24 h of anoxic conditions (Shelp et al. 1999) and also in radish leaves within 4 h (Streeter and Thompson 1972). Accumulation of this hypotensive compound under anaerobic conditions also occurred in tea leaves, glutamic acid again being the source of GABA (Tsushida and Murai 1987).

Mechanical Stress in Regulation of GABA Synthesis

Soybean (*Glycine max* [L.] Meff) leaves contain low levels (0.05 µmole g^{-1} fresh weight) of GABA but the concentration of this non-protein amino acid reportedly increased to 1 to 2 µmole g^{-1} fresh weight within 5 min of the transfer of plants (or detached leaves) to lower temperature from 33 to 22°C. Maximum accumulation of GABA was nearly attained within 2 min of detachment of the leaflets, leading to a 20-fold increase within 5 min. A parallel decrease occurred in the concentration of glutamate. Accumulation of GABA was also triggered by mechanical damage to the soybean leaves, but when subjected to a gradual reduction in temperature (2°C/min), only a small increase in GABA occurred in the leaves. Even when the soybean leaves were placed in liquid N$_2$ before detachment, a low level of GABA was always detected (approximately 0.05

mmol/g fresh weight). Here GABA accumulation in response either to a low-temperature shock or mechanical stimulation or damage possibly results from an altered intracellular compartmentation of glutamate or due to increase in GAD activity mediated by increases in levels of cytosolic H$^+$ or Ca^{2+} (Wallace et al. 1984).

While the biochemical basis for GABA accumulation in wounded leaves is well founded, the underlying mechanisms for wounding-induced GABA accumulation in systemic leaves remain unclear. In recent investigations on *Arabidopsis thaliana*, knock-out mutant lines pop2-5, unable to degrade GABA, and tpc1-2, lacking a wounding-induced systemic cytosolic calcium elevation, were employed for a comprehensive investigation of systemic GABA accumulation. A wounding-induced systemic GABA accumulation was detected in tpc1-2 plants demonstrating that an increased calcium level was not involved. Similarly, after both mechanical wounding and *Spodoptera littoralis* (cotton leaf worm) feeding, GABA accumulation in pop2-5 plants was significantly higher in local and systemic leaves, compared to the wild-type plants. Consequently, larvae feeding on these GABA-enriched mutant plants decreased significantly. Upon exogenous application of a D$_2$-labeled GABA to wounded leaves of pop2-5 plants, its uptake but no translocation to unwounded leaves was detected. In contrast, an accumulation of endogenous GABA triggered by a *de novo* synthesis of GABA was observed in vascular connected systemic untreated leaves. These findings suggested that systemic accumulation of GABA upon wounding does not depend on the translocation of GABA from wounded to systemic leaves or on an increase in cytosolic calcium (as had been propounded earlier). These studies have proposed that GABA might also have a role as a signaling molecule, which is activated upon wounding. The signal responsible for this observation still remains to be elucidated, but electrophysiological reactions might represent conceivable signaling candidates (Scholz et al. 2017).

Heavy Metal Stress in Regulation of GABA Synthesis

Potted plants of *Sesamum indicum* L. cultivated variety Cumhuriyet treated with 100 μg g⁻¹ selenium solutions showed increased GABA accumulation. There was 35% increase in GABA level on day 7 and 53% on day 21 (end of experimentation period). This is perhaps the only report on increased GABA levels under heavy metal stress (Bor et al. 2009).

Conclusion

More studies are warranted to elucidate the physiological roles of GABA in plants. GABA cannot be merely treated as a metabolite in plants in the context of its response to stresses. Exploitation of *Arabidopsis* functional genomic tools has established the notion that GABA has a new role of a signaling molecule in plants. GABA probably plays a dual role both as a signaling molecule and as a metabolite in plants, similar to those played by other metabolites, such as glutamate and sugars. Cellular studies with the help of functional genomic tools (e.g. mutants) and other whole-genome approaches would possibly uncouple the signaling and metabolic roles of GABA, and identify the molecular components and their modes of action (Bouché and Fromm 2004; Fait et al. 2006). An integration of the roles of GABA in both biotic and abiotic stress including its novel role as a plant neurotransmitter would perhaps unravel the plethora of roles of this ubiquitous molecule in plants.

REFERENCES

Abeles, F. B. (1973). Regulation of ethylene production by internal, environmental and stress factors In: *Ethylene in Plant Biology* (p. 83), Academic Press, New York.

Abeles F. B., Morgan P. and Saltveit M. (1992). *Ethlene in Plant Biology*, 2nd ed. Academic Press, New York.

Aurisano, N., Bertani, A. and Reggiani, R. (1995). Anaerobic accumulation of 4-aminobutyrate in rice seedlings; causes and significance. *Phytochemistry*, 38, 1147–1150.

Bai, Q., Yang, R., Zhang, L. and Gu, Z. (2013). Salt stress induces accumulation of γ-aminobutyric acid in germinated foxtail millet (*Setaria italica* L.). *Cereal Chemistry*, 90, 145–149.

Baum, G., Chen, Y., Arazi, T., Takatsuji, H. and Fromm, H. (1993). A plant glutamate decarboxylase containing a calmodulin binding domain. Cloning, sequence, and functional analysis. *Journal of Biological Chemistry*, 268, 19610–19617.

Baum, G., Lev-Yadun, S., Fridmann, Y., Arazi, T., Katsnelson, H., Zik, M. and Fromm, H. (1996). Calmodulin binding to glutamate decarboxylase is required for regulation of glutamate and GABA metabolism and normal development in plants. *The EMBO Journal*, 15, 2988–2996.

Bolarín, M. C., Santa-Cruz, A., Cayuela, E. and Pérez-Alfocea, F. (1995). Short-term solute changes in leaves and roots of cultivated and wild tomato seedlings under salinity. *Journal of Plant Physiology*, 147, 463–468.

Bor, M., Seckin, B., Ozgur, R., Yılmaz, O., Ozdemir, F. and Turkan, I. (2009). Comparative effects of drought, salt, heavy metal and heat stresses on gamma-aminobutyric acid levels of sesame (*Sesamum indicum* L.). *Acta Physiologiae Plantarum*, 31, 655–659.

Bouché, N., Fait, A., Bouchez, D., Møller, S. G. and Fromm, H. (2003). Mitochondrial succinic-semialdehyde dehydrogenase of the γ-aminobutyrate shunt is required to restrict levels of reactive oxygen intermediates in plants. *Proceedings of the National Academy of Sciences*, 100, 6843–6848.

Bouché, N. and Fromm, H. (2004). GABA in plants: just a metabolite? *Trends in Plant Science*, 9, 110–115.

Cholewa, E., Cholewinski, A. J., Shelp, B. J., Snedden, W. A. and Bown, A. W. (1997). Cold-shock-stimulated γ-aminobutyric acid synthesis is mediated by an increase in cytosolic Ca²⁺, not by an increase in cytosolic H⁺. *Canadian Journal of Botany*, 75, 375–382.

Crawford, L. A., Bown, A. W., Breitkreuz, K. E. and Guinel, F. C. (1994). The synthesis of [gamma]-aminobutyric acid in response to treatments reducing cytosolic pH. *Plant Physiology*, 104, 865–871.

Du, H., Wang, Z., Yu, W., Liu, Y. and Huang, B. (2011). Differential metabolic responses of perennial grass *Cynodon transvaalensis×Cynodon dactylon* (C₄) and *Poa pratensis* (C₃) to heat stress. *Physiologia Plantarum*, 141, 251–264.

Fait, A., Yellin, A. and Fromm, H. (2005). GABA shunt deficiencies and accumulation of reactive oxygen intermediates: insight from *Arabidopsis* mutants. *FEBS Letters*, 579, 415–420.

Fait, A., Yellin, A. and Fromm, H. (2006). GABA and GHB neurotransmitters in plants and animals neuronal aspects of plant life. In: *Communication in Plants* (pp. 171–185), Baluška, F., Mancuso, S. and Volkmann D. (Eds.), Springer-Verlag, Germany.

Gill, S. S. and Tuteja, N. (2010). Reactive oxygen species and antioxidant machinery in abiotic stress tolerance in crop plants. *Plant Physiology and Biochemistry*, 48, 909–30.

Jaffe, M. J. (1973). Thigmomorphogenesis: The response of plant growth and development to mechanical stimulation: With special reference to *Bryoria dioria*. *Planta*, 114, 143–157.

Jaffe, M. J. and Biro, R. (1979). Thigmomorphogenesis, the effect of mechanical perturbation on the growth of plants, with special reference to anatomical changes, the role of ethylene and interaction with other environmental stresses. In: *Stress Physiology in Crop Plants* (pp. 25–69), Mussell, H. and Staples, R. C. (Eds.), John Wiley and Sons, New York.

Kinnersley, A. M. and Turano, F. J. (2000). Gamma aminobutyric acid (GABA) and plant responses to stress. *Critical Reviews in Plant Sciences*, 19, 479–509.

Kornberg, H. L. (1966). Anaplerotic sequences and their role in metabolism. *Essays in Biochemistry*, 2, 1–31.

Li, Z., Yu, J., Peng, Y. and Huang, B. (2016). Metabolic pathways regulated by γ-aminobutyric acid (GABA) contributing to heat tolerance in creeping bentgrass (*Agrostis stolonifera*). *Scientific Reports*, 6, 1–16.

Mayer, R. R., Cherry, J. H. and Rhodes, D. (1990). Effects of heat shock on amino acid metabolism of cowpea cells. *Plant Physiology*, 94, 796–810.

Morgan, P. W. and Drew, M. C. (1997). Ethylene and plant responses to stress. *Physiologia Plantarum*, 100, 620–630.

Nayyar, H., Kaur, R., Kaur, S. and Singh, R. (2014). γ-Aminobutyric acid (GABA) imparts partial protection from heat stress injury to rice seedlings by improving leaf turgor and upregulating osmoprotectants and antioxidants. *Journal of Plant Growth Regulation*, 33, 408–419.

Palanivelu, R., Brass, L., Edlund, A. F. and Preuss, D. (2003). Pollen tube growth and guidance is regulated by POP2, an *Arabidopsis* gene that controls GABA levels. *Cell*, 114, 47–59.

Raggi, V. (1994). Changes in free amino acids and osmotic adjustment in leaves of water-stressed bean. *Physiologia Plantarum*, 91, 427–434.

Ramakrishna, A. and Ravishankar, G. A. (2011). Influence of abiotic stress signals on secondary metabolites in plants. *Plant Signaling and Behavior*, 6, 1720–1731.

Ramakrishna, A. and Ravishankar, G. A. (2013) Role of plant metabolites in abiotic stress tolerance under changing climatic conditions with special reference to secondary compounds. In: *Climate Change and Plant Abiotic Stress Tolerance* (pp. 705–726), Wiley-VCH Verlag GmbH & Co. KGaA, Weinheim, Germany.

Ramputh, A. I. and Bown, A. W. (1996). Rapid [gamma]-aminobutyric acid synthesis and the inhibition of the growth and development of oblique-banded leaf-roller larvae. *Plant Physiology*, 111, 1349–1352.

Recht, L., Töpfer, N., Batushansky, A., Sikron, N., Gibon, Y., Fait, A. and Zarka, A. (2014). Metabolite profiling and integrative modeling reveal metabolic constraints for carbon partitioning under nitrogen starvation in the green algae *Haematococcus pluvialis*. *Journal of Biological Chemistry*, 289, 30387–30403.

Renault, H., El Amrani, A., Berger, A., Mouille, G., Soubigou-Taconnat, L., Bouchereau, A. and Deleu, C. (2013). γ-Aminobutyric acid transaminase deficiency impairs central carbon metabolism and leads to cell wall defects during salt stress in *Arabidopsis* roots. *Plant, Cell and Environment*, 36, 1009–1018.

Scholz, S. S., Malabarba, J., Reichelt, M., Heyer, M., Ludewig, F. and Mithöfer, A. (2017). Evidence for GABA-induced systemic GABA accumulation in *Arabidopsis* upon wounding. *Frontiers in Plant Science*, 8, 1–9.

Shelp, B. J., Bown, A. W. and McLean, M. D. (1999). Metabolism and functions of gamma-aminobutyric acid. *Trends in Plant Science*, 4, 446–452.

Snedden, W. A. and Fromm, H. (1999). Regulation of the γ-aminobutyrate-synthesizing enzyme, glutamate decarboxylase, by calcium-calmodulin: a mechanism for rapid activation in response to stress. In: *Plant Responses to Environmental Stresses: From Phytohormones to Genome Reorganization* (pp. 549–574), Lerner, H. R. (Ed.), Marcel Dekker, New York.

Streeter, J. G. and Thompson, J. F. (1972). Anaerobic accumulation of gamma-aminobutyric acid and alanine in radish leaves (*Raphanus sativus*, L.). *Plant Physiology*, 49, 572–578.

Thompson, J. F., Stewart, C. R. and Morris, C. J. (1966). Changes in amino acid content of excised leaves during incubation I. The effect of water content of leaves and atmospheric oxygen level. *Plant Physiology*, 41, 1578–1584.

Tsushida, T. and Murai, T. (1987). Conversion of glutamic acid to γ-aminobutyric acid in tea leaves under anaerobic conditions. *Agricultural and Biological Chemistry*, 51, 2865–2871.

Wallace, W., Secor, J. and Schrader, L. E. (1984). Rapid accumulation of γ-aminobutyric acid and alanine in soybean leaves in response to an abrupt transfer to lower temperature, darkness, or mechanical manipulation. *Plant Physiology*, 75, 170–175.

Woodrow, P., Ciarmiello, L. F., Annunziata, M. G., Pacifico, S., Iannuzzi, F., Mirto, A. and Carillo, P. (2017). Durum wheat seedling responses to simultaneous high light and salinity involve a fine reconfiguration of amino acids and carbohydrate metabolism. *Physiologia Plantarum*, 159, 290–312.

Section III

Role of Specialized Proteins During Abiotic Stress Management

22

MicroRNAs: Emerging Roles in Abiotic Stresses and Metabolic Processes

Susana S. Araújo, Carolina Gomes, Jorge A.P. Paiva, Alma Balestrazzi, and Anca Macovei

CONTENTS

Abbreviations

ABA	Abscisic acid
AGO1	Argonaute 1
ARF	Auxin Responsive Factor
DCL	Dicer-like
GA	Gibberellic acid
miRNA	MicroRNA
RNA	Ribonucleic acid
ROS	Reactive oxygen species
TFs	Transcription factors

Introduction

To flourish as sessile organisms, plants require ideal environmental conditions (temperature, soil, air, light, etc.) while changes in several parameters can affect the species reproduction and survival. Nowadays, the predicted increase in temperature, alteration of rain patterns, and extreme weather events across the globe continuously impact plant growth. Adaptation to high stress habitats is a key component in the evolution of plants with resistance to abiotic stress factors (Ramakrishna and Ravishankar, 2011, 2013). This involves a combination of phenotypic plasticity and genetic adaptations, mediated through the activity of multiple genes. Among the many mechanisms responsible for stress adaptation, transcriptional and post-transcriptional regulation mediated by specific transcription factors (TFs) and microRNAs (miRNAs), respectively, represent the major families of gene regulators (Samad et al., 2017).

MiRNAs are small (18–22 nucleotides), non-coding, single-stranded ribonucleic acids (RNAs) that specifically bind to target endogenous mRNAs and repress their translation or induce their cleavage (Bartel, 2004; Borges and Martienssen, 2015; Djami-Tchatchou et al., 2017). They can trigger translational repression or gene silencing by binding to complementary sequences on target mRNA transcripts, thereby controlling the regulation of their target genes at the post-transcriptional level (Jones-Rhoades et al., 2006; Yang et al., 2007). MiRNAs control many biological processes, such as plant development, differentiation, signal transduction, or stress responses (Sunkar et al., 2004; Lin et al., 2009). In plants, miRNAs are processed from stem-loop regions of long primary transcripts by a Dicer-like (DCL) enzyme and are loaded into silencing complexes, where they direct cleavage of complementary mRNAs or inhibit their translation (Jones-Rhoades et al., 2006). Comprehensive reviews describing the major steps in miRNA discovery, biogenesis, and activity can be found in Jones-Rhoades and Bartel (2004) or Voinnet (2009). As miRNA target specific genes based on sequence complementarity, many online tools and databases were built to provide solid platforms for the collection, standardization, and browsing of publicly available data to support miRNA research. For example, miRBase (http://www.mirbase.org) is the public repository for all published miRNA sequences and associated annotations (Kozomara and Griffiths-Jones, 2011). Presently, miRBase gathers 28,645 miRNA entry sequences and respective annotations (miRBase, release 21, accessed on the 20th January 2017). Of them, 8496 mature miRNAs sequences can be retrieved from miRBase for Viridiplantae. Just as interesting is PASmir (http://pcsb.ahau.edu.cn:8080/PASmiR), which is another comprehensive literature-curated and web-accessible repository for miRNAs involved in plant response to abiotic stresses (Zhang et al., 2013). Presently, PASmir gathers information from more than 1400 miRNA-abiotic stress regulatory entries identified in more than 40 plant species.

This chapter aims to review the current literature discussing the implication of miRNAs in a number of adaptive processes, namely their roles in the regulation of abiotic stresses, secondary metabolite production, and seed pre-germinative metabolism.

MiRNAs: Master-Payers in Abiotic Stress Responses

Plants exposed to stress use multiple gene regulatory mechanisms, including post-transcriptional regulation of gene expression to restore and re-establish cellular homeostasis (Sunkar

et al., 2012; Kawaguchi et al., 2004). Mounting evidence has shown that miRNAs are decisive post-transcriptional gene-expression regulators that modulate plant development and the responses to biotic or abiotic stresses (Shriram et al., 2016). In plants, most of the miRNA targets are mRNAs that code for TFs (Jones-Rhoades et al., 2006). Consequently, this also highlights the important role that post-transcriptional regulation mediated by miRNA plays in controlling numerous processes in plants, including the responses to abiotic stresses. Numerous studies have shown that plants up- or down-regulate the expression of miRNAs in response to environmental stresses, but also that abiotic stresses lead to the synthesis of new miRNAs (Khraiwesh et al., 2012).

Over 200 research articles report studies dealing with miRNA identification, profiling, or validation in response to abiotic stresses were retrieved from PubMed at the National Centre for Biotechnology Information (NCBI, http://www.ncbi.nlm.nih.gov/pubmed, on 10th January 2017) (Figure 22.1a). The current number of research articles devoted to the study of miRNAs in the modulation of abiotic stress responses in plants has increased approximately fivefold from 2004. Based on the overall tendency, it is tempting to speculate that the number of studies on the topic is expected to increase in the next few years. This could be supported by the release of several plant reference genomes coupled to the general decrease in the costs to perform high-throughput sequencing. As depicted in Figure 22.1b, numerous conserved and novel miRNAs have been identified and implicated in the regulation of drought (Bakhshi et al., 2016; Ferreira et al., 2012), cold stress (Maeda et al., 2016; Song et al., 2017), nutrient deprivation (Zhao et al., 2012; Panda and Sunkar, 2015), and trace element toxicity, among others (Huang et al., 2016; Cakir et al., 2016). One interesting aspect is that the majority of studies are conducted to investigate the role of miRNAs in response to multiple stresses (Ragupathy et al., 2016; Chen et al., 2017). This last aspect is not surprising since a single miRNA can regulate the expression of more than 100 different transcripts (Felekkis et al., 2010), and a cross-talk among regulatory networks in response

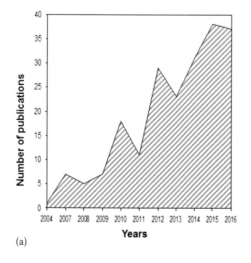

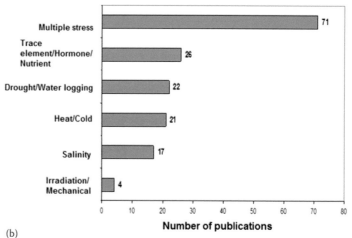

(a)   (b)

FIGURE 22.1 Number of research scientific manuscripts on the miRNA and abiotic stress responses in plants available at NCBI between 2004 and 2016. (a) Chronological evolution of the number of publications describing identification, profiling, or validation of abiotic stress-induced miRNAs. (b) Number of publications in which miRNA expression was validated in each abiotic stress type. The query was performed on 10th January 2017, at PUBMED using the following settings: miRNA AND plant AND abiotic stress NOT "review" [Publication Type].

to multiple stresses has been widely reported (Nakashima et al., 2014; Niu et al., 2016). Importantly, this also highlights the high potential of modulating miRNA expression as a strategy to generate multiple stress resistant plants to face the current challenges of changing environments. More details about the studies conducted on different types of abiotic stress are provided in the next sections.

Drought Stress

Drought, or water deficit, plays a major role in determining crop productivity (Boyer, 1982). It is generally well reported that molecular changes take place rapidly after plants sense water depravation (Trindade et al., 2010). Beside transcriptional regulation, a growing body of research indicates that miRNA-mediated post-transcriptional regulation has been implicated in the modulation of the expression of target gene response to water deprivation (Sunkar et al., 2012; Ferdous et al., 2015). Research studies were conducted to identify drought-responsive miRNAs in both crop and model species, such as *Medicago truncatula* (Trindade et al., 2010), *Nicotiana tabacum* (Chen et al., 2017), *Zea mays* (Wang et al., 2014), *Brassica napus* (Jian et al., 2016), and *Solanum tuberosum* (Yang et al., 2014).

Katiyar et al. (2015) conducted a study in sorghum (*Sorghum bicolor*) to identify genotype-specific and drought-responsive small interfering RNAs, including miRNAs. Sorghum is a major staple food crop of millions of people, especially in semi-arid tropics regions. To accomplish their goal, small RNA libraries were constructed using RNA samples of two genotypes, namely M35-1 (drought tolerant) and C43 (drought susceptible), grown under control and drought stress conditions. In this study, 97 conserved and 526 novel miRNAs, representing 472 unique miRNA families, were identified by sequencing the small RNA libraries using a Solexa deep sequencing technology combined with bioinformatics analysis. Of the 96 drought-regulated miRNAs, 63 miRNAs showed opposite regulation in drought tolerant and drought sensitive genotypes (Katiyar et al., 2015). For example, miR160 that targets several Auxin Responsive Factor (ARF) family members, including ARF10, ARF16, and ARF17, showed a differential regulation depending on the genotype studied. While miRNA160 was up-regulated in M35-1, the same miRNA was down-regulated in C43, suggesting it may contribute, at least in part, to differential drought tolerance in the studied sorghum genotypes.

Drought is one of the main abiotic stresses that dramatically hampers rice production. In this context, Chung et al. (2016) investigated how post-transcriptional regulatory mechanisms modulate rice responses under well-watered and drought conditions. RNA-sequencing (RNA-seq) transcript profiling was used to evaluate the levels of non-coding RNAs, including pre-miRNAs. This study identified 66 drought-responsive miRNA precursors, assigned to 29 miRNA families. One interesting aspect was the fact that members of the same miRNA family are functionally diverse during drought responses (Chung et al., 2016). For example, pre-miR399k and pre-miR399d were up-regulated by drought stress, while pre-miR399e and pre-miR399i were down-regulated. Parallel Analysis of RNA Ends (PARE), also known as RNA degradome analysis, was used to validate the miRNA-mediated target cleavages. This technique was particularly useful to validate that miR171f, miR156, miR530, and miR819i, have a large number of putative target genes in rice, as these targets were up- or down-regulated in an opposite manner to the change in miRNA expression in response to drought (Chung et al., 2016). Such results represent a valuable resource for investigating how the extensive set of non-coding RNAs interact during drought stress.

The differential expression of miRNAs between two sugarcane cultivars with distinct drought tolerance under simulated drought treatments was investigated by Gentile et al. (2015). The sugarcane cultivars RB867515 (higher drought tolerance-HT) and RB855536 (lower drought tolerance-LT) were used in this study. Differential miRNA expression depends not only on the cultivar studied, but also on the duration of the drought treatment. For example, three miRNAs (ssp-miR164, ssp-miR397, and ssp-miR528) were up-regulated in the RB867515 (HT), while in the RB855536 (LT) cultivar, four miRNAs (ssp-miR164, ssp-miR394, ssp-miR399-seq1, and ssp-miR1432) were down-regulated. Only ssp-miR397 displayed the same pattern in both cultivars as it was induced after 2 days of water stress. A NAC (NAM/ATAF/CUC) TF was found among the targets of ssp-miR164. A previous study has described that miR164 regulates the expression of 5 NAC TFs in *Arabidopsis* (Guo et al., 2005), with important roles in development, growth, and drought stress responses (Ooka et al., 2003).

Salinity Stress

Salinity affects crop productivity and almost all aspects of plant development from germination to reproductive development (for review see Park et al., 2016). Soil salinity imposes ion toxicity, osmotic stress, nutrient (N, Ca, K, P, Fe, and Zn) deficiency, and oxidative stress on plants, thus limiting water up-take from soil (Shrivastava and Kumar, 2015). Numerous studies are providing mounting evidence that miRNAs play a major role in the molecular regulation of salt stress responses and adaptation (Macovei and Tuteja, 2012; Mangrauthia et al., 2013; Mittal et al., 2016).

Zhuang et al. (2014) conducted a study in the salt tolerant *Solanum linnaeanum* to investigate the involvement of miRNAs in the response to NaCl treatments. Among the 98 conserved miRNAs identified, only 14 miRNAs were found to be differentially regulated in response to the applied salt stress. For example, miR164c, sli-miR166d, and sli-miR397a were up-regulated under salt stress. These miRNAs are predicted to target lipase-related genes, the DNA repair Rad4 gene family, and laccase encoding genes, respectively. Unlike the targets of up-regulated miRNAs in *S. linnaeanum*, some of the down-regulated miRNAs target mRNAs of TFs, indicating an upstream regulation of miRNAs during the response to salt stress. For example, the sli-miR171 family targets the scarecrow TF gene family. Nevertheless, more studies are needed to link the potential function of these potential miRNAs targets to the response to salt stress, which would be relevant for developing strategies for the genetic improvement of the *Solanaceae* crops.

Salinity strongly influences miRNA biogenesis by enhancing or repressing the transcription of specific MIR genes (Parreira et al., 2016). However, this could also affect the activities of the proteins responsible for miRNA processing, maturation, and target cleavage (Capitão et al., 2011). To extend the knowledge on the mechanisms underlying the regulation of miRNA expression in plants, Dolata et al. (2016) investigated the influence of salinity stress on the miRNA in *Arabidopsis* by using a transcriptome approach. To accomplish this goal, the authors used mirEX, a high-throughput real-time quantitative PCR (qRT-PCR)-based platform, which allows the simultaneous examination of the accumulation of all known *Arabidopsis* miRNA precursors (pri-miRNAs) in response to salt stress. The study was also conducted in *Arabidopsis* mutants containing a mutated version of the argonaute1 (AGO1) gene to access the overall importance of AGO1 in general miRNA stabilization. Overall profiling results showed that 40% of all *Arabidopsis* pri-miRNAs were changed during salt stress. In some cases, the increased levels of *MIR* gene transcription resulted in higher miRNA production, thus leading to more efficient cleavage or translation inhibition of the target mRNAs. Other interesting findings of this study are related to the stress experiments performed in the *ago1-11* mutant. The lack of AGO1 led to a dramatic decrease in miRNA161 and miRNA173 levels in response to salt stress, suggesting that AGO1 bound to and stabilized both of these miRNAs, leading to effective cleavage of their target sequences. The results demonstrated that plant AGO1 is also involved in the co-transcriptional regulation of specific miRNAs that are responsive to salt stress (Dolata et al., 2016).

Extreme Temperatures

Temperature is a primary factor affecting the rate of plant development and growth, with each species having a specific temperature range (Hatfield and Prueger, 2015). Extreme weather events, such as waves of frost or heat, are predicted to increase under future climate scenarios, posing a significant challenge for grain production (Barlow et al., 2015). Recent advances have revealed key physiologic, metabolic, and regulatory processes that play a major role in temperature stresses. For example, impaired pollen development under high temperature conditions has been implicated in reduced yields across a large number of crop systems, including tomato (Frank et al., 2009). Some TFs that play a role in the impairment of pollen development during heat stresses were identified (Frank et al., 2009). Gao et al. (2016) conducted a study to identify miRNAs and their targets in response to high temperatures in tomato. Small-RNA and degradome libraries were constructed from the leaves of a heat-tolerant tomato at normal (26/18°C), moderate (33/33°C), and acute (40/40°C) temperatures. Among the different miRNAs identified, 96 and 150 miRNAs were responsive to the moderately and acutely elevated temperature, respectively. The targets for some conserved miRNAs and 8 novel miRNAs were also identified. For example, spi-miR6300_gma targeted the *Hsp70* gene in tomato plants at the moderately elevated temperature, while spi-miR166c-3p and spi-miR166g-3p_osa targeted the *Hsp60-3A* gene in tomato plants at the acutely elevated temperature. Such results suggest that these miRNAs could play important roles in tomato responses to elevated temperatures by regulating the transcript abundances of heat shock proteins (Gao et al., 2016).

When considering cold stress, the highly conserved miR394 targets the leaf curling responsiveness (LCR) gene, encoding an F-box protein (SKP1-Cullin/CDC53-F-box) (Song et al., 2012). A recent study conducted in *Arabidopsis* showed that miR394a-overexpressing plants (*35S::MIR394a*) and *lrc* loss-of-function mutants exhibited more tolerance to low temperature (4°C to −11°C) when compared to wild type plants (Song et al., 2016). The authors performed another experiment in which they expressed a miR394 cleavage resistance version of *35S::m5LCR* aiming to corroborate if the down-regulation of LCR is mediated by miR394 and linked to the physiological response observed. The *35S::m5LCR* plants showed cold hypersensitive phenotype, indicating that LCR plays a negative role in response to low temperature (Song et al., 2016). Altogether the results of this study demonstrate the vast potentiality of miRNAs modulation to improve a plant's performance under cold stress.

Heavy Metals Toxicity

Metals, generally called heavy metals, are naturally present in the soil. Among these metals, there are some essential micronutrients, like copper (Cu), zinc (Zn), iron (Fe), and manganese (Mn) that are needed for normal growth and development, which, however, are toxic in high amounts (Gielen et al., 2012). Other metals like cadmium (Cd), lead (Pb), aluminum (Al), or mercury (Hg) are described as non-essential elements, being toxic even at low concentrations. Geological and anthropogenic activities have increased the concentration of these elements to amounts that are nowadays harmful to both plants and animals (Chibuike and Obiora, 2014). Research conducted in both model and crop plants, like *Arabidopsis* (Kayıhan et al., 2016), rice (Sharma et al., 2015), or radish (Liu et al., 2015), showed that heavy metals toxicity results in a relevant inhibition of plant growth and developmental processes. Importantly, contaminated plants also pose a serious threat to human and animal health when consumed (Sharma et al., 2015).

Numerous review manuscripts describe a pivotal role of miRNAs in regulating heavy metal toxicity in crops (Noman and Aqeel, 2017; Gielen et al., 2012; Yang and Chen, 2013). Chromium (Cr) has been described as one of the most hazardous and widespread environmental pollutants (Shanker et al., 2009). The molecular mechanisms associated with Cr toxicity in plants are poorly documented when compared to the number of studies conducted in other heavy metals (Sharmin et al., 2012). To address this open question, Liu et al. (2015) conducted a study in radish, an important root vegetable crop, to identify Cr-responsive miRNAs and validate the targeted transcripts for Cr stress-regulated miRNAs. Two small RNA libraries were constructed and sequenced from Cr-free (CK) and Cr-treated (200 mg L^{-1} CrVI) roots. The results showed that 54 known and 16 novel miRNAs were significantly differentially expressed under Cr stress. As an example, miR156/157, miR164, and miR854 were found to target squamosa promoter-binding-like proteins (SPLs), ARF16, and WRKY DNA-binding protein 26 (WRKY26), respectively. These genes encode for different TF families that might potentially regulate heavy metal transcriptional processes in plants. The expression patterns of some Cr-responsive miRNAs and their targets (e.g. miRNA520, which targets the TF TCP15) were validated by qRT-PCR, showing an approximate negative correlation. The outcomes of this study provided novel insights into miRNA-mediated regulatory mechanisms underlying plant response to Cr stress in root vegetable crops.

The exposition to arsenic (As) through the consumption of contaminated rice poses a serious issue in terms of food safety for humans (McCarty et al., 2011). Although several studies have been conducted to investigate the mechanisms of As uptake and accumulation in rice, the role of miRNAs in regulating these mechanisms is being also investigated. Sharma et al. (2015) carried out an miRNA profiling study in contrasting As-accumulating rice accessions (high As-accumulating genotype-HARG versus low As-accumulating genotype-LARG) using miRNA Array. Several differentially expressed miRNAs were identified in response to arsenite (AsIII) and arsenate (AsV) exposure. The expression of the miR396, miR399, miR408, miR528, miR1861, miR2102, and miR2907 families was up-regulated in response to As exposure in both cultivars. Similarly, down-regulation of miR164, miR171, miR395, miR529, miR820, miR1432, and miR1846 was observed in both cultivars under the same experimental conditions. The qRT-PCR study conducted corroborated the differences among genotypes, in which miR171g, miR529b, miR820a, and miR1432 expression were repressed in HARG but were significantly enhanced in the LARG genotype in response to AsIII stress. These results suggest a miRNA-dependent regulatory mechanism during As exposure in rice, which is dependent on the genotype/accessions.

MiRNAs Involved in Post-Transcriptional Regulation of Secondary Metabolites

Despite the major role of plant secondary metabolites in plant development and defense, and their importance for human health, the understanding of miRNA-mediated regulation of flavonoids, lignins, terpenes and terpenoids, and alkaloids biosynthetic pathways is still very limited (see review by Wagner and Kroumova, 2008; Gupta et al., 2017). Here, we will present some recent examples of the identification and characterization of miRNAs involved in the regulation of different classes of secondary metabolites, potential biotechnological targets to modulate the accumulation of different secondary metabolites.

Flavonoids

Flavonoids are a group of polyphenolic compounds and one of the most widely distributed groups of plant secondary metabolites. These phenylpropanoid pathway-derived compounds share a basic structural unit, two rings interconnected by a central ring. Within the flavonoid group of compounds, subgroup classification is based on the oxidation level of the central ring. Flavonoid subgroups include flavonols, flavones, isoflavones, anthocyanins, proanthocyanidins (PAs), and phlobaphene pigments. In plants, flavonoids constitute a highly diverse group with more than 6000 flavonoids having been identified. Due to this structural diversity, flavonoids play a wide assortment of physiological roles in plants, participating in the mediation of plant environmental responses, including the coordination of plant–microbe interactions, protection against diverse biotic and abiotic stresses, attraction by pollinators, and attenuation of photo-oxidative damage (Sharma et al., 2016). Here we will be focusing on the two major subgroups of flavonoids: anthocyanins and flavonols.

Anthocyanins are the most widely distributed water-soluble plant pigments. Besides being responsible for a range of colors in fruits and vegetables, anthocyanins also play pivotal roles in plant–environment interactions, such as the protection against different stresses, including salt, drought, heavy metal, high light stress, and pathogenic agents (Winkel-Shirley, 2001). Anthocyanin biosynthesis pathway comprises of three reaction steps of the general phenylpropanoid pathway that lead to the production of p-coumaroyl CoA, which is considered a major branch-point metabolite between flavonoid and monolignol production (Schilmiller et al., 2009). After the synthesis of p-coumaroyl CoA, the specific anthocyanin biosynthesis pathway starts with the synthesis of chalcone by chalcone synthase (CHS), then proceeds along a single reaction path to anthocyanins. The gene coding for the enzymes of anthocyanins biosynthetic pathway and its biological functions have been extensively characterized in some plant species (Tanaka et al., 2008; Shi and Xie, 2014). Anthocyanin biosynthesis-related functional genes have been shown to be regulated by several TFs, including members of the MYB, bHLH, WRKY, and WD-repeat families, which might regulate gene expression in a separate or cooperative manner. Despite this extensive knowledge regarding anthocyanin biosynthesis-related genes, molecular mechanisms of anthocyanin biosynthesis regulation during plant developmental stages and in response to different environmental stimuli is not yet fully understood. Several studies have been focused on the miRNAs-mediated regulation of *MYB* genes. In *Arabidopsis*, the overexpression of miR828 resulted in plants with lower anthocyanins content (Yang et al., 2013). In these plants, the expression of different *MYB* transcripts (*MYB75*, *MYB90*, and *MYB113*) that regulate the anthocyanin biosynthesis were reduced, as well as the genes directly involved in the biosynthesis of anthocyanins, including *PAL*, *CHS*, *CHI*, *F3H*, *F30H*, *DFR*, and *LDOX*. In tomato (*Solanum lycopersicum*), miR858 was identified as a negative regulator of anthocyanin biosynthesis, mediating the cleavage of *SlMYB7-like* and *SlMYB48-like* transcripts (Jia et al., 2015). The blockage of miR858 by target mimic resulted in an increased expression of several genes of the anthocyanin biosynthetic pathway (e.g. *PAL*, *CHS*, *DFR*, *ANS*, and *3GT*). An increased expression of the *SlMYB48* transcript was also observed, suggesting the involvement of this TF in the regulation of anthocyanin biosynthesis. Contrary to that observed in tomato, a recent study in *Arabidopsis* (Wang et al., 2016) showed that the accumulation of miR858a leads to the increase of anthocyanins content by transcriptional repression of the *AtMYBL2*, a key negative regulator of anthocyanin biosynthesis. In *Arabidopsis* seedlings, light-mediated regulation of miR858a occurs under control of ELONGATED HYPOCOTYL 5 (*HY5*) (Wang et al., 2016). Hence, under high light or other environmental stresses, the biosynthesis of anthocyanins might be modulated through a HY5-miR858a-MYBL2 regulatory loop, integrating transcriptional and post-transcriptional regulation of anthocyanin biosynthesis. More recently, a general model of miRNA-mediated regulation of anthocyanin biosynthesis in Chinese radish (*Raphanus sativus* L.) was proposed (Sun et al., 2017) based on the genome-wide identification of miRNA combined with target prediction and expression validation by qRT-PCR. This model opened new insights for exploring the miRNAs-based regulation of anthocyanins by integrating different regulatory mechanisms that directly or indirectly regulate the transcription of anthocyanin biosynthetic genes (*PAL, C4H, 4CL, CHS, CHI, F3H, DFR, ANS*, and *UFGT*). This includes the MYB-bHLH-WDR (MBW) complex, targeted by miR156, miR528, miR582a, and WRKY, and MADS-box, targeted by miR419 and miR773a.

Flavonols differ from anthocyanins in the degree of oxidation of the central pyran ring and derive from a different branch of the flavonoids biosynthetic pathway. In plants, these compounds have important physiological functions including the regulation of auxin transporters and play a role in the control of plant growth and development (Nguyen et al., 2013). Furthermore, it was reported that flavonols have protective functions under drought stress after the accumulation of such compounds were observed (Gao et al., 2016). Flavonol biosynthesis is a highly-regulated process dependent on tissue and developmental stages, both in normal and stressful environmental conditions. The regulation of flavonol biosythesis is thought to be mainly modulated by the MYB group of TFs (Gao et al., 2016; Sharma et al., 2016), but the global regulatory network of this biosynthetic pathway is not yet fully understood. Therefore, recent studies have focused on the miRNA post-transcriptional regulatory role on flavonol biosynthesis. Such studies have been performed in trees like *Ginkgo biloba* and *Ammopiptanthus mongolicus*, or in the model plant *Arabidopsis*. The study of Wang et al. (2015) used the Illumina next-generation sequencing HiSeqTM 2000 platform to identify and quantify the profiles of small RNAs potentially related to secondary metabolism in *Ginkgo biloba* leaves. In this study, the authors identified several miRNAs targeting flavonoid biosynthesis-related proteins, proving the involvement of miRNAs in the regulation of this metabolic pathway. Further, carotenoid biosynthesis, terpenoid backbone biosynthesis, and diterpenoid biosynthesis-related proteins were also among the predicted miRNA targets, forecasting the widespread regulatory role of miRNAs on secondary metabolism in *G. biloba* leaves. Another tree species, *Ammopiptanthus mongolicus*, known to grow in particularly harsh environments under extreme drought and freezing conditions in desert regions, has recently been used as a model to study the role of miRNAs on secondary metabolism regulation in response to drought stress (Gao et al., 2016). In this report, it was shown that miR858 participates in drought stress response by up-regulating MYB TF genes involved in the regulation of flavonol biosynthesis. Conversely, in a functional study conducted in *Arabidopsis thaliana* (Sharma et al., 2016) it was demonstrated that miR858a down-regulates flavonol-specific MYBs involved in regulating early steps of the flavonoid biosynthetic pathway. The authors also concluded that the higher expression of MYBs in miR858a-silenced lines leads to the redirection of the metabolic flux towards the synthesis of flavonoids at the cost of lignin synthesis, showing that miR858 can be a potential regulator of the phenylpropanoid pathway in *Arabidopsis*.

Lignin

Lignin is the second most abundant biopolymer, after cellulose, and is mainly found in the xylem secondary cell walls, although it can be found in other specialized cells like stone cells of fresh fruits. Lignin is considered one of the most complex polyphenolic heteropolymers. It derives from the oxidative polymerization of three p-hydroxycinammyl alcohol precursors (monolignols: p-coumaryl, coniferyl, and sinapyl alcohols) giving rise to p-hydroxyphenyl (H), guaiacyl (G), and syringyl (S) units (Boerjan et al., 2003; Vanholme et al., 2010). Lignin confers mechanical rigidity, imperviousness, water conductivity, and resistance to biodegradation, playing an important role in defense responses to disease and pests (Bhuiyan et al., 2009). However, lignin recalcitrance to biodegradation is considered one of the most important issues not only in the paper and pulp industry (Boudet et al., 2003), but also for the digestibility of forage crops (Zhao and Dixon, 2014). Thus, understanding the molecular regulatory mechanisms, including post-transcriptional miRNA-mediated regulation of the lignin biosynthesis pathway, is of foremost importance to produce plant biomass with reduced content and/or more extractible lignin.

Next generation sequencing (NGS) platforms, such as Illumina platforms, are particularly useful for miRNAome characterization and identification of miRNA putatively involved in the regulation of different metabolic processes. For example, to study the role of miRNA in fruit development and fruit quality, Wu et al. (2014) generated and sequenced six small RNA libraries corresponding to six development stages of pear (*Pyrus bretschneideri*). Pear is of particular interest as a model in this study field due to the presence of lignified stone cells. The analysis of miRNAome modifications along with different developmental stages allowed the identification of a large number of known and novel miRNAs, confirming the important role of miRNAs during fruit development. More interestingly, among the identified miRNAs, 11 miRNA were found to be putatively implicated in the regulation of the lignin biosynthesis pathway; namely, hydroxycinnamoyl transferase (*HCT*), encoding the enzyme and catalyzing the conversion of p-coumaroyl-CoA (C3'H) into caffeoyl-Coa (PCC) and feruloyl (FC), is putatively targeted by miR3711, miR419, and miR5260; the *4CL* is putatively targeted by miR396b, and the peroxidase (*POD*) is putatively targeted by miR5021. In addition, several laccases-blue copper oxidases involved in the polymerization of coniferyl alcohol (Ranocha et al., 1999), were found to be putatively targeted by miR397a, in accordance to previous observations in *Arabidopsis* (Wang et al., 2014) and in poplar (Lu et al., 2013). In poplar, overexpression of Ptr-MIR397 resulted in a reduction of ~40% of activity, whereas the expression of all monolignol genes remained stable. For both plants, the modulation of laccases by members of the miR397 family resulted in plants with reduced lignin content.

Known to alter plant architecture and biomass productivity, the miR156 family has a potential role in the regulation of lignin composition and content in lignocellulosic biomass. This has been demonstrated in forest trees, such as poplar, and in forage crops, such as alfalfa (*Medicago sativa*). In poplar, Rubinelli et al. (2013) overexpressed the miRNA Corngrass1 (Cg1) from maize, belonging to the miR156 family, under the control of the cauliflower mosaic virus 35S promoter, and found a significant reduction (up to 30%) of the lignin content, followed by modifications in plant architecture. A decrease of syringyl to guaiacyl ratio was also observed in plants harboring the 35S:Cg1 plants when compared to wild type plants. More recently, Aung et al. (2015) cloned and overexpressed miR156 in alfalfa. The miR156-overexpressing plants showed improved forage quality and biomass production. This enhanced biomass production resulted from delayed flowering and consequent prolongation of the vegetative stage. Furthermore, two miR156-overexpressing lines (A11 and A17) showing a 10% reduction in lignin content were identified, however, no significant effect on lignin composition in terms of the ratio of syringyl-like (S) to guaiacyl-like (G) units was observed. When the transcriptomes of these two overexpressed miR156 lines were compared to that of wild type plants, the expression of flavonoid biosynthesis and lignin catabolism pathway related genes correlated well with the variation of lignin content and composition in these plants (Gao et al., 2016).

Terpenes and Terpenoids

Terpenes and terpenoids are the most numerous and diverse secondary metabolites, including plant hormones, sterols, carotenoids, rubber, among others (Zwenger and Basu, 2008). Despite their vast diversity, all members of these secondary metabolites share the same basic unit, isoprene, a simple hydrocarbon molecule. Terpenes contain only the hydrocarbon structure, whereas in terpenoids the basic hydrocarbon structures are modified, for example, by the addition of oxygen. Due to their vast importance for plant development and for pharmaceutical applications, the molecular mechanisms involved in the regulation of these secondary metabolite pathways are the object of several studies to elucidate the biosynthetic pathways and the regulation of these natural products.

The regulatory function of miR5021 in terpenoid backbone biosynthesis has been reported in different plant species, indicating a key role in the modulation of isopentenyl diphosphate (*IPP*) and dimethyl allyl diphosphate (*DMAPP*) biosynthesis, the common precursors for all downstream end terpenoids. In *Xanthium strumarium*, Fan et al. (2015) predicted that miR5021 mediates the post-transcriptional regulation of a number of genes, among them are *HMGR*, a key regulatory enzyme that controls the amount of isoprenoids (Choi et al., 1992), as well as *IDS* and *IDI*, two genes involved in the terpenoid backbone biosynthesis. More recently, Singh et al. (2016) reported that miR5021 may regulate the expression of gene encoding enzymes involved in 2-C-methyl-D-erythritol 4-phosphate (MEP)/1-deoxy-D-xylulose 5-phosphate pathway (DOXP) in mint (*Mentha spp.*). The same authors reported the potential role of miR414 in the regulation of sesquiterpenoids and triterpenoid biosynthesis by modulating the expression of a homologous gene of TSP21. In *Xanthium strumarium*, a medicinal plant that synthetizes terpenoids with pharmaceutical interest, the downstream terpenoid biosynthesis pathway is responsible for the biosynthesis of mono-, sesqui-, di-, and tri-terpenoids (e.g. xanthatin). Fan et al. (2015) reported that germacrene A oxidase (*GAO*) transcripts were potentially targeted by miR6435. GAO is a key P450 enzyme (downstream isoprenoid pathway) involved in the biosynthesis of xanthanolides (Eljounaidi et al., 2014). Lycopene and beta-carotene are the major carotenoids present in tomato fruit. Here, miR172 plays an important role in regulating the content of a major carotenoid, trans-lycopene. On the other hand, the higher expression of miR395, which targets zeaxanthin epoxidase (*ZEP*) transcript during the ripe stage, suggests a longer retention of lycopene in ripe tomatoes (Koul et al., 2016). In the medicinal herb *Picrorhiza kurroa*, Vashisht et al. (2015) reported the implication of miR4995 in the regulation of picroside I, a secondary metabolite with important pharmaceutical activity.

Alkaloids

Alkaloids are a diverse nitrogen-containing group of secondary metabolites. These compounds are widespread among the plant kingdom; it is estimated that over 20% of plant species produce alkaloids, with roughly 12000 different plant alkaloids discovered. In nature, alkaloids generally act as defensive agents against herbivores and pathogens. Many alkaloids are pharmaceutically active, possessing a wide range of medicinal applications (e.g. antitussives, purgatives, sedatives, and anticancer drugs) (Carocho and Ferreira, 2013). Contrasting with other groups of secondary metabolites that derive from the same biosynthetic pathway, alkaloids are synthesized through different metabolic pathways. Therefore, considerable efforts have been undertaken to describe alkaloid-related biosynthetic pathways. Despite these advances, the post-transcriptional regulation of alkaloid biosynthesis through miRNAs is rather incomplete. Few studies have been performed in alkaloid producing plants, such as *Nicotiana tabacum*, *Catharanthus roseus*, and *Papaver somniferum* (opium poppy). In tobacco, miR164 and its target, the NAC TF gene *NtNAC-R1* which mediates auxin signaling to promote lateral root development (Fu et al., 2013), were shown to modulate nicotine content upon topping. *NtNAC-R1* is upregulated by the increase of IAA content and the decrease of miR164 in tobacco roots after topping results in an increase of the lateral roots as well as nicotine content. Li et al. (2015) identified a novel tobacco miRNA, (nta)-miRX27, which targets the *QUINOLINATE PHOSPHORIBOSYLTRANSFERASE2 (QPT2)* gene. To clarify the functional role of this miRNA, a series of functional studies were performed, including the overexpression and silencing of miRX27, which revealed that this miRNA plays an important regulatory role in nicotine biosynthesis following topping. Most remarkably, the authors demonstrated that in topping-treated tobacco plants, enhanced nicotine biosynthesis is achieved by nta-eTMX27-mediated miRX27 inhibition, thus constituting the first report of a miRNA-eTM regulatory module regulating secondary metabolite biosynthesis in plants. Some studies on the regulatory role of miRNAs have also been performed in the medicinal plant *Catharanthus roseus*, which is an important source of alkaloids, particularly terpenoid indole alkaloids (TIA). Pani and Mahapatra (2013) conducted a computational study to identify miRNAs and their targets involved in TIA biosynthesis. As referred before for the miR5021, this miRNA has a potential role in terpenoid backbone biosynthesis by targeting the gene encoding for the genarylgenaryl-diphosphate synthase enzyme. More recently, Shen et al. (2017) observed that MeJA-induced miRNAs, specifically miR160, act as positive regulators of the TIA biosynthesis pathway by the repression of ARFs, thus attenuating the auxin-mediated repression of the TIA pathway.

MiRNAs Implication in Seed Pre-Germinative Metabolism

An intricated network of metabolic processes, e.g. *de-novo* synthesis of nucleic acids and proteins (Rajjou et al., 2006; Stasolla et al., 2001), ATP production (Czarna et al., 2016), and lipid biosynthesis (Penfield, 2007; Borek et al., 2015), have been revealed during the transition from quiescent dry seeds to the proliferating state of germinating seeds (Galland et al., 2014; Han et al., 2015; Macovei et al., 2017). This is entailed as the so-called 'pre-germinative metabolism', where the seed's response to these changes plays a crucial role in preserving seed vigour, a complex physiological trait controlled by genetic and environmental factors (Rajjou and Debeaujon, 2008; Rajjou et al., 2008; Balestrazzi et al., 2010; Macovei et al., 2011a,b). Seeds face a panoply of stresses during maturation, and post-dispersal storage, as well as upon the start of the germination process, when water up-take is associated with ROS (Reactive Oxygen Species) accumulation and extensive oxidative injury to cellular macromolecules (Kranner et al., 2010; Ventura et al., 2012; Paparella et al., 2015). Involvement of miRNAs in the different aspects of seed biology, from embryo maturation to germination, have recently started to be documented (see reviews by Martin et al., 2010; Das et al., 2015; Liu and El-Kassaby, 2017).

Hence, seed development, as well as the transition phase from dry seed to seedling development, are accompanied by extensive genetic and epigenetic regulatory mechanisms, and miRNAs have a role to play during these complex processes (Figure 22.2). When considering the epigenetic changes, it has been hypothesized that extensive epigenetic regulation marks the transition from dormant-to-germinated embryos. Within this context, it was demonstrated that certain miRNAs of 24-nt-long derived from genomic repeats, transposons, and intergenic regions, can act as heterochromatin siRNAs (Pang et al., 2009; Morin et al., 2008). In plants, the canonical 21-nt miRNAs are predominant, but longer miRNAs (24-nt) can also be found and these are believed to have a conserved role in the modulation of DNA methylation (Wu et al., 2010). High-throughput sequencing of small RNA

libraries from *Pinus contorta*, *Oryza sativa*, and *Gossypium hirsutum* revealed a higher peak of 24-nt miRNAs in dormant embryos while the peak decreased in germinating embryos. The observed differences in the distribution of miRNAs with 21-nt and 24-nt could be attributed to changes in the activity levels of RNA-dependent RNA polymerase2 (RDR2) and/or RDR6 and are regulated by hormones (Zhang et al., 2013).

The involvement of miRNAs in seed germination started to be illustrated during the characterization of mutants defective in functions related to their biogenesis pathway, such as *dcl1*, *hyl1* (*HYPONASTIC LEAVES*), *hen1* (*HUA ENHANCER*), and *ago1* (*ARGONAUTE1*), that revealed severe defects in seed development (Willmann et al., 2011). MiRNA-mediated repression of specific genes can be removed by adding silent mutations within the nucleotide sequence of the target mRNA, complementary to the miRNA itself. Such an approach disclosed plant phenotypes associated with defective cellular and molecular pathways controlled by specific miRNAs and represents a helpful tool for a better understanding of miRNAs roles *in vivo*. Several miRNAs have been detected in imbibed seeds, among which there are conserved miRNA families targeting components of specific phytohormone networks and/or phytohormone cross-talk. Some miRNA families (miR156, miR159, miR164, miR166, miR168, miR169, miR172, miR393, miR394, and miR397) are down-regulated during seed imbibition, whereas there are other miRNA families (miR398, miR408, miR528, and miR529) that appear to be up-regulated (Li et al., 2013).

Representative examples of miRNAs involved in essential phases of seed development and seed germination, like embryogenesis, dormancy, embryo-to-seedling transition, stress, DNA repair, and radicle protrusion, are evidenced in Figure 22.2 and discussed in the subsequent sections.

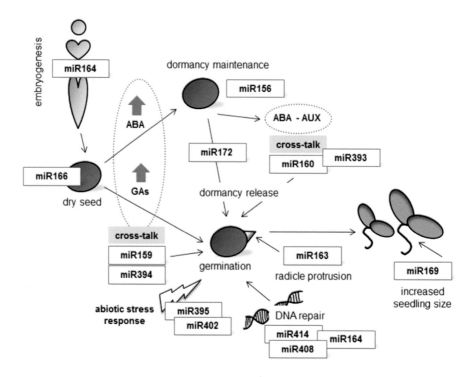

FIGURE 22.2 The main miRNA families involved in different aspects of seed germination, including phytohormone cross-talk. ABA, Abscisic acid; GAs, Gibberellins; AUX, auxin.

Embryogenesis

MiRNAs play essential roles during embryo formation as well as during an embryo's developmental timing (Vashisht and Nodine, 2014). An interesting case is miR164 targeting the *NAC* (NA M/ATAF/CUC) genes, whose expression is dependent on TCP (*TEOSINTE BRANCHED1, CYCLOIDEA, PROLIFERATING CELL NUCLEAR ANTIGEN FACTOR1*) TFs. Liu and El-Kassaby (2017) showed that miR164b accumulates when the transition from the heart to torpedo embryo takes place. Using a genome-wide transcript profiling approach, Nodine and Bartel (2010) found that 50 putative miRNA targets were up-regulated in early *Arabidopsis dcl1* mutant embryos when compared to wild type embryos. These include TFs required for proper cotyledon formation during embryo development, like the *SQUAMOSA PROMOTER-BINDING PROTEIN-LIKE* (*SPL10* and *SPL11*), targeted by miRNA156 which is active as early as the 8-cell stage. Another recent study evidenced that 30 known and 10 novel miRNA families were differentially expressed during pod development in peanut (Gao et al., 2017). Among these, miR156/157 (targeting the *SPL* gene family), miR164 (targeting the *NAC* gene family), miR167 and miR1088 (targeting the Pentatricopeptide repeat protein-*PPRP*), miR172 (targeting the Ethylene-responsive TF *AP2*), and miR396 (targeting the growth regulating factor *GRF*), are actively modulated during early pod development.

Crossover from Seed Dormancy to Seed Germination

Seed dormancy is determined by a proper hormonal balance, with the abscisic acid (ABA) and gibberellic acids (GAs) (Nambara et al., 2010). While ABA is abundant in dormant seeds, it decreases when dormancy is released, and the GAs increase during germination. An ABA-dependent accumulation of miR159 during *Arabidopsis* seed germination was reported by Reyes and Chua (2007). Overexpression of miR159 in *Arabidopsis* triggered the down-regulation of *AtMYB33* and *AtMYB101* genes, coding for TFs that are positive regulators of ABA-mediated pathways, and caused a decline in ABA hypersensitivity (Reyes and Chua, 2007).

A role in the control of ABA- and auxin-mediated signalling/cross-talk has been ascribed to miR393 (Curaba et al., 2014). Treatments with ABA trigger miR393 biosynthesis, impairing auxin perception. In *Arabidopsis*, this is achieved by the down-regulation of gene encoding members of the *TRANSPORT INHIBITOR RESPONSE1* (*TIR1*)/*AUXIN SIGNALLING F-BOX* (*AFB*) clade of auxin receptor (*TAAR*). Similarly, miR393 overexpression in rice leads to reduced drought tolerance due to the reduced expression of *OsTIR1* and *OsAFB2* genes (Curaba et al., 2014). A similar pattern was observed by Hu et al. (2016) in sacred lotus seeds where two *NnTIR1* gene homologues were detected as targets of miR393 in a degradome dataset. These studies set the stage for the hypothesis that miR393-mediated down-regulation of components in auxin signal transduction may represent a crucial step for decreasing ABA sensitivity in mature seeds, thus triggering germination.

A number of miRNA families (miR160, miR164, and miR166) act on auxin-mediated signalling pathways during embryo and seed development, thus playing a crucial role in maintaining or breaking dormancy, and promoting the embryo-to-seedling transition (Martin et al., 2010). In *Arabidopsis* seeds, miR160 targets the TFs *ARF10*, *ARF16*, and *ARF17* that specifically bind the promoter regions of auxin-regulated genes (Liu et al., 2007). When silent mutations were introduced in the *AtARF10* sequence complementary to miR160, overaccumulation of the target TF was triggered in the mutant (*mARF10*) lines. This caused impairment of the *Arabidopsis* developmental process, resulting in anomalous leaf and flower morphologies while no effects were observed in terms of seed germination performance. However, *mARF10* seeds showed an increased sensitivity to ABA, revealing a potential cross-talk between the two phytohormones during seed germination (Liu et al., 2007; Nonogaki et al., 2008). The expression of miR160 was higher in germinated embryos than in dormant embryos, while the target *ARF10* mRNA decreased in germinated compared to dormant embryos, in agreement with the miR160-mediated negative regulation of the *ARF10* gene. The study by Liu et al. (2016) reinforced the interaction of miR160 with *AtARF10* and *AtARR15* (*ARABIDOPSIS RESPONSE REGULATOR 15*) and demonstrated that this miRNA acts as a repressor of callus induction in *Arabidopsis* by controlling the auxin-cytokinin cross-talk. In sacred lotus (*Nelumbo nucifera*), miR160 was found to target *NnARF17* and the other three members of this family and was proven to be essential in promoting germination by decreasing ABA levels (Hu et al., 2016).

An interesting case concerning the involvement of miRNAs in embryo-to-seedling transition consists of the activity of miR156 and miR172. In *Arabidopsis* mature embryos, enhanced accumulation of miR156 concomitant with a reduced level of miR172 affects the developmental process, maintaining seed dormancy (Martin et al., 2010). The *DOG1* (*DELAY OF GERMINATION 1*) gene, a key player in the regulation of seed dormancy in response to temperature, is discussed in cooperation with miR157 and miR172 (Huo et al., 2016). In lettuce (*Lactuca sativa* L.), down-regulation of the *LsDOG1* gene expression facilitates seed germination at a high temperature in association with reduced miR156 levels and increased miR172 accumulation. Indeed, the expression of genes involved in miR156 and miR172 processing was lowered in *Arabidopsis* and lettuce dry seeds of *dog1* mutants (Huo et al., 2016). This finding demonstrates the essential role played by miR156 and miR172 in sensing and integrating environmental changes in the context of seed germination. This work contributed by disclosing the molecular link between dormancy and embryo-to-seedling transition resulting from the interaction between DOG1, miR156, and miR172.

Information concerning the roles of miRNAs in germination is scanty in gymnosperms, which display significant differences in tissue anatomy, cell biology, and molecular features during embryogenesis when compared to angiosperms (Cairney and Pullman, 2007). The miRNAs involved in the transition from dormant-to-germinated embryos in larches (*Larix leptolepis* L.) included miR159a, whose expression was significantly higher in germinated than in dormant embryos and concomitant with the abundance of the *LlMYB33* transcript (Zhang et al., 2013).

Early Seed Germination

The role of miRNAs in the early stage of germination was investigated by Wang et al. (2011) in maize (*Zea mays* L.) using two independent small RNA libraries from seeds collected at 24 h of imbibition. The screening allowed for the retrieval of several miRNA conserved families, among them miR159 targeting the *ZmMYB33* and *ZmMYB101* mRNAs. He et al. (2015) reported on the composition and expression profiles of miRNAs in dry and imbibed rice seeds (embryos collected at 0, 12, and 24 h of imbibition). The miR159 family was one of the most represented in imbibed rice seeds, and particularly abundant was miR159f. The role of miR159 in the cross-talk between ABA and GA during germination involves the cleavage of *OsGAMYB* (the ortholog of *Arabidopsis MYB33*) and *OsGAMYB-like* mRNAs (Tsuji et al., 2006). The GAMYB proteins participate in the GA-mediated signalling in cereal aleurone cells (Millar and Gubler, 2005). *OsGAMYB* and *OsGAMYB-like* genes and miR159 are co-expressed in rice anthers, whereas microarray analysis showed that the *OsGAMYB* gene is crucial for the control of GA-mediated gene expression in rice aleurone cells and promote storage hydrolysis during seed germination (Jossier et al., 2009).

Along with miR159, the large miR166 family (14 members) was present in RNA libraries constructed from dry and imbibed maize seed embryos. The miR166 family targets homeodomain-leucine zipper (HD-ZIP) TFs involved in the control of leaf morphogenesis, vascular development, lateral organ polarity, and meristem formation (Kim et al., 2005). miR166 negatively regulates genes belonging to the HD-ZIP Class III family by cleaving their mRNAs, promoting the asymmetric development of leaves in maize (Juarez et al., 2004). miR166 was significantly down-regulated in imbibed seeds while it was extremely abundant in dry seeds. Evidently, high levels of miR166 are needed in dry seeds to repress the activity of HD-ZIP Class III TFs (Wang et al., 2011; Li et al., 2013).

The highly conserved miR394 family is present in *Arabidopsis* with two members, miR394a and miR394b, both targeting the same *leaf curling responsiveness (LCR)* gene, which encodes an F-box protein (*SKP1-Cullin/CDC53-F*-box). Song et al. (2013) showed that the *LCR* gene knockdown by means of miR394 over-expression as well as or knockout *lcr* mutants caused in ABA hypersensitivity and impaired seed germination.

The miR169 family includes stress-responsive miRNAs up-regulated in the presence of drought, salinity, low temperatures, and aluminium in plants, and modulates the expression of *NFY* genes coding for the nuclear factor Y (Li et al., 2008). Besides this, it is known that in *Arabidopsis*-imbibed seeds ABA-mediated gene expression involves several *NFY* genes (e.g. *LEC1*, encoding the subunit B of NFY), which interact with seed-specific ABRE-binding factors (Warpeha et al., 2007). A similar role was suggested for the nuclear factor Y subunit A (*NFY-A*) genes, such as the *AtNFYA5* gene in *Arabidopsis*, and *NnNFYA2*, *NnNFYA3*, and *NnNFYA10* genes in sacred lotus seeds (Hu et al., 2016). Other roles have been evidenced for miR169 by Kumar et al. (2013), who investigated the influence of gold nanoparticles on miRNA expression in *Arabidopsis* seeds and showed that a reduced expression of miR169 resulted into a significant increase in seedling size.

Stress-Related Germination

The miR395 family includes six components that can act both as positive or negative regulators of seed germination under abiotic stress (Kim et al., 2010a). The target genes are *ATP sulphurylase* (*APS1*), *APS3*, *APS4*, and *sulfate/selenate transporters* (*SULTR*) responsible for sulfate up-take and assimilation. Overexpression of miR395c impairs germination under excess salt or water deficit while overexpression of miR395e positively influences germination (Kim et al., 2010a). The same authors reported that the overexpression of miR402 accelerated *Arabidopsis* seed germination under salt stress, dehydration, and cold stress conditions (Kim et al., 2010b). As for seedling growth, this was promoted only under salt stress. miR402 targets the *ROS1* (*REPRESSOR OF SILENCING 1*)-like gene, encoding a putative DNA glycosylase, also known as *DML3* (*DEMETER-LIKE 3*), involved in DNA demethylation (Ortega-Galisteo et al., 2008). *AtDML3* gene expression was down-regulated in *Arabidopsis* plants over-expressing miR402.

High-throughput sequencing of small RNA libraries from rapeseed (*Brassica napus* L.) seeds confirmed the role of miR159 and relative targets, namely the *B. napus* homologs of *AtMYB101*, in germination control under stress (Jian et al., 2016). This study reported that miR156 was the most abundant in *B. napus* seeds imbibed in the presence/absence of salt, and under drought treatments (Jian et al., 2016). In a different work (Li et al., 2016), over-expression of miR172c in soybean (*Glycine max* L.) resulted in enhanced tolerance to water shortage associated with increased ABA sensitivity. The correlation between the highly conserved miR172 family and abiotic stress tolerance has been documented also for wheat (*Triticum* spp.) and rice (Gupta et al., 2014; Zhou et al., 2010). Different to what was observed by Li et al. (2013) in maize, miR156 was down-regulated in sacred lotus at 72 h of germination (Hu et al., 2016). This study performed a small RNA degradome analysis in *N. nucifera*, highlighting that the highly conserved miR159 participates in the GA-mediated aleurone vacuolation during seed germination (Hu et al., 2016). The *N. nucifera* miR159 was found to target three *NnGAMYB* genes whose expression was highly up-regulated at 36 h during seed germination.

DNA Damage Response During Seed Germination

Genome integrity must be preserved in the embryo to enhance seed vigour and ensure a successful germination and seedling establishment. Once DNA damage has been properly repaired, the embryo cells resume cell cycle progression and undergo DNA replication, but when repair mechanisms are defective, oxidative injury leads to cell death (Waterworth et al., 2010, 2011; Kranner et al., 2010; Balestrazzi et al., 2011; Ventura et al., 2012). DNA repair is a key component of the 'pre-germinative metabolism'. All DNA repair pathways are activated during the early phase of seed imbibition to maintain genome integrity (Macovei et al., 2011a,b; Balestrazzi et al., 2010, 2011, 2012; Chen et al., 2012; Cordoba-Canero et al., 2014). The Transcription Coupled-Nucleotide-Excision Repair (TC-NER), a sub-pathway which specifically recognises and removes lesions from the transcribed strands of transcriptionally active genes, has been investigated during rice seed germination by Macovei et al. (2014) who demonstrated that DNA helicases are required to facilitate proper unwinding of the DNA duplex during transcription. MiRNAs targeting helicases in the context of the DNA damage response and repair in rice seeds/seedlings were also investigated (Macovei and Tuteja, 2013). The expression patterns of rice miR414, miR164e, and miR408, and their targeted helicase genes *OsABP* (ATP-Binding Protein), *OsDBH* (DEAD-Box Helicase), and *OsDSHCT* (DOB1/SK12/helY-like DEAD-box Helicase) were monitored in seeds treated with both low dose rate (LDR) and high dose rate (HDR) γ-rays. The expression of DNA helicase genes was always negatively correlated with the miRNAs expression. Interestingly, in irradiated rice seeds, there was a significantly higher induction of miRNA expression in response to LDR as compared with HDR (Macovei and Tuteja, 2013). This work has provided the first evidence for the possible role played *in planta* by miRNA targeting proteins involved in DNA repair and their contribution to the overall plant response to ionizing radiation.

Radicle Protrusion

Light-dependent induction of miR163, predominantly expressed in radicles, has been observed during seed germination, concomitant with the down-regulation of its target gene *PXMT1* (1,7-paraxanthine methyltransferase *1*), encoding a putative member of the SABATH (SALICYLIC ACID/BENZOIC ACID/THEOBROMINE) group of methyltransferases (Chung et al., 2016). This catalyses the AdoMet-dependent methylation of natural plant metabolites, including phytohormones and signalling molecules (Zhao et al., 2008). Both *miR163* mutant and transgenic *Arabidopsis* lines overexpressing the *AtPXMT1* gene revealed delayed radicle emergence during seed germination, shorter primary roots, and an increased number of lateral roots in young seedlings. It has been hypothesized that the substrate of PXMT1 methyltransferase may be necessary for root elongation or that accumulation of the product of the reaction catalysed by this enzyme may inhibit root elongation (Chung et al., 2016).

In addition to their essential roles in the temporal control of developmental processes and dormancy-release, miR156 and miR172 also play important functions during the post-germinative developmental phases (Martin et al., 2010). MiRNA156 targets SPL gene members, but when a mutated *SPL13* gene is not cleaved by miR156, this is overexpressed at the post-germination phase and results in delayed leaflet development in cotyledon-stage seedlings (Martin et al., 2010). Thus, the miR156-mediated repression of the *SPL13* gene impairs the transition to the vegetative-leaf stage characterized by autotrophic growth. The same response has been observed in rice (Wu et al., 2009).

Conclusion

While plants are sessile in nature, their response to environmental stress is quite dynamic, reflected by a multitude of internal and external signals that trigger appropriate responses which allow the plant to continue to develop under different types of conditions. Within this context, miRNAs play a major role in facilitating plant adaptation responses at a post-transcriptional level. Constant investigations into the miRNA-mediated post-transcriptional regulation can give rise to countless opportunities to advance research and develop crops that are more resilient to climatic changes with increased productivity and high seed vigor. As shown in the examples cited in this chapter, by tackling specific metabolic pathways, plants with increased production of secondary metabolites, useful for pharmaceutical and nutraceutical purposes, can be produced. Moreover, these secondary metabolites are also among the main signals of plant defense. Hence, due to all their actions, the field of miRNA investigation will continue to produce more ideas for research in the near future.

Acknowledgments

S.S.A. acknowledges the financial support of Fundação para a Ciência e a Tecnologia (Lisbon, Portugal) through the research unit "GREEN-IT: Bioresources for Sustainability" (UID/Multi/04551/2013) and postdoctoral grant (SFRH/BPD/108032/2015). C.G. acknowledges the PhD scholarship in the frame of project SONATA BIS "PurpleWalls" (ref. UMO-2015/18/E/NZ2/00694) funded by NCN (Poland). J.A.P.P. acknowledges the European Union's Seventh Framework Programme for research, technological development and demonstration (EU FP7 Agreement no. 621321) and the Polish financial sources for education (2015–2019) allocated to project np (W26/7.PR/2015). A.B. acknowledges Regione Lombardia D.G. Istruzione, Formazione e Lavoro, Struttura Asse V—Interregionalità e Transnazionalità POR FSE 2007–2013, Project ID 46547514 'Advanced Priming Technologies for the Lombardy Agro-Seed Industry-PRIMTECH' (Action 2).

REFERENCES

Aung B, Gruber MY, Hannoufa A. The MicroRNA156 system: A tool in plant biotechnology. *Biocatal. Agric. Biotechnol.* 2015;4:432–442.

Bakhshi BE, Mohseni Fard E, Nikpay N, Ebrahimi MA, Bihamta MR, Mardi M, Salekdeh GH. MicroRNA signatures of drought signaling in rice root. *PLoS ONE* 2016;11:e0156814.

Balestrazzi A, Confalonieri M, Donà M, Carbonera D. Genotoxic stress, DNA repair, and crop productivity. In: Tuteja N, Gill SS (eds.). *Crop Improvement Under Adverse Conditions.* Berlin: Springer-Verlag, 2012; pp. 153–169.

Balestrazzi A, Confalonieri M, Macovei A, Carbonera D. Seed imbibition in *Medicago truncatula* Gaertn.: Expression profiles of DNA repair genes in relation to PEG-mediated stress. *J. Plant Physiol.* 2010;168:706–713.

Balestrazzi A, Confalonieri M, Macovei A, Donà M, Carbonera D. Genotoxic stress and DNA repair in plants: Emerging functions and tools for improving crop productivity. *Plant Cell Rep.* 2011;30:287–295.

Barlow KM, Christy BP, O'Leary GJ, Riffkin PA, Nuttall JG. Simulating the impact of extreme heat and frost events on wheat crop production: A review. Field Crops Res. 2015;171:109–119.

Bartel DP. MicroRNAs: Genomics, biogenesis, mechanism, and function. *Cell* 2004;116:281–297.

Bhuiyan NH, Selvaraj G, Wei Y, King J. Role of lignification in plant defense. *Plant Signal. Behav.* 2009;4:158–159.

Boerjan W, Ralph J, Baucher M. Lignin biosynthesis. *Annu. Rev. Plant Biol.* 2003;54:519–546.

Borek S, Ratajczak W, Ratajczak L. Regulation of storage lipid metabolism in developing and germinating lupin (*Lupinus* spp.) seeds. *Acta Physiol. Plant.* 2015;37:119.

Borges F, Martienssen RA. The expanding world of small RNAs in plants. *Nat. Rev. Mol. Cell Biol.* 2015;16:727–741.

Boudet AM, Kajita S, Grima-Pettenati J, Goffner D. Lignins and lignocellulosics: A better control of synthesis for new and improved uses. *Trends Plant Sci.* 2003;8:576–581.

Boyer JS. Plant productivity and environment. *Science* 1982; 218:443–448.

Cairney J, Pullman GS. The cellular and molecular biology of conifer embryogenesis. *New Phytol.* 2007;176:511–536.

Cakir OB, Candar-Cakir B, Zhang B. Small RNA and degradome sequencing reveals important microRNA function in *Astragalus chrysochlorus* response to selenium stimuli. *Plant Biotechnol. J.* 2016;14:543–556.

Capitão C, Paiva JAP, Santos DM, Fevereiro P. In *Medicago truncatula*, water deficit modulates the transcript accumulation of components of small RNA pathways. *BMC Plant Biol.* 2011;11:79.

Carocho M, Ferreira IC. The role of phenolic compounds in the fight against cancer – A review. *Anticancer Agents Med. Chem.* 2013;13:1236–1258.

Chen H, Chu P, Zhou Y, Li Y, Liu J, Ding Y, Tsang EW, Jiang L, Wu K, Huang S. Overexpression of *AtOGG1*, a DNA glycosylase/ AP lyase, enhances seed longevity and abiotic stress tolerance in *Arabidopsis. J. Exp. Bot.* 2012;63:4107–4112.

Chen Q, Li M, Zhang Z, Tie W, Chen X, Jin L, Zhai N, Zheng Q, Wang R, Xu G, Zhang H, Liu P, Zhou H. Integrated mRNA and microRNA analysis identifies genes and small miRNA molecules associated with transcriptional and post-transcriptional-level responses to both drought stress and re-watering treatment in tobacco. *BMC Genomics* 2017;18:62.

Chibuike GU, Obiora SC. Heavy metal polluted soils: Effect on plants and bioremediation methods. *Appl. Environ. Soil Sci.* 2014;2014:1–12.

Choi D, Ward BL, Bostock RM. Differential induction and suppression of potato 3-hydroxy-3-methylglutaryl coenzyme A reductase genes in response to *Phytophthora infestans* and to its elicitor arachidonic acid. *Plant Cell* 1992;4:1333–1344.

Chung PJ, Jung H, Jeong DH, Ha SH, Do Choi Y, Kim JK. Transcriptome profiling of drought responsive noncoding RNAs and their target genes in rice. BMC Genomics 2016; 17:563.

Chung PJ, Park BS, Wang H, Liu J. Light-inducible MiR163 targets *PXMT1* transcripts to promote seed germination and primary root elongation in *Arabidopsis. Plant Physiol.* 2016;170:1772–1782.

Cordoba-Canero D, Roldan-Arjona T, Ariza RR. *Arabidopsis* ZDP DNA 30-phosphatase and ARP endonuclease function in 8-oxoG repair initiated by FPG and OGG1 DNA glycosylases. *Plant J.* 2014;79:824–834.

Curaba J, Singh MB, Bhalla PL. miRNAs in the crosstalk between phytohormone signalling pathways. *J. Exp. Bot.* 2014;65:1425–1438.

Czarna M, Kolodziejczak M, Janska H. Mitochondrial proteome studies in seeds during germination. *Proteomes* 2016;4:19.

Das SS, Karmakar P, Nandi AK, Sanan-Mishra N. Small RNA mediated regulation of seed germination. *Front. Plant Sci.* 2015;6:828.

Djami-Tchatchou AT, Sanan-Mishra N, Ntushelo K, Dubery IA. Functional roles of microRNAs in agronomically important plants-potential as targets for crop improvement and protection. *Front. Plant Sci.* 2017;8:378.

Dolata J, Bajczyk M, Bielewicz D, Niedojadlo K, Niedojadlo J, Pietrykowska H, Walczak W, Szweykowska-Kulinska Z, Jarmolowski A. Salt stress reveals a new role for ARGONAUTE1 in miRNA biogenesis at the transcriptional and posttranscriptional levels. *Plant Physiol.* 2016;172:297–312.

Eljounaidi K, Cankar K, Comino C, Moglia A, Hehn A, Bourgaud F, Bouwmeester H, Menin B, Lanteri S, Beekwilder J. Cytochrome P450s from *Cynara cardunculus* L. CYP71AV9 and CYP71BL5, catalyze distinct hydroxylations in the sesquiterpene lactone biosynthetic pathway. *Plant Sci.* 2014;223:59–68.

Fan R, Li Y, Li C, Zhang Y. Differential microRNA analysis of glandular trichomes and young leaves in *Xanthium strumarium* L. reveals their putative roles in regulating terpenoid biosynthesis. *PLoS ONE* 2015;10:e0139002.

Felekkis K, Touvana E, Stefanou C, Deltas C. microRNAs: A newly described class of encoded molecules that play a role in health and disease. *Hippokratia* 2010;14:236–240.

Ferdous J, Hussain SS, Shi BJ. Role of microRNAs in plant drought tolerance. *Plant Biotechnol. J.* 2015;13:293–305.

Ferreira TH, Gentile A, Vilela RD, Costa GGL, Dias LI, Endres L, Menossi M. microRNAs associated with drought response in the bioenergy crop sugarcane (*Saccharum* Spp.). *PLoS ONE* 2012;7:e46703.

Frank G, Pressman E, Ophir R, Althan L, Shaked R, Freedman M, Shen S, Firon N. Transcriptional profiling of maturing tomato (*Solanum lycopersicum* L.) microspores reveals the involvement of heat shock proteins, ROS scavengers, hormones, and sugars in the heat stress response. *J. Exp. Bot.* 2009;60:3891–3908.

Fu Y, Guo H, Cheng Z, Wang R, Li G, Huo G, Liu W. NtNAC-R1, a novel NAC transcription factor gene in tobacco roots, responds to mechanical damage of shoot meristem. *Plant Physiol. Biochem.* 2013;69:74–81.

Galland M, Huguet R, Arc E, Cueff G, Job D, Rajjou L. Dynamic proteomics emphasizes the importance of selective mRNA translation and protein turnover during *Arabidopsis* seed germination. *Mol. Cell. Proteomics* 2014;13:252–268.

Gao C, Wang P, Zhao S, Zhao C, Xia H, Hou L, Ju Z, Zhang Y, Li C, Wang X. Small RNA profiling and degradome analysis reveal regulation of microRNA in peanut embryogenesis and early pod development. *BMC Genomics* 2017;18:220.

Gao F, Wang N, Li HY, Liu JS, Fu CX, Xiao ZH, Wei CX, Lu XD, Feng JC, Zhou YJ. Identification of drought-responsive microRNAs and their targets in *Ammopiptanthus mongolicus* by using high-throughput sequencing. *Sci. Rep.* 2016;6:34601.

Gao R, Austin RS, Amyot L, Hannoufa A. Comparative transcriptome investigation of global gene expression changes caused by miR156 overexpression in *Medicago sativa. BMC Genomics* 2016;17:658.

Gentile A, Dias LI, Mattos RS, Ferreira TH, Menossi M. MicroRNAs and drought responses in sugarcane. *Front Plant Sci.* 2015;23(6):58. doi: 10.3389/fpls.2015.00058.

Gielen H, Remans T, Vangronsveld J, Cuypers A. MicroRNAs in metal stress: Specific roles or secondary responses? *Int. J. Mol. Sci.* 2012;13:15826–15847.

Guo HS, Xie Q, Fei JF, Chua NH. MicroRNA directs mRNA cleavage of the transcription factor NAC1 to downregulate auxin signals for Arabidopsis lateral root development. *Plant Cell* 2005;17:1376–1386.

Gupta OP, Karkute SG, Banerjee S, Meena NL, Dahuja A. Contemporary understanding of miRNA-based regulation of secondary metabolites biosynthesis in plants. *Front. Plant Sci.* 2017;8:374.

Gupta OP, Meena NL, Sharma I, Sharma P. Differential regulation of microRNAs in response to osmotic, salt and cold stresses in wheat. *Mol. Biol. Rep.* 2014;41:4623–4629.

Han C, Yang P. Studies on the molecular mechanisms of seed germination. *Proteomics* 2015;15:1671–1679.

Hatfield JL, Prueger JH. Temperature extremes: Effect on plant growth and development. *Weather Clim. Extremes* 2015;10:4–10.

He D, Wang Q, Wang K, Yang P. Genome-wide dissection of the microRNA expression profile in rice embryo during early stages of seed germination. *PloS ONE* 2015;10:e0145424.

Hu J, Jin J, Qian Q, Huang K, Ding Y. Small RNA and degradome profiling reveals miRNA regulation in the seed germination of ancient eudicot *Nelumbo nucifera*. *BMC Genomics* 2016;17:684.

Huang JH, Qi YP, Wen SX, Guo P, Chen XM, Chen LS. Illumina microRNA profiles reveal the involvement of miR397a in citrus adaptation to long-term boron toxicity via modulating secondary cell-wall biosynthesis. *Sci. Rep.* 2016;6:22900.

Huo H, Wei S, Bradford KJ. DELAY OF GERMINATION1 (DOG1) regulates both seed dormancy and flowering time through microRNA pathways. *Proc. Natl. Acad. Sci. USA* 2016;E2199–E2206.

Jia XY, Shen J, Liu H, Li F, Ding N, Gao CY, Pattanaik S, Patra B, Li RZ, Yuan L. Small tandem target mimic-mediated blockage of microRNA858 induces anthocyanin accumulation in tomato. *Planta* 2015;242:283–293.

Jian H, Wang J, Wang T, Wei L, Li J, Liu L. Identification of rapeseed microRNAs involved in early stage seed germination under salt and drought stresses. *Front. Plant Sci.* 2016;7:658.

Jones-Rhoades MW, Bartel DP. Computational identification of plant microRNAs and their targets, including a stress-induced miRNA. *Mol. Cell* 2004;14:787–799.

Jones-Rhoades MW, Bartel DP, Bartel B. MicroRNAs and their regulatory roles in plants. *Annu. Rev. Plant Biol.* 2006;57:19–53.

Jossier M, Bouly JP, Meimoun P, Arjmand A, Lessard P, Hawley S, Grahame HD, Thomas M. SnRK1 (SNF1-related kinase1) has a central role in sugar and ABA signalling in *Arabidopsis thaliana*. *Plant J.* 2009;59:316–328.

Juarez MT, Kui JS, Thomas J, Heller BA, Timmermans MC. microRNA-mediated repression of rolled leaf1 specifies maize leaf polarity. *Nature* 2004;428:84–88.

Katiyar A, Smita S, Muthusamy SK, Chinnusamy V, Pandey DM, Bansal KC. Identification of novel drought-responsive microRNAs and trans-acting siRNAs from *Sorghum bicolor* (L.) Moench by high-throughput sequencing analysis. *Front. Plant Sci.* 2015;6:506.

Kawaguchi R, Girke T, Bray EA, Bailey-Serres J. Differential mRNA translation contributes to gene regulation under non-stress and dehydration stress conditions in *Arabidopsis thaliana*. *Plant J.* 2004;38:823–839.

Kayıhan DS, Kayıhan C, Çiftçi YÖ. Excess boron responsive regulations of antioxidative mechanism at physio-biochemical and molecular levels in *Arabidopsis thaliana*. *Plant Physiol. Biochem.* 2016;109:337–345.

Khraiwesh B, Zhu JK, Zhu J. Role of miRNAs and siRNAs in biotic and abiotic stress responses of plants. *BBA* 2012;1819:137–148.

Kim J, Jung JH, Reyes JL, Kim YS, Chung KS, Kim JA, Lee M, Lee Y, Narry Kim V, Chua NH, Park CM. microRNA-directed cleavage of ATHB15 mRNA regulates vascular development in Arabidopsis inflorescence stems. *Plant J.* 2005;42:84–94.

Kim JY, Lee HJ, Jung HJ, Maruyama K, Suzuki N, Kang H. Overexpression of microRNA395c or 395e affects differently the seed germination of *Arabidopsis thaliana* under stress conditions. *Planta* 2010a;232:1447–1454.

Kim JY, Kwak KJ, Jung HJ, Lee HJ, Kang H. MicroRNA402 affects seed germination of *Arabidopsis thaliana* under stress conditions via targeting DEMETER-LIKE protein 3 mRNA. *Plant Cell Physiol.* 2010b;51:1079–1083.

Koul A, Yogindran S, Sharma D, Kaul S, Rajam MV, Dhar MK. Carotenoid profiling, in silico analysis and transcript profiling of miRNAs targeting carotenoid biosynthetic pathway genes in different developmental tissues of tomato. *Plant Physiol. Biochem.* 2016;108:412–421.

Kozomara A, Griffiths-Jones S. miRBase: Integrating microRNA annotation and deep-sequencing data. *Nucl. Acids Res.* 2011;39:D152–D157.

Kranner I, Beckett RP, Minibayeva FV, Seal CE. What is stress? Concepts, definitions and applications in seed science. *New Phytol.* 2010;188:655–673.

Kumar V, Guleria P, Kumar V, Yadav SK. Gold nanoparticle exposure induces growth and yield enhancement in *Arabidopsis thaliana*. *Sci. Total Environ.* 2013;461–462:462–468.

Li D, Wang L, Liu X, Cui D, Chen T, Zhang H, Jiang C, Xu C, Li P, Li S, Zhao L, Chen H. Deep sequencing of maize small RNAs reveals a diverse set of miroRNA in dry and imbibed seeds. *PLoS ONE* 2013;8:e55107.

Li FF, Wang WD, Zhao N, Xiao BG, Cao PJ, Wu XF, Ye CY, Shen EH, Qiu J, Zhu QH, Xie JH, Zhou XP, Fan LJ. Regulation of nicotine biosynthesis by an endogenous target mimicry of microRNA in tobacco. *Plant Physiol.* 2015;169:1062–1071.

Li S, Castillo-González C, Yu B, Zhang X. The functions of plant small RNAs in development and in stress responses. *Plant J.* 2016;doi:10.1111/tpj.13444.

Li WX, Oono Y, Zhu J, He XJ, Wu JM, Iida K, Lu XY, Cui X, Jin H, Zhu JK. The *Arabidopsis* NFYA5 transcription factor is regulated transcriptionally and posttranscriptionally to promote drought resistance. *Plant Cell* 2008;20:2238–2251.

Lin HJ, Zhang ZM, Shen YO, Gao SB, Pan GT. Review of plant miRNAs in environmental stressed conditions. *Res. J. Agric. Biol. Sci.* 2009;5:803–814.

Liu PP, Montgomery TA, Fahlgren N, Kasschau KD, Nonogaki H, Carrington JC. Repression of AUXIN RESPONSE FACTOR10 by microRNA160 is critical for seed germination and post-germination stages. *Plant J.* 2007;52:133–146.

Liu W, Xu L, Wang Y, Shen H, Zhu X, Zhang K, Chen Y, Yu R, Limera C, Liu L. Transcriptome-wide analysis of chromium-stress responsive microRNAs to explore miRNA-mediated regulatory networks in radish (*Raphanus sativus* L.). *Sci. Rep.* 2015;5:14024.

Liu Y, El-Kassaby YA. Regulatory crosstalk between microRNAs and hormone signaling cascades controls the variation on seed dormancy phenotype at *Arabidopsis thaliana* seed set. *Plant Cell Rep.* 2017;36(5):705–717. doi:10.1007/s00299-017-2111-6.

Liu Z, Li J, Wang L, Li Q, Lu Q, Yu Y, Li S, Bai MY, Hu Y, Xiang F. Repression of callus initiation by the miRNA-directed interaction of auxin-cytokinin in *Arabidopsis thaliana. Plant J.* 2016;87:391–402.

Lu S, Li Q, Wei H, Chang MJ, Tunlaya-Anukit S, Kim H, Liu J, Song J, Sun YH, Yuan L, Yeh TF, Peszlen I, Ralph J, Sederoff RR, Chiang VL. Ptr-miR397a is a negative regulator of laccase genes affecting lignin content in *Populus trichocarpa. Proc. Natl. Acad. Sci. USA* 2013;110:10848–10853.

Macovei A, Balestrazzi A, Confalonieri M, Faè M, Carbonera D. New insights on the barrel medic *MtOGG1* and *MtFPG* functions in relation to oxidative stress response *in planta* and during seed imbibition. *Plant Physiol. Biochem.* 2011a;49:1040–1050.

Macovei A, Balestrazzi A, Confalonieri M, Buttafava A, Carbonera D. The TFIIS and TFIIS-like genes from *Medicago truncatula* are involved in oxidative stress response. *Gene* 2011b;470:20–30.

Macovei A, Garg B, Raikwar S, Balestrazzi A, Carbonera D, Buttafava A, Bremont JF, Gill SS, Tuteja N. Synergistic exposure of rice seeds to different doses of γ-ray and salinity stress resulted in increased antioxidant enzyme activities and gene-specific modulation of TC-NER pathway. *BioMed Res. Int.* 2014;2014:676934.

Macovei A, Pagano A, Leonetti P, Carbonera D, Balestrazzi A, Araújo SS. Systems biology and genome-wide approaches to unveil the molecular players involved in the pre-germinative metabolism: Implications on seed technology traits. *Plant Cell Rep.* 2017;36(5):669–688. doi:10.1007/s00299-016-2060-5.

Macovei A, Tuteja N. Different expression of miRNAs targeting helicases in rice in response to low and high dose rate γ-ray treatments. *Plant Signal. Behav.* 2013;8:e25128.

Macovei A, Tuteja N. microRNAs targeting DEAD-box helicases are involved in salinity stress response in rice (*Oryza sativa* L.). *BMC Plant Biol.* 2012;12:183.

Maeda S, Sakazono S, Masako-Suzuki H, Taguchi M, Yamamura K, Nagano K, Endo T, Saeki K, Osaka M, Nabemoto M, Ito K, Kudo T, Kobayashi M, Kawagishi M, Fujita K, Nanjo H, Shindo T, Yano K, Suzuki G, Suwabe K, Watanabe M. Comparative analysis of microRNA profiles of rice anthers between cool-sensitive and cool-tolerant cultivars under cool-temperature stress. *Genes Genet. Syst.* 2016;91:97–109.

Mangrauthia SK, Agarwal S, Sailaja B, Madhav MS, Voleti SR. MicroRNAs and their role in salt stress response in plants. In: *Salt Stress in Plants.* New York, NY: Springer New York, 2013; pp. 15–46.

Martin RC, Liu PP, Goloviznina NA, Nonogaki H. microRNA, seeds, and Darwin?: Diverse function of miRNA in seed biology and plant responses to stress. *J. Exp. Bot.* 2010;61:2229–2234.

McCarty KM, Hanh HT, Kim KW. Arsenic geochemistry and human health in South East Asia. *Rev. Environ. Health* 2011;26:71–78.

Millar AA, Gubler F. The Arabidopsis *GAMYB-like* genes, *MYB33* and *MYB65*, are microRNA-regulated genes that redundantly facilitate anther development. *Plant Cell* 2005;17:705–721.

Min Yang Z, Chen J. A potential role of microRNAs in plant response to metal toxicity. *Metallomics* 2013;5:1184.

Mittal D, Sharma N, Sharma V, Sopory SK, Sanan-Mishra N. Role of microRNAs in rice plant under salt stress. *Ann. Appl. Biol.* 2016;168:2–18.

Morin RD, Aksay G, Dolgosheina E, Ebhardt HA, Magrini V, Mardis ER, Sahinalp SC, Unrau PJ. Comparative analysis of the small RNA transcriptomes of *Pinus contorta* and *Oryza sativa. Genome Res.* 2008;18:571–584.

Nakashima K, Yamaguchi-Shinozaki K, Shinozaki K. The transcriptional regulatory network in the drought response and its crosstalk in abiotic stress responses including drought, cold, and heat. *Front. Plant Sci.* 2014;5:170.

Nambara E, Okamoto M, Tatematsu K, Yano R, Seo M, Kamiya Y. Abscisic acid and the control of seed dormancy and germination. *Seed Sci. Res.* 2010;20:55.

Nguyen HN, Kim JH, Hyun WY, Nguyen NT, Hong SW, Lee H. TTG1-mediated flavonols biosynthesis alleviates root growth inhibition in response to ABA. *Plant Cell Rep.* 2013;32:503–514.

Niu J, Wang J, Hu H, Chen Y, An J, Cai J, Sun R, Sheng Z, Liu X, Lin S. Cross-talk between freezing response and signaling for regulatory transcriptions of MIR475b and its targets by miR475b promoter in *Populus suaveolens. Sci Rep.* 2016;6:20648.

Nodine MD, Bartel DP. MicroRNAs prevent precocious gene expression and enable pattern formation during plant embryogenesis. *Genes Dev.* 2010;24:2678–2692.

Noman A, Aqeel M. miRNA-based heavy metal homeostasis and plant growth. *Environ. Sci. Pollut. Res.* 2017;24(11):10068–10082. doi:10.1007/s11356-017-8593-5.

Nonogaki H. Repression of transcription factors by microRNA during seed germination and postgermination. *Plant Signal. Behav.* 2008;3:65–67.

Ooka H, Satoh K, Doi K, Nagata T, Otomo Y, Murakami K, Matsubara K, Osato N, Kawai J, Carninci P, Havashizaki Y, Kojima K, Takahara Y, Kikuchi S. Comprehensive analysis of NAC family genes in *Oryza sativa* and *Arabidopsis thaliana. DNA Res.* 2003;10:239–247.

Ortega-Galisteo AP, Morales-Ruiz T, Ariza RR, Roldán-Arjona T. *Arabidopsis* DEMETER-LIKE proteins DML2 and DML3 are required for appropriate distribution of DNA methylation marks. *Plant Mol. Biol.* 2008;67:671–681.

Panda SK, Sunkar R. Nutrient- and other stress-responsive microRNAs in plants: Role for thiol-based redox signaling. *Plant Signal. Behav.* 2015;10:e1010916.

Pang M, Woodward AW, Agarwal V, Guan X, Ha M, Ramachandran V, Chen X, Triplett BA, Stelly DM, Chen ZJ. Genome wide analysis reveals rapid and dynamic changes in miRNA and siRNA sequence and expression during ovule and fiber development in allotetraploid cotton (*Gossypium hirsutum* L.). *Genome Biol.* 2009;10:R122.

Pani A, Mahapatra RK. Computational identification of microRNAs and their targets in *Catharanthus roseus* expressed sequence tags. *Genomics Data* 2013;1:2–6.

Paparella S, Araujo SS, Rossi G, Wijayasinghe M, Carbonera D, Balestrazzi A. Seed priming: State of the art and new perspectives. *Plant Cell Rep.* 2015;34:1281–1293.

Park HJ, Kim WY, Yun DJ. A new insight of salt stress signaling in plant. *Mol. Cells* 2016;39:447–459.

Parreira JR, Branco D, Almeida AM, Czubacka A, Agacka-Mołdoch M, Paiva JAP, Tavares-Cadete F, Araújo SS. Systems biology approaches to improve drought stress tolerance in plants: State of the art and future challenges. In: Hossain MA, Wani SH, Bhattacharjee S, Burritt DJ, Tran LSP (eds.). *Drought*

Stress Tolerance in Plants Vol 2: Molecular and Genetic Perspectives. Cham, Switzerland: Springer International Publishing, 2016; pp. 433–471.

Penfield S, Pinfield-Wells H, Graham IA. Lipid metabolism in seed dormancy. In: Bradford KJ, Nonogaki H (eds.). *Annual Plant Reviews Volume 27: Seed Development, Dormancy and Germination.* Oxford, UK: Blackwell Publishing Ltd, 2007.

Ragupathy R, Ravichandran S, Mahdi MSR, Huang D, Reimer E, Domaratzki M, Cloutier S. Deep sequencing of wheat sRNA transcriptome reveals distinct temporal expression pattern of miRNAs in response to heat, light and UV. *Sci. Rep.* 2016;6:39373.

Rajjou L, Balghazi M, Huguet R, Robin C, Moreau A, Job C, Job D. Proteomic investigation of the effect of salicylic acid on *Arabidopsis* seed germination and establishment of early defense mechanisms. *Plant Physiol.* 2006;141:910–923.

Rajjou L, Belghazi M, Huguet R, Ogè L, Bourdais G, Bove J, Collet B, Godin B, Granier F, Boutin JP, Job D, Jullien M, Grappin P. Protein repair L-isoaspartyl methyltransferase 1 is involved in both seed longevity and germination vigor in *Arabidopsis.* *Plant Cell* 2008;20:3022–3037.

Rajjou L, Debeaujon I. Seed longevity: Survival and maintenance of high germination ability of dry seeds. *CR Biol.* 2008;331:796–805.

Ramakrishna A, Ravishankar GA. Influence of abiotic stress signals on secondary metabolites in plants. *Plant Signal. Behav.* 2011;6:1720–1731.

Ramakrishna A, Ravishankar GA. Role of plant metabolites in abiotic stress tolerance under changing climatic conditions with special reference to secondary compounds. In: Tuteja N, Gill SS (eds.). *Climate Change and Abiotic Stress Tolerance.* Weinheim: Wiley-VCH, 2013. ISBN 978-3-527-33491-9.

Ranocha P, McDougall G, Hawkins S, Sterjiades R, Borderies G, Stewart D, Cabanes-Macheteau M, Boudet AM, Goffner D. Biochemical characterization, molecular cloning and expression of laccases – a divergent gene family – in poplar. *Eur. J. Biochem.* 1999;259:485–495.

Reyes LJ, Chua NH. ABA induction of miR159 controls transcript levels of two MYB factors during *Arabidopsis* seed germination. *Plant J.* 2007;49:592–606.

Rubinelli PM, Chuck G, Li X, Meilan R. Constitutive expression of the Corngrass1 microRNA in poplar affects plant architecture and stem lignin content and composition. *Biomass Bioenerg.* 2013;54:312–321.

Samad AFA, Sajad M, Nazaruddin N, Fauzi IA, Murad AMA, Zainal Z, Ismail I. MicroRNA and transcription factors: Key players in plants regulatory network. *Front. Plant Sci.* 2017;8:565. doi:10.3389/fpls.2017.00565.

Schilmiller AL, Stout J, Weng JK, Humphreys J, Ruegger MO, Chapple C. Mutations in the cinnamate 4-hydroxylase gene impact metabolism, growth and development in *Arabidopsis.* *Plant J.* 2009;60:771–782.

Shanker AK, Djanaguiraman M, Venkateswarlu B. Chromium interactions in plants: Current status and future strategies. *Metallomics* 2009;1:375.

Sharmin SA, Alam I, Kim KH, Kim YG, Kim PJ, Bahk JD, Lee BH. Chromium-induced physiological and proteomic alterations in roots of *Miscanthus sinensis.* *Plant Sci.* 2012;187:113–126.

Sharma D, Tiwari M, Lakhwani D, Tripathi RD, Trivedi PK. Differential expression of microRNAs by arsenate and arsenite stress in natural accessions of rice. *Metallomics* 2015;7:174–187.

Sharma D, Tiwari M, Pandey A, Bhatia C, Sharma A, Trivedi PK. MicroRNA858 is a potential regulator of phenylpropanoid pathway and plant development. *Plant Physiol.* 2016;171:944–959.

Shen EM, Singh SK, Ghosh JS, Patra B, Paul P, Yuan L, Pattanaik S. The miRNAome of *Catharanthus roseus*: Identification, expression analysis, and potential roles of microRNAs in regulation of terpenoid indole alkaloid biosynthesis. *Sci. Rep.* 2017;7:43027.

Shi MZ, Xie DY. Biosynthesis and metabolic engineering of anthocyanins in *Arabidopsis thaliana.* *Recent Pat. Biotechnol.* 2014;8:47–60.

Shriram V, Kumar V, Devarumath RM, Khare TS, Wani SH. MicroRNAs as potential targets for abiotic stress tolerance in plants. *Front. Plant Sci.* 2016;7:817.

Shrivastava P, Kumar R. Soil salinity: A serious environmental issue and plant growth promoting bacteria as one of the tools for its alleviation. *Saudi J. Biol. Sci.* 2015;22:123–131.

Singh N, Srivastava S, Shasany AK, Sharma A. Identification of miRNAs and their targets involved in the secondary metabolic pathways of *Mentha* spp. *Comput. Biol. Chem.* 2016;64:154–162.

Song G, Zhang R, Zhang S, Li Y, Gao J, Han X, Chen M, Wang J, Li W, Li G. Response of microRNAs to cold treatment in the young spikes of common wheat. *BMC Genomics* 2017;18:212.

Song JB, Gao S, Sun D, Li H, Shu XX, Yang ZM. miR394 and LCR are involved in *Arabidopsis* salt and drought stress responses in an abscisic acid-dependent manner. *BMC Plant Biol.* 2013;13:210.

Song JB, Gao S, Wang Y, Li BW, Zhang YL, Yang ZM. miR394 and its target gene LCR are involved in cold stress response in *Arabidopsis.* *Plant Gene* 2016;5:56–64.

Song JB, Huang SQ, Dalmay T, Yang ZM. Regulation of leaf morphology by microRNA394 and its target LEAF CURLING RESPONSIVENESS. *Plant Cell Physiol.* 2012;53:1283–1294.

Stasolla C, Loukanina N, Ashihara H, Yeung EC, Trevor AT. Changes in pyrimidine nucleotide biosynthesis during germination of white spruce (*Picea glauca*) somatic embryos. *In Vitro Cell Dev. Biol. Plant* 2001;37:285–292.

Sun Y, Qiu Y, Duan M, Wang J, Zhang X, Wang H, Song J, Li X. Identification of anthocyanin biosynthesis related microRNAs in a distinctive Chinese radish (*Raphanus sativus* L.) by high-throughput sequencing. *MGG* 2017;292:215–229.

Sunkar R, Li YF, Jagadeeswaran G. Functions of microRNAs in plant stress responses. *Trends Plant Sci.* 2012;17:196–203.

Sunkar R, Zhu JK. Novel and stress-regulated microRNAs and other small RNAs from *Arabidopsis.* *Plant Cell* 2004;16:2001–2019.

Tanaka Y, Sasaki N, Ohmiya A. Biosynthesis of plant pigments: Anthocyanins, betalains and carotenoids. *Plant J.* 2008;54:733–749.

Trindade I, Capitão C, Dalmay T, Fevereiro MP, Dos Santos DM. miR398 and miR408 are up-regulated in response to water deficit in *Medicago truncatula.* *Planta* 2010;231:705–716.

Tsuji H, Aya K, Ueguchi-Tanaka M, Shimada Y, Nakazono M, Watanabe R, Nishizawa NK, Gomi K, Shimada A, Kitano H, Ashikari M, Matsuoka M. GAMYB controls different sets of genes and is differentially regulated by microRNA in aleurone cells and anthers. *Plant J.* 2006;47:427–444.

Vanholme R, Ralph J, Akiyama T, Lu F, Pazo JR, Kim H, Christensen JH, Van Reusel B, Storme V, De Rycke R, Rohde A, Morreel K, Boerjan W. Engineering traditional monolignols out of lignin by concomitant up-regulation of F5H1 and down-regulation of COMT in *Arabidopsis*. *Plant J.* 2010;64:885–897.

Vashisht I, Mishra P, Pal T, Chanumolu S, Singh TR, Chauhan RS. Mining NGS transcriptomes for miRNAs and dissecting their role in regulating growth, development, and secondary metabolites production in different organs of a medicinal herb, *Picrorhiza kurroa*. *Planta* 2015;241:255–1268.

Vashisht D, Nodine MD. MicroRNA functions in plant embryos. *Biochem. Soc. Trans.* 2014;42(2):352–357.

Ventura L, Donà M, Macovei A, Carbonera D, Buttafava A, Mondoni A, Rossi G, Balestrazzi A. Understanding the molecular pathways associated with seed vigor. *Plant Physiol. Biochem.* 2012;60:196–206.

Voinnet O. Origin, biogenesis, and activity of plant microRNAs. *Cell* 2009;136:669–687.

Wagner GJ, Kroumova AB. The use of RNAi to elucidate and manipulate secondary metabolite synthesis in plants. In: Ying SY (ed.). *Current Perspectives in microRNAs (miRNA)*. Dordrecht, Netherlands: Springer, 2008; pp. 431–459.

Wang CY, Zhang S, Yu Y, Luo YC, Liu Q, Ju C, Zhang YC, Qu LH, Lucas WJ, Wang X, Chen YQ. MiR397b regulates both lignin content and seed number in *Arabidopsis* via modulating a laccase involved in lignin biosynthesis. *Plant Biotechnol. J.* 2014;12:1132–1142.

Wang L, Liu H, Li D, Chen H. Identification and characterization of maize microRNAs involved in the very early stage of seed germination. *BMC Genomics* 2011;12:154.

Wang L, Zhao JG, Zhang M, Li WX, Luo KG, Lu ZG, Zhang CQ, Jin B. Identification and characterization of microRNA expression in *Ginkgo biloba* L. leaves. *Tree Genet. Genomes* 2015;11:76.

Wang YG, An M, Zhou SF, She YH, Li WC, Fu FL. Expression profile of maize microRNAs corresponding to their target genes under drought stress. *Biochem. Genet.* 2014;52:474–493.

Wang Y, Wang Y, Song Z, Zhang H. Repression of MYBL2 by both microRNA858a and HY5 leads to the activation of anthocyanin biosynthetic pathway in *Arabidopsis*. *Mol. Plant* 2016;9:1395–1405.

Warpeha KM, Upadhyay S, Yeh J, Adamiak J, Hawkins SI, Lapik YR, Anderson MB, Kaufman LS. The GCR1, GPA1, PRN1, NF-Y signal chain mediates both blue light and abscisic acid responses in *Arabidopsis*. *Plant Physiol.* 2007;143:1590–1600.

Waterworth WM, Drury GE, Bray CM, West CE. Repairing breaks in the plant genome: the importance of keeping it together. *New Phytol.* 2011;192(4):805–822. doi: 10.1111/j.1469-8137.2011.03926.x.

Waterworth WM, Drury GE, Bray CM, West CE. Repairing breaks in the plant genome: the importance of keeping it together. *New Phytol.* 2011;192(4):805–22. doi: 10.1111/j.1469-8137.2011.03926.x.

Waterworth WM, Masnavi G, Bhardwaj RM, Jiang Q, Bray CM, West CE. A plant DNA ligase is an important determinant of seed longevity. *Plant J.* 2010;63:848–860.

Willmann MR, Mehalick AJ, Packer RL, Jenik PD. MicroRNAs regulate the timing of embryo maturation in Arabidopsis. *Plant Physiol.* 2011;155:1871–1884.

Winkel-Shirley B. Flavonoid biosynthesis. A colorful model for genetics, biochemistry, cell biology, and biotechnology. *Plant Physiol.* 2001;126:485–493.

Wu J, Wang D, Liu Y, Wang L, Qiao X, Zhang S. Identification of miRNAs involved in pear fruit development and quality. *BMC Genomics* 2014;15:953.

Wu L, Zhang QQ, Zhou HY, Ni FR, Wu XY, Qi YJ. Rice microRNA effector complexes and targets. *Plant Cell* 2009;21:3421–3435.

Wu L, Zhou H, Zhang Q, Zhang J, Ni F, Liu C, Qi Y. DNA methylation mediated by a microRNA pathway. *Mol. Cell.* 2010;38:465–475.

Yang F, Cai J, Yang Y, Liu Z. Overexpression of microRNA828 reduces anthocyanin accumulation in *Arabidopsis*. *PCTOC* 2013;115:159–167.

Yang J, Zhang N, Mi X, Wu L, Ma R, Zhu X, Yao L, Jin X, Si H, Wang D. Identification of miR159s and their target genes and expression analysis under drought stress in potato. *Comput. Biol. Chem.* 2014;53:204–213.

Yang T, Xue L, An L. Functional diversity of miRNA in plants. *Plant Sci.* 2007;172:423–432.

Zhang J, Zhang S, Han S, Li X, Tong Z, Qi L. Deciphering small noncoding RNAs during the transition from dormant embryo to germinated embryo in larches (*Larix leptolepis*). *PLoS ONE* 2013;8:e81452.

Zhang S, Yue Y, Sheng L, Wu Y, Fan G, Li A, Hu X, Shangguan M, Wei C. PASmiR: A literature-curated database for miRNA molecular regulation in plant response to abiotic stress. *BMC Plant Biol.* 2013;13:33.

Zhao M, Tai H, Sun S, Zhang F, Xu Y, Li WX. Cloning and characterization of maize miRNAs involved in responses to nitrogen deficiency. *PLoS ONE* 2012;7:e29669.

Zhao N, Ferrer JL, Ross J, Guan J, Yang Y, Pichersky E, Noel JP, Chen F. Structural biochemical, and phylogenetic analyses suggest that indole-3-acetic acid methyltransferase is an evolutionarily ancient member of the SABATH family. *Plant Physiol.* 2008;146:455–467.

Zhao Q, Dixon RA. Altering the cell wall and its impact on plant disease: From forage to bioenergy. *Ann. Rev. Phytopathol.* 2014;52:69–91.

Zhou L, Liu Y, Liu Z, Kong D, Duan M, Luo L. Genome wide identification and analysis of drought-responsive microRNAs in *Oryza sativa*. *J. Exp. Bot.* 2010;61:4157–4168.

Zhuang Y, Zhou XH, Liu J. Conserved miRNAs and their response to salt stress in wild eggplant *Solanum linnaeanum* roots. *Int. J. Mol. Sci.* 2014;15:839–849.

Zwenger S, Basu C. Plant terpenoids: Applications and future potentials. *Biotechnol. Mol. Biol. Rev.* 2008;3:1–7.

23

Current Understanding of Regulation of GBF3 Under Abiotic and Biotic Stresses and Its Potential Role in Combined Stress Tolerance

Sandeep Kumar Dixit, Aarti Gupta, and Muthappa Senthil-Kumar

CONTENTS

Abbreviations

ABA	Abscisic acid
ABRE	ABA-responsive element
Adh	Alcohol dehydrogenase
AREB1	ABA-responsive element binding protein 1
bZIP	Basic leucine zipper
CK	Casein Kinase
esk1	*eskimo1*
GBFs	G-box binding factors
NPR1	Non-expressor of PR-1
PRD	Proline-rich domain
TuMV	Turnip mosaic virus

Introduction

The basic leucine zipper (bZIP) family is one of the largest families of transcription factors that regulate myriad plant processes, including abiotic and biotic stress responses in plants (Riechmann et al., 2000). The bZIP family proteins are characterized by having a basic amino acid-rich region and a leucine zipper region (Jakoby et al., 2002). Based on the sequence similarity of the basic region and the presence of conserved motifs, bZIP proteins from *Arabidopsis thaliana* have been categorized into 10 groups (A, B, C, D, E, F, G, H, I and S) (Jakoby et al., 2002). G-Box binding factors (GBFs) are important bZIP family members, which are spread over group A, G and S. GBFs bind to the specific, palindromic, hexameric G-box element (CACGTG), noted within the promoter of various stress-responsive genes (Menkens et al., 1995). The flanking sequence of the G-box element dictates interaction with a specific GBF (Williams et al., 1992).

In *A. thaliana*, six members of GBF family, viz., *AtGBF1*, *AtGBF2*, *AtGBF3*, *AtGBF4*, *AtGBF5* and *AtGBF6* are reported

in TAIR database (TAIR 10; https://www.arabidopsis.org). These GBFs are distributed over different chromosomes with *AtGBF1*, *AtGBF2* and *AtGBF6* being located at the chromosome number four, *AtGBF3* and *AtGBF5* at chromosome number two and *AtGBF4* located at chromosome number one. AtGBF2 and AtGBF3 show 50% homology in their amino acid sequence (www.ncbi.nlm.nih.gov/). AtGBF4 shows 45% similarity with AtGBF1 and AtGBF3 and 37% similarity with AtGBF6. AtGBF5 is more similar to AtGBF6 with 46% amino acid sequence similarity (www.ncbi.nlm.nih.gov/). Additionally, amino acid sequence analysis of different AtGBFs revealed the presence of a conserved basic leucine zipper domain at C-terminus that binds to the G-box element. AtGBF1, AtGBF2 and AtGBF3 bear a proline-rich domain at N-terminus, which, however, is absent in AtGBF4, AtGBF5 and AtGBF6. AtGBF4 contains an acidic amino acid-rich region in place of a proline-rich region (Menkens and Cashmore, 1994).

Dimerization is a central trait of GBF binding to DNA where they can either form a homo- or hetero-dimer. AtGBF1, AtGBF2 and AtGBF3 can form homo- as well as hetero-dimer (Schindler et al., 1992). Reportedly, hetero-dimerization increases the response specificity to signals by increasing the number of GBF interactions with different cis-elements (Schindler et al., 1992; Siberil et al., 2001). AtGBF4, however, cannot form homo-dimer because negatively charged residues in the leucine zipper create electrostatic repulsion (Menkens and Cashmore, 1994). GBF1 is involved in regulating different target genes' expression both positively and negatively and accordingly influence different plant processes. For example, GBF1 plays a positive role in leaf senescence by negatively regulating catalase 2 expression (Smykowski et al., 2010).

Expression of different GBFs is induced in response to various environmental cues. *AtGBF2* is induced by heat stress (Kant et al., 2008). *OsABF1*, a rice homolog of *AtGBF4*, is induced under anoxia, salinity, drought, oxidative stress, cold and abscisic acid (ABA) in rice seedlings (Hossainet al., 2010). *AtGBF5* is induced by hypo-osmolarity (Satoh et al., 2004). *AtGBF1* expresses in light and dark grown leaves and roots (Schindler et al., 1992). AtGBF2 localizes in both the cytoplasm and the nucleus and is largely light-driven, where blue light increased AtGBF2 localization in the nucleus (Terzaghi et al., 1997). As per the public transcriptomic database, *AtGBF4* expresses in leaves, roots, flowers and siliques and *AtGBF5* mainly expresses in roots and carpels (www.bar.utoronto.ca/efp/cgi-bin/efpWeb.cgi). *AtGBF6* is highly expressed in stems and flowers (Rook et al., 1998b). *AtGBF6* expression is induced by light and is suppressed under high sucrose concentrations, reflecting its role in the maintenance of carbohydrates partitioning (Rook et al., 1998a).

AtGBF3 expresses mainly in etiolated leaves and in roots (Schindler et al., 1992). Besides, literature reports indicate that *AtGBF3* transcripts accumulate to high levels in dry seed, stamens, mature pollen and siliques (www.bar.utoronto.ca/efp/cgi-bin/efpWeb.cgi). Importantly, among different members of *AtGBFs,* literature suggests that *AtGBF3* expression increases under abiotic stresses like drought, cold, heat and salt stress and biotic stress like *Pseudomonas syringae* and *Turnip mosaic virus* (TuMV) (Chen et al., 2002;Xin et al., 2007; Prasch and Sonnewald, 2013). Thus, AtGBF3 might play a role in multiple and combined abiotic and biotic stress tolerance mechanisms, which however have not been characterized so far. Taking cues from different literature sources, we put forth the current understanding of the role of GBF3 in multiple and combined abiotic and biotic stresses in this chapter.

Structural and Functional Traits of AtGBF3

AtGBF3 gene spans 3213 bp region with 11 exons and 11 introns (Figure 23.1a, TAIR 10; https://www.arabidopsis.org). As per the NCBI database, it contains 12 exons and 11 introns. However, the 5′ untranslated region (UTR) interrupted with an intron is considered as the first exon and the 3′ UTR region along with the 11th exon is also considered as an exon. Two splice variants have been reported for *AtGBF3* (TAIR 10; https://www.arabidopsis.org). The two transcript variants (AtGBF3.1 and AtGBF3.2) are the result of alternative splicing within the 8th exon. Nucleotide sequence comparison between the two splice variants shows that *AtGBF3.2* lacks 69 nucleotides and this corresponds to 23 amino acids. Promoter analysis by Plant care and PlantPAN2 database (http://bioinformatics.psb.ugent.be/webtools/plantcare/html/, http://plantpan2.itps.ncku.edu.tw/) reveals that the *AtGBF3* promoter contains a G-box element, MYB binding cis-elements, abscisic acid response elements (ABRE) and TGA-box elements (Figure 23.1b).

AtGBF3 protein contains a proline-rich domain at the N-terminal and a basic leucine zipper domain is present at C-terminus (Figure 23.1c) (Lu et al., 1996). PRD may play a role in transcriptional regulation. (Tamai et al., 2002) by regulating downstream genes. The basic region, rich in basic amino acids, is involved in DNA sequence recognition and cognate binding. The helical leucine zipper domain contains periodic repetition of leucine residues. The two helical leucine zipper domains of two GBFs interact (either in GBF3 homo-dimer or hetero-dimer with other GBFs) by hydrophobic interaction and form a parallel

α helical coil structure. As stated earlier, this dimerization is essential for DNA binding (Alber, 1992; Pabo and Sauer, 1992).

Phylogenetic analysis of AtGBF3 based on amino acid sequence homology reveals that GBF3 is present across monocots and dicots (Figure 23.1d). Moreover, the three domains (viz., PRD, basic region and leucine zipper domain) were also found to be conserved across different plant species (Figure 23.1e).

The basic amino acid region of AtGBF3 contains a nuclear localization signal (NLS), so it is expected that AtGBF3 in localized in the nucleus (Terzaghi et al., 1997). However, no direct evidence of localization of AtGBF3 is reported so far. *In silico* analysis reveals that GBF3 localizes in the nucleus and the cytoplasm (http://bar.utoronto.ca/cell_efp/cgi-bin/cell_efp.cgi). Additionally, immunoblot-based localization studies with polyclonal anti-GBF1 antibodies (which can also bind with GBF2 and GBF3 as GBF1) reveal that most of the AtGBFs are mainly localized in the cytoplasm (Terzaghi et al., 1997). The AtGBF1 has a putative Casein Kinase II (CKII) phosphorylation site between amino acids 112 and 164. The CK II phosphorylation of GBF1 enhances its binding to DNA (Klimczak et al., 1992, 1995). Deletion of the phosphorylation site spanning amino acid 112 to 164 from AtGBF1 showed increased nuclear localization. So it is possible that the phosphorylation state of GBF1 is involved in its localization. Several CK II phosphorylation sites are present in AtGBF3, presenting the possibility that the nuclear import of AtGBF3 may be regulated by kinase activity (Terzaghi et al., 1997). So it is possible that GBF3 also localizes in both the cytoplasm and the nucleus.

AtGBF3 Mediates Environmental Stress Responses

The expression pattern of AtGBF3 documented in public microarray data reveals that *AtGBF3* expression increases during different abiotic stresses like drought, osmotic, cold, salt and heat stress and biotic stresses like biotrophic and necrotrophic pathogens (www.bar.utoronto.ca/efp/cgi-bin/efpWeb.cgi; https://genevestigator.com/gv/). Reportedly, exogenous ABA induces *AtGBF3* in *A. thaliana* cell culture (Lu et al., 1996). ABA is an important plant hormone which integrates different abiotic and biotic stress signals. *In silico* promoter analysis reveals that the *AtGBF3* promoter contains many ABA-responsive elements (ABRE; ACGTGGC), an MYB binding site (CAACTG), many G-box (CACGTG) and TGA2/3 binding sites (TGACG) (http://bioinformatics.psb.ugent.be/webtools/plantcare/html/; http://plantpan2.itps.ncku.edu.tw/) (Figure 23.1b). The ABA-responsive element binding protein 1 (AREB1) is a transcription factor, strongly induced by ABA, drought and high salinity and binds to the ABRE present in the target genes (Fujita et al., 2005). The presence of two ABREs in the promoter of a target gene is essential for induction of target genes by AREB1 (Choi et al., 2000; Uno et al., 2000). Transcriptomic analysis of the *AtAREB1* over-expression line showed that *AtGBF3* was up-regulated in such plants (Fujita et al., 2005). This evidence suggests that abiotic stresses like drought, salt and ABA induces *AREB1* and AREB1 in turn activates *AtGBF3*. Additionally, the *GBF3* promoter contains an MYB binding site (Figure 23.1b) and, under drought stress, At*GBF3* expression may also be induced by MYB (e.g., MYB2). *ESK1* encodes a negative regulator of cold acclimation and *esk1* mutant plants show freezing tolerance. Transcriptomic analysis of *A. thaliana* esk1 mutants

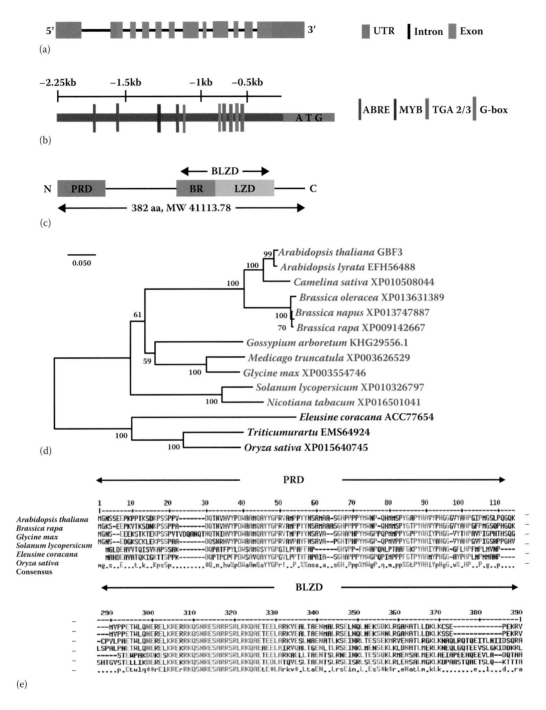

FIGURE 23.1 Illustration of *AtGBF3* gene structure and protein structure predicted by *in silico* analysis. (a) *AtGBF3* gene structure depicts 11 exons, 11 introns, 5′ UTR interrupted by intron and a 3′ UTR. (b) *AtGBF3* promoter composition with stress-related motifs is shown here. A 2.25 kb long region upstream of the *AtGBF3* translation start site was queried in PlantCARE (http://bioinformatics.psb.ugent.be/webtools/plantcare/html/) and PlantPAN2 (plantpan2.itps.ncku.edu.tw/). The figure shows few cis-acting elements specific to *A. thaliana*; ABA response element (ABRE), Myb binding site, TGA2 or 3 binding site and the G-box element. (c) Schematic diagram of AtGBF3 protein, harbouring proline-rich domain (PR, blue box), basic amino acid-rich region (BR, green box) and leucine zipper domain (LZD, orange box) is shown here (https://www.ebi.ac.uk/interpro/). AtGBF3 constitute 382 amino acid (aa) and molecular weight (MW) of AtGBF3 is 41113.78. (d) Phylogenetic relationship of AtGBF3 orthologues is represented here in the form of dendrogram. The AtGBF3 homologues in other plant species were searched based on input query as 'GBF3' and also based on the AtGBF3 amino acid sequence blast in the NCBI protein database. The accession numbers are presented for each. The tree was constructed by the Neighbour-Joining method using MEGA7 software, where the number at each node indicates the bootstrap score showing the closeness of proteins. (e) The proline-rich domain (PRD) and the basic leucine zipper domain (BLZD) are shown in *Arabidopsis thaliana, Arabidopsis lyrata, Brassica rapa, Solanum lycopersicum, Eleusine coracana* and *Oryza sativa*. The multiple sequence alignment of AtGBF3 with its orthologues was performed using multAlin *(http://multalin. toulouse.inra.fr/multalin/)*.

revealed that *GBF3* was up-regulated in these mutants and suggests GBF3 plays a role in cold tolerance (Xin et al., 2007). Taken together, GBF3 expression is induced under abiotic stresses.

GBF3 expression has been reported to be induced by the exogenous salicylic acid (SA) (Thibaud-Nissen et al., 2006). GBF3 is induced by *P. syringae* pv. *tomato* DC3000 (avirulent) (Chen et al., 2002). Expression of *GBF3* was suppressed in *npr1* mutants under *P. syringae* pv. *maculicola* ES4326 infection (Chen et al., 2002). NPR1 interacts with TGA2 and TGA3 transcription factors (Zhou et al., 2000). *In silico* promoter analysis revealed that the *AtGBF3* promoter contains TGA2 or TGA3 binding sites. This evidence suggests that GBF3 functions downstream of SA and NPR1 and TGA2/3 and mediates pathogen resistance.

Under abiotic stresses, expressed GBF3 protein regulates the G-box element containing stress-responsive genes for example, *Adh* (Lu et al., 1996). bHLH106 is a transcription factor that can bind to the E-box and G-box and has a role in salt stress tolerance. Transcriptomic analysis of the *AtbHLH106* over-expression line suggests that many stress-responsive genes that have a G-box element in their promoter are regulated by AtbHLH106

(Ahmad et al., 2015). Based on the presence of a G-box element in the promoter of *CytochromeP450*, it could also be one of the downstream components of the GBF3 regulation. Moreover, bioinformatics analysis shows that GBF3 interacts with a number of proteins like GBF6, GBF4, GPRI1 bZIP16, bZIP68, ABF3, bZIP, GN and ERD5 (STRING, http://string-db.org/). GBF3 can also interact with bZIP to form hetero-dimer and regulate downstream gene expression. GPRI1 interacts with the proline-rich region of GBF3 and this interaction may activate target genes (Tamai et al., 2002). The role of ERD5 is well-documented in drought stress (Bhaskara et al., 2015) and bacterial defence (Cecchini et al., 2011; Fabro et al., 2016), but its interaction and role with GBF3 is still unknown. This evidence strongly points towards a role for GBF3 in regulating multiple abiotic and biotic stresses.

Possible Role in Combined Stress

A transcriptomic study on combined stress showed that AtGBF3 was up-regulated under individual as well as combined heat,

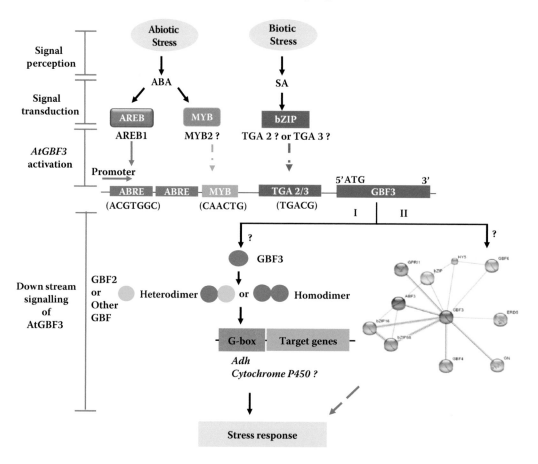

FIGURE 23.2 Proposed Model for role of GBF3 in abiotic and biotic stress. We hypothesize that under abiotic stress like drought, ABA accumulation increases, increased ABA level induces *AREB1* or ABA accumulation may also induce *MYB (MYB2)*. AREB1 protein induces *AtGBF3*. *GBF3* may also be induced by MYB (MYB2). Under biotic stress like *Pseudomonas syringae* pv. *tomato* Salicylic acid (SA) level increases, increased SA may induce *TGA2* or *TGA3* (*bZIP* family member). TGA2 or TGA3 protein may induce *GBF3*. The expressed GBF3 protein induces other stress-responsive genes like *Adh* (Lu et al., 1996) containing a G-box element in their promoter. Further, on the basis of *in silico* analysis, it potentially binds to the G-box element found in the promoter region of stress-responsive genes like *Cytochrome P450* (I). Additionally, STRING-based analysis (http://string-db.org/) of GBF3 shows that elongated hypocotyl 5 (HY5), GBF6, bZIP68, bZIP16, GBF4, ABA-responsive elements binding factor 3 (ABF3), GBFs proline-rich region interacting factor 1 (GPRI1), GN and early-responsive to dehydration (ERD5) are interacting partners of GBF3. ERD5 is one of the interacting partners of GBF3, the role of ERD5 has been proven in drought stress (Bhaskara et al., 2015) and defence response (Fabro et al., 2016; Cecchini et al., 2011) (II). However, the mechanism of interaction is still unknown.

drought and TuMV stress (Prasch and Sonnewald, 2013). Under drought stress, ABA accumulation induces *AREB1* and expressed AREB1 induces GBF3. As stated earlier, GBF3 may be involved in drought tolerance by regulating stress-responsive genes. Under pathogen stress, NPR1 and TGA2/3 may induce GBF3 in SA mediated signalling (Figure 23.2). It was reported that high levels of SA accumulate under combined virus and drought stress (Xu et al., 2008). Henceforth, we hypothesize that under combined drought and pathogen stress, SA accumulation increases and induces GBF3, which, in turn, regulates the expression of genes involved in combined stress tolerance.

Conclusion

GBF3 binds to the G-box element present in the promoter of stress-responsive genes. Transcriptomic analysis reveals that GBF3 is induced under different individual as well as combined abiotic and biotic stress. GBF3 is a downstream target of ABA signalling under drought stress. GBF3 is also a downstream component of SA signalling under pathogen stress. So GBF3 may be an important gene for future studies on stress tolerance.

Acknowledgments

Projects at MS-K Lab are supported by National Institute of Plant Genome Research core funding and the DBT-Ramalingaswami re-entry fellowship grant (BT/RLF/re-entry/23/2012). SKD acknowledges DBT-Junior Research Fellowship (DBT/2014/NIPGR/255) and AG acknowledges the SERB National Post-Doctoral Fellowship (N-PDF/2015/000116).

REFERENCES

Ahmad A, Niwa Y, Goto S, Ogawa T, Shimizu M, Suzuki A, Kobayashi K, Kobayashi H. bHLH106 integrates functions of multiple genes through their G-Box to confer salt tolerance on Arabidopsis. *PloS One.* 2015;10:e0126872.

Alber T. Structure of the leucine zipper. *Curr. Opin. Genet. Dev.* 1992;2:205–210.

Bhaskara GB, Yang TH, Verslues PE. Dynamic proline metabolism: Importance and regulation in water limited environments. *Front. Plant Sci.* 2015;6:484.

Boratyn GM, Camacho C, Cooper PS, Coulouris G, Fong A, Ma N, Madden TL, Matten WT, McGinnis SD, Merezhuk Y, Raytselis Y, Sayers EW, Tao T, Ye J, Zaretskaya I. BLAST: a more efficient report with usability improvements. *Nucleic Acids Res.* 2013;41:W29–W33.

Cecchini NM, Monteoliva MI, Alvarez ME. Proline dehydrogenase contributes to pathogen defense in Arabidopsis. *Plant Physiol.* 2011;155:1947–1959.

Chen W, Provart NJ, Glazebrook J, Katagiri F, Chang HS, Eulgem T, Mauch F, Luan S, Zou G, Whitham SA, Budworth PR. Expression profile matrix of Arabidopsis transcription factor genes suggests their putative functions in response to environmental stresses. *Plant Cell* 2002;14:559–574.

Choi HI, Hong JH, Ha JO, Kang JY, Kim SY. ABFs, a family of ABA-responsive element binding factors. *J. Biol. Chem.* 2000;275:1723–1730.

Chow CN, Zheng HQ, Wu NY, Chien CH, Huang HD, Lee TY, Chiang-Hsieh YF, Hou PF, Yang TY, Chang WC. PlantPAN 2.0: an update of plant promoter analysis navigator for reconstructing transcriptional regulatory networks in plants. *Nucleic Acids Res.* 2015:gkv1035v1–gkv1035.

Corpet F. Multiple sequence alignment with hierarchical clustering. *Nucleic Acids Res.* 1988;16:10881–10890.

Fabro G, Rizzi YS, Alvarez ME. Arabidopsis proline dehydrogenase contributes to flagellin mediated PAMP-triggered immunity by affecting RBOHD. *Mol. Plant Microbe Interact.* 2016;29:620–628.

Fujita Y, Fujita M, Satoh R, Maruyama K, Parvez MM, Seki M, Hiratsu K, Ohme-Takagi M, Shinozaki K, Yamaguchi-Shinozaki K. AREB1 is a transcription activator of novel ABRE-dependent ABA signaling that enhances drought stress tolerance in Arabidopsis. *Plant Cell* 2005;17:3470–3488.

Hossain MA, Lee Y, Cho JI, Ahn CH, Lee SK, Jeon JS, Kang H, Lee CH, An G, Park PB. The bZIP transcription factor OsABF1 is an ABA responsive element binding factor that enhances abiotic stress signaling in rice. *Plant Mol. Biol.* 2010;72:557–566.

Huala E, Dickerman AW, Garcia-Hernandez M, Weems D, Reiser L, LaFond F, Hanley D, Kiphart D, Zhuang J, Huang W, Mueller L, Bhattacharyya D, Bhaya D, Sobral B, Beavis B, Somerville C, Rhee S. The *Arabidopsis* Information Resource (TAIR): a comprehensive database and web-based information retrieval, analysis, and visualization system for a model plant. *Nucleic Acids Res.* 2001;29:102–105.

Jakoby M, Weisshaar B, Dröge-Laser W, Vicente-Carbajosa J, Tiedemann J, Kroj T, Parcy F. bZIP transcription factors in Arabidopsis. *Trends Plant Sci.* 2002;7:106–111.

Kant P, Gordon M, Kant S, Zolla G, Davydov O, Heimer YM, Chalifa-Caspi VE, Shaked R, Barak S. Functional-genomics-based identification of genes that regulate Arabidopsis responses to multiple abiotic stresses. *Plant Cell Environ.* 2008;31:697–714.

Klimczak LJ, Collinge MA, Farini D, Giuliano G, Walker JC, Cashmore AR. Reconstitution of Arabidopsis casein kinase II from recombinant subunits and phosphorylation of transcription factor GBF1. *Plant Cell* 1995;7:105–115.

Klimczak LJ, Schindler U, Cashmore AR. DNA binding activity of the Arabidopsis G-box binding factor GBF1 is stimulated by phosphorylation by casein kinase II from broccoli. *Plant Cell* 1992;4:87–98.

Kumar S, Stecher G, Tamura K. MEGA7: Molecular Evolutionary Genetics Analysis Version 7.0 for Bigger Datasets. *Mol Biol Evol.* 2016;33:1870–1874.

Lescot M, Dehais P, Thijs G, Marchal K, Moreau Y, Van de Peer Y, Rouze P, Rombauts S. PlantCARE, a database of plant cis-acting regulatory elements and a portal to tools for in silico analysis of promoter sequences. *Nucleic Acids Res.* 2002;30:325–327.

Lu G, Paul AL, McCarty DR, Ferl RJ. Transcription factor veracity: Is GBF3 responsible for ABA-regulated expression of Arabidopsis Adh?. *Plant Cell* 1996;8:847–857.

Menkens AE, Cashmore AR. Isolation and characterization of a fourth *Arabidopsisthaliana* G-box-binding factor, which has similarities to Fos oncoprotein. *Proc. Natl. Acad. Sci. USA* 1994;91:2522–2526.

Menkens AE, Schindler U, Cashmore AR. The G-box: A ubiquitous regulatory DNA element in plants bound by the GBF family of bZIP proteins. *Trends Biochem. Sci.* 1995;20:506–510.

Pabo CO, Sauer RT. Transcription factors: Structural families and principles of DNA recognition. *Annu. Rev. Biochem.* 1992;61:1053–1095.

Prasch CM, Sonnewald U. Simultaneous application of heat, drought, and virus to Arabidopsis plants reveals significant shifts in signaling networks. *Plant Physiol.* 2013;162:1849–1866.

Riechmann JL, Heard J, Martin G, Reuber L, Jiang CZ, Keddie J, Adam L, Pineda O, Ratcliffe OJ, Samaha RR, Creelman R. Arabidopsis transcription factors: Genome-wide comparative analysis among eukaryotes. *Science* 2000;290:2105–2110.

Rook F, Gerrits N, Kortstee A, Van Kampen M, Borrias M, Weisbeek P, Smeekens S. Sucrose-specific signalling represses translation of the Arabidopsis ATB2 bZIP transcription factor gene. *Plant J.* 1998a;15:253–263.

Rook F, Weisbeek P, Smeekens S. The light-regulated Arabidopsis bZIP transcription factor gene *ATB2* encodes a protein with an unusually long leucine zipper domain. *Plant Mol Biol.* 1998b;37:171–178.

Satoh R, Fujita Y, Nakashima K, Shinozaki K, Yamaguchi-Shinozaki K. A novel subgroup of bZIP proteins functions as transcriptional activators in hypoosmolarity-responsive expression of the *ProDH* gene in Arabidopsis. *Plant Cell Physiol.* 2004;45:309–317.

Schimid M, Davison TS, Henz SR, Pape UJ, Demar M, Vingron M, Scholkopf B, Weigel D, Lohmann JU. A gene expression map of *Arabidopsis thaliana* development. *Nat Genet.* 2005;37:501–506.

Schindler U, Menkens AE, Beckmann H, Ecker JR, Cashmore AR. Heterodimerization between light-regulated and ubiquitously expressed Arabidopsis GBF bZIP proteins. *EMBO J.* 1992;11:1261–1273.

Siberil Y, Doireau P, Gantet P. Plant bZIP G-box binding factors. *Eur. J. Biochem.* 2001;268:5655–5666.

Smykowski A, Zimmermann P, Zentgraf U. G-Box binding factor1 reduces CATALASE2 expression and regulates the onset of leaf senescence in Arabidopsis. *Plant Physiol.* 2010;153:1321–1331.

Szklarczyk D, Franceschini A, Wyder S, Forslund K, Heller D, Huerta-Cepas J, Simonovic M, Roth A, Santos A, Tsafou KP, Kuhn M, Bork P, Jensen LJ, Mering CV. STRING v10: protein-protein interaction networks, integrated over the tree of life. *Nucleic Acids Res.* 2015;43:D447–D452.

Tamai H, Iwabuchi M, Meshi T. Arabidopsis GARP transcriptional activators interact with the Pro-rich activation domain shared by G-box-binding bZIP factors. *Plant Cell Physiol.* 2002;43:99–107.

Terzaghi WB, Bertekap RL, Cashmore AR. Intracellular localization of GBF proteins and blue light-induced import of GBF2 fusion proteins into the nucleus of cultured Arabidopsis and soybean cells. *Plant J.* 1997;11:967–982.

Thibaud-Nissen F, Wu H, Richmond T, Redman JC, Johnson C, Green R, Arias J, Town CD. Development of Arabidopsis whole-genome microarrays and their application to the discovery of binding sites for the TGA2 transcription factor in salicylic acid-treated plants. *Plant J.* 2006; 47:152–162.

Uno Y, Furihata T, Abe H, Yoshida R, Shinozaki K, Yamaguchi-Shinozaki K. Arabidopsis basic leucine zipper transcription factors involved in an abscisic acid-dependent signal transduction pathway under drought and high-salinity conditions. *Proc. Natl. Acad. Sci. USA* 2000;97:11632–11637.

Williams ME, Foster R, Chua NH. Sequences flanking the hexameric G-box core CACGTG affect the specificity of protein binding. *Plant Cell* 1992;4:485–496.

Xin Z, Mandaokar A, Chen J, Last RL, Browse J. Arabidopsis *ESK1* encodes a novel regulator of freezing tolerance. *Plant J.* 2007;49(5):786–799.

Xu P, Chen F, Mannas JP, Feldman T, Sumner LW, Roossinck MJ. Virus infection improves drought tolerance. *New Phytol.* 2008;180:911–921.

Zdobnov EM, Apweiler R. InterProScan—an integration platform for the signature-recognition methods in InterPro. *Bioinformatics.* 2001;17:847–848.

Zhou JM, Trifa Y, Silva H, Pontier D, Lam E, Shah J, Klessig DF. NPR1 differentially interacts with members of the TGA/OBF family of transcription factors that bind an element of the *PR1* gene required for induction by salicylic acid. *Mol. Plant Microbe Interact.* 2000;13:191–202.

Zimmermann P, Hirsch-Hoffmann M, Hennig L, Gruissem W. GENEVESTIGATOR. Arabidopsis microarray database and analysis toolbox. *Plant Physiol.* 2004;136:2621–2632.

24

microRNAs: Key Modulators of Drought Stress Responses in Plants

A. Thilagavathy, Kavya Naik, and V.R. Devaraj

CONTENTS

Abbreviations

ABA	Abscisic acid
ARF	Auxin Response Factor
CSD	Cu/Zn-superoxide dismutase
DREB	Dehydration-Responsive Element Binding protein
GRF	Growth-Regulating Factor
miRNA, miR	microRNA
PTGS	Post-transcriptional gene silencing
RISC	RNA-induced silencing complex
RT-qPCR	Reverse Transcription quantitative PCR
TF	Transcription Factor
TIR	Transport Inhibitor Response

Introduction

From the agricultural perspective, drought is described in terms of its effect on crop yield under water deficit conditions imposed by environmental factors. Physiological adjustments to abiotic stress in plants involve regulation of gene expression at various layers from transcription to post-transcriptional regulation. miRNAs have emerged as remarkable players in the regulation of gene expression at the post-transcriptional level. These 20-22 nt endogenous regulatory RNAs identified first in *C.elegans* by Lee et al. (1993) were so named by Gary Ruvkun in 2001 (Muljo et al., 2010). Ever since the report of a plant miRNA from *A. thaliana*, a large number of miRNAs have been identified from diverse sources as indicated by (28,645) entries in miRBase; a miRNA database current release-21, http://www.mirbase.org/in.

Biogenesis of miRNAs in Plants

Biochemistry of miRNA biogenesis and the mechanism of action in gene regulation have been discussed in recent reviews (Ferdous et al., 2015; Pashkovskiy and Ryazansky, 2013; Ding et al., 2013). Briefly, from the plant biochemistry perspective, miRNAs are expressed as long primary miRNA (pri-miRNA) generated from MIR genes by the activity of RNA polymerase II (Lee et al., 2003). These pri-miRNAs are processed in the nucleus by RNase III-like enzyme, Dicer-like 1 (DCL1), and hyponastic leaves *HYL1* protein to stem-loop intermediates called miRNA precursors or pre-miRNAs (Bartel, 2004; Kurihara and Watanabe, 2004). The pre-miRNAs are later processed into mature miRNA duplexes. The mature miRNA duplex containing two strands, miRNA and miRNA* with 2 nucleotide overhang at 3' end, is methylated by a methyltransferase, HEN1 (Hua Enhancer 1) found in plants. Then, the duplex is exported from the nucleus to the cytoplasm by *HASTY* (Park et al., 2002). Upon entry into the cytoplasm, a helicase unwinds the miRNA-miRNA* duplex, exposing the mature miRNA that is incorporated into the RNA-induced silencing complex (RISC) containing *ARGONAUTE1*, which directs the RISC to regulate gene expression either by mRNA cleavage, or by translational repression (Bartel, 2004) (Figure 24.1). While perfect base-pairing between miRNA and mRNA target causes mRNA cleavage, imperfect base-pairing leads to translational repression. Usually, plant miRNA mediates target mRNA cleavage (Voinnet, 2009). The recent identification of a new miRNA-mediated regulation known as miRNA-mediated decay involving removal of poly (A) tail of the target mRNA, thus making it unstable (Rüegger and Grobhans, 2012), adds to the diversity of miRNA mediated post-transcriptional gene regulation.

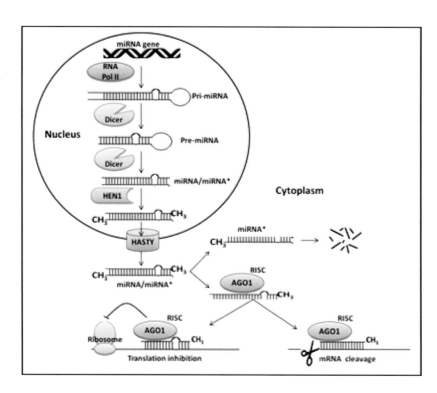

FIGURE 24.1 Biogenesis of miRNAs and their mode of action in plants.

Identification and Characterization of Drought-Responsive miRNAs

Well-designed experimental and computational approaches have led to reliable methods of miRNA isolation from plant sources. Research on plant miRNAs gained momentum around 2002, when the first miRNA of plant origin was identified in *A. thaliana* (Llave et al., 2002; Park et al., 2002). Since then, the list of plant miRNAs and their target genes are constantly being expanded as they are identified and characterized from various plants species including rice, wheat, and maize etc. Plant miRNAs were initially identified by cloning and sequencing of small RNA population (Llave et al., 2002; Park et al., 2002). Recently, high throughput sequencing technology/Next-Generation Sequencing has emerged to be an effective tool for miRNA discovery and profiling (Wang et al., 2011). Generation of immense data from high throughput sequencing poses a challenge to the efficacy and reliability of newly discovered miRNAs. Further, challenges can be faced in distinguishing non-homozygous sequence variants from sequencing errors in the case of plants which lack genome sequences for the purpose of mapping/aligning. Apart from this, several plant miRNAs have also been identified by computational approaches from genome sequences based on the conservation of the small RNA sequence in the context of a fold-back precursor in different species (Jones-Rhoades and Bartel, 2004). Northern blotting, stem-loop real-time RT-PCR, and microarray technologies are the most commonly used techniques to characterize miRNAs. With the advent of stem-loop primers, RT-qPCR is being used as a robust and potential method for detecting expression of miRNA molecules accurately despite their small size. It is also employed to validate the identified

miRNAs. Microarrays are the preferred platform for genome-wide miRNA expression analysis.

A number of computational tools and databases have been developed to facilitate the identification and prediction of miRNA and pre-miRNA. A computer-based homology search provides robust strategies for the discovery and identification of mature sequences of miRNA families that are widely conserved across the plant lineages, as demonstrated by studies on mosses, monocots, and gymnosperms (Zhang et al., 2006; Axtell and Bartel, 2005). Computational identification exploits the paramount features of miRNA genes; a characteristic stem-loop pre-miRNA structure, and a high degree of conservation of mature sequences between related genomes (Dezulian et al., 2006; Huang et al., 2007). Though several computational tools have been developed for identification of plant miRNAs and their target prediction, a curated database of stress-related miRNAs is a necessity to build a cohesive database system.

Prediction of targets for identified miRNAs is a crucial step in exploring the post-transcriptional regulation of gene expression. Experimental approaches developed to identify targets of miRNA include RT-PCR using miRNAs as endogenous primers, labeled miRNA pull-down assay (LAMP), and RNA ligase mediated amplification of cDNA ends (RLM-RACE) followed by sequencing (Martinez-Sanchez and Murphy, 2013). Computational methods have been effectively used to complement experimental approaches to identify the plant miRNA targets, owing to the perfect complementarity between miRNAs and mRNA targets. The last decade has witnessed the development of several bioinformatic tools to predict miRNA targets. Computational tools such as; psRNATarget, Targetfinder, Target scan, Tapirhybrid, Tapirfasta, p-TAREF, patscan, miranda, and Target-align are a few notable ones (Table 24.1).

TABLE 24.1

List of Databases/Repositories and Major Tools Available for Plant miRNAs, Their Target Identification/Prediction*

Name of the Database/Resource/Repository/Tool	Description	Web Link	References
BioVLAB-MMIA-NGS	MiRNA and mRNA integrated analysis using high-throughput sequencing data coupled with bioinformatics tools	http://epigenomics.snu.ac.kr/biovlab_mmia_ngs/	Chae et al. (2015)
C-mii	A tool for plant miR and target identification	http://www.biotec.or.th/isl/c-mii	Numnark et al. (2012)
DMD	A dietary miRNA database from 15 dietary plant and animal species	http://sbbi.unl.edu/dmd/	Chiang et al. (2015)
MFSN	A tool for prediction of plant miRNA functions based on a functional similarity network (MFSN) through the application of transductive multi-label classification (TRAM) to the MFSN		Meng et al. (2016)
miPEPs	MiRNAs Encode Peptides is a tool for the functional analysis of plant miRNA family members		Couzigou et al. (2015)
miRA	Plant miRNA identification tool especially for organisms without existing miRNA annotation. It is also useful for identifying species-specific miRNAs	https://github.com/mhuttner/miRA	Evers et al. (2015)
miRanalyzer	miR detection and analysis tool for next-generation sequencing experiments	http://bioinfo5.ugr.es/miRanalyzer/miRanalyzer.php	Hackenberg et al. (2011)
miRBase	Searchable database of published miR sequences and annotation	http://www.mirbase.org	Kozomara and Griffiths-Jones (2014)
miRDeep-P	A computational tool for analyzing the miRtranscriptome in plants	http://faculty.virginia.edu/lilab/miRDP/	Yang and Li (2011)
miRge	A fast multiplexed method of processing sRNA-sequence data to determine miRNA entropy and identify differential production of miRNA isomiRs	http://atlas.pathology.jhu.edu/baras/miRge.html.	Baras et al. (2015)
miRPlant	An Integrated Tool for Identification of Plant MiR from RNA Sequencing Data	http://www.australianprostatecentre.org/research/software/mirplant	An et al. (2014)
miRTarBase	The experimentally validated miR-target interactions Database	http://mirtarbase.mbc.nctu.edu.tw/index.php	Hsu et al. (2011)
MTide	An integrated tool for the identification of miR-target interaction in plants	http://bis.zju.edu.cn/MTide/	Zhang et al. (2014)
mirTool	A comprehensive web server providing detailed annotation information for known miRs and predicting novel miRs that have not been characterized before	http://centre.bioinformatics.zj.cn/mirtools/	Wu et al. (2013)
miTRATA	A tool for miRNA truncation and tailing analysis	https://wasabi.dbi.udel.edu/~apps/ta/	Patel et al. (2016)
PASmiR	A literature-curated database for miR molecular regulation in plant response to abiotic stress		Zhang et al. (2013)
PlantMirnaT	A miRNA-mRNA integrated analysis system	https://sites.google.com/site/biohealthinformaticslab/Resources	Rhee et al. (2015)
PlanTE-MIR	Database for transposable element-related plant microRNAs	http://bioinfo-tool.cp.utfpr.edu.br/plantemirdb/	Lorenzetti et al. (2016)
PmiRKB	Plant miR Knowledge Base. Four major functional modules, SNPs, Pri-miRs, MiR-Tar and Self-reg, are provided	http://bis.zju.edu.cn/pmirkb/	Meng et al. (2011)
PMRD	Plant miRNA Database	http://bioinformatics.cau.edu.cn/PMRD/	Zhang et al. (2010)
PNRD	It is an updated version of PMRD	http://structuralbiology.cau.edu.cn/PNRD/index.php	Yi et al. (2015)
P-SAMS	A Plant Small RNA Maker Site is a web tool for artificial miRNAs and synthetic trans-acting small interfering RNAs	http://p-sams.carringtonlab.org	Fahlgren et al. (2016)

(Continued)

TABLE 24.1 (CONTINUED)

List of Databases/Repositories and Major Tools Available for Plant miRNAs, Their Target Identification/Prediction*

Name of the Database/Resource/ Repository/Tool	Description	Web Link	References
Semirna	Searching for plant miRNAs using target sequences	http://www.bioinfocabd.upo.es/semirna/	Muñoz-Mérida et al., 2012
sRNAtoolbox	A set of tools for expression profiling and analysis of sRNA bench results	http://bioinfo5.ugr.es/srnatoolbox	Rueda et al. (2015)
TAPIR	Target prediction for Plant miRs	http://bioinformatics.psb.ugent.be/webtools/tapir/	Bonnet et al. (2010)
WMP	Database for abiotic stress responsive miRNAs in Wheat	http://wheat.bioinfo.uqam.ca	Remita et al. (2015)

* Modified from: Shriram et al. (2016).

Drought Stress Physiology of Plants

Drought is a widespread adverse environmental condition which is likely to be frequent in the future, due to climate change with an estimated 4°C warming, reducing renewable surface water and groundwater resources significantly in most tropical and subtropical regions. Since water accounts for 80%–90% of the fresh weight of most herbaceous plant structures, and over 50% of the fresh weight of woody plants, the amount of water available to plants is crucial (Kramer and Boyer, 1995). Absorption and transport of water by roots are entwined with numerous physiological processes in plants, including the acquisition of nutrients, translocation, assimilation, and photosynthesis. Also, roots transport a huge amount of water to the leaves to maintain the turgor, which sustains the leaves in the extended form for adequate absorption and assimilation of CO_2 (Hsiao and Xu, 2000). Therefore, decreasing water content during drought is accompanied by loss of turgor and wilting, cessation of cell enlargement, closure of stomata, reduction in photosynthesis, and interference with many other basic metabolic processes (Neumann, 2008).

Detrimental effects of drought stress are circumvented by physiological and metabolic adjustments, and by altering gene expression. Genes induced during drought stress are classified into two groups. The first group comprises of genes whose products are components that protect the cell against drought stress directly, such as osmolytes, antioxidants, antioxidant enzymes, chaperones, LEA (late embryogenesis abundant) proteins, water channel proteins, and transporters. The second group comprises of genes coding protein components involved in the signal transduction, such as protein kinases and transcription factors that regulate gene expression in drought stress response (Shinozaki and Yamaguchi-Shinozaki, 2006). These regulations involve proteins, especially transcription factors (TFs), as major regulators and protein-coding genes as targets of regulation at the transcriptional level. A better understanding of post-transcriptional regulations has led to the identification of miRNAs as post-transcriptional regulators (Lee et al., 1993). miRNAs are recognized as important modulators in drought avoidance and tolerance by controlling the expression of drought-responsive genes (Covarrubias and Reyes, 2010; Shuai et al., 2013). miRNAs can be placed at the center of the gene regulatory network as most of them target genes encoding TFs, which, in turn, regulate large numbers of genes associated with the molecular events

of drought response. An miRNA controlled regulatory network would provide an additional layer of regulation of genes at a post-transcriptional level for abiotic-stress response in plants (Shuai et al., 2013).

miRNAs and Drought Stress

miRNAs have been identified from a number of plant species, such as *Arabidopsis* (Sunkar and Zhu, 2004), rice (Zhou et al., 2010), cowpea (Barrera-Figueroa et al., 2011), tobacco (Frazier et al., 2011), *Glycine max* (Kulcheski et al., 2011), and *Phaseolus vulgaris* (Arenas-Huertero et al., 2009), etc. While drought-induced miRNAs are believed to down-regulate their target mRNAs coding for negative functional proteins, drought-repressed miRNAs are known to cause accumulation of their target mRNAs, which may code for positive functional protein in drought stress response. This is in line with results of both induction as well as repression of miRNA expression under drought stress (Chae et al., 2015).

The most extensively studied model plant, *Arabidopsis*, has been shown to respond to drought stress by up-regulating miR156, miR159, miR167, miR168, miR171, miR172, miR319, miR393, miR394a, miR395c, miR395e, miR396, and miR397, and down-regulating miR161, miR168a, miR168b, miR169, miR171a, miR319c, and miR474 (Liu et al., 2008; Zhou et al., 2010). Generally, levels of miRNAs expressed in response to applied stress like drought is species-specific (Barrera-Figueroa et al., 2011). For instance, miR156 that is up-regulated under drought in *Arabidopsis*, barley, and *Triticum dicoccoides* (Eldem et al., 2012; Kantar et al., 2010, 2011; Sun et al., 2012), is found to be down-regulated in rice and maize (Wei et al., 2009; Zhou et al., 2010). Likewise, the level of miR166 was up-regulated in *Medicago truncatula* (Boualem et al., 2008; Trindade et al., 2010), and down-regulated in *Oryza sativa* and *Triticum dicoccoides* (Zhou et al., 2010; Kantar et al., 2011). The developmental stage of the plant and the degree of drought has been shown to alter the miRNA expression pattern, wherein contrasting patterns of miRNA expression have been noted in a tissue-specific and stress-specific manner. Thus, more often, the same miRNA families show differential expression under drought stress, as seen in the case of rice, wherein miR319 family was both up and down-regulated (Zhou et al., 2010). Similarly, levels of miR398a/b in *M. truncatula* under drought condition set up by Trindade et al.

(2010) was up-regulated, while under drought stress imposed by Wang et al. (2011) was down-regulated. In some cases, the expression of the same miRNA in the same plant species under drought conditions may exhibit spatio-temporal differential expression pattern. With the identification of increasing numbers of miRNAs under drought stress, validation of their targets by various experimental and computational methods would provide better insight into the physiological and biochemical role of miR-NAs under stress (Lu et al., 2005).

Complex post-transcriptional regulation of genes by stress-specific and spatio-temporal variation in miRNAs in a species-dependent manner would be better appreciated by annotating and validating targets of various families of miRNAs. Such annotation and validation would help in explaining physiological variations already recognized, and those yet to be explored under drought stress. Attempts have been made in networking/linking the predicted physiological roles of these annotated targets. Such networking may either correlate with well-understood physiological events under drought stress, or may lead to unidentified physiological patterns with newer players (Table 24.2).

miRNAs in Early Drought Responses

As key molecules that regulate gene expression, miRNAs are believed to influence every stage of plant development and growth. Under drought stress, miRNA may influence the drought response in plants from the very onset of drought signaling (Wan et al., 2017). Physiological alteration, such as the reduction of turgor resulting from a water deficit, reduced photosynthetic activity, and consequent reduction in plant growth are thought to be coordinated in order to balance stress response activation and plant development (Fan et al., 1994). As an adaptive response during drought, extraction of water from deep soil is facilitated by suppressing the growth of lateral roots, thereby redirecting resources towards the generation of deeper roots, which are more efficient in fetching water. miRNAs are recognized as important players in the regulation of the complex network involved in reprogramming of plant cell growth during drought (Mallory and Vaucheret, 2006). Functional analyses of conserved miRNAs have revealed that miRNAs regulate various aspects of developmental programs, including auxin signaling, leaf development, and lateral root formation (Jones-Rhoades et al., 2006). Phytohormone, the auxin that plays a central role in the plant growth response signaling pathway, is found to be predominantly regulated by miRNAs (Figure 24.2). TIR1 and GRR1, which are closely related F-box proteins and are known to attack auxin repressors, AUX/IAA proteins, causing proteolysis, are targets of miR393. This proteolysis of Aux/IAA repressor releases the Auxin Response Transcription Factors (ARFs) from repressor ARF complexes facilitating their binding to respective response elements and causes expression of auxin-responsive genes which regulate root development (Guilfoyle and Hagen, 2007). Interestingly, apart from the auxin receptor mRNAs, even the *ARF* mRNAs are subjected to miRNA-mediated regulation directly or indirectly, as inferred from pleiotropic regulation of *ARF10*, *ARF16*, and *ARF17* regulating root growth, which are targets of miR160. This effect of miR160 has also been reported in *A. thaliana*, wherein over-expression of miR160 caused

abnormal root development and growth (Wang et al., 2005). Similarly, levels of ARF 6 and ARF 8, major root growth and gynoecium development regulators are shown to be regulated by miR167 (Lima et al., 2012). ARF3 and ARF4 are also regulated indirectly by miRNAs such as miR390 via cleavage of ta-siRNAs which target mRNAs of ARF3 and ARF4. Highly conserved among plant species, these miRNAs are involved in the modulation of auxin signaling early in the development of plants (Yoon et al., 2010).

Besides influencing auxin signaling, miRNAs have been shown to play a vital role in regulating transcription factors related to growth. TFs, No Apical Meristem (*NAM-NAC*), and cup-shaped cotyledon (*CUC*), which are important in root and shoot development, respectively, are reported to be regulated by miR164 (Hasson et al., 2011). This has been substantiated from athMIR164 mutants with more roots, thus indicating the diverse functional role of the miR164 family (Guo et al., 2005). Another family of miRNA that is involved in cell development and expressed in response to drought stress is miR166. mRNAs of Class III homeodomain-leucine zipper (*HD-Zip III*) transcription factors have been identified as targets of miR166 (Trindade et al., 2010). This apart, over-expression of miR166 in *M. truncatula* has been associated with a reduced number of lateral roots (Boualem et al., 2008). Growth-regulating factors (*GRF*) have been shown to play a crucial role in co-ordination of cell division and differentiation during leaf development in *Arabidopsis* (Jones-Rhoades and Bartel, 2004; Wang et al., 2011) and *Brachypodium* (Bertolini et al., 2013). These GRF mRNA levels have been shown to be regulated in a countervailing manner to levels of a conserved miRNA, miR396. miRNAs have also been found to be associated with senescence as noticed in *A. thaliana*, wherein TFs called teosinte branched/cycloidea proteins (TCP), which regulate jasmonic acid biosynthesis involved in development and leaf senescence, are targets of miR319 (Lima et al., 2012). The transcription factor, TCP was shown to bind to the promoter region of miR164 and cause the inhibition of cell growth (Martin-Trillo and Cubas, 2010). TCP has also been shown to regulate expression of miR396, thus causing growth inhibition by lowering the levels of GRF transcripts (Rodriguez et al., 2010). The regulatory role of miR319 at this stage seems to be more indirect and substantiates miRNA-mediated regulation of miRNAs. The inverse relationship between the miR319 and miR164/miR396 expression level is also observed in *Brachypodium*, supporting the intriguing regulatory network between miR164, miR396, and miR319 (Bertolini et al., 2013).

miRNAs in Drought Signaling

Drought treatment evokes plant responses through an accurate signal perception and transduction, mastered in the due course of evolution. Classically, two transcriptional regulatory networks induced by drought have been well-characterized; an ABA-dependent and another DREB-(dehydration-responsive element binding protein) mediated ABA-independent pathway. Various classes of TFs involved in these pathways are found to be ideal targets of miRNAs, contributing to the post-transcriptional regulation of drought stress response in favor of resistance or avoidance (Zhang, 2015).

TABLE 24.2

Drought Resistance Induced by Expression Pattern of Major miRNAs Under Drought Stress and Physiological Functions of Their Predicted Targets*

miRNA Family	Target	Species	Drought Resistance Mechanism	Response Under Drought	Reference
miR159	MYB	*Arabidopsis thaliana*	ABA signaling and osmotic stress tolerance	Up-regulated	Abe et al. (2003) and Reyes and Chua (2007)
miR160	ARF	*Arabidopsis thaliana*	ABA response		Liu et al. (2007)
miR164	NAC-TF	*Hordeum vulgare*			Kantar et al. (2010)
	MDR-like ABC Transporter	*Oryza sativa*			Zhou et al. (2010)
miR166	HD-Zip	*Medicago truncatula*	Root and nodule development	Up-regulated	Boualem et al. (2008) and Trindade et al. (2010)
		Oryza sativa		Down-regulated	Zhou et al. (2010)
		Triticum dicoccoides		Down-regulated	Kantar et al. (2011)
miR167	ARF	*Arabidopsis thaliana*	ABA response	Up-regulated	Liu et al. (2008)
		Oryza sativa			Liu et al. (2009)
	PLD	*Zea mays*	ABA response and controlling stomatal movement	Down-regulated	Wei et al. (2009)
miR168	AGO	*Arabidopsis thaliana*	miRNA processing	Up-regulated	Liu et al. (2008)
		Oryza sativa		Down-regulated	Zhou et al. (2010)
miR169	NFYA	*Arabidopsis thaliana*	ABA response and controlling stomatal aperture	Down-regulated	Li et al. (2008)
		Medicago truncatula		Down-regulated	Wang et al. (2011)
		Oryza sativa		Up-regulated	Zhao et al. (2007)
		Solanum lycopersicum		Up-regulated	Zhang et al. (2011)
miR319	TCP	*Arabidopsis thaliana*	Jasmonic acid biosynthesis	Down-regulated	Rodriguez et al. (2010)
miR390	TAS3-ARF	*Vigna unguiculata*	Auxin signaling and lateral root development	Up-regulated	Barrera-Figueroa et al. (2011)
miR393	TIR1	*Arabidopsis thaliana*	Auxin signaling	Up-regulated	Sunkar and Zhu (2004) and Liu et al. (2008)
		Oryza sativa		Up-regulated	Zhao et al. (2007)
miR396	GRF	*Arabidopsis thaliana*	Leaf development	Up-regulated	Liu et al. (2008)
		Nicotiana tabacum		Up-regulated	Yang and Yu (2009)
		Oryza sativa		Down-regulated	Zhou et al. (2010)
miR397	β-Fructofuranosidase	*Oryza sativa*	CO_2 fixation	Down-regulated	Zhou et al. (2010)
	Laccase	*Arabidopsis thaliana*	Unknown	Up-regulated	Sunkar and Zhu (2004)
miR398	CSD	*Medicago truncatula*	ROS detoxification	Up-regulated	Trindade et al. (2010)
	COX	*Triticum dicoccoides*	Respiration pathway	Up-regulated	Kantar et al. (2011)
miR 408	PSK	*Oryza sativa*	Cell-to-cell communication	Down-regulated	Khraiwesh et al. (2012)
	PSKR	*P. trichocarpa*	Sulfate stress	Up-regulated	Trindade et al. (2010)
miR474	PDH	*Zea mays*	Proline accumulation	Up-regulated	Wei et al. (2009)
	PPR	*Oryza sativa*	Controlling organelle gene expression	Up-regulated	Zhou et al. (2010)
		Medicago truncatula		Up-regulated	Trindade et al. (2010)
miR528	POD	*Zea mays*	ROS detoxification	Down-regulated	Wei et al. (2009)
miR858	MYB2	*A. mongolicus*	Flavonoids biosynthesis	Down-regulated	Gao et al. (2016)

* Modified from: Ding et al. (2013).

ABA, being a key regulator of water status and stomatal movement, is positively regulated by MYB101 and MYB33. The stability of these TFs has been found to be regulated by over-expressed miR159, by suppressing their transcripts. This has been demonstrated in transgenic plants that are rendered hypo-sensitive to ABA. The same has also been corroborated from transgenic plants over-expressing cleavage-resistant MYB101

and MYB33 transcripts, which are hypersensitive to ABA (Reyes and Chua, 2007).

ABA-mediated guard cell response to drought is regulated by another signaling pathway component, Phospholipase-D (PLD), which directs stomatal movement (Xue et al., 2005). This PLD transcript has been predicted to be a target for miR167 family, which is down-regulated under drought stress, leading to the

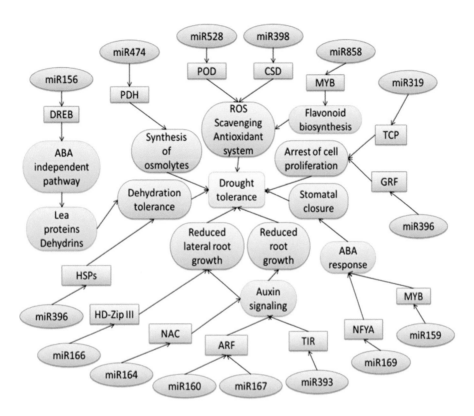

FIGURE 24.2 Probable regulatory network of miRNAs and their target genes in drought response. ARF, auxin response factor; CSD, Cu/Zn-superoxide dismutase; DREB, Dehydration Response Element binding proteins; GRF, growth-regulating factor; HD-Zip, homeodomain-leucine zipper; NFY, nuclear factor Y; PDH, proline dehydrogenase; POD, peroxidase; TIR, transport inhibitor response; HSPs, Heat shockproteins; TCP, Teosinte cycloidea proteins. Modified from Ding et al. (2013).

accumulation of PLD, and drought tolerance via stomatal closure in maize (Wei et al., 2009). Similarly, ABA-mediated drought tolerance has been associated with regulation of another TF, NFYA5 responsible for responding to environmental cues, like drought, and associated oxidative stress by associating with a number of drought-responsive genes, such as peroxidases (POD) and glutathione transferases (GT), as ascertained by microarray methods (Kumimoto et al., 2008). NFYA5 transcripts have been predicted to be targets of miR169a as reported by Li et al. (2008) who demonstrated induction of NFYA5 under down-regulation of miR169a in *A. thaliana* treated with ABA and drought stress.

Drought-response signaling via the ABA-independent pathway involves DREB TFs binding DRE (Dehydration Response Elements) to influence various drought-responsive genes. Although DREB-mediated drought response is not fully elucidated yet, enhanced drought tolerance through the DREB-mediated signaling pathway is considered to involve transcriptional activation of genes such as LEA proteins, thought to protect macromolecules such as enzymes from dehydration, and dehydrins that help in maintaining the original cell volume (Hanin et al., 2011). Prediction of DREB TFs as targets of miR156 and miR168 in *Jatropha* (Galli et al., 2014), further supports the role of miRNAs in drought tolerance by regulating key players of drought tolerance at the transcriptional level.

A group of proteins with WD motif repeats are shown to function in signal transduction and transcriptional control. The WD motifs act as sites for protein-protein interaction and serve as platforms for the assembly of protein complexes/mediators (Bertolini,

2013). The transcripts of these WD motif proteins have been predicted to be targets of two miRNAs: miRCB23a and miRCB23b from *Brachypodium*. Similarly, miR1876 and miR399j have been shown to be targeting WD motif proteins as a part of the drought signaling pathway. WD40 repeat-containing protein (DWD) and CBL interacting protein kinase (CIPK), respectively, are targets of miR1876 and miR399j. While DWD has been reported to play a negative role in ABA signaling, CIPK is reported to be involved in Ca^{2+} signaling. EF-hand family proteins which play a pivotal role as Ca^{2+} sensors like CDPK have been shown to be regulated by miRNAs under drought stress. For example, miR1318/1432 have been shown to be down-regulated under drought stress, leading to increased activity of the Ca^{2+} sensors, followed by increased expression of stress-responsive genes (Bakhshi et al., 2016).

Two drought-responsive miRNAs, miR408 and miR164a/b/f are implicated in drought stress signaling in rice through regulation of phytosulfokine (PSK) and phytosulfokine receptor (PSKR) co-operation. PSKR receptor kinase, binding to its signaling molecule PSK, essential for cell-to-cell communication, probably becomes highly active under drought stress, due to down-regulation of both miR408 and miR164a/b/f (Bakhshi et al., 2016).

miRNA-Induced Physiological Responses under Drought

Regulation of a number of well-characterized players involved in physiological adjustments, which help plants to resist effects of drought and other abiotic stress, are being linked to miRNAs. Osmotic adjustment is a basic mechanism by which cell turgidity is maintained to stabilize structural integrity of proteins and other macromolecules during drought stress. Plants cope with drought stress through osmotic adjustment, achieved by the accumulation of various osmoprotectants or osmolytes such as proline, glycine betaine, sugar, and sugar alcohol such as mannitol. Proline displays a multifunctional role in defense mechanisms, as an osmolyte and a stress-related signaling molecule (Nanjo et al., 1999). Under drought stress, proline accumulation is achieved by its biosynthesis and preventing degradation of pre-formed proline. In higher plants, proline degradation to glutamic acid is catalyzed in mitochondria by proline dehydrogenase (PDH) and P5C-dehydrogenase (P5CDH). As PDH is the first enzyme in the proline oxidation pathway, suppression of its gene leads to proline accumulation (Reddy et al., 2004; Nanjo et al., 2003). Post-transcriptional regulation of PDH mRNA by miRNA under drought has been found to be associated with elevated accumulation of proline. This is exemplified by elevated levels of miR474 in maize (Wei et al., 2009), rice (Zhou et al., 2010), and *M. truncatula* (Trindade et al., 2010).

Heat shock proteins (HSPs) are major players in drought response at physiological levels. By acting as molecular chaperones, they contribute to balance/restoration of protein homeostasis (Guan et al., 2013). The recent identification of a conserved miRNA, miR396d-e, and three novel miRNAs; sbi-miR-26, sbi-miR-85a-k, and sbi-miR-336, which target HSP mRNAs demonstrates post-transcriptional gene regulation of HSPs by miRNAs in *Sorghum bicolor*. In line with these observations, Katiyar et al. (2015) have reported down-regulation of these families of

miRNA causing elevated levels of HSP and resultant drought tolerance.

It is a well-established fact that drought induces oxidative stress by the generation of reactive oxygen species (ROS) in different cellular compartments, chloroplasts, mitochondria, and peroxisomes leading to irreversible damage to the plants (Cruz, 2008; Gill and Tuteja, 2010). At a physiological level, various antioxidants and antioxidant enzymes have been proved to be major players in drought responses (Mittler, 2002). The first cellular defense mechanism against oxidative stress, scavenging the (ROS) by converting superoxide radicals to H_2O_2 and O_2, is catalyzed by superoxide dismutase (SOD) (Fridovich, 1995). Cu/Zn–SOD, a type of copper enzyme encoded by two closely related genes, cytosolic *CSD1* and plastidic *CSD2*, is known to be targeted by miR398. Detoxification of superoxide radicals, as a physiological drought tolerance mechanism, is due to the post-transcriptional induction of the two genes mediated by down-regulation of miR398 (Sunkar et al., 2006). Decreased levels of miR398 in drought-stressed barley, contributing to higher transcript levels of *CSD1* and *CSD2*, demonstrates the ability of a single miRNA to have multiple targets (Ferdous et al., 2017).

Peroxidase (POD), another key enzyme that is known to confer antioxidant defense and regulate stomatal movement, is also reported to be the target of a miRNA, miR528. Accumulation of *POD* transcripts in drought-stressed maize, validated by qRT-PCR confirmed down-regulation of miR528 levels, and accounted for the removal of excessive peroxide radicals (Wei et al., 2009). Apart from antioxidant enzymes, enzymes of flavonoid biosynthesis, which contribute to drought tolerance via excessive flavonoid synthesis, have also been identified as

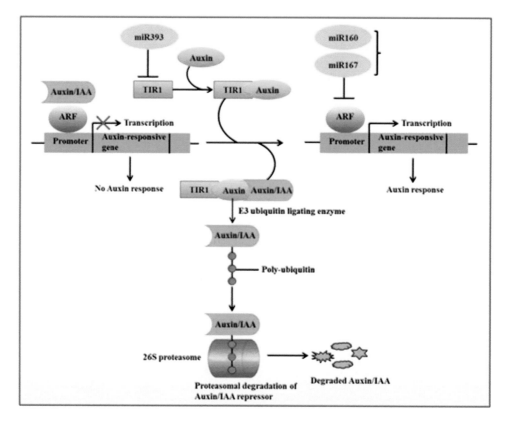

FIGURE 24.3 miRNA-mediated gene regulation upon auxin perception and signaling under normal growth and stress condition. Source: Kumar (2014).

targets of miRNA mediated post-transcriptional gene regulation. Demonstration of up-regulated MYB genes, which positively regulate flavonoid biosynthetic genes in *A. mongolicus* (Gao et al., 2016), coinciding with decreased levels of miR858 under drought stress, substantiated the role of miRNA in this physiological adjustment (Figure 24.3).

Auto-Regulation of miRNA Expression

miRNA-mediated responses occur via modulation of self-expression as well as that of their targets under stressed conditions. Further, in response to stress, these target miRNAs, in turn, can modify the expression of other protein-coding genes. miRNA biogenesis plays a crucial role in development and, hence, understanding the key molecules which are regulated to establish a homeostasis is critical. Two miRNAs and their target genes central to miRNA biogenesis and function are miR162 that targets DCL1 (Dicer like1) and miR168 that targets AGO1 (Argonaute1) (Jones-Rhoades et al., 2006). DCL1, an integral component of miRNA biogenesis pathway, functions by cleaving primary miRNAs (pri-miRNAs) and precursor miRNAs (pre-miRNAs) to form mature miRNA (Kurihara and Watanabe, 2004). AGO1, the catalytic subunit of the RISC, is responsible for post-transcriptional gene silencing. Expression of AGO1 is controlled through a feedback loop involving miR168 that ensures the maintenance of AGO1 homeostasis, crucial for the correct functioning of the miRNA pathway. The increased accumulation of mRNAs targeted for cleavage by miRNAs in AGO1 mutants of *A. thaliana* can be considered to propose that AGO1 acts within the miRNA pathway, probably within the miRNA programmed RISC, wherein a lack of AGO1 destabilizes few miRNAs (Lima et al., 2012). There are also reports suggesting the necessity of regulation of *AGO1* transcript by miR168 for proper plant development, illustrating the importance of feedback control by this miRNA (Vaucheret et al., 2004). The fact that DCL1 and AGO1 are also miRNA targets, suggests that plant miRNAs tune their own biogenesis and function.

Conclusions

Population growth, urbanization, competition for fresh water between agricultural lands and cities, global warming, and environmental changes due to changing lifestyles demand the immediate development of efficient strategies by which food production can be increased. The biochemical and molecular basis of plant tolerance to drought has been intensely investigated over the past decades. These investigations have identified numerous genes that are induced under drought and salt stress, with the anticipation that over-expression of these stress-responsive genes would improve plant stress tolerance. However, such transgenics have provided little improvements in stress tolerance, largely because of complicated genetic interactions underlying plant stress tolerance, which are still poorly understood. Therefore, exploration of the post-transcriptional regulatory system through which group of genes can be manipulated by miRNAs appears to be promising. Although an increasing number of conserved and novel miRNAs have been identified and cataloged, the functions of the majority of them are obscured. Thus, the priority would be an annotation of their targets, rather than mere identification. Since much of the data pertaining to the miRNA and their targets have originated via bio-informatics methods, an experimental demonstration of the proposed regulatory outcome becomes inevitable to employ miRNAs as drought tolerant modulators in transgenic crops.

Acknowledgments

The authors thank DST for Financial assistance under WOS-A, Inspire, SERB, and FIST schemes.

REFERENCES

Abe H, Urao T, Ito T, Seki M, Shinozaki K, Yamaguchi-Shinozaki K. Arabidopsis AtMYC2 (bHLH) and AtMYB2 (MYB) function as transcriptional activators in abscisic acid signaling. *Plant Cell* 2003; 15: 63–78.

An J, Lai J, Sajjanhar A, Lehman, ML, Nelson CC. miRPlant: An integrated tool for identification of plant miRNA from RNA sequencing data. *BMC Bioinf.* 2014; 15: 275. doi:10.1186/1471-2105-15-275.

Arenas-Huertero C, Perez B, Rabanal F, Blanco-Melo D, De la Rosa C, Estrada-Navarrete G, Sanchez F, Covarrubias AA, Reyes JL. Conserved and novel miRNAs in the legume *Phaseolus vulgaris* in response to stress. *Plant Mol. Biol.* 2009; 70: 385–401.

Axtell MJ, Bartel DP. Antiquity of microRNAs and their targets in land plants. *Plant Cell* 2005; 17: 1658–1673.

Bakhshi B, Fard EM, Nikpay N, Ebrahimi MA, Bihamta MR, Mardi M, Salekdeh GH. MicroRNA signatures of drought signaling in rice root. *PLoS ONE* 2016; 11(6): e0156814. doi:10.1371/journal.pone.0156814.

Baras AS, Mitchell CJ, Myers JR, Gupta S, Weng LC, Ashton JM, Cornish TC, Pandey A, Halushka MK. miRge—A multiplexed method of processing small RNA-seq data to determine microRNA entropy. *PLoS ONE* 2015; 10: e0143066. doi:10.1371/journal.pone.0143066.

Barrera-Figueroa BE, Gao L, Diop NN, Wu Z, Ehlers JD, Roberts PA, Close TJ, Zhu JK, Liu R. Identification and comparative analysis of drought-associated microRNAs in two cowpea genotypes. *BMC Plant Biol.* 2011; 11(1): 127.

Bartel D. MicroRNAs: Genomics, biogenesis, mechanism, and function. *Cell* 2004; 116: 281–297.

Bertolini E, Verelstb W, Hornerd DS, Gianfranceschid L, Piccolod V, Inzéb D, Pèa ME, Micaa E. Addressing the role of microRNAs in reprogramming leaf growth during drought stress in Brachypodium distachyon. *Mol. Plant* 2013; 6(2): 423–443.

Bonnet E, He Y, Billiau K, Van de Peer Y. TAPIR, a web server for the prediction of plant micro RNA targets, including target mimics. *Bioinformatics* 2010; 26: 1566–1568. doi:10.1093/bioinformatics/btq233.

Boualem A, Laporte P, Jovanovic M, Laffont C, Plet J, Combier JP, Niebel A, Crespi M, Frugier F. MicroRNA166 controls root and nodule development in Medicagotruncatula. *Plant J.* 2008; 54: 876–887. doi:10.1111/j.1365-313X.2008.03448.x.

Chae H, Rhee S, Nephew KP, Kim S. BioVLAB-MMIA-NGS: microRNA-mRNA integrated analysis using high through-put sequencing data. *Bioinformatics* 2015; 31: 265–267. doi:10.1093/bioinformatics/btu614.

Chiang K, Shu J, Zempleni J, Cui J. Dietary microRNA database (DMD): An archive database and analytic tool for food-borne microRNAs. *PLoS ONE* 2015; 10: e0128089. doi:10.1371/journal.pone.0128089.

Couzigou JM, Lauressergues D, Bécard G, Combier JP. miRNA-encoded peptides (miPEPs): A new tool to analyze the roles of miRNAs in plant biology. *RNA Biol.* 2015; 12: 1178–1180. doi:10.1080/15476286.2015.10 94601.

Covarrubias AA, Reyes JL. Post-transcriptional gene regulation of salinity and drought responses by plant microRNAs. *Plant Cell Environ.* 2010; 33(4): 481–489.

Cruz CMH. Drought stress and reactive oxygen species: Production, scavenging and signaling. *Plant Signal. Behav.* 2008; 3: 156–165.

Dezulian T, Remmert M, Palatnik JF, Weigel D, Huson DH. Identification of plant microRNA homologs. *Bioinformatics* 2006; 22(3): 359–360.

Ding Y, Tao Y, Zhu C. Emerging roles of microRNAs in the media-tion of drought stress response in plants. *J. Exp. Bot.* 2013; 64: 3077–3086.

Eldem V, Akcay UC, Ozhuner E, Bakir Y, Uranbey S, Unver T. Genome-wide identification of miRNAs responsive to drought in peach (*Prunus persica*) by high-throughput deep sequenc-ing. *PLoS ONE* 2012; 7: e50298.

Evers M, Huttner M, Dueck A, Meister G, Engelmann JC. miRA: Adaptable novel miRNA identification in plants using small RNA sequencing data. *BMC Bioinf.* 2015; 16: 370. doi:10.1186/s12859-015-0798-3.

Fahlgren N, Hill ST, Carrington JC, Carbonell A. P-SAMS: A web site for plant artificial microRNA and synthetic trans-acting small interfering RNA design. *Bioinformatics* 2016; 32: 157–158. doi:10.1093/bioinformatics/btv534.

Fan S, Blake TJ, Blumwald E. The relative contribution of elastic and osmotic adjustments to turgor maintenance of woody spe-cies. *Physiol. Plant.* 1994; 90(2): 408–413.

Ferdous J, Carlos J, Sanchez-Ferrero JC, Langridge P, Milne L, Chowdhury J, Brien C, Tricker PJ. Differential expression of microRNAs and potential targets under drought stress in bar-ley. *Plant Cell Environ.* 2017; 40: 11–24.

Ferdous J, Hussain SS, Shi BJ. Role of microRNAs in plant drought tolerance. *Plant Biotechnol. J.* 2015; 13(3): 293–305.

Frazier TP, Sun G, Burklew CE, Zhang BH. Salt and drought stresses induce the aberrant expression of microRNA genes in tobacco. *Mol. Biotechnol.* 2011; 49: 159–165.

Fridovich I. Superoxide radical and superoxide dismutases. *Annu. Rev. Biochem.* 1995; 64: 97–112.

Galli V, Guzman F, de Oliveira LFV, Loss-Morais G, Körbes AP, Silva SD, Margis-Pinheiro MM, Margis R. Identifying microRNAs and transcript targets in Jatropha seeds. *PLoS ONE* 2014; 9(2): e83727. doi:10.1371/journal.pone.0083727.

Gao F, Wang N, Li H, Liu J, Fu C, Xiao Z, Wei C, Lu X, Feng J, Zhou Y. Identification of drought responsive microRNAs and their targets in *Ammopiptanthus mongolicus* by using highthroughput sequencing. *Sci. Rep.* 2016; 6: 34601.

Gill SS, Tuteja N. Reactive oxygen species and antioxidant machin-ery in abiotic stress tolerance in crop plants. *Plant Physiol. Biochem.* 2010; 48: 909–930.

Guan Q, Lu X, Zeng H, Zhang Y, Zhu J. Heat stress induc-tion of miR398 triggers a regulatory loop that is critical for thermotolerance in Arabidopsis. *Plant J.* 2013; 74(5): 840–851.

Guilfoyle TJ, Hagen G. Auxin response factors. *Curr. Opin. Plant Biol.* 2007; 10: 453–460. doi:10.1016/j.pbi.2007.08.014.

Guo HS, Xie Q, Fei JF, Chua NH. MicroRNA directs mRNA cleav-age of the transcription factor NAC1 to downregulate auxin signals for arabidopsis lateral root development. *Plant Cell* 2005; 17: 1376–1386.

Hackenberg M, Rodríguez-Ezpeleta N, Aransay AM. miRanalyzer: An update on the detection and analysis of microRNAs in high-throughput sequencing experiments. *Nucl. Acids Res.* 2011; 39: W132–W138. doi:10.1093/nar/gkr247.

Hanin M, Brini F, Ebel C, Masmoudi K. Plant dehydrins and stress tolerance. *Plant Signal. Behav.* 2011; 6: 10, 1–7.

Hasson A, Plessis A, Blein T, Adroher B, Grigg S, Tsiantis M, Boudaoud A, Damerval C, Laufs P. Evolution and diverse roles of the CUP-SHAPED COTYLEDON genes in *Arabidopsis* leaf development. *Plant Cell* 2011; 23: 54–68.

Hsiao TC, Xu LK. Sensitivity of growth of roots versus leaves to water stress: Biophysical analysis and relation to water trans-port. *J. Exp. Bot.* 2000; 51: 1596–1616.

Hsu SD, Lin FM, Wu WY, Liang C, Huang WC, Chan WL, Tsai WT, et al. miRTarBase: A database curates experimentally validated microRNA-target interactions. *Nucl. Acids Res.* 2011; 39: D163–D169. doi:10.1093/nar/gkq1107.

Huang TH, Fan B, Rothschild MF, Hu ZL, Li K, Zhao SH. MiRFinder: An improved approach and software implemen-tation for genome-wide fast microRNA precursor scans. *BMC Bioinf.* 2007; 8: 341.

Jones-Rhoades MW, Bartel DP. Computational identification of plant microRNAs and their targets, including a stress-induced miRNA. *Mol. Cell* 2004; 14: 787–799.

Jones-Rhoades MW, Bartel DP, Bartel B. MicroRNAs and their regulatory roles in plants. *Annu. Rev. Plant Biol.* 2006; 57: 19–53.

Kantar M, Lucas SJ, Budak H. miRNA expression patterns of *Triticum dicoccoides* in response to shock drought stress. *Planta* 2011; 233: 471–484.

Kantar M, Unver T, Budak H. Regulation of barley miRNAs upon dehydration stress correlated with target gene expression. *Funct. Integr. Genom.* 2010; 10: 493–507.

Katiyar A, Smita S, Muthusamy SK, Chinnusamy V, Pandey DM, Bansal KC. Identification of novel drought-responsive microR-NAs and trans-acting siRNAs from Sorghum bicolor(L.) Moench by high-throughput sequencing analysis. *Front. Plant Sci.* 2015; 6: 506. doi:10.3389/fpls.2015.00506.

Khraiwesh B, Zhu JK, Zhu JH. Role of miRNAs and siRNAs in biotic and abiotic stress responses of plants. *Biochim. Biophys. Acta* 2012; 1819: 137–148.

Kozomara A, Griffiths-Jones S. miRBase: Annotating high confi-dence microRNAs using deep sequencing data. *Nucl. Acids Res.* 2014; 42: D68–D73. doi:10.1093/nar/gkt1181.

Kramer PJ, Boyer JS. *Water Relations of Plants and Soils.* 1st edi-tion, Academic Press, London, 1995.

Kulcheski FR, de Oliveira LF, Molina LG, Almerão MP, Rodrigues FA, Marcolino J, Barbosa JF, et al. Identification of novel soy-bean microRNAs involved in abiotic and biotic stresses. *BMC Genom.* 2011; 12: 307.

Kumar R. Role of microRNAs in biotic and abiotic stress responses in crop plants. *Appl. Biochem. Biotechnol.* 2014; 174(1): 93–11.

Kumimoto RW, Adam L, Hymus GJ, Repetti PP, Reuber TL, Marion CM, Hempel FD, Ratcliffe OJ. The nuclear factor Y subunits NF-YB2 and NF-YB3 play additive roles in the promotion of flowering by inductive long-day photoperiods in Arabidopsis. *Planta* 2008; 228: 709–723.

Kurihara Y, Watanabe Y. Arabidopsis micro-RNA biogenesis through Dicer-like 1 protein functions. *Proc. Natl. Acad. Sci. USA* 2004; 101(34): 12753–12758.

Lee R, Feinbaum R, Ambros V. The *C. Elegans* heterochronic gene lin-4 encodes small RNAs with antisense complementarity to lin-14. *Cell* 1993; 75(5): 843–854.

Lee Y, Ahn C, Han J, Choi H, Kim J, Yim J, Lee J, Provost P, Radmark O, Kim S, Kim VN. The nuclear RNase III Drosha initiates microRNA processing. *Nature* 2003; 425(6956): 415–419.

Li Y, Wan L, Bi S, Wan Z, Li Z, Cao J, Tong Z, Xu H, He F, Li X. Identification of drought responsive microRNAs from roots and leaves of Alfalfa by High-Throughput sequencing. *Genes* 2017; 8: 119.

Liu HH, Tian X, Li YJ, Wu CA, Zheng CC. Microarray-based analysis of stress-regulated microRNAs in *Arabidopsis thaliana*. *RNA* 2008; 14(5): 836–843.

Liu PP, Montgomery TA, Fahlgren N, Kasschau KD., Nonogaki H, Carrington JC. Repression of AUXIN RESPONSE FACTOR10 by microRNA160 is critical for seed germination and post-germination stages. *Plant J.* 2007; 52: 133–146.

Liu Q, Zhang YC, Wang CY, Luo YC, Huang QJ, Chen SY, Zhou H, Qu LH, Chen YQ. Expression analysis of phytohormone regulated microRNAs in rice, implying their regulation roles in plant hormone signaling. *FEBS Lett.* 2009; 583: 723–728.

Lima JC, Loss-Morais G, Margis R. MicroRNAs play critical roles during plant development and in response to abiotic stresses. *Genet. Mol. Biol.* 2012; 35(4) (suppl): 1069–1077.

Llave C, Xie Z, Kasschau KD, Carrington JC. Cleavage of Scarecrow-like mRNA targets directed by a class of *Arabidopsis* miRNA. *Science* 2002; 297: 2053–2056.

Lorenzetti APR, de Antonio GYA, Paschoal AR, Domingues DS. PlanTE-MIR DB: A database for transposable element-related microRNAs in plant genomes. *Funct. Integr. Genom.* 2016; 16: 235–242. doi:10.1007/s10142-016-0480-5.

Lu S, Sun YH, Shi R, Clark C, Li L, Chiang VL. Novel and mechanical stress-responsive MicroRNAs in *Populus trichocarpa* that are absent from *Arabidopsis*. *Plant Cell* 2005; 17: 2186–2203.

Mallory AC, Vaucheret H. Functions of microRNAs and related small RNAs in plants. *Nat. Genet.* 2006; 38: S31–S36. doi:101038/ng1791.

Martin-Trillo M, Cubas P. TCP genes: A family snapshot ten years later. *Trends Plant Sci.* 2010; 15: 31–39.

Martinez-Sanchez A, Murphy CL. MicroRNA target identification experimental approaches. *Biology* 2013; 2: 189–205.

Meng J, Shi GL, Luan YS. Plant miRNA function prediction based on functional similarity network and transductive multi-label classification algorithm. *Neuro Comput.* 2016; 179: 283–289. doi:10.1016/j.neucom.2015.12.011.

Meng Y, Gou L, Chen D, Mao C, Jin Y, Wu P, et al. PmiRKB: a plant microRNA knowledge base. *Nucleic Acids Res.* 2011; 39: D181–D187. doi:10.1093/nar/gkq721

Mittler R. Oxidative stress, antioxidants and stress tolerance. *Trends Plant Sci.* 2002; 7: 405–410.

Muljo SA, Kanellopoulou C, Aravind L. MicroRNA targeting in mammalian Genomes: Genes and mechanisms. *Wiley Interdiscip. Rev. Syst. Biol. Med.* 2010; 2(2): 148–161.

Muñoz-Mérida A, Perkins JR, Viguera E, Thode G, Bejarano ER, Pérez-Pulido AJ. Semirna: Searching for plant miRNAs using target sequences. *Omics* 2012; 16: 168–177. doi:10.1089/omi.2011.0115.

Nanjo T, Fujita M, Seki M. Kato T, Tabata S, Shinozaki K. Toxicity of free proline revealed in an Arabidopsis T-DNA-tagged mutant deficient in proline dehydrogenase. *Plant Cell Physiol.* 2003; 44: 541–548.

Nanjo T, Kobayashi M, Yoshiba Y, Sanada Y, Wada K, Tsukaya H, Kakubari Y, Yamaguchi-Shinozaki K, Shinozaki K. Biological functions of proline in morphogenesis and osmotolerance revealed in antisense transgenic Arabidopsis thaliana. *Plant J.* 1999; 18: 185–193.

Neumann PM. Coping mechanisms for crop plants in drought-prone environments. *Ann. Bot.* 2008; 101(7): 901–907.

Numnark S, Mhuantong W, Ingsriswang S, Wichadakul D. C-mii: A tool for plant miRNA and target identification. *BMC Genom.* 2012; 13: S16. doi:10.1186/1471-2164-13-S7-S16.

Park W, Li J, Song R, Messing J, Chen X. CARPEL FACTORY, a Dicer homolog, and HEN1, a novel protein, act in microRNA metabolism in *Arabidopsis thaliana*. *Curr. Biol.* 2002; 12: 1484–1495.

Pashkovskiy PP, Ryazansky SS. Biogenesis, evolution, and functions of plant microRNAs. *Biochemistry* 2013; 78(6): 627–637.

Patel P, Ramachandruni SD, Kakrana A, Nakano M, Meyers BC. miTRATA: A web-based tool for microRNA truncation and tailing analysis. *Bioinformatics* 2016; 32: 450–452. doi:10.1093/bioinformatics/btv583.

Reddy AR, Chaitanya KV, Vivekanandan M. Drought induced responses of photosynthesis and antioxidant metabolism in higher plants. *J. Plant Physiol.* 2004; 161: 1189–1202.

Remita MA, Lord E, Agharbaoui Z, Leclercq M, Badawi M, Makarenkov V, Sarhan F, Diallo AB. WMP: A novel comprehensive wheat miRNA database, including related bioinformatics software. BioRxiv. 2015; 024893. doi:10.1101/024893.

Reyes JL, Chua NH. ABA induction of miR159 controls transcript levels of two MYB factors during Arabidopsis seed germination. *Plant J.* 2007; 49: 592–606.

Rhee S, Chae H, Kim S. PlantMirnaT: miRNA and mRNA integrated analysis fully utilizing characteristics of plant sequencing data. *Methods* 2015; 83: 80–87. doi:10.1016/j.ymeth.2015.04.003.

Rodriguez RE, Mecchia MA, Debernardi JM, Schommer C, Weigel D, Palatnik JF. Control of cell proliferation in Arabidopsis thaliana bymicroRNA miR396. *Development* 2010; 137: 103–112.

Rueda A, Barturen G, Lebrón R, Gómez-Martín C, Alganza Á, Oliver JL, Hackenberg M. sRNAtoolbox: An integrated collection of small RNA research tools. *Nucl. Acids Res.* 2015; 43: W467–W473. doi:10.1093/nar/gkv555.

Rüegger S, Grobhans H. MicroRNA turnover: When, how, and why. *Trends Biochem. Sci.* 2012; 37(10): 436–446.

Shinozaki K, Yamaguchi-Shinozaki K. Gene networks involved in drought stress response and tolerance. *J. Exp. Bot.* 2006; 58(2): 221–227.

Shriram V, Kumar V, Devarumath RM, Khare TS, Wani SH. MicroRNAs as potential targets for abiotic stress tolerance in plants. *Front. Plant Sci.* 2016; 7: 817. doi:10.3389/fpls.2016.00817.

Shuai P, Liang D, Zhang Z, Yin W, Xia X. Identification of drought-responsive and novel *Populus trichocarpa* microRNAs by high-throughput sequencing and their targets using degradome analysis. *BMC Genom.* 2013; 14: 233.

Sun G. MicroRNAs and their diverse functions in plants. *Plant Mol. Biol.* 2012; 80: 17–36. doi:10.1007/s11103-011-9817-6.

Sunkar R, Kapoor A, Zhu JK. Posttranscriptional induction of two Cu/Zn superoxide dismutase genes in Arabidopsis is mediated by down-regulation of miR398 and important for oxidative stress tolerance. *Plant Cell* 2006; 18: 2051–2065.

Sunkar R, Zhu JK. Novel and stress-regulated microRNAs and other small RNAs from *Arabidopsis. Plant Cell* 2004; 16: 2001–2019.

Trindade I, Capitao C, Dalmay T, Fevereiro MP, Santos DM. miR398 and miR408 are upregulated in response to water deficit in Medicagotruncatula. *Planta* 2010; 231: 705–716.

Vaucheret H, Vazquez F, Crete P, Bartel DP. The action of ARGONAUTE1 in the miRNA pathway and its regulation by the miRNA pathway are crucial for plant development. *Genes Dev.* 2004; 18: 1187–1197.

Voinnet O. Origin, biogenesis, and activity of plant microRNAs. *Cell* 2009; 136: 669–687.

Wang JW, Wang LJ, Mao YB, Cai WJ, Xue HW, Chen XY. Control of root cap formation by microRNA-targeted auxin response factors in Arabidopsis. *Plant Cell* 2005; 17: 2204–2216.

Wang T, Chen L, Zhao M, Tian Q, Zhang WH. Identification of drought-responsive microRNAs in *Medicago truncatula* by genome-wide high throughput sequencing. *BMC Genom.* 2011; 12: 367.

Wei L, Zhang D, Xiang F, Zhang Z. Differentially expressed miRNAs potentially involved in the regulation of defense mechanism to drought stress in maize seedlings. *Int. J. Plant Sci.* 2009; 170: 979–989.

Xue C, Li F, He T, Liu G P, Li Y, Zhang X. Classification of real and pseudo microRNA precursors using local structure-sequence features and support vector machine. *BMC Bioinf.* 2005; 6: 310.

Yang F, Yu D. Overexpression of Arabidopsis miR396 enhances drought tolerance in transgenic tobacco plants. *Acta Bot. Yunnan.* 2009; 31: 421–426.

Yang X, Li L. miRDeep-P: A computational tool for analyzing the microRNA transcriptome in plants. *Bioinformatics* 2011; 27: 2614–2615. doi:10.1093/bioinformatics/btr430.

Yi X, Zhang Z, Ling Y, Xu W, Su Z. PNRD: A plant non-coding RNA database. *Nucl. Acids Res.* 2015; 43: D982–D989. doi:10.1093/nar/gku1162.

Yoon EK, Yang JH, Lim J, Kim SH, Kim SK, Lee WS. Auxin regulation of the microRNA390-dependent transacting small interfering RNA pathway in *Arabidopsis* lateral root development. *Nucl. Acids Res.* 2010; 38: 1382–1391.

Zhang B. MicroRNA: a new target for improving plant tolerance to abiotic stress. *J. Exp. Bot.* 2015; 66: 1749–1761. doi:10.1093/jxb/erv013.

Zhang B, Pan X, Cannon CH, Cobb GP, Anderson TA. Conservation and divergence of plant microRNA genes. *Plant J.* 2006; 46: 243–259.

Zhang S, Yue Y, Sheng L, Wu Y, Fan G, Li A, Hu X, ShangGuan M, Wei C. *PASmiR*: A literature-curated database form iRNA molecular regulation in plant response to abiotic stress. *BMC Plant Biol.* 2013; 13: 33. doi:10.1186/1471-2229-13-33.

Zhang X, Zou Z, Gong P, Zhang J, Ziaf K, Li H, Xiao F, Ye Z. Overexpression of microRNA169 confers enhanced drought tolerance to tomato. *Biotechnol. Lett.* 2011; 33: 403–409.

Zhang Z, Jiang L, Wang J, Chen M. MTide: An integrated tool for the identification of miRNA-target interaction in plants. *Bioinformatics* 2014; 31: 290–291. doi:10.1093/bioinformatics/btu633.

Zhang Z, Yu J, Li D, Zhang Z, Liu F, Zhou X, Wang T, Ling Y, Su Z. PMRD: Plant microRNA database. *Nucl. Acids Res.* 2010; 38: D806–D813. doi:10.1093/nar/gkp818.

Zhao B, Liang R, Ge L, Li W, Xiao H, Lin H, Ruan K, Jin Y. Identification of drought-induced microRNAs in rice. *Biochem. Biophys. Res. Commun.* 2007; 354: 585–590.

Zhou LG, Liu YH, Liu ZC, Kong DY, Duan M, Luo LJ. Genome-wide identification and analysis of drought-responsive microRNAs in *Oryza sativa. J. Exp. Bot.* 2010; 61: 4157–4168.

25

Proteomics of Salinity Stress: Opportunities and Challenges

Shweta Jha

CONTENTS

Abbreviations

2-D DIGE	Two-dimensional differential gel electrophoresis
2-DE	Two-dimensional gel electrophoresis
CBL	Calcineurin B-like protein
CBS	Cystathionine β-synthase
CIPK	CBL-interacting protein kinase
ESI MS/MS	Electron spray ionization coupled with tandem mass spectrometry
ICAT	Isotope-coded affinity tags
iTRAQ	Isobaric tags for relative and absolute quantitation
LC MS/MS	Liquid chromatography coupled with tandem mass spectrometry
LEA	Late embryogenesis abundant
MALDI-TOF	Matrix-associated laser desorption ionization-time of flight
NGS	Next generation sequencing
Q-TOF	Quadruple time of flight
SILAC	Stable isotope labeling of amino acids

Introduction

Abiotic stresses such as high or low temperature, salinity, water deficit, drought, flooding, etc. cause extensive loss to crop production worldwide and comprise major challenges for world food security, especially in developing countries. Soil salinity is one of the most serious environmental stresses in arid and semi-arid regions, which poses significant threats to plant growth and productivity. According to the Food and Agriculture Organization of the United Nations (FAO), over 6–20% of the total land in the world is affected by salinity, and by 2050 this loss of arable land is expected to increase by up to 50% due to increased salinization, which may have a devastating global impact.

High salinity causes accumulation of toxic ions (Na^+ and Cl^-) within the cell (ion toxicity) as well as a reduction in the external soil water potential, resulting in water deficit and hyper-osmotic stress followed by oxidative damage in plants. Most plants under salt stress conditions exhibit a decrease in photosynthesis rate and wilting, resulting in reduced plant growth or even death (Munns, 2002). This grave situation has stimulated interest in plant engineering for salt tolerance. To cope with salinity stress, plants have evolved complex mechanisms that facilitate cell metabolism under stressed conditions and include morphological, developmental, physiological, and biochemical strategies. Examples include selective ion uptake or exclusion; sequestration of toxic ions into vacuoles; synthesis and accumulation of antioxidative enzymes/antioxidants and osmolytes such as proline, mannitol, sucrose, trehalose, and glycine betaine; and adjustment of energy metabolism and photosynthesis (Munns and Tester, 2008). This active reversible acclimation to stress conditions occurs via profound changes in gene expression, which result in alteration of the plant transcriptome, proteome, and metabolome.

Knowledge of plants' responses and adaptation at cellular and molecular levels to adverse environmental conditions is necessary for effective engineering of plants for stress tolerance. Plant responses to abiotic stress and stress signaling pathways have been studied for a long time (Hanin et al., 2016). Recently, considerable progress in determining the mechanisms conferring stress tolerance has been achieved by the use of modern genetic approaches. Some important genes encoding proteins for osmolyte synthesis, ion channels, signaling factors, antioxidative enzymes, transcription factors, metabolic enzymes, redox regulatory proteins, etc. have been cloned and characterized, which revealed the basic functions of these genes in plants' response and adaptation to salinity (Munns, 2005; Hanin et al., 2016). Examples include stress associated proteins SAP1/11 (Mukhopadhyaya et al., 2004; Tyagi et al., 2014); eukaryotic translation initiation factor eIF4A (Bhadra Rao et al., 2017); small GTP-binding protein Rab7 (Agarwal et al., 2008); voltage dependent anion channel VDAC (Desai et al., 2006); transcription factors DREB2A (Agarwal et al., 2010), DREB1B (Gutha et al., 2008), and NAC1 (Ramegowda et al., 2012); glyoxalase pathway enzymes gly I and II (Singla-Pareek et al., 2003; Ghosh et al., 2014); LEA protein gene HVA1 (Xu et al., 1996) and Rab16A (Ganguly et al., 2012); CCCH-tandem zinc finger protein-TZF1 (Jan et al., 2013); CBS domain containing protein (Singh et al., 2012); vacuolar Na^+/H^+ antiporter (Apse et al., 1999); annexin (Jami et al., 2008); CBL-CIPK signaling components (Mahajan et al., 2006); RING-H2 finger protein gene BIRF1 (Liu et al., 2008); mannose-1-phosphate guanyl transferase gene MPG1 (Kumar et al., 2012); and many more. However, the molecular mechanism for regulation of salinity stress responses in plants still remains unclear (Haak et al., 2017).

Role of "Omics" in Salinity Stress Signaling

High-throughput "omics" technologies (genomics, transcriptomics, proteomics, metabolomics, etc.) may provide useful information related to salinity stress tolerance in plants. The analysis of plant samples using "omics" methods results in the production of a large amount of data. For example, high-throughput transcriptomics studies have provided an immense amount of data on gene expression at the mRNA level. The transcriptome of *Thellungiella salsuginea/halophila* (a salt-tolerant relative of *Arabidopsis*) was studied using microarray for the comparison of salt response in glycophytes and halophytes. More than 194 transcripts in *Arabidopsis* and at least 2300 expressed sequence tags (ESTs) in *Thellungiella* have been shown to be significantly altered under salinity stress (Wong et al., 2006; Zhang et al., 2008). Transcriptomics studies have also been performed in the halophytic plants *Suaeda salsa* (Zhang et al., 2001), *Aeluropus littoralis* (Zouari et al., 2007), *Salicornia brachiata* (Jha et al., 2009), and *Festuca rubra* (Diedhiou et al., 2009), revealing important information about differentially expressed genes and factors providing tolerance to these species under high salinity conditions.

Proteomics as an Emerging Tool for Stress Signaling Studies

The data obtained from transcriptomics studies provide a global view of stress-responsive genes in different plants. However, due to post-transcriptional and post-translational modifications, the changes in gene expression at transcript level usually do not correlate with the changes at protein level. Proteins are direct effectors of plant stress response, and unlike transcripts, they are directly involved in a wide array of cellular responses associated with stress acclimation. The changes in protein composition under stress are closely interrelated with plant developmental and physiological responses determining stress tolerance and exert a crucial impact on the plasma membrane, the cytoplasm, and the cytoskeleton, as well as other cellular structures. Thus, it is crucial to investigate the changes in the plant proteome under stress conditions.

The proteomics of plant abiotic stress has great potential to determine the key processes involved in stress response and stress tolerance acquisition in plants and can be used as an essential approach in the post-genomic era to study global protein expression. Analyses of abiotic stress-responsive proteomes in plants have yielded important information for understanding the complex mechanisms of the plant stress response (Zhang et al., 2012). Comparative proteomics has been successfully applied for the systematic analysis of total proteins in various plant species under a wide range of abiotic stresses, including salt stress (Pang et al., 2010), drought (Alvarez et al., 2008), high or low temperature (Majoul et al., 2004; Gao et al., 2009), ultraviolet radiation (Willey et al., 2009), heavy metals (Hu et al., 2003), and herbicides (Castro et al., 2005). Recently, some comprehensive reviews focused on salinity response in plants have been published (Agarwal et al., 2009; Zhang et al., 2012; Hanin et al., 2016). Plant response to salinity has been investigated at the proteome level in a wide range of plants, including model plants, agricultural and economic crops, wild species, arid plants, halophytic species, forage plants, and tree species (Table 25.1). More than 2000 salt-responsive proteins have been identified from whole seedlings or different plant organs/tissues such as shoots, roots, leaves, hypocotyls, and seeds, revealing their functions mainly in signaling, regulation of gene expression, enzymes for energy metabolism, redox homeostasis, synthesis of osmoprotectants, activation of antioxidant machinery, defense proteins, and secondary metabolism.

In the field of plant abiotic stress research, the most common strategy is the comparison of proteomes isolated from unstressed (control) plants and the corresponding proteomes from plants treated with one kind of stress or a combination of stresses. Other strategies include the comparison of proteomes from different cultivars or genotypes exhibiting contrasting tolerance to a specific environmental stress. Differential expression protein profiling represents the core of comparative proteomics and is used to determine changes in protein composition at global level on exposure to stress, followed by relative quantitation and identification of differentially expressed proteins. So far, a few studies have compared proteome responses to salt stress in related plant species with contrasting salinity tolerance, such as *A. thaliana* and *T. salsuginea* (Pang et al., 2010), rice and its salt-tolerant wild relative *Porteresia coarctata* (Sengupta and Majumder, 2009), and common bread wheat and a *Triticum aestivum/Thinopyrum ponticum* introgression hybrid (Peng et al., 2009). Comparative proteomic analysis has enabled researchers to identify differentially expressed proteins in genetically related plant species revealing differential stress tolerance. Moreover, proteomic studies on halophytic plants have shown that the mechanism of salt ion exclusion is associated with enhanced expression of several plasma membrane–associated ion transporters such as SOS1, NhaA (Na^+/H^+ antiporter), VDAC (voltage-dependent anion channel), and calcium-mediated sensors (Askari et al., 2006; Barkla et al., 2009; Li et al., 2011; Wang et al., 2015).

Several gel-based or gel-free techniques are available for the differential analysis of protein expression, such as two-dimensional electrophoresis (2-DE) or liquid chromatography (LC) coupled with tandem mass spectrometry (MS/MS), Differential gel electrophoresis (2-D DIGE), isotope-coded affinity tags (ICAT), stable isotope labeling of amino acids (SILAC), isobaric tags for relative and absolute quantitation (iTRAQ), etc. Among these techniques, 2-DE coupled with MS/MS has been

TABLE 25.1

Summary of Published Research on Plant Proteome Analyses in Response to Salinity Stress

Plant	Organ/Tissue	Salt Concentration	Exposure Time	Proteomic Approach	No. of Identified Protein	Reference
Model Plants						
Arabidopsis thaliana	Root	150 mM	6 h or 48 h	2-DE[a], LC-MS/MS[b]	86	Jiang et al. (2007)
	Leaf	50, 150 mM	5 d	2-DE[a], nanoESI MS/MS[c]; MALDI-TOF/TOF[d]	79	Pang et al. (2010)
	Microsomal fraction	150 mM	5 d	iTRAQ[e]; 2-D LC-MS/MS[b]	152	Pang et al. (2010)
Thellungiella salsuginea/halophila	Leaf	50, 150 mM	5 d	2-DE[a], nanoESI MS/MS[c]; MALDI-TOF/TOF[d]	32	Pang et al. (2010)
	Microsomal fraction	150 mM	5 d	iTRAQ[e]; 2-D LC-MS/MS[b]	93	Pang et al. (2010)
Nicotiana tabacum	Leaf	150 mM	2 d	2-DE[a]; MALDI-TOF/TOF[d]	18	Razavizadeh et al. (2009)
Medicago sativa, Medicago truncatula	Root	300 mM	8 h	2-DE[a]; MALDI-TOF/TOF[d]	60 (*M. sativa*), 26 (*M. truncatula*)	Long (2016)
Crop Plants						
Oryza sativa	Root	150 mM	48 h	2-DE[a]; MALDI-TOF/TOF[d]	34	Cheng et al. (2009)
	Leaf	200 mM	72 h	2-DE[a]; MALDI-TOF/TOF[d]	16	Sengupta and Majumder (2009)
Porteresia coarctata	Leaf	200, 400 mM	72 h	2-DE[a]; MALDI-TOF/TOF[d]	16	Sengupta and Majumder (2009)
Triticum aestivum	Root; leaf	200 mM	24 h	2-DE[a]; MALDI-TOF/TOF[d]	86 (root); 58 (leaf)	Peng et al. (2009)
	Shoot mitochondria	200 mM	4 d	2-DE[a]; nanoLC MS/MS[b]	68	Jacoby et al. (2010)
Somatic hybrid of *Triticum aestivum* and *Thinopyrum ponticum*	Root; leaf	200 mM	24 h	2-DE[a]; MALDI-TOF/TOF[d]	86 (root); 58 (leaf)	Peng et al. (2009)
Hordeum vulgare	Leaf	300 mM	24 h	2-DE[a]; MALDI-TOF/TOF[d]	22	Rasoulnia et al. (2010)
	Leaf	300 mM	3 weeks	2-DE[a]; MALDI-TOF/TOF[d]	44	Fatehi et al. (2012)
Setaria italica	Seedling	100, 150, 200 mM	7 d	2-DE[a]; MALDI-TOF/TOF[d]	29	Veeranagamallaiah et al. (2008)
Sorghum bicolor	Leaf	100 mM	14 d	2-DE[a]; MALDI-TOF/TOF[d]	55	Ngara et al. (2012)
Plants of Economic Use						
Agrostis stolonifera	Root, leaf	10 ds m[-1]	28 days	2-D DIGE[f]; MALDI-TOF/TOF[d]	106 (leaf); 24 (root)	Xu et al. (2010)
Brassica napus	Leaf	200 mM	24, 48, 72 h	2-DE[a]; MALDI-TOF/TOF[d]	42	Jia (2015)
Glycine max	Hypocotyl, root	100 mM	3 d	2-DE[a]; ESI-Q/TOF-MS/MS[g]	7	Aghaei et al. (2009)
	Leaf, hypocotyl, root	40 mM	1 week	2-DE[a]; MALDI-TOF/TOF[d]	19 (leaf), 22 (hypocotyl), 14 (root)	Sobhanian et al. (2010b)
Pisum sativum	Root	75, 150 mM	6 weeks/7 d	2-DE[a]; ESI-Q-TOF MS/MS[g]	35	Kav et al. (2004)

(Continued)

TABLE 25.1 (CONTINUED)

Summary of Published Research on Plant Proteome Analyses in Response to Salinity Stress

Plant	Organ/Tissue	Salt Concentration	Exposure Time	Proteomic Approach	No. of Identified Protein	Reference
Solanum lycopersicum	Root	100 mM	14 d	2-DE[a]; nanoLC MS/MS[b]	80	Manaa et al. (2011)
Solanum tuberosum	Shoot	90 mM	4 weeks	2-DE[a]; Cleveland peptide mapping; protein sequencing by Edman degradation method	39	Aghaei et al. (2008)
Vitis vinifera	Shoot tips	55, 250 mM	8, 16 d	2-DE[a]; MALDI-TOF/TOF[d]	191	Vincent et al. (2007)
Gossypium hirsutum	Root	200 mM	24 h	iTRAQ[e]; Q-TOF MS/MS[g]	128	Li (2015)
Cucumis sativus	Root	84 mM	3 d	2-DE[a]; MALDI-TOF/TOF[d]	33	Shao (2016)
Beta vulgaris	Root, shoot	125 mM	7 d	2-DE[a]; nanoESI-MS/MS[c]	3 (root); 6 (shoot)	Wakeel et al. (2011)
Citrus aurantium	Leaf	150 mM	16 d	2-DE[a]; nanoESI-MS/MS[c]	85	Tanou et al. (2009)
Populus cathayana	Leaf	75, 150 mM	4 weeks	2-DE[a]; ESI-Q-TOF MS/MS[g]	73	Chen et al. (2011)
Halophytic Plants						
Suaeda aegyptiaca	Leaf	150, 300, 450, 600 mM	30 d	2-DE[a]; LC-MS/MS[b]	27	Askari et al. (2006)
Suaeda salsa	Leaf	100, 200 mM	3 weeks	2-DE[a]; MALDI-TOF/TOF[d]	57	Li et al. (2011)
Mesembryanthemum crystallinum	Leaf Tonoplast	200 mM	7 d	2-D DIGE[f]; nanoESI-MS/MS[c]	4	Barkla et al. (2009)
Aster tripolium	Leaf	375 mM	4 weeks	2-DE[a]; MALDI-TOF/TOF[d]	5	Geissler et al. (2010)
Puccinellia tenuiflora	Leaf	50, 150 mM	7 d	2-DE[a]; ESI-Q-TOF MS/MS[g]	93	Yu et al. (2011)
Aeluropus lagopoides	Shoot	450 mM	10 d	2-DE[a]; MALDI-TOF/TOF[d]	83	Sobhanian et al. (2010a)
Halogeton glomeratus	Leaf	200 mM	24 h, 72 h, 7 d	2-DE[a]; MALDI-TOF/TOF[d]	49	Wang (2015)
Tangut Nitraria	Leaf	500 mM	1, 3, 5, 7 d	iTRAQ[e]; triple Q-TOF MS[g]	71	Cheng et al. (2015)
Bruguiera gymnorhiza	Root, leaf	500 mM	1, 3, 6, 12, 24 h and 3, 6, 12 d	2-DE[a]; protein sequencing; LC-MS/MS[b]	3	Tada and Kashimura (2009)
Kandelia candel	Leaf	150, 300, 450, 600 mM	3 d	2-DE[a]; MALDI-TOF/TOF[d]	48	Wang (2014)

[a] Two-dimensional gel electrophoresis.
[b] Liquid chromatography coupled with tandem mass spectrometry.
[c] Electron spray ionization coupled with tandem mass spectrometry.
[d] Matrix-associated laser desorption ionization-time of flight/time of flight.
[e] Isobaric tag for relative and absolute quantitation.
[f] Two-dimensional differential gel electrophoresis.
[g] Quadruple time of flight-mass spectrometry.

extensively used to study the effect of salt stress on protein expression using leaves, roots, or shoots of many plant species (Table 25.1). Proteomics studies could thus significantly contribute to the identification and characterization of key plant proteins imparting stress tolerance, which can further be used as potential biomarkers for a specific kind of abiotic stress. Recently, proteomics studies dealing with plant response to abiotic stress have been focused mainly on the description of quantitative changes in plant proteomes or a specific subcellular proteome (e.g., the nuclear, mitochondrial, plastid, or secretory proteome). Taken together, these studies may provide the framework for a better understanding of the mechanisms that govern plant responses to salinity stress and are useful for the identification of novel candidate genes that can be used further for the improvement of stress tolerance in sensitive crop plant species through genetic engineering.

Conclusion and Future Prospects

As suggested earlier, proteomics studies can significantly contribute to the understanding of physiological mechanisms underlying plant stress tolerance, which can further be used to describe stress tolerance at the genotype level. So far, however, we have important gaps in our knowledge regarding the regulation of plant response to abiotic stress, as this regulation takes place at multiple transcriptional, post-transcriptional, post-translational, and epigenetic levels (Haak et al., 2017). By a combination of proteomics and physiological approaches, several stress acclimation strategies have been described by the aforementioned studies. However, most results showed the previously characterized salt-induced proteins rather than providing new mechanistic insights into salinity tolerance. Quite a lot of information is available on changes in cellular metabolism as well as stress-responsive proteins in the plant proteome, while much less is known about regulatory proteins involved in stress signaling and regulation of gene expression (mainly transcription factors), membrane proteins, and transporters, due to low levels in the cell or difficulties in purification. In the near future, the emergence of new, powerful proteomic techniques and the selection of specific developmental stages/cell/tissues or subcellular organelles, especially laser capture–mediated micro-dissection (LCM)-mediated single cell isolation, may allow the cell-specific expression, protein enrichment, and successful detection of low-abundance proteins. In addition, studies of post-translational modifications (PTMs) such as glycosylation, phosphorylation, and ubiquitination, protein–protein interactions (interactomics), and redox proteomics are also needed (Gong et al., 2015). These approaches would contribute to a detailed functional characterization of proteins and a better understanding of the mechanisms of plant stress tolerance.

The majority of these studies have focused on model plant species with sequenced genomes, such as *Arabidopsis* and rice. These are mainly glycophytes and do not generally possess critical defenses against salinity. Moreover, extrapolation of the results of salinity tolerance in model plants to cereal crops is difficult due to species-specific differences in salt tolerance mechanisms. Crops grown in arid and semi-arid regions are often resistant to several abiotic stresses and thus, can be valuable donor genotypes for stress-tolerant genes. Since genome sequences of many crop species are still lacking, protein databases available for model plants are currently being used as reference databases for the analysis of data obtained from other plant species. However, further improvements are expected with the advent of next generation sequencing (NGS) and an increase in the number of genome sequences available for different plant species. In addition, more studies on halophytic plants may provide valuable tools for the identification of the basic mechanisms of plant salinity tolerance.

Abiotic stress proteomic studies primarily focus on quantitative changes in protein abundance between stressed and unstressed plants in related genotypes of contrasting tolerance. However, a mere differential abundance of proteins does not give much information; therefore, the results should be validated and combined with those from transcriptomics and metabolomics. The integration of the huge amount of data coming from high-throughput "omics" approaches is a challenging task but is crucial for the prediction of accurate function of biological systems under changing environmental conditions. Translational proteomics may thus lead toward the application of this knowledge to the improvement of crop plants for abiotic stress tolerance by identifying novel candidate proteins. But, the experimental validation of the function of these unique identified stress proteins (by gene silencing or overexpression) remains of the utmost importance. The data obtained under controlled laboratory conditions need to be further investigated in field conditions, which may exhibit a true response to a combination of stresses under varying environmental conditions at the whole-plant level, and is of vital importance for application in crop breeding programs.

Acknowledgments

The author thankfully acknowledges the Department of Science and Technology, Government of India, for DST-SERB Young Scientist grant (SB/YS/LS-39/2014) and University Grants Commission, Government of India for UGC-BSR Start-up research grant to SJ (F.30-50/2014/BSR); DST-FIST program and UGC-SAPII CAS program in Department of Botany, J.N.V. University, Jodhpur.

The author declares no financial or commercial conflict of interest.

REFERENCES

Agarwal P, Agarwal PK, Joshi AJ, Sopory SK, Reddy MK. Overexpression of PgDREB2A transcription factor enhances abiotic stress tolerance and activates downstream stress-responsive genes. *Mol. Biol. Rep.* 2010; 37(2):1125–1135. doi:10.1007/s11033-009-9885-8.

Agarwal PK, Agarwal P, Jain P, Jha B, Reddy MK, Sopory SK. Constitutive overexpression of a stress-inducible small GTP-binding protein PgRab7 from *Pennisetum glaucum* enhances abiotic stress tolerance in transgenic tobacco. *Plant Cell Rep.* 2008; 27:105–115. doi:10.1007/s00299-007-0446-0.

Aghaei K, Ehsanpour AA, Komatsu S. Proteome analysis of potato under salt stress. *J. Proteome Res.* 2008; 7(11):4858–4868. doi:10.1021/pr800460y.

Aghaei K, Ehsanpour AA, Shah AH, Komatsu S. Proteome analysis of soybean hypocotyl and root under salt stress. *Amino Acids* 2009; 36(1):91–98. doi:10.1007/s00726-008-0036-7.

Agrawal GK, Jwa NS, Rakwal R. Rice proteomics: Ending phase I and the beginning of phase II. *Proteomics* 2009; 9:935–963. doi:10.1002/pmic.200800594.

Alvarez S, Marsh EL, Schroeder SG, Schachtman DP. Metabolomic and proteomic changes in the xylem sap of maize under drought. *Plant Cell Environ.* 2008; 31(3):325–340. doi:10.1111/j.1365-3040.2007.01770.x.

Apse MP, Aharon GS, Snedden WA, Blumwald E. Salt tolerance conferred by overexpression of a vacuolar Na+/H+ antiport in *Arabidopsis. Science* 1999; 285(5431):1256–1258. doi:10.1126/science.285.5431.1256.

Askari H, Edqvist J, Hajheidari M, Kafi M, Salekdeh GH. Effects of salinity levels on proteome of *Suaeda aegyptiaca* leaves. *Proteomics* 2006; 6(8):2542–2554. doi:10.1002/pmic.200500328.

Barkla BJ, Vera-Estrella R, Hernandez-Coronado M, Pantoja O. Quantitative proteomics of the tonoplast reveals a role for glycolytic enzymes in salt tolerance. *Plant Cell* 2009; 21(12):4044–4058. doi:10.1105/tpc.109.069211.

Bhadra Rao TSR, Naresh JV, Reddy PS, Reddy MK, Mallikarjuna G. Expression of *Pennisetum glaucum* Eukaryotic Translational Initiation Factor 4A (PgEIF4a) confers improved drought, salinity, and oxidative stress tolerance in groundnut. *Front. Plant Sci.* 2017; 8(453):1–15. doi:10.3389/fpls.2017.00453.

Castro AJ, Carapito C, Zorn N, Magne C, Leize E, Van Dorsselaer A, Clement C. Proteomic analysis of grapevine (*Vitis vinifera* L.) tissues subjected to herbicide stress. *J. Exp. Bot.* 2005; 56(421):2783–2795. doi:10.1093/jxb/eri271.

Chen FG, Zhang S, Jiang H, Ma WJ, Korpelainen H, Li C. Comparative proteomics analysis of salt response reveals sex-related photosynthetic inhibition by salinity in *Populus cathayana* cuttings. *J. Proteome Res.* 2011; 10(9):3944–3958. doi:10.1021/pr200535r.

Cheng T, Chen J, Zhang J, Shi S, Zhou Y, Lu L, Wang P, et al. Physiological and proteomic analyses of leaves from the halophyte *Tangut Nitraria* reveals diverse response pathways critical for high salinity tolerance. *Front. Plant Sci.* 2015; 6:30. doi:10.3389/fpls.2015.00030.

Cheng YW, Qi YC, Zhu Q, Chen X, Wang N, Zhao X, Chen HY, Cui XJ, Xu LL, Zhang W. New changes in the plasma membrane-associated proteome of rice roots under salt stress. *Proteomics* 2009; 9(11):3100–3114. doi:10.1002/pmic.200800340.

Desai MK, Mishra RN, Verma D, Nair S, Sopory SK, Reddy MK. Structural and functional analysis of a salt stress inducible gene encoding voltage dependent anion channel (VDAC) from pearl millet (*Pennisetum glaucum*). *Plant Physiol. Biochem.* 2006; 44:483–493. doi:10.1016/j.plaphy.2006.08.008.

Diedhiou CJ, Popova OV, Golldack D. Transcript profiling of the salt-tolerant *Festuca rubra* ssp. *litoralis* reveals a regulatory network controlling salt acclimatization. *J. Plant Physiol.* 2009; 166(7):697–711. doi:10.1016/j.jplph.2008.09.015.

Fatehi F, Hosseinzadeh A, Alizadeh H, Brimavandi T, Struik PC. The proteome response of salt-resistant and salt-sensitive barley genotypes to long-term salinity stress. *Mol. Biol. Rep.* 2012; 39(5):6387–6397. doi:10.1007/s11033-012-1460-z.

Ganguly M, Datta K, Roychoudhury A, Gayen D, Sengupta DN, Datta SK. Overexpression of Rab16A gene in indica rice variety for generating enhanced salt tolerance. *Plant Signal. Behav.* 2012; 7(4):502–509. doi:10.4161/psb.19646.

Gao F, Zhou Y, Zhu W, Li X, Fan L, Zhang G. Proteomic analysis of cold stress-responsive proteins in *Thellungiella* rosette leaves. *Planta* 2009; 230(5):1033–1046; PMID: 19705148. doi:10.1007/s00425-009-1003-6.

Geissler N, Hussin S, Koyro HW. Elevated atmospheric CO_2 concentration enhances salinity tolerance in *Aster tripolium* L. *Planta* 2010; 231(3):583–594. doi:10.1007/s00425-009-1064-6.

Ghosh A, Pareek A, Sopory SK, Singla-Pareek SL. A glutathione responsive rice glyoxalase II, OsGLYII-2, functions in salinity adaptation by maintaining better photosynthesis efficiency and anti-oxidant pool. *Plant J.* 2014; 80(1):93–105. doi:10.1111/tpj.12621.

Gong F, Hu X, Wang W. Proteomic analysis of crop plants under abiotic stress conditions: Where to focus our research? *Front. Plant Sci.* 2015; 6(418):1–5. doi:10.3389/fpls.2015.00418.

Gutha LR, Reddy AR. Rice DREB1B promoter shows distinct stress-specific responses, and the overexpression of cDNA in tobacco confers improved abiotic and biotic stress tolerance. *Plant Mol. Biol.* 2008; 68(6):533–555. doi:10.1007/s11103-008-9391-8.

Haak DC, Fukao T, Grene R, Hua Z, Ivanov R, Perrella G, Li S. Multilevel regulation of abiotic stress responses in plants. *Front. Plant Sci.* 2017; 8(1564):1–24. doi:10.3389/fpls.2017.01564.

Hanin M, Ebel C, Ngom M, Laplaze L, Masmoudi K. New insights on plant salt tolerance mechanisms and their potential use for breeding. *Front. Plant Sci.* 2016; 7:1787. doi:10.3389/fpls.2016.01787.

Hu Y, Wang G, Chen GY, Fu X, Yao SQ. Proteome analysis of *Saccharomyces cerevisiae* under metal stress by two-dimensional differential gel electrophoresis. *Electrophoresis* 2003; 24(9):1458–1470. doi:10.1002/elps.200390188.

Jacoby RP, Millar AH, Taylor NL. Wheat mitochondrial proteomes provide new links between antioxidant defense and plant salinity tolerance. *J. Proteome Res.* 2010; 9(12):6595–6604. doi:10.1021/pr1007834.

Jami SK, Clark GB, Turlapati SA, Handley C, Roux SJ, Kirti PB. Ectopic expression of an annexin from *Brassica juncea* confers tolerance to abiotic and biotic stress treatments in transgenic tobacco. *Plant Physiol. Biochem.* 2008; 46(12):1019–1030. doi:10.1016/j.plaphy.2008.07.006.

Jan A, Maruyama K, Todaka D, Kidokoro S, Abo M, Yoshimura E, Shinozaki K, Nakashima K, Yamaguchi-Shinozaki K. OsTZF1, a CCCH-tandem zinc finger protein, confers delayed senescence and stress tolerance in rice by regulating stress-related genes. *Plant Physiol.* 2013; 161(3):1202–1216. doi:10.1104/pp.112.205385.

Jha B, Agarwal PK, Reddy PS, Lal S, Sopory SK, Reddy MK. Identification of salt-induced genes from *Salicornia brachiata*, an extreme halophyte through expressed sequence tags analysis. *Genes Genet. Syst.* 2009; 84(2):111–120. doi:10.1266/ggs.84.111.

Jia H, Shao M, He Y, Guan R, Chu P, Jiang H. Proteome dynamics and physiological responses to short-term salt stress in *Brassica napus* leaves. *PLoS ONE* 2015; 10(12):e0144808. doi:10.1371/journal.pone.0144808.

Jiang Y, Yang B, Harris NS, Deyholos MK. Comparative proteomic analysis of NaCl stress-responsive proteins in Arabidopsis roots. *J. Exp. Bot.* 2007; 58(13):3591–3607. doi:10.1093/jxb/erm207.

Kav NNV, Srivastava S, Goonewardende L, Blade SF. Proteome-level changes in the roots of *Pisum sativum* in response to salinity. *Ann. Appl. Biol.* 2004; 145:217–230. doi:10.1111/j.1744-7348.2004.tb00378.x.

Kumar R, Mustafiz A, Sahoo KK, Sharma V, Samanta S, Sopory SK, Pareek A, Singla-Pareek SL. Functional screening of cDNA library from a salt tolerant rice genotype Pokkali identifies mannose-1-phosphate guanyl transferase gene (OsMPG1) as a key member of salinity stress response. *Plant Mol. Biol.* 2012; 79(6):555–568. doi:10.1007/s11103-012-9928-8.

Li W, Zhang C, Lu Q, Wen X, Lu C. The combined effect of salt stress and heat shock on proteome profiling in *Suaeda salsa*. *J. Plant Physiol.* 2011; 168(15):1743–1752. doi:10.1016/j.jplph.2011.03.018.

Li W, Zhao F, Fang W, Xie D, Hou J, Yang X, Zhao Y, Tang Z, Nie L, Lv S. Identification of early salt stress responsive proteins in seedling roots of upland cotton (*Gossypium hirsutum* L.) employing iTRAQ-based proteomic technique. *Front. Plant Sci.* 2015; 6:732. doi:10.3389/fpls.2015.00732.

Liu H, Zhang H, Yang Y, Li G, Yang Y, Wang X, Basnayake BM, Li D, Song F. Functional analysis reveals pleiotropic effects of rice RING-H2 finger protein gene OsBIRF1 on regulation of growth and defense responses against abiotic and biotic stresses. *Plant Mol. Biol.* 2008; 68(1–2):17–30. doi:10.1007/s11103-008-9349-x.

Long R, Li M, Zhang T, Kang J, Sun Y, Cong L, Gao Y, Liu F, Yang Q. Comparative proteomic analysis reveals differential root proteins in *Medicago sativa* and *Medicago truncatula* in response to salt stress. *Front. Plant Sci.* 2016; 7:424. doi:10.3389/fpls.2016.00424.

Mahajan S, Sopory SK, Tuteja N. Cloning and characterization of CBL-CIPK signalling components from a legume (*Pisum sativum*). *FEBS J.* 2006; 273(5):907–925. doi:10.1111/j.1742-4658.2006.05111.x.

Majoul T, Bancel E, Triboi E, Ben Hamida J, Branlard G. Proteomic analysis of the effect of heat stress on hexaploid wheat grain: Characterization of heat-responsive proteins from nonprolamins fraction. *Proteomics* 2004; 4(2):505–513. doi:10.1002/pmic.200300570.

Manaa A, Ben Ahmed H, Valot B, Bouchet JP, Aschi-Smiti S, Causse M, Faurobert M. Salt and genotype impact on plant physiology and root proteome variations in tomato. *J. Exp. Bot.* 2011; 62(8):2797–2813. doi:10.1093/jxb/erq460.

Mukhopadhyay A, Vij S, Tyagi AK. Overexpression of a zinc-finger protein gene from rice confers tolerance to cold, dehydration, and salt stress in transgenic tobacco. *Proc. Natl. Acad. Sci. USA* 2004; 101:6309–6314. doi:10.1073/pnas.0401572101.

Munns R. Comparative physiology of salt and water stress. *Plant Cell Environ.* 2002; 25(2):239–250. doi:10.1046/j.0016-8025.2001.00808.x.

Munns R. Genes and salt tolerance: Bringing them together. *Tansley Rev. New Phytol.* 2005; 167(3):645–663. doi:10.1111/j.1469-8137.2005.01487.x.

Munns R, Tester M. Mechanisms of salinity tolerance. *Annu. Rev. Plant Biol.* 2008; 59:651–681. doi:10.1146/annurev.arplant.59.032607.092911.

Ngara R, Ndimba R, Borch-Jensen J, Jensen ON, Ndimba B. Identification and profiling of salinity stress-responsive proteins in *Sorghum bicolor* seedlings. *J. Proteomics* 2012; 75:4139–4150. doi:10.1016/j.jprot.2012.05.038.

Pang Q, Chen S, Dai S, Chen Y, Wang Y, Yan X. Comparative proteomics of salt tolerance in *Arabidopsis thaliana* and *Thellungiella halophila*. *J. Proteome Res.* 2010; 9(5):2584–2599. doi: 10.1021/pr100034f.

Peng ZY, Wang MC, Li F, Lv HJ, Li CL, Xia GM. A proteomic study of the response to salinity and drought stress in an introgression strain of bread wheat. *Mol. Cell Proteomics* 2009; 8(12):2676–2686. doi:10.1074/mcp.M900052-MCP200.

Ramegowda V, Senthil-Kumar M, Nataraja KN, Reddy MK, Mysore KS, Udayakumar M. Expression of a finger millet transcription factor, EcNAC1, in tobacco confers abiotic stress-tolerance. *PLoS ONE* 2012; 7(7):e40397. doi:10.1371/journal.pone.0040397.

Rasoulnia A, Bihamta M, Peyghambari S, Alizadeh H, Rahnama A. Proteomic response of barley leaves to salinity. *Mol. Biol. Rep.* 2010; 38(8):5055–5063. doi:10.1007/s11033-010-0651-8.

Razavizadeh R, Ehsanpour AA, Ahsan N, Komatsu S. Proteome analysis of tobacco leaves under salt stress. *Peptides* 2009; 30(9):1651–1659. doi:10.1016/j.peptides.2009.06.023.

Sengupta S, Majumder AL. Insight into the salt tolerance factors of a wild halophytic rice, *Porteresia coarctata*: A physiological and proteomic approach. *Planta* 2009; 229:911–929. doi:10.1007/s00425-008-0878-y.

Shao QS, Shu S, Du J, Yuan YH, Xing WW, Guo SR, Sun J. Proteome analysis of roots in cucumber seedlings under iso-osmotic NaCl and Ca(NO$_3$)$_2$ stresses. *Plant Mol. Biol. Rep.* 2016; 34:303–317. doi:10.1007/s11105-015-0916-4.

Singh AK, Kumar R, Pareek A, Sopory SK, Singla-Pareek SL. Overexpression of rice CBS domain containing protein improves salinity, oxidative, and heavy metal tolerance in transgenic tobacco. *Mol. Biotechnol.* 2012; 52(3):205–216. doi:10.1007/s12033-011-9487-2.

Singla-Pareek SL, Reddy MK, Sopory SK. Genetic engineering of the glyoxalase pathway in tobacco leads to enhanced salinity tolerance. *Proc. Natl. Acad. Sci. USA* 2003; 100(25):14672–14677. doi: 10.1073_pnas.2034667100.

Sobhanian H, Motamed N, Jazii FR, Nakamura T, Komatsu S. Salt stress induced differential proteome and metabolome response in the shoots of *Aeluropus lagopoides* (Poaceae), a halophyte C4 plant. *J. Proteome Res.* 2010a; 9(6):2882–2897. doi:10.1021/pr900974k.

Sobhanian H, Razavizadeh R, Nanjo Y, Ehsanpour AA, Jazii FR, Motamed N, Komatsu S. Proteome analysis of soybean leaves, hypocotyls and roots under salt stress. *Proteome Sci.* 2010b; 8:19. doi:10.1186/1477-5956-8-19.

Tada Y, Kashimura T. Proteomic analysis of salt-responsive proteins in the mangrove plant, *Bruguiera gymnorhiza*. *Plant Cell Physiol.* 2009; 50(3):439–446. doi:10.1093/pcp/pcp002.

Tanou G, Job C, Rajjou L, Arc E, Belghazi M, Diamantidis G, Molassiotis A, Job D. Proteomics reveals the overlapping roles of hydrogen peroxide and nitric oxide in the acclimation of citrus plants to salinity. *Plant J.* 2009; 60(5):795–804. doi:10.1111/j.1365-313X.2009.04000.x.

Tyagi H, Jha S, Sharma M, Giri J, Tyagi AK. Rice SAPs are responsive to multiple biotic stresses and overexpression of OsSAP1, an A20/AN1 zinc-finger protein, enhances the basal resistance against pathogen infection in tobacco. *Plant Sci.* 2014; 225:68–76. doi:10.1016/j.plantsci.2014.05.016.

Veeranagamallaiah G, Jyothsnakumari G, Thippeswamy M, Chandra Obul Reddy P, Surabhi G-K, Sriranganayakulu G, Mahesh Y, Rajasekhar B, Madhurarekha C, Sudhakar C. Proteomic analysis of salt stress responses in foxtail millet (*Setaria italica* L. cv. Prasad) seedling. *Plant Sci.* 2008; 175(5):631–641. doi:10.1016/j.plantsci.2008.06.017.

Vincent D, Ergul A, Bohlman MC, Tattersall EA, Tillett RL, Wheatley MD, Woolsey R, et al. Proteomic analysis reveals differences between *Vitis vinifera* L. cv. Chardonnay and cv. Cabernet Sauvignon and their responses to water deficit and salinity. *J. Exp. Bot.* 2007; 58(7):1873–1892. doi:10.1093/jxb/erm012.

Wakeel A, Asif AR, Pitann B, Schubert S. Proteome analysis of sugar beet (*Beta vulgaris* L.) elucidates constitutive adaptation during the first phase of salt stress. *J. Plant Physiol.* 2011; 168(6):519–526. doi:10.1016/j.jplph.2010.08.016.

Wang J, Meng Y, Li B, Ma X, Lai Y, Si E, Yang K, et al. Physiological and proteomic analyses of salt stress response in the halophyte *Halogeton glomeratus. Plant Cell Environ.* 2015; 38: 655–669. doi:10.1111/pce.12428.

Wang L, Liu X, Liang M, Tan F, Liang W, Chen Y, Lin Y, Huang L, Xing J, Chen W. Proteomic analysis of salt-responsive proteins in the leaves of Mangrove *Kandelia candel* during short-term stress. *PLoS ONE* 2014; 9(1):e83141. doi:10.1371/journal.pone.0083141.

Willey NJ, Heinekamp YJ, Burridge A. Genomic and proteomic analyses of plant response to radiation in the environment—an abiotic stress context. *Radioprotection* 2009; 44(5):887–890. doi:10.1051/radiopro/20095158.

Wong CE, Li Y, Labbe A, Guevara D, Nuin P, Whitty B, Diaz C, et al. Transcriptional profiling implicates novel interactions between abiotic stress and hormonal responses in *Thellungiella*, a close relative of *Arabidopsis. Plant Physiol.* 2006; 140(4):1437–1450. doi:10.1104/pp.105.070508.

Xu CP, Sibicky T, Huang BR. Protein profile analysis of salt-responsive proteins in leaves and roots in two cultivars of creeping bentgrass differing in salinity tolerance. *Plant Cell Rep.* 2010; 29(6):595–615. doi:10.1007/s00299-010-0847-3.

Xu D, Duan X, Wang B, Hong B, Ho T, Wu R. Expression of a late embryogenesis abundant protein gene, HVA1, from barley confers tolerance to water deficit and salt stress in transgenic rice. *Plant Physiol.* 1996; 110(1):249–257. doi:10.1104/pp.110.1.249.

Yu J, Chen S, Zhao Q, Wang T, Yang C, Diaz C, Sun G, Dai S. Physiological and proteomic analysis of salinity tolerance in *Puccinellia tenuiflora. J. Proteome Res.* 2011; 10(9):3852–3870. doi:10.1021/pr101102p.

Zhang H, Han B, Wang T, Chen S, Li H, Zhang Y, Dai S. Mechanisms of plant salt response: Insights from proteomics. *J. Proteome Res.* 2012; 11(1):49–67. doi:10.1021/pr200861w.

Zhang L, Ma XL, Zhang Q, Ma CL, Wang PP, Sun YF, Zhao YX, Zhang H. Expressed sequence tags from a NaCl treated *Suaeda salsa* cDNA library. *Gene* 2001; 267(2):193–200. doi:10.1016/S0378-1119(01)00403-6.

Zhang YY, Lai JB, Sun SH, Li Y, Liu YY, Liang LM, Chen MS, Xie Q. Comparison analysis of transcripts from the halophyte *Thellungiella halophila. J. Integr. Plant Biol.* 2008; 50(10):1327–1335. doi:10.1111/j.1744-7909.2008.00740.x.

Zouari N, Ben Saad R, Legavre T, Azaza J, Sabau X, Jaoua M, Masmoudi K, Hassairi A. Identification and sequencing of ESTs from the halophyte grass *Aeluropus littoralis. Gene* 2007; 404(1–2):61–69. doi:10.1016/j.gene.2007.08.021.

Section IV

Role of Signaling Molecules Under Abiotic Stress Management

26

Signaling Molecules and Their Involvement in Abiotic and Biotic Stress Response Crosstalk in Plants

V.R. Devaraj and Myrene Dsouza

CONTENTS

Abbreviations

ABA	Abscisic acid
ABRE ABA	responsive elements
APX	Ascorbate peroxidase
BR	Brassinosteroids
CAT	Catalase
CK	Cytokinin
DAG	Diacyl glycerol
ET	Ethylene
GA	Gibberellic acid
GR	Glutathione reductase
IAA	Indole-3-acetic acid
JA	Jasmonic acid
MAPK	Mitogen active protein kinase
NO	Nitric oxide
PA	Phosphatidic acid
PAMP	Pathogen associated molecular patterns
PR	Pathogenesis-related
PYR	Pyrabactin resistance protein (PYR)
RBOH	Respiratory burst oxidase homolog protein
ROS	Reactive oxygen species
RWR	Rapid wound-responsive protein
SA	Salicylic acid
SAR	Systemic acquired resistance
TF	Transcription factors
WIPK	Wounding-induced protein kinase

Introduction

Higher plants are sessile organisms that perceive environmental cues such as light and chemical signals and respond by changing their morphologies (Ramakrishna and Ravishankar, 2011). Several genes are responsible for determining the position and fate of the cell. These genes encode a plethora of signaling molecules that are responsible for updating spatial and temporal conditions yet providing the stability of morphogenesis needed to maintain the overall plant structure. Nonetheless, some cells respond plastically to environmental conditions in complex mechanisms involving robust and dynamic networks. The control circuitry is mostly modular, with both specific and shared control elements responding to negative feedback and feed-forward control.

In general, signal transduction in plant cells commences with receptor activation and generation of second messengers, thereby converting the primary external signal into intracellular signals. The second messengers then orchestrate biochemical and physiological responses via downstream pathways. Most often, reversible protein phosphorylation takes place to enable signal propagation, providing stability and resilience to the response. A signal transduction pathway is thus a delicate cooperative process involving receptors/sensors, second messengers, phosphoprotein cascades, transcription factors (TFs), and stress-responsive genes. In addition, signal transduction is underlined by distinct spatial and temporal dimensions, the former being satisfied by placing receptors, channels, G proteins, and kinases in specific membranes either permanently or ephemerally. The investigation of temporal control by microarray analysis is still in its infancy. A thorough understanding of signaling-dependent change in response to time is needed.

While trying to comprehend signaling mechanisms in a single cell, the response of the whole plant must not be neglected. Complex coordination mechanisms involving proteins, mRNA, growth regulators, phytohormones, sugars, and nodulation factors exist between cells and tissues. Cell-to-cell communication through plasmodesmata has been well documented (Kitagawa and Jackson, 2017). However, even more crucial than intercellular communication is the elucidation of signaling and highly tuned coordination between cellular organelles; for example, between the chloroplast and the nucleus. In this chapter, an exploration of signaling molecules will be undertaken to understand how plants cope with the various stresses encountered during their lifespan.

Second Messengers

Several second messengers are known to be active participants in stress signal transduction. These are invariably small intracellular signaling molecules or ions, normally located in the cytoplasm and causing the activation of various kinases to regulate stress-responsive enzyme activities. The second messengers are usually reactive oxygen species (ROS), lipid-derived phosphoinositides, cyclic nucleotides, phytohormones, growth regulators, and TFs.

Reactive Oxygen Species in Signal Transduction

The most prominent discovery from research on the response of plants to environmental stress is increased ROS levels and the resistance or susceptibility of the plant to antioxidants and to the ROS scavenging enzymes produced. On the one hand, ROS are highly toxic due to their tendency to form harmful by-products, resulting in damage to cellular biomolecules and ultimately causing cell death. On the other hand, ROS serve as biological stimuli to activate and regulate various genetic stress-response processes through alteration of gene expression and signal transduction pathways (Gill and Tuteja, 2010; Benderradji et al., 2011). The basic ROS cycle (Figure 26.1) responsible for the modulation of cellular levels of ROS during normal metabolism includes the key ROS scavenging enzymes: ascorbate peroxidase (APX), glutathione reductase (GR), and catalase (CAT). The extended ROS cycle, on the other hand, is operational in plants subjected to biotic or abiotic stresses (Figure 26.1) and comprises heat shock proteins (HSPs), pathogenesis related proteins (PR), chalcone synthase (CHS), phenylalanine ammonia-lyase (PAL),

cytochrome P450, and so on. As ROS are short-lived, sensing is efficiently done via at least three different mechanisms that are specific in terms of targets and perception mechanisms: unidentified receptor proteins; redox sensitive compounds, including TFs (such as NPR1 or HSFs); and direct inhibition of phosphatases by ROS.

ROS must be used and/or interfere with other signaling pathways or molecules to affect plant growth and cell metabolism. ROS target enzymes such as guanylyl cyclase (Vranova and Atichartpongkul, 2002), phospholipase C (Foyer and Noctor, 2003), phospholipase A2 (Moller, 2001), and phospholipase D (PLD) (Rasmusson and Soole, 2004). ROS generated by stress have also been documented to alter the redox status of the cell by shifting the glutathione/glutathione disulfide (GSH/GSSG) ratio, causing the oxidation of cysteine on key redox sensors such as ribonucleotide reductase, thioredoxin reductase, and some TFs (Xiong and Zhu, 2002). Furthermore, ROS such as H_2O_2 enhance the gene expression of oxygen scavenging enzymes and induce HSPs (Lee et al., 2000).

Ion channels, especially calcium channels, possess ubiquitous physiological functions. These channels are largely activated by ROS generated by respiratory burst oxidase homologs (RBOHs), resulting in an increase in cytosolic Ca^{2+} and further downstream signaling involving calmodulin (Bowler and Fluhr, 2000) (Figure 26.2), activation of G-protein (Baxter-Burrell et al., 2002), and activation of phospholipid signaling (Hirt et al., 2011). Ca^{2+} increase has been reported by some researchers to be upstream of ROS production (Bhattacharjee, 2005), while in other cases, Ca^{2+} elevation is downstream of ROS production (Bowler and Fluhr, 2000).

Plant stress responses are regulated by multiple signaling pathways that activate gene transcription and its downstream machinery. The MAPK cascade linking different receptors to cellular and nuclear targets can be activated by accumulation of H_2O_2 or may trigger an H_2O_2-induced oxidative burst (Petrov and Van Breusegem, 2012). A hypothetical model depicting some of the players involved in this pathway is shown in Figure 26.2. These include MAP kinase kinase kinase (MAPKKK), AtANP1, the MAPKs, AtMPK3/6, Ntp46MAPK, and calmodulin. A two-component histidine kinase or a receptor-like protein kinase is responsible for sensing H_2O_2. Calmodulin and an MAPK cascade are then activated, resulting in the induction/activation/suppression of several TFs and possible cross-talk with the pathogen-response signal transduction pathway (Figure 26.2). Positive feedback by protein phosphorylation through MAPK cascades regulates Ca^{2+} and ROS via the activation of RbohD and RbohF (Kimura et al., 2012). Microarray studies have examined responsive *cis*-elements in TF genes regulated by ROS (Gadjev et al., 2006), such as those belonging to the WRKY, Zat, RAV, GRAS, and Myb families (Desikan et al., 2001). The induction of different sets of genes by the same TF is made possible by the interaction of specific cysteinyl residues with various types of ROS (Delaunay et al., 2002). TFs that are well characterized to be regulated by ROS are WRKY, DREBA, AP-1, NRF2, CREB, HSF1, HIF-1, TP53, NF-κB, NOTCH, SP1, and SCREB-1 (Marinho et al., 2014).

Water loss triggers ABA synthesis, leading to the induction of many, but not all, drought-inducible genes. The existence of both ABA-independent and ABA-dependent signal transduction

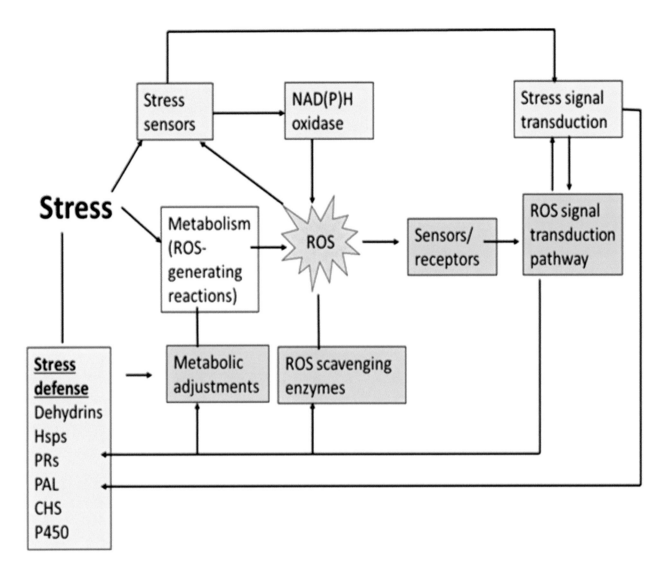

FIGURE 26.1 ROS cycle. Basic ROS cycle (green boxes): modulation of cellular ROS levels during normal metabolism with key ROS scavenging enzymes, ascorbate peroxidase (APX) and catalase (CAT). Extended ROS cycle (yellow boxes): operates in plants under biotic or abiotic stresses. CHS: chalcone synthase; HSPs: heat shock proteins; P450: cytochrome P450; PAL: phenylalanine ammonia lyase; PR: pathogenesis related proteins. (Modified from http://biology.unt.edu/ros/rosmetabolismdoc.htm)

cascades has been demonstrated. H_2O_2, largely produced by NADPH oxidase, is a prerequisite for ABA-induced stomatal closure (Zhang et al., 2001), causing a reduction in CO_2 availability for photosynthesis and ROS formation due to misdirection of electrons in the photosystem. Mutations in genes encoding catalytic subunits of NADPH oxidase result in impairment of H_2O_2 production as well as the activation of guard cell Ca^{2+} channels and stomatal closure (Kwak et al., 2003). Suzuki et al. (2013) demonstrated temporal-spatial interaction between ROS and ABA, resulting in systemic acquired acclimation to environmental challenges in plants. Apart from H_2O_2, impaired synthesis of NO, as seen in loss-of-function mutations in *Arabidopsis* NO synthases (AtNOS1), also weakens ABA-induced stomatal closure (Guo et al., 2002). NO produced concomitantly with ROS under various stresses as a defense response (Del Río, 2015) could have either toxic or protective effects, depending on its concentration, combination with ROS compounds, and its subcellular localization (Correa-Aragunde et al., 2015). Subjecting citrus plants to salinity stress after pre-treatment with H_2O_2 or

NO has suggested an overlap between H_2O_2 and NO signaling pathways in acclimation to salinity (Tanou et al., 2010). Several reports have also assigned an antioxidant role to NO during various stresses by modulation of superoxide formation and inhibition of lipid peroxidation (Correa-Aragunde et al., 2015).

Temporal and Spatial Synchronization of ROS with Other Signals

Any change in environmental conditions requires rapid transduction of the signal to initiate plant response. ROS production is known to be biphasic, occurring in an ROS wave, that is, an initial burst within local tissue producing $O_2^{\bullet-}$ within 3 min of stress induction, followed by the release of $O_2^{\bullet-}$ and H_2O_2 into systemic tissues about 6 h from induction (Suzuki et al., 2013). Experiments using inhibitors of NADPH oxidase found that they reduced the synthesis of ROS, resulting in suppression of $O_2^{\bullet-}$ and wound response proteins at later stages. Therefore, an early burst of ROS is required for the later phase of ROS production

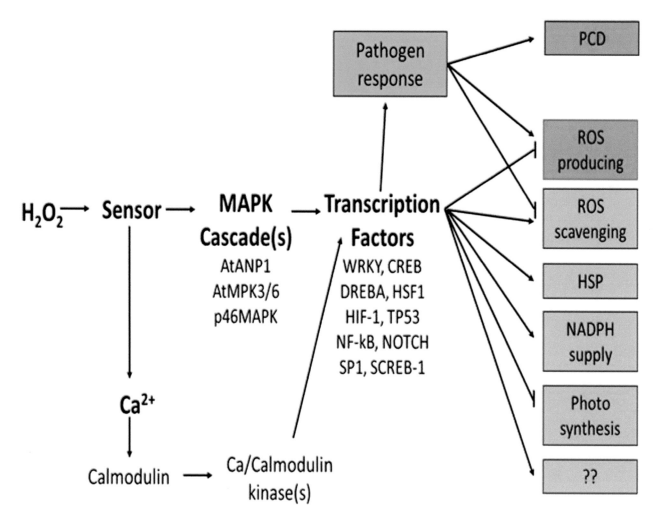

FIGURE 26.2 Hypothetical model depicting some of the players involved in the ROS signaling pathway. (Modified from http://biology.unt.edu/ros/ros-metabolismdoc.htm)

and other downstream signaling processes in response to stimuli. The early signaling events include ion fluxes across the plasma membrane, increased cytosolic Ca^{2+} levels, activation of MAPKs, and production of ROS (Finka et al., 2012). For example, abiotic stress such as heat causes transient opening of Ca^{2+} channels (Finka et al., 2012) and ROS production within 10 and 3 min, respectively (Konigshofer et al., 2008). In addition, early defense responses to a pathogen elicitor in *Arabidopsis* activated the MAPK cascade, resulting in induction of membrane protein phosphorylation, calmodulins, protein kinases, protein phosphatases, and proteins associated with auxin signaling within 5–10 min (Benschop et al., 2007). Mechanical wounding in *Arabidopsis* also resulted in elevated levels of jasmonic acid (JA) within 30 s to 5 min (Glauser et al., 2009). Suzuki and Mittler (2012) implicated the integration of JA and H_2O_2 produced in peroxisomes to electric signals generated under wounding stress.

Transcriptome analysis by a number of researchers has shown similar signals occurring in both local and systemic tissues subjected to abiotic stress (Suzuki et al., 2013). Proof of the ability of plants to spatially distribute metabolites in response to stress was obtained by metabolic profiling of secondary metabolites in pathogen-infected and uninfected leaf tissues. Linking of this ability to levels of ROS in the cell was made possible by studies

on mutants lacking CAT2. Emphasis was laid on the ROS wave generated by the pathogen infestation and its propagation from local to systemic cells via cell-to-cell communication. In addition, APX1 expression too occurs in a wave-like pattern that correlates positively with levels of H_2O_2 but negatively with non-phytochemical quenching (Karpinski et al., 2013). Studies of systemic acquired resistance in plants have demonstrated a number of different signaling molecules transported from local tissue to systemic tissue (Suzuki et al., 2013; Shah and Zeier, 2013). Spatial coordination can be further appreciated in long-distance signaling occurring during root herbivory, resulting in an induction of ABA accumulation in leaves (Soler et al., 2013). The distribution of RBOH proteins from roots to reproductive tissues to establish the role of ROS signaling in mediating this connection has been elucidated (Suzuki et al., 2013). Thus, integration of ROS signals and metabolic cues could be a promising subject for future studies.

Integration of H_2O_2 and Ca^{2+} Signals with Phytohormones

Plants regulate their growth and development in response to various internal and external stimuli by a diverse group of endogenous signaling molecules, the phytohormones. Even in small quantities, these play fundamental roles in plant acclimatization by mediating growth, development, source/sink transitions, and nutrient allocation (Fahad et al., 2015) and in plant defenses against biotic and biotic stresses (Taiz and Zeiger, 2010). Evidence of hormonal regulation of plant stress signal perception and integration for adaptive responses has increased in the past decade (Pareek et al., 2010). Phytohormones are translocated throughout the plant body via the xylem or phloem; they move short distances between cells or are maintained in their site of synthesis, exerting influence on target cells by binding to plasma membrane receptors or intracellular receptors. The downstream effects of hormonal signaling include alterations in gene expression patterns and in some cases non-genomic responses (Fraire-Velázquez et al., 2011). Phytohormones are subjected to positive or negative feedback and are largely involved in cross-talk. Independent hormone cascades tend to converge at certain points, forming a signaling network and thus enabling cross-talk. The fluctuation of phytohormone levels is responsible for fine tuning cellular dynamics as a means of regulating growth responses under stress. The possible roles of phytohormones in abiotic stress tolerance and cross-talk between phytohormone signaling cascades are illustrated in Figure 26.3.

ABA, the primary hormone involved in the perception of abiotic stress, is expressed quickly in responses to drought, osmotic and cold stress, and wounding (Bari and Jones, 2009). On the other hand, salicylic acid (SA), JA, and ethylene (ET) are more important in biotic stress conditions (Bari and Jones, 2009; Luna et al., 2011). Considerable evidence of cross-talk by ABA, SA, JA, and ET with auxins, gibberellic acids (GAs), and cytokinins (CKs) in regulating plant defense response has been reported (Bari and Jones, 2009). ABA acts antagonistically with SA/JA/ET, increasing the susceptibility of the plant to pathogens (Rejeb et al., 2014). For example, ET-regulated stomatal closure mediated by H_2O_2 production inhibits ABA-induced stomatal closure. This indicates that ET formed as a result of other stress conditions could interfere with ABA-regulated stomatal closure and may serve as a feedback mechanism to permit CO_2 influx for photosynthesis even under extreme stress conditions. Long-term physiological responses to abiotic stresses result in alteration of ABA-mediated gene regulation of TFs that bind to ABA-responsive elements (ABREs) on ABA-regulated genes (Zhu et al., 2007), such as activation of the ABA-responsive TFs ABF1 and ABF4 on their phosphorylation by the ABA-inducible kinase CPK4 or CPK11. Drought-response gene expression overlapped

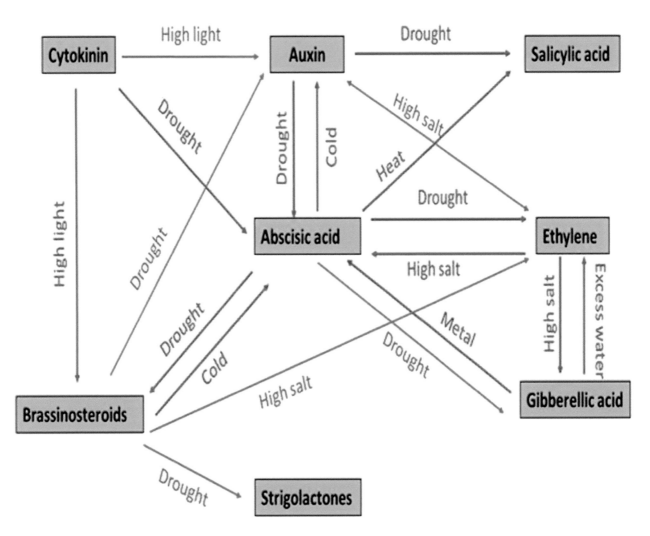

FIGURE 26.3 The possible roles of phytohormones in abiotic stress tolerance and cross-talk between phytohormone signaling. (Wani, S.H. et al., *Crop J.*, 4, 162, 2016. With permission.)

with the expression profiles of genes regulated by hormones such as JA, auxin, CK, and ET (Huang et al., 2008). It was also found that ABA appears to inhibit ET production (Harrison, 2012). In addition, ABA and ET independently increase H_2O_2 levels, which then stimulate NO production and stomatal closure (Wilkinson and Davies, 2009). CKs work antagonistically with ABA (Pospíšilová, 2003); decreased CK levels and accumulation of ABA lead to an increased ABA/CK ratio that enhances apical dominance and closure of stomata under drought (O'Brien and Benkova, 2013). Figure 26.4 illustrates hormonal cross-talk between abiotic and biotic stress.

It has been proposed that JA induced by herbivory interacts with ABA-regulated stomatal closure by increasing Ca^{2+} influx and/or by activating H_2O_2/NO signaling. Ca^{2+} influx into the cytoplasm in turn stimulates calcium-dependent protein kinase (CDPK) production and activates slow-type anion channels, interacting with ABA-induced stomatal closure, NO, ROS, ET, and JA (Harrison, 2012). Similarly, upregulation of JA pathway genes has been reported under salinity stress (Walia et al., 2006).

Cross-tolerance signaling seen in a signal peptide system coupled with JA was reported in wound-induced salt-stress adaptation in tomato (Capiati et al., 2006).

GA, involved in cross-talk of hormonal interactions in signaling pathways, was found in close functional association with auxins and brassinosteroids (BRs) (Stamm and Kumar, 2013). Cross-talk between GA and JA is mediated through the DELLA and JAZ proteins directly interacting with each other (Riemann et al., 2015), the latter also competing for binding to MYC2, a TF resulting in downregulation of JA-responsive genes (Hou et al., 2010). The DELLA gene, RGL3, is transcriptionally upregulated by JA signaling, and the promotor of RGL3 is a target of MYC2 (Wild et al., 2012). RGL3 binds to JAZ1, resulting in JA-mediated degradation and release of MYC2, enabling induction of RGL3, which in turn binds to JAZ8, enhancing MYC2-dependent JA responses (Wild et al., 2012). GA thus links the two pathways and determines the strength of JA response.

Research by Tiryaki and Staswick (2002) demonstrated that JA- and auxin-triggered signaling acts both antagonistically

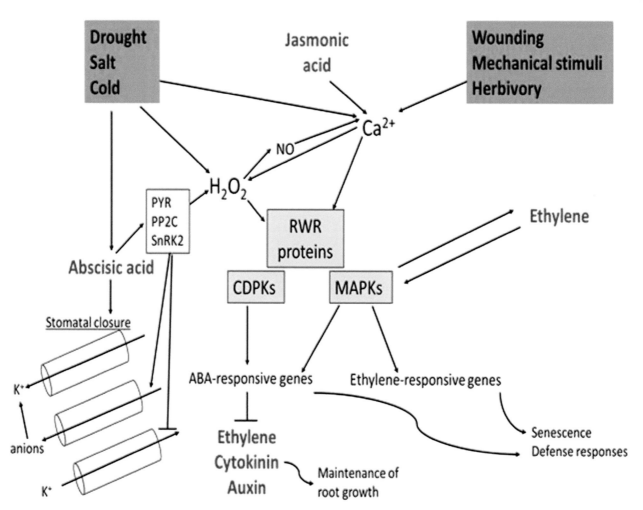

FIGURE 26.4 Phytohormone cross-talk integrating elements involved in the response of abiotic and biotic events. Abiotic stress triggers an increase in ABA concentration in guard cells resulting in activation of a phosphatase (PP2C), pyrabactin resistance protein (PYR), and SnRK2 with consequent inactivation of the K^+ import channel and the activation of an anion efflux channel. SnRK2 also activates H_2O_2 production, resulting in NO synthesis, which triggers the influx of Ca^{2+} from the vacuole into the cytoplasm. H_2O_2 simultaneously activates expression of rapid wound-responsive (RWR) proteins, which along with Ca^{2+}, triggers CDPKs to stimulate anion efflux and inhibit K^+ influx. The resulting ion loss causes water efflux, loss of turgor, and stomatal closure. Stress-induced JA production interacts with ABA-mediated stomatal closure by stimulating the influx of extracellular Ca^{2+} and/or by activating H_2O_2/ NO signaling. Ethylene acts as a negative regulator of ABA.

and synergistically. This convergence is made possible by the GH3 family of acyl acid amido synthetases, which catalyze amino acid conjugation of both JA and indole-3-acetic acid (IAA), leading to molecular activation and inactivation, respectively (Khan and Stone, 2007). JA and auxin cross-talk during lateral root formation involving the ethylene response factor109 (ERF109) has been reported (Cai et al., 2014). Drought-induced biosynthesis of JA in roots and repression of CK biosynthesis and signaling leads to growth and expansion of the root system, aiding in the development of drought tolerance (Werner et al., 2010). Thus, the CK and JA signaling pathways are linked antagonistically. Some researchers have reported restoration of reduced IAA levels under salinity stress by treatment with SA, suggesting that hormonal balance and cross-talk are critical to signal perception, transduction, and mediation of stress response (Fahad et al., 2015).

BRs are involved in the ROS-dependent stress response, causing enhanced tolerance to oxidative stress accompanied by induction of H_2O_2 production in the apoplast and expression of RBOH, MAPK1, and MAPK3 (Xia et al., 2015). BRs are, however, not directly involved in long-distance signaling but affect the distribution of auxins and polyamines (Li et al., 2013). BRs inhibit root growth by negatively regulating JA-induced synthesis of anthocyanins needed for signaling by implicating WD-repeat/Myb/bHLH transcriptional complexes (Qi et al., 2011). Moreover, BR-mediated signaling is regulated by ABA, and in turn, ABA inhibits BR-induced responses under abiotic stress. BRs regulate strigolactones (SLs), a group of carotenoid-derived compounds produced and exuded in small amounts in roots in response to nutritional conditions triggered by drought (Koltai and Beveridge, 2013). SLs also act as signaling molecules by stimulating nodulation in the legume–rhizobium interaction process (Foo and Davies, 2011). They find application in agriculture for various purposes, including as inducers of suicidal seed germination of parasitic plants (Vurro and Yoneyama et al., 2012). In summary, ABA, BR, SA, JA, and NO induce stomatal closure, while CK and IAA promote stomatal opening.

Role of Lipid-Derived Signal Molecules

Besides the structural role of modulating membrane fluidity, phospholipids also serve as mitigators to reduce the impact of stress and are directly involved in intracellular signal transduction (Okazaki et al., 2013). Such signaling lipids are low in abundance, are synthesized quickly from pre-existing lipids, are turned over rapidly, and possess lipid-binding domains for activation of enzymes and recruitment of proteins to the membrane. Signaling lipids include lysophospholipids, fatty acids, phosphatidic acids, inositol phosphate, diacylglycerol, oxylipin, sphingolipids, and N-acylethanolamine (Okazaki and Saito, 2014). An overview of phospholipid signaling pathways in plants is illustrated in Figure 26.5. In addition to lipids, lipid hydrolytic enzymes such as phospholipases and esterases, which have different substrate preferences, distribution, and mode of induction on stress treatment, are important. The fact that a variety of lipo-oxygenases and kinases use lipids as substrates is suggestive of a lipid-mediated signaling system characterized by flexibility and complexity, enabling adaptation to environmental stresses.

Free Fatty Acids and Lysophospholipids

Glycerolipids, the major lipid component of biological membranes, are hydrolyzed by phospholipase A (PLA), yielding two important signal mediators, lysophospholipids and free fatty acids (FFAs). FFAs regulate salt, drought, and heavy metal tolerance, wound-induced responses, and defenses against insect/herbivore feeding in plants (Kachroo and Kachroo, 2009). Oleic acid controls NO-mediated defense signaling by modulating the levels of NO-associated protein (Mandal et al., 2012). Azelaic acid, a derivative of FFAs, is a known defense signal in pathogen-induced systemic acquired resistance (SAR) for accumulating SA (Jung et al., 2009). The role of oxidized products of FFAs, oxylipins, in response to drought and pathogen infection and in organ development has been demonstrated (Yang et al., 2012). It has been recently shown that linolenic acid, the precursor of JA, induces several antioxidant systems, such as galactinol synthase or Met sulfoxide reductase enzymes (Mata-Pérez et al., 2015). Nitro-fatty acids (NO_2-FAs) formed as a product of the reaction between reactive nitrogen species and unsaturated fatty acids are found to occur freely, esterified to complex lipids in hydrophobic compartments, and adducted with proteins (Rudolph et al., 2009). These serve as potent NO donors in signal transduction cascades and in posttranslational modification of proteins involved in signaling (Geisler and Rudolph, 2012).

Lysophospholipids, normally occurring in trace quantities in plant tissues, show a substantial increase, accompanied by a transient increase in PLA activity, under responses to abiotic stress and plant–environment interactions (Welti et al., 2002). Yeast-derived elicitor induces transient accumulation of lysophosphatidylcholine (lysoPC) in Californian poppy, leading to phytoalexin production (Viehweger et al., 2002). The application of root extract of mycorrhizal plants containing lysoPC induces the phosphate transporter, which is normally induced by arbuscular mycorrhizal (AM) fungal colonization of cortex cells (Drissner et al., 2007). Lysophosphatidylethanolamine acts as a signal transducer in potato roots colonized by AM fungi (Drissner et al., 2007). The transcription factor MYB30 is known to physically interact with PLA2-a after triggering its translocation from cytoplasmic vesicles to the nucleus (Froidure et al., 2010), indicating a rather complex regulation by gene silencing.

Phosphatidic Acid

PA plays a dual role as a metabolic intermediate for membrane lipid biosynthesis as well as a signaling molecule. The former role is well elucidated in root hair development (Ohashi et al., 2003), as a precursor for phospho- and galactolipids (Arisz, 2010), and in lipid remodeling as a consequence of phosphorus starvation (Misson et al., 2005). Its role in signal transduction has been reported in the SA-mediated signaling cascade (Kalachova et al., 2013), transduction of H_2O_2 signals through interaction with cytosolic glyceraldehyde-3-phosphate dehydrogenase (Guo et al., 2012a), suppression of cell death induced by hydrogen peroxide (Zhang et al., 2003), response to ABA (Guo et al., 2012b), nitrogen signaling (Hong et al., 2009), ROS production by NADPH oxidase activity regulation (Zhang et al., 2009), low-temperature stress response (Arisz et al., 2013), salt-stress response (McLoughlin et al., 2013),

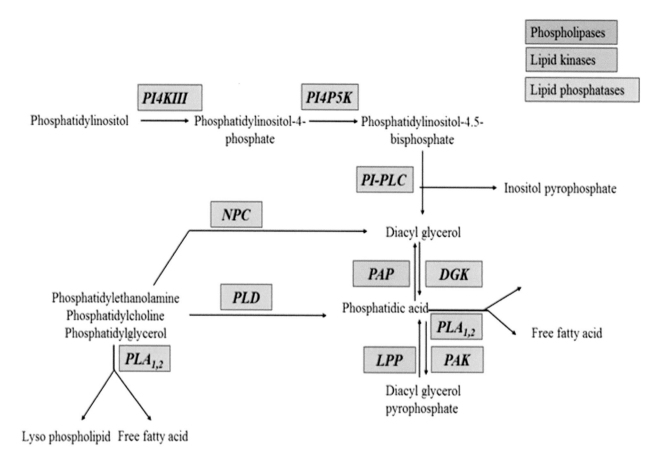

FIGURE 26.5 Phospholipid signaling pathways in plants. DGK: diacylglycerol kinase; LPP: lipid phosphate phosphatase; NPC: non-specific phospholipase C; PAK: phosphatidic acid kinase; PAP: phosphatidic acid phosphatase; PI4K: phosphatidylinositol 4-kinase; PI4P5K: phosphatidylinositol-4-phosphate 5-kinase; PI-PLC: phosphatidylinositol-specific phospholipase C; PLA: phospholipase A; PLD: phospholipase D. (Modified from Ruelland, E. et al., *Environ. Exp. Bot.*, 114, 129, 2015.)

microtubule organization in response to salt stress (Zhang et al., 2012), and drought (Uraji et al., 2012). PA regulation of cytoskeletal dynamics include its binding to microtubule bundling protein MAP65-1 (Zhang et al., 2012) and capping protein (Li et al., 2013) to promote stress tolerance in *A. thaliana* (Pleskot et al., 2013).

In addition to interaction with other signaling lipids, responses of PA are also integrated with other cellular signals, including Ca²⁺ (Li et al., 2009), ROS (Zhang et al., 2009), NO (Raho et al., 2011), ABA, and cytoskeleton (Lanteri et al., 2008), enabling the fulfilment of its multitasking role. PA regulates ABA-mediated stomatal closure by stimulating the production of ROS through NADPH oxidase (Zhang et al., 2009) while simultaneously inhibiting ABI1 protein phosphatase activity (Mishra et al., 2006). It activates the kinase of long-chain bases of sphingolipid (SPHK), which, along with PLDα1, codependently amplifies response to ABA (Guo et al., 2012b). PA is also able to bind to CTR1 kinase in *A. thaliana*, resulting in inhibition of its activity and contributing to the activation of the ET signaling pathway in the absence of ET (Testerink et al., 2007). PA binding of a trigalactosyldiacylglycerol 4, a lipid trafficking protein, has been recently reported (Wang et al., 2013b), indicating a potential role for PA in shifting main metabolic fluxes in response to environmental stimuli.

Inositol Polyphosphate (IP3) and Diacylglycerol (DAG)

Phosphoinositides comprise a divergent group of phospholipids derived from phosphatidylinositol. Isoforms of PLD bind polyphosphoinositides, resulting in their activation. This is considered as an important link between two branches of lipid signaling. Phosphatidylinositol-4,5-bisphosphate (PIP2) is rapidly hydrolyzed on stress by phospholipase C (PLC) to form inositol triphosphate. and DAG, which can be metabolized into other signaling molecules such as inositol hexaphosphate (IP6) and PA (Munnik and Vermeer, 2010) that potentiate JA and IAA signaling (Sheard et al., 2010). In addition, DAG can also be supplied from other phospholipids by non-specific PLC (Pokotylo et al., 2014). IP3-mediated signaling in response to environmental stimuli occurs under extreme temperatures (Delage et al., 2013), salinity (Munnik and Nielsen, 2011), drought, and ABA (DeWald et al., 2001). Normal physiological functions of IP3 include stomatal guard-cell closure, pollen-tube elongation, and pollen dormancy.

The involvement of DAG as a lipid messenger is still controversial, as its exact function in plant cells remains elusive. DAG produced by non-specific phospholipase 4 (NPC4) is converted to PA, triggering ABA response and plant tolerance to osmotic stress

(Peters et al., 2010). Disruption of NPC4 gene in *Arabidopsis* decreases DAG levels, compromising the plant's response to ABA and causing hyperosmotic stress effects (Peters et al., 2010). Scarce experimental evidence, however, does support the role of DAG as a signaling molecule. The involvement of PLC/DAG in PA formation is shown under aluminum stress in *Coffea arabica* suspension cells (Ramos-Diaz et al., 2007). Accumulation of DAG at the tip of the tobacco pollen tube is essential for tube tip growth through signaling (Helling et al., 2006). Phosphate starvation upregulates *AtNPC4* and *AtNPC5*, producing DAG and accumulating inorganic phosphate (Gaude et al., 2008). The fact that PA, and not DAG, is the active signaling molecule can also be concluded from a reverse genetic approach. However, the content of DAG within plant cells is known to fluctuate in response to a variety of developmental and environmental cues.

Oxylipins

Oxylipins are formed by the action of PLAs on linolenic acid in membrane lipids. The most extensively researched oxylipins involved in the activation of defense responses through the expression of genes are JA and its immediate precursor 12-oxo-phytodienoic acid (OPDA) (Block et al., 2005). JA-induced biosynthesis of poisonous compounds serves as a plant defense against necrotrophic pathogens. It is involved in signaling in response to UV damage (Demkura et al., 2010) and ozone-induced hypersensitive cell death (Mackerness et al., 1999). JA induces a set of Coronatine-insensitive1 (COI1)-dependent genes, while OPDA has been found to induce a set of largely COI1-independent genes (Taki et al., 2005). Signaling causes interaction of the F-box ubiquitin ligase COI1 with JAZ transcriptional repressors, leading to degradation of these repressors of downstream JA-induced genes, many of which are dependent on the key transcription factor MYC2/JIN1 (Chini et al., 2007). OPDA, on the other hand, regulates stress-responsive cellular redox homeostasis via the activation of thiol metabolism in chloroplasts (Park et al., 2013). In *Capsicum chinense* J. cells, JA was reported to stimulate PI-PLC and PLD activities, as measured *in vitro*, as well as that of PI kinases, in a biphasic dose-dependent manner (Munoz-Sánchez et al., 2012). Other types of oxylipins, such as azelaic acid, prime plants to accumulate SA, a defense signal for SAR against pathogen infection (Jung et al., 2009). Phytoprostanes, jasmonate-like products formed by nonenzymatic lipid peroxidation, activate the stress response genes, thereby enhancing protection from subsequent oxidative stress (Loeffler et al., 2005).

Sphingolipids

Several reports describe the importance of sphingolipids in biotic response and the defense against bacteria and fungi pathogens by inducing programmed cell death (PCD) (Markham et al., 2013). With respect to signaling, sphingolipid-mediated enhanced cell death is mediated by a mitogen activated protein kinase (MPK6; Saucedo-Garcia et al., 2011) and a Ca^{2+}-dependent protein kinase regulated by 14-3-3-proteins (Lachaud et al., 2013). Evidence suggests that free long chain base phosphates (LCB) are a potential trigger signal for PCD and are implicated in cold-stress response (Guillas et al., 2013). LCB kinase activity is stimulated by ABA and PA (Guo et al., 2012b).

This suggests a link between PLD-dependent PA signaling and sphingolipid-dependent signaling.

Role of Phosphoproteins in Stress Signaling

In most signaling pathways, reversible protein phosphorylation is the crux for the relay of signals within cells. Therefore, controlling the phosphorylation states of these proteins by protein kinases and phosphatases plays a fundamental role in coordinating the activities of signal transduction pathways.

Mitogen Activated Protein Kinase (MAPK)

The major pathway involved in abiotic stress response in plants is invariably the MAPK cascade, linking several cellular responses to external stimuli. Briefly, the constitution of this cascade is MAP kinase kinase kinases (MAP3Ks/MAPKKKs/MEKKs), MAP kinase kinases (MAP2Ks/MAPKKs/MEKs/MKKs), and MAP kinases (MAPKs/MPKs) (Mishra et al., 2006). Stimulation of the plasma membrane under stress causes activation of MAP3Ks or MAP kinase kinase kinase kinases (MAP4Ks), serving as adapters to connect upstream signaling steps to the core MAPK cascade. The MAP3Ks are serine/threonine kinases phosphorylating two amino acids in the $S/T-X_{3-5}-S/T$ motif of the MAP2K activation loop, while MAP2Ks phosphorylate MAPKs on threonine and tyrosine residues at a conserved T-X-Y motif. MAPKs are serine/threonine kinases able to phosphorylate a wide range of substrates, including other kinases, cytoskeleton-associated proteins, and/or TFs. Mediation for the formation and integrity of the MAPK cascade is made possible by scaffold proteins, shared docking domains, and adaptor or anchoring proteins (Takekawa et al., 2005). After signal completion, MAPK phosphatases (MKPs) are involved in the time-dependent shutdown of the pathway (Ulm et al., 2002). The entire cascade is regulated by various mechanisms, such as transcriptional and translational regulation, and posttranscriptional regulation such as protein–protein interactions. Through extensive research, it is known that MAP3K/MAP2K/MAPK signaling modules control diverse cellular functions through the formation of interconnected networks within cells. These include normal cellular functions such as cell division, development, hormone signaling and synthesis, and response to abiotic stress (extreme temperature, drought, salinity, wounding, UV radiation, ozone, heavy metals) and biotic stress events.

MAPK Activation in Abiotic Stress Signaling

Extensive evidence indicates that plants rapidly activate MAPKs when exposed to multiple abiotic stress stimuli (Ligterink and Hirt, 2001). Many stress-responsive genes expressed under drought stress play important roles in drought resistance. Involvement of ABA in the expression and activation of p44MKK4 (MAP kinase kinase) under drought was elaborately shown by in-gel kinase assays and experiments on immunoprecipitation (Jonak et al., 1996). In addition, increased tolerance to dehydration stress was demonstrated by overexpression of DSM1 (a putative rice MAPKKK gene in rice) (Ning et al., 2010). Work on yeast two-hybrid analyses in *Arabidopsis* subjected to salinity stress demonstrated at least two MAPK cascades involving protein–protein interactions between MEKK1 and MKK2/MEK1 (MAPKKs), between MKK2/MEK1 and MPK4 (a MAPK), and between MPK4 and MEKK1 (Ichimura et al., 1998), while a combination of salinity and cold stress involved MEKK1 as an upstream activator of MKK2 (Teige et al., 2004). For high temperature stress, AtMEKK1 and AtMPK3 are transcriptionally upregulated, while AtMPK4, AtMKK2, and AtMPK6 are activated in response to cold stress (Teige et al., 2004) in *Arabidopsis* (Figure 26.6). *Arabidopsis* plants lacking MPK3 and MPK6 exhibited hypersensitivity to ozone (Miles et al., 2005). Ozone treatment was also found to activate ROS-dependent MAPKs, leading to Ca^{2+} influx and the activation of an upstream membrane localized component and a cognate MAPK kinase (Hamel et al., 2005). Rice plants stressed with Cd exhibited activation of *OsMSRMK2*, a novel MAPK gene (Agarwal et al., 2002), while As(III) treatment resulted in the overexpression of OsMPK3, OsMPK4, and OsMKK4 (Gupta et al., 2009) (Figure 26.6). Exposure of *Medicago* seedlings to excess Cu or Cd ions resulted in a complex activation pattern of four distinct MAPKs: SIMK,

MMK2, MMK3, and SAMK (stress activated MAPK) (Jonak et al., 2004). The first report of activation of a wound-induced MAPK cascade, WIPK (wound induced MAP kinase), involved studies on tobacco (Seo et al., 1995). AtMPK4 and AtMPK6 in *Arabidopsis* (Ichimura et al., 2000) and NtMPK4 in tobacco revealed wound-induced activation along with two other wound-responsive tobacco MAPKs, WIPK and salicylic acid-induced protein kinase (SIPK) (Gomi et al., 2005). Other plants with activation of MAPKs on wounding include tomato (Xing et al., 2001), soya bean (Lee et al., 2001), and cotton (Wang et al., 2007). All the described research implies a means of achieving abiotic tolerance via genetic manipulation.

MAPK Activation in Pathogen Defense

Recognition of pathogen attack by plants occurs non-specifically through pathogen associated molecular patterns (PAMPs) via cell surface–located pathogen-recognition receptors, resulting in the induction of PAMP-triggered immunity via convergent signaling pathways (Jones and Dangl, 2006). The MAPK cascade

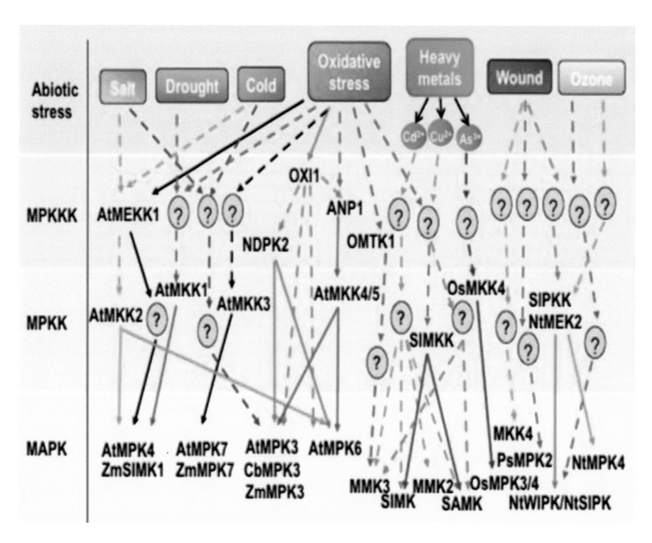

FIGURE 26.6 Schematic representation of cross-talk among different plant MAP kinase signaling components. The scheme of general signal transduction is shown on the left. The homologs in *Arabidopsis* (At), tobacco (Nt), maize (Zm), pea (Ps), and *Chorispora bungeana* (Cb) are shown. Solid arrows show proved pathways; dashed arrows indicate postulated pathways; question marks indicate unknown cascade components. (Sinha, A.K. et al., *Trends Plant Sci.*, 18, 149, 2011. With permission.)

bridges signal transduction from receptors to several downstream components. For example, flg22, derived from flagellin, triggers rapid activation of MPK3, MPK4, and MPK6 (Droillard et al., 2004). *hrp* (hypersensitive response and pathogenicity)-encoded harpins are known to activate MPK4 and MPK6, followed by induction of pathogenesis-related (PR) genes (Desikan et al., 2001) and production of proteins possessing antimicrobial activities. In addition to their role in abiotic stresses and biotic defenses, both MPK3 and MPK6 are regulators of a variety of processes, including abscission and stomatal development (Colcombet and Hirt, 2008). Modules involving the MAPKKs MKK1/MKK4/MKK5/MKK9 and the MAPKs MPK3/MPK6 are clearly implicated in different defense strategies (Suarez-Rodriguez et al., 2007). MPK3 and MPK6 are implicated in the synthesis of the phytoalexin camalexin, produced in defense against the fungal pathogen *Botrytis cinerea* (Ren et al., 2008).

Regulation of stomatal opening is not only important in abiotic stress tolerance but is also required to restrict pathogen invasion via the innate immune response. MPK3/MPK6 are the key players in stomatal development and dynamics (Wang et al., 2007). The pathogen *Xanthomonas campestris* pv. Campestris (Xcc) excretes a cell–cell signal-regulated virulence factor, which reverses stomatal closure induced by ABA produced under drought (Gudesblat et al., 2007). This complements infection by *Pseudomonas syringae* in tomato (Pst) mutants deficient in the production of coronatine, a toxin required to overcome stomatal defense.

Another member of the MAP kinase family, SIPK, is activated in multiple signaling pathways as a response to various pathogen-associated stimuli by elicitors from *Phytophthora* sp, bacterial harpin and tobacco mosaic virus (TMV) (Zhang et al., 1998). Posttranslational regulation of SIPK occurs via dual phosphorylation at threonine and tyrosine residues in the conserved motif of TEY present between sub-domains VII and VIII. WIPK is activated under wounding stress (Zhang et al., 1998) in response to pathogen elicitors such as cell wall–derived carbohydrate elicitor from *Phytophthora* spp as well as TMV. Transgenic tobacco possessing depressed WIPK activity exhibits increased SA production and thus, provokes the induction of acidic PR1 and PR2 genes on wounding (Ling, 2015).

Cross-talk of Genetic Pathways in Response to Abiotic and Biotic Stress

The existence of a complex network permitting plants to protect themselves against environmental conditions and invading pathogens has been demonstrated by many researchers in the past two decades (Chen et al., 2006). It was found that *Arabidopsis* induced five cytochrome P450 genes (CYP81D11, CYP710A1, CYP81D8, CYP71B6, and CYP76C2) from among 29 genes when subjected to biotic and abiotic stress. A common characteristic shared by these induced genes is the presence of *cis*-acting elements in regulatory regions of the gene; W-box (DNA binding sites for WRKY TFs), P-box (a positive *cis*-acting regulator of pathogen defense), and MYB recognition sites are common. In *Arabidopsis*, genes encoding TFs triggered by cross-talk between signaling pathways for biotic and abiotic stress include

DEAR1 (transcriptional repressor of DREB protein mediating plant defense and freezing stress responses) (Tsutsui et al., 2009), BOS1 encoding a transcription factor regulating *Botrytis cinerea* infection as well as in response to water deficit, increased salinity and oxidative stress (Mengiste et al., 2003), and SlERF5 induction of the dehydration-responsive genes through the ABA-mediated abiotic stress response (Chuang et al., 2010).

Pathogen invasion activates the phenylpropanoid pathway to stimulate the production of secondary metabolites such as lignin, flavonoids, and phytoalexins. CHS, the key enzyme in the pathway, is also induced in response to abiotic stress. Thus, molecular events required to achieve more direct action against invading microorganisms are achieved through phytoalexins, reinforcement of the cell wall with lignin, and so on through cross-talk between the biotic and abiotic arms of signaling. In both cases, increase in cytoplasmic Ca^{2+}, ROS production, induction of WRKY, and PR gene expression (Huang et al., 2010) was found to occur. Even though cross-talk occurs, its level is determined by pathogen identity in the case of biotic stress and level of the stressing factor in the case of abiotic stress (Fraire-Velázquez et al., 2011). For instance, it was found that the genes upregulated in chickpea in response to pathogen attack were more similar to those induced by high salinity than to those upregulated in response to cold or drought conditions (Fraire-Velázquez et al., 2011). The authors further found that the SUMO gene and the EF-hand calcium-binding protein gene were responsive to pathogens as well as to various abiotic stress events in the *Phaseolus vulgaris/Colletotrichum lindemuthianum* pathosystem. These two genes were also differentially regulated transcriptionally, as demonstrated by temporal and spatial studies on the specific mRNA levels (Fraire-Velázquez et al., 2011). SUMO targets up to five WRKY transcription factors (WRKY3, WRKY4, WRKY6, WRKY33, and WRKY72), indicating the convergence of resistance protein signaling and SUMO conjugation at transcription complexes. It was previously shown that SUMO conjugation transforms transcription activators into repressors, thereby suppressing defense signaling in non-infected plants (Burg and Takken, 2010). SUMO and the EF-hand calcium-binding protein genes also exhibit differential regulation by the daily photoperiod and circadian rhythm, wherein similar kinetics was demonstrated in the dark period but not in the light period (Alvarado-Gutiérrez et al., 2008).

Conclusion

As crop growth and crop yields are affected by environmental and biological cues, there is an immense need for understanding the mechanisms of homeostatic control in the plant. Combinatorial stress alters the plant phenotype such that the expression of defense is largely affected by the type of abiotic stress and the pathogens involved. The complex responses thus exhibited are initiated by the interplay of specific signaling pathways involved in abiotic and biotic stress, resulting in the accumulation of diverse signaling molecules that contribute to cross-tolerance. Perception of stress triggers the activation of complex signal transduction cascades that interact with the baseline pathways to perceive, amplify, and transmit stress signals and trigger stress responses. ROS, TFs, phytohormones, Ca^{2+}

and Ca^{2+}-regulated proteins, and MAPK cascades are credited with altering the plant's physiological state and re-establishing homeostasis. ROS signaling is integrated with hormone signaling networks, while the latter are known to induce changes in cytosolic Ca^{2+}. Specific TFs, including WRKYs, ATAF1 and 2, MYC2, RD2, BOS1, OsNAC6, and OsMPK5 kinase, are common to multiple networks or involved in cross-talk between stress signaling pathways regulated by ABA, SA, JA, and ET as well as ROS signaling.

Although much has been achieved in the context of plant stress signaling pathways, efforts should be made with transcriptome and proteome analyses, the sequencing of entire genomes, bioinformatic analyses, and functional studies, to get more insights into the molecular mechanism(s) underlying sensing and signal transduction. Although complete genome sequences are available for several crop plants, in comparison, protein and metabolite databases are still incomplete, which complicates the task of integrating all observations. Minor differences exist between plant species and even between different cultivars. Additionally, plant responses are often organ dependent, and results obtained with whole plants may be misleading.

REFERENCES

Agrawal, G.K., Rakwal, R., Iwahashi, H. 2002. Isolation of novel rice (*Oryza sativa* L.) multiple stress responsive MAP kinase gene, OsMSRMK2, whose mRNA accumulates rapidly in response to environmental cues. *Biochem. Biophys. Res. Commun.* 294: 1009–1016.

Alvarado-Gutiérrez, A., Del Real-Monroy, M., Rodríguez-Guerra, R., Almanza-Sánchez, L., Lozoya-Gloria, E., Fraire-Velázquez, S. 2008. A *Phaseolus vulgaris* EF-hand calcium-binding domain is induced early in the defense response against *Colletotrichum lindemuthianum* and by abiotic stress: Sequences shared between interacting partners. *Physiol. Mol. Plant Pathol.* 72: 111–121.

Arisz, S.A. 2010. Plant phosphatidic acid metabolism in response to environmental stress. PhD thesis, University of Amsterdam, Amsterdam.

Arisz, S.A., van Wijk, R., Roels, W., Zhu, J.K., Haring, M.A., Munnik, T. 2013. Rapid phosphatidic acid accumulation in response to low temperature stress in *Arabidopsis* is generated through diacylglycerol kinase. *Front. Plant Sci.* 4: 1.

Bari, R., Jones, J.D.G. 2009. Role of plant hormones in plant defence responses. *Plant Mol. Biol.* 69: 473–488.

Baxter, A., Mittler, R., Suzuki, N. 2014. ROS as key players in plant stress signaling. *J. Exp. Bot.* 65: 1229–1240.

Baxter-Burrell, A., Yang, Z., Springer, P.S., Bailey-Serres, J. 2002. RopGAP4-dependent Rop GTPase rheostat control of Arabidopsis oxygen deprivation tolerance. *Science* 296: 2026–2028.

Benderradji, L., Brini, F., Amar, S.B., Kellou, K., Azaza, J., Masmoudi, K., et al. 2011. Sodium transport in the seedlings of two bread wheat (*Triticum aestivum* L.) genotypes showing contrasting salt stress tolerance. *Aust. J. Crop Sci.* 5: 233–241.

Benschop, J.J., Mohammed, S., O'Flaherty, M., Heck, A.J., Slijper, M., Menke, F.L. 2007. Quantitative phosphoproteomics of early elicitor signaling in *Arabidopsis*. *Mol. Cell. Proteomics* 6: 1198–1214.

Bhattacharjee S. 2005. Reactive oxygen species and oxidative burst: Roles in stress, senescence and signal transduction in plants. *Curr. Sci.* 89(7): 1113–1121.

Block, A., Schmelz, E., Jones, J.B., Klee, H.J. 2005. Coronatine and salicylic acid: The battle between *Arabidopsis* and *Pseudomonas* for phytohormone control. *Mol. Plant Pathol.* 6: 79–83.

Bowler, C., Fluhr, R. 2000. The role of calcium and activated oxygens as signals for controlling cross-tolerance. *Trends Plant Sci.* 5: 241–246.

Burg, H.A., Takken, F.L. 2010. SUMO-, MAPK-, and resistance protein-signaling converge at transcription complexes that regulate plant innate immunity. *Plant Signal. Behav.* 5: 1597–1601.

Cai, X.T., Xu, P., Zhao, P.X., Liu, R., Yu, L.H., Xiang, C.B. 2014. *Arabidopsis* ERF109 mediates cross-talk between jasmonic acid and auxin biosynthesis during lateral root formation. *Nat. Commun.* 5: 5833.

Capiati, D.A., Pais, S.M., Tellez-Inon, M.T. 2006. Wounding increases salt tolerance in tomato plants: Evidence on the participation of calmodulin-like activities in cross-tolerance signalling. *J. Exp. Bot.* 57(10): 2391–2400.

Chen, F., Li, Q., Sun, L., He, Z. 2006. The rice 14-3-3 gene family and its involvement in responses to biotic and abiotic stress. *DNA Res.* 13: 53–63.

Chini, A., Fonseca, S., Fernandez, G., Adie, B., Chico, J.M., Lorenzo, O., et al. 2007. The JAZ family of repressors is the missing link in jasmonate signalling. *Nature*: 448, 666–671.

Chuang, H., Harnrak, A., Chen, Y., Hsu, C. 2010. A harpin-induced ethylene-responsive factor regulates plant growth and responses to biotic and abiotic stresses. *Biophys. Res. Commun.* 402: 410–420.

Colcombet, J., Hirt, H. 2008. *Arabidopsis* MAPKs: A complex signaling network involved in multiple biological processes. *Biochem. J.* 413: 217–226.

Correa-Aragunde, N., Foresi, N., Lamattina, L. 2015. Nitric oxide is a ubiquitous signal for maintaining redox balance in plant cells: Regulation of ascorbate peroxidase as a case study. *J. Exp. Bot.* 66: 2913–2921.

Del Río, L.A. 2015. ROS and RNS in plant physiology: An overview. *J. Exp. Bot.* 66: 2827–2837.

Delage, E., Puyaubert, J., Zachowski, A., Ruelland, E. 2013. Signal transduction pathways involving phosphatidylinositol 4-phosphate and phosphatidylinositol 4,5-bisphosphate: Convergences and divergences among eukaryotic kingdoms. *Prog. Lipid Res.* 52: 1–14.

Delaunay, A., Pflieger, D., Barrault, M.B., Vinh, J., Toledano, M.B. 2002. A thiol peroxidase is an H_2O_2 receptor and redox-transducer in gene activation. *Cell* 111: 471–481.

Demkura, P.V., Abdala, G., Baldwin, I.T., Ballare, C.L. 2010. Jasmonate-dependent and -independent pathways mediate specific effects of solar ultraviolet B radiation on leaf phenolics and antiherbivore defense. *Plant Physiol.* 152: 1084–1095.

Desikan, R., A-H-Mackerness, S., Hancock, J.T., Neill, S.J. 2001. Regulation of the *Arabidopsis* transcriptome by oxidative stress. *Plant Physiol.* 127: 159–172.

DeWald, D.B., Torabinejad, J., Jones, C.A., Shope, J.C., Cangelosi, A.R., Thompson, J.E., et al. 2001. Rapid accumulation of phosphatidylinositol 4,5-bisphosphate and inositol 1,4,5-trisphosphate correlates with calcium mobilization in salt-stressed arabidopsis. *Plant Physiol.* 126: 759–769.

Drissner, D., Kunze, G., Callewaert, N., Gehrig, P., Tamasloukht, M., Boller, T., et al. 2007. Lyso-phosphatidylcholine is a signal in the arbuscular mycorrhizal symbiosis. *Science* 318: 265–268.

Droillard, M.J., Boudsocq, M., Barbier-Brygoo, H., Lauriere, C. 2004. Involvement of MPK4 in osmotic stress response pathways in cell suspensions and plantlets of *Arabidopsis thaliana*: Activation by hypoosmolarity and negative role in hyperosmolarity tolerance. *FEBS Lett.* 574: 42–48.

Fahad, S., Hussain, S., Bano, A., Saud, S., Hassan, S., Shan, D., et al. 2015. Potential role of phytohormones and plant growth-promoting rhizobacteria in abiotic stresses: Consequences for changing environment. *Environ. Sci. Pollut. Res.* 22: 4907–4921.

Fahad, S., Hussain, S., Matloob, A., Khan, F.A., Khaliq, A., Saud, S., et al. 2015. Phytohormones and plant responses to salinity stress: A review. *Plant Growth Regul.* 75: 391–404.

Finka, A., Cuendet, A.F., Maathuis, F.J., Saidi, Y., Goloubinoff, P. 2012. Plasma membrane cyclic nucleotide gated calcium channels control land plant thermal sensing and acquired thermotolerance. *Plant Cell* 24: 3333–3348.

Foo, E., Davies, N.W. 2011. Strigolactones promote nodulation in pea. *Planta* 234: 1073–1081.

Foyer, C.H., Noctor, G. 2003. Redox sensing and signalling associated with reactive oxygen in chloroplasts, peroxisomes and mitochondria. *Physiol. Planta.* 119(3): 355–364.

Fraire-Velázquez, S., Rodríguez-Guerra, R., Sánchez-Calderón, L. 2011. Abiotic and biotic stress response crosstalk in plants. In: *Abiotic Stress Response in Plants – Physiological, Biochemical and Genetic Perspectives*, Arun Shanker and B. Venkateswarlu (Eds.), pp. 3–25. ISBN 978-953-307-672-0.

Froidure, S., Canonne, J., Daniel, X., Jauneau, A., Briere, C., Roby, D., Rivas, S. 2010. AtsPLA2-alpha nuclear relocalization by the *Arabidopsis* transcription factor AtMYB30 leads to repression of the plant defense response. *Proc. Natl. Acad. Sci. USA* 107: 15281–15286.

Gadjev, I., Vanderauwera, S., Gechev, T.S., Laloi, C., Minkov, I.N., Shulaev, V., et al. 2006. Transcriptomic footprints disclose specificity of reactive oxygen species signaling in *Arabidopsis*. *Plant Physiol.* 141: 436–445.

Gaude, N., Nakamura, Y., Scheible, W.R., Ohta, H., Dormann, P. 2008. Phospholipase C5 (NPC5) is involved in galactolipid accumulation during phosphate limitation in leaves of *Arabidopsis*. *Plant J.* 56: 28–39.

Geisler, A.C., Rudolph, T.K. 2012. Nitroalkylation—A redox sensitive signaling pathway. *Biochim. Biophys. Acta* 1820: 777–784.

Gill, S.S., Tuteja, N. 2010. Reactive oxygen species and antioxidant machinery in abiotic stress tolerance in crop plants, *Plant Physiol. Biochem.* 48: 909–930.

Glauser, G., Dubugnon, L., Mousavi, S.A., Rudaz, S., Wolfender, J.L., Farmer, E.E. 2009. Velocity estimates for signal propagation leading to systemic jasmonic acid accumulation in wounded *Arabidopsis*. *J. Biol. Chem.* 284: 34506–34513.

Gomi, K., Ogawa, D., Katou, S., Kamada, H., Nakajima, N., Saji, H., et al. 2005. A mitogen-activated protein kinase NtMPK4 activated by SIPKK is required for jasmonic acid signaling and involved in ozone tolerance via stomatal movement in tobacco. *Plant Cell Physiol.* 46: 1902–1914.

Gudesblat, G.E., Iusem, N.D., Morris, P.C. 2007. Guard cell-specific inhibition of *Arabidopsis* MPK3 expression causes abnormal stomatal responses to abscisic acid and hydrogen peroxide. *New Phytol.* 173: 713–721.

Guillas, I., Guellim, A., Reze, N., Baudouin, E. 2013. Long chain base changes triggered by a short exposure of *Arabidopsis* to low temperature are altered by AHb1 non-symbiotic haemoglobin overexpression. *Plant Physiol. Biochem.* 63: 191–195.

Guo, L., Devaiah, S.P., Narasimhan, R., Pan, X., Zhang, Y., Zhang, W., et al. 2012a. Cytosolic glyceraldehyde-3-phosphate dehydrogenases interact with phospholipase D delta to transduce hydrogen peroxide signals in the *Arabidopsis* response to stress. *Plant Cell* 24: 2200–2212.

Guo, L., Mishra, G., Markham, J.E., Li, M., Tawfall, A., Welti, R., et al. 2012b. Connections between sphingosine kinase and phospholipase D in the abscisic acid signaling pathway in *Arabidopsis*. *J. Biol. Chem.* 287: 8286–8296.

Guo, Y., Xiong, L., Ishitani, M., Zhu, J.K. 2002. An *Arabidopsis* mutation in translation elongation factor 2 causes superinduction of CBF/DREB1 transcription factor genes but blocks the induction of their downstream targets under low temperatures. *Proc. Natl. Acad. Sci. USA* 99: 7786–7791.

Gupta, M., Sharma, P., Sarin, N.B., Sinha, A.K. 2009. Differential response of arsenic stress in two varieties of *Brassica juncea* L. *Chemosphere* 74: 1201–1208.

Hamel, L.P., Miles, G.P., Samuel, M.A., Ellis, B.E., Seguin, A., Beaudoin, N. 2005. Activation of stress-responsive mitogen activated protein kinase pathways in hybrid poplar (*Populus trichocarpa* × *Populus deltoides*). *Tree Physiol.* 25: 277–288.

Harrison, M.A. 2012. Cross-talk between phytohormone signaling pathways under both optimal and stressful environmental conditions. In: *Phytohormones and Abiotic Stress Tolerance in Plants*, N.A. Khan, R. Nazar, N. Iqbal, N.A. Anjum (Eds.), Springer-Verlag, Berlin/Heidelberg, pp. 49–76.

Helling, D., Possart, A., Cottier, S., Klahre, U., Kost, B. 2006. Pollen tube tip growth depends on plasma membrane polarization mediated by tobacco PLC3 activity and endocytic membrane recycling. *Plant Cell* 18: 3519–3534.

Hirt, H., Garcia, A.V., Oelmüller, R. 2011. AGC kinases in plant development and defense. *Plant Signal. Behav.* 6: 1030–1033.

Hong, Y., Devaiah, S.P., Bahn, S.C., Thamasandra, B.N., Li, M., Welti, R., et al. 2009. Phospholipase D epsilon and phosphatidic acid enhance *Arabidopsis* nitrogen signaling and growth. *Plant J.* 58: 376–387.

Hou, X., Lee, L.Y., Xia, K., Yan, Y., Yu, H. 2010. DELLAs modulate jasmonate signaling via competitive binding to JAZs. *Dev. Cell* 19: 884–894.

Huang, D., Wu, W., Abrams, S.R., Adrian, J., Cutler, A.J. 2008. The relationship of drought-related gene expression in *Arabidopsis thaliana* to hormonal and environmental factors. *J. Exp. Bot.* 59: 2991–3007.

Huang, X., Li, J., Bao, F., Zhang, X., Yang, S. 2010. A gain-of-function mutation in the *Arabidopsis* disease resistance gene RPP4 confers sensitivity to low temperature. *Plant Physiol.* 154: 796–809.

Ichimura, K., Mizoguchi, T., Irie, K., Morris, P., Giraudat, J., Matsumoto, K., et al. 1998. Isolation of ATMEKK1 (a MAP Kinase kinase kinase)-interacting proteins and analysis of a MAP kinase cascade in *Arabidopsis*. *Biochem. Biophys. Res. Commun.* 253: 532–543.

Ichimura, K., Mizoguchi, T., Yoshida, R., Yuasa, T., Shinozaki, K. 2000. Various abiotic stresses rapidly activate *Arabidopsis* MAP kinases ATMPK4 and ATMPK6. *Plant J.* 24: 655–665.

Jonak, C., Kiegerl, S., Ligterink, W., Barker, P.J., Huskisson, N.S., Hirt, H. 1996. Stress signaling in plants: A mitogen-activated protein kinase pathway is activated by cold and drought. *Proc. Natl. Acad. Sci. USA* 93: 11274–11279.

Jonak, C., Nakagami, H., Hirt, H. 2004. Heavy metal stress. Activation of distinct mitogen-activated protein kinase pathways by copper and cadmium. *Plant Physiol.* 136: 3276–3283.

Jones, J.D., Dangl, J.L. 2006. The plant immune system. *Nature* 444: 323–329.

Jung, H.W., Tschaplinski, T.J., Wang, L., Glazebrook, J., Greenberg, J.T. 2009. Priming in systemic plant immunity. *Science* 324: 89–91.

Kachroo, A., Kachroo, P. 2009. Fatty acid-derived signals in plant defense. *Annu. Rev. Phytopathol.* 47: 153–176.

Kalachova, T., Iakovenko, O., Kretinin, S., Kravets, V. 2013. Involvement of phospholipase D and NADPH-oxidase in salicylic acid signaling cascade. *Plant Physiol. Biochem.* 66: 127–133.

Karpinski, S., Szechynska-Hebda, M., Wituszynska, W., Burdiak, P. 2013. Light acclimation, retrograde signalling, cell death and immune defences in plants. *Plant Cell Environ.* 36: 736–744.

Khan, S., Stone, J.M. 2007. *Arabidopsis thaliana* GH3.9 in auxin and jasmonate crosstalk. *Plant Signal. Behav.* 2: 483–485.

Kimura, S., Kaya, H., Kawarazaki, T., Hiraoka, G., Senzaki, E., Michikawa, M., et al. 2012. Protein phosphorylation is a prerequisite for the Ca^{2+}-dependent activation of *Arabidopsis* NADPH oxidases and may function as a trigger for the positive feedback regulation of Ca^{2+} and reactive oxygen species. *Biochim. Biophys. Acta.* 1823: 398–405.

Kitagawa, M., Jackson D. 2017. Plasmodesmata-mediated cell-to-cell communication in the shoot apical meristem: How stem cells talk. *Plants* 6(12): 1–14.

Koltai, H., Beveridge, C.A. 2013. Strigolactones and the coordinated development of shoot and root. In: *Long-Distance Systemic Signaling and Communication in Plants*, F. Baluska (Ed.), Springer, Berlin, pp. 189–204.

Konigshofer, H., Tromballa, H.W., Loppert, H.G. 2008. Early events in signalling high-temperature stress in tobacco BY2 cells involve alterations in membrane fluidity and enhanced hydrogen peroxide production. *Plant Cell Environ.* 31: 1771–1780.

Kwak, J.M., Mori, I.C., Pei, Z.M., Leonhardt, N., Torres, M.A., Dangl, J.L., et al. 2003. NADPH oxidase AtrbohD and AtrbohF genes function in ROS-dependent ABA signaling in *Arabidopsis*. *EMBO J.* 22: 2623–2633.

Lachaud, C., Prigent, E., Thuleau, P., Grat, S., Da Silva, D., Brière, C., et al. 2013. 14-3-3-Regulated $Ca2^+$-dependent protein kinase CPK3 is required for sphingolipid-induced cell death in *Arabidopsis. Cell Death Differ.* 20: 209–217.

Lanteri, M.L., Laxalt, A.M., Lamattina, L. 2008. Nitric oxide triggers phosphatidic acid accumulation via phospholipase D during auxin induced adventitious root formation in cucumber. *Plant Physiol.* 147: 188–198.

Lee, B.H., Won, S.H., Lee, H.S., Miyao, M., Chung, W.I., Kim, I.J. 2000. Expression of the chloroplast-localized small heat shock protein by oxidative stress in rice. *Gene* 245(2): 283–290.

Lee, S., Hirt, H., Lee, Y. 2001. Phosphatidic acid activates a wound-activated MAPK in *Glycine max. Plant J.* 26: 479–486.

Li, M., Hong, Y., Wang, X. 2009. Phospholipase D- and phosphatidic acid-mediated signaling in plants. *Biochim. Biophys. Acta* 1791: 927–935.

Li, P., Chen, L., Zhou, Y., Xia, X., Shi, K., Chen, Z., et al. 2013. Brassinosteroids-induced systemic stress tolerance was associated with increased transcripts of several defence-related genes in the phloem in *Cucumis sativus. PLoS One* 8: 66582.

Ligterink, W., Hirt, H. 2001. Mitogen-activated protein (MAP) kinase pathways in plants: Versatile signaling tools. *Int. Rev. Cytol.* 201: 209–275.

Ling, H.H. 2015. Functional roles of plant protein kinases in signal transduction pathways during abiotic and biotic stress. *J. Biodivers. Biopros. Dev.* 2: 147.

Loeffler, C., Berger, S., Guy, A., Durand, T., Bringmann, G., Dreyer, M., et al. 2005. B1- phytoprostanes trigger plant defense and detoxification responses. *Plant Physiol.* 137: 328–340.

Luna, E., Pastor, V., Robert, J., Flors, V., Mauch-Mani, B., Ton, J. 2011. Callose deposition: A multifaceted plant defense response. *Mol. Plant Microbe Interact.* 24: 183–193.

Mackerness, S.A.H., Surplus, S.L., Blake, P., John, C.F., Buchanan-Wollaston, V., Jordan, B.R., et al. 1999. Ultraviolet-B-induced stress and changes in gene expression in *Arabidopsis thaliana*: Role of signaling pathways controlled by jasmonic acid, ethylene and reactive oxygen species. *Plant Cell Environ.* 22: 1413–1423.

Mandal, M.K., Chandra-Shekara, A.C., Jeong, R.D., Yu, K., Zhu, S., Chanda, B., et al. 2012. Oleic acid-dependent modulation of NITRIC OXIDE ASSOCIATED1 protein levels regulates nitric oxide-mediated defense signaling in *Arabidopsis. Plant Cell* 24: 1654–1674.

Marinho, H.S., Real, C., Cyrne, L., Soares, H., Antunes, F. 2014. Hydrogen peroxide sensing, signaling and regulation of transcription factors. *Redox Biol.* 2: 35–62.

Markham, J.E., Lynch, D.V., Napier, J.A., Dunn, T.M., Cahoon, E.B. 2013. Plant sphingolipids: Function follows form. *Curr. Opin. Plant Biol.* 16: 350–357.

Mata-Pérez, C., Sánchez-Calvo, B., Begara-Morales, J.C., Luque, F., Jiménez-Ruiz, J., Padilla, M.N., et al. 2015. Transcriptomic profiling of linolenic acid-responsive genes in ROS signaling from RNA-seq data in *Arabidopsis. Front. Plant Sci.* 6: 122.

McLoughlin, F., Arisz, S.A., Dekker, H.L., Kramer, G., De Koster, C.G., Haring, M.A., et al. 2013. Identification of novel candidate phosphatidic acid-binding proteins involved in the salt-stress response of *Arabidopsis thaliana* roots. *Biochem. J.* 450: 573–581.

Mengiste, T., Chen, X., Salmeron, J., Dietrich, R. 2003. The BOS1 gene encodes an R2R3MYB transcription factor protein that is required for biotic and abiotic stress responses in arabidopsis. *The Plant Cell.* Online.

Miles, G.P., Samuel, M.A., Zhang, Y., Ellis, B.E. 2005. RNA interference-based (RNAi) suppression of AtMPK6, an *Arabidopsis* mitogen-activated protein kinase, results in hypersensitivity to ozone and misregulation of AtMPK3. *Environ. Pollut.* 138: 230–237.

Mishra, G., Zhang, W., Deng, F., Zhao, J., Wang, X. 2006. A bifurcating pathway directs abscisic acid effects on stomatal closure and opening in *Arabidopsis. Science* 312: 264–266.

Misson, J., Raghothama, K.G., Jain, A., Jouhet, J., Block, M.A., Bligny, R., et al. 2005. A genome-wide transcriptional analysis using *Arabidopsis thaliana* Affymetrix gene chips determined plant responses to phosphate deprivation. *Proc. Natl. Acad. Sci. USA* 102: 11934–11939.

Moller, I.M. 2001. Plant mitochondria and oxidative stress: Electron transport, NADPH turnover, and metabolism of reactive oxygen species. *Ann. Rev. Plant* Physiol. Plant Mol. Biol. 52(1): 561–591.

Munoz-Sánchez, J.A., Altúzar-Molina, A., Teresa Hernández-Sotomayor, S.M. 2012. Phospholipase signaling is modified differentially by phytoregulators in *Capsicum chinense* J. cells. *Plant Signal. Behav*. 7: 1103–1105.

Munnik T, Nielsen E. 2011. Green light for polyphosphoinositide signals in plants. *Curr. Opin. Plant Biol.* 14: 489–497.

Munnik, T., Vermeer, J.E.M. 2010. Osmotic stress-induced phosphoinositide and inositol phosphate signalling in plants. *Plant Cell Environ.* 33(4): 655–659.

Ning, J., Li, X., Hicks, L.M., Xiong, L. 2010. A Raf-like MAPKKK gene DSM1 mediates drought resistance through reactive oxygen species scavenging in rice. *Plant Physiol.* 152: 876–890.

O'Brien, J.A., Benkova, E. 2013. Cytokinin cross-talking during biotic and abiotic stress responses. *Front. Plant Sci.* 4: 451.

Ohashi, Y., Oka, A., Rodrigues-Pousada, R., Possenti, M., Ruberti, I., Morelli, G., et al. 2003. Modulation of phospholipid signaling by GLABRA2 in root-hair pattern formation. *Science* 300: 1427–1430.

Okazaki, Y., Saito, K. 2014. Roles of lipids as signaling molecules and mitigators during stress response in plants. *Plant J.* 79: 584–596.

Okazaki, Y., Otsuki, H., Narisawa, T., Kobayashi, M., Sawai, S., Kamide, Y., et al. 2013. A new class of plant lipid is essential for protection against phosphorus depletion. *Nat. Commun.* 4: 1510.

Pareek, A., Sopory, S.K., Bohnert, H.J., Govindjee, M.A.L. (Eds.). 2010. *Abiotic Stress Adaptation in Plants.* Springer, Dordrecht.

Park, S.W., Li, W., Viehhauser, A., He, B., Kim, S., Nilsson, A.K., et al. 2013. Cyclophilin 20-3 relays a 12-oxo-phytodienoic acid signal during stress responsive regulation of cellular redox homeostasis. *Proc. Natl. Acad. Sci. USA* 110: 9559–9564.

Peters, C., Li, M., Narasimhan, R., Roth, M., Welti, R., Wang, X. 2010. Nonspecific phospholipase C NPC4 promotes responses to abscisic acid and tolerance to hyperosmotic stress in *Arabidopsis. Plant Cell* 22: 2642–2659.

Petrov, V.D., Van Breusegem, F. 2012. Hydrogen peroxide—a central hub for information flow in plant cells. *AoB Plants* 2012:ls014. doi:10.1093/aobpla/pls014.

Pleskot, R., Li, J., Zársky, V., Potocky, M., Staiger, C.J. 2013. Regulation of cytoskeletal dynamics by phospholipase D and phosphatidic acid. *Trends Plant Sci.* 18: 496–504.

Pokotylo, I., Kolesnikov, Y., Kravets, V., Zachowski, A., Ruelland, E. 2014. Plant phosphoinositide-dependent phospholipases C: Variations around a canonical theme. *Biochimie* 96: 144–157.

Pospíšilová, J. 2003. Participation of phytohormones in the stomatal regulation of gas exchange during water stress. *Biol. Plant.* 46: 491–506.

Qi, T., Song, S., Ren, Q., Wu, D., Huang, H., Chen, Y., et al. 2011. The jasmonate-ZIM-domain proteins interact with the WD-Repeat/bHLH/MYB complexes to regulate Jasmonate-mediated anthocyanin accumulation and trichome initiation in *Arabidopsis thaliana. Plant Cell* 23: 1795–1814.

Raho, N., Ramirez, L., Lanteri, M.L., Gonorazky, G., Lamattina, L., Ten, H.A., et al. 2011. Phosphatidic acid production in chitosan elicited tomato cells, via both phospholipase D and phospholipase C/diacylglycerol kinase, requires nitric oxide. *J. Plant Physiol.* 168: 534–539.

Ramakrishna, A., Ravishankar, G.A. 2011. Influence of abiotic stress singles on secondary metabolites in plants. *Plant Signal Behav* 6:1720–1731.

Ramos-Diaz, A., Brito-Argaez, L., Munnik, T., Hernandez-Sotomayor, S.M. 2007. Aluminum inhibits phosphatidic acid formation by blocking phospholipase C pathway. *Planta* 225: 393–401.

Rasmusson, A.G., Soole, K.L., Elthon, T.E. 2004. Alternative NAD (P) H dehydrogenases of plant mitochondria. *Annu. Rev. Plant Biol.* 55(1): 23–39.

Rejeb, I.B., Pastor, V., Mauch-Mani, B. 2014. Plant responses to simultaneous biotic and abiotic stress: Molecular mechanisms. *Plants.* 3: 458–475.

Ren, D., Liu, Y., Yang, K.Y., Han, L., Mao, G., Glazebrook, J., et al. 2008. A fungal-responsive MAPK cascade regulates phytoalexin biosynthesis in *Arabidopsis. Proc. Natl. Acad. Sci. USA* 105: 5638–5643.

Riemann, M., Dhakarey, R., Hazman, M., Miro, B., Kohli, A., Nick, P. 2015. Exploring jasmonates in the hormonal network of drought and salinity responses. *Front. Plant Sci.* 6: 1077.

Rudolph, V., Schopfer, F.J., Khoo, N.K., Rudolph, T.K., Cole, M.P., Woodcock, S.R., et al. 2009. Nitro-fatty acid metabolome: Saturation, desaturation, β-oxidation, and protein adduction. *J. Biol. Chem.* 284: 1461–1473.

Ruelland, E., Kravets, V., Derevyanchuk, M., Martinec, J., Zachowski, A., Pokotylo, I. 2015. Role of phospholipid signalling in plant environmental responses. *Environ. Exp. Bot.* 114: 129–143. doi:10.1016/j.envexpbot.2014.08.009.

Saucedo-Garcia, M., Guevara-Garcia, A., Gonzalez-Solis, A., Cruz-Garcia, F., Vazquez-Santana, S., Markham, J.E., et al. 2011. MPK6, sphinganine and the LCB2a gene from serine palmitoyltransferase are required in the signaling pathway that mediates cell death induced by long chain bases in *Arabidopsis. New Phytol.* 191(4): 943–957.

Seo, S., Okamoto, M., Seto, H., Ishizuka, K., Sano, H., Ohashi, Y. 1995. Tobacco MAP kinase: A possible mediator in wound signal transduction pathways. *Science* 270: 1988–1992.

Shah, J., Zeier, J. 2013. Long-distance communication and signal amplification in systemic acquired resistance. *Front. Plant. Sci.* 4: 30.

Sheard, L.B., Tan, X., Mao, H., Withers, J., Ben-Nissan, G., Hinds, T.R., et al. 2010. Jasmonate perception by inositol-phosphate-potentiated COI1-JAZ co-receptor. *Nature* 468: 400–405.

Sinha, A.K., Jaggi, M., Raghuram, B., Tuteja, N. 2011. Mitogen-activated protein kinase signaling in plants under abiotic stress. *Plant Signal. Behav.* 6(2): 196–203.

Soler, R., Erb, M., Kaplan, I. 2013. Long distance root–shoot signalling in plant–insect community interactions. *Trends Plant Sci.* 18: 149–156.

Stamm, P., Kumar, P.P. 2013. Auxin and gibberellin responsive *Arabidopsis* SMALL AUXIN UP RNA36 regulates hypocotyl elongation in the light. *Plant Cell Rep.* 32: 759–769.

Suarez-Rodriguez, M.C., Adams-Phillips, L., Liu, Y., Wang, H., Su, S.H., Jester, P.J., et al. 2007. MEKK1 is required for flg22-induced MPK4 activation in *Arabidopsis* plants. *Plant Physiol.* 143: 661–669.

Suzuki, N., Miller, G., Salazar, C., Mondal, H.A., Shulaev, E., Cortes, D.F., et al. 2013. Temporal-spatial interaction between reactive oxygen species and abscisic acid regulates rapid systemic acclimation in plants. *Plant. Cell.* 25: 3553–3569.

Suzuki, N., Mittler, R. 2012. Reactive oxygen species-dependent wound responses in animals and plants. *Free Radic. Biol. Med.* 53: 2269–2276.

Taiz, L., Zeiger, E. 2010. *Plant Physiology*, 5th edition. Sinauer Associates, USA.

Takekawa, M., Tatebayashi, K., Saito, H. 2005. Conserved docking site is essential for activation of mammalian MAP kinase kinases by specific MAP kinase kinase kinases. *Mol. Cell.* 5(18): 295–306.

Taki, N., Sasaki-Sekimoto, Y., Obayashi, T., Kikuta, A., Kobayashi, K., Ainai, T., et al. 2005. 12-oxo-phytodienoic acid triggers expression of a distinct set of genes and plays a role in wound-induced gene expression in *Arabidopsis*. *Plant Physiol.* 139: 1268–1283.

Tanou, G., Job, C., Belghazi, M., Molassiotis, A., Diamantidis, G., Job, D. 2010. Proteomic signatures uncover hydrogen peroxide and nitric oxide cross-talk signaling network in citrus plants. *J. Proteome Res.* 9: 5994–6006.

Teige, M., Scheikl, E., Eulgem, T., Doczi, R., Ichimura, K., Shinozaki, K., et al. 2004. The MKK2 pathway mediates cold and salt stress signaling in *Arabidopsis*. *Mol. Cell.* 15: 141–152.

Testerink, C., Larsen, P.B., van der Does, D., van Himbergen, J.A.J., Munnik, T. 2007. Phosphatidic acid binds to and inhibits the activity of *Arabidopsis* CTR1. *J. Exp. Bot.* 58: 3905–3914.

Tiryaki, I., Staswick, P.E. 2002. An *Arabidopsis* mutant defective in jasmonate response is allelic to the auxin-signaling mutant axr1. *Plant Physiol.* 130: 887–894.

Tsutsui, T., Kato, W., Asada, Y., Sako, K., Sato, T., Sonoda, Y., et al. 2009. DEAR1, a transcriptional repressor of DREB protein that mediates plant defense and freezing stress responses in *Arabidopsis*. *J. Plant Res.* 122: 633–643.

Ulm, R., Ichimura, K., Mizoguchi, T., Peck, S.C., Zhu, T., Wang, X., et al. 2002. Distinct regulation of salinity and genotoxic stress responses by *Arabidopsis* MAP kinase phosphatase 1. *EMBO J.* 21: 6483–6493.

Uraji, M., Katagiri, T., Okuma, E., Ye, W., Hossain, M.A., Masuda, C., et al. 2012. Cooperative function of PLDdelta and PLDalpha1 in abscisic acid-induced stomatal closure in *Arabidopsis*. *Plant Physiol.* 159: 450–460.

Viewweger, K., Dordschbal, B., Roos, W. 2002. Elicitor-activated phospholipase A(2) generates lysophosphatidylcholines that mobilize the vacuolar H(+) pool for pH signaling via the activation of Na (+)-dependent proton fluxes. *Plant Cell* 14: 1509–1525.

Vranová, E., Atichartpongkul, S., Villarroel, R., Van Montagu, M., Inzé, D., Van Camp, W. 2002. Comprehensive analysis of gene expression in *Nicotiana tabacum* leaves acclimated to oxidative stress. *Proc. Natl. Acad. Sci. USA* 99: 10870–10875.

Vurro, M., Yoneyama, K. 2012. Strigolactones—intriguing biologically active compounds: Perspectives for deciphering their biological role and for proposing practical application, *Pest Manage. Sci.* 68: 664–668.

Walia, H., Wilson, C., Wahid, A., Condamine, P., Cui, X., Close, T.J. 2006. Expression analysis of barley (*Hordeum vulgare* L.) during salinity stress, *Funct. Integr. Genom.* 6: 143–156.

Wang, M., Zhang, Y., Wang, J., Wu, X., Guo, X. 2007. A novel MAP kinase gene in cotton (*Gossypium hirsutum* L.), GhMAPK, is involved in response to diverse environmental stresses. *J. Biochem. Mol. Biol.* 40: 325–332.

Wani, S.H., Kumar, V., Shriram, V., Sah, S.K. 2016. Phytohormones and their metabolic engineering for abiotic stress tolerance in crop plants. *Crop J.* 4(3): 162–176.

Welti, R., Li, W., Li, M., Sang, Y., Biesiada, H., Zhou, H.E., et al. 2002. Profiling membrane lipids in plant stress responses. Role of phospholipase D alpha in freezing-induced lipid changes in *Arabidopsis*. *J. Biol. Chem.* 277: 31994–32002.

Werner, T., Nehnevajova, E., Kollmer, I., Novak, O., Strnad, M., Kramer, U., et al. 2010. Root-specific reduction of cytokinin causes enhanced root growth, drought tolerance, and leaf mineral enrichment in *Arabidopsis* and tobacco. *Plant Cell* 22: 3905–3920.

Wild, M., Davière, J.M., Cheminant, S., Regnault, T., Baumberger, N., Heintz, D., et al. 2012. The *Arabidopsis* DELLARGA-LIKE3 is a direct target of MYC2 and modulates jasmonate signaling responses. *Plant Cell* 24: 3307–3319.

Wilkinson, S., Davies, W.J. 2009. Ozone suppresses soil drying and abscisic acid (ABA)-induced stomatal closure via an ethylene-dependent mechanism. *Plant Cell Environ.* 32: 949–959.

Xia, X.J., Zhou, Y.H., Shi, K., Zhou, J., Foyer, C.H., Yu, J.Q. 2015. Interplay between reactive oxygen species and hormones in the control of plant development and stress tolerance. *J. Exp. Bot.* 66(10): 2839–2856.

Xing, T., Malik, K., Martin, T., Miki, B.L. 2001. Activation of tomato PR and wound-related genes by a mutagenized tomato MAP kinase kinase through divergent pathways. *Plant Mol. Biol.* 46: 109–120.

Xiong, L., Zhu, J-K. 2002. Molecular and genetic aspects of plant responses to osmotic stress. *Plant Cell Environ.* 25(2): 131–139.

Yang, W.Y., Zheng, Y., Bahn, S.C., Pan, X.Q., Li, M.Y., Vu, H.S., et al. 2012. The patatin-containing phospholipase A PLAII alpha modulates oxylipin formation and water loss in *Arabidopsis thaliana*. *Mol. Plant.* 5: 452–460.

Zhang, Q., Lin, F., Mao, T., Nie, J., Yan, M., Yuan, M., et al. 2012. Phosphatidic acid regulates microtubule organization by interacting with MAP65-1 in response to salt stress in *Arabidopsis*. *Plant Cell* 24: 4555–4576.

Zhang, S., Du, H., Klessig, D.F. 1998. Activation of the tobacco SIP kinase by both a cell wall-derived carbohydrate elicitor and purified proteinaceous elicitins from *Phytophthora* spp. *Plant Cell* 10: 435–450.

Zhang, W., Wang, C., Qin, C., Wood, T., Olafsdottir, G., Welti, R., et al. 2003. The oleate-stimulated phospholipase D, PLDdelta, and phosphatidic acid decrease H_2O_2-induced cell death in *Arabidopsis*. *Plant Cell* 15: 2285–2295.

Zhang, X., Zhang, L., Dong, F., Gao, J., Galbraith, D.W. 2001. Hydrogen peroxide is involved in abscisic acid-induced stomatal closure in *Vicia faba*. *Plant Physiol.* 126(4): 1438–1448.

Zhang, Y., Zhu, H., Zhang, Q., Li, M., Yan, M., Wang, R., et al. 2009. Phospholipase D alpha1 and phosphatidic acid regulate NADPH oxidase activity and production of reactive oxygen species in ABA-mediated stomatal closure in *Arabidopsis*. *Plant Cell* 21: 2357–2377.

Zhu, S.Y., Yu, X.C., Wang, X.Y., Zhao, R., Li, Y., Fan, R.C., et al. 2007. Two calcium-dependent protein kinases, CPK4 and CPK11, regulate abscisic acid signal transduction in *Arabidopsis*. *Plant Cell* 19: 3019–3036.

27

Current Understanding of the Role of Jasmonic Acid during Photoinhibition in Plants

Ruquia Mushtaq, Sarvajeet Singh Gill, Shruti Kaushik, Anil K. Singh, Akula Ramakrishna, and Geetika Sirhindi

CONTENTS

Abbreviations

13-HPOT	(13S)-hydroperoxyoctadecatrienoic acid		**JA**	Jasmonic acid
$^1O_2{}^*$	Singlet oxygen		**KAT**	l-3-ketoacyl-CoA-thiolase
ABA	Abscisic acid		**LOX**	1,3-Lipoxygenase
AIF	Anther Indehiscence Factor		**LOX**	lipoxygenase
AOC	Allene oxide cyclase		**Me-JA**	methyl jasmonate
AOS	Allene oxide synthase		**MFPs**	multifunctional proteins
BADH	betaine aldehyde dehydrogenase		*NLS*	nuclear localization signal
CaMV	Cauliflower mosaic virus		**NPQ**	non-photochemical quenching
DAD1	Defective in anther dehiscence 1		$O_2{}^-$	superoxide
DAF	DAD1-ACTIVATING FACTOR		**OEC**	oxygen-evolving complex
DALL	DAD1-like LIPASES		**OPDA**	12-oxo- phytodienoic acid
DES	divinylether synthase		**OPR3**	12-oxo-phytodienic acid reduc- tase3
DGDG	digalactosyldiacylglycerol		**PI**	Photoinhibition
ETC	electron transport chain		**PLA1**	phospholipase A1
GLA1	galactolipase A1		**PPFD**	Photosynthetic photon flux density
GLRs	glutamate receptor-like genes		**ROS**	Reactive oxygen species
HPL	hydroperoxide lyase		**TFs**	transcription factors
			α-LeA	α-linolenic acid

Introduction

Photoinhibition (PI) is a broad term that initially described a decline in photosynthetic viability of oxygen evolving photosynthetic organisms due to excessive illumination. 100 years ago, the phenomenon of photoinhibition was described by Ewart (1896) during his examination of the effects of external factors on this intricate process. Rabinowitch (1951) further examined this interaction between the external environment and internal factors that inhibits photosynthesis. PI has great importance in the field of plant physiology as it solely affects the quality as well as the quantity of photosynthesis, which essentially determines the plant productivity. Light is the most influential factor of the environment in terms of size and speed, and thus it is very interesting to study the growth and development of plants under challenging environmental conditions.

On the basis of the quality and quantity of light responsible for triggering the phenomenon of PI, it can be classified into (i) dynamic photoinhibition, which is observed in plants exposed to moderately excessive light, and (ii) chronic photoinhibition, triggered by exposure to highly excessive light conditions (Osmond, 1994). The plants show that dynamic PI decreases quantum yield while the maximum photosynthetic rate remains unchanged. This type of PI is caused by diversification in the absorbed light, towards a decline in quantum yield instigated by heat dissipation. Chronic PI happens mostly in plants exposed to any type of stress conditions, and this is associated with the damage of D1 protein of PSII RC (Figure 27.1). The number of photons that fall on a leaf from sunlight determines the photosynthetic reactions (Demmig-Adams and Adams, 2000). This photosynthetic photon flux density (PPFD) changes over the course of day as the quality and quantity of light that the leaf receives during day changes. The photons that fall on the leaf belong to a different wavelength of the sunlight. However, the light-capturing complexes present in inner thylakoid membrane of chloroplast are only designed to capture photons of red and blue wavelengths.

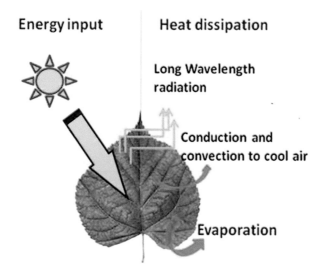

FIGURE 27.1 Various pathways leaf follow to protect from excess excitation energy to avoid damage.

The energy that photons carry is most damaging at the molecular level, and this damage is more pronounced under unfavorable conditions causing photoinhibition (Powles, 1984; Krause, 1988). Hence, photoinhibition is a state of physiological stress caused by any unfavorable environmental conditions in which the regulated process of light capturing, dissipations, and use for photochemistry is disturbed.

Regulation of Photoinhibition

Photosynthetic organisms employ complex regulatory mechanisms to avoid photoinhibition, including photo-damage repair of the D1 protein of PSII. However, the two photosystems are equally driven by the excitation energy of photons and this is ensured by the tightly regulated mechanism of energy flow in the antenna systems surrounding the two photosystems. Despite the photosynthetic machinery's three lines of defense, including protective and scavenging mechanisms, damage can still occur to the photosynthetic machinery, leading to photoinhibition. Thus, additional mechanisms are required which manage the repair and scavenging of the system.

In this category, the first role that appears is that of accessory pigments which are the part of antenna complexes surrounding the two photosystems. For example, *carotenoids* which, in addition to their role as accessory pigments, play a vital role in photoprotection. The excited state of chlorophyll molecules, mainly *chlorophyll a*, is called a quenched state if it transfers this energy for photochemistry. But if this excited state of *Chl* molecules is not quenched then their energy reacts with the molecular oxygen leading to the formation of singlet oxygen ($^1O_2^*$) which is extremely reactive. The production of the reactive oxygen species (ROS) damages cellular components, especially lipids of the chloroplast related to membrane degeneration (Gill and Tuteja, 2010). Carotenoids play a protective role here, like a rapid task force, and quench the energy from the excited chlorophyll molecule to get excited. On the other hand, carotenoids do not have the sufficient energy to form singlet oxygen; hence they dwindle to back to their ground state by dissipating the energy in the form of heat and thus help by mitigating the photo-inhibitory decline in photosynthesis.

Another major process involved in regulating photoinhibition is non-photochemical quenching (NPQ), which includes quenching of overexcited chlorophyll molecules known as chlorophyll fluorescence. This quenching is also achieved by the dissipation of excitation energy in the form of heat. Although the molecular mechanism involved in NPQ is not well understood, the evidence available suggests that different underlying mechanisms for the quenching process to protect plants from photoinhibition. In this series, the aggregation of antenna complexes and the pH of the thylakoid lumen are important factors in regulating PI. Three different carotenoids, namely violaxanthin, antheraxanthin, and zeaxanthin, are well reported to be involved in regulating NPQ. These three carotenoids operate in a cyclic manner and this cycle is known as the xanthophyll cycle. Out of the three carotenoids, zeaxanthin is reported to be the most effective component of this cycle in the dissipation of heat and regulating PI (Pfündel and Bilger, 1994).

The stacking arrangement of thylakoid membranes also permits portioning of energy between two photosystems and thus helps to manage and regulate photoinhibition. Out of the two photosystems, PSII is the most susceptible component to damage by light. Powles (1984) explained the unique susceptibility of PSII to photoinhibition and also proposed a mechanism used by PSII for photo-damage protection. On other hand, PSI is very vulnerable to damage, particularly that which is caused by ROS produced by overexcited chlorophyll molecules. PSI is associated with very strong reductant ferredoxin which can easily reduce molecular oxygen to superoxide (O_2^-). This reduction works in competition with the normal reduction of $NADP^+$ by the electron flow in the thylakoid membrane. Superoxide is one member of ROS series, which when induced by PSI can be easily eliminated by the action of antioxidant enzymes present in the lumen of the thylakoid membrane.

Mechanisms of Photoinhibition and Its Protection

Asada (1999) proposed an overall picture of the regulation of photons captured by the antenna complexes particularly those surrounding PSII to protect photosynthetic systems from photo-damage (Figure 27.2). PI is a process resulting from the energy absorbed by the photosynthetic pigments and channeled to PSII. This was suggested on the basis of observations that the action spectrum for photosynthesis follows the spectrum of PI. Another point of support for this assumption is the decline in PSII photochemical efficiency which is the initial symptom of PI. PSII plays a central role in PI which involves light impairment and the protection of PSII.

There are different mechanisms by which higher plants, unicellular alga, or other photosynthetic organisms reduce or manage PI. The mechanisms involved in the management of PI in nature are an array of responses towards the quantity of excess light which can operate on a range of time scales. Reduction in the maximum quantum yields for CO_2 uptake (ϕ) and O_2 evolution (ϕO) and prolonged exposure to excessive light causing a decline in the rate of light-saturated photosynthesis (A_{sat}) are a few functional consequences of photoinhibition (Long et al., 1983; Powles et al.,1983; Sassenrath et al., 1990).

Prolonged exposure to excitation energy of high-intensity results in damage to the photosynthetic system and PSII reaction center and the main target in this is the D1 protein. Damaged D1 should be removed from the reaction center and replaced by the newly synthesized protein and it is the only component of PSII that needs to be synthesized. However, it doesn't mean that other components of the PSII reaction center are not damaged during PI but it is thought that they are recycled whereas D1 is not. On the other hand, the PSI reaction center is associated with a strong electron acceptor ferredoxin hence PSI is a very good reductant and is protected from PI. The superoxides produced at PSI, due to the oxidation-reduction reaction of Fd and $NADP^+$, can be eliminated by the actions of a series of enzymes including SOD and APOX. The reaction center of PSII which is a multiprotein complex present in plants and cyanobacteria plays an important role in PI. The PSII reaction center contains several transition metal co-factors; including haems, non-haem iron, and an oxygen-evolving manganese cluster containing four Mn ions. PSII is vulnerable to PI under visible and UV light (Aro et al., 1993; Melis, 1999). In PI, PS II lost its electron transfer activity and the repair of PSII *in vivo* requires both degradation and re-synthesis of the D1 protein.

When plants are exposed to environmental stresses and the availability of CO_2 within the leaf (*Ci*) is restricted or the synthesis of ATP is impaired, the activity of the Calvin cycle is reduced, but the PSII remains active. In such a situation, the concentration of the final electron acceptor $NADP^+$ is generally very low which is a consequence of an excess of excitation energy in the photosystems. Excitation energy may be dissipated from the system by various methods, one of which is non-photochemical quenching, potentially through the xanthophyll cycle. Some other alternative

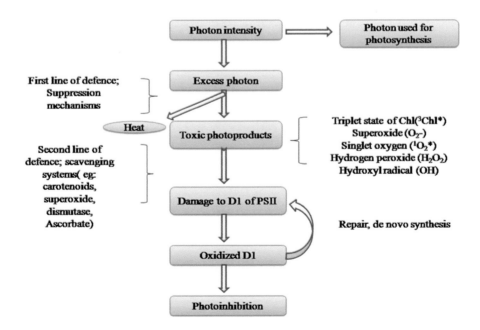

FIGURE 27.2 Overall picture of the regulation of photon capture and the protection and repair of photo-damage.

processes like photorespiration metabolism may also be involved in dissipating the excitation energy and protecting the plant from photoinhibition (Niyogi, 2000).

However, if the energy is not dissipated then electrons accumulating in the electron transport chain initiate Mehler's Reaction by transferring electrons to oxygen and producing ROS such as superoxide. These ROS have very high reactive potential and are thus responsible for damaging many cellular components such as proteins, DNA, and lipids as well as constituting oxidative stress. These ROS also inactivate the photochemical reactions occurring in the PSII reaction center and thus initiate photoinhibition (Figure 27.3). The ROS generated by different stress factors are liable for inhibition of the PSII repair mechanism by inhibiting the transcription and translation of *psbA* genes which encode the D1 protein of the PSII reaction center (Nishiyama et al., 2001, 2004; Allakhverdiev et al., 2002). Most of the photosynthetic genes are regulated through redox reaction (Pfannschmidt, 2003).

Various abiotic factors causing environmental stresses are responsible for deactivating PSII reaction centers by inhibiting the repairing mechanisms for photo-damage rather than by directly attacking it (Murata et al., 2007). Regulation of the expression of genes involved in photosynthesis is linked with the redox state which is in turn regulated through kinase activity responsible for controlling the activity of plastid transcription factors (Baginsky et al., 1999). Abiotic stresses are also reported to have negative effects on CO_2 diffusion, ATP synthesis, and reductant status in addition to a reduction in the activities of enzymes involved in the Calvin cycle, including the key enzyme Rubisco thus limiting CO_2 assimilation.

Photoinhibition Regulating Factors

Inhibition of various activities of photosynthesis metabolism is regulated not only by light quality and quantity but also by factors such as CO_2 concentration, temperature, humidity, and internally by metabolites and other factors (Edwards and Walker, 1983;

Sharkey, 1985). Internal factors also included the cross-talk of signaling pathways involving various hormones which are working as signaling molecules such as ABA, ethylene, polyamines, salicylic acid, brassinosteroids, and jasmonates.

Photoinhibition triggered by the over-reduction of the electron transport chain (ETC) at the thylakoid membrane of the chloroplast is the product of various adverse environmental conditions (Foyer and Noctor, 2005; Nishiyama et al., 2006; Takahashi and Murata, 2008; Rochaix, 2011; Foyer et al., 2012; Kangasjarvi et al., 2012; Nishiyama and Murata, 2014). Plants have several mechanisms to overcome this problem such as reducing the rate of electron transport by converting the excessively absorbed light into thermal energy. The dissipation of excess excitation energy as heat is called non-photochemical quenching (NPQ) of chlorophyll fluorescence (Rochaix, 2011; Tikkanen et al., 2011; Nath et al., 2013; Spetea et al., 2014).

Stress signals are perceived by the plants through receptor proteins present at the plasma membrane that trigger cascades of molecules involved in transducing the signals to the regulatory systems via secondary messengers and various protein kinases pathways (Schmutz et al., 2010; Choudhary et al., 2012; Le et al., 2012). The signal transduction pathways are composed of various components, including phytohormones, transcription factors (TFs), mitogen-activated protein kinases, protein kinases, and phosphatases which eventually regulate the expression of genes involved in stress tolerance and/or photosynthesis or photoinhibition regulation (Foyer and Shigeoka, 2011; Puranik et al., 2012; Osakabe et al., 2014).

Jasmonates in Photoinhibition Protection

It is now well established that signaling molecules such as jasmonates, single-handedly or in cross-talk with other phytohormones, protect the plant against the photoinhibition of photosynthesis (Sharma et al., 2013; Sirhindi et al., 2015, 2016). Other oxylipin producing branches such as hydroperoxide lyase branch (*HPL*) are found to play a very protective role against

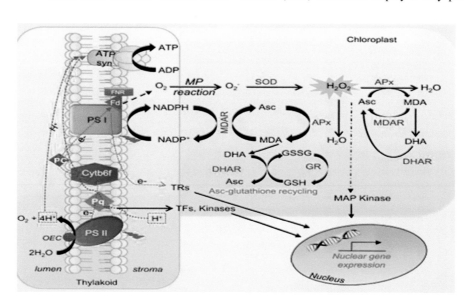

FIGURE 27.3 Production of reactive oxygen species during photosynthetic redox reactions in the thylakoid membrane of chloroplast.

the photoinhibition of photosynthetic machinery (Savchenko et al., 2017). In their study, it has been reported that overexpressing lines of *HPL* in *Arabidopsis* intact leaves exhibit enhanced adaptability towards high light intensity which is evidenced from a lower rate constant of photoinhibition during high light treatment and a higher rate constant of recovery after photoinhibition of PSII. They also reported the suppression of photochemical activity of PSII in intact leaves and isolated thylakoid membranes by exogenous application of methyl jasmonate (Me-JA). Another factor supporting the possible role of jasmonate in protecting against photoinhibition is its site of biosynthesis which occurs in the lipids present in the thylakoid membrane. Various biosynthetic pathways of jasmonates have been explored till date and were most recently all reviewed by Wasternack and Strnad in 2016.

Occurrence and Biosynthesis of Jasmonates

After the initial identification of Me-JA from flowering plants (Demole et al., 1962) and of JA in the culture medium of the fungus *Lasiodiplodia theobromae* (Aldridge et al., 1971), numerous JA compounds were later reported in other plants (Meyer et al., 1984). High levels were found in fruits. Successive screenings revealed occurrences in all organs carrying plastids. The basal level was found to increase rapidly upon wounding or in response to other environmental stimuli. Recently, 12-oxo-phytodienoic acid (OPDA) but not JA has been identified in *Marchantia polymorpha*, the most basal lineage of extant land plants (Yamamoto et al., 2015). In *Physcomitrella patens* OPDA was detected but JA was not detected (Stumpe et al., 2010). Green and brown algae contain several lipoxygenase (LOX) pathway intermediates including JAs.

In the 1980s, the biosynthetic pathway of JA from α-LeA esterified in chloroplast membranes was established. Originally, a sequence of a LOX, a hydroperoxide cyclase, a reductase, and β-oxidation of the carboxylic acid side chain was proposed (Vick and Zimmerman, 1983). Afterward, the single step of hydroperoxide cyclase was revealed to be a two-step reaction of membrane-associated allene oxide synthase (AOS), whose highly unstable product is cyclized by an allene oxide cyclase (AOC) to OPDA. Besides the enzymatic reactions, the unstable epoxide was hydrolyzing to α- and γ-ketols and non-enzymatic cyclization to racemic OPDA (Brash et al., 1988). All enzymes of OPDA formation are located in chloroplasts and the second half of JA biosynthesis occurs in the peroxisomes. Here, OPDA reductase reduced the OPDA followed by enzymes of β-oxidation, the acyl-CoA oxidase (ACX), the multifunctional proteins (MFPs), and the l-3-ketoacyl-CoA-thiolase (KAT) upon activation of co-esters by fatty acid CoA-synthetases and 4-coumaroyl: CoA-ligases.

JA biosynthesis has been frequently reviewed (Browse, 2009a,b; Schaller and Stintzi, 2009; Acosta and Farmer, 2010; Wasternack and Kombrink, 2010; Kombrink, 2012; Wasternack and Hause, 2013) including mechanistic explanations of enzyme crystallization (Lee and Black, 2005). These studies reported the involvement of different pathways for biosynthesis of different forms of jasmonates.

Galactolipases Involved in JA Biosynthesis

JA biosynthesis is introduced by the release of α-LeA from galactolipids of chloroplast membranes by *phospholipase A1* (*PLA1*) which hydrolyzes with *sn-1* specifically. For JA formation the flower-specific protein *DEFECTIVE IN ANTHER DEHISCENCE 1* (*DAD1*), a *PLA1*, is absolutely required (Ishiguro et al., 2001). *DAD1* is activated by the homoeotic protein *AGAMOUS*, thereby affecting late stamen development (Ito et al., 2007). Besides, *DAD1* and most of the subsequently acting genes involved in JA biosynthesis [*AOS*, *AOC*, and *12-oxo-phytodienic acid reduc-tase3* (*OPR3*)] are repressed by the *NAC*-like gene *ANTHER INDEHISCENCE FACTOR* (*AIF*) (Shih et al., 2014). A *galactolipase A1* (*GLA1*) of *Nicotiana attenuata* has drawn attention to for forming JA in its leaves and roots but not during infection with *Phytophothora parasitica* (Bonaventure et al., 2011). More studies exposed that, upon *Phythophtora* infection, the *GLA1 lipase* is involved in the generation of 9-OH-18:2 and other C18 and C19 oxylipins (Schuck et al., 2014). An interesting link between the biosynthesis of the lipid digalacto-syldiacylglycerol (DGDG) and JA biosynthesis was described recently (Lin et al., 2016).

1,3-Lipoxygenase (LOX)

The lipoxygenase occurrence and mechanism have been repeatedly reviewed (Andreou and Feussner, 2009; Andreou et al., 2009). Among the seven different branches of the *LOX* pathway, only the *AOS* branch leads to JA formation (58Feussner and Wasternack, 2002). The introduction of molecular oxygen in position C-13 of α-LeA is the ultimate reaction in JA formation. All 13-LOXs known to date are involved in wound-induced JA formation with partially specific activity: (i) wounding (LOX2); (ii) lipid peroxidation (LOX2); (iii) early wound responses of xylem cells (LOX6); (iv) wound responses in vascular tissue (LOX3, LOX4); (v) natural and dark-induced senescence (LOX2); and (vi) flower development and fertility (LOX3, LOX4). Meanwhile, some new data on 13-LOXs of *Arabidopsis* has been accumulated. *AtLOX6* was revealed to be exclusively required for JA accumulation in roots under abiotic and biotic stress and the generation of the basal level of OPDA (Grebner et al., 2013). An association of *AtLOX3* with the salinity stress response has been detected (Ding et al., 2016). *AtLOX3* and *AtLOX4* have different activities and it was identified that *AtLOX4* has a major role in defense control against root-knot nematodes and cyst nematodes(Ozalvo et al., 2014). The generation of a variety of *13-LOX* mutants carrying a JA reporter system improved our knowledge on intercellular JA transport in the axial (root to shoot) and radial (cell to cell) directions (Gasperini et al., 2015). This biosensor for JA was considered with a specific *Jas* motif of *JA ZIM-domain9* (*JAZ9*) and the *VENUS* alternate of the yellow fluorescent protein carrying a nuclear localization signal (*NLS*) and the *Cauliflower Mosaic Virus* (*CaMV*) 35S promoter (Larrieu et al., 2015). Along with the six tomato *LOXs* (*TomloxA–F*), *TomloxC*, and *TomloxD* are chloroplast-localized *13-LOX* enzymes. *TomloxC* is essential for C5 flavor volatiles without any important defensive function (Shen et al., 2014).

Genetic and biochemical evidence obtained from characterizing wound response *spr8* mutants showed that *TomloxD* is involved in JA generation and consequently linked to all defense reactions against herbivores (Yan et al., 2013). This mutant carries a point mutation in the catalytic domain of *TomloxD*. *13-LOXs* were reported for the first time in the plastids of barley using an immunological approach (Feussner et al., 1995), as well as by immunogold labeling (Bachmann et al., 2002). The current localization of a *13-LOX* of barley lacks novelty and sufficient quality in the activity data (Springer et al., 2016).

Allene Oxide Synthase (AOS)

AOS is a member of the *CYP74* gene family, collectively working with hydroperoxide lyase (*HPL*) and divinylether synthase (*DES*) (58Feussner and Wasternack and Hause, 2002; Hughes et al., 2009). The cloning, mechanism, characterization, and function of *AOS* in stress responses and development have been reviewed (Brash, 2009; Schaller and Stintzi, 2009; Kombrink, 2012). In many plants, *AOSs* arise in gene families, except *Arabidopsis*, where a single-gene copy allowed the generation of JA- and OPDA-deficient mutant plants (Park et al., 2002). *AOS1* in rice has a chloroplast targeting sequence (Haga and Ino, 2004), and the enzyme has been effectively used without a chloroplast target signal in a combined *AOS/AOC* enzymatic assay as a functional *AOS* (Riemann et al., 2013). In rice, mutant analysis revealed an exciting cross-talk between the AOS and the HPL branch (Liu et al., 2012).

Allene Oxide Cyclase (AOC)

AOCs occur in small gene families that have been cloned and characterized, with mechanistic studies from several plant species (Hofmann and Pollmann, 2008; Schaller and Stintzi, 2009; Kombrink, 2012) and the crystal structures available for *AtAOC2* (Hofmann et al., 2006) and *PpAOC1* and *PpAOC2* (Neumann et al., 2012). Lately, an *AOC* of the liverwort *Marchantia polymorpha* was cloned (Yamamoto et al., 2015). The *M. polymorpha* accumulates OPDA, not JA, suggesting a signal property of OPDA (Yamamoto et al., 2015). Commonly, overexpression of *AOS* or *AOC* did not lead to elevated JA levels due to lack of substrate α-LeA which is generated only in the case of external stimuli such as wounding (Laudert et al., 2000; Stenzel et al., 2003). Constitutive overexpression of AOC in *Salvia miltiorrhiza* and in wheat led to elevated JA levels and JA responses in the corresponding plants (Gu et al., 2012; Zhao et al., 2014). In rice *AOC* mutants (*cpm2*, *hebiba*) demonstrated the role of JA in defense against the blast fungus *Magnaporthe oryzae* (Riemann et al., 2013). Both of these mutants are impaired in the *AOC* function and exhibited a deficiency in OPDA accompanied by increasing salt tolerance and reactive oxygen species (ROS)-scavenging activity (Hazman et al., 2015). In the case of *Arabidopsis*, the four members of the *AOC* gene family have tissue- and organ-specific promoter activities including *in vivo* heteromerization (Stenzel et al., 2012). Further inspection of heteromeric pairs of AOCs is in line with enzyme activity control via heteromerization (Otto et al., 2016).

OPDA Reductase (OPR3)

Within the chloroplast OPDA has formed and been transported across two membranes. Whereas, no transporter for the inner and outer envelope of the chloroplast has been identified until now, import into peroxisomes seems to occur by *COMATOSE* exhibiting *PXA1* activity or by the anion trapping (Theodoulou et al., 2005). The OPRs take place in small gene families and only the *OPR3* of *Arabidopsis* and its homologs are involved in JA biosynthesis. In the case of rice, the homolog of *AtOPR3* is *OsOPR7* (Tani et al., 2008). *OPRs* have been cloned, localized, biochemically characterized, and functionally analyzed from the different plant species (Schaller and Stintzi, 2009; Kombrink, 2012; Wasternack and Hause, 2013). In rice, 10 members of the gene family were relatively characterized in terms of expression, structure, and function which are divided into five sub-groups, and *OPR7* has been identified as a JA-forming enzyme (Li et al., 2011). In *Arabidopsis*, six members of the gene family and *OPR3* are exclusively involved in JA formation. The JA-independent direct role of *OPR3* has been identified recently in the primary root growth under P deficiency (Zheng et al., 2016). Even though primary root growth inhibition by P deficiency is a well-known phenomenon, the underlying mechanism was unknown. By suppression of root tip growth at the transcriptional level under P deficiency a novel function of *AtOPR3* has been shown, including a cross-talk with the ethylene and gibberellin (GA) signaling pathways (Zheng et al., 2016).

β-Oxidation of the Pentenyl Side Chain (*ACX, MFP,* and *KAT*)

In the final steps of auxin biosynthesis, the pentenyl side chain of the JA is shortened by the fatty acid β-oxidation machinery (Hu et al., 2012). This was revealed by labelling experiments (Miersch and Wasternack, 2000), as well as biochemical and genetic data for the enzymes involved, for example, acyl-CoA oxidase (Li et al., 2005; Schilmiller et al., 2007), l-3-ketoacyl CoA thiolase (Castillo et al., 2004), and 4-coumarate:CoA ligase-like enzymes (Schneider et al., 2005; Koo et al., 2006). Further evidence was collected regarding diminished JA formation in the *pex* mutants and the affected components of the peroxisomal import complex (Hu et al., 2012; Kombrink, 2012; Wasternack, 2014). Along with the *pex* mutants, *PEX13* interacts with *DAYU/ABERRANT PEROXISOME MORPHOLOGY9* (*DAU*) which is significant in pollen maturation as well as JA formation (Li et al., 2014). In between the evolutionarily conserved *KAT1* and *KAT5* and *Arabidopsis* inflorescence development, an interesting link was shown suggesting the requirement of JA for normal inflorescence development (Wiszniewski et al., 2014). This is suggestive of the flower inflorescence initially observed in the mutant *aim1*, which is affected in the *MFP* protein (Richmond and Bleecker, 1999).

Intracellular Localization of *13-LOXs, AOS,* and *AOCs*

Enzymes for OPDA formation such as 13-LOXs, AOS, and AOCs carry a target sequence for chloroplast import, with a few exceptions, but all of them have been localized in the chloroplasts by import studies and immune-cytochemistry, including electron microscopy (Farmaki et al., 2007; Wasternack, 2007; Schaller et al., 2008). The chloroplast is the site of photoinhibition, therefore jasmonates may successfully regulate photoinhibition, which allows the plant to sustain under abiotic or biotic stress, thus maintaining a functioning metabolome. However, there is still a lack of understanding of the chloroplast targeting sequence of *AOS1* of barley and *AOS1* of rice, and both of them localize in the chloroplasts (Maucher et al., 2000; Yoeun et al., 2013). AOS is found in lipid-rich plastoglobuli, where high concentrations of quinone and α-tocopherol content has been found which become stable the highly unstable epoxide formed in the *AOS* reaction (Ytterberg et al., 2006). For tomato, *in vitro* import studies divulge the import of *AOS* to the inner envelope membrane (Froehlich et al., 2001) which is the site of biosynthesis of lipid oxides, acting as a precursor for jasmonates biosynthesis. The localization of AOS in the inner envelope of *Arabidopsis* with the rhomboid-like protease *AtRBL8/9* leads to regulation of AOS (Knopf et al., 2012).

The three main components of the regulation of JA biosynthesis have been repeatedly reviewed and these are (i) upon external stimuli there is requirement to release α-LeA as the substrate; (ii) a positive feedback loop in the expression of JA biosynthesis genes by JA; and (iii) tissue specificity (Wasternack, 2007; Schaller and Stintzi, 2009; Wasternack and Hause, 2013). Through concurrent activity between the *AOS* and *HPL* additional regulatory factors may occur and through negative regulators *JAZs* other branches of the *LOX* pathway may occur, including heteromerization (e.g. *AOCs* of *Arabidopsis*) (Otto et al., 2016), post-translational control (e.g. *OPR3*), the mitogen-activated protein kinase cascade, and Ca2+ signaling (Wasternack and Hause, 2013). In JA formation recent data argues against a positive feedback loop but still shows regulation in the expression of JA biosynthesis genes, thereby suggesting the involvement of post-translational modifications (Scholz et al., 2015).

JA Metabolism

Twelve metabolic pathways are known today that convert JA into active and inactive compounds, or compounds which are partially active (Koo and Howe, 2012; Wasternack and Hause, 2013; Heitz et al., 2016). Whereas, homeostasis provides the majority of JA signaling among different JA-Ile derivatives, some reactions also lead to compounds with activity in specific reactions of various stress responses and development such as leaf movement. The leaf is the site of light perception leading to photoinhibition, and thus it may possible that JA could regulate photoinhibition through either regulating the activity levels of various stress response genes, such as those involved in ROS-scavenging, including SOD, CAT, APOX, GR, etc., or by regulating the expression of stress-related genes.

Conjugation

In JA perception, the formation of *JA-Ile* by jasmonoyl-isoleucine synthetase (*JAR1*), which is a member of the *GH3* gene family (AtGH3.11) (Staswick and Tiryaki, 2004), is the ultimate step. The *GH3* protein family comprises the acyl acid-amido synthetases with auxin, JA, or benzoate as substrates (Westfall et al., 2013). The role of JA-Ile as the most biologically active JA compound has become obvious through reports on the JA-Ile-mediated promotion of the *SCF*COI1–*JAZ* co-receptor complex formation (Thines et al., 2007; Katsir et al., 2008; Melotto et al., 2008) and the properties of JA-Ile as the ligand of the complex (Yan et al., 2009; Sheard et al., 2010).(Fonseca et al., 2009). This development has been used to design the ligand-based antagonist of JA-Ile by modifying coronatine (Monte et al., 2014) or for the generation of inhibitors of *JAR1* such as *jarin1* by chemical screening (Meesters et al., 2014). Crystallization of *JAR1* led to the explanation of why the epimer (+)-7-iso-JA-Ile is formed (Westfall et al., 2012). Even though (+)-7-iso-JA-Ile is exclusively formed upon wounding, tomato *JAR1-RNAi* lines showed only 50% to 75% of down-regulation, suggesting the occurrence of another JA-conjugating enzyme other than *JAR1* (Suza et al., 2010). In the use of *jar1* mutants this has to be taken into account. In rice, *OsJAR1*, exhibiting a similar substrate specificity to *AtJAR1*, this is the most important JA-conjugating enzyme in wounding response (Fukumoto et al., 2013; Svyatyna et al., 2014), seed development under field conditions (Fukumoto et al., 2013), as well as the floret opening, and also anther dehiscence (Xiao et al., 2014). *OsJAR1* activity is also required in photomorphogenesis but acts partially redundantly to the *OsJAR2* (Svyatyna et al., 2014). *OsJAR2* cannot complement the two major functions of *OsJAR1* (Fukumoto et al., 2013). The level of JA-Ile plays an important regulatory role for all the JA-dependent processes and this is sustained by *JAR1* activity and different metabolic reactions such as hydroxylation, carboxylation, and hydrolysis of conjugates.

Recently, endogenous bioactive JA derivatives (+)-7-iso-JA-Ala, (+)-7-iso-JA-VAL, (+)-7-iso-JA-Leu, and (+)-7-iso-JA-Met have also been identified via the corresponding coronafacic acid conjugates as constant molecular mimics (Yan et al., 2016). Distinct binding is demonstrated by these conjugates in the *COI1–JAZ* interaction assay.

Role of Jasmonates in Growth and Development

The life cycle of a plant initiates in the form of seed germination which is a complex phenomenon regulated by various external and internal factors. The germination process starts with the emergence of the root beyond the seed coat which requires the interaction of diverse phytohormones (Bewlay and Black, 1994; Kucera et al., 2005). The literature available regarding the involvement of JAs in seed germination, explores the inhibitory side of various JAs rather than its promoting potential. Linkies and Leubner-Metzger (2012) reviewed the role of Me-JA on germination and seedling growth while illustrating its inhibitory potential. Dave et al. (2011) evidenced that OPDA, the precursor of JA biosynthesis, is the inhibitory compound and acts together with ABA in a *COI1* independent manner. Poonam et al. (2013) first reported the promoting effect of JA on germination of *C. cajan* seeds. The promoting effect of Me-JA on the germination of a number of dormant seeds including apple, pear, two species of *Acer*, and Douglas fir (*Pseudotsuga menziesii*) has also been reported (Daletskaya and Sembdner, 1989;

Berestetzky et al., 1991; Ranjan and Lewak, 1992; Jarvis et al., 1996; Yildiz et al., 2007).

The inhibitory effect of JA and its derivatives on germination in maize seed may be due to a decrease in the concentration or activity of α-amylase and a reduction in ethylene production was suggested (Norastehnia et al., 2007). Dave and Graham (2012) revealed that cis-OPDA is much more effective than JA in inhibiting seed germination in *A. thaliana*. Exogenous application of JA/Me-JA causing the inhibition of seed germination has been reported in angiosperm *Solanum lycopersicum*, *Brassica napus*, *Linum usitatissimum*, *Lupinus luteus*, *Z. mays*, *Amarantus caudatus*, cocklebur, sunflower, tobacco, rapeseed, and flax (Corbineau et al., 1988; Wilen et al., 1991; Kepczynski and Bialecka, 1994; Wilen et al., 1994; Nojavan and Ishizawa, 1998; Preston et al., 2002; Bialecka and Kepczynski, 2003; Norastehnia et al., 2007; Miersch et al., 2008; Oh et al., 2009; Zalewski et al., 2010).

Creelman and Mullet (1997) avowed that JA and Me-JA inhibit germination of non-dormant seeds while stimulating germination in dormant seeds. Yang et al. (2012) reported faster germination in rice plants where *COI1* gene was silenced, when compared to wild type rice plants. Lee and Back (2005) illustrated the overexpression of pepper *MAPKinase*, which is responsible for the higher accumulation of JA in transgenic rice causing inhibition in seed germination. Kaur et al. (2013) investigated the role of JA in the germination percentage and growth of *B. napus* under normal and salt stress conditions and reported that exogenous application of different concentrations of JA inhibited seed germination when compared to the control. Antagonistic reports are given on *C. cajan* seed germination by exogenous application of different concentrations of JA where about 22% enhancement in the rate of seed germination was observed (Poonam et al., 2013; Sirhindi et al., 2015).

It is a general observation that the effect of JAs on plant growth in terms of root and shoot length is inhibitory most of the time. Zhang and Turner (2008) clarified that the inhibition of cell mitosis in presence of a wound-induced accumulation of endogenous JAs might be the cause of the suppression of root and shoot growth in plants. Previously, Swiatek et al. (2003) also demonstrated the inhibition of cell elongation and cell division in tobacco and *Arabidopsis* roots. It might be the reason that exogenous application of Me-JA leads to blockage of the G_1/S and G_2/M transitions in the cell cycle of cultured tobacco *BY-2* cells responsible for the reduction in the inhibition of root growth (Swiatek et al., 2002). Similar results were conveyed by Pauwels et al. (2008) and Patil et al. (2014) who elucidated that Me-JA in μM concentration, when added to *A. thaliana* suspension cultures, repressed the activation of M-phase genes, arresting cells in the G_2 phase and thus slowed down cell cycle progression, impaired the G_1/S transition and decreased the number of actively dividing cells in cultured *Taxus* cells.

Hanaka et al. (2015) reported no influence of exogenous application of Me-JA on net root growth of *Phaseolus coccineus* plants exposed to excess copper (Cu^{2+}) but significant reduction in net root growth was described in *P. coccineus* plants without Cu^{2+}. Samet et al. (2012) studied the influence of JA alone or in combination with salicylic acid (SA) and yeast extract on hairy root cultures of *Atropa belladonna* and found that 25 μM JA enhanced the root growth in these plants. Monzón et al. (2012)

investigated the interaction of JA with auxin to modulate the growth of primary and lateral roots and its effect on the root architecture of *Helianthus annuus* seedlings and showed that JA determines sunflower root architecture and reduces the length of both primary and lateral roots as well as the number of lateral roots. This is due to the fact that the addition of JA to the growth medium reduced the root cell elongation and cell division.

Sheteawi (2007) illustrated the protective role of exogenous application of JA for improving growth in salt-stressed soybean plants. Lorenzo et al. (2004) described a reduction in the rate of growth inhibition in roots and leaves by JA application in JA signaling mutants such as *jin1/myc2*, *jin 4/jar1* and *jai3*. Wasternack and Hause (2013) reviewed the inhibitory effect of JA on root growth and pointed out that JA's induced root growth inhibition should be analyzed in relation to other factors responsible for controlling the process of root development which is a complex phenomenon (Petricka et al., 2012; Ubeda-Tomás et al., 2012).

In most of the findings, JA's induced inhibition of root growth seems to occur due to the modulation in the effect of auxin signaling in root growth and development. Péret et al. (2009) and Petricka et al. (2012) reported cross-talk between JA and auxins in *Arabidopsis* lateral root formation suggesting that JA plays a role in lateral root formation. Similar results were obtained on lateral root emergence in response to JA in *Arabidopsis* by Stenzel et al. (2012). Adventitious root formation is also a complex phenomenon that is regulated by a number of environmental factors along with the endogenous level of hormones. JA played an antagonistic role with respect to auxins in adventitious root formation in which auxins play a positive role by inducing adventitious root formation when there is a negative *coi1* and *myc2* dependent regulation via JA-Ile homeostasis. The mutants of *jar1*, *coi1-16*, *myc2*, *myc3*, and *myc4* which were impaired in JA perception and signaling showed more formation of adventitious roots than the wild type (Gutierrez et al., 2012).

Tuber formation is a type of vegetative reproduction which is initiated by swelling and occurs mainly as a result of the expansion of cells (Takahashi et al., 1994). Takahashi et al. (1995) reported the swelling of tuber cells and an increase in sucrose content on JA application which resembles the process of tuber formation. Wasternack and Hause (2002, 2013) reviewed the tuber-inducing activities of JAs, particularly 12-OH-JA (TA) and its glucoside (TAG). Nam et al. (2008) suggested that the accumulation of JA, TA, and TAG at low temperature favors tuber induction. However, Miersch et al. (2008) argued against the specific role of JA in tuber formation on the basis of findings by Gidda et al. (2003) who reported the occurrence of TA, 12-HSO4–JA, and 12-O-GLC-JA in different non-tuber bearing plant species. Noir et al. (2013) studied the regulation of leaf growth in *Arabidopsis* mutants altered in JA synthesis and perception, *allene oxide synthase*, and *coi1 1-16B* and reported that Me-JA treatment delays the switch from the mitotic cell cycle to the endo-reduplication cycle which accompanies cell expansion, in a *coi1-* dependent manner and inhibits the mitotic cycle itself, arresting cells in the G_1 phase prior to the S phase transition. A number of studies (Castro et al., 1999; Koda, 1997; Takahashi et al., 1994, 1995) reviewed the role of JA in controlling bulb and tuber formation by stimulating cell division and promoting cell expansion which takes place by the disruption of the cell microtubule network allowing the cells to expand radially (Abe et al., 1990; Matsuki et al., 1992; Shibaoka, 1994). Kolomiets et al. (2001) documented that tuber development in potato was regulated by lipoxygenase (*LOXs*) which produces hydroperoxide fatty acids involved in the formation of JA and related compounds. Similarly, Sohn et al. (2011) suggested the role of Jasmonic acid carboxyl methyl transferase (*JMT*) in increasing tuber yield and the size of transgenic potatoes. Overexpression of *JMT* mRNA led to the higher accumulation of JA, Me-JA and TA.

Boughton et al. (2005) investigated the effect of Me-JA in tomato and found that Me-JA induced and enhanced the density of glandular trichomes. War et al. (2013) revealed that JA pre-treatment resulted in an increase in trichome density in groundnut plants which caused anti-xenosis for oviposition by *Helicoverpa armigera* and induced resistance. Kobayashi et al. (2010) detailed that Me-JA alters trichome density on leaf surfaces of Rhodes grass (*Chloris gayana* Kunth) by affecting leaf area and trichome initiation. Yoshida et al. (2009) reported the enhanced expression of *GLABRA3* (*GL3*) in JA treatment which promoted trichome formation in response to JA but in a dose-dependent manner. Van Schie et al. (2007) avowed that the increased expression of *Le-MTS1* by JA increases the tomato linalool synthase in trichome. Qi et al. (2011) talked about the interaction of jasmonate *ZIM* domain proteins with *WD-Repeat/bHLH/MYB* complexes to regulate jasmonate-mediated trichome initiation and anthocyanin formation in *A. thaliana*.

Besides these JA also plays important roles in plant reproduction including male fertility, sex determination, and seed maturation (Yuan and Zhang, 2015). Goetz et al. (2012) documented that the prevailing occurrence of OPDA compared to JA, showed its role during flower and seed development in tomato. The presence of 12-OH-JA, 12-HSO₄, and 12-O-Glc-JA in some flower organs showed that JA is present as a storage product in flowers (Miersch et al., 2008). The jasmonate *ZIM* domain protein NaJAZ was found to regulate JA levels and flower abscission in *Nicotiana attenuate* plants (Oh et al., 2013). Pak et al. (2009) conveyed that Me-JA treatments promoted flowering time and affected the development of floral organs in oilseed rape (*B. napus* L.). They also proved the role of floral identity genes *BnAP1, BnAP2, BnAP3, BnAG1,* and *BnPI3* in different kinds of floral abnormalities in *B. napus*. Cai et al. (2014) detailed the role of JA in spikelet morphogenesis in rice (*Oryza sativa*) and suggested the role of *EG1* and *EG2/OsJAZ1* in spikelet development. Shih et al. (2014) reported that the *NAC*-like protein named Anther Indehiscence Factor (*AIF*) is involved in anther dehiscence by suppressing the expression of JA biosynthetic genes. Kim (2014) reviewed the role of JA and ethylene in flower development and floral organ abscission. These hormones regulated the timing of floral organ abscission both independently and interdependently. Oh et al. (2013) documented the role of *NaJAZd* to counteract flower abscission, by regulating JA and JA-Ile level and/or expression of *NaMYB305* genes in the flowers of *N. attenuate*. In spite of floral abscission, Me-JA had shown its remarkable potential as an abscission agent in several fruits such as citrus, California table olives (*Olea europaea* L.), and table and wine grapes (*Vitis vinifera* L.) (Hartmond et al., 2000; Kender et al., 2001; Fidelibus et al., 2007; Gonzalez-Herranz et al., 2009). Me-JA was also effective in reducing fruit detachment force (*FDF*) and inducing fruit abscission in grapes (Fidelibus et al., 2007; Gonzalez-Hurranz et al., 2009). Burns et al. (2003) reported that coronatine plays a functionally similar role to JA's induced fruit abscission in citrus. Exogenous application of JA's induced senescence-like phenotypes such as yellowing of leaves, induction of senescence-associated genes, and an increase in the levels of endogenous JA which was also observed during senescence. (He et al., 2002; Seltmann et al., 2010). Wu et al. (2007) reported the role of *HDA6* for jasmonate responses, senescence and flower development. Besseau et al. (2012) investigated that the JA-linked *TFs*, specifically *WRKY53*, *WRKY54*, and *WRKY70*, are involved in the regulatory network that modulates the onset and progression of leaf senescence, possibly through an interaction with *WRKY30*.

Role of Jasmonates in Photoinhibition

Me-JA stimulated production of reactive oxygen species resulted in an alteration of the mitochondrial dynamics as well as the collapse of photosynthetic system (Zhang et al., 2008). Ananiev et al. (2004) avowed that Me-JA inhibited chlorophyll synthesis in excised cotyledons of *Cucurbita pepo* (*zucchini*). Czerpak et al. (2006) in contrast reported that JA induced an accumulation of chlorophyll in *Chlorella vulgaris*. Jamalomidi et al. (2013) observed the enhancement of chlorophyll *a* and *b* on Me-JA treatments. Kaur et al. (2013) reported an increase in the total chlorophyll in JA treated *B. napus* seedlings. Poonam et al. (2013) and Sirhindi et al. (2015) also reported an increase in the total chlorophyll in pigeon pea. Beale et al. (1978) discussed that the stimulatory effect of JA on photosynthetic pigment could be due to its stronger effect on δ-*ALA* (aminolevulinic acid) which is a rate-limiting step in the chlorophyll biosynthetic pathway. Sorial and Gendy (2010) recorded an increase in chlorophyll concentration in sweet basil plants. Ueda and Saniewski (2006) showed that Me-JA treatment induced chlorophyll formation in basal parts of the tulip bulb as a result of the regulation of sugar metabolism. It seems that the expression of the genes involved in chlorophyll synthesis increases in the presence of Me-JA (Wasternack and Hause, 2002). Barley plant, on the other hand, showed decreased chlorophyll content, net photosynthetic, and transpiration rates when treated with Me-JA (Hristova and Popova, 2002). Me-JA induced the gene expression of some key enzymes involved in chlorophyll biosynthesis through the formation of 5-aminolevulinic acid. In addition, carotenoid pigment, which has a protective role in photosynthetic excitation energy transfer to the photosystems and protecting the chloroplast membrane through dispelling a lot of energy, was also enhanced in JA-treated *C. cajan* plants (Poonam et al., 2013).

In addition to this, Abbaspour and Rezaei (2014) reported increases in the Hill reaction, chlorophyll *a*, chlorophyll *b*, chlorophyll *a* + *b*, β-carotene, and xanthophylls after foliar application of JA and SA in *Dracocephalum moldavica* L. under normal irrigation conditions. Jung et al. (2004) documented chlorophyll reduction along with a decrease in photosynthetic efficiency in *A. thaliana* upon exogenous application of Me-JA and norflurazon. A similar decrease in F_v/F_m was observed by Hanaka et al. (2015) in the application of Me-JA in *P. coccineus* plants. Me-JA stimulated anthocyanin formation in *Kalanchoe blossfeldiana* (Saniewski et al., 2003) and showed that photosynthetic products such as sugar molecules important for anthocyanin formation were ameliorated in Me-JA treated plants. Skrzypek et al. (2005) also suggested that Me-JA treatment changed sugar metabolism which led to ethylene formation. Babst et al. (2005) reported that JA encouraged an alteration in the carbon transport and partitioning in populous.

Sorial and Gendy (2010) reported an increase in the concentration of osmolytes such as proline, total soluble sugars, and total amino acid in sweet basil plants. Gao et al. (2004) showed that JA application leads to the accumulation of glycine-betaine in *Pyrus bretschneideri* plants with or without drought stress. JA induces the betaine level in the plant by increasing its biosynthesis via up-regulation of betaine aldehyde dehydrogenase (*BADH*) gene expression and increasing its activity. Ali et al. (2007) explained the increase in proline content in Me-JA and SA treated *Panax ginseng* bioreactor root suspension cultures. Similarly, an increase in proline content was observed by Jamalomidi et al. (2013) in Me-JA treated seeds of *Cocker 347* cultivar of *N. tobacum*. Kaur et al. (2013) reported an increase in the total soluble sugars under JA treatment in *B. napus* seedlings and concluded that this accumulation of soluble sugars may be due to the hydrolysis of starch to a higher level in JA treated plants (Fischer and Holl, 1991).

Jasmonates are well reported to modulate the Calvin cycle products of the plants under normal as well as stress conditions which are directly or indirectly involved in protection from photoinhibition. Sugar is the primary metabolic product while proline is the secondary metabolic product, and both act as osmoprotectants to maintai homeostasis. Cellular status of the cell plays an important role in the working properties of the photoinhibition regulating components. JA along with SA play an important role in the regulation of proline and sugar production, stabilizing cellular membranes and maintaining turgor pressure in chamomile plants (Nazarli et al., 2014). Total soluble sugar and vitamin C content of tomato fruit were significantly affected by Me-JA which contributed to the significant increase in tomato fruit yield (Kazemi, 2014). Geransayeh et al. (2015) documented that Me-JA treated strawberry (*Fragaria ananassa* L. cv. *Gaviota*) fruits showed a decrease in fresh weight, pH, TSS, and higher vitamin C, anthocyanin, calcium, and pectin. Wang et al. (2015) reported increased flavonoid content on Me-JA elicitation in the cell suspension culture of *Hypericum perforatum*. Abdelgawad et al. (2014) also reported an increase in total carbohydrates, total soluble sugars, polysaccharide, as well as free amino acids, and proline in Me-JA treated seedlings of maize. On the other hand, exogenous application of different concentrations of JA reduced starch concentration in poplar trees, sugar in tulip stems, and amino acids in cabbage (Babst et al., 2005; Skrzypek et al., 2005). However, there are antagonistic reports about the protecting role of JA in photosynthesis and thus in photoprotection of plants from photoinhibition. Wang et al. (2014) reported that starch concentration was significantly lowered in *N. tobacum* by JA's signaling. Whereas, Hanik et al. (2010) reported that JA application induced amino acids in the leaves of the tobacco plant.

Sirhindi et al. (2016) evaluated the effects of JA on photosynthetic pigments, which act as the primary protector of plants from photoinhibition, in soybean (*Glycine max* L.) plants subjected to Ni toxicity, and the effect of JA on antioxidant enzyme activity and gene expression in soybean (*Glycine max* L.) plants subjected to nickel (Ni) stress. In this study they avowed that JA helps in restoring the chlorophyll fluorescence disturbed by Ni stress along with an increase in proline, glycinebetaine, total protein, and total soluble sugar (TSS) by 33.09%, 51.26%, 22.58%, and 49.15%, respectively, over the control. The addition of JA to Ni-stressed plants is reported to enhance the chlorophyll content to significant levels over the control as well as stressed plants. Supplementation of JA minimizes the accumulation of H_2O_2, MDA, and NADPH oxidase, which helps in the stabilization of biomolecules and thus helps in photoinhibition protection as the second level of defense. The activity of superoxide dismutase (SOD), peroxidase (POD), catalase (CAT), and ascorbate peroxidase (APX) is increased by the application of JA which is also shown to enhance the expression of the genes of the same. These results signified that the co-application of JA helps seedlings combat the detrimental effects of Ni through enhanced osmolytes, the activity of antioxidant enzymes, and gene expression which are involved at different levels in photoinhibitory protection mechanisms.

Conclusions

In the last two decades, there has been an exponential increase in the data regarding jasmonate behavior and understanding with respect to plant growth and development under normal and stress conditions. Most recently identification of JA receptors along with their mode of action through proteasomal degradation of receptors was a breakthrough in our understanding of JA-dependent processes by identification of so-called JAZ-proteins. However, we are still at the first step of finding the answer to the question of "How JA signaling is switched on or switched off?". The application of JA mitigates the harmful effects of abiotic stress by protecting the photosynthetic machinery through the modulation of compatible solutes, antioxidant defense system, and expression of various antioxidants as SOD, POD, APX, and CAT genes. Now it has been established that jasmonates play a very important central role in stress-management of plants. Further understanding through transcriptome and next-generation sequencing tools will give insights into the mechanism of activity of JA signaling components including its cross-talk to other plant hormones. These future studies will address the similarities and divergences in the jasmonate-dependent regulatory network in response to different stressors and also towards the natural variegation in the role of jasmonate in photoinhibition protection in plants under different stress conditions.

REFERENCES

Abbaspour, S. H. and H. Rezaei. 2014. Effects of salicylic acid and jasmonic acid on hill reaction and photosynthetic pigment (*Dracocephalum Moldavica L.*) in different levels of drought. *International Journal of Advanced Biological and Biomedical Research* 2 (12): 2850–2859.

Abdelgawad, Z. A., A. A. Khalafaallah, and M. M. Abdallah. 2014. Impact of methyl jasmonate on antioxidant activity and some biochemical aspects of maize plant grown under water stress condition. *Agriculture Science* 5(October): 1077–1088.

Abe, M., H. Shiboaka, and H. Yamane. 1990. Cell cycle-dependent disruption of microtubules by methyl jasmonate in tobacco BY-2 cells. *Protoplasma* 156:1–8.

Acosta, I. F. and E. E. Farmer. 2010. Jasmonates. *The Arabidopsis Book* 8: e0129.

Aldridge, D. C., S. Galt, and D. Giles. 1971. Metabolites of Lasiodiplodia theobromae. Journal of the Chemical Society. (C), 1623–1627.

Ali, M. B., E. J. Hahn, and K. Y. Paek. 2007. Methyl jasmonate and salicylic acid induced oxidative stress and accumulation of phenolics in Panax ginseng bioreactor root suspension cultures. *Molecules.* 12(3): 607–621.

Allakhverdiev, S. I., Y. Nishiyama, S. Miyairi, et al. 2002. Salt stress inhibits the repair of photodamaged photosystem II by suppressing the transcription and translation of psbA genes in Synechocystis. *Plant Physiology* 130: 1443–1453.

Ananiev, E. D., K. Ananieva, and I. Todorov. 2004. Effects of methyl ester of jasmonic acid, abscisic acid and benzyladenine on chlorophyll synthesis in excised cotyledons of Cucurbita pepo (Zucchini). *Bulgarian Journal of Plant Physiology* 30: 51–63.

Andreou, A. and I. Feussner. 2009. Lipoxygenases structure and reaction mechanism. Phytochemistry 70: 1504–1510.

Andreou, A., F. Brodhun, and I. Feussner. 2009. Biosynthesis of oxylipins in non-mammals. *Progress in Lipid Research* 48:148–170.

Aro, E. M., I. Virgin, and B. Andersson.1993. Photoinhibition of photosystem II. Inactivation, protein damage and turnover. Biochimica et Biophysica Acta 1143: 113–134.

Asada, K. 1999. The water-water cycle in chloroplasts: Scavenging of active oxygen species and dissipation of excess photons. *Annual Review of Plant Biology* 50: 601–639.

Babst, B. A., R. A. Ferrieri, D. W. Gray, et al. 2005. Jasmonic acid induced rapid changes in carbon transport and partitioning in Populus. *New Phytologist* 167: 63–72.

Bachmann, A., B. Hause, H. Maucher, et al. 2002. Jasmonate-induced lipid peroxidation in barley leaves initiated by distinct 13-LOX forms of chloroplasts. *Journal of Biological Chemistry* 383: 1645–1657.

Baginsky, S., K. Tiller, T. Pfannschmidt, et al. 1999. PTK, the chloroplast RNA polymerase-associated protein kinase from mustard (Sinapis alba), mediates redox control of plastid in vitro transcription. *Plant Molecular Biology* 39: 1013–1023.

Beale, S. I. 1978. δ-Aminolevulinic acid in plants: Its bio-synthesis, regulation and role in plastid development. *Annual Review Plant Physiology* 29: 95–120.

Berestetzky, V., W. Dathe, T. Daletskaya, et al. 1991. Jasmonic acid in seed dormancy of Acer tataricum. *Biochemie und Physiologie der Pflanzen* 187(1): 13–19.

Besseau, S., J. Li, and E. T. Palva. 2012. WRKY54 and WRKY70 cooperate as negative regulators of leaf senescence in Arabidopsis thaliana. *Journal of Experimental Botany* 63: 2667–2679.

Bewlay, J. D. and M. Black. 1994. *Seeds: Physiology of developmental and germination.* 2nd edition, Plenum press, New York.

Bialecka, B. and J. Kepczynski. 2003. Endogenous ethylene and reversing methyl jasmonate inhibition of *Amaranthus caudatus* seed germination by benzyladenine or gibberellin. *Plant Growth Regulation* 41: 7–12.

Bonaventure, G., S. Schuck, and I. T. Baldwin. 2011. Revealing complexity and specificity in the activation of lipase-mediated oxylipin biosynthesis: A specific role of the Nicotiana attenuata GLA1 lipase in the activation of jasmonic acid biosynthesis in leaves and roots. *Plant Cell Environment* 34: 1507–1520.

Boughton, A. J., K. Hoover, and G. W. Felton. 2005. Methyl jasmonate application induces increased densities of glandular trichomes on tomato (*Lycopersicon esculentum*). *Journal of Chemical Ecology* 31:2211–2216.

Brash, A. R. 2009. Mechanistic aspects of CYP74 allene oxide synthases and related cytochrome P450 enzymes. *Phytochemistry* 70: 1522–1531.

Brash, A. R., S. W. Baertschi, C. D. Ingram, et al. 1988. Isolation and characterization of natural allene oxides: Unstable intermediates in the metabolism of lipids hydroxides. *Proceedings of the National Academy of Sciences of the United States of America* 85: 3382–3386.

Browse, J. 2009a. Jasmonate passes muster: A receptor and targets for the defense hormone, *Annual Review in Plant Biology* 60: 183–205.

Browse, J. 2009b. Jasmonate: Preventing the maize tassel from getting in touch with his feminine side. *Science Signalling* 2(59): 9.

Burns, J. K., L. V. Pozo, C. R. Arias, et al. 2003. Coronatine and abscission in citrus. *Journal of the American Society for Horticultural Science* 128: 309–315.

Cai, Q., Z. Yuan, M. Chen, et al. 2014. Jasmonic acid regulates spikelet development in rice. *Nature Communication* 5: 3476.

Castillo, M. C., C. Martínez, A. Buchala, et al. 2004. Gene-specific involvement of beta-oxidation in wound-activated responses in Arabidopsis. *Plant Physiology* 135: 85–94.

Castro, G., T. Kraus, and G. Abdala. 1999. Endogenous jasmonic acid and radial cell expansion in buds of potato tubers. *Journal of Plant Physiology* 155:706–10.

Choudhary, S.P., J. Q. Yu, K. Yamaguchi-Shinozaki, et al. 2012. Benefits of brassinosteroid crosstalk. *Trends in Plant Science* 17:594–605.

Corbineau, F., R. M. Rudnicki, and D. Come. 1988. The effects of methyl jasmonate on sunflower (*Helianthus annuus L.*) seed germination and seedling development. *Plant Growth Regulation* 7:157–169.

Czerpak, R., A. Piotrowska, and K. Szleska. 2006. Jasmonic acid affects changes in the growth and some components content in alga *Chlorella vulgaris*. *Acta Physiologeiae Plantarum* 28:195–203.

Daletskaya, T. V. and G. Sembdner. 1989. Effect of jasmonic acid on germination of nondormant and dormant seeds. *Fiziologija Rastenij* 36: 1118–1123.

Dave, A., M. L. Hernández, Z. He, et al. 2011. 12-Oxo-phytodienoic acid accumulation during seed development represses seed germination in Arabidopsis. *The Plant Cell* 23:583–599.

Demmig-Adams, B. and Adams, W. W. III. 2000. Photosynthesis: Harvesting sunlight safely. *Nature* 403: 371–374.

Demole, E., E. Lederer, and D. Mercier. 1962. Isolement et détermination de la structure du jasmonate de méthyle, constituant odorant charactéristique de lèssence de jasmin. *Helvetica Chimica Acta* 45: 675–685.

Ding, H., J. Lai, Q. Wu, et al. 2016. Jasmonate complements the function of Arabidopsis lipoxygenase3 in salinity stress response. *Plant Science* 244: 1–7.

Edwards, G. E. and D. A. Walker. 1983. *C3, C4: Mechanisms, and Cellular and Environmental Regulation of Photosynthesis.* Blackwell, London.

Ewart, A. J. 1896. On assimilatory inhibition in plants. *The Botanical Journal of the Linnean Society* 31: 364–461.

Farmaki, T., M. Sanmartín, P. Jiménez, et al. 2007. Differential distribution of the lipoxygenase pathway enzymes within potato chloroplasts. *Journal of Experimental Botany* 58: 555–568.

Feussner, I. and C. Wasternack. 2002. The lipoxygenase pathway. Annual Review of Plant Biology 53: 275–297.

Feussner, I., B. Hause, K. Vörös, et al. 1995. Jasmonate-induced lipoxygenase forms are localized in chloroplasts of barley leaves (*Hordeum vulgare* cv. Salome). *The Plant Journal* 7: 949–957.

Fidelibus, M. W., K. A. Cathline, and J. K. Burns. 2007. Potential abscission agents for raisin table and wine grapes. *HortScience* 42:1626–1630.

Fischer, C. and W. Holl. 1991. Food reserves in Scots pine (L.) I. Seasonal changes in the carbohydrate and fat reserves of pine needles. *Tree* 5: 187–195.

Fonseca, S., A. Chini, M. Hamberg, et al. 2009. ± 7-iso-Jasmonoyl-L-isoleucine is the endogenous bioactive jasmonate. *Nature Chemical Biology*. 5:344–350.

Foyer, C. H., J. Neukermans, G. Queval, et al. 2012. Harbinson Photosynthetic control of electron transport and the regulation of gene expression. *Journal of Experimental Botany* 63:1637–1661.

Foyer, C. H. and G. Noctor. 2005. Redox homeostasis and antioxidant signaling: A metabolic interface between stress perception and physiological responses. *Plant Cell* 17: 1866–1875.

Foyer, C. H. and S. Shigeoka. 2011. Understanding oxidative stress and antioxidant functions to enhance photosynthesis. *Plant Physiology*. 155: 93–100.

Froehlich, J. E., A. Itoh, and G. A. Howe. 2001. Tomato allene oxide synthase and fatty acid hydroperoxide lyase, two cytochrome P450s involved in oxylipin metabolism, are targeted to different membranes of chloroplast envelope. *Plant Physiology* 125: 306–317.

Fukumoto, K., K. Alamgir, Y. Yamashita, et al. 2013. Response of rice to insect elicitors and the role of OsJAR1 in wound and herbivory-induced JA-Ile accumulation. *Journal of Integrative Plant Biology* 55: 775–784.

Gao, X., X. Wang, Y. Lu, et al. 2004. Jasmonic acid is involved in the water stress induced betaine accumulation in pear leaves. *Plant Cell Environment* 27: 5497–5507.

Gasperini, D., A. Chauvin, I. F. Acosta, et al. 2015. Axial and radial oxylipin transport. *Plant Physiology* 169: 2244–2254.

Geransayeh, M., S. Sepahvand, V. Abdossi, et al. 2015. Effect of methyl jasmonate treatment on decay, post-harvest life and quality of Strawberry (*Fragaria ananassa* L. cv. Gaviota) fruit. *International Journal of Current Science* 15(E): 123–131.

Gidda, S., O. Miersch, A. Levitin, et al. 2003. Biochemical and molecular characterization of a hydroxyjasmonate sulfotransferase from Arabidopsis thaliana. *Journal of Biological Chemistry* 278: 17895–17900.

Gill, S. S., N. Tuteja. 2010. Reactive oxygen species and antioxidant machinery in abiotic stress tolerance in crop plants. *Plant Physiology and Biochemistry* 48(12): 909–930.

Goetz, S., A. Hellwege, and I. Stenzel. 2012. Role of cis-12-oxo-phytodienoic acid in tomato embryo development. *Plant Physiology* 158: 1715–1727.

Gonzalez-Herranz, R., K. A. Cathline, M. W. Fidelibus, et al. 2009. Potential of methyl jasmonate as a harvest aid for thompson seedless grapes: Concentration and time needed for consistent berry loosening. *HortScience* 44:1330–1333.

Grebner, W., N. E. Stingl, A. Oenel, et al. 2013. Lipoxygenase6-dependent oxylipin synthesis in roots is required for abiotic and biotic stress resistance of Arabidopsis. *Plant Physiology* 161: 2159–2170.

Gu, X. C., J. F. Chen, Y. Xiao, et al. 2012. Overexpression of allene oxide cyclase promoted tanshinone/phenolic acid production in *Salvia miltiorrhiza*. *Plant Cell Reports* 31: 2247–2259.

Gutierrez, L., G. Mongelard, K. Flokova, et al. 2012. Auxin controls Arabidopsis adventitious root initiation by regulating jasmonic acid homeostasis. *The Plant Cell*. 24(6): 2515–2527.

Haga, K. and M. Iino. 2004. Phytochrome-mediated transcriptional up-regulation of ALLENE OXIDE SYNTHASE in rice seedlings. Plant and Cell Physiology 45: 119–128.

Hanaka, A., W. Maksymiec, and W. Bednarek. 2015. The effect of methyl jasmonate on selected physiological parameters of copper-treated *Phaseolus coccineus* plants. *Plant Growth Regulation* 77(2): 167–177.

Hanik, N., S. Gómez, M. Best, et al. 2010. Partitioning of new carbon as and supl; and supl; C in Nicotiana tabacum reveals insight into methyl jasmonate induced changes in metabolism. *Journal of Chemical Ecology* 36: 1058–1067.

Hartmond, U., R. Yuan, J. K. Burns, et al. 2000. Citrus fruit abscission induced by methyl-jasmonate. *Journal of the American Society for Horticultural Science* 125(5): 547–552.

He, Y., H. Fukushige, D. F. Hildebrand, et al. 2002. Evidence supporting a role of jasmonic acid in Arabidopsis leaf senescence. *Plant Physiology* 128: 876–884.

Heitz, T., E. Smirnova, E. Widemann, et al. 2016. The rise and fall of jasmonate biological activities. In: *Lipids in plant and algae development*, Nakamura, Y., Li-Beisson, Y., eds. Cham: Springer International Publishing, 405–426.

Hofmann, E. and S. Pollmann. 2008. Molecular mechanism of enzymatic allene oxide cyclization in plants. *Plant Physiology and Biochemistry* 46: 302–308.

Hofmann, E., P. Zerbe, and F. Schaller. 2006. The crystal structure of *Arabidopsis thaliana* allene oxide cyclase: Insights into the oxylipin cyclization reaction. *The Plant Cell* 18: 3201–3217.

Hristova, V. A. and L. P. Popova. 2002. Treatment with methyl jasmonate alleviates the effects of paraquat on photosynthesis in barley plants. *Photosynthetica* 40: 567–574.

Hu, J., A. Baker, B. Bartel, et al. 2012. Plant peroxisomes: Biogenesis and function. *The Plant Cell* 24: 2279–2303.

Hughes, R. K., S. De Domenico, and A. Santino. 2009. Plant cytochrome CYP74 family: Biochemical features, endocellular localisation, activation mechanism in plant defence and improvements for industrial applications. *ChemBioChem* 10: 1122–1133.

Ishiguro, S., A. Kawai-Oda, and K. Ueda. 2001. The DEFECTIVE IN ANTHER DEHISCENCE1 gene encodes a novel phospholipase A1 catalyzing the initial step of jasmonic acid biosynthesis, which synchronizes pollen maturation, anther dehiscence, and flower opening in Arabidopsis. *Plant Cell* 13:2191–2209.

Ito, T., K. H. Ng, T. S. Lim, et al. 2007. The homeotic protein AGAMOUS controls late stamen development by regulating a jasmonate biosynthetic gene in Arabidopsis. *The Plant Cell* 19: 3516–3529.

Jamalomidi, F., J. Sarmad, and M. Jamalomidi. 2013. Changes caused by Methyl Jasmonate in Cocker 347- a cultivar of tobacco (Nicotine tabacum L.) under salinity stress. *International Research Journal of Applied and Basic Sciences*. 4(5): 1139–1145.

Jarvis, S. B., M. A. Taylor M. R. MacLeod, et al. 1996. Cloning and characterisation of the cDNA clones of three genes that are differentially expressed during dormancy-breakage in the seeds of Douglas fir (*Pseudotsuga menziesii*). *Journal of Plant Physiology* 147(5): 559–566.

Jung, S. 2004. Effect of chlorophyll reduction in Arabidopsis thaliana by methyl jasmonate or norflurazon on antioxidant systems. *Journal of Plant Physiology and Biochemistry* 42: 231–255.

Kangasjarvi, S., J. Neukermans, S. Li, et al. 2012. Photosynthesis, photorespiration, and light signalling in defence responses. *Journal of Experimental Botany* 63:1619–1636.

Katsir, L., A. L. Schilmiller, P. E. Staswick, et al. 2008. COI1 is a critical component of a receptor for jasmonate and the bacterial virulence factor coronatine. Proceedings of the National Academy of Sciences, USA 105: 7100–7105.

Kaur, H., P. Sharma, and G. Sirhindi. 2013. Sugar accumulation and its regulation by jasmonic acid in *Brassica napus L.* under salt stress. *Journal of Plant Physiology and Biochemistry* 9:53–64.

Kazemi, M. 2014. Effect of foliar application with salicylic acid and methyl jasmonate on growth, flowering, yield and fruit quality of tomato. *Bulletin of Environment, Pharmacology and Life Sciences* 3:154–158.

Kender, W. J., U. Hartmond, J. K. Burns, et al. 2001. Methyl jasmonate and CMN-pyrazole applied alone and in combination can cause mature orange abscission. *Scientia Horticulturae* 88:107–120.

Kepczynski, J. and B. Bialecka. 1994. Stimulatory effect of ethephon, ACC, gibberellin A-3 and A-4+7 on germination of methyl jasmonate inhibited *Amaranthus caudatus L.* seeds. *Plant Growth Regulation* 14: 211–216

Kim, J. 2014. Four shades of detachment: Regulation of floral organ abscission. *Plant Signaling and Behaviour* 9:e976154.

Knopf, R. R., A. Feder, K. Mayer, et al. 2012. Rhomboid proteins in the chloroplast envelope affect the level of allene oxide synthase in Arabidopsis thaliana. *The Plant Journal* 72: 559–571.

Kobayashi, H., M. Yanaka, and T. M. Ikeda. 2010. Exogenous methyl jasmonate alters trichome density on leaf surfaces of Rhodes grass (Chloris gayana Kunth). *Journal of Plant Growth Regulation* 29:506–511

Koda, Y. 1997. Possible involvement of jasmonates in various morphogenic events. *Physiologia Plantarum* 100: 639–646.

Koeduka, T., K. Ishizaki, C. M. Mwenda, et al. 2015. Biochemical characterization of allene oxide synthases from the liverwort Marchantia polymorpha and green microalgae *Klebsormidium flaccidum* provides insight into the evolutionary divergence of the plant CYP74 family. *Planta* 242: 1175–1186.

Kolomiets, M. V., D. J. Hannapel, H. Chen., et al. 2001. Lipoxygenase is involved in the control of potato tuber development. *The Plant Cell* 13(3): 613–626.

Kombrink, E. 2012. Chemical and genetic exploration of jasmonate biosynthesis and signaling paths. Planta 236: 1351–1366.

Koo, A. J., H. S. Chung, and Y. Kobayashi. 2006. Identification of a peroxisomal acyl-activating enzyme involved in the biosynthesis of jasmonic acid in Arabidopsis. *Journal of Biological Chemistry* 281: 33511–33520.

Koo, A. J. and G. A. Howe. 2012. Catabolism and deactivation of the lipidderived hormone jasmonoyl-isoleucine. Frontiers in Plant Science 3: 19.

Krause, G. H. 1988. Photoinhibition of photosynthesis. an evaluation of damaging and protective mechanisms. *Plant Physiology* 74: 566–574.

Kucera, B., M. A. Cohn, G. Leubner-Metzger. 2005. Plant hormone interactions during seed dormancy release and germination. *Seed Science Research* 15:281–307.

Larrieu, A., A. Champion, J. Legrand, et al. 2015. A fluorescent hormone biosensor reveals the dynamics of jasmonate signalling in plants. *Nature Communications* 6: 6043.

Laudert, D., F. Schaller, and E. W. Weiler. 2000. Transgenic *Nicotiana tabacum* and *Arabidopsis thaliana* plants overexpressing allene oxide synthase. *Planta* 211: 163–165.

Le, D. T., R. Nishiyama, Y. Watanabe, et al. 2012. Identification and expression analysis of cytokinin metabolic genes in soybean under normal and drought conditions in relation to cytokinin levels. *PLoS One* 7: 42411.

Lee, D. E. and K. Back. 2005. Ectopic expression of MAP kinase inhibits germination and seedling growth in transgenic rice. *Plant Growth Regulation* 45:251–257.

Li, C., A. L. Schilmiller, G. Liu, et al. 2005. Role of β-oxidation in jasmonate biosynthesis and systemic wound signaling in tomato. *The Plant Cell* 17: 971–986.

Li, W., F. Zhou, B. Liu, et al. 2011. Comparative characterization, expression pattern and function analysis of the 12-oxo-phytodienoic acid reductase gene family in rice. *Plant Cell Reports* 30: 981–995.

Li, X-R., H. J. Li, L. Yuan, et al. 2014. Arabidopsis DAYU/ABERRANT PEROXISOME MORPHOLOGY9 is a key regulator of peroxisome biogenesis and plays critical roles during pollen maturation and germination in planta. *The Plant Cell* 26: 619–635.

Lin, Y. T., L. J. Chen, C. Herrfurth, et al. 2016. Reduced biosynthesis of digalactosyldiacylglycerol, a major chloroplast membrane lipid, leads to oxylipin overproduction and phloem cap lignification in Arabidopsis. *The Plant Cell* 28: 219–232.

Linkies, A. G. and G. Leubner-Metzger. 2012. Beyond gibberellins and abscisic acid: How ethylene and jasmonates control seed germination. *Plant Cell Reports* 31: 253–270.

Liu, X., F. Li, J. Tang, et al. 2012. Activation of the jasmonic acid pathway by depletion of the hydroperoxide lyase OsHPL3 reveals crosstalk between the HPL and AOS branches of the oxylipin pathway in rice. *PLoS One* 7: e50089.

Long, S. P., T. M. East, and N.R. Baker. 1983. Chilling damage to photosynthesis in young Zea mays. I. Effects of light and temperature variation on photosynthetic CO_2 assimilation. *Journal of Experimental Botany*. 34, 177–188.

Lorenzo, O., J. M. Chico, and J. J. Sa. 2004. JASMONATE-INSENSITIVE1encodes a MYC transcription factor essential to discriminate between different jasmonate regulated defense responses in Arabidopsis. *Plant Cell*. 16(7):1938–1950.

Matsuki, T., H. Tazaki, T. Fujimori, et al. 1992. The influences of jasmonic acid methyl-ester on microtubules in potato cells and formation of potato-tubers. *Bioscience Biotechnology and Biochemistry* 56 (8):1329–1330.

Maucher, H., B. Hause, and I. Feussner. 2000. Allene oxide synthases of barley (*Hordeum vulgare* cv. Salome): Tissue specific regulation in seedling development. *The Plant Journal* 21: 199–213.

Meesters, C., T. Mönig, J. Oeljeklaus, et al. 2014. A chemical inhibitor of jasmonate signaling targets JAR1 in *Arabidopsis thaliana*. *Nature Chemical Biology* 10: 830–836.

Melis, A. 1999. Photosystem II damage and repair cycle in chloroplasts: What modulates the rate of photodamage in vivo? *Trends in Plant Sciences* 4:130–135.

Melotto, M., C. Mecey, Y. Niu, et al. 2008. A critical role of two positively charged amino acids in the Jas motif of Arabodopsis JAZ proteins in mediating coronatine- and jasmonoyl isoleucine-dependent interactions with the COI1 F-box protein. *The Plant Journal* 55: 979–988.

Meyer, A., O. Miersch, and C. Biittner. 1984. Occurrence of the plant growth regulator jasmonic acid in plants. *Journal of Plant Growth Regulation* 3: 1–8.

Miersch, O., J. Neumerkel, M. Dippe, et al. 2008. Hydroxylated jasmonates are commonly occurring metabolites of jasmonic acid and contribute to a partial switch-off in jasmonate signaling. *New Phytologist* 177: 114–127.

Miersch, O. and C. Wasternack. 2000. Octadecanoid and jasmonate signaling in tomato (Lycopersicon esculentum Mill.) leaves: endogenous jasmonates do not induce jasmonate biosynthesis. Biological Chemistry 381: 715–722.

Monte, I., M. Hamberg, A. Chini, et al. 2014. Rational design of a ligand-based antagonist of jasmonate perception. *Nature Chemical Biology* 10: 671–676.

Monzón, G. C., M. Pinedo, L. Lamattina, et al. 2012. Sunflower root growth regulation: The role of jasmonic acid and its relation with auxins. *Plant Growth Regulation* 66(2): 129–136.

Murata, N., S. Takahashi, Y. Nishiyama. et al. 2007. Photoinhibition of photosystem II under environmental stress. *Biochimica et Biophysica Acta* 1767: 414–421.

Nath, K., A. Jajoo, R. S. Poudyal, et al. 2013. Towards a critical understanding of the photosystem II repair mechanism and its regulation during stress conditions. *FEBS Letters* 587: 3372–3381.

Nazarli, H., A. Ahmadi, and J. Hadian. 2014. Salicylic acid and methyl jasmonate enhance drought tolerance in chamomile plants. *Journal of HerbMed Pharmacology* 3(2):87–92.

Neumann, P., F. Brodhun, K. Sauer, et al. 2012. Crystal structures of Physcomitrella patens AOC1 and AOC2: Insights into the enzyme mechanism and differences in substrate specificity. *Plant Physiology* 160: 1251–1266.

Nishiyama, Y. and N. Murata. 2014. Revised scheme for the mechanism of photoinhibition and its application to enhance the abiotic stress tolerance of the photosynthetic machinery. *Applied Microbiology and Biotechnology* 98: 8777–8796.

Nishiyama, Y., S. I. Allakhverdiev, N. Murata, et al. 2006. A new paradigm for the action of reactive oxygen species in the photoinhibition of photosystem II. *Biochimica et Biophysica Acta* 1757: 742–749.

Nishiyama, Y., S. I. Allakhverdiev, H. Yamamoto, et al. 2004. Singlet oxygen inhibits the repair of photosystem II by suppressing the translation elongation of the D1 protein in Synechocystis sp. PCC 6803. *Biochemistry* 43: 11321–11330.

Nishiyama, Y., H. Yamamoto, S. I. Allakhverdiev, et al. 2001. Oxidative stress inhibits the repair of photodamage to the photosynthetic machinery. *EMBO Journal* 20: 5587–5594.

Niyogi, K. K. 2000. Safety valves for photosynthesis. Current Opinion in Plant Biology 3: 455–460.

Noir, S., M. Bömer, N. Takahashi, et al. 2013. Jasmonate controls leaf growth by repressing cell proliferation and the onset of endoreduplication while maintaining a potential stand-by mode. *Plant Physiology* 161:1930–1951.

Nojavan-Asghari, M. and K. Ishizawa. 1998. Inhibitory effects of methyl jasmonate on the germination and ethylene production in cocklebur seeds. *Journal of Plant Growth Regulation* 17:13–18.

Norastehnia, A., R. H. Sajedi, and M. Nojavan-Asghari. 2007. Inhibitory effects of methyl jasmonate on seed germination in maize (Zea Mays): Effect on α-amylase activity and ethylene production. *General and Applied Plant Physiology* 33: 13–23.

Oh, E., H. Kang, S. Yamaguchi, et al. 2009. Genome-wide analysis of genes targeted by PHYTOCHROME INTERACTING FACTOR 3-LIKE5 during seed germination in Arabidopsis. *The Plant Cell* 21:403–19.

Oh, Y., I. T. Baldwin, and I. Galis. 2013. A Jasmonate ZIM-Domain protein NaJAZd regulates floral jasmonic acid levels and counteracts flower abscission in Nicotiana attenuata plants. *PLoS One* 8(2): 1–11.

Osakabe, Y., K. Yamaguchi-Shinozaki, K. Shinozaki, et al. 2014. ABA control of plant macroelement membrane transport systems in response to water deficit and high salinity. *New Phytologist* 202:35–49.

Osmond, C. B. 1994. What is photoinhibition? Some insights from comparisons of sun and shade plants. In: *Photoinhibition of Photosynthesis: From Molecular Mechanisms to the Field*, Baker, N.R., Bowyer, J.R. eds., pp. 1–24. Bios Scientific Publishers, Oxford.

Otto, M., C. Naumann, W. Brandt, et al. 2016. Activity regulation by heteromerization of arabidopsis allene oxide cyclase family members. *Plants* 5: 3.

Ozalvo, R., J. Cabrera, C. Escobar, et al. 2014. Two closely related members of Arabidopsis 13-lipoxygenases (13-LOXs), LOX3 and LOX4, reveal distinct functions in response to plant–parasitic nematode infection. *Molecular Plant Pathology* 15: 319–332.

Pak, H., Y. Guo, M. Chen, et al. 2009. The effect of exogenous methyl jasmonate on the flowering time, floral organ morphology, and transcript levels of a group of genes implicated in the development of oilseed rape flowers (*Brassica napus L.*). *Planta* 231: 79–91.

Patil, R. A., S. K. Lenka, J. Normanly, et al. 2014. Methyl jasmonate represses growth and affects cell cycle progression in cultured taxus cells. *Plant Cell Reports*. 33(9):1479–92.

Pauwels, L., K. Morreel, E. De Witte, et al. 2008. Mapping methyl jasmonate-mediated transcriptional reprogramming of metabolism and cell cycle progression in cultured Arabidopsis cells. *Proceedings of the National Academy of Sciences of the USA* 105: 1380–1385.

Péret, B., A. Larrieu, and M. J. Bennett. 2009. Lateral root emergence: A difficult birth. *Journal of Experimental Botany* 60:3637–43

Petricka, J. J., C. M. Winter, and P. N. Benfey. 2012. Control of Arabidopsis root development. *Annual Review of Plant Biology*. 63:563–590.

Pfannschmidt, T. 2003. Chloroplast redox signals: how photosynthesis con- trols its own genes. Trends in Plant Science 8: 33–41.

Pfündel, E. E. and W. Bilger. 1994. Regulation and possible function of the violaxanthin cycle. *Photosynthesis Research* 42: 89–109.

Poonam, S., H. Kaur, and S. Geetika. 2013. Effect of Jasmonic acid on photosynthetic pigments and stress markers in *Cajanus cajan* (L.) Millsp. seedlings under copper stress. *American Journal of Plant Sciences* 4:817–823.

Powles, S. B. 1984. Photoinhibition of photosynthesis induced by visible light. *Annual Review of Plant Physiology* 35: 15–44.

Powles, S. B., J. A. Berry, and O. Bjorkman. 1983. Interaction between light and chilling temperature on the inhibition of photosynthesis in chilling-sensitive plants. *Plant, Cell and Environment* 6: 117–123.

Preston, C. A., H. Betts, and I. T. Baldwin. 2002. Methyl jasmonate as an allelopathic agent: Sagebrush inhibits germination of a neighboring tobacco, Nicotiana attenuata. *Journal of Chemical Ecology* 28: 2343–2369.

Puranik, S., P. P. Sahu, P. S. Srivastava, et al. 2012. NAC proteins: Regulation and role in stress tolerance. *Trends in Plant Sciences* 17: 369–381.

Qi, T., S. Song, Q. Ren, et al. 2011. The jasmonate-ZIM-domain proteins interact with the WD-repeat/bHLH/MYB complexes to regulate jasmonate-mediated anthocyanin accumulation and trichome initiation in *Arabidopsis thaliana. Plant Cell* 23: 1795–1814.

Rabinowitch, E. I. 1951. *Photosynthesis.* Volume II, Part 1. Interscience Publishers, New York.

Ranjan, R. and S. Lewak. 1992. Jasmonic acid promotes germination and lipase activity in nonstratified apple embryos. *Physiologia Plantarum* 86: 335–39.

Richmond, T. and A. Bleecker. 1999. A defect in β-oxidation causes abnormal inflorescence development in Arabidopsis. The Plant Cell 11: 1911–1923.

Riemann, M., K. Haga, T. Shimizu, et al. 2013. Identification of rice allene oxide cyclase mutants and the function of jasmonate for defence against *Magnaporthe oryzae. The Plant Journal* 74: 226–238.

Rochaix, J. D. 2011. Regulation of photosynthetic electron transport. *Biochimica et Biophysica Acta* 1807: 375–383.

Samet, A. E., K. Piri, M. Kayhanfar, et al. 2012. Influence of jasmonic acids, yeast extract and salicylic acid on growth and accumulation of hyosciamine and scopolamine in hairy root cultures of *Atropa belladonna* L. *International Journal of Agriculture: Research and Review* 2(4): 403–409.

Saniewski, M., M. Horbowicz, J. Puchalski, et al. 2003. Methyl jasmonate stimulates the formation and accumulation of anthocyanins in *Kalanchoe blossfeldiana. Acta Physiologie Plantarum* 25: 143–149.

Sassenrath, G. F., D. R. Ort, and A. R. Jr. Portis. 1990. Impaired reductive activation of stromal bisphosphatases in tomato leaves following low-temperature exposure at high light. *Archives of Biochemistry and Biophysics* 282: 302–308.

Savchenko, T., D. Yanykin, A. Khorobrykh, et al. 2017. The hydroperoxide lyase branch of the oxylipin pathway protects against photoinhibition of photosynthesis. *Planta* 245: 1179–1192.

Schaller, A. and A. Stintzi. 2009. Enzymes in jasmonate biosynthesis—Structure, function, regulation. *Phytochemistry* 70: 1532–1538.

Schaller, F., P. Zerbe, S. Reinbothe, et al. 2008. The allene oxide cyclase family of Arabidopsis thaliana: Localization and cyclization. *FEBS Journal* 275: 2428–2441.

Schilmiller, A. L., A. J. K. Koo, and G. A. Howe. 2007. Functional diversification of acyl-coenzyme A oxidases in jasmonic acid biosynthesis and action. *Plant Physiology* 143: 812–824.

Schmutz, J., S. B. Cannon, J. Schlueter, et al. 2010. Genome sequence of the palaeopolyploid soybean. *Nature* 463:178–183.

Schneider, K., L. Kienow, E. Schmelzer, et al. 2005. A new type of peroxisomal acyl-coenzyme A synthetase from *Arabidopsis thaliana* has the catalytic capacity to activate biosynthetic precursors of jasmonic acid. *Journal of Biological Chemistry* 280: 13962–13972.

Scholz, S. S., M. Reichelt, W. Boland. 2015. Additional evidence against jasmonate-induced jasmonate induction hypothesis. *Plant Science* 239: 9–14.

Schuck, S., M. Kallenbach, I. T. Baldwin, et al. 2014. Bonaventure G. The Nicotiana attenuata GLA1 lipase controls the accumulation of Phytophthora parasitica-induced oxylipins and defensive secondary metabolites. *Plant Cell Environment* 37(7): 1703–1715.

Seltmann, M. A., N. E. Stingl, J. K. Lautenschlaeger, et al. 2010. Differential impact of lipoxygenase 2 and jasmonates on natural and stress-induced senescence in Arabidopsis. *Plant Physiology* 152: 1940–1950.

Sharkey, T. D. 1985. Photosynthesis in intact leaves of C3 plants: Physics, physiology and rate limitations. *Botanical Review* 51: 53–105.

Sheard, L. B., X. Tan, H. Mao, et al. 2010. Jasmonate perception by inositol-phosphate-potentiated COI1–JAZ co-receptor. *Nature* 468: 400–405.

Sheteawi, S. 2007. Improving growth and yield of salt-stressed soybean by exogenous application of jasmonic acid and ascobin. *International Journal of Agriculture and Biology* 9(3): 473–478.

Shibaoka, H. 1994. Plant hormone-induced changes in the orientation of cortical microtubules: Alteration of the cross-linking between microtubules and the plasma membrane. *Annual Review Plant Physiology* 45: 527–544

Shih, C. F., W. H. Hsu, Y. J. Peng, et al. 2014. The NAC-like gene ANTHER INDEHISCENCE FACTOR acts as a repressor that controls anther dehiscence by regulating genes in the jasmonate biosynthesis pathway in Arabidopsis. *Journal of Experimental Botany* 65: 621–639.

Sirhindi, G., M. A. Mir, E. F. Abd-Allah, et al. 2016. Jasmonic acid modulates the physio-biochemical attributes, antioxidant enzyme activity, and gene expression in *Glycine max* under nickel toxicity. *Frontiers in Plant Science* 7: 591.

Sirhindi, G., P. Sharma, A. Singh, et al. 2015. Alteration in photosynthetic pigments, osmolytes and antioxidants in imparting copper stress tolerance by exogenous jasmonic acid treatment in *Cajanus cajan. International Journal of Plant Physiology and Biochemistry* 7: 30–39.

Skrzypek, E., K. Miyamoto, M. Saniewski, et al. 2005. Jasmonates are essential factors inducing gummosis in tulips: Mode of action of jasmonates focusing on sugar metabolism. *Journal of Plant Physiology* 162: 495–505.

Sohn, H. B., H. Y. Lee, J. S. Seo, et al. 2011. Overexpression of jasmonic acid carboxyl methyltransferase increases tuber yield and size in transgenic potato. *Plant Biotechnology Reports* 5(1): 27–34.

Sorial, M. E. and A. A. Gendy. 2010. Response of sweet basil to jasmonic acid application in relation to different water supplies. *Bioscience Research* 7: 39–47.

Spetea, C., E. Rintamäki, B. Schoefs, et al. 2014. Changing the light environment: Chloroplast signalling and response mechanisms. Philosophical transactions of the *Royal Society of London. Series B, Biological sciences Royal Society* 369: 20130220.

Springer, A., C. Kang, S. Rustgi, et al. 2016. Programmed chloroplast destruction during leaf senescence involves 13-lipoxygenase (13-LOX). *Proceedings of the National Academy of Sciences, USA* 113: 3383–3388.

Staswick, P. E. and I. Tiryaki. 2004. The oxylipin signal jasmonic acid is activated by an enzyme that conjugates it to isoleucine in Arabidopsis. The Plant Cell 16: 2117–2127.

Stenzel, I., B. Hause, H. Maucher, et al. 2003. Allene oxide cyclase dependence of the wound response and vascular bundle-specific generation of jasmonates in tomato—amplification in wound signaling. *The Plant Journal* 33: 577–589.

Stenzel, I., M. Otto, and C. Delker. 2012. Allene oxide cyclase (AOC) gene family members of *Arabidopsis thaliana*: Tissue- and organ-specific promoter activities and in vivo heteromerization. *Journal of Experimental Botany* 63: 6125–6138.

Stumpe, M., C. Göbel, and B. Faltin. 2010. The moss Physcomitrella patens contains cyclopentenones but no jasmonates: Mutations in allene oxide cyclase lead to reduced fertility and altered sporophyte morphology. *New Phytologists* 188:740–749.

Suza, W. P., M. L. Rowe, and M. Hamberg. 2010. A tomato enzyme synthesizes (+)-7-iso-jasmonoyl-l-isoleucine in wounded leaves. *Planta* 231:717–728.

Svyatyna, K., Y. Jikumaru., R. Brendel, et al. 2014. Light induces jasmonate– isoleucine conjugation via OsJAR1-dependent and -independent pathways in rice. *Plant, Cell and Environment* 37:827–839.

Swiatek, A., A. Azmi, E. Witters, et al. 2003. Stress messengers jasmonic acid and abscisic acid negatively regulate plant cell cycle. *Bulgarian Journal of Plant Physiology* 29: 172–178.

Swiatek, A., M. Lenjou, D. Van Bockstaele, et al. 2002. Differential effect of jasmonic acid and abscisic acid on cell cycle progression in tobacco BY-2 cells. *Plant Physiology* 128:201–211.

Takahashi, K., K. Fujino., Y. Kikuta, et al. 1994. Expansion of potato cells.

Takahashi, S. and N. Murata. 2008. How do environmental stresses accelerate photoinhibition? *Trends in Plant Science* 13:178–182.

Takahashi, T., A. Gasch, N. Nishizawa, et al. 1995. The DIMINUTO gene of Arabidopsis is involved in regulating cell elongation. *Genes and Development* 9:97–107.

Tani, T., H. Sobajima, K. Okada, et al. 2008. Identification of the OsOPR7 gene encoding 12-oxophytodienoate reductase involved in the biosynthesis of jasmonic acid in rice. *Planta* 227:517–526.

Theodoulou, F. L., K. Job, S. P. Slocombe, et al. 2005. Jasmonic acid levels are reduced in COMATOSE ATP-binding cassette transporter mutants. Implications for transport of jasmonate precursors into peroxisomes. Plant Physiology 137: 835–840.

Thines, B., L. Katsir, M. Melotto, et al. 2007. JAZ repressor proteins are targets of the SCF(COI1) complex during jasmonate signalling. *Nature* 448:661–665.

Tikkanen, M., M. Grieco, E. Aro, et al. 2011. Novel insights into plant light-harvesting complex II phosphorylation and 'state transitions'. *Trends in Plant Sciences* 16:126–131.

Ubeda-Tomás, S., G. T. S. Beemster, and M. J. Bennett. 2012. Hormonal regulation of root growth: Integrating local activities into global behaviour. *Trends in Plant Science* 17(6):326–331.

Ueda, J. and M. Saniewski. 2006. Methyl jasmonate-induced stimulation of chlorophyll formation in the basal part of tulip bulbs kept under natural light conditions. *Journal of Fruit and Ornamental Plant Research* 14:199–210.

Van Schie, C. C. N., M. A. Haring, and R. C. Schuurink. 2007. Tomato linalool synthase is induced in trichomes by jasmonic acid. *Plant Molecular Biology* 94: 251–263.

Vick, B. A. and D. C. Zimmerman. 1983. The biosynthesis of jasmonic acid: a physiological role for plant lipoxygenase. Biochem Biophys Res Commun. 111(2):470–477.

Wang, J., J. Qian, L. Yao, et al. 2015. Enhanced production of flavonoids by methyl jasmonate elicitation in cell suspension culture of Hypericum perforatum. Bioresources and Bioprocessing 2(1):1–5.

Wang, L., X. Yang, Z. Ren, et al. 2014. Alleviation of photosynthetic inhibition in copper-stressed tomatoes through rebalance of ion content by exogenous nitric oxide. *Turkish Journal of Botany* 38:1–13.

War, A. R., B. Hussain, and H. C. Sharma. 2013. Induced resistance in groundnut by jasmonic acid and salicylic acid through alteration of trichome density and oviposition by *Helicoverpa armigera* (Lepidoptera: Noctuidae). *AoB Plants* 1–6.

Wasternack, C. 2007. Jasmonates: An update on biosynthesis, signal transduction and action in plant stress response, growth and development. *Annals of Botany* 100:681–697.

Wasternack, C. 2014. Jasmonates in plant growth and stress responses. In: *Phytohormones: A Window to Metabolism, Signaling and Biotechnological Applications*, Tran, L.-S., Pal, S., eds., pp. 221–263. Springer, New York.

Wasternack, C. and B. Hause. 2002. Jasmonates and octadecanoids—Signals in plant stress response and development. In: *Progress in Nucleic Acid Research and Molecular Biology*, Moldave, K., ed., pp. 165–222. Academic Press, New York.

Wasternack, C. and B. Hause. 2013. Jasmonates: Biosynthesis, perception, signal transduction and action in plant stress response, growth and development. *Annals of Botany* 111(6):1021–58.

Wasternack, C. and E. Kombrink. 2010. Jasmonates: Structural requirements for lipid-derived signals active in plant stress responses and development. *ACS Chemical Biology* 5:63–77.

Wasternack, C. and M. Strnad. 2016. Jasmonate signaling in plant stress responses and development—active and inactive compounds. *New Biotechnology* 33:604–613.

Westfall, C. S., A. M. Muehler, and J. M. Jez. 2013. Enzyme action in the regulation of plant hormone responses. Journal of Biological Chemistry 288: 19304–19311.

Westfall, C. S., C. Zubieta, J. Herrmann, et al. 2012. Structural basis for prereceptor modulation of plant hormones by GH3 proteins. *Science* 336:1708–1711.

Wiszniewski, A. A., J. D. Bussell, R. L. Long, et al. 2014. Knockout of the two evolutionarily conserved peroxisomal 3-ketoacyl-CoA thiolases in Arabidopsis recapitulates the abnormal inflorescence meristem 1 phenotype. *Journal of Experimental Botany* 65:6723–6733.

Wu, M., A. Neilson, A. L. Swift, et al. 2007. Multiparameter metabolic analysis reveals a close link between attenuated mitochondrial bioenergetic function and enhanced glycolysis dependency in human tumor cells. *American Journal Physiology and Cell Physiology* 292:C125–136.

Xiao, Y., Y. Chen, T. Charnikhova, et al. 2014. OsJAR1 is required for JA-regulated floret opening and anther dehiscence in rice. *Plant Molecular Biology* 86:19–33.

Yamamoto Y., J. Ohshika, and T. Takahashi. 2015. Functional analysis of allene oxide cyclase, MpAOC, in the liverwort Marchantia polymorpha. *Phytochemistry* 116:48–56.

Yan, J., S. Li, M. Gu, et al. 2016. Endogenous bioactive jasmonate is composed of a set of (+)-7-iso-JA amino acid conjugates. *Plant Physiology* 172(4):2154–2164. DOI:10.1104/pp.16.00906.

Yan, J., C. Zhang., M. Gu, et al. 2009. The Arabidopsis CORONATINE INSENSITIVE1 protein is a jasmonate receptor. *The Plant Cell* 21: 2220–2236.

Yan, L., Q. Zhai, J. Wei, et al. 2013. Role of tomato lipoxygenase D in wound-induced jasmonate biosynthesis and plant immunity to insect herbivores. *PLoS Genetics* 9:e1003964.

Yang, D. L., J. Yao, C. S. Mei, et al. 2012. Plant hormone jasmonate prioritizes defense over growth by interfering with gibberellin signaling cascade. *Proceedings of the National Academy of Sciences of the USA* 109:E1192–E1200.

Yıldız, K., C. Yazıcı, and F. Muradoğlu. 2007. Effect of jasmonic acid on germination dormant and nondormant apple seeds. Asian Journal of Chemistry 19 (2): 1098–1102.

Yoeun, S., R. Rakwal, and O. Han. 2013. Dual positional substrate specificity of rice allene oxide synthase-1: Insight into mechanism of inhibition by type II ligand imidazole. *BMB Reports* 46:151–156.

Yoshida, Y., R. Sano, T. Wada, et al. 2009. Jasmonic acid control of GLABRA3 links inducible defense and trichome patterning in Arabidopsis. *Development* 136:1039–1048.

Ytterberg, A. J., J. B. Peltier, and K. J. van Wijk. 2006. Protein profiling of plastoglobules in chloroplasts and chromoplasts. A surprising site for differential accumulation of metabolic enzymes. *Plant Physiology* 140: 984–997.

Yuan, Z. and D. Zhang. 2015. Roles of jasmonate signalling in plant inflorescence and flower development. *Current Opinion in Plant Biology* 27:44–51.

Zalewski, K., B. Nitkiewicz, L. B. Lahuta, et al. 2010. Effect of jasmonic acid-methyl ester on the composition of carbohydrates and germination of yellow lupine (*Lupinus luteus L.*) seeds. *Journal of Plant Physiology* 167(12):967–973.

Zhang, Y. and J. G. Turner. 2008. Wound-induced endogenous jasmonates stunt plant growth by inhibiting mitosis. *PLos OnE* 3:1–9.

Zhao, Y., W. Dong, N. Zhang, et al. 2014. A wheat allene oxide cyclase gene enhances salinity tolerance via jasmonate signaling. *Plant Physiology* 164:1068–1076.

Zheng, H., X. Pan, Y. Deng, et al. 2016. AtOPR3 specifically inhibits primary root growth in Arabidopsis under phosphate deficiency. *Scientific Reports* 6:24778.

28

Current Scenario of NO (S-Nitrosylation) Signaling in Cold Stress

Yaiphabi Sougrakpam, Priyanka Babuta, and Renu Deswal

CONTENTS

Abbreviations

APX	Ascorbate peroxidase
CAT	Catalase
CBF	C-repeat binding factor
DHAR	Dehydroascorbate reductase
GSNO	S-nitrosoglutathione
GSNOR	S-nitrosoglutathione reductase
NO	Nitric oxide
NOS	Nitric oxide synthase
NR	Nitrate reductase
NTR	NADPH-dependent thioredoxin reductase
RNS	Reactive nitrogen species
ROS	Reactive oxygen species
SNO	S-nitrosothiol
SOD	Superoxide dismutase
Trx	Thioredoxin
TrxR	Thioredoxin reductase

Introduction

Abiotic stress affects the growth and development of plants, ultimately leading to yield loss. To shield themselves from adverse environmental conditions, plants have developed complex recognition and response mechanism(s) (Bita and Gerats, 2013). The first response of plants to combat any stress conditions is the production of reactive oxygen species (ROS) (hydrogen peroxide [H_2O_2]) and reactive nitrogen species (RNS) (mainly nitric oxide [NO]) (Jaspers and Kangasjarvi, 2010). Under adverse environmental conditions, ROS (H_2O_2) accumulation is triggered, contributing to oxidative stress, which causes damage to DNA, proteins, lipids, and carbohydrates, leading to cell death (Neill et al., 2002; Gill et al., 2010; Choudhury et al., 2017). Apart from inducing cellular damage, ROS also acts as a signaling molecule regulating the expression of various stress-responsive genes (Qiao et al., 2014). NO, a redox-active molecule, was shown to have both a positive as well as a negative impact on the stress response, depending on its concentration and localization in cellular compartments (Romero-Puertas et al., 2013; Fancy et al., 2017). It maintains the cellular homeostasis by activating the antioxidant machinery, leading to downregulation of ROS (Qiao et al., 2014). In the present chapter, the role of NO in low-temperature stress is discussed to understand the regulatory mechanisms.

Sources of NO in Plants

NO is generated in plants by different pathways broadly categorized as enzymatic (oxidative and reductive) and non-enzymatic (Gupta et al., 2011). In the oxidative pathway, NO generation is mediated by nitric oxide synthase (NOS)-like enzyme via the oxidation of L-arginine (Corpas et al., 2009). Polyamines also contribute to NO production; however, the mechanism is not yet elucidated (Tun et al., 2006). Hydroxylamine-mediated NO production occurs by reaction with superoxide to form NO

(Vetrovsky et al., 1996) or is catalyzed by hydroxylamine oxidoreductase (Hooper and Terry, 1979). The reductive pathway involves the reduction of nitrite, using NAD(P)H as electron donor, by nitrate reductase (NR) (Neill et al., 2008) and membrane-bound nitrite:NO reductase (Ni:NOR) (Stohr et al., 2001). The reduction of nitrite to NO is also mediated by xanthine oxidoreductase (XOR, a peroxisomal enzyme) under anaerobic conditions using xanthine or NADH as reducing agent (Godber et al., 2000; Corpas et al., 2004). It can also be generated from nitrite by mitochondrial cytochrome c oxidase and/or reductase with NAD(P)H as electron donor via ubiquinone and the mitochondrial electron transport chain in the inner mitochondrial membrane (Tielens et al., 2002; Stoimenova et al., 2007). Nonenzymatically, NO is generated from nitrite under acidic pH in the presence of reductants such as ascorbic acid (Bethke et al., 2004). Light-mediated conversion of nitrite to NO by carotenoids has also been reported (Cooney et al., 1994).

NO Homoeostasis in Plants

NO, being a small, highly labile, and diffusible free radical, is involved in various growth processes, from germination of seed to senescence (Ya'acov et al., 1998; Besson-Bard et al., 2008; Baudouin, 2011; Qiau et al., 2014). Under physiological conditions, NO undergoes secondary reactions with oxygen to form nitrogen dioxide (NO_2), dinitrogen trioxide (N_2O_3), or peroxynitrite ($ONOO^-$) (Broniowska and Hogg, 2012). NO-responsive biological functions are mediated by post-translational modifications (PTMs) of proteins: namely, tyrosine nitration, metal nitrosylation, and S-nitrosylation (Baudouin, 2011; Kovacs and Lindermyer, 2013). In tyrosine nitration, $ONOO^-$ reacts with tyrosine residues of the target proteins, forming 3-nitrotyrosine and modulating their activity (Radi, 2004; Baudouin, 2011). In metal nitrosylation, NO reacts with transition metals of metalloproteins, forming a metal-nitrosyl complex, which transfers NO to a cysteine of the same protein. S-nitrosylation is the most widely studied NO-based PTM, whereby –SH in cysteine is modified to –SNO. In biological systems, S-nitrosylation occurs either by reaction of NO with thiyl (RS·, thiolate) radicals; or between N_2O_3 and thiolate anion, where N_2O_3 is the donor of NO^+; or by trans-nitrosylation, whereby NO from low-molecular weight thiols such as S-nitrosoglutathione (GSNO) or nitrosocysteine (CysNO) or from one protein is transferred to another protein (Kovacs and Lindermyer, 2013; Zaffaginini et al., 2016). In biological systems, S-nitrosylation is mainly mediated by GSNO, a low–molecular weight S-nitrosothiol that is a mobile reservoir of NO (Thalineau et al., 2016). S-nitrosylation mediates its effects via changing protein activity, translocation, and conformation.

A high concentration of NO leads to over-accumulation of RNS, which causes nitrosative stress (Neill et al., 2003, 2008; Ziogas et al., 2013; Wilson et al., 2008). Protein denitrosylation (cellular catabolism of SNO) is an important molecular mechanism involved in the regulation of NO-based signaling. It is important for maintaining the level of SNO, conferring cellular protection, and alleviating the symptoms of pathological conditions linked to nitrosative stress (Benhar et al., 2010). Denitrosylation can be mediated by either non-enzymatic (heat, light, reducing agents,

nucleophilic compounds, and transition metals) or enzymatic (S-nitrosoglutathione reductase [GSNOR] and thioredoxin-thioredoxin reductase [Trx-TrxR] system) pathways (Kovacs and Lindermyer, 2013). The enzymatic pathway of denitrosylation is widely studied in both animal and plant systems.

GSNOR, a zinc-dependent dehydrogenase belonging to the Class III alcohol dehydrogenase (ADH3; EC 1.1.1.1) family, was initially identified as glutathione-dependent formaldehyde dehydrogenase (GS-FDH; EC 1.2.1.1) due to its role in the formaldehyde detoxification pathway. The NADH-dependent GSNOR-catalyzed reaction results in the depletion of GSNO levels and reduces the formation of S-nitrosothiols (SNOs) by trans-nitrosylation regulating protein nitrosylation (Sakamoto et al., 2002).

Thioredoxins (Trx), small ubiquitous redox proteins, also play an important role in NO-based signaling and in ameliorating nitrosative stress (Hashemy et al., 2008; Benhar et al., 2008, 2009; Sengupta et al., 2012; Kneeshaw et al., 2014). For Trx to function as a denitrosylase, the combination of the Trx-TrxR system is required. TrxR, also known as NADPH-dependent thioredoxin reductase (NTR), is a well-characterized principal protein disulfide reductase that maintains cellular redox equilibrium. On the basis of compartmentalization, NTR is classified into three classes: NTRA (cytosolic), NTRB (mitochondrial); and NTRC (plastidial) (Serrato et al., 2004; Reichheld et al., 2005). It is the balance of S-nitrosylation/denitrosylation that carries forward NO-mediated cold stress signaling.

Different Routes of NO Production Lead to Nitrosative Burst during Cold Stress

NO production follows different routes during cold stress. NR-mediated NO production has been observed in *Arabidopsis thaliana* (Zhao et al., 2009; Cantrel, 2011), *Capsicum annuum* (Airaki et al., 2012), and *Brassica juncea* seedling (Sehrawat et al., 2013), while NO production by an NOS-like enzyme was also reported in *Chorispora bungeana* (Liu et al., 2010), *Solanum lycopersicum* fruit (Zhao et al., 2011), *C. annuum* (Airaki et al., 2012), and *B. juncea* (Sehrawat et al., 2013). Non-enzymatic NO production was observed in the apoplast of *B. juncea* (Sehrawat and Deswal, 2014). Therefore, both NR and NOS-like enzyme seem to contribute to NO production in cold stress. In addition, site-specific non-enzymatic NO production was observed in the apoplast, the acidic compartment of a cell (Sehrawat and Deswal, 2014).

NO production via different routes leads to the accumulation of SNOs, stable metabolites of NO that participate in storage and transport. These are generated by the action of nitrosonium (NO+) with thiols in protein/peptide cysteine residues and cysteines (Carver et al., 2005). Glutathione (GSH) combines with NO to form GSNO, a low–molecular weight thiol (Lima et al., 2009; Corpas et al., 2008). The SNO content was reported to increase five-fold in *Pisum sativum* (Corpas et al., 2008), 1.4-fold in apoplast of *B. juncea* (Sehrawat and Deswal, 2014), and by 12% in *Arabidopsis* plantlets (Puyaubert et al., 2014) during low-temperature conditions. The enhanced SNO content drives NO-mediated signaling by either nitration, glutathionylation, or S-nitrosylation of proteins during cold stress. Of these, S-nitrosylation is the most investigated PTM.

S-Nitrosoproteome Analysis during Cold Stress

The protein complement of S-nitrosylated proteins, called the *S-nitrosoproteome*, was analyzed under cold stress in *Brassica* seedlings as well as in sub-cellular compartments. In *B. juncea* seedlings, cold stress induced differential S-nitrosylation of proteins in crude extract, RuBisCO depletome, and apoplast (Abat and Deswal, 2009a; Sehrawat et al., 2013; Sehrawat and Deswal, 2014). S-nitrosylated proteins belonged to different functional categories: photosynthesis, stress/redox/signaling, metabolism, and cell wall modifying. Stress/signaling/redox-related proteins constituted the major (49%) category of S-nitrosylated proteins in crude as well as in apoplast of *B. juncea*, followed by metabolism (24%), photosynthesis (15%), and cell wall–modifying proteins (12%) (Figure 28.1). The study of the S-nitrosoproteome of different cellular compartments would help to increase the S-nitrosoproteome pool and also to discover new roles of NO-mediated S-nitrosylation in plants (Table 28.1).

Various photosynthetic proteins, such as beta-carbonic anhydrase (CA), oxygen evolving enhancer protein, a 23 kDa polypeptide involved in photosystem II (PSII), RuBisCO, photosystem I (PSI) reaction center, and sedoheptulose-bisphosphatase were S-nitrosylated during cold stress. NO negatively regulates the enzymes involved in the photosynthetic machinery. Inhibition of CA, an enzyme involved in carboxylation and decarboxylation reactions in photosynthesis and respiration, by S-nitrosylation was reported (Moroney et al., 2001; Sharma et al., 2016). Also, the CA activity of AtSABP3 (*Arabidopsis thaliana* salicylic acid-binding protein 3) was also inhibited by S-nitrosylation at Cys-280 (Wang et al., 2009). Cold induced S-nitrosylation of

CA, sedoheptulose-bisphosphatase, and light-dependent PSI and PSII was reported in *B. juncea* (Sehrawat et al., 2013; Sehrawat and Deswal, 2014). Large and small subunits of RuBisCO exhibited differential S-nitrosylation during cold stress, causing 39% reduction in its activity (Abat and Deswal, 2009a). Thus, it can be concluded that the negative impact of S-nitrosylation on the photosynthetic machinery might contribute to yield loss during cold stress.

We have suggested earlier that glycolysis is heavily regulated by nitrosylation, as 90% of the glycolytic enzymes are S-nitrosylated (Abat and Deswal, 2009b). Cold stress differentially modulates the glycolytic cycle. Fructose-1-6-bisphosphate aldolase was positively regulated in *B. juncea* seedlings (Sehrawat et al., 2013), while glyceraldehyde-3-phosphate dehydrogenase, which possesses cysteine in its active site, making it susceptible to oxidative protein modification, was negatively regulated in *A. thaliana* (Lindermayr et al., 2005; Holtgrefe et al., 2008).

A very interesting observation was modification of the Brassicaceae specific glucosinolate pathway. Glucosinolates (secondary metabolites) are modulated under both biotic and abiotic stress (Martinez-Ballesta et al., 2013). Cold stress modulates S-nitrosylation of glucosinolate hydrolytic proteins, myrosinase, and epithiospecifier proteins. The exogenous application of isothiocyanates (a by-product of glucosinolate hydrolysis) confers heat tolerance on *Arabidopsis* by increasing the expression of heat shock proteins (HSPs) and GST and the accumulation of H_2O_2 (Hara et al., 2010, 2013; Khokon et al., 2011). This suggests the role of the pathway in stress tolerance. As S-nitrosylation of myrosinase has not been investigated yet, our group is trying to decipher this interesting aspect of myrosinase regulation by S-nitrosylation regarding glucosinolate metabolism.

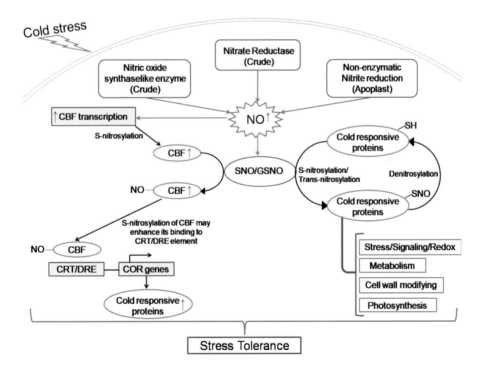

FIGURE 28.1 Schematic representation of NO-mediated cold stress tolerance. Cold stress induces the accumulation of NO via different routes, which increases the SNO/GSNO content in a cell. This leads to S-nitrosylation/trans-nitrosylation of cold responsive proteins categorized as stress/signaling/redox, photosynthetic, cell wall modifying, and metabolic. CBF, a transcription factor, is regulated by NO at transcriptional and post-translational level, ultimately enhancing COR gene expression. Both the pathways contribute to stress tolerance in plants.

TABLE 28.1

Identified S-Nitrosylated Targets of *Brassica juncea* Seedling Classified in Functional Categories Indicating Major Metabolic Signaling in Cold Stress (from Different Compartments)

Sl. No.	Protein	S-Nitrosylation	Cellular Compartment	Activity	Functional Category
1	RuBisCO large subunit	Upregulated	Total proteome	Inhibit	Photosynthesis
2	RuBisCO small subunit	Upregulated/ downregulated	Total proteome	Inhibit	Photosynthesis
2	Sedoheptulose-bisphosphatase	Upregulated	RuBisCO depletome	Inhibit	Photosynthesis
3	Beta-carbonic anhydrase	Upregulated	RuBisCO depletome	–	Photosynthesis
4	Oxygen-evolving enhancer protein 2	Upregulated	RuBisCO depletome	–	Photosynthesis
5	Major latex protein–like protein	Upregulated	Total proteome	–	Stress/signaling/redox
6	Putative cysteine-rich repeat secretory protein	Downregulated	Total proteome	–	Stress/signaling/redox
7	Putative lactoylglutathione lyase	Upregulated	RuBisCO depletome, apoplast	–	Stress/signaling/redox
8	PpX-GppA phosphatase protein	Upregulated	Total proteome	–	Stress/signaling/redox
9	Daikon cysteine protease RD21	Upregulated	RuBisCO depletome	–	Stress/signaling/redox
10	Vacuolar calcium binding protein	Downregulated	RuBisCO depletome	–	Stress/signaling/redox
11	Fe-SOD	Upregulated	RuBisCO depletome	–	Stress/signaling/redox
12	Glutathione S-transferase	Upregulated	RuBisCO depletome, apoplast	Induce	Stress/signaling/redox
13	Chaperonin 10	Upregulated	RuBisCO depletome	–	Stress/signaling/redox
14	Epithiospecifier protein	Upregulated	RuBisCO depletome	–	Stress/signaling/redox
15	MLP-like protein 328	Upregulated	RuBisCO depletome	–	Stress/signaling/redox
16	Myrosinase	Downregulated	Apoplast	–	Stress/signaling/redox
17	Germin-like protein	Upregulated	Apoplast	–	Stress/signaling/redox
18	Dehydroascorbate reductase (DHAR)	Upregulated	Apoplast	Induce	Stress/signaling/redox
19	Cu/Zn-SOD	Upregulated	Apoplast	Induce	Stress/signaling/redox
20	Soluble inorganic pyrophosphatase 1	Downregulated	RuBisCO depletome	–	Metabolism
21	Putative fructose-bisphosphate aldolase	Upregulated	RuBisCO depletome	Induce	Metabolism
22	Fructose-bisphosphate aldolase, class I	Upregulated	RuBisCO depletome	Induce	Metabolism
23	Glyceraldehyde-3-phosphate dehydrogenase	Upregulated	RuBisCO depletome	Inhibit	Metabolism
24	Fructose-1,6-bisphosphate aldolase	Upregulated	Total proteome, RuBisCO depletome	Induce	Metabolism
25	Aspartyl protease family protein	Upregulated	Apoplast	–	Stress/signaling/redox
26	GDSL lipase, GLIP	Upregulated	Apoplast	–	Cell wall modifying
27	Acyl-binding/lipid transfer protein	Upregulated	Apoplast	–	Cell wall modifying
28	1.7S seed storage protein	Downregulated	Apoplast	–	Metabolism
29	Photosystem I reaction center subunit II	Upregulated	Apoplast	–	Metabolism
30	Cruciferin precursor	Downregulated	Apoplast	–	Metabolism
31	Cruciferin cru4 subunit	Downregulated	Apoplast	–	Metabolism
32	FASCICLIN-like arabinogalactan	Upregulated	Apoplast	–	Cell wall modifying
33	Alpha-L-arabinofuranosidase	Downregulated	Apoplast	–	Cell wall modifying

Gly I (lactoylglutathione lyase), a stress marker enzyme, was found to be S-nitrosylated in the RuBisCO depletome (Sehrawat et al., 2013) and apoplast (Sehrawat and Deswal, 2014) of *B. juncea* under cold stress. Gly I detoxifies methylglyoxal, a highly toxic glycolytic by-product that reacts with proteins and DNA, thereby causing inactivation (Korithoski et al., 2007). Reversible inactivation of Gly I activity was observed in human endothelial cells exposed to GNSO along with a shift in pI from 5.0 to 5.2. Inhibition of activity was reversed by dithiothreitol treatment (Mitsumoto et al., 1999). Depletion of reduced glutathione (GSH) inhibits the inactivation of Gly I activity by GSNO, showing the requirement for GSH for its inactivation by NO (Mitsumoto et al., 2000). Gly I activity was completely inhibited by glutathionylation and by treatment with oxidized

glutathione (GSSG) (Birkenmeier et al., 2010). A nitrated form of Gly I has been reported in citrus during salt stress; however, the effect of nitration has not been reported (Tanou et al., 2012). Therefore, Gly I seems to be a PTM cross-talk point for NO signaling. Moreover, it seems to be PTM cross-talk point/node in NO signaling.

Cold stress also mediates the S-nitrosylation of proteins involved in signaling (PpX-GppA phosphatase, soluble inorganic pyrophosphatase, major latex proteins such as protein 328, and vacuolar calcium binding protein), proteins involved in seed germination and storage (germin-like protein, 1.7S seed storage protein, and cruciferin), proteins involved in translation, folding, and degradation (chaperonin and daikon cysteine protease RD21), and cell wall–modifying proteins (acyl-binding/lipid transfer protein isoform III, GDSL lipase, GLIP, and FASCILIN-like arabinogalactan protein 8). However, the effect of S-nitrosylation on these proteins has not been investigated yet. Thus, cold stress affects the S-nitrosylation of proteins involved in important biological functions by either inducing or inhibiting their activity, thereby promoting the cold stress signaling response by modulating major metabolic pathways. The "enzymatic antioxidant machinery" is one such group of proteins tightly regulated by nitrosylation.

NO Modulates Antioxidant Machinery in Cold Stress

Redox signaling is emerging as an intrinsic mechanism in providing abiotic stress tolerance to plants (Fancy et al., 2017). Abiotic stress leads to the production of ROS, causing oxidative damage to plants. NO has been reported to induce the activity of antioxidant enzymes to protect plants from oxidative damage. Interestingly, S-nitrosylation of antioxidant scavenging enzymes Fe-superoxide dismutase (Fe-SOD), Cu/Zn-SOD, dehydroxyascorbate reductase (DHAR), and glutathione-S-transferase (GST) has been reported (Sehrawat et al., 2013; Sehrawat and Deswal, 2014). SOD converts superoxide radical to H_2O_2, providing protection against oxidative stress (Bowler et al., 1991). During cold stress, S-nitrosylation led to a 50% increase in the activity of Fe-SOD in the RuBisCO depletome (Sehrawat et al., 2013). DHAR, an important antioxidant involved in the ascorbate-glutathione cycle, is positively regulated, with a 67.2% increase in its activity, while GST, another antioxidant, showed a two-fold increase in its activity in *B. juncea* apoplast (Sehrawat and Deswal, 2014). Cold stress–induced oxidative damage is prevented by increasing activity of these antioxidant enzymes via S-nitrosylation. This seems to be an adaptive mechanism for combating oxidative stress. Regulation of activity of the antioxidant enzymes by other NO-based PTMs may not be ruled out, as DHAR is also S-glutathionylated and tyrosine nitrated (Sehrawat and Deswal, 2014), suggesting the existence of NO-based PTM cross-talk in cold stress signaling. NO has been reported to increase the activity of SOD, ascorbate peroxidase (APX), peroxidase (POD), catalase (CAT), and monodehydroascorbate reductase (MDAR) in *C. bungeana* (Liu et al., 2010), *S. lycopersicum* fruit (Zhao et al., 2011), *C. annuum* (Airaki et al., 2012), and *Triticum aestivum* (Bavita et al., 2012). Increased levels of antioxidants such as ascorbate (Asc) and GSH were also observed in these systems. APX, MDHAR, DHAR, and GR are enzymes involved in the Asada–Halliwell pathway, which is responsible for maintaining cellular redox homeostasis (Noctor et al., 1998; Sharma et al., 2016). GSH protects cells from toxic effects of ROS and also provides a reducing environment. It is involved in the synthesis of proteins and nucleic acids and the detoxification of free radicals and peroxides (Meister, 1988). Therefore, NO provides protection from cold stress–induced oxidative damage by regulating the Asada–Halliwell pathway enzymes and other enzymes involved in detoxification of ROS.

NO-Induced CBF Expression and PTM (S-Nitrosylation) Provides Cold Tolerance

Nitrosylation targets also include transcription factors such as non-expressor of pathogenesis related protein 1 (NPR1). Investigations by our group have shown for the first time the involvement of NO in modulating C-repeat binding factor (CBF) and hence, cold stress signaling. S-nitrosylation could also modulate transcription factors that have an important role in providing abiotic stress tolerance in plants. ICE-CBF (Inducer of CBF expression-C-repeat binding factor)–dependent cold stress signaling is widely studied in cold acclimation. CBF was reported to be regulated by NO transcriptionally and post-translationally (via S-nitrosylation) in both fruit and seedlings (Zhao et al., 2011; Kashyap et al., 2015). Besides enhancing CBF expression, NO treatment also increased proline content and reduced malondialdehyde (MDA) content and ion leakage, thus imparting cold stress tolerance (Zhao et al., 2011; Kashyap et al., 2015). S-nitrosylation may enhance binding of CBF to C-repeat/Dehydration responsive element (CRT/DRE) in the promoter region of cold responsive (COR) genes, inducing their expression. Thus, transcriptional and post-translational regulation of CBF by NO may contribute to cold stress tolerance in plants (Figure 28.1).

Denitrosylases and Cold Stress

The role of GSNOR and Trx-TrxR in multiple biotic and abiotic stress conditions has been demonstrated in plants (Barroso et al., 2006; Corpas et al., 2008; Dai et al., 2011; Salgado et al., 2013). The heat tolerance of knockdown mutants of GSNOR (hot5) and NTRC (hot1) was compromised, suggesting their role in the thermotolerance mechanism (Lee et al., 2008; Chae et al., 2013; Serrato et al., 2004). Cold stress induces the accumulation of GSNOR and NO content in *P. sativum*, *Helianthus annuus*, and *C. annuum* (Corpas et al., 2008; Chaki et al., 2011; Airaki et al., 2012). Cheng et al. (2015) also reported accumulation of GSNOR during cold stress in *Populus trichocarpa* but suppressed it after prolonged cold stress. Similar results were also observed in pepper, where GSNOR activity increased by 32% and then declined after low temperature treatment for 1 and 2 days, respectively, suggesting the existence of a feedback mechanism (Airaki et al., 2012). Contradictory results were reported in citrus, in which downregulation of GSNOR was observed during cold stress at both transcript and protein level (Ziogas et al., 2013). It is reported that the acclimatization of pepper to low temperature is linked with enhancement of GSNOR activity and a decrease in NO content (Airaki et al., 2011).

Moon et al. (2015) observed that over-expression of AtNTRC also confers cold and freezing stress tolerance and also showed cryoprotective activity for malate dehydrogenase and lactic dehydrogenase. Conversely, expression of *S. lycopersicum* NTR was increased in response to oxidative and salinity stress but not under cold stress (Dai et al., 2011). Cold stress also induces the expression of thioredoxin, which is suggested to activate GSNOR through denitrosylation, in turn alleviating RNS accumulation. Also, the NO scavenger 2-(4-carboxyphenyl)-4,4,5,5-tetramethylimidazoline-1-oxyl-3-oxide (cPTIO) and a GSNOR inhibitor

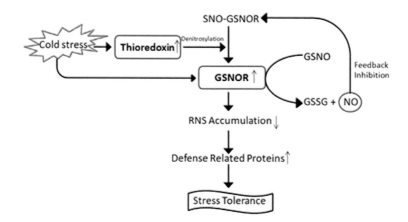

FIGURE 28.2 Schematic representation of the role of denitrosylases in cold stress tolerance. Trx denitrosylate GSNOR, resulting in the activation of the enzyme, which is regulated by a feedback inhibition mechanism. Accumulation of Trx and GSNOR in response to cold stress alleviates RNS accumulation. This, in turn, activates the defense response machinery, leading to stress tolerance in plants.

(N6022 or dodecanoic acid) mediated the accumulation of RNS-suppressed cold-induced defense-related proteins (Cheng et al., 2015). This suggests a crucial role of Trx and GSNOR in modulating plants' sensitivity to cold stress (Figure 28.2). How denitrosylation contributes to stress tolerance needs to be dissected. It would be a balance between nitrosylation/denitrosylation that decides the up-/downregulation of the targets leading to the continuation of the signaling pathway.

Recent Advancements in NO Research

Putative NO Receptor Hints at NO Perception by N-End Rule Pathway

Recently, Gibbs et al. (2014) discovered group VII ethylene response factors (ERFs) as a putative NO receptor in plants. NO perception occurs via proteolysis of transcriptional regulators by the N-end pathway of proteolytic degradation of a protein, depending on its N-terminal residue. Group VII ERFs have a conserved Met-Cys (MC) at the N-terminus, which is destabilized by the presence of an oxygen molecule, causing cysteine oxidation with subsequent removal of the N-terminus by methionine aminopeptidase (MAP) activity. Both NO and oxygen cause destabilization of group VII ERFs. The exact mechanism of NO perception via the N-end rule pathway is not known; however, it has been proposed that N-terminal cysteine (Nt-Cys) first undergoes S-nitrosylation, followed by oxidation to produce Cys sulfinic and Cys sulfonic acids, followed by arginylation of these residues. The arginine residue is recognized by N-recognins, causing its ubiquitination and leading to proteosomal degradation. Thus, the NO signal seems to be mediated by proteolytic degradation of its receptor.

NO as a Priming Agent

Agriculture suffers from severe yield loss due to different abiotic stress conditions. Therefore, studies in the last few decades have focused on improving the stress tolerance of crop plants (Antoniou et al., 2016). It has been shown that treatment of plants with certain chemical compounds prior to stress conditions was able to increase the stress tolerance of plants (Savvides et al., 2016). NO, along with H_2O_2, hydrogen sulfide (H_2S), melatonin, and polyamines, also plays a role in abiotic stress tolerance in plants (Antoniou et al., 2016; Savvides et al., 2016). These molecules have been used for priming plants. Exposure to low concentrations of these molecules leads to an increase in endogenous levels without any inhibitory effect, along with a faster and stronger response to stress (Savvides et al., 2016). Increased transcript level of antioxidant enzymes, enhanced antioxidant capacity, post-translational modification of proteins, and reduced MDA and electrolytic leakage are the indicators of the primed state of the plant (Antoniou et al., 2016; Savvides et al., 2016). The low cost and low concentration requirement indicate that this is an economical alternative; however, factors such as mode and time of application, as well as environmental effects, need to be studied in detail before its adoption in agriculture to prevent yield loss.

Conclusion and Future Directions

Advancements in NO research have proved that S-nitrosylation is a fundamental post-translational modification in plants, linked with the regulation of numerous targets in multiple physiological processes, including abiotic stress. In response to low-temperature stress, NO production is triggered by both oxidative and reductive pathways, modulating the S-nitrosoproteome of various compartments of a cell. Cold stress–induced S-nitrosylation targets are grouped into the categories of stress and redox related, metabolic, photosynthetic, and cell wall modifying. The cold stress–induced oxidative burst is scavenged by NO-mediated enhancement of antioxidant enzymes (Fe-SOD, Cu/Zn-SOD, DHAR, APX, CAT, and MDHAR) and accumulation of antioxidants (Asc and GSH). Negative regulation of S-nitrosylation in the photosynthetic and glycolytic pathways during cold might reduce the primary productivity of plants, leading to yield loss. S-nitrosylation of glucosinolate hydrolytic enzymes (myrosinase and epithiospecifier proteins) by S-nitrosylation during cold stress suggests the involvement of secondary metabolites in NO-mediated stress tolerance. In addition to the S-nitrosoproteome, transcription factors are also regulated by NO-based PTM in response to low-temperature stress. CBF, a transcription factor in the ICE-CBF-dependent cold stress signaling cascade, is induced transcriptionally and post-translationally. S-nitrosylated CBF binds to the CRT/DRE element, which in turn, enhances the expression of COR genes, leading to cold acclimation in plants. Excessive removal of RNS by positive modulation of denitrosylases (GSNOR and Trx-TrxR) during cold stress suggests a pivotal role of nitrosylation/denitrosylation homeostasis in regulating cold stress tolerance in plants. Recent advancements in the discovery of putative NO receptors and NO priming agents in plants may help in better understanding the NO signaling cascade in physiological response(s). These insights might further help in exploring the complex network of cross-talk between NO and stress tolerance, ultimately leading to the design of improved strategies to combat adverse stress conditions in agricultural fields.

Acknowledgement

YS and PB thank the University grant commission (UGC) for a Junior Research Fellowship (JRF).

REFERENCES

Abat JK, Deswal R. Differential modulation of S-nitrosoproteome of *Brassica juncea* by low temperature: Change in S-nitrosylation of Rubisco is responsible for the inactivation of its carboxylase activity. *Proteomics* 2009a;9(18):4368–4380.

Abat JK, Deswal R. Post translational modifications of proteins by nitric oxide—a new tool for metabolome regulation. In: Hayat S, Mori M, Pichtel J, Ahmad A, editors. *Nitric Oxide in Plant Physiology.* Germany:Wiley-VCH; 2009b, 189–200.

Airaki M, Leterrier M, Mateos RM, Valderrama R, Chaki M, Barroso JB, et al. Metabolism of reactive oxygen species and reactive nitrogen species in pepper (*Capsicum annuum* L.) plants under low temperature stress. *Plant Cell Environ.* 2012;35(2):281–295.

Airaki M, Sánchez-Moreno L, Leterrier M, Barroso JB, Palma JM, Corpas FJ. Detection and quantification of S-nitrosoglutathione (GSNO) in pepper (*Capsicum annuum* L.) plant organs by LC-ES/MS. *Plant Cell Physiol.* 2011;52(11):2006–2015.

Antoniou C, Savvides A, Christou A, Fotopoulos V. Unravelling chemical priming machinery in plants: The role of reactive oxygen–nitrogen–sulfur species in abiotic stress tolerance enhancement. *Curr. Opin. Plant Biol.* 2016;33:101–107.

Barroso JB, Corpas FJ, Carreras A, Rodríguez-Serrano M, Esteban FJ, Fernández-Ocaña A, et al. Localization of S-nitrosoglutathione and expression of S-nitrosoglutathione reductase in pea plants under cadmium stress. *J. Exp. Bot.* 2006;57(8):1785–1793.

Baudouin E. The language of nitric oxide signalling. *Plant Biol.* 2011;13(2):233–242.

Bavita A, Shashi B, Navtej SB. Nitric oxide alleviates oxidative damage induced by high temperature stress in wheat. *Indian J. Exp. Biol.* 2012;50:372–378.

Benhar M, Forrester MT, Hess DT, Stamler JS. Regulated protein denitrosylation by cytosolic and mitochondrial thioredoxins. *Science* 2008;320:1050–1055.

Benhar M, Forrester MT, Stamler JS. Protein denitrosylation: Enzymatic mechanisms and cellular functions. *Nat. Rev. Mol. Cell Biol.* 2009;10:1–12.

Benhar M, Thompson JW, Moseley MA, Stamler JS. Identification of S-nitrosylated targets of thioredoxin using a quantitative proteomics approach. *Biochemistry* 2010;49(32):6963–6969.

Besson-Bard A, Courtois C, Gauthier A, Dahan J, Dobrowolska G, Jeandroz S, et al. Nitric oxide in plants: Production and cross-talk with Ca^{2+} signaling. *Mol. Plant* 2008;1(2):218–228.

Bethke PC. Apoplastic synthesis of nitric oxide by plant tissues. *Plant Cell* 2004;16(2):332–341.

Birkenmeier G, Stegemann C, Hoffmann R, Günther R, Huse K, Birkemeyer C. Posttranslational modification of human glyoxalase 1 indicates redox-dependent regulation. *PLoS One* 2010;5(4):e10399.

Bita CE, Gerats T. Plant tolerance to high temperature in a changing environment: Scientific fundamentals and production of heat stress-tolerant crops. *Front. Plant Sci.* 2013;4:1–18.

Bowler C, Slooten L, Vandenbranden S, De Rycke R, Botterman J, Sybesma C, et al. Manganese superoxide dismutase can reduce cellular damage mediated by oxygen radicals in transgenic plants. *EMBO J.* 1991;10(7):1723–1732.

Broniowska KA, Hogg N. The chemical biology of S-nitrosothiols. *Antioxid. Redox Signal.* 2012;17(7): 969–980

Cantrel C, Vazquez T, Puyaubert J, Rezé N, Lesch M, Kaiser WM, et al. Nitric oxide participates in cold-responsive phosphosphingolipid formation and gene expression in *Arabidopsis thaliana*. *New Phytol.* 2011;189(2):415–427.

Carver J, Doctor A, Gaston B. S-Nitrosothiol formation. *Methods Enzymol.* 2005;396(5):95–105.

Chae HB, Moon JC, Shin MR, Chi YH, Jung YJ, Lee SY, et al. Thioredoxin reductase type C (NTRC) orchestrates enhanced thermotolerance to *Arabidopsis* by its redox-dependent holdase chaperone function. *Mol. Plant* 2013;6(2):323–336.

Chaki M, Valderrama R, Fernández-Ocaña AM, Carreras A, Gómez-Rodríguez MV, Pedrajas JR, et al. Mechanical wounding induces a nitrosative stress by down-regulation of GSNO reductase and an increase in S-nitrosothiols in sunflower (*Helianthus annuus*) seedlings. *J. Exp. Bot.* 2011;62(6):1803–1813..

Cheng T, Chen J, Ef AA, Wang P, Wang G, Hu X, et al. Quantitative proteomics analysis reveals that S-nitrosoglutathione reductase (GSNOR) and nitric oxide signaling enhance poplar defense against chilling stress. *Planta* 2015;242(6):1361–1390.

Choudhury FK, Rivero RM, Blumwald E, Mittler R. Reactive oxygen species, abiotic stress and stress combination. *Plant J.* 2017;90(5):856–867.

Cooney RV, Harwood PJ, Custer LJ, Franke AA. Light-mediated conversion of nitrogen dioxide to nitric oxide by carotenoids. *Environ. Health Perspect.* 1994;102(5):460–462.

Corpas FJ, Barroso JB, Carreras A, Quirós M, León AM, Romero-Puertas MC, et al. Cellular and subcellular localization of endogenous nitric oxide in young and senescent pea plants. *Plant Physiol.* 2004;136(1):2722–2733.

Corpas FJ, Chaki M, Fernández-Ocaña A, Valderrama R, Palma JM, Carreras A, et al. Metabolism of reactive nitrogen species in pea plants under abiotic stress conditions. *Plant Cell Physiol.* 2008;49(11):1711–1722.

Corpas FJ, Hayashi M, Mano S, Nishimura M, Barroso JB. Peroxisomes are required for in vivo nitric oxide accumulation in the cytosol following salinity stress of *Arabidopsis* plants. *Plant Physiol.* 2009;151(4):2083–2094.

Dai C, Wang MH. Isolation and characterization of thioredoxin and NADPH-dependent thioredoxin reductase from tomato (*Solanum lycopersicum*). *BMB Rep.* 2011;44(10):692–697.

Fancy NN, Bahlmann AK, Loake GJ. Nitric oxide function in plant abiotic stress. *Plant Cell Environ.* 2017;40(4):462–472.

Gibbs DJ, Md Isa N, Movahedi M, Lozano-Juste J, Mendiondo GM, Berckhan S, et al. Nitric oxide sensing in plants is mediated by proteolytic control of Group VII ERF transcription factors. *Mol. Cell* 2014;53(3):369–379.

Gill SS, Tuteja N. Reactive oxygen species and antioxidant machinery in abiotic stress tolerance in crop plants. *Plant Physiol. Biochem.* 2010;48(12):909–930.

Godber BLJ, Doel JJ, Sapkota GP, Blake DR, Stevens CR, Eisenthal R, et al. Reduction of nitrite to nitric oxide catalyzed by xanthine oxidoreductase. *J. Biol. Chem.* 2000;275(11):7757–7763.

Gupta KJ, Fernie AR, Kaiser WM, van Dongen JT. On the origins of nitric oxide. *Trends Plant Sci.* 2011;16(3):160–168.

Hara M, Harazaki A, Tabata K. Administration of isothiocyanates enhances heat tolerance in *Arabidopsis thaliana*. *Plant Growth Regul.* 2013;69(1):71–77.

Hara M, Yatsuzuka Y, Tabata K, Kuboi T. Exogenously applied isothiocyanates enhance glutathione S-transferase expression in *Arabidopsis* but act as herbicides at higher concentrations. *J. Plant Physiol.* 2010;167(8):643–649.

Hashemy SI, Holmgren A. Regulation of the catalytic activity and structure of human thioredoxin 1 via oxidation and S-nitrosylation of cysteine residues. *J. Biol. Chem.* 2008;283(32): 21890–21898.

Holtgrefe S, Gohlke J, Starmann J, Druce S, Klocke S, Altmann B, et al. Regulation of plant cytosolic glyceraldehyde 3-phosphate dehydrogenase isoforms by thiol modifications. *Physiol. Plant.* 2008;133(2):211–228.

Hooper AB, Terry KR. Hydroxylamine oxidoreductase of *Nitrosomonas*. Production of nitric oxide from hydroxylamine. *BBA Enzymol.* 1979;571(1):12–20.

Jaspers P, Kangasjärvi J. Reactive oxygen species in abiotic stress signaling. *Physiol. Plant.* 2010;138(4):405–413.

Kashyap P, Sehrawat A, Deswal R. Nitric oxide modulates *Lycopersicon esculentum* C-repeat binding factor 1 (LeCBF1) transcriptionally as well as post-translationally by nitrosylation. *Plant Physiol. Biochem.* 2015;96:115–123.

Khokon MAR, Jahan MS, Rahman T, Hossain MA, Muroyama D, Minami I, et al. Allyl isothiocyanate (AITC) induces stomatal closure in *Arabidopsis*. *Plant Cell Environ.* 2011;34(11): 1900–1906.

Kneeshaw S, Gelineau S, Tada Y, Loake GJ, Spoel SH. Selective protein denitrosylation activity of thioredoxin-h5 modulates plant immunity. *Mol. Cell* 2014;56(1):153–162.

Korithoski B, Lévesque CM, Cvitkovitch DG. Involvement of the detoxifying enzyme lactoylglutathione lyase in *Streptococcus mutans* aciduricity. *J. Bacteriol.* 2007;189(21):7586–7592.

Kovacs I, Lindermayr C. Nitric oxide-based protein modification: Formation and site-specificity of protein S-nitrosylation. *Front. Plant Sci.* 2013;4:1–10.

Lee U, Wie C, Fernandez BO, Feelisch M, Vierling E. Modulation of nitrosative stress by S-nitrosoglutathione reductase is critical for thermotolerance and plant growth in *Arabidopsis*. *Plant Cell* 2008;20(3):786–802.

Lima B, Lam GKW, Xie L, Diesen DL, Villamizar N, Nienaber J, et al. Endogenous S-nitrosothiols protect against myocardial injury. *PNAS* 2009;106(15):6297–6302.

Lindermayr C, Saalbach G, Durner J. Proteomic identification of S-nitrosylated proteins. *Plant Physiol.* 2005;137:921–930.

Liu Y, Jiang H, Zhao Z, An L. Nitric oxide synthase like activity-dependent nitric oxide production protects against chilling-induced oxidative damage in *Chorispora bungeana* suspension cultured cells. *Plant Physiol. Biochem.* 2010;48(12):936–944.

Martínez-Ballesta MDC, Moreno DA, Carvajal M. The physiological importance of glucosinolates on plant response to abiotic stress in Brassica. *Int. J. Mol. Sci.* 2013; 14(6):11607–11625.

Meister A. Glutathione metabolism and its selective modification. *J. Biol. Chem.* 1988;263(33):17205–17208.

Mitsumoto A, Kim KR, Oshima G, Kunimoto M, Okawa K, Iwamatsu A, et al. Glyoxalase I is a novel nitric-oxide-responsive protein. *Biochem. J.* 1999;344:837–844.

Mitsumoto A, Kim KR, Oshima G, Kunimoto M, Okawa K, Iwamatsu A, et al. Nitric oxide inactivates glyoxalase I in cooperation with glutathione. *J. Biochem.* 2000;128(4): 647–654.

Moon JC, Lee S, Shin SY, Chae HB, Jung YJ, Jung HS, et al. Overexpression of *Arabidopsis* NADPH-dependent thioredoxin reductase C (AtNTRC) confers freezing and cold shock tolerance to plants. *Biochem. Biophys. Res. Commun.* 2015;463(4):1225–1229.

Moroney JV, Bartlett SG, Samuelsson G. Carbonic anhydrases in plants and algae: Invited review. *Plant Cell Environ.* 2001;24(2):141–153.

Neill S, Bright J, Desikan R, Hancock J, Harrison J, Wilson I. Nitric oxide evolution and perception. *J. Exp. Bot.* 2008;59(1):25–35.

Neill SJ, Desikan R, Clarke A, Hurst RD, Hancock JT, Lane C, et al. Hydrogen peroxide and nitric oxide as signalling molecules in plants. *J. Exp. Bot.* 2002;53(372):1237–1247.

Neill SJ, Desikan R, Hancock JT. Nitric oxide signalling in plants. *Bot. Rev.* 2003;159:11–35.

Noctor G, Foyer CH. Ascorbate and glutathione: Keeping active oxygen under control. *Annu. Rev. Plant Biol.* 1998;49(1): 249–279.

Puyaubert J, Fares A, Rézé N, Peltier J, Baudouin E. Identification of endogenously S-nitrosylated proteins in *Arabidopsis* plantlets : Effect of cold stress on cysteine nitrosylation level. *Plant Sci.* 2014;215–216:150–156.

Qiao W, Li C, Fan LM. Cross-talk between nitric oxide and hydrogen peroxide in plant responses to abiotic stresses. *Environ. Exp. Bot.* 2014;100:84–93.

Radi R. Nitric oxide, oxidants, and protein tyrosine nitration. *Proc. Natl. Acad. Sci. USA* 2004;101(12):4003–4008.

Reichheld JP, Meyer E, Khafif M, Bonnard G, Meyer Y. AtNTRB is the major mitochondrial thioredoxin reductase in *Arabidopsis thaliana*. *FEBS Lett.* 2005;579(2):337–342.

Romero-Puertas MC, Rodríguez-Serrano M, Sandalio LM. Protein S-nitrosylation in plants under abiotic stress: An overview. *Front. Plant Sci.* 2013;4:1–6.

Sakamoto A, Ueda M, Morikawa H. *Arabidopsis* glutathione-dependent formaldehyde dehydrogenase is an S-nitrosoglutathione reductase. *FEBS Lett.* 2002;515(1–3):20–24.

Salgado I, Carmen Martínez M, Oliveira HC, Frungillo L. Nitric oxide signaling and homeostasis in plants: A focus on nitrate reductase and S-nitrosoglutathione reductase in stress-related responses. *Rev. Bras. Bot.* 2013;36(2):89–98.

Savvides A, Ali S, Tester M, Fotopoulos V. Chemical priming of plants against multiple abiotic stresses: Mission possible? *Trends Plant Sci.* 2016;21(4):329–340.

Sehrawat A, Abat JK, Deswal R. RuBisCO depletion improved proteome coverage of cold responsive S-nitrosylated targets in *Brassica juncea. Front. Plant Sci.* 2013;4:1–14.

Sehrawat A, Deswal R. S-nitrosylation analysis in *Brassica juncea* apoplast highlights the importance of nitric oxide in cold-stress signaling. *J. Proteome Res.* 2014;13(5):2599–2619.

Sengupta R, Holmgren A. The role of thioredoxin in the regulation of cellular processes by S-nitrosylation. *Biochim. Biophys. Acta* 2012;1820(6):689–700.

Serrato AJ, Pérez-Ruiz JM, Spínola MC, Cejudo FJ. A novel NADPH thioredoxin reductase, localised in the chloroplast, which deficiency causes hypersensitivity to abiotic stress in *Arabidopsis thaliana. J. Biol. Chem.* 2004;279(42):43821–43827.

Sharma S, Sehrawat A, Deswal R. Asada-Halliwell pathway maintains redox status in *Dioscorea alata* tuber which helps in germination. *Plant Sci.* 2016;250:20–29.

Stöhr C, Strube F, Marx G, Ullrich WR, Rockel P. A plasma membrane-bound enzyme of tobacco roots catalyses the formation of nitric oxide from nitrite. *Planta* 2001;212(5):835–841.

Stoimenova M, Igamberdiev AU, Gupta KJ, Hill RD. Nitrite-driven anaerobic ATP synthesis in barley and rice root mitochondria. *Planta* 2007;226(2):465–474.

Tanou G, Filippou P, Belghazi M, Job D, Diamantidis G, Fotopoulos V, et al. Oxidative and nitrosative-based signaling and associated post-translational modifications orchestrate the acclimation of citrus plants to salinity stress. *Plant J.* 2012;72(4): 585–599.

Thalineau E, Truong H-N, Berger A, Fournier C, Boscari A, Wendehenne D, et al. Cross-regulation between N metabolism and nitric oxide (NO) signaling during plant immunity. *Front. Plant Sci.* 2016;7:1–14.

Tielens AGM, Rotte C, Van Hellemond JJ, Martin W. Mitochondria as we don't know them. *Trends Biochem. Sci.* 2002;27(11):564–572.

Tun NN, Santa-Catarina C, Begum T, Silveira V, Handro W, Segal Floh EI, et al. Polyamines induce rapid biosynthesis of nitric oxide (NO) in *Arabidopsis thaliana* seedlings. *Plant Cell Physiol.* 2006;47(3):346–354.

Vetrovsky P, Stoclet JC, Entlicher G. Possible mechanism of nitric oxide production from N(G)-hydroxy-L-arginine or hydroxylamine by superoxide ion. *Int. J. Biochem. Cell Biol.* 1996;28(12):1311–13418.

Wang YQ, Feechan A, Yun BW, Shafiei R, Hofmann A, Taylor P, et al. S-nitrosylation of AtSABP3 antagonizes the expression of plant immunity. *J. Biol. Chem.* 2009;284(4): 2131–2137.

Wilson ID, Neill SJ, Hancock JT. Nitric oxide synthesis and signalling in plants. *Plant Cell Environ.* 2008;31(5):622–631.

Ya'acov YL, Wills RB, Ku VV. Evidence for the function of the free radical gas—nitric oxide (NO·)—as an endogenous maturation and senescence regulating factor in higher plants. *Plant Phys. Biochem.* 1998;36(11):825–833.

Zaffagnini M, De Mia M, Morisse S, Di Giacinto N, Marchand CH, Maes A, et al. Protein S-nitrosylation in photosynthetic organisms: A comprehensive overview with future perspectives. *Biochim. Biophys. Acta* 2016;1864(8):952–966.

Zhao M-G, Chen L, Zhang L-L, Zhang W-H. Nitric reductase-dependent nitric oxide production is involved in cold acclimation and freezing tolerance in *Arabidopsis. Plant Physiol.* 2009;151(2):755–767.

Zhao R, Sheng J, Lv S, Zheng Y, Zhang J, Yu M, et al. Nitric oxide participates in the regulation of LeCBF1 gene expression and improves cold tolerance in harvested tomato fruit. *Postharvest Biol. Technol.* 2011;62(2):121–126.

Ziogas V, Tanou G, Filippou P, Diamantidis G, Vasilakakis M, Fotopoulos V, et al. Nitrosative responses in citrus plants exposed to six abiotic stress conditions. *Plant Physiol. Biochem.* 2013;68:118–126.

29

Physiological Roles of Brassinosteroids in Conferring Temperature and Salt Stress Tolerance in Plants

Geetika Sirhindi, Renu Bhardwaj, Manish Kumar, Sandeep Kumar,
Neha Dogra, Harpreet Sekhon, Shruti Kaushik, and Isha Madaan

CONTENTS

Introduction

Brassinosteroids (BRs) belong to the sixth class of Plant Growth Regulators (PGRs), a type of plant steroid hormone that regulates growth and development. Structurally BRs show similarity to cholesterol-derived animal steroid hormones and insect ecdysteroids. They are involved in an array of plant activities such as cell expansion, vascular differentiation, pollen tube formation, and other significant growth-promoting activities, along with biotic and abiotic stress tolerance during the life of the plant. Pollen extracts were tested from around 60 different plant species (Mandava and Mitchell, 1971), and the unexpected response was a combination of elongation with swelling and curvature of the treated internode was produced by pollens of rape (*Brassica napus* L.) and alder tree (*Alnus glutinosa* L.). Later it was discerned that rape pollens contained a new group of lipidic plant hormones, which they called *brassins* after the name of the plant from which they were extracted (Mitchell et al., 1970). *Brassins* occurred in fractions as an active constituent mainly having glucosyl esters of fatty acids (Mandava et al., 1973). Further work revealed that although elongation was promoted by these esters, they were not able to reproduce all of the responses as any of the established PGRs have (Grove et al., 1979). To date, more than 70 types of BRs from different plant species have been identified and characterized as having a pleiotropic effect on plant growth and development.

Intensive research conducted on BRs revealed that they elicit a broad spectrum of physiological and morphological responses in plants, including stem elongation, leaf bending, epinasty, induction of ethylene biosynthesis, proton pump activation and synthesis of nucleic acids, proteins, regulation of carbohydrate assimilation, allocation, and regulation of photosynthesis (Zhu et al., 2017). Brassinosteroids also respond to various biotic and abiotic stresses (Clouse and Sasse, 1998; Li and Chory, 1999;

Sreeramulu et al., 2013; Sharma et al., 2015; Sirhindi et al., 2016). Anti-stress properties of different active forms of BRs have been suggested by various studies as salt stress (Ali et al., 2008; Kaur and Patti, 2017), cold stress (Hu et al., 2008; Kumar et al., 2010, 2012), heat stress (Ogweno et al., 2008; Kumar et al., 2015), and heavy metal stress (Rady, 2011). Because of their different types of roles, extensive research has been conducted to promote BRs as an essential plant growth regulator for modern agriculture (Ikekawa and Zhao, 1991; Divi and Krishna, 2009). For various biological functions in plants, the maintenance and regulation of the endogenous level of BRs is crucial (Tanaka et al., 2005). BR biosynthesis, transport, and degradation are critical components of BR homeostasis and for maintaining the endogenous level of BR in plants. It has been observed that extreme dwarfism, altered leaf morphology, abnormal vascular development, delayed flowering and senescence, and reduced male fertility have been exhibited by BR-deficient mutants (Clouse, 2011).

Biosynthesis and Metabolism of BRs

BRs are unlike any other phytohormones in the respect that they are not subject to active transport within the plant and, thus, their levels inside the plant are determined by the balance between local biosynthetic and interactive reactions (Hategan et al., 2011). BRs can be classified as C_{27}, C_{28}, or C_{29} according to the number of carbons, the orientation of the side chain, and the present active chemical group (Vardhini, 2013; Vardhini et al. 2013). Although more than 70 different compounds related to BRs have been identified from different plant species (Haubrick and Assmann, 2006), brassinolide (BL), 28-homobrassinolide (28-HomoBL), and 24-epibrassinolide (24-EpiBL) are the only three bioactive BRs which are widely used in most physiological and experimental studies, as reported by Vardhini et al. (2006, 2015) (Figure 29.1).

BRs biosynthesis occurs from intricate network pathways and is mostly modulated by the transcriptional regulation of BRs biosynthetic genes (Chung and Choe, 2013; Vriet et al., 2013). Campesterol is a precursor for the synthesis of the most active form of BRs i.e. brassinolides (BL). Firstly, campesterol is converted to campestenal which was initially believed to branch into two parallel pathways, as early and late C-6 oxidation pathways which involved a chain of reductions, hydroxylations, epimerizations, and oxidations which eventually converge to castesterone (Figure 29.2). This castesterone then leads to the formation of BLs (Fujioka et al., 1998; Marco et al., 2002). Later it was revealed that the BRs biosynthetic pathway is a triterpenoid pathway (Choe, 2007; Chung and Choe, 2013). Mevalonic acid serves as a precursor of the triterpenoid pathway. Where it is condensed and transformed to 2, 3–oxidosqualene which then further undergoes modification to form major plant sterols like sitosterols and campesterols. Campesterol can be modified by two different enzymes: a C-22 hydroxylase *dwarf4/CYP90B1 (DWF4)* and a C-3 hydrogenase constitutive photomorphogenesis *dwarf/ CYP90A1 (CPD)*, depending upon the availability of substrate and enzymes. Because *DWF4* can act in multiple biosynthetic intermediates, including campesterol and campestenol, the pathway branches to a third pathway, the early C-22 hydroxylation pathway (Choe et al., 2001; Fujioka and Yokota, 2003).

CPD forms several other branches in which a C-23 hydroxylase metabolizes campesterol, and other intermediates can be shown by LC-MS and in various genetic studies (Ohnishi et al., 2012). In the C-6 oxidation pathway, the intermediates formed in the above-mentioned reactions are further modified and later merge into a certain degree of crosstalk between the parallel pathways manifesting complex networking of BRs biosynthesis. Later, several key genes involved in BRs biosynthesis such as a *5α-reductase* known as *de-etiolated-2 (DET2)*, *CPD*, a C-3 oxidase, *DWF4*, a C-22 hydroxylase, *rotundifolia3/CYP90C1 (ROT3)*, and *CYP90D1*,

C-23 hydroxylases have been characterized to advance the perspective of BRs biosynthesis (Wang et al., 2012a; Vriet et al., 2013). Interestingly, BRs biosynthesis culminates at castesterone in rice because monocots like rice and maize lack the enzyme (*CYP85A2*) which is responsible for C-6 oxidation reaction (Kim et al., 2008).

In contrast to other hormones, BRs are not transported long distances but are used in proximity with synthesizing cells. However, intracellular transport occurs either passively or actively, from their site of synthesis in ER (Endoplasmic Reticulum) to the plasma membrane where its perception occurs (Symons et al., 2008). Because of their crosstalk with other hormones like auxins, BRs are able to exert a long-distance effect (Symons et al., 2008; Vriet et al., 2013). Some carrier mechanisms are suggested to be mediated for the short distance transport of BRs, like BR conjugates formed by the binding of BR to fatty acids or glucose or short distance transport can also be mediated through specific proteinaceous transporters (Fujioka and Yokota, 2003; Symons et al., 2008). The spatial and temporal regulation of BRs homeostasis at the tissue or at the cellular level is extremely crucial for normal growth and development when there is the absence of a mode for long-distance transport of BRs (Symons et al., 2008).

BRs biosynthesis undergoes a two-way regulation mechanism: first, at the endogenous level of BRs by modulating the expression of biosynthetic genes and secondly by inactivating bioactive BRs. Most of the genes, such as *DET2*, *DWF4*, *CPD*, *BR6ox1*, and *ROT3*, are feedback regulators of BR-specific biosynthesis (Sun et al., 2010; Yu et al., 2011). Moreover, the BR signaling mutant *Brassinosteroid Insensitive 1 (bri1)* shows considerable accumulation of endogenous BR as its feedback regulation requires an intact BR perception and signaling pathway (Sun et al., 2010). At the level of transcription, regulation is mediated by two major BR signaling transcription factors *brassinazole-resistant 1 (BZR1)* and *BRI1-EMS-suppressor1 (BZR2/BES1)* as well as by several other novel transcription factors (*CESTA, RAVL1, TCP1*). These have been lately identified to regulate the expression of key BR biosynthetic genes such as *CPD* and *DWF4* (Je et al., 2010; Sun

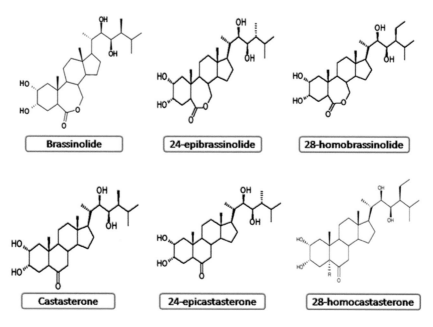

FIGURE 29.1 Structure of commonly used bioactive BRs (From Vardhini and Anjum, 2015).

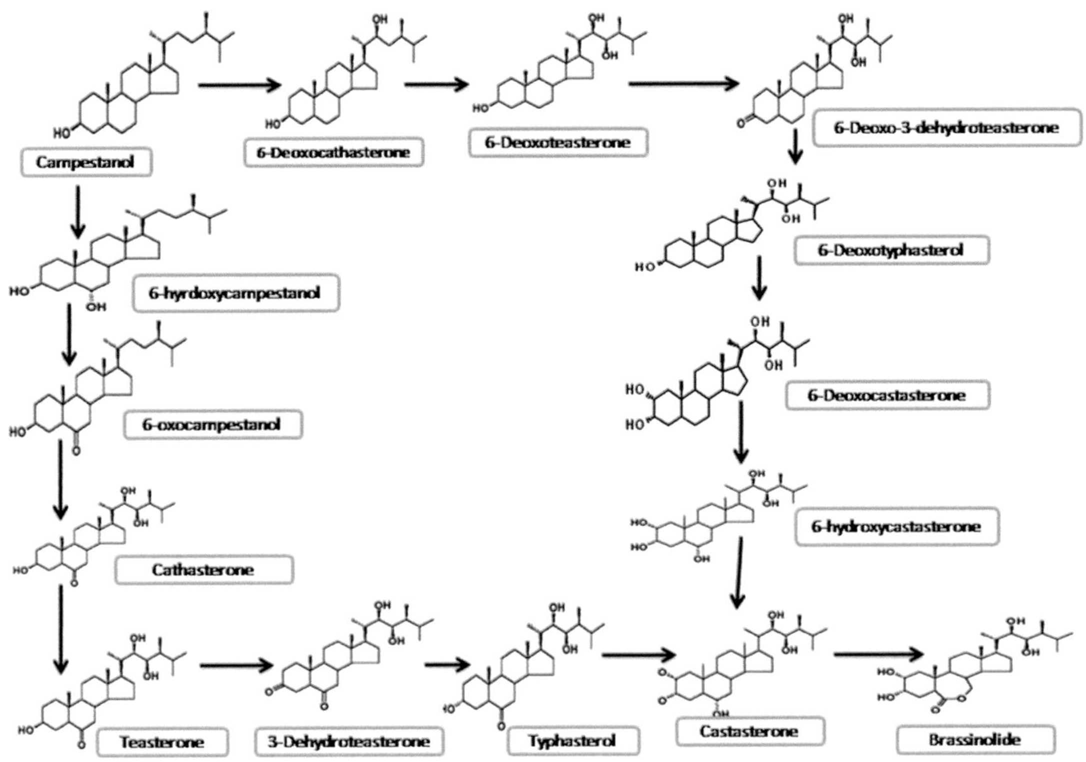

FIGURE 29.2 Biosynthesis of Brassinolide from campestanol (Marco et al., 2002).

et al., 2010; Poppenberger et al., 2011; Guo et al., 2013). In the positive regulation of BR signaling, some *bHLH*, like transcription factors such as *CESTA* and *TCP* are involved (Poppenberger et al., 2011; Wang et al., 2012; Guo et al., 2013).

In maintaining the optimum levels of bioactive BRs in the cell, BRs catabolism/metabolism involves various processes like acylation, sulphonation, glycosylation, etc., which play a crucial role. BR novel genes *brassinosteroid inactivator1 (BIA1)* and *abnormal shoot-1 (abs-1)*, belonging to the family of *BAHD acyltransferases*, were found to be involved (Roh et al., 2012; Wang et al., 2012) in BR biosynthesis and activity and to maintain the optimum level of bioactive BRs. Inactivation of these novel genes is done by acylation. The inactivation of another *BAHD acyltransferases pizza (PIZ)* is also done by acylation that plays a redundant role with *BIA1* (Schneider et al., 2012). Similarly, differential specificities to castesterone are possessed by *bri1-5 enhanced 1 (BEN1)* and *Brassica napus sulfotransferase 3 (BNST3)* and by various mechanisms involved in the reduction and sulfonation of inactive BL and in the active BR (Marsolais et al., 2007; Yuan et al., 2007). The 23 *O*-glycosylation of CS and BL as part of the inactivation process are catalyzed by a set of *glycotransferases* enzymes *UGT73C6* and its close homolog *UGT73C5* (Poppenberger et al., 2005). Recent studies have also shown that BR biosynthesis can be regulated by external stimuli like salt and temperature stress (Maharjan and Choe, 2011; Sharma et al., 2013).

Role of BRs under Temperature and Salt Stress

Plants can cope with the high-temperature stress through physiological, morpho-anatomical, and biochemical alterations

(Figure 29.3). Under high-temperature stress, plants accumulate the compatible osmolytes which help to increase the retention of water in plants for better stomatal regulation and increased photosynthetic rate. Plants also exhibit some morpho-anatomical alterations to cope with high-temperature stress which includes a reduction in cell size, closure of stomata, increased stomatal and trichomes density, and greater xylem vessels (Hasanuzzaman et al., 2013). The third mechanism to cope with high temperature is biochemical alteration. The plants increase the stress-related proteins which enhance the activities of antioxidants like superoxide dismutase (SOD), catalase (CAT), and peroxidase (POD) in the plant cells. These antioxidants scavenge the reactive oxygen species (ROS), reduce the photo-oxidation of the plasma membrane, and maintain the integrity of the chloroplast membrane, thus increasing the photosynthetic rate (Tables 29.1 and 29.2).

High-temperature stress-induced changes in water relation (Cabanero et al., 2004; Morales et al., 2003; Simoes-Araujo et al., 2003) and accumulation of compatible osmolytes (Hare et al., 1998; Sakamoto and Murata, 2002). High-temperature stress can also cause decrease in photosynthesis (Sharkova, 2001; Wise et al., 2004) and hormonal changes (Maestri et al., 2002). High-temperature stress ($\leq 40°C$) can cause scorching of leaves and twigs, sunburns on leaves, branches and stems, leaf senescence and abscission, shoot and root growth inhibition, fruit discoloration, damage, and reduced yield in plants (Vollenweider and Gunthardt-Goerg, 2005). In barley and radish, super-optimal temperatures can cause various damages like inhibition of seed germination and seedling growth (Cavusoglu and Kabar, 2007), retardation of flowering (Nakano et al., 2013), failure of fertilization, reduction of fruit growth, and sugar accumulation (Lu et al., 2010). Patel and Franklin (2009) reported that high-temperature

FIGURE 29.3   Effect of high temperature stress on plants (Hasanuzzaman et al., 2013).

TABLE 29.1

Effect of BRs on Various Parameters Under the Influence of Different Temperature Stresses

Plant Species	BRs analogous	Mode of Application	Stress type	Parameters studied	Reference
Vigna radiata	10^{-6}, 10^{-8}, 10^{-10} M of 28-HBL	Seedling treatment	Drought stress	Proline, carbohydrate, nitrogen, phosphorus and potassium content	Nasser and Mohammed (2014)
Brassica juncea	HBL, 0.01 μM	Foliar spray	Drought stress	Photosynthesis and antioxidant systems	Fariduddin et al. (2009)
Brassica juncea	28-HBL, 10^{-8} M	Seedling treatment	High temperature stress (40°C)	Antioxidant systems and PPO	Sirhindi et al. (2009)
Cucumis sativus	28-HBL, 10^{-8}, 10^{-6} M	Foliar spray	Chilling temperature (10/8°C, 5/3°C)	Chlorophyll content, Photosynthetic rate, Efficiency of PSII and Antioxidant systems	Fariduddin et al. (2011)
Brassica juncea	28-HBL (10^{-9} M)	Seed treatment	Temperature stress (4°C and 44°C)	Growth, stress markers & antioxidant enzymes	Sirhindi et al. (2014)
Brassica juncea	28-homoBL (10^{-6}, 10^{-9}, 10^{-12} M)	Seed treatment	Extreme temperature stress (4 and 44°C)	Growth and photosynthesis	Kaur et al. (2014)
Vigna radiata	28-HBL, 10^{-6}, 10^{-8}, 10^{-10} M	Seed treatment	High temperature	Antioxidant enzymes	Kumar et al. (2014)
Brassica juncea	28-HBL, 0.01 μM	Seed treatment	High temperature (30°C or 40°C)	Growth, chlorophyll, photosynthesis, photosystem II	Fariduddin et al. (2014)
Brassica juncea	24-EBL, 10^{-8} M	Seedling treatment	Chilling stress (4°C)	Antioxidant systems	Kumar et al. (2010)
Brassica napus	24-EBL, 0.05, 1.00 μM	Cotyledns treatment	Cold stress	Membrane permeability	Janeczko et al. (2007)
Chorispora bungeana	24-EBL, 0.05 mgL⁻¹	Cultured cells	Chilling stress	Reactive oxygen species levels and lipid peroxidation, antioxidant enzymes	Liu et al. (2009)
Cucumis sativus	24-EBL, 0.1 μM	Seedling treatment	Chilling stress	Electron transport rate, NADPH oxidase, H_2O_2 levels	Xia et al. (2009)
Robinia pseudoacacia	BL, 0–0.4 mg L⁻¹	Soaking treatment	Water stress	Activities of POD, SOD and CAT along with decreased rate of transpiration, Stomatal conductance and MDA	Li et al. (2008)
Lycopersicon esculentum	1 μM of EBR	Spraying treatment	Drought stress	Relative water content, stomatal conductance, photosynthetic rate, IntercellularCO₂ concentration, lipid peroxidation, antioxidant enzymes	Yuan et al. (2010)

TABLE 29.2

Effect of BRs on Various Parameters Under the Influence of Different Stresses

Plant Species	BRs Analogous	Mode of Application	Salt Stress	Parameters Studied	Reference
Brassica juncea	28-HBL, 10^{-10}, 10^{-8} and 10^{-6} M	Seedling treatment	Salt stress	Chlorophyll content, nitrate reductase, carbonic anhydrase activity, MDA content and proline content	Hayat et al. (2007)
Brassica juncea	28-HBL, 10^{-10}, 10^{-8} and 10^{-6} M	Pre-imbibitions	Salt stress	Photosynthetic rate, carbonic anhydrase, nitrate reductase and plant yield	Alyemeni et al. (2013)
Zea mays	28-HBL, 10^{-8}, 10^{-6} and 10^{-4} M	Seed soaking treatment	Salt stress	Antioxidant systems, protein and MDA content	Arora et al. (2008)
Brassica juncea	28-HBL, 10^{-10} or 10^{-8} M	Seed soaking	Salt stress	Carbonic anhydrase, nitrate reductase, nodule number and nodule fresh, dry weight	Hayat et al. (2007)
Brassia juncea	28-HB, 10^{-12}, 10^{-9} and 10^{-6} M	Seed treatment	Salt stress	Plant grwth, carbohydrates, total soluble sugars and glycine betaine	Kaur et al. (2015)
Brassica juncea	28-HBL, 10^{-10}, 10^{-8} and 10^{-6} M	Seed priming	Salt stress	Chlorophyll content, carbonic anhydrase, chlorophyll content and seed yield	Hayat et al. (2007)
Pusa Basmati	28-HBL, 10^{-7} M	Seed treatment	Salt stress	Antioxidant enzyme activities, and their gene expression analysis	Sharma et al. (2015)
Hordeum vulgare	HBR 0.5 and 1 μM	Seed treatment	Salt stress	Root germination, cell division, root length and shoot length and antioxidant systems	Marakali et al. (2014)
Cicer arietinum	28-HBL, 10^{-10} and 10^{-8} M	Pre-sowing soaking	Salt stress	Nitrate reductase, carbonic anhydrase activities	Ali et al. (2007)
Brassica juncea	28-HBL, 10^{-10}, 10^{-8} and 10^{-6} M	Seedling treatment	Salt stress	Nitrate reductase carbonic anhydrase activities, chlorophyll content, photosynthesis	Hayat et al. (2007)
Solanum melongena	24-EpiBL, 50, 100, 200 and 400 nM	Seed treatment	Salt stress	Plant height, shoot, root fresh mass, gas exchange parameters and chlorophyll pigments and chlorophyll fluorescence parameters	Wu et al. (2012)
Solanum tuberosum	0, 0.01, 0.1, 1, 10, and 100 g/L BL	In vitro treatment	Salt stress	Root elongation, lateral root development, relative biomass, root shoot ratio, Na+ and K+ ions estimation	Yueqing et al. (2016)
Cucumis sativus	Brz a μM, EpiBL, 0.1 μM	Seed treatment and spraying	Salt stress	Chlorophyll fluorescence, istochemical staining of $O_2{}^{\cdot-}$ and H_2O_2, MDA content, Antioxidants enzymes	Xia et al. (2009)

stress in developing shoots caused a severe reduction in the first internode length, resulting in the premature death of plants. Similarly, sugarcane plants grown under high temperature exhibited smaller internodes, increased tillering, early senescence, and reduced total biomass (Cardozo and Sentelhas, 2013). In rice, anthesis and fertilization and to some extent microsporogenesis (booting) are the most susceptible stages to high-temperature stress (Farrell et al., 2006; Satake and Yoshida, 1978). High-temperature stress-induced spikelet sterility was linked to decreased anther dehiscence, poor shedding of pollen, poor germination of pollen grains on the stigma, and decreased elongation of pollen tubes in rice (Prasad et al., 2006).

BRs alleviated heat-induced inhibition of photosynthesis, increased carboxylation efficiency, and enhanced antioxidant systems in *Lycopersicum esculentum* (Nogues, 2008; Ogweno et al., 2008). Mazzora et al. (2001) reported that BRs ameliorate the negative impact of different temperatures and enhanced physiological, molecular, and antioxidant enzyme activities in tomato. Yadava et al. (2016) avowed that, in *Zea mays* plants, exogenous application of 1 μM epibrassinolide (epiBL) significantly altered the activities of SOD, CAT, and APOX. It is possible that the induction of thermotolerance by the external application of 24-epiBL may be directly related to the role of 24-EpiBL in modulating the antioxidant enzyme activities. BR protects plants from their germination until maturity by

upregulation and downregulation of various non-enzymatic and enzymatic activities at the cellular level (Sirhindi et al., 2009).

Accumulation of ROS as a result of high-temperature stress is a major cause of loss of crop productivity worldwide (Mittler, 2002; Apel and Hirt, 2004; Mahajan and Tuteja, 2005; Tuteja, 2007, 2010; Khan and Singh, 2008). In wheat (*Triticum aestivum* L.), high-temperature stress during reproductive development is a primary constraint to production. Among the various methods to induce high-temperature stress tolerance in plants, foliar application, or pre-sowing seed treatment with low concentrations of inorganic salts, osmoprotectants, signaling molecules (e.g. growth hormones), and oxidants (e.g., H_2O_2), as well as the preconditioning of plants are common approaches (Wahid et al., 2007). Turf grass leaves manifested higher thermostability, lower lipid peroxidation product malondialdehyde (MDA), and lower damage to chloroplast upon exposure to high-temperature stress in heat-acclimated as compared to non-acclimated plants (Xu et al., 2006). Pre-sowing hardening of the seed at a high temperature (42°C) resulted in tolerance to overheating and dehydration, higher levels of water-soluble proteins, and lower amounts of amide-N in leaves compared to non-hardened plants in pearl millet (Tikhomirov et al., 1987).

Kolupaev et al. (2005) reported that exogenous application of Ca^{2+} promoted heat tolerance in plants by stimulating the activities of guaiacol peroxidase, SOD, and catalase, which could be the reason for the induction of heat tolerance. Glycine betaine and polyamines are low-molecular-weight organic compounds that have been successfully applied to induce heat tolerance in various plant species. Wahid and Shabbir (2005) reported that barley seeds pre-treated with glycine betaine led to plants with lower membrane damage, better photosynthetic rate, improved leaf water potential, and greater shoot dry mass, compared to untreated seeds. Under heat stress, Ca^{2+} was required for the maintenance of antioxidant activity and not for osmotic adjustment in some cool-season grasses (Jiang and Haung, 2001).

Cold temperature stress (0°C–10°C) has a broad spectrum of effects on cellular components and metabolic processes of plants. Cold temperature imposes stress of variable severity depending on the intensity and duration of the stress (Figure 29.4). Several studies indicate that the membrane systems of the cell are the primary site of freezing injury in plants (Levitt, 1980) and freeze-induced membrane damage results primarily from the severe dehydration associated with freezing (Steponkus et al., 1993). When temperatures drop below 0°C, ice formation is generally initiated in the intercellular spaces whereas the extracellular fluid has a higher freezing point (lower solute concentration) than the intracellular fluid (Jan et al., 2009). Multiple forms of membrane damage can occur as a consequence of freeze-induced cellular dehydration including expansion-induced lysis, lamellar to hexagonal II phase transitions, and fracture jump lesions (Steponkus et al., 1993). It is well established that freeze-induced production of reactive oxygen species contributes to membrane damage and that intercellular ice can form adhesions with the cell wall and cell membranes and cause cell ruptures (Baek and Skinner, 2012). There is evidence that protein denaturation occurs in plants at low temperatures (Sanghera, 2011), which could result in potential damage.

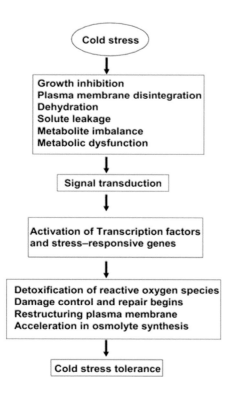

FIGURE 29.4 Effect of cold temperature stress on plants (Yadav, 2010).

To protect themselves from the lethal effects of low-temperature stress, plants adopt a strategy which mainly includes the balance of water potential required for normal metabolic machinery function in maintaining the temperature of the cell along with other factors. BRs are also reported to play an important role in protecting plants from temperature stress (Yadav, 2010). Fariduddin et al. (2011), while studying the deleterious effects of various stresses with or without 28-homoBL in *Cucumis sativus*, also provided evidence that exogenous application of 28-homoBL improved chilling tolerance in *C. sativus* L. Janeczko et al. (2011) reported a similar type of tolerant potential of 28-homoBL in cold-treated rape leaves. Xia et al. (2009) also showed the ameliorative potential of BRs in *C. sativus* towards chilling stress, but it was 0.1 µM 24-epiBL which showed this potential. BRs enhanced the overall growth of the plant after direct sowing of paddy in submerged pots at low temperature (Bajguz and Hayat, 2008). Kaur et al. (2014) avowed that 28-homoBL showed great potential in protecting the PSII from oxidative damage caused by 4°C in *B. juncea* L. They further admitted that 28-homoBL modulates the carbon skeleton of *B. juncea* exposed to 4°C temperature and thus protect the plants from the harmful effects of low temperature. Chilling stress can be ameliorated by the exogenous application of different concentrations of 24-epiBL and 0.05 mg/L^{-1} concentration was found to be the most effective in cultured cells of *Chorispora bungeana* (Liu et al., 2009). 24-epiBL performed this management of chilling tolerance in *C. bungeana* by tight regulation of reactive oxygen species and lipid peroxidation through alleviation enzyme activities. Kagale et al. (2007) also reported that BRs treatment enhanced tolerance to drought and cold stress in *A. thaliana* and *B. napus* seedlings.

Brassinosteroids can regulate cellulose biosynthesis by controlling the expression of cellulose synthase (CESA) genes (Liqiong et al., 2011) that are considered to be the principal regulator in increasing carbohydrate and sugar accumulation subjected to temperature conditions. Exogenous application of 28-homoBL lowered the toxicity that occurred due to temperature stress, both 4°C and 44°C, in *B. juncea* seedlings by ameliorating the proline content and the activities of various antioxidant enzymes along with protecting the total proline content of the plants grown under stress. 28-homoBL also cut down the MDA level in *B. juncea* plants which was otherwise shotout where exposure of temperature stress was given to plants. On the perspective of chilling/freezing temperature in rice, 24-epiBL treatment reduced electrolyte leakage during chilling at 1–5°C and showed an increase in the level of ATP and proline and enhanced resistance which was attributed to BR-induced effects on membrane stability and osmoregulation (Wang and Zeng, 1993). Kumar et al. (2014) studied the role of 28-homobrassinolide in the induction of growth and antioxidants such as SOD, CAT, POD, APOX, DHAR, and MDHAR of *Brassica juncea* cv. *RLM-619* under temperature stress. BR also accelerated the recovery of photosynthetic apparatus from cold stress by balancing the electron partitioning carboxylation and homeostasis in cucumbers (Jiang et al., 2013). Wang et al. (2012) also reported that BL mitigated the negative effect of chilling injury in green bell peppers during storage, thus efficiently increasing their shelf life.

Plant chilling and heat injury inhibit growth by affecting some essential metabolic enzymes from photosynthesis and respiration. BRs application induces synthesis of heat-shock proteins, antioxidants, and the expression of cold-related genes. The study on the effect of 24-EBL on *Brassica napus* and tomato seedlings under thermal stress showed that 24-EBL treatment significantly increased thermotolerance capacity and the accumulation of specific heat-shock proteins during stress in EBL-treated plants (Dhaubhadel et al., 2002). Ogweno et al. (2008) found that when exogenously 24-EBL concentrations (0.01, 0.1, and 1.0 mg L^{-1}) were applied on tomato (*Lycopersicon esculentum* Mill. cv. 9021) plants exposed to high temperature (40/30°C), net photosynthetic rate, stomatal conductance, and maximum carboxylation rate of rubisco (ribulose 1,5-bisphosphate carboxylase oxygenase) were decreased, while activities of antioxidant enzymes such as super-oxide dismutase, ascorbate peroxidase, guaiacol peroxidase, and catalase increased during heat treatments, and these increases proved to be more significant in EBL-treated plants (0.1 mg L^{-1} EBL). EBL application reduced malonaldehyde contents and significantly increased shoot weight following heat stress; thus, it can be concluded that EBR could alleviate the detrimental effects of high temperatures on plant growth by increasing car-boxylation efficiency and enhancing antioxidant enzyme systems in leaves of tomato. Dhaubhadel (2002) reported on the protec-tive role of BRs in tomato and *Brassica napus* seedlings grow-ing under low and high temperature conditions. Singh and Shono (2005) found that tomato plants (*Lycopersicon esculentum* Mill.) treated with 24-EBL were more tolerant to high-temperature stress than untreated plants. It was determined that mitochon-drial small heat-shock proteins (Mt-sHSP) did not accumulate in EBL-treated plants (1 µM) at 25°C, although at 38°C this HSPs accumulation was many-fold higher in EBL-treated than in untreated plants. EBL possibly induced thermotolerance in tomato plants through induction of sHSPs, and these plants had better photosynthetic efficiency. Exposure to 45°C for 3 hours completely killed more than 90% of untreated plants, while 1 µM EBR application was found to be the most effective for survival of tomato plants at lethal temperatures. Nassar (2004) studied the effect of HBL on in vitro growth of apical meristems and heat tolerance of banana shoots. They confirmed that HBL may have a role in thermotolerance by the regulation of stem elonga-tion and protecting the membrane against heat stress. Similarly, Wilen et al. (1995) also reported that EBL markedly enhanced cell viability of brome grass cell suspension culture following exposure to high-temperature stress by the accumulation of a set of heat-shock proteins of 90KD.

Huang et al. (2006) discovered that growth of mung bean epi-cotyls (*Vigna radiata* L.) was suppressed by chilling treatment which was recovered after treatment with 24-EBL (10 µM), 17 proteins downregulated by chilling were upregulated and these upregulated proteins were involved in methionine assimilation, ATP synthesis, cell wall construction, and other stress responses. Kagale et al. (2007) found that transcripts of cold-related genes accumulated to higher levels in 24-EBL-treated plants of *Brassica napus* and *Arabidopsis thaliana* which were grown on a nutrient solution containing 1 µM 24-EBL and exposed to cold stress (2°C) for 3 days. Kumar et al. (2010) reported that exoge-nous application of H$_2$O$_2$ to *Brassica juncea* L. seedlings adapted them to tolerate chilling stress, which further enhanced to higher levels when supplemented with brassinosteroids. Janeckzo et al. (2007) investigated the effect of 24-EBL on the cold resistance of rape seedlings (*Brassica napus* L. cv. *Lycosmos*). They verified

that at 2°C, BR injection into cotyledons (0.05 and 1.00 µM) or primary leaves (1.0 µM) abolished the effect of cold on perme-ability as plants submitted to cold treatment without BR injection elevated the membrane permeability. Liu et al. (2009) studied the effect of 24-EBL on suspension cultured cells of *Chorispora bungeana* under low temperature (4°C and 0°C) for 5 days. They observed that 24-EBL-treated cells (0.05 mg.L^{-1}) exhibited higher viability after exposure to low temperature compared to untreated control plants. Under chilling stress, reactive oxygen species (ROS) levels and lipid peroxidation were increased in cultured cells which were significantly inhibited by EBL treat-ment. Activities of antioxidant enzymes such as ascorbate per-oxidase, catalase, peroxidase, and superoxide dismutase were increased during chilling treatments, and this enhancement was more significant in EBL-applied suspension cells. The EBL treatment also greatly enhanced contents of ascorbic acid and reduced glutathione under chilling stress. From these results, it was concluded that EBL could play positive roles in the allevia-tion of oxidative damage caused by ROS overproduction through enhancing antioxidant defense system, resulting in improving the tolerance of *C. bungeana* suspension cultures to chilling stress. BRs may reduce chilling injury of plant cell membranes due to lipid peroxidation, therefore protecting the structural integrity of the membranes and resulting in the enhancement of chilling tol-erance. Xia et al. (2009) verified the effects of 24-EBL (0.1 µM) in cucumber (*Cucumis sativus* L. cv. *Jinyan No.4*) seedlings sub-mitted to chilling stress. Chilling stress (8°C) caused a signifi-cant reduction in the electron transport rate while EBL treatment alleviated chilling stress and enhanced the electron transport rate. In one of the earlier studies, the oxidative-related genes encoding monodehydroascorbate reductase and thioredoxin h, the cold-responsive gene *COR-47* and *COR-78*, and heat stress-related genes *HSP-83*, *HSP-70*, *HSF-3*, *Hsc70-3*, *Hsc70-G7*, have been identified by microarray analysis of either brassinosteroids-deficient or brassinosteroid (BRs)-treated plants by Müssig et al. (2002).

Several adaptive strategies in response to various abiotic stresses such as salt, water, cold, and heat stress are exhibited by plants, which ultimately affect plant growth and yield (McCue and Hanson, 1990). It is well documented that, besides other stresses, exposure to saline conditions limits plant growth and productivity (Abbas et al., 2010). It is one of the most brutal envi-ronmental factors and soil salinity is a complex phenotypic and physiological phenomenon in plants. It imposes ionic imbalance, hyperionic, and hyperosmotic stress. This salt stress disrupts the overall metabolic activities and thus limits the productivity of crop plants worldwide (Munns and Tester, 2008). Photosynthesis can be restricted by salt stress by decreasing green pigments (Sudhir and Murthy, 2004), and this can suppress rubisco activ-ity (Soussi et al., 1998)and reduce stomatal conductance, thus affecting internal CO$_2$ availability (Bethkey and Drew, 1992). Total plant growth can be reduced by a decrease in root and shoot growth in a saline environment (Sehrawat et al., 2013a,b,c). This inhibition in growth may be due to the diversion of energy from growth to maintenance under salt stress (Greenway and Gibbs, 2003). Reactive oxygen species (ROS) can be increased by salt stress, which then accelerates the toxic reactions like lipid peroxi-dation, protein degradation, and DNA mutation (McCord, 2000). To alleviate these oxidative effects, different kinds of antioxidants like superoxide dismutase (SOD), catalase (CAT), and peroxidase (POD) are generated by plants (Noreen and Ashraf, 2009).

To overcome the threats generated by salinity, different types of mechanical, biological, and chemical approaches are being practiced in different parts of the world. To cope with the toxic effects of salinity, many compounds are being used, including ascorbic acid (Khafagy et al., 2009), glycinebetaine (Arafa et al., 2009; Abbas et al., 2010), silicon (Ashraf et al., 2010), proline (Hoque et al., 2007), etc. However, the use of BRs is the most prominent (Rao et al., 2002). BRs are reported to enhance abiotic stress tolerance by two mechanisms, through membrane stability and/or osmotic adjustment (Wang and Zeng, 1993). In mitigating reactive oxygen species (ROS) produced in response to salt stress BRs play a key role (Bajguz and Hayat, 2009).

Recent studies demonstrate that, at various levels with the signaling components of other phytohormones, BRs interact and regulate processes like plant growth, development, and stress responses (Hu and Yu, 2014; Tong et al., 2014; Chaiwanon and Wang, 2015; Divi et al., 2015; Yuan et al., 2015). BR signaling involves its perception by the cell membrane receptor followed by the activation of a cascade of phosphorylation events (Figure 29.5) to relay the signal to the downstream partners resulting in the BR-induced gene expression (Belkhadir and Jaillais, 2015; Saini et al., 2015). Various biological approaches such as mutant screening, microarray, proteomics, protein–protein interaction studies, and bioinformatics played a vital role in the identification and characterization of various components involved in BR signaling (Divi et al., 2010, 2015).

Recent molecular studies have demonstrated the positive role of exogenous application of BR or endogenous BR contents in salt stress tolerance (Krishna et al., 2003; Koh et al., 2007; Manavalan et al., 2012). Enhancing BR signaling activity in *Arabidopsis* led to the increase in salt stress tolerances, but BR-defective mutants showed hyper salt stress responses (Krishna et al., 2003). Potential applications of BRs in agriculture to improve crop yield under salinity stress are well documented (Wu et al., 2008; Divi and Krishna, 2009; Hao et al., 2013). Similar to dicot plants, the positive roles of BRs in the salt stress responses in rice have been reported. A loss of functional rice *gsk1* mutant, an orthologue of a major BR negative regulator *BIN2*, increased salt stress tolerance compared to wild type (Koh et al., 2007). These findings suggested that adverse effects in rice could be reduced by activated BR signaling pathways under salt stress. Consistently, BRs treatment could remove the salinity-induced inhibition of seed germination and seedling growth in rice. Moreover, increasing endogenous BR levels by the gene silencing-mediated disruption *SQS* (*Squalene synthase*) gene that is involved in critical functions for BRs biosynthesis led to an increase in salt stress tolerance (Manavalan et al., 2012). Together, these imply the positive roles of BRs in plant salt and drought stress tolerances. Recently, there have been studies of the molecular action mechanisms in various species explaining the involvement of BRs in salt stress. Interestingly, through *BIN2*-mediated inactivation of *YDA* signaling cascades, BRs signaling pathways are found to play an important role in the stomata development (Kim et al., 2012), which suggested that water loss could be controlled by activated

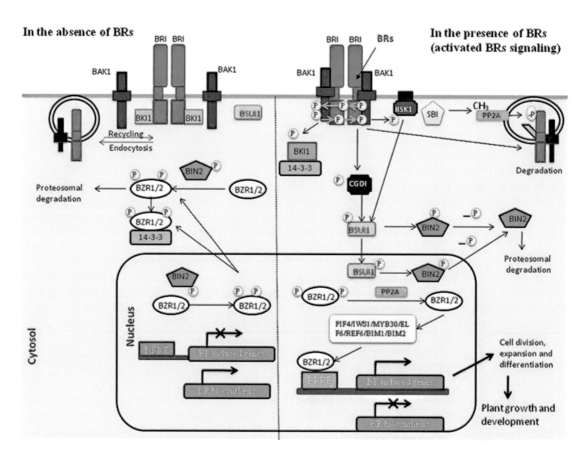

FIGURE 29.5 Brassinosteroids signaling (From Saini, 2015).

BRs under high salt stress conditions by reducing stomatal density in plants.

BR-enhanced stress tolerance has been correlated with stress-related marker gene expression and crosstalk with other hormones including auxin, ABA, and salicylic acid (SA) (Dhaubhadel et al., 1999; Divi et al., 2010). It is likely that the stress tolerance ability of BRs is connected partially to its interaction with the other stress hormones (Divi and Krishna, 2009). The crosstalk between BRs and ABA is important in salt stress response. Under salt stress condition, the accumulation of ABA rapidly induced seedling growth arrest by activating *ABI5* (*ABA INSENSITIVE5*), which enabled the post germinated seedlings to cope with various unfavorable conditions (Finkelstein et al., 2002; Lopez-Molina et al., 2002). Recent molecular studies suggested that these BRs act as central positive regulators in early seedling development by attenuating the signaling outputs of ABA signals (Zhang et al., 2009; Ryu and Hwang, 2013).

In early seedling development, the antagonistic interactions between the BR and ABA signaling pathways have long been reported, and the underlying the molecular mechanisms have been investigated recently (Steber and McCourt, 2001; Zhang et al., 2009). The phosphorylation of *BES1* enhanced by the *ABI2* (*ABA INSENSITIVE2*)-mediated ABA signaling pathway, indicated that ABA inhibits the primary signaling outputs of BR by activating *BIN2* kinase (Zhang et al., 2009). The exogenous BR application or overexpression of BR biosynthetic *DWF4* suppresses the ABA-mediated inhibition of early seedling development (Steber, 2001; Zhang et al., 2009). Moreover, the BR-defective mutants display ABA and salt hypersensitive phenotypes during the early seedling development (Steber and McCourt, 2001; Zhang et al., 2009) suggesting that BRs have similar positive roles with GA in early seedling development in salt stress conditions (Figure 29.6).

The role of BRs and associated compounds in the modulation of various components of antioxidant defense systems in salinity-exposed plants has been extensively reported (Nunez et al., 2003; Özdemir et al., 2004; Song et al., 2006; Shahbaz and Ashraf, 2007; Zhang et al., 2007; Ali et al., 2008b; Arora et al., 2008a; El-Khallal et al., 2009; Hayat et al., 2010b; Rady, 2011; Vardhini, 2011; Ding et al., 2012; El-Mashad and Mohamed, 2012; Abbas et al., 2013; Fariduddin et al., 2013b; Lu and Yang, 2013; Sharma et al., 2013b). The negative impact of salt stress in *Zea mays* is mitigated by BL by inducing the activities of different antioxidant enzymes (El-Khallal et al., 2009). Applications of 28-homoBL (10^{-7}, 10^{-9}, and 10^{-11} M) for 7 days improved the seedling growth and lipid peroxidation and this was done via elevating antioxidative enzyme activities (SOD, CAT, GR, APX, and GPX) in the seedlings of *Zea mays* (var. *Partap-1*) subjected to salt (25, 50, 75, and 100 mM NaCl) stress (Arora et al., 2008a). 24-EpiBL applied as foliar spray could alleviate the adverse effects of salt on two hexaploid wheat (*Triticum aestivum*) cultivars that are S-24 (salt tolerant) and MH-97 (moderately salt sensitive), grown in saline conditions (150 mM of NaCl) by enhancing the activity of POD and CAT (Shahbaz and Ashraf, 2007). BL treatment increased the activities of CAT, SOD, and GR; this reduced the activities of POD and PPO of two varieties of the sorghum plants (*CSH-5* and *CSH-6*) grown in two saline experimental sites of Karaikal (Varchikudy and Mallavur), thus indicating its ability to counteract the negative impact of saline stress (Vardhini, 2011). Exogenous BL (0. 01 mgL^{-1}) decreased the salt stress index, MDA, mortality rate and electrolyte leakage by enhancing the activities of SOD, POD, and CAT in *Cucumis sativus* seedlings (Song et al., 2006). Against salt stress the

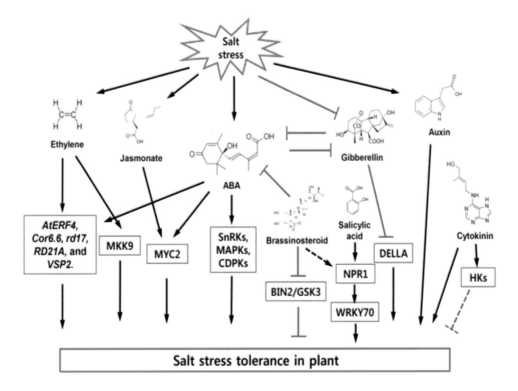

FIGURE 29.6 Role of various plant growth regulators in salt stress tolerance in plants (From Hojin et al., 2015).

exogenous BR (0.005, 0.01, 0.05, 0.1, and 0.2 mgL^{-1}) protected *Cucumis sativus* seedlings by elevating the activity of SOD, POD, and CAT, and that lowered the salt-injured index (40.2%) and also increased the contents of free-proline, soluble sugars (Shang et al., 2006). Application of epiBL to salinity-exposed *Cucumis sativus* seedlings decreased leaf superoxide anion production rate, H_2O_2, MDA, cell membrane permeability, and improved seedling growth as a result of increased the activities of SOD, POD, and CAT (Lu and Yang, 2013). Applications of epiBL to the Cu+NaCl (150 mM)-stressed seeds of two cultivars (Rocket and Jumbo) of *Cucumis sativus* plant enhanced the activities of various antioxidant enzymes viz., CAT, POD, SOD, that eventually improved the growth, carbonic anhydrase activity, and photosynthetic efficiency (Fariduddin et al., 2013b). Seed priming with 5.0 µM L^{-1} BL was reported to improve the seed germination as well as seedling growth of 3 lucerne (*Medicago sativa* L.) varieties Victoria, Golden Empress, and Victor by significantly increasing the activities of POD, SOD, and CAT under salt stress (13.6 dSm^{-1} NaCl solution) (Zhang et al., 2007).

Salinity (120 mM NaCl)-exposed *IR-28 Oryza sativa* seedlings when supplemented with 24-EpiBL, showed considerable alleviation of the oxidative damage along with improving seedling growth by increasing APX activity and by reducing lipid peroxidation (Özdemir et al., 2004). A polyhydroxylated spirostanic brassinosteroid analog (BB-16; 0.001 or 0.01 mg dm–3) application to salinity (75 m NaCl) exposed *O. sativa* seedlings showed significant increases in the activities of CAT, SOD, and GR (Nunez et al., 2003). Exogenous application of 24-epiBL to *Oryza sativa* var. *Pusa Basmati-1*, grown under salt stress conditions exhibited improvement in growth, levels of protein, proline contents, and antioxidant enzyme activities through the expression of various BRs (*OsBRI1*, *OsDWF4*) and salt (*SalT*)-responsive genes (Sharma et al., 2013b). Eggplant seedlings, exposed to 90 mM NaCl with 0, 0.025, 0.05, 0.10, and 0.20 mg dm^{-3} of epiBL for 10 days, exhibited decreased electrolyte leakage, superoxide production, MDA, and H_2O_2 probably as a result of increased activities of SOD, GPX, CAT, and APX enzymes and the contents of non-enzymatic antioxidants such as AsA and GSH (Ding et al., 2012). 24-epiBL decreased the adverse effects of salinity stress on two varieties of pepper (*Capsicum annuum*) arguably by increasing the activities of antioxidative enzymes and the contents of proline, total anthocyanins, and minerals (Abbas et al., 2013). Supplementation of 24-epiBL or 28-homoBL to NaCl-exposed *Brassica juncea* detoxified the stress generated by NaCl by enhancing antioxidative enzymes and the level of proline (Ali et al., 2008b; Sirhindi et al., 2015; Kaur and Patti, 2017). Supplementation of *Vigna radiata* plants with 28-homoBL detoxified the stress generated by NaCl by elevating the activities of antioxidative enzymes and also proline content that in turn improved the MSI, leaf water potential (ψ) (Hayat et al., 2010b). Spraying 24-epiBL at 5 µM concentration to NaCl-exposed *Phaseolus vulgaris* improved the MSI, RLWC as a result of significant elevations in the activities of antioxidative enzymes and proline content. Imbibition of 24-epiBL to pea (*Pisum sativum* L.) seeds, subjected to sodium chloride stress significantly elevated the activity of SOD, POD, and CAT enzymes that helped plants to improve fresh as well as dry biomass, seedling height, photosynthetic rate, stomatal conductance, and total chlorophyll

content (Shahid et al., 2011). Treatment with 0.05 ppm BL as a foliar spray mitigated salt stress impacts in cowpea (*Vigna sinensis*) by inducing the activities of various antioxidant enzymes such as SOD, POD, PPO, and GR and the contents of AsA (El-Mashad and Mohamed, 2012).

Rattan et al. (2012) studied the role of brassinosteroids (BRs) in osmoprotectant accumulation in response to NaCl salt stress (0, 40, 60, 80, 100 mM), which enables the plants to grow under stress conditions by influencing various physiological and morphological processes of 30- and 60-day old *Zea mays* var. *DKC 9106* plants. Serna et al. (2015) carried out a study to analyze the effects of brassinosteroids on alleviating salt stress in lettuce plants that were separated into shoot and root after 5 days of brassinosteroid application and a reduction in weight due to NaCl was observed. When 1 µM BRs application was given, it resulted in polyamine and ethylene synthesis which protected the plant from salinity. Application of 28-homoBL enhanced the superoxide dismutase, catalase, guaiacol peroxidase, ascorbate peroxidase, and glutathione reductase activities in salt-stressed plants of *Brassica juncea* L. cv *RLC 1*, which was found to provide tolerance against NaCl salt stress (Kaur et al., 2015). 28-homoBL diminishes the effects of salt stress by regulating the efficiency of photosynthesis and seed yield in *Brassica juncea* L. (Alyemeni et al., 2013).

A foliar spray of 24-EBL (0.0125, 0.025, and 0.0375 mg.L^{-1}) to wheat cultivars (S-24 and MH-97) increased plant biomass and leaf area per plant of both cultivars under non-saline conditions. However, under salinity stress (150 mM NaCl), improvement in growth due to exogenous EBL was observed only in S-24 (salt tolerant cultivar). Photosynthetic rate was reduced due to salt stress in both cultivars, but this inhibitory effect was ameliorated significantly by the exogenous application of EBL. The most effective dose in improving growth of both cultivars due to EBL spray under non-saline or saline conditions was found to be 0.025 mg.L^{-1}. EBL-induced increase in growth was associated with improved photosynthetic capacity. Wu et al. (2012) studied the effect of 24-epibrassinolide on photosynthesis in seedlings of eggplant (*Solanum melongena* L.) under salt stress. They concluded that the application of 100 nM epibrassinolide concentration was most effective for salt-stressed plants in which significant increases in chlorophyll content, net photosynthetic rate, stomatal conductance, intercellular CO_2 concentration, maximal quantum efficiency of photosystem II (PSII) photochemistry, PSII maximum efficiency, photochemical quenching coefficient, PSII operating efficiency, and the fraction of light absorbed (utilized in PSII photochemistry) while nonphotochemical quenching was reduced in 24-EBL-treated plants.

They further found that higher concentrations of EBL (200 and 400 nM) reduced its effect or even caused a decline in photosynthetic capacity as compared to plants that received high salinity treatment alone. They concluded that salinity-affected gas exchange and chlorophyll fluorescence parameters and the decreased growth of eggplant seedlings, and EBL, especially 100 nM, alleviated the detrimental effects of salinity by improving the photosynthetic efficiency. Ali et al. (2007) evaluated the effect of a pre-sowing soaking treatment of an aqueous solution of 10^{-10} or 10^{-8} M of 28-HBL under NaCl (1 or 10 mM) in seeds of chickpea (*Cice rarietinum* L. cv.

KPG-59). It was found that the plants grown from seeds soaked in HBR (10^{-8} M) possessed higher leaf nitrate reductase and carbonic anhydrase activities, more dry mass, higher nodule number, and more nodule fresh and dry mass, compared with control distilled water-treated plants. Hayat et al. (2007) verified the effect of 28-HBL (10^{-10}, 10^{-8}, and 10^{-6} M) on salinity-induced changes in *Brassica juncea*. Plants that received only NaCl (50, 100, or 150 mM) treatment exhibited decreases in nitrate reductase and carbonic anhydrase activities, chlorophyll content, and photosynthetic rate, 60 days after sowing. Subsequent treatment with HBR significantly increased all of these parameters. 10^{-8} M concentration of HBR generated the best response and also overcame the detrimental effects when NaCl concentration was 50 mM. The HBR concentration of 10^{-8} M along with the NaCl concentration of 150 mM resulted in increased concentration of tissue proline concentration compared to the other treatments.

BRs and Oxidative Stress Tolerance

One of the major consequences of various abiotic and biotic stresses is oxidative damage, which is induced by the excessive production of ROS, such as superoxide radical ($O_2{}^-$), hydrogen peroxide (H_2O_2), hydroxyl radical (OH•), and alkoxyl radical (RO•). In plants, ROS are continuously produced predominantly in the chloroplast, mitochondria, and peroxisomes. Production and removal of ROS have been balanced. However, the production and scavenging of ROS might be disturbed by the number of biotic and abiotic factors (Apel and Hirt, 2004). These ROS have the capacity to initiate lipid peroxidation and degrade proteins, lipids, and nucleic acids (McCord, 2000; Halliwell and Gutteridge, 1999). Ultimately these ROS might lead to the death of the plant cell by enhancing the expression of the ROS-dependent and cell death-related genes.

Brassinosteroids have also been shown to protect plants from both biotic and abiotic stresses (Krishna, 2003) along with having a role in plant growth and development. From all the BRs isolated from different plant species, 28-homobrassinolide (HBL) and 24-epibrassinolide (EBL) are the most active compounds for exogenous applications. These compounds have been used in plant tissue culture applications, leading to increases in the freezing and thermotolerance of cell suspension (Wilen et al., 1995). Ahmad et al. (2012) determined that BR improved seed germination and early development of tomato (*Solanum lycopersicum* L. cv. *Hezuo903*) seedling under phenanthrene (PHE) stress by increasing the activities of antioxidant enzymes over PHE alone, but decreased the MDA contents both in shoot and root of young tomato seedlings. 1.0 nM EBR was the most effective followed by 100 and 0.01 nM for the improvement of germination and seedling growth under PHE stress in tomato. Xia et al. (2006) showed that 24-EBL alleviated the pesticide-induced depression of photosynthesis in *Cucumis sativus* L. by increasing CO_2 assimilation capacity and activities of antioxidant enzymes. Kitanaga et al. (2006) reported that BRs involved in plant defense, by regulating the low-molecular-weight, basic cysteine-rich antimicrobial protein named theionine protein. Transcripts of theionine genes encoding antimicrobial peptides were present at a high level in rice coleoptiles just after germination and

decreased to an undetectable level after about 3 days but this decline was suppressed by co-treatment with gibberellic acid and brassinosteroids. Ding et al. (2009a,b) evaluated the potential of root and foliar applications of 24-EBL in reducing *Fusarium* wilt and their influence on antioxidant and phenolic metabolism in roots of cucumber plants (*Cucumis sativus* L. cv. *Jinyan No. 4*).

EBL treatments significantly reduced pathogen-induced accumulation of reactive oxygen species, flavonoids and phenolic compounds on one hand while increased activities of defense-related and ROS-scavenging enzymes on the other hand as well as phenylalanine ammonia lyase and polyphenol oxidase. Arora et al. (2008) demonstrated that the application of 28-HBL to Cu-stressed maize seedlings significantly enhanced the tolerance towards oxidative stress generated by Cu. Alam et al. (2007) concluded that spraying treatment of 28-HBL (10–8 M) to plants of *Brassica juncea* L. cv. *T59* neutralized the toxic effect of nickel (50 or 100 µM) by increasing net photosynthetic rate, content of chlorophyll, and the activities of nitrate reductase, carbonic anhydrase, and antioxidant enzymes. It was reported that BRs enhanced the level of antioxidants under cadmium, temperature, and salt stress in *Brassica juncea* (Sirhindi et al., 2015; Kaur et al., 2017). Activities of antioxidant enzymes, such as catalase, peroxidase, and superoxide dismutase, and the contents of proline increased against the control by the exogenous application of 28-homobrassinolide, which made the plants tolerant to heavy metal stress.

Ramakrishna and Rao (2012) studied that 24-EBL treatment protected radish (*Raphanus sativus* L.) seedlings from zinc-induced oxidative stress by enhancing the antioxidative system. They concluded that application of 24-epibrassinolide had a protective role in lipid peroxidation, protein oxidation, and membrane integrity in radish seedlings and thus significantly alleviated zinc-induced oxidative stress. Anuradha and Rao (2007) found that the supplementation of 24-EBL to radish seedlings (*Raphanus sativus* L.) reduced lead toxicity and enhanced the growth by ameliorating the activities of antioxidant enzymes viz. catalase, ascorbate peroxidase, guaiacol peroxidase, superoxide dismutase, and total glutathione content showed an increase in their level of activity or concentration in BR-treated Pb-stressed seedlings as compared to the control while peroxidase activity was reduced in such treated seedlings.

Rady et al. (2011) studied the effect of 24-EBL on growth, yield, and antioxidant systems in bean (*Phaseolus vulgaris* L.) plants under salinity stress and cadmium stress. From their observations it was found that 24-EBL treatment detoxified the stress caused by salinity and cadmium by increasing antioxidant systems, proline, and relative water content and protected the photosynthetic machinery and plant growth as compared to control and stressed plants. The effects of 24-EBL on seedling growth, antioxidative systems, lipid peroxidation, proline, and soluble protein content were investigated in seedlings of the salt-sensitive rice (*Oryza sativa* L.) cultivar *IR-28* (Özdemir et al., 2004). Seed application of 24-EBL (3 µM) improved seedling growth, alleviated the lipid damage, and decreased proline accumulation caused by salt stress (120 mM) in salt-sensitive rice varieties. Zhang et al. (2007) tested the seeds of three lucerne (*Medicago sativa* L.) varieties (cv. Victor, Victoria, and Golden Empress) to investigate the effects of seed priming with BL and found improvement in salt tolerance of lucerne seedlings. This was supported by increased germination ability, root length, root vigour, root dry weight, and shoot fresh and dry weight under high level of salt stress (13.6 dSm^{-1} NaCl solution) which was attributed to an increase in peroxidase, catalase, and superoxide dismutase activities and lowered malondialdehyde, reflecting the level of lipid peroxidation in lucerne seedlings under salt stress.

Li et al. (2008) demonstrated that soaking roots with BL (0–0.4 mgL^{-1}) prior to planting increased the activities of POD, SOD, and CAT, and decreased the rate of transpiration, stomatal conductance, and MDA as compared with water-stressed control DW plants of *Robinia pseudoacacia*, which would be the possible mechanism in making the plant drought tolerant. Vardhini (2011) found enhanced activities of four oxidizing enzymes SOD, GR, IAAO, and PPO while the activity of two hydrolyzing enzymes protease and ribonuclease was decreased by BRs treatments in seedlings of sorghum under PEG-induced water stress as compared to control seedlings. Zhang et al. (2008) found that the foliar spray of EBL (0.1 mg L^{-1}) enhanced drought tolerance, minimized the yield loss of soybean (*Glycine max* L.), and increased the concentration of soluble sugars and proline, and the activities of POD and SOD in soybean leaves under drought stress conditions could be the possible mechanism for this. EBL treatment decreased malondialdehyde concentrations and electrical conductivity of leaves under drought stress.

24-EBL induced the lowering of ROS levels, MDA, and carbonyl levels by increasing the activities of ROS-scavenging enzymes and decreasing the activities of lipoxygenase and NADPH oxidase. Xia et al. (2009) reported that reactive oxygen species were involved in brassinosteroid-induced stress tolerance in cucumber. They found that BR levels positively correlated with the tolerance to photo-oxidative and cold stresses and resistance to Cucumber Mosaic Virus (CMV). They also showed that BR treatment enhanced NADPH oxidase activity and elevated H$_2$O$_2$ levels in apoplast. H$_2$O$_2$ levels were elevated after BR treatment in cucumber seedlings which was accompanied by increased tolerance to oxidative stress. Similar results were observed by Xia et al. (2011), who reported that 24-EBL treatments repressed the expression of BR biosynthetic genes but elevated the production of H$_2$O$_2$. They further found that EBL-induced systemic induction of BR biosynthetic genes was mediated by systemically elevated H$_2$O$_2$. Cui et al. (2011) examined the involvement of NO• and H$_2$O$_2$ in BR-induced tolerance in cucumber (*Cucumis sativus* L. cv. *Jinyan No.4*) plant. They provided evidence that BR-induced H$_2$O$_2$ was necessary for BR-induced NO• production. They further found that BRs treatment increased NO• production in cucumber leaves, which is involved in EBR-induced stress tolerance by the induction of antioxidant genes, which in turn leads to increased activities of the corresponding antioxidant enzymes for mitigating oxidative stress due to various stress conditions.

Role of BRs in Regulation of Antioxidant Defense System

In addition to regulating growth activities, BRs are also reported to confer resistance to plants as well as animals against a plethora of abiotic and biotic stresses. The various stresses against which BRs have been found to be effective include thermal stress, drought stress, heavy metals stress, salt stress, infection, pesticides, and even viruses (Dhaubhadel et al., 1999, 2002; Wachsman et al., 2004; Krishna, 2003; Haubrick and Assmann, 2006; Bhardwaj et al., 2007; Kumar et al., 2009, 2010, 2012; Sirhindi et al., 2015; Kaur et al., 2017). BRs had been reported to provide resistance against these stresses by regulating the activities of various antioxidative enzymes [superoxide dismutase (SOD), guaiacol peroxidase (POD), ascorbate peroxidase (APOX), catalase (CAT), and glutathione reductase (GR)], antioxidants (α-tocopherol, ascorbic acid, glutathione, etc.), and the level of lipid peroxidation (Asada and Takahashi, 1987; Mazorra et al., 2002; Özdemir et al., 2004; Hayat et al., 2007, 2007; Arora et al., 2008; Sirhindi et al., 2014).

The ROS are removed by an army of enzymatic and non-enzymatic antioxidative defense systems. Non-enzymatic antioxidative systems include various antioxidants like ascorbate, reduced glutathione (GSH), tocopherol, flavonoids, alkaloids, and carotenoids. Ascorbate and α-tocopherol are extremely effective antioxidants as they are relatively poor electron donors and effectively scavenge OH•, O$_2$•, and singlet oxygen. Carotenoids protect chlorophyll by absorbing the excess of excitation energy and by quenching singlet oxygen. Enzymatic antioxidative systems consist of various antioxidative enzymes like SOD, POD, APOX, CAT, GR, monodehydroascorbate reductase (MDHAR), and dehydroascorbate reductase (DHAR). SOD act as the first line of defense against ROS, dismutating O$_2$•$^-$ to H$_2$O$_2$. Subsequently, CAT, APOX, and POD detoxify the H$_2$O$_2$ to H$_2$O. H$_2$O$_2$ is also converted into water by the ascorbate glutathione cycle. APOX detoxify the H$_2$O$_2$ to H$_2$O by using ascorbate as a reducing agent which is oxidized into monodehydroascorbate. This can later be regenerated into ascorbate with the help of MDHAR using NADPH as a reducing power. Monodehydroascorbate can be spontaneously dismutated into dehydroascorbate. Then, ascorbate regeneration is mediated by DHAR driven by the oxidation of reduced glutathione (GSH) to oxidized glutathione (GSSG). Finally, GR regenerates GSH from GSSG using NADPH as a reducing agent. The extent of oxidative stress in a cell is determined by the amount of O$_2$•$^-$, H$_2$O$_2$, and OH•.

Therefore, the balance of antioxidative enzymes activities will be vital for suppressing the toxic level of ROS. All components of enzymatic and non-enzymatic defense systems act in a co-ordinated manner and constitute the pathway called the "Asada-Halliwell pathway" (Asada and Takahashi, 1987; Noctor and Foyer, 1998; Arora et al., 2002). Several plant hormones are implicated in modulating the plant responses to oxidative stress, including ethylene (Vahala et al., 2003), abscisic acid (Kovtun et al., 2000), salicylic acid (SA) (Metwally et al., 2003; Vahala et al., 2003), auxin (Kovtun et al., 2000), and BRs (Almeida et al., 2005; Mazorra et al., 2002). Conversely, Cao et al. (2005) found in an experiment that the BR-deficient det2 mutant showed an enhanced resistance to general oxidative stress. A constitutive increase in SOD activity and increased transcript levels of the CAT gene were detected in det2 plants. A similar phenomenon was also observed in an SA-deficient transgenic line expressing a salicylate hydroxylase gene (Borsani et al., 2001). Goda et al. (2002) also demonstrated that ATPA2 and ATP24a genes encoding peroxidases were constitutively upregulated in the det2 mutant. Furthermore, the oxidative stress-related genes encoding monodehydroascorbate reductase and thioredoxin h, the cold and drought stress response genes COR47 and COR78, and the heat stress-related genes HSP83, HSP70, HSF3, Hsc70-3, and Hsc70-G7 have been identified by microarray analysis of either

BR-deficient or BR-treated plants (Müssig et al., 2002). As per previous reports, it was observed that BRs have a capacity to ameliorate the detrimental effects of various abiotic and biotic stresses by regulating the activities of antioxidative enzymes and antioxidants. Various reports are available which further confirm the potential to regulate antioxidative defense systems. Özdemir et al. (2004) had reported that EBL ameliorates the salt stress in rice by regulating the activities of POD, CAT, APOX, and SOD. Almeida et al. (2005) had found that homobrassinolide treatments alleviate the negative effects of H_2O_2 on the leaf structures and allow better recovery of the cell component. It was reported that BR enhanced the level of antioxidative enzymes (catalase, peroxidase, and superoxide dismutase) under cadmium stress in *Brassica juncea* and the content of proline increased against the control plant by 28-HOMOBL applications (Hayat et al., 2007). Further, Zhang et al. (2007a) reported that seed priming of lucerne (*Medicago sativa* L.) with BL significantly enhanced the activities of antioxidant enzymes (POD, SOD, and CAT) and reduced the MDA content of seedlings exposed to salt stress thereby ameliorating salt stress. Foliar application of EBL wheat plants and *B. juncea* plants successfully ameliorated the salinity-induced growth inhibition by enhancing activities of antioxidative enzymes (POD, CAT, SOD and GR) (Ali et al., 2008a). In addition to this, EBL pretreatment to *Cucumis sativus* L. and *Vicia faba* also helped to alleviate the pesticide and herbicide (terbutryn)-induced depression of photosynthesis by increasing activities of antioxidative enzymes (Xia et al., 2006; Piñol and Simón, 2009). 28-homoBL also ameliorated the salinity stress and heavy metal stress in *Zea mays* and *B. juncea* by increasing the activities of antioxidative enzymes and by reducing the level of MDA (Sharma et al., 2007; Bhardwaj et al., 2007b; Arora et al., 2008a,b).

To regulate ROS production and mitigation homeostasis, cells ameliorate antioxidant defense systems which are composed of various antioxidant enzymes such as superoxide dismutase (SOD), catalase (CAT), glutathione peroxidase (GPOD), ascorbate peroxidase (APOX), and non-enzymatic antioxidants such as ascorbic acid, α-tocopherol, carotenoids, glutathione, etc. Clouse (1997) showed that exogenous application of BRs modified antioxidant defense systems under different stress conditions to mitigate overproducing ROS under such stress (Kaur et al., 2015; Sirhindi et al., 2014). Anuradha and Rao (2007) affirmed the enhanced activity of CAT but a reduction in the activity of peroxidases and APOX under cadmium stress in radish plants by exogenous application of BRs. Sirhindi et al. (2009) reported a significant increase in activities of SOD, CAT, and polyphenol oxidase (PPO) by exogenous application of BRs to *B. juncea* plants grown under natural field conditions. While the activity of APOX showed this enhancement to non-significant level. The difference in alteration of antioxidant enzyme activities in salt-sensitive rice cultivar suggested that the higher salt stress resistance in sensitive seedlings induced by 24-epiBL may be due to the maintenance of a higher activity of APX under salt stress conditions. Goda et al. (2002) demonstrated constituent upregulation in *det2* mutant of *A. thaliana*. These *det2* mutant plants showed enhanced oxidative stress resistance which was correlated with a constituent increase in SOD activity and the transcript level of CAT (Cao et al., 2004). Oxidative stress-related genes encoding MDHAR and thioredoxin, along with heat-shock *HSP83*,

HSP70, *HSF3*, *Hsc70*, and *Hsc70-G7*, and cold stress-related genes *COR47* and *COR78* by microarray analysis in BR-deficient and BR-treated plants are identified by Müssig et al. (2002).

Conclusion

The description and characterization of various isomeric forms of brassinosteroids applications to plants raised under stress or unstressed, help in understanding the biosynthesis, regulation, and signaling pathways. Brassinosteroids are not involved in active transportation, hence their mechanism of action involves crosstalk with other hormones such as auxins, gibberellins, ABA, SA, and JA to give the effective response. Responses related to biotic and abiotic stresses as well as plant growth and development-related processes are influenced by exogenous application of BRs but the response of BRs depend on the type of BR used as well as the dose and time of application. Out of most commonly use BRs, 28-HBR has more potential in protecting plants exposed to temperature and/or salt stress although 24-EBR also has a similar impact. In conclusion, it is evident that brassinosteroids which were previously known for their association with normal growth and development amelioration in plants now demonstrate multiple functions in protecting crops from various abiotic and biotic stresses and have an important role in temperature and salt stress management in plants.

Acknowledgment

The authors extend their appreciation to the University Grants Commission, New Delhi for supporting the project in form of financial grants vide F.No. 42-920/2013 (SR) and F. No. 33-194/2007 (SR). The authors are also grateful to the Head of the Department of Botany, Punjabi University, Patiala for providing laboratory facilities.

REFERENCES

Abbas S, Latif HH, Elsherbiny EA. Effect of 24-epibrassinolide on the physiological and genetic changes on two varieties of pepper under salt stress conditions. *Pak. J. Bot.* 2013;45:1273–1284.

Abbas W, Ashraf M, Akram NA. Alleviation of salt induced adverse effects in eggplant (Solanum melongena) by glycinebetaine and sugarbeat extract. *Sci. Hort.* 2010;125:188–195.

Ahmad P, Bhardwaj R, Tuteja N. Plant signaling under abiotic stress environment. In: *Environmental Adaptations and Stress Tolerance of Plants in the Era of Climate Change,.* Ahmad P, Prasad MNV. (Eds.). New York: Springer, 2012; 297–323.

Alam MM, Hayat S, Ali B, Ahmad A. Effect of 28 homo-brassinolide on nickel induced changes in *Brassica juncea. Photosynthetica* 2007;45:139–142.

Ali B, Hassan SA, Hayat S, Hayat Q, Yadav S, Fariduddin Q, Ahmad A. Role for brassinosteroids in the amelioration of aluminium stress through antioxidant system in mung bean (*Vigna radiata* L. *Wilczek*). *Environ. Exp. Bot.* 2008a;62:153–159.

Ali B, Hayat S, Ahmad A. 28 homo-brassinolide ameliorates the saline stress in chickpea (*Cicer arietinum*). *Environ. Exp. Bot.* 2008b;59:217–223.

Ali G, Hadi F, Ali Z, Tariq M, Khan MA. Callus induction and *in vitro* complete plant regeneration of different cultivars of tobacco (*Nicotiana tabacum* L.) on media of different hormonal concentrations. *Biotechnology* 2007;4:561–566.

Almeida JM, Fidalgo F, Confraria A, Santos A, Pires H, Santos I. Effect of hydrogen peroxide on catalase gene expression, isoform activities and levels in leaves of potato sprayed with homobrassinolide and ultrastructural changes in mesophyll cells. *Funct. Plant Biol.* 2005;32:707–720.

Alyemeni MN, Hayat S, Wijaya L, Anaji A. Foliar application of 28-homobrassinolide mitigates salinity stress by increasing the efficiency of photosynthesis in *Brassica juncea*. *Acta Bot. Brasilica* 2013;27:502–505.

Anuradha S, Rao SSR. Effect of 24-epibrassinolide on the growth and antioxidant enzyme activities in radish seedlings under lead toxicity. *Indian J. Plant Physiol.* 2007;12:396–400.

Anuradha S, Rao SSR. The effect of brassinosteroids on radish (*Raphanus sativus* L.) seedlings growing under cadmium stress. *Plant Soil Environ.* 2003;53:465–472.

Apel K, Hirt H. Reactive oxygen species: Metabolism, oxidative stress, and signal transduction. *Annu. Rev. Plant Biol.* 2004;55:373–399.

Arafa AA, Khafagy MA, El-Banna MF. The effect of glycinebetaine or ascorbic acid on grain germination and leaf structure of sorghum plants grown under salinity stress. *Aust. J. Crop Sci.* 2009;3:294–304.

Arora A, Sairam RK, Srivastave GC. Oxidative stress and antioxidative systems in plants. *Curr. Sci.* 2002;82:1227–1238.

Arora N, Bhardwaj R, Sharma P, Arora HK. 28-homobrassinolide alleviates oxidative stress in salt treated maize (*Zea mays* L.) plants. *Braz. J. Plant Physiol.* 2008;20:153–157.

Asada K, Takahashi M. Production and scavenging of active oxygen in photosynthesis. In: *Photoinhibition: Topics in Photosynthesis*. Kyle DJ, Osmond CB, Arnten CJ. (Eds.). Amsterdam: Elsevier, 1987; 9: 227–287.

Ashraf M, Akram NA, Arteca RN, Fooled MR. The Physiological biochemical and molecular roles of brassinosteroids and salicylic acid in plant processes and salt tolerance. *Crit. Rev. Plant Sci.* 2010;29:162–190.

Athar HUR, Ashraf M. Modulation of growth, photosynthetic capacity and Ali Q, water relations in salt stressed wheat plants by exogenously applied 24-epibrassinolide. *Plant Growth Regul.* 2008;56:107–116.

Baek KH, Skinner DZ. Production of reactive oxygen species by freezing stress and the protective roles of antioxidant enzymes in plants. *J. Agric. Chem. Environ.* 2012;1:34–40.

Bajguz A. Brassinosteroid enhanced the level of abscisic acid in Chlorella vulgaris subjected to short-term heat stress. *J. Plant Physiol.* 2009;166:882–886.

Bajguz A, Hayat S. Effects of Brassinosteroids on the plant responses to environmental stresses. *Plant Physiol. Biochem.* 2008;8(47):1–8.

Belkhadir Y, Jaillais Y. The molecular circuitry of brassinosteroid signaling. *New Phytol.* 2015;206:522–540.

Bethkey PC, Drew MC. Stomatal and non-stomatal components to inhibition of photosynthesis in leaves of Capsicum annum during progressive exposure to NaCl. *Plant Physiol.* 1992;99(1):219–26.

Bhardwaj R, Kaur S, Nagar PK, Arora HK. Isolation and characterization of brassinosteroids from immature seeds of Camellia sinensis (O) Kuntze. *Plant Grow Regul.* 2007;53(1):1–5.

Borsani O, Valpuestan V, Botella MA. Evidence for a role of salicylic acid in the oxidative damage generated by NaCl and osmotic stress in Arabidopsis seedlings. *Plant Physiol.* 2001;126:1024–1030.

Cabanero FJ, Martinez V, Carvajal M. Does calcium determine water uptake under saline conditions in pepper plants or is it water flux which determines calcium uptake. *Plant Sci.* 2004;166:443–450.

Cao JJ, Chow J, Lee S, Li Y, Chen S, An Z. Characterization and source apportionment of atmospheric organic and elemental carbon during fall and winter of 2003 in Xi'an, China. *Atmos. Chem. Phys.* 2005;5:127–137.

Cao L, Leers-Sucheta S, Azhar S. Aging alters the functional expression of enzymatic and non-enzymatic anti-oxidant defence systems in testicular rat Leydig cells. *J. Steroid Biochem. Mol. Biol.* 2004;88:61–67.

Cardozo NP, Sentelhas PC. Climatic effect on sugarcane ripening under the influence of cultivars and crop age. *Sci. Agric.* 2013;70(6):449–456.

Cavusoglu K, Kabar K. Comparative effects on some plant growth regulator on the germination of barley and radish seeds under high tempera-ture stress. *EurAsian J. BioSci.* 2007;1:1–10.

Chaiwanon J, Wang ZY. Spatiotemporal brassinosteroid signaling and antagonism with auxin pattern stem cell dynamics in Arabidopsis roots. *Curr. Biol.* 2015;25:1031–1042.

Choe S. Brassinosteroid biosynthesis and metabolism. In: *Plant Hormones:Biosynthesis,SignalTransduction,Action*.Dordrecht: Kluwer Academic Publishers, 2007; 156–178.

Choe S, Fujioka S, Noguchi T, Takatsuto S, Yoshida S, Feldmann KA. Overexpression of DWARF4 in the brassinosteroid biosynthetic pathway results in increased vegetative growth and seed yield in Arabidopsis. *Plant J.* 2001;26:573–582.

Chung Y, Choe S. The regulation of brassinosteroid biosynthesis in Arabidopsis. *Crit. Rev. Plant Sci.* 2013;32:396–410.

Clouse SD. Brassinosteroid signal transduction: From receptor kinase activation to transcriptional networks regulating plant development. *Plant Cell* 2011;23:1219–1230.

Clouse SD. Molecular genetic analysis of brassinosteroid action. *Physiol. Plant* 1997;100:702–709.

Clouse SD, Sasse JM. Brassinosteroids: Essential regulators of plant growth and development. *Annu. Rev. Plant Physiol. Mol. Biol.* 1998;49:427–451.

Cui JX, Zhou YH, Ding JG, Xia XJ, Shi K, Chen SC, Asami T, Chen Z, Yu JQ. Role of nitric oxide in hydrogen peroxidedependent induction of abiotic stress tolerance by brassinosteroids in cucumber. *Plant Cell Environ.* 2011;34:347–358.

Dhaubhadel S, Browning KS, Gallie DR, Krishna P. Brassinosteroid function to protect the translational machinery and heat shock protein synthesis following thermal stress. *Plant J.* 2002;29(6):681–691.

Dhaubhadel S, Chaudhary S, Dobinson KF, Krishna P. Treatment with 24-epibrassinolide, a brassinosteroid, increases the basic thermotolerance of Brassica napus and tomato seedlings. *Plant Mol. Biol.* 1999;40:333–342.

Ding HD, Zhu XH, Zhu ZW, Yang SJ, Zha DS, Wu XX. Amelioration of salt induced oxidative stress in eggplant by application of 24-epibrassinolide. *Biol. Plant.* 2012;56:767–770.

Ding J, Shi K, Zhou YH, Yu JQ. Effects of root and foliar applications of 24-epibrassinolide on fusarium wilt and antioxidant metabolism in cucumber roots. *Hort. Sci.* 2009a;44(5):1340–1349.

Ding J, Shi K, Zhou YH, Yu JQ. Microbial community responses associated with the development of Fusarium oxysporium sp. Cucumerinum after 24-epibrassinolide applications to shoots and roots in cucumber. *Eur. J. Plant Pathol.* 2009b;124:141–150.

Divi UK, Krishna P. Brassinosteroid: A biotechnological target for enhancing crop yield and stress tolerance. *New Biotech.* 2009;26:131–136.

Divi UK, Krishna P. Overexpression of the Brassinosteroid biosynthetic gene AtDWF4 in Arabidopsis seeds overcomes Abscisic acid- induced inhibition of germination and increases cold tolerance in transgenic seedling. *J. Plant Growth Regul.* 2010;29:385–393.

Divi UK, Rahman T, Krishna P. Brassinosteroid-mediated stress tolerance in Arabidopsis shows interactions with abscisic acid, ethylene and salicylic acid pathways. *BMC Plant Biol.* 2010;10:151.

Divi UK, Rahman T, Krishna P. Gene expression and functional analyses in brassinosteroid-mediated stress tolerance. *Plant Biotechnol.* 2015;14:419–432.

El-Khallal SM, Hathout TA, Ashour AERA, Kerrit AAA. Brassinolide and salicylic acid induced antioxidant enzymes, hormonal balance and protein profile of maize plants grown under salt stress. *Res. J. Agric. Biol. Sci.* 2009;5:391–402.

El-Mashad AA, Mohamed HI. Brassinolide aaleviates salt stress and increase antioxidant activity of cowpea plants (Vigna sinensis). *Protoplasma* 2012;249:625–635.

Fariduddin Q, Khanam S, Hasan SA, Ali B, Hayat SA, Ahmad A. Effect of 28- homobrassinolide on the drought stress-induced changes in photosynthesis and antioxidant system of Brassica juncea L. *Acta Physiol. Plant* 2009;31:889–897.

Fariduddin Q, Mir BA, Yusuf M, Ahmad A. Comparative roles of brassinosteroids and polyamines in salt stress tolerance. *Acta Physiol. Plant* 2013;35:2037–2053.

Fariduddin Q, Yusuf M, Begum M, Ahmad A. 28- homobrassinolide protect photosynthetic machinery in Indian mustard high temperature stress. *J. Stress Physiol. Biochem.* 2014;10:181–194.

Fariduddin Q, Yusuf M, Chalkoo S, Hayat S, Ahmad A. 28- homobrassinolide improves growth and photosynthesis in Cucumis sativus L. through an enchanced antioxidant system in the presence of chilling stress. *Photosynthetica* 2011;49:55–64.

Farrell TC, Fox KM, Williams RL, Fukai S. Genotypic variation for cold tolerance during reproductive development in rice: Screening with cold air and cold water. *Field Crops Res.* 2006;98:178–194.

Finkelstein RR, Gampala SSL, Rock CD. Abscisic acid signaling in seeds and seedlings. *Plant Cell* 2002;14:S15–S45.

Fujioka S, Noguchi T, Yokota T, Takatsuto S, Yoshida S. Brassinosteroids in Arabidopsis thaliana. *Phytochemistry* 1998;48:595–599.

Fujioka S, Yokota T. Biosynthesis and metabolism of brassinosteroids. *Annu. Rev. Plant Biol.* 2003;54:137–164.

Goda H, Shimada Y, Asami T, Fujioka S, Yoshida S. Microarray analysis of brassinosteroid regulated genes in Arabidopsis. *Plant Physiol* 2002;130:1319–1334.

Greenway H, Gibbs J. Mechanisms of anoxia tolerance in plants. II. Energy requirements for maintenance and energy distribution to essential processes. *Funct. Plant Biol.* 2003;30:999–1036.

Grove MD, Spencer GF, Rohwedder WK, Mandava N, Worley JF, JDW, Steffens GL, Flippen-Anderson JL, Carter Cook J. Brassinolide, a plant growth-promoting steroid isolated from *Brassica napus* pollen. *Nature* 1979;281:216–217.

Guo H, Li L, Aluru M, Aluru S,Yin Y. Mechanisms and networks for brassinosteroid regulated gene expression. *Curr. Opin. Plant Biol.* 2013;16:545–553.

Halliwell B, Gutteridge JMC. Free radicals in biology and medicine. *Oxford University Press, Oxford* 1999;1–25.

Hao JJ, Yin YH, Fei SZ. Brassinosteroid signaling network: Implications on yield and stress tolerance. *Plant Cell Rep.* 2013;32:1017–1030.

Hare PD, Cress WA, Staden J. The involvement of cytokinins in plant responses to environmental stress. *Plant Growth Regul.* 1998;23:79–103.

Hasanuzzaman M, Nahar K, Fujita M. Extreme temperatures, oxidative stress and antioxidant defense in plants. *Abiotic Stress—Plant Responses and Applications in Agriculture* 2013;10:169–205.

Hasanuzzaman M, Nahar K, Fujita M. Regulatory role of polyamines in growth, development and abiotic stress tolerance in plants. In: *Plant Adaptation to Environmental Change: Significance of Amino Acids and Their Derivatives.* Anjum NA, Gill SS, Gill R. (Eds.). Boston: CABI, 2014; 157–193.

Hategan L, Godza B, Szekeres M. Regulation of brassinosteroid metabolism. In: *Brassinosteroids: A Class of Plant Hormone.* Hayat S, Ahmed A. (Eds.) New York: Springer, 2011; 57–81.

Haubrick LL, Assmann SM. Brassinosteroids and plant function: some clues, more puzzles. Plant Cell Environ. 2006;29:446–457.

Hayat S, Ali B, Hassan SA, Ahmad A. Effects of 28- homobrassinolide on salinity-induced changes in Brassica juncea. *Turk. J. Biol.* 2007;31:141–146.

Hayat S, Hasan SA, Hayat Q, Ahmad A. Brassinosteroids protect Lycopersicon esculentum from cadmium toxicity applied as shotgun approach. *Protoplasma* 2010a;239:3–14.

Hayat S, Hasan SA, Yusuf M, Hayat Q, Ahmad A. Effect of 28-homobrassinolide on photosynthesis, fluorescence and antioxidant system in the presence or absence of salinity and temperature in Vigna radiate. *Environ. Exp. Bot.* 2010b;69:105–112.

Hojin R, Yong GC. Plant hormones in salt stress tolerance. *J. Plant Biol.* 2015;58:147–155.

Hoque MA, Banu MNA, Okuma E. Exogenous proline and glycinebetaine increase NaCl-induced ascorbate-glutathione cycle enzyme activities, and proline improves salt tolerance more than glycinebetaine in tobacco Bright Yellow-2 suspension-cultured cells. *J. Plant Physiol.* 200b;164:1457–1468.

Hoque MA, Okuma E, Banu MNA, Nakamura Y, Shimoishi Y, Murata Y. Exogenous proline mitigates the Detrimental effects of salt stress more than exogenous betaine by increasing antioxidant enzyme activities. *J. Plant Physiol.* 2007a;164:553–561.

Hu H, You J, Fang Y, Zhu X, Qi Z, Xiong L. Characterization of transcription factor gene SNAC2 conferring cold and salt tolerance in rice. *Plant Mol. Biol.* 2008;67:169–181.

Hu Y, Yu D. BRASSINOSTEROID INSENSITIVE2 interacts with ABSCISIC ACID INSENSITIVE5 to mediate the antagonism of brassinosteroids to abscisic acid during seed germination in Arabidopsis. *Plant Cell* 2014;26:4394–4408.

Huang B, Chu CH, Chen SL, Juan HF, Chen YM. A proteomics study of the mung bean epicotyl regulated by brassinosteroids under conditions of chilling stress. *Cell Mol Biol Lett.* 2006;11(2):264–78.

Ikekawa N, Zhao YJ. Application of 24-epibrassinolide in agriculture. In: Cutler HG, Yokota T, Adam G, eds. Brassinosteroids. Chemistry, bioactivity, and applications. ACS Symposium Series, 474. *Washington: American Chemical Society* 1991;474:280–291.

Jan N, Hussain MU, Andrabi KI. Cold resistance in plants: A mystery unresolved. *Electron. J. Biotechnol.* 2009;15:12.

Janeczko A, Gullner G, Skoczowski A, Dubert F, Barna B. Effects of brassinosteroid infiltration prior to cold treatment on ion leakage and pigment contents in rape leaves. *Biol. Plant.* 2007;51:355–358.

Janeczko A, Oklestkova J, Pociecha E, Koscielniak J, Mirek M. Physiological effects and transport of 24-epibrassinolide in heat-stressed barley. *Acta Physiol. Plant* 2011;233:1249–1259.

Je BI, Piao HL, Park SJ, Park SH, Kim CM, Xuan YH. RAV-Like1 maintains brassinosteroid homeostasis via the coordinated activation of BRI1 and biosynthetic genes in rice. *Plant Cell* 2010;22(6):1777–1791.

Jiang Y, Huang B. Drought and heat stress injury to two cool season turfgrasses in relation to antioxidant metabolism and lipid peroxidation. *Crop Sci.* 2001;41:436–442.

Jiang YP, Huang LF, Cheng F, Zhou YH, Xia XJ, Mao WH, Shi K, Yu JQ. Brassinosteroids accelerate recovery of photosynthetic apparatus from cold stress by balancing the electron partitioning, carboxylation and redox homeostasis in cucumber. *Physiol. Plant.* 2013;148:133–145.

Kagale S, Divi UK, Krochko JE, Keller WA, Krishna P. Brassinosteroids confers tolerance in Arabidopsis thaliana and Brassica napus to a range of abiotic stresses. *Planta* 2007;225(2):353–364.

Kaur H, Sirhindi G, Bhardwaj R. Influence of 28-homobrassinolide on photochemical efficiency in *Brassica juncea* under dual stress of extreme temperatures and salt. *Canadian Journal of Pure and Applied Sciences* 2017;11:4205–4213.

Kaur H, Sirhindi G, Bhardwaj R. Aiternation of antioxidant machinery by 28-homobrassinolide in Brassica juncea L. under salt stress. *Adv. Appl. Sci. Res.* 2015;6:166–172.

Kaur H, Sirhindi G, Bhardwaj R, Sharma P, Mudasir M. 28-homobrassinolide modulate antenna complexes and carbon skeleton of Brassica juncea L. under temperature stress. *J. Stress Physiol. Biochem.* 2014;10:186–196.

Kaur N, Patti PK. Integrating classical with emerging concepts for better understanding of salinity stress tolerance mechanisms in rice. *Environ. Sci.* 2017;5:42.

Khafagy MA, Arafa AA, El-Banna MF. Glycinebetaine and ascorbic acid can alleviate the harmful effects of NaCl salinity in sweet pepper. *Aust. J. Crop Sci.* 2009;3:257–267.

Khan NA, Singh S. *Abiotic Stress and Responses*. New Delhi: IK International Publishing House Private. Limited, 2008.

Kim BK, Fujioka S, Takatsuto S, Tsujimoto M, Choe S. Castasterone is a likely end product of brassinosteroid biosynthetic pathway in rice. *Biochem. Biophys. Res. Commun.* 2008;374:614–619.

Kim TW, Michniewicz M, Bergmann DC, Wang ZY. Brassinosteroid regulates stomatal development by GSK3-mediated inhibition of a MAPK pathway. *Nature* 2012;482:419–422.

Kitanaga Y, Jian C, Hasegawa M, Yazaki J, Kishimoto N, Kikuchi S. Sequential regulation of gibberellin, brassinosteroid, and jasmonic acid biosynthesis occurs in rice coleoptiles to control the transcript levels of anti-microbial thionin genes. *Biosci. Biotechnol. Biochem.* 2006;70:2410–2419.

Koh S, Lee SC, Kim MK, Koh JH, Lee SG, Choe S, Kim SR. T-DNA tagged knockout mutation of rice OsGSK1, an orthologue of Arabidopsis BIN2, with enhanced tolerance to various abiotic stresses. *Plant Mol. Biol.* 2007;65:453–466.

Kolupaev Y, Akinina G, Mokrousov A. Induction of heat tolerance in wheat coleoptiles by calcium ions and its relation to oxidative stress. *Russian J. Plant Physiol.* 2005;52:199–204.

Kovtun Y, Chiu WL, Tena SJ. Fuctional analysis of oxidative stress-activated mitogen-activated protein kinase cascade in plants. *Proc. Natl. Acad. Sci. USA* 2000;97:2940–2945.

Krishna P. Brassinosteroid-mediated stress responses. *J. Plant Growth Regul.* 2003;22:289–2997.

Kumar A, D'Souza SS, Tickoo S, Salimath BP, Singh HB. Antiangiogenic and proapoptotic activities of allyl isothiocyanate inhibit ascites tumor growth in vivo. *Integr. Cancer Ther.* 2009;8:75–87.

Kumar M, Sirhindi G, Bhrdwaj R, Kumar S, Jain G. Effect of exogenous H_2O_2 on antioixidant enzymes of *Brassica juncea* L. seedling in relation to 24-epibrassinolide under chilling stress. *Ind. J. Biochem. Biophys.* 2010;47(6):378–382.

Kumar S, Dubey RS, Tripathi RD, Chakrabarty D, Trivedi PK. Omics and biotechnology of arsenic stress and detoxification in plants: Current updates and prospective. *Environ. Int.* 2015;74:221–230.

Kumar S, Singh D, Dutta M. Quality characteristics in rapeseed-mustard and role of some antinutritional factors in plant defense: Future strategies. *J. Oilseed Brassica* 2014;5:87–95.

Kumar S, Sirhindi G, Bhardwaj R. 28- homobrassinolide induced exaggerated growth, biochemical molecular aspects of Brassica juncea L. RLM-619 seedling under high temperature stress. *Plant Biochem. Physiol.* 2014;2:127.

Kumar S, Sirhindi G, Bhardwaj R, Kimar M, Arora P. Role of 24-epibrassinolide in amelioration of high temperature stress through antioxidant defence system in Brassica juncea L. *Plant Stress* 2012;55–58.

Levitt J. *Responses of plant to environmental stress Volume 1:Chilling, Freezing, and High Temperature Stress.*, London: Academic Press. 1980.

Li JM, Chory J. Brassinosteroid actions in plants. *J. Exp. Bot.* 1999;50:275–282.

Li KR, Wang HH, Han G, Wang QJ, Fan J. Effect of brassinolide on the survival, growth and drought resistance of Robinia pseudoacacia seedling under water stress. *New Forests* 2008;35:255–266.

Li QF, Wang C, Jiang L, Li S, Sun SS, He JX. An interaction between BZR1 and DELLAs mediates direct signaling crosstalk between brassinosteroids and gibberellins in *Arabidopsis*. *Sci. Signal.* 2012;5:ra72.

Liqiong X, Yang C, Wang X. Brassinosteroids can regulate cellulose biosynthesis by controlling the expression of CESA genes in Arabidopsis. *J. Exp. Bot.* 2011;62(13):4495–4506.

Liu ZJ, Zhang XL, Bai JG, Suo BX, Xu PL, Wang L. Exogenous paraquat changes antioxidant enzymes activities and lipid peroxidation in drought stressed cucumber leaves. *Sci. Hort.* 2009;121:138–143.

Lopez-Molina L, Mongrand B, McLachlin DT, Chait BT, Chua NH. ABI5 acts downstream of ABI3 to execute an ABA dependent growth arrest during germination. *Plant J.* 2002;32:317–328.

Lu XM, Yang W. Alleviation effects of brassinolide on cucumber seedlings under NaCl stress. *Ying Yong Sheng Tai Xue Bao* 2013;24:1409–1414.

Lu Y, Zhao L, Wang B. From virtual community members to c2c e-commerce buyers: trust in virtual communities and its effect on consumers' purchase intention. *Electronic Commerce Research & Applications* 2010;94:346–360.

Maestri E, Klueva N, Perrotta C, Gulli M, Nguyen HT, Marmiroli N. Molecular genetics of heat tolerance and heat shock proteins in cereals. *Plant Mol. Biol.* 2002;48:667–681.

Mahajan S, Tuteja N. Cold, salinity and drought stresses: An overview. *Arch. Biochem. Biophys.* 2005;444:139–158.

Maharjan PM, Choe S. High temperature stimulates DWARF4 (DWF4) expression to increase hypocotyle elongation in Arabidopsis. *J. Plant Biol.* 2011;54:425–429.

Manavalan LP, Chen X, Clarke J, Salmeron J, Nguyen HT. RNAi-mediated disruption of squalene synthase improves drought tolerance and yield in rice. *J. Exp. Bot.* 2012;63:163–175.

Mandava NB, Mitchell JW. Structural elucidation of brassins. *Chem. Ind.* 1972;930–932.

Mandava NB, Sidwell BA, Mitchell JW, Worley JF. Production of brassins from rape pollen: A convenient preparative method. *Ind. Eng. Chem. Prod. Res. Dev.* 1973;12:138–139.

Marakali S, Temel A, Nermin G. Salt stress and homobrassinosteroid interactions during germination in barley roots. *Not. Bot. Horti Agrobot. Cluj-Napoca* 2014;42(2):446–452.

Marco A, Zullo T, Adam G. Brassinosteroid phytohormones - structure, bioactivity and applications. *Brazilian Journal of Plant Physiology* 2002;14:143–181.

Marsolais F, Boyd J, Paredes Y, Schinas AM, Garcia M, Elzein S. Molecular and biochemical characterization of two brassinosteroid sulfotransferases from Arabidopsis, AtST4a (At2g14920) and AtST1 (At2g03760). *Planta* 2007;225:1233–1244.

Mazorra LM, Nunez M, Hechavarria M, Coll F, Sanchez-Blanco MJ. Influence of brassinosteroids on antioxidant enzymes activity in tomato under different temperatures. *Biol. Plant.* 2002;45:593–596.

McCord JM. The evolution of free redicals and oxidative stress. *Am. J. Med.* 2000;108:652–659.

McCue AM, Hanson A. Drought and salt tolerance: toward understanding and application. *Trends Biotechnol.* 1990;8:358–362.

Metwally A, Finkermeier I, Georgi M, Dietz KJ. Salicylic acid alleviates the cadmium toxicity in barley seedlings. *Plant Physiol.* 2003;132:272–281.

Mitchell JW, Mandava NB, Worley JF, Plimmer JR, Smith MV. Brassins a new family of plant hormones from rape pollen. *Nature* 1970;225:1065–1066.

Mitchell R, Stocklin W, Stefanovi M, Geissman T. Chap-arrolide and castelanolide, New bitter principles from castela nicholsoni. *Phytochemistry* 1971;10:411–417.

Mittler R. Oxidative stress, antioxidants and stress tolerance. *Trends Plant Sci.* 2002;7:405–410.

Morales D, Rodrigues P, Dell Amico J, Nicolas E, Torrecillas A, Sanchez-Blanco MJ. High temperature preconditioning and thermal shock imposition affects water relations, gas exchange and root hydraulic conductivity in tomato. *Biol. Plant.* 2003;47:203–208.

Munns R, Tester M. Mechanisms of salinity tolerance. *Annu. Rev. Plant Biol.* 2008;59:651–681.

Müssig C, Fischer S, Altman T. Brassinosteroid regulated gene exression. *Plant Physiol.* 2002;129:1241–1251.

Nakano Y, Higuchi Y, Sumitomo K, Hisamatsu T. Flowering retardation by high temperature in chrysanthemums: Involvement of FLOWERING LOCUS T- like 3 gene repression. *J. Exp. Bot.* 2013;64:909–920.

Nassar AH. Effect of homobrassinolide on in vitro growth of apical meristems and heat tolerance of banana shoots. *Int. J. Agric. Boil.* 2004;771–775.

Nasser AM, Mohammed AS. Effect of 28- homobrassinolide on the drought induced changes in the seeds of Vigna radiate. *Legume Res.* 2014;37:515–519.

Noctor G, Foyer C. Ascorbate and glutathione: Keeping active oxygen under control. *Annu. Rev. Plant Physiol. Plant Mol. Biol.* 1998;49:249–279.

Nogues S. Brassinosteroids alleviate heat induced inhibition of photosynthesis by increasing carboxylation efficiency and enhancing antioxidant systems in Lycopersicon esculentum. *J. Plant Growth Regul.* 2008;27:49–57.

Noreen Z, Ashraf M. Assessment of variation in antioxidative defense system in salt- treated pea (Pisum sativum) cultivars and its putative use as salinity tolerance markers. *J. Plant Physiol.* 2009;166:1764–1774.

Nunez M, Mazzafera P, Mazorra LM, Siqueira WJ, Zullo MAT. Influence of brassinosteroid analogue on antioxidant enzymes in rice grown in culture medium with NaCl. *Biol Plant.* 2003;47:67–70.

Ogweno JO, Song XS, Shi K, Hu WH, Mao WH, Zhou YH, Yu JQ, Nogues S. Brassinosteroids Allivate Heat-induced inhibition of photosynthesis by increasing carboxylation efficiency and enhancing antioxidant system in Lycopersicon esculentum. *J. Plant Growth Regul.* 2008;27:49–57.

Ohnishi ST, Godza B, Watanabe B, Fujioka S, Hategan L, Ide K, Shibata K, Yokota T, Szkeres M, Mizutani M. CYP90A1/CPD, a brassinosteroid biosynthetic cytochrome P450 of Arabidopsis, catalyzes C-3 oxidation. *J. Biol. Chem.* 2012;287:31551–31560.

Özdemir F, Bor M, Demiral T, Turkan I. Effect of 24-epibrassinolide on seed germination, seedling growth, lipid peroxidation, proline content and antioxidant system of rice (Oryza sativa L.) under salinity stress. *Plant Growth Regul.* 2004;41:1–9.

Patel D, Franklin KA. Temperature regulation of plant architecture. *Plant Signal. Behav.* 2009;4:577–579.

Piñol R, Simón E. Effect of 24-epibrassinolide on chlorophyll fluorescence and photosynthetic CO_2 assimilation in Vicia faba plants treated with the photosynthesis-inhibiting herbicide terbutryn. *J. Plant Growth Regul.* 2009;28:97–105.

Poppenberger B, Fujioka S, Soeno K, George GL, Vaistij FE, Hiranuma S. The UGT73C5 of Arabidopsis thaliana glucosylates brassinosteroids. *Proc. Natl. Acad. Sci. USA* 2005;102:15253–15258.

Poppenberger B, Rozhon W, Khan M, Husar S, Adam G, Luschnig C. CESTA, a positive regulator of brassinosteroid biosynthesis. *EMBO J.* 2011;30:1149–1161.

Prasad PVV, Boote LH, Allen JR, Sheehy JE, Thomas JMG. Species, ecotype and cultivar differences in spikelet fertility and harvest index of rice in response to high temperature stress. *Field Crops Res.* 2006;95:398–411.

Rady MM. Effect of 24-epibrassinolide on growth, yield, antioxidant system and cadmium content of bean (Phaseolus vulgaris L.) plants under salinity and cadmium stress. *Sci. Hort.* 2011;129(2):232–237.

Ramakrishna A, Ravishankar GA. Influence of abiotic stress signals on secondary metabolites in plants. *Plant Signal. Behav.* 2011;6:1720–1731.

Ramakrishna B, Rao SSR. 24-epibrassinolide alleviated zinc-induced oxidative stress in radish (Raphanus sativus L.) seedlings by enhancing antioxidative system. *Plant Growth Regul.* 2012;68:249–259.

Rao AAR, Vardhini BV, Sujatha E, Anuradha S. Brassinosteroids a new class of phytohormones. *Curr. Sci.* 2002;82:1239–1245.

Rattan A, Kapoor N, Bhardwaj R. Role of brassinosteroids in osmolytes accumulation under salinity stress in Zea mays plants. *Int. J. Sci. Res.* 2012;3:1822–1827.

Roh H, Jeong CW, Fujioka S, Kim YK, Lee S, Ahn JH. Genetic evidence for the reduction of brassinosteroid levels by a BAHD acyltransferase-like protein in Arabidopsis. *Plant Physiol.* 2012;159:696–709.

Ryu H, Hwang I. Brassinosteroids in plant developmental signaling networks. *J. Plant Biol.* 2013;56:267–273.

Saini S, Sharma I, Pati PK. Versatile roles of brassinosteroid in plants in the context of its homoeostasis, signaling and cross-talks. *Front. Plant Sci.* 2015;6:950.

Sakamoto A, Murata N. The role of glycine betaine in the protection of plants from stress: Clues from transgenic plants. *Plant Cell Environ.* 2002;25:163–171.

Sanghera GS, Wani SH, Hussain W, Singh NB. Engineering cold stress tolerance in crop plants. *Curr. Genom.* 2011;12(1):30–43.

Satake T, Yoshida S. High temperature induced sterility in Indica rice at flowering. *Jpn. J. Crop Sci.* 1978;47:6–17.

Schneider K, Breuer C, Kawamura A, Jikumaru Y, Hanada A, Fujioka S, Ichikawa T, Kondou Y, Matsui M, Kamiya Y, Yamaguchi S, Sugimoto K. Arabidopsis pizza has capacity to acylate brassinosteroids. *PLoS One* 2012;7:46805.

Sehrawat N, Bhat KV, Sairam RK, Jaiwal PK. Screening of mungbean (Vigna radiata L. Wilczek) genotypes for salt tolerance. *Int. J. Plant Anim. Environ. Sci.* 2013a;4:36–43.

Sehrawat N, Bhat KV, Sairam RK, Tomooka N, Kaga A, Shu Y, Jaiwal PK. Diversity analysis and confirmation of intra-specific hybrids for salt tolerance in mungbean (Vigna radiata L. Wilczek). *Int. J. Integr. Biol.* 2013b;14:65–73.

Sehrawat N, Jaiwal PK, Yadav M, Bhat KV, Sairam RK. Salinity stress restraining mungbean (Vigna radiata L. Wilczek) production: Gateway for genetic improvement. *Int. J. Agric. Crop Sci.* 2013c;6:505–509.

Serna M, Yamilet CB, Pedro J, Zapataa AB, Maria TPA, Asunction AA. Brassinosteroid analogue prevented the effect of salt stress on ethylene synthesis and polyamines in lettuce plants. *Sci. Hort.* 2015;185:105–112.

Shahbaz M, Ashraf M. Influence of Exogenous application of Brassinosteroids on growth and mineral nutrients of wheat (Triticum aestivum L.) under saline conditions. *Pak. J. Bot.* 2007;39(2):513–522.

Shahbaz M, Nasreen S, Afza T. Environmental consequences of economic growth and foreign direct investment: evidence from panel data analysis. *Bull. Energy Econ.* 2014;2:14–27.

Shahid MA, Pervez MA, Balal RM, Mattson, Rashid A, Ahmad R, Ayyub CM, Abbas T. Brassinosteroid (24-epibrassinolide) enhances growth and alleviates the deleterious effects induced by salt stress in pea (Pisum sativum L.). *Austral. J. Crop Sci.* 2011;5(5):500–510.

Shang Q, Song S, Zhang Z, Guo S. Exogenous brassinosteroid induced salt resistance of cucumber (Cucumis sativus L.) seedlings. *Sci. Agric. Sin.* 2006;39:1872–1877.

Sharkova VE. The effect of heat shock on the capacity of wheat plants to restore their photosynthesis electron transport after photoinhibition or repeated heating. *Russian J. Plant Physiol.* 2001;48:793–797.

Sharma RK, Agrawal M, Marshall FM. Heavy metals contamination of soil and vegetables in suburban areas of Varanasi, India. *Ecotox. Environ. Saf.* 2007;66:258–266.

Sharma I, Bhardwaj R, Pati PK. Exogenous application of 28-Homobrassinolide modulates the dynamics of salt and pesticides induced stress responses in an elite rice variety Pusa Basmati-1. *J. Plant Growth Regul.* 2015;34:509–518.

Sharma I, Ching E, Saini S, Bhardwaj R, Pati PK. Exogenous application of brassinosteroid offers tolerance to salinity by altering stress responses in rice variety Pusa Basmati-1. *Plant Physiol. Biochem.* 2013;69:17–26.

Simoes-Araujo JL, Rumjanek NG, Margis-Pinheiro M. Small heat shock proteins genes are differentially expressed in distinct varieties of common bean. *Brazilian J. Plant Physiol.* 2003;15:33–11.

Singh I, Shono M. Physiological and molecular effects of 24-epi-brassinolide, a brassinosteroid on thermotolerance of tomato. *Plant Growth Regul.* 2005;47:111–119.

Sirhindi G, Kaur H, Bhardwaj R, Kaur NP, Sharma P. Thermo protective role of 28- homobrassinolide in Brassica juncea plants. *Am. J. Plant Sci.* 2014;5:48074–48079.

Sirhindi G, Kumar S, Bhardwaj R, Kumar M. Effect of 24-epibrassinolide and 28-homobrassinolide on the growth and antioxidant enzymes activities in the seedling of Brassica juncea L. *Physiol. Mol. Biol. Plants* 2009;15:335–341.

Sirhindi G, Mir MA, Abd-Allah EF, Ahmad P, Gucel S. Jasmonic acid modulates the physio-biochemical attributes, antioxidant enzyme activity, and gene expression in glycine max under nickel toxicity. *Front. Plant Sci.* 2016;7:591.

Sirhindi G, Mir MA, Sharma P, Gill SS, Kaur H, Mushtaq R. Modulatory role of jasmonic acid on photosynthetic pigments, antioxidants and stress markers of Glycine max L. under nickel stress. Physiol. *Mol. Biol. Plants.* 2015;4:559–565.

Song WJ, Zhou WJ, Jin ZL, Zhang D, Takeyuchi Y, Joel DM. Growth regulators restore germination of Orobanche seeds that are conditioned under water stress and suboptimal temperature. *Aust. J. Agric. Res.* 2006;57:1195–1201.

Soussi M, Dcana A, Lluch C. Effect of salt stress on growth, photosynthesis and nitrogen fixation in chickpea (Cicer arietinum L.). *J. Exp. Bot.* 1998;49:1329–1327.

Sreeramulu S, Mostizky Y, Sunitha S, Shani E, Nahum H, Salomon D. BSKs are partially redundant positive regulators of brassinosteroid signaling in Arabidopsis. *Plant J.* 2013;74:905–919.

Steber CM, McCourt P. A role for brassinosteroids in germination in Arabidopsis. *Plant Physiol.* 2001;125:763–769.

Steponkus PL, Uemura M, Webb MS. A contrast of the cryostability of the plasma membrane of winter rye and spring oat-two species that widely differ in their freezing tolerance and plasma membrane lipid composition. In: *Advances in Low Temperature Biology.* Steponkus PL (Ed.). London: JAI Press, 1993; 2: 211–312.

Sudhir P, Murthy SDS. Effects of salt stress on basic processes of photosynthesis. *Photosynthetica* 2004;42:481–486.

Sun Y, Fan XY, Cao DM, Tang W, He K, Zhu JY. Integration of brassinosteroid signal transduction with the transcription network for plant growth regulation in Arabidopsis. *Dev. Cell* 2010;19:765–777.

Symons GM, Ross JJ, Jager CE, Reid JB. Brassinosteroid transport. *J. Exp. Bot.* 2008;59:17–24.

Tanaka K, Asami T, Yoshida S, Nakamura Y, Matsuo T, Okamoto S. Brassinosteroid homeostasis in Arabidopsis is ensured by feedback expressions of multiple genes involved in its metabolism. *Plant Physiol.* 2005;138:1117–1125.

Tikhomirov AA, Zolotukhin IG, Lisovskii GM, Sidko FY. Specificity of responses to the spectral composition of the PAR in plants of different species under artificial illumination. *Soviet Plant Physiol.* 1987;34:624–633.

Tong H, Xiao Y, Liu D, Gao S, Liu L, Yin Y. Brassinosteroid regulates cell elongation by modulating gibberellin metabolism in rice. *Plant Cell* 2014;26:4376–4393.

Tuteja N. Mechanism of high salinity tolerance in plants. *Methods Enzymol.* 2007; 428: 419–218.

Tuteja N. Cold, salt and drought stress. In: *Plant Stress Biology: From Genomics Towards System Biology.* Heribert, H. (Ed.). Weinheim, Germany: Wiley–Blackwell, 2010; 137–159.

Vahala J, Keinanen M, Schutzendubel A, Polle A, Kangasjarvi J. Differential effects of elevated ozone on two hybrid aspen genotypes predisposed to chronic ozone fumigation role of ethylene and salicylic acid. *Plant Physiol.* 2003;132:196–205.

Vardhini BV. Comparative study of Sorghum vulgare Pers. Grown in two experimental sites by brassinolide application at vegetative, flowering and grain filling stage. *Proc. Andhra Pradesh Akad. Sci.* 2013;15:75–79.

Vardhini BV. Studies on the effect of brassinolide on the antioxidative system of two varieties of sorghum grown in saline soils of Karaikal. *Asian Aust. J. Plant Sci. Biotechnol.* 2011;5:31–34.

Vardhini BV, Anjum NA. Brassinosteroids make plant life easier under abiotic stresses mainly by modulating major components of antioxidant defense system. *Front. Environ. Sci.* 2015;21:70–75.

Vardhini BV, Anjum NA, Gill SS, Gill R. Brassinosteroids role for amino acids, peptides and amines modulation in stressed plants are view. In: *Plant Adaptation to Environmental Change: Significance of Amino Acids and their Derivatives.* Wallingford, CT: CAB International, 2013; 300–316.

Vardhini BV, Anuradha S, Rao SSR. Brassinosteroids a great potential to improve crop productivity. *Indian J. Plant Physiol.* 2006;11:1–12.

Vollenweider P, Gunthardt-Goerg MS. Diagnosis of abiotic and biotic stress factors using the visible symptoms in foliage. *Environ. Pollut.* 2005;137:455–465.

Vriet C, Russinova E, Reuzeau C. From squalene to brassinolide: The steroid metabolic and signaling pathways across the plant kingdom. *Mol. Plant* 2013;6:1738–1757.

Wachsman MB, Ramirez JA, Talarico LB, Galagovsky LR, Coto CE. Anti viral activity of natural and synthetic brassinosteroids. *Curr. Med. Chem. Anti Infective Agents* 2004;3:163–179.

Wahid A, Gelani S, Ashraf M, Foolad MR. Heat tolerance in plant: An overview. *Environ. Exp. Bot.* 2007;61:199–223.

Wahid A, Shabbir A. Induction of heat stress tolerance in Barley seedlings by pre sowing seed treatment with glycinebetaine. *Plant Growth Regul.* 2005;46:133–141.

Wang B, Zeng G. Effect of 24-epibrassinolide on the resistance of rice seedlings to chilling injury. *Zhiwa Shengi Xuebao* 1993;19:53–60.

Wang H, Nagegowda DA, Rawat R, Bouvier-Nave P, Guo D, Bach TJ. Overexpression of Brassica juncea wild-type and mutant HMG-CoA synthase 1 in Arabidopsis upregulates genes in sterol biosynthesis and enhances sterol production and stress tolerance. *Plant Biotechnol. J.* 2012;10:31–42.

Wilen RW, Sacco M, Gusta LV, Krishna P. Effects of 24-epibrassinolide on freezing and thermo tolerance of brome grass (Bromusinermis) cell cultures. *Physiol. Plant.* 1995;95:195–202.

Wise R, Olson A, Schrader S, Sharkey T. Electron transport is the functional limitation of photosynthesis in field-grown pima cotton plants at high temperature. *Plant Cell Environ.* 2004;27:717–724.

Wu L, Zhang Z, Zhang H, Wang XC, Huang R. Transcriptional modulation of ethylene response factor protein JERF3 in the oxidative stress response enhances tolerance of tobacco seedlings to salt, drought, and freezing. *Plant Physiol.* 2008;148:1953–1963.

Wu X, Oh MH, Kim HS, Schwartz D, Imai BS, Yau PM. Transphosphorylation of E. coli proteins during production of recombinant protein kinases provides a robust system to characterize kinase specificity. *Front. Plant Sci.* 2012;3:262.

Xia XJ, Huang YY, Wang L, Huang LF, Yu YL, Zhou YH, Yu JQ. Pesticides –induced depression of photosynthesis was alleviated by 24- epibrassinolide pretreatment in Cucumis sativus L. *Pestic. Biochem. Physiol.* 2006;86:42–48.

Xia XJ, Wang YJ, Zhou YH, Tao Y, Mao WH, Shi K, Asami T, Chen Z, Yu JQ. Reactive oxygen species are involve in brassinosteroid- induced stress tolerance in cucumber. *Plant Physiol.* 2009;150:801–814.

Xia XJ, Zhou YH, Ding J, Shi K, Asami T, Chen Z. Induction of systemic stress tolerance by brassinosteroid in Cucumis sativus. *New Phytol.* 2011;191:706–720.

Xiao-Jian X, Yan J, Yan-Hong Z,Yuan T, Wei-Hua M, Kai S, Tadao A, Zhixiang C, Jing Q. Reactive oxygen species are involved in brassinosteroid-induced stress tolerance in cucumber. *Plant Physiol.* 2009;150(2):801–804.

Xu S, Li J, Zhang X, Wei H, Cui L. Effect of heat acclimation pretreatment on changes of membrane lipid peroxidation, antioxidant metabolites, and ultrastructure of chloroplast in two cool-season turf grass species under heat stress. *Environ. Exp. Bot.* 2006;56:274–285.

Yadav SK. Cold stress tolerance mechanisms in plants. *Agron. Suatain. Dev.* 2010;30:515–527.

Yadava P, Kaushal J, Gautam A, Parmar H, Singh I. Physiological and biochemical effect of 24-epibrassinolide on heat-stress adaptation in maize. *Nat. Sci.* 2016;59:171–179.

Yu X, Li L, Zola J, Aluru M, Ye H, Foudree A. A brassinosteroid transcriptional network revealed by genome-wide identification of BESI target genes in Arabidopsis thaliana. *Plant J.* 2011;65:634–646.

Yuan JS, Yong JS, Ming TW, Xu YL, Xiang G, Rong H, Jun M. Effect of seed priming in germination and seedling growth under water stress in rice. *Acta Agron. Sin.* 2010;36:1931–1940.

Yuan LB, Peng ZH, Zhi TT, Zho Z, Liu Y, Zhu Q. Brassinosteroid enhances cytokinin-induced anthocyanin biosynthesis in Arabidopsis seedlings. *Biol. Plant.* 2015;59:99–105.

Yuan T, Fujioka S, Takatsuto S, Matsumoto S, Gou X, He K. BEN1, a gene encoding a dihydroflavonol 4-reductase (DFR)-like protein, regulates the levels of brassinosteroids in Arabidopsis thaliana. *Plant J.* 2007;51:220–233.

Yueqing H, Shitou X, Yi S, Huiqun W, Weigui L, Shengying S, Langtao X. Brassinolide increases potato root growth in vitro in a dose-dependent way and alleviates salinity stress. *Bio Med Res. Int.* 2016; 16: 1–11.

Zhang F, Wang Y, Yang Y, Wu D, Wang H, Liu J. Involvement of hydrogen peroxide and nitric oxide in salt resistance in the calluses from Populus euphratica. *Plant Cell Environ.* 2007;7:775–785.

Zhang HX, Xia Y, Wang GP. Excess copper induces accumulation of hydrogen peroxide and increases lipid peroxidation and total activity of copper-zinc superoxide dismutase in roots of Elsholtzia haichowensis. *Planta* 2008;227:465–475.

Zhang SS, Cai ZY, Wang XL. The primary signaling outputs of brassinosteroids are regulated by abscisic acid signaling. *Proc. Natl. Acad. Sci. USA* 2009;106:4543–4548.

Zhu J, Liu Z, Brady EC, Otto-Bliesner BL, Marcott SA, Zhang J, Wang X, Nusbaumer J, Wong TE, Jahn A, Noone D. Investigating the direct meltwater effect in terrestrial oxygen-isotope paleoclimate records using an isotope-enabled Earth System Model. *Geophys. Res. Lett.* 2017;44:12501–12510.

Section V

Biotechnological Applications to Improve the Plant Metabolic Pathways Towards Better Adaptations

30

Genetic Engineering Approaches for Abiotic Stress Tolerance in Broccoli: Recent Progress

Pankaj Kumar, Ajay Kumar Thakur, and Dinesh Kumar Srivastava

CONTENTS

Abbreviations

KEGG	Kyoto Encyclopedia of Genes and Genomes
LEAP	Late embryogenesis abundant proteins
MALDI-TOF MS	Matrix-assisted laser desorption ionization time-of-flight mass spectrometry
miRNAs	MicroRNAs
PF2D	Phase fractionation
QTL	Quantitative trait locus
RT-PCR	Real-time reverse transcription polymerase chain reaction
sRNAs	small RNAs
TFs	Transcription factors

Introduction

Vegetables are an important source of nutrition to human health, as they provide minerals, micronutrients, vitamins, antioxidants, phytosterols, and dietary fiber. Vegetable cultivation is a significant part of the agricultural economy of nations, especially in the developing world. However, their productivity and quality are seriously hindered by a number of biotic and abiotic stresses, and post-harvest storage constraints. Lack of appropriate processing facilities also leads to qualitative and quantitative losses (Srivastava et al. 2016). A major challenge for plant scientists in the 21st century is to ensure global food security by using various scientific interventions to develop stable crops that are tolerant to multiple stresses, thus improving yields particularly in areas with adverse environmental conditions. In the past four decades, conventional breeding has played a significant role in the improvement of vegetable yield, quality, post-harvest life, and resistance to biotic and abiotic stresses. However, there are many constraints in conventional breeding, which can only be overcome by modern biotechnological tools.

Broccoli is a plant of the cabbage family, Brassicaceae (formerly Cruciferae). It is classified as the *Italica* cultivar group of the species *Brassica oleracea*. Other cultivars of this group of *B. oleracea* include cabbage (*Capitata* group), cauliflower (*Botrytis* group), kale and collard greens (*Acephala* group), kohlrabi (*Gongylodes* group), and brussels sprouts (*Gemmifera* group). The Chinese broccoli (*Alboglabra* group) is also a cultivar group of *B. oleracea*. Common varieties are Calabrese and Purple sprouting broccoli. Broccoli possesses abundant fleshy green flower heads arranged in a tree-like fashion on branches sprouting from a thick, edible stalk. The large mass of flower heads is surrounded by leaves. Broccoli most resembles its close relative cauliflower but is green rather than white. It is a cool weather crop that does poorly in hot summer weather. It requires fertile soil with good moisture supply. Broccoli is a relatively new crop in North, South, and Central America, Northern Europe, and Asia. In broccoli, purple and green variants occur, though, in the United States, it is only the green types that are grown. The green sprouting broccolis are classified according to their maturity such as early, medium, and late cultivars. In India, sprouting broccoli is hardly a commercial crop. A large number of F_1 hybrid cultivars are being marketed by the different seed companies in Japan, United States, and Europe. All *B. oleracea* are cross-pollinating plants and pollination occurs via insect vectors, the most common of which is the honeybee. Broccoli has a genetic characteristic of self-incompatibility, which encourages cross pollination resulting in higher levels of variability. Feher (1985) described nutritional quality and suitability for the export of the cultivars Kayak, Corvett and Coaster. All had good organoleptic properties, high protein, vitamin C, and mineral contents.

Broccoli (*B. oleracea* L. var. *italica*) is an important, nutritionally rich vegetable crop high in calcium, antioxidants, vitamin A, vitamin K, β-carotene, riboflavin, iron (Vallejo et al. 2003; Kumar and Srivastava 2016a,b), and selenium content (Finley et al. 2001; Finley 2003). This crop is not only economically important but also provides nutrients that have anti-cancerous properties contributed by sulforaphane glucosinolate (Keck et al. 2003) and quinone reductase glutathione S-transferase (Zhang et al. 1992; Fahey et al. 1997). However, this crop is amenable to a number of biotic and abiotic stresses. Conventional breeding, so far, has failed to address these issues due to the lack of resistance sources in the germplasm. Genetic manipulation using genetic engineering techniques has become the method of choice for broccoli improvement. Recent advancements in molecular biology and genetic transformation have made it possible to identify, isolate, and transfer desirable genes from any living organism to plants. Genetic engineering consists of isolating a gene of interest, ligating that gene with a desirable vector to form the recombinant-DNA molecule, and then transferring that gene into the plant genome to create a new function. In contrast to conventional breeding, which involves the random mixing of tens of thousands of genes present in both the resistant and susceptible plants, recombinant DNA technology allows the transfer of only the desirable genes to the susceptible plants and the preservation of valuable economic traits. However, a pre-requisite for transferring genes into plants is the availability of efficient regeneration and transformation methods. Plant regeneration studies in broccoli were successfully reported by various researchers using different explants such as peduncle (Christey and Earle 1991), anther (Chang et al. 1996), protoplasts (Kaur et al. 2006), hypocotyl (Zhong and Li 1993; Puddephat et al. 2001; Kim and Botella 2002; Ravanfar et al. 2009; Huang et al. 2011; Kumar and Srivastava 2015a; Kumar et al. 2015a), leaf tissue (Robertson and Earle 1986; Cao and Earle 2003; Farzinebrahimi et al. 2012; Kumar and Srivastava 2015a), cotyledon (Qin et al. 2006; Ravanfar et al. 2011, 2014; Kumar and Srivastava 2015a,b), and petiole (Kumar et al. 2015a,b). Metz et al. (1995a,b) were the first to report on *Agrobacterium*-mediated gene transfer studies in broccoli, which was followed by other works (Puddephat et al. 1996, 2001; Chen et al. 2004, 2007; Huang et al. 2005; Higgins et al. 2006; Bhalla and Singh 2008; Kumar et al. 2017). Highly

efficient plant regeneration and stable genetic transformation protocol of the broccoli are dependent on genotype and need to be established for each cultivar. This chapter addresses different aspects of the genetic improvement of broccoli for abiotic stress tolerance using plant genetic engineering techniques to provide insight towards its potential applications.

Broccoli: Abiotic stresses

Abiotic stresses such as heat, drought, and salinity are the major environmental constraints affecting the production and productivity of almost all vegetable crops. Conventional plant breeding has not been that successful in addressing abiotic stress mitigation at present. The reason might be that the traits are controlled by a number of genes present at a quantitative trait locus (QTL). To combat the negative effects of various abiotic stresses, it is a pre-requisite to identify potential candidate genes or QTLs (gene networks) associated with broad-spectrum multiple abiotic stress tolerance. Various abiotic stresses including drought, high temperature, salinity, frost, flood, etc. adversely affect overall crop growth and productivity by affecting the vegetative and reproductive stages of growth and development (Figure 30.1). These stresses generally trigger a series of physiological, biochemical, and molecular changes in the plants which result in damage to the cellular machinery (Figure 30.2). These abiotic stresses lead to a disruption in cellular osmotic balance leading to dysfunctional homeostasis and ion distribution and oxidative stress, which lead to the denaturation of the integral proteins of plants (Gill et al. 2016). Plants respond to such stresses in a variety of mechanisms which trigger the cell signaling process, transcriptional controls, and production of a number of stress-related tolerant proteins, antioxidants, and osmotic solutes to maintain homeostasis and to protect and repair the damaged integral proteins (Ramakrishna and Ravishankar 2011, 2013). Generally, stress-sensitive plants are unable to synthesize such compounds under stress conditions

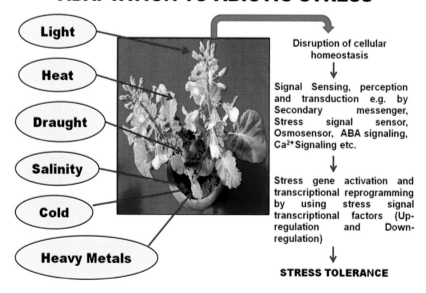

FIGURE 30.1 Adaptation to abiotic stress in broccoli.

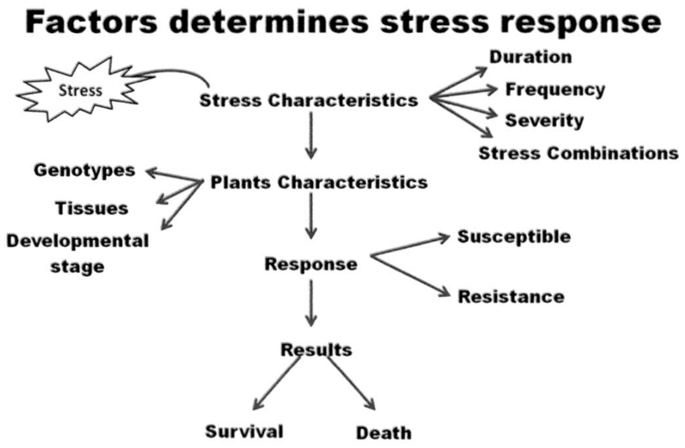

Factors determines stress response

FIGURE 30.2 Factors determine stress response in broccoli.

and thus are rendered liable to various stresses which hamper their overall growth. A number of genes have been identified in a number of plants/organisms, closely or distantly related, which code for the synthesis of these stress protecting compounds and thus can be targeted for genetic transformation into sensitive genotypes. Such genes have been classified into three categories: (1) genes which code for the synthesis of various osmolytes such as mannitol, glycine betaine, proline, heat shock proteins, etc.; (2) genes responsible for ion and water uptake and transport like aquaporins and ion transporter, etc., and (3) genes regulating transcriptional controls and signal transduction mechanism, for example *MAPK*, *DREBl*, etc.

Change in the variability of rainfall and temperature adversely affects the yield and nutritional quality of crops (Porter and Semenov 2005). Climate change will also influence the habitat range of pests and pathogens, with the increasing temperature facilitating pathogen spread (Bale et al. 2002; Luck et al. 2011; Madgwick et al. 2011; Nicol et al. 2011). Crop plants are therefore likely to encounter different environmental stresses, which, when occurring simultaneously, can have severe consequences. The changing climatic conditions, combined with an increasing pressure on global food productivity due to population increase, results in a demand for stress-tolerant crop varieties through the genetic manipulation of transcription factors (TFs), late embryogenesis abundant (LEA) proteins, antioxidant proteins, etc. (Umezawa et al. 2006; Bhatnagar-Mathur et al. 2008; Takeda and Matsuoka 2008; Newton et al. 2011). Stress-induced gene expression has been broadly categorized into three groups: (1) genes encoding proteins with known enzymatic or structural functions, (2) proteins with as yet unknown functions, and (3) regulatory proteins. Initially, attempts made by researchers to develop transgenics (mainly tobacco) for abiotic stress tolerance involved "single action genes", i.e. genes responsible for the modification of a single metabolite that confers increased tolerance to salt or drought stress (Bhatnagar-Mathur et al. 2008; Gill et al. 2016). Stress-induced proteins with known functions such as water channel proteins, key enzymes for osmolyte (proline, betaine, sugars such as trehalose, and polyamines) biosynthesis, detoxification enzymes, and transport proteins were the initial targets of plant transformation. However, that approach has not factored in the fact that abiotic stress tolerance is likely to involve

coordinated actions of many genes at a time, and that single-gene tolerance is unlikely to be sustainable. Therefore, a second "wave" of transformation attempts provides better integration of genetic engineering and plant physiology techniques to transform plants with the third category of stress-induced genes, namely, regulatory proteins. Through these proteins, many genes involved in stress response can be simultaneously regulated by a single gene encoding stress-inducible TF (Kasuga et al. 1999), thus offering the possibility of enhancing multiple stress tolerance (Bhatnagar-Mathur et al. 2008). Further, genetic engineering allows timing control, tissue-specificity, and expression level of the introduced genes for their optimal function. This is an important consideration if the action of a given gene or TF is desired only at a specific time, in a specific organ, or under specific conditions of stress. The basic findings and characterization of stress-inducible promoters have led to a major shift in the paradigm for genetically engineering stress-tolerant crops in recent years (Gill et al. 2016). With an increasing number of stresses, inducible gene/TF availability, genetic transformation, and molecular breeding gained a momentum shift for the incorporation of abiotic stress tolerance. Different crops respond to various stresses by highly complex coordinated mechanisms that involve changes at the physiological, cellular, and transcriptome levels. Evaluating the transgenic plants under stress conditions and understanding the physiological effect of the inserted desired genes at the whole plant level remain major challenges to overcome (Atkinson and Urwin 2012; Kayum et al. 2016).

Broccoli (*Brassica oleracea* var. *italica*) cultivars are commonly grown in cold climate regions and are regarded as cool-season vegetable crops (Suri et al. 2005). Under optimal temperatures of 18–25°C, this crop grows with normal flower buds and forms one large central head. However, when the temperature exceeds 30°C, heads become loose and branchy, tend to develop leaf-like structures, and initial floret development is disrupted (Bjorkman and Pearson 1998). The cultivar, which could also be grown on the highlands of the tropics, responds adversely to extreme temperatures and high humidity in the lowland; thus, gene transformation is essential for improving its tolerance against heat stress (Ravanfar and Aziz 2014). It was also reported that, when plants are subjected to high temperature and waterlogging, there is a reduction in chlorophyll content and water in the leaves but the level of hydrogen peroxide in the cells increases, resulting in reduced plant growth and development (Pucciariello et al. 2012).

Transgenic Approaches for Abiotic Stress Tolerance

Transgenic technologies involve different gene transfer methods for modern crop improvement programs. Differentially expressed abiotic stress tolerant gene(s)/TFs can be manipulated into novel combinations with desired traits, expressed ectopically and transferred to different crop species in which they do not naturally occur or vary. Hence, the ability to transform the vegetable crops with the gene(s) from any biological source (plant, animal, microbial) is an enormously prevailing tool for molecular plant breeding. Transgenic vegetables can be used as sources of new cultivars and their germplasm as new sources of variation in breeding programs. They are extremely useful as proof-of-concept tools to dissect and characterize the activity and interplay of gene networks for abiotic stress resistance/tolerance (Bhatnagar-Mathur et al. 2008; Gill et al. 2016).

Genetic improvement programs could be a reliable approach to tackle the broccoli cultivation problems. Genetic transformation is effective and less time-consuming compared to other methods of genetic improvement including conventional breeding (Cardoza and Stewart 2003; Hong et al. 2009). The most extensively applied technique for genetic transformation of *Brassica* spp, including cabbage and broccoli, is the *Agrobacterium tumefaciens*-mediated method (Cardoza and Stewart 2003). *Agrobacterium rhizogenes* has also been applied for the transformation of several members of the *Brassica* family (Henzi et al. 2000). However, in *A. rhizogenes*-mediated transformation, the production of transgenic plants with abnormal phenotypes after regeneration from the hairy root cultures is usually observed, which is a disadvantage of this method (Puddephat et al. 2001; Young et al. 2003). Different important factors that affect the transformation efficiency need to be optimized for achieving a high frequency of *Agrobacterium*-mediated transformation (Opabode 2006; Srivastava et al. 2016). The factors include pretreatment of explants (Hiei et al. 2006), bacterial concentration (Rafat et al. 2010) and immersion time (Xing et al. 2007), cocultivation condition (Vasudevan et al. 2007) as well as bacterial strain, plant genotype, and explant type (Kumar and Srivastava 2015a).

Heat shock proteins (HSPs) are functionally related proteins also called molecular chaperones. Their expression is increased when cells are exposed to elevated temperatures or other stresses (Narberhaus 2010; Omar et al. 2011). Among different HSPs, HSP101 appears to play a key role in conferring acquired thermotolerance (Waters et al. 1996; Torok et al. 2001). The production of heat-tolerant transgenic broccoli via *Agrobacterium*-mediated transformation with *Arabidopsis thaliana HSP 101* (*AtHSP101*), cDNA established the crucial role of *AtHSP101* in the heat tolerance characteristics. Ravanfar and Aziz (2014) reported heat stress tolerant broccoli cv. Green Marval with 5% transformation efficiency based on the positive PCR results using the optimized procedure. The expression of the luciferase reporter gene in the transformed cells and the transcription of *AtHSP101* using real-time reverse transcription polymerase chain reaction (RT-PCR) further confirmed the transgenic status of the regenerated plants. Transgenic *Brassica* plants have also shown improved tolerance to dehydration, as well as to other types of stresses (salt, heavy metals, and hydrogen peroxide) (Vinocur and Altman 2005). There is very limited literature available on abiotic stress tolerance efforts in broccoli. A strategy is proposed to genetically improve broccoli by the transfer of different abiotic stress tolerant genes characterized in other model crops, particularly in other vegetable *brassicas* using plant genetic engineering techniques.

Recent Molecular Advancement for Understanding Abiotic Stress Tolerance in Broccoli

The cellular metabolic components including calcium, reactive oxygen species (ROS), protein kinase, protein phosphatase etc. are activated and promote signal transduction pathways inside cells during stress and convey the abiotic stress signals that are necessary to regulate TFs (Gill and Tuteja 2010). The TFs control the downstream genes that regulate the abiotic stress through activating their corresponding *cis*-acting elements. Plant cells receive the external signals of each stress and consequently switch on the response to produce particular molecules to overcome such stress. Some molecules are involved in the defense response to the specific stress that contributes to the protection of the plant and expression of its resistance/tolerance to that particular stress. Molecular engineering is a new approach that is used to modify a gene coding and introduce specific genes in crop plants, which lead to increased tolerance against abiotic stresses (Gill et al. 2016). We need to understand the functions or mechanisms of the stress responsive candidate genes. Thereafter, introgression of those candidate genes through conventional breeding, marker-assisted back crossing, or even, in urgent situations, genetic transformation, will lead to a developed resistance or tolerance in the cultivar of broccoli.

Differentially expressed genes have been identified that activate against various abiotic stresses including heat, flood, drought, cold, and intense light (Kreps et al. 2002; Seki et al. 2002; Rizhsky et al. 2002). However, very few were reported for the simultaneous occurrence of multiple abiotic stresses, as opposed to individual stresses, which damage crop production (Rizhsky et al. 2002; Mittler and Blumwald 2010; Atkinson and Urwin 2012). Therefore, uncovering the physiological mechanisms whereby plants can withstand combined waterlogging and high temperature stress is greatly desired. A number of NAC TFs are involved in various abiotic stresses affecting Chinese cabbage (Liu et al. 2014). Different brassinazole-resistant (BZR) TFs and the longevity assurance gene of *B. rapa* showed responsiveness against cold, drought, and ABA stresses (Ahmed et al. 2012; Saha et al. 2015). Kayum et al. (2016) and Wang et al. (2015) showed that *BrMYB210*, *BrMYB137*, *BrMYB88*, *BrMYB154*, and *BrMYB222* significantly responded against cold and osmotic stresses. Ahmed et al. (2015) showed that 12 *BoCRGs* TFs were expressed differentially after cold stress treatment in two contrasting cabbage lines, and *BoCRG54, 56, 59, 62, 70, 72*, and *99* were predicted to be involved in cold regulatory pathways.

MicroRNAs (miRNAs) are known as a new class of endogenous regulators of a wide range of plant physiological processes, which act by regulating gene expression post-transcriptionally. The *Brassica* vegetable, broccoli growth and yield are also affected by salt stress. Tian et al. (2014) identified and characterized miRNAs related to salt stress in broccoli using high-throughput sequencing and bioinformatics analysis. The differential regulation of miRNAs between control and salt-stressed broccoli indicated that miRNAs play an integral role in the regulation of responses to salt stress. Differential miRNA expression was confirmed by RT-PCR analysis and miRNA target prediction was undertaken using the Kyoto Encyclopedia of Genes and Genomes (KEGG) Orthology (KO) database and Gene Ontology (GO)-enrichment analyses. Two libraries of small (or short) RNAs (sRNAs) were constructed and sequenced by high-throughput Solexa sequencing and a total of 24,511,963 and 21,034,728 clean reads, representing 9,861,236 (40.23%) and 8,574,665 (40.76%) unique reads, were reported for control and salt-stressed broccoli, respectively. Furthermore, 42 putative known and 39 putative candidate miRNAs that were differentially expressed between control and salt-stressed broccoli were revealed by their read counts and confirmed by the use of stem-loop RT-PCR. Among these, the putative conserved miRNAs, miR393 and miR855, and two putative candidate miRNAs, miR3 and miR34, were the most strongly down-regulated when broccoli was salt-stressed, whereas the putative conserved miRNA, miR396a, and the putative candidate miRNA, miR37, were the most up-regulated. Finally, analysis of the predicted gene targets of miRNAs using the GO and KO databases indicated that a range of metabolic and other cellular functions known to be associated with salt stress were up-regulated in broccoli treated with salt.

Proteomic analysis is a powerful approach for revealing differentially expressed proteins under given conditions. Liu et al. (2013) identified a number of differentially expressed proteins using proteome analysis in broccoli florets that were treated with N6-benzylaminopurine, revealing a complex network that provides comprehensive information on post-harvest yellowing response mechanisms (Liu et al. 2011). The production of broccoli (*Brassica oleracea* L. var. *italica*) is largely reduced by waterlogging and high temperature stresses. Lin et al. (2015) carried out proteomic analysis in broccoli (*Brassica oleracea* L. var. *italica*) under high temperature and waterlogging stresses using two-dimensional liquid phase fractionation (PF2D) and matrix-assisted laser desorption ionization time-of-flight mass spectrometry (MALDI-TOF MS) and identified 31 differentially expressed proteins from heat-tolerant and heat-susceptible broccoli cultivars TSS-AVRDC-2 and B-75, respectively, under high temperature and/or waterlogging stresses. They further cloned the stress-responsive Rubisco genes and studied their transcript levels under stress conditions. From results, it was revealed that the broccoli cultivar TSS-AVRDC-2 exhibited significantly higher chlorophyll content, lower stomatal conductance, and better H_2O_2 scavenging under stress in comparison to the broccoli cultivar B-75. Two-dimensional liquid PF2D analyses revealed that Rubisco proteins in both cultivars were regulated under stressing treatments, and that TSS-AVRDC-2 had higher levels of both Rubisco large and small subunit transcripts than the cultivar B-75 when subjected to high temperature and/or waterlogging. Finally, it was concluded that higher levels of Rubisco proteins in TSS-AVRDC-2 could lead to increased carbon fixation efficiency to provide sufficient energy to enable stress tolerance under waterlogging at 40°C. This provides a basis for understanding broccoli metabolic pathways and their cross-talk under stress.

Conclusion

With the advent of modern molecular biotechnology, newer tools permitting gene transfer across the species, such as transgenics, have opened an avenue for solving the age-old problems of global hunger due to increasing population, continuous depletion of natural resources, and erratic changes in global climate. Conventional agricultural practices alone are unable to sustain the quality and quantity of the agricultural produce. The use of transgenes to improve abiotic stress tolerance in crops remains an attractive option by a thorough understanding of the physiological processes in response to different abiotic stresses, choice of a given promoter or TF to be used during transformation. So far, comparatively lesser efforts have been directed to the genetic improvement of broccoli for the incorporation of an abiotic stress tolerant trait as compared to the other aspects including biotic stresses (Kumar and Srivastava 2016a). Huge research efforts are needed to identify particular gene/gene networks and TFs/miRNAs/siRNAs that enable the expression of a number of genes involved in imparting heat or water logging stress tolerance in broccoli using a number of '*omics*' technologies. Further, there is a need to address various regulatory obstacles for the commercial release of various transgenic crops in the country so that the real benefit of this wonderful technology may reach to the consumers, the end users.

REFERENCES

Ahmed, N. U., Jung, H. J., Park, J. I., Cho, Y. G., Hur, Y., and Nou, I. S. 2015. Identification and expression analysis of cold and freezing stress responsive genes of *Brassica oleracea*. *Gene* 554:215–223.

Ahmed, N. U., Park, J. I., Jung, H. J., Lee, I. H., Song, I. J., Yang, S. Y., et al. 2012. Identification and characterization of longevity assurance gene related to stress resistance in *Brassica*. *African Journal of Biotechnology* 11:12721–12727.

Atkinson, N. J., and Urwin, P. E. 2012. The interaction of plant biotic and abiotic stresses: From genes to the field. *Journal of Experimental Botany* 63(10):3523–3543.

Bale, J. S., Masters, G. J., Hodkinson, I. D., Awmack, C., Bezemer, T. M., Brown, V. K., et al. 2002. Herbivory in global climate change research: Direct effects of rising temperature on insect herbivores. *Global Change Biology* 8:1–16.

Bhalla, P. L., and Singh, M. B. 2008. *Agrobacterium*-mediated transformation of *Brassica napus* and *Brassica oleracea*. *Nature Protocols* 3:181–189.

Bhatnagar-Mathur, P., Vadez, V., and Sharma, K. K. 2008. Transgenic approaches for abiotic stress tolerance in plants: Retrospect and prospects. *Plant Cell Reports* 27:411–424.

Bjorkman, T., and Pearson, K. J. 1998. High temperature arrest of inflorescence development in broccoli (*Brassica oleracea* L. var. *italica*). *Journal of Experimental Botany* 49: 101–106.

Cao, J., and Earle, E. D. 2003. Transgene expression in broccoli (*Brassica oleracea* var. *italica*) clones propagated *in vitro* via leaf explants. *Plant Cell Reports* 21:789–796.

Cardoza, V., and Stewart, C. N. Jr. 2003. Increased Agrobacterium mediated transformation and rooting efficiencies in canola (*Brassica napus* L.) from hypocotyl explants. *Plant Cell Reports* 21:599–604.

Chang, Y. M., Liou, P. C., and Hsiao, C. H. 1996. Anther culture of cabbage (*Brassica oleracea* L. var. *capitata*) and broccoli (*B. oleracea* L. var. *italica*). *Journal of Agriculture Research in China* 45:35–46.

Chen, L. O., Chin, H. L., Kelkarc, S. M., Chang, Y. M., and Shawa, J. F. 2007. Transgenic broccoli (*Brassica oleracea* L. var. *italica*) with antisense chlorophyllase (*BOCLH1*) delays postharvest yellowing. *Plant Science* 174:25–31.

Chen, L. O., Hwang, J. Y., Wang, Y. H., Chen, Y. T., and Shaw, J. 2004. Ethylene insensitive and post-harvest yellowing retardation in mutant ethylene response sensor gene transformed broccoli. *Molecular Breeding* 14:199–213.

Christey, M. C., and Earle, E. D. 1991. Regeneration of *Brassica oleracea* from peduncle explants. *Horticulture Science* 26:1069–1072.

Fahey, J. W., Zhang, Y., and Talalay, P. 1997. Broccoli sprouts: An exceptionally rich source of inducers of enzymes that protect against chemical carcinogens. *Proceeding of National Academy of Sciences USA* 99:10367–10372.

Farzinebrahimi, R., Taha, R. M., Fadainasab, M., and Mokhtar, S. 2012. *In vitro* plant regeneration, antioxidant and antibacterial studies on broccoli, *Brassica oleracea* var. *italica*. *Pakistan Journal of Botany* 44:2117–2122.

Feher. 1985. Sprouting broccoli. In *Vegetable Crops*, ed. Bose, T. K., Kabir, J., Maity, T. K., Parthasarathy, V. A., and Som, M. G., 411–419. Naya Prakash, Kolkata.

Finley, J. W. 2003. Reduction of cancer risk by consumption of selenium-enriched plants: Enrichment of broccoli with selenium increases the anticarcinogenic properties of broccoli. *Journal of Medicinal Food* 6:19–26.

Finley, J. W., Ip, C., Lisk, D. J., Davis, C. D., Hintze, K. G., and Whanger, P. D. 2001. Cancer-protective properties of high-selenium broccoli. *Journal of Agriculture and Food Chemistry* 49:2679–2683.

Gill, S. S., Anjum, N. A., Gill, R., and Tuteja, N. 2016. Abiotic stress signaling in plants—An overview. In *Abiotic Stress Response in Plants*, ed. Tuteja, N., and Gill, S. S., 3–12. Wiley-VCH Verlag GmbH & Co. KGaA, Weinheim.

Gill, S. S., and Tuteja, N. 2010. Reactive oxygen species and antioxidant machinery in abiotic stress tolerance in crop plants. *Plant Physiology and Biochemistry* 48:909–930.

Henzi, M. X., Christey, M. C., and McNeil, D. L. 2000. Factors that influence *Agrobacterium rhizogenes*-mediated transformation of broccoli (*Brassica oleracea* L. var. *italica*). *Plant Cell Reports* 19:994–999.

Hiei, Y., Ishida, Y., Kasaoka, K., and Komari, T. 2006. Improved frequency of transformation in rice and maize by treatment of immature embryos with centrifugation and heat prior to infection with *Agrobacterium tumefaciens*. *Plant Cell Tissue and Organ Culture* 87:233–243.

Higgins, J. D., Newbury, H. J., Barbara, D. J., Muthumeenakshi, S., and Puddephat, I. J. 2006. Production of marker-free genetically engineered broccoli with sense and antisense *ACC synthase 1* and *ACC oxidase 1* and *2* to extend shelf life. *Molecular Breeding* 17:7–20.

Hong, Z. J., Yan, Z. H., Guo, T. N., Fang, T. Y., Yun, Z. X., and Xiu, L. Y. 2009. Several methods to detect the inheritance and resistance to the Diamondback Moth in transgenic Chinese cabbage. *African Journal of Biotechnology* 8(12):2887–2892.

Huang, K., Jiashu, C., Xiaolin, Y., Wanzhi, Y., Gang, L., and Xiang, X. 2005. Plant male sterility induced by antigene *CYP86MF* in *Brassica oleracea* L. var. *italica*. *Agriculture Science China* 4:806–810.

Huang, K., Qiuyun, W., Juncleng, L., and Zheng, J. 2011. Optimization of plant regeneration from broccoli. *African Journal of Biotechnology* 10:4081–4085.

Kasuga, M., Liu, Q., Miura, S., Yamaguchi-Shinozaki, K., and Shinozaki, K. 1999. Improving plant drought, salt, and freezing tolerance by gene transfer of a single stress inducible transcription factor. *Nature Biotechnology* 17:287–291.

Kaur, N., Vyvadilova, M., Klima, M., and Bechyne, M. 2006. A simple procedure for mesophyll protoplast culture and plant regeneration in *Brassica oleracea* L. and *Brassica napus* L. *Czech Journal of Genetics and Plant Breeding* 3:103–110.

Kayum, M. A., Park, J. I., Ahmed, N. U., Saha, G., Chung, M. Y., Kang, J. G., et al. 2016. Alfin-like transcription factor family: Characterization and expression profiling against stresses in *Brassica oleracea*. *Acta Physiologia Plantarum* 38:127.

Kayum, M. D., Kim, H. T., Nath, U. K., Park, J. I., Kho, K. H., Cho, Y. G., et al. 2016. Research on biotic and abiotic stress related genes exploration and prediction in *Brassica rapa* and *B. oleracea*: A review. *Plant Breeding and Biotechnology* 4(2):135–144.

Keck, A. S., Qiao, Q., and Jeffery, E. H. 2003. Food matrix effects on bioactivity of broccoli-derived sulforaphane in liver and colon of f344 rats. *Journal of Agriculture and Food Chemistry* 51:3320–3327.

Kim, J. H., and Botella, J. R. 2002. Callus induction and plant regeneration from broccoli (*Brassica oleracea* var. *italica*) for transformation. *Journal of Plant Biology* 45:177–181.

Kreps, J. A., Wu, Y. J., Chang, H. S., Zhu, T., Wang, X., and Harper, J. F. 2002. Transcriptome changes for *Arabidopsis* in response to salt, osmotic, and cold stress. *Plant Physiology* 130:2129–2141.

Kumar, P., Gambhir, G., Gaur, A., and Srivastava, D. K. 2015a. Molecular analysis of genetic stability in *in vitro* regenerated plants of broccoli (*Brassica oleracea* L. var. *italica*). *Current Science* 109(8):1470–1475.

Kumar, P., Gaur, A., and Srivastava, D. K. 2015b. Morphogenic response of leaf and petiole explants of broccoli using thidiazuron. *Journal of Crop Improvement* 29:432–446.

Kumar, P., Gaur, A., and Srivastava, D. K. 2017. *Agrobacterium*-mediated insect resistance gene (*cry1Aa*) transfer studies pertaining to antibiotic sensitivity on cultured tissues of broccoli (*Brassica oleracia* L. var. *italica*): An important vegetable crop. *International Journal of Vegetable Science* 23(6):523–535. doi:10.1080/19315260.2017.1334734.

Kumar, P., and Srivastava, D. K. 2015a. Effect of potent cytokinin thidiazuron (TDZ) on *in vitro* morphogenic potential of broccoli (*Brassica oleracea* L. var. *italica*), an important vegetable crop. *Indian Journal of Plant Physiology* 20(4):317–323.

Kumar, P., and Srivastava, D. K. 2015b. High frequency organogenesis in hypocotyl, cotyledon, leaf and petiole explants of broccoli (*Brassica oleracea* L. var. *italica*), an important vegetable crop. *Physiology and Molecular Biology of Plants* 21(2):279–285.

Kumar, P., and Srivastava, D. K. 2016a. Biotechnological advancement in genetic improvement of broccoli (*Brassica oleracea* L. var. *italica*), an important vegetable crop. *Biotechnology Letters* 38(7):1049–1063.

Kumar, P., and Srivastava, D. K. 2016b. Biotechnological application in *in vitro* plant regeneration studies of Broccoli (*Brassica oleracea* l. var. *italica*), an important vegetable crop. *Biotechnology Letters* 38(4):561–571.

Lin, H. H., Lin, K. H., Chen, S. C., Shen, Y. H., and Lo, H. F. 2015. Proteomic analysis of broccoli (*Brassica oleracea*) under high temperature and waterlogging stresses. *Botanical Studies* 56:18.

Liu, M. S., Li, H. C., Chang, Y. M., Wu, M. T, and Chen, L. F. 2011. Proteomic analysis of stress-related proteins in transgenic broccoli harboring a gene for cytokinin production during postharvest senescence. *Plant Science* 181:288–299.

Liu, M. S., Li, H. C., Lai, Y. M., Lo, H. F., and Chen, L. F. 2013. Proteomics and transcriptomics of broccoli subjected to exogenously supplied and transgenic senescence induced cytokinin for amelioration of postharvest yellowing. *Journal of Proteomics* 20:133–144.

Liu, T., Song, X., Duan, W., Huang, Z., Liu, G., Li, Y., et al. 2014. Genome-wide analysis and expression patterns of NAC transcription factor family under different developmental stages and abiotic stresses in Chinese cabbage. *Plant Molecular Biology Reports* 32:1041–1056.

Luck, J., Spackman, M., Freeman, A., Trebicki, P., Griffiths, W., Finlay, K., and Chakraborty, S. 2011. Climate change and diseases of food crops. *Plant Pathology* 60:113–121.

Madgwick, J. W., West, J. S., White, R. P., Semenov, M. A., Townsend, J. A., Turner, J. A., et al. 2011. Impacts of climate change on wheat anthesis and fusarium ear blight in the UK. *European Journal of Plant Pathology* 130:117–131.

Metz, T. D., Dixit, T. R., and Earle, E. D. 1995a. *Agrobacterium tumefaciens* mediated transformation of broccoli (*Brassica oleracea* var. *italica*) and cabbage (*B. oleracea* var. *capitata*). *Plant Cell Reports* 15:287–292.

Metz, T. D., Roush, R. T., Tang, J. D., Shelton, A. M., and Earle, E. D. 1995b. Transgenic broccoli expressing a *Bacillus thuringiensis* insecticidal crystal protein: Implications for pest resistance management strategies. *Molecular Breeding* 4:309–317.

Mittler, R., and Blumwald, E. 2010. Genetic engineering for modern agriculture: Challenges and perspectives. *Annual Review in Plant Biology* 61:443–462.

Narberhaus, F. 2010. Translational control of bacterial heat shock and virulence genes by temperature-sensing mRNAs. *RNA Biology* 7:84–89.

Newton, A. C., Johnson, S. N., and Gregory, P. J. 2011. Implications of climate change for diseases, crop yields and food security. *Euphytica* 179:3–18.

Nicol, J. M., Turner, S. J., Coyne, D. L., Den, N. L., Hockland, S., and Tahna, M. Z. 2011. Current nematode threats to world agriculture. In *Genomics and Molecular Genetics of Plant–nematode Interactions*, ed. Jones, J., Gheysen, G., and Fenoll, C., 21–44. Springer, London.

Omar, S. A., Fu, Q. T., Chen, M. S., Wang, G. J., Song, S. Q., Elsheery, N. I., et al. 2011. Identification and expression analysis of two small heat shock protein cDNAs from developing seeds of biodiesel feedstock plant *Jatropha curcas*. *Plant Science* 181:632–637.

Opabode, J. T. 2006. *Agrobacterium*-mediated transformation of plants: Emerging factors that influence efficiency. *Biotechnology and Molecular Biology Review* 1:12–20.

Porter, J. R., and Semenov, M. A. 2005. Crop responses to climatic variation. *Philosophical Transactions of the Royal Society B: Biological Sciences* 360:2021–2035.

Pucciariello, C., Parlanti, S., Banti, V., Novi, G., and Perata, P. 2012. Reactive oxygen species-driven transcription in *Arabidopsis* under oxygen deprivation. *Plant Physiology* 159(1): 184–196.

Puddephat, I. J., Riggs, T. J., and Fenning, T. M. 1996. Transformation of *Brassica oleracea* L.: A critical review. *Molecular Breeding* 2(3):185–210.

Puddephat, I. J., Robinson, H. T., Fenning, T. M., Barbara, D. J., Morton, A., and Pink, D. A. C. 2001. Recovery of phenotypically normal transgenic plants of *Brassica oleracea* L. var. *italica* upon *Agrobacterium rhizogenes*-mediated co-transformation and selection of transformed hairy roots by GUS assay. *Molecular Breeding* 7:229–242.

Qin, Y., Li, H. L., and Guo, Y. D. 2006. High frequency embryogenesis, regeneration of broccoli (Brassica oleracea var. italica) and analysis of genetic stability by RAPD. *Scientia Horticulturae* 111:203–208.

Rafat, A., Aziz, M. A., Rashid, A. A., Abdullah, S. N. A., Kamaladini, H., Sirchi, M. H. T., et al. 2010. Optimization of *Agrobacterium tumefaciens*-mediated transformation and shoot regeneration after co-cultivation of cabbage (*Brassica oleracea* subsp. *capitata* cv. KY Cross) with *AtHSP101* gene. *Scientia Horticulture* 124:1–8.

Ramakrishna, A., and Ravishankar, G. A. 2011. Influence of abiotic stress signals on secondary metabolites in plants. *Plant Signaling and Behavior* 6:1720–1731.

Ramakrishna, A., and Ravishankar, G. A. 2013. Role of plant metabolites in abiotic stress tolerance under changing climatic conditions with special reference to secondary compounds. In *Climate Change and Plant Abiotic Stress Tolerance*, 705–726, Wiley-VCH Verlag GmbH & Co. KGaA, Weinheim.

Ravanfar, S. A., and Aziz, M. A. 2014. Shoot tip regeneration and optimization of *Agrobacterium tumefaciens*-mediated transformation of Broccoli (*Brassica oleracea* var. *italica*) cv. Green Marvel. *Plant Biotechnology* 9:27–36.

Ravanfar, S. A., Aziz, M. A., Kadir, M. A., Rashid, A. A., and Haddadi, F. 2011. *In vitro* shoot regeneration and acclimatization of *Brassica oleracea* var. *italica* cv. Green marvel. *African Journal of Biotechnology* 10:5614–5619.

Ravanfar, S. A., Aziz, M. A., Kadir, M. A., Rashid, A. A., and Sirchi, M. H. T. 2009. Plant regeneration of *Brassica oleracea* var. *italica* (broccoli) cv. Green marvel was affected by plant growth regulators. *African Journal of Biotechnology* 8:2523–2528.

Ravanfar, S. A., Aziz, M. A., Rashid, A. A., and Shahida, S. 2014. *In vitro* adventitious shoot regeneration from cotyledon explant of *Brassica oleracea* subsp. *italica* and *Brassica oleracia* subsp. *capitata* using TDZ and NAA. *Pakistan Journal of Botany* 46:329–335.

Rizhsky, L., Liang, H. J., and Mittler, R. 2002. The combined effect of drought stress and heat shock on gene expression in tobacco. *Plant Physiology* 130:1143–1151.

Robertson, D., and Earle, E. D. 1986. Plant regeneration from leaf protoplasts of *Brassica oleracea* L. var. *italica*. *Plant Cell Reports* 5:61–64.

Saha, G., Park, J. I., Jung, H. J., Ahmed, N. U., Kayum, M. A., Kang, J. G., et al. 2015. Molecular characterization of BZR transcription factor family and abiotic stress induced expression profiling in *Brassica rapa*. *Plant Physiology and Biochemistry* 92:92–104.

Seki, M., Narusaka, M., Ishida, J., Nanjo, T., Fujita, M., Oono, Y., et al. 2002. Monitoring the expression profiles of 7000 *Arabidopsis* genes under drought, cold and high-salinity stresses using a full-length cDNA microarray. *The Plant Journal* 31:279–292.

Srivastava, D. K., Kumar, P., Sharma, S., Gaur, A., and Gambhir, G. 2016. Genetic engineering for insect resistance in economically important vegetable crops. In *Plant Tissue Culture: Propagation, Conservation and Crop*, ed. Ahmad, N., and Anis, M., 343–378, Springer Publishing House, Singapore.

Suri, S. S., Saini, A. R. K., and Ramawat, K. G. 2005. High Frequency Regeneration and *Agrobacterium tumefaciens*-mediated Transformation of Broccoli (*Brassica oleracea* var. *italica*). *European Journal of Horticulture Science* 70:71–78.

Takeda, S., and Matsuoka, M. 2008. Genetic approaches to crop improvement: Responding to environmental and population changes. *Nature Reviews Genetics* 9:444–457.

Tian, Y., Tian, Y., Luo, X., Zhou, T., Huang, Z., Liu, Y., et al. 2014. Identification and characterization of microRNAs related to salt stress in broccoli, using high-throughput sequencing and bioinformatics analysis. *BMC Plant Biology* 14:226.

Torok, Z., Goloubinoff, P., Horvath, I., Tsvetkova, N. M., Glatz, A., Balogh, G., et al. 2001. Synechocystis *HSP17* is an amphitropic protein that stabilizes heat-stressed membranes and binds denatured proteins for subsequent chaperone mediated refolding. *Proceeding of National of Academy Science USA* 98:3098–3103.

Umezawa, T., Fujita, M., Fujita, Y., Yamaguchi-Shinozaki, K., and Shinozaki, K. 2006. Engineering drought tolerance in plants: Discovering and tailoring genes to unlock the future. *Current Opinion in Biotechnology* 17:113–122.

Vallejo, F., Garcia-viguera, C., and Tomas-barberan, F. A. 2003. Changes in broccoli (*Brassica oleracea* var. *italica*) health promoting compounds with inflorescence development. *Journal of Agriculture and Food Chemistry* 51:3776–3782.

Vasudevan, A., Selvaraj, N., Ganapathi, A., and Choi, C. W. 2007. *Agrobacterium*-mediated genetic transformation in cucumber (*Cucumis sativus* L.). *American Journal of Biotechnology and Biochemistry* 3:24–32.

Vinocur, B., and Altman, A. 2005. Recent advances in engineering plant tolerance to abiotic stress: Achievements and limitations. *Current Opinion in Biotechnology* 16:123–132.

Wang, Z., Tang, J., Hu, R., Wu, P., Hou, X. L., Song, X.M., et al. 2015. Genome-wide analysis of the *R2R3-MYB* transcription factor genes in Chinese cabbage (*Brassica rapa* ssp. *pekinensis*) reveals their stress and hormone responsive patterns. *BMC Genomics* 16:17.

Waters, E. R., Lee, G. J., and Vierling, E. 1996. Evolution, structure and function of the small heat shock proteins in plants. *Journal of Experimental Botany* 47:325–338.

Xing, Y., Yang, Q., Ji, Q., Luo, Y., Zhang, Y., Gu, K., et al. 2007. Optimization of *Agrobacterium*-mediated transformation parameters for sweet potato embryogenic callus using β-glucuronidase (GUS) as a reporter. *African Journal of Biotechnology* 6:2578–2584.

Young, J. M., Kerr, A., and Sawada, H. 2003. Genus Agrobacterium. In *Bergey's Manual of Systematic Bacteriology*, 2nd ed. Springer, New York.

Zhang, Y. S., Talalay, P., Cho, C. G., and Posner, G. 1992. A major inducer of anticarcinogenic protective enzymes from broccoli: Isolation and elucidation of structure. *Proceeding of National Academy of Sciences USA* 89:2399–2403.

Zhong, Z. X., and Li, X. 1993. Plant regeneration from hypocotyl protoplasts culture of *Brasscia oleracea* L. var. *italica*. *Acta Agriculture Shanghai* 9:13–18.

31

Impact of Abiotic Stresses on Metabolic Adaptation in Opium Poppy (Papaver somniferum L.)

Ankesh Pandey, Satya N. Jena, and Sudhir Shukla

CONTENTS

Abbreviations

6OMT	6-*O*-Methyltransferase
7OMT	7-*O*-Methyltransferase
ABA	Abscisic Acid
AOX1a	Alternate Oxidase
AP2	APETALA2
At	*Arabidopsis thaliana*
BBE	Berberine Bridge Enzyme
bHLH	Basic Helix Loop Helix
BIA	Benzylisoquinoline Alkaloids
BZIP	Basic Zipper
CAP2	Single AP2 Domain Containing Transcription Activator from Chickpea
COR	Codeinone Reductase
Cr	*Catharanthus roseus*
DCL	Dicer Like
DREB	Dehydration Responsive Element Binding
DRE/CRT	Dehydration Responsive Element/C-Repeat
ELIPs	Early Light-Induced Proteins
ERD	Early Responsive To Dehydration
EREBP	Ethylene-Responsive Element Binding Proteins
ERF	ETS2 Repressor Factor
FAO	Food and Agriculture Organization
GABA	Gamma Aminobutyric Acid
GC-MS	Gas Chromatography-Mass Spectroscopy
GSH	Glutathione (Reduced)
HSPs	Heat Shock Proteins
L-DOPA	L-Dihydroxy Phenylalanine
LEA	Late Embryogenesis Abundant
MADS	SMCM1, AGAMOUS, DEFICIENS, SRF
MeJA	Methyl Jasmonate
MIPS	Major Intrinsic Proteins
MLPs	Major Latex Protein
MYB	Myeloblastosis
NAC	NAM, ATAF, CUC
NADPH	Nicotinamide Adenine Dinucleotide Phosphate (Reduced)
NMR	Nuclear Magnetic Resonance
OPP	Oxidative Pentose Phosphate
P5CS1	Pyrroline-5-Carboxylate Synthase
PAL	Phenylalanine Ammonia Lyase
PDF1.2	PLANT DEFENSIN1.2
PR	Pathogenesis Related
Ps	*Papaver somnifera*
R	Resistance
RNS	Reactive Nitrogen Species
ROS	Reactive Oxygen Species
SA	Salicylic Acid
T6ODM	Thebaine 6-*O*-Demethylase
TCA	Tri-Carboxylic Acid
TYDC	Tyrosine/DOPA Decarboxylase
UAS	Upstream Activating Sequence
UGTs	UDP-Glucuronosyltransferases
VIGS	Virus-Induced Gene Silencing
WRKY	W-Tryptophan, R-Arginine, K-Lysine, Y-Tyrosine

Introduction

Plants are exposed to various types of stresses in adverse environmental conditions that compel them to alter their normal metabolic activities for their survival. Stresses are the factors which alter the usual vital activities of a plant such as metabolism, adaptive functions, and growth. Metabolic pathways are interlinked directly and indirectly with the growth and development of a plant during its life cycle, however, its adaptive features to cope with stress are performed by the secondary metabolism. Stresses are broadly categorized as factors associated with living beings, i.e., biotic factors and the factors associated with nonbiological parameters known as abiotic factors. Abiotic factors are major environmental factors affecting the whole life cycle of the plant. Major abiotic factors are light, temperature, soil, water, and chemicals, etc. The relationships among these abiotic factors may be additive, antagonistic, or agonistic, for example, an increase in light intensity accelerates photosynthesis as well as increases water absorption by roots in plants. Soil property is another factor related to water holding capacity, which ultimately indicates the water availability for the plant. Certain chemicals like osmolytes are crucial for the plant to survive in a high salt environment. These chemicals are necessary for maintaining osmotic balance of the cells. Temperature and photosynthesis are positively co-related at the sub-optimum level, which has a direct influence on the rate of transpiration as well as water absorption by the root from the soil.

Abiotic stresses influence the production and accumulation of the secondary metabolites in plants as per their requirement to defend them against the different stresses. However, the concentration of secondary metabolites fluctuates in plants depending on the type of abiotic stress, which has been proven by several studies conducted on various plants. Abiotic stresses are known to activate a multigame response by dynamic changes in various primary and secondary metabolite content (Ramakrishna and Ravishankar, 2011, 2013). The precise analysis of proteome and metabolome is indispensable to integrate the fundamentals of stress biochemistry and physiology. Despite these factors, geographical locations and environmental conditions can also affect the metabolite biosynthesis. In some geographical areas or climatic conditions, plants adapt themselves better than in other conditions. Optimum climatic conditions, which support the growth of plants, are an estimation of the concentration of secondary metabolites. But, the same plant can show variation in concentration of secondary metabolites in different geographical regions that depend upon the availability of favorable or unfavorable climatic conditions to the plants.

Opium poppy (*Papaver somniferum*), a medicinally important plant, is a well characterized and established crop cultivated by only licensed cultivators due to its narcotic nature. *Papaver somniferum* belongs to the family Papaveraceae. This family is prominently distributed in north temperate to subtropical regions. They also grow in habitats in southern montane Africa and eastern Australia, but the highest diversity has been seen in western North America and temperate Eurasia. There are 42 genera and 775 species altogether. Common genera distributed in the US and Canada are *Argemone, Papaver, Eschscholzia,* and *Corydalis.* The plants are mostly annual or perennial herbs but can be shrubs

or trees (Christenhusz and Byng, 2016). The plants of the *Papaver somniferum* have specialized lactiferous cells, prominently in its capsules, which accumulate opium latex. The latex contains more than 80 secondary metabolites (alkaloids); the five most common of which are morphine, codeine, thebaine, narcotine, and papaverine (Rastogi et al., 2012). The content of alkaloids varies according to environment/climatic conditions, especially due to various biotic and abiotic stresses (Verma and Shukla, 2015). This chapter deals with the influence of various abiotic stresses on metabolic adaptation in *Papaver somniferum* L.

Benzylisoquinoline Alkaloid (BIA) Biosynthetic Pathway in Opium

The Benzylisoquinoline class of compounds represents the major alkaloids of *Papaver someniferum* that includes papaverine, noscapine, codeine, morphine, berberine, protopine, etc. Although these alkaloids have a common precursor (tyrosine), they are regulated at multilevel points under their modifications and substitution. Nowadays, people are confident at targeting a particular gene of the pathway leading to a diversion of metabolic flux for the accumulation of desired alkaloids (Gomez et al., 2007; Leonard et al., 2009; Allen et al., 2004; Beaudoin and Facchini, 2014). Though in recent years the metabolic pathway for the biosynthesis of the Benzylisoquinoline class of compounds has been better understood (Figure 31.1), there are still several regulatory mechanisms running in the background of the biosynthesis that are unknown. The challenge of solving the enigma that is linked to the yield and accumulation of certain alkaloids under altered growth conditions during stress still needs to be explored.

The detailed physiological role of BIA biosynthetic enzymes has been investigated using a variety of gene suppression and over-expression methods in *Papaver somniferum* (Wijekoon and Facchini, 2012; Dang and Facchini, 2012; Desgagné-Penix and Facchini, 2012; Alagoz et al., 2016; Boke et al., 2015). The latex of poppy capsules is the main source of BIAs, but the differential accumulation of alkaloids is found throughout the plant. Induction of chemical elicitors, such as methyl jasmonate (MeJa), induces stress-related transcripts in *Papaver somniferum*, which is complemented by the responses to stimuli. Such types of studies on MeJa treatment can provide a better understanding of the metabolic pathways involved in the molecular mechanism of *Papaver somniferum* during stresses against defense (Gurkok et al., 2015).

Metabolic Adaptation Induced by Abiotic Stresses

The metabolic diversities, which observed in plants are due to continuous evolutionary processes, may vary from species to species and inter-generic climatic conditions, and metabolic adaptation needs the fine tuning of primary and secondary metabolism. Metabolism is the central hub for signaling, physiological regulation, and defense responses. At the same time, biosynthesis, concentration, transport, and storage of primary and secondary metabolites are affected by abiotic stresses. However, time-series experiments showed that metabolic activities are quicker than transcriptional activities in response to stresses. With the

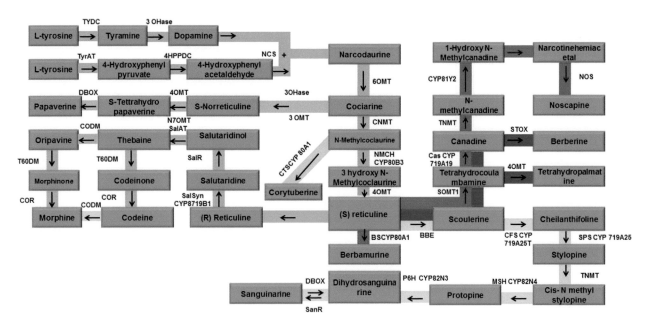

FIGURE 31.1 BIA biosynthetic pathways for BIA in opium poppy and other plant species. *TYDC*-tyrosine/DOPA decarboxylase, *3OHase*-tyrosine/ tyramine 3-hydroxylase, *4HPPDC*-4-hydroxyphenylpuruvate decarboxylase, *NCS*-norcoclaurine synthase, *6OMT*-norcoclaurine 6-*O*-methyltransferase, *CNMT*-coclaurine *N*-Methyltransferase, *NMCH*-*N*-methylcoclaurine3'-hydroxylase, *4'OMT2*-3'-hydroxyl-*N*-methylcoclaurine 4'-*O*-methyltransferase (isoform 2), *BBE*-berberine bridge enzyme, *SOMT1*-scoulerine 9-*O*-methyltransferase, *CAS*-canadine synthase, *TNMT*-tetrahydroprotoberberine*N*-methyltransferase, *CYP82Y1*-*N*-methylcanadine 1-hydroxylase, *NOS*-noscapine synthase, *STOX* -(S)-tetrahydroxyprotoberbineoxidase, *CoOMT*-columbamine*O*-methyltransferase, *CFS*-cheilanthifoline synthase, *SPS*-stylopine synthase, *MSH*-*N*-methylstylopine 14-hydroxylase, *P6H*-protopine 6-hydroxylase, *DBOX*-dihydrosanguinarine oxidase, *SanR*-sanguinarinereductase, *SalSyn*-salutaridine synthase, *SalR*-salutaridine reductase, *SalAT*-salutaridinol 7-*O*-acetyltransferase, *T6ODM*-Thebaine, 6-*O*-demethylase, *COR*-codeinone reductase, *CODM*-codeine *O*-demethylase, *N7OMT*-norreticuline 7-*O*-methyltransferase, *3'OHase*-uncharacterized 3'-hydroxylase, *3'OMT*-uncharacterized 3'-*O*-methyltransferase.

purpose of understanding simultaneous metabolic responses and alteration in transcriptional profiles in response to a specific abiotic stress, integrative and comprehensive approaches are needed. Metabolic profiling and the establishment of their co-relation with gene activity might increase our knowledge of how complex metabolic networks interact with each other and how they are dynamically regulated under stress adaptation and tolerance processes. The integration of the collected metabolic data concerning abiotic stress response to *Papaver somniferum* and related plant species could be helpful for the identification of tolerance-related traits that might be transferred to valuable crop species (Fraire and Balderas, 2013). Several studies have been done to mark these adaptive variations in many crops, but very limited attempts have been made in *Papaver somniferum*. This chapter deals with the impact of various abiotic factors in the metabolic adaptation by *Papaver somniferum* (Figure 31.2).

Metabolic Adaptation Induced by Salt Stresses

As per the estimation of the United Nations' Food and Agriculture Organization (FAO), salinity is a major abiotic factor for more than 20% of irrigated land (Rodziewicz et al., 2014). Considering plant ability to cope up with salt stress, it is found that plants which can tolerate salt stress are called halophyte while the plants which cannot tolerate salinity and eventually die are known as glycophyte. The majority of crop species belong to the latter category. The source of salinity in a plant may be through soil or water. Of the many salts, NaCl is a major contributor to salinity.

It causes a reduction in soil porosity, which leads to restrictions in water, nutrient, and oxygen availability to the plant. High salt concentration induces abscisic acid (ABA) synthesis, which is transported to guard cells, and stimulates the stomatal closure, ultimately causing a reduction in photosynthesis, photoinhibition, and oxidative stress. The inhibition of cell expansion results in immediate plant growth inhibition and senescence. Salinity is sensed by the plant as the driving force for adaptation and speciation. To cope with salt stress, plants evolve different strategies including lowering the rate of photosynthesis, stomatal conductance, and transpiration. Sodium ions compete with similar potassium ions and inhibit potassium uptake by the root. The exclusion of potassium results in growth inhibition because this ion is involved in enzyme activities, maintaining membrane potential, and cell turgor (Erxleben et al., 2012).

The metabolic agitation in plants to salinity involves both primary and secondary metabolism. For example, 29 proteins were significantly up- or down-regulated due to NaCl stress, which regulates the primary metabolic activities in foxtail millet (cv. Prasad) (Veeranagamallaiah et al., 2008). Studies in model plants and crops have established that the physiology of plants in salt stress conditions exhibit complex metabolic responses including different well-coordinated mechanisms, salt-dose dependence, and time course changes.

Time-course metabolite changes in the cell cultures of *A. thaliana* exposed to salt stress revealed that glycerol and inositol remain abundant 24 hrs after salt stress exposure, but lactate and sucrose are accumulated after 48 hrs. There are two types of responses that have been observed by this study; first

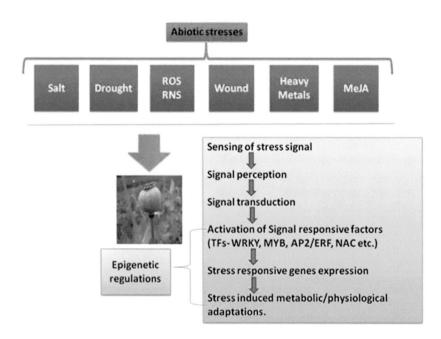

FIGURE 31.2 Impact of abiotic stresses and major adaptations in *Papaver somniferum L.*

are short-term responses, such as the methylation cycle, glycine betaine biosynthesis, and the phenyl-propanoid pathway, and second are long-term responses like glycolysis, sucrose metabolism, and a diminished methylation processes. Long-term salt exposure also causes a reduction in the metabolites (Kim et al., 2007). In a separate study, dose-dependent treatment of a tobacco plant showed that treatment with 50 mMNaCl for one day induced accumulation of sucrose, glucose, and fructose in reducing order, through gluconeogenesis. However, high stress (500 mMNaCl) for another day results in the induction of proline and a higher elevation of the sucrose level as compared to the lower dose; at the same time, a reduction of glucose and fructose pool, as well as transamination-related metabolites (asparagine, glutamine, and gamma-aminobutyric acid (GABA)), is observed. It is validated that sugar and proline synthesis show metabolic mechanisms for the control of salt stress as short-term responses. The proline continues to be observed at high levels at later stages (under high salt stress at 500 mMNaCl), and the sucrose decreases (though it remains high compared to the control). Despite the reductions in glucose, fructose, glutamine, GABA, malate, fumarate, choline, uridine, hypoxanthine, nicotine, N-methylnicotinamide, and formate, enhanced levels of asparagine, valine, isoleucine, tryptophan, myo-inositol, uracil, and allantoin, have been reported (Zhang et al., 2011).

Plant physiology about salt stress has undergone complex metabolic responses with time. The time and dose-dependent studies revealed differential tissue responses as a biological unit; even in some cases, the responses are contrasting. These studies prove that the metabolic plant responses to salinity stress are not ubiquitous; they have wide variability depending on the genus and species, and even the cultivar under consideration. Differential metabolic transcriptions have an intimate correlation with genetic backgrounds. Although the study of salt-induced metabolic adaptations in *Papaver somniferum* has not been done so far, despite this, salt stress–induced metabolic changes have been studied in model plants such as tobacco and *A. thaliana*, which

can shed some light on the probable changes in *Papaver somniferum* during salt stresses.

Drought stress is a significant abiotic stress that directly affects plant growth and development. Drought stress is a physiological condition defined by the reduction of available water in the soil to a critical level, and the atmospheric condition adds to the continuous loss of water. Drought stress tolerance is exhibited by all plants, but their mechanisms and levels of resistance are species specific. Drought stresses influence changes in the ratio of chlorophyll "a" and "b," and carotenoids. In a study, the reduction in chlorophyll content under drought stress was observed in *Gossypium* and *Catharanthus roseus*. Anthocyanins are accumulated under drought stress, and in cold climates anthocyanins are usually more resistant to drought. For example, a purple cultivar of chili resists water stress more efficiently than a green cultivar. The *Hypericum brasiliense* plants grown under drought stress have a high concentration of phenolics in comparison to the control plants. Stressed plants remain smaller in size, but a 10% increase in the total content of phenolic compounds was found in the stressed plants of *Hypericum*. However, flavonoids have protective functions during drought stress (De and Mazzafera, 2005).

When water deficit is sensed, signal transduction takes place, which involves ABA as a central component. ABA accumulates in plant tissues under drought and promotes transpiration reduction via stomatal closure. The expression of several stress-responsive genes is governed by ABA, including late embryogenesis abundant (LEA) proteins, which eventually leads to the induction of drought stress tolerance in plants (Aroca et al., 2008). The burning questions are: How does drought stress disturb metabolism in crop plants? and How can biotechnology contribute to the support of crop breeding for more sustainable agriculture? The probable answers to these questions are briefly elaborated. The central mechanism by which plants cope with water deficit is the osmotic adjustment. Differential metabolic adaptations were observed in drought-tolerant (NA5009RG) and drought-sensitive (DM50048) soybean cultivars when analyzed by nuclear magnetic resonance (1H NMR). During water stress, surprisingly, there was an absence of the accumulation of common osmo-protectants, such as proline, soluble sugars, such as sucrose or myo-inositol, organic acids, and other amino acids (except for aspartate)observed in the leaves of either genotype. However, the active accumulation of compatible solutes is essential for maintaining a positive cell turgor. Levels of 2-oxoglutaric acid, pinitol, and allantoin are altered differentially, despite common metabolites in the genotypes when drought is enforced, suggesting their functions as osmo-protectants (Silventeet al., 2012). In a particular study, an amino acids pool including proline, tryptophan, leucine, isoleucine, and valine enhances under drought stress during the analysis of three different cultivars of wheat (*Triticumaestivum*) for 103 metabolites through gas chromatography-mass spectroscopy (GC-MS) (Bowne et al., 2012).

Metabolic adjustments in plants to cope with these stress conditions are transient and might be dependent on the severity of stress. To validate this, changes in the concentrations of 28 metabolites were observed in a metabolic analysis by GC-MS in drought-exposed plants of maize (*Zea mays*). An accumulation of soluble carbohydrates, proline and eight other amino acids, shikimate, aconitase, and serine and glycine was justified by the decrement of leaf starch, malate, fumarate, 2-oxoglutarate, and seven amino acids during drought treatment. Moreover, water potential became more negative during the 8 to 10 day period, and the changes in some metabolites were quite dynamic, validating their dependence on the stress severity of the stress exposed (Sicher and Barnaby, 2012). Studies related to the impact of drought stress on the metabolism of *Papaver somniferum* have not been done so far, so in this chapter a prediction for the drought tolerance mechanism in *Papaver somniferum* has been tried using some model plants.

Metabolic Adaptations Induced by Oxidative Stress

It was observed that a high concentration of reactive oxygen species (ROS) causes damage to protein structures, and oxidation of DNA and lipids. It affects cellular integrity, and eventually, cell death occurs. The stress-induced imbalance between ROS generation and scavenging in both biotic and abiotic types of stresses have been reported (Kar, 2011). Normal cellular metabolic activity causes ROS generation under steady growth conditions. Thus, cells sense this uncontrolled elevation of ROS and implement their own signaling mechanism to activate protective responses (Moller and Sweetlove, 2010). Metabolic responses initiate efficient mechanisms for the reduction of toxic concentrations of ROS. The antioxidant system consists of protective enzymes (e.g. superoxide dismutase, peroxidase, catalase, reductase, and redoxin) and radical scavenger metabolites (mainly glutathione (reduced) (GSH) and ascorbate). In *A. thaliana*, methyl viologen induces oxidative stress that causes the down-regulation of photosynthesis-related genes and the inhibition of starch and sucrose synthesis pathways which ultimately activate the catabolic pathways. These metabolic adjustments avert the wasted energy, which is utilized for non-defensive processes, and also shift carbon reserves as emergency relief, i.e. the accumulation

of the protein structure-stabilizer molecule as maltose (Fraire et al., 2013).

The early (0.5 hrs) response, which was observed in the culture as well as in the plant system, was due to the inhibition of the tri-carboxylic acid cycle (TCA) through the accumulation of pyruvate and citrate that was finally adjusted by a decrement of malate, succinate, and fumarate pools. Inhibition of the TCA cycle concomitantly causes a decrement in the glutamate and aspartate pool due to the inhibition of the synthesis of TCA-linked precursor's 2-oxoglutarate and oxaloacetate, respectively. Metabolic redistribution of carbon flux from glycolysis to the oxidative pentose phosphate (OPP) pathway for the reduction of glycolytic pools of glucose-6 phosphate and fructose 6-P, and the rise in the OPP pathway intermediates, ribulose 5-phosphate and ribose 5-phosphate, has been observed. Increased carbon flux through the OPP pathway might be giving reducing power (via nicotinamide adenine dinucleotide phosphate (reduced) (NADPH)) for antioxidant activity when oxidative stress decreases the level of radical scavenger reductants GSH, ascorbate, and NADPH. After 2 and 6 hrs of stress progression, differential metabolic adjustments in response to oxidative stress were observed in roots as well as in cell suspension cultures. The concentrations of these metabolites reduced throughout the time course, indicating higher oxidative stress in cell cultures, whereas in roots, pools of TCA cycle intermediates and amino acids were recovered. GABA, aromatic amino acids (tryptophan, phenylalanine, and tyrosine), proline and other amino acids, and metabolites significantly altered in roots after the six-hour treatment (Lehmann et al., 2009). It is concluded that plants tune their primary as well as secondary metabolism in response to elevated ROS. These types of studies in *A. thaliana* could provide a framework to design further specified study in *Papaver somniferum*, which will have great impact on developing a salt tolerance strategy.

Metabolic Adaptations Induced by Heavy Metal Stress

The production of secondary metabolites can be altered as a metabolic adaptation through some metal ions like lanthanum, europium, silver, and cadmium. Nickel (Ni) is a trace element which is an essential factor of the urease enzyme and is needed for plant developmental processes. Ni stress causes the severe reduction in anthocyanin level and the inhibition of plant growth because trace metals bind and inhibit phenylalanine ammonia lyase (PAL) activity to limit anthocyanin biosynthesis (Hawrylak, 2008). Similarly, copper stimulates the production of betalains in *Beta vulgaris*. The hairy root culture of *Brugmansia candida* expressed that $AgNO_3$ or $CdCl_2$ are other very promising elicitors for tropane alkaloids, such as scopolamine and hyoscyamine. The content of cadmium and zinc remain lesser in capsules in comparison to seeds of *Papaver somniferum*, which made up 2.5% to 12.9% of above-ground biomass, but 15% to 42% of the total Cd remains stored in seeds, which signifies a preferential translocation of Cd into seeds. However, the content of zinc remains much higher than the cadmium in both capsules and seeds (Lachman et al., 2006). The heroin obtained from the opium of poppy contains copper and iron that increases the undesired toxicity

(Infante et al., 1999). Following the activation of the biosynthesis of narcotine in seeds of all the varieties, and capsules of only the Opal variety, as well as morphine in capsules of cv. Opal, they have been found to have high cadmium. Cadmium is toxic to plants as well as inducing an abiotic stress response. *Papaver somniferum* plants respond to this stress through over-expressing alkaloids, prominently narcotine, morphine, and papaverine, in the capsules of cv. Opal, whereas higher Zn content reduces narcotine in the seeds of all varieties, codeine in the Opal variety capsules, and papaverine and narcotine in capsules of all varieties. Zinc is a micro biogenic element for the poppy plant, and its higher content could be correlated with the lower content of major alkaloids, especially to codeine, papaverine, and narcotine, in capsules. Cd and Zn may be related through their ability to form chelate complexes with hydroxyl groups via the alkaloid in poppy capsules and seeds (the molecule of morphine contains two hydroxyl groups). The constitutive presence of morphine together with bis-morphine and other metabolites involved in poppy alkaloid biosynthesis could be involved in inducing the defense system (Morimoto et al., 2001; Szabó et al., 2003). It has been validated that narcotine-Cd, narcotine-Zn in seeds, and morphine-Cd (esp. in cv. Opal) in capsules are positively correlated linearly. Cadmium in higher concentrations caused abiotic stress and increased the content of alkaloids as a defense system in poppy. The positive relation between narcotine and zinc has been reported in the seeds, while the negative relation between papaverine-Zn and codeine-Zn (noticeably in cv. Opal) prevail in the capsules. This fact could be related to the positive role of zinc with the protein and saccharide metabolism and growth in poppy. Further investigations have reported that high levels of Ni, Zn, Cd, and Se provide effective protection against biotic stresses in *Papaver somniferum*. These relations between heavy metals, especially of Cadmium and Zn, act as inducers for the secondary metabolite mediated defense mechanisms in *Papaver somniferum*, which need to be investigated in the future to reveal heavy metal–dependent biosynthetic substitutions for metabolic adaptations.

Metabolic Adaptation with MeJA as Abiotic Elicitor

The alkaloids in the latex which oozes out from the capsules of *Papaver somniferum* remain abundant in terms of concentration as well as in diversity in the capsules of the plant (Weid et al., 2004). It has been studied that the alkaloid level induces after the treatment of MeJa for 3 hrs and increases continuously to 12 hrs after treatment. Out of these induced alkaloids, morphine and noscapine are abundant in capsule tissues under un-induced conditions which show a range shift in the metabolites during stress. The bis-morphine is rapidly produced through morphine in the alkaloid biosynthetic pathway of poppy, which increases its cellular concentration upon mechanical damage (Morimoto et al., 2001). The treatment of MeJa leads to the gradual induction of morphine accumulation in response to stress as morphine and noscapine are bi-routed from (S)-reticuline. Despite morphine and noscapine existing at all time points, it has been suggested that genes expressed in the BIA pathway upstream to (S)-reticuline might be the key points in noscapine accumulation

rather than the downstream genes. Instead of these, there might be unknown steps and/or regulators for noscapine biosynthesis (Maldonado, 2014). A large number of stress-related genes have been identified in response to MeJa treatment in *Papaver somniferum*. Moreover, differentially expressed transcripts and alkaloid concentrations consistently reached their highest level at 12 hrs after the treatment of MeJa. Major latex protein (MLP) is one of the proteins which is expressed abundantly under MeJA exposure, however the functional characterization of MLP is not completely known, but the *Papaver somniferum* transcriptome is confirmed to be affected by MeJa treatment. There is no induction of MLP transcripts after the elicitation of the cell culture of *Papaver somniferum* from 0 to 80 hrs in contrast to some findings which indicated alterations at all times (Facchini and Park, 2003). Moreover, heat shock proteins (HSPs) are involved in protecting plants against stress by re-establishing normal protein configuration, thus helping to maintain cellular homeostasis. The family members of Hsp90 remain involved in protein folding, but they also remain involved in cellular signal regulation and seem to be necessary for the functionality of resistance (R) protein in *A. thaliana* (Hubert et al., 2003). HSP70 is another most up-regulated protein in response to both low- and high-elicitor treatment in root cell cultures of the California poppy (Angelova et al., 2010). However, earlier it was thought that elicitor treatment did not considerably up-regulate HSP70 protein level in cell cultures of *Papaver somniferum* (Zulak et al., 2009). It was noticed that HSP 70 and HSP 90 might be working in co-ordination to promote protection and maintain the steady state in *Papaver somniferum* within 3 hrs. Beside this, one of the pathogenesis related (PR) family gene encoding chitinases has the ability to degrade fungal cell wall and is able to make the plant resistant to fungal pathogens. Chitinase is induced after 3 hrs treatment of MeJa as an early responsive gene. Furthermore uridine diphosphate glycosyl transferases (UGTs) is involved in the transfer of glycosyl residues from activated nucleotide sugars to a wide range of acceptor molecules, such as phytoalexins, cell wall precursors, and plant hormones like salicyclic acid (SA) (Bricchi et al., 2012). To date, no data is available about whether UGT 74e2 plays any crucial part in alkaloid biosynthesis. However, findings observed by Bricchi et al. (2012) indicated that UGT 74e2 transcript levels altered in the time courses of the MeJa treatment. Although poppy culture cells are a good system to observe stress responses on elicitor treatments, in our opinion a whole plant study would be beneficial *in vivo* to investigate elicitor effects on metabolic adaptations in plants.

MeJa-Induced KEGG Pathways

Amino acids cysteine (Cys) and methionine (Met) are essential for living beings because of their crucial function in primary and secondary metabolism. Cys and Met metabolism–based transcripts can be molded by MeJa treatment. Moreover, it was noticed that exogenous treatment of 50 μM MeJa induces only 6-O-Methyltransferase (6OMT), 7-O-methyltransferase (7OMT), thebaine 6-O-demethylase (T6ODM), berberine bridge enzyme (BBE), and codeinone reductase (COR) transcripts in poppy seedlings (Mishra et al., 2013). Further, it has been reported that the expressions of these enzymes are spatial and temporal, although cellular or tissue-restricted localization of morphine-based alkaloid biosynthesis is not completely understood (Huang and Kutchan, 2000). BIA pathway specific differentially expressed genes are regulated in elicitor-treated *Papaver somniferum* at different time points. Treatment with low-elicitor stimulates a pool of genes required for the phyto-alexin response. In a study, it was observed that highly up-regulated proteins remain linked to the alkaloid biosynthesis. The proteins with a sharp increase in the expression remain identical among different conditions such as different doses of elicitor treatment (initiating hypersensitive responses), mRNA degradation, cell death, and artificial shifts in cytosolic pH which stimulate the alkaloid biosynthesis instead of cell death (Angelova et al., 2010).

Co-expressed genes are crucial as these are committed for the up-regulation of alkaloid biosynthesis. A rapid down-regulation of translatable mRNAs have been observed which clearly indicate the existence of an additional defense response which does not need an elicitor for the alkaloid overproduction. The decay of these mRNAs might reflect the onset of a hypersensitive response but does not confirm the final stage of apoptotic cell death since the compromised cells (showing alkaloid-stained nuclei and an absence of viability probes) increase only 18 hrs after elicitor contact. It was seen that only the rate-limiting enzyme 3′-hydroxy-N-methylcoclaurine 4′-O-methyltransferase (4′OMT) of benzophenanthridine group is highly up-regulated. Other gene-encoding biosynthetic enzymes are activated in root-derived cell cultures, which was confirmed by gene expression analysis. An example for these genes is BBE, whose induction after elicitor treatment has been demonstrated (Viehwegeret al., 2006). Moreover, it was suggested that the strong overexpression of 4′-OMT might be due to its involvement in the rate-limiting step (synthesis of reticuline), which attributes the biosynthetic bottleneck to an upstream enzyme, (S)-norcoclaurine 6OMT (Inui et al., 2007). A strong increase in the alkaloid accumulation was observed in the cultured cells of *Eschscholzia* after overexpressing the homologous 6OMT gene from *Coptis japonica*, while no response to elicitor contact was shown in the plant system (Angelovaet al., 2010).

However, the biosynthetic flow of metabolites is not restricted due to low/no activity of 6OMT. What the overexpression of 4′-OMT induces through elicitor is partially compensated for by 6OMT as this enzyme also has 6OMT activity. The limitation imparted due to 6OMT activity has been tested in the biosynthesis of alkaloids in *Papaver somniferum*, where it was the only enzyme of alkaloid biosynthesis undergoing strong up-regulation in pathogen-treated cell cultures (Zulak et al., 2009). Highly up-regulated sets of proteins via each signal path for different components of plant-defense responses need to be studied thoroughly. However, the synergistic induction and their linkage to a secondary biosynthesis pathway of *Papaver somniferum* have not yet been reported. The protein HSP70 is a member of the HSP family that helps in the recovery of cells from different kinds of stresses. Plant cyclophilins, as found in animals, undertake protein-folding catalysis to co-perform with stress responses and developmental processes (Romano et al., 2004). They might play a role in mRNA processing and protein degradation, as well as signal transduction. It is well known that cyclophilins interact with HSP 70 chaperones (Rassow et al., 1995). Based on the particular functionality of the stress-protective proteins, their

synergistic up-regulation and their co-expression with the rate-limiting biosynthetic enzymes of alkaloid biosynthesis constitute a special cellular 'emergency kit' which determine the overproduction of the benzophenanthridines class of secondary metabolites. However, half an hour of low-dose treatment results from transcriptional change as a response to mild oxidative stress. Moreover, alkaloid content is also responsible for generating oxidative and other kinds of stress within the producing cell. The final biosynthetic step is the oxidation of dihydrobenzophenanthridines, catalyzed by a flavoprotein oxidase. This step requires molecular oxygen to produce H_2O_2; an oxidative stress response may be elicited by this membrane-permeable H_2O_2. Eventually, H_2O_2 may intercalate dsDNA and also may inhibit membrane-bound enzymes (Angelova et al., 2010).

Alkaloid biosynthesis requires effective compartmentation and exportation to avoid contact with DNA and cytoplasm. The extracellular benzophenanthridines easily re-enter the cell and reduce dihydro-alkaloids (less toxic) by NADPH-dependent cytosolic and enzyme sanguinarine reductase at above-threshold concentration. However, this mechanism is not efficient, especially in aged cultures and under conditions of energy (NADPH) limitations where alkaloid-damaged cells are prominent (Weiss et al., 2006; Vogel et al., 2010). The protective proteins are co-elicited with alkaloid biosynthesis to avoid self-intoxication, but high doses of externally added alkaloids rarely show an adverse effect in elicitor-treated cells compared to non-elicited cell suspensions. However, low-elicitor treatment does not cause significant cytotoxicity as discussed earlier, even at a high rate of alkaloid production. So, there must be a well-coordinated expression of biosynthetic and stress-defensive genes triggered by the elicitor. For example, in *Papaver somniferum*, a closely related species to *Eschscholzia*, the pathogen induces the up-regulation of several isoforms of HSP 70 and other chaperones with rate-limiting biosynthetic enzymes. Transcription factors (TFs) of myeloblastosis (MYB) and APETALA2 (AP2)/ ETS2 repressor factor (ERF) families target indole alkaloid biosynthetic genes, most probably with the induction of HSPs as found in transcriptional profiling of wound/jasmonate-responsive genes (Cheong et al., 2002). Further transcripts may appear from genome-wide analyses of ROS-induced gene expression, hormonal effects at the link point of secondary metabolism, as well as by the characterization of organelle-specific proteomes concerning the intracellular compartmentalization of secondary biosynthesis (Angelova et al., 2010). Exact mechanisms of the underlying signal transfer are yet unknown, and co-regulation of secondary biosynthesis and stress-protective proteins in plants have also not yet been investigated at the protein level. However, some TFs can be molded or regulated genetically, as well as epigenetically, to provide reconstitution of metabolic response under abiotic stresses which need to be revisited.

Trans-Factors Induced with Response to Abiotic Stresses

The regulation of TFs for stress adaptations is crucial for biotechnological engineering to develop stress tolerance in plant species. The TFs play a crucial role in water stress signaling, and some of them constitute pivotal roles in the signaling webs.

The chief TFs in this network are MYB, basic helix loops helix (bHLH), basic zipper (bZIP), ERF, NAM, ATAF, CUC (NAC), and W-Tryptophan, R-Arginine, K-Lysine, Y-Tyrosine (WRKY). The 43 drought, cold, or high-salinity stress–induced TF genes have been identified based on the transcriptome analyses in *Arabidopsis* using 7,000 genes. Out of these putative genes, six were dehydration responsive element binding (DREBs), two ERFs, ten zinc finger–containing factors, three MYBs, two bHLHs, four bZIPs, five NACs, and four WRKYs (Tripathi et al., 2014). A model was proposed, which elaborated the primary perception of signaling consequences during the drought which was finally received by TFs including SMCM1, AGAMOUS, DEFICIENS, SRF (MADS)-box, NAC, and WRKY proteins (Gechev et al., 2013). Early light-induced proteins (ELIPs), LEAs, HSPs, PR proteins, regulation of photosynthetic genes, ROS-scavenging metabolites, sucrose accumulation, and synthesis of GABA are the ultimate outcomes of stress signal perception. AP2/ERFs are involved in the regulatory framework of metabolism, development, and stress responses in plants (Zhu et al., 2014). They have a unique feature to bind multiple *cis*-elements present in stress-responsive gene promoters (Park et al., 2001). AP2/ERF has been isolated from *Papaver somniferum* and named PsAP2. The PsAP2 has shown maximum homology with AP2/ERF protein detected from *Populus trichocarpa*. It keeps significant sequence homology with ethylene-responsive element binding proteins (EREBP)/AP2 domain with other proteins; otherwise, amino acid sequences outside the DNA binding domain remain quite unique. The expression of PsAP2 remains highest in the flower, followed by the capsule and root, in comparison to leaf. Besides stress response, many AP2/ERFs are involved in plant developmental responses also. Hence, based on the PsAP2 tissue-specific and basal level expression, it is assumed that AP2/ERFs can have some direct or indirect roles in normal plant growth. Stress-induced transcriptomics have shown that the PsAP2 transcript is induced by various stresses in early stages and the elicitor-induced study indicated its role in the first line of regulation in defense. With the induced expression of PsAP2 in response to wounding, ethylene and MeJA seemed much higher than dehydration, salt, and ABA (Mishra et al., 2015). This indicated the probable role of PsAP2 in the regulation of multiple simultaneous responses in plants. In another separate study, it was observed that PsAP2 probably takes part in the regulation of multiple simultaneous stress responses in plants. Unlike TSI1 from tobacco and ERF1 from *Arabidopsis*, PsAP2 shows a delay in expression with response to salt and dehydration stress (Cheng et al., 2013). The action of the PsAP2 transcript was different from TSI1 or AtERF1 because PsAP2 showed induced expression after the 5th hour of MeJA exposure, ABA, and ethylene treatment. Basically, PsAP2 encodes a conserved AP2/EREBP DNA binding domain which binds either dehydration responsive element/c-repeat (DRE/CRT) or GCC box elements. The DRE and GCC boxes are linked with stress-responsive gene expressions in plants, that is why PsAP2 over-expressed lines respond better to different stresses. There is no difference in the affinity of the binding of PsAP2 towards both the DRE/CRT and GCC box *cis*-element as it might be PsAP2 that regulates both biotic as well as abiotic stresses. The specific and preferential binding affinity, as shown by PsAP2 for DRE/CRT with the core element ACCGAC, and the GCC box element with GCCGCC,

was confirmed by gel retardation assays, which suggested that PsAP2 is novel in comparison to other characterized AP2/ERF proteins. Many AP2/ERF proteins have been characterized in response to stresses in plants that showed an induced level of stress-related marker genes like early responsive to dehydration (ERD), PLANT DEFENSIN1.2 (PDF1.2), LEA, and pyrroline-5-carboxylate synthase (P5CS1). To investigate the possible mechanism of stress tolerance involving PsAP2, transgenic line analysis of 26 stress-inducible marker genes of tobacco has been done. Despite tobacco transgenic plants expressing PsAP2, they did not induce the expression of ERD's, which contains the DRE/CRT box, whereas it was shown in a single AP2 domain containing transcription activator from chickpea (CAP2) (an ERF from chickpea) under normal growth conditions. Out of the 26 transcripts which have been analyzed for their expression, only *Nicotiana tabacum* alternate oxidase (AOX1a) and major intrinsic proteins (MIPS) can express constitutively upon induction in PsAP2-overexpressing transgenic tobacco plants. There might be a species-specific evolution of trans-regulation (Shukla et al., 2006).

Alternative oxidase in mitochondria is maintaining both the stress-induced ROS and reactive nitrogen species (RNS) level. The induced expression of MIPS and NtAOX1a in PsAP2 overexpressing transgenic tobacco plants indicates an increase in antioxidant activity, which in turn provides the better response of transgenic lines under different stress conditions. Through the antioxidant enzyme system, AOX1a, MIPS, and PsAP2 might regulate the level of RNS and ROS under stress conditions. In a particular study, PsAP2-VIGS (virus-induced gene silencing) poppy plants exhibited a reduction in the AOX1a transcript level with reduced antioxidant properties, indicating a similar mode of action of PsAP2 in poppy as observed in transgenic tobacco. NtAOX1a and MIPS are induced to express in PsAP2 transgenic tobacco plants, and the specific activation of the NtAOX1a promoter recently opens up a chapter to investigate a new possible phenomenon and the role of the AP2/ERF protein in stress response regulation (Mishra et al., 2015).

Abiotic stresses are perceived by signal molecules and stimulate a subset of genes, which include many TFs like WRKY, ERF, NAC, and MADS. WRKY TFs are the most studied TFs because of their involvement in diverse biotic/abiotic stress responses as well as in developmental/physiological activities displayed by plants. WRKY genes in crop improvement are most promising because the WRKY gene can improve metabolic responses to multiple stresses (Jiang et al., 2015). Stress-induced expression of 'Aquaporin (GO238746)', 'LEA (GO238788)', and 'Glutamate decarboxylase (GO238819)' may be linked to the denovo synthesis of ABA after wounding (Shukla et al., 2006). Three novel TFs, first the WRKY domain containing TF (GT617707) and then two different NAC domain-containing TFs (GO238759; GO238737) were further identified with wounding response in *Papaver somniferum*. Conserved wound inducible W-box motif is detected, and the up-regulation of stress-specific metabolites with their respective transcripts leads to the validation of WRKY as a regulator of BIAs pathway. PsWRKY proteins showed maximum percent identity of 51% with CrWRKY (*Catharanthus roseus*), and MtWRKY (*Medicago truncatula*) was 47% with SPF1 *Ipomoea batatas*. CrWRKY and PsWRKY overexpress in response to MeJa, which suggests the role of these factors in abiotic stress responses (Suttipanta et al., 2011). PsWRKY induces expression against cold, dehydration, wounding, and salt treatment while CrWRKY responds differently. The PsWRKY expression is documented mainly for the capsule, followed by the root, and the straw, while CrWRKY expresses mostly in root tissues, followed by fruit, and leaves in *Papaver somniferum*. Both TFs have differential expression patterns and are functionally regulated by different pathways. Recombinant PsWRKY bind specifically with the W-box element present in the promoter region of BIAs transcripts. Involvement of PsWRKY in the regulation of BIAs pathway has been validated. The yeast one-hybrid assay and protoplast transient analysis have confirmed that PsWRKY can activate the tyrosine/DOPA decarboxylase (TYDC) 59 upstream activating sequence (UAS). *In vivo* study through interactome completely infer of its role in the regulation of BIAs pathway. Wound-induced expressions of MYB TFs remain involved in the regulation of the expression of flavonoid genes, which are the component of stress tolerance mechanisms adapted by plants. The wound-responsive genes are assorted according to their functions like metabolism, cellular defense, cellular localization, BIAs pathway enzymes, transcription, and unclassified. However, an unclassified group of ESTs is most abundant, and most of them are still unknown in *Papaver somniferum* though they have the potential for applicable investigation (Mishra et al., 2013).

Epigenetic processes are the new and cryptic aspect in plant metabolic adaptations in response to environmental stresses. Salinity-induced phosphorylation of histone H3 and acetylation of histone H4 in *A. thaliana* and tobacco has been identified at chromatin level while drought-induced acetylation and tri-methylation of histone H3 drought-responsive genes of *A. thaliana* has also been reported. Inheritable modifications of stress adaptations, such as salt stress, including altered genomic DNA methylation, as well as dicer like (DCL) proteins, suggest the possible role of small RNA pathways in epigenetic regulations. In barley, polycomb protein expression and histone methylation function are affected by ABA, which indicate hormone-mediated pathways in epigenetic modifications (Golldack et al., 2011). Application of an epigenetic framework to improve stress-responsive modulation of TFs seems like the future perspective for the engineering of plant tolerance to abiotic stresses. Despite the importance of revealing fundamental epigenetic mechanisms in model plants, it is necessary to investigate relevant insensitive and naturally tolerant species with respect to epigenetic changes. A coordinative study with metabolomics, transcriptomics, and epigenetics might be a better approach to link plant metabolic adaptations in response to environmental stresses.

Conclusion

Metabolic adaptations rely on environmental influences for a genus of particular species and their evolution under abiotic as well as biotic stresses. Plants share some common mechanisms for metabolic adaptation during abiotic stresses; instead of such consensus, genus-specific mechanisms have also been observed, which are well adapted during the course of evolution.

In developing countries, where adequate facilities of irrigation and anti-stress machinery are still in the natal phase, superior yield cannot be expected; however, it is possible to derive some strategic mechanisms which can cope with these abiotic stresses. Although intensive studies on the impact of abiotic stress on metabolic adaptation have been carried out on model plants, as well as on closely related plants like *Papaver somniferum*, independent study, especially on *Papaver somniferum*, is completely lacking and needs the proper attention of the scientific community.

REFERENCES

Alagoz Y, Gurkok T, Zhang B, Unver T. Manipulating the biosynthesis of bioactive compound alkaloids for next-generation metabolic engineering in opium poppy using CRISPR-Cas 9 genome editing technology. *Scientific Reports* 2016;6:30910.

Allen RS, Millgate AG, Chitty JA, Thisleton J, Miller JA, Fist AJ, et al. RNAi-mediated replacement of morphine with the nonnarcotic alkaloid reticuline in opium poppy. *Nature Biotechnology* 2004;22(12):1559–1566.

Angelova S, Buchheim M, Frowitter D, Schierhorn A, Roos W. Overproduction of alkaloid phytoalexins in California poppy cells is associated with the co-expression of biosynthetic and stress-protective enzymes. *Molecular Plant* 2010;3(5):927–939.

Aroca R, Vernieri P, Ruiz-Lozano JM. Mycorrhizal and non-mycorrhizal *Lactuca sativa* plants exhibit contrasting responses to exogenous ABA during drought stress and recovery. *Journal of Experimental Botany* 2008;59(8):2029–2041.

Beaudoin G A, Facchini PJ. Benzylisoquinoline alkaloid biosynthesis in opium poppy. *Planta* 2014;240(1):19–32.

Boke H, Ozhuner E, Turktas M, Parmaksiz I, Ozcan S, Unver T. Regulation of the alkaloid biosynthesis by miRNA in opium poppy. *Plant Biotechnology Journal* 2015;13(3):409–420.

Bowne JB, Erwin TA, Juttner J, Schnurbusch T, Langridge P, Bacic A, et al. Drought responses of leaf tissues from wheat cultivars of differing drought tolerance at the metabolite level. *Molecular Plant* 2012;5(2):418–429.

Bricchi I, Bertea CM, Occhipinti A, Paponov IA, Maffei ME. Dynamics of membrane potential variation and gene expression induced by *Spodoptera littoralis*, *Myzus persicae*, and *Pseudomonas syringae* in *Arabidopsis*. *PLoS One* 2012;7(10):e46673.

Cheng M-C, Liao P-M, Kuo W-W, Lin T-P. The *Arabidopsis* ETHYLENE RESPONSE FACTOR1 regulates abiotic stress-responsive gene expression by binding to different cis-acting elements in response to different stress signals. *Plant Physiology* 2013;162(3):1566–1582.

Cheong YH, Chang H-S, Gupta R, Wang X, Zhu T, Luan S. Transcriptional profiling reveals novel interactions between wounding, pathogen, abiotic stress, and hormonal responses in *Arabidopsis*. *Plant Physiology* 2002;129(2):661–677.

Christenhusz MJ, Byng JW. The number of known plants species in the world and its annual increase. *Phytotaxa* 2016;261(3):201–217.

Dang T-TT, Facchini PJ. Characterization of three *O*-Methyltransferases involved in noscapine biosynthesis in opium poppy. *Plant Physiology* 2012;159(2):618–631. doi:10.1104/pp.112.194886

de Abreu IN, Mazzafera P. Effect of water and temperature stress on the content of active constituents of *Hypericum brasiliense* Choisy. *Plant Physiology and Biochemistry* 2005;43(3):241–248.

Desgagné-Penix I, Facchini PJ. Systematic silencing of benzyl-isoquinoline alkaloid biosynthetic genes reveals the major route to papaverine in opium poppy. *The Plant Journal* 2012;72(2):331–344.

Erxleben A, Gessler A, Vervliet-Scheebaum M, Reski R. Metabolite profiling of the moss *Physcomitrella patens* reveals evolutionary conservation of osmoprotective substances. *Plant Cell Reports* 2012;31(2):427–436.

Facchini PJ, Park S-U. Developmental and inducible accumulation of gene transcripts involved in alkaloid biosynthesis in opium poppy. *Phytochemistry* 2003;64(1):177–186.

Fraire-Velázquez S, Balderas-Hernández VE. *Abiotic Stress in Plants and Metabolic Responses*. INTECH Open Access Publisher, Croatia, 2013; pp. 25–48.

Gechev TS, Benina M, Obata T, Tohge T, Sujeeth N, Minkov I, et al. Molecular mechanisms of desiccation tolerance in the resurrection glacial relic *Haberlea rhodopensis*. *Cellular and Molecular Life Sciences* 2013;70(4):689–709.

Golldack D, Lüking I, Yang O. Plant tolerance to drought and salinity: Stress regulating transcription factors and their functional significance in the cellular transcriptional network. *Plant Cell Reports* 2011;30(8):1383–1391.

Gómez-Galera S, Pelacho AM, Gené A, Capell T, Christou P. The genetic manipulation of medicinal and aromatic plants. *Plant Cell Reports* 2007;26(10):1689–1715.

Gurkok T, Turktas M, Parmaksiz I, Unver T. Transcriptome profiling of alkaloid biosynthesis in elicitor induced opium poppy. *Plant Molecular Biology Reporter* 2015;33(3):673–688.

Hawrylak-Nowak B. Changes in anthocyanin content as indicator of maize sensitivity to selenium. *Journal of Plant Nutrition* 2008;31(7):1232–1242.

Huang F-C, Kutchan TM. Distribution of morphinan and benzo [c] phenanthridine alkaloid gene transcript accumulation in *Papaver somniferum*. *Phytochemistry* 2000;53(5):555–564.

Hubert DA, Tornero P, Belkhadir Y, Krishna P, Takahashi A, Shirasu K, et al. Cytosolic HSP90 associates with and modulates the *Arabidopsis* RPM1 disease resistance protein. *The EMBO Journal* 2003;22(21):5679–5689. doi:10.1093/emboj/cdg547.

Infante F, Dominguez E, Trujillo D, Luna A. Metal contamination in illicit samples of heroin. *Journal of Forensic Science* 1999;44(1):110–113.

Inui T, Tamura K-i, Fujii N, Morishige T, Sato F. Overexpression of *Coptis japonica* norcoclaurine 6-O-methyltransferase overcomes the rate-limiting step in benzylisoquinoline alkaloid biosynthesis in cultured *Eschscholzia californica*. *Plant and Cell Physiology* 2007;48(2):252–262.

Jiang W, Wu J, Zhang Y, Yin L, Lu J. Isolation of a WRKY30 gene from *Muscadinia rotundifolia* (Michx) and validation of its function under biotic and abiotic stresses. *Protoplasma* 2015;252(5):1361–1374.

Kar RK. Plant responses to water stress: Role of reactive oxygen species. *Plant Signaling and Behavior* 2011;6(11):1741–1745.

Kim JK, Bamba T, Harada K, Fukusaki E, Kobayashi A. Time-course metabolic profiling in *Arabidopsis thaliana* cell cultures after salt stress treatment. *Journal of Experimental Botany* 2007;58(3):415–424.

Lachman J, Hejtmankova A, Miholova D, Kolihová D, Tluka P. Relations among alkaloids, cadmium and zinc contents in opium poppy (*Papaver somniferum* L.). *Plant Soil and Environment* 2006;52(6):282.

Lehmann M, Schwarzländer M, Obata T, Sirikantaramas S, Burow M, Olsen CE, et al. The metabolic response of *Arabidopsis* roots to oxidative stress is distinct from that of heterotrophic cells in culture and highlights a complex relationship between the levels of transcripts, metabolites, and flux. *Molecular Plant* 2009;2(3):390–406.

Leonard E, Runguphan W, O'Connor S, Prather KJ. Opportunities in metabolic engineering to facilitate scalable alkaloid production. *Nature Chemical Biology* 2009;5(5):292–300.

Maldonado-Bonilla LD. Composition and function of P bodies in *Arabidopsis thaliana*. *Frontiers in Plant Science* 2014;5:201.

Mishra S, Phukan UJ, Tripathi V, Singh DK, Luqman S, Shukla RK. PsAP2 an AP2/ERF family transcription factor from *Papaver somniferum* enhances abiotic and biotic stress tolerance in transgenic tobacco. *Plant Molecular Biology* 2015;89(1–2):173–186.

Mishra S, Triptahi V, Singh S, Phukan UJ, Gupta M, Shanker K, et al. Wound induced tanscriptional regulation of benzylisoquinoline pathway and characterization of wound inducible PsWRKY transcription factor from *Papaver somniferum*. *PLoS One* 2013;8(1):e52784.

Møller IM, Sweetlove LJ. ROS signalling–specificity is required. *Trends in Plant Science* 2010;15(7):370–374.

Morimoto S, Suemori K, Moriwaki J, Taura F, Tanaka H, Aso M, et al. Morphine metabolism in the opium poppy and its possible physiological function biochemical characterization of the morphine metabolite, bismorphine. *Journal of Biological Chemistry* 2001;276(41):38179–38184.

Park JM, Park C-J, Lee S-B, Ham B-K, Shin R, Paek K-H. Overexpression of the tobacco Tsi1 gene encoding an EREBP/AP2–type transcription factor enhances resistance against pathogen attack and osmotic stress in tobacco. *The Plant Cell* 2001;13(5):1035–1046.

Ramakrishna A, Ravishankar GA. Influence of abiotic stress signals on secondary metabolites in plants. *Plant Signaling and Behavior* 2011;6:1720–1731.

Ramakrishna A, Ravishankar GA. Role of plant metabolites in abiotic stress tolerance under changing climatic conditions with special reference to secondary compounds. In: *Climate Change and Plant Abiotic Stress Tolerance*, Tuteja, N., and Gill, S. S.(ed.). Wiley-VCH Verlag GmbH & Co. KGaA, Weinheim, Germany, 2013; pp. 705–726.

Rassow J, Mohrs K, Koidl S, Barthelmess IB, Pfanner N, Tropschug M. Cyclophilin 20 is involved in mitochondrial protein folding in cooperation with molecular chaperones Hsp70 and Hsp60. *Molecular and Cellular Biology* 1995;15(5):2654–2662.

Rastogi A, Mishra BK, Shukla S. Opium poppy: An ancient medicinal plant. In: *Breeding of Field Crops*, Bharadwaj DN (ed.). Agrobios Publications, Jodhpur, India, 2012; pp. 865–903.

Rodziewicz P, Swarcewicz B, Chmielewska K, Wojakowska A, Stobiecki M. Influence of abiotic stresses on plant proteome and metabolome changes. *Acta Physiologiae Plantarum* 2014;36(1):1–19.

Romano PG, Horton P, Gray JE. The *Arabidopsis* cyclophilin gene family. *Plant Physiology*. 2004;134(4):1268–1282.

Shukla RK, Raha S, Tripathi V, Chattopadhyay D. Expression of CAP2, an APETALA2-family transcription factor from chickpea, enhances growth and tolerance to dehydration and salt stress in transgenic tobacco. *Plant Physiology* 2006;142(1):113–123.

Sicher RC, Barnaby JY. Impact of carbon dioxide enrichment on the responses of maize leaf transcripts and metabolites to water stress. *Physiologia Plantarum* 2012;144(3):238–253.

Silvente S, Sobolev AP, Lara M. Metabolite adjustments in drought tolerant and sensitive soybean genotypes in response to water stress. *PLoS One* 2012;7(6):e38554.

Suttipanta N, Pattanaik S, Kulshrestha M, Patra B, Singh SK, Yuan L. The transcription factor CrWRKY1 positively regulates the terpenoid indole alkaloid biosynthesis in *Catharanthus roseus*. *Plant Physiology* 2011;157(4):2081–2093.

Szabó B, Tyihák E, Szabó G, Botz L. Mycotoxin and drought stress induced change of alkaloid content of *Papaver somniferum* plantlets. *Acta Botanica Hungarica* 2003;45(3–4):409–417.

Tripathi P, Rabara RC, Rushton PJ. A systems biology perspective on the role of WRKY transcription factors in drought responses in plants. *Planta* 2014;239(2):255–266.

Veeranagamallaiah G, Jyothsnakumari G, Thippeswamy M, Reddy PCO, Surabhi G-K, Sriranganayakulu G, et al. Proteomic analysis of salt stress responses in foxtail millet (*Setaria italica* L. cv. Prasad) seedlings. *Plant Science* 2008;175(5):631–641.

Verma N, Shukla, S. Impact of various factors responsible for fluctuation in plant secondary metabolite. *Journal of Applied Research on Medicinal and Aromatic Plants* 2015;2:105–113.

Viehweger K, Schwartze W, Schumann B, Lein W, Roos W. The Gα protein controls a pH-dependent signal path to the induction of phytoalexin biosynthesis in *Eschscholzia californica*. *The Plant Cell* 2006;18(6):1510–1523.

Vogel M, Lawson M, Sippl W, Conrad U, Roos W. Sanguinarine reductase -structure and mechanism of an enzyme of alkaloid detoxication. *Journal of Biological Chemistry* 2010;285(24):jbc.M109.088989.

Weid M, Ziegler J, Kutchan TM. The roles of latex and the vascular bundle in morphine biosynthesis in the opium poppy, *Papaver somniferum*. *Proceedings of the National Academy of Sciences of the United States of America* 2004;101(38):13957–13962.

Weiss D, Baumert A, Vogel M, Roos W. Sanguinarine reductase, a key enzyme of benzophenanthridine detoxification. *Plant, Cell and Environment* 2006;29(2):291–302.

Wijekoon CP, Facchini PJ. Systematic knockdown of morphine pathway enzymes in opium poppy using virus-induced gene silencing. *The Plant Journal* 2012;69(6):1052–1063.

Zhang J, Zhang Y, Du Y, Chen S, Tang H. Dynamic metabonomic responses of tobacco (*Nicotiana tabacum*) plants to salt stress. *Journal of Proteome Research* 2011;10(4):1904–1914.

Zhu X, Qi L, Liu X, Cai S, Xu H, Huang R, et al. The wheat ethylene response factor transcription factor pathogen-induced ERF1 mediates host responses to both the necrotrophic pathogen *Rhizoctonia cerealis* and freezing stresses. *Plant Physiology* 2014;164(3):1499–1514.

Zulak KG, Lippert DN, Kuzyk MA, Domanski D, Chou T, Borchers CH, et al. Targeted proteomics using selected reaction monitoring reveals the induction of specific terpene synthases in a multi-level study of methyl jasmonate-treated Norway spruce (*Picea abies*). *The Plant Journal* 2009;60(6):1015–1030.

32

In Vitro *Selection and Genetic Engineering for Abiotic Stress Tolerant Plants and Underlying Mechanism*

Srinath Rao and H. Sandhya

CONTENTS

Abbreviations

APX	ascorbate peroxidase
BADH	betaine aldehyde dehydrogenase
CAT	catalase
CMO	choline monooxygenase
COX	choline oxidase
GB	glycine betaine
P5CS	pyrroline-5-carboxylate synthetase
PEG	poly ethylene glycol
PPO	polyphenol oxidase
ROS	reactive oxygen species
SOD	superoxide dismutase

Introduction

An external factor incurring negative influence on the plant can be defined as stress. The major stresses that affect plant growth are water deficiency, chilling and freezing, heat and salinity stress. Drought and salinity are expected to cause serious salinization of more than 50% of all arable land by the year 2050, therefore salinity (mostly sodium chloride (NaCl)) stresses pose a major threat to crop productivity. The production and productivity of several crops are adversely affected by abiotic stresses, mainly salinity and drought. Damage caused by these stresses are responsible for enormous economic losses worldwide. Soil salinity has existed long before humans and agriculture. But due to agricultural practices, such as irrigation and poor drainage systems, this problem has been aggravated. Salinity in agricultural fields is thus a severe constraint to crop growth and productivity in many regions and the situation has become a global concern. It is estimated that 20% of irrigated land in the world is affected by salinity (Yamaguchi and Blumwald, 2005).

Moreover, it has been predicted that increasing salinization will reduce the land available for cultivation by 30% within the next 25 years and up to 50% by the year 2050 (Wang et al., 2003). Several physiological and biochemical processes are influenced due the presence of excess salt in soil or in solutions. It interferes with normal physiology, resulting in problems such as ion imbalance, ion toxicity, mineral deficiency, osmotic stress and oxidative stress; these conditions ultimately interact with several cellular components, including DNA, lipids, pigments and proteins in plants (Zhu, 2002; Ramakrishna and Ravishankar, 2011, 2013), which in turn will severely affect the growth and development of the majority of crop plants. It has become a global challenge to protect crops against salinity-induced damage. High salinity (e.g., increased concentrations of Na^+ and Cl^- in the soil solution) causes osmotic/ionic stress (Hasegawa et al., 2000). Development of salt-tolerant crops that can tolerate high levels of salinity in the soils would be a practical solution for such a problem (Yamaguchi and Blumwald, 2005). Traditional breeding technologies and proper management strategies continue to play a vital role in crop improvement. In conventional breeding programs, favorable genes of interest from wild but inter-crossing genera and species, are introduced into commercial and food crops to induce stress tolerance. However, conventional breeding methods have little success and have failed to provide desirable results (Purohit et al., 1998). Therefore, biotechnological tools are to be deployed to address the critical problems of crop improvement for sustainable agriculture. The advent of genetic engineering approaches has enabled researchers to overcome such problems (Kavi Kishor et al., 1995; Bakhsh et al., 2009; Khan et al., 2010). Genetic engineering for developing stress-tolerant plants, based on the introgression of genes that are known to be involved in stress response and putative tolerance, is to be introduced to improve crop varieties. In this direction, genetic transformation is a widely used procedure for introducing genes from distant gene pools into many plant species for the development of stress-tolerant plants, and considerable efforts have been made to produce stress-tolerant plants using this technique (Borsani et al., 2003; Yamaguchi and Blumwald, 2005). However, the major limiting factors in utilizing this technique are low transformation frequency, consequent reduction of gene expression and silencing of the transgene (Mondal et al., 1997). Nonetheless, this technique holds promise in crop improvement. Tissue culture technique has emerged as a feasible and cost-effective alternative tool for developing stress-tolerant plants. This technique can be performed under controlled conditions with limited space and time (Sakhanokho and Kelley, 2009), and has the potential for the selection of stress-tolerant variants using a low-cost laboratory set up.

Development of Abiotic Stress-Tolerant Plants through *In Vitro* Selection

The productivity of several commercial crops is limited by major abiotic stresses including salinity, drought, water logging, heat, frost and mineral toxicities. Successful application of biotechnology to improve crop growth and production will require knowledge of good biological information regarding the target species as well as the mechanisms underlying resistance/tolerance to these stresses (Dita et al., 2006). Development of abiotic stress–tolerant plants, especially for salt and drought conditions using plant breeding techniques, is a difficult task.

Conventional breeding has contributed significantly to the development of drought-tolerant high-yielding crop varieties for centuries; however, the development of new crop cultivars has been relatively slow mainly due to the limitation of fertility barriers. Tissue culture techniques are good tools for studying the effect of salinity and investigating the mechanism of salt tolerance at the unorganized cellular or organized tissue level, and may provide information on the potential for physiological, biochemical and growth responses to salt stress at different levels of tissue organization. Additionally, *in vitro* studies allow relatively faster results in a shorter time and controlled environment as compared to *ex vivo* conditions (Zhang et al., 2004; Perez-Tornero et al., 2009) and the inferences obtained from *in vitro* cultures under salt stress may be directly applicable at the whole-plant level (Cano et al., 1998; Basu et al., 2002; Bracci et al., 2008). The importance of plant tissue culture in the improvement of salt tolerance in plants has been long observed (Nabors et al., 1980; Dix, 1993; Hasegawa et al., 1994; Tal, 1994). In recent years, tissue culture techniques have been used as a useful tool to elucidate the mechanism involved in salt tolerance by using *in vitro*–selected salt-tolerant cell lines (Davenport et al., 2003; Gu et al. 2004; Lutts et al., 2004; Naik and Harinath, 1988; Rao & Patil. 1999; Venkataiah. et al., 2004). In addition, these lines have been used to regenerate salt-tolerant plants (Reddy and Vaidyanath, 1986; Naik and Harinath, 1988; Chen et al., 2001; Jaiwal and Singh, 2001; Miki et al., 2001; Ochatt et al., 1999; Rao and Krupanidhi, 1996; Rao and Patil, 1999; Shankhdhar et al., 2000; Sajid and Aftab, 2014).

In vitro culture of plant cells, tissues or organs on a medium containing selective agents offers the best opportunity to select and regenerate plants with desirable characteristics. In this technique, the explants are exposed to a broad range of selective agents added to the culture medium. The selecting agents usually employed for *in vitro* selection include NaCl (for salt-tolerance), poly ethylene glycol (PEG) or mannitol (for drought tolerance). In this technique, explants are exposed to a saline environment through the culture medium and only those explants that are capable of sustaining such environments survive in the long-term cultures and regenerate plants which are selected. During the culture of explants on the medium with a selective agent, only those explants capable of sustaining such environments survive in the long run and are selected. Two types of selection methods have been suggested: (a) stepwise long-term treatment, in which cultures are exposed to stress with a gradual increase in concentrations of the selecting agent and (b) shock treatment, in which cultures are directly subjected to a shock of high concentration and only those which can tolerate that level are selected and further sub-cultured. These methods are based on the induction of genetic variation among cells, tissues and/or organs in cultured and regenerated plants (Mohamed et al., 2000). The tissue culture induces variation in regenerated plants, known as somaclonal variations (Larkin and Scowcroft, 1981), which can result in genetically stable variations, useful in crop improvement. The advantage of *in vitro* selection is that it can considerably shorten the time for the selection of desirable traits under selection pressure with minimal environmental interaction, and can complement field selection (Jain, 2001). *In vitro* selection techniques are widely used for the development of such tolerant lines (Hossain et al., 2007). Since the first report in *Nicotiana sylvestris* (Zenk, 1974), many attempts have been made to produce salt/drought-tolerant plants using *in vitro* techniques (Nabors et al., 1980; Watad et al., 1983; Kumar and Sharma, 1989; Vajrabhaya et al., 1989; Binh et al., 1992; Karadimova and Djambova, 1993; Olmos et al., 1994; Tal, 1994; Barakat and Abdel-Latif, 1996; Patnaik and Debata, 1997a,b; Singh et al., 2003; Zair et al., 2003; Gandonou et al., 2006; Hossain et al., 2007). Benavides et al. (2000) raised potato plants from salt-tolerant callus lines. Alvarez et al. (2003) obtained salt-tolerant sunflower plants through *in vitro* techniques. Shanthi et al. (2010) selected callus of rice varieties CSR 10, TRY(R) 2 and TRY1 tolerant to NaCl and regenerated plants from the selected callus. Plants were regenerated successfully through shoot organogenesis of a NaCl-selected callus line in *Tanacetum cinerariaefolium* (Abdi et al., 2011). This is done using a number of systems (callus, suspension cultures, somatic embryos, shoot cultures, etc.) which are screened for variation in their ability to tolerate relatively high levels of NaCl or PEG in media (Vajrabhaya et al., 1989; Woodward and Bennett, 2005; Abdi et al., 2011; Musa, 2011; Rao and Jabeen, 2013). In the majority of salinity studies, the salt used is NaCl. However, several researchers during *in vitro* screening have compared the response of other Cl^- and SO_4^- salts including KCl, $MgSO_4$ and Na_2SO_4. Different responses were found in *Nicotiana tabacum* when growth on seawater, synthetic seawater, mannitol, NaCl and other Cl^- and SO_4^- salts was compared (Chen et al., 1980). Use of multiple salts as a selection pressure will parallel the salinity under field conditions and may be a better choice. Callus line selection procedure is likely to play an important role in the recovery of stable variant plants with improved salt tolerance (Winicov, 1996). Experiment results obtained from these studies strongly indicate that stepwise increase in NaCl concentration, from a relatively low level to cytotoxic level, is a better way to isolate the NaCl-tolerant callus line since direct transfer of callus to the highest selection procedure was found to be lethal. Though this technique offers many advantages, development of stress-tolerant plants through *in vitro* selection has also some limitations like the loss of regeneration during selection and reduced growth of plants. Reduction in growth is a very common phenomenon of salt-stressed plants, which has also been observed in cultured cells, tissues or organs on a medium supplemented with NaCl/PEG. However, slower growth is an adaptive feature for plant survival under stress (Queiros et al., 2007) Other drawbacks are the lack of correlation between the mechanisms of tolerance operating in a cultured cell, tissue or organ and those of the whole plants, and the phenomenon of epigenetic adaptation (Tal, 1994). During *in vitro* selection, non-tolerant cells sometimes undergo an epigenetic adaptation, i.e. stable epigenetic alterations that are inherited only through mitosis and not through meiosis, to the specific selective agents and thus obscure the selection of rare mutants with true tolerance, i.e. meiotically inherited (Tal, 1994). Numerous studies document that altered DNA methylation is a major cause of epigenetic modifications, which are sometimes observed in tissue-cultured experiments (Guo et al., 2007; Li et al., 2007; Gao et al., 2010). Many authors have suggested that the problem of epigenetic adaptation during *in vitro* selection can be overcome by the use of short-term or one-step selection that may prevent the development of epigenetically adapted cells (Dix, 1993; Chandler and Vasil, 1984; McHugen and Swartz, 1984; Tal, 1994).

Development of Abiotic Stress-Tolerant Plants through Genetic Engineering

To transfer a trait within a species into the high yielding locally adapted cultivars through conventional breeding requires five to six generations and a large number of progenies planted to be able to select the plants with the appropriate combination of traits. The improved lines that have developed have to then go through a set of multi-location tests before a variety can be identified for cultivation by farmers. This process takes a minimum of 7–10 years. However, genetic transformation provides access to genes from other species, which can be used for producing transgenic crops. In this technique, introgression of genes that are known to be involved in stress response and putative tolerance are introduced for improving crop varieties, mainly through *Agrobacterium tumifaciens*–mediated genetic transformation. This procedure enables the introduction of genes from distant gene pools into many plant species for the development of stress-tolerant plants, and considerable efforts have been made to produce stress-tolerant plants using this technique. The genes of interest can be transferred into the target crops/cultivars in a single event, and it takes 5–6 years to develop cultivars with stable gene expression. The lines thus produced can be released for cultivation by the farmers or used as donor parents in conventional plant breeding. Abiotic stress–related genes are being transferred into crop plants using gene transfer techniques (Ashraf et al., 2008). Importance is given to introduce gene encoding compatible organic osmolytes, heat shock proteins, late embryogenesis

abundant proteins and transcription factors responsible for activating gene expression (Ashraf et al., 2010). Many genes play an important role in the synthesis of osmoprotectants in stress-tolerant plants like proline, glycine betaine (GB) and polyamines, mannitol, trehalose and galactinol, which accumulate during osmotic adjustment (Vincour and Altman, 2005). It is now well documented that these organic solutes play an important role in the induction of drought tolerance in crop plants (Ashraf and Foolad, 2007). GB, which has a substantial role in stress tolerance, is richly synthesized under dehydration stress (Mansour, 2000; Mohanty et al., 2002; Yang et al., 2003; Ashraf and Foolad, 2007). The important enzymes that are required in the production of GB in higher plants are choline monooxygenase (CMO) and betaine aldehyde dehydrogenase (BADH). Shen et al. (2002) and Zhang et al. (2009) reported tobacco transgenic lines overexpressing CMO. Quan et al. (2004) reported that an increased GB accumulation under water-deficient conditions ultimately leads to enhanced drought stress in the genetically transformed maize inbred line DH4866 with the *E. coli* betA gene encoding choline dehydrogenase. Transgenic plants were reported to have a higher level of GB and showed more drought tolerance than control plants. Zhang et al. (2009) introduced AhCMO into a Chinese cotton cultivar SM3 via *Agrobacterium*-mediated transformation; they observed that transgenic lines of cotton have 26% and 131% more GB when compared to control plants and these transgenic plants showed improved salt tolerance. Rice has also been engineered to accumulate more levels of GB with improved tolerance against drought and salt stresses (Su et al., 2006; Mohanty et al., 2002). Li et al. (2011) reported that transgenic rice plants overexpressing *AtNHX5* showed not only high salt tolerance, but also high drought tolerance. Shiraswa et al. (2006) and Ahmad et al. (2008) reported the development of transgenic potato lines with improved tolerance against oxidative, drought and salt stresses using the choline oxidase gene (COX). *Brassica* and tomato have also been engineered, using the same approach, to encode resistance against chilling, drought and salt stresses respectively (Parsad et al., 2000; Park et al., 2007). Genetic transformation with GB synthesizing enzymes' gene(s) in naturally non-accumulating plants has resulted in enhanced tolerance against a variety of abiotic stresses (Ahmad et al., 2014). Ahmad et al. (2008) developed transgenic potato plants expressing the GB synthesizing choline oxidase (codA) gene from *Arthrobacter globiformis*, targeted to chloroplasts. They observed that these transgenic plants showed stable integration, and expressed the chimeric gene, resulting in an accumulation of GB. Further, they reported that the transgenic plants exhibited tolerance to different abiotic stress in *in vitro* conditions. GB synthesizing transgenic crop plants such as cotton (Zhang et al., 2009), rice (Mohanty et al., 2002), tomato (Zhou et al., 2007) and wheat (He et al., 2010) not only exhibited tolerance to salt and drought stress, but they showed enhanced yield as compared to NT plants under stress conditions.

Proline is another significant organic osmolyte which plays an important role in stress tolerance. A key enzyme for the proline biosynthesis is pyrroline-5-carboxylate synthetase (*P5CS*). The genes encoding this enzyme have been engineered in a number of crops like rice by Zhu et al. (1998), wheat by Sawahel and Hassan (2002), carrot by Han and Hwang (2003), petunia by Ronde et al. (2004), soybean by Yamada et al. (2005), potato by Sayari et al. (2005) and tobacco by Kavi Kishor et al. (1995) and Gubis et al. (2007). All the transgenic lines that developed showed increased accumulation of proline and thus enhanced drought tolerance. Expression of the *P5CS* transgene in chick pea resulted in the overproduction of the *P5CS* enzyme in transgenic chickpea plants as well as an increase in proline accumulation, and stress tolerance (Ghanti et al., 2011). Saijo et al. (2000) demonstrated that the overexpression of a single Ca2+-dependent protein kinase conferred cold, salt and drought tolerance in rice plants. Prasad et al. (2000) reported that a bacterial codA gene when transformed into *Brassica juncea* enhances its tolerance to salt.

Role of Osmolytes in Abiotic Stress Tolerance

Varieties of physiological responses have been evolved by salt-tolerant plants that confer tolerance to salinity stress (Yang et al., 2010). These plants accumulate osmolytes such as glycine, betaine and proline and soluble sugars that maintain the osmotic balance disrupted by the presence of ions in the vacuole (McCue and Hanson, 1990; Wang et al., 2004). El Sayed and El Sayed (2011) reported that in pea, the NaCl resistant callus showed greater accumulation of proline than the sensitive callus. Similar observations have been made for several other plant species (Dix and Pearce, 1981; Watad et al., 1983; Pandey and Ganapathy, 1984; Ahmad et al., 2008) where the sharp increase in the accumulation of proline was observed in stress-tolerant callus lines. Salt-selected cell lines accumulate proline in cytoplasm to a greater extent than the non-selected cell lines of lucerne (Chaudhary et al., 1997). Proline content in salt-tolerant plants is higher than that of salt-sensitive ones (Lin et al., 2002) and Ashraf and Harris (2004) opined that proline may afford much protection against salt stress in tolerant plants. Niknam et al. (2011) reported that in *A. sordidum*, a higher accumulation of proline was found in the *A. glandulosum* callus cultures that were tolerant to salt stress than callus cultures that were salt sensitive. Shah et al. (2012) reported a 17-fold increase in the proline content of 20% PEG-selected calli than the non-selected calli of rice. Increase in the proline accumulation on exposing calli to PEG is reported in *Sorghum* (Bhaskaran et al., 1985) and rice (Aqeel-Ahmad et al., 2007). Proline is involved in the protection of enzymes (Solomon et al., 1994) and cellular structures (Van Resenburg et al., 1993) and acts as a free radical scavenger (Tripathi and Gaur, 2004). GB, a ziwitterionic, fully N-methyl substituted glycine derivative, has been detected in a wide variety of microorganisms, higher plants and animals (Mohanty et al., 2002). Several species come from diverse taxonomic backgrounds, like spinach, sugar beet and wheat synthesize GB (Jones and Storey, 1981). On the contrary, many other economically important crop plants like rice, potato and tomato do not synthesize GB; therefore, they are called GB non-accumulators. The role of GB in protecting enzymes in stress has previously been reported in different studies. GB accumulation prevents membrane damage from a variety of environmental stresses (Deshnium et al., 1997; Chen et al., 2000) via direct membrane stabilization (Rudolph et al., 1986) and the maintenance of the water shell that surrounds the surface-exposed membrane proteins (Coughlan et al., 1982).

GB is a potent compatible compound which does not interfere in the normal metabolism of plants. Natural accumulators of GB (spinach, sugar beet and mangroves) are tolerant to different stresses (Ashraf, 2004; Mohanty et al., 2002). It has been reported that exogenous application of GB enhanced the activity of various antioxidant enzymes, such as superoxide dismutase (SOD), ascorbate peroxidase (APX) and catalase (CAT), during stress conditions (Banu et al., 2009; Sairam et al., 2002). Additionally, *in vivo* synthesis of GB, by the overexpression of GB synthesizing enzymes in GB non-accumulator plants, has resulted in GB synthesis. Due to its diverse positive effects to ameliorate stresses other than an osmolyte, transgenic GB synthesizing plants showed enhanced tolerance to various abiotic stresses (Ahmad et al., 2010; Park et al., 2007; Yang et al., 2007). GB is well known to function like a chaperone to protect enzymatic activities during stress, as it has been reported to protect Rubisco and malate dehydrogenase activities during salt stress (Aran et al., 2986). It is also involved in helping to refold and stabilize proteins in stressful environments (Diamant et al., 2003).

Role of Antioxidant Enzymes in Abiotic Stress

Many environmental stresses, including drought and salinity, cause oxidative stress; several cytotoxic reactive oxygen species (ROS) are continuously generated in the mitochondria, peroxisome and cytoplasm. As a result of accumulated ROS, such as superoxide radical ($O_2\bullet^-$), hydrogen peroxide (H_2O_2), hydroxyl radical (OH•) and singlet oxygen $^1O_2$, damage is caused to lipids, nucleic acids and proteins (Froyer and Noctor, 2005; Turkan and Demiral, 2009; Gill and Tuteja, 2010). Plant cells have developed complex antioxidant defense systems in the form of antioxidant enzymes, such as SOD, CAT, peroxidase (POX), glutathione reductase and polyphenol oxidase (PPO), and non-enzymatic antioxidants, such as ascorbate and glutathione, protect plant cells and subcellular systems from the effects of ROS (Foyer et al., 1991; Bartels and Sunkar, 2005). SOD acts as the first line of defense converting O_2^- to H_2O_2, followed by the detoxification of H_2O_2 by APX, GPX and CAT. The peroxidase enzyme decomposes H_2O_2 by the oxidation of co-substrates, such as phenolic compounds and/or antioxidants (Gaspar et al., 1991; Gill and Tuteja, 2010). Many reports demonstrate a strong relationship between antioxidant enzyme activity and abiotic stress tolerance in plants (Bor et al., 2003; Demiral and Turkan, 2005). CAT enzyme plays an important role in the acquisition of tolerance to oxidative stress in the adaptive response of cells (Abassi et al., 1998). The SOD enzyme converts the superoxide ion produced under stressed conditions to hydrogen peroxide and molecular oxygen, thus playing an important role in defense mechanisms (Harinasut et al., 2003). Gosset et al. (1996) reported that in the NaCl-tolerant cell line grown on 150 mM NaCl, CAT and peroxidase activities were significantly greater than in the control callus. Munir and Aftab (2013) observed that the sugarcane plants regenerated from salt-treated callus cultures generally had elevated levels of antioxidant enzymes as compared to the control. The activities of SOD, CAT, guaiacol peroxidase (GPX), APX and glutathione reductase were higher in callus cultures of *A. glandulosum* under salt stress (Niknam et al., 2011). Sreenivasulu et al. (1999) reported that in salt-tolerant *Setaria italica*, POX activity was found to be higher to protect plants against the stress. Shalata and Tal (1998) reported that in salt-tolerant tomato, CAT was an effective antioxidant enzyme. Munir and Aftab (2013) reported maximum value for peroxidase activity in the sugarcane plants regenerated from 80 mM NaCl-treated callus cultures. Rao and Jabeen (2013) reported that in the PEG tolerant callus line of sugarcane the activity of POX, APX and SOD enzymes was greater in selected calli than in non-selected callus line. PEG induced an increase in the activity of CAT, APX and SOD in sugarcane callus cultures (Patade et al., 2011, 2012). Sen and Alikamanoglu (2011) reported that in *in vitro*–raised seedlings of wheat under NaCl stress there was an increase in the enzyme activities of SOD, CAT and POD, with respect to the control. The higher activity of antioxidant enzymes could be the reason for better growth of selected calli than the non-selected calli on PEG-supplemented medium (Rao and Jabeen, 2013).

Conclusion

The development of salt-tolerant crops that can tolerate high levels of salinity in soils would be a practical solution of such problems. Different strategies are in progress for the development of stress-tolerant plants. *In vitro* selection procedure and *Agrobacterium*-mediated transformation techniques are widely used for the development of such tolerant lines. Transgenic plants containing genes for the overexpression of proline, GB and antioxidant enzymes can survive better in stressful conditions than the normal plants. Callus lines selected for NaCl and drought tolerance (Using PEG), and also plants regenerated from such tolerance, maintain higher activity of antioxidant enzymes and increased accumulation of proline and GB for better growth and survival under stressful environments.

REFERENCES

Abassi NA, Kushad MM, Endress AG. Active oxygen-scavenging enzymes activities in developing apple flowers and fruits. *Sci. Hort.* 1998; 74: 183–194.

Abdi G, Hedayat M, Khush-Khui M. Development of NaCl-tolerant line in *Tanacetum cinerariaefolium* (Trevir.) Schultz-Bip. through shoot organogenesis of selected callus line. *J. Biol. Environ. Sci.* 2011; 5(15): 111–119.

Ahmad R, Hussain J, Jamil M, Kim MD, Kwak SS, Maroof M, Shah1 el-Hendawy SE, Al-Suhaibani NA, Rehman SU. Glycinebetaine synthesizing transgenic potato plants exhibit enhanced tolerance to salt and cold stresses. *Pak. J. Bot.* 2014; 46(6): 1987–1993.

Ahmad R, Kim MD, Back KH, Kim HS, Lee HS, Kwon SY, Murata N, Chung WI, Kwak SS. Stress-induced expression of choline oxidase in potato plant chloroplasts confers enhanced tolerance to oxidative, salt, and drought stresses. *Plant Cell Rep.* 2008; 27: 687–698.

Ahmad MSA, Ashraf M, Ali Q. Soil salinity as a selection pressure is a key determinant for the evolution of salt tolerance in Blue Panicgrass (Panicum antidotale Retz.). *Flora (Jena).* 2010; 205 (1): 37–45.

Ahmad P, Prasad MNV. (2004). Abiotic stress responses in plants. *J. Plant Physiol.* 168: 807–815.

Alvarez I, Tomaro LM, Bernavides PM. Changes in polyamines, proline and ethylene in sunflower calluses treated with NaCl. *Plant Cell Tissue Organ Cult.* 2003; 74(1): 51–59.

Ansari SR, Frooqi AHA, Sharma S. (1998). Interspecific variation in sodium and potassium ion accumulation and essential oil metabolism in three Cymbopogon species raised under sodium chloride stress. *J. Essen. Oil Res.* 10: 413–418.

Aqeel-Ahmad MS, Javed F, Ashraf M. Iso-osmotic effect of NaCl and PEG on growth, cations and free proline accumulation in callus tissue of two indica rice (*Oryza sativa* L.) genotypes. *J. Plant Growth Regul.* 2007; 53: 53–63.

Ashraf M. Inducing drought tolerance in plants: recent advances. *Biotechnol. Adv.* 2010; 28: 169–183.

Ashraf M, Athar HR, Harris PJC, Kwon TR. Some prospective strategies for improving crop salt tolerance. *Adv. Agron.* 2008; 97: 45–110.

Ashraf M, Foolad MR. Roles of glycinebetaine and proline in improving plant abiotic stress resistance. *Environ. Exp. Bot.* 2007; 59: 206–216.

Ashraf M, Harris PJC. Potential biochemical indicators of salinity tolerance in plants. *Plant Sci.* 2004; 166(1): 3–16.

Bakhsh A, Rao AQ, Shahid AA, Husnain T, Riazuddin S. Insect resistance and risk assessment studies in advance lines of Bt cotton harboring Cry1Ac and Cry2A genes. *AEJAES.* 2009; 6: 1–11.

Banu MNA, Hoque MA, Watanabe-Sugimoto M, Matsuoka K, Nakamura Y, Yasuaki S, Murata Y. Proline and glycinebetaine induce antioxid ant defense gene expression and suppress cell death in cultured tobacco cells unde r salt stress. *J. Plant Physiol.* 2009; 166: 146–156.

Barakat MN, Abdel-Latif, TH. *In vitro* selection of wheat callus tolerance high levels of salt and plant regeneration. *Euphytica* 1996; 91: 127–140.

Bartels D, Sunkar R. Drought and salt tolerance in plants. *Crit. Rev. Plant Sci.* 2005; 24: 23–58.

Basu S, Gangopadhyay G, Mukherjee BB. Salt tolerance in rice *in vitro*: implication of accumulation of Na$^+$, K$^+$ and proline. *Plant Cell Tissue Organ Cult.* 2002; 69: 55–64.

Benavides PM, Marconi LP, Gallego MS, Comba EM, Tomaro LM. Relationship between antioxidant defense systems and salt-tolerance in *Solanum tuberosum. Aust. J. Plant Physiol.* 2000; 27: 273–278.

Bhaskaran S, Smith RH, Newton RJ. Physiological changes in cultured *Sorgham* cells in response to induced water stress 1. Free proline. *Plant Physiol.* 1985; 79: 239–248.

Binh DQ, Heszky LE, Gyulai G, Csillag A. Plant regeneration of NaCl-pretreated cells from long-term suspension culture of rice (*Oryza sativa* L.) in high saline conditions. *Plant Cell Tissue Organ Cult.* 1992; 29: 75–82.

Bor M, Ozdemir F, Turkan I. The effect of salt stress on lipid peroxidation and antioxidants in leaves of sugar beet (*Beta vulgaris* L.) and wild beet (*Beta maritima* L.). *Plant Sci.* 2003; 16: 77–84.

Borsani O, Valpuesta V, Botella MA. Developing salt tolerant plants in a new century: a molecular biology approach. *Plant Cell Tissue Organ Cult.* 2003; 73: 101–115.

Boyer J.S. Plant productivity and environment. *Science* 1982, *218*, 443–448.

Boyer J.S., and Knipling E.B. Isopiestic technique for measuring leaf water potentials with a thermocouple psychrometer. *Proc Natl Acad Sci USA* 1965, *54*, 1044–1051.

Bracci T, Minnocci A, Sebastiani L. *In vitro* olive (*Olea europaea* L.) cvs Frantoio and Moraiolo microshoot tolerance to NaCl. *Plant Biosyst*. 2008; 142: 563–571.

Cano EA, Perez-Alfocea F, Moreno V, Caro M, Bolarin MC. Evaluation of salt tolerance in cultivated and wild tomato species through *in vitro* shoot apex culture. *Plant Cell Tissue Organ Cult*. 1998; 53: 19–26.

Cerekovic N., Fatnassi N., Santino A., Poltronieri P. Differences in physiological, biochemical, and molecular mechanisms of water stress tolerance in chickpea varieties. Chapter 6. In: *Approaches in Enhancing Abiotic Stress Tolerance to Plants* (Hasanuzzaman M., Fujita M., Oku H., Islam T.; Eds). Taylor and Francis, CRC in press, 2019. Catalog n. K345347.

Chandler SF, Vasil IK. Selection and characterization of NaC1 tolerant cells from embryogenic cultures of *Pennisetam purpureum* Schum. (Napir grass). *Plant Sci. Lett*. 1984; 37: 157–164.

Chaudhary MT, Merrett MJ, Wainwright SJ. Growth, ion content and proline accumulation in NaCl-selected and non-selected cell lines of lucerne cultured on sodium and potassium salts. *Plant Sci*. 1997; 127: 71–79.

Chen R, Gyokusen K, Saito A. Selection, regeneration and protein profile characteristics of NaCl-Tolerant callus of *Robinia pseudoacaia* L. *J. For. Res*. 2001; 6: 43–48.

Chen WP, Li PH, Chen THH. Glycine betaine increases chilling tolerance and reduces chilling-induced lipid peroxidation in Zea mays L. *Plant Cell Environment* 2000; 23: 609–618.

Chen Y, Zahavi E, Barak P, Ummiel N. Effects of salinity stresses on tobacco. I. The growth of N. tabacum callus cultures under seawater, NaCl, and mannitol stresses. *Z. Pflanzenphysiol*. 1980; 98: 141–153.

Cossett DR, Banks SW, Millhollon EP, Lucas MC. Antioxidant response to NaCl stress in a control and an NaC1-tolerant cotton cell line grown in the presence of paraquat, buthionine sulfoximine, and exogenous glutathione. *Plant Physiol*. 1996; 11(2): 803–809.

Coughlan SJ, Heber U. The role of glycine betaine in the protection of spinach thylakoids against freezing stress. *Planta*. 1982; 156: 62–69.

Croteau R, Kutchan T, Lewis N. (2000) Natural products (secondary metabolites). In: *Biochemistry and Molecular Biology of Plants*, B Buchanan, W Gruissem, R Joneas, (eds.). American Society of Plant Biologists, Rockville, MD, pp. 1250–1268.

Davenport SB, Gallego SM, Benavides MP, Tomaro ML. Behavior of antioxidant defense system in the adaptive response to salt stress in *Helianthus annuus* L. cells. *J. Plant Growth Regul*. 2003; 40: 81–88.

De Ronde JA, Cress WA, Kruger GHJ, Strasser RJ, VanStaden J. Photosynthetic response of transgenic soybean plants, containing an *Arabidopsis P5VR* gene, during heat and drought stress. *Plant Physiol*. 2004; 161: 1211–1224.

Demiral T, Turkan I. Comparative lipid peroxidation, antioxidant defense systems and proline content in roots of two rice cultivars differing in salt tolerance. *Environ. Exp. Bot*. 2005; 53: 247–257.

Deshnium P, Gombos Z, Nishiyama Y, Murata N. The action in vivo of glycine betaine in enhancement of tolerance of Synechococcus sp. strain PCC 7942 to low temperature. *J. Bacteriol*. 1997; 179, 339–344.

Dita, MA, Rispai N, Prats E, Rubiales D, Singh KB. Biotechnology approaches to overcome biotic and abiotic stress constraints in legumes. *Euphytica* 2006; 147: 1–24.

Dix PJ. The role of mutant cell lines in studies on environmental stress tolerance: an assessment. *Plant J*. 1993; 3: 309–313.

Dix PJ, Pearce RS. Proline accumulation in NaCl resistant and sensitive cell lines of *Nicotiana sylvestris*. *Z. Pflanzenphysiol*. 1981; 102: 243–248.

El Sayed H, El Sayed A. Isolation and characterization of NaCl resistant callus culture of field pea (*Pisum sativum*, L.) to salinity. *Agric. Biol. J. N. Am*. 2011; 2(6): 964–973.

Fatima S, Farooqi AHA, Sangwan RS. (2006) Water stress mediated modulation in essential oil, proline and polypeptide profile in palmarosa and citronella java. In: *Physiology and Molecular Biology of Plants*. Lucknow, India: Central Institue of Medicinal and Aromatic Plants, CIMAP.

Foyer CH, Lelandais M, Galap C, Kunert KJ. Effect of elevated cytosolic glutathione reductase activity on the cellular glutathione pool and photosynthesis in leaves under normal and stress conditions. *Plant Physiol*. 1991; 97: 863–872.

Foyer CH, Noctor G. Oxidant and antioxidant signalling in plants: a reevaluation of the concept of oxidative stress in a physiological context. *Plant Cell Environ*. 2005; 28: 1056–1071.

Gandonou CB, Errabii T, Abrini J, Idaomar M, Senhaji NS. Selection of callus cultures of sugarcane (*Saccharum* sp.) tolerant to NaCl and their response to salt tolerance. *Plant Cell Tissue Organ Cult*. 2006; 87: 9–16.

Gao X, Yang D, Cao D, Ao M, Sui X, Wang Q, Kimatu JN, Wang L. *In vitro* micropropagation of *Freesia hybrida* and the assessment of genetic and epigenetic stability in regenerated plantlets. *J. Plant Growth Regul*. 2010; 29: 257–267.

Gaspar T, Penel C, Hagega D, Greppin H. Peroxidases in plant growth, differentiation and development processes. In: *Biochemical, Molecular and Physiological Aspects of Plant Peroxidases* (eds. J. Lobarzewsky, H. Greppin, C. Penel, and Th. Gaspar). University of Geneve, Switzerland, 1991, pp. 249–280.

Ghanti SKK, Sujata KG, Vijay Kumar BM, Karba NN, Reddy KJ, Srinath Rao M, Kavi Kishor PB. Heterologous expression of *P5CS* gene in chickpea enhances salt tolerance without affecting yield. *Biol. Plant*. 2011; 55 (4): 634–640.

Gill SS, Tuteja N. Reactive oxygen species and antioxidant machinery in abiotic stress tolerance in crop plants. *Plant Physiol. Biochem*. 2010; 48: 909–930.

Gu R, Liu Q, Pie D, Jiang X. Understanding saline and osmotic tolerance of *Populus euphratica* suspended cells. *Plant Cell Tissue Organ Cult*. 2004; 78: 261–265.

Gubis JR, Vankova V, Cervena M, Dragunova M, Hudcovicova T, Dokupil L, Jurekova Z. Transformed tobacco plants with increased tolerance to drought. *S. Afr. J. Bot*. 2007; 73(4): 505–511.

Guo WL, Wu R, Zhang YF, Liu XM, Wang HY, Gong L, Zhang ZH, Liu B. Tissue culture-induced locus-specific alteration in DNA methylation and its correlation with genetic variation in *Codonopsis lanceolata* Benth. et Hook. f. *Plant Cell Rep*. 2007; 26: 1297–1307.

Han KH, Hwang CH. Salt tolerance enhanced by transformation of a *P5CS* gene in carrot. *J. Plant Biotechnol*. 2003; 5: 149–153.

Harinasut P, Poonsopa D, Roengmongkol K, Charoensataporn R. Salinity effects on antioxidant enzymes in mulberry cultivar. *Sci. Asia* 2003; 29: 109–113.

Hasegawa PM, Bressan RA, Nelson DE, Samaras Y, Rhodes D. Tissue culture in the improvement of salt tolerance in plants. In: AR Yeo and TJ Flowers (eds) *Monographs on Theoretical and Applied Genetics: Breeding Plants with Resistance to Problem Soils*, p 83. 1994; Springer-Verlag, Berlin.

Hasegawa PM, Bressan RA, Zhu JK, Bohnert HJ. Plant cellular and molecular responses to high salinity. *Annu. Rev. Plant Physiol. Plant Mol. Biol.* 2000; 51: 463–499.

He C, Yang A, Zhang W, Gao Q, Zhang J. Improved salt tolerance of transgenic wheat by introducing betA gene for glycine betaine synthesis. *Plant Cell Tissue Organ Cult.* 2010; 101: 65–78.

Hossain Z, Mandal AK, Datta SK, Biswas, AK. Development of NaCl tolerant line in *Chrysanthemum morifolium* Ramat. through shoot organogenesis of selected callus line. *J. Biotechnol.* 2007; 129: 658–667.

Jain M. Tissue culture-derived variation in crop improvement. *Euphytica* 2001; 118: 153–166.

Jaiswal R, Singh NP. Plant regeneration from NaCl tolerant callus/ cell lines of chickpea. *ICPN* 2001; 8: 21–23.

Jones RG, Storey R. Betaines. The unusually strong stabilizing effect of glycine betaine on the structure and function of the oxygen-evolving PSII comp. In: Paleg LG and Aspinal D (eds) The Physiology and Biochemistry of Draught Resistance in Plants, pp 171–204. 1981; Academic Press, Sydney, Australia.

Karadimova M, Djambova G. Increased NaCl-tolerance in wheat (*Triticum aestivum* L. and *T. durum* desf.) through *in vitro* selection. *In Vitro Cell Dev. Biol. Plant* 1993; 29: 180–182.

Kavi Kishore PB, Hong Z, Miao G, Hu CAA, Verma DPS. Over expression of Pyrroline 5 carboxylate synthetase increases proline over production and confers osmotolerance in transgenic plants. *Plant Physiol.* 1995; 108: 1387–1394.

Khan A, Iqbal I, Shah A, Humera N, Farooq A, Ibrahim M. Alleviation of adverse effects of salt stress in brassica (*Brassica campestris*) by pre-sowing seed treatment with ascorbic acid. *AEJAES.* 2010; 7: 557–560.

Kumar V, Sharma DK. Isolation and characterization of sodium chloride resistantcallus culture of *Vigna radiata* (L.) Wilczek var. radiata. *J. Exp. Bot.* 1989; 40: 143–147.

Larkin PJ, Scowcroft SC. Somaclonal variation – A novel source of variability from cell culture for plant improvement. *Theor. Appl. Genet.* 1981; 60: 197–214.

Li M, Lin X, Li H, Pan X, Wu G. Overexpression of *AtNHX5* improves tolerance to both salt and water stress in rice (*Oryza sativa* L.). *Plant Cell Tissue Organ Cult.* 2011; 107(2): 283–293.

Li X, Yu X, Wang N, Feng Q, Dong Z, Liu L, Shen J, Liu B. Genetic and epigenetic instabilities induced by tissue culture in wild barley (*Hordeum brevisubulatum* (Trin.) Link). *Plant Cell Tissue Organ Cult.* 2007; 90: 153–168.

Lin CC, Hsu YT, Kao CH. The effect of NaCl on proline accumulation in rice leaves. *J. Plant Growth Regul.* 2002; 36: 275–285.

Lutts S, Almansouri M, Kinet JM. Salinity and water stress have contrasting effects on the relationship between growth and cell viability during and after stress exposure in durum wheat callus. *Plant Sci.* 2004; 167: 9–18.

Mansour MMF. Nitrogen containing compounds and adaptation of plants to salinity stress. *Biol. Plant* 2000; 43: 491–500.

McCue KF, Hanson AD. Drought and salt tolerance: towards understanding and application. *Trend Biotechnol.* 1990; 8: 358–362.

Mc-Hughen A, Swartz M. A tissue culture derived salt-tolerant line of flax (*Linum usitatissimum*). *Plant Physiol.* 1984; 117: 109–118.

Miki Y, Hashiba M, Hisajima S. Establishment of salt stress tolerant rice plants through step-up NaCl treatment *in vitro*. *Biol. Plant.* 2001; 44: 391–395.

Mohamed MAH, Harris PJC, Henderson J. *In vitro* selection and characterization of a drought tolerant clone of *Tagetes minuta*. *Plant Sci.* 2000; 159: 213–222.

Mohanty A, Kathuria H, Ferjani A, Sakamoto A, Mohanty P, Murata N, Tyagi A. Transgenics of an elite rice variety pusa basmati 1 harbouring the codA gene are highly tolerant to salt stress. *Theor. Appl. Genet.* 2002; 106: 51–57.

Mondal TK, Kundu PK, Ahuja PS. Gene silencing: a problem in transgenic research. *Curr. Sci.* 1997; 72: 699–700.

Munir N, Aftab F. Changes in activities of antioxidant enzymes in response to NaCl stress in callus cultures and regenerated plants of sugarcane. *J. Anim. Plant Sci.* 2013; 23(1): 203–209.

Musa Y. The use of polyethylene glycol (PEG) as selection agent of callus and plantlets of some sugarcane varieties for drought tolerance. *J. Agrivigor.* 2011; 10(2): 130–140.

Nabors MW, Gibbs GE, Bernstein CS, Meis ME. NaCl-tolerant tobacco plants from cultured cells. *Z. Pflanzenphysiol.* 1980; 97: 13–17.

Naik GR, Harinath K. Redifferentiation of NaCl tolerant sugarcane plants from callus derived resistant lines. *Curr. Sci.* 1988; 57(8): 432–433.

Niknam V, Meratan AA, Ghaffari SM. The effect of salt stress on lipid peroxidation and antioxidative enzymes in callus of two *Acanthophyllum* species. *In Vitro Cell Dev. Biol. Plant* 2011; 47(2): 297–308.

Nizam J, Khan IA, Farook SA. Ukaz Publication, Hyderabad, pp. 149–154.

Ochatt SJ, Marconi PL, Radice S, Arnozis PA, Caso OH. *In vitro* recurrent selection of potato: production and characterization of salt tolerant cell lines and plants. *Plant Cell Tissue Organ Cult.* 1999; 55: 1–8.

Olmos E, Hernandez JA, Sevilla F, Hellin E. Induction of several antioxidant enzymes in the selection of a salt-tolerant cell line of *Pisum sativum*. *Plant Physiol.* 1994; 144: 594–598.

Pandey R, Ganapathy PS. Isolation of NaCl tolerant callus line of *Cicer arietinum* L Cv BG-203. *Plant Cell Rep.* 1984; 3: 45–47.

Park EJ, Jeknic Z, Pino MT, Murata N, Chen TH. Glycinebetaine accumulation in chloroplasts is more effective than that in cytosol in protecting transgenic tomato plants against abiotic stress. *Plant Cell Environ.* 2007; 30: 994–1005.

Parsad VS, Sharmila KP, Kumar PA, Saradhi PP. Transformation of *Brassica juncea* (L.) Czern with a bacterial codA gene enhances its tolerance to salt stress. *Mol. Breed.* 2000; 6: 489–499.

Patade VY, Bhargava S, Suprasanna P. Effects of NaCl and iso-osmotic PEG stress on growth, osmolytes accumulation and antioxidant defense in cultured sugarcane cells. *Plant Cell Tissue Organ Cult.* 2012; 108(2): 279–286.

Patade VY, Bhargava S, Suprasanna P. Salt and drought tolerance of sugarcane under iso-osmotic salt and water stress: growth osmolyte accumulation and antioxidant defense. *J. Plant Interact.* 2011; 6(4): 275–282.

Patil P, Rao S. Selection and characterization of NaCl tolerant callus cultures of *Vigna radiate* (L.) Wilczek. In: *Plant Tissue Culture and Biotechnology Emerging Trends* (ed. P.B.K. Kishor). Universities Press, Hyderabad, India, 1999.

Patnaik J, Debata BK. *In vitro* selection of NaCl tolerant callus lines of *Cymbopogon martinii* (Roxb.) Wats. *Plant Sci.* 1997a; 124: 203–210.

Patnaik J, Debata BK. Regeneration of plantlets from NaCl tolerant callus lines of *Cymbopogon martinii* (Roxb.) Wats. *Plant Sci.* 1997b; 128: 67–74.

Perez-Tornero O, Tallón CI, Porras I, Navarro JM. Physiological and growth changes in micropropagated Citrus macrophylla explants due to salinity. *J Plant Physiol.* 2009; 166(17): 1923–1933.

Purohit M, Srivastava S, Srivastava PS. Stress tolerant plants through tissue culture. In: *Plant Tissue Culture and Molecular Biology: Application and Prospects* (ed. P.S. Srivastava). Narosa Publishing House, New Delhi, 1998, pp.554–578.

Quan R, Shang M, Zhang H, Zhao Y, Zhang J. Engineering of enhanced glycine betaine synthesis improves draught tolerance in maize. *Plant Biotech. J.* 2004; 2: 477–486.

Queiros F, Fidalgo F, Santos I, Salema R. *In vitro* selection of salt tolerant cell lines in *Solanum tuberosum* L. *Biol. Plant.* 2007; 51: 728–734.

Ramakrishna A, Ravishankar GA. Influence of abiotic stress signals on secondary metabolites in plants. *Plant Sig. Behav.* 2011; 6: 1720–1731.

Ramakrishna A, Ravishankar GA. Role of plant metabolites in abiotic stress tolerance under changing climatic conditions with special reference to secondary compounds. In: *Climate Change and Plant Abiotic Stress Tolerance* (eds. N. Tuteja, and S.S. Gill). Wiley-VCH Verlag GmbH & Co. KGaA, Weinheim, Germany, 2013, pp. 705–726.

Rao S, Krupanidhi. Regeneration of plantlets from sodium chloride (tolerant) cell lines in pigeon pea (*Cajanus cajan* (L.) Millsp.) 1996. In: Role of Biotechnology in Pulse Crops. Eds. Rao S, Jabeen FTZ. *In vitro* selection and characterization of polyethylene glycol (PEG) tolerant callus lines and regeneration of plantlets from the selected callus lines in sugarcane (*Saccharum officinarum* L.). *Physiol. Mol. Biol. Plants* 2013; 19: 261–268.

Reddy PJ, Vaidyanath K. *In vitro* characterization of salt stress effects and the selection of salt-tolerant plants in rice (*Oryza sativa* L.). *Theor. Appl. Genet.* 1986; 71: 757–760.

Saijo Y, Hata S, Kyozuka J, Shimamoto K, Izui K. Overexpression of a single Ca2+ dependent protein kinase confers both cold and salt/drought tolerance on rice plants. *Plant J.* 2000; 23: 319–327.

Sairam RK, Rao KV, Srivastava GC. Differential response of wheat genotypes to longterm salinity stress in relation to oxidative stress, antioxidant activity and osmolytes concentration. *Plant Science* 2002; 163: 1037–1046.

Sajid ZA, Faheem Aftab F. Plant regeneration from *in vitro*-selected salt tolerant callus cultures of *Solanum tuberosum* L. *Pak. J. Bot.* 2014; 46(4): 1507–1514.

Sakhanokho HF, Kelley RY. Influence of salicylic acid on *in vitro* propagation and salt tolerance in *Hibiscus acetosella* and *Hibiscus moscheutos* (cv 'Luna Red'). *Afr. J. Biotechnol.* 2009; 8(8): 1474–1481.

Sassi A.S., Aydi S., Gonzalez E.M., Abdelly C. Osmotic stress affects water relations, growth, and nitrogen fixation in *Phaseolus vulgaris* plants. *Acta Physiol Plant* 2008, *30*, 441–449.

Sawahel WA, Hassan AH. Generation of transgenic wheat plants producing high levels of the osmoprotectant proline. *Biotechnol. Lett.* 2002; 24: 721–725.

Sayari AH, Bouzid RG, Bidani A, Jaoua L, Savoure A, Jaoua S. Over expression of Δ1-pyrroline-5-carboxylate synthetase increases proline production and confers salt tolerance in transgenic potato plants. *Plant Sci.* 2005; 169: 746–752.

Sen A, Alikamanoglu S. Effect of salt stress on Growth parameters and Antioxidant enzymes of different wheat (*Triticum aestivum* L.) varieties on *in vitro* tissue culture. *Fresenius Environ. Bull.* 2011; 20(2a): 489–495.

Shah AH, Shah SH, Ahmad H, Swati ZA. Adaptation to polyethylene stress maintains totipotency of cell lines of Oryza sativa L. CV Swat-1 for a longer period. *Pak. J. Bot.* 2012; 44(1): 313–316.

Shalata A, Tal M. The effect of salt stress on lipid peroxidation and antioxidants in the of the cultivated tomato and its wild salt tolerant relative *Lycopersicon pennellii*. *Physiol. Plant.* 1998; 104: 169–174.

Shankhdhar D, Shankhdhar SC, Mani SC, Pant RC. *In vitro* selection for salt tolerance in rice. *Biol. Plantarum* 2000; 43(3): 477–480.

Shanthi P, Jebaraj S, Geetha S. *In vitro* screening for salt tolerance in Rice (*Oryza sativa*). *Electron. J. Plant Breed.* 2010; 1(4): 1208–1212.

Shen YG, Du BX, Zhang WK, Zhang JS, Chen SY. AhCMO, regulated by stresses in triplex hortensis, can improve drought tolerance in transgenic tobacco. *Theor. Appl. Genet.* 2002; 105: 815–821.

Shao HB, Chu LY, Jaleel CA, Manivannan P, Panneerselvam R, Shao MA. (2009). Understanding water deficit stress-induced changes in the basic metabolism of higher plants-biotechnologically and sustainably improving agriculture and the ecoenvironment in arid regions of the globe. *Crit. Rev. Biotechnol.* 29: 131-151

Shirasawa K, Takabe T, Takabe T, Kishitani S. Accumulation of glycine betaine in rice plants that overexpress choline mono-oxygenase from spinach and evaluation of their tolerance to abiotic stress. *Ann. Bot.* 2006; 98: 565–571.

Singh M, Jaiswal U, Jaiswal VS. *In vitro* selection of NaCl-tolerant callus line and regeneration of plantlets in a bamboo (*Dendrocalamus strictus* Nees.) *In Vitro Cell Dev. Biol. Plant* 2003; 39: 229–233.

Solomon A, Beer S, Waisel Y, Jones GP, Poleg LG. Effect of NaCl on the carboxylating activity of Rubisco from *Tamarix jordanis* in the presence and absence of proline related compatible solutes. *Physiol. Plant* 1994; 90: 189–204.

Sreenivasulu N, Ramanjulu S, Rmachandra-Kini K, Prakash HS, Shekar-Shetty H, Savithri HS, Sudhakar C. Total peroxidase activity and peroxidase isoforms as modified by salt stress in two cultivars of fox-tail millet with differential salt tolerance. *Plant Sci.* 1999; 141: 1–9.

Su J, Hirji R, Zhang L, He C, Selvaraj G, Wu R. Evaluation of the stress-inducible production of choline oxidase in transgenic rice as a strategy for producing the stress protectant glycine betaine. *J. Exp. Bot.* 2006; 57(5): 1129–1135.

Tal M. *In vitro* selection for salt tolerance in crop plants: theoretical and practical considerations. *In Vitro Cell Dev. Biol. Plant* 1994; 30: 175–180.

Tripathi B N, Gaur J P. Relationship between copper- and zinc-induced oxidative stress and praline accumulation in *Scenedesmus* sp. *Planta* 2004; 219: 397–404.

Turkan I, Demiral T. Recent developments in understanding salinity tolerance. *Environ. Exp. Bot.* 2009; 67: 2–9.

Tuteja N, Gill SS, Tuteja R. (2011). Plant response to abiotic stress: Shedding light on salt, drought, cold and heavy metal stress. *Omics and Plant Abiotic Stress Tolerance* 1: 39-64.

Vajrabhaya M, Thanapaisal T, Vajrabhaya T. Development of salt tolerant lines of KDML and LPT rice cultivars through tissue culture. *Plant Cell Rep.* 1989; 8: 411–414.

Van Resenburg L, Kruger GHJ, Kruger H. Proline accumulation as drought tolerance selection criterion: its relationship to membrane integrity and chloroplast ultrastructure in *Nicotiana tabacum* L. *Plant Physiol.* 1993; 141: 188–194.

Venkataiah P, Christopher T, Subhash K. Selection and characterization of sodium chloride and mannitol tolerant callus lines of red pepper (*Capsicum annuum* L.). *Plant Physiol.* 2004; 9(2): 158–163.

Vinocur B, Altman A. Recent advances in engineering plant tolerance to abiotic stress: achievements and limitations. *Curr. Opin. Biotechnol.* 2005; 16: 123–132.

Wang B, Luttge U, Ratajczak R. Specific regulation of SOD isoforms by NaCl and osmotic stress in leaves of the C3 halophyte *Suaeda salsa* L. *Plant Physiol.* 2004; 161: 285–293.

Wang W, Vinocur B, Altman A. Plant responses to drought, salinity and extreme temperatures: towards genetic engineering for stress tolerance. *Planta* 2003; 218: 1–14.

Watad AEA, Reinhold L, Lerner HR. Comparison between a stable NaCl selected *Nicotiana* cell line and the wild type. *Plant Physiol.* 1983; 73: 624–629.

Winicov I. Characterization of rice (*Oryza sativa* L.) plants regeneration salt-tolerant cell lines. *Plant Sci.* 1996; 113: 105–111.

Woodward AJ, Bennett IJ. The effect of salt stress and abscisic acid on proline production, chlorophyll content and growth of in vitro propagated shoots of *Eucalyptus camaldulensis*. *Plant Cell Tissue Organ Cult.* 2005; 82: 189–200.

Yamada M, Morishita H, Urano K, Shiozaki N, Yamaguchi-Shinozaki K, Shinozaki K, Yoshiba Y. Effects of free proline accumulation in petunias under drought stress. *J. Exp. Bot.* 2005; 56: 1975–1981.

Yamaguchi T, Blumwald E. Developing salt-tolerant crop plants: challenges and opportunities. *Trends Plant Sci.* 2005; 10: 615–620.

Yang P, Wen LL, Zhao C, Zhao B, Yangdong G. Cloning and functional identification of ProDH gene from broccoli. *Journal of Guangxi Agricultural and Biological Science* 2010; 29: 206–214.

Yang WJ, Rich PJ, Axtell JD, Wood KV, Bonham CC, Ejeta G, Mickelbart V, Rhodes D. Genotypic variation for glycine betaine in *Sorghum*. *Crop Sci.* 2003; 43: 162–169.

Yang X, Wen X, Gong H, Lu Q, Yang Z, Tang Y, Liang Z, Lu C. Genetic engineering of the biosynthesis of glycine betaine enhances thermo tolerance of photosystem II in tobacco plants. *Planta* 2007; 225: 719–733.

Zair I, Chlyah A, Sabounji K, Tittahsen M, Chlyah H. Salt tolerance improvement in some wheat cultivars after application of *in vitro* selection pressure. *Plant Cell Tissue Organ Cult.* 2003; 73: 237–244.

Zenk MH. Haploids in physiological and biochemical research. In: *Haploids in Higher Plants* (ed. K.J. Kasha), Canada University Guelph Press, Ontario, Canada, 1974.

Zhang F, Yang YL, He WL, Zhao X, Zhang LX. Effects of salinity on growth and compatible solutes of callus induced from *Populus euphratica*. *In Vitro Cell Dev. Biol. Plant* 2004; 40: 491–494.

Zhang H, Dong H, Li W, Sun Y, Chen S, Kong X. Increased glycine betaine synthesis and salinity tolerance in AhCMO transgenic cotton lines. *Mol. Breed.* 2009; 23: 289–298.

Zhou SF, Chen XY, Xue XN, Zhang XG, Li YX. Physiological and growth responses of tomato progenies harboring the betaine aldehyde dehydrogenase gene to salt stress. *J. Integr. Plant. Biol.* 2007; 49: 628–637.

Zhu B, Su J, Chang M, Verma DPS, Fan YL, Wu R. Over expression of a ∧1 pyrroline 5-carboxylate synthetase gene and analysis of tolerance to water and salt stress in transgenic rice. *Plant Sci.* 1998; 139: 41–48.

Zhu JK. Salt and drought stress signal transduction in plants. *Annu. Rev. Plant Physiol. Plant Mol. Biol.* 2002; 53: 247–273.

33

Plant–Environment Interaction: Influence of Abiotic Stress on Plant Essential Oil Yield and Quality

Marine Hussain, Barbi Gogoi, Babita Joshi, Bitupon Borah, Lucy Lalthafamkimi, and Brijmohan Singh Bhau

CONTENTS

Abbreviations

ABA	Abscisic acid
Apx1	Ascorbate peroxidase
ATAF	Arabidopsis transcription activation factor
AtMGL	Methionine gamma lyase
AtRALFL8	Rapid alkalinization factor-like8
AZI1	Azelaic acid induced1
bZIPs	Basic Leucine Zipper
CAT	Catalase
CBF/DREB	Cold-binding factor/dehydration responsive element binding
EO	Essential oil
ERFs	Ethylene-responsive element-binding factors
FAs	Fatty acids
FTA	Farnesyl transferase
GC	Gas chromatography
GPX	Glutathione peroxidase
LC	Liquid chromatography
LOX	Lysyl oxidase
MAPK	Mitogen-activated protein kinase
MEKK1	Mitogen-activated protein kinase kinase kinase 1
miRNA	MicroRNAs
MPK	Myotonin protein kinase
MS	Mass spectrometry
NMR	Nuclear magnetic resonance spectroscopy
QTL	Quantitative trait loci
ROS	Reactive oxygen species
SOD	Superoxide dismutase
TFs	Transcription factors

Introduction

Nature possesses a unique resource of high phytochemical diversity , which has a wide range of biological and therapeutic properties. Since ancient times, aromas and flavors have been associated with individual lifestyles. Aromatic plants are the basis of these aromas and known to release aromatic or volatile substances in the form of essential oils (EOs). There are 17,500 aromatic plant species among higher plants and approximately 3,000 EOs are known, out of which 300 are supposed to be commercially important for the pharmaceutical, cosmetics and perfume industries (Bhattacharjee, 2005). Some of the major families bearing EO plants are Apiaceae, Asteraceae, Combretaceae, Geraniaceae, Graminae, Lamiaceae, Myrtaceae, Meliaceae, Piperaceae, Rutaceae, Verbenaceae and Zingiberaceae (Bedi et al., 2008). The production and utilization of EOs have increased all over the world with an annual production of EO at about 40,000–60,000 tons per annum in conjunction with a market value of 700 million

US$ (Djilani et al., 2012). India ranks second in the world trade of EOs (Rao et al., 2005). Recently, a market research study by Grand View Research estimates that the global EOs market is expected to reach $11.67 billion by 2022 (Yeager, 2017).

Oil production in plants depends on various biotic and abiotic factors to which the plants are subjected to during their growth. Phytochemicals that occur naturally in therapeutic plants have well-built defense mechanisms against various diseases and environmental stresses (Nostro et al., 2000). Phytochemicals in plants are classified as primary and secondary metabolites (Geissman et al., 1969; Mann et al., 1978) where primary constituents are carbohydrates, proteins and fats, which are required for the normal metabolic functioning of plants, and secondary metabolites are referred to as a plant's natural products, *viz.*, terpenes and terpenoids (25,000 types), alkaloids (12,000 types) and phenolic compounds (8,000 types) (Croteau et al., 2000). Natural volatile complexes of secondary metabolites, such as EOs, are characterized by a strong odor and possess lower density than that of water. EOs are oily aromatic liquids extracted from aromatic plant materials and could be biosynthesized in different plant organs as secondary metabolites (Asbahani et al., 2015). Chemical constituents of EO mainly comprise of a complex mixture of monoterpenes, diterpenes and sesquiterpenes. Some of the volatile terpenoids like hemiterpenoids (C_5), monoterpenoids (C_{10}), sesquiterpenoids (C_{15}) and diterpenoids (C_{20}) play a key role in the interaction between plants and insect herbivores or pollinators, and also implicate general defense or stress responses (Dudareva et al., 2004; Pichersky et al., 2002, 2006). The significance of EOs are their flavor concentrations, their similarity to their corresponding sources and their natural antioxidants and antimicrobial agents, as found in citrus fruits (Somesh et al., 2015). Moreover, extensive use of EO in perfumery, aromatherapy, cosmetics, incense, medicine, household cleaning products, food flavoring and drinks has designated them as key commodities in the fragrance and food industries. Additionally, EOs can be used as herbicides, pesticides and anticancer compounds due to their biological activities (Mahmoud et al., 2002; Abrahim et al., 2003).

Production of secondary metabolites is a significant part of the defense response to stress conditions. Therefore, the accumulation of secondary metabolites is primarily related to membrane lipid protection from oxidative stress and reactive oxygen species (ROS), which are considered the mediators in the biosynthesis of particular secondary metabolites (Zhu et al., 2009; Ramakrishna and Ravishankar, 2011, 2013). EO production in plants are exaggerated by a wide range of both abiotic and biotic stresses including genetic variation, plant ecotype or variety, plant nutrition, application of fertilizers, geographic location of the plants, surrounding climate, seasonal variations, stress during growth or maturity and also post-harvest drying and storage (Chaves et al., 2002; Jaleel et al., 2007, 2008). These environmental stresses trigger a wide variety of plant responses, ranging from altered gene expression and cellular metabolism to changes in growth rate and crop yield (Bhatt et al., 2005; Jaleel et al., 2007; Farooq et al., 2008). Among the abiotic stresses, drought is the most vital one which affects plants periodically in some growth stages, or enduringly for the full life cycle (Jaleel et al., 2009). Drought stress is the result of water discrepancy in soil, and atmospheric conditions increase water loss through evapotranspiration (Reddy et al.,

2008). In higher plants, the oxygen toxicity is more serious under water deficit conditions. Water stress causes stomatal closure, which reduces the CO_2/O_2 ratio in leaves and inhibits photosynthesis (Melzer et al., 1987; Jason et al., 2004; Moussa, 2006). These circumstances boost the rate of ROS like superoxide radical (O^{2-}) hydrogen peroxide (H_2O_2), and hydroxyl radical (OH) predominantly in chloroplast and mitochondria (Mittler, 2002; Neill et al., 2002), using enhanced leakage of electrons to oxygen. Therefore, plants defend the cellular and sub-cellular system from the cytotoxic effects of active oxygen radicals with several anti-oxidative enzymes such as superoxide dimutase (SOD), POX and catalase (CAT) as well as metabolites like glutathione, ascorbic acid, tocopherol and carotenoids (Alscher et al., 2002). Additionally, the stress induces disintegration in free radicals, which could also be partially related to the activity of lipoxygenase that converts C18:2 and C18:3 to the corresponding hydroxyl peroxides (Bell et al., 1991). On the other hand, many researchers have revealed that drought also enhances the number of secondary metabolites in a wide diversity of plant species, such as *Rehmannia glutinosa* (Gaertn.) DC. (Chung et al., 2006). It increases the EO percentage of medicinal and aromatic plants, because, during stress, metabolites produced in the plants increase and prevent oxidization in the cells. The interaction between the oil percentage and shoot yield is considered an important component of the EO content, and exerting stress results in an increase in the EO percentage but a decrease in shoot yield, and therefore a reduction in the content of EO (Aliabadi et al., 2009). Overall, the percentage of EO was increased under water stress but the content of EO was decreased under this condition.

In recent decades, the demand for EO has increased rapidly due to the increasing per capita consumption of oil in commercial and therapeutic applications. Therefore, increase in oil yields is a promising option to fill the gap in the scarcity of oil production. Traditionally, EO extraction from plant materials is carried out by distillation with water or steam. Other methods of oil extraction that exist in the perfumery and food industries include cold pressing or scarification, effleurage, solvent extraction and upper critical carbon dioxide extraction. Subsequently, biotechnology also influences the breeding or development of EO plants and could theoretically produce plants containing oil of a required composition. Genetic and metabolic regulation can determine the proportion of oil in different species. Further, oil yield is a quantitative trait controlled by many loci and is influenced by the environment (Rahman et al., 2013). The identification of genes that are involved in lipid biosynthesis and its regulation generates transgenic oil plants, which are capable of producing the desired oil characteristics with the help of advance molecular biology tools and transformation methods (Maheshwar et al., 2014). These emerging interests in enhancing the oil yields of plants have resulted in the discovery of several molecular regulators of oil biosynthesis in *Arabidopsis* and other plant species. Predominantly, the molecular regulators of oil content include gene encoding transcription factors (TFs) and enzymes involved with fatty acids (FAs) and triacylglyceride (TAG) synthesis and carbon lux. In addition, recent advancements in the high-throughput technologies like genomics, transcriptomics and metabolomics have led to the understanding of genes, gene networks, regulatory factors and their interactions that govern plant oil biosynthesis that allow the devising of novel strategies for the genetic engineering of plants (Sanghera et al., 2011; Kashyap et al., 2011).

This book chapter provides an account of the studies on some common abiotic stresses to which EO plants are exposed during their growth period and their influence on the quality and quantity of oil, and also the strategies for the improvement in abiotic stresses by omics technology. Enhancing EO productivity is an important challenge, and understanding the role played by stress may offer significant advantages to the EO farming and processing industry. Scientific evaluation of the data on many important but unexplored EO plants will also help in mitigating, ameliorating and minimizing the harmful effects caused by stress. Further, rapid developments in high-throughput technologies like genomics, proteomics and metabolomics, along with new genetic engineering strategies such as the altered expression of TFs, introduction of novel genes like microRNA (miRNA) and multigene engineering, will hope to contour EO in a way to convene the potential EO desires of the world.

Impact of Abiotic Stress on Oil Yield

Abiotic stress can be defined as any factor endeavored by the environment on the optimal functioning of an organism. It is an altered physiological condition caused by external elements that tend to disrupt the equilibrium. Abiotic stresses cause changes in the soil–plant–atmosphere continuum, which is responsible for changes in yield and growth of major plants in different parts of the world. Therefore, the subject of abiotic stress response in plants – metabolism, productivity and sustainability – is gaining considerable significance in the contemporary world (Ahmad and Prasad, 2004). The term stress is used with various meanings; the physiological definition and appropriate term as responses in different situations. Stress, being a constraint or a highly unpredictable fluctuation imposed on regular metabolic patterns, can cause injury, disease or aberrant physiology. As a natural part of every ecosystem, abiotic stresses affect organisms in a variety of ways. These effects may be beneficial or detrimental, depending on the type of stress triggering on a plant species. Plants growing in natural habitats are often challenged by multiple unfavorable environmental conditions – abiotic factors, causing abiotic stresses – that play a major role in determining the productivity of crop yields in such a way that growth, development and reproduction of the yield of crops becomes compromised. The growth, yield and EO composition of the majority of oil yielding plants have been affected by various environmental constraints. Changes in both EO yield and their compositions (quality) have been reported to be influenced by environmental conditions. Some studies have shown a decrease in EO yields and changes in their compositions under abiotic stress (Dow et al., 1981). On the other hand, there are reports in support of the stimulating effects of abiotic stresses on the synthesis of secondary metabolites in plants (Hendawy et al., 2005).

Environmental stress regulates a wide variety of plant responses, ranging from altered gene expression and cellular metabolism to changes in growth rate and crop yield (Jaleel et al., 2007). It is well known that environmental abiotic stresses, such as temperature, humidity, light intensity, the supply of water, minerals, heavy metals, salinity ozone, UV-B radiation, etc., significantly affect plant survival, growth and development, and thus decrease plant quality, yield and biomass production (Assche et al., 1990; Wang et al., 2003; Camejo et al., 2005). The major abiotic stresses such as drought, flood, high salinity, cold, heat, soil humidity, etc., influence the survival, biomass production and yields of staple food crops up to 70% (Foy, 1988; Kaur et al., 2008); hence, they threaten food security worldwide. Dehydration stress imparted by drought, salinity and temperature severity is the most prevalent abiotic stress that limits plant growth and productivity (Hikosaka et al., 1999; Jaleel et al., 2009; Cao et al., 2011). In this regard, this content will focus on the effects and roles of different abiotic stress on the yield and quality of EOs in oil yielding plants.

Effects Due to Water Stress

Water is one of the most important factors of all forms of life since it is necessary for life to thrive. Plants are living organisms and require water in order to grow well and reproduce. Water participates directly or indirectly in all metabolic processes of all living organisms. It is an important climatic factor which affects or determines plant growth and development. Its availability, or scarcity, can mean a successful harvest, diminution in yield or total failure. However, response of plant against water is different depending on the plant species. Most plants are mesophytes (adapted to conditions with moderate supply of water), hydrophytes (require watery habitats) and xerophytes (adapted to dry conditions). Shao et al. (2009) observed that plants experience water stress either when the water supply to their roots becomes limiting or when the transpiration rate becomes intense. A plant requires a certain amount of water for its optimal survival; too much water (flooding stress) can cause plant cells to swell and burst, whereas too little water (drought stress) can cause the plant to dry up, a condition called desiccation. Hence, the effects or role of water on plants can be briefly divided into the following broad categories.

Drought Stress

Drought stress is one of the most destructive environmental stresses affecting agricultural productivity around the world and may result in considerable yield reductions (Boyer, 1982; Ludlow et al., 1990). It is an abiotic stress that affects plant morphology, physiology and biochemistry, causing a significant reduction in agricultural production (Hsiao, 1973). Drought stress limits the production of 25% of the world's land (Delfine et al., 2005). Although the effects of drought on many plants have been widely investigated, less is known about the biosynthesis and accumulation of oil in aromatic plants under water deficit conditions (Sangwan et al., 1994; Rahmani et al., 2008). Water deficit in plants may lead to physiological disorders, such as a reduction in photosynthesis and transpiration, which may cause changes in the yield and composition of EO in aromatic plants. A number of studies indicate drastic reductions in grain yield as a result of water deficit during the reproductive period of coriander (Aliabadi et al., 2008) but an increase in the content of EOs. Similar results were found by Mohamed et al. (2002) in Mexican marigold. Consequently, drought stress reduced the vegetative growth period and the plant moved to the flowering stage. Thus, quantity characteristics of oil decreased sorely under drought conditions. Sangwan et al. (1994) studied the lemongrass species, *Cymbopogon nardus* and *C. pendulus*, that were exposed to water stress. They observed that water deficit reduced plant height, leaf length, leaf area, fresh and dry weight and leaf moisture content. According to their studies of the lemongrass species, the major oil *viz.*, geraniol and citral, increased in both the *Cymbopogon* species. The effect of drought on EO was studied in excised leaves of palmarosa (*Cymbopogon martinii* var. *motia*) and citronella java (*C. winterianus*) and it was found that the EO percentage was increased under water stress and EO content was decreased under this condition (Fatima et al., 2006). Drought stress has adverse effects on growth and development of oil yielding plants which in turn affects the percentage or quality of EOs in them.

The content of EOs and their composition are also affected by drought stress and other abiotic factors including genetic makeup (Muzik et al., 1989) and cultivation conditions, such as climate, habitat, harvesting time, environmental stresses and the use of mineral nutrients (Min et al., 2005; Stutte, 2006). Similarly, Aliabadi et al. (2009) studied the influence of water deficit stress on the EO of balm and they found that EO yield was reduced under water deficit stress but EO percentage was increased under stress. Also, the study for effects of drought on EO was carried out in two species of an herb plant that is *Ocimum basilicum* L. and *Ocimum americanum* L. (Khalid, 2006). For both species, EO percentage and the main constituents of EO were increased. Few studies suggested that the reduction in leaf area due to drought might result in a higher density of the leaf oil glands, leading to an elevated amount of oil accumulation (Charles et al., 1990; Simon et al., 1992). However, experiments with *Cymbopogons* revealed that water stress could alter the oil biogenetic capacity of plants that occurred without any change in the number of oil glands in the leaves as a result of short term stress conditions (Sangwan et al., 1994). According to Rahmani et al. (2008), drought stress had a significant effect on oil yield and oil percentage of *calendula*. The yield of flower and EO in coriander were achieved under non-stress conditions, but the highest percentage of oil was achieved under water stress conditions (Aliabadi et al., 2008). Singh et al. (2000) reported that drought decreased the oil yield of rosemary (*Rosmarinus officinalis* L.). Pirzad et al. (2006) also reported that the yield of EO in *Matricaria chamomilla* L. was decreased due to water deficiency. In contrast, water stress caused an increase in the oil production of thyme (Aziz et al., 2008). Khalid et al. (2006) evaluated the influence of water stress on the yield of EO of two species of *Ocimum basilicum* L. (sweet basil) and *Ocimum americanum* L. (American basil). For both species under water stress, EO percentage and the main constituents of EO increased. Studies on *Salvia officinalis* were conducted and it was found that water deficit on FA contents, EO yield and its composition and FAs in the aerial parts are widely affected (Bettaieb et al., 2008). Moderate water deficit increased the yield of EO significantly. The main EO constituents (camphor, thujone and 1, 8 cineole) were also increased significantly under moderate water condition. According to Ozkan et al. (2013), growth and yield in cultivars of *Sesamum indicum* were adversely affected under drought stress whereas the oil yield was not much affected. However, the percentage of major FA compositions was drastically affected by the limited water supply.

Drought stress increases the EO percentage of more medicinal and aromatic plants because, in the case of stress, more metabolites are produced in the plants and substances prevent oxidization in the cells. EO content was reduced under drought stress; the interaction between the amount of EO percentage and flower yield is considered important as they are two components of the EO content.

Flooding Stress

Flooding is one of the most harmful abiotic stresses caused by heavy rains, excessive irrigation and low infiltration rate of soils, and its extensive presence severely reduces the productivity of crops in major growing regions in the world. Excess of water may also influence the growth and development of plants, which in turn may affect their secondary metabolite production. Flooding enforces severe pressure on plants, principally because excess water in their surroundings can deprive them of certain basic needs, notably of oxygen and carbon dioxide and light for photosynthesis (Michael et al., 2009). It is one of the major abiotic influences on species distribution and agricultural productivity worldwide. Temporal flooding is a common environmental stress for terrestrial plants and most widely affects crop plants. According to Wu et al. (2007), flood stress plays a major role in the seed germination of soybean, which in turn effects its growth and oil production.

Salinity Stress

Salinity is one of the most atrocious environmental factors limiting the productivity of crop plants because most of crop plants are sensitive to salinity caused by high concentrations of salts in the soil, and the areas of land affected by it are increasing day by day. Soil has abundant salt, which is one of the major criteria for global food crop production, thereby sustaining the capacity of agriculture to maintain the rapidly increasing human population (Flowers, 2004). Excess of salt may affect plant growth adversely, and it has been reported that about 20% of all cultivated land is salt-affected, thereby reducing yield below the genetic potential (Munns, 2008; Flowers, 2004). Soil salinity has become a key dilemma because it alters irrigation water quality. Salinity has highly adverse impacts on the growth and productivity of oil yielding plants. Increase in the number of soluble salts in the soil leads to osmotic stress, specific ion toxicity and ionic imbalances (Munns, 2008; Chartzoulakis et al., 2002; Zhu, 2003; Arshi et al., 2002,2004 respectively) leading to plant death or yield losses both in conventional crop species and medicinal plants. A preliminary study indicated that some oil yielding plants might be suitable in salt affected soils. Few species of poaceae, which are the storehouse of many EOs like palmarosa (*Cymbopogon martinii*) and lemongrass (*Cymbopogon flexuosus*), are also reported to withstand salinity to a great extent. Many studies have been done concerning salt tolerance of conventional crops, but reports on volatile oil plants are scarce (Kumar et al., 1994). According to Kumar et al. (1995), *C. flexuosus* suffered less from the reduced shoot and root yield. Increase in salt stress resulted in a reduction of both shoot and root yield of citronella, lemongrass and veti-ver. Ansari et al. (1998) studied the effect of salts in three *Cymbopogon* grasses, *viz.*, *C. winterianus*, *C. flexuosus* and *C. martini*, and found that salinity resulted in retardation of plant growth and significant decline in content and yield of EO in the *Cymbopogon* species. However the composition, proportions of citral and geraniol increased under salt stress in the oil of lemongrass and palmarosa, respectively. Some studies showed that a decrease in EO yield under salinity is related to an alteration in EO composition (Dow et al., 1981). According to Neffati et al. (2008), EO content was increased in coriander (*Coriandrum sativum*) at 25 and 50 mM NaCl, but a decrease under high salinity. Dorman et al. (2000) found that salt stress affects growth, mineral nutrition and yield and composition of EOs in marjoram (*Origanum majorana L.*). EO yield reduced due to salt stress in *Trachyspermum ammi* (Ashraf et al., 2006). In several medicinal plants, the quality and quantity of EOs have been reported to decrease due to salinity, e.g., *Mentha piperita* (Tabatabaie et al., 2007), *Thymus maroccanus* (Belaqziz et al., 2009), basil (Said-Al Ahl et al., 2011) and apple mint (Aziz et al., 2008).

Strategies to Improve Abiotic Stress and Oil Yield

Improvement of abiotic stress tolerance in plants would be an advantage to the growers, increasing growth and yield, in the presence of cold, drought, flood, salt stress, heavy metals, acid rain, pollution, heat, UV stress and other abiotic stresses. Oil yield is a complicated trait involving the interaction of many biochemical pathways and interacting factors on a molecular basis. The emergence of the recent "omics" technologies, such as genomics, proteomics and metabolomics, is now allowing researchers to establish the genetics behind abiotic stress responses. These omics technologies allow tracking of the factors affecting plant growth and yield and provide the data that can be directly used to explore the complex interaction between the plants, its metabolism and also the stress caused by environmental threats. Plant responses to abiotic stress are mediated via extreme alteration in gene expression which results in changes in the composition of plant transcriptome, proteome and metabolome (Munns et al., 2008). Many strategies have been undertaken to improve abiotic stress, of which some were successful and others are on the verge of the meeting point.

Genomics

The knowledge of the molecular mechanisms underlying the responses of plants to environmental stresses is still rather limited, but an increasing number of genes have been identified in recent years that mediate these responses. Some of these genes are induced by stress stimuli and encode products that confer tolerance to adverse conditions. Some of the genes are listed in Table 33.1.

Many genes linked to different pathways and processes, such as stress interpretation and signaling, contributing to molecular, biochemical, cellular, physiological and morphological adaptations are differentially regulated in response to plant stress (Perez-Alfocea et al., 2011). Stress responsive genes include those that pacify the effect of the stress and lead to adjustment of the cellular environment and plant tolerance. Numerous investigations show that plant defense response genes are transcriptionally activated by different types of abiotic stress. It has been characterized that the selection of specific defense genes, in the response against environmental factors, suggests the existence of a complex signaling network that allows the plant to recognize and protect itself against different abiotic stress (Jaspers et al., 2010). Many approaches have been undertaken to improve abiotic stress tolerance in genetic backgrounds including screening of diverse genetic resources, wide crossing and subsequent frequent backcrossing, identification and selection of the major modifying

TABLE 33.1

Genes Are Induced by Stress Stimuli and Encode Products that Confer Tolerance to Adverse Conditions

Genes	Stress	References
Mitogen-activated protein kinases (MAPK)	Abiotic and biotic stresses	Pitzschke et al. (2009)
HVA1	Salinity and drought	Fu et al. (2007)
Glycerol-3-phosphateacyltransferase gene	Cold	Yan et al. (2008)
Lysyl oxidase (LOX)	Drought	Yang et al. (2012)
Arabidopsis transcription activation factor (ATAF)	Drought, salinity, cold,	Christianson et al. (2010)
Basic Leucine Zipper (bZIPs) family (e.g.,ABF1, ABF2)	Drought, temperature, salinity	Abe et al. (2005)
Glutathione peroxidase (GPX),	Drought, temperature, salinity	Tuteja et al. (2011)
Farnesyl transferase (FTA)	Drought	Wang et al. (2005)
SOD	Drought, salinity, cold	Tuteja et al. (2011)
catalase (CAT)	Low temperature	Tuteja et al. (2011)
Ascorbate peroxidase (Apx1)	Drought, heat stress, cold	Koussevitzky et al. (2008)
Mitogen-activated protein kinase kinase kinase 1 (MEKK1) and ANP1	Environmental stress	Nakagami et al. (2006)
14.3.3 gene family (GF14b, GF14c)	Salinity, drought,	Chen et al. (2006)
Cold-binding factor/dehydration responsive element binding (CBF/DREB) families (CBF1, CBF2, DREB2A)	Drought, cold, salinity	Agarwal et al. (2010)
Myotonin protein kinase 3 (MPK3), MPK4 and MPK6	Abiotic stress	Pitzschke et al. (2009)
Rapid alkalinization factor-like8 (AtRALFL8)	Water deficit	Atkinson et al. (2013)
Azelaic acid induced1 (AZI1)	Water deficit	Atkinson et al. (2013)
Methionine gamma lyase (*AtMGL*)	Drought	Atkinson et al. (2013)

genes through linkage mapping and quantitative trait loci (QTL) analysis, the production and screening of mutant populations and the transgenic introduction of novel genes. With the advancement of QTL mapping, new breeding approaches such as marker-assisted selection have been emerging (Peleman et al., 2003). It has also been suggested that marker-assisted selection will enhance oil product security under a changing climate and will be integrated into the development of stress tolerant *Brassicca* crops (Zhang et al., 2014). Genomic approaches help in providing deep insight into the mechanisms of established environment–microbe interactions as reported that Apx1, a gene coding for cytosolic ascorbate peroxidase 1, is specifically required for tolerance to drought and heat stress in *Arabidopsis* (Koussevitzky et al., 2008). Moreover, the DREB family, DREB2A protein, has also been used to develop genetically modified drought-tolerant plants. In *Arabidopsis*, the over expression of a constitutively active (CA) DREB2A form resulted in significant tolerance to drought and heat stress (Sakuma et al., 2006). However, knowledge of plant-environment interactions in the genomic level is scarce in oil yielding plants and to uncover the exact mechanism for abiotic stress tolerance, much better research is required.

Proteomics

The Plant adaptation to abiotic stress conditions is resolved through deep alteration in gene expression, which results in a variation in the composition of plant transcriptome. Since proteins are directly involved in plant stress response, proteomics studies can significantly contribute to unfolding the possible relationships between protein abundance and plant stress acclimatization. Several studies have already proven that the changes in gene expression at transcript level do not often correspond with the changes at the protein level (Bogeat-Triboulot et al., 2007). The investigation of variation in plant proteome is highly significant since proteins are direct effectors of plant stress response. Proteins not only include enzymes catalyzing changes in metabolite levels, but also include components of the transcription and translation system (Bogeat-Triboulot et al., 2007). Salt stress extremely affects the production of EOs and the compounds of medicinal and aromatic plants, thus the investigation on the mechanisms of salt tolerance in these plants draws great attention (Aghaei et al., 2013). Establishment of salt-tolerant plants leads to increased production of raw materials for drugs, flavors, fragrances and oil globally (Aghaei et al., 2013). In order to increase the production of a special compound in these types of plants, it is necessary to know which protein or proteins are engaged in the biosynthetic pathway, and, as a consequence, proteomics becomes a powerful approach.

The most important families of plant TFs linked to plant stress responses are mentioned below.

Ethylene-Responsive Element-Binding Factors (ERFs)

This protein family has been associated with a wide array of stresses; the RNA levels of specific ERF genes are managed by cold, drought and other environmental conditions (Onate-Sanchez et al., 2002). ERF proteins are shown to function as either activators or repressors of transcription, which is of great importance in all processes interconnected with plant responses

to adverse growing conditions due to both biotic and abiotic factors (Fujimoto et al., 2000). It has been reported that ERF proteins from one plant species function in other plant species, enhancing their potential utility in increasing the stress tolerance of plants (Nakano et al., 2006; Wu et al., 2007). However, the basic over expression of ERF genes generally causes detrimental effects. To conquer this problem, the use of stress-inducible promoters to regulate the expression of the ERF genes has been successfully used (Mizoi et al., 2012).

MYC Proteins

They are involved in the plant' response to adverse environmental conditions. This TF family plays a major role in abscisic acid (ABA) dependent stress response (Agarwal et al., 2010).

MYB Proteins

They are the major factors in regulatory networks controlling development (Mandaokar Browse, 2009), metabolism (Lepiniec et al., 2006) and responses to abiotic stresses (Dubos et al., 2010). Since the *Arabidopsis* genome sequence was reported, an important amount of data has been compiled on the roles of MYB TFs in plants, and some members of this family are involved in these responses. The AtMYB96 acts through the ABA signaling cascade to control water stress and disease resistance (Seo et al., 2010). AtMYB33 and AtMYB101 are indulged in ABA-mediated responses to environmental signals. AtMYB15 is also involved in cold stress tolerance (Agarwal et al., 2006). AtMYB108 in both biotic and abiotic stress responses (Zhang et al., 2011). The elucidation of MYB protein function and regulation that is possible in *Arabidopsis* will allow predicting the contributions of MYB proteins to the responses of abiotic stress conditions in other plant species.

Therefore, studies of plant responses to different abiotic stress conditions at protein level can significantly contribute to our understanding of physiological and transcriptional mechanisms underlying plant stress tolerance. Proteomics studies could therefore lead to the identification of potential protein markers whose changes in abundance can be related to quantitative changes in some physiological parameters related to stress tolerance (Nakashima et al., 2011).

Metabolomics

Metabolites reflect the assimilation of gene expression, protein interaction and other different regulatory processes and are therefore closer to the phenotype than mRNA transcripts or proteins alone (Arbona et al., 2013). It has been successfully applied to the study of the molecular phenotypes of plants in response to abiotic stress in order to find particular patterns associated with stress tolerance and its production. These studies have highlighted the crucial involvement of primary metabolites: sugars, amino acids and Krebs cycle intermediates as direct markers of photosynthetic dysfunction as well as effectors of osmotic readjustment (Castellana et al., 2010). Conversely, secondary metabolites are more specific of genera and species and respond to specific stress conditions as antioxidants, ROS scavengers, coenzymes, UV and excess radiation screen and also as regulatory molecules (Kirakosyan et al., 2004). Metabolomics technique have been successful in evaluating stress responses in barley (Widodo et al., 2009), *Citrus* (Djoukeng et al., 2008), *Medicago truncatula* (Broeckling et al., 2005) and *Arabidopsis thaliana* (Fukushima et al., 2011). This technological tool includes different approaches, namely, targeted analysis, metabolic fingerprinting and metabolite profiling. Targeted analysis is used to examine the concentration of a limited number of known metabolites precisely, by using either gas chromatography (GC) or liquid chromatography (LC) coupled to mass spectrometry (MS) or nuclear magnetic resonance spectroscopy (NMR); it is the most developed approach used in metabolomics (Djoukeng et al., 2008). Metabolic fingerprinting uses signals from hundreds to thousands of metabolites for rapid sample classification through statistical analysis (Chatterjee et al., 2010). Metabolite profiling attempts to identify and measure a specific class or classes of chemically related metabolites that often share chemical properties that facilitate simultaneous analysis (Seger et al., 2007). Like other functional genomics research, metabolomics produces large amounts of data. Handling, processing and analyzing this data is a distinctive challenge for researchers and requires specialized mathematical, statistical and bioinformatics tools (Shulaev, 2006). Further developments in this area require improvements in both analytical science and bioinformatics.

Involvement of Novel Gene miRNAs in Plant Stress Tolerance

The miRNAs are endogenous, small 21–24 nucleotide, single stranded, non-protein coding RNAs that have come out as an important regulator of gene expression (Bartel, 2004). Discovery and functional association of miRNAs have led to a wide research area in the world of non-coding RNAs (Reinhart et al., 2002). Targets of miRNA comprise of TFs or other regulatory proteins that function in plant development or signal transduction. Researches on miRNAs have suggested an association between miRNAs and plant stress responses (Patade et al., 2010). Several miRNAs are either up- or down-regulated by abiotic stresses, which suggests that being indulged in stress-responsive gene expression and stress adaptation affects a variety of cellular and physiological processes (Sunkar et al., 2004; Shukla et al., 2008). Sunkar et al. in 2004 identified abiotic stress-regulated miRNAs and reported differential expression of some of the identified miRNAs in *Arabidopsis* seedlings exposed to drought, salinity or cold stress (Sunkar et al., 2004). In order to resolve the function of miRNA, Zhao et al. in 2007 studied transcript expression profiles of miRNAs in rice (*O. sativa*) under drought stress (Zhao et al., 2007). Later, he developed a computational transcriptome-based approach to elucidate stress inducible miRNAs in plants. Interestingly, the promoter analysis of the miRNA genes revealed the presence of many known stress-responsive cis-regulatory elements (Zhou et al., 2008). Very limited miRNA researches have been carried out in oil yielding plants. Soybean, an oil yielding plant, is sensitive to environmental stresses such as drought, which is the major abiotic stress for its productivity worldwide. It was reported that small RNAs interfere with gene regulation in roots under water deficit stress (Kulcheski et al., 2011). In wild soybean, a number of miRNAs were also detected to play important regulatory roles in aluminum stress response

(Ying et al., 2012). Switch grass, a biofuel plant, is also affected by drought. It is reported that miR156 and miR162 plays a role in the adaption of switch grass to drought and salinity stress (Sun et al., 2012).

MiRNA technology uses deep sequencing technologies and other expression analyses, such as quantitative real-time PCR. In the future, more function and expression studies will be necessary in order to demonstrate the common miRNA mediated regulatory mechanisms that regulate tolerance to different abiotic stresses. The use of artificial miRNAs, as well as over expression and knockout/down of both miRNAs and their targets, will be the best techniques for identifying the specific roles of individual miRNAs in response to environmental stresses.

Conclusion

Increased protection of plants against abiotic stresses encompasses a complex regulatory network controlling morphological, physiological, biochemical and molecular changes. And to understand how plants respond to stress, it must be considered that they are subjected to different abiotic stress conditions. This preliminary consideration is essential to understanding the performance and productivity of plants under stress and also to identifying strategies to improve the effect of tolerance. As the growth, yield and EO composition of the majority of oil yielding plants have been affected by the adverse conditions, it is necessary to understand the underlying stress mechanisms. Enhancing oil productivity and understanding the role played by stress is a big challenge to the scientific community. The integration of the omics technology is likely to enable researchers to reconstruct the whole cascade of cellular events leading to rapid responses and adaptation to the various abiotic stimuli. However, more focused omics-based research data generation following integrated approaches encompassing genomics, proteomics and metabolomics, and the involvement of novel genes like miRNA studies on specific plant–abiotic stress systems, will be needed to resolve many facts behind the precise mechanisms of stress tolerance in plants.

REFERENCES

Abe M, Kobayashi Y, Yamamoto S. (2005). FD, a bZIP protein mediating signals from the floral pathway integrator FT at the shoot apex. *Science*. 309(5737): 1052–1056.

Abrahim D, Francischini AC, Pergo EM, Kelmer-Bracht AM, Ishii-Iwamoto EL. (2003). Effects of α-pinene on the mitochondrial respiration of maize seedlings. *Plant Physiol. Biochem.* 41: 985–991.

Agarwal M, Hao Y, Kapoor A. (2006). A R2R3 type MYB transcription factor is involved in the cold regulation of CBF genes and in acquired freezing tolerance. *J. Biol. Chem.* 281(49): 37636–37645.

Agarwal PK, Jha B. (2010). Transcription factors in plants and ABA dependent and independent abiotic stress signalling. *Biol. Plant.* 54(2): 201–212.

Aghaei K, Komatsu S. (2013) Crop and medicinal plants proteomics in response to salt stress. *Front. Plant Sci.* 4: 8.

Ahmad P, Prasad MNV. (2004). Abiotic stress responses in plants. *J. Plant Physiol.* 168: 807–815.

Aliabadi FH, Lebaschi MH, Hamidi A. (2008). Effects of arbuscular mycorrhizal fungi, phosphorus and water stress on quantity and quality characteristics of coriander. *Appl. Sci.* 2(2): 55–59.

Aliabadi FH, Valadabadi SAR, Daneshian J, Khalvati MA. (2009). Evaluation changing of essential oil of balm (*Melissa officinalis* L.) under water deficit stress conditions. *J. Med. Plant. Res.* 3(5): 329–333.

Alscher RG, Erturk N, Heath LS. (2002). Role of superoxide dismutase (SODs) in controlling oxidative stress in plants. *Exp. Bot.* 53: 133–141.

Ansari SR, Frooqi AHA, Sharma S. (1998). Interspecific variation in sodium and potassium ion accumulation and essential oil metabolism in three Cymbopogon species raised under sodium chloride stress. *J. Essen. Oil Res.* 10: 413–418.

Arbona V, Manzi M, de Ollas C, Gomez-Cadenas A. (2013). Metabolomics as a tool to investigate abiotic stress tolerance in plants. *Int. J. Mol. Sci.* 14: 4885–4911.

Arshi A, Abdin MZ, Iqbal M. (2002). Growth and metabolism of senna as affected by salt stress. *Biol. Plant.* 45: 295–298.

Arshi A, Abdin MZ, Iqbal M. (2004). Changes in biochemical status and growth performance of Senna (*Cassia angustifolia* Vahl.) grown under salt stress. *Phytomorphology* 54: 109–124.

Asbahani AE, Miladi K, Badri W, Sala M, Addi EHA, Casabianca H, Mousadik., AE, Hartmann D, Jilale A, Renaud FNR, Elaissari A. (2015). Essential oils from extraction to encapsulation. *Int. J. Pharm.* 483(1–2): 220–243.

Ashraf M, Orooj A. (2006). Salt stress effects on growth, ion accumulation and seed oil concentration in an arid zone traditional medicinal plant ajwain (*Trachyspermum ammi* L. Sprague). *J. Arid. Environ.* 64(2): 209–220.

Assche FV, Clijsters H. (1990). Effects of metals on enzyme activity in plants. *Plant Cell Environ.* 13(3): 195–206.

Atkinson NJ, Lilley CJ, Peter E. (2013). Identification of genes involved in the response of *arabidopsis* to simultaneous biotic and abiotic stresses. *Plant Physiol.* 162(4): 2028–2041.

Aziz EE, Al-Amier H, Craker LE. (2008). Influence of salt stress on growth and essential oil production in peppermint, pennyroyal and apple mint. *J. Herbs Spices Med. Plants* 14: 77–87.

Bartel D. (2004). MicroRNAs: genomics, biogenesis, mechanism, and function. *Cell* 116: 281–297.

Bedi S, Tanuja., Vyas SP. (2008). *A Handbook of Aromatic and Essential Oil Plants: Cultivation, Chemistry, Processing and Uses,* AGROBIOS Publishers, Jodhpur, India.

Belaqziz R, Romane A, Abbad A. (2009). Salt stress effects on germination, growth and essential oil content of an endemic thyme species in Morocco *Thymus maroccanus* Ball. *J. Appl. Sci. Res.* 5: 858–863.

Bell E, Mullet JE. (1991). Lipoxygenase gene expression is modulated in plants by water deficit, wounding and methyl jasmonate. *Mol. Gen. Genet.* 230: 456–462.

Bettaieb I, Zakhama N, Wannes WA, Kchouk ME, Marzouk B. (2008). Water deficit effects on *Salvia officinalis* fatty acids and essential oils composition. *Sci. Hortic.* 120: 271–275.

Bhatt RM, Rao SNK. (2005). Influence of pod load response of okra to water stress. *Ind. J. Plant Physiol.* 10: 54–59.

Bhattacharjee SK. (2005). *A Handbook of Aromatic Plants,* Pointer Publishers Vyas-Building, Jaipur, India.

Bogeat-Triboulot MB, Brosche M, Renaut J. (2007). Gradual soil water depletion results in reversible changes of gene expression, protein profiles, ecophysiology, and growth performance in *Populus euphratica*, a poplar growing in arid regions. *Plant Physiol.* 143(2): 876–892.

Boyer JS. (1982). Plant productivity and environment. *Science* 218: 443–448.

Broeckling CD, Huhman DV, Farag MA. (2005). Metabolic profiling of *Medicago truncatula* cell cultures reveals the effects of biotic and abiotic elicitors on metabolism. *J. Exp. Bot.* 56(410): 323–336.

Camejo D, Rodriguez P, Morales MA, Dell Amico JM, Torrecillas A, Alarcon JJ. (2005). High temperature effects on photosynthetic activity of two tomato cultivars with different heat susceptibility. *J. Plant Physiol.* 162: 281–289.

Cao HX, Sun CX, Shao HB, Lei XT. (2011). Effects of low temperature and drought on the physiological and growth changes in oil palm seedlings. *Afr. J. Biotechnol.* 10: 2630–2637.

Castellana N, Bafna V. (2010). Proteogenomics to discover the full coding content of genomes: a computational perspective. *J. Proteomics.* 73: 2124–2135.

Charles DJ, Joly RJ, Simon JE. (1990). Effect of osmotic stress on the essential oil content and comparison of peppermint. *Phytochemistry* 2: 2837–2840.

Chartzoulakis K, Loupassaki M, Bertaki M, Androulakis I. (2002). Effects of NaCl salinity on growth, ion content and CO2 assimilation rate of six olive cultivars. *Sci. Hortic.* 96: 235–247.

Chatterjee S, Srivastava S, Khalid A. (2010). Comprehensive metabolic fingerprinting of *Withania somnifera* leaf and root extracts. *Phytochemistry* 71(10): 1085–1094.

Chaves MM, Pereira JS, Maroco ML, Rodriques CPP, Ricardo ML, Osorio I, Carvatho TF, Pinheiro C. (2002). How plants cope with water stress in the field photosynthesis and growth. *Ann. Bot.* 89: 907–916.

Chen F, Li Q, Sun L, He Z. (2006). The rice 14-3-3 gene family and its involvement in responses to biotic and abiotic stress. *DNA Res.* 13(2): 53–63.

Christianson JA, Dennis ES, Llewellyn DJ, Wilson IW. (2010). ATAF NAC transcription factors: regulators of plant stress signaling. *Plant Signal. Behav.* 5(4): 428–432.

Chung IM, Kim JJ, Lim JD, Yu CY, Kim SH, Hahn SJ. (2006). Comparison of resveratrol SOD activity, phenolic compounds and free amino acids in *Rehmannia glutinosa* under temperature and water stress. *Environ. Exp. Bot.* 56: 44–53.

Croteau R. (2000). Kutchan T, Lewis N. (2000) Natural products (secondary metabolites). In: *Biochemistry and Molecular Biology of Plants,* B Buchanan, W Gruissem, R Joneas, (eds.). American Society of Plant Biologists, Rockville, MD, pp. 1250–1268.

Delfine S, Loreto F, Pinell P, Tognetti R, Alvino A. (2005). Isoprenoids content and photosynthetic limitations in rosemary and spearmint plants under water stress. *Agric. Ecosyst. Environ.* 106: 243–252.

Djilani A, Dicko A. (2012). *The Therapeutic Benefits of Essential Oils, Nutrition, Well Being and Health,* Jaouad Bouayed (ed.). ISBN: 978-953-51-0125.

Djoukeng JD, Arbona V, Argamasilla R, Gomez C. (2008). Flavonoid profiling in leaves of citrus genotypes under different environmental situations. *J. Agric. Food Chem.* 56(23): 11087–11097.

Dorman HJD, Figueiredo AC, Barraso G, Deans SG. (2000). *In vitro* evaluation of antioxidant activity of essential oils and their components. *Flavour. Fragr. J.* 15: 12–16.

Dow AI, Cline TA, Horning EV. (1981). Salt tolerance studies on irrigated mint. Bulletin of Agriculture Research Center, Washington State University, Pullman. 906: 11.

Dubos C, Stracke R, Grotewold E, Weisshaar B, Martin C, Lepiniec L. (2010). MYB transcription factors in *Arabidopsis*. *Trends. Plant Sci.* 15(10): 573–581.

Dudareva N, Pichersky E, Gershenzon J. (2004). Biochemistry of plant volatiles. *Plant Physiol.* 135: 1893–1902.

Farooq M, Basra SMA, Wahid A, Cheema ZA, Cheema MA, Khaliq A. (2008). Physiological role of exogenously applied glycine-betaine in improving drought tolerance of fine grain aromatic rice (*Oryza sativa* L.). *J. Agron. Crop Sci.* 194: 325–333.

Fatima S, Farooqi AHA, Sangwan RS. (2006). Water stress mediated modulation in essential oil, proline and polypeptide profile in palmarosa and citronella java. In: *Physiology and Molecular Biology of Plants*. Lucknow, India: Central Institue of Medicinal and Aromatic Plants, CIMAP.

Flowers TJ. (2004). Improving crop salt tolerance. *J. Exp. Bot.* 55(396): 307–319.

Foy CD. (1988). Plant adaptation to acid, aluminum-toxic soils. *Commun. Soil Sci. Plant Anal.* 19: 959–987.

Fu D, Huang B, Xiao Y, Muthukrishnan S, Liang GH. (2007). Overexpression of barley hva1 gene in creeping bentgrass for improving drought tolerance. *Plant Cell Rep.* 26(4): 467–477.

Fujimoto SY, Ohta M, Usui A, Shinshi H, Ohme-Takagi M. (2000). *Arabidopsis* ethylene-responsive element binding factors act as transcriptional activators or repressors of GCC box-mediated gene expression. *Plant Cell* 12(3): 393–404.

Fukushima A, Kusano M, Redestig H, Arita M, Saito K. (2011). Metabolomic correlation-network modules in *Arabidopsis* based on a graph-clustering approach. *BMC Syst. Biol.* 5(1): 1–12.

Geissman TA, Crout DHG. (1969). Organic Chemistry of Secondary Plant Metabolism. Freeman. San Francisco., T. Robinson. (1969). The Organic Constituents of Higher Plants. Cordus, Amherst, Mass ed., M. Luckner. (1983). Secondary Metabolism in Plants and Animals. Academic Press. New York, 1972.

Hendawy SF, Khalid KA. (2005). Response of sage (*Salvia officinalis* L.) plants to zinc application under different salinity levels. *J. Appl. Sci. Res.* 1: 147–155.

Hikosaka K, Muradami A, Hirose T. (1999). Balancing carboxylation and regeneration of ribulose- 1, 5-bisphosphate in leaf photosynthesis: temperature acclimation of an evergreen tree, *Quereus myrsinaefolia*. *Plant Cell Environ.* 22: 841–849.

Hsiao TC. (1973). Plant responses to water stress. *Annu. Rev. Plant Physiol.* 24: 519–570.

Jaleel CA, Gopi R, Panneerselvam R. (2008). Growth and photosynthetic pigments responses of two varieties of *Catharanthus roseus* to triadimefon treatment. *C. R. Biol.* 331: 272–277.

Jaleel CA, Gopi R, Panneerselvam R. (2009). Alterations in non-enzymatic antioxidant components of *Catharanthus roseus* exposed to paclobutrazol, gibberellic acid and *Pseudomonas fluorescens*. *Plant Omics* 2(1): 30–40.

Jaleel CA, Gopi R, Sankar B, Gomathinayagam M, Panneerselvam M. (2008). Differential responses in water use efficiency in two varieties of *Catharanthus roseus* under drought stress. *C. R. Biol.* 331: 42–47.

Jaleel CA, Manivannan P, Kishorekumar AB, Sankar R, Gopi R, Somasundaram PR. (2007). Alterations in osmoregulation, antioxidant enzymes and indole alkaloid levels in *Catharanthus roseus* exposed to water deficit. *Colloids. Surf. B. Biointerfaces.* 59: 150–157.

Jaleel CA, Manivannan P, Sankar B, Kishorekumar A, Gopi R, Somasundaram R, Panneerselvam R. (2007). Water deficit stress mitigation by calcium chloride in *Catharanthus* roseus; effects on oxidative stress, proline metabolism and indole alkaloid accumulation. *Colloids. Surf. B. Biointerfaces.* 60: 110–116.

Jason JG, Thomas GR, Mason-Pharr D. (2004). Heat and drought influence photosynthesis, water relations and soluble carbohydrates of two ecotypes of redbud (*Cercis Canadensis*). *J. Am. Soc. Hort. Sci.* 129(4): 497–502.

Jaspers P, Kangasjarvi J. (2010). Reactive oxygen species in abiotic stress signaling. *Physiol. Plantarum.* 138(4): 405–413.

Kashyap PL, Sanghera GS, Wani SH, Shai W, Kumar S. (2011). Genes of microorganisms paving way to tailor next generation fungal disease resistant crop plants. *Not. Sci. Biol.* 3: 147–157.

Kaur SK, Nayyar GH. (2008). Exogenous application of abscisic acid improves cold tolerance in chickpea (*Cicer arietinum* L.). *J. Agron. Crop Sci.* 194: 449–456.

Khalid KA. (2006). Influence of water stress on growth, essential oil, and chemical composition of herbs (*Ocimum* sp.). *Int. Agrophys.* 20(4): 289–296.

Kirakosyan A, Kaufman P, Warber S, Zick S, Aaronson K, Bolling S, Chul Chang S. (2004). Applied environmental stresses to enhance the levels of polyphenolics in leaves of hawthorn plants. *Physiol. Plant.* 121: 182–186.

Koussevitzky S, Suzuki N, Huntington S, Armijo L, Sha W, Cortes D. (2008). Ascorbate peroxidase 1 plays a key role in the response of *Arabidopsis thaliana* to stress combination. *J. Biol. Chem.* 283: 34197–34203.

Kulcheski FR, de Oliveira LFV, Molina LG, Almerao MP, Rodrigues FA, Marcolino J. (2011). Identification of novel soybean microRNAs involved in abiotic and biotic stresses. *BMC Genom.* 12 (307): 1–17.

Kumar A, Abrol YP. (1994). Effect of gypsum on five tropical grasses grown in normal and extensive sodic soil. *Exp. Agr.* 19: 169–177.

Kumar AA, Gill KS. (1995). Performance of aromatic grasses under saline and sodic stress condition. Salt tolerance of aromatic grasses. *Indian Perfum.* 39: 39–44.

Lepiniec L, Debeaujon I, Routaboul JM. (2006). Genetics and biochemistry of seed flavonoids. *Annu. Rev. Plant Biol.* 57: 405–430.

Ludlow MM, Muchow RC. (1990). A critical evaluation of the traits for improving crop yield in water limited environments. *Adv. Agron.* 43: 107–153.

Maheshwar P, Kovalchuk I. (2014). Genetic engineering of oilseed crops. *Biocatal. Agric. Biotechnol.* 3: 31–37.

Mahmoud SS, Croteau RB. (2002). Strategies for transgenic manipulation of monoterpene biosynthesis in plants. *Trends Plant Sci.* 7: 366–373.

Mandaokar A, Browse J. (2009). MYB108 acts together with MYB24 to regulate jasmonate-mediated stamen maturation in *Arabidopsis*. *Plant Physiol.* 2(149): 851–862.

Mann J. (1978). Secondary metabolism. Bell EA, Charlwood BV, Eds., *Encylopedia of Plant Physiology*. Springer-Verlag, New York, 1980. vol. 8.J. Meinwald., G.D. Prestwich., K. Nakanishi., I. Rubo., *Science* 199: 1167.

Melzer E, O'Leary M. (1987). Anapleurotic CO_2 fixation by phosphoenolpyruvate carboxylase in C3 plants. *Plant Physiol.* 84: 58–60.

Michael BJ, Kimiharu I, Osamu I. (2009). Evolution and mechanisms of plant tolerance to flooding stress. *Ann. Bot.* 103: 137–142.

Min SY, Tawaha ARM, Lee KD. (2005). Effects of ammonium concentration on the yield, mineral content and active terpene components of *Chrysanthemum coronarium* L. in a hydroponic system. *Res. J. Agric. Biol. Sci.* 1: 170–175.

Mittler R. (2002). Oxidative stress, antioxidants and stress tolerance. *Trends Plant Sci.* 7(9): 405–410.

Mizoi J, Shinozaki K, Yamaguchi-Shinozaki K. (2012). AP2/ERF family transcription factors in plant abiotic stress responses. *Biochim. Biophys. Acta* 1819(2): 86–96.

Mohamed MAH, Harris PJC, Henderson J, Senatore F. (2002). Effect of drought stress on the yield and composition of volatile oils of drought-tolerant and non-drought-tolerant clones of *Tagetes minuta*. *Planta Med.* 68(5): 472–474.

Moussa HR. (2006). Influence of exogenous application of silicon on physiological response of salt-stressed maize (*Zea mays L.*). *IJAB.* 3(8): 293–297.

Munns R, Tester M. (2008). Mechanisms of salinity tolerance. *Annu. Rev. Plant Biol.* 59: 651–681.

Muzik RM, Pregitzer KS, Hanover JW. (1989). Changes in terpene production following nitrogen fertilization of grand fir *Abies grandis* (Dougl.) Lindl. seedlings. *Oecologia* 80: 485–489.

Nakagami H, Soukupova H, Schikora A, Zarsky Z, Hirt H. (2006). A mitogen-activated protein kinase kinase kinase mediates reactive oxygen species homeostasis in *Arabidopsis*. *J. Biol. Chem.* 281(50): 38697–38704.

Nakano T, Suzuki K, Fujimura T, Shinshi H. (2006). Genome wide analysis of the ERF gene family in *Arabidopsis* and rice. *Plant Physiol.* 140(2): 411–432.

Nakashima K, Takasaki H, Mizoi J, Shinozaki K, Yamaguchi-Shinozaki K. (2011). NAC transcription factors in plant biotic and abiotic stress responses. *Biochim. Biophys. Acta* 1819(2): 97–103.

Neffati M, Marzouk B. (2008). Changes in essential oil and fatty acid composition in coriander (*Coriandrum sativum* L) leaves under saline conditions. *Ind. Crops Prod.* 28: 137–142.

Neill S, Desikan R, Clarke S, Hurs RD, Hancock JT. (2002). Hydrogen peroxide and nitric oxide as signaling molecules in plants. *J. Exp. Bot.* 53: 1237–1247.

Nostro A, Germano MP, Dangelo V, Marino A, Cannatelli MA. (2000). Extraction methods and bioautography for evaluation of medicinal plant antimicrobial activity. *Lett. Appl. Microbiol.* 30: 379–384.

Onate-Sanchez L, Singh KB. (2002). Identification of *Arabidopsis* ethylene-responsive element binding factors with distinct induction kinetics after pathogen infection. *Plant Physiol.* 128(4): 1313–1322.

Ozkan A, Kulak M. (2013). Effects of water stress on growth, oil yield, fatty acid composition and mineral content of *Sesamum indicum*. *J. Anim. Plant Sci.* 23(6): 1686–1690.

Patade VY, Suprasanna P. (2010). Short-term salt and PEG stresses regulate expression of MicroRNA, *miR159* in sugarcane leaves. *J. Crop Sci. Biotechnol.* 13(3): 177–182.

Peleman JD, van der Voort JR. (2003). Breeding by design. *Trends Plant Sci.* 8(7): 330–334.

Perez-Alfocea F, Ghanem ME, Gomez-Cadenas A, Dodd IC. (2011). Omics of root-to-shoot signaling under salt stress and water deficit. *OMICS.* 15(12): 893–901.

Pichersky E, Gershenzon J. (2002). The formation and function of plant volatiles: perfumes for pollinator attraction and defense. *Curr. Opin. Plant Biol.* 5: 237–243.

Pichersky E, Noel JP, Dudareva N. (2006). Biosynthesis of plant volatiles: nature's diversity and ingenuity. *Science* 311: 808–811.

Pirzad A, Alyari H, Shakiba MR, Zehtab-Salmasi S, Mohammadi A. (2006). Essential oil content and composition of German chamomile (*Matricaria chamomilla* L.) at different irrigation regimes. *J. Agron.* 5(3): 451–455.

Pitzschke A, Schikora A, Hirt H. (2009). MAPK cascade signaling networks in plant defense. *Curr. Opin. Plant Biol.* 12(4): 421–426.

Qiao-Ying Z, Yang CY, Ma QB, Li XP, Dong WW, Nian H. (2012). Identification of wild soybean miRNAs and their target genes responsive to aluminum stress. *BMC Plant Biol.* 12: 182.

Rahman H, Harwood JL, Weselake R. (2013). Increasing seed oil content in Brassica species through breeding and biotechnology. *Lipid Technol.* 25: 182–185.

Rahmani N, Farahani H, Valadabadi SAR. (2008). Effects of nitrogen on oil yield and its component of Calendula (*Calendula officinalis* L.) in drought stress conditions. *Abstracts Book of The World Congress on Medicinal and Aromatic Plants*, South Africa, p. 36.

Ramakrishna A and Ravishankar GA. (2011). Influence of abiotic stress signals on secondary metabolites in plants. *Plant Signal. Behav.* 6: 1720–1731.

Ramakrishna A, Ravishankar GA. (2013). Role of plant metabolites in abiotic stress tolerance under changing climatic conditions with special reference to secondary compounds. In: *Climate Change and Abiotic Stress Tolerance*, N. Tuteja and S.S. Gill (eds.). Wiley-VCH, Weinheim. ISBN 978-3-527-33491-9.

Rao BRR, Kaul PN, Syamasundar KV, Ramesh S. (2005). Chemical profiles of primary and secondary essential oils of palmarosa (*Cymbopogon martini* (Roxb.) Wats var. motia Burk.). *Ind. Crops Prod.* 21: 121–127.

Reddy PCO, Sairanganayakulu G, Thippeswamy M, Reddy PS, Reddy MK, Sudhakar CH. (2008). Antioxidant enzyme activities and gene expression patterns in leaves of Kentucky blue grass in response to drought and post-drought recovery. *Plant Sci.* 175: 372–384.

Reinhart BJ, Weinstein EG, Rhoades MW, Bartel B, Bartel DP. (2002). MicroRNAs in plants. *Genes Dev.* 16: 1616–1626.

Said-Al Ahl HAH, Omer EA. (2011). Medicinal and aromatic plants production under salt stress. *Herba Pol.* 57: 72–87.

Sakuma Y, Maruyama K, Qin F, Osakabe Y, Shinozaki K, Yamaguchi-Shinozaki K. (2006). Dual function of an *Arabidopsis* transcription factor DREB2A in water- stress-responsive and heat-stress–responsive gene expression. *Proc. Natl. Acad. Sci. U.S.A.* 103: 18822–18827.

Sanghera GS, Kashyap PL, Singh G, da Silva JAT. (2011). Transgenics: fast track to plant stress amelioration. *Transgenic Plant. J.* 5: 1–26.

Sangwan NS, Farooqi AH, Sangwan RS. (1994). Effects of drought stress on growth and essential oil metabolism in lemongrasses. *New Phytol.* 128: 173–179.

Seger C, Sturm S. (2007). Analytical aspects of plant metabolite profiling platforms: current standings and future aims. *J. Proteome Res.* 6(2): 480–497.

Seo PJ, Park CM. (2010). MYB 96-mediated abscisic acid signals induce pathogen resistance response by promoting salicylic acid biosynthesis in *Arabidopsis. New Phytol.* 186(2): 471–483.

Shao HB, Chu LY, Jaleel CA, Manivannan P, Panneerselvam R, Shao MA. (2009). Understanding water deficit stress-induced changes in the basic metabolism of higher plants-biotechnologically and sustainably improving agriculture and the ecoenvironment in arid regions of the globe. *Crit. Rev. Biotechnol.* 29: 131–151.

Shukla LI, Chinnusamy V, Sunkar R. (2008). The role of microRNAs and other endogenous small RNAs in plant stress responses. *Biochim. Biophys Acta.* 1779: 743–748.

Shulaev V. (2006). Metabolomics technology and bioinformatics. *Brief Bioinform.* 7(2): 128–139.

Simon JE, Bubenheim DR, Joly RJ, Charles DJ. (1992). Water stress induced alterations in essential oil content and composition of sweet basil. *JEOR.* 4: 71–75.

Singh M, Ramsh S. (2000). Effect of irrigation and nitrogen on herbage, oil yield and water use efficiency in rosemary grown under semi-arid and tropical conditions. *J. Med. Aromat Plant Sci.* 22: 659–662.

Somesh M, Rupali S, Swati S, Jose M, Manish M. (2015). *In vitro* comparative study on antimicrobial activity of five extract of few citrus fruit: peel & pulp vs Gentamicin. *AJBAS* 9(1): 165–173.

Stutte GW. (2006). Process and product: recirculation hydroponics and bioactive compounds in a controlled environment. *HortScience* 41: 526–530.

Sun G, Stewart C, Jr N, Xiao P, Zhang B. (2012). MicroRNA expression analysis in the cellulosic biofuel crop Switchgrass (*Panicum virgatum*) under abiotic stress. *PLoS ONE* 7: e32017.

Sunkar R, Zhu JK. (2004). Novel and stress-regulated microRNAs and other small RNA from *Arabidopsis. Plant Cell* 16: 2001–2019.

Tabatabaie SJ, Nazari, J. (2007). Influence of nutrient concentration and NaCl salinity on growth, photosynthesis and essential oil content of peppermint and lemon verbena. *Turk J. Agric. For.* 31: 245–253.

Tuteja N, Gill SS, Tuteja R. (2011). Plant response to abiotic stress: Shedding light on salt, drought, cold and heavy metal stress. *Omics and Plant Abiotic Stress Tolerance* 1: 39–64.

Wang W, Vinocur B, Altman A. (2003). Plant responses to drought, salinity and extreme temperatures: towards genetic engineering for stress tolerance. *Planta* 218: 1–14.

Wang Y, Ying J, Kuzma M, Chalifoux M, Sample A, McArthur C, Uchacz T, Sarvas C, Wan J, Dennis DT, McCourt P, Huang Y. (2005). Molecular tailoring of farnesylation for plant drought tolerance and yield protection. *Plant J.* 43: 413–424.

Widodo., Patterson JH, Newbigin E, Tester M, Bacic A, Roessner U. (2009). Metabolic responses to salt stress of barley (*Hordeum vulgare* L.) cultivars, Sahara and Clipper, which differ in salinity tolerance. *J. Expt. Bot.* 60(14): 4089–4103.

Wu L, Chen X, Ren H. (2007). ERF protein JERF1 that transcriptionally modulates the expression of abscisic acid biosynthesis-related gene enhances the tolerance under salinity and cold in tobacco. *Planta* 226(4): 815–825.

Yan QC, Kuo MS, Li S. (2008). AGPAT6 is a novel microsomal glycerol-3-phosphate acyltransferase. *J. Biol. Chem.* 283(15): 10048–10057.

Yang XY, Jiang W.J, Yu HJ. (2012). The expression profiling of the lipoxygenase (LOX) family genes during fruit development, abiotic stress and hormonal treatments in cucumber (*Cucumis sativus* L.). *Int. J. Mol. Sci.* 13: 2481–2500.

Yeager N. (2017). The Environmental Impact of Essential Oils. http://www.earthisland.org.

Yoshimura K, Miyao K, Gaber A, Takeda T, Kanaboshi H, Miyasaka H, Shigeoka S. (2004). Enhancement of stress tolerance in transgenic tobacco plants overexpressing Chlamydomonas peroxidase in chloroplasts or cytosol. *Plant J.* 37: 21–33.

Zhang L, Zhao G, Jia J, Liu X. (2011). Molecular characterization of 60 isolated wheat MYB genes and analysis of their expression during abiotic stress. *J. Expt. Bot.* 63(1): 203–214.

Zhang X, Long W, Zou1 X, Li1 F, Nishio T. (2014). Recent progress in drought and salt tolerance studies in Brassica crops. *Breed. Sci.* 64: 60–73.

Zhao BT, Liang RQ, Ge LF, Li W, Xiao HS, Lin HX, Ruan KC, Jin YX. (2007). Identification of drought induced microRNAs in rice. *Biochem. Biophys. Res. Commun.* 354: 585–590.

Zhou X, Wang G, Sutoh K, Zhu JK, Zhang W. (2008). Identification of cold-inducible microRNAs in plants by transcriptome analysis. *Biochim. Biophys Acta.* 1779: 780–788.

Zhu JK. (2003). Regulation of ion homeostasis under salt stress. *Curr. Opin. Plant Biol.* 6: 441–445.

Zhu Z, Liang Z, Han R. (2009). Saikosaponin accumulation and antioxidative protection in drought-stressed *Bupleurum chinense* DC plants. *Environ. Exp. Bot.* 66: 326–333.

34

Differences in Adaptation to Water Stress in Stress Sensitive and Resistant Varieties of Kabuli and Desi Type Chickpea

Nadia Fatnassi, Ralph Horres, Natasa Cerekovic, Angelo Santino, and Palmiro Poltronieri

CONTENTS

Abbreviations

13-AOS	allene oxide synthase
ACX	acyl CoA oxidase
AOC	allene oxide cyclase
AOX	alternative oxidase
APX	ascorbate peroxidase
CAT	catalase
DHAR	dehydroascorbate reductase
DPB	dry plant biomass
GPX	glutathione peroxidase
GR	glutathione reductase
GST	glutathione-S-transferase
JA	jasmonic acid
KAT2	ketoacyl-CoA thiolase
MPF	multifunctional protein
OPDA	12-oxophytodienoic acid
PEG	polyethylene glycol
RDW	root dry weight
RSR	root to shoot ratio
RWC	relative water content
SDW	shoot dry weight
SOD	superoxide dismutase
WUE	water use efficiency

Introduction

Chickpea (*Cicer arietinum* L), the second most important grain crop in the world, is widely grown across the Mediterranean basin, East Africa, India, the Americas, and Australia. Most chickpea producing areas are in the arid and semi-arid zones, and approximately 90% of the world's chickpea is grown under rain-fed conditions. Drought is one of the major constraints for crop productivity and is typical of the post-rainy season in the semi-arid tropical regions. Drought is determined by the rainfall and the evaporative demand before and during the crop season, and by soil characteristics; under terminal drought, early crop duration and the yield potential were shown to contribute to crop yield in dry climates, with differences between Desi type (small-seeded) and Kabuli type (large-seeded) chickpea varieties (Nayyar et al., 2006a). The seed composition is slightly different in the two types, as shown by Singh et al., analysing BG-1053, L-551 for the Kabuli and GPF-2, PBG-1, PDG-4 for the Desi type (2008). Desi type varieties are more tolerant to drought stress than Kabuli type. Presently, a selected germplasm for water stress tolerance is available for both types (Krishnamurthy).

Facing off stressful conditions imposed by their environment that could affect their growth and their development throughout their life cycle, plants must be able to perceive, process, and translate different stimuli into adaptive responses. To support food availability and security, knowledge about plant stress response is vital for the development of breeding and biotechnological strategies to improve stress tolerance in crops and, consequently, crop yields.

Understanding the plant-coordinated responses involves fine description of the mechanisms occurring at the cellular and molecular level. These mechanisms involve numerous components that are organised into complex transduction pathways and networks, from signal perception to physiological responses. The major challenges of plant signalling are understanding what kind of signals cells receive, how these signals are recognized, and how cells respond spatially and temporally to these signals to programme a specific response at the organism level. Furthermore, signal transduction cascades involve a large array of molecular and cellular processes not restricted to a particular stimulus.

This has been illustrated in several studies showing cross-talk between hormones and signalling pathways (Poltronieri et al., 2011). This review intends to provide a synthesis of recent knowledge on the signalling pathways induced by environmental challenges such as drought, high salinity, temperature fluctuations, cold, heat, nutrient deprivation, CO_2, and osmotic stress. Individual researches have focused on the components of signalling cascades (receptors/sensors, Ca2+, MAP kinases, nitric oxide, reactive oxygen species, ion fluxes) (Poltronieri et al., 2011, 2015) and their role under different stress conditions, changes in intracellular compartmentalization, phytohormone retrograde signalling, source-to-sink transport of sugars through the phloem (Lemoine et al., 2013) and other compounds (lipids, hormones, peptides, RNAs) moving through xylem or phloem. The collected information may provide a link between the molecular cell signal transduction cascades and the plant response at the whole organism level. Finally, "omics" techniques such as proteomics (Pandey et al., 2008), genomics (Varshney et al., 2009), transcriptomics (Garg et al., 2011, 2015; Hiremath et al., 2011; Jain and Chattopadhyay, 2010; Agarwal et a., 2012), systems biology, and network modelling have been recently added to the tools used in deciphering the complexity of cell signalling transduction and the triggering of gene expression (upregulation or downregulation), focusing on the recruitment of transcription factors (Nguyen et al., 2015; Ramalingam et al., 2015; Anbazhagan et al., 2014), regulation of chromatin by epigenetic mechanisms, and post-transcriptional control by small RNAs (Poltronieri et al., 2013).

Several characters have been shown to be important in drought stress tolerance, among them are early flowering and cropping, root depth, conservative pattern of water use (Zaman-Allah et al., 2011a, 2011b), and induction of genes (catalase, ascorbate peroxidase, etc.) conferring protection from reactive oxygen species and pro-oxidative states (Maccarrone et al., 1995; Foyer, 2005; Molina et al., 2011; Poltronieri and Miwa, 2015). Exploitation of a combination of useful traits in chickpeas requires the formulation of criteria for genetic improvement in breeding for drought tolerance (Hamwieh et al., 2013). One of the parameters nowadays used for enabling cropping systems under water shortage is the establishment of short duration chickpea cropping, enabling high yields even in more arid southern regions in India.

Among the physiological parameters (Gupta et al., 2010; Mir et al., 2012) used by researchers to compare different varieties and their ability to cope with abiotic stress are the following: dry plant biomass (DPB), shoot dry weight (SDW) (Purushothaman et al., 2016), root dry weight (RDW), root to shoot ratio (RSR), carbon allocation (Hasibeder et al., 2015), relative water content (RWC), water use efficiency (WUE) (Siddique et al., 2001) estimated by the ratio of net CO_2 assimilation (A) and transpiration (E) (A/E) (Gonzalez et al., 1995), leaf gas exchange rates (Leport et al., 1998, 1999; Basu et al., 2007), leaf water potential (Boyer and Knipling, 1965; Karamanos and Papatheohari, 1999), stomatal conductance (Farquhar and Sharkey, 1982), reduction in transpiration in response to high vapour pressure deficit (VPD) (Purushothaman et al., 2015), leaf electrolyte leakage, photosynthesis rates, levels of chlorophyll (Nayyar et al., 2006a), the partitioning coefficient (Krishnamurthy et al., 2013), pod production (Leport et al., 2006; Fang et al., 2010), seed filling (Davies et al., 2000; Behboudian et al., 2001; Turner et al., 2005; Palta et al., 2005; Awasthi et al., 2014), and yield (Behboudian et al., 2001; Zaman-Allah et al., 2011b;

Ulemale et al., 2013; Yaqoob et al., 2013). Other parameters used by researchers included estimation of abscisic acid (Liu et al., 2005; Mantri et al., 2007), redox potential (Foyer, 2005), levels of proline (Kavikishore et al., 1995; Mafakheri et al., 2010; Mohammadi et al., 2011; Mathur et al., 2009; Szabados and Savouré, 2009), glycine betaine (Chen and Murata, 2002; Ashraf and Foolad, 2007), polyamines, polyols, and sugar osmolytes (Poltronieri et al., 2011).

In studies on nodules (Gonzalez et al., 1995; Ramos et al., 1999; Galvez et al., 2005; Mhadhbi et al., 2008; Larrainzar et al., 2007; Sohrabi et al., 2012; Kurdali et al., 2013; Esfahani et al., 2014b), several parameters have been analysed, such as fresh and dry nodule biomass, decline in N_2ase activity as a measure of metabolic limitation of bacteroids, O_2 consumption, leghemoglobin content, energy charge and respiratory capacity, sucrose synthase activity and sucrose, ATP, and organic acids content (Talbi et al., 2012). As shown by Labidi et al.(2009) and Naya et al. (2007), symbiotic nitrogen fixation (SNF) is sensitive to osmotic stress, while *Rhizobium etli* CFNX713 strain, overexpressing *cbb*3 oxidase with higher respiratory capacity, was shown to improve energy change in nodules (Talbi et al., 2012).

Recently the drought response in Desi lines in field conditions was assessed, measuring the canopy temperature depression (CTD) using Thermal infrared thermography and the partitioning coefficient in elite varieties under water stress (Kashiwagi et al., 2008; Krishnamurthy et al., 2013; Purushothaman and Krishnamurthy, 2014; Purushothaman et al., 2015).

In a recent report by researchers at ICRISAT, genotype ICCV 03408 with high CTD and high yield in terminal drought conditions has been identified as a drought-tolerant line, while lines ICCV03104, ICCV00202, ICCV01102, ICCV10112, and ICCV04106 with high CTD values have been identified as donors for drought tolerance (Bharadwaj et al., 2015).

Chickpea growth and development are dependent on the plant–root system, due to its role in water and mineral uptake, nodulation, and symbiotic relationships (Mhadhbi et al., 2008; Esfahani et al., 2014a,b). A strong root system allows the capture of more soil water during the growth period (Kashiwagi et al., 2006). Plants spend energy on root production in search of water and/or reducing water loss. This parameter has been considered a criterion of adaptation to drought (Sassi et al., 2010), although some findings showed no clear correlation between root traits and water extraction ability. However, if water is unavailable deeper in the soil profile, longer roots may reduce shoot dry weight and harvest index by allowing the preferential partitioning of carbohydrates to roots at the expense of shoots (Pace et al., 1999). Moreover, the ability of plants to maintain root growth while under osmotic stress is crucial for nodulation. Drought impairs the development of root hairs and the site of entry of rhizobia into the host, resulting in poor or no nodulation.

The existence of a large diversity for root biomass, root prolificacy, and rooting depth in chickpea mini-core germplasm accessions (Kashiwagi et al., 2005) prompted new efforts of germplasm improvement through selection of characteristics such as enhanced absorption.

Among the varieties most studied, ICC 4958 is drought tolerant, but salt sensitive. Varieties with a "deep root system" such as ICC 4958 grow well, their accumulation of seed mass after flowering is faster, and they accumulate a large seed mass before the soil moisture recedes. Varieties such as ICC 4958 have been used in breeding programmes for crossing with other drought-tolerant varieties to produce élite hybrids. In this way, many pedigree hybrids were produced using ICC 4958 as parent variety.

INRAT-73 (also known as Beja 1) is salt and drought tolerant, and Amdoun 1 is drought sensitive (Labidi et al., 2009). In their study, change in root to shoot ratio, loss of chlorophylls, and nodule mortality were considered an indicator of stress tolerance, with Beja and Kesseb as the varieties with the best performances. However, RSR may be negatively affected in case the shoot biomass is reduced.

The genomes of Desi-type ICC 4958 (Parween et al., 2015) is available, as well as the Kabuli CDC Frontier variety (Hiremath et al., 2011), whose sequence was compared with another 26 Kabuli genomes, 5 wild chickpea genomes, and 58 Desi-type genomes.

The identification of markers linked to salinity tolerance is underway at ICRISAT, a study that will aid in the Marker Assisted Selection (MAS) and Marker Assisted Backcross (MABC) programmes. Under salt stress, tolerant chickpea varieties such as Pusa 362, Pusa 1103, and Pusa 72 outperformed other varieties, showing a reduction of 35% in shoot growth and 5% in root growth, while sensitive plants showed 73% in shoot and 51% in root growth compared to the unstressed conditions (Kumar et al., 2015).

Terminal drought reduces the chickpea growth duration, especially the reproductive phase (Krishnamurthy et al., 2013; Pushpavalli et al., 2014; Farooq et al., 2016), with effects more pronounced in Kabuli type varieties. Studies in recent years (Nayyar et al., 2006b; Leport et al., 1999; Purushothaman et al., 2014) showed a greater inhibition of yield in Kabuli compared with Desi following water stress and attributed this to a less effective remobilisation of the assimilates towards developing seeds under stress conditions. Nayyar used rhizobium-inoculated Desi-type GPF2, and Kabuli-type L550 varieties water stressed at the reproductive stage (plants with 7–9 pods), monitoring the water stress for 14 days. A greater stress tolerance in the Desi type indicated a greater capacity of the Desi type to deal with oxidative stress, ascribed to its superior ability to maintain better water status, which resulted in lesser oxidative damage. In addition, they conducted osmotic stress studies by subjecting both types of chickpea to similar levels of polyethylene glycol-induced water stress and to10 μmol/L abscisic acid (ABA), involved in oxidative injury, demonstrating that Kabuli-type shoots were inhibited more than Desi-type plants (Nayyar et al., 2006a).

Chickpea Improvement by Plant Biotechnology for Drought Stress Tolerance

Several genes coding for signalling pathways and transcription factors providing increased drought tolerance have been overexpressed in chickpea or used to transform this species, among them are *Dehydration Responsive Element Binding protein 1A* (*DREB1A*), a TF related to drought and cold tolerance (Anbazhagan et al., 2014) and *pyrroline-5-carboxylate synthetase* (*P5CS*), which increases proline production (Mathur et al., 2009). It is envisaged that further advancements will bring novel varieties obtained by traditional gene modification methods (Agrobacterium transformation, cisgenesis, CRISPR/cas genome editing) as well as genome and marker-assisted breeding

methods (Kumar et al., 2008; Varhsney et al., 2009; Mir et al., 2012; Kumar et al., 2015).

Gene Expression and Transcriptomic Studies in Chickpea under Water and Salt Stress

In recent years, several groups have described tissue-specific gene expression, and transcriptomic data in various chickpea varieties subjected to water and salt stress challenges. Agarwal performed a comparative analysis of Kabuli chickpea transcriptome with Desi and wild chickpea, collecting root and shoot tissue samples from 15-day-old seedlings (Agarwal et al., 2012). Previously, several approaches have been used in the identification of expressed gene sequences. For instance, Expression Sequence Tags (ESTs) have been retrieved from chickpeas under various stress conditions (Jayashree et al., 2005; Deokar et al., 2011). ESTs were classified and grouped from abiotic stressed chickpea plants (Varshney et al., 2009) using ICC 4958 (drought tolerant) and ICC 1882 (sensitive) varieties. Other groups contributed by performing tissue-specific transcriptome analysis in chickpea with Massively Parallel Pyrosequencing (Garg et al., 2011). With the availability of Next Generation Sequencing platforms, ICC 4958 was used by many researchers for transcriptomic studies and tissue-specific gene expression analysis (Garg et al., 2011; Varshney et al., 2009). Other varieties, such as Kabuli chickpea ICCV2, were also used for gene expression studies, determining the root, shoot and floral bud transcriptomes (Agarwal et al., 2012).

Ramalingam et al. analysed JA 11, an elite drought tolerant cultivar and an introgression line (JG 11+) developed by crossing ICC 4958 and JG 11 (ICC 4958×JG 11) (2015), using greenhouse conditions. Slow drought stress was imposed on the four genotypes, and roots were collected when the transpiration ratio decreased tenfold.

ICC 4958 was used to study root transcripts under drought stress (Hiremath et al., 2011) compared to transcripts in the sensitive variety ICC 1882. Hiremath et al. (2011) used 22 different tissues of the ICC 4958 variety, representing different developmental stages as well as drought and salinity stressed roots, harvested at up to several days after stress onset.

Some studies have analysed the changes in gene expression during prolonged water stress; the longest period of drought treatment was up to 12 days, in chickpea lines ICCV2 and PUSABGD72 (Jain and Chattopadhyay, 2010). The researchers studied the drought-tolerant PUSABGD72 and sensitive ICCV2 varieties (Jain and Chattopadhyay, 2010). As for drought treatment, soil-grown 12 day-old plants were subjected to progressive drought by withholding water for 3, 6, and 12 days respectively. The soil moisture content decreased from approximately 50% to approximately 15% at day 12.

Previously, Mantri performed transcripts profiling using abiotic stress-tolerant and sensitive chickpea varieties (Mantri et al., 2007), but used only flower and leaf tissues, due to the poor quality of root RNA. Mantri showed that drought tolerant-2 (BG 362) had twice the number of repressed transcripts than tolerant-1 (BG 1103), drought susceptible-1 (Kaniva) and susceptible-2 (Genesis 508), determining the differentially expressed (DE) transcripts affected in response to drought stress (desiccation), and, in parallel, determining the chickpea varieties gene expression profiles in response to cold (Sonali, ILC 01276, Amethyst, DOOEN varieties) and high salinity (250 mM NaCl) (CPI 060546, CPI 60527, ICC 06474, ICC 08161 varieties). The water content in each pot at the start of drought treatment was estimated to be 30% of the initial pot weight. In the subsequent days, the control pots were maintained at 80% water content and the treatment pots were allowed to lose 5–10% of their water content per day and any extra water lost (>10%) was replenished. The leaf, root, and flower/bud tissues were collected separately when the treatment pots reached 30% water content, indicative of a drought or high water deficit condition. As for salt treatment, leaf/shoot and root tissues were used from plants at 24 and 48 h post-treatment.

In recent years, RNA sequencing and SuperSAGE (Serial Analysis of Gene Expression) combined with Next Generation Sequencing was performed on RNA transcripts in Beja 1 nodules (Afonso-Grunz et al., 2014), in studies on chickpea roots and nodules during drought stress (Molina et al., 2008), and on roots and nodules during salt stress response (Molina et al., 2011; Kahl et al., 2011).

Molina et al., using the Desi-type water stress tolerant ILC 588 variety, subjected to 2 h of water stress, studied gene expression changes showing the main deregulated transcripts in roots (Molina et al., 2008). In a subsequent study on chickpea roots and nodules in response to salt stress, using ICC 4958, ICC 6098, Beja 1, and Amdoun 1, varieties tolerant and sensitive to salt stress, respectively, several isoforms of *lipoxygenase* (*LOX*) were identified and other genes in the oxylipin biosynthesis pathway were shown to be deregulated. Two *LOX* sequences were found deregulated in drought stress as well as under salt stress; *STCa-24417*, in particular, was upregulated several folds at 2, 8, 24 and 72 h of salt stress (Molina et al., 2011), in roots and nodules. In their study, Beja 1 (salt tolerant), Amdoun 1 (salt sensitive), ICC 4958 (salt sensitive), and ICC 6098 (salt weakly tolerant) varieties were used (Kahl et al., 2011). More recently, a study on root and nodule transcriptome was performed in Beja under salt conditions (Afonso-Grunz et al., 2014).

To confirm the SuperSAGE data, specific RT-PCR primers (Taqman probes) were designed, using the identified isoforms and splicing variants as a template, and the expression of transcripts was analysed by RT-PCR. Jasmonate and oxylipin biosynthesis, the reactive oxygen species (ROS) scavenging, and oxidative stress response were among the main pathways involved in water stress response. The identified transcripts are thought to protect cells from water deficit by producing important metabolic proteins and also regulating genes for signal transduction under water stress response. Considering the ROS response, Molina et al. (2011) identified several transcripts changing in expression levels, such as *ascorbate peroxidase* (*APX*), *alternative oxidase* (*AOX*), *superoxide dismutase* (*SOD*), *catalase* (*CAT*), *dehydroascorbate reductase* (*DHAR*), *glutathione peroxidase* (*GPX*), *glutathione reductase* (*GR*), and *glutathione-S-transferase* (*GST*). This data confirmed that the main early salt stress response in chickpea is the production and management of reactive oxygen species (ROS).

The genes involved in jasmonic acid (JA) synthesis codify for a plastidial 13-LOX, that produces 13(*S*)-hydroperoxy fatty acids from polyunsaturated fatty acids, and allene oxide synthase (13-AOS) and allene oxide cyclase (AOC) that act sequentially to produce 12-oxophytodienoic acid (OPDA). OPDA is then reduced by OPDA Reductase into 3-oxo-2(2'-pentenyl)-cyclopentane-1-octanoic acid. This undergoes three cycles of beta oxidation by acyl CoA oxidase (ACX), producing OPC:6, that is processed by a multifunctional protein (MPF), involved in the synthesis of OPC:4CoA, and by the ketoacyl-CoA thiolase (KAT2) that produces JA-CoA and finally JA.

Following Molina et al. (2011), De Domenico et al. monitored the expression of two *lipoxygenase* (*LOX*) transcripts and other genes involved in jasmonate (JA) synthesis in roots of ICC 4958 subjected to water stress up to 72 h, compared to the expression of the same genes in roots of the drought sensitive ICC 1882 (2012). This study differed from previous studies as the response to water stress was analysed using chickpea plants not inoculated with *Mesorhizobium ciceri 835*. The PCR analysis made use of the Taqman probes and primers designed and tested by Molina et al. (2011). De Domenico et al. monitored the expression of oxylipin biosynthesis genes (2 *LOX*s, 2 *AOS*s, 2 *AOC*s, 2 *HPL*s, and *OPDR*) in roots of ICC 4958 and ICC 1882 plants (2012).

The study performed HPLC profiling of oxylipins and intermediate compounds (such as OPDA, Jasmonic acid, and Jasmonate-Isoleucine) on ICC 4958 and ICC 1882 for all the stress treatment periods. The differential synthesis of JA, at early stages of treatment in the resistant variety, was in agreement with the timing of induction of oxylipin biosynthesis genes in ICC 4958.

Starting from the second hour of water stress in ICC 4958, the level of 13(*S*)-hydroperoxy fatty acid, the product of 13(*S*)LOX, and specific substrate for 13-AOS, started to increase. This result confirmed that the increase in expression of LOX1 (Lipoxy7252, StCa-24417) corresponded to the enzyme with 13(*S*) activity, different from other LOX enzymes with 9-13(*S*) or 9(*S*) activity in the synthesis pathway of other types of oxylipins.

In subsequent studies, experiments were conducted to monitor prolonged exposure (for 7 and 10 days) to osmotic stress and drought stress, using various tissues (nodules, roots) and cultivars, such as the Kabuli type Amdoun 1 and Beja 1, and the Desi type ICC 4958 and ICC 6098.

Osmotic Stress Response: Differences in Cultivars Contrasting in the Stress Response

Several groups studied osmotic stress response, monitoring plants subjected to controlled negative water potentials, by application of polyethylene glycol (PEG) and mannitol solutions to the roots (Basu et al., 2007; Mhadhbi et al., 2008; Varshney et al., 2009). Induced water deficit by non-electrolytes as neutral osmotically active compounds, showed similar values to that observed in the field, also permitting vigour evaluation. Mannitol is preferred in the osmotic stress challenge to avoid damaging the plants (Mhadhbi et al., 2008; Sassi et al., 2008). Osmotic stress shares features and similarities in the physiological responses with other abiotic stresses, such as salt and drought.

In a study, we evaluated differences in physiological parameters during response and adaptation to osmotic stress for 7 and 10 days, in the Kabuli type cultivars Amdoun 1 and Beja 1/INRAT-93, and in the Desi type ICC 6098, showing the improved ability of the resistant varieties to maintain plant biomass, relative water content, photosynthetic activity, and net assimilation of CO_2 (Gonzalez et al., 1995). The three varieties were hydroponically grown and challenged by 50 mM mannitol stress. Plantlets were monitored in physiological parameters, such as growth performance and plant biomass (PDW); water distribution (RWC); gas exchanges; symbiotic nitrogen fixation, evaluated as nodule biomass; and nodule functionality, evaluated as leghemoglobin content.

To monitor plants subjected to controlled negative water potentials, the application of polyethylene glycol (PEG) and mannitol solutions to the roots has been used often. Induced water deficit by non-electrolytes as neutral osmotically active compounds, showed similar values to that observed in the field, also permitting vigour evaluation. Recent studies showed that PEG application had deleterious effects on plants when subjected to it for a long period of time (Mhadhbi et al., 2008), whereas mannitol seems to be better suited for the generation of osmotic stress without damaging the plants (Mhadhbi et al., 2008). Osmotic stress shares features and similarities in the physiological responses with other abiotic stresses, such as salt and drought. However, chickpea varieties have a gradient of tolerance that may differ from stress to stress. For instance, ICC6098 is considered weakly drought tolerant and salt sensitive, Amdoun 1 is classified as weakly tolerant to drought (Labidi et al., 2009), and Beja 1 as salt and drought tolerant. Under osmotic stress conditions, ICC6098 showed the biggest DW equal to 3.16 g, compared to 1.60 g in Amdoun 1. After 14 days, the osmotic stress caused a decrease in plant biomass (20%), relative water content (23%), photosynthetic activity (41%), and nodule biomass (30%) in all lines, while it did not decrease nodule functionality measured as leghemoglobin (Lb) content. Using Real Time PCR, we monitored a selected set of genes in roots and nodules of Amdoun 1 and ICC 6098.

Among the findings, an increase in the root to shoot ratio (RSR) was observed in Amdoun 1 under osmotic stress (mannitol 50 mM) after 10 days, compared to control plants, possibly due to a preferential biomass allocation to roots (Cerekovic et al., in press), while, at the same time point a reduction in RDW was found in ICC 6098, compared to its control, because of the high RDW content in control plants. Small changes in leaf relative water content (RWC) were found after 10 days exposure to osmotic stress in both Amdoun 1 and ICC 6098. Mannitol treatment lowered RWC to 45% in Amdoun 1, to 40% in Beja 1, and to 30% in ICC 6098, compared to their controls. The osmotic

stress decreased nodule FW more severely in Amdoun 1 than in ICC 6098. The decrease of nodule FW was put in relation to a reduction of individual nodule FW and numbers. Nodule efficiency, assessed by the leghaemoglobin content in nodules, was not severely affected by 10 days osmotic stress in both Amdoun 1 and ICC 6098 varieties. Prolonged exposure to mannitol stress reduced Lb content in Amdoun 1 to the greatest extent. Data on performance under osmotic stress and assessed water use efficiency, plant shoot and root dry weight, nodule biomass, and leghaemoglobin content was not so different in the cultivars analysed.

Water Stress Response

In another study, we evaluated differences in physiological parameters during response and adaptation to water stress for 7 and 10 days, in the Kabuli type Amdoun 1 and Beja 1/INRAT-93, and in the Desi type ICC 4958 and ICC 6098, with plants grown in pots and inoculated with rhizobia. After 7 days, moderate drought stress affected plant growth and nodule fresh weight only marginally. After 10 days (severe drought), the total vegetative dry mass was more reduced in the Kabuli type than in the Desi type. Drought treatments significantly reduced nodule fresh mass at all stages of growth in Amdoun 1 and ICC 6098, whereas ICC 4958 and Beja 1 showed only a slight decrease in nodular biomass with drought treatment. Relative Water Content (RWC) decreased in all cultivars. In Amdoun 1 RWC showed a decrease of about 25% after 7 days and 30% after 10 days, while in Beja 1 it remained around 17%–18%. In ICC 6098 and ICC 4958 varieties, RWC decreased about 14% after 7 days, and about 22% after 10 days. Changes in physiological parameters (PDW, RWC, and NFW) are shown in Table 34.1.

Differential Expression of Lipoxygenase Genes

We analysed the differential gene expression of two *LOX* genes involved in Jasmonate production in nodules, and analysed the differences in the time of induction and level of expression in the four varieties.

In Amdoun 1 plants subjected to osmotic stress, in the nodules, the two LOX genes (Lipoxy7252 and Lip478466) were

TABLE 34.1

Plant Growth (Plant Dry Weight), RWC (Relative Water Content) and Nodule Fresh Weight (NFW) of Four Cultivars of Chickpea and Effect of Moderate and Prolonged Drought and After Rewatering

Treatment	Moderate Drought			Severe Drought			Re-watering		
	PDW	RWC	NFW	PDW	RWC	NFW	PDW	RWC	NFW
Amdoun control	0.73 ± 0.02^a	86 ± 3.2^a	0.5 ± 0.13^a	0.96 ± 0.05^a	88 ± 1.13^a	0.4 ± 0.08^a	1.10 ± 0.08^a	80 ± 5^a	0.44 ± 0.08^a
Amdoun stress	0.54 ± 0.07^b	66 ± 3^b	0.32 ± 0.01^b	0.78 ± 0.09^b	63 ± 1.2^c	0.2 ± 0.04^c	0.89 ± 0.009^b	78 ± 7^b	0.4 ± 0.01^a
Beja 1 control	0.54 ± 0.08^b	82 ± 4^a	0.63 ± 0.08^a	0.71 ± 0.11^b	85 ± 3^a	0.42 ± 0.11^a	0.72 ± 0.008^b	85 ± 6^a	0.5 ± 0.012^a
Beja 1 stress	0.52 ± 0.06^b	68 ± 4^b	0.38 ± 0.06^c	0.69 ± 0.03^b	70 ± 4^c	0.1 ± 0.03^d	0.74 ± 0.03^b	84 ± 10^a	0.2 ± 0.07^d
ICC 4958 control	0.6 ± 0.15^{ab}	83 ± 4^a	0.5 ± 0.15^b	0.87 ± 0.01^a	87 ± 2^a	0.4 ± 0.08^a	1 ± 0.01^a	76 ± 11^b	0.5 ± 0.01^a
ICC 4958 stress	0.51 ± 0.08^b	72.3 ± 3^{ab}	0.3 ± 0.04^b	0.79 ± 0.08^b	68 ± 7^a	0.2 ± 0.04^c	0.83 ± 0.012^b	75 ± 8^b	0.25 ± 0.012^c
ICC 6098 control	0.67 ± 0.11^b	82.5 ± 6^a	0.7 ± 0.14^a	0.84 ± 0.01^a	87 ± 3^a	0.42 ± 0.11^a	0.86 ± 0.001^b	80 ± 7^{ab}	0.52 ± 0.008^a
ICC 6098 stress	0.62 ± 0.01^b	71 ± 2^b	0.4 ± 0.05^c	0.69 ± 0.08^b	66 ± 4^c	0.1 ± 0.03^d	1.47 ± 0.03^a	78 ± 4^b	0.43 ± 0.011^b

Notes: Values are means ± standard errors ($n=8$). For each parameter different letters indicate significant differences ($P<0.05$).

upregulated at 7 days (4- and 5-fold, respectively), and remained upregulated to these values at 10 days. In Amdoun 1 roots, only Lipoxy7252 was upregulated 2.5-fold, and only in the first period, at 7 days. In ICC 6098 nodules, Lipoxy7252 was upregulated from basal levels up to 9.2-fold after 10 days, while in roots increased only slightly at 10 days, but was upregulated up to 8.9-fold at rewatering. Lip478466 remained stable during all the periods of stress.

During drought stress, in Amdoun 1 nodules the Lipoxy7252 gene was 6-fold upregulated, remaining at this level in the second time point, while Lip578466 was 5-fold upregulated and remained at similar levels (4 fold) in the second time point. In Beja 1 nodules, Lipoxy7252 increased up to 6-fold in the first time point and up to 17-fold after 10 days, while Lip578466 was found slightly upregulated only in the first time point and decreased to basal levels at the second time point (Figures 34.1 and 34.2).

In ICC 6098 nodules, Lipoxy7252 was upregulated up to 3-fold during the first time point of drought stress and up to 9-fold at 10 days, while Lip478466 remained unchanged at the first time point, and slightly downregulated after 10 days. In ICC 4958 nodules, Lip478466 was 2-fold upregulated after 7 days, and 10-fold induced after 10 days, while Lipoxy7252 remained 3-fold upregulated for all the period. Therefore, Desi-type

varieties showed a contrasting trend; ICC 4958 maintained lower levels of Lipoxy7252 at both time points, while the expression of Lip478466 increased during prolonged stress. In ICC 6098 and Beja 1, in agreement with their higher drought tolerance, Lipoxy7252 was upregulated in the second time point, with an increase of 9.3-fold and 17.5-fold in ICC 6098 and Beja 1, respectively.

Conclusion

There is a need for selection programmes for stress-resistant parent varieties on the basis of biomarker genes showing high or rapid inducible expression, to be used in breeding programs to select the pedigrees for crossing, and monitor the ability of hybrids to tolerate better abiotic stress challenges.

We showed variety-dependent performances under osmotic and drought stress, and assessed water use efficiency, plant shoot and root dry weight, nodule biomass, and leghemoglobin content in treated and control plants. The tolerance of Desi type ICC 6098 may be related to its superior ability to maintain the plant water status for longer periods, which results in lesser oxidative damage. These results confirmed the higher tolerance of Desi type varieties to osmotic stress and suggest that this may be

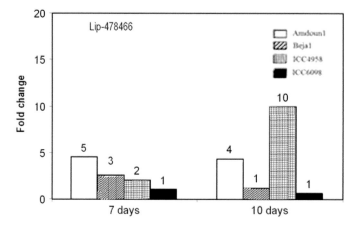

FIGURE 34.1 Lipoxygenase *Lip478466* expression changes after water stress in Amdoun 1, Beja 1, ICC 4958 and ICC 6098 after 7 and 10 days.

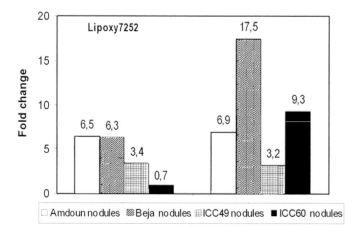

FIGURE 34.2 Lipoxygenase *Lipoxy7252* expression changes after water stress in Amdoun 1, Beja 1, ICC 4958 and ICC 6098 after 7 and 10 days.

linked to Desi plants ability to maintain the functional integrity of the photosynthetic system and to develop an abundant nodule network, which in turn favours the rate of symbiotic nitrogen fixation.

REFERENCES

Afonso-Grunz F., Molina C., Hoffmeier K., Rycak L., Kudapa H., Varshney R.K., Drevon J.J., Winter P., and Kahl G. Genome-based analysis of the transcriptome from mature chickpea root nodules. *Front. Plant Sci.* 2014, *5*, 235.

Agarwal G., Jhanwar S., Priya P., Singh V.K., Saxena M.S., Parida S.K., Garg R., Tyagi A.K., and Jain M. Comparative analysis of Kabuli chickpea transcriptome with Desi and wild chickpea provides a rich resource for development of functional markers. *PLoS One* 2012, *7*(12), e52443.

Anbazhagan K., Bhatnagar-Mathur P., Valdez V., Dumbala S.R., Kishor P.B., and Sharma K.K. DREB1A overexpression in transgenic chickpea alters key traits influencing plant water budget across water regimes. *Plant Cell Rep.* 2014, *34*, 199–210.

Ashraf M. and Foolad M.R. Roles of glycine betaine and proline in improving plant abiotic stress tolerance. *Environ. Exp. Bot.* 2007, *59*, 206–216.

Awasthi R., Kaushal N., Vadez V., Turner N.C., Berger J., Siddique K.H.M., and Nayyar H. Individual and combined effects of transient drought and heat stress on carbon assimilation and seed filling in chickpea. *Funct. Plant Biol.* 2014, *41*, 1148–1167.

Basu P. S., Berger J.D., Turner N.C., Chaturvedi S.K., Ali M., and Siddique K.H.M. Osmotic adjustment of chickpea (*Cicer arietinum*) is not associated with changes in carbohydrate composition or leaf gas exchange under drought. *Ann. Appl. Biol.* 2007, *150*, 217–225.

Behboudian M.H., Ma Q., Turner N.C., and Palta J.A. Reactions of chickpea to water stress: yield and seed composition. *J. Sci. Food Agric.* 2001, *81*, 1288–1291.

Bharadwaj C., Kumar S., Singhal T., Kumar T., Singh P., Tripathi S., Pal M., Hegde V.S., Jain P.K., Chauhan S., Kumar Verma A., Roorkiwal M., Gaur P.M., and Varshney R.K. Genomic approaches for breeding drought-tolerant chickpea. Conference Paper, NGG-P11. 5th International Conference on Next Generation Genomics and Integrated Breeding for Crop Improvement (ICRISAT), Patancheru, India, February 18–20, 2015.

Boyer J.S. Plant productivity and environment. Science 1982, 218, 443–448.

Boyer J.S., and Knipling E.B. Isopiestic technique for measuring leaf water potentials with a thermocouple psychrometer. *Proc Natl Acad Sci USA* 1965, 54, 1044–1051.

Cerekovic N., Fatnassi N., Horres R., Santino A., and Poltronieri P. Differences in physiology response and adaptation to osmotic and drought stress in Kabuli-type Amdoun-1 and Beja 1, and Desi-type ICC 4958 and ICC 6098, chickpea varieties. *Sci. China Life Sci.*, in press.

Cerekovic N., Fatnassi N., Santino A., Poltronieri P. Differences in physiological, biochemical, and molecular mechanisms of water stress tolerance in chickpea varieties. Chapter 6. In: *Approaches in Enhancing Abiotic Stress Tolerance to Plants* (Hasanuzzaman M., Fujita M., Oku H., Islam T.; Eds). Taylor and Francis, CRC in press, 2019. Catalog n. K345347.

Chen T.H. and Murata N. Enhancement of tolerance of abiotic stress by metabolic engineering of betaines and other compatible solutes. *Curr. Opin. Plant Biol.* 2002, *5*, 250–257.

Davies S.L., Turner N.C., Palta J.A., Siddique K.H.M., and Plummer J.A. Remobilisation of carbon and nitrogen supports seed filling in chickpea subjected to water deficits. *Aust. J. Agric. Res.* 2000, *51*, 855–866.

De Domenico S., Bonsegna S., Horres R., Pastor V., Taurino M., Poltronieri P., Imtiaz M., Kahl G., Flors V., Winter P., and Santino A. Transcriptomic analysis of oxylipin biosynthesis genes and chemical profiling reveal an early induction of jasmonates in chickpea roots under drought stress. *Plant Physiol. Biochem.* 2012, *62*, 115–122.

Deokar A.A., Kondawar V., Jain P.K., Karuppayil S.M., and Raju N.L. Comparative analysis of expressed sequence tags (ESTs) between drought-tolerant and susceptible genotypes of chickpea under terminal drought stress. *BMC Plant Biol.* 2011, *11*, 70.

Esfahani M.N., Sulieman S., Schulze J., Yamaguchi-Shinozaki K., Shinozaki K., and Tran L.S. Mechanisms of physiological adjustment of N_2 fixation in chickpea (*Cicer arientum* L.) during early stages of water deficit: Single or multi-factor controls? *Plant J.* 2014a, *79*, 964–980. doi: 1111/tpj.12599.

Esfahani M.N., Sulieman S., Schulze J., Yamaguchi-Shinozaki K., Shinozaki K., and Tran L.S. Approaches for enhancement of N_2 fixation efficiency of chickpea (*Cicer arietinum* L.) under limiting nitrogen conditions. *Plant Biotechnol. J.* 2014b, *12*(3), 387–397.

Fang X., Turner N.C., Yan G., Li F., and Siddique K.H.M. Flower numbers, pod production, pollen viability, and pistil function are reduced and flower and pod abortion increased in chickpea (*Cicer arietinum* L.) under terminal drought. *J. Exp. Bot.* 2010, *61*, 335–345.

Farooq M., Gogoi N., Barthakur S., Baroowa B., Bharadwaj N., Alghamdi S.S., and Siddique K.H.M. Drought stress in grain legumes during reproduction and grain filling. *J. Agron. Crop Sci.* 2016, *203*(2), 81–102. doi:10.1111/jac.12169

Farquhar G.D. and Sharkey T.D. Stomatal conductance and photosynthesis. *Annu. Rev. Plant Physiol.* 1982, *33*, 317–345.

Foyer H.C. Redox homeostasis and antioxidant signaling: A metabolic interface between stress perception and physiological responses. *Plant Cell* 2005, 17, 1866–1875.

Galvez L., Gonzalez E.M., and Arrese-Igor C. Evidence for carbon flux shortage and strong carbon/nitrogen interactions in pea nodules at early stages of water stress. *J. Exp. Bot.* 2005, *56*, 2551–2561.

Garg R., Batthacharjee A., and Jain M. Genomic scale transcriptomic insights into molecular aspects of abiotic stress response in chickpea. *Plant Mol. Biol. Rep.* 2015, *33*, 388–400.

Garg R., Patel R.K., Jhanwar S., Priya P., Bhattacharjee A., Yadav G., Bhatia S., Chattopadhyay D., Tyagi A.K., and Jain M. Gene discovery and tissue-specific transcriptome analysis in chickpea with Massively Parallel Pyrosequencing and web resource development. *Plant Physiol.* 2011, *156*, 1661–1678.

Gonzalez E.M., Gordon A.J., James C.L., and Arrese-Igor C. The role of sucrose synthase in the response of soybean nodules to drought. *J. Exp. Bot.* 1995, *46*, 1515–1523.

Gupta V., Bhatia S., Mohanty N.A., Sethy N., and Tripathy B.C. Comparative analysis of photosynthetic and biochemical characteristics of Desi and Kabuli gene pools of chickpea (*Cicer arietinum* L.). *Int. J. Gen. Eng. Biotechnol.* 2010, *1*, 65–76.

Hamwieh A., Imtiaz M., and Malhotra R.S. Multi-environment QTL analyses for drought-related traits in a recombinant inbred population of chickpea (*Cicer arietinum* L.). *Theor. Appl. Genet.* 2013, *126*, 1025–1038.

Hasibeder R., Fuchslueger L., Richter A., and Bahn M. Summer drought alters carbon allocation to roots and root respiration in mountain grassland. *New Phytol.* 2015, *205*(3), 1117–1127.

Hiremath P.J., Farmer A., Cannon S.B., Woodward J., Kudapa H., Tuteja R., Kumar A., Bhanuprakash A., Mulaosmanovic B., Gujaria N., Krishnamurthy L., Gaur P.M., Kavikishor. P.B., Shah T., Srinivasan R., Lohse M., Xiao Y., Town C.D., Cook D.R., May G.D., and Varshney R.K. Large-scale transcriptome analysis in chickpea (*Cicer arietinum* L.), an orphan legume crop of the semi-arid tropics of Asia and Africa. *Plant Biotechnol. J.* 2011, *9*, 922–931.

Jain D. and Chattopadhyay D. Analysis of gene expression in response to water deficit of chickpea (*Cicer arietinum* L.) varieties differing in drought tolerance. *BMC Plant Biol.* 2010, *10*, 24.

Jayashree B., Buhariwalla H.K., Shinde S., and Crouch J.H. A legume genomics resource: the chickpea root expressed sequence tag database. *Electron. J. Biotechnol.* 2005, *8*, 128–133.

Kahl G., Molina C.M., and Winter P. Functional genomics-transcriptomics for legumes: background, tools and insights. In: Perez de la Vega M., Torres A.M., Cubero J.I., and Kole C., Eds. *Genetics, Genomics and Breeding of Cool Season Grain Legumes*, pp. 237–284. CRC Press, Boca Raton, FL, 2011.

Karamanos A.J. and Papatheohari A.Y. Assessment of drought resistance of crop genotypes by means of the water potential index. *Crop Sci.* 1999, *39*, 1792–1797.

Kashiwagi J., Krishnamurthy L., Crouch J.H., and Serraj R. Variability of root characteristics and their contribution to seed yield in chickpea (*Cicer arietinum* L.) under terminal drought stress. *Field Crop Res.* 2006, *95*, 171–181.

Kashiwagi J., Krishnamurthy L., Upadhyaya, H.D., and Gaur, PM. Rapid screening technique for canopy temperature status and its relevance to drought tolerance improvement in chickpea. *J. SAT Agric. Res.* 2008, *6*, 1–4.

Kashiwagi J., Krishnamurthy L., Upadhyaya H. D., Krishna H., Chandra S., Vadez V., and Serraj R. Genetic variability of drought-avoidance root traits in the mini-core germplasm collection of chickpea (*Cicer arietinum* L.). *Euphytica* 2005, *146*, 213–222.

Kavikishore P.B., Hong Z., Miao G.H., Hu C.A.A., and Verma D.P.S. Overexpression of Dl-pyrroline-5-carboxylate synthetase increases proline production and confers osmotolerance in transgenic plants. *Plant Physiol.* 1995, *108*, 1387–1394.

Krishnamurthy L., Kashiwagi J., Gaur P.M., Upadhyaya H.D., and Vadez V. Sources of tolerance to terminal drought in the chickpea (*Cicer arietinum* L.) minicore germplasm. *Field Crops Res.* 2010, *119*, 322–330.

Krishnamurthy L., Kashiwagi J., Upadhyaya H.D., Gowda C.L.L., Gaur P.M., Singh S., Purushothaman R., and Varshney R.K. Partitioning coefficient: A trait that contributes to drought tolerance in chickpea. *Field Crops Res.* 2013, *149*, 354–365.

Kumar A., Bernier J., Verulkar S., Lafitte H.R., and Atlin G.N. Breeding for drought tolerance: Direct selection for yield, response to selection and use of drought-tolerant donors in upland and lowland-adapted populations. *Field Crops Res.* 2008, *107*, 221–231.

Kumar T., Bharadwaj C., Rizvi A.H., Sarker A., Tripathi S., Alam A., and Chauhan S.K. Chickpea landraces: A valuable and divergent source for drought tolerance. *Int. J. Trop. Agric.* 2015, *33*(3), 1–6.

Kurdali F., Al-Chammaa M., and Mouasess A. Growth and nitrogen fixation in silicon and/or potassium fed chickpeas grown under drought and well watered conditions. *J. Stress Physiol. Biochem.* 2013, *9*, 385–406.

Labidi N., Mahmoudi H., Dorsaf M., Slama I., and Abdelly C. Assessment of intervarietal differences in drought tolerance in chickpea using both nodule and plant traits as indicators. *J. Plant Breed. Crop Sci.* 2009, *1*(4), 080–086.

Larrainzar E., Wienkoop S., Scherling C., Kempa S., Ladrera R., Arres-Igor C., and Gonzalez E.M. *Medicago truncatula* root nodule proteome analysis reveals differential plant and bacteroid responses to drought stress. *Plant Physiol.* 2007, *144*, 1495–1507.

Lemoine R., La Camera S., Atanassova R., Dédaldéchamp F., Allario T., Pourtau N., Bonnemain J.-L., Laloi M., Coutos-Thévenot P., Maurousset L., Faucher M., Girousse C., Lemonnier P., Parrilla J., and Durand M. Source-to-sink transport of sugar and regulation by environmental factors. *Front Plant Sci.* 2013, *4*, 272.

Leport L., Turner N.C., Davies S.L., and Siddique K.H.M. Variation in pod production and abortion among chickpea cultivars under terminal drought. *Eur. J. Agron.* 2006, *24*, 236–246.

Leport L., Turner, N.C., French R.J., Barr M.D., Duda R., Davies S.L., Tennant D., and Siddique K.H.M. Physiological responses of chickpea genotypes to terminal drought in a Mediterranean-type environment. *Eur. J. Agron.* 1999, *11*, 279–291.

Leport L., Turner N.C., French R.J., Tennant D., Thomson B.D., and Siddique K.H.M. Water relations, gas-exchange, and growth of cool-season grain legumes in a Mediterranean-type environment. *Eur. J. Agron.* 1998, *9*, 295–303.

Liu F., Jensen C.R., and Andersen M.N. A review of drought adaptation in crop plants: changes in vegetative and reproductive physiology induced by ABA-based chemical signals. *Aust. J. Agric. Res.* 2005, *56*, 1245–1252.

Maccarrone M., Veldink G.A., Agro A.F., and Vliegenthart J.F. Modulation of soybean lipoxygenase expression and membrane oxidation. by water deficit. *FEBS Lett.* 1995, *371*, 223–226.

Mafakheri A., Siosemardeh A., Bahramnejad B., Struik P.C., and Sohrabi Y. Effect of drought stress on yield, proline and chlorophyll contents in three chickpea cultivars. *Aust. J. Crop Sci.* 2010, *4*, 580–585.

Mantri N.L., Ford R., Coram T.E., and Pang E.C.K. Transcriptional profiling of chickpea genes differentially regulated in response to high-salinity, cold and drought. *BMC Genomics* 2007, *8*,303.

Mathur P.B., Vadez V., Jyotsna Devi M., Lavanya M., Vani G., and Sharma K.K. Genetic engineering of chickpea (*Cicer arietinum* L.) with the P5CSF129A gene for osmoregulation with implications on drought tolerance. *Mol. Breed.* 2009, *23*, 591–606.

Mhadhbi H., Jebara M., Zitoun A., Limam F., and Aouani M.E. Symbiotic effectiveness and response to mannitol-mediated osmotic stress of various chickpea–rhizobia associations. *World J. Microbiol. Biotechnol.* 2008, *24*, 1027–1035.

Mir R.R., Zaman-Allah M., Sreenivasulu N., Trethowan R., and Varshney R.K. Integrated genomics, physiology and breeding approaches for improving drought tolerance in crops. *Theor. Appl. Genet.* 2012, *125*, 625–645.

Mohammadi A., Habibi D., Rohami M., and Mafakheri S. Effect of drought stress on antioxidant enzymes activity of some chickpea cultivars. *Am.-Eurasian J. Agric. Environ. Sci.* 2011, *11*, 782–785.

Molina C., Rotter B., Horres R., Udupa S.M, Besser B., Bellarmino L., Baum M., Matsumura H., Terauchi R., Kahl G., and Winter P. SuperSAGE: The drought stress-responsive transcriptome of chickpea roots. *BMC Genomics* 2008, *9*, 553.

Molina C., Zaman-Allah M., Khan F., Fatnassi N., Horres R., Rotter B., Steinhauer D., Amenc L., Drevon J.J., Winter P., and Kahl G. The salt-responsive transcriptome of chickpea roots and nodules via deepSuperSAGE. *BMC Plant Biol.* 2011, *11*, 31.

Naya L., Ladrera R., Ramos J., Gonzalez E.M., Arrese-Igor C., Minchin F.R., and Becana M. The responses of carbon metabolism and antioxidant defenses of alfalfa nodules to drought stress and to subsequent recovery of plants. *Plant Physiol.* 2007, *144*, 1104–1114.

Nayyar H., Singh S., Kaur S., and Kumar S., Upadhyaya H.D. Differential sensitivity of *Macrocarpa* and *Microcarpa* types of chickpea (*Cicer arietinum* L.) to water stress: association of contrasting stress response with oxidative injury. *J. Integr. Plant Biol.* 2006a, *48*, 1318–1329.

Nayyar H., Kaur S., Singh S., and Upadhyaya H.D. Differential sensitivity of Desi (small-seeded) and Kabuli (large-seeded) chickpea genotypes to water stress during seed filling: Effects on accumulation of seed reserves and yield. *J. Sci. Food Agric.* 2006b, *86*(13), 2076–2082.

Nguyen K.H., Ha C.V., Watanabe Y., Tran U.T., Nasr Esfahani M., Nguyen D.V., and Tran L.S. Correlation between differential drought tolerability of two contrasting drought-responsive chickpea cultivars and differential expression of a subset of CaNAC genes under normal and dehydration conditions. *Front. Plant Sci.* 2015, *6*, 449.

Pace P.F., Crale H.T., El-Halawany S.H.M., Cothren J.T., and Senseman S.A. Drought induced changes in shoot and root growth of young cotton plants. *J. Cotton Sci.* 1999, *3*, 183–187.

Palta J.A., Nandwal A.S., Kumari S., and Turner N.C. Foliar nitrogen applications increase the seed yield and protein content in chickpea (*Cicer arietinum* L.) subject to terminal drought. *Aust. J. Agric. Res.* 2005, *56*, 105–112.

Pandey A., Chakraborty S., Datta A., and Chakraborty N. Proteomics approach to identify dehydration responsive nuclear proteins from chickpea (*Cicer arietinum* L). *Mol. Cell. Proteomics* 2008, *7*, 88–107.

Parween S., Nawaz K., Roy R., Pole A.K., Venkata Suresh B., Misra G., Jain M., Yadav G., Parida S.K., Tyagi A.K., Bhatia S., and Chattopadhyay D. An advanced draft genome assembly of a desi type chickpea (*Cicer arietinum* L.). *Sci. Rep.* 2015, *5*, 12806. doi:10.1038/srep12806.

Poltronieri P., Bonsegna S., De Domenico S., and Santino A. Molecular mechanisms of abiotic stress response in plants. *Field Veg. Crops Res./Ratarstvo Povrtartsvo* 2011, *48*, 15–24.

Poltronieri P., and Miwa M. PARP proteins, NAD, epigenetics, and antioxidative response in abiotic stress. In: Poltronieri P. and Hong Y. Eds. *Applied Plant Genomics and Biotechnology*, pp. 237–252, Chapter 15. Elsevier, Oxford, UK, 2015.

Poltronieri P, Taurino M, Bonsegna S, De Domenico S, Santino A. Monitoring the activation of jasmonate biosynthesis genes for selection of chickpea hybrids tolerant to drought stress. In:

Chakraborty U. and Chakraborty B. Eds. *Abiotic Stresses in Crop Plants*, pp. 54–70, Chapter 4. CABI, Wallingford, UK, 2015.

Poltronieri P., Taurino M., De Domenico S., Bonsegna S., and Santino A. Activation of the jasmonate biosynthesis pathway in roots in drought stress. In: Tuteja N. and Gill S.S. Eds. *Climate Change and Abiotic Stress Tolerance*, pp. 325–342. Wiley-VCH GmbH & Co. KGaA, Weinheim, Germany, 2013.

Purushothaman R. and Krishnamurthy L. Timing of sampling for the canopy temperature depression can be critical for the best differentiation of drought tolerance in chickpea. *J. SAT Agric. Res.* 2014, *12*, 1–8.

Purushothaman R., Krishnamurthy L., Upadhyaya H.D., Vadez V., and Varshney R.K. Shoot traits and their relevance in terminal drought tolerance of chickpea (*Cicer arietinum* L.). *Field Crops Res.* 2016, *197*, 10–27. doi:10.1016/j.fcr.2016.07.016.

Purushothaman R., Thudi, M., Krishnamurthy L., Upadhyaya, H.D., Kashiwagi, J., Gowda C.L.L., and Varshney R.K. Association of mid-reproductive stage canopy temperature depression with the molecular markers and grain yields of chickpea (*Cicer arietinum* L.) germplasm under terminal drought. *Field Crops Res.* 2015, *174*, 1–11.

Purushothaman R., Upadhyaya H.D., Gaur P.M., Gowda C.L.L., and Krishnamurthy L. Kabuli and desi chickpeas differ in their requirement for reproductive duration. *Field Crops Res.* 2014, *163*, 4–31.

Pushpavalli R., Zaman-Allah M., Turner N.C., Baddam R., Rao M.V., and Vadez V. Higher flower and seed number leads to higher yield under water stress conditions imposed during reproduction in chickpea. *Funct. Plant Biol.* 2014, *42*, 162–174.

Ramalingam A., Kudapa H., Pazhamala L.T., Garg V., and Varshney R.K. Gene expression and yeast two-hybrid studies of 1R-MYB Transcription Factor mediating drought stress response in chickpea (*Cicer arietinum* L.). *Front. Plant Sci.* 2015, *6*, 1117.

Ramos M.L.G., Gordon A.J., Minchen F.R., Sprent J.I., and Parsons R. Effect of water stress on nodule physiology and biochemistry of a drought tolerant cultivar of common bean (*Phaseolus vulgaris* L.). *Ann. Bot.* 1999, *83*, 57–63.

Sassi A.S., Aydi S., Gonzalez E.M., Abdelly C. Osmotic stress affects water relations, growth, and nitrogen fixation in *Phaseolus vulgaris* plants. *Acta Physiol Plant* 2008, 30, 441–449.

Sassi S., Aydi S., Gonzalez E.M., Arrese-Igor C., and Abdelly C. Understanding osmotic stress tolerance in leaves and nodules of two *Phaseolus vulgaris* cultivars with contrasting drought tolerance. *Symbiosis* 2010, *52*, 1–10.

Siddique K.H.M., Regan K.L., Tennant G., and Thomson B.D. Water use and water use efficiency of cool season grain legumes in low rainfall Mediterranean-type environments. *Eur. J. Agron.* 2001, *15*, 267–280.

Singh G.D., Wani A.A., Kaur D., and Sogi D.S. Characterisation and functional properties of proteins of some Indian chickpea (*Cicer arietinum*) cultivars. *J. Sci. Food Agric.* 2008, *88*, 778–786.

Sohrabi Y., Heidari G., Weisany W., Ghasemi Golezani K., and Mohammadi K. Some physiological responses of chickpea cultivars to arbuscular mycorrhiza under drought stress. *Russ. J. Plant Physiol.* 2012, *59*, 708–716.

Szabados L. and Savouré A. Proline: a multifunctional amino acid. *Trends Plant Sci.* 2009, *15*, 89–97.

Talbi C., Sanchez C., Hidalgo-Garcia A., Gonzalez E.M., Arrese-Igor C., Girard L., Bedmar E.J., and Delgado M.J. Enhanced expression of *Rhizobium etli cbb3* oxidase improves drought tolerance of common bean symbiotic nitrogen fixation. *J. Exp. Bot.* 2012, *63*, 5035–5043.

Turner N.C., Davies S.L., Plummer J.A., and Siddique K.H.M. Seed filling in grain legumes (pulses) under water deficits with emphasis on chickpea (*Cicer arietinum* L.). *Adv. Agron.* 2005, *87*, 211–250.

Ulemale C.S., Mate S.N., and Deshmukh D.V. Physiological indices for drought tolerance in chickpea (*Cicer arietinum* L.). *World J. Agric. Sci.* 2013, 9, 123.d–131.d.

Varshney R.K., Hiremath P.J., Lekha P., Kashiwagi J., Balaji J., Deokar A.A., Vadez V., Xiao Y., Srinivasan R., Gaur P.M., Siddique K.H.M., Town C.D., and Hoisington D.A. A comprehensive resource of drought- and salinity responsive ESTs for gene discovery and marker development in chickpea (*Cicer arietinum* L.). *BMC Genomics* 2009, *10*, 523.

Yaqoob M., Hollington P.A., Mahar A.B., and Gurmani Z.A. Yield performance and responses studies of chickpea (*Cicer arietinum* L.) genotypes under drought stress. *Emirates J. Food Agric.* 2013, *25*, 117–123.

Zaman-Allah M., Jenkinson D.M., and Vadez V. A conservative pattern of water use, rather than deep or profuse rooting, is critical for the terminal drought tolerance of chickpea. *J. Exp. Bot.* 2011a, *62*, 4239–4252.

Zaman-Allah M., Jenkinson D.M., and Vadez V. Chickpea genotypes contrasting for seed yield under terminal drought stress in the field differ for traits related to the control of water use. *Funct. Plant Biol.* 2011b, *38*, 270–281.

Index